Kingdoms, Empires, and Domains

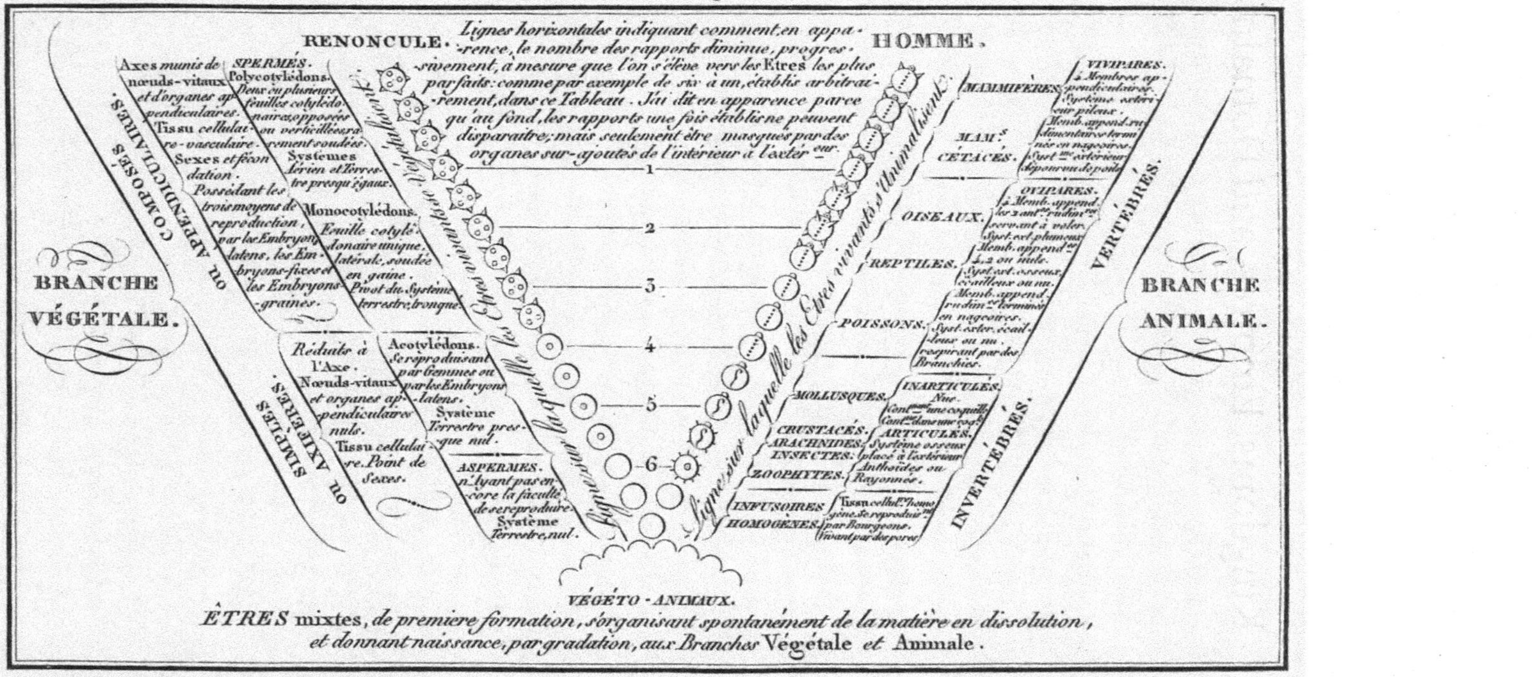

Kingdoms, Empires, and Domains

The History of High-Level Biological Classification

MARK A. RAGAN

OXFORD
UNIVERSITY PRESS

Oxford University Press is a department of the University of Oxford. It furthers
the University's objective of excellence in research, scholarship, and education
by publishing worldwide. Oxford is a registered trade mark of Oxford University
Press in the UK and certain other countries.

Published in the United States of America by Oxford University Press
198 Madison Avenue, New York, NY 10016, United States of America.

© Oxford University Press 2023

All rights reserved. No part of this publication may be reproduced, stored in
a retrieval system, or transmitted, in any form or by any means, without the
prior permission in writing of Oxford University Press, or as expressly permitted
by law, by license, or under terms agreed with the appropriate reproduction
rights organization. Inquiries concerning reproduction outside the scope of the
above should be sent to the Rights Department, Oxford University Press, at the
address above.

You must not circulate this work in any other form
and you must impose this same condition on any acquirer.

Library of Congress Cataloging-in-Publication Data
Names: Ragan, Mark A., author.
Title: Kingdoms, empires, and domains : the history of high-level
biological classification / Mark A. Ragan.
Other titles: History of high-level biological classification
Description: New York, NY : Oxford University Press, [2023] |
Includes bibliographical references and index.
Identifiers: LCCN 2022039010 (print) | LCCN 2022039011 (ebook) |
ISBN 9780197643037 (hardback) | ISBN 9780197643051 (epub)
Subjects: LCSH: Biology—Classification—History. |
Biology—Nomenclature—History. | Classification of sciences—History.
Classification: LCC QH83 .R343 2023 (print) | LCC QH83 (ebook) |
DDC 578.01/2—dc23/eng/20221005
LC record available at https://lccn.loc.gov/2022039010
LC ebook record available at https://lccn.loc.gov/2022039011

DOI: 10.1093/oso/9780197643037.001.0001

Printed by Integrated Books International, United States of America

Fellow Labourers! The Great Vintage & Harvest is now upon Earth.
The whole extent of the Globe is explored. Every scatter'd Atom
Of Human Intellect now is flocking to the sound of the Trumpet.
All the Wisdom which was hidden in caves & dens from ancient
Time, is now sought out from Animal & Vegetable & Mineral. . . .

So spake Ololon in reminiscence astonish'd, but they
Could not behold Golgonooza without passing the Polypus,
A wondrous journey not passable by Immortal feet, & none
But the Divine Saviour can pass it without annihilation.
For Golgonooza cannot be seen till having pass'd the Polypus
It is viewed on all sides round by a Four-fold Vision,
Or till you become Mortal & Vegetable in Sexuality
Then you behold its mighty Spires & Domes of ivory & gold . . .

—William Blake, *Milton* (1804)

[Blake, *Milton*. In: Maclagan & Russell (1907), at 25:17–21 and 35:18–25
(original pagination)]

Contents

List of illustrations	xiii
Preface	xv
Acknowledgement: copyrighted material	xix

1. The earliest Nature	1
Primitive concepts of natural entities	2
Early figurative art	2
Symbolic language and folk taxonomies	4
Creation myths	6
Animal deities and anthropomorphized plants	6
Transformation and metamorphosis	8
Metempsychosis, reincarnation, and anamnesis	10
Transmutation and transubstantiation	12
2. Eastern Nature	14
The Indian subcontinent	14
Buddhism	16
China: the common tradition	17
China: Confucianism	19
China: Taoism	20
China: Mohism	21
Japan	22
The "three kingdoms of nature" are not rooted in prehistory	23
3. Philosophical Nature	24
Hellenic philosophical traditions before Socrates	25
The Pythagoreans	26
The Eleatics	28
The atomists	28
Empedocles	29
Diogenes of Apollonia	30
Socrates	30
Plato	31
Aristotle	34
Theophrastus	42
Stoic and later triadic divisions of soul or beings	43
Scepticism	44
Envoi	46
4. Utilitarian Nature	48
Lucretius	48
Seneca	49

viii CONTENTS

Pliny the Elder	50
Herbals and pharmacopœias	52
Early medical texts	53
Bestiaries	55
Summary	57

5. Neoplatonic Nature   59
Philo of Alexandria	60
Calvenus Taurus	63
Plotinus	63
Porphyry and Anatolius	64
Iamblichus and Dexippus	65
Themistius	67
Athens and Alexandria	68
Ammonius Hermiæ and John Philoponus	69
Elias and David	71
Summary: philosophical themes within Neoplatonism	73

6. Christian Nature   75
Early theologians and polemicists	76
Origen	77
Nemesius	79
The Cappadocian Fathers	80
Augustine	83
Pseudo-Dionysius	85
Boëthius	87
John of Damascus	87
Summary	89

7. Islamic and Jewish Nature   91
Islam and the translation of Hellenic philosophy into Arabic	93
Arabic natural history, an-Naẓẓām, and al-Jāḥiẓ	96
Al-Kindī, al-Fārābī, and al-Masʿūdī	97
The Ikhwān al-Ṣafā	98
Al-Bīrūnī, Ibn Sīnā, al-Ghazālī, Ibn Rushd, and al-Abharī	100
Niẓamī Arūzī, al-Qazwīnī, and later authors	103
Ṣufiyya	106
The Jewish philosophical tradition: Ibn Daud and Maimonides	107
Kabbalah	108
Duran, Alemanno, and Albotini	109
The rediscovery of Aristotle's natural history	111

8. Monastic and scholastic Nature   114
Cassiodorus to Hrabanus Maurus	115
Eriugena	116
Anselm, Peter Abelard, and Peter Lombard	118
Adelard and Berachya	119
Hildegard and Marius	120
The School of Chartres	121
Bernard Silvestris and John Blund	122
Robert Grosseteste	123
Thomas of Cantimpré, Bartholomæus Anglicus, and Vincent of Beauvais	124

CONTENTS ix

	Albertus Magnus and Thomas Aquinas	125
	Bonaventure and Dante	128
	The Fourteenth century	130
	Nicholas of Cusa	132
	From scholasticism to humanism	133
9.	Nature's mystic book	136
	Oracles and mysteries	136
	Thrice-great Hermes	137
	Universal truths and hidden meanings	138
	Gnostic texts	139
	Macrocosm and microcosm	140
	Alchemy	142
	The *Kitāb Sirr al-khalīqa*	143
	The *Sirr al-asrār* or *Secretum secretorum*	144
	Magic	146
	From Jābir to the Renaissance	147
	Three Renaissance humanists: Ficino, Pico, and Agrippa	149
	Paracelsus and the alchemists	153
	Bruno, Fludd, and the nature-mystics	157
	Summary and questions	160
10.	Allegory, myth, and superstition	162
	Allegory	163
	Beings with exaggerated features	164
	Chimæras: the *borametz*	164
	Active transformation: the barnacle-goose-tree	167
	Return from the dead	170
	Monsters and marvels	170
	Ancients and Moderns	171
11.	The return of the zoophytes	173
	Dictionaries	173
	Guillaume Budé: Roman law (1508)	175
	Otto Brunfels: *materia medica* (1534)	176
	François Rabelais: literature in the vernacular (1546)	177
	Jean Bodin: political theory (1576)	178
	Jacopo Zabarella: Aristotelian logic (1606)	180
	Johann Thomas Freig: Ramist natural history (1579)	181
	Robert Burton: English vernacular (1621)	183
	Juan Eusebio Nieremberg: baroque nature (1635)	183
	David Person: rare and excellent matters (1635)	185
	Henry More: the Spirit of Nature (1682)	186
	Concluding comments	187
12.	Plants and animals	188
	Herbals (from 1475)	189
	The rise of scientific botany 1: 1490–1580	190
	Andrea Cesalpino	193
	The rise of scientific botany 2: 1580–1680	194
	Medieval and early Renaissance animal-books	197

The rise of scientific zoology 1: 1520–1550	198
The rise of scientific zoology 2: the momentous 1550s	199
The rise of scientific zoology 3: the encyclopædists 1560–1660	203
The rise of scientific zoology 4: curiosities and specialization	205
Zoophyta: a fourth division of nature?	207
Plants and animals in 1680	208

13. The most wretched creatures — 210

Multiple worlds	210
Invisible airborne seeds	212
Leibniz and monads	213
Leeuwenhoek and Joblot: little animals observed	215
Buffon, Needham, and Spallanzani: spontaneous generation	216
A class of their own?	219
Summary: one hundred years of little animals	221

14. Continuity in the living world — 223

The Great Chain under attack	226
Richard Bradley: *A philosophical account*	227
Corals: an ancient enigma resolved	231
Hydra: a new enigma	233
Charles Bonnet: the canonical Great Chain of Being	236
The Great Chain after 1780	240

15. Classifying God's handiwork — 243

Magnol and Tournefort	244
Ray and natural theology	245
Linnæus	248
What, then, are fungi?	257
Adanson, Scopoli, and de Jussieu	259
Zoophyta as animals	261
Summary	263

16. Beyond the end of the Chain — 265

Nature as a map	266
Nature as a network	268
Nature as a polygon or Easter egg	269
Nature as a branched tree	276
Nature as a spiral	280
Nature as a circle	281
Quinarian nature	282
Summary	286

17. From *histoire naturelle* to *anatomie* and *morphologie* — 289

Denis Diderot and Jean Le Rond d'Alembert	290
Louis-Jean-Marie Daubenton	291
Jean-Baptiste Lamarck	293
Georges Cuvier	298
Étienne Geoffroy Saint-Hilaire	300
Félix Vicq-d'Azyr: *le règne vivant*	302
Jean Guillaume Bruguière: a new arrangement of Vermes	303
Julien-Joseph Virey: *evolution* along parallel chains	303

CONTENTS   xi

Pierre-Jean-François Turpin: *végéto-animaux*   305
Henri Marie Ducrotay de Blainville: infusoria as an appendage   306
Henri Milne-Edwards: embryology and classification   307
Jean-Baptiste Bory de Saint-Vincent: Règne Psychodiaire   307
Summary: France   309

18. *Naturphilosophie*, polygastric animalcules, and cells   311
Johann Gottfried Herder   312
Johann Wolfgang von Goethe   312
Immanuel Kant: transcendental idealism   315
German Romanticism   316
Naturphilosophie   317
Lorenz Oken   318
Gottfried Reinhold Treviranus   322
Alexander von Humboldt   322
Karl Ernst von Baer   324
Christian Gottfried Ehrenberg   325
Cell theory   327
The last *Naturphilosoph*: Carl Gustav Carus   331
Summary: Germany   333

19. Green matter, zoospores, and diatoms   335
Simple animals, simple plants   335
How, then, do algae reproduce?   336
Case study 1: Priestley's green matter   338
Case study 2: zoospores   341
Case study 3: metamorphosis   344
Benjamin Gaillon   346
Friedrich Traugott Kützing   347
Case study 4: diatoms and desmids   350
Summary   352

20. Temples of Nature   355
Britain: three Linnæan kingdoms   356
Erasmus Darwin   359
Natural theology   361
Richard Owen   364
Vestiges of the natural history of Creation   368
Charles Darwin   369
John Hogg   370
Thomas B. Wilson and John Cassin   373
Popular natural histories in Victorian Britain   375
Summary: Britain   377

21. Ernst Haeckel and Protista   379
Die Radiolarien (1862)   382
Generelle Morphologie (1866)   383
New classes of Protista   388
Sponges and gastraea theory   390
Monera, protozoa, and protophyta   392
Das Protistenreich (1878)   395

xii CONTENTS

Protists and Histones	396
Four kingdoms of life	397
The protozoological tradition	400
The phycological tradition	403
The bacteriological tradition	405
The protistological tradition	407
Summary: Haeckel and Protista	409

22. Beyond three kingdoms — 411

Kingdoms and superkingdoms	412
Four kingdoms (Copeland, 1938–1956)	417
Five kingdoms (Whittaker, 1969)	417
Other high-level proposals to 1975	419
The rise of cellular ultrastructure	421
Eukaryogenesis 1: *Natura facit saltum*	421
Eukaryogenesis 2: science may discover ten	426
Summary	430

23. Genes, genomes, and domains — 432

Introduction: the molecular basis of heredity	432
Molecular phylogenetics before sequences	433
The ribosomal RNA Tree of Life	435
The molecular consensus erodes	438
Thinking laterally about genomes	439
Genomes and pan-genomes	440
Genomes from the environment	441
Retrospective: the domains of life	443
Last words on kingdoms, empires, and domains	448

Appendix: Victorian popular natural histories	451
Acronyms	453
Notes	455
References	637
Index of Persons	777
Index of Subjects	803

Illustrations

Figures

6.1. Anonymous Twelfth-century manuscript, souls ascending from earth to God.	86
9.1. Cosmology of alchemy, according to the *Kitāb Sirr al-khalīqa*.	145
9.2. Pico della Mirandola's three worlds, as modified by Le Fèvre de La Boderie.	151
9.3. The Fountain of Mercury, from *Rosarium philosophorum*.	155
9.4. Kirchweger, *Aurea catena Homeri*.	160
11.1. Freig, division of corporeal bodies following the method of Ramus.	182
16.1. Linnæus (Giseke), genealogical-geographical table of the affinities of plants.	267
16.2. Buffon, genealogical table of the different races of dogs.	270
16.3. Goldfuss, animal-globe or Easter egg.	271
16.4. Von Baer, schematic representation of animals.	273
16.5. Eichwald, table of transitions among animals.	274
16.6. Adrien de Jussieu, map of genera of Rutaceæ.	275
16.7. Duchesne, genealogy of strawberries.	277
16.8. Eichwald, tree of animal life.	279
16.9. Horaninov, schematic diagram of nature.	280
16.10. Fischer von Waldheim, organized nature as two *cercles de mouvement*.	282
16.11. Macleay, general view of organized matter.	283
16.12. Macleay, quinarian arrangement of the animal kingdom.	285
20.1. Hogg, diagram of natural bodies, or of the four Kingdoms of Nature.	373
20.2. Hitchcock, palæontological chart showing plants and animals as branched series.	374
21.1. Haeckel, phylogenetic table of stem-relationships of the phyla of the Animal Kingdom.	394
21.2. Haeckel, *Stammbaum* of the organic world.	398
21.3. Bütschli, *Stammbaum* of organisms.	402
21.4. Klebs, relationships among the lower organisms.	405
21.5. Entz, relationships in the organic world.	408
22.1. Whittaker, four kingdoms of organisms.	418
22.2. Whittaker, five-kingdom system based on three levels of organization.	419
22.3. Margulis, modification of Whittaker's five-kingdom scheme.	420
22.4. Mereschkowsky, representation of the organic world as two *Stämme*.	425

xiv LIST OF ILLUSTRATIONS

23.1. Margoliash, Fitch, and Dickerson, statistical phylogenetic tree of cytochromes *c*. 434

23.2. Woese, universal phylogenetic tree determined from rRNA sequence comparison. 438

23.3. Woese, Kandler, and Wheelis, universal phylogenetic tree in rooted form. 438

Table

15.1. Major editions of *Systema naturæ*. 250

Preface

Animals, vegetables, minerals: for twenty-six centuries, these three great groups have encompassed the scope of material bodies on earth. Philosophers, poets, naturalists, even children recognize their members: things sensate but mortal; living but insensate; and organized but nonliving. *Homo sapiens*, the one animal capable of pride, sometimes accords himself a fourth.

Yet as many of us are aware—perhaps dimly from some long-ago biology lecture—certain microscopic beings are motile like animals, yet pigmented like plants. In 1866, citing little precedent, Ernst Haeckel established for these and other simple creatures a third major group of living beings: the protists. Why did he take this step where others had not? Or perhaps there *had* been other supernumerary kingdoms of life, now—or even in Haeckel's day—lost to the mists of time. And this odd word *zoophyte*, tucked away at the back of the dictionary: are zoophytes somehow both plant *and* animal? Are (some) protists zoophytes?

As a young scientist interested in the diversity of life, I read Haeckel's *Generelle Morphologie* and pondered these questions. Later, as a member, then President, of the International Society for Evolutionary Protistology (ISEP), I had occasion to investigate the history of Third Kingdoms of Life. I was scarcely prepared for what I found. Plants and animals have *not*, in fact, been recognized as the highest-level taxa of living beings since time immemorial. Zoophytes were first considered intermediate between plants and animals not in Eighteenth-century Europe, but in Fourth-century Syria, and eventually found a place in a classification of animals that was falsely attributed to Aristotle. The most-inclusive taxa began to be called *kingdoms* only in 1604—first by alchemists, not botanists or zoologists. Since then, more than fifty taxa at kingdom level or above been proposed by philosophers, scholars, or naturalists including Linnæus. Haeckel's Protista has itself been pronounced dead more than once.

I invite you, dear reader, to join me on this eclectic tour of the history of the idea that there is more to the living world than plants and animals. We shall meet aboriginals and alchemists, philosophers and mystics, saints and heretics, physicians and poets, soldiers and diplomats, sea-captains and stonemasons (well, one stonemason), botanists, zoologists, Haeckel himself, and other protistologists. Our story unfolds across twenty-three chapters, each describing a theme or historical episode. The narrative is broadly chronological, although certain chapters run in parallel and are best read together. I rely heavily on primary texts and authoritative translations, but draw on secondary literature where appropriate.

xvi PREFACE

My fascination with algae and protozoa can be traced to a microscope given me at age twelve by my parents; then classes and an honours paper with David J. Chapman at the University of Chicago. James S. Craigie supervised my PhD project, on brown seaweeds, at Dalhousie University. Over succeeding decades, research collaborations brought encounters with amazing creatures: diatoms, sponges, red algae, *Spirogyra*, dinoflagellates, charophytes, *Cryptococcus*, chytrids, ichthyosporeans, ciliates, labyrinthulids, diplomonads, *Sulfolobus*, various bacteria, dinoflagellates again, corals, and *Chromera*, in approximately that order. I outlined this book (1997) and drafted two early chapters (1998–2000) while at National Research Council (NRC) Canada, then took it up again in early 2018 upon retirement from the University of Queensland. In the interim, I read widely across the history of high-level biological classification, with special focus on organisms that do not fit comfortably within the canonical kingdoms of living things.

I did not make this journey alone. First and foremost, I thank my wife, Chikako, and our children, Alicia and Wesley, for their love and forbearance. Colleagues in my research institutions, in ISEP, in the Evolutionary Biology Program of the Canadian Institute for Advanced Research, in the Reef Future Genomics 2020 consortium organized by the Great Barrier Reef Foundation, and many others brought inspiration and wondrous organisms. Marvin Green, Dave Chapman, Emanuel Margoliash, Jim Craigie, Arne Jensen, Jack McLachlan, Carolyn Bird, Max Taylor, Lynn Margulis, John Corliss, Tom Cavalier-Smith, Lynn Rothschild, Miklós Müller, Ford Doolittle, Carl Woese, and John Mattick have been teachers, mentors, colleagues, and friends.

I could not have begun this book without a firm grounding in Latin, for which I thank Agnes Landis. My first steps into the early literature were facilitated by Gina Douglas (Linnean Society of London), Tony Swann (Wheldon & Wesley), and colleagues at the Natural History Museum (London). Staff at the NRC Canada National Science Library and the University of Queensland Library tracked down obscure books and articles. Online resources, notably the Biodiversity Heritage Library and the Internet Archive, opened an unimaginable world of early texts. J. Patrick Atherton KHS, Denis Brosnan, Anitra Laycock, Georges Merinfeld, Amin R. Mohamed, and Atefeh Taherian Fard provided expert translations. William Chittick, Ford Doolittle, Ernst Mayr, Minaka Nobuhiro, Marie-Odile Soyer-Gobillard, and David M. Williams helped me understand specific issues or viewpoints. Nick Hamilton photographed Figure 9.1.

I owe a debt of gratitude to my editors at Oxford University Press, Jeremy Lewis and Michelle Kelley, for their belief in this project and unfailingly wise advice. Suganya Elango and her team at Newgen Knowledge Works brought my dream to reality on the printed (and electronic) page.

Finally, dear reader, let us return to the theme of texts. Primary texts—or more typically, their instantiations in print—form the bedrock of any deep history of ideas. We accept each text as a gift handed down to us, through the years, from its author. Texts set out an argument, reveal assumptions, and open a window into the author's world. Sometimes we glimpse the author's personality: even on the printed page, we do not mistake Tertullian for Augustine, nor Erasmus Darwin for Gilbert White. Moreover, a printed work bears evidence of skill and craftsmanship: of its author to be sure, but also of its illustrator, editor, printer, and bookbinder. Institutions and individuals have preserved books through natural disasters, wars, and periods of intolerance. For all these reasons—commitment to historicity, gratitude to our antecessors, respect for the printed word—I present texts as I find them. My translations are literal rather than literary. Items in the reference list hew closely to their original form, with minimal stylistic concessions for capitalization, italicization, and forms of diacritical marks. Nor have I retrofit italics onto scientific names or embedded titles. Even so, we shall often enough find ourselves in contested areas of orthography, for which I accept full responsibility.

Mark Ragan
Brisbane, Australia

Acknowledgement: copyrighted material

The following illustrations and textual matter are presented with the written permission of the copyright owners. Formal statements of license or permission are presented in the captions (for Figures) and endnotes (for textual material).

Figure 9.1, from HR Turner, *Science in medieval Islam* (1995), page 193, by permission of University of Texas Press. It builds on a hand-drawn figure reproduced in U Weisser, *Buch über das Geheimnis der Schöpfung und die Darstellung der Natur (Buch der Ursachen) von Pseudo-Apollonius von Tyana* (Aleppo, 1979), at page 371 in the Arabic section, but was substantially adapted by Michael Graham, and translated by Michele de Angelis, for the University of Texas publication.

Figure 9.3, from the 1992 facsimile, edited by J Telle, of the anonymous *Rosarium Philosophorum* (Frankfurt, 1550). Facsimile originally published by VCH Verlagsgesellschaft mbH (Weinheim). Republished by permission of John Wiley & Sons Ltd.

Figure 16.2, from WS Macleay, *Horæ entomologicæ* Part II (1821), at page 318, as re-drawn by Michal De-Medonsa in A Novick, *Journal of the History of Biology* (2016). Republished by permission of Springer Nature.

Figure 22.1, from RH Whittaker, *Quarterly Review of Biology* (1959), at page 217. Republished by permission of University of Chicago Press.

Figure 22.2, from RH Whittaker, *Science* (1969), at page 157. Republished by permission of American Association for the Advancement of Science.

Figure 22.3, from L Margulis, *Evolution* (1971), at page 244. Republished by permission of John Wiley and Sons.

Figure 23.2, from CR Woese, *Microbiological Reviews* (1987), at page 231. Republished by permission of American Society for Microbiology.

Epigraph introducing Chapter 2, from the poem "Yr awdil vraith" ("Diversified song") attributed to Gwion, son of Gwreang (*floruit* Sixth century CE), from the late Sixteenth-century Peniardd manuscripts, as interpreted by Robert Graves in *The white goddess. A historical grammar of poetic myth*. London: Faber & Faber (1961), at page 154. Quoted by permission of Carcanet Press Ltd.

Quotation from Surapala in Chapter 2, from the *Vrikshayurveda* by Surapala, as translated by N Sadhale, *Asian Agri-History Foundation Bulletin No. 1* (1998), at page 58, by permission of Asian Agri-History Foundation.

XX ACKNOWLEDGEMENT: COPYRIGHTED MATERIAL

Epigraph introducing Chapter 7, from the poem "I died a mineral" by *Jalāl ad-Dīn Rūmī* in the translation of AJ Arberry, from *Routledge Revivals: Classical Persian Literature* (1958), at page 241, by permission of Taylor & Francis Ltd UK.

Other images have been sourced from publications that are in the public domain (or otherwise free from copyright) owing to their date of publication and/or the year of their author's death. Many of these can be accessed online thanks to significant investment of the sponsoring institution, and (in many cases) the Biodiversity Heritage Library, the Internet Archive, or the Wikimedia Foundation. Permission has been obtained where required, and conditions attached to their reuse have been honoured. The institutions are acknowledged in the figure captions. Particular thanks are in order for the following:

Frontispiece, from PJF Turpin, in JLM Poiret & PJF Turpin, *Leçons de flore. Cours complet de botanique* (1820) provided by Real Jardín Botánico de Madrid.

Figure 9.2, from a particularly fine copy of N Le Fèvre de La Boderie, in the introduction to *L'harmonie dv monde, divisee en trois cantiqves* (1579), with permission of Niedersächsisch Staats- und Universitätsbibliothek (SUB) Göttingen.

I also acknowledge private individuals who have made images available, but wish to remain anonymous. In some cases, electronic means have been sparingly used to lighten the background, sharpen the contrast, and/or eliminate fold marks, foxing, dust spots or shadows. Figure 16.4 has been redrawn; Figure 21.1 has been redrawn, and translated in part; and the labels in Figure 21.5 have been translated.

1

The earliest Nature

We have been transported one million years back in time. From our vantage point, the panorama of the vast African plain unfolds before us; morning mists rise as the sun begins its ascent. Herds of zebra graze silently, alert for lions. A small hunting party of hominids, armed with simple clubs and poles, stalks the herd. A kilometre away another party gathers grasses and seeds, watchful too for the lion. Distinguishing carnivore from herbivore was then, as now, a key survival skill; had these early hominids not recognized the distinction and adjusted their behaviour correspondingly, there might be no *Homo sapiens* today. For a hundred thousand generations hunters and gatherers, carnivores and herbivores were linked in this way. Long before there was symbolic language, or indeed the prerequisite cranial capacity, our distant ancestors must have been strongly selected to recognize and distinguish broad classes of objects in their environment: those immobile, green, and cellulosic, and those motile and suffused with blood.

We follow our ancestors back to their camp, watch them re-enact the hunt, gesticulate, dance. We cannot know precisely what they imagine themselves to be doing: teaching and learning, celebrating, appeasing the spirits? Whatever their intent, their abstracted or ritualized behaviour conveys selective advantage by bringing past success to bear on present and anticipated future needs: roots, seeds, meat, skins. Even without symbolic language, each member of the group shares, as common points of reference, the host of agents in their environment—the sun, moon, and stars; day and night; earth, sky, and horizon; wind and rain; drought and wildfire. Why should we imagine that they linked *animal* more strongly with *plant* than with *rainy season* or *hunting spirit of our clan*?

We are still painting with too broad a brush. These distant ancestors of ours were dramatically successful. Whether primarily by selection or through their own emerging perspicacity, they must have reacted differently to carnivores than to herbivores, differently to fruit trees than to grasses or fungi. Even birds and insects avoid poisonous plants. Was not the lion-hunting party constituted differently than the zebra-hunting party, the root diggers differently than the fruit gatherers? There is no a priori reason to believe that *animal*, or *plant*, elicited from them any common activity or social organization. Even if our ancestors conceptualized herbivores and carnivores as a single class, did this class include birds, fish, snakes, insects, worms? Do these too graze in fine herds in the morning mists, or stalk their prey in the tall

Kingdoms, Empires, and Domains. Mark A. Ragan, Oxford University Press. © Oxford University Press 2023.
DOI: 10.1093/oso/9780197643037.003.0001

2 KINGDOMS, EMPIRES, AND DOMAINS

grass? Do their carcasses permeate the camp with the smell of roasting flesh, fill the belly, increase fertility, appease the spirits?

Primitive concepts of natural entities

As we shall see in subsequent chapters, for century upon century all manner of authorities—philosophers, naturalists, *littérateurs*—have juxtaposed the terms *animal* and *vegetable* (or *animal, vegetable,* and *mineral*; or these plus *man*) with confidence often verging on nonchalance. One might almost be forgiven for concluding that the concepts and distinctions implied by each of these words alone, and by their juxtaposition, must somehow be self-evident. But is this, in fact, the case? Have high-level concepts of *animal* and *vegetable*[1] truly been with us since the beginning?

The question has received surprisingly little attention. In *Geschichte des Ursprungs der Eintheilung der Naturkörper in drey Reiche,*[2] Johann Beseke identified Emmanuel König as the first (in *Regnum animale,* 1682) to refer to minerals, vegetables, and animals as *the three kingdoms of nature*.[3] In fact (as we shall see in due course), the first such description dates to the first decade or two of the Seventeenth century. The three-volume *Histoire naturelle générale des règnes organiques* (1854–1862) by Isidore Geoffroy Saint-Hilaire[4] offers far more detail, not all of it accurate. More recently, Susannah Gibson introduced her delightful account of mid-Eighteenth-century natural history with a brief synopsis of the views of Aristotle, Pliny, Linnæus, and others on animals, vegetables, and minerals.[5] More typically, however, treatments such as Samuel Haughton's *The three kingdoms of nature* (1868) offer no historical context whatsoever.

How, then, were animals and plants conceptualized long ago? The very nature of this question requires us to look far beyond written sources. Our argument has multiple interdependent strands: none is entirely unproblematic or fully definitive, but most lead in the same direction. In this chapter we consider the earliest possible lines of evidence, including from cave drawings and other early figurative art, comparative linguistics, folk taxonomies, creation myths, totemism, animal deities, and anthropomorphized plants, up to the earliest written records. We shall find only hints of evidence for significant antiquity of *animal* and *vegetable* as static high-level concepts. Instead, we uncover a diverse, fluid conceptual landscape in which every boundary is transgressed again and again: a landscape not of fixity, but of metamorphosis. But let us not get ahead of our story.

Early figurative art

Three million years ago, australopithecines wielded stones as tools; carved tools had appeared by half a million years later. Inevitably, little can be inferred about their probable use, or the mindset of those who used them. Flint-tipped spears obviously useful for hunting appeared very much later, towards the end of the Upper Palæolithic, an era we can glimpse *via* figurative art—etchings, paintings, and carvings, sometimes in combination—across more than one hundred present-day countries.

Abstract representations in red ochre at Blombos Cave, South Africa, date to 75000 BCE.[6] The earliest known figurative art comes to us from the Upper Palæolithic: cave art depicting a native pig (Sulawesi, at least 43500 BCE); hunting scenes, including humanoid images with facial and other bodily features of animals (Sulawesi, 41900 BCE); a standing human figure with the head of a lion, carved in mammoth ivory (Swabia, 38000 BCE); and engravings of aurochs, ibex, horses, and mammoths (France, 36000 BCE). Modern humans carved fertility-figures from mammoth tusk at 40000–35000 BCE, and fashioned others in ceramic around 24000 BCE. A long-eared owl is engraved at Chauvet (31000–28000 BCE), and other birds (perhaps geese) at Cussac (around 25000 BCE) and Roucadour (around 24000 BCE) in France. Images of fish, crocodiles, turtles, wallabies, emus, and other animals are common in early Australian rock art.[7]

Representational art flourished in Europe during 17500–9000 BCE. Best known are the magnificent cave drawings in southwestern Europe including at Lascaux (around 17000 BCE) and Altamira (14800–13100 BCE), with powerful horses, aurochs, bison, and large animals depicted by the thousands. We may speculate on what these drawings meant to their creators, but they do not constitute a zoology in any modern sense. The cave floor at Lascaux is littered with reindeer bones, for example, but no reindeer are depicted on the walls—although they are at Le Placard (17500 BCE) and Les Combarelles (12000 BCE). Images of snakes, insects, birds, or plants are uncommon. At Lascaux in particular, images of different animals are not distributed at random across the walls and ceilings, but instead are sometimes associated spatially in ways that may reflect shared ecosystems, seasonality, or "male" and "female" animal types.[8]

Human images are infrequent in European cave art of this era, and unlike the anatomically accurate animals, are greatly simplified. Some are associated with the hunt, while others probably had a ritualistic or shamanistic significance. In particular, "sorcerer" figures combine human with animal features: a humanoid torso with antlers (Trois Frères, 14000–13000 BCE), a spear-hunter with a bird-like head or mask (Lascaux), bison-headed (Gabillou) and seal-headed men (Cosquer, 17000 BCE). At Lascaux we also find a two-horned chimæra with "the body of a woolly rhino, the shoulder of a bison, the head of a lion and the tail of a horse".[9] Humans become the centre of attention in the later Spanish Levantine rock art (8000–3500 BCE). We see them hunting with bow and arrow, engaging in battle, dancing, tending domesticated animals, and collecting honey. Birds, fish, spiders, and bees are occasionally depicted, plants rarely so. At several sites, women are depicted working in fields, holding digging sticks, poles, or plants.[10] A famous mural at Selva Pascuala shows mushrooms, perhaps the psychoactive *Psilocybe*, and (if so) may point to its ritual use.[11]

4   KINGDOMS, EMPIRES, AND DOMAINS

Symbolic language and folk taxonomies

Our capacity for language is one of the more-credible reasons why we humans deem ourselves superior to other earthly beings. The genes that enable us to vocalize had counterparts in the genomes of Neandertals and Denisovans.[12] In light of research with chimps, great apes, and marine mammals, our superiority is perhaps less clear-cut than we once proudly assumed. Even so, human language is surely unequalled in descriptive power, nuance, and abstraction. But the challenge our ancestors faced was great indeed: to conceptualize and express the richness of nature and natural phenomena.

Adam gave names to cattle, fowls, and beasts—names that refer, of course, not to individuals but to groups: naming is classifying.[13] Egyptian papyri refer to oxen, cattle, goats, asses, lions, hounds, crocodiles, wild fowl, corn, sycamores, shrubs, thorns, hedges, and melons.[14] The Egyptian *Book of the dead* adds the ape, ram, hippopotamus, antelope, cat, panther, dogs, jackal, serpent, phœnix (*bennu* bird), ostrich, duck, heron, hawk and other birds, fish, tortoise, scarab, scorpion, worms, palm tree, lotus, barley, figs, orchards, and grass.[15] From cuneiform characters on clay tablets we know that 5000 years ago, Babylonians distinguished at least thirty kinds of fish, hundreds of terrestrial animals, and at least 250 kinds of plants. Lists of plants and medicines in the royal library at Nineveh ran to more than 400 names or expressions.[16]

Primary types could be aggregated into more-inclusive groups. The *Book of the dead* refers to *beasts* and *beasts of the field*.[17] According to Reginald Thompson, in the Ras Shamra (Ugarit) texts (13000–11000 BCE) *šam-mu* referred to plants in general, whereas in Assyrian texts it could mean *plant, grass, vegetable*, or *mineral drug*.[18] In old Sumer, names for animals were written with a cuneiform prefix that could place them within a more-inclusive group.[19] Some Babylonian tablets distinguish fish from other water-dwellers; on others, plants are arranged into trees, herbs, spices and drugs, and cereals. Large quadrupeds (horses, camels, asses) are sometimes listed separately from small ones (dogs, hyenas, lions), and fruit trees are grouped by shape of their fruit.[20] *Genesis* refers to quadrupeds, birds, fishes, and reptiles, as do the Babylonian and Hebrew narratives of the Great Flood. The earliest known epic is that of Gilgamesh, the legendary fifth ruler in the first postdiluvian dynasty of Uruk in Mesopotamia, during the Fourth millennium BCE.[21] The story is replete with names of animals, plants, and precious stones encountered by Gilgamesh and Enkidu. Several passages speak of aggregate groups: "all living creatures born of the flesh", "all the trees of the forest", "all long-tailed creatures", "the beasts we hunted", and "small creatures of the pastures".[22]

A system of aggregated and subdivided groups is the beginning of a hierarchy. Only slightly later than Gilgamesh is an alabaster vase from Warka, now in the Iraq Museum in Baghdad.[23] This meter-high vase bears three registers of decoration: the top shows the goddess Inanna, the middle a procession of porters bearing offerings, the bottom a row of sheep or goats above, stalks and heads of grain beneath. Is this a statement that god, man, animal and plant are arranged in hierarchical order and comprehend all life? Or is it something simpler, the record of a harvest ritual or feast day?

Around the world, native tongues are often rich in words for plants and animals. As often as not, the named groups correspond acceptably well to modern taxa.[24] However, folk taxonomies are limited to local flora and fauna, tend to be shallow (that is, have few intermediate ranks), and as pre-theoretical systems often deal poorly, or not at all, with organisms that are anomalous or share features of different groups, such as bats. Aggregate groups can be broad (bird, fish, tree) or narrow, and to outsiders may seem arbitrary. Moreover, folk taxonomies tend not to have a name for the group—*animal* or *plant*—at the very top of the hierarchy. In the anthropological literature, this group is called the *unique beginner*.[25] In the Mayan (Tzeltal) language, which has a unique beginner (*canbalam$_1$*) for *animal*, the term excludes man; different informants variously included or excluded bats and/or armadillos.[26]

Of course, absence of a standard term does not necessary imply absence of the corresponding concept. Tzeltal lacks a word for *plant*, but numerous expressions contrast plants with members of other domains. Plants "don't move" (*ma šhihik*), whereas animals do; they "don't walk" (*ma šbenik*), "are planted in the earth" (*ʔay c'unulik ta lum*), and "possess roots" (*ʔay yisimik*). Names for types of plants are joined with the numerical classifier *tekh*, whereas animal names are used with the numerical classifier *hoht*, and names of humans with *tul*.[27] Variations on these themes can be multiplied. Early explorers recorded no words for *animal, plant,* or *vegetable* in Aztec, Algonquin, Cree, or Yoruba.[28] Dawson[29] reported the same for three Australian aboriginal languages, although each had inclusive words for *bird* and *insect*, and distinguished *freshwater fish* from *saltwater fish*. Greenland Eskimos have general terms for *plants, fish,* and *sea animals*, while their word for *living* can also mean *an animal*.[30] Adam named cattle, fowls, and beasts, but it is not recorded that he coined the terms *animal* and *plant*. On the other hand, Gilgamesh's *all living creatures born of the flesh* surely comes close to being a unique beginner.

Finally, folk taxonomies often lump small, superficially undifferentiated beings into a catch-all group of "bugs", whether because such creatures are difficult to examine, intrude little into human lives and livelihood, or (following Atran) fall far short of the glorious standard of comparison, man himself.[31] The rich vocabulary Australian aborigines deploy for insects,[29] many of which serve as sources of food, provides a telling counterexample. But lumping residual forms under a single name conflates heterogeneous types, while leaving the boundaries of *animal* or *plant* vague or unexamined.

Perhaps our propensity to aggregate and subdivide arises ultimately from the hierarchical nature of human society;[32] in any event, taxonomy is "everywhere as old as language".[33] Beyond this it is risky to generalize: textual materials are fragmentary, languages evolve and die, and we—as anthropologists, archæologists, biologists, or translators—necessarily seek, and find, reference points in our own societies, time, and languages. Perhaps the scribes of Sumer or Thebes considered it

Creation myths

By their nature, creation myths are stories of transformation. A recurrent theme from many lands and cultures is that of the Earth Mother: humans were fashioned from earth, clay, dust, or rock, or arose from a spring.[34] Maimonides connected the name of the first man, Adam, to *adamah*, "earth".[35] A goddess shaped Enkidu, Gilgamesh's companion, from clay,[36] and Prometheus fashioned man and woman from clay. The Qumran scrolls declare man to be "moulded clay" and "a creature of clay".[37] Enoch dreamt not only of man but also of diverse animals and birds, springing up from the earth.[38]

Other creation myths link man with animals. Storyteller "Jim" Mulluk of the Ngulugwongga people in northern Australia begins the story of Manark the mother kangaroo and Memembel the porpoise as follows: "In the Dreamtime when animals were the people, the Porpoise was a woman called Memembel." Similarly, the story of Micherin, the dingo: "A long time ago when animals were people the Dingo was a black fellow"; and the story of Murroo, the dugong: "In the Dreamtime when animals and trees were people, the Dugong used to be a beautiful girl."[39]

A close relationship between mankind and other earthly beings lies at the heart of totemism, "that curious system of superstition which unites by a mystic bond a group of human kinfolk to a species of animals or plants".[40] We take the word *totam* from Ojibwa, but totemism is, or was, more widespread, not only among North American native societies but also in Australia, Africa, and elsewhere.[41] The totem itself has been interpreted as an imagined intermediate between the group and its ancestor, although totems include not only mammals but also fish, salamanders, caterpillars, plants, seeds, fire, wind, stars, planets, and the sea.[42] In other traditions, our link with creation or the heavens is instead represented as a tree, seen not as a totem (a symbolic ancestor or intermediate) but instead as an organic bridge between earth and heaven.[43]

Whatever the nature of these imagined relationships—and we would be foolhardy to generalize too aggressively in the face of such diversity—it is clear that even in conceptualizing ourselves, human imagination has not been overly constrained by the boundaries of what have been recognized as the animal, vegetable, and mineral kingdoms.

Animal deities and anthropomorphized plants

Diverse societies have depicted gods, deities, and other sacred figures as animal or part-animal. Ancient Egypt was notably well-supplied with animal gods including Horus, with the face of a hawk; Anubis, with the head of a canine; Sekmet, with the head of a lion; and Thoth, with the face of a heron.[44] An Eighth-century BCE relief from ancient Khorsabad depicts a Babylonian water-god, man above the waist, fish below. Indians had Ganesha, the elephant God; Cambodians Garuda, a winged monster with the face of a bird and the feet of a beast; Cretans the Minotaur, Greeks Pan, and so on. The winged Lion of St Mark graces Chartres Cathedral.

Hybrid creatures inhabited not only the sacred world but also the profane.[45] Gilgamesh encountered two "awful guardians" at the mountain of the sun, part man, part lion, and with the tail of a scorpion.[46] The sphinx of Thebes dates to about 1400 BCE; winged and sometimes human-headed bulls adorned walls and gates in imperial Assyria and Persia. The phœnix served both pagan and Christian myth, appearing in the *Physiologus* and in authors as diverse as Ovid, Tertullian, Tacitus, Celsus, Philostratus, and St Ambrose.[47] Chinese ritualistic bronze vessels dating to the Second millennium BCE depict composite animals, and ancient Chinese texts described hybrid beasts "unhampered by conventional genetic laws";[48] to this day the merlion, with the head of a lion and the body of a fish, symbolizes the city-state of Singapore. Mythology and literature are replete with griffins, manticores, ægopitheci, arctopitheci, leontopitheci, cynocephali, centaurs, fauns, satyrs, silenians, sea-wolves, sea monsters, eals, gorgons, sirens, harpies, mermaids, lamias, tritons, nereids, winged serpents, yales (centicores), and other chimæric beings. Other creatures, although not hybrids per se, certainly lay outside the standard boundaries of humanity: lares, genii, wood nymphs, foliots, airies, Robin Goodfellows, and trolls.[49]

Hybridization could draw upon not only animals and man but also the physical world. According to the *Iliad*, the west wind Zephyrus fathered wind-swift mares. Astolpho's horse Rabican was born of "a close union of the wind and flame, and, nourished not by hay or heartening corn, fed on pure air". The wind is personified in Bengalese tradition, the sky in many cultures, and signs of the zodiac are even today depicted as animals. Herodotus relates that Egyptians believed fire to be *thurion emphochon*, a live beast.[50]

Although mankind has seen fit to worship an appalling diversity of zoomorphs, we seem to have drawn the line at plant deities.[51] Nonetheless, a herb conveyed immortality to Gilgamesh,[52] and mushrooms have been held sacred in various religions. Many early systems of medicine were centred on a fundamental sympathy between medicament and patient; diverse plants were considered to share these sympathies, while minerals were largely excluded from the Western pharmacopœia until much later.[53]

The most human-like of plants was undoubtedly the mandrake (genus *Mandragora*), which was widely held to occur in two forms, male and female. Various myths, some rather gruesome, speak of its growth from human sperm or urine spilled upon the ground. Often explicitly described as a hybrid be-tween person and plant, mandrakes were reputed to be so "full of animal life and

8   KINGDOMS, EMPIRES, AND DOMAINS

consciousness that they shriek when torn out of the earth". They could bring fertility not only to women but to female elephants as well.[54]

Physical objects have also been deified, personified, or otherwise assigned properties more usually associated with animals. In the Neoplatonic tradition, heavenly bodies were beings or souls. Saint Paul called the sun, moon, and stars *creatures*, and in this was followed by Origen, Jerome, and other Fathers of the Church.[55] Giordano Bruno later described the heavenly bodies as animals possessing not only sensation but also intelligence—indeed, intelligence perhaps greater than our own.[56] The Fourth-century BCE *Epinomis*, earlier misattributed to Plato, posits that there are five principal types of creatures, each associated with an element: the visible gods (*i.e.* the stars) associated with fire; creatures of the earth (men, animals, and vegetables) linked through legs, belly, or roots to the earth; *daimones* of the æther; a race of æry beings; and one of watery beings, perhaps the nymphs.[57] In many cultures, anthropomorphic concepts of sexuality, fecundity, birth, and death have been projected onto the earth, and expressed as various rites and mysteries. Some involved mining and metallurgy—for example, the Malaysian belief that ore bodies are alive and can actively elude miners precisely as animals can elude hunters.[58] Others surrounded agriculture, horticulture, and the fertility of the earth.[59]

Transformation and metamorphosis

Our early hominid ancestors must have noticed not only regularities among individuals—the basis of classification—but also how these individuals change. Wind and clouds turn to rain, plants sprout up from bare earth, tadpoles become frogs, grubs and caterpillars are reborn as beetles and butterflies, and all flesh journeys "from dust to dust". The Neolithic horses and deer at Lascaux are painted with seasonal coats corresponding to the mating and rutting seasons. It is hardly a surprise that from every age, land, and people come abstracted tales of transformation and metamorphosis. Perhaps more surprising is the frequency with which these changes transgress every boundary, including those that might separate animals, vegetables, and minerals.

A Neolithic mural at Çatal Höyük (6500–5650 BCE), in present-day Anatolia, depicts the fertility goddess having given birth to the head of a bull (the male god).[60] Dumuzi, primary consort of the Sumerian fertility-goddess Inanna, was transformed into a snake.[61] In the Egyptian *Book of the dead*, the deceased can be transformed into a golden hawk, a lily, lotus, phœnix, heron, serpent, crocodile—indeed into whatsoever form he pleases.[62] Lot's wife became a pillar of salt, and Aaron's rod a serpent.[63] Athena turned the Gorgon Medusa into a winged monster with reptile locks. Empedocles famously stated that "I have been ere now a boy and a girl, a bush and a bird and leaping dumb fish in the sea."[64] In *The Papyrus of Ani* the deceased states that his physical body does not decay after death, but is instead

THE EARLIEST NATURE 9

transformed into a spiritual body symbolized by the *bennu* (phœnix). Budge translates the hieroglyphs "I grew [germinated] in the form of plants."[65] Although hardly a taxonomic statement, this does imply a class of *things that germinate*.

In *Metamorphoses*, Ovid announced his "intention . . . to tell of bodies changed to different forms" and went on to relate more than one hundred transformations.[66] Some sixty-five of these speak of transformations among animals, men, demigods, and gods, with man-to-bird being a favourite; sixteen describe the change of animal, man, or god into vegetables or vegetable products. There are twenty metamorphoses from animals into minerals, earth, or waters; six apotheoses of animal or god into a star; the wooden ships of Troy are changed to sea-nymphs; and three minerals are made animal (as is the entire city of Ardea). Wood is changed to stone, seaweed to coral, and the spear of Romulus into a tree.[67] In *The golden ass*, Apuleius tells of a man who was first transformed by witches into an animal, then returned to human form after an ecstatic vision of the goddess Isis.[68]

Writing in the Seventh century CE, Isidore of Seville allegorized classical mythological monsters, and was noncommittal on many reported transformations of men into animals. Other transformations—including that of Diomedes's companions into birds—he accepted as not mere "fabled fictions" but proven by "historical evidence". He accepted too that bees can arise from cattle, locusts from mules, and (citing Ovid) scorpions from crabs.[69] Isidore was made a saint (indeed, a Doctor of the Church), but the Church remained cautious about transformation, on the one hand celebrating the rite of transubstantiation (see below) while on the other condemning, or worse, witchcraft and most forms of magic. Indeed, the canon law collated by Burchard of Worms in about the year 1020 expressly rejected the possibility that people could be transformed into animals[70]—a move that neither prevented the persecution of witches, nor diminished the appeal of tales of magic.[71] Much later, Thomas Browne rejected all reports of human metamorphosis except that of Lot's wife.[72]

Transformations involving non-human animals were, of course, quite another matter. Belief that flies, bees, wasps, scorpions, and the like were generated from animal carcasses seems to have been widespread in the ancient world; we return to this idea in subsequent chapters. Such beliefs persisted undiminished during the Middle Ages. Petrus Bonus, for example, wrote in about 1330 CE that "nature generates frogs in the clouds, or by means of putrefaction in dust moistened with rain, by the ultimate disposition of kindred substances . . . . The decomposition of a basilisk generates scorpions. In the dead body of a calf are generated bees, wasps in the carcass of an ass, beetles in the flesh of a horse, and locusts in that of a mule."[73] Even in Linnæus's time, many Norwegians apparently believed that rats fell from the clouds.[74]

Plants could undergo transformations as well. Theophrastus discussed— sceptically except for wheat and barley—the apparently common belief that crop plants could degenerate into weeds. In the Second century CE, Galen's father, Nicon, conducted experiments that confirmed this changeability. Scaliger claimed to have witnessed the transmutation of wheat into barley. Likewise, the ivy *Hedera helix* was widely thought to transmute at maturity into a *Cissus* bush.[75] Similar folk beliefs persisted into the Twentieth century.[76] Minerals, particularly ores, could of course be forced to undergo transmutation: such was the basis of smelting and, later, alchemy.[77]

A case might be made that transformationist ideas underlie the beginnings of scientific explanation. Thales of Miletus, considered the first man of science, believed that there was but one substance, water, which took on variety of states. Anaximenes and Diogenes held that there was only air, but it could also exist as fire, water, and earth. Heraclitus and Hippasus believed the same for fire. Whatever the identity of the fundamental substance, it could take on other forms and appearances.[78]

Belief in transformation and metamorphosis extended far beyond Mediterranean shores. The *Nabatæan agriculture* relates ancient examples of transmutation among water, soil, animals, vegetables, and man.[79] The Icelandic sagas relate that sorcerers could turn men into animals; similar beliefs are known from Siberia and other northern regions.[80] Witches turned themselves into animals from Lapland to India, Alaska to Paraguay.[81] The Mi'kmaq creator-god Glooskap turned his evil brother into a mountain.[82] Wang Ch'ung wrote in *Lun hêng* that frogs could be transformed into quails, sparrows into clams.[83] The Taoist *Chuang tzu* opens with the metamorphosis of the fishlike *khun* into a birdlike *phêng*.[84] The legendary farmer Hou Chi was born from a footprint, the legendary ruler Ch'i was born from a stone, and the *Chin kuang ming ching* records that flowing water can become fish.[85] In old Japan, a badger transformed itself into a woman.[86] Metamorphosis is a recurrent theme throughout the *Thousand and one nights*; it was often in animal form that *djinns* approached humans.[87] The *Quran* reports that Allāh cursed idolaters, transforming them into monkeys and swine.[88] The *Kitāb Ahbar az-zaman* of pseudo-Mas'udi mentions a crab-like fish that, removed from water, turns to stone.[89] Even today, transformation remains a staple of popular culture, from Pinocchio to Kafka to Spider-Man.[90]

A recurrent variant is lycanthropy, the transformation of a man into a wolf or other animal. Wolves are favoured throughout much of Europe, while the fox, bear, serpent, tiger, hyena, lion, alligator, leopard, or jaguar have been preferred elsewhere. Such transformations were usually associated with demonic or satanic agency, although gods, apostles, and saints have sometimes been credited with such powers. Triumphant Gilgamesh rejected the advances of Ishtar, queen of heaven, by reminding her of previous lovers she had turned into animals, including a shepherd she had made into a wolf. Jove himself, disguised as a wanderer, transformed evil king Lycaon into a wolf.[91]

Metempsychosis, reincarnation, and anamnesis

Not only can the body be transformed but also, perhaps more importantly, the soul can associate itself with different bodies. Indeed—whether because supposed evidence was more abundant, or disproof more difficult—belief in the transmigration of souls (metempsychosis) has been particularly widespread.[92] According to Herodotus, the belief originated in Egypt:

> The Egyptians . . . were also the first to broach the opinion, that the soul of man is immortal, and that, when the body dies, it enters into the form of an animal which is born at the moment, thence passing on from one animal into another, until it has circled through the forms of all the creatures which tenant the earth, the water, and the air, after which it enters again into a human frame, and is born anew. The whole period of the transmigration is (they say) 3000 years.[93]

Pythagoras taught that human souls could migrate into the bodies of animals. Life was breath, and the portion of the air that animates the body (*i.e.* the soul) was returned to the air when the body dies; thus any creature (human or beast) that breathes can inhale a soul.[94] Pythagoras forbad the killing of animals for food. As related by his admiring if not always accurate biographers, Pythagoras justified this injunction in different ways, including reference to his belief that a human soul might be resident in the animal. By some accounts, his injunction against eating certain fish seemed to be weaker. Plants, of course, do not breathe, but it is unclear to what extent Pythagoras distinguished clearly between plants and animals. Iamblichus, however, noted that Pythagoreans were enjoined from eating mallow because "it is the first sign of the sympathy between heavenly and earthly beings".[95] Heracleides of Pontus and Empedocles used the terms *zoa* and *emphycha* to include plants, and Heracleides is said to have believed that his soul could migrate into plants as well as into animals.[96]

In *Phaedrus*,[97] Plato notes that soul can pass from man to beast, or *vice versa*, in relation to the moral character of its former host. The soul cannot typically return to its source in less than ten thousand years, although the guileless and true soul of the philosopher, and that of the lover who is not devoid of philosophy, can complete its cycle in only three thousand. Later Platonists and the Neoplatonists accepted metempsychosis, although disagreeing over details.[98] If indeed souls can be exchanged between men and animals, then perhaps men and animals are essentially homogeneous.[99] Christian writers did not welcome this conclusion; Tertullian, in particular, devoted eight chapters of his *De anima* to an attack on metempsychosis.[100,101] Christian Neoplatonists of the Fifteenth and Sixteenth centuries typically interpreted the ancients' words as allegory, or accepted that human souls could sometimes become trapped or imprisoned in the bodies of animals; others rejected metempsychosis altogether. But for Giovanni Pico della Mirandola, metempsychosis represented the mutability of man's essence, his potential to become a plant, animal, celestial being or angel, or to be unified with God.[102]

Metempsychosis appears in some, although not all, branches of the Jewish kabbalistic tradition. For example, in 1585 Rabbi Barukh Abraham da Spoleto of Modena claimed that after bodily death, a sinful soul would migrate to the body

12   KINGDOMS, EMPIRES, AND DOMAINS

of an animal. Abraham Yagel disagreed, instead accepting only that there could be superficial physiognomic changes that fell short of actual transmigration. But he admitted that a human soul could be imprisoned "in the body of a beast or animal, fowl or snake, tree or grass or mineral" for up to twelve months, and noted that "all the children of Israel acknowledge that metempsychosis and migration of the soul [exist] in the order of divine justice".[103]

Transmigration of souls plays a role in other traditions too. In the first half of the Eleventh century, Abū ʾl-Rayḥān Muḥammad ibn Aḥmad al-Bīrūnī reported that Hindus believed Nature to be pervaded by *purusha*, a spirit that flows through matter, through the different shapes of plants and animals, "ascending through them according the rising degrees of their animation. This finally leads to the process of common metempsychosis." Al-Bīrūnī went on to report that his fellow countryman Abu-Yaʿkub believed that metempsychosis instead "proceeds in one and the same species, never crossing the limits and passing into another species".[104] Belief in *karma*, a metempsychosis after which the soul experiences happiness or misery in successive rebirths, predates Buddhism.[105] Buddhist *karma* differs in that happiness or misery is a consequence of one's morals or ethics, not of whether one has or has not performed certain rituals or sacrifices. *Karma* transcended the human form: "grass, shrub was I, worm, tree, full many a sort of beast, bird, snake, stone, man and demon.... In every species born, Great Lord!"[106]

Indeed, the supposed continuity of the soul has often been contrasted with the transience of the bodies through which it passes in succession. Passage into a new body is described as *reincarnation*. Sometimes the soul has, or imparts to its new bodily host, a recollection of its previous incarnations (*anamnesis*).[94,107] According to John Dillon, every Platonist and Pythagorean believed in reincarnation. He quotes Albinus:

> There is no other way in which learning could come about than as a result of the recollection of things that one had known in a previous existence. For if we formed general concepts from a survey of individual objects, (a) how could we make a comprehensive survey of particular objects, seeing that they are infinite; or (b) how could we form such concepts on the basis of only a few? For we would be deceived, as for instance if we concluded that only what breathes is an animal.[108]

Thus, for Albinus, there are beings of the category *animal* that nevertheless do not breathe.

Transmutation and transubstantiation

Neoplatonists often distinguished between *transmutation* and *transubstantiation*. In the former, the accidents (*e.g.* shape, colour, taste) change, but substance (the *materia prima*) remains unchanged. A leaden colour could be transmuted to a golden colour and, depending on one's understanding of what gold "is", something that acquires the qualities of gold (particularly its most important property, colour) could in some sense actually *be* gold.[109] By contrast, transubstantiation is a change of substance without a concomitant change of accidents:[110] after consecration, the bread and wine still look, feel, and taste like bread and wine, but the material within the form is no longer the material of bread and wine, but rather that of the body of Christ. Bread and wine are of course derived from plants, but were believed to become flesh, as God had taken on the body of man.[111] Transubstantiation in this specific sense became an article of faith for Catholics in 1215.[112]

* * *

In this chapter we have found little evidence that *animal* and *vegetable* were well-delimited unique beginners in the earliest conceptualizations of the natural world. Instead, we have seen that humans, animals, plants, and physical objects were thought able to form hybrids, change into one another, exchange souls, and take on each others' substance or properties. Under such circumstances, little or no separate intermediate territory can exist, and the issue of intermediate organisms could scarcely arise.

Our focus, in the remainder of this book, will be on the philosophical, religious, and scientific traditions of Europe and the Mediterranean region over the past twenty-six centuries. Less is known of historical ideas about animals, vegetables, and minerals outside the classical and European spheres of intellectual influence. It would be immensely interesting to search for ideas of Third Kingdoms of Life in other traditions; but this will have to await another author. Before we turn our focus to the Mediterranean, however, let us briefly examine some relevant ideas in Indian Subcontinental, Buddhist, Confucian, Taoist, Mohist, Japanese, and Manichæan traditions.

2

Eastern Nature

Solomon obtained
In Babel's tower
All the sciences
   Of Asia's land.
                    —Robert Graves, *Diversified song*[1]

From the dawn of European civilization, poets and sages reflecting on the origin of practical and occult wisdom have turned their thoughts to the Middle East, India, and beyond. Astrology, arts of healing, the Eleusinian and Orphic mysteries, alchemy, and much else has been traced (rightly or wrongly) to Sumer, the Indus valley, or Cathay. To be sure, cultural and intellectual exchanges have occurred between Asia and Europe; some are attested historically, others through linguistics or genetics. We have already noted the (supposed) Egyptian origin of the notion of metempsychosis, and will encounter others in due course. Within the broader landscape, however, let us ask specifically: were concepts of animals, vegetables, and minerals established in eastern cultures, philosophies, or religions prior to Western contact? Did such ideas flow in one direction or the other? What, indeed, can be said about ancient Asian ideas of nature?

The Indian subcontinent

Little is known of the earliest civilizations in the Indian subcontinent. Archeological, linguistic, and genetic evidence indicates that human migrations into the Indus River basin from Persia (about 4000 BCE) and from the Caspian area (about 2000 BCE) were followed by mixing between northern and southern populations.[2] As a consequence, Sanskrit and Hindi are part of the Indo-European language family, whereas the Dravidian languages of southern India (Tamil and Telugu) are not. In the Indus basin, substantial cities at Harappa and Mohenjo-Daro grew up from about 2600 BCE and flourished until about 1900 BCE. Among the carved seals at Harappa are those depicting composite animals, *e.g.* of a bull and an elephant, or

Kingdoms, Empires, and Domains. Mark A. Ragan, Oxford University Press. © Oxford University Press 2023.
DOI: 10.1093/oso/9780197643037.003.0002

a human and a tiger. Neolithic rock art in west-central India likewise depicted not only animals and humans, but also human figures with animal heads.[3]

The Vedas, ancient wisdom-texts revered in the Hindu tradition, were at first transmitted orally in a now-lost language which gave rise to Sanskrit. The earliest texts constitute the *Rigveda*, composed around 1500–1000 BCE. We may discern in its mantras the stirrings of enquiry about the origin of the universe and its relation to the human body. Several types of plants (trees, herbs, shrubs, creepers, grasses) and animals (serpents, fish, birds, wild beasts, cattle) are mentioned, although not in the creation mantras per se.[4] Words translated as *animals* and *plants* appear by themselves, and in limiting constructions such as *creatures of the air, and animals both wild and tame* or *plants and trees*.

Later Vedic works introduce the individual soul and its endless journey (*samsara*) through cycles of birth, misery, and death. According to two early Upanishads (around 900 BCE), in the beginning the world was "only water"; the earth, atmosphere, sky, gods, men, beasts, birds, grass, trees, animals, worms, flies, and ants are all "just water solidified".[5] Another extensively analogizes man and tree, with skin as bark, blood as sap, and so forth.[6] Plants possess a soul, eat and drink, experience pleasure and pain, and "retain their consciousness inward".[7] The *Mahabharata*, an early Sanskrit-language epic, goes even farther, concluding that plants and trees possess life, perceive touch, have ears and can hear, possess vision, smell and taste, feel pleasure and pain, and require sleep.[8]

The *Vrikshayurveda* is a text on gardening and plant life. A Sixth-century CE text mentions the existence of such a body of teaching, and a Thirteenth-century CE compilation draws on the same tradition; but a comprehensive Veda was unknown until the early 1990s, when a manuscript fitting this description was discovered in the Bodleian Library, Oxford. It is not possible to date the manuscript more precisely than 1000–1400 CE; its author, Surapala, was said to be a prominent physician.[9] Its 325 verses discuss trees and gardens, soil and groundwater, seed, planting, protection and nourishment of plants, plant diseases and their treatment, and land use; some 170 species of plants are mentioned. Verse 45 identifies the four types of plants as "*vanaspati, druma, lata*, and *gulma*";[10] yet the term translated here as *plants* does not include trees, which are extensively discussed elsewhere in the *Vrikshayurveda*. At least six verses give recipes for changing one species to another; thus

> A seed of any variety, freely rubbed with the bark of mango creeper, jasmine, *dhataki*, and *madhavi* mixed with the milk of a she-goat and then sown in a pit, filled with soil dug up from around the roots of trees belonging to different species and thereafter sufficiently sprinkled with the powder of sesame and barley, and (the seed so sown) watered with curd and milk grows into the respective creeper.[11]

Other traditions survive in the Subcontinent, among which Jainism is notable here for its teaching that life forms can be ranked in ascending order by the

16 KINGDOMS, EMPIRES, AND DOMAINS

number of senses they possess. All physical bodies, even rocks and lumps of soil, have life, breath, and the sense of touch. Molluscs add the sense of taste, crawling insects a further sense of smell, and flying insects the sense of sight; higher animals can also hear, and some can think.[12] A life force (*jīva*) resides in each being; when its body dies, the *jīva* is reborn into a higher form, or descends to a lower one, depending on the *karma* it has attracted. In the so-called *prākrit* texts, composed in Middle Indo-Aryan languages other than Sanskrit,[13] beings are likewise ranked by how many senses they exhibit, although details (*e.g.* the number of senses) often differ. Terms translated as *animal* and *plant* are prominent in the earliest chapters of the first book of the *Ākārānga sūtra*,[14] which dates to the Fourth or Fifth century BCE.

The extent to which the development of Subcontinental natural-historical thought has been influenced by external sources has been, and remains, a contentious issue. It seems unlikely that primary evidence will be found that predates the migration of 4000 BCE. Evidence from rock art, stone and brass artefacts, the Vedic oral tradition, and perhaps other domains situates certain ideas in the Indus basin and/or farther south well before Darius I conquered Sind (late Sixth century BCE). If (as has been claimed since antiquity) Alexander sought out Brahmin philosophers for discussion and debate during his occupation (327–325 BCE),[15] a degree of Hellenic influence cannot be ruled out. Obstacles to further research include those associated with any oral tradition (*e.g.* meanings hidden in the play of sounds); words whose meaning is no longer known (as in the *Rigveda*); the lack of resilient media for written records (tree bark was commonly used); and difficulties in dating the texts that have survived.

Buddhism

Gautama Siddhārtha, the Buddha, lived from 563 to 483 BCE[16] in the northeast of the Subcontinent. Over the subsequent dozen centuries Buddhism split into several movements and schools, some of which enjoyed state support—notably, in the extensive Maurya Empire under Ashoka (ruled *ca* 268–232 BCE). Two monks, Dharmaraksa and Kasyapa Matanga, were invited to China by Han Emperor Mingdi in about 67 CE, and over the next millennium thousands of Buddhist texts were translated into Chinese.[17]

The Buddha presented his teaching as focusing on practises (the Eightfold Path) that allow the devotee to escape from the endless cycle of rebirth and suffering. His philosophy was based on the attainment of knowledge, and respect for reason and truth; in practise, however, these factors were more than counterbalanced by Buddhism's emphasis on the transitory nature of existence, the need for self-annihilation as a prerequisite for spiritual advancement, and a general distraction from the immediate surroundings. Moreover, some things, including the possibility of a vital principle, were considered unknowable. In such a tradition, some branches of enquiry might flourish (*e.g.* atomic theory, arithmetic, and medicine), but Buddhism was not generally conducive to natural history, including what today would be known as biology.[18] Within the Madhyamkia school of Buddhism in particular, there arose the idea that since everything is in perpetual change, nothing is real and all is illusion—a notion obviously inimical to modern scientific enquiry.[19]

China: the common tradition

The Fifth century BCE was a chaotic period in Chinese history, one of wars and social upheaval. Three major philosophical traditions arose during this period: Confucianism, Taoism, and Mohism, developing within a broadly common heritage and sharing, to varying extents, certain fundamental concepts. Perhaps the most fundamental shared concept is that of *tao*, later given literary form by Laozi[20] (see below). *Tao*, or *the way*, is the source of all things, the force by which the universe arose from chaos; it gives life to all things, and encompasses the distribution of powers in nature according to which heaven rules earth, seasons follow one another, day alternates with night, and growth leads to decay. One attains *tao* by harmonizing one's life with nature, as heaven unites itself with earth to complete the seasonal cycle.[21] *Tao* was seen as complete, all-embracing, the whole; nothing lacks *tao*. Indeed, a famous passage from the *Zhuangzi* emphasizes that *tao* is found in animals (represented by the ant), vegetables (wild grasses) and minerals (an earthenware tile).[22]

Equally prominent in early Chinese thought is the concept of a duality or polarity of power, *yin* and *yang*. It appears in the earliest Chinese medical text, the *Huang ti su wên nei ching*: "the principle of Yin and Yang is the basis of the entire universe. It is the principle of everything in creation. . . . Heaven was created by an accumulation of Yang; the Earth was created by an accumulation of Yin. . . . Yang ascends to Heaven; Yin descends to earth."[23] The interplay of *yin* and *yang* was said to have given rise to the *wu hsing*, or five fundamental elements: water, fire, metal, wood, and earth. This idea, credited to Zou Yan (about 350–270 BCE),[24] was further developed by the so-called Naturalist philosophers. Zou taught that each element dominates another in a continuing cycle called *mutual* [or *cyclical*] *conquest*: wood overcomes earth, metal overcomes wood, fire overcomes metal, water overcomes fire, and earth overcomes water. Mutual conquest came to be widely accepted, and was later formalized in the short treatise *Wu ti tê* (*The virtues by which the Five Emperors ruled*) that may date to the late Third century BCE. Although the Mohists (see below) argued that the five elements do not perpetually overcome each other, the idea was nonetheless largely taken over into Neo-Confucianism.[25]

There gradually evolved a series of symbolic associations among these five elements and "every conceivable category of things in the universe which it was possible to classify in fives".[26] Associations with classes of living animals (*zhong*) were primarily developed by a group of agricultural Naturalists called the Yue Ling group; the *Yue ling* (*Monthly ordinances*), a lengthy section of the *Liji* (*Book of rites*), is also found in other books in complete (in the *Da dai li ji*) or abridged form (*Huai nan zi*). *Scaly* (fishes) became associated with wood, *feathered* (birds) with fire, *naked* (man) with earth, *hairy* (mammals) with metal, and *shell-covered* (invertebrates) with water. According to the *Da dai li ji*: "birds and fishes are born under the sign of the Yin, but they belong to the Yang. This is why birds and fishes both lay eggs. Fishes swim in the waters, birds fly among the clouds. But in winter, the swallows and starlings go down into the sea and change into mussels."[27] The *Da dai li ji* refers to worms that "live in the earth, have no hearts and do not breathe" as animals; silkworms too were considered animals. In all, there were said to be 360 kinds of each of the five types of animals.

Another tradition in early China sought to organize knowledge of natural history as the basis for a *materia medica*. The earliest known *Pen tsao* texts date to the Fifth century BCE, although a link is sometimes traced back to the legendary "heavenly husbandman" Shen Nong, who not only taught the Chinese people to cultivate farmland and domesticate animals, but also founded traditional medicine by tasting hundreds of plants and recording their effects on his body.[28] The 278 extant *Pen tsao* texts[29] describe a broad spectrum of natural sources. The early *Shen Nong pen tsao ching* arranged these from minerals through plants to animals, perhaps reflecting the influence of Han *fang shih* or alchemists,[30] while the later *Pen tsao ching chi chu* of Thao Hung-Ching (about 500 CE) grouped them instead into "natural categories" (*wu lei*): minerals, herbs and trees, fruits and vegetables, cereals, insects and animals, and "things that have a name but are not used in medicine".[31] The richly illustrated *Hsin hsiu pen tsao* (659 CE), regarded as the first national pharmacopœia, recognized nine categories, breaking out birds, mammals, fishes, and invertebrates.[32]

Joseph Needham and colleagues record the usage of *plants* either by itself, or in the combination *vegetable drugs*, in *Pen tsao* texts from the Fourth through Seventh centuries CE.[33] However, it is sometimes unclear whether the authors of *Science and Civilization in China* are claiming that the primary texts actually employed characters that correspond to modern concepts of *animal*, *vegetable*, or *mineral*, or whether these ideas were superimposed by later commentators or academics, or indeed by the authors of *SCC* themselves.[34]

Finally, external ideas influenced Chinese thought as well. Zhang Qian journeyed far into central Asia as an emissary of the Han court in the Second century BCE.[35] As mentioned above, Buddhism appeared in China by the middle of the First century CE, well after the establishment of the three major philosophical schools discussed below, presenting a challenge to both Confucianism and Taoism that was still being played out more than a millennium later.[36] The *zhi* (or *shi*) associated with animal life (see below) may be Buddhist-inspired: the name of the Gautama's clan, Sakyamuni, can be rendered as *Shi ga mou ni*.[37]

An analogy between man and tree begins the famous reply of Hui Neng (later to become the sixth and last Zen patriarch of China) in a poetry contest with Shen Xiu:[38]

Shen Xiu:	*Hui Neng:*
Our body is the Bodhi tree;	There is no Bodhi tree
Our mind is a mirror bright.	Nor stand of mirror bright.
Carefully we wipe them clean	Since all is void, all empty,
And let no dust alight.	Where can the dust alight?

Other outside influences spread into China from the Second century BCE, with the opening of what would become known as the Silk Road.[39] Over time, this trade route brought to China not only exotic agricultural plants, goods, and technologies but also (in addition to Buddhism) Nestorianism, Manichæism, Zoroastrianism, and Islam. Contact with the Byzantine empire—hence with Hellenic, Neoplatonic, and other ideas from the Mediterranean area—was extensive by the Fourth century CE, nearly a millennium before the journeys of Marco Polo.

China: Confucianism

Kōng Fūzi (Confucius) was born in the State of Lu, in what is now Shandong Province, in 551 BCE, and died in 479 BCE. Although he was a rationalist and accepted the existence of a moral order in the universe, his teachings focused on man and human society, particularly in submission to an authoritarian state, and Confucianism became a conservative force opposing a scientific approach to nature and the development of technology.[40] His teachings are preserved in the *Lún yǔ* (*Conversations and discourses*), better known as the *Analects*.[41] These state that "we become largely acquainted with the names of birds, beasts and plants" from the Book of Poetry.[42] This is not quite "animals and vegetables": *beasts* presumably excluded birds, and the pictograms translated *plants* are those for *grasses* and *trees*.[43] A similar formulation is given in the Confucian *Chung yung*.[44]

In the *Xunzi* (*Book of Master Xun*), Xun Kuang arranged these groups in ascending order: "water and fire have subtle spirits [*qi*] but not life [*seng*]. Plants and trees have life [*seng*] but not perception [*zhi*]; birds and animals have perception [*zhi*] but not a sense of justice [*yi*]. Man has spirits, life, and perception, and in addition the sense of justice; therefore he is the noblest of earthly beings."[45] This startlingly Aristotelian formulation was written in the mid-Third century BCE, a century after Aristotle but a century before the Silk Road was established. Similar concepts appear in later Confucian teachings,[46] including those of Zhu Xi, the greatest of the Five Philosophers and the twelfth of the Twelve Confucian Sages:

20 KINGDOMS, EMPIRES, AND DOMAINS

Someone said: Birds and beasts, as well as men, all have perception and vitality [*zhi-jue*], though with different degrees of penetration. Is there perception and vitality also in the vegetable kingdom?

[Zhu Xi] answered: There is. Take the case of a plant; when watered, its flowers shed forth glory; when pinched, it withers and droops. Can it be said to be without perception and vitality? Chou Tun-I refrained from clearing away the grasses from in front of his window, because, he said, "their vital impulse is just like my own". In this he attributed perception and vitality to plants. But the vitality of the animals is not on the same plane as man's vitality, nor is that of plants on the same level as that of animals.[47]

China: Taoism

Taoism was elaborated by philosophers who, unlike Confucius, shunned the courts of feudal princes during the period of wars, instead withdrawing to the forests or mountains to meditate. Thus, while Confucianism focused on one's obligations to authority, Taoism emphasized the unity of nature and man's subservience to natural forces. Taoism was distrustful of reason and logic, and showed more influence of ancient Chinese mysticism, shamanism, and alchemy than did Confucianism; but at the same time, it was more empirical, observational, and anti-authoritarian. According to Feng Yu-Lan, Taoism was "the only system of mysticism which the world has ever seen which was not profoundly anti-scientific".[48] The most important Taoist work is the *Tao te ching* (*Canon of the virtue*), attributed to the perhaps fictitious Laozi but probably compiled in the late Fourth century BCE.[49] Next in importance is the *Zhuangzi* of Zhuang Zhou, according to which the *tao* "produces vital energy, and this gives birth to [organic] forms; all the myriad things [reproduce their kind] shape giving rise to shape".[50] Nonetheless, Taoists came to deny the fixity of biological species:

All *zhong* [species] contain (certain) *ji* [germs]. These germs, when in water, become *jue* [an otherwise unspecified minute organism]. In a place bordering upon water and land they become (lichens or algae, like what we call the) "clothes of frogs and oysters". On the bank they become *ling-xi* [perhaps a certain plant]. Reaching the fertile soil the *ling-xi* become *wu-zu* [a kind of plant]. The roots of this give rise to the *ji-cao* [unknown, currently used to denote a beetle larva]; the leaves become *hu-die* [a butterfly] or *xu* [a kind of crab]. The *hu-die* later changes into an insect, born in the chimney-corner, which has the appearance of newly formed skin. Its name is *zhu-duo* [identity unknown]. After a thousand days the *zhu-duo* becomes a bird called *gan-yu-gu* [precise identity unknown]; the saliva of which becomes the *si-mi* [apparently a type of insect]. The *si-mi* becomes a *shi-xi* [wine-fly], and from this in turn comes the *yi-lu* [apparently another type of insect]. The *huang-guang* [yet another type of insect?] are produced from the *jiu-you* [apparently a type of insect]. *Mou-nei* [mosquitoes] are produced from rotting *huan* [a kind of insect]. *Yang-xi* [another type of insect?] paired with the *bu-xun-ru-zhu* [an insect parasitic on bamboos?] produces the *jing-ning* [unknown, perhaps related to dragonfly or cicada?], which produces the *zheng* [at least later meant leopard], which [ultimately] produces the horse, which [ultimately] produces man. Man again goes back into the germs. All things come from the germs and return to the germs.[51]

Although empirical and focused on nature, Taoism was unsystematic both conceptually and methodologically, and Taoist philosophers did not develop a hierarchical classification of what they observed in nature. Discussions touching on mineral, vegetable and animal life were admixed with mysticism, as in the early *Guan zi*:

> The earth is the origin of all things, the root and garden of all life . . . water is the blood and breath of the earth, flowing and communicating [within its body] as if in sinews and veins . . . [water] is collected in the heavens and on earth, and stored up in all things. It is produced amidst metal and stone, and collected in all living beings. It is thus mysterious and magical. Being collected in herbs and trees, their roots grow in measured increase, their flowers in due profusion blossom, and their fruits get measured ripeness. [Being collected in] birds and animals, they get their form and flesh, their feathers and furs, their clearly marked fibres and veins.[52]

But often the mystical component dominated, not only in the early writings but perhaps even more so during the Taoist renaissance of the Sixth to Tenth centuries CE. Thus

> If you know that the Tao, which is formless, can change the things that have form, you can change the bodies of birds and animals. If you can attain the purity of the Tao, you can never be implicated in things; your body will feel light, and you will be able to ride on the phoenix and the crane. . . . Being is Non-Being and Non-Being is Being; if you know this, you can control the ghosts and demons. The Real is Empty and the Empty is Real.[53]

China: Mohism

Unlike the Confucians or the Taoists—indeed, uniquely in early China—the Mohists may be considered an organized school of thought. Mohism was based on the so-called *Canon* of Mozi,[54] a remarkable and difficult work that emphasizes disputational skills, geometry, mechanics (balance, support, pulleys, walls), and optics. Internal evidence suggests that the school was based in the artisan class including builders, engineers, and specialists in military fortification. Biological

22 KINGDOMS, EMPIRES, AND DOMAINS

issues feature only peripherally among the fifty-three surviving chapters of the *Canon*. Frogs and rats are stated to undergo metamorphosis.[55] Perhaps more important is a treatment of classification, but characteristically, its text is difficult and the meaning remains unclear.[56]

Japan

Chinese texts and teachings spread into Japan during the Fifth and Sixth centuries CE. The slow grafting of Buddhist and Confucian ideas onto existing native beliefs yielded, by the early Ninth century CE, the syncretist *riōbu shintō* in which native deities were considered to be transmigrations of Buddhist divinities. Shintō, at least in this syncretized form, imposes "no clear lines between natural objects, such as rocks, trees, waterfalls, and mountains, and living creatures of all sorts, vegetable or animal, and humans, or between humans and gods."[57]

In 731 CE a Japanese scholar, Tanabe Fubito, copied the *Hsin hsiu pen tsao* (above) and carried it back to Japan. Derivative Japanese texts in natural history, called *honzō*, followed in which inorganic and organic beings were arranged as in the *Pen tsao* originals,[58] *i.e.* into six to nine groups. The Buddhist monk Kūkai used *plant* (*shokubutsu*) and *animal* (*dobutsu*) in his *Henzyō hokki shourei shu* (815 CE), but these terms do not occur in the *Honzō wamyo* (about 923 CE) or the *Wamyō ruishusho* (930 CE), encyclopædic works based closely on the *Hsin hsiu pen tsao*, nor indeed elsewhere in Japan until the Eighteenth century.[59] Ueno[60] dates the scientific study of natural history in Japan from 1613, when the fifty-two-volume *Pen tsao kang mu* (*Honzō kōmoku*) was brought over from China. Nonetheless Japanese encyclopædias and pharmacopœias of that period continued to follow Chinese texts closely. Although the *Yamato-honzō* (*Medicinal natural history of Japan*) of Kaibara Ekiken (1709) explicitly broke with the *Pen tsao* tradition, even in the "last great work of Japanese natural history", the *Honzō kouoku keimou* of Ono Ranzan (1803, revised 1847), classification still followed the *Pen tsao kang mu* of 1590.[61]

Otherwise, apart from a modest level of trade through ports in Kyūshū from the mid-Sixteenth century, Japan remained isolated until the 1850s.[62] In the decades immediately prior to the treaties of 1858, Udagawa Yōan published a small botanical text in the form of a Buddhist sutra (*Botanikakyō*, 1822) in which he divided living things into those with voluntary locomotion, and those without it; both were said to be "essentially of one nature". His *Shokugaku keigen* (1833), regarded as the first Western-style botany book in Japan, described three kingdoms (animals, plants, minerals) and generally followed the Tenth edition of Linnæus's *Systema naturae*. The Linnæan *Sōmokuzusetsu* of Iinuma Yokusai (1856) was "the first comprehensive scientific monograph that appeared in Japan".[63]

The "three kingdoms of nature" are not rooted in prehistory

So far, we have uncovered only fleeting hints that concepts equivalent to *animal*, *vegetable*, or *mineral* were recognized anywhere in the world until perhaps five centuries before the Common Era; have come down to us via an uninterrupted historical tradition; or (except in parts of China and India) encompassed the natural world in a comprehensive, mutually exclusive manner. Instead, from all corners of the world we encounter ancient beliefs in affinities between people and natural entities; ensouled stars and planets; hybrid gods and heroes; transformation, metamorphosis, metempsychosis, and reincarnation. It was a fluid conceptual landscape in which every boundary could be, and often was, transgressed.

Taxonomy may indeed be as old as language,[64] but classification at the highest level appears to be a relatively recent development. Recognition and clear delineation of *animal*, *vegetable*, and *mineral* seems to have been no more necessary for the engineers of Ashur, the priests of Thebes or the philosophers of Lu than for Neolithic hunter-gatherers.

Our investigations in upcoming chapters will be based largely on textual sources, not only within established philosophical traditions such as Neoplatonism but also in those of alchemy, religion, mysticism, literature, and poetry. But we should not imagine that mythologies, folk classifications, and traditional ways of thought disappeared. Some were taken over into the new religious and philosophical systems, while others persisted as fable, folk belief, or superstition.[65] Totemism, for example, has long since disappeared from most modern societies, but even today there is widespread belief, on ethical or religious grounds, that man is the brother of all creatures and bears special responsibility for their protection and wellbeing.[66]

Hereafter we shall focus almost exclusively on ideas about plants, animals, and intermediate beings as understood around the Mediterranean, and in Europe. We begin with the very foundations of Western thought: the Hellenic philosophical tradition—that is, in Greece and nearby lands—from about 600 BCE up to the capture of Athens by Roman armies in 86 BCE.

3
Philosophical Nature

> The safest general characterization of the European philosophical tradition is that it consists of a series of footnotes to Plato. I do not mean the systematic scheme of thought which scholars have doubtfully extracted from his writings. I allude to the wealth of general ideas scattered through them. His personal endowments, his wide opportunities for experience at a great period of civilization, his inheritance of an intellectual tradition not yet stiffened by excessive systematization, have made his writing an inexhaustible mine of suggestion.
>
> —Alfred North Whitehead, *Process and reality*[1]

Many of the philosophical foundations of Western thought—and not of Western thought alone—date to Fourth-century BCE Hellas. This was the age of Xenophon, Alexander, Theophrastus, Archytas, Heracleides, Pyrrho, and Epicurus—and of Plato and Aristotle. As Whitehead observed, one can scarcely exaggerate Plato's influence on the subsequent development of ideas central to almost every area of philosophical enquiry—form, idea, knowledge, virtue, justice, happiness, order, law, the state, soul, the cosmos, and many others.

Whatever claim might be made for Plato's contribution to European philosophy in general, applies with even greater justification for Aristotle's contribution to the philosophical basis of natural history. As we shall see, Aristotle did much to develop the conceptual framework within which the concepts *animal*, *vegetable*, and *mineral* can be delineated, juxtaposed, and required to comprehend all beings lower than Man. A key part of this framework was his recognition of continuity in nature—an idea that, recast as the Great Chain of Being, was to guide Western understanding of nature for the next twenty-two centuries. We shall also investigate some key differences between Platonic and Aristotelian doctrine regarding nature and the universe—differences that, despite centuries of attempts to unify the two philosophies under one or another form of Neoplatonism, returned to the fore time and time again.

A broader reading of fragments attributed to philosophers outside Plato's Academy and Aristotle's Peripatetic School suggests, however, that reasonably modern ideas of *animal* and *vegetable* were already afoot in the Hellenic world

Kingdoms, Empires, and Domains. Mark A. Ragan, Oxford University Press. © Oxford University Press 2023.
DOI: 10.1093/oso/9780197643037.003.0003

of the Fourth century BCE. Caution is required, as dating and attribution can be problematic. But even as we rightly credit Aristotle with precisely defining these concepts and applying them to actual beings, we must avoid overpersonalizing what may have been just another facet of the broad-based transition from a mythological to a materialistic world-view that marked Ionian thought in the Sixth to Fourth centuries BCE.[2]

We begin this chapter by examining Hellenic philosophical traditions in this revolutionary period. Although primary textual evidence is often fragmentary, we nonetheless see that this was a time in which philosophers travelled from settlement to settlement, region to region. The identifiable (and self-identified) schools of thought did not develop in isolation, and their respective influences became to some extent interwoven. This cross-fertilization extended beyond Hellas to intellectual traditions in Phœnicia, Egypt, and perhaps beyond; we must avoid delineating "Hellenic" philosophy too narrowly. Nor is only Western philosophy indebted to these fourth-century intellectual revolutionaries. Aristotle's student Alexander marched his armies into India, where he engaged learned Brahmins in philosophical debate.[3] Less than two hundred years after Aristotle's death, caravans plied the Silk Road—carrying not only goods but concepts, ideas, and ways of thought.

This chapter does not presume to survey the entire philosophy, nor even the entire natural philosophy, of the Hellenes. Instead, we focus on specific concepts that would later prove fundamental to ideas of a Third Kingdom of Life. Importantly, we shall see that, despite numerous claims to the contrary, neither Plato nor Aristotle proposed, or came close to proposing, such a Third Kingdom.

Hellenic philosophical traditions before Socrates

"Sing, goddess . . ."—these immortal words, echoing down to us through almost thirty centuries, remind us that the *Iliad* was oral tradition.[4] Papyrus was introduced into Hellas only during the mid- to late-Seventh century BCE, and several hundred years were to elapse before writing came into common use.[5] It is to the final years of purely oral tradition that we can trace the first stirrings of philosophical thought about Nature. Thales of Miletus,[6] first among the legendary Seven Wise Men of ancient Ionia, is reputed to have been the first to reflect on the primal substance of the world, deducing that water must be the most fundamental substance. According to Diogenes Laërtius, the poet Chœrilus credited Thales as the first to teach that soul—that which distinguishes the living from the dead—is immortal.[7]

Orpheus was the Father of Songs,[8] and the lost Orphic hymns related the origins and exploits of the gods and their interactions with mortals—themes common to Near Eastern traditions too. Fragments of texts from the late Sixth century BCE reveal that Heraclitus of Ephesus, Parmenides of Elea, Alcmaeon of Croton, Xenophanes of Colophon, and others were pondering the nature (*physis*) and ordering (*kosmos* or *diakosmos*) of the physical world and its relation to a cosmic deity.[9] Tradition has it that Anaximander of Miletus[10] was the first to write a treatise on nature (*Peri physeos*), of which only a single fragment remains. Later authors credit Anaximander as the first to conceptualize the soul as air or breath, and to propose that the first animals were "born in moisture", enclosed in a thorny bark; Censorinus calls these "fish or living things very similar to fish".[11] The analogy between animals and trees (thorny bark) catches our eye, but we do not have it in Anaximander's own words; there is no hint of an evolutionary scale, as each genus arose independently from the slime. Anaximander and Empedocles[12] recognized earth, water, air, and fire as the four elements, among which Thales accorded precedence to water, Anaximenes to air, Xenophanes to earth and water, and Heraclitus to fire.[13]

The word ζωή (*zoë*), meaning *life* in the sense of "an ensemble of habitual acts", came into use at about this time. According to Hall,[14] the root ζω had been in use by the time of Homer and marked the distinction between living beings and the lifeless, but Homer used it only as a participle and verb, not as a substantive (noun). It appeared as a substantive in the Seventh century BCE[15] with the cosmos, stars, gods, and humans all considered to partake of ζωή. The term seems not to have been used for the group of all animals until about 550 BCE.

During the half-dozen generations between Thales and Plato, philosophical foundations were set down not only for natural history, but more broadly for logic, ethics, politics, metaphysics, mathematics, geometry, astronomy, and other areas. Within this intellectual ferment there developed philosophical schools, some formal, others less so. Three of these are of special importance to our argument: the Pythagoreans, the Eleatics, and the atomists.

The Pythagoreans

Pythagoras of Samos—mathematician, philosopher, mystic—flourished in the Sixth century BCE.[16] According to Iamblichus, Pythagoras was the first to call himself a *philosopher*.[17] Pythagoras was reputed to have spent many years in the eastern Mediterranean lands and Egypt, where he was inducted into their mystery rites. He later settled in what is now southern Italy, then a region of Greek colonies. His followers dressed simply and abstained from meat, fish, beans, and wine.[18] As they grew more numerous, they began to suffer persecution, indeed sometimes martyrdom. In the early- to mid-Fifth century BCE the Pythagorean school divided, one branch continuing as a religious brotherhood, the other developing "a quasi-scientific doctrine of number"[19] based on an emanationist[20] cosmology:

> The principle of all things is the Monad; from this Monad there comes into existence the Indefinite Dyad as matter for the Monad, which is cause. From the Monad and the Indefinite Dyad arise the numbers; from numbers, points; from these, lines; from these, plane figures, from plane figures, solids; from solid figures

there arise sensible bodies, the elements of which are four, fire, water, earth and air. These elements interchange and turn into one another completely and combine to produce a cosmos animate, intelligent, and spherical, with the earth at its centre, the earth itself too being spherical and inhabited round about.[21]

The philosophical branch of Pythagoreanism came to posit not only the primacy of number, but also the existence of mathematical harmony throughout nature; a hierarchically ordered cosmos of which man is a microcosm; continuity and kinship among all living things; the body-soul duality; and metempsychosis.[22] It is unclear what Pythagoras himself may have taught about corporeal beings. If we can take the above passage from Alexander Polyhistor at face value, Pythagoras taught that the entire cosmos was animate and intelligent. Although we might connect his injunction against killing animals or fish for food (Chapter 1) to belief in the transmigration of souls,[23] beans and wine were forbidden too, perhaps for different reasons. Pythagoreans were enjoined against eating sea-anemones;[24] perhaps they were considered animals or fish.

Outstanding among the philosophical Pythagoreans was Philolaus of Croton, a contemporary of Socrates.[25] His *Peri physeos* is preserved in some two dozen fragments, about half of which are considered genuine.[26] One fragment, preserved in the anonymous Second-century *Theologumena arithmeticae*, reads:

> And there are four principles of the rational animal, just as Philolaus says in *On Nature*: brain, heart, navel, genitals. The head [is the seat] of intellect, the heart of life and sensation, the navel of rooting and first growth, the genitals of the sowing of seed and generation. The brain [contains] the origin of man, the heart the origin of animals, the navel the origin of plants, the genitals the origin of all (living things). For all things both flourish and grow from seed.[27]

Here we see a class of "all living things" divided, comprehensively it seems, into three partitions—man, animals, and plants. Attribution to Philolaus presents several difficulties: the distinction between intellect and sense-perception, generally attributed to Democritus, appears here suspiciously well-developed; and plant-embryo analogies related to the navel are found in Democritus and the Hippocratic corpus,[28] not to mention Aristotle.[29] But the distinction differs in detail from Aristotle's (see below), and inclusion of the fourth (comprehensive) grouping can be seen as a characteristically Pythagorean touch. If this fragment is correctly attributed to Philolaus, a tripartite ordered division of living things into man, animals and vegetables predated Aristotle by perhaps two generations.

A *Peri physeos* attributed to Ocellus Lucanus,[30] a very early Pythagorean, refers to the generation of man, the other animals, and plants. If the attribution is correct, the trichotomy *man, animal, plant* was recognized in the Pythagorean tradition two or three generations earlier than Philolaus; but as transmitted by Stobaeus a millennium later, the work is suspiciously Peripatetic. Hippodamus of Miletus,

28 KINGDOMS, EMPIRES, AND DOMAINS

a contemporary of Philolaus, refers to animals in a broad sense encompassing the gods, man, and beasts.[31]

Even in Plato's day, a wise philosopher might be circumspect in seeming to promote Pythagorean doctrines.[32] Even so, Plato allowed Socrates to praise the wisdom of Pythagoras and the lifestyle of his followers. Plato's pupils Aristotle, Speusippus, and Xenocrates debated what, if anything, their master's teachings owed at a deeper level to Pythagoras. Discussion usually focuses on *Phaedo* and *Timaeus*, the latter being highly relevant here because it contains Plato's only sustained consideration of nature.[33] We return to *Timaeus* momentarily.

The last Pythagorean settlement in Hellas disappeared in about the late Fourth century BCE, and from then until the so-called Neopythagorean revival of the First century CE we have mostly Pythagorean pseudepigrapha of unknown provenance. Pythagoreanism was revived under Nicomachus of Gerasa and Numenius of Apamea, although in increasingly formal harmonization with Platonic teachings. By the time of Plotinus, Neopythagoreanism had been subsumed into the broad stream of Neoplatonism.[34]

The Eleatics

The Eleatic philosophers Parmenides, Zeno, and Melissus[35] are best known for having explored the paradoxes of time, space, continuity, divisibility, and infinity. Their teachings emphasized logic and metaphysics, even when discussing nature and the cosmos:

> Nothing that is inanimate and irrational can give birth to an animate and rational being; but the world gives birth to animate and rational beings; therefore the world is animate and rational.[36]

In this, and in regarding the cosmos as a sphere, they agreed with the Pythagoreans. But on many other issues the Eleatics developed distinctive ideas, for example that the essence of being is unity and immutability, hence that motion, change, generation, and plurality are illusion. Being arose neither from being nor from non-being; in fact, it never came into being at all, but simply *is*. What we perceive by sense-perception is not core reality, but instead a mere name or convention, devoid of meaning. These ideas threw Hellenic philosophy into a great crisis.

The atomists

Leucippus and Democritus probably intended their atomic theories to address the philosophical crises of the Eleatic paradoxes.[37] Like the Eleatics, the atomists accepted that matter is indestructible; but unlike the Eleatics, they did not believe change to be mere illusion. Instead, they interpreted change as the rearrangement of indestructible particles. Of Leucippus only a few fragments survive, among them his statement of the principle of causation: "nothing happens in vain, but everything from reason (*logos*) and by necessity".[38] We know from Aristotle and other ancients, however, that Leucippus was the first to consider that substance was not infinitely divisible. Democritus travelled to Egypt, Babylon, Persia, and perhaps India, and upon his return wrote extensively on mathematics, astronomy, medicine, ethics, and other topics. It is unclear to what extent his thinking was shaped by Eastern ideas, *e.g.* of unity and duality. But following Leucippus, he resolved the Eleatic paradoxes by positing a limit to divisibility: *átomon*, the undivided. "By convention sweet and by convention bitter, by convention hot, by convention cold, by convention colour; but in reality [nothing but] atoms and void."[39] To critics this left little room for soul or spirit, although Democritus seems to have considered that the entire universe was infused with a *psychē* composed of the lightest, most-ethereal and most-spherical atoms.[40] Thus the universe was ensouled, although in a very different sense than supposed by Pythagoras or Parmenides, or indeed by most later philosophers.

Atomism was rejected by almost all subsequent schools of Hellenic philosophy, notably including Plato and Aristotle. It was likewise rejected, much later, by the fathers of the Christian church and by Philo,[41] but was taken up in the generation after Aristotle by Epicurus,[42] who argued that the world had come into being through chance combinations of atoms. In this he was followed by Lucretius, whom we shall meet in the next chapter.[43]

Empedocles

Many of the numerous fragments attributed to Empedocles[44] have been assigned to his poem *Peri physeos*. Understanding Empedocles has not been made easier by the fact that he wrote in verse and used an "exotic vocabulary and complex style". Like Parmenides, he viewed the objects of sense-perception as mere conventions:

> When there has been a mixture in the shape of a man which comes to the air, or the shape of the species of wild animals, or of plants, or of birds, then people say that this is to be born, and when they separate they call this again ill-fated death; these terms are not right, but I follow the custom and use them myself.[45]

Empedocles addressed the Eleatic challenge by considering change to be the reorganization of the four eternal elements (earth, air, fire, water); hence nothing actually comes into being *de novo*. Empedocles repeatedly referred to seven types of beings, and appears to arrange them into four groups, albeit not in any obvious order. He referred to the elements as the "roots" (*rizomata*) of all things, and states that from these roots spring

30 KINGDOMS, EMPIRES, AND DOMAINS

trees . . . and men and women, and animals and birds and water-nourished fish, and long-lived gods too, highest in honour.[46]

No obvious correlation exists between these types or groups and the four elements. Whether as part of his belief in reincarnation or by poetic license, Empedocles drew many analogies between plants, and animals or man: humans are "plants that grow from 'shoots'", olive trees bear "eggs", and "hair, leaves, the close-packed feathers of birds and scales on strong limbs—as the same they grow".[47] Indeed, "all things have intelligence and a share of thought"[48]—an idea with which Pythagoras would have been comfortable.

Diogenes of Apollonia

At one level, Diogenes of Apollonia also distinguished vegetables from animals, although with ambiguous wording that seems to leave room for the existence other earthly beings:

> It seems to me, to state it comprehensively, that all existing things change from the same thing and are the same thing. And that is quite clear; for if the things that now exist in this universe—earth and water and air and fire and the other things which appear as existing in this universe—if any of these were different from the others (different in its proper nature) and were not the same as they changed in many ways and altered, they could in no way mingle with one another, nor would advantage and harm come to one from another, nor would plants grow from the earth, nor animals, nor anything else be born, if things were not so put together as to be the same. But all these things, being alterations from the same thing, become different at different times and return to the same thing.[49]

At another level, of course, this passage states that earth, plants, animals, and "anything else" are all variants of the same thing, temporarily distinguishable but soon to return to a state of identity.

Socrates

> Socrates first called philosophy down from the sky, set it in cities and even introduced it into homes, and compelled it to consider life and morals, good and evil.[50]

Socrates[51] is considered the founder of moral philosophy. Not coincidentally, he may also have been the first to assert a divine purpose in creation. Aristotle credited him with introducing inductive reasoning and universal definition.[52] In arguing that only mind, not sensory perception, can grasp true ideals, Socrates can be seen as reacting against both the Eleatics and the atomists. But by focusing on soul (which partakes of divine nature) and mind, not body, Socrates diverted attention from the study of actual plants and animals. The so-called Socratic schools—the Megarians and Cynics—focused on logic and morality respectively; neither is known to have developed an analysis of natural beings.

Plato

To understand Plato[53] we must begin with his theory of *intelligible forms*. Plato argued that objects in the world around us are as shadows, fleeting, changeable, devoid of reality. Groups of like individuals do, however, partake in a common quality, the universal *form* or *idea* (*eidos*) pertinent to that group. Reality is resident in this transcendent form, not in its imperfect instantiations. The forms are in fact the *only* reality, and can be apprehended only mentally, not by sense-perception. The visible cosmos is a model of these forms, and the forms in turn are modelled on the sovereign Idea (the Good, the One, the Beautiful) that is God.[54] This deep idealism represents a departure from both Parmenides and the atomists.

Reality for Plato thus lay not in the individual, but in the corresponding idea. Valuing mathematics as a path to understanding, he scorned the arts and trades as "base and mechanical".[55] His cosmology was complex and intricately wrought; not for nothing was what we today call *physics* known for so many centuries as *natural philosophy*. But Plato showed only passing interest either in the diversity of earthly minerals, plants, or animals, or in the corresponding universal forms. His world-view is set out in one of his final works, the *Timaeus*.[56] From our vantage point the cosmology of *Timaeus* is remote and unfamiliar, but this dialogue was the "single most authoritative model of the cosmic order" through at least the Twelfth century, reaching a medieval audience mostly unable to read Greek via a Fourth-century Latin translation by Calcidius. *Timaeus* was in fact the only Platonic work known in most of Europe until *Meno* and *Phaedo* were translated in about 1156.[57]

In the dialogue Timaeus, "our best astronomer" and having studied the nature of the cosmos, describes to Socrates and his guests[58] how the universe began, down to the creation of mankind. Timaeus relates that the Demiurge, desirous that all things should be good, fashioned the world in his own image. The Demiurge put intelligence in soul, and soul in body; and the universe, the fairest and best among God's creation, is therefore endowed with soul and intelligence. It is in fact a living creature, perfect, and contains within itself all others.[59] To this spherical creature the Demiurge assigned circular motion.[60] From the divisible and the indivisible he made an essence, and by mixing these three he created soul[61] and united it centre-to-centre with body. Time too he created as a (necessarily imperfect) model of eternity, and appointed the stars to mark it out.[62]

32 KINGDOMS, EMPIRES, AND DOMAINS

Up to this point there had been only the singular Living Creature, but after the birth of time he fashioned four classes of creature within it, so that the universe would not be imperfect.[63] He first created the heavenly gods (including the visible stars) and enjoined them to continue his work according to the cosmic nature and the laws of destiny. He entrusted to each a soul constituted similarly as the world-soul, albeit diluted "to a second and third degree of purity", to be implanted into a human body. The soul that lived righteously would, upon death of the host body, return to its star to enjoy there a blessed existence. But souls that failed to live well would instead pass into a woman's body; and if it even there continued in evil, it would eventually pass into the body of an appropriate beast.[64]

Almost the entire remainder of the dialogue is devoted to a consideration of the soul and body of man, the nature of each element, space, motion, properties of bodies, sensations, passions, desire, ageing, death, disease, and the benefits of physical exercise.[65] Only at the very end does Timaeus return to the three remaining classes of creatures: birds, created of innocent light-minded men; wild beasts, fashioned from those who have ignored philosophy; and the race of fishes, shellfish, and other aquatic animals, from those so ignorant and impure as to be no longer worthy even of respiring air. Beasts are of three kinds, their attachment to the earth increasing with their foolishness: quadrupeds, polypods, and reptiles.[66] There is thus an order among creatures, from the heavenly gods downward through man, woman, birds, quadrupeds, polypods, reptiles, and finally fish and aquatic animals.

What, specifically, is the basis of this order among beings? A soul can move upward (back towards the heavenly gods) or downward (into animals in increasingly direct contact with the earth) in respect of how successfully (or not) it has achieved a harmonious balance among reason, passion, and appetite.[67] The soul is in constant motion, indeed self-moved,[68] but motion characterizes even the inharmonious and unrighteous soul: that is, the Platonic gradation among beings is not based on the presence or degree of self-motion. Or to put it the other way around: self-motion is clearly the purest form of motion, but Plato does not rank beings by their purity of motion.

The Demiurge intended the three classes of creatures (together with the heavenly gods and stars) to make the universe complete and perfect.[63] Why then did he make plants and minerals? Plants appear to have been an afterthought:

And when all the limbs and parts of the mortal living creature had been naturally joined together, it was so that of necessity its life consisted in fire and air; and because of this it wasted away when dissolved by these elements or left empty thereby; wherefore the Gods contrived succour for the creature. Blending it with other shapes and senses they engendered a substance akin to that of man, so as to form another living creature: such are the cultivated trees and plants and seeds which have been trained by husbandry and are now domesticated amongst us; but formerly the wild kinds only existed, these being older than the cultivated kinds. For everything, in fact, which partakes of life may justly and with perfect truth be termed a living creature. Certainly that creature which we are now describing partakes of the third kind of soul, which is seated, as we affirm, between the midriff and the navel, and which shares not at all in opinion and reasoning and mind but in sensation, pleasant and painful, together with desires. For inasmuch as it continues wholly passive and does not turn within itself around itself, repelling motion from without and using its own native motion, it is not endowed by the original constitution with a natural capacity for discerning or reflecting upon any of its own experiences. Wherefore it lives indeed and is not other than a living creature, but it remains stationary and rooted down owing to its being deprived of the power of self-movement.[69]

"Trees and plants and seeds" are thus living creatures, made of the material to which soul is bonded; but unlike animals, in which are found all three faculties of the soul (reason, passion, appetite), they partake only of the appetitive faculty,[70] and human soul cannot pass into them.[71] Timaeus does not use an inclusive term here to encompass "trees and plants and seeds"; elsewhere he employs the word *phyton*,[72] but it is unclear whether he means it to encompass other organisms that are today considered *plants*. Stones and minerals are not bodies into which individual souls can pass, being mere manifestations of combinations of the four elements.[73]

Thus in *Timaeus*, living beings are those that have a soul bonded to their "matter":[74] heavenly gods, men, women, birds, quadrupeds, polypods, reptiles, fish, shellfish, trees, plants, and seeds, indeed the cosmos as a whole. In man, soul is of three parts: the logical, the spirited, and the appetitive,[75] each centred in a different part of the body. Plants share only the lowest mortal soul; stones and minerals are not creatures and are not alive. Plato certainly put important concepts into play (including a hierarchy of beings), but he did not develop a technical vocabulary for them; nor did he formally delineate kingdoms either of beings or of their transcendent forms.

Plato's universe is one of motion: elements are ceaselessly transmuted into one another, souls migrate upwards or downwards, bodies age and die. But the transcendent ideas—the only reality—remain intact, and do not intergrade into one another. Any given horse might partake more or less fully in the idea *horse*, but it could not partake in the idea *lion*, much less in the ideas *oyster* or *myrtle*. In *Phaedrus*, Plato has Socrates characterize stories of chimæras, gorgons, and winged steeds as frivolous allegories,[76] thereby leaving little room for ambiguous or intermediate beings.[77]

Before leaving Plato, we must note two further points. One is methodological; we return to it in a moment. The other is one of those fecund "general ideas" to which Whitehead referred: that of *macrocosm* and *microcosm*.

Like the universe as a whole, man has body and soul. He is a compound of the mortal and the immortal, the base and the heavenly; his soul emanates directly from the body of the universe. The fire in man, small and weak and mean, is nourished and generated and ruled by the fire in the universe, wonderful in quantity and

34 KINGDOMS, EMPIRES, AND DOMAINS

beauty.[78] Indeed, in every way man is the microcosm, the universe the macrocosm. This idea, in all its mystical beauty, became a defining feature of Platonism (and even more so of Neoplatonism). It unlocked a Pandora's Box of hidden correspondences that could be imagined to mediate every facet of human existence. The doctrine of macrocosm and microcosm took root in astrology, numerology, divination, dream interpretation, fortune-telling, magic, alchemy, and just about every other form of occult practise known to gullible mankind for more than two thousand years.[79] As we shall see, the idea that the earthly and heavenly realms manifest occult symmetries based on the number *three* (the triangle; the composition of the soul from divisible, indivisible, and their essence) lay just beneath the surface as subsequent generations of philosophers, naturalists, and mystics sought to classify the living world.

We also find in Plato a conscious recognition of the power of classification, both bottom-up and top-down. The former is operative when, as part of the process of mentally apprehending a specific transcendent form, we discover its corresponding class. Many aspects of his theory of intelligible forms remained problematic, however, and at times Plato seemed unable to convince the sceptic, or indeed even himself.[80] His top-down application of logical division as a way of knowing met with more success, as in the famous example of the *aspaleintos*, or angler.[81]

Angling is an art, argues the Eleatic stranger. Of the two kinds of art (creative and acquisitive), it is acquisitive. The acquisitive is of two kinds, that proceeding by voluntary exchange and that proceeding by coercion; angling is of the latter. Coercion may be open (fighting) or secret (hunting). Hunting may be of the lifeless or of the living (animal hunting). Animal-hunting may be of land animals or swimming animals, and swimming animals may be winged (fowling) or native to the water (in which case the art is called fishing). Of fishing, there is fishing by use of enclosures, and fishing by use of hooks and tridents. The latter is divided into night-fishing and day-fishing. Day-fishing is further divided into barbing, where the blow is downwards, and angling, where the fish is hooked from below.

This example yields us a small bonus. The Eleatic stranger says that "hunting of the lifeless" has no special name, except "some kinds of diving and the like, which are of little importance". Mair[82] argues that Plato "without doubt" has in mind σπογγοφηρική, sponge-cutting. If so, then for Plato the sponge is lifeless.[83] Aristotle will express a different opinion.

Aristotle

Aristotle,[84] like Plato, based his philosophical system on a theory of forms (ideas). For both, form is reality; but whereas Platonic form is external and transcendent, Aristotelian form inheres in material things, imbuing them with reality.[85] Thus it is precisely in the world of natural things where matter, which by itself is mere potentiality, is spontaneously shaped and controlled, to greater or lesser extents, by form. Aristotelian *nature* is both this intrinsic spontaneity, and the receptiveness of matter thereto.[86] We should not be surprised at Aristotle's great interest in animals and other natural things: by studying their generation, structures, growth, and development, we begin to understand form.

Aristotle rejected the atomism of Leucippus and Democritus; indeed, his *Physica* presents the classical argument for continuity in space and time.[87] Physical dimension, hence motion, is irreducibly continuous. In place of atoms, he proposed four essential qualities of matter (hot, cold, wet, dry) which he supposed to unite in various binary combinations to produce the four elements: earth, air, fire, and water.[88] Matter itself is formless, potential without reality. In inanimate things (*e.g.* minerals) form imbues but does not govern matter, whereas in living things— vegetables, animals (beasts), and man—form governs matter. Form in living beings is *soul*,[89] and *De anima* is Aristotle's enquiry into its nature. Let us follow his argument for a moment:

> We resume our inquiry from a fresh starting-point by calling attention to the fact that what has soul in it differs from what has not, in that the former displays life. Now this word has more than one sense, and provided any one alone of these is found in a thing we say that thing is living. Living, that is, may mean thinking or perception or local movement and rest, or movement in the sense of nutrition, decay and growth. Hence we think of plants also as living, for they are observed to possess in themselves an originative power through which they increase or decrease in all spatial directions; they grow up *and* down, and everything that grows increases its bulk alike in both directions or indeed in all, and continues to live so long as it can absorb nutriment.
>
> This power of self-nutrition can be isolated from the other powers mentioned, but not they from it—in mortal beings at least. The fact is obvious in plants; for it is the only psychic[90] power they possess.
>
> This is the originative power the possession of which leads us to speak of things as *living* at all, but it is the possession of sensation that leads us for the first time to speak of living things as animals; for even those beings which possess no power of local movement but do possess the power of sensation we call animals and not merely living things.
>
> The primary form of sense is touch, which belongs to all animals. Just as the power of self-nutrition can be isolated from touch and sensation generally, so touch can be isolated from all other forms of sense. (By the power of self-nutrition we mean that departmental power of the soul which is common to plants and animals: all animals whatsoever are observed to have the sense of touch.) What the explanation of these two facts is, we must discuss later.[91] At the moment we must confine ourselves to saying that soul is the source of these phenomena and is characterized by them, viz. by the powers of self-nutrition, sensation, thinking, and motivity.[92]

36 KINGDOMS, EMPIRES, AND DOMAINS

After discussing whether these powers are complete souls, or parts of soul, and speculating on whether the mind might be a "widely different kind of soul, differing as what is eternal from what is perishable",[93] Aristotle continues:

> Further, some animals possess all these parts of soul, some certain of them only, others one only (this is what enables us to classify animals); the cause must be considered later. A similar arrangement is found also within the field of the senses; some classes of animals have all the senses, some only certain of them, others only one, the most indispensable, touch.[94]
>
> Of the psychic powers above enumerated some kinds of living things, as we have said, possess all, some less than all, others one only. Those we have mentioned are the nutritive, the appetitive, the sensory, the locomotive, and the power of thinking. Plants have none but the first, the nutritive, while another order of living things has this *plus* the sensory. If any order of living things has the sensory, it must also have the appetitive; for appetite is the genus of which desire, passion, and wish are the species; now all animals have one sense at least, viz. touch, and whatever has a sense has the capacity for pleasure and pain and therefore has pleasant and painful objects present to it, and wherever these are present, there is desire, for desire is just appetition of what is pleasant. Further, all animals have the sense for food (for touch is the sense for food); the food of all living things consists of what is dry, most, hot, cold, and these are the qualities apprehended by touch; all other sensible qualities are apprehended by touch only indirectly.[95]

Thus, earthly things are alive by virtue of being ensouled.[96] Soul (*psyche*) gives rise to and is characterized by specific characteristics or powers, of which Aristotle lists five: the nutritive (originative), sensory, appetitive, locomotive, and ratiocinative. The nutritive power has as its end the generation of a like being, so "the first soul ought to be named the reproductive soul";[97] indeed, "nutrition and reproduction are due to one and the same psychic power".[98] Soul as inherent in plants—vegetative soul—is thus characterized by this nutritive and generative power, and by it alone. Animals possess, in superaddition, the appetitive power and at least the most basic sensory power, touch; these three powers characterize animate soul, inherent in all animals.[99] In rational soul, inherent in man, all five powers are at their fullest extent.[100] There are three and only three "order(s) of living things"—vegetables, animals, and man—because the properties of their souls are distributed in a superadditive way.[101] Minerals, although organized beings, are not ensouled.[102]

Animals (beasts), moreover, differ among themselves according to the extent to which they possess the sensory and locomotive powers of animate soul. All animals possess the sense of touch, including the sense for food; taste, hearing, vision, and smell are distributed in more-complex ways. But not all animals are equally self-motile—indeed, some are quite immobile.[103] Among the five powers of soul, locomotion is unique in failing to map exactly with the boundary delineating animals from vegetables. As we shall see, later authors often misunderstood Aristotle on this point, incorrectly claiming that he distinguished animal from vegetable based on the ability to self-move, locomote or move "willingly" or "purposefully".[104]

It is true that in *Historia animalium*, Aristotle explicitly ranked animals by movement.[105] He also grouped or ranked them by number and type of feet,[106] means of bearing their young,[107] "amount of vitality",[108] goal-directed activity,[109] life in water or on land, gregarious or solitary lifestyle (and the former by presence, as among bees and cranes, or absence of a "ruler"), diet, habitat, seasonality, lifespan, susceptibility to domestication and disease, and other traits.[110] In another passage[111] he considered man to be more perfect because man "alone partakes of the divine, or at any rate partakes of it in a fuller measure than the rest". Every *genos* is divisible into contraries (*eide*), but not every pair of contraries can together constitute a *genos*; and differentiæ (*diaphores*) are not unique to specific genera, but often can be applied quite widely across animals.[112] Aristotle thus rejected Plato's dichotomous classification because it can fail to capture essences, and can break apart natural groups.[113] Whether for these reasons alone, or because he also appreciated that knowledge of animal structure was incomplete, Aristotle did not present a unitary classification of animals.[114] This did not stop innumerable later authors[115] from offering (usually without appropriate *caveat*) "Aristotelian" classifications, many of which included a "Zoophyta" intermediate between vegetables and animals. We return below to this important matter.

Although Aristotle's genera are incommensurable, they might nonetheless be compared indirectly by means of analogy, for example between vegetables and animals:

> And animals seem literally to be like divided plants, as though one should separate and divide them, when they bear seed, into the male and female existing in them.[116]
>
> Plants are the reverse of animals in this respect . . . there is a correspondence between the roots in a plant and what is called the mouth in animals, by means of which they take in their food, whether the source of supply be the earth or each other's bodies.[117]

By analogizing animals with plants Aristotle clearly intends to emphasize the vegetative, *i.e.* nutritive and generative, soul inherent in them.[118] Indeed there are animals, and parts of animals, that in these respects differ little from true plants:

> Now it is by sense-perception that an animal differs from those organisms which have only life. But since, if it is a living animal, it must also live; therefore, when it is necessary for it to accomplish the function of that which has life, it unites and copulates, becoming like a plant. . . . Testaceous animals, being intermediate between animals and plants, perform the function of neither class as belonging to both. As plants they have no sexes, and one does not generate in another; as animals they do not bear fruit from themselves like plants; but they are formed and

38 KINGDOMS, EMPIRES, AND DOMAINS

generated from a liquid and earthy concretion. However, we must speak later of the generation of these animals.[119]

For nobody would put down the unfertilized embryo as soulless or in every sense bereft of life (since both the semen and the embryo of an animal have every bit as much life as a plant), and it is productive up to a certain point. That then they possess the nutritive soul is plain. . . . As they develop they also acquire the sensitive soul in virtue of which an animal is an animal.[120]

And regarding the animal embryo: "for at first all such embryos seem to live the life of a plant", and the embryo "grows by means of the umbilicus in the same way as a plant by its root".[121] Indeed, animals "actually live the life of a plant" when they sleep.[122]

With these analogies, Aristotle poses two possible middle grounds between animals and plants. One might be delimited by absence of positive characters: plants lack separate sexes and do not bear the young internally, whereas no animal bears its young externally as a tree bears fruit. Testacea arise from neither fruit nor embryo, but from a liquid and earthy concretion. Another middle ground might be temporal: animals pass through a vegetable-like (embryonic) phase before taking on or exhibiting the animate powers. As we shall see in a moment, Aristotle follows neither tack; analogy has its limits. He instead returns to his case against atomism, finding continuity not only in pure space or motion, but now in inherent form, in the world of beings.

Aristotle's continuum among beings

Aristotle begins Book VIII of *Historia animalium* by comparing the temperaments of animals and humans, and the psychological habits of children and adults. Occasionally, he says, we see in animals a natural potentiality towards human qualities, and in children adumbrations of adults' psychological habits. Then follows the classic formulation of *synecheia*, the principle of continuity:

Nature proceeds little by little from things lifeless to animal life in such a way that it is impossible to determine the exact line of demarcation, nor on which side thereof an intermediate form should lie. Thus, next after lifeless things in the upward scale comes the plant, and of plants one will differ from another as to its amount of apparent vitality; and, in a word, the whole genus of plants, whilst it is devoid of life as compared with an animal, is endowed with life as compared with other corporeal entities. Indeed, as we have just remarked, there is observed in plants a continuous scale of ascent towards the animal. So, in the sea, there are certain objects concerning which one would be at a loss to determine whether they be animal or vegetable. For instance, certain of these objects are fairly rooted, and in several cases perish if detached; thus the pinna[123] is rooted to a particular spot, and the solen (or razor-shell) cannot survive withdrawal from its burrow. Indeed, broadly speaking, the entire genus of testaceans have a resemblance to vegetables, if they be contrasted with such animals as are capable of progression.[124]

In regard to sensibility, some animals give no indication whatsoever of it, whilst others indicate it but indistinctly. Further, the substance of some of these intermediate creatures is fleshlike, as is the case with the so-called tethya (or ascidians) and the acalephae (or sea-anemones); but the sponge is in every respect like a vegetable. And so throughout the entire animal scale there is a graduated differentiation in amount of vitality and in capacity for motion.[125]

These passages are as puzzling as they are famous, for as we have seen, Aristotle elsewhere carefully distinguished animal from plant with respect to the sensory power, especially the sense of touch.[126] He discusses these ambiguous or intermediate creatures further:

The Ascidians differ but slightly from plants, and yet have more of an animal nature than the sponges, which are virtually plants and nothing more.[127] For nature passes from lifeless objects to animals in such unbroken sequence, interposing between them beings which live and yet are not animals, that scarcely any difference seems to exist between two neighbouring groups owing to their close proximity.

A sponge, then, as already said, in these respects completely resembles a plant, that throughout its life it is attached to a rock, and that when separated from this it dies. Slightly different from the sponges are the so-called Holothurias and the sea-lungs, as also sundry other sea-animals that resemble them. For these are free and unattached. Yet they have no feeling, and their life is simply that of a plant separated from the ground. For even among land-plants there are some that are independent of the soil, and that spring up and grow, either upon other plants, or even entirely free. Such, for example, is the plant which is found on Parnassus, and which some call the Epipetrum.[128] This you may hang up on a peg and it will yet live for a considerable time. Sometimes it is a matter of doubt whether a given organism should be classed with plants or with animals. The Ascidians, for instance, and the like so far resemble plants as that they never live free and unattached, but, on the other hand, inasmuch as they have a certain flesh-like substance, they must be supposed to possess some degree of sensibility.

An Ascidian has a body divided by a single septum and with two orifices, one where it takes in the fluid matter that ministers to its nutrition, the other where it discharges the surplus of unused juice, for it has no visible residual substance, such as have the other Testacea. This is itself a very strong justification for considering an Ascidian, and anything else there may be among animals that resembles it, to be of a vegetable character; for plants also never have any residuum. Across the middle of the body of these Ascidians there runs a thin transverse partition, and here it is that we may reasonably suppose the part on which life depends to be situated.

40 KINGDOMS, EMPIRES, AND DOMAINS

The Acalephae, or Sea-nettles, as they are variously called, are not Testacea at all, but lie outside the recognized groups. Their constitution, like that of the Ascidians, approximates them on one side to plants, on the other to animals. For seeing that some of them can detach themselves and can fasten upon their food, and that they are sensible of objects which come into contact with them, they must be considered to have an animal nature. The like conclusion follows from their using the asperity of their bodies as a protection against enemies. But, on the other hand, they are closely allied to plants, firstly by the imperfection of their structure, secondly by their being able to attach themselves to the rocks, which they do with great rapidity, and lastly by their having no visible residuum notwithstanding that they possess a mouth.[129]

These acalephae (sea-anemones) are sensible, but closely allied to plants in their lack of complex structure, fixity to the substratum and absence of undigested food.[130] The description of these beings "approximat(ing) on one side to plants, on the other to animals"[131] turns up again, as we shall see, in the early Seventeenth century. Sponges are even more plant-like but in the final analysis, they too possess the sensory power:

It is said that the sponge is sensitive; and as a proof of this statement they say that if the sponge is made aware of an attempt being made to pluck it from its place of attachment it draws itself together, and it becomes a difficult task to detach it. It makes a similar contractile movement in windy and boisterous weather, obviously with the object of tightening its hold. Some persons express doubts as to the truth of this assertion; as, for instance, the people of Torone.[132]

The borderland between animal and vegetable

Aristotle was interested in the borderland between animals and plants—marine invertebrates—because unlike previous philosophers, he believed it necessary to learn from as many beings as possible. Folk taxonomies largely ignored these organisms, or dealt with them in an unnatural way. By contrast, by including man among the animal, and reasoning by analogy where necessary, Aristotle might bring these organisms, and indeed plants, into a common system.[133]

As we have seen, Aristotle described sea-anemones as simultaneously "approximating" to plants on one side, to animals on the other. For this he seized on an established word, ἐπαμφοτερίξειν, to "dualize" or "hold the middle ground between two groups",[134] and by careful explanation crafted it into a precise technical term. Some animals (the seal, dolphin, sea turtle, and others) are terrestrial but not purely so, while also aquatic but not purely so, by virtue of whether they depend on air or water for respiration, thermoregulation, and/or food.[135] Similarly bats, apes, the ostrich, hermit crabs, and other beings can by some criteria be grouped one way, by other criteria another. Elsewhere he stated that all testaceans (not acelephs alone) hold a middle ground between animals and vegetables.[136] Aristotle's usage differs from that of Plato, for whom ἐπαμφοτερίξειν denoted ambiguity or equivocation,[137] or a readiness to waver between opposite sides or opinions.[138] Early Latin translations and later naturalists attempted to capture Aristotle's concept with the adjective *anceps*—two-faced, or facing both ways.[131,139]

By some criteria these sponges, sea-anemones, or testaceans might be grouped with the vegetables. They nonetheless are (or might be supposed to be) sensate, and consequently partake of animate soul—that is, they are animals, however attenuated their sensory power. These "certain objects in the sea" approximate, or show a face to, beings immediately below them in the continuum, which happen to be vegetables, and equally to those immediately above, which happen to be undoubted animals. Like seals and bats they are dualizers, but only vis-à-vis their immediate neighbours in the continuum of beings. This is surely what it means to ascend little by little without precise lines of demarcation. Rather than disrupting existing taxa, or allowing sponges, anemones, and testaceans to be members of two taxa or none at all, Aristotle accepted their anomalies and two-facedness, and retained them as animals.[140]

The idea that certain beings might be ambiguously vegetable *or* animal, or part-vegetable and part-animal, has undoubted poetic appeal. However, these fantasies require Aristotle's word ἐπαμφοτερίξειν to carry the meaning it bore in Plato's dialogues, not the meaning Aristotle actually constructed for it. To be sure, Fourth-century BCE Greek did not yet offer a highly developed technical vocabulary for either philosophy or natural history.[141] For better or worse, his ideas would soon be known predominantly through the filter of Neoplatonism, so it was perhaps inevitable that these "certain objects in the sea" would become plant-animals, *zoophytes*. But nowhere in Aristotle's extant works does any variant of the term *zoophyte* appear.[142]

Much of the remainder of this book is devoted to this fascinating concept *zoophyta* (or, as a formal taxon, *Zoophyta*), from its origin (or origins) through philosophical traditions in the Near East, Egypt, and Byzantium into Greece, Italy, and France. As we shall see, at an early date *zoophyta* became falsely associated with Aristotelian natural philosophy—an association that persisted into the Twentieth century, as historians of science continued to purvey the myth of an Aristotelian Zoophyta.[143]

De plantis

So far, we have shared Aristotle's perspective on these "certain objects in the sea" as he looked downward from undoubtedly sensate beings. It would be most interesting to share his perspective from below, looking upward from among the vegetables as they ascend little by little towards the animate. Aristotle may or may not have written a work *De plantis*[144] but if so, it has long been lost. A book of that name once included in the Aristotelian corpus[145] was probably written instead by

42 KINGDOMS, EMPIRES, AND DOMAINS

Nicolaus of Damascus in the First century BCE.[146] This *De plantis*, which may have been intended as a commentary on Aristotle's original, presents an uneven mixture of more-or-less Aristotelian formulations and descriptive observation, together with (in the Arabic version) quite un-Aristotelian comments on how heavenly bodies influence the growth of plants. The text begins:

> Life is found in animals and plants; but while in animals it is clearly manifest, in plants it is hidden and less evident.[147]

The author takes issue with Plato, who said that plants have sensation and desire; and even more with Anaxagoras, Democritus, and Empedocles, who said that they have intellect and intelligence.[148] The vegetable soul encompasses nutrition and growth, but is imperfect, lacking sensation and locomotion.[149] The fact that plants are alive does not make them animals; just as some animals lack foresight and intelligence (but are still animals), plants lack sensation but are still ensouled beings.[150] "Sea-shells" dualize as in Aristotle:

> It is wrong for us to suppose any intermediate state between the animate and the inanimate. We know that sea-shells are animals which lack foresight and intelligence and are at once plants and animals. The only reason, therefore, for their being called animals is that they have sensation . . .[151]
>
> . . . there is a mean between life and the inanimate, because the inanimate is that which has no soul nor any portion of it. But a plant is not one of those things which entirely lack a soul, because there is some portion of a soul in it; and it is not an animal, because there is no sensation in it, and plants pass one by one gradually from life into death . . . a plant is imperfect, and, whereas an animal has definite limbs, a plant is indefinite in a form, and a plant derives its own particular nature from the motion which it possesses in itself.[152]

The author of *De plantis* divides plants into trees, bushes, herbs, and vegetables,[153] and considers fungi and mushrooms to be plants that lack branches and leaves.[154] A number of features differentiate animate from inanimate beings, *e.g.* stones fail to release water upon heating, and do not grow.[155] An interesting contrast is drawn between plants and earthenware pottery, but the treatment tends to be superficial, with "greenness" said to be the most common characteristic of plant life.[156] Disappointingly, we learn little about the ascent towards the animate; the author does not tell us which plants most closely approximate to sea-shells, nor in what ways they do so.

Theophrastus

Theophrastus of Eresos succeeded Aristotle as head of the Peripatetic School.[157] A prolific author on diverse topics,[158] he is best known today as the father of philosophical botany. His *Historia plantarum* and *De causis plantarum* were translated into Latin in the mid-Fifteenth century, and served (often word-for-word) as the basis of botanical texts for the subsequent hundred years.[159] Although Aristotle certainly pointed the way, Theophrastus was the first to assemble "everything that vegetates", including seaweeds and fungi, into "plants".[160]

Historia plantarum begins with the contrast between plants and animals. Plants lack ἤθη and πρᾶξεις—"conduct" and "activities".[161] Even fundamental concepts such as "part", he wrote, may not be strictly comparable between the two groups: fruits are reckoned to be parts of plants, whereas animals' young are not considered parts of the parent; and parts of plants may be of indeterminate number.[162] Although some internal parts of plants have no special technical names, they are referred to by imperfect analogy with parts of animals: thus muscle, veins, and flesh (wood). Bark and core are "properly so called", and some (perhaps less careful than Theophrastus himself) call the innermost part of the woody stem its "heart" or "marrow".[163]

Theophrastus arranged "all or nearly all" plants into trees, shrubs, half-shrubs (or "undershrubs"), and herbs.[164] Features such as root, stem, branch, twig, leaf, flower or fruit, bark, core, and fibres or veins "belong to a plant's essential nature", but pertain especially to trees, the most perfect among plants. Fungi and truffles lack every one these features, but he showed no reluctance in considering them plants as well.[165] Theophrastus described various fungi, lichens, and seaweeds with sufficient accuracy to allow them to be identified today, and for the most part discussed aquatic vascular plants separately from algae.[166] He may have considered corals among the "well known" aquatic plants that "turn to stone";[167] this ancient myth persisted well into the Eighteenth century. But "sponges and what are called *aplysiai* and such-like growths are of a different character".[168] Unfortunately, Theophrastus does not elaborate upon this assertion.

Theophrastus provided the earliest description of a "sensitive plant", probably *Mimosa polyacantha*.[169] He considered it—along with palms, gum acacia, olive- and plum-trees, and aquatic plants—one of the "most conspicuous things" peculiar to Egypt, but did not accord it any special proximity to animals. Theophrastus maintained trees to be the most perfect plants; their perfection was based on the presence of essential characters (root, stem, etc.) and was not based on a continuum extending from stones to man.

Stoic and later triadic divisions of soul or beings

Aristotle considered minerals, plants, and animals to be what they are, with characteristic powers and limitations, as a consequence of the type of soul they possess. Although soul is not corporeal it nonetheless is a kind of nature, inhering in natural things. For the Stoics, however, all was material.

44 KINGDOMS, EMPIRES, AND DOMAINS

Zeno of Cition[170] divided knowledge into physics, logic, and ethics. Like Heraclitus, he accepted the pre-eminence of fire, and Stoic physics emphasized *pneuma*, the fiery soul that infuses and holds together the form and qualities of the cosmos. Yet fire was an element: for the Stoics everything, even souls and gods, was material, indeed corporeal. Thus, in their different ways, six major Hellenic philosophical schools—Pythagoreans, Eleatics, atomists, Platonists, Aristotelians, and Stoics—accepted that the cosmos is ensouled. Opinions differed, of course, on what this soul actually "is": the alternative duality within the One; the incorporeal within each and every body, indeed within the cosmos; the lightest and most ethereal atoms; transcendent idea, inherent form, or fiery *pneuma*.

The concept of *pneuma* was further developed by Chrysippus of Soloi.[171] For Chrysippus, *pneuma* permeates the cosmos,[172] binding in different grades or dispositions to different types of objects and thereby bringing about their physical properties and degree of orderly organization according to the level of tension (*tonos*) present. *Pneuma* which sustains structure in stones and other inanimate objects is called tenor or state (*hexis*). Plants have *hexis* plus *physis* (nature), a higher level of *pneuma* that governs their growth, nourishment, and reproduction. Animals have *hexis* (governing the bones), *physis* (governing the hair and fingernails) plus *psyche* (soul), a disposition of *pneuma* that allows animals to perceive the world and respond with actions. Rational beings—humans and gods—have these three plus *pneuma* in its greatest tension, *logos* or *nous*.[173]

The earlier Stoics engaged in lively disputation with the Peripatetics, but over time the two schools converged to some extent. Posidonius allowed a role for spirit in the human soul, although not (as claimed by Galen) the full Platonic trichotomy of body, soul, and spirit.[174] He distinguished animals, vegetables, and minerals, but recognized only the former two as unified bodies.[175] Like Plato, Posidonius attributed appetite, desire, and sensation to plants, these plus passion to animals,[176] and considered the world to be a rational animal. He viewed the universe as "a continuous ascending scale of being, in which the lower and simpler subserves the needs of the higher and more complex, until the crown of all is reached in man, whose possession of reason causes him to project beyond the sensible or phenomenal universe altogether, so as to have entry into the world of intelligibles."[177]

Unlike Pythagoreanism and atomism, which were driven underground, Stoicism remained as an ethical and political force to the end of paganism. In 176 CE Roman Emperor Marcus Aurelius, a Stoic himself, established chairs of Stoic, Epicurean, Academic, and Peripatetic philosophy in Athens. Stoic ideas were occasionally cited in regard to the composition of plants and animals.[178]

Scepticism

We conclude this chapter with what may or may not be the first appearance of the word *zoophyta*. Our story begins with Pyrrho of Elis, who is said to have accompanied Alexander to India and studied with wise men there. A man of imperturbable disposition,[179] he is regarded as the first Sceptical philosopher. Scepticism eventually split into two schools, both at first associated with Plato's Academy. Adherents of one aspired to a life based on reason, and engaged other philosophers on questions of truth and belief; by contrast, Aenesidemus of Knossos sought to develop a Scepticism that could bring equanimity in life following the example of Pyrrho. In *Pyrrhonist discourses*, Aenesidemus describes how this can be done: by setting competing appearances and thoughts in opposition to each other in one's mind, a balance (*isostheneia*) can be achieved among them, allowing one to suspend judgement and thereby achieve tranquility. Pyrrhonian Scepticism became popular among the so-called empirical school of medicine. Pyrrho left no writings and the *Discourses* have been lost, but a description of this way of thought has been preserved in *Outlines of Pyrrhonism*, written in Greek by the physician Sextus Empiricus around 200 CE.[180]

According to Sextus, there are ten modes of thought (*tropes*) by which appearances and thoughts can be put into opposition.[181] None of this would be relevant to our question, were it not for how Sextus chose to illustrate the first trope, *arguments based on the differences between kinds of animals*.[182] He tells us that because different kinds of animals (with their different sensory facilities) can be inferred to perceive a thing differently, there is no privileged viewpoint: no decision can be made, so judgement should be held in abeyance.[183] He illustrates this by naming diverse animals, beginning with some produced via what would later be called spontaneous generation:

> Thus, as to origin, some animals are produced without sexual union, others by coition. And of those produced without coition, some come from fire, like the animalcules which appear in furnaces, others from putrid water, like gnats; others from wine when it turns sour, like ants; others from earth, like grasshoppers; others from marsh, like frogs; others from mud, like worms; others from asses, like beetles; others from greens, like caterpillars; others from fruits, like the gall-insects in wild figs; others from rotting animals, as bees from bulls and wasps from horses.[184]

Animalcules is not the only possible reading for the beings which appear in furnaces. Greek manuscripts of the *Outlines* had become available in Europe by the early Fourteenth century[185] but as Latin was much more-widely understood, Henri Estienne made a Latin translation for the first printed edition (1562), employing the term *zwophyta* in the passage above.[186] A visually more-attractive edition soon followed (1569) from the press of Christophe Plantin, using Estienne's translation but with *zoophyta* in place of *zwophyta*. The bilingual Chouët edition (1621) paired Estienne's translation (again with *zoophyta*) with Greek text reading ζωόφυτα. Working directly from Greek manuscripts, Johann Fabricius produced a bilingual edition (1718) which followed Chouët with ζωόφυτα but replaced *zoophyta* with *animalia*; an "emended edition" (1842) did likewise. The Greek-only edition of Bekker (1842) gave ζωῦφια [*animalcules*] rather than ζωόφυτα,[187] as did Mutschmann (1912), who however acknowledged ζωόφυτα as the consensus reading of the five Greek manuscripts dating from 1465 to 1542. In 1888 a Thirteenth-century Latin manuscript of the *Outlines* was discovered in Paris, reading simply *aîa* [*animals*].[188] Notably, the well-regarded *Dictionarivm Graecvm* (1524) of Giovanni Crastone gave ζώδιον, ζωῦφιον for *animal parvum*.[189]

A contemporary of Sextus, Diogenes Laërtius, paraphrased Aenesidemus's first trope using the unusual word πυρίβια (*fire-beings*).[190] In any case, it is unclear why such a fire-being would be regarded as an animal-plant. Today it is accepted that Sextus wrote of animalcules, not of zoophytes: Pappenheim translates the word as *Thierchen*, Patrick as *little animals*, and Annas and Barnes as *little creatures*.[191] But what exactly are the animalcules that appear in furnaces? Fortunately, other authors provide critical detail. Pliny reports that

> in the copper foundries of Cyprus even in the middle of the fire there flies a creature with wings and four legs, of the size of a rather large fly; it is called the *pyrallis* or by some the *pyrotocon*. As long as it is in the fire it lives, but when it leaves it on a rather long flight it dies off.[192]

Likewise Cicero, Apuleius and Aelian wrote of winged animals that are born of the fire, flutter about in glowing furnaces, and die if they get too far from the heat;[193] Philo and Aelian called them πυριγόνα,[194] and Porta employed the same word (*pyrigones*) in retelling the story some fourteen centuries later.[195] Given that fish are born in water, beasts in or upon earth, and (perhaps) spirits or dæmons in air, it made sense that some being was engendered in the remaining element, fire. No description was ever forthcoming, however, and no such insect is known to modern science. Perhaps Sextus Empiricus actually wrote ζωόπυρα and was ascribing a brief life to the glowing sparks that rise above a flame. It would be ironic if these were not animal-plants, but mineral.[196]

Envoi

Aristotle's Peripatetic school flourished for less than a century after his death. After Theophrastus its leadership fell to Strato of Lampsacus, who tried to reconcile the views of Aristotle and the atomists. Thereafter it went into decline, and although names are recorded for head scholars over another 150 years,[197] "by the middle of the third century, its work was done".[198] Aristotle's manuscripts passed to Theophrastus and thence to his nephew Neleus, who is said to have buried them in a cave in Asia Minor; they were later unearthed and taken to Rome, where in about 70 BCE Andronicus of Rhodes drew together and edited what became the Aristotelian corpus.[199] Copies of manuscripts circulated in Alexandria, Syria, and the Middle East, but the Roman empire entrenched Latin throughout western Europe, and Latin translations of Aristotle's works began to appear only in the early Thirteenth century CE.[200]

Plato's Academy remained active longer, but evolved in a Sceptical direction. Relations with the Lyceum were not always warm, but by the time of Antiochus the two schools were seen as variants on the same Platonic tradition.[201] At about this time, Athens became entangled in the first war between Rome and Pontus. In 88 BCE Antiochus and other philosophers fled to Rome ahead of the advance of Mithridates,[202] and the destruction of the Academy by Sulla in 86 BCE marked the end of the New Academy. The centre of Platonism subsequently moved to Alexandria, where Heraclitus of Tyre was established by 76 BCE. It is there that we shall pick up the story in Chapter 5.

But before we do so, we must acknowledge that many texts from the period we considered in this chapter—from Pythagoras to Sextus Empiricus—were not expressly philosophic. These years saw the first popular natural history (Pliny the Elder); specialist works on hunting, fowling, fishing (Oppian), and agriculture (Cato the Elder, Varro, Columella); herbals and *materia medica* (Dioscorides, Pamphilius[203]); and books of medicine (Hippocrates, Celsus, Galen). Before we pick up the (neo)Platonic tradition, let us examine how animals, vegetables, and minerals were treated in these parallel, more-utilitarian traditions.

4

Utilitarian Nature

> [The books of Theophrastus] are not so well adapted to those who wish to
> tend land as to those who wish to attend the schools of the philosophers.
> —Marcus Terentius Varro, *Rerum rusticarum*[1]

We now turn to utilitarian descriptions of nature in Classical times. In Hellas, and even more in Rome,[2] there appeared a proliferation of works intended not as philosophy, poetry, or drama, but instead dealt with practical arts including agriculture, horticulture, medicine, geography, and a host of other technical matters—topics of immediate interest to a growing empire.

In the previous chapter we established that by the Fourth century BCE—indeed, probably earlier—terms meaning *animal* and *plant* were in use to denote broadly inclusive groups of beings.[3] Even so, in *De agri cultura*—the oldest surviving text in Latin prose—Cato the Elder[4] avoided these terms altogether. In *Rerum rusticarum* Varro[5] wrote instead of *iumentum* (beasts of burden), *pecudes* or *pecudes culturae* (cattle), or *bestia* (wild animals), or employed constructions such as *inclusa animalia quae pascantur* (animals enclosed for pasture-feeding). Columella took a different approach in *De re rustica*, using *animal* to refer not only to livestock and man but also, on multiple occasions, to insects.[6] In *De arboribus* Columella spoke of "worms and other animals".[7]

In *De natura deorum* Cicero divided living things into animals and vegetables.[8] He cited Chrysippus to the effect that corn and fruits were created for the sake of animals, and animals for the sake of man—although he used these examples to argue for the perfection of the world, not for a hierarchy in nature.[9] Acknowledging fire, air, water, and earth as the four elements, Cicero apportioned animals among their realms: "animals again are divided into those that live on land and those that live in the water, while a third class are amphibious and dwell in both regions, and there are also some that are believed to be born from fire, and are occasionally seen fluttering about in glowing furnaces."[10]

Lucretius

Lucretius is known today only through his didactic poem *De rerum natura*. Written in the mid-First century BCE, this Epicurean work was well received in

Kingdoms, Empires, and Domains. Mark A. Ragan, Oxford University Press. © Oxford University Press 2023.
DOI: 10.1093/oso/9780197643037.003.0004

classical antiquity, although later disparaged by the early Christian Church Fathers. Manuscripts circulated in Europe until the Ninth century but thereafter the work lay forgotten, perhaps because of its anti-religious tenor and argument against the existence of an immortal soul.[11] A manuscript was famously discovered in Germany by Gian Francesco Poggio Bracciolini in 1417,[12] and the work soon regained its (much) earlier popularity.[13]

In *De rerum natura* Lucretius argued that the world came into existence not by action of the gods, but through the association of elemental particles. These atoms are not only primal bodies but primordial germs, procreant, the seeds of things.[14] Once the world itself had arisen from chaos, first there arose grasses, shrubs, and trees, then birds, beasts (first as monsters and prodigies, and forms unable to survive or leave progeny; later as the mammals we see today), and finally man.[15] It is not difficult to interpret this temporal series as a broad-brush scale of nature; yet even today, according to Lucretius, animals can spring forth from the earth.[16] He nonetheless denied the possible existence, now or in times past, of centaurs, fishmen, or any other chimæra in which all parts of the hybrid body are well integrated and strong.[17]

Once rediscovered and disseminated in print, *De rerum natura* was to influence Renaissance and later thought. Its materialist physics based on random associations of atoms eliminated a role for a divine creative force, whether Neoplatonic or Christian. Ada Palmer documents how *Epicurean* became a term of abuse in the Fifteenth and Sixteenth centuries, not least between Catholics and Protestants. Luminaries as diverse as Popes Sixtus IV and Pius II, Montaigne, Ben Jonson, and Thomas Jefferson owned copies; Marsilio Ficino wrote a commentary on it (before destroying his commentary later), while Niccolò Machiavelli's handwritten transcript is preserved in the Vatican Library.[18]

Seneca

By contrast, the works of Seneca—tragedies, essays, letters, and the *Naturales quaestiones*—have remained in high regard from the time of his suicide (on orders of Roman emperor Nero) to the present day. In *Naturales quæstiones*, written about 62–64 CE, Seneca considered the nature of meteors, thunder, lightning, earthquakes, winds, and the River Nile, and offered thoughts on ethics and morals guided by Stoicism. He mentioned animals and plants by way of example, at one point contrasting "vegetable life" with "animal life".[19] Nils von Hofsten[20] claimed that the *Naturales quæstiones* influenced the young Linnæus, whose "general concept of Nature as a wonderful harmony manifesting a divine purpose is on the whole very often a reflection of ideas in Stoic philosophy."

Seneca examined the living world more deeply in his *Epistulae morales ad Lucilium*. In Epistle 58, "On being", he illustrates Aristotle's use of *genus* and *species*. Man, horse, and dog are species of the genus animal; animals, which possess life

50 KINGDOMS, EMPIRES, AND DOMAINS

and mind, and plants and trees, which possess only the former, are species of the animate; and the animate, together with the inanimate (*e.g.* rocks), constitute substance. Stoics, he wrote, would place above this another genus, "things which exist", divisible into the substantial and the insubstantial; the latter would include such "figments of unsound reasoning" as centaurs and giants. He followed with an explanation of Plato's sixfold division of "things which exist".[21] Seneca is not concerned to enumerate the beings that fall within *animal* or *plant*, although among the former he mentions "tiny animals" that can sting, presumably insects.[22]

In his concluding Epistle, Seneca reiterated that there exist four natures: those of the tree, animal, man, and God. The last two of these share the power of reason, but one is mortal and the other immortal; it is in the latter where Nature perfects the Good.[23]

Pliny the Elder

Pliny's *Naturalis historia* is the largest single book to have survived from Roman times, and for Europeans served as the main source of knowledge about the living world for more than 1500 years. Pliny himself was an *equite*, a commander of military forces and the imperial fleet, Procurator of two small Roman provinces, and a secretary of state to Emperor Vespasian.[24,25] Members of the equestrian class were often interested in the physical and natural world, and some wrote on technical topics.[25] Pliny himself authored numerous books, among which only the *Naturalis historia* has survived. He perished in 79 CE while attempting to rescue friends during the eruption of Vesuvius.

In his dedication of *Naturalis historia* to Emperor Titus (son of Vespasian), Pliny lamented that his subject matter—"the nature of things, *i.e.* life"—was sterile and hardly a beaten path to authorship, as neither Roman nor Greek had heretofore taken on the entire subject. He would nonetheless attempt to "give novelty to the old, authority to the new, a shine to the worn-out, illumination to the obscure, agreeability to the disdained, credibility to the doubtful, true nature to all and to each its own nature."[26]

Naturalis historia is a compilation, by Pliny's own estimate bringing together 20,000 facts from 200 books of 100 authors.[27] He advised that it is better consulted than read front-to-back and to this end provided a *summarium*, rather like a table of contents.[28] Although not a work of philosophy, its contents are set out in a considered order: first the universe, heavenly bodies, and elements; the geography of the known world; the history of man, his inventions and marvels; land and aquatic animals, birds and insects; trees, plants, fruits, and flowers; wine-making; medical botany, diseases and medicines made from trees, plants, and animals; magic, superstitions, and astrology; minerals including gold and silver; pigments, sculpture, and painting; gems and precious stones. Without fail, each new topic led Pliny into unexpected, often fascinating digressions.

UTILITARIAN NATURE   51

Pliny began his history of animals with man, followed by land animals in descending order based on size (and to some extent geographical range). Next came aquatic animals again from large to small, including the octopus, cuttlefish, squid, crabs, sea-snails, sea-urchins, and various shellfish.[29] Then after a series of anecdotes to show that fish and sea-worms must be sensate,[30] he arrived at the borderland between plants and animals:

> For my part, I adjudge that those [creatures] possess a sense-perception which has the nature of neither the animal nor the vegetable, but is of a certain third nature which draws upon them both [tertiam quandam ex utroque naturam habent]—I mean the sea anemones and sponges.[31]

Not unlike Aristotle's ἐπαμφοτερίξειν, Pliny's word *utroque* means *looking to* or *drawing upon both sides*. But unlike Aristotle, Pliny found "a certain third nature" in sea-anemones and sponges. He did not give it a name, nor had he encountered one in hundreds of Greek and Latin books; but his own words seem to call out for the term *zoophyte*. Pliny did not, however, accept that all aquatic creatures are equally sensate. Sponges exhibit intelligence, because when they become aware of the presence of a sponge-cutter they contract, making it much more difficult to tear them loose; they do the same when pounded by the surf. Like some plants, sponges can grow back from roots left behind on the rock, and in some cases the colour of blood remains.[32] Oysters, sea cucumbers, sea-lungs, and starfish, on the other hand, lack sense-perception and have only the nature of a bush.[33]

Pliny discussed "marine shrubs and trees" alongside terrestrial shrubs;[34] although he did not realize it, many of these are corals, or perhaps seaweeds. He reckoned mushrooms and fungi among plants;[35] likewise, he interspersed remedies that use seaweeds or fungi with those based on plants.[36] Corals (recognized as such) are animals: in Book 32 he reported they have the form of a shrub, and he referred to the "berries" of one, but the surrounding sections deal entirely with fishes and other marine animals.[37]

Sponges, though, are problematic. We have seen that he considers them to be of "a certain third nature" that partakes of both animal and plant natures.[31,38] He related that some people believe there to be male and female sponges,[39] although this would not disqualify them from being plants, as Pliny believed all plants to be of both sexes.[40] But he placed his discussion of remedies derived from sponges at the very end of Book 31, along with the medicinal properties of waters and salts; he turned to "remedies derived from the marine and aquatic animals" only in Book 32. On the evidence of how the *Naturalis historia* is divided into books, sponges are the step just beneath living beings.[41]

Not unintentionally and more than any other work from antiquity, the *Naturalis historia* was all things to all people: for Alberti a history of painting,[42] for Shakespeare a mine of fable and nature-lore,[43] for Diderot a well-styled and subtly

52   KINGDOMS, EMPIRES, AND DOMAINS

subversive panorama by a fellow *philosophe*.[44] Sometimes modern and lofty,[45] at other points uncritical and credulous, in the *Naturalis historia* Pliny had indeed "taken the measure of Nature".[46]

Herbals and pharmacopœias

Another important tradition with roots in this era is the plant-book. The very fact that a book might be wholly given over to the use of plants in agriculture, horticulture, or medicine tells us that by the last century or two BCE, it was widely accepted that plants constitute a distinct high-level group of beings—a unique beginner. However, there was not yet consensus about what was or was not a plant. Here we examine what beings were included or excluded from these earliest plant-books, and why. Some authors discuss plants in an order that implies boundaries or draws attention to ambiguous cases; others simply list them by size, or alphabetically.

We have already met the father of philosophical botany, Theophrastus. Although not the first to write on plants,[47] he was the first to enquire systematically into their nature and interrelationships of their external and internal parts. By nature, plants have roots, stems, branches, twigs, leaves, flowers, fruits, bark, core, fibres, and veins; this is true above all for trees, which are thus the standard of comparison. Yet none of these features is common to all plants, as evidenced by fungi and truffles.[48] In *Historia plantarum* Theophrastus thus proceeded from trees, the most perfect of plants, to shrubs and half-shrubs and finally to the grasses. At Book 4 he digressed to consider trees and shrubs of other lands, and aquatic and marine plants including seaweeds; he failed to return to fungi and truffles.[49] Sponges and similar growths, however, are "of a different character"—whether he meant animals or stones, he does not say.[50] In another work, *De lapidibus*, he reported that coral is "like a stone" but "grows in the sea" and "is a subject for another enquiry".[51]

Historia plantarum is often cited as a source of the herbal tradition; or indeed by virtue of its Book 9, which deals with medicinal properties of herbs, as a herbal in its own right.[52,53] There is some doubt as to the authorship of Book 9; Arber[53] suspects that it was compiled posthumously, as much to cast ridicule on the self-serving mystique around herb-gatherers and root-diggers[54] as to guide actual practise. Book 9 does not describe plants sufficiently for identification, nor is it arranged by disease, or in any consistent manner whatsoever; it is difficult to see how it alone might have been used as a working herbal.

Other plant-books were in circulation from the Fourth century BCE,[55] but the strongest claim to being the first herbal goes to a late Second- or early First-century BCE work by Cratevas, personal physician to Mithridates VI of Pontus.[56] In this work, or perhaps in a later or popular edition of it, plants were presented in alphabetical order; notably, text describing how each plant could be identified was replaced by a colour painting of the plant, while its medicinal properties were given below.[57] Pliny rightly complained that plants change their appearance by season, so no single image would suffice.[58]

The plant-book of Cratevus was a primary source for the *De materia medica* of Dioscorides.[59] Written around 70 CE, *De materia medica* remained in high esteem through the Renaissance and beyond; and not only in Europe, but also in the Middle East and Islamic lands. Its contents are arranged in a curious manner, neither alphabetic nor Theophrastian: aromatics and oils in Book 1; products of animals and herbs in Book 2; roots, juices, and other herbs in Books 3 and 4; then wines and extracts from vines and minerals in Book 5.

Dioscorides led off Book 2 with seventy-four living creatures, among them the jellyfish;[60] the sea-slug and other molluscs; the ray, eel, sea-horse, and other fish; and the cuttlefish. Although he did not say as much, we may suppose that with this grouping, he considers the jellyfish to be an animal. Fungi, a lichen, and possibly seaweeds are among herbs in Book 4.[61] The truffle, a "round root", is placed among pot herbs.[62] On the other hand, Dioscorides waits until Book 5, in the section on "metallic stones", to describe sponges and a coral that some call *lithodendron* (stone-tree)—although he allows that the latter "seems to be a sea plant".[63]

Pamphilius, a Greek physician who practised in Rome towards the end of the First century CE, compiled a dictionary of plants in which entries were arranged alphabetically.[64] *Herbarium*, an illustrated herbal attributed to Apuleius Platonicis and based on the *De materia medica* of Dioscorides and Pliny's *Naturalis historia*, circulated widely in Europe from about the Fifth century CE. The *Alphabet of Galen*, a *materia medica* set out in alphabetical order and perhaps dating to the same period,[65] likewise was known in Europe until about the Thirteenth century; despite superficial similarities, it is not related to the *De materia medica* of Dioscorides, nor does it draw on Pliny. Eventually printed in about 1481 and in 1490 respectively, these latter two works helped to extend the tradition of plant-books into central Europe. We take up this story again in Chapter 12.

The poems and fragments of Nicander fit broadly into this tradition. His *Georgica*, now mostly lost, warns that few mushrooms are safe to eat.[66] Two of his poems are extant: *Alexipharmica*, on various poisons particularly of plants, and *Theriaca*, on venomous animals including insects, spiders, and the sting ray. Fungi (an "evil ferment") are again mentioned in the former. Both poems describe remedies, usually to be prepared from plants.[67] No order is apparent among the organisms, poisons, and remedies.

Early medical texts

Long before Classical times, herbs were used to treat wounds and disease. In many lands—not only the East—healing might also require visiting a temple for dream interpretation, incantation, or magic. In early Rome, for example, a cult grew up around Febris, goddess of (malarial) fevers; three temples were dedicated to her

54 KINGDOMS, EMPIRES, AND DOMAINS

in Rome alone. Regimens of diet and exercise, studies of anatomy and physiology, and classification of symptoms became part of medicine only later. Hippocrates[68] helped to move medicine beyond theurgy, not least by promoting the ancient idea[69] that illness is caused by disharmony among opposing qualities, the four humours;[70] the physician should help nature return them to equality (*isonomia*), by gentle means if possible.

The *Hippocratic corpus*, a body of works associated with Hippocrates or in the spirit of his teaching, was mostly assembled during the Third century BCE.[71] Celsus seems to have drawn on other Greek sources from the latter part of this period for his *De medicina*, originally part of an encyclopædia that has otherwise been lost.[72] In it Celsus describes foods and drugs from animals and plants,[73] at one point referring to "hard-fibred fish of the intermediate class, oysters, scallops, the shellfish murex and purpura, [and] snails".[74] Sponges appear only as a medium for the application of liquids or steam to the body, or for soaking up blood.

Until the mid-Sixteenth century, however, the prince of classical medicine was Galen. Schooled in philosophy, on the urging of his father he became a physician (indeed, the most famous in Rome under Emperor Marcus Aurelius) and taken as a whole, his work might be called philosophical medicine. In *De placitis Hippocratis et Platonicis* he argued strongly against Chrysippus, who taught that the soul is unitary and is ruled from the heart;[75] Galen agreed instead with Plato that the soul is tripartite, and with Posidonius that nonhuman animals are endowed with its appetitive and passionate parts. However, those animals that "are not easily moved and are attached like plants to rock" possess only its appetitive facility.[76,77] The version of Stoicism then ascendant in Rome[78] held that the affairs of man were determined by external causes including those in the heavens; Galen embraced external causality but rejected its astrological component, arguing instead that macrocosm and microcosm (the human body and all its parts) were perfect, reflecting the perfection of the Demiurge. The compatibility of this view with monotheistic religions including Christianity, which had begun to make inroads into Rome, doubtlessly contributed to his subsequent esteem.[79]

Like Celsus before him, Galen grouped drugs by the type of being from which they are produced: plants and herbs, earths, stones and metals, or animals.[80] Fungi are plants, although "altogether cold and moist".[81] Oysters, cuttlefish, snails, crabs, fish, the sea-horse, an insect, salamanders, and sea-dragons are animals, while alcimonium[82] and sponges, like brines, salts, and sea-foam, are not.

In classical times books, like great statesmen, were models to copy and perhaps improve upon. Texts could be (and often were) appropriated without acknowledgement, passages inserted from other sources, words changed or removed; plagiarism was the sincerest form of respect. Copies were made on demand, with scribes paid per word; and the Roman Empire carried Latin, not Greek, into the reaches of Europe. Given that Galen's output may have exceeded three or four million words,[83] it is not surprising that his books—increasingly written in codices rather than on scrolls[84]—were abridged, excerpted, combined, translated, and retranslated to meet the growing demand, including from physicians remote from the main centres of scholarship.[85] From the Thirteenth century until authentic texts began to be published in the Sixteenth, Galenic medicine was best known through a compendium known as the *Articella*.[86] In the meantime, the *De medicina* of Celsus was rescued from obscurity, and became the first medical work of classical times to be printed using moveable type (1478).

The new availability of authentic Galen was significant for reasons beyond a much-improved standard of medicine and medical education. Galenic medicine became intertwined with Aristotelian natural history and method at Padua, which for two centuries from 1400 was arguably the leading centre of learning in Europe.[87] More broadly, the Galenic ideal of the "thinking, learning practitioner"[88] accorded well with the humanistic movement. The printing house of Aldus Manutius in Venice published scholarly editions of Aristotle and Galen, as indeed of many other Greek and Roman authors. By the close of the Sixteenth century Galen—in his own way a Platonist—was part of Renaissance Aristotelianism.

Bestiaries

The earliest animal-books likewise date to the Fifth century BCE. Herodicus of Selymbia, a teacher of Hippocrates known for his advocacy of exercise and massage in the maintenance of good health, is credited with *Peri diatios* (*On regimen*). It considers only animals that form part of the human diet, and classifies them variously by habitat, domestication, and whether they are carnivores or herbivores.

As we have seen, Aristotle did not classify animals into a single unitary hierarchy. Over the centuries various authors constructed so-called Aristotelian systems based on his works, for example dividing animals into blooded or bloodless, then the former into four subgroups and the latter into five including ὀστρακόδερμα ("mussels, sea-snails, ascidia, holothuria, actinia, sponges").[89] But Aristotle did not flesh these groups out, so to speak, with extensive lists of actual species.[90] In the early Second century BCE Aristophanes of Byzantium made an epitome of Aristotle's zoological works, listing species by name with each name followed by information about that animal's parts, generation, lifespan, and character[91]—intentionally sweeping away any philosophical order (*e.g.* by type of soul) for a purely utilitarian reason, as explained by Aristophanes himself:

> This I tried to do, in order that you need not go through Aristotle's treatise on animals which is divided into many parts, but you can have the entire enquiry about each single animal brought together.[92]

Unhelpfully for our purpose here, Aristophanes intentionally "omit[ted] the genus of animals which produce grubs, as they are small and weak and not worth observation".[93]

56 KINGDOMS, EMPIRES, AND DOMAINS

Early sources mention other works, now mostly lost, on aquatic animals, fishes, birds, and beasts, and on hunting, fowling, and fishing.[94] Oppian of Anazarbus wrote didactic poems on each of the latter three topics; only *Halieutica*, the poem on fishing, survives in original form. Cuttlefish are mentioned in passing as bloodless and boneless; bivalve molluscs are produced from slime and have no sexes, and some tiny fishes have no blood and no parents.[95] *Cynegetica*, a similar poem on hunting now attributed to a different Oppian (of Apamea), vividly describes the risks of sponge-diving; sponges "and other things that grow on the rocks" are said to have breath, *i.e.* sense-perception.[96]

With the exception of Seneca, our authors up to this point have shared a common intent: to describe plants or animals themselves, their capture, and/or their use in agriculture or medicine. Knowledge was of course imperfect, reports were often secondhand, and readers might enjoy the occasional tall tale, as in Pliny. But a very different, much older tradition existed in parallel, in which animals serve to illustrate human character: the cunning fox, the industrious ant, the brave lion, the loyal dog. Many of these stories were successively compiled and embellished, culminating by the late First century CE in the lexicon of Pamphilius.[97] It was inevitable that these two traditions would intersect, as they did in the *De natura animalium* of Aelian.

A Stoic and rhetorician known as "the honey-voiced", Aelian was a member of the same circle around the imperial family as Galen, Philostratus, and Oppian of Anazarbus.[98] His *De natura animalium* presents some two hundred animals in apparently random order,[99] each with a description that often involves a colourful folk-tale or wondrous "fact" in the service of moral instruction: the mother dolphin sacrifices her life for her young; bees protect and are deferential to their king; young vipers avenge the death of their father. Admittedly, little moral guidance might be gained from the sponge:

> The sponge is directed by a small animal resembling a spider rather than a crab. For the Sponge is no lifeless or bloodless object engendered by the sea, but clings to rocks like other creatures and has a certain power of movement in itself, though it needs, as you might say, someone to remind it that it is a living creature, for owing to some natural porosity it remains motionless and at rest, until something encounters its pores; then the spider-like creature pricks it, and it seizes what has fallen in and makes a meal. But when a man approaches to cut it off, the Sponge is pricked by the animal that lives in it, shudders, and contracts, and the trouble and labour that this causes to the fisherman is considerable, and no mistake.[100]

At least the zoology in *De natura animalium* was of high quality: Aelian relied on the best sources. By contrast, the *Physiologus*—the archetypal bestiary—made little pretense of accuracy. Instead, its purpose was to guide the reader towards knowledge of the Christian God using an iconography based on animals.[101] Written in Greek, probably in the cultural melting pot that was Second- or Third-century CE Alexandria, *Physiologus* drew on multiple sources: legends from Egypt, the Middle

East, and India; folklore common to the Eastern Mediterranean; Herodotus, Pliny, and Aelian.[101,102] The word *Physiologus* refers to the author and, according to Curley, identified him as "one who interpreted metaphysically, morally and mystically the transcendent significance of the natural world";[101] only later did the work itself become known as the *Physiologus*.

Moralistic animal-tales are older than written history, but what set the *Physiologus* apart from Aesop[103] or Aelian was its thick layer of Christian allegory, as in the phœnix immolating itself then rising from its ashes on the third day. Early versions included forty-nine subjects (in no obvious order) including five or six stones, two trees, and the Prophet Amos. As the *Physiologus* was translated—first into Ethiopian, Arabic, Armenian, Syriac, and Latin, eventually into all the main European languages—then copied and revised, further subjects were added including the mandrake and coral. Its base of pagan legends made some Christians uncomfortable,[104] but its stories enlivened many a sermon and, graced by illustrations that could appeal even to the illiterate, it became one of the most widely circulated books of the Middle Ages.

One standard story was that of the salamander. From time immemorial it has been believed that salamanders live in the fire, and indeed can extinguish a fire simply by crawling into it. We find this legend not only in *Physiologus* but also in Aristotle,[105] Theophrastus, Nicander, Seneca, Pliny, Dioscorides, Galen, Tertullian, Aelian, Isidore, Albert, Gessner, Cellini, da Vinci, Porta, Bacon, Browne, and Topsell; in the *Talmud*, the *Suda*, and the *Hortus sanitatis*; and in stained-glass windows, bestiaries, and books of hermeticism and alchemy.[106] Even today, the solidified material that remains in the hearth of a blast furnace is called a *salamander*.[107] For *Physiologus*, the salamander allegorized the three men who survived being cast into Nebuchadnezzar's fiery furnace.[108]

The oyster[109] is as close as *Physiologus* ventures to the boundary between animals and another group of beings. However, its relentless use of allegories in which animals, trees, and stones display human or divine traits is a throwback to a prephilosophical, pre-utilitarian conception of nature. Unfortunately, this would be well-suited to the dawning Middle Ages, during which the study of plants and animals largely died out in Europe.

Summary

By the end of the Classical era, words meaning *animal*, *vegetable*, and *mineral* were in widespread use well beyond the schools of the philosophers, and (allowing for the state of knowledge) meant more-or-less what they mean today. A variety of opinions persisted on some borderland organisms, especially sponges and corals, but without involving an intermediate genus or mentioning *zoophyta*. This era saw the rise of books on practical matters including animals, plants, and medicine; the first encyclopædia; the establishment of herbal, pharmacopœial, and bestiary

58   KINGDOMS, EMPIRES, AND DOMAINS

traditions; and the entanglement of the latter in Christian allegory. The subsequent development of ideas on natural history, whether philosophical, practical, religious, or occult, would draw heavily on the common if diverse heritage of Neoplatonism. In Chapter 5 we introduce Neoplatonism, focusing on how Aristotle's souls were progressively fashioned into a Great Chain of Being, and on the consequent implications for understanding, delineating, and classifying plants and animals.

5

Neoplatonic Nature

> There does not exist, nor will there ever exist, any treatise of mine dealing
> with [Idea, the Real that lies behind all existence]. For it does not at all
> admit of verbal expression like other studies, but, as a result of continued
> application to the subject itself and communion therewith, it is brought to
> birth in the soul on a sudden, as light that is kindled by a leaping spark, and
> thereafter it nourishes itself.
>
> —*Epistle 7* (attributed to Plato)[1]

In Chapter 3 we followed Plato's Academy and Aristotle's Peripatetic school up until
the sack of Athens by the Roman consul Sulla in 86 BCE. The Academy had remained
active after Plato's death in 348 BCE, developing in a Pythagorean (mathematical)
direction that can already be discerned in the later teachings of Plato himself[2] while
also embracing Aristotelian, Sceptical, and Stoic ideas regarding the cosmological
First Principle, logic, and ethics.[3] By contrast, Aristotle's primary texts disappeared
within a generation after his death in 332 BCE, and remained unavailable for more
than 250 years.

The relative longevity of the Academy mattered: these were years of Greek cultural hegemony. In the wake of Alexander, Hellenic cities and colonies were established from Pergamon and Antioch eastward to Bactria and beyond. Educated
men[4] in these lands spoke Greek, studied Plato, and lived by Stoic principles. This
continued to be the case even after Rome sent its armies into Egypt, Asia Minor,
and Judæa. But this is only part of the story. Throughout the Empire local religions
survived or even spread, their deities finding counterparts in the Roman pantheon,
while new mystery-religions arose promising salvation and eternal life after death.[5]
These were variously tolerated, co-opted, ignored, or driven underground by the
imperial authorities and their satraps; one such new religion, Christianity, eventually won favour even at the seat of empire.

Among the myriad areas that Plato had addressed in his dialogues, metaphysics
and cosmology proved particularly fertile. Later philosophers sought to reconstruct
a unified metaphysics that seamlessly wove together not only the creation myth in
Timaeus and the theory of forms in *Parmenides*,[6] but also Pythagorean number-
symbolism. For some Platonists this required a four-level metaphysics;[7] others

Kingdoms, Empires, and Domains. Mark A. Ragan, Oxford University Press. © Oxford University Press 2023.
DOI: 10.1093/oso/9780197643037.003.0005

60 KINGDOMS, EMPIRES, AND DOMAINS

preferred three levels. Great effort was expended to situate God (or for Christians, the Trinity), the beginning (if any) of time, the origin of the soul, and the source of evil at one or another of these levels. These are questions of theology. For others— and not only philosophers—it was of equal or greater urgency to put this metaphysics into practise: what rituals enable an individual human soul to know, partake of, or ascend to God? This is no longer theology, but theurgy.

In addressing these exalted issues, middle and Neoplatonists sometimes made reference to animals and plants, for example in describing how the universe (and with it, the hierarchy of souls) had come into being, or—ironically perhaps—in commentary on Aristotle as a first step towards the study of Plato. Although the philosophers we meet in this chapter came from different backgrounds—pagan, Jewish, or Christian—they lived and worked in the multicultural, multiconfessional milieu dominated by Hellenic culture and, later, the Roman army. In Chapter 6 we follow our question into Christian theology, scriptural exegesis, and apologetics, and in Chapter 7 into Islamic and Jewish religious traditions. First, however, we focus on the Neoplatonic philosophical tradition in its own right—and finally encounter *zoophyta*.

Philo of Alexandria

As the armies of Mithridates VI approached Greece in 88 BCE many philosophers fled to Rome, among them the head of the Academy, Philo of Larisa, and his pupil Antiochus. There Antiochus was befriended by the aristocrat Lucullus, and when Lucullus was appointed a quæstor under Sulla in Asia, Antiochus accompanied him. As related by Cicero,[8] at one point in 87 BCE Antiochus was in Alexandria (where Sulla was raising a fleet) in the company of the Platonic philosopher Heraclitus, when texts reached them in which Philo of Larisa argued for the "essential unity"[9] of Plato (read in a way that emphasized the Sceptical tendencies he was said to share with Socrates), Aristotle, and the Stoics. Antiochus became indignant, for although he accepted a fundamental unity among Plato, Aristotle, and the Stoics, he rejected a place for Scepticism among them.

It is telling that this anecdote is set in Alexandria. Founded by Alexander the Great in 332 BCE—the year of Aristotle's death—Alexandria had been a seat of learning since the beginning of the Ptolemaic dynasty.[10] The second city of the Roman empire, with a population reckoned at 500,000,[11] it was a mixing pot of Hellenic, Roman, Egyptian, Jewish, other Middle Eastern, and even Indian ideas.[12]

This admixture is seen in the works of Philo of Alexandria.[13] Scion of a prominent Jewish family, Philo received a Hellenic education and named his sons in honour of the Roman emperor. He had learned from Stoic readings of Homer "what philosophic truths could be concealed behind battles and fornications, shipwrecks and homecomings, and it must have suddenly struck him that this was just what was going on in the Pentateuch".[14] Interpreting the tenets of Judaism through the lens of a Platonic philosophy with Pythagorean, Aristotelian, and Stoic characteristics, he claimed that Moses had been the teacher of Pythagoras, just as Pythagoras had (supposedly) been the teacher of Plato.[15] Faith and reason are mutually consistent; any apparent inconsistency was simply a signal from Moses to the initiate that a particular passage is to be read allegorically.[16]

Philo taught that there were two souls, one rational and the other irrational.[17] The essence of the former is *pneuma*, while that of the latter is blood. In his *Questions and solutions on Genesis* Philo draws an Aristotelian distinction among the nutritive, sense-perceptive, and rational parts of the soul,[18] but in *The immutability of God* the division is Stoic:

> Among the various kinds of bodies, the Creator has bound some by means of cohesion [*hexis*], others by growth [*physis*], others by soul [*psyche*], and others yet by rational soul [*logike psyche*]. Thus, in stones and timber that has been detached from its organic growth, he made cohesion a truly powerful bond.... Growth God allotted to plants, constituting it a blend of many faculties, nutritive, transformative, and augmentative.[19]

A central challenge facing Jewish (and later, Christian and Islamic) philosophers was reconciling the creation of the world as revealed by scripture, with the eternal cosmos of Plato and Aristotle. With *De opificio mundi* Philo initiated the hexæmeral tradition,[20] in which the *Genesis* story of six days of creation is explored in a philosophical manner. For Philo, the God of the Old Testament created the Platonic form of the world on the first day, and the sensible world itself on days two through six. Philo seems comfortable with the terms *animal* and *plant*, although without greatly exploring their limits. On the third day, plants were created perfect: fruit trees already laden with fruit, ready for the "immediate use and enjoyment of the animals that were forthwith to come into being".[21] Then on the fifth and sixth days,

> To crown all he made man, in what way I will say presently, when I have first pointed out the exceeding beauty of the chain of sequence which Moses has employed in setting forth the beginning of life. For of the forms of animal life, the least elaborately wrought has been allotted to the race of fish; that worked out in greatest detail and best in all respects to mankind; that which lies between these two to creatures that tread the earth and travel in the air. For the principle of life in these is endowed with perceptions keener than that in fishes, but less keen that that in men. Wherefore, of the creatures that have life, fishes were the first which he brought into being, creatures in whose being the body predominates over the soul or life-principle. They are in a way animals and not animals; lifeless beings with the power of movement. The seed of the principle of life has been sown in them adventitiously, with a view only to the perpetuation of their bodies, just as salt (we are told) is added to flesh that it may not easily decay. After the fishes He

made the birds and land-creatures; for, when we come to these, we find them with keener senses and manifesting by their structure far more clearly all the qualities proper to beings endowed with the life-principle. To crown all, as we have said before, he made man, and bestowed on him mind par excellence, life-principle of the life-principle itself, like the pupil in the eye.[22]

Characteristically, Philo infuses the creation story with numerology. Animals were created beginning on the fifth day, for example, and have five senses (sight, hearing, taste, smell, touch).[23] In an extended allegory on husbandry (*De plantatione*), Philo refers to the world itself as a plant set in the universe, and the faculties as having been planted in man. Animals are plants too: the roots of the latter, directed downward, are analogous with the head of an animal, positioned upward. But whereas the eyes of animals face the earth, the eyes of man incline upward towards heaven, for man "is a plant not earthly but heavenly".[24] He identifies the power of locomotion as the defining feature of animals, with terrestrial animals having been created to belong naturally to the earth, swimming animals to water, winged creatures to air, and the fire-born (πυρίγονα) to the fire.[25]

Although later translators have rendered πυρίγονα correctly, Philo has nevertheless been incorrectly credited with introducing the word ζωόφυτα in a work called *Peri khosmos*, later translated by Guillaume Budé as *De mundo*. The sentence in question reads "Again, all animals that swim and zoophytes are allotted to the water, and all terrestrial animals and plants to the land."[26] However, *De mundo* is clearly an epitome—Boëthius, who lived five centuries after Philo, is cited—and not an original work by Philo.[27]

Philo also wrote a *De animalibus*, although it focuses entirely on the question of whether animals have reason. Arguing for the affirmative, Philo's nephew Alexander[28] recites the usual stories of the industrious ant, the wondrous skill of the spider, the motherly attention of the swallow to her young, and many others. In refutation, Philo argues that all these admirable qualities are involuntary, according to nature and without foresight; we might just as well accept that vines show maternal attention to grapes.[29] He concludes that we must "stop criticizing nature and committing sacrilege", as it is "the height of injustice" to grant equality to brute animals.[30]

The cross-fertilization of Hellenic and Eastern ideas continued after Philo. Ammonius, an Egyptian, revived Platonism in Athens around 66 or 67 CE;[31] both he and his student Plutarch engaged in Pythagorean number-mysticism, while Plutarch became a priest at the temple of Apollo in Delphi and wrote favourably on Eastern religion, finding useful similarities between *e.g.* Eros and the Sun, and Aphrodite and the Moon.[32] Plutarch accepted that the Supreme Being is eternal, but read *Timaeus* literally to say that because soul is prior to body and initiates change and motion, the cosmos must therefore have come into being.[33]

Calvenus Taurus

In 176 BCE, Emperor Marcus Aurelius endowed chairs of Platonic, Aristotelian, Stoic, and Epicurean philosophy at Athens. One of the leading Athenian Platonists at about this time, Calvenus Taurus, was prominent in a movement to disentangle the teachings of Plato from those of Aristotle and the Stoics.[34] With the eternity or creation of the cosmos of prime concern, Taurus sought to re-establish a conventional interpretation of *Timaeus* in which Plato's description of a temporal creation is only for "clarity of instruction".[35] Specifically he argued that *genetos*, which is normally translated as *created*, should rather be considered to mean *subject to coming into being*. In fact, Taurus gave four possible senses of *genetos*, the second of which is particularly interesting:

> That is also called "created" which is in theory composite, even if it has not in fact been combined. Thus the *mesê*[36] is a "combination" of the *netê* and the *hypatê*; for even if it has not been combined from these two, its value is seen to be an equal proportion between the one and the other; and the same thing goes for flowers and animals. In the cosmos, then, there is seen to be combination and mixture, so that we can by subtracting and separating off from it the various qualities analyse it into its primary substratum.[37]

So far as we know, Taurus did not follow up on "flowers and animals", although their *mesê* would be the two-faced beings—sponges, corals, and the like.

Plotinus

Returning to Alexandria, we next encounter Ammonius Saccas. Different reports cast him as a lapsed Christian[38] and/or of Northern Indian ancestry.[39] By about 225 CE he was teaching in Alexandria, where his pupils included Cassius Longinus,[40] Origen Adamantius,[41] and Plotinus. According to Hierocles, Ammonius taught that behind the teachings of Plato and Aristotle lay a single unitary philosophy.[42] After eleven years of study under Ammonius, Plotinus joined Roman Emperor Gordian III on his ill-fated campaign against Persia. Thereafter he moved to Rome, where he attracted sponsors and students. His teachings, particularly the essays edited by his student Porphyry as the *Enneads*, are regarded as the foundation of the grand synthesis of Hellenic (and some Eastern) philosophy, mysticism, and religion now known as Neoplatonism.

For Neoplatonists, all that exists, whether immaterial (ideas, numbers) or material, has emerged from an eternal, divine First Principle (the One). This emergence or *emanation* has proceeded (and continues to proceed) through a succession of stages, each of which serves as the generative principle of the next; conversely, everything that exists was necessarily prefigured within a prior stage. For Plotinus, the

64   KINGDOMS, EMPIRES, AND DOMAINS

One is threefold and gave rise to *nous*, which in turn gave rise to *psyche*, and it to Soul. Lower yet is nature (*physis*), which brought about matter, the final step in the chain of causation.[43] It is but a small step from this all-inclusive causal hierarchy to macrocosm and microcosm, to a linear chain of being, indeed to sympathies between stars and men, augury, and fortune-telling. Thus for Plotinus, Soul

> takes fulness by looking to its source; but it generates its image by adopting another, a downward, movement. This image of Soul is Sense and Nature, the vegetal principle . . . . In the case of soul entering some vegetal form, what is there is one phase, the more rebellious and less intellectual, outgone to that extreme; in a soul entering an animal, the faculty of sensation has been dominant and brought it there; in soul entering man, the movement outward has either been wholly of its reasoning part or has come from [*psyche*] in the sense that the soul, possessing that principle as immanent to its being, has an inborn desire of intellectual activity and of movement in general.[44]

Plotinus taught that animals and plants share in reason, soul, and life.[45] *Life* "embraces many forms which shade down from primal to secondary and so on, all massed under the common term—life of plant and life of animal—each phase brighter or dimmer than its next."[46] But this is not the gradation of Aristotle's *Historia animalium*: for Plotinus, soul is fundamentally the same in plant, animal or man, even if its reasoning and intellectual principles may be attenuated. As a consequence, souls can migrate from man not only into the bodies of lower animals, but even into trees and plants.[47]

Porphyry and Anatolius

Porphyry[48] studied under Longinus in Athens, then from 262 CE under Plotinus in Rome. He argued on ethical grounds against killing animals and eating flesh,[49] and sought to spread distrust of Christianity, arguing that the Old Testament prophesies (particularly those attributed to Daniel) were actually written long after the fact.[50] He was sympathetic towards the mystical side of Neoplatonism,[51] but in *Epistle to Anebo*[52] cautioned against animal sacrifice, divination, and theurgy. He may also have argued against Aristotle's definition of the soul.[53] Ironically for a Platonist, Porphyry became best known for his introduction (*Isagoge*) to Aristotelian logic.[54]

In *Isagoge*, Porphyry defines, compares, and contrasts the five categories of essence: genus, species, difference, property, and accident. According to Aristotle, these *cinque voces* exhaust all possible relations between a predicate and its subject.[55] Porphyry asked (but did not answer) whether these exist solely in our minds (a position later called nominalism), or have a substantial existence of their own (realism); and within the latter, whether they might exist apart from sensible phenomena (Plato's transcendence of the idea) or *in sensibilibus posita et circa hæc consistentia* (Aristotle's doctrine of the essence as inseparable from the individual thing).[56]

Porphyry's explication of the species under the genus *substance*, distinguished by application of *differentiæ*, was later immortalized in the so-called Tree of Porphyry.[57] It shows substance divided into the species corporeal and incorporeal; corporeal substance (body) into animate and inanimate; animate corporeal substance into plants and animals; animals into rational and irrational; and rational animals into mortals (men) and immortal. Aristotle did not pursue division to the level of individuals: *man*, not Socrates, is the *infima species*. Since animate-inanimate and sensitive-insensitive divide *substance*, animate-sensitive completes the genus *animal*, and animate-insensitive completes the genus *plant*.[58] On its face, this leaves no room for a third genus of equal status.

Isagoge was intentionally brief, so it is no surprise that Porphyry provided no detail in its pages on what might be included in or excluded from *animal* and *plant*. In his surviving commentary *On Aristotle's Categories*, however, he includes sea-urchins, sea-anemones, oysters, and crabs among animals although (in his estimation) they lack heads.[59]

Anatolius was the second-ranking philosopher in Porphyry's school.[60] Passages from his neo-Pythagorean *On the decad* are preserved in *The theology of arithmetic*,[61] an anonymous compilation attributed to the school of Iamblichus (see below). According to Anatolius "the triad, the first odd number, is called perfect by some, because it is the first number to signify the totality—beginning, middle and end. When people exalt extraordinary events, they derive words from the triad and talk of *thrice blessed, thrice fortunate*."[62] The text goes on to enumerate four distinct senses, four kinds of plants (trees, shrubs, vegetation, herbs), and three (terrestrial, ærial, and aquatic) or five (living in fire, ærial, terrestrial, aquatic, and amphibious) types of living creatures. The number five is associated with addition and increase, *i.e.* the vegetative part of the soul, and "the general structure of plants is five-fold—root, stem, bark, leaf and fruit".[63]

Iamblichus and Dexippus

Iamblichus[64] studied under Anatolius, and possibly under Porphyry himself, before returning to Syria in the final decade of the Third century.[65] The following century would see the social status of paganism decline as Christianity spread. From persecution under Diocletian (303–313 CE), Christians were protected under Constantine (312–337 CE) through the Edict of Milan (313 CE); Julian "the Apostate" (361–363 CE) sought to revive paganism, in part by weakening Christianity by extending tolerance to Christian dissidents and Jews; but after Julian died in battle, Jovian re-established Christianity as the state religion. Theodosius (379–395 CE) banned sacrifices to the gods, while Christian mobs vandalized temples. As the common people across the Empire abandoned the old ways for the mystery of salvation on offer from Christian bishops, the leaders of paganism sought to revive traditional belief and practise. For this they turned to the teachings of Iamblichus, regarded by Julian—himself a philosopher—as the equal of Plato.

The plan was not without merit. Iamblichus had sought to strip away what he saw as the Gnosticism of Plotinus and Porphyry, to recover the true insights of Pythagoras and Plato, even—especially—those kindled by a leaping spark in the soul. In his purified Neoplatonism, matter was no longer evil, soul was fully engaged with matter, and the soul of man could be free to achieve union with the gods.[66] Iamblichus populated the upper realms of the universe with principles, and gods in groups of three and seven; below these came archangels, angels, dæmons, heroes, sublunary and hylic archons, then souls, joining God to man in "one continuous link from highest to lowest and mak[ing] indivisible the community of the universe".[67] He wrote a lengthy commentary (now lost) on the *Chaldæan Oracles*,[68] and under the transparent pseudonym of an Egyptian high priest wrote *De mysteriis*,[69] a defence of cult and ritual in response to Porphyry's *Epistle to Anebo*. Indeed, Iamblichus was known for his commitment to theurgical *praxis*.[70] Nonetheless Iamblichus shared an almost Aristotelian concept of animals and plants: man is the wisest of animals,[71] the "best and most worthy of all animals on the earth".[72] "Man deprived of sense and intellect together is reduced to the condition of a plant; deprived of intellect alone he becomes a brute; deprived of irrationality but yet remaining in the possession of intellect he becomes similar to God."[73]

Like Porphyry, Iamblichus wrote a commentary on Aristotle's *Categories*; and like the *Ad Gedalium* from which it reportedly borrowed *verbatim*, it has been lost.[74] We do, however, have a commentary on the *Categories* by his disciple Dexippus,[75] set out as a dialogue between the master (Dexippus) and a student, Seleucus. At one point they have established that differentiæ are neither substance nor accident. Seleucus then asks what differentiæ can be, as all things are necessarily one or the other. Dexippus begins by referring to the experts on substance, the Stoics, who say that "the passage, in nature, from one genus to another is inexpressible, because the intermediate stages are imperceptible to us, as for instance potentiality in the case of substances." Or (continues Dexippus) something else might be intermediate, as between an actual and a potential man.[76] In the same way, differentiæ might provide "a sort of common bond" between quality and substance. For example,

> Nature does not pass directly from opposites to opposites, as for instance from animals to plants, but she has also contrived a type of life median between the two, in the shape of the class of "zoophytes", which links the two extremes, and completes and binds each to the other. And in this case also, between the different genera, as it were, of substance and accident there will be a mediating entity, according to some as partaking of both of these, while according to others as distinct from both of them.[77]

Thus, as the mediating entity, zoophytes would be regarded by some as partaking in the natures of both animals and plants, but by others as a distinct genus. Dexippus does not tease this apart further, but instead puts forward a third solution which Dillon[78] identifies as likely Iamblichus's own opinion on the matter. Two centuries later Simplicius of Cilicia (below) covered the same ground in his commentary on the *Categories*,[79] attributing the argument followed by Dexippus (above) to Iamblichus and reproducing the passage quoted above ("Nature does not pass directly from opposites to opposites . . .") *verbatim*. This raises the distinct possibility that Iamblichus may have used the term *zoophyte* in his lost commentary.[80] In the 1549 Latin translation of Dexippus,[81] ζωόφυτον was left in Greek.

Themistius

Themistius[82] was based in Constantinople, outside the established schools in Athens, Alexandria, and Rome. Although respectful of Plato, alone among philosophers of the time his paraphrases of the books of Aristotle were free of heavy-handed Neoplatonism,[83] for example regarding the soul:

> Anything with a soul is distinguished from anything without a soul by being alive, and while the soul has numerous capacities, which we have often listed, we see [only] one adequate for just being alive, namely the "vegetative", whose three ways of functioning are causing nourishment, causing growth, and finally reproducing. This is why all plants are also thought to be alive, for they clearly have in themselves the sort of capacity and source through which they move to opposite locations at the same time.[84]

Themistius uses the term ζωόφυτον seven times in his paraphrase on *De anima*:

(1) Aristotle criticizes the Orphic poems, and other philosophers, for holding that the soul is borne upon the wind and enters the body during the act of breathing: this cannot be so, as plants and some animals do not breathe. Themistius identifies these non-breathing animals καί τὰ καλούμενα ζώφυτα καί τὰ ἔντομα, the "so-called zoophyta and insects".[85]

(2 and 3) Aristotle states that while possession of the generative power allows us to speak of a thing as living, possession of sensation lets us refer to a living thing as an animal; and that all animals have the sense of touch. In his paraphrase Themistius refers to living things that do not move from place to place "yet have some [degree of] perception, like testacea, [whether these be] 'zoophytes' or even animals". Similarly, as the nutritive capacity can be separated from sense-perception, so can touch be separated from the remaining senses, as in zoophytes.[86]

68 KINGDOMS, EMPIRES, AND DOMAINS

(4 and 5) A little later, Aristotle notes that some kinds of living things possess the nutritive, appetitive, sensory, and locomotive powers and can think, while other beings possess fewer of these, or only one. Plants have only the nutritive power, "while another order of living beings has this *plus* the sensory" and, necessarily, the appetitive. All animals have at least the sense of touch, which they need to recognize food. Themistius reviews these capacities of soul in order: nutrition, sense-perception, desire, movement from place to place, and discursive thinking. Plants have the capacity for nutrition; zoophytes add only touch and perceive nutriment, employing touch rather than taste.[87]

(6) Aristotle enumerates the senses as sight, hearing, smell, taste, and touch, and says that if sensory organs are found in an animal it must possess all these senses, unless it is imperfect or has been damaged. Themistius again notes that the so-called zoophytes possess only the sense of touch.[88]

(7) Considering simple and compound sensation, Aristotle argues that taste, unlike the sensing of sound, colour, or odour, is a sort of touch, and is related to nutriment; without taste and touch "it is impossible for an animal to be". In paraphrasing this passage, Themistius comments that this is why the so-called zoophytes "share slightly in an animal's existence". By contrast, the other senses are "possessed not just by any [animal] species but [only] by those that are more developed, and thus have a capacity for forward movement".[89]

Thus, for Themistius *zoophytes* are animals, albeit animals with the fewest possible senses. Only in the second of these seven instances does his choice of words admit that zoophytes might be distinct from animals, and nowhere does he set out zoophytes as a genus intermediate between, and at the same rank as, plants and animals.

Themistius's paraphrase of *De anima* became an important path for the transmission of Aristotle's teaching on the soul into the Arabic world and, later, into medieval Europe. Its translation into Arabic by Isḥāq ibn Hunain[90] was known to Ibn Sīnā (Avicenna) and Ibn Rushd (Averroës), while its translation into Latin by William of Moerbeke[91] was used by Aquinas. The *editio princeps* was published by Aldus in 1534.[92] Could this tradition have been responsible for the myth that *zoophyte* is Aristotelian?

Athens and Alexandria

Neoplatonism enjoyed a revival in the early Fifth century[93] with the Athenian school following Iamblichus in an idealistic and religious direction, while the Alexandrian school "prudently and diligently applied itself to textual matters and exegesis".[94] In Athens the revival was led by Plutarch son of Nestorias, whose only known contribution to the matter at hand was the dubious statement that "Plato, Anaxagoras, and Democritus suppose plants to be terrestrial animals", together with a reference to forests (presumably of corals) in the Red Sea.[95] He was succeeded by Syrianus and then Proclus, whose voluminous works included attacks on Christians for accepting that the world had been created. The Athenian school was ultimately closed by Justinian in 529 CE as part of his consolidation of Orthodoxy.

Simplicius (above) studied in Alexandria under Ammonius Hermiæ, whom we shall meet soon, and in Athens under Damascius.[96] On the closure of the Athenian school Simplicius, together with Damascius and five other scholars, sought refuge at the court of Khosrow I of Persia.[97] In addition to his use of *zoophyton* following Dexippus (Iamblichus) in his commentary on the *Categories*,[77] Simplicius makes use of the term four more times: in the introduction to his commentary on Aristotle's *Physics* ("and the animate ones are the animals and the plants and the zoophytes"), and within that commentary in reference to plants and many kinds of animals that lack organs for locomotion; in his commentary on *De anima*, regarding sensory organs of perfect and imperfect animals; and in his commentary on *De caelo*, regarding animals in which there is no distinction between left and right.[98]

Platonism had long been established in Alexandria[99] when Hermias arrived from Athens and set up a school of Neoplatonism.[100] Despite considerable ebb and flow of political power in the city and society, the school flourished in its focus on textual exegesis—not least of Aristotle and Porphyry—for nearly two centuries; pagan heads of the school (among them Ammonius Hermiæ and Olympiodorus) largely accommodated Christian philosophers, and vice versa (Elias, David).[101] Notably a consensus emerged that although the universe is eternal, God nonetheless is "causally responsible for [its] existence".[102]

Ammonius Hermiæ and John Philoponus

Although a Neoplatonist, Ammonius[103] is remembered for his commentaries on Aristotle; or more precisely, for those written by his students based on their notes of his lectures.[104] A commentary on the *Categories* transmitted by an anonymous student, unlike that of Iamblichus transmitted by Dexippus, does not treat transitional or intermediate beings. The term *zoophyte* does, however, appear in two passages[105] in a commentary on Porphyry's *Isagoge* attributed to Ammonius but likely of mixed authorship.[106] It records that being is divided into the corporeal and the incorporeal; corporeal being into living and non-living; and living being (the ensouled, *emphychon*) into animal, plant, and zoophyte:

> for the plant has only three powers, nutrition, growth and reproduction and the animal has these [three powers] and sensory power and locomotion from place to place; and so the zoophyte is in the middle [in between] both of these, for it has the

70 KINGDOMS, EMPIRES, AND DOMAINS

three powers and the sense of touch, not, however, changing from place to place, these are the oysters and the sponges; for they grow naturally on rocks.[107]

Another of his students was John Philoponus,[108] whose notes on Ammonius's lectures on *De anima* make the distinction that although zoophytes (*e.g.* sponges) move, their motion is a sort of contraction and expansion rather than from place to place (locomotion).[109] Later he presents arguments for and against the viewpoint that embryos are animals, but states that such arguments may be unnecessary:

> For even if something is nourished through itself, it is not therefore an animal. A living thing and an animal are not the same thing. We will say it is alive and a life, but not also an animal, since plants also are nourished through themselves, but are not animals. Indeed, even the very sharing in sense does not thereupon make an animal. Things, at least, that share only in the sense of touch are not animals, for zoophytes share in touch. Besides, what is most proper to animals is that they change their places as wholes, something that does not belong to embryos. For they are attached to the mother and bound with her like a part belonging to her, and they are changed [in respect of place] like zoophytes. . . . Besides, if creation advances step by step from the less perfect to the more perfect, and the superior souls do not otherwise come along unless the more deficient have been present in advance, and the order is first the inanimate, then the vegetative life, then that of zoophytes, then that of the non-rational, and lastly that of the rational.[110]

Philoponus concludes that "to use sense and change [place]" are proper to animals; embryos are not animals, and neither are zoophytes. He makes the point even more forcefully in his commentary on Aristotle's *Analytica priora*, stating that "it is not possible for the same thing to be both animal and zoophyte, or that the zoophyte be at the same time a plant".[111] Philoponus (Ammonius) concedes that Aristotle calls both animals-that-locomote and zoophytes by the common term *animals*, since both are "animate perceiving substances", unlike plants or heavenly bodies.[112] Sponges have touch alone, and "animals with shells" have touch and "a sort of taste"; but zoophytes lack a rational soul,[113] the deliberative imagination that accompanies sense-perception,[114] and "the more perfect senses" of sight, hearing, and smell.[115]

With these interpretations, Ammonius and (perhaps even more) Philoponus recognized zoophytes as a genus intermediate between, but distinct from, plants and animals. Philoponus was familiar with the works of Themistius,[116] who as we have seen was more cautious in this regard. Philoponus is regarded as having progressively replaced "Aristotelian science with rival theories, which were taken up at first by the Arabs and came fully into their own in the West only in the Sixteenth century";[117] to these we may add the recognition of Zoophyta.

Before leaving Philoponus, we might remark upon his further thoughts about the vegetative and irrational souls. We are simultaneously men, animals, and animate beings (*i.e.* plants). "As men we possess the rational faculties . . . as animals the irrational, as animate beings the vegetative."[118] Bodily we are likewise tripartite: Philoponus speaks of our terrestrial body, pneumatic body, and luminous body. Our vegetative soul is inseparable from our terrestrial body and perishes when the terrestrial body does, although its faculty of growth may persist briefly, *e.g.* causing hair and nails to grow after death.[119] Our irrational soul is separable from our terrestrial body, although inseparable from the pneumatic body; it is through this pneumatic body that the irrational soul is subject to punishment after death. After purification, the irrational soul can set aside the pneumatic body, but "some other body heavenly and consequently eternal, which they call of the nature of light or starry" clings to it.[120] But if the pneumatic body has been "thickened by an evil way of life", it may be fed (*i.e.* retain something of the vegetative life) by vapours, and appear around tombs for some time. Only our rational soul is separable from body, and therefore immortal.

Elias and David

The second successor to Ammonius as head of the Neoplatonic school at Alexandria was Olympiodorus,[121] who argued for the eternity of the world (and pagan beliefs more generally) even as Christianity spread among the populace.[122] The last pagan head philosopher in Alexandria, he was succeeded by his student Elias,[122,123] a regional prefect of the Byzantine Empire and highly regarded by Justinian. Elias taught the works of Aristotle as preliminary to Plato,[124] although none of his Platonic scholarship survives. Together with his successors David and Stephanus, Elias is credited with a systematic analysis of the exact wording of Aristotelian texts, not just of their general intent.[125] In his commentary on Porphyry's *Isagoge*, Elias follows Philoponus in dividing earthly beings into animal, zoophyte, and plant.[126] The faculties of zoophytes are composite:

> it is called a plant, on which account the philosopher Plotinus said concerning those devoting time to these plant powers such as to eating and to communing, and despising rational speculation, that they are in danger of being turned into trees. But if in addition to these [plant] powers it were to take up both motion and sensation, it is an animal. But if it were to take up sensation alone, but not movement, this makes a zoophyte, as in the case of the oyster and things of that sort. For these sense, but do not move. For when it feeds it contracts [its] stalks, and in good water it unfolds [them] but it does not move, since it is attached to the rocks. Supposing there to be motion for these [zoophytes], [because] they stretch out and draw in, it is of their parts [*i.e.* not the whole organism]. And just as the name of the zoophyte is a composite, thusly also is the fact [that its powers are a composite of animal and plant powers]. For it has from the animal sensation, and from the plant lack of movement.[127]

72 KINGDOMS, EMPIRES, AND DOMAINS

Like Elias, David[122] studied under Olympiodorus. In his two surviving works,[128] *Prologomena philosophiae* and a lengthy commentary on Porphyry's *Isagoge*, he follows Philoponus and Elias in recognizing three distinct genera of living beings:

> for indeed first is natural reason and second the mathematical and third the theological. But otherwise genera are never divided into three, so that the animals are divided into the rational and the non-rational and again, the colours into white and black. And so how can we speak of the division of genus into species, by which these are three? For there is natural reason, the mathematical, and the theological. And yet in these opposing species one never participates in the other, such that the rational never participates in the non-rational, nor the non-rational in the rational. But if the mathematical participates in both natural reason and the theological, so that of the theological some is in the material and some in things immaterial, it is clear that the division is not of the sort that is from genus into species. And concerning this some say, "look at the ensouled genus which is divided into three species, such that there are animals, the zoophytes, and the plants, and not only is it divided into three species, but indeed one of these species participates in the others; for indeed the zoophyte participates in both the animal and the plant, as indeed the name makes clear." We can say concerning this, that it is possible to show that division holds badly. Again, those responding say, "look at the genus rhetoric which is divided into three species, the judicial, the deliberative, and the panegyric."[129]

And later:

> And when the Platonists take up these kinds of divisions . . . well, we will lay to rest speculation in a few things concerning the animal and the zoophyte and the plant for those seeking this, in the second place preparing to teach these things in response to the Platonists. Indeed, the ensouled is divided into animal, zoophyte, plant. And the plant is that having three powers, reproduction, growth, nutrition. And well the plant is made. For [the plant] being beneficial for us, will indeed be fit for our nature. For if we remain children, it is not possible to bear another. But since indeed growth requires nutrition, necessarily nutrition was given to us. And so, if some sensation will be associated with these three powers, it makes the zoophyte. That the zoophyte has sensation is clear from the fact that it likes to unfold in the waters, but when distressed the stalks are contracted. And if for these there is movement, it becomes the animal. And it must be said that where there is movement, there also is sensation, but not that where there is sensation there is also movement, so that sensation being inferior to movements is subordinated to movement. But perhaps there will be a difficulty, someone saying, "from what is it clear that movement entails sensation?" From what is it clear? From the fact that nature makes nothing in vain. . . . We hold hence by deduction that such a definition is not sound for the animal, for indeed it includes the zoophyte. For we say

the animal to be ensouled perceptive substance, and such are the zoophytes. So it is necessary to add self-movement; this is ensouled, perceptive, self-moving substance. And I mean movement from place to place.[130]

Three other Greek-language texts potentially dating to this period likewise distinguish zoophytes from plants and animals. An anonymous *Prolegomena* very similar to the one by David likewise argues that the division of beings into animals, plants, and zoophytes is not an example of division of genera into species, but should instead be likened to the division of rhetoric into panegyric, deliberative, and judicial.[131] The brief *Ten universal assertions* falsely ascribed to Archytas[132] contains similar concepts and wording. Its modern editor considers its tripartite distinction among plants, zoophytes, and animals to be a misreading arising from erroneous punctuation,[133] but (as we have seen) the three-way division is well attested from Ammonius onward. The third text, a Seventh-century Alexandrian commentary on Hippocrates by one Joannes,[134] recalls Philoponus in drawing this three-way distinction within a discussion of the embryo.

Summary: philosophical themes within Neoplatonism

From Plotinus onward, Neoplatonists elaborated upon teachings of Plato, particularly in *Timaeus*, to construct an elaborate speculative framework encompassing all that can be said or imagined to exist. From the First Principle, the unknowable One, emerged its perfect image, the Demiurge (*nous*), and from it a succession of stages, each the generative principle for the next: the World Soul (*psyche*); soul itself; multiple ranks of gods, angels, and the like; and individual souls that are (Iamblichus) or are not (Plotinus) joined to matter. The faculties of soul were progressively attenuated as the unfolding of the universe reached men, animals, zoophytes (if distinct from animals), and plants. This stood in contrast to the progressive superaddition of souls taught by Aristotle and Philo, and to the much-older ideas that man had been formed from clay, or that the first animals were born in moisture. Yet while the philosophical arm of Neoplatonism emphasized a downward chain of causation, Neoplatonic theurgy aimed to help the soul ascend towards the One, with the mediation of the gods and heavenly bodies. Christian mystics too sought the ascent of the soul, as we shall see in Chapter 6.

If man, like the universe, has arisen by emanation, we might find in him a reflection of the whole. As mentioned in Chapter 3, this notion is deeply embedded in Platonism,[135] even if Plato does not use the words *macrocosm* or *microcosm*. Surprisingly, the terms appear (once) in Aristotle: "now if this can occur in an animal, why should not the same be true also of the universe as a whole? If it can occur in a small world it could also occur in a great one; and if it can occur in the world, it could also occur in the infinite."[136] Ideas of macrocosm and microcosm also became established among the Stoics[137] and were spoken of positively by Galen[138] and

74 KINGDOMS, EMPIRES, AND DOMAINS

Macrobius.[139] But it was in Neoplatonism that macrocosm and microcosm attained its richest philosophical development, and it was through the Neoplatonic tradition that the concept was transmitted to Arabic and later European natural history.

In this chapter we encountered the views of late Platonic and Neoplatonic philosophers, and commentators, on animals, zoophytes, and plants. Some were pagan, others Jewish or Christian; our focus was on how they delineated animals or plants, and their openness to a third or intermediate genus. In the next two chapters we shift our focus to how natural groups were described within Christian, Islamic, and Jewish religious traditions per se. For Christianity and Islam in particular, the interface with Neoplatonism was important, and we will briefly meet some of the same teachers and written works. Chapter 6 will take us from the First century CE (the Apostle Paul) into the Sixth century (Boëthius) in the Western Roman empire, and into the Eighth century (John of Damascus) in the East. In those years even as today, religion cast a wide net; words and ideas in religious texts reflected and shaped how everyday people thought about living organisms far more than those in the books of the philosophers.

6

Christian Nature

> What indeed has Athens to do with Jerusalem? What concord is there be-
> tween the Academy and the Church?
> —Tertullian, *The prescription against heretics*[1]

In Chapter 5 we focused on the period from the destruction of the Academy through the rise of Neoplatonism in the Third and Fourth centuries CE. These were eventful years as well for a certain upstart religion that originated in Roman Judæa. Jesus of Nazareth, son of a carpenter, was baptized by an itinerant prophet; spent forty days in the desert, where he was tempted by Satan; declared himself the Son of God and the long-awaited Jewish Messiah; and attracted large crowds and performed miracles, before falling afoul of Jewish and Roman authorities. After his crucifixion in 33 CE his disciples carried his teachings as far afield as Athens and Rome.

Unlike most other religions, the mysteries of Christianity were open to all: no training in philosophy was required. This accessibility facilitated its rapid spread and made it threatening to pagan authorities. A particularly severe persecution under Diocletian (303–311 CE) was relieved by edicts of toleration.[2] Constantine went farther, building churches and exempting bishops from taxation, although without abandoning paganism. He summoned the First Council of Nicæa (325 CE) and enjoined the three hundred assembled bishops to reach consensus on certain points of doctrine and practise that were sowing disunity. Later, on his deathbed Constantine was baptized by Bishop Eusebius of Nicomedia.[3] In 380 CE the Edict of Thessalonica established Nicene Christianity as the state religion across the Empire.

Christianity faced important challenges in these years apart from its relation with the state. In the absence of settled doctrine, all manner of sects grew up within or attached themselves to the church, so heresies had to be confronted. Innumerable points of theology remained to be settled including the unity and persons of God, the creation or eternity of the world, the origin of evil, and the nature of the human soul. Such matters occasionally touched on natural history, for example in the Six Days of Creation, or whether soul could pass from man into lower animals. But in general, Christian theologians and polemicists were uninterested in natural history, except as allegory or to illustrate a point of doctrine.

Kingdoms, Empires, and Domains. Mark A. Ragan, Oxford University Press. © Oxford University Press 2023.
DOI: 10.1093/oso/9780197643037.003.0006

Early theologians and polemicists

One of the most-active early apostles was Paul (later Saint Paul) of Tarsus in Cilicia.[4] He had been educated in Jerusalem under a respected rabbi[5] and his letters, now preserved as books of the Christian New Testament, show both the Stoicism expected of Roman citizens of the time,[6] and some knowledge of Plato.[7] Nonetheless he resolved to preach to the untutored, and avoided philosophical arguments.[8] The so-called Apostolic Fathers[9] followed Paul in eschewing philosophy, although Clement of Rome used *Demiurge* and *Creator* interchangeably.[10]

Nevertheless, attempts were soon underway to reconcile Christianity with philosophy, in particular Platonism. The next generation of Fathers were converts from paganism and had studied philosophy;[11] some cited pagan philosophy alongside Christian theology, for example to argue against polytheism or, a little later, against the Gnostics.[12] Superficially at least, it was easy to find common ground between Platonism (in particular) and Christian faith, for example in speaking of God as Logos, or interpreting the three persons of the Godhead as hypostases. Justin Martyr considered philosophy to have been "God's revelation to the Greeks".[13] He wrote that Plato had learned divine truths from Moses,[14] and drew parallels between Socrates and Jesus.[15] Justin did not accept that souls could transmigrate,[16] and distinguished herbs from beasts in respect of dietary laws.[17]

Tertullian was the most anti-philosophic of the early Christian writers. Born in Carthage, he was well-educated in both Latin and Greek, and cited Plato, the Stoics, the Hermetic corpus, and medical treatises. While working in Rome he converted to Christianity, and upon his return to Carthage became associated with a Montanist community there.[18] He railed against "stupid curiosity on natural objects",[19] yet wrote on soul in plants and animals, following Pliny in arguing that insects must be able to breathe, because soul is in the life-breath and insects are alive.[20] He accepted that plants have soul too, indeed that trees have wisdom and vines climb to avoid being trampled upon.[21]

Clement of Alexandria studied under Pantænus, a former Stoic who headed the catechetical school at Alexandria. Like Justin, this Clement believed that the Greeks had learned from Moses, and he adduced support for tenets of his faith from Homer, Pythagoras, Socrates, Plato, and others.[22] Indeed, he went so far as to claim that it was by divine inspiration that pagan philosophers and poets had sometimes grasped the truth.[23] Even so, he cautioned his fellow believers to "be not much with a strange woman", *i.e.* philosophy.[24] Like Justin he rejected the idea that souls can transmigrate;[25] rather, he presented the Christian life as a stepwise progression of the soul from faith to knowledge then upward through the seven heavens[26] to union with God.[27] Clement did not belabour these heavens as either pagan metaphysics or a Christian celestial hierarchy, but it is fascinating to read (a century before Iamblichus, perhaps two centuries before Pseudo-Dionysius):

For on one original first Principle, which acts according to [the Father's] will, the first and the second and the third depend. Then at the highest extremity of the visible world is the blessed band of angels; and down to ourselves there are ranged, some under others, those who, from One and by One, both are saved and save.[28]

The *Recognitions*, falsely attributed to an earlier Clement, probably dates to about this period.[29] Its eighth book reviews Hellenic theories of the origin of the world including those of the Pythagoreans, Epicurus, Plato, and Aristotle, concluding that "the director of such order is the very wisdom of God".[30] Exceptional cases of spontaneous generation remind us of the ongoing activity of the Creator:

> But lest this [the orderly generation of animals from their parents] should seem, as some think, to be done by a certain order of nature, and not by the appointment of the Creator, He has, as a proof and indication of His providence, ordained a few animals to preserve their stock on the earth in an exceptional way: for example, the crow conceives through the mouth, and the weasel brings forth through the ear; and some birds, such as hens, sometimes produce eggs conceived of wind or dust; other animals convert the male into the female, and change their sex every year, as hares and hyænas, which they call monsters; others spring from the earth, and get their bodies from it, as moles; others from ashes, as vipers; others from putrifying flesh, as wasps from horse-flesh, bees from ox-flesh; others from cow-dung, as beetles; others from herbs, as the scorpion from the basil; and again, herbs from animals, as parsley and asparagus from the horn of the stag or the she-goat.[31]

Origen

Origen Adamantius was probably born in Alexandria, son of one Leonides, a teacher of Greek rhetoric or grammar and a convert to Christianity.[32] In an eventful life, Origen succeeded Clement of Alexandria as head of the catechetical school, attended lectures by Ammonius Saccas, founded the Christian school in Cæsarea and, under the patronage of a wealthy friend, produced a prodigious body of theological treatises covering nearly all the Christian sacred texts. Porphyry may have studied under him. Later, Origen's teachings became caught up in the battles for theological purity that began after 375 CE and culminated in the systematic destruction of his works after the Second Council of Constantinople in 553 CE. As a consequence, few of his writings survive in their original Greek, although some were preserved in Latin translations of variable trustworthiness.[33]

Origen was the first to set out Christian theology in a systematic manner, "bringing philosophy into union with religion".[34] Like Justin and Clement, he imagined that Plato had learned "the Jewish mysteries" while in Egypt,[35] and actively defended Plato (and in his view, therewith orthodoxy) against the anti-Christian

78 KINGDOMS, EMPIRES, AND DOMAINS

polemic *On the true doctrine* by the Greek philosopher Celsus.[36] Particularly as Paul and the earliest Fathers had avoided philosophical arguments, Christianity was often looked down upon as a religion for the uneducated and unsophisticated;[37] Origen's *Contra Celsum* dramatically reversed such opinion, and remained highly influential for generations.[38]

For Origen, not only animals and man but also the sun, moon and stars, angels, principalities and powers, the Son of God, and the Holy Spirit, as well as Satan and his angels, are living rational beings with free will.[39] Indeed "the whole world also ought to be regarded as some huge and immense animal, which is kept together by the power and reason of God as by one soul".[40] Origen accepted that souls can migrate,[41] and asserted that God is absent from animals, pieces of wood, and stones.[42] He accepted a hierarchy of beings ordered by the nature of their movement: the lowest are "things held together by their form" such as quarry stones and pieces of timber that are moved only by an external force or during decay; above these are inanimate beings such as plants, moved by their inherent nature or soul "out of themselves"; then animate creatures moved "from within themselves", for example a spider driven by an internal impulse ("phantasm") to spin a web, or a bee to produce wax; and finally rational creatures, for which movement and the mover are inseparable.[43] Origen notes in passing that metals, fire, and fountains of water might be included in this second group,[44] while hunting dogs and war-horses seem almost to have the faculty of reason.[45] On the other hand, only the impious could believe that animals could be useful in augury or divination.[46]

In *Phaedrus*, Plato had Socrates argue that the soul is perpetually self-moved.[47] Yet only rational beings are self-moved. One understands why Origen asked whether there might be two souls, one heavenly and the other of the flesh.[48] Moses himself taught that blood is the soul of all flesh.[49] What then about "bees, wasps and ants, and those other things which are in the waters, oysters and cockles, and all others which are without blood, and [yet] are most clearly shown to be living things"? Origen explains that they contain a liquid analogous to blood; for "colour is a thing of no importance, provided the substance be endowed with life".[50] Elsewhere, though, he dismisses such animals "of subordinate rank" as scarcely worthy of enquiry among the diversity of God's creation.[51]

Origen expanded Plato's microcosm to encompass a hierarchy of earthly bodies, and claimed the combination for Christian theology. In *Commentary on the Song of Songs* he wrote that according to the Apostle Paul,

> this visible world may teach about the invisible, and that earth may contain certain
> patterns of things heavenly, so that we may rise from lower to higher things, and
> out of those we see on earth perceive and know those which are in the heavens. As
> a certain likeness of those, the Creator has given a likeness of creatures which are
> on earth by which the differences might be more easily gathered and perceived.
> And perhaps just as God made man in his own image and likeness, so also did he
> make the remaining creatures after certain other heavenly images as a likeness.

And perhaps every single thing on earth has something of an image and likeness in heavenly things, to such a degree that even the grain of mustard which is the smallest of all seeds may have something of an image and likeness in heaven.[52]

This may extend to innumerable worlds existing one after the other.[53] Origen's extensive use of allegory, and his arguments from pagan philosophy—for example in casting the Trinity as an emanation and Christ as the Demiurge[54]—greatly concerned some Church authorities, even during his lifetime. Although persecuted for his faith,[55] he was posthumously pronounced a heretic (twice) and eventually excommunicated. Even so, his arguments remained influential within the Church well into the Middle Ages.

Nemesius

By the Fourth century, Christianity was on the march across the eastern Mediterranean lands, its main enemy not Julian's soft Hellenism but rather its own fractious disunity. Regional blocs and factions disagreed—sometimes violently—over theology and doctrine, canon and liturgy. Through the efforts of Origen, Athanasius,[56] and others, a somewhat Platonized Christianity took hold in Cappadocia, whereas a more Aristotelian version was typical of Antioch.[57]

On the nature of man by Nemesius, Bishop of Emesa[58,59] is thought to date to about 392 CE. The treatise begins with a standard Platonic concept of the incorporeal self-moving soul, then follows Galen in localizing the various human faculties to different parts of the body. Like the universe itself, soul is "disposed in a series of ascending grades with, at the bottom, an unconscious and uncontrollable irrational life-urge, and at the top, the detached and rational activity of the human spirit contemplating Good and the world of intelligibles".[60] The book then takes on a more Christian tone as he considers pain, pleasure, free will, and issues of faith. According to Nemesius,

There is no so marked difference between inanimate things and plants, but for the self-nutrient faculty of the latter. Likewise plants are not so different from irrational, but sentient, animals, nor are these, in turn, in total contrast with the rational creatures. One order is not unrelated to another, nor do they lack palpable and natural bonds of union. For example, while some inherent power makes one kind of stone differ from another, the lodestone seems to stand out, in comparison with other stones, by its celebrated power of first attracting, and then holding, iron to itself, as if it would feed upon it.

Again, when the Creator passed in turn from the creation of plants to that of animals, we may suppose that he did not, so to say, leap from the one order to the next, and suddenly make creatures endowed with the powers of locomotion and sensation. Rather, he advanced towards this end by slow degrees and seemly

80   KINGDOMS, EMPIRES, AND DOMAINS

moderation. He framed the marine animals called *pinna* and sea-nettle to have all the appearance of sensitive plants. Like plants he fixed them to the bed of the sea as if with roots. He surrounded them with shells as trees grow bark, rendering them stationary like plants. Nevertheless, he implanted in them the sense common to the whole animal creation, the sense of touch. They are thus like plants in being rooted and stationary, and like animals in their possession of feeling or perception. Aristotle observed that sponges, in like manner, though they grow on the rocks, close and open, or, rather, spread themselves out, as in self-defense, when they perceive anything approaching. For this reason the scientists of ancient days used to call them, and all such creatures, Zoophytes. After the *pinna* and such like creatures, God made next the animals with but a very limited range of movement, yet able to move themselves from one place to another. Such are most of the shellfish, and earthworms. Next he endowed particular species with more of this or that faculty, such as sentience or locomotion, until he reached the highest types of animal. By that, I mean those animals which possess all the senses, and are capable of unrestricted movement. And when God passed from the irrational animals to create a rational living creature, man, he did not introduce this rational creature abruptly, but led up to it, by the development, in certain animals, of instinctive intelligence, of devices and clever tricks for self-preservation, which make them appear almost rational. Only after them did God bring forth man, the truly rational living creature.[61]

Here the quartet *stones, plants, animals, man* is clearly on display. However, "scientists of ancient days" had not, in fact, used *zoophytes* in reference to organisms sharing features of both plants and animals. As we saw in Chapter 5, *zoophyte* is first attested in works by Dexippus and Themistius that antedate *On the nature of man* by perhaps four decades. Alfanus translated *On the nature of man* into Latin in the mid- to late Eleventh century, as did Burgundio of Pisa in the late Twelfth, bringing zoophytes to the attention of Albertus Magnus, Peter Lombard, Thomas Aquinas, and John of Salisbury, among others.[59,62] Nemesius used the example of zoophytes—animate beings devoid of blood—to argue against the idea that soul is blood, or indeed anything material.[63] On the question of transmigration he seems to side with Iamblichus, accepting that souls can migrate only among men, or among beasts, but not between man and beast (much less plants).[64]

The Cappadocian Fathers

Basil, his younger brother Gregory of Nyssa, and their friend Gregory of Nazianzus[65] played important roles in the establishment of Christianity in Anatolia, a region visited several times by the Apostle Paul. From wealthy, influential Christian families, their lives and ministries intertwined with issues critical to the early Church including asceticism and monasticism, ministry to the poor, the ecclesiastical role

CHRISTIAN NATURE   81

of women, and development of doctrinal orthodoxy. They were deeply involved in Church affairs including synods, councils, and (in the case of Basil) resistance against Emperor Julian "the Apostate", whom the former had come to know while they were fellow students in Athens.[66] Gregory of Nyssa, renowned as a theologian and humanist, drew pagan philosophy more closely into Christian theology.

Their thoughts on natural history are found largely, although not exclusively, in their works on the six days of creation. The account in *Genesis* 1 raises several issues, not least that *Genesis* 2:4 states that God created the earth and the heavens in one day. Philo had argued that that "six days" meant only that creation was orderly and perfect, the number six being perfect and "most suitable to productivity".[67] One might also ask how there could be light, how grass and trees could flourish and three days could pass, before God made the sun and moon.[68] Origen had argued that the first three days could not have been solar days, and that the entire account should be read allegorically.[69] Basil rejected an allegorical interpretation, arguing that light was created not on Day One but on "one day", a unique day without end.[70] Metaphysical questions remained as well, including how the immaterial (God) could bring about the material (the world). Gregory of Nyssa's solution was that the visible world and everything within it are projections from the mind of God;[71] more precisely, that God set in motion a process within which the world unfolded in orderly succession.[72]

Basil's *Hexaemeron*[73] took the form of nine homilies in which he drew upon Plato, Aristotle, and Philo, and on Aelian and the *Physiologus* for anecdotes about animals.[74] As for the late creation of the sun, light was already available for the growth of plants; the sun became the "material vehicle" of that light, useful for sailors and travellers. Further, its late creation was a signal from God that the sun was not to be worshipped.[75] Basil accepted that "frogs, gnats and flies" are produced from water, then continued "for though plants and trees be said to live, seeing that they share the power of being nourished and growing; nevertheless they are neither living beings, nor have they life".[76] He distinguished "sea nettles, cockles and all hard-shelled creatures" from fish without explicitly calling them animals,[77] and asked "how is it that coral, a stone so much esteemed, is a plant in the midst of the sea, and when once exposed to the air becomes hard as a rock?"[78] He allowed the bat to be "at the same time quadruped and fowl".[79]

Gregory of Nyssa dealt with the Creation in three works: *De opificio hominis, Liber in Hexaemeron,* and *Homiliæ in verba "faciamus hominem"*, the former intended as a supplement to the *Hexaemeron* of his brother Basil. In it, Gregory is admirably explicit that the three divisions of soul are superadditive:

for in the rational are included the others also, while in the sensitive there also surely exists the vegetative form, and that again is conceived only in connection with what is material: thus we may suppose that nature makes an ascent as it were by steps—I mean the various properties of life—from the lower to the perfect form.[80]

82 KINGDOMS, EMPIRES, AND DOMAINS

But, he continued,

> Let no one suppose on this account that in the compound nature of man there are three souls welded together, contemplated each in its own limits, so that one should think man's nature to be a sort of conglomeration of several souls. The true and perfect soul is naturally one, the intellectual and immaterial.[81]

Gregory was too careful a theologian to draw a superficial parallel with the *nous*, *logos*, and *psyche* within us, much less with the divine Trinity of whom they are a reflection; but it is fair to point out that in each case, three are united as one.

In Gregory's *De anima et resurrectione* the Teacher counsels that "it has been said by wise men that man is a little world in himself and contains all the elements which go to complete the universe." Gregory was receptive to using the microcosm-macrocosm analogy to "infer a wisdom transcending the universe from the skillful and artistic designs observable in this harmonized fabric of physical nature", although he doubted that it would be useful in informing us about the soul.[82]

Gregory of Nazianzus was famed for his orations, which however rarely touch on points of natural history. He affirmed that after God brought about the angels and heavenly powers, he then "gave being to the world of thought"; and when this was in good order, he "conceives a second world, material and visible".[83] In reciting differences among animals, he asked "how is it that some are crawling things, and others upright; some attached to one spot, some amphibious".[84] One imagines that those "attached to one spot" include sea-anemones and sponges; if so, he considered them to be animals.

Ambrose[85] produced the first Latin *Hexameron* in 387 CE, about seventeen years after that of Basil. It often follows Basil's closely—the explanation about "one day", anecdotes of fish, the oyster, and the coral—supplemented with more than one hundred reminiscences, spiritual lessons from animals, and local references sourced from Virgil (the *Georgics* and *Eclogues*), Cicero, and the *Aeneid*.[86] Basil's *Hexaemeron* was translated into Latin in about 440 CE by Eustathius Afer, and an abridgement forms the basis of the Anglo-Saxon *Hexameron* attributed to Ælfric, Abbot of St Albans.[87] By contrast, Gregory of Nyssa's *Liber in Hexaemeron* was left out of the main collections of his writings, and remains little-known even today.[88]

Although these hexæmeral works had little to say about the boundary between plants and animals, they embedded into an unimpeachably orthodox Christian tradition the Aristotelian concept of superadditive vegetative, sensible, and rational souls. By and large, these Fathers treated philosophy as a useful handmaiden to theology,[89] and their description of the persons of God, the days of Creation, macrocosm-microcosm, and soul remained broadly Neoplatonic. Transmigration of souls was downplayed and restricted in scope, while a limited role for spontaneous generation was a reminder of divine power. Combined with the teaching of Nemesius that nature steps little by little from the motionless and insensate upward

CHRISTIAN NATURE   83

towards rational man (with zoophytes as a bonus), works from this period allowed a valuable core of classical natural history to go forward as Christian belief.

Augustine

The story of Augustine is well known: born in Roman North Africa to a Christian mother and pagan father, he received a classical education. While a teacher of Latin grammar in Carthage he left Christianity to follow the Manichæan religion, but became disillusioned and later, in Rome, read Plotinus and passed through "a period of attachment to Neoplatonism".[90] Later, while the Imperial orator at Milan, Augustine was brought back to the faith by Ambrose, Bishop of Milan, and was baptized in 387 CE. He later returned to North Africa, became Bishop of Hippo, and took up a monastic life.

Augustine credited "some books of the Platonists" for helping him come to terms with the immateriality of God.[91] He writes respectfully of Plato,[92] and explains that an instantaneous Creation was followed by a progressive unfolding in the visible world of *rationes seminales* which even today continue to bring about new forms;[93] the six days are figurative and point to deeper spiritual truths.[94] His powerful and moving description of his discovery of "the immutable and true eternity of truth above my changing mind"[95] seems structured on *Timaeus*.

Augustine could be critical of the value of natural history,[96] but conceded that ignorance of the natural world "makes figurative expressions obscure, as when we do not know the nature of the animals, or minerals, or plants, which are frequently referred to in Scripture".[97] He was a keen observer of nature, willing to contradict Pliny where necessary.[98] Augustine accepted the division into stones, vegetables, animals, and man (and celestial beings):

> when [those who think about God] go on to look into the nature of the life itself, if they find it mere nutritive life, without sensibility, such as that of plants, they consider it inferior to sentient life, such as that of cattle; and above this, again, they place intelligent life, such as that of men. And, perceiving that even this is subject to change, they are compelled to place above it, again, that unchangeable life which is not at one time foolish, at another time wise, but on the contrary is wisdom itself.[99]

Or in a more-inclusive if less-Aristotelian formulation: the life of trees nourishes and conserves, that of animals also has sensation, that of man adds intelligence, while angels do not need nutriment but only maintain, feel, and understand.[100] He made sport of the Manichæan belief that there was a degree of sentience in all beings, and that soul could migrate among man, animals, and plants:

84   KINGDOMS, EMPIRES, AND DOMAINS

I think it right to refer here to the authority of Scripture, because we cannot here enter on a profound discussion about the soul of animals, or the kind of life in trees . . . [I ask you] to tell me first what harm is done to a tree, I say not by plucking a leaf or an apple,—for which, however, one of you would be condemned at once as having abused the symbol, if he did it intentionally, and not accidentally,—but if you tear it up by the root. For the soul in trees, which, according to you, is a rational soul, is, in your theory, freed from bondage when the tree is cut down,—a bondage, too, where it suffered great misery and got no profit. For it is well known that you, in the words of your founder, threaten as a great, though not the greatest punishment, the change from a man to a tree; and it is not probable that the soul in a tree can grow in wisdom as it does in a man. There is the best reason for not killing a man, in case you should kill one whose wisdom or virtue might be of use to many, or one who might have attained to wisdom, whether by the advice of another without himself, or by divine illumination in his own mind. And the more wisdom the soul has when it leaves the body, the more profitable is its departure, as we know both from well-grounded reasoning and from wide-spread belief. Thus to cut down a tree is to set free the soul from a body in which it makes no progress in wisdom. You—the holy men, I mean—ought to be mainly occupied in cutting down trees, and in leading the souls thus emancipated to better things by prayers and psalms. Or can this be done only with the souls which you take into your belly, instead of aiding them by your understanding?[101]

At least to make a rhetorical point, Augustine followed Varro[102] in assigning an insentient life to our bones and hair:

Let not our religion be the worship of the life that trees live, for it is not sentient life. It is of the kind that goes on in the rhythm of our bodies, the sort of life that our bones and hair have, and our hair can be cut without our feeling anything. Sentient life is better than this, and yet we must not worship such life as beasts have.[103]

He included snails[104] and shellfish[105] among the animals; it is not possible to judge from his (in)famous metaphor of Creation as a sponge in the vast sea of God's Being[106] whether he considered sponges to be animals. Finally, Augustine found stepwise ascents, descents, and hierarchies throughout nature, the grandest of these extending from God to forms, spiritual formed, spiritual unformed, material formed and material unformed, and finally to nothing.[107] For Augustine these hierarchies are not mere philosophical constructs; he is seeking a path of spiritual ascent to God.[108] In Chapter 14 we will return to visions of intellectual, spiritual, or mystical ascent from earthly bodies to the divine, and their implications for boundaries between natural types.

We began these paragraphs on Augustine with a précis of his life. He departed Italy for North Africa in 388 CE; Rome was sacked by the Senones in about 390 CE, and by the Visigoths in 410 CE. The latter event in particular was widely, if prematurely, viewed as the end of Western civilization, and prompted Augustine's call for Christians to focus on the body of the elect (the City of God) rather than on earthly power.[109] Even so, Augustine would become the most important early link between the Western Europe that would emerge, and its Hellenic intellectual roots.[110] Like John the Baptist, the Platonists prepared the way (the immaterial One, the hypostases, the soul in its material body); the incarnate Christ and God's all-permeating love supplied the missing elements, allowing the individual soul to ascend to God. Yet this spiritual benefit came at a cost, as

> the omnipotence of God was declared to manifest itself equally in the flowers of spring and the budding of Aaron's rod, in the wine harvested from the vineyard and the miracle of Cana, in the infants born every day and the resurrection of a dead person . . . [thanks to Augustine] we find the explanation of phenomena by their immediate causes disappearing in favor of interpretations that, quite legitimately of course, could be given these phenomena in sacral or poetic terms within an overall reference to their supreme destiny.[111]

Pseudo-Dionysius

The Neoplatonic works falsely attributed to Dionysius, a First-century Athenian jurist who became the first Bishop of Athens,[112] are thought to date to the early Sixth century.[113] The writer embraces the elaborate celestial hierarchy that, following Iamblichus and Proclus, had come to encumber Neoplatonism, recasting it in Christian terms.[114] Guided by the "material figures and bodily compositions" presented to us in "the Divine Institution of sacred Rites"[115] as symbols of the immaterial, the faithful can be borne "as far as our capacity permits" towards the Immaterial.[116] In *On the celestial hierarchy* Pseudo-Dionysius is far more interested in the three successive triads of angelic types than in animals or plants, although he associates the usual moral characters with "Divine representations of the Heavenly Minds through wild beasts".[117] In *On the divine names*, however, he ranks the powers that stream out from the Super-Essential Mystery as bestowing Godhead, Being, Life, or Wisdom in that order;[118] and in another passage establishes the superadditive hierarchy

> Good ⊃ existence ⊃ life ⊃ sentience ⊃ reason ⊃ intellect.

Accordingly, all that derives from the providence of God must be Good, whether existent or not. Qualities (such as good and evil), stones, plants, animals, souls, and angels also exist. Plants, animals, souls, and angels are Good, exist, and have life. Animals, souls and angels have these, plus sensation; souls and angels have reason as well; and angels all these plus intellect.[119] The closer anything approaches to God, the more it partakes of divinity, even as the above particulars

86 KINGDOMS, EMPIRES, AND DOMAINS

are progressively cast aside: God is beyond all quality, materiality, reason, or intellect.[120] We have just travelled the *via negativa*, the philosophical basis of Medieval scholasticism.[121]

Christianity offered a fertile soil for ideas of ascent from the material to the divine. Attention focused on (re)unification of the soul with God, but vegetables and animals sometimes received a mention as stations along the hierarchy, as in Figure 6.1. We return to specific metaphors of hierarchy and ascent in Chapter 14.

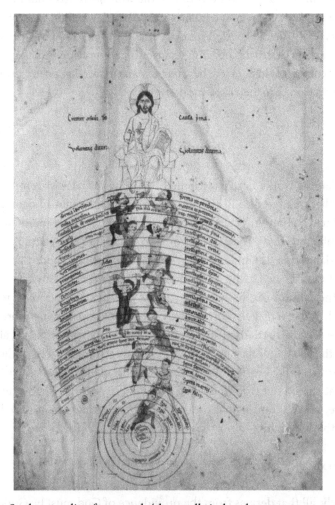

Figure 6.1. Souls ascending from earth (the small circle at bottom, *terra centrum mundi*) to God (the image of Christ at top, with inscriptions *Creator omnium Deus*, *Causa prima*, and *Voluntas divina*). In the arc labelled *Natura principium corporis* we find *Anima vegetabilis*, *Anima animalis*, *Anima rationalis*, and *Anima celestis*. Anonymous Twelfth-century manuscript Latin 3236A, folio 90r (Bibliothèque nationale de France): BnF Banque d'images.

Boëthius

Anicius Boëthius, a Roman consul, served as head of government services under Theoderic, the Ostrogoth king who ruled Italy from 493 to 526 CE.[122] A Nicene Christian, Boëthius wrote on theological issues and against heterodoxy. Somewhat unusually for a Roman of the day, he was fluent in Greek[123] and had begun to translate the works of Aristotle into Latin.[124] He also translated, and composed two commentaries upon, Porphyry's *Isagoge*. In the second of these commentaries, he presents the three Aristotelian souls:

> There is a triple power of the soul to be found in animated bodies. Of these, one power supports the life for the body, that it may arise by birth and subsist by nourishment; another lends judgment to perception; the third is the foundation for the strength of the mind and for reason. Of these, it is the function of the first to be at hand for creating, nourishing, and sustaining bodies, but it will exercise no judgment of reason or of sense. This power is possessed by herbs and trees and anything that is fixed, rooted to the earth. But the second is composite and conjoined: taking over to itself the first and making it part of itself, it is further able to form a varied and multiform judgment of things. For every animal who has the power of sense, is also born, and nourished, and sustained.[125]

The words "anything that is fixed, rooted to the earth" catch the eye. Plants are fixed to the earth through roots, but what of sponges? Pliny spoke of their "roots" (*radices*);[126] Boëthius used the derivative adverb (*radicitus*).

Boëthius has long been best known for his *Philosophiæ consolationis*, written in prison as he awaited an undeserved death. Hugely popular in Europe throughout the Middle Ages, it dealt movingly with good, evil, happiness, and the vicissitudes of fate in a way that was neither ponderously philosophic nor overtly Christian. Its so-called philosopher's prayer "O qui perpetua mundum . . ."[127] has been called "a brilliant distillation of the cosmology of Plato's *Timaeus*",[128] while the poem on the ascent of the soul is broadly Hellenic, although not specifically Neoplatonic.[129] Nearly eight centuries later Dante (likewise guided by a female figure) met the soul of Boëthius, alongside those of Dionysius and Augustine, in the fourth heavenly sphere—that of the wise.[130]

John of Damascus

John of Damascus[131] is considered the last Greek Church Father, and the last major link between Athens and Jerusalem. Muslim Arab forces took Damascus from the Byzantines during 634–636 CE, and a quarter century later the city became the capital of the Umayyad caliphate. Until about 685 CE, however, Greek remained the predominant language of the political administration, and Christians (including

88 KINGDOMS, EMPIRES, AND DOMAINS

John's grandfather and father) served in important roles. It is unclear whether John did likewise, but at some point he left Damascus to become a monk at Mar Saba, near Jerusalem.

His major work, *Peri gnoseos* (*Fons scientiæ*, or *Fount of knowledge*), consists of three sections. In the first section, known as the *Dialectica*, John reviews Hellenic philosophy, particularly Aristotelian logic. He makes it clear that knowledge is to be preferred over ignorance, and claims that the best contributions of the philosophers were revealed to them by God. The second section presents 103 heresies (among which John included teachings of the Pythagoreans, Platonists, Stoics, and Epicureans, as well as Judaism and Islam), while the final section systematically covers orthodox theology. The latter incorporates an account of the creation[132] that goes well beyond *Genesis*, for example by including the elements, planets, zodiac, Galenic humours, and an overview of geography.[133] In this third section, known as *De fide orthodoxa*, John states that

> One should note that man has something in common with inanimate things, that he shares life with the rational living beings, and that he shares understanding with the rational. In common with inanimate things, he has his body and its composition from the four elements. In common with the plants, he has these same things plus the power of assimilating nourishment, of growing and of semination of generation. In common with the brute beasts, he has all these plus appetite—that is to say, anger and desire—sensation, and spontaneous movement.
>
> Now, the senses are five; namely, sight, hearing, smell, taste, and touch. Belonging to spontaneous movement are the power of moving from place to place, that of moving the entire body, and that of speech and breathing—for in us we have the power either to do these things or not to do them.
>
> Through his power of reason man is akin to the incorporeal and intellectual natures, reasoning, thinking, judging each thing, and pursuing the virtues, particularly the acme of the virtues which is religion. For this reason, man is also a microcosm.[134]

But let us return to the *Dialectica*, "the first example of a manual of philosophy especially composed as an aid to the study of theology".[135] John restates the familiar formulation that stones are inanimate; plants animate but insensate; beasts animate, sentient, and irrational; and man animate, sentient, and rational.[136] In illustrating genera and species in nature, he introduces a new species:

> *Substance* is the most general genus. It is divided into *corporeal* and *incorporeal*.
> The *corporeal* is divided into *animate* and *inanimate*.
> The *animate* is divided into *sentient*, or *animal*, *zoophyte*, and *non-sentient*, or *plant*.
> The *animal* is divided into *rational* and *irrational*.

The *rational* is divided into *mortal* and *immortal*.

The *mortal* is divided into man, ox, horse, dog, and the like.[137]

Thus, *animate* contains three species: animal, zoophyte, and plant. But are zoophytes sensate or insensate? In the Latin translation by Grosseteste this appears in two slightly different formulations, first "animatum [dividitur] in sensibile animal et insensibile et zoophyton et plantam", then "animatum [dividitur in] sensibile animal et insensibile, zoophyton et plantam."[138] The Latin editions published at Basel in 1548 and 1559 follow:

The animate [is divided into] the sentient animal & the insentient, zoophyte, & plant.[139]

Right or wrong, the latter formulations are unambiguous: zoophytes are neither animal nor plant, but like plants are insentient. John explains this middle ground elsewhere:

It is called *animate* if it is nourished, grows, and begets offspring similar to itself. And this is further divided into three. Into animal; and the animal has not only these three, it has movements to change from place to place. Into plant, in which there are only the three. Into zoophyte. The zoophyte is between animal and plant. For the plant has the three powers and the animal, in addition to these three, movement from place to place. This [zoophyte] belongs in between. For indeed it moves itself but not that [movement] from place to place, but that [movement] of contraction, if anything should touch it.[140]

This is the same argument we encountered from Ammonius Hermiæ and John Philoponus in Chapter 5. According to John of Damascus, for a living being to be an animal (and not merely a zoophyte) it must be sentient and have the power of locomotion. Zoophytes are not animals; they exhibit contractile motion, but cannot move their entire body from one place to another. John implicitly acknowledges that zoophytes have the sense of touch, as that is what prompts them to contract; but he fails to pursue the argument to its Aristotelian conclusion that touch is the most widely shared and indispensable sense.[141] Nor is he interested to distinguish (as did Philoponus) between sponges, which have touch alone, and animals with shells, which additionally have "a sort of taste".[142] Such is philosophy in the service of theology.

Summary

Well before John turned his back on Damascus, the eastern Hellenic philosophical tradition had "all but vanished".[143] Justinian persecuted pagan teachers, and in 529 CE closed the Neoplatonic Academy in Athens; seven scholars including Damascius

90 KINGDOMS, EMPIRES, AND DOMAINS

and Simplicius fled to protection under the Sasanian king Khosrow I, although they were soon allowed to return, albeit not to teach.[144] In a later round of repression (562 CE) Justinian ordered pagan books to be burnt in Constantinople. Even so, some philosophical schools seem to have continued. Arab forces conquered Alexandria in 641 CE, but its Neoplatonic school remained active until perhaps 720 CE, when its philosophers moved to Antioch and then to Harran.[145]

Platonism enjoyed a modest renaissance in Constantinople under Basil II.[146] The *Quæstionum naturalium* attributed to Michael Psellus[147] is almost entirely concerned with the physical universe, heavenly bodies, and earth. His student John of Italy taught on Plato, Porphyry, and Iamblichus, and for his effort was condemned for heresy.[148] Other philosophers gathered around the court of Anna Comnena, daughter of Emperor Alexius I Comnenus; one of these, Michael of Ephesus, wrote a commentary on the *De motu animalium* and *De incessu animalium*.[149] Neither of these Aristotelian works, however, considers the borderland between animals and plants.

The natural history we encountered in this chapter was not written for its own sake: the authors were engaged in theology or apologetics. Yet the creation of grass, herbs, fruit trees, birds, sea creatures, beasts, cattle, and creeping things after their kind could scarcely be avoided, and some combination of Neoplatonic philosophy and allegorical interpretation could harmonize the accounts in *Genesis* 1 and 2. Particularly in Alexandria and what would become the Eastern church, Hellenic philosophy was accepted as a handmaiden to theology; with exceptions, the Latin Fathers were less welcoming.[150] Even so, animals and plants could offer spiritual lessons to the faithful.

The Aristotelian hierarchy of plants, animals, and man, and its Neoplatonic extension to multiple ranks of angels, were easily taken over into Christian theology;[151] Nemesius and John of Damascus found room for zoophytes. Microcosm and macrocosm proved compatible too, implicitly if not always explicitly. The ancient idea of metempsychosis was, however, a casualty. The few biblical references to transmigration could be explained as allegory,[152] and the Fathers generally denied or downplayed the possibility, or limited its scope. Tellingly, one of the charges of heresy against John of Italy was that he taught that souls could transmigrate.[148]

In the next chapter we consider how these same issues fared in the other eastern Mediterranean religious traditions, particularly Judaism and Islam. Although both are Abrahamic, one is ancient and inwardly focused, the other relatively recent and open to all who accept submission to Allāh. Consistent with the historical circumstances of its appearance, Islamic philosophy and culture was significantly influenced by Neoplatonism, and we shall examine the passage of Hellenic learning into (and subsequently out from) the Islamic world, with a focus on works that deal with animals, plants, and the boundary between them.

Judaism, Christianity, and Islam did not, however, have the field all to themselves. We begin Chapter 7 by considering the broader religious landscape in the Middle East in the centuries preceding John of Damascus.

7

Islamic and Jewish Nature

> I died as a mineral and became a plant,
> I died as a plant and rose to animal,
> I died as animal and I was Man.
> Why should I fear? When was I less by dying?
> Yet once more I shall die as Man to soar
> With angels blest . . .
> —Jalāl ad-Dīn Rūmī, *I died a mineral*[1]

In this chapter we examine how animals and plants were delineated in Islam and Judaism, and ask whether these religious traditions were receptive to the idea of intermediate organisms. We have already encountered points of contact between Hellenic philosophy and Middle Eastern thought: Pythagoras and Alexander, Porphyry and Iamblichus, Philo of Alexandria, Paul of Tarsus, and John of Damascus. But the story is broader than this, and begins much earlier. Let us start at the beginning.

Agriculture, a momentous step beyond hunter-and-gatherer livelihoods, arose first in the Fertile Crescent. As nomads settled and took up farming, the land could support a greater density of people, who in turn became tied to their fields, pastures, and water sources. Cities grew up, and with them new occupations in construction, trade, and civil administration.[2] Kingdoms and empires rose and fell, and with them their gods. The ancient Sumerian, Akkadian, Canaanite, and Hittite religious were polytheistic, although one god (the one associated with a particular city) was often considered principal. In the *Enuma Elish* creation story from Babylon, Marduk created mankind from the blood of the defeated god Quingu, while Asarre was "the giver of arable land who established plough-land / The creator of barley and flax, who made plant life grow."[3]

By contrast, Zoroastrianism is monotheistic, with a cosmology based around the contest between good and evil. The *Gathas*,[4] hymns ascribed to Zarathustra (Zoroaster), very occasionally refer to plants, but offer no clues as to what organisms might be included.

Kingdoms, Empires, and Domains. Mark A. Ragan, Oxford University Press. © Oxford University Press 2023.
DOI: 10.1093/oso/9780197643037.003.0007

In Canaan, tribes claiming descent from Abraham's grandson Jacob[5] adopted the god Yahweh from the ancient Canaanite religion, and over time became recognized as a distinct ethnic group, the Israelites.[6] Among them two main religious traditions developed, Samaritan and Jewish, with separate temples, priesthoods, and versions of the Pentateuch.[7] In 111–110 BCE, forces led by the Maccabean high priest John Hyrkanus conquered Samaria and destroyed the Samaritan temple. Later, under Roman rule, the Samaritan tradition enjoyed a brief renaissance under Marquah, whose prayers and hymns were collected in the *Defter*, or Samaritan *Book of common prayer*, and the *Memar* (*Teaching*).[8] According to Marquah, God created ten things ex nihilo: light and darkness, the four seasons, and the four elements fire, air or wind, earth, and water.[9] Each of the elements was divided four ways at creation, resulting in four kinds of earthly beings: fish, animals, birds, and man.

We have already encountered the Jewish creation story, *Genesis*, which became part of the Christian Old Testament. In addition, the Jewish tradition incorporates *midrashim* (rabbinical commentaries) that offer textual exegesis or ethical guidance. In *Genesis Rabbah*[10] we find several versions of the order of creation,[11] an opinion that souls and bodies were created separately,[12] and a comment that all things were created on the first day but were revealed only later, for example as the plants sprouted above the surface of the earth, or as the clouds parted to reveal the sun and moon.[13] A *midrash* that even fleas, gnats, and flies were "included in the creation of the world"[14] might cast doubt on spontaneous generation, whilst other *midrashim* imply that heavenly bodies look after earthly matters, even the growth of herbs and the ripening of fruit.[15] Most interestingly perhaps, man is an intermediate, "created with four attributes of the higher beings (*i.e.* angels) and four attributes of the lower beings (*i.e.* beasts)".[16]

The Roman-Byzantine and Sasanian empires dominated the Middle East during late antiquity,[17] but it would be an oversimplification to paint the religious landscape in those years solely in three primary colours—Jewish, Zoroastrian, and Christian—against a monochrome pagan background. The *Talmud* was completed in Baghdad, perhaps around 550 CE.[18] Zoroastrianism had fallen into decline by the Third century CE, and Shāpūr I[19] sought to bolster its appeal by incorporating Aristotelian and Platonic concepts.[20] In this he was followed by Khosrow I,[21] to whom the Athenian Neoplatonists fled after Justinian closed the Academy in 529 CE. The Christian church split following a dispute between Cyril, Patriarch of Alexandria, and Nestorius, Patriarch of Constantinople, over the person(s) of Christ. After the First Council of Ephesus (431 CE) deposed Nestorius, his followers took refuge in the Sasanian empire, where they joined local Christians who for political reasons had already sought to distance themselves from Byzantium.[22] These Nestorians brought their Syriac translations of Hellenic philosophy, and it was via these texts that Hellenic philosophy later began to pass into Arabic. Nor was paganism of a piece, encompassing as it did Hellenic philosophies from Stoicism to theurgical Neoplatonism, as well as diverse local gods and beliefs.

Further complicating this landscape were syncretic religions. The most successful was Manichæism, which combined elements of Christianity and Buddhism within an overall cosmology in which good (light) was opposed to evil (darkness).[23] The significance of light in Manichæism was rooted in the much older Iranian cosmology of the Avesta.[24] In contrast to plants and animals, which contain a mixture of light and darkness, stand the infinite lights. A progression or hierarchy of lights is apparent in the heavens: the faint stars, brighter moon, bright sun, and *anagrān* ("infinite lights", *i.e.* Paradise). In the Avesta and the Pahlavi texts, heavenly bodies are paired with living beings: the stars with plants, the moon with animals, the sun with man, infinite lights with God (or the gods). This implies a hierarchy of plants, animals, man, and God(s).[25]

Manichæism spread westward to Spain and eastward to Mongolia in the Third to Seventh centuries, and briefly counted Augustine as an adherent;[26] we have already seen his hostile comments on the Manichæan doctrine of transmigration of souls among plants, animals, and man (Chapter 6). A Turkish Manichæan prayer recognized five kinds of living beings: man, quadrupeds, and animals that fly, live in water, and creep upon the ground.[27] Other syncretic religions incorporated the so-called *Chaldæan Oracles*, which date to the Second century CE, or writings attributed to Hermes Trismegistus; we return to these in Chapter 9.

Manichæism can be considered Gnostic, in the sense that it promised salvation through special knowledge rather than by faith.[28] Gnostic beliefs were widespread during late antiquity, as evidenced by the considerable effort that the early Church invested in opposing them. Some were concerned primarily with heavenly matters, whereas others proposed highly detailed alternative creation stories; we return to the latter in Chapter 9 as well.

Islam and the translation of Hellenic philosophy into Arabic

Little is known of natural-historical thought in the Arabian Peninsula before the Seventh century CE. The nomadic tribes were polytheistic, while sizeable Christian and Jewish minorities could be found in some cities and regions.[29] These tribes tended to be fractious, and sharp rivalries existed among cities. This changed dramatically within a single decade, from 622 to 632 CE. Muḥammad, an orphan, was accompanying a trading caravan when a Christian hermit foretold that he would become a prophet of God. Later, while secluded for contemplation and prayer in a cave near Mecca, he received revelations from the Angel Gabriel, and in 610–613 CE began preaching that God is unitary and man should submit to His will.[30] In 622 Muḥammad and his followers decamped from Mecca to Yathrib,[31] where he instituted an inclusive civic constitution. His forces captured Mecca in 629 and went on to claim other cities, destroying images of the traditional gods. A prolonged war (602–628) between the Sasanian and Byzantine empires had weakened both, and

94 KINGDOMS, EMPIRES, AND DOMAINS

Muslim armies soon took advantage, capturing Damascus by 636 and Alexandria in 642 CE.

Ongoing revelations to Muḥammad were transcribed and became the verses (*surahs*) of the *Quran*. Surah 13 describes the creation of the heavens, sun, moon, earth, mountains, rivers, trees, and plants;[32] this is generally interpreted as creation ex nihilo.[33] Surah 16 provides further detail: God created men, cattle, horses, and mules and asses, trees, crops, olives, palms, vines, fruit, and things of diverse hues.[34] Surah 55 stipulates that "the heavens and the earth, and what between them is" were created in six days.[35] Various animals, plants, stones, and coral[36] are mentioned in other surahs, although not for the purpose of botanical or zoological classification. The *Quran* allows "the game of the sea and the food of it" to be eaten;[37] this is sometimes interpreted as excluding crustaceans and shellfish.[38]

Prior to the engagement with Hellenic philosophy, a system of thought known as ʿilm al-kalām (*science of discourse*), often shortened to *kalām* (*word* or *speech*), developed within Islam.[39] The term denotes reasoned argumentation beyond simple appeal to tradition, and although primarily applied to defence of the faith, it could also be used in other areas of learning, as in *al-kalām al-tabīʿī* (physical *kalām*, *i.e.* discourse about nature). By the Eighth century CE two points of disagreement had arisen within theological *kalām*;[40] one side emerged as the Muʿtazilite school. Muʿtazilites were favourably disposed towards logic and reason, and by the reign of ʿAbbāsid Caliph al-Māʾmūn[41] their Baghdad branch had adopted further logical tools from Aristotle.[42]

The expanding Arab caliphate often showed tolerance towards other monotheistic religions, allowing officials to remain in their posts and continue working in the local language, as at Damascus from 636 to about 685 CE.[43] Such environments were conducive to encounters with ideas from Neoplatonism, Stoicism, and the teachings of Aristotle.[44] Al-Māʾmūn took this further, seeking out manuscripts of Hellenic philosophy, recruiting scholars (including many Nestorian Christians) and eventually declaring Muʿtazilism the official creed. Nor did the philosophical trend in Islam look solely to Constantinople: Aristotelian, Neoplatonic, and Patristic ideas had earlier reached the Sasanian court (above), and "the procession of Hellenizing scholars and apprentices came to Baghdad from quite another direction, from towns in the Iranian highlands and the border of the eastern steppe".[45] The last remnants of the Neoplatonic school of Alexandria, including the philosopher al-Fārābī, are said to have reached Baghdad from Marw and Harran around 900 CE.[46]

Between 750 and 1050 CE the entire Aristotelian corpus (except the *Politics*), plus "the pseudepigraphs, the commentaries, introductions, anthologies, epitomes, and glosses" were rendered into Arabic.[47] So too were works by Hippocrates, Plato, Theophrastus, Euclid, Ptolemy, Galen, and others, often via a Syriac copy rather than directly from Greek. These efforts were organized under Caliph al-Manṣūr[48] and reached their apex under al-Māʾmūn, and coincided with the introduction of papermaking from China.[49] But the great translation extended well beyond the court

ISLAMIC AND JEWISH NATURE 95

library: Christian scholars across the Muslim lands translated the Greek Church Fathers, Philoponus[50] and many others, efforts that continued until well after the time of Maimonides.[51] Aristotle's treatises on animals (*Historia animalium, De partibus animalium*, and *De generatione animalium*) were translated by 850 CE and were combined into a single *Kitāb al-Ḥayawān* attributed to Yaḥyā ibn al-Baṭrīḳ.

Did these translations include works that employ the term *zoophyte*?[52] The *De natura hominis* of Nemesius was translated into Arabic perhaps four times, including by the Nestorian Christian physician Ḥunain ibn 'Isḥāq,[53] head of Mā'mūn's House of Wisdom.[54] His son Isḥāq ibn Ḥunain translated Themistius's paraphrase of Aristotle's *De anima*,[55] bringing *zoophyte* into Arabic as a prolix expression that means "objects between plant and animal".[56] One Antonius of Antioch translated the *Dialectica* by John of Damascus at the end of the Tenth century.[57] Philoponus's commentary on the *Analytica priora* was translated, at least in part.[58] Simplicius, on the other hand, appears to have been poorly represented.[59,60]

Porphyry's *Isagoge* was accorded multiple translations, although as we have seen, *zoophyte* was to be found in commentaries, not in the work itself. Ibn al-Ṭayyib's commentary on Porphyry's *Isagoge* bears substantial similarity, in part, to those attributed to Elias and David.[61] In Ibn al-Ṭayyib's commentary we read:

> You must know that animate is divided into animal, plant, and vegetative animal. Animal is such as man; plant is such as tree and vegetative animal is such as snail and sponge. For, in terms of the meaning of plant, these two (*i.e.* animal and plant) have vegetative soul, for both of them feed, grow and reproduce their kind. And, in terms of the meaning of animal, they have a feeling for likes and dislikes and also they do not lack movement in a place.[62] Lest someone might doubt the fact that genus is not divided into three species, we say that animate is divided into animal and not-animal, and animal is divided into vegetative and perfect.[63]

We mentioned above that al-Mā'mūn employed Nestorian Christians as translators. One of these was Ayyūb ar-Ruhāwi, or Job of Edessa, a prolific translator of Galen. Ḥunain ibn 'Isḥāq himself used several of Job's Syriac translations in bringing Galen into Arabic. About 817 CE Job wrote (in Syriac) an encyclopædic treatise dealing with the cosmos, soul, natural history, medicine, mathematics, and music. The *Book of treasures* is not purely a compilation; in it Job explains that individual bodies (not simply the earth as a whole) are composed of the four elements. All of nature arose in a "three-fold order", with three physical dimensions; past, present, and future; and beginning, middle, and end. Correspondingly, animals arose "in the high sphere, plants in the low sphere, and animal-plants in the middle sphere, because the latter possess affinity with the moving animals". The animal-plants establish an equilibrium between animals, which came into existence nearer the "light elements" (fire and air), and trees, which "stuck to the earth" because the "heavy elements" (earth and water) predominate in them. The animal-plant "feels and shrinks from fear like animals, and contains also something resembling fleshy

96 KINGDOMS, EMPIRES, AND DOMAINS

matter, but is fixed in the earth, and like trees does not move from place to place."[64] Each of these genera is itself threefold: terrestrial, aquatic, and aerial animals; trees, shrubs, and herbs; shellfish (oysters), sponges, and molluscs.[65]

With the *Book of treasures*, the concept *zoophyte* (although not the term itself) reached Islamic natural philosophy beyond the translations per se, and as a third genus on equal footing with plants and animals. By contrast, Ibn al-Ṭayyib's later commentary on the *Isagoge* drew "vegetative animals" closer to animals than to plants. Let us now ask whether either of these ideas gained traction among Muslim thinkers.

Arabic natural history, an-Naẓẓām, and al-Jāḥiẓ

No Arabic scientific botanical or zoological tradition predated the Muslim encounter with Hellenic philosophy and natural history;[66,67] nor did the arrival of Islam bring about a flowering of interest in natural history in its own right. However, Muslims (like Jews and Christians) could study nature to discover "signs or tokens of the glory of God",[68] and works in natural history were written expressly for that purpose.[69] Nasr notes that "the phenomena of Nature, the events taking place within the soul of man, and the verses of the *Quran* are all called *āyāt*, the human soul and Nature being respectively the microcosmic and macrocosmic counterparts of the celestial archetypes contained in the Divine Word."[70]

The Arabic language accorded one word for wild animals, two or three for other sorts of animals, and four words for domestic animals including the camel.[67] Islam superposed a complicated set of distinctions for dietary purposes. There also existed a tradition of associating human qualities with animals, particularly in proverbs.[68] Individual animals are described in books from about the Eighth century CE,[71] including *One thousand and one nights*.[72] Plants were arranged according to the season in which they flower; their utility as food, medicine, or dyestuff; into "rough" versus "delicate" herbs, or into "sweet" versus "salty and bitter" pasture plants.[73] Some later authors followed Theophrastus in classifying plants by overall form, for example in the *Kitāb al-Nabāt* (*Book of plants*) by Abū Ḥanīfa ad-Dīnawari, which includes a chapter on mushrooms "and similar plants".[74]

Muslim writers generally denied the possibility of spontaneous generation, although exceptions were made for flies, lice, ticks, scorpions, and frogs. On the other hand, it was widely believed that certain animals (notably the partridge) and date palms were engendered by the wind.[75] Mu'tazilite philosopher Ibrāhīm an-Naẓẓām wrote that God created all earthly things at once but hid away some things inside others, such that they could later be revealed in temporal progression.[76]

An-Naẓẓām's student al-Jāḥiẓ, renowned for his works of prose, authored a gigantic *Kitāb al-Ḥayawān* (*Book of animals*)[77] with the aim of demonstrating the wisdom of the Creator. Its 397 animals are presented in no discernable order, but it is not a simple bestiary; instead, it interweaves information from lexicographic, literary, moralistic, natural historical, and religious traditions, and explores logic, classification, and terminology. Al-Jāḥiẓ divides matter into that which does and does not grow, the growing into animals and plants, and animals according to whether they walk, fly, swim, or creep.[78] He finds, however, that these classes immediately throw up anomalies: the ostrich is excluded from the birds while the bat included, and insects end up in multiple groups. All flying animals walk, whereas few walking animals can fly. Similar problems arise if we divide animals according to other criteria, for example whether they are predator or prey.

Substantial parts of the *Kitāb al-Ḥayawān* are structured around debates, for example between proponents of the dog and of the rooster. The latter hold the dog unworthy because it combines attributes from distinct categories of animals, namely predators and prey. This allows al-Jāḥiẓ to argue that we must study material creation as we find it, including "intercategory" creatures.[79] Given that al-Jāḥiẓ recognizes different levels of inclusivity—breeds, kinds, and groups or classes of kinds—*intercategory* encompasses different cases, from hybrids to dualizers, the two-faced beings we encountered in Chapter 3. The Arabic version of Aristotle's animal books renders ἐπαμφοτερίζειν as *mushtarak* ("shared"): the seal (to take one example) is *mushtarak* because it shares some features with aquatic animals, others with terrestrial animals.[80] In the dog-rooster debate, al-Jāḥiẓ refuses to concede that shared features exclude an animal from its group: the dog remains a predator, even if (like prey) it is friendly to humans.

Al-Kindī, al-Fārābī, and al-Masʿūdī

Al-Kindī is identified with the birth of *falsafah* (philosophy and logic) as opposed to *kalām*. A native of Baghdad, he became prominent in the House of Wisdom at the time of al-Māʾmūn[41] and his successor al-Muʿtasim.[81] Many of his writings have been lost, but those that have survived reveal extensive knowledge of Aristotle, and familiarity with other philosophers including John Philoponus, Elias, and David.[82] He speaks of three faculties of soul (growth, animality, rationality) and follows "the ancient sages who did not speak our language" in describing man as a small universe, analogizing our bones, veins, organs, and so on with corresponding features of the earth and atmosphere.[83]

In the Arabic philosophical tradition, al-Fārābī is considered the "second master" after Aristotle. Building on the first master, in *Mabādiʾ ārāʾ ahl al-madīnat al-fāḍila* (*Principles of the views of the citizens of the best state*, or more succinctly *On the perfect state*) al-Fārābī presents an overview of *falsafah* and accords it priority over *kalām*. Material natural bodies have arisen from the four elements through a four-stage process of mixing, in which the mixed bodies successively interact with the elements and with each other, and are acted upon by the celestial bodies, until the mixed bodies cannot become any more dissimilar from the elements:

Some bodies arise from the first mixture, and some from the second, and some from the third, and some from the last mixture. The minerals arise as the result of a mixture which is nearer to the elements and is less complex, and their distance from the elements is less in rank. The plants arise as the result of a more complex mixture than theirs (*i.e.* the minerals), and they are a further stage removed from the elements. The animals which lack speech and thought arise as the result of a mixture which is more complex than that of the plants. Man alone arises as the result of the last mixture.[84]

According to al-Fārābī, the soul possesses five faculties: the nutritive, sensitive, appetitive, imaginative, and rational. The nutritive faculty is found in plants, animals, and man. Animals and man share the five external senses (touch, taste, smell, hearing, sight) plus an internal appetitive faculty that governs desire for objects perceived through these senses. The imaginative faculty retains and combines the resulting impressions, while the rational faculty, found only in man, rules the others.[85] He does not mention intermediate or two-faced organisms.

Al-Masʿūdī is best known as a historian and geographer ("the Muslim Herodotus").[86] His *Murūj al-dhahab wa maʿādin al-jawāhir* (*Meadows of gold and mines of gems*) and *Kitab at-Tanbīh wa-l-ʿishrāf* (*Book of notification and review*) were intended as précis or extensions of his *Kitab Akhbār al-zamān* (*Book of the history of the ages*),[87] now lost. He began the *Murūj* by recounting the Creation in a partly allegorical manner with details not elsewhere attested, for example Adam carrying a grain of wheat and cuttings from thirty fruit trees to earth from Paradise.[88] No detailed order of living creation can be deduced from this account; animals are mentioned only obliquely. Al-Masʿūdī referred several times to the trio animals, plants, and minerals,[89] and to six categories of bodies: celestial, earthly, man, brute animal, plant, and inert or mineral.[90] Oysters are animals.[91] In the *Tanbīh* he explained how "the celestial rotations added to the essential qualities act upon the elements from the heavens down to the earth and cause the special developments of minerals, plants, and animals", and continued:

> According to that arrangement, all the worlds hold together as in a chain, they are united actually and potentially, all of them bear the stamp of the divine art, the mark of wisdom, the evident signs of supreme power.[92]

The Ikhwān al-Ṣafā

The Ikhwān al-Ṣafā (Brethren of Purity)[93] were an "esoteric fraternity"[94] of intellectuals active in Tenth-century Iraq. They read the *Quran* allegorically, found elements of truth within most religions, and drew from the Pythagorean tradition and Neoplatonism. Their encyclopædia of mathematics, logic, natural sciences, metaphysics, mysticism, astrology, and magic (their eponymous *Rasāʾil*, or *Epistles*) is not purely a compendium of natural history, but embeds a spiritual dimension. They describe man as a microcosmos, and conversely the world as "a macroanthropos endowed with a body, a soul, a life, and knowledge";[95] heavenly bodies directly affect each and every earthly being and event.[96] They recognize three (mineral, plant, animal) or four genera (these plus man) of sublunar beings.

The Brethren assert that "the lowest plants verge into the highest mineral gems, and the highest plants into the lowest animals . . . the highest animals verge with the lowest rank of human beings; and the highest rank of humans, with the lowest of the angels."[97] They refer to this hierarchy as one of "nobility" or "purity",[98] but its basis is the universal soul as found in each genus of beings. The four elements are continually transformed into one another.[99] Minerals are simple mixtures of the elements, "the lowest things that come to be". Plants are nourished by the elements and grow in three dimensions; animals are nourished, grow, have locomotion and sensitivity; man has all of these, plus reason and discernment. Indeed, this reflects the temporal order in which minerals, plants, animals, and man came into being.[100] Angels are the animal soul, directed by the rational soul.[101]

Minerals arise by the coagulation, in different (sometimes harmonic) proportions, of substances variously described as moistures, dampness, juices, waters, and vapours, or as earthen, watery, and ærial, together with heat.[102] The basest minerals—those closest to the earth—are gypsum, acidic salts, and alums. Metals are produced from mercury and sulphur via a second stage of mixing and natural refinement. The soul, spirit, and body of gold—that is, its ærial, watery and earthen components—exist in perfect balance, while light from heavenly bodies guides the formation of sapphire and ruby; these are the noblest minerals.[103]

Plants—"every body that comes out from earth, is nourished, and grows"—are of three kinds: trees, crops, and herbs. Unlike minerals, plants cannot be transformed into one another. The vegetative soul brings about seven biological functions: attraction, retention, digestion, expulsion, nutrition, formation, and accretion.[104] The Brethren's description of the mineral-plant interface is difficult to understand, but involves "green manure" (moss?) and truffles, one a "vegetal mineral", the other a "mineral plant".[105] Other vegetal degrees rank above these, and share with animals the sense of touch.[106] The plant adjacent to the animal degree is the date palm, an "animal plant, as some of its actions and states are distinct from the states of plants, even though its body is a plant" and "a plant with regard to its body, but an animal with regard to its soul". Parasitic plants, which draw nutrition not from the earth but (like a worm) from trees, likewise exhibit the actions of the animal soul.[107] Coral is a plant.[108]

The Brethren classify animals in several ways, employing diverse criteria. With regard to the faculties of soul they recognize five stages: worms that breed in clay (the lowest, "almost on a par with plants", with only the sense of touch); worms that crawl on leaves, with taste and touch; animals which live in the deep sea and other dark places, which additionally have the sense of smell; insects, possessing these plus hearing; and perfect animals, which also have sight.[109] Worms in general are

100   KINGDOMS, EMPIRES, AND DOMAINS

"vegetal animals" because their body "grows as some plants do, but it stands autonomously, and because of the fact that it moves its body with a voluntary movement, it is an animal." More specifically, the lowest animal is the "cane worm", perhaps referring to the snail.[110]

Later they arrange animals by habitat, diet, and size, and take a step towards ranking animals *within* each class: the lion is king of the carnivores, the mythical Simurgh the king of the birds, the bee the king of the swarming creatures, and so on.[111] As "the animal level is a place of origin for virtues and source for noble traits, these cannot be contained by one animal species only." Thus, the ape is nearest the human level in bodily form, but the horse by its dispositions, the parrot and nightingale by their voices, the bee by its "subtle deeds", and the elephant by its acumen.[112] The Brethren recognize that some species exist at the intersection of discrete types, for example the ostrich, "a cross between a bird and a beast",[113] but do not hold them up as intermediary links in the ladder of nature. Repeating the time-honoured legend of how pearls are born, the Brethren state that the oyster is the "basest and weakest of sea animals" yet "the greatest in soul".[114]

Their terminology "vegetal mineral", "mineral plant", "vegetal animal", and "animal plant" recalls "zoophyte", but such beings nonetheless remain minerals, plants, or animals and are not assigned to a new intermediate genus. The Brethren also argue that during development, the human embryo is successively under the influence of the vegetable (the first four months) and animal souls (month five until birth), with the human soul taking control only thereafter.[115]

Finally, *Epistle* 22 is notable for its message that animals have intrinsic worth, apart from their utility to man. Animals were created by Allāh, and praise Him as they are able.[116] And "there is no animal kind, no species or individual among them, great or small, that does not have a band of angels charged by God with overseeing its growth, preservation, and welfare, at every stage."[117]

Al-Bīrūnī, Ibn Sīnā, al-Ghazālī, Ibn Rushd, and al-Abharī

Al-Bīrūnī is highly regarded for his writings on astronomy, geography, physics, and mineralogy, and on history and anthropology particularly of India. So far as is known, his interest in natural history did not extend far beyond pharmacology and medicine, while his philosophical writings have been lost.[118] According to Nasr, al-Bīrūnī accepted that minerals, plants, and animals "comprise the totality of creatures possessing physical existence here on earth", and that man has migrated through the plant and animal realms and therefore bears within himself the nature of these creatures. He believed that minerals live and grow in the earth, and if removed must be used in accordance with God's laws.[119] In *Kitāb al-Tafhīm li-awa'īl ṣinā'at al-tanjīm* (*Book of instruction in the elements of the art of astrology*) he referred to corals as "precious stones from water",[120] whereas in *Kitāb al-Jamāhir fī ma'rifah al-jawāhir'* (*Book most comprehensive in knowledge of precious stones*) he credited corals and sponges with sense-perception and contraction—that is, they are animals.[121] In *Kitāb al-Muḥaṣṣaṣ* (*Book of customs*) his contemporary Ibn Sīda included fungi among a miscellany of plants and plant products.[122]

Falsafah, and medieval philosophy more broadly, reached its peak in the works of Ibn Sīnā (Avicenna), a Persian-speaking native of Bukhara.[123] A polymath of prodigious learning, Ibn Sīnā was a physician, advisor, philosopher, and poet. His *al-Qānūn fī al-ṭibb* (*Canon of medicine*) remained in use for six centuries after his death,[124] while his *Kitāb al-Šifaʾ* (*Book of healing*)[125] is a masterclass in Aristotelian philosophy from the *Isagoge* and *Organon* through physics, mathematics, metaphysics, politics, law, and ethics. The section on physics was further subdivided into expositions on nature *per se* (Aristotle's *Physica*), the heavens (*De caelo*), coming to be and passing away (*De generatione et corruptione*), mineralogy (*Meteorology* 4), meteorology (*Meteorology* 1–3), the soul (*De anima*), plants (the pseudo-Aristotelian *De plantis*), and animals (*Historia animalium, De partibus animalium* and *De generatione animalium*).[126]

For Ibn Sīnā, composite bodies (those produced by mixtures of the four elements) are of three types: mineral, plant, and animal. Minerals lack a soul; the vegetative soul has nutritive, growth-promoting and propagative powers; animals have these plus external and internal sensation, and volition. This is much as in al-Fārābī (above). Of the five senses, only touch and taste are necessary to support movement-at-will and to perceive food. Man is unitary and possesses not only the vegetative and animal souls but also a rational soul, which brings the ability to reason and to receive divine guidance. Ibn Sīnā associates the human soul with the faculty of speech.[127]

Motion can be uniform or non-uniform, voluntary or involuntary. The motion of plants is non-uniform and involuntary, while that of animals is non-uniform and voluntary.[128] Even a shellfish opens and closes its shells to seek nutrition, and if turned over will move until it has righted itself.[129] Ibn Sīnā—who may never have seen an ocean—does not consider sponges or sea-anemones.[130] But he has a surprise for us: in the section of *al-Šifaʾ* dealing with plants, he claims that because plants do not move voluntarily to seek food, they are not alive.[131] Living organisms need not be able to locomote, he writes, pointing again to the mollusc flexing its shells in situ. But plants cannot do even this; only the animal soul prepares the body for the functions of life. With this he broke not only from his earlier opinion, but from the *De plantis* as well, where shellfish were "at once plants and animals".[132]

Later, Ibn Sīnā renounced his *al-Šifaʾ* as having been written for "common people . . . devoid of understanding", announcing that he would henceforth write only for those who meditate deeply on these matters.[133] Nasr interprets this as a promise from Ibn Sīnā to pull aside the supposed veil behind which the Pythagoreans had hidden the Egyptian, Orphic and Babylonian mysteries, and thereby provide illumination rather than mere knowledge.[134]

For Aristotle, the study of plants and animals helps us understand not only man, but also deeper matters including causation and soul (life). We observe their form,

102 KINGDOMS, EMPIRES, AND DOMAINS

parts, movement, and generation, and draw conclusions about the number and powers of the senses. We marvel, moreover, at their fitness-for-purpose and beauty, and reflect on the natural and good.[135] Ibn Sīnā constructed a comprehensive philosophy, grounded in his faith, that integrated substantial portions of the Aristotelian tradition gently infused with Neoplatonic cosmology. Plants and animals (although not zoophytes) find a place, but his *falsafah* emphasizes logic and metaphysics and is focused on man. Unlike Heraclitus, Ibn Sīnā neither acknowledges the gods in the kitchen of natural history, nor bids us much of a welcome there.

With Ibn Sīnā, the golden age of *falsafah* came to an end.[136] No person or treatise was solely responsible, but if we had to associate one thinker with its demise a leading candidate would be al-Ghazālī. Al-Ghazālī divided "the seekers of knowledge" into *mutakallimūn* (orthodox scholastics), *bāṭinyya* (Ismāʿīlīs), *falāsifa*, and *ṣufiyya*.[137] In *Tahāfut al-falāsifa* (*Incoherence of the philosophers*) he prosecuted twenty charges against the *falāsifa* in general and Ibn Sīnā in particular,[138] finding them guilty on seventeen counts of heresy and three of *kufr*, infidelity to Islam. The three beliefs he singled out as blasphemous are the eternity of the world,[139] secondary causality, and denial that bodies will be resurrected on the day of final judgement.[140] Regarding causation, for al-Ghazālī each and every event, no matter how inconsequential, reveals the immediate and present hand of God, who is not constrained to act in any particular way but "does not deceive humans and does not lead them astray".[141]

The *Tahāfut* brought several consequences, probably unintended. As pointed out by Marmura, al-Ghazālī explained philosophical ideas so clearly that later scholars of *kalām* were obliged to engage them.[142] And although his verdicts were clear, they were limited: he spared logic, physics, and astronomy, and insisted that Muslims should study medicine and mathematics to "protect the interests of their community".[143] Indeed, does not opportunity to witness a sign of the hand of God validate observation-based enquiry into nature?

Al-Ghazālī is credited with two works on natural history. In *al-Ḥikma fi 'l-makhlūḳāt illah* (*Wisdom in God's creation*) he "describes creation, from the heavens down to the plants" but excluded minerals.[144] The *al-Hikma* is problematic, as al-Ghazālī appears to have copied it nearly verbatim from an earlier source, the *Kitāb al-Dalāʾil* incorrectly attributed to al-Jāḥiẓ.[145] In *Maqāṣid al-falāsifa* (*Doctrines of the philosophers*),[146] al-Ghazālī reviews the philosophers' teachings on the four elements and their mixtures; minerals; the souls of vegetables and animals, their powers, and external and internal senses; the soul of man; and angels. Of this, little if anything deviates from al-Fārābī or Ibn Sīnā. He concludes by describing steps or grades leading from man, through the prophets, finally joining to the angels.[147] A perhaps insoluble puzzle remains: did al-Ghazālī actually espouse these doctrines, or was he only positioning them for his later assault in the *Tahāfut*?[148]

Despite al-Ghazālī's criticism, *falsafah* persisted for some time, notably in al-Andalus, where we encounter Ibn Rushd. He is best known for his commentaries on the texts of Aristotle (as opposed to summaries such as the *al-Šifaʾ*) and for his defence of *falsafah* against al-Ghazālī. This defence, the *Tahāfut al-tahāfut* (*Incoherence of the incoherence*), was translated into Latin in 1328 and became an important path by which knowledge of Aristotle reached European scholars including Albertus Magnus and Nicholas of Autrecourt.[149] While his positions on being, causality, and intellect, and his argument that religion is not a branch of knowledge but rather a "personal and inward power",[150] were influential (if controversial) in the later European Middle Ages, here we focus on his natural history. Ibn Rushd accepted the tripartite division into minerals, plants, and animals,[151] and argued that plants and animals appear to have been designed by God.[152] Commenting on Aristotle's *De anima*, he asserts that "many sensitive and animate things" including sea sponges and many shellfish do not move with respect to place, yet "are called *animals*, not merely *living*".[153] Likewise, plants are alive.[154] However, the term (and concept) *zoophyte* is absent in Ibn Rushd, including at passages where it occurs in Ammonius or Themistius.[155]

Two less well-known works of this period likewise imply a hierarchy of beings. The *Nuzhat-nāma-i ʿAlāʾī* (*Book of refreshment dedicated to ʿAlāʾ*) of Šahmardān describes man, quadrupeds, birds, insects and fish, plants, and minerals in that order. Fruit trees are presented first within the plants, and metals first within the minerals.[156] Later al-Abharī wrote a *Hidāyat al-ḥikmah* (*Guide to philosophy*)[157] that embraced logic, physics, and metaphysics. A noted astronomer and astrologer, he held that stars are alive and have volition.[158] Within the section on physics al-Abharī considered the corporeal world in successive chapters on minerals, plants, animals, and man.

Niẓamī Arūzī, al-Qazwīnī, and later authors

Niẓamī Arūzī, an astronomer, physician, and court poet to the Ghaznavid empire, is best known today for his *Čahār maqāla* (*Four discourses*),[159] written between 1150 and 1160 CE, which describes a continuous scale of nature linking the inorganic world with the angelic. When God "desired to produce in this world minerals, plants, animals and men, He created the stars, and in particular the sun and moon". After these interacted to produce the earth's atmosphere and geological features, God created "for that substance wherefrom the plants were made manifest" four forces (attractive, retentive, assimilative, and expulsive) and three faculties (nutritive, diffusive, and reproductive).[160]

> So this Kingdom rose superior to the inorganic world in these several ways which have been mentioned; and the far-reaching Wisdom of the Creator so ordained that these Kingdoms should be connected one with another successively and continuously, so that in the inorganic world the first material, which was clay, underwent a process of evolution and became higher in organisation until it grew

104 KINGDOMS, EMPIRES, AND DOMAINS

to coral, which is the ultimate term of the inorganic world and is connected with
the most primitive stage of plant-life. And the most primitive thing in the vegetable kingdom is the thorn, and the most highly developed the date-palm and
the grape, which resemble the animal kingdom in that the former needs the male
to fertilise it and so that it may bear fruit, while the latter flees from its foe. For
the vine flees from the bind-weed, a plant which, when it twists round the vine,
causes it to shrivel up, wherefore the vine flees from it. In the vegetable kingdom
there is nothing higher than the date-palm and the vine, inasmuch as they have
assimilated themselves to that which is superior to their own kingdom, and have
subtly overstepped the limits of their own world, and evolved themselves in a
higher direction.[161]

Now when this kingdom had attained perfection, and the influence of the
"Fathers" [the seven planets] of the upper world had reacted on the "Mothers"
[the four elements] below, and the interspace between the air and the fire in its
turn became involved, a finer offspring resulted and the manifestation of the animal world took place. This, bringing with it the faculties already possessed by
the vegetable kingdom, added thereunto two others, one the faculty of discovery,
which is called the "Perceptive Faculty", whereby the animal discerns things; the
second the power of voluntary movement, by the help of which the animal moves,
approaching that which is congenial to it and retreating from that which is offen-
sive, which is called the "Motor Faculty".[162]

Arūzī then describes the ten branches of the perceptive faculty, namely five external senses (touch, taste, sight, hearing, smell) and five internal: the "composite
sense", imagination, and imaginative (in animals) or cognitive (in man), appre-
hensive, and retentive faculties. These are the servants of the animal soul. Every
animal that possesses the perceptive and motor faculties, and the ten subordinate
faculties, is "perfect". Animals may be defective, for example by lacking eyes (the
ant or snake) or ears (the deaf adder). No animal is more defective than the maggot,
while the highest is the satyr (nasnās),

> a creature inhabiting the plains of Turkistán, of erect carriage and vertical stature,
> with wide flat nails. It cherishes a great affection for men; wherever it sees men, it
> halts on their path and examines them attentively; and when it finds a solitary man,
> it carries him off, and it is even said that it will conceive from him. This, after man-
> kind, is the highest of animals, inasmuch as in several respects it resembles man; first
> in its erect stature; secondly in the breadth of its nails; and thirdly in the hair of its
> head.[163]

Arūzī goes on to distinguish a hierarchy within mankind: wild men of
wastelands and mountains; those who dwell in towns and cities; and those who
reflect on "the real essences of things"—the philosophers and (above them)
prophets. The king is lieutenant to the imām, the imām to the prophet, and

the prophet to God. The prophet inherently knows everything, can speak of past and future events without recourse to analogic reasoning, and is the link between man and the angels.[164] Although in translating this passage Browne refers to the mineral, plant, and animal "kingdoms", Arūzī's word *'ālam* is better translated as *world*.[165]

At about the same time, and likewise in the eastern reaches of Islam, Zakarīyā al-Qazwīnī wrote *'Ajā'ib al-makhlūqāt wa-ġarā'ib al-mawjūdāt* (*Marvels of creation and miraculous aspects of creatures*), which came to be known simply as his *Cosmology*. He reviews the origin of minerals, plants, and animals, and ranks them in ascending series: the lowest minerals are joined to earth or water, the highest to the plants. The lowest plants are joined to the minerals, the highest to the animals; and similarly, the lowest animals are joined to the plants, the highest to man. The lowest men are joined to animals, highest to angelic souls.[166] According to Carus, al-Qazwīnī distinguishes animals from plants by their sensory ability and mobility. The animal closest to plants, the snail, has only the sense of touch, while apes are closest to man in both form and soul.[167] He recognizes seven types of animals: man; *jinn*, dæmons, and spirits; horses and beasts of burden; cattle and ungulates; beasts of prey; birds; and vermin, snakes, snails, worms, insects, spiders, and the like. Aquatic animals are treated separately, in the section on seas and ocean, where they are arranged alphabetically.[168]

There followed in the Islamic world a series of descriptive books on natural history that drew on the *Kitāb al-Ḥayawān* of al-Jāhiz, the *'Ajā'ib al-makhlūqāt* of Zakarīyā al-Qazwīnī, and/or specialized treatises, for example on desert animals or horses. The *Nuzhat al-qulūb* (*Delights of the heart*) of Mustaufī al-Qazwīnī contains sections on minerals, plants, and animals. The latter are classified as of the land, sea, or air, and land animals are subdivided into domestic, wild, beasts of prey, poisonous and creeping, and a curious group of mythical animals.[169] The wide-ranging *Muqaddimah* (*Prolegomena*) of Ibn Khaldūn, written in 1377 CE, not only identifies the organisms at the interface between plants and animals, and between animals and man, but also explains how the major groups are "connected":

> One should then look at the world of creation. It started out from the minerals and progressed, in an ingenious, gradual manner, to plants and animals. The last stage of minerals is connected with the first stage of plants, such as herbs and seedless plants. The last stage of plants, such as palms and vines, is connected with the first stage of animals, such as snails and shellfish which have only the power of touch. The word "connection" with regard to these created things means that the last stage of each group is fully prepared to become the first stage of the next group.
>
> The animal world then widens, its species become numerous, and in a gradual process of creation, it finally leads to man, who is able to think and to reflect. The higher stage of man is reached from the world of the monkeys, in which both

106 KINGDOMS, EMPIRES, AND DOMAINS

sagacity and perception are found, but which has not reached the stage of actual reflection and thinking.[170]

Ṣufiyya

The hierarchy *minerals, plants, animals* seems to have been passed uneventfully into the Islamic religious tradition *per se*. The relevant literature has been little explored from a natural-historical perspective, and works of important authors—indeed of entire schools—remain unavailable in any European language. A happy if limited exception concerns the last of al-Ghazālī's four classes of "seekers of knowledge", the *ṣufiyya*. Sufism emphasizes the internal dimensions of faith, and includes a reverence for Muḥammad as a spiritual example and guide, the first Perfect Man. Ibn al-ʿArabī developed a cosmology in which the seven spheres and the four elements ("fathers" and "mothers" respectively, as in Arūzī) give rise to *al-muwalladāt al-thalātha*—the three progeny or children,[171] namely minerals, plants, and animals (al-ʿArabī occasionally added a fourth, the *jinn*). In the *Miftāḥ al-ghayb* (*Key to the unseen*) of al-Qūnawī[172] we learn about the spiritual ascent of a Perfect Man:

> If he is a Perfect Man, when he enters the world of the three kingdoms [*al-muwalladāt al-thalātha*], his journey will be unitary. In other words, the first plant, for example, within which he becomes manifest, will be free of all corrupting impediments until it reaches its full growth at its own level. Or rather, he would normally become manifest in the most perfect species of plant that exists in the place appropriate to his spiritual reality and his station, or in the place which is the residence of his parents. Then God will send to the plant whomsoever He will. This person will pick it, for example, and cause it to reach the parents . . . they will eat that plant's form in the time appropriate to his level and to the level of the Command within which he is included. . . . Then that plant will be transformed into digested food, then blood, then sperm. It will become connected to the bodies of the parents in a manner that causes it to rise from the level of the plant kingdom and the mineral kingdom to the level of the animal kingdom. Finally, the material of his form will become entified and transferred from the loins to the womb. This is the first all-comprehensive entification which he undergoes . . . . In the speed of his transferal from the mineral to the plant kingdom, you can behold the speed of his transferal from the mineral to the plant kingdom. The levels are interrelated; no barriers separate them except *barzakhs* conceived by the mind.[173]

The Bektashi order of Sufis, founded in 1501, was influential in the Balkans during the middle and later Ottoman period. Their beliefs encompass a Neoplatonic emanation through intelligences, souls, and the celestial spheres, ending in the four elements; and a cycle of existence in which body "takes on more and more aspects of Real Being until in the Perfect Man there is a complete return to the Godhead".

The return ascent traverses the three progeny and culminates in the Perfect Man, a microcosm of the universe.[174] As transcribed from a Bektashi text by Seyit Ali Riza Baba,[175] "substance of coral" is positioned between minerals and seedless plants; the date palm is the highest plant; and ape and monkey are the animals closest to the "stage of man".

The Perfect Man			The First Intelligence	
Stage of Man			The Nine Intelligences	
			The Nine Souls	
Ape, Monkey			The Great Sphere	
			Sphere of the Constellations	
Kinds of Animals				
	ARC		Sphere of Saturn	
Palm of the Date	**OF**		Sphere of Jupiter	
Fruit Trees	**ASCENT**		Sphere of Mars	
		ARC	Sphere of the Sun	
Plants with seed		**OF**	Sphere of Venus	
		DESCENT	Sphere of Mercury	
Plants without seed			Sphere of the Moon	
Substance of Coral			Condition of Heat	
Minerals, stones			Condition of Cold	
			Condition of Moisture	
Minerals completed			Condition of Dryness	
			Fire	
			Air	
Volatile toughness			Water	
			Earth	

Examples could be proliferated,[176] but it is clear that a hierarchy of animals, vegetables, and minerals was broadly established in the Islamic tradition by no later than the Tenth century CE, although without recognition of *zoophyta*.

The Jewish philosophical tradition: Ibn Daud and Maimonides

As mentioned above, the *Talmud* (a collection of Jewish oral law, ethical precepts, history, and other rabbinical teachings) was completed around 550 CE.[18] According to Maimonides, an oral philosophical tradition once existed in Judaism alongside the oral law, but was lost during the Babylonian captivity, leaving only traces in the *Talmud* and *midrashim*.[177] When eastern Judaism encountered Hellenic philosophy, it did so in the context of *kalām*, specifically that of the Islamic *mutakallimūn* (who themselves had imported arguments from Philoponus and others against Aristotle). By contrast, Maimonides presented western Judaism—he was writing from al-Andalus—as unsullied by *kalām*.[178]

108 KINGDOMS, EMPIRES, AND DOMAINS

His fellow countryman Ibn Daud was one of the first Jewish philosophers to attempt to "hold two lamps"—that is, to reconcile Aristotelian philosophy, as presented by al-Fārābī and Ibn Sīnā, with the *Talmud*.[179] Ibn Daud's *Sefer ha-emunah ha-ramah* (*Book of exalted faith*)[180] is structured as three sections, on physics and metaphysics, Jewish religion, and moral philosophy respectively. He accepts that plants, animal, and man are ensouled. Plants possess three major (nutritive, augmentative, and procreative) and four subsidiary forces (attractive, retentive, assimilative, and expulsive); animals have twelve more, namely the five external senses, five internal senses, involuntary motion, and locomotion. Within the plant world there are "almost uncountably many gradations". Animals range from "the monkeys, which are close to the nature of man, down to the coral-trees, where plant- and animal-life are touching". Man possesses these nineteen forces of the soul, plus a rational mind.[181] Souls do not transmigrate.[182]

Ibn Daud's lamps were soon outshone by those of Moses Maimonides. In *Moreh nevukhim* (*The guide for the perplexed*) he sought to harmonize the Jewish religion with Aristotelian philosophy.[183] For Maimonides, the cosmos is a single living being, with the earth in the centre and celestial spheres above. As the heart rules the human body by its constant motion, so the outermost sphere rules the rest of the universe. The spheres communicate four forces to the sublunary world. One force governs the mixture and composition of the elements, and the formation of minerals; there are separate forces for vegetative and vital functions; and another force "endows rational beings with intellect".[184,185] Maimonides finds in the three sons of Adam an allusion to the three souls of man: "first, the animal element (Abel) becomes extinct; then the vegetable elements (Cain) are dissolved; only the third element, the intellect (Seth), survives, and forms the basis of mankind."[186] Within this symbolic reading there is no place for beings intermediate between plants and animals. Maimonides notes that some animals arise spontaneously, for example in rotting fruit; they do not reproduce sexually, and are "accidental species" that "take origin merely the general nature of transient things". He is clear, however, that these are animals.[187]

Kabbalah

We consider Kabbalah here (rather than in Chapter 9) because its major texts emerged in western Judaism during the lifetime of Maimonides (*Sefer ha-bahir*) or soon thereafter (*Sefer ha-zohar*) and influenced scholarship not only of Torah, but more broadly. An esoteric strand had long existed within the Judaic religious tradition, occupied with finding hidden truths in (for example) the vision of the Prophet Ezekiel.[188] *Kabbalah* ("tradition") builds on this strand, but refers more specifically to beliefs based on three premises: that beyond the creator of *Genesis* is an unknowable infinite God; emanated attributes (*sefirot*) of this infinite God gave rise to the universe, without in any way diminishing the infinite; and these *sefirot* form a bridge between the finite universe and the infinite God.[189] These tenets, which diverge markedly from rabbinical Judaism, might inform meditation, scriptural exegesis, or indeed theurgical practise.

The *Sefer yetzirah* (*Book of creation*) is the oldest book of Jewish mysticism. Versions of different length have survived, perhaps reflecting an origin among "hidden scrolls", personal records of material that was otherwise transmitted in a secret oral tradition. Opinions on the date it was committed to writing span more than one thousand years; indeed, different parts may have originated at different times.[190] The book is replete with number- and letter-symbolism, some shared with the *Memar* of Marqah,[191] and references to man as microcosm. God "made sevens beloved under all the heavens".[192] In his commentary on the *Sefer yetzirah* written about 1285 CE, Rabbi Abraham Abulafia identified seven levels in creation: form, matter, combination, mineral, vegetable, animal, and man. As the seventh level, man is the most beloved by God.[193]

Medieval kabbalah is thought to have arisen in Languedoc, in those years a hotbed of dualistic beliefs,[194] nucleated by direct revelations by the Prophet Elijah to distinguished members within the rabbinic community over several decades from the 1170s, and by the appearance of the book *Bahir* (*Bright*).[195] Opinions differ on whether kabbalah was a reaction against the Aristotelianism of Maimonides, who made no secret of his opposition to mysticism; his *Guide* appeared in Hebrew translation in 1204 CE, the year of his death. *Bahir* appears to have been edited between 1160 and 1180 CE in Provence by a group of anonymous scholars working with much older material originating from European Hasidic and eastern Gnostic traditions. Its appearance put Jewish mysticism into "unavoidable competition with the rabbinic and philosophic forms of this same medieval Judaism".[196]

The *Zohar* (*Splendour*), written by Rabbi Moses de León in the last decade or two of the Thirteenth century,[197] describes the universe as having come into existence following a series of emanations within the infinite, giving rise to the sefirot. These sefirot, ten in number, correspond to the names of God and to stages (in some cases, specific days) of creation, and indicate the path by which the divine reaches down to man and its reflection might return. They also signify the unfolding of the universe, which nonetheless (in some interpretations) might still be said to have been created ex nihilo.[198] Every blade of grass has its own star, and every star a designated higher being to represent before the Holy One.[199] Even so, the *Zohar* does not link any particular emanated attribute to animals, plants, or minerals.

Duran, Alemanno, and Albotini

Despite kabbalah (and mysticism more generally), the Aristotelianism of Maimonides largely prevailed in natural-historical thought in western Judaism. Beginning in the Thirteenth century, Hebrew-language encyclopædias by Yehudah ben Schlomoh, Ibn Falaquera, Gershon ben Shlomoh, and others paraphrased

110 KINGDOMS, EMPIRES, AND DOMAINS

Aristotelian natural history as transmitted by Ibn Rushd, Ibn Sīnā, al-Fārābī, and Themistius.[200] In his encyclopædic *Magen avot* (*Shield of the Fathers*), Rabbi Simeon Duran wrote:

> Out of the commingling of the elements come four [forms]. The first form is the form that becomes inanimate: the parts which are inanimate [bodies] and the minerals which are dead bodies; they have no movement except the movement from above downward when they have come out of their place due to a compulsion, and by the mystery of the compulsive [force] they return to their natural place, and this is their nature, as I have written above. And in the totality of this form are all the metals according to their kinds, and all the stones according to their kinds, and they all agree in this form, and differ [only] in their quality according to the mixture of their elements. And all of these were created on the third day, when the earth became revealed.
>
> The second form is the form of vegetation, for in it was added to the first form the form of sprouting and growth, and of producing seed that remains in it. This is the life-force to preserve the species, and this is the form which is called soul, for there are in it movements that change, downward and upward, for the roots move downward and branches upward, and the roots sense the sweetness of water and move toward it, and move away from the bitter water.

He goes on to present animals as the "third form" and man as the fourth, "speaking living form".[201] Yohanan Alemanno is noted for having introduced Pico della Mirandola[202] to kabbalah. In *Sha'ar ha-ḥesheq* (*Book of the gate of desire*)[203] he attempts to "organise all the sciences into one organic whole under the aegis of religion".[204] Alemanno was convinced that the adept could perform amazing transformations:

> And you, if you would know the relations of the metals which affect each other, you could transmute the whole world into silver and gold and precious stones and pearls. And likewise with the relations of the living things and the plants you could perform miraculous things never imagined by the ancients. And if, in addition, you would know the science of the relations of the stars and their representations on earth and all that which they pour out [*i.e.*, emanate] day after day, nothing would be too miraculous for you in judging the future and the past and the present, and all that the Lord has wrought, to find its solution in the [things] inanimate, vegetative, speechless living, and the living which has the power of speech.[205]

Rabbi Judah Albotini wrote commentaries on religious tracts by Maimonides, and the kabbalistic *Sullam ha-aliyyah* (*The ladder of ascent*):

> Indeed, our holy rabbis, peace be upon them, the prophets, the sons of the prophets, the Tana'im, and the 'Amora'im, who had in their hands the true Tradition from Moses, our Rabbi, peace be upon him, together with the wise men of the Kabbala

of recent generations and even the ancient wise men of research of the nations [of the world] who inclined toward the wisdom of the Sages of Israel in [these] matters—all agree that the worlds which encompass all created beings are three [in number]: the world of the separated Intelligences with its ten steps and hosts; second, the world of the spheres with its ten heavens, the stars, the constellations, and their hosts; [and] third, the world of the lower beings with its four elements; inanimate, vegetable, living, and human beings and their progeny. And the Master of all rules over all of them.[206]

All philosophic and religious worldviews eventually came up against early modern science. In Judaism this encounter was personified by Abraham Yagel, a kabbalist and "practicing magus"[207] who advocated the religious value of understanding the natural world. He sought to reformulate the kabbalistic tradition into the language of science, while preserving a legitimate role for the magical arts.[208] We explore these magical arts further in Chapter 9.

The rediscovery of Aristotle's natural history

We conclude this chapter with the European rediscovery of Aristotle. Building on Augustine's philosophy of reason, during the first half of the Twelfth century scholars discovered Aristotle's logic, theory of science, and method in his *Analytica*; then in the second half of the century, his fundamental concepts and principles of nature in the *Physica*.[209] Their debt to their Islamic counterparts can scarcely be overstated. Yet the Islamic and Jewish traditions were not mere chains of transmission connecting Athens to Paris, Oxford, and Padua via Baghdad, Cairo, and Toledo; they were, and remain, important in their own right and as part of our shared intellectual and cultural heritage. Moreover, as we have seen, some Hellenic and Roman thought passed into Europe directly, or through Constantinople: in the Eighth century Alcuin in the court of Charlemagne drew on *De interpretatione*, Themistius's gloss on the *Categoriae*, and Porphyry's *Isagoge* for his textbook of logic, *De dialectica*.[210] By legend, the medical school at Salerno was founded, perhaps in the Ninth century, by four physicians: one Arab, one Jew, one Greek, and one Latin.[211]

Even so, the Rediscovery was important in *its* own right: by the end of the Thirteenth century nearly the entire Aristotelian corpus had been rendered into Latin, as had texts by Archimedes, al-Fārābī, Galen, Hippocrates, Ibn Rushd, Ibn Sīnā, Ptolemy, and others, along with commentaries.[212] This time there was a receptive and increasingly sophisticated audience—not least in the universities, which attracted scholars from across Europe. Against this context, let us once more take stock of Latin translations of texts bearing ideas of a third or intermediate group of organisms, and/or the term *zoophyta*:

Themistius. On 22 November 1267 William of Moerbeke completed his translation of *In Aristotelis De anima* by Themistius,[213] preserving all seven instances of *zoophyte*.[214] A 1480 Latin translation by Barbaro uses *zoophyta* only once, to mean a third group between plants and animals.[215] The (Greek) *editio princeps* was published by Aldus in 1534.[216] The Arabic translation by Isḥāq ibn Ḥunain[55] was known to Ibn Sīnā and Ibn Rushd, but seems not to have been brought into Latin.

Although Themistius did not refer to zoophytes in his paraphrase of Aristotle's *Physica*,[217] the term nonetheless found its way into the Barbaro translation[218] because Barbaro appropriated parts of Simplicius's commentary on the *Physica* (see below), including as a prologue.[219] In the hands of Barbaro, Simplicius recognized three genera of beings: animals, vegetables, and "those that have something of both natures, called Zoophyta".[220] The Latin terminology used to position zoophytes (*ancipiti, ambigus, amborum*) is noteworthy.

Simplicius. A single fragment is known of a translation, perhaps by Grosseteste, of Simplicius's commentary on the *Physica*.[221] Much later (1480), Barbaro independently translated parts of this commentary for use in his interpretation of the paraphrase by Themistius (above). Latin editions of Simplicius on the *Physica* appeared in 1544, 1551, and 1558. Likewise, a single fragment has been preserved of Grosseteste's translation of Simplicius's commentary on *De caelo*.[222] A later translation by Moerbeke (1271) was published, with emendations, in 1540 and 1544.[223] By contrast, no Latin manuscript is known for the commentary on *De anima* attributed to Simplicius; the *editio princeps* is Aldine (1527) and the first Latin edition dates to 1543, although there are hints of an earlier edition (1480s).[224]

Philoponus. In 1268 Moerbeke translated chapters 4–8 of Book 3 of Philoponus on Ammonius's lectures on *De anima* from a Greek manuscript that has since been lost, but must have differed very greatly from all manuscripts of Book 3 now known. This lends credence to the idea that the latter are from another hand, perhaps that of Stephanus.[225] Even so, *zoophyton* appears in all three books of "Philoponus" on the *De anima*.[226] By contrast, Philoponus on Aristotle's *Analytica priora* was first brought into Latin not long before it was published in 1541. The *editio princeps* appeared in 1536; a partial Arabic translation[58] appears never to have been brought into Latin.

Nemesius. The Greek text referred to as *De natura hominis* was rendered into Latin (as *Premnon physicon*, or *Key to nature*) by Archbishop Alfanus of Salerno[227] prior to 1085 CE. A more-literal translation was made by Burgundio of Pisa after 1160, dedicated to emperor Frederic Barbarossa.[228] The Alfanus translation was read by Albertus Magnus and John of Salisbury, while the Burgundio translation was known to Peter Lombard and Aquinas.[229] Other translations appeared from 1512.[230]

De natura hominis was taken twice into Arabic, first in abbreviated form, then later at full length (as *Kitāb fī Ṭabīʿat al-insān*) by Ḥunain ibn ʾIsḥāq.[53] The shorter version was subsequently incorporated, along with extended excerpts from *Elements of theology* by Proclus and the first known instance of the *Tabula smaragdina* (*Emerald tablet*), into the alchemical treatise *Kitāb Sirr al-khalīqa wa ṣanʿat al-ṭabīʿa* (*Book of the secret of creation and art of nature*) falsely attributed to Apollonius of Tyana.[231] The *Sirr al-khalīqa* is ordered into treatises on creation, celestial bodies, minerals, plants, animals, and man. It exists in two recensions, with *Ṭabīʿat* in only the longer, although some brief parallels can be found in the shorter recension.[232] The shorter version of *al-Khalīqa* was brought into Latin (as *De secretis natura*) by Hugo de Santalla in the early Twelfth century;[233] we return to it in Chapter 9.

John of Damascus. The *Dialectica* was translated into Old Slavonic early in the Tenth century, and into Arabic by Antonius of Antioch[57] in the second half of the Tenth century. Most early Latin translations of *De fide orthodoxa* also included an abbreviated translation of the *Dialectica*, perhaps by Burgundio.[234] Grosseteste then translated the *De fide*, along with some or all of the *Dialectica*, between 1235 and 1253,[235] preserving *zoophyton*: "*animatum [dividitur] in sensibile animal et insensible et zoophyton et plantam*."[236] Other translations into Latin followed.[237] The Arabic version seems not to have been taken into Latin.

No record has come down to us of any pre-Sixteenth century Latin translation of three other works mentioning zoophytes: *In Aristotelis Categoriae* by Dexippus,[238] *In Porphyrii Isagogen* by "Ammonius Hermiae",[239,240] and the epitome *Peri khosmos* (*De mundo*) misattributed to Philo.[241] *In Porphyrii Isagogen* attributed to Elias, *In Porphyrii Isagogen* attributed to David, and Ibn al-Ṭayyib's commentary on Porphyry's *Isagoge* were never translated into Latin.

Aristotle's *De anima* was translated into Latin by James of Venice before 1150; Michael Scot made another translation, and rendered Ibn Rushd's *Long commentary* into Latin, between 1220 and 1235, while Moerbeke translated Themistius's paraphrase in 1267 and revised James's translation before 1268. Scot translated Aristotle's *Historia animalium*, *De partibus animalium* and *De generatione animalium* by about 1220, and the animal books of Ibn Sīnā's *Kitāb al-Šifaʾ* and perhaps the epitome of Ibn Rushd by 1232. Peter Gallego also translated the epitome between 1250 and 1267, while Moerbeke translated the *Historia*, *De partibus*, *De generatione*, *De progressu* and *De motu animalium* directly from Greek in 1260.[242]

And so passed Aristotle's works, and at slight remove zoophytes, into the European Middle Ages. Next, we consider the *status quo ante* in Europe, and the impact of the Islamic philosophical tradition and the rediscovery of Aristotle on medieval natural history, with particular reference to ideas of intermediate types of organisms.

8

Monastic and scholastic Nature

Nothing can exist in the world except animals, vegetables, and minerals.
—Marius, *On the elements*[1]

The end of classical antiquity and the dawn of the European Middle Ages can be dated with surprising precision. With the execution of Boëthius in 524 CE and closure of the Athenian schools in 529 CE the Hellenic philosophical spirit, already much attenuated, finally succumbed to an unforgiving Latinate Christianity. Yet 529 CE also saw the foundation of the monastery at Monte Cassino by Benedict of Nursia, while nine years later Cassiodorus—the successor to Boëthius as *magister officiorum* to Ostrogoth King Theoderic—retired to his estate at Vivarium, where he not only followed Benedict's example in founding a monastery but also equipped it with a library and school.[2] Benedict's *Regula* guiding the design and operation of monastic communities would underpin one of the key institutions of the ensuing Middle Ages.

Monastic communities appeared first in Egypt,[3] then in other Eastern Mediterranean lands and Europe.[4] The Sixth through Eleventh centuries brought a proliferation of monasteries across Europe, with thousands of Benedictine monasteries established from Barcelos (in present-day Portugal) to Kraków. Isolation, plague, and plundering Vikings often made their existence precarious, but monasteries preserved religious education, literacy, and access to rudimentary healthcare. Except in Ireland, they did not, however, preserve skills in Greek. Already by the time of Boëthius few Western scholars could read or even transcribe Greek, and as the eastern and western confessions became increasingly estranged from each other, contact was lost not only with the texts of Plato and Aristotle but also with the subsequent Alexandrian tradition.[5] Thus when Isidore, Bede, and Hrabanus Maurus attempted to rebuild classical learning across western Europe they necessarily drew on Latin sources, notably Virgil, Pliny, Boëthius, Jerome, and Augustine.

During the 780s Charlemagne reformed his own palace school; recruited Alcuin, the most highly renowned scholar in Europe, as its head; and directed bishops and monasteries to provide a liberal education to young men.[6] Momentum was soon lost, however, and by the mid-Ninth century Strabo despaired that "in our own

Kingdoms, Empires, and Domains. Mark A. Ragan, Oxford University Press. © Oxford University Press 2023.
DOI: 10.1093/oso/9780197643037.003.0008

MONASTIC AND SCHOLASTIC NATURE   115

time the thirst for knowledge is disappearing again: the light of wisdom is less and less sought after and is now becoming rare again in most men's minds".[7]

Only in the late Eleventh and Twelfth centuries did the dynamic change. Papal authority began to be restored under Pope Gregory VII.[8] Crusades were mounted, and the first Christian religious orders were founded.[9] Education and scholarship shifted from monasteries to urban cathedral schools, and the University of Paris emerged from the Parisian cathedral school around 1150 CE. The relationship between university and civil authorities was sometimes fraught; the University of Paris became locked in "angry feuds" with the municipality, at one point (1229 CE) prompting the scholars to flee.[10] The friars established schools in their convents and became "formidable rivals" to the University. The Franciscans attracted Alexander of Hales, Jean de la Rochelle, and Bonaventure, while the Dominicans countered with Albertus Magnus, Thomas Aquinas, and Duns Scotus. Meanwhile the rediscovery of Aristotle was underway in present-day Spain (Chapter 7).

Whether in universities or the convent schools, scholarship was confined almost exclusively to the Church[11] with the unsurprising aim of joining faith and reason; the approach that developed, particularly its reliance on inference, finely argued distinctions, and resolution set out as a *disputatio* between master and student, became known as scholasticism. For our purposes here, we need not engage deeply with the issues in play including the extent of the power of God, the nature of the Trinity, transubstantiation (whether substance can change without its accidents changing also), and the reality of universals (realism versus nominalism).[12] Studies of logic (dialectic) took precedence until the mid-Twelfth century.[13]

Cassiodorus to Hrabanus Maurus

We have already met Cassiodorus. His *De institutionibus divinarum litterarum* sets out a plan of study for monks (not only the scriptures but also histories, commentaries, and the medical works of Galen and Hippocrates), while his *De artibus ac disciplinis liberalium litterarum*[14] specifies the curriculum for monastery schools: religious studies, the *trivium* of arts (grammar, rhetoric, and dialectic), and the *quadrivium* of mathematics (arithmetic, music, geometry, and astronomy).[15] His plan for dialectic begins with Porphyry's *Isagoge* and Aristotle's *Categoriae*. He wrote a *De anima*,[16] but unlike that of Aristotle, it deals only with the human soul and does not mention a superaddition of vegetative, animate, and rational powers. In *Variarum* Cassiodorus points to the salamander as abiding in flames, as an example of how the sinful will suffer the fires of Hell.[17] On a more practical note, he may have translated works of Dioscorides into Latin.[18]

Archbishop Isidore of Seville considered himself a Spaniard, not a Roman. At the Fourth Synodal Council of Toledo (from 633 CE) he ordered his bishops to set up schools to educate the clergy in liberal arts as well as religion.[19] Isidore is best known for his *Etymologiæ*, a dictionary and encyclopædia of word origins that may

116   KINGDOMS, EMPIRES, AND DOMAINS

have been intended for the newly literate Visigothic ruling class as much as for the clergy.[20] Etymology was part of the art of grammar,[21] and while Isidore's word origins are often fanciful, he intended them to aid in the understanding and correct use of words, rather than as moral or religious allegory.[22] His entries preserve folklore and passages from many ancient works that have otherwise since been lost. However, *Etymologiæ* fails to advance natural history beyond its state in Roman times. Animals (Book 12) are considered separately from stones and metals (Book 16), while plants make up the great majority of "rural matters" (Book 17). The dragon, basilisk, two-headed *amphisbæna*, and many-headed hydra are serpents, while the hippopotamus, whale, frogs, and sponges are types of fish.[23] Bees can arise from cattle, locusts from mules, and (on the authority of Ovid) scorpions from crabs.[24] Coral is a "red gem" and is considered with gemstones.[25] There is no listing for *zoophyte*.

Isidore also wrote a *De natura rerum* that considered the heavens and earth, day and night, months and years, solstice and equinox, the earth, planets, sun, moon, stars, eclipses, lightning, clouds, snow, wind (including pestilences, which are due to corrupted air), the ocean, seas, rivers, earthquakes, and the Mount Ætna volcano.[26] Bede, an English Benedictine monk, likewise wrote a *De natura rerum* that draws heavily on Isidore and Pliny, and shows evidence of some personal observation.[27] Neither work treats living organisms.

We have also briefly met Alcuin, recruited from York to head Charlemagne's palace school (and teach Charlemagne himself). In his written work, Alcuin refers occasionally to plants and animals ("trees, herbs, animals, all living things, men likewise, are called substances").[28] But living organisms merit no place in his outline of knowledge.[29] A few decades later, Archbishop Hrabanus Maurus[30] based his encyclopædic *De universo*[31] on Isidore's *Etymologiae*, including its medical portion "almost *in toto*".[32] Likewise his books on animals, stones and metals, and agriculture borrow heavily from the corresponding sections of Isidore.[33] Sponges are treated in Book 8, with fish. Hrabanus also wrote a brief *De anima* which, like that of Cassiodorus, focuses only on the human soul.[34]

Eriugena

The Ninth century brought a very different voice to Carolingian Europe: John the Gael, known as Eriugena.[35] An Irish philosopher,[36] he wrote commentaries on the idiosyncratic *De nuptiis Philologiæ et Mercurii* (*On the Wedding of Philology and Mercury*) by Martianus Capella,[37] and translated from Greek into Latin the *De hominis opificio* by Gregory of Nyssa, the *Ambigua ad Iohannem* of Maximus the Confessor, and the known works of Pseudo-Dionysius. He held that (Neoplatonic) philosophy was in full accord with (mystical) Christianity, indeed that the liberal arts "constitute an independent way to salvation" and that "no one enters into heaven except through philosophy".[38]

While at the palace school of Charles the Bald[39] Eriugena wrote *Periphyseon*, also known as *De divisione naturæ*. For Eriugena, *natura* is the totality of things, those that exist and those that do not.[40] Its divisions are four: God, who creates and is not created; the primordial causes, which are created and themselves create; temporal effects, which are created but do not create; and that which cannot be, which perforce is neither created nor creates.[41] The latter three arise in continual emanation from God, to whom the soul will return. He treats these temporal effects—from the intelligible and celestial essences down to the lowest level of visible creation, that of bodies—in Book 3 of *Periphyseon*. Much of his argument focuses on the meaning of creation ex nihilo,[42] but he eventually reaches animals, plants, and (citing Basil and Augustine) the six days of creation.[43]

Eriugena makes it clear that "all bodies which are naturally constituted are governed by some species of life", by which (following Plato and early Augustine) he includes the celestial and heavenly spheres, animals, plants, trees, and "even the bodies which appear to our senses as dead".[44] Indeed, (unnamed) philosophers have not unreasonably regarded plants and trees to be "animals fixed in place . . . for they are animate bodies which increase".[45] He then sets out a position that owes much to Aristotle:

> Now, the irrational life is divided into that which participates in sense and that which is without sense; and the one is distributed among all animals which possess the power of perceiving, the other among [matters] which lack all sense, the kind of life which is held to rule plants and trees, and below which reason finds no kind of life at all.
>
> Thus by four differentiations created life is brought together into four species: the intellectual in angels, [the rational in men, the sensitive in beasts, the insensitive] in plants and in the other bodies, in which only the form shows a trace of life, as are the four elements of the world whether as simple in themselves or as composite: earth, I mean, water and air and ether. And this is why man is not inappropriately called the workshop of all creatures since in him the universal creature is contained. [For] he has intellect like an angel, reason like a man, sense like an [irrational] animal, life like a plant, and subsists in body and soul: [there is no creature that he is without]. [For] outside these you (will) find no creature.[46]

Moreover, animals and man constitute a single species, hence animals possess an immortal soul.[47] Even so, man is more than a rational animal: he is a mirror of God.

Eriugena's *Periphyseon* makes for heavy reading, and was not immediately influential. Its arguments were nonetheless censured by various Church councils, and later by Popes Honorius III (1225) and Gregory XIII (1585), and it ended up on the *Index librorum prohibitorum*. It was better-known through the *Clavis physicæ*, a non-technical summary by Honorius, a student of Anselm.[48]

118 KINGDOMS, EMPIRES, AND DOMAINS

Anselm, Peter Abelard, and Peter Lombard

The Twelfth century brought stirrings of vitality to a moribund Europe. A warm climate[49] facilitated the spread of agriculture as large tracts of forest were cleared and towns established. Political consolidation and the spread of Christianity brought an end to Viking raids. Public baths were commonplace. Windmills, the keel and rudder, the magnetic compass, the astrolabe, and mechanical clocks were introduced, while new engineering technologies brought Gothic architecture and the bridge at Avignon. Latin classics, poetry, and jurisprudence were revived even as vernacular languages emerged and new visual arts appeared. The Hanseatic League was founded, and new trade routes developed. Travellers and Crusaders returned from Eastern Mediterranean lands with texts of Plato, Aristotle, and other Hellenic philosophers, while papermaking reached Spain from Islamic lands, and the Retranslation got underway in earnest. Universities emerged at Oxford, Cambridge, and Paris. On the other hand, disease, poverty, and violence remained the lot of many, while the Twelfth century also brought armed conflicts, persecution of Jews, and the Inquisition.[50]

Anselm of Canterbury is a key figure in the Twelfth-century intellectual renaissance through his concern with "the relations between God and man, the presence of the divine in nature . . . [and the] implications of the cosmic and psychological renewal effected by the Incarnation."[51] Anselm taught that reason must be subordinate to faith, and encouraged his readers to begin their intellectual journey with Augustine.[52] With his focus on issues such as the existence of God, Anselm has little to say about natural history. He does, however, accept a gradation in nature:

> if anyone considers the natures of things, he cannot help perceiving that they are not all of equal excellence but that some of them differ by an inequality of gradation. For if anyone doubts that a horse is by nature better than a tree and that a man is more excellent than a horse, then surely this [person] ought not to be called a man.[53]

Anselm defines *animal* as *living substance capable of perception*, and man as a rational animal.[54] He counterposes stone, animal, and man (although not trees or plants) in *De veritate*.[55]

For Peter Abelard, the path to truth begins with doubt, not faith. His *Sic et non*, in which he brought together the conflicting opinions of the Fathers on matters of doctrine, was seen as promoting scepticism towards authority.[56] Despite his statement to Héloïse that "he would not wish to be a philosopher if it were necessary to contradict St Paul" nor "wish to be an Aristotelian if that meant it were necessary to separate himself from Christ", his nemesis Bernard of Clairvaux replied that "Abelard sweats blood and water to make Plato a Christian, but proves only that he (Abelard) is himself a pagan."[57] Abelard made important contributions to our understanding of the relationship between individuals and categories. Like Eriugena and Anselm, he illustrated his arguments using the categories stones, animals, and man (although so far as I can determine, not plants).[58]

Peter Lombard, a professor at the cathedral school of Notre Dame, was briefly Bishop of Paris. His *Liber quatuor sententiarum* (*Four books of sentences*) brings together the opinions of Augustine, various Fathers, and later masters on the mystery of the Trinity, creation, incarnation of the Word, and the doctrine of signs.[59] Within the book on creation is a hexæmeron[60] that breaks no new philosophical ground. He is reluctant to commit to whether "plants, trees, and perhaps animals" were created at one and the same time, and if so whether only formally, or materially as well.[61] He follows Augustine in allowing that certain small animals, particularly those that arise from the bodies of dead animals, may be said not to have been created on the Fifth or Sixth day; but "those which are born from the earth or from waters, or from those things which come from the germinating earth, may not incongruously be said to have been created at that time".[62]

Adelard and Berachya

Adelard was born in Bath,[63] studied at Tours, and began teaching in Laon before travelling to Salerno, Sicily, Greece, and the eastern Mediterranean, probably including Palestine. His first major work, *De eodem et diverso* (*On the same and the different*), was in the spirit of Boëthius: Philosophy is accompanied by the seven liberal Arts, who guide the soul in its earthly journey.[64] He translated Euclid's *Elements* from Arabic, and wrote treatises on hawkery and on the astrolabe. His *Quæstiones naturales*[65] is more relevant to our topic at hand. Presented as a dialogue with his nephew, *Quæstiones* purports to set out what Adelard has learned from the Arabs[66]—although he mentions no Islamic scholars or books by name, and employs no Arabic terms. He refers repeatedly to creation and the Creator,[67] describes the four Galenic humours,[68] and quotes from the *Premnon physicon* of Alfanus.[69] The questions themselves proceed from the lowest things to the highest[70]—from plants "fixed in the ground by their roots" through trees, brute animals, man, the earth, atmosphere, planets, and the heavenly spheres, and conclude by acknowledging "the incomposite, formless, unchangeable, infinite God from whom is the cause of all things".[71,72]

In the late Twelfth century, the Jewish moralist Berachya ha-Naqdan[73] adapted Adelard's *Quæstiones naturales* as *Dodi ve-nechdi* (*Uncle and nephew*).[74] Although Berachya sought to "redeem" the *Quæstiones* by emphasizing spiritual rather than scientific and philosophical issues,[75] his text nonetheless progresses from the earth and sea through plants, trees, animals, and man to atmospheric phenomena and the heavenly bodies. Just as the ground brings forth herbs and the waters fish, the air brings forth "a creation called *Lilith*, which inhabits ruins", while fire gives rise to "a bird called *Salamander*".[76] In Scripture the word *soul* is used only for man, but he concedes that even "dumb animals that move about" have a soul even if it cannot,

120 KINGDOMS, EMPIRES, AND DOMAINS

after death, rise up to God. Soul has three parts: *nephesh*, associated with generation, growth, and "longing for food and cohabitation", located in the liver; *neshama*
(breath), located in the forehead; and *ruach* (spirit).[77]

Hildegard and Marius

Hildegard, Abbess of the Benedictine house at Bingen,[78] represents the final phase
of monastic writings before the rise of universities. She kept a herbal garden, and
in addition to liturgical and mystical works wrote *Liber subtilitatum diuersarum
naturarum creaturarum* (*Book of the intricacies of the diverse natures of creatures*).
A few decades after her death the *Liber subtilitatum* was split into separate texts,
Liber simplicis medicinæ and *Liber compositæ medicinæ*, better known as *Physica*
and *Causæ et curæ* respectively.[79] The *Physica*[80] is set out as nine books: on plants
(including mushrooms, honey, sugar, milk, butter, salt, vinegar, birds' eggs, resins,
and sulphur), the elements (including the earth, sea, and rivers), trees (rose hips,
thorns, the grapevine, berries, wood smoke, moss), precious stones (magnets,
pearls), fish (whale, seal), birds (insects, bats), terrestrial animals (unicorn, flea,
ant), reptiles (dragon, snakes, spiders, scorpions, basilisk, earthworms, snails,
slugs), and metals. She does not mention the oyster, mussel, coral, or sponge.

Hildegard begins *Causæ et curæ* with the creation of the angels, soul, elements,
sun, stars, planets, moon, and winds. As man is a small world, the heavenly bodies
are deeply relevant to the causes of human disorders and can guide their treatment,
for example in the use of extracts from plants whose shape, taste or colour correspond to, or oppose, the specific qualities that are out of balance. Her understanding
of physiology was generally Galenic, although she referred to qualities (*tepidum,
spuma, humidum, siccum*) as humours[81] and emphasized the salubrious effects of
viriditas, the vitality that plants draw from the earth. Hildegard referred to *viriditas*
in different ways: as a quality (the colour green), a substance (the juice or sap of
plants), or a metaphor for "positive spiritual action—integrity, truth, penance, abstinence".[82] She referred to human reproductive anatomy in botanical terms: the
penis is a stem (*stirps*), the menstrual flow is (one form of) *viriditas*.[83]

By contrast, little is known of Marius, author of *De elementis* (*On the elements*). As
it has come down to us the treatise is incomplete, but was probably written during
the 1160s.[84] Notably, Marius refers to minerals (*congelata*) rather than "stones and
metals", and plants (*virentia*) rather than "herbs and trees". Unlike Hildegard, he
treats minerals, plants, and animals in successive sections, and in several places explicitly juxtaposes the three terms.[85] Plants are growing things that do not move
from place to place; they have appetitive, digestive, and retentive powers; and after
a plant begins to grow it also exhibits an expulsive power.[86] Unlike plants, animals
have sensibility and can move from place to place, although "many animals lack
transitive motion, such as oysters and things similar to them".[87] Yet "if any one

MONASTIC AND SCHOLASTIC NATURE   121

defining an animal should say that it is a thing which grows and declines, is mobile and sensible, he would not deviate from the truth of the matter".[88]

[Student:] Now I know very well what the differences and similarities are between animals and plants. Now I ask whether or not there is anything in addition to sensibility and transitive motion which might to any extent be added to a plant, by which something different from the plant or animal might be brought about. I do not question that oysters, which grow like plants and do not move from place to place, nevertheless have feeling like animals.[89] And there are other things intermediate between mineral and plants, such as coral, which is heavy, solid, cold, dry, and flat. We know that this, while it is in its native place under the ocean waves, grows without having branches or roots. But I am asking about a thing which is not an animal, not a plant, and not intermediate between these.

[Master:] . . . nothing can be added to a plant except motion and rest. If you add rest, when the motion ceases it will be like a mineral. But if you add motion, it will either be of doing something, of decaying, of growing larger or smaller, or even of changing into something else. . . . But this cannot be, for the power of this motion cannot change one species into another, but only one quality into another . . . never can the plants be changed as a result of changes of their qualities. Therefore, nothing by which something can be brought about different from an animal or plant, can be added to plants except sensibility and transitive motion. And nothing can exist in the world except animals, vegetables, and minerals. You should know, therefore, that this is the foundation of this whole art.[90]

This remarkable conclusion could scarcely have been asserted with such confidence before the Twelfth century. The treatise (as we have it) concludes with a summary by the student: "Man is composed of the four elements. I also know that he is, in a certain way, similar to minerals. For if he is lifted up, he will fall to the earth like a mineral. And after death he can be counted among the minerals. He is also similar to plants, for he grows like a plant. That he is an animal, no one denies. He is also similar to the angels, for he is rational like an angel." The master adds "he is also called a small world (*minor mundus*) by philosophers", and closes with a benediction.[91]

The School of Chartres

During the Twelfth century the cathedral school at Chartres, in Paris, attracted some of the leading scholars of the age, among them Bernard and Thierry of Chartres, Gilbert of Poitiers, William of Conches, John of Salisbury, and Alain de Lille. The school first came into prominence after 990 CE under Fulbert, a student of Gerbert of Aurillac (later Pope Sylvester II),[92,93] and by the Twelfth century was notable for its emphasis on the quadrivium and natural philosophy rather than

122 KINGDOMS, EMPIRES, AND DOMAINS

grammar, rhetoric, and dialectic.[94] Scholasticism ruled, but the intellectual culture at Chartres has been described as rational, Platonic, and humanistic.[95]

Chartrian philosophers wrote on cosmology, the elements, universals, and "the *causae* operative in created life".[96] Thierry of Chartres, in his gloss on *Genesis*, limited the hand of God to creation of the elements themselves: *ita igitur ignis est quasi artifex et efficiens causa.*[97] Hugh of St Victor (citing Valerius Soranus) identified this creative fire with nature,[98] and in *De planctu naturæ* Alain de Lille personified Nature and relates a dream in which she descends to earth to admonish and reform degenerate mankind.[99] John of Salisbury encouraged his readers to consult the *Premnon physicon*, "a book which discusses the soul most fully",[100] so presumably he encountered the term *zoophyta*. Man remained a microcosm, but step by step, nature was disentangled from the mind of God and set on her way as an autonomous if mysterious lifegiving principle.[101] Even so, William of Conches[102] and Hugh of St Victor[103] among others retained the three superadditive types of soul and the five senses.

When rationalist push inevitably came to theological shove, William recanted quickly if perhaps strategically.[104] By the third quarter of the Century the fortunes of Chartres—or at least its Platonic strain—were in decline, hemmed in on one side by religious tradition, on the other by increased specialization in the sciences, and by the books of Aristotle that were becoming available in Latin translation.[105]

Bernard Silvestris and John Blund

Little is known of Bernard Silvestris; a poet and teacher associated with Tours, he may have known Adelard of Bath, and was presumably influenced by Archbishop Hildebert, "the only earlier medieval poet in whose work the personified goddess Natura tended to be more than a rhetorical flourish, carrying for Hildebert, as later for Bernard, the sense of both a creatrix in the macrocosm and an inner law in the microcosm."[106] Silvestris wrote *Mathematicus*, a poem which has nothing whatsoever to do with mathematics per se[107], and *Cosmographia*[108], an allegorical poem dedicated to Thierry of Chartres. In *Cosmographia*, the world has become weary and run-down; the cosmic cycle of birth and renewal must be restarted. "Gods and goddesses of the classical pantheon flit across the stage as through the writer were a pagan";[109] the elements, soul (Anima), and world-soul (Endelichia) play their roles, and the firmament, the earth, animals, and man come forth. Sea-urchins are fish, as are oysters;[110] plants and animals are mentioned in abundance, without hint of intermediate types. His *Cosmographia* was a success: later in the century Giraldus Cambrensis plagiarized it for his poem *De mundi creatione*, and in due course Alexander Neckam, Boccaccio, Chaucer, and others cited the *Cosmographia* or borrowed its themes.[111]

Among the scholars we encounter in this chapter, John Blund was the first to make use of the new translations of Aristotle and Ibn Sīnā. Blund lectured on

Aristotle at Oxford and Paris, and—perhaps when the University of Paris was dispersed in 1229—returned to England, where he briefly served as Archbishop of Canterbury, and then became Chancellor of York Cathedral. His *Tractatus de anima*[112] appears to have been written around 1200 CE, before the Condemnation of 1210.[113] Blund accepts the three superadditive souls[114] and, following Ibn Sīnā, argues that in man they are parts of a single soul.[115] The sensitive soul is responsible for voluntary movement.[116] The rational soul brings man together with the *angelis* and *intelligentiis*.[117] Blund discusses most of the passages in Aristotle's *De anima* at which Themistius used the term *zoophyte*, but gives no indication of intermediate organisms.[118]

Robert Grosseteste

As we have seen, the translation of works from Arabic, and later from Greek, into Latin gathered pace during the Twelfth century. Several of the translators were notable philosophers in their own right, including Domingo Gundisalvo and Michael Scot. Of these, by far the most interesting in the present context is Robert Grosseteste, Bishop of Lincoln.[119] His *Hexæmeron*, written in the 1230s,[120] provided far more detail on plants and animals than had, for instance, those of Ambrose or Augustine, and for the first time found a place for Aristotle's dualizers in the *Genesis* story.

As was standard for hexæmeral literature, Grosseteste began at the Beginning and worked his way Day by Day through Creation. He asserts that herbs differ from trees in that the former were made to be eaten by animals, whereas trees exist to provide fruit to man and animals—although "the tender shoots of some trees are like herbs and are often eaten by animals".[121] Plants have "something analogous to heart" where the vegetative soul is located, and "rotatory motions" that attract nourishment through the root, retain and digest it, and expel that which is superfluous.[122] The vegetative soul is not "living soul"; plants live "a kind of dead life" inferior to the "living life" of sentient beings.[123]

He quotes Aristotle that "in the sea there are some things about which there is a doubt whether they are animals and plants. They are joined to the places where they are: and many of them die if you take them somewhere else."[124] Grosseteste returns to this point:

> The first power of the soul is the vegetative power which can be called animation; secondly, there is the sensitive power, which is what is understood here by the word "life"; and thirdly there is the power of local movement. For not all sentient beings have the power of processive change with regard to place, e.g. many shellfish and the creatures that are attached to stones.[125]

He goes on to consider the "creeping things" brought forth in the waters:

124   KINGDOMS, EMPIRES, AND DOMAINS

> Perhaps the word "creep" includes the motion of dilation and contraction. . . . Or maybe these water creatures that have no processive motion are referred to as plants and things with sense: i.e. as a mid-point between vegetables and sensible things. They share with sensible things touch and taste, and the motion of contraction and dilation. But they share with plants immobility as regards place, and some of them share fixed attachment. Hence they are a sort of mid-point, as the philosophers say, between vegetables and things with sense. And so perhaps they are left by the Lawmaker [Moses] to be understood as included in virtue of the inclusion of the extremes.[126]

Alternatively, their power of creeping may be imperfect, as the ostrich is included among the birds and "has some kind of imperfect power of flying",[126] and "as halfway between the plants and the animals are the immobile water animals, so midway between the water creatures and the walking land animals are the creeping land animals".[127]

Thus "following the very words of Scripture" we can "make a first general division of animals" into the sentient-but-unable-to-move-from-place-to-place, and the sentient-and-mobile. The former are "a kind of mid-way point between plants, which have only vegetative activity, and animals, which, without qualification, have sense activity and move." This is demanded by the order of nature. Moreover, "it is necessary that the sentient creatures which are immobile would be water creatures. Since they cannot move locally to get their nourishment, they have to live in an element which brings nourishment to them." And as they cannot flee danger, "nature has armed them with hard shells against things that might damage them".[128]

Time and again Grosseteste offers alternative readings—literal, philosophical, allegorical—and invites the reader to decide. He declines to be drawn on whether the six Days are actual units of time,[129] or on whether "the things that are born from corruptions and purgings or exhalations or the decay of dead bodies" are natural species, or "a sort of degeneration from perfect species". But of what perfect species, he asks, is the louse, or the worm, a degeneration?[130] Grosseteste respects the reader, but his *Hexæmeron* is a work of faith. He cannot resist moral lessons, and following Augustine "and many others", accepts that salamanders live in the fire.[131]

Thomas of Cantimpré, Bartholomæus Anglicus, and Vincent of Beauvais

Between 1228 and 1244 Thomas of Cantimpré, a Dominican and student of Albertus Magnus (below), composed a book of nature (*De natura rerum*)[132] that covered many of the same areas, but from a much less philosophical (or theological) perspective. Its twenty books treat the human body, soul, strange races of men, quadrupeds, birds, sea monsters, fish, serpents, an eclectic group of "worms" (including insects, mice, frogs, and tortoises), ordinary trees, aromatic trees, aromatic herbs, fountains and rivers, precious stones, the seven metals,[133] the seven regions of the atmosphere, the seven planets, the seven *passionibus aeris*, the four elements, and the *ornatu celi*. Sponges, starfish, and the octopus are deemed fish, while dragons and salamanders are registered as serpents. Within each chapter the arrangement is largely alphabetical. Later Thomas's *De natura rerum* was set to verse by Jacob van Maerlant as *Naturen bloeme*,[134] and served as the model for Conrad von Megenberg's *Buch der Natur*.[135]

Bartholomæus Anglicus, a Franciscan, taught theology at Paris,[136] perhaps also at the time of Albert. His *De proprietatibus rerum* (*On the properties of things*), written around mid-century, sought to "explain the allusions to natural objects met with in the Scriptures . . . for the benefit of the village preaching friar".[137] An immediate success, *De proprietatibus* was translated into French, Spanish, Dutch, and English, and after the invention of moveable type went through numerous printed editions. Its nineteen books deal with the Trinity, angels, soul (including its vegetable, sensible, and rational powers),[138] the elements, the human body, the ages and conditions of man, diseases and infirmities, the heavens, time, forms of the elements, atmospheric phenomena, birds, waters and fish, the earth, countries and regions, stones, trees and herbs, beasts, and colours. The entries within some chapters (including those on stones, trees and herbs, and beasts) are again alphabetical.[139] Following Pliny and Isidore, Bartholomew places shellfish, crabs, crocodiles, beavers, hippopotami, and whales among fish.[140] Coral is a tree so long as it is immersed in water, but in air turns into a stone,[141] while insects, spiders, mermaids,[142] and worms are animals. Salamanders quench a fire, and their spittle causes the hair to fall out.[143] Bartholomew spares us neither tall tales recycled from Pliny, nor moral lessons from the *Physiologus*.

Vincent of Beauvais, a Dominican of Albert's generation, is known for his ambitious *Speculum maius* (*Great mirror*), intended to compile everything known about nature, philosophy, the practical arts, history, and morals. Arranged to parallel the Bible from *Genesis* to *Revelation*, the three volumes attributable to Vincent himself cite more than 450 authors over 9885 chapters.[144] He attributes to Aristotle the statement that marine sponges are bodies intermediate between plants and animals because they have roots like plants yet perceive touch like animals, and draws further information from Pliny, Ibn Sīnā, Dioscorides, and others.[145] He credits to Ambrose and Aristotle the belief that corals are marine herbs that in the air are transformed to stone,[146] although elsewhere he considers them marine trees[147] or groups them with stones.[148] In none of these three highly detailed works, explicitly dependent on sixteen centuries of authority, do we find the term *zoophyta*.

Albertus Magnus and Thomas Aquinas

More than comprehensively than anyone before or since, Albert,[149] a Dominican friar and bishop, welcomed the Aristotelian philosophical and natural-historical

126 KINGDOMS, EMPIRES, AND DOMAINS

tradition into Christianity. He wrote paraphrases of most of Aristotle's works—not to mention those of Porphyry, Boëthius, and others—and "where he felt that Aristotle should have produced a work, but it was missing, Albert produced the work himself".[150] In decades of trekking across central Europe on Church business[151] Albert had opportunity to observe the local flora and fauna, and indeed made empirical investigations of his own. He intended his *De mineralibus, De vegetabilibus*, and *De animalibus* as steps towards commenting on all of human knowledge.[152]

In Albert's works we again find a hierarchy of vegetative, sensible, rational, and intellective souls[153] and a progression in nature, which "never causes genera to be distant without creating some medium between them, since nature only passes from extreme to extreme through a medium".[154] Plants are intermediate between the non-living and living, as they are alive compared to minerals but non-living compared to animals. More to the point, fungi are intermediate between minerals and plants;[155] indeed, perhaps *fungi, tuberes*, and *boleti* are not plants at all, but instead "plant-like things that are exhaled and evaporated from other plants".[156] Plants (excluding fungi) can be divided into herbs, bushy herbs, bushes, and trees,[157] among which the tree is the "most animate", and from whence there is "no great distance" to animals.[158]

The marine sponge is "like an intermediary between plants and animals . . . for it moves by expansion and contraction just like an animal and yet has leaves in the manner of a plant, which we have seen with our own eyes in the sea."[159] There are many gradations in animality, and in the sea "there are certain animals concerning which there is doubt whether they are animals or plants" including shellfish, the sea-squirt (*tycho*),[158] and sea-nettles (*stincus*).[160] Some animals arise from putrefaction, or are asexual, or lack movement from place to place, or vary collectively or individually *e.g.* in colour, or hatch from eggs; yet if they are sensate, they are animal.[161] Like its archetypes sixteen centuries earlier, Albert's *De animalibus*[162] classifies animals according to diverse criteria including the element in which they live, and their mode of locomotion. Indeed, by some criteria (he asserts) a child can be considered intermediate between brute animals and humans. Albert is careful to explain the precise manner in which fungi, sponges, or children might be intermediates: as between potency and act, and in combining properties of both extremes, but not as an external negation.[163]

Albert was "the first Scholastic who reproduced the whole philosophy of Aristotle in systematic order . . . [and] remodeled it to meet the requirements of ecclesiastical dogma", but his student Thomas Aquinas fitted it all together more perfectly.[164] Aquinas began his education at the Benedictine monastery at Monte Cassino, where his uncle was abbot, but he later escaped to join the Dominicans. He met Albert at Paris, followed him to Köln, then taught theology at Paris and was appointed Papal theologian at the Vatican. If contemporary reports are to be believed, Aquinas walked away from his final project—the *Summa theologica*—following a series of ecstatic visions.

MONASTIC AND SCHOLASTIC NATURE   127

Aquinas's philosophy, like that of Albert, is grounded foursquare in Aristotle. Aquinas wrote a dozen commentaries on Aristotle, from *De interpretatione* to the *Politica* and *Metaphysica*, although not on the books on animals. Aquinas was not reluctant to apply reason to matters claimed by the Church, for example the existence of God. But he argued that some of the higher mysteries—the Trinity, the Incarnation, creation of the world in time—cannot be demonstrated *propter quid*, and must be approached from a position of faith.[165] In another innovation, he transformed the Aristotelian notion of being-as-form (*forma* or *modus*) to that of being-as-the-act-of-existing (*viventus*).[166]

Aquinas recognizes the vegetative, animate, and rational types of soul,[167] but also refers to grades or steps of life (*in gradibus viventium*).[168] Life reveals itself in four modes: as intellective, as sensitive, as the cause of spatial motion and rest, and as the cause of the motions of taking nourishment, decay, and growth. The life of plants entails only the latter three motions, whereas other living things "also have sensation, but are always fixed to one place—such are the inferior animals like shellfish (*ostreae*)".[169] All animals share the sense of touch; many inferior animals possess only this one sense.[170] In Chapter 5 we saw that Themistius, commenting on Aristotle's *De anima*, used *zoophyta* seven times to refer to sensate beings that do not move from place to place. In the corresponding passages of his own commentary, Aquinas sometimes refers to "plants and things resembling plants [that] do not move locally at all but are fixed in one place"[171] or to *ostreae* (above), but shows no awareness of the term *zoophyta*.[172]

Aquinas returns on multiple occasions to the gradient in nature:

Following a diversity of natures, one finds a diverse manner of emanation in things, and, the higher a nature is, the more intimate to the nature is that which flows from it. For, in all things, inanimate bodies have the lowest place. There can be no emanations in these except by the action of some one upon another one. For this is the way in which fire is generated by fire . . .

Among animate bodies the next place is held by the plants, and in these the emanation does proceed somewhat from what is within: to the extent, namely, that the internal humor of the plant is converted into seed and that the seed committed to the soil grows into a plant. Here, then, one has already found the first grade of life, for living things are those which move themselves to action, but those which can move only things external to them are entirely devoid of life. . . . The life of plants is nevertheless imperfect; this is because, although the emanation in plants proceeds from what is within, what comes forth little by little in the emanation is, at the end, found to be entirely external. For the humor first emerging from the tree becomes a blossom, and at length a fruit distinct from the tree's bark, yet still fastened to it. But, when the fruit is perfected, it is separate from the tree altogether; it falls to the ground and its seeding power produces another plant. If one also considers this carefully, he will see that originally this emanation comes from

128 KINGDOMS, EMPIRES, AND DOMAINS

what is external, for the internal humor of the tree is taken through the roots from the soil from which the plant receives nourishment.

Beyond plants one finds a higher grade of life: that of the sensitive soul. Its emanation may have an external beginning, but has an internal termination, and, the more fully the emanation proceeds, the more it reaches what is within. For the exterior sensible impresses its form on the exterior senses; from these it proceeds to the imagination and, further, to the storehouse of the memory. Nevertheless, in each step of this emanation the principle and the term refer to different things; no sensitive power reflects upon itself. This grade of life, then, is higher than the life of plants—higher to the extent that its operation takes place within the principles which are within; it is, nevertheless, not an entirely perfect life, since the emanation is always from some first to some second.

That, then, is the supreme and perfect grade of life which is in the intellect, for the intellect reflects upon itself and the intellect can understand itself. But even in the intellectual life one finds diverse grades. For the human intellect, although it can know itself, does indeed take the first beginning of its knowledge from without, because it cannot understand without a phantasm. . . . There is, therefore, a more perfect intellectual life in the angels. In them the intellect does not proceed to self-knowledge from anything exterior, but knows itself through itself. Nonetheless, it is not the ultimate perfection to which their life belongs. The reason is this: Although the intention understood is entirely intrinsic to them, the very intention understood is not their substance, for in them understanding is not identified with being. . . . Therefore, the ultimate perfection of life belongs to God, in whom understanding is not other than being . . . the intention understood in God must be the divine essence itself.[173]

Aquinas makes it clear that "nature is found to ascend by degrees through diverse species" *within* groups as well: among elements, earth is lowest, while fire is most noble. Gold is the most-perfect mineral, trees the most-perfect plants, and man the most-perfect animal. "Moreover, certain animals are more like plants, that is the immobile ones which have touch only. Similarly, certain plants are more like inanimate bodies, as is clear from what the Philosopher says in the book *De plantis.*"[174] This ascent is gradual and unbroken, like the series of numbers increasing one by one.[175]

Bonaventure and Dante

As we have seen, the translation of philosophical texts from Arabic and Greek gathered pace in the Twelfth century. Aristotle became available in Latin between about 1210 and 1225, revealing a nearly forgotten body of thought in which the natural world was not simply a source of wonder and allegory, and which offered logical tools that could be applied to problems in theology. Among these were texts

MONASTIC AND SCHOLASTIC NATURE   129

misattributed to Aristotle, for example the *Liber de causis* and a *Theologia*. Church authorities reacted with an abundance of caution; in 1210 they banned the books and commentaries on natural philosophy, then relaxed the prohibition slightly in 1215 to allow lectures on the *Organon*. By 1231 all but Aristotle's *Physica* had been cleared, and in 1254 it too found a place in the curriculum set out for the Faculty of Arts at the University of Paris.[176,177]

As a student and lecturer in theology at Paris, Bonaventure[178] was familiar with the logical works of Aristotle, but believed deeply that philosophy, indeed all academic and practical arts, merely led back to divine revelation.[179] He became head of the Franciscan order in 1257. Following a mystical vision in 1259 he wrote the work for which he became best known, *Itinerarium mentis in Deum* (*Journey of the mind to God*). Guided by Augustine, Bonaventure began his journey in the physical world.[180] From Bonaventure's earlier commentary on Peter Lombard's *Sentences* we know that he accepted that the soul was vegetative, sensible, and rational,[181] and that God had ornamented the world with minerals, plants, and animals.[182] Bonaventure wrote that the world (the macrocosm) enters our soul (the microcosm) through the portals of the five senses[183] and is a "ladder by which we may ascend to God".[184] Accordingly, he "read *Genesis* spiritually, distinguishing seven levels of 'vision' corresponding to the seven days of creation".[180,185]

By 1267 Bonaventure was speaking out against a strain of Aristotelianism championed by Siger of Brabant, in which philosophy was held to be autonomous from faith, and which looked to Ibn Rushd on hot-button issues including the eternity of the world and the relationship between body and soul. The protagonists could scarcely have failed to notice that—as we saw in Chapter 7—Ibn Rushd had pushed back with some success against the faith-based conservatism of al-Ghāzalī. In 1270 Bonaventure's position received the imprimatur of the Church. Bishop Tempier of Paris[186] formally condemned thirteen erroneous doctrines, excommunicating any who might knowingly teach or uphold them. To be sure, Aquinas had argued against these doctrines on philosophical grounds. Bonaventure himself was made a cardinal in 1273, then died in 1274, while these disputes were still simmering. In 1277, following an intervention from Pope John XXI, Tempier approved a much broader list of 219 propositions for condemnation, many of which now appear—indeed, were seen at the time—to be completely orthodox. Among the supposed errors were positions held by Aquinas, and Thomism fell under something of a cloud while a mystically inclined neo-Augustinian movement inspired by Bonaventure became ascendant.[177,187] More broadly, the Church became wary of pagan learning[188] and remained so for centuries.

The philosophical and theological drama that played out at Paris was watched closely in the major centres of Christendom, and not only inside the Church. In his *Commedia* (1308–1321), Dante Alighieri refers more than 650 times to the *Summa theologica* of Aquinas.[189] In *Paradisio*, Beatrice reminds Dante that God created the elements, and plant and animal bodies, only indirectly, and indirectly draws out their souls; but He created the angels directly, and directly inspires human life.[190]

130 KINGDOMS, EMPIRES, AND DOMAINS

Even so, Dante "explicitly defends the superiority of a contemplative path to salvation".[191] Near the end of his *itinerarium* Dante encounters both Aquinas and Bonaventure in the Heaven of the Sun, each in exalted company.[192] We shall return to allegorical or mystical journeys, ascents, and ladders in Chapter 14.

The Fourteenth century

The Fourteenth century brought new challenges. The three-century climatic warm period came to an end; glaciers advanced, crop yields plummeted, and the Great Famine of 1315–1317 was the first of many to visit Europe.[193] One in three Europeans succumbed to the Black Death (1347–1351), and prolonged conflict over the crown of France (1337–1453) further depopulated France and the Low Countries. The Papacy moved from Avignon back to Rome, precipitating the so-called Western Schism (1378–1417) that drew in not only clerical factions but also secular powers. Rivalry sharpened between Dominicans and Franciscans.[194] Mystical sects appeared,[195] while John Wyclif was dismissed from the University of Oxford for agitating against the Pope, clergy, monasteries, and Church doctrines including transubstantiation.[196] Secular voices included not only Dante but also Jean Buridan and Geoffroy Chaucer.

Although often regarded today as an idealist or mystic, Meister Eckhart was a senior official in the Dominican order and, like Aquinas a half-century earlier, occupied one of the Dominican chairs of theology at the University of Paris. Eckhart held that texts can be read not only literally (*e.g.* Augustine's *Literal commentary on Genesis*) but also as parables of two kinds: one, highly granular, in which individual words hold a parallel meaning; the other in which "the whole parable is the likeness and expresses the whole matter of which it is a parable".[197] His readings of the creation story are correspondingly complex, pursuing multiple interpretations in parallel and drawing on a vast range of Christian, Islamic, Jewish, and classical sources.[198] For example, regarding the fruit trees created on the third Day, his argument veers suddenly from grades of life and the dominion of man, to Jesus as the son of a carpenter, and heaven as the garden (orchard) of paradise.[199] Even so, Eckhart accepted a hierarchy of perfection in which grasses, herbs, and trees constitute the first level; reptiles, fish, and birds the second; beasts the third; and man the fourth. He emphasized that plants, inferior animals, and beasts receive the spirit of life only indirectly, from the earth or the waters and thereafter by procreation within their kinds; into man alone did God directly breathe the breath of life.[200]

Within the universe there exist four levels—existence, life, human intellect, and angelic intellect[201]—among which "nature does not make a leap, but progresses in an orderly fashion, or in a progressive series absenting itself gradually and by the smallest steps possible".[202] In a different formulation, Eckhart identified six powers of the soul; their functions correspond to those of plants, beasts, and man, although he discussed them only in the human context.[203] He pointed to touch as the foundation of all senses, and found it more perfect in man than in any other earthly animal.[204]

A different strain of Fourteenth-century thought is evident in the writings of the so-called *moderni*, beginning with William of Ockham, a Franciscan. Ockham held that the truth of Scripture is absolute, neither requiring nor admitting proof by reason. Similar positions were taken by Jean Buridan and Nicole Oresme. This disentanglement of philosophy from theology, coupled with the strong nominalist position that only individuals are real—hence the natural world should be observed at firsthand—can be seen as marking the downfall of scholasticism.[205] Ockham wrote detailed commentaries on Aristotle's books of logic and the *Physica*, and a small *Summa* of his natural philosophy. He enquired in particular depth into issues of infinity and continuity, interpreting Aristotle to claim that natural things (for example, continua) consist of infinitely many parts in actuality, not just potentially.[206] However, Ockham did not extend his investigation to living things. He tirelessly held up stones, plants, animals, and man as examples of genera that are distinct from one another,[207] but was disinterested in the diversity of their species, and did not explore intermediate forms.

Jean Buridan, a student of Ockham, taught in the Faculty of Arts at the University of Paris independently of the religious orders. Like Ockham, he held that doctrine is not susceptible to reason. Buridan commented on Aristotle, and is best known for recasting the *Organon* to make it more accessible.[208] He accepted a hierarchy of stones, plants, animal, and man,[209] but whereas Aristotle discussed soul in its own right—as the first actuality of a body that is potentially alive[210]—for Buridan, soul is defined by physical body. Thereafter, at the University of Paris "considerations of the soul's ultimate nature migrated to the faculty of theology". Buridan's student Nicole Oresme[211] later downplayed soul per se even further.[212]

The *moderni* moved beyond Aristotle in other ways as well. Buridan introduced the notion of *impetus* to analyze projectile motion. Oresme worked in analytical geometry, introduced rectangular coordinates, and wrote on thresholds, limits, and infinite series. Building on his interest in music, he described what would later be called overtones, and developed methods to calculate mathematical series with irrational exponents.[213] His insights into continuity render his views on continuity in living nature particularly interesting. Oresme points out that there is a greater difference between a man at birth and at maturity "than between a pig and a dog at birth, or between an ass and a horse or mule, or a crow and an eagle, or between a wolf and a dog, all of which are of different species".[214] He continues:

> It must then be supposed that nature always tends toward the more perfect if it can; or it must be imagined that there is an order in the successive generation of species or individuals. This is evident from Aristotle, *De animalibus*, Book 7, Chapter 1. For example, first comes the sperm. . . . Then the spirit and sperm are received in a suitable place and the spirit begins to work on the sperm. Whether that action is finished or unfinished, there is a change to another species and something

132 KINGDOMS, EMPIRES, AND DOMAINS

else comes to be as a certain kind of life which is nourished and grows and lives like a plant. Then the spirit in question works some more on that material and that material which up to then was only living is disposed in such a way that it becomes such as to have sensation and the property or power that distinguishes animal from plant. And the said spirit or formative power has such a power that it shapes and colours and gives the same kind of being, if it can, as that from which it was separated . . .

As an example and illustration, say that in the generation of a human there is the following sequence: first, the sperm; then, something fungus-like; third, something like an animal but unformed as it were, as Aristotle reports in *De animalibus*, Book 7, something about which there is doubt whether it be plant or animal etc.; fourth, something monkey-like; fifth, something like a pygmy; sixth, a completed human being etc. Then I say, if the said spirit and the material are potent, they will make all these things in order; and if not, accordingly as they will be more or less potent they will depart more or less from the completed [creature], and there will remain the dispositions of one somehow similar to it and in the place immediately next to it [in the order of generation]. For example, a fungus has certain dispositions which the preceding species has and also some which the subsequent species has. And so the animal mentioned in the third stage will have some dispositions of a fungus and some of a monkey etc.; and similarly for the monkey with respect to etc.[215]

This idea that "first we live the life of a plant"—which, as we have seen, can be traced through Philoponus (Chapter 5) and Aristotle to Democritus and the Hippocratic corpus—was taken up later by Thomas Browne in *Religio medici*.[216]

Nicholas of Cusa

Nicholas of Cusa may not have been the last scholastic, but "after the beginning of the 15th century Scholasticism was divorced from the spirit of the time".[217] Faith, tradition, and mysticism increasingly took one road, reason and empiricism another. Nicholas, a cardinal and papal legate, argued that human knowledge is conjecture, a representation in our mind of images perceived only in part and from our individual perspective.[218] God—infinite, transcendent, unknowable—can be apprehended only via "mystical intuition".[219] Even so, Nicholas recognized a hierarchy in nature, with stones, salts, and metals constituting a first level, and above them vegetable, sensitive, and intellectual spirits.[220] He argued that

all things are distinguished from one another by degrees, so that no thing coincides with another. Accordingly, no contracted thing[221] can participate precisely in the degree of contraction of another thing, so that, necessarily, any given thing is comparatively greater or lesser than any other given thing. Therefore, all contracted things exist between a maximum and a minimum . . .

The first general contraction of the universe is through a plurality of genera, which must differ by degrees. However, genera exist only contractedly in species; and species exist only in individuals, which alone exist actually. Therefore, just as in accordance with the nature of contracted things the individual is positable only within the limit of its species, so too no individual can attain to the limit of its genus and of the universe. Indeed, among many individual things of the same species, there must be a difference of degrees of perfection. . . .

However, the union of all things is through God, so that although all things are different, they are united. Accordingly, among genera, which contract the one universe, there is such a union of a lower [genus] and a higher [genus] that the two coincide in a third [genus] in between. And among the different species there is such an order of combination that the highest species of the one genus coincides with the lowest [species] of the immediately higher [genus], so that there is one continuous and perfect universe. However, every union is by degrees; and we do not arrive at a maximum union, because that is God. Therefore, the different species of a lower and a higher genus are not united in something indivisible which does not admit of greater and lesser degree; rather, [they are united] in a third species, whose individuals differ by degrees, so that no one [of them] participates equally in both [the higher and the lower species], as if this individual were a composite of these [two species]. Instead, [the individual of the third species] contracts, in its own degree, the one nature of its own species. As related to the other species this [third] species is seen to be composed of the lower and of the higher [species], though not equally, since no thing can be composed of precise equals; and this third species, which falls between the other two, necessarily has a preponderant conformity to one of them—i.e., to the higher or to the lower. In the books of the philosophers examples of this are found with regard to oysters, sea mussels, and other things.[222]

From scholasticism to humanism

By the mid-Fifteenth century Scholasticism had largely run its course, leaving intact many ancient ideas directly relevant to the borderland between plants and animals. There was consensus that the universe had emanated (indeed, is still emanating) within or from God, and includes a hierarchy of the heavenly spheres, man, beasts, plants, and stones. All but stones are ensouled, although Aquinas emphasized grades or steps of life, whereas nominalists focused on individuals. The developing human embryo likewise passes through a series of stages. Ockham and Nicholas of Cusa asked what it means for a species to be intermediate in such a series, and accepted oysters and the like, although not corals, as time-honoured examples. Yet with few exceptions (Albert and perhaps Bede), there was little critical observation

134 KINGDOMS, EMPIRES, AND DOMAINS

of nature. New mathematical tools and radical possibilities relating to the physical world (multiple worlds, the void) appeared, while the word *zoophyta* languished in dusty manuscripts, seemingly forgotten.

With resolution of the Western Schism (1417), power in the Church shifted for some decades to a succession of councils. Meanwhile Byzantium, greatly reduced by war and plague, watched the rise of the Ottomans with growing alarm. The time seemed right to heal the Great Schism between Eastern and Western churches. In 1438 Byzantine emperor John VIII Palaiologos joined the Council of Ferrara,[223] where reconciliation was on the agenda. Escorted by Papal envoy Nicholas of Cusa, John was accompanied by numerous dignitaries including Metropolitan Bessarion of Nicæa[224] and the Neoplatonist Georgius Gemistus Pletho.[225] While the delegates debated points of doctrine, humanists flocked to Pletho's authoritative lectures on Plato, whose actual texts had not been read in the West for many centuries.

While in Florence, Pletho wrote *Peri ōn Aristoteles pros Platōna diapheretai*, or *Wherein Aristotle disagrees with Plato*.[226] This tract, commonly known as *De differentiis*, praises the wisdom and piety of Plato and identifies points on which the two philosophers disagree. One such point concerns Aristotle's treatment of equivocal beings: it is difficult to reconcile ζωοφύτων with Plato's strictly binary division of genera into species.[227] A later translator did not bring ζωοφύτων directly into Latin, but referred instead to things *distinguens ipsum a genere illorum quæ tertiam quamdam, non animalium neque fruticum, habere naturam dicuntur*.[228]

Another Byzantine churchman who took part in the Council of Ferrara was John Eugenikos,[229] a friend of Pletho. In a brief historical item on Trebizond, Eugenikos wrote: "and so the mainland and sea are here in open conflict, by projecting forward [προαλλομένην] the whole multitude of birds and land animals and wild beasts, these and fish and ζωοφύτων".[230]

The Council's resolutions were ultimately rejected by the Greek church, and Constantinople fell to Mehmed II in 1453. Nonetheless Cosimo de' Medici[231] agreed to sponsor a new forum, the Accademia Platonica, whose aim was to translate and study authentic texts of Plato and other Hellenes. Its leader was Marsilio Ficino, and illustrious members included Giovanni Pico della Mirandola. We shall meet several of these early Renaissance humanists again in Chapter 9.

Other scholars moved from East to West at about this time. George of Trebizond,[232] an Aristotelian, wrote an introduction to dialectics (*Isagoge dialectica*) that has been called "the first humanistic textbook in logic".[233] Nicholas of Cusa commissioned him to translate the *Parmenides*.[234] Theodorus Gaza[235] made technically precise translations of Aristotle, Theophrastus, and Aelian, and became professor of Greek at the new University of Ferrara. Cardinal Bessarion assembled a great library of Greek and Byzantine texts,[236] and eventually weighed in decisively on the side of Plato, ending the controversy among partisans of Plato and Aristotle that Pletho had set off.[237]

Further threads extended forward from the scholastics. Although the cloud eventually lifted from Thomism, both Meister Eckhart and William of Ockham were investigated for heresy. Eckhart was posthumously found guilty, while William of Ockham became convinced that the Pope himself was the heretic, and fled into exile.[238] In due course new universities appeared across Europe, some influenced by the philosophy of Duns Scotus, others by Ockham—both of whom held theology and philosophy to be distinct undertakings. At Erfurt, where Eckhart had first joined the Dominicans, the new university grew strongly Ockhamist; in 1501 one of the incoming students was a seventeen-year-old by the name of Martin Luther.[239]

Philosophy and organized religion enjoyed no monopoly of perspective on the natural world. Beyond reason and faith lay a badlands of spurious texts, esoteric traditions, occult beliefs, and misguided practise that, to varying extents, drew on more-accepted modes of enquiry and on each other. We have already considered theurgical strains of Neoplatonism (Chapter 5) and Kabbalah (Chapter 7), and touched on Hermeticism. We now ask more broadly whether intermediate genera found a place in mysticism, the occult, alchemy, and magic.

9

Nature's mystic book

Thrice happy he who, not mistook,
Hath read in Natures mystick Book.
                    —Andrew Marvell, *Upon Appleton House*[1]

In the previous four chapters we considered the origin and transmission of ideas about plants, animals, and intermediate organisms within philosophical and religious traditions up to the Fifteenth century. Although diverse, these traditions tended to share a common feature: the support of an institution—the state, church, or university (or in the case of scholasticism, all three). Yet their writ ran only so far into society more broadly. Ficino, Pico, and Agrippa pursued their scholarship largely outside the royal court, cathedral, or university rooms.[2] Farther beyond official faith and reason lay a wild terrain of notions about the natural world. In one direction we find myths and tales of wondrous beings; in Chapter 10 we shall venture there, in search of intermediate or transgressive organisms. In other directions we find the perennial belief that the world is not as it seems: natural things are in secret communication with each other, the stars direct our fate, and it is possible to call on hidden powers for good or evil. Sages and prophets of all ages have discovered keys to this occult knowledge, and their writings contain coded messages that we too could see and follow, if only our eyes were truly open.

In this chapter we explore the world of occult writings and practise including Hermeticism, alchemy, and magic. A recurrent theme is that of transformation.

Oracles and mysteries

Humankind has long imagined itself worthy of a place amongst the grandeur of the heavens. From time to time, individuals have claimed privileged access to the mysteries of the universe, in particular how mortal man can achieve union with God in the afterlife. Mystery cults grew up, including the Eleusinian mysteries perhaps dating to the Mycenæan period in ancient Greece,[3] and the later Orphic mysteries involving cycles of rebirth and metempsychosis.[4] Many accounts identify Egypt as

Kingdoms, Empires, and Domains. Mark A. Ragan, Oxford University Press. © Oxford University Press 2023.
DOI: 10.1093/oso/9780197643037.003.0009

the primordial source of occult knowledge. Herodotus wrote that belief in metempsychosis arose first in Egypt,[5] and both Pythagoras and Plato were said to have spent time there. Iamblichus defended religious rites under the pseudonym of an Egyptian high priest.[6]

Farther east, the *Chaldæan Oracles*[7] were held up as another wellspring of occult knowledge. Once attributed to Zoroaster, they were actually written in the Second century CE by one Julianus[8] and incorporate "a good deal of contemporary Platonic-Pythagorean doctrine"[9] along with a Gnostic view of the world as evil.[10] The *Oracles* refer in passing to animals and plants[11] without defining or delineating either.

Thrice-great Hermes

The greatest of Egyptian sages was Hermes Trismegistus, "master of all arts and sciences, perfect in all crafts, Ruler of the Three Worlds, Scribe of the Gods, and Keeper of the Books of Life".[12] It is not known whether this Hermes was an actual historical person, or the collective literary identity of a sect or cult.[13] Clement of Alexandria[14] reports that forty-two books of Hermes were carried in ceremonial processions.[15] It is unlikely that any of these books has survived—although this being the occult, we find the obligatory reports of copies hidden in the desert, their location known to only a few initiates.[16] The Neoplatonists of Harran and the Ikhwān al-Safā identified Hermes as the prophet Idrīs (Enoch).[17] Others identified him with Moses, such that "the entire Hermetic philosophy [became] an esoteric revelation made to Moses on Sinai along with the exoteric doctrine accessible in Scripture".[18] A belief that Aristotle translated the books of Hermes from Egyptian into Greek as the basis of his own natural philosophy seems to have been current in Thirteenth-century Damascus.[19]

Attribution of the Hermetic texts becomes yet further entangled. As early as 172 BCE Thoth, the Egyptian god of wisdom, was referred to as "three times great".[20] The Greeks equated Thoth with their god Hermes, as did the Romans with Mercury.[21] Iamblichus refers to the author Hermes as a god "common to all priests".[22] During the Second and Third centuries CE hundreds[23] of texts dealing with "astrology and the occult sciences, with the secret virtues of plants and stones and the sympathetic magic based on knowledge of such virtues, with the making of talismans for drawing down the powers of the stars, and so on" were written (in Greek) under the name Hermes Trismegistus. At about the same time a much smaller group of texts, likewise attributed to Hermes Trismegistus but focused instead on religious experience and related philosophical issues (the cosmos, gods, soul, mind, good and evil), emerged and by the Eleventh century were seen as a distinct Hermetic corpus.[24] Fourteen of these books were translated into Latin in 1463 by Marsilio Ficino at the behest of Cosimo de' Medici, and form the core of the *Corpus Hermeticum* as recognized today.[25] A related treatise, the *Asclepius*, is almost always bound with the core *Corpus Hermeticum*.[26] Its text, as we have it, dates to at least the Fourth century, and was known to Augustine (who attacked its teaching that men might invite

138 KINGDOMS, EMPIRES, AND DOMAINS

spirits to animate idols that might then work good or harm).[27] An earlier Greek *Asclepius*, cited by the early Christian apologist Lactantius,[28] has been lost.

Individual species of plants and animals, and groups including trees, fruits, flowers, shrubs, beasts, and birds, are mentioned throughout the *Corpus Hermeticum* and *Asclepius*, although not in a way that reflects any particular underlying natural order. Living things encompass the mortal and immortal, the rational and irrational, the ensouled and soulless.[29] The latter (plants) are as inverted animals: "of all these kinds, the ensouled have roots reaching them from on high to below, but living things without soul branch from a root that grows from beneath to above."[30] Man, mortal in body but immortal in essence,[31] is unique in combining sensation with consciousness, reason, and knowledge.[32] The cosmos too is living, "has its own sensation and understanding",[33] and is subject to passions.[34]

Another text attributed to Hermes is the *Smaragdina tabula* or *Emerald tablet*,[35,36] which we briefly encountered in Chapter 7 bound into the *Kitāb Sirr al-khalīqa wa ṣanʿat al-ṭabīʿa* (*Book of the secret of creation and art of nature*) attributed to the alchemist Jābir ibn Ḥayyān.[37] According to a treatise falsely attributed to Albertus Magnus, the original text—inscribed in Phœnician on a tablet of emerald—had been taken from the very hands of the dead Hermes by Sarah, wife of Abraham, and was later recovered from the tomb of Hermes by Alexander the Great.[38] More likely, the *Emerald tablet* dates to the early Christian era; Latin translations (from Arabic) circulated from the Twelfth century.[36] According to Ficino, Hermes was called thrice-great because he was "the greatest philosopher and the greatest priest and the greatest king".[39] The *Emerald tablet* itself calls him three times great because he "had the three parts of the philosophy of the entire world". These three might be alchemy, astrology, and theurgy;[40] creation, preservation, and destruction;[41] or the "Three Kingdoms (as he calls them,) *viz. Mineral, Vegetable, Animal*".[42] We return below to the supposed evidence behind this claim.

Alchemists revere the *Emerald tablet* as the foundational text of their art. Its cryptic words link macrocosm and microcosm, or perhaps astrology and alchemy ("that which is below is like that which is above, and that which is above is like that which is below"); teach that all things arise from One; hint at a cosmology ("the Sun is its father, the Moon its mother"), and the path of soul from earth to heaven and back; and encourage the reader to separate earth from fire. It does not explicitly mention minerals, metals, plants, animals, or man, although of course these fall under "all things" and can be read-in as desired. The claim that Hermes was thrice-great because he was master of the mineral, vegetable, and animal kingdoms (above) was not based on the *Emerald tablet* itself, but more likely on the *Kitāb Sirr al-khalīqa*, to which we return below.

Universal truths and hidden meanings

Ficino, head of the Florentine Accademia Platonica, famously imagined a direct line of philosophical authority extending from Hermes through Orpheus,

Aglaophemus, Pythagoras, and Philolaus to Plato.[43,44] Lactantius (above) had welcomed Hermes Trismegistus as a "holy Gentile prophet",[45] as the *Corpus Hermeticum* refers repeatedly to God as *Father*, and two passages identify "the lightgiving word" or the Demiurge as His son.[46] According to Ficino, the Orphic story that Minerva (Athena) was born from the forehead of Jupiter (Zeus) tells us that wisdom comes from God.[47] An unbroken esoteric tradition beginning with Hermes would thus draw Platonism closer to Christianity, and reinforce the fertile if optimistic ideas of a *prisca theologia*, a common thread of divine truth running through religions from all nations and times,[39] and a *philosophia perennis*, a core set of truths underlying seemingly diverse philosophical systems.[44]

In reality, philosophies and religions often competed keenly with one another, so it must be supposed that this core set of universal truths was well-concealed. Truths may lie hidden in names, words, letters or numbers: the number *three*, for example, symbolized harmony, completeness, or resurrection.[48] They might be found through allegorical and other non-literal readings, or in correlations between macrocosm and microcosm. Needless to say, the Jewish and Christian scriptures and the works of Plato were held to be particularly rich treasure-houses of occult meaning. God is infinitely knowing, so we should never take His word in sacred texts purely at face value.[49] Other texts might be intentionally obscure. It could be significant—indeed necessary—that there be three types of soul, or three main groups of natural things.

Gnostic texts

In Chapter 7 we briefly encountered Gnosticism, a blanket term for syncretic religions that emphasize special knowledge (*gnosis*) rather than faith. Common around the eastern Mediterranean and Middle East for several centuries from about 100 CE, Gnostic belief systems typically shared core concepts including an unknowable One identified with light, and serial emanations from the One, giving rise to multiple ranks of supramundane beings including Sophia (Wisdom), from whom flows a demiurge who in turn creates the physical universe. These supramundane beings often bore Hellenic or Christian names and might be paired male-with-female, for example Christ with Sophia.[50] By contrast, materiality (*e.g.* of the world itself, and of the body) was evil, and individuals may seek emancipation from it through *gnosis*. At least a dozen major currents in Gnosticism have been recognized, differing not only geographically but also in their relative use of concepts from Babylonian, Christian, Egyptian, Hermetic, Jewish, Persian, or Platonic sources and in the number, identification, and pairings of supramundane beings. Given the syncretic nature of these teachings, it is not difficult to find what appear to be Gnostic influences in philosophic and religious figures from Paul of Tarsus[51] to the Prophet Muḥammad[52] and al-Ghazāli.[53]

Gnostics believed that they had access to knowledge which the Hebrews had long thought was beyond human understanding, but "which according to Paul, God

140 KINGDOMS, EMPIRES, AND DOMAINS

through the Holy Spirit causes to be revealed to every true believer in Christ and, according to Plato, reveals itself to every true philosopher".[54] In the Valentinian Gnostic tradition, such knowledge included "esoteric teachings which originate from Jesus that were passed on in secret" to his Disciples and eventually to Paul.[55] Extant texts[56] refer in passing to animals, beasts, cattle, reptiles, birds, fish, trees, herbs, and flowers, but offer no systematic overview of the natural world, and do not indicate intermediate status for any type of organism.

Macrocosm and microcosm

The idea that man is a reflection in miniature of the universe—a "little world" epitomizing, in a deep sense, the "great world" of the cosmos—can be traced to Anaximenes[57] and may be implicit in the association of gods with heavenly bodies, not to mention outright sun- and star-worship dating back to Babylon and Assyria. The term μικρῷ κόσμῳ appears first in Aristotle,[58] but the analogy itself, in political and ethical as well as cosmological contexts, was developed by Plato[59] and was thence extended, often under cover of his authority, by Hellenic, Islamic, Jewish, and Christian philosophers. Thus, for Galen and Ibn Sīnā the four humours correspond to the four elements, and "the seven cervical and the twelve dorsal vertebrae correspond to the seven planets and the twelve signs of the Zodiac, as well as to the days of the week and the months of the year; and the total number of discs of the vertebrae, which they consider to be twenty-eight, to the letters of the Arabic alphabet and the stations of the moon."[60] Even the empirically minded al-Rāzī accepted a medical role for precious stones, which in Arabic lore draw their colours from heavenly bodies.[61]

Plato had Timaeus say that the Demiurge assigned each soul to its own star.[62] From this remarkably durable idea is it but a short step to the notion that the heavens influence or determine one's personal fate—which might be reassuring, neutral, or terrifying, depending on one's philosophy or religion.[63] Indeed, however far a soul might journey—into a wild beast, for example—it could still look to its star. Plato restricted soul to humans and animals, although as we saw in Chapter 1, others considered that it could migrate to plants, or even become briefly imprisoned in minerals. So, do beings intermediate between plants and animals have or retain their own star? Remarkably, this question was addressed in some detail by the Jesuit intellectual and mystic Juan Eusebio Nieremberg[64] in his *Historia naturale maxime peregrinae* (1635). In a chapter bearing the title "Whether celestial bodies might have a middle [state of] perfection between plants and animals" he argues:

> But is it possible for another nature, extraneous to both, to intervene in the middle, between the plants and the beasts: the sponge truly, and the *vertibula* or *hard-skinned*, as well as the *tubera*, as Theodorus[65] translates it into Latin, take a middle path, particularly the sponge, so that both kinds of being, the tethya and

holothurians, somehow are seen to be a sponge. Surely there is no middle that takes on either of these [categories]? [No,] because, to me, the heavenly bodies do not favour such a kind of being, as the philosophy of Moses demonstrates; and if his philosophy shows this, then that of God does so all the more. . . . It was with the greatest scrupulousness that Moses cultivated scientific method in his story of the generation of the universe, by the perpetual ascent of the natural world, progressing from the more-imperfect to the more-excellent, from the primary elements, then to other things continuously right on to man, by uninterrupted ascent. Since he described the creation [*conditionem*] of the stars as in-between [those of] plants and animals, he seems—by [recognizing] a mid-way perfection between both—to have done them honour. God has created nothing rashly, nothing by accident; Moses has narrated nothing without great deliberation, nothing without a measure of mystery; so that I wonder, with how much learning and experience of nature, that so excellent a Philosophy presents itself to a person who is considering its words and its main points, as well as the overall arrangement of all things.

I ask you, although there had been an opportune moment (or perhaps necessity) for making stars on the first day [of creation], so that a foreign light would not be wandering about without a home, rejected without a cause, did it exist, without mystery, until the fourth day? No, he [God] wanted to uphold an inviolable method, so that his law and its individual adornment are maintained in their untroubled perfection. I hear you say that the heavens are not alive. And surely you have not heard from Moses that plants live? I don't think that lives and souls instituted [*inaugurarunt*] the name *animals* in the writings of Moses. If it be so, even with the applause of the Stoics, without doubt only a minority of stars are endowed with a sentient nature. In addition, we may grant life to the plants, something perhaps, it seems, not to be denied to the stars.

Movement that involves a spontaneous and perpetual outcome is especially what defines life: it corresponds with the stars in a quite excellent and eternal fashion. Of their own selves, as I have said, they are moved; indeed, in this respect alone am I able to have said this. In one instance, quite obviously and evidently, they certainly move in a limited orbit; in another, they move rather steadily but not to a point of rest, as the elements are, which are therefore inanimate: for truly, that thing lives whose perfection is to move itself. This is remote from [the situation with the] elements: their perfection is rest, or an aspiration to quietude: for them, motion proves an unsuitable situation. The stars, nevertheless, detest leisure the most of all, therefore some sort of life is what defines them. Truly then, if there is localised motion, it is action pertaining to life, such that it is highly satisfactory to most people; a sort of life belongs to those stars, a life midway between the vegetative and the sentient, more of this and less of that. It required, as an ornamentation of nature, that that life-action, of progression in a local place, be allocated to its own separate classification, just as for life's other functions. Where mind is absent, there are sensation and intuition; where sight is absent, there is hearing; olfaction is present, when there is no hearing; taste, when there is no olfaction; the sense of touch, where not accompanied by taste. Wherefore [the capability of] progressive movement will exist appropriately in some other nature, where there is no sense of touch; likewise in its nourishment too it merits the class of inanimate things.[66]

We shall return to Nieremberg in other contexts in the next two chapters.

Alchemy

From Asia and the Middle East to the Americas, prehistoric peoples fashioned tools and weapons from meteoric iron: AN.BAR, the Sumerian word for iron, combines pictograms for *sky* and *fire*.[67] Beginning about seven millennia ago, our ancestors learned to transform lumps of earth into shiny metal—first lead and tin, then more interestingly copper, bronze, and iron.[68] The knowledge and technology needed to bring these transformations about was often maintained within an exclusive group such as a tribe, caste, or guild.[69]

Human ambition was not sated by making iron kettles. By combining primary elements into a more-perfect product man became Demiurge, and mystico-religious interpretations were there for the taking. What was smelting, but the fiery emergence of the noble from the base? The production of an alloy, but the mutual dissolution of two substances into a transcendent unity? Or refining, but perfection through a progression of new appearances (colours)? Even so, man set his sights higher. From the earliest times, it had been accepted that minerals were engendered and grew inside the earth.[70] Indeed, left to themselves, ores might continue to perfect themselves and become that noblest of metals, gold.[71] Surely man could hasten the process along.[72]

Alchemy is a body of prescientific theory and practise that had as its goal the perfection of physical nature. Transmutation of base metals into gold drew the most attention, but a truly perfect substance—the *lapis philosophorum*, or philosopher's stone—might work its power beyond the metals, for example by curing disease, restoring youth, and imparting immortality. The *Emerald tablet* was perennially imagined to be an occult recipe for preparing or discovering the philosopher's stone. What distinguished alchemy from other similarly intentioned practises—for example the use of magic amulets, incantations, or yoga—was its underlying theory.

Aristotle held that the four "so-called elements" share pairs of contrary properties, albeit in different combinations: earth is cold and dry, air hot and moist, fire hot and dry, water cold and moist. These terms signify more than their face value: *hot*, for example, "is that which associates things of the same kind", whereas *cold* brings together "homogeneous and heterogeneous things alike". These properties themselves are immutable, but simple bodies corresponding to the elements can and do change into each other.[73] Furthermore, the heat of the sun brings about two types of exhalation from the earth; one, involving the transition from water to air, is the

NATURE'S MYSTIC BOOK   143

material cause of minerals, while the other, implicating the transition from earth to fire, is the efficient cause of metals.[74]

For al-Fārābī, alchemy is "the science of converting realities into another species".[75] He surely had in mind the teachings of Jābir ibn Ḥayyān, mentioned above. Jābir explained that minerals and metals do not arise from the exhalations per se; instead, the smoky (earth-to-fire) exhalation leads to sulphur, the vaporous (water-to-air) exhalation produces mercury, and it is *their* combination in different ratios and states of purity that generates the metals. If the sulphur and mercury are pure, and are unified in the most-harmonious ratio, the metal produced is gold; impurities or an inharmonious combination instead yield (for example) silver, copper, iron, or lead. Thus (for example) lead is but impure and inharmonious gold, and an alchemist might perfect lead to gold by cleansing it of its fundamental impurities and harmonizing its proportions of sulphur and mercury.[76]

Sulphur-mercury theory as introduced by Jābir may have had some empirical basis,[77] but alchemical sulphur and mercury were not elements in a modern sense. Instead, like Aristotle's contrary properties, they are bearers of further virtues: sulphur is hot, dry, and readily combustible, mercury cold, moist (indeed liquid), volatile, and metallic. Nine hundred years before the emergence of modern chemistry,[78] sulphur-mercury theory offered a unitary basis for mining, metallurgy, and alchemy. Jābir's acolytes further encumbered it (and alchemy more generally) with Pythagorean, Neoplatonic, Hebrew, Gnostic, and Christian concepts and imagery, not least those of macrocosm-microcosm. Sulphur became male, active, yellow like the sun, symbolic of soul; mercury female, receptive, silver-coloured like the moon, symbolic of spirit. Farther into the labyrinth of symbolism, sulphur and mercury are joined in marriage; the forge and crucible bring the suffering, death, and resurrection of matter.[79] Alchemy aims to verify that "all things arise from the One"; alchemists purify not mere metal, but "the divine essence from the bonds of materiality".[80] Gold is the "living God".[81]

The *Kitāb Sirr al-khalīqa*

As mentioned above, we encountered *Kitāb Sirr al-khalīqa wa ṣanʿat al-ṭabīʿa* (*Book of the secret of creation and art of nature*) in Chapter 7, in connection with the *De natura hominis* by Nemesius of Emesa. According to Jābir, the *Sirr al-khalīqa* was written by Apollonius of Tyana based on the *Kitāb al-ʿIlal* (*Book of causes*) of Hermes Trismegistus. Curiously, Apollonius too is associated with Emesa, now known as Homs.[82] More likely, however, the *Sirr al-khalīqa* originated as a Fifth- or Sixth-century CE Greek treatise. A copy was later translated into Syriac by a monk named Sagiyūs, who added lengthy arguments against various Christian heresies; then in the Ninth or Tenth century his text was in turn translated, with further redactions, into Arabic.[83] The *Sirr al-khalīqa* is known in two recensions, each of which includes the *Emerald tablet*.

144   KINGDOMS, EMPIRES, AND DOMAINS

The *Sirr al-khalīqa* is arranged in six books: on the Creator, celestial bodies, minerals, plants, animals, and man. Briefly, the word of the Creator brought about the causes, light, action, movement, and heat. The heavens began to move before they were fully formed, admixing the coarse and subtle parts of matter and bringing forth the substances of the three major groups: first of minerals, which are lifeless because there were as yet no stars to ensoul them; then of plants; and finally of animals, which "unite body and soul, that is, life and movement, because they are the product of the action of the three celestial powers: the heavens, the stars, and the rotation of the celestial bodies. Plants, which have as principles of their formation only the heavens and the stars, are only animated bodies and are deprived of movement. Minerals are purely bodily substances, also devoid of movement and life, because they owe their formation only to the celestial spheres in their state of inertia, without the concurrence of any other power."[84] Sulphur and mercury arise from subterranean exhalations, each under the domination of a sign of the zodiac. The metals show different degrees of perfection, but all are "corruptions of gold".[85]

The minerals, plants, and animals are referred to as *mawālīd*, which as we saw in Chapter 7 means "progeny" or "children" rather than "kingdoms".[86] Intermediating these groups are transitional stages of aquatic beings. Between minerals and plants, corals have the body of a mineral but the spirit of a plant, while between plants and animals, mussels and sponges have the body of an animal and the spirit of a plant. Corals originate from fire and earth through the mediation of water. Mussels and sponges originate from water mixed with air and earth; fixed to the earth but endowed with the sense of taste, they combine life and lifelessness (Figure 9.1).[87] These organisms are not imperfect plants or animals, but comprise separate (if unnamed) stages.

The shorter recension was translated into Latin by Hugo de Santalla, probably between 1145 and 1151 CE, as *Liber de secretis naturae*.[88] The natural genera appear in the *Liber* as *animalium, germinantium et minerarium*,[89] as do the *corallus*, *ostrea*, and *conchilibus* between the animate and inanimate.[90] No natural genus is called a *kingdom*. Haq has noted that Arabic alchemy differed from the earlier Hellenic version in utilizing animal and plant materials as well as minerals.[91]

The *Sirr al-asrār* or *Secretum secretorum*

At about the same time another mysterious book made its way from the eastern Mediterranean, via the Caliphate, to Latin Europe: the *Sirr al-asrār*, or *Secret of secrets*. It is written as a letter of advice from Aristotle to Alexander the Great, and presents itself as having been translated from Greek through Syriac to Arabic by Ibn al-Baṭrīk, who as we have seen (Chapters 3 and 7) was credited with the translation of Aristotle's books of animals. It claims to contain "prohibited and profound mysteries" and, lest they "fall into the hands of wicked and tyrannical men, who might discover what God did not deem them worthy to understand", should be read

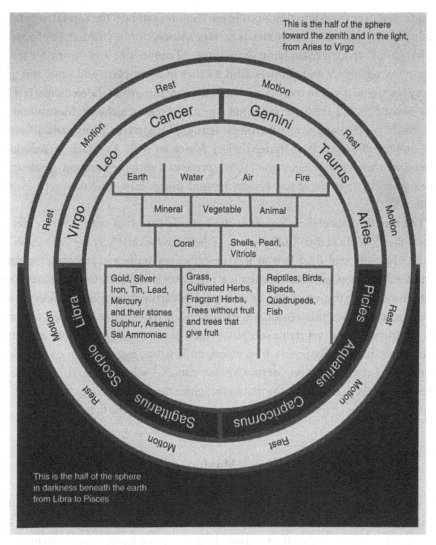

Figure 9.1. Cosmology of alchemy, according to the *Kitāb Sirr al-khalīqa* attributed to Apollonius of Tyana. From Weisser (1979:371), as adapted by Michael Graham in HR Turner, *Science in medieval Islam* (1995:193) by permission of University of Texas Press.

on two levels, one "evident and apparent", the other "secret and mysterious".[92] No Greek or Syriac version is known, and the first (partial) translation from Arabic into Latin was made about 1120[93] or 1135–1142.[94] By that time two manuscript traditions had emerged in Islamic lands, one in the East and the other in the West; the former typically included the *Emerald tablet*.[95] A second Latin translation was made by Philip of Tripoli around 1243, and a subsequent recension (1259) served as the basis for the English version by Roger Bacon *circa* 1270.[96]

In *Secretum secretorum* substance is demarcated into mineral, vegetable, and animal.[97] In the Latin these are not called kingdoms, and no intermediate groups are

146   KINGDOMS, EMPIRES, AND DOMAINS

mentioned. Some versions describe the philosopher's stone ("the animal, vegetable, and mineral stone, the stone which is neither a stone, nor has the nature of a stone, although it is created resembling some stones of mountains and mines, for it also resembles vegetables and animals. And it exists in every place and time, and with every man. And it has all colours, and in it there are present all the elements. It is the microcosm"[98]), but the *Secretum* is not a practical text of alchemy. More attention is devoted to physiognomy, numerology, astrology, augury, and talismans, although the author offers practical (indeed wise) advice as well. A sequel, in Ashmole's *Theatrum chemicum Britannicum* (1652), presents a supposed second epistle from Alexander to Aristotle; it too divides natural things into animal, vegetable, and mineral.[99]

The pseudo-Aristotelian *Secretum* has become confused with a very different work, the alchemical text *Kitāb al-Asrār* (*Liber secretorum*)[100] by the Persian physician and alchemist al-Rāzī. In it, al-Rāzī describes apparatus and protocols for distillation, sublimation, and other alchemical procedures, and classifies substances (not organisms) as earthy, vegetable, or mineral. His earthy substances are four spirits, seven fusible bodies, thirteen stones, six vitriols, six types of borax, and eleven salts. He mentions a single vegetable substance (the ash of a marsh plant) and ten animal substances: hair, skulls, brains, bile, blood, milk, urine, eggs, mother of pearl, and horn. There are, in addition, "derivative" substances—salts and alloys.[101] Al-Rāzī is sometimes said to have been the first to recognize three great groups in nature;[102] as this and previous chapters amply demonstrate, he was not the first.

Magic

In *The golden bough*, Frazer argued that science and magic are sisters: both assume that "in nature one event follows another necessarily and invariably without the intervention of any spiritual or personal agency"; magic, however, misconceives the nature of the laws that govern this regularity. By contrast, religion involves belief that one or more superhuman beings govern the world and upon supplication might intervene to change the course of events. Insofar as magic deals with spirits, "it treats them exactly in the same fashion as it treats inanimate agents, that is, it constrains or coerces instead of conciliating or propitiating them as religion would do".[103]

Perhaps the headline misconception of regularity in nature is the so-called doctrine of signatures. In Hildegard's *Causæ et curæ* (Chapter 8) we encountered the medicinal use of extracts from plants whose shape, taste, or colour correspond to, or oppose, the humours that, in disease, are supposedly out of balance. By the late Middle Ages, it was widely held that the world is replete with signs that mark occult connections within macrocosm and microcosm. Thus, the kernel of the walnut, which bears "the very figure of the Brain", has power to "comfort the brain and head mightily".[104] Paracelsus believed that God places signatures in nature for the use of man.[105] Jakob Böhme went farther: "the whole outward visible World with all its Being is a Signature, or Figure of the inward spiritual World", and "every Thing which is generated out of the internal has its Signature". His *Signatura rerum* describes a tangle of signatures and antipathies linking the human body with the planets, animals, plants, minerals, metals, elements, sulphur, mercury, salt, and events and prophecies from the Bible, and indicating good or evil, health or disease.[106] Most of these signatures reached across, or indeed beyond, the main natural genera.

In this context, it should be no surprise that ingredients of magical potions—wool of bat, eye of newt—might bear the names of transgressive organisms.[107] Bats dualize, with four legs like beasts yet wings like birds. The newt (salamander) lives in the fire. The mandrake-root is formed in the image of man.

Nor was all magic providential: sorcery, witchcraft and other forms of "black" magic were repressed with varying degrees of zeal throughout the Medieval period.[108] Pagan societies that opposed magic did so because it was practised in secret and was assumed to be a threat. The Church opposed it because it involved consorting with demons, *i.e.* spirits who set themselves up as gods; and invoking false gods was evil, whether secretive or open. "In short, the pagan definition of magic had a moral and a theological dimension but was grounded in social concerns; the Christian definition had a moral and a social dimension but was explicitly centered on theological concerns."[109] Those sympathetic to magic countered by ascribing occult powers to esteemed figures from antiquity onward including Moses, the Patriarchs, Hermes, Pythagoras, Socrates, Plato, Aristotle, Porphyry, Virgil, Albertus Magnus, Thomas Aquinas, Roger Bacon, Jeanne d'Arc, Martin Luther, and Oliver Cromwell.[110]

From Jābir to the Renaissance

As we have seen, Jābir was familiar with the *Sirr al-khalīqa* and thus with its four groups: minerals, plants, animals, and man. Because everything that affects minerals, plants, and animals is due to the four elements, "he who succeeds in manipulating the elements in the three kingdoms succeeds, by the same act, in acquiring knowledge of all things, and in understanding the science of creation and the art of Nature."[111] In *Kitāb al-Aḥjār* (*Book of stones*) he outlines elaborate correspondences between the metals and Arabic letters and words, musical modes, and numerical ratios.[112] His claim that "the shape of animals arise[s] out of straight lines, that of stones out of curves, and that of plants out of the combination of straight lines and curves"[113] offers little opportunity for intermediate groups. Elsewhere, Jābir offers practical recipes for alchemical transformations, including (in *Kitāb al-Tajmī*, or *Book of assemblage*) for the artificial generation of humans.[114]

148 KINGDOMS, EMPIRES, AND DOMAINS

A century later, Muḥammad al-Rāzī brought a rationalist spirit to alchemy and medicine. He favoured observation over divine revelation, and wrote a rebuttal to the *De mysteriis* of Iamblichus. In the same vein, he had little to say about the origin of metals from sulphur and mercury.[115] He introduced salinity and oiliness as new physical qualities, and in a step towards chemistry rejected the symbolic dimension of alchemy altogether.[116] Al-Bīrūnī accepted the sulphur-mercury origin of metals, and believed that metals have a life in the earth and eventually become perfected as gold, but rejected alchemy as a form of witchcraft.[117] Ibn Sīnā agreed, and found opportunities in his writings to ridicule alchemists.[118]

The Ikhwān al-Ṣafā wrote that sulphur and mercury arising from subterranean vapours are the material cause of minerals,[119] and imply that minerals do not grow further.[120] This, however, is not the entire story: in another sense minerals "have a life of their own. They grow like fruits and have love, desire, hatred, and repulsion just as animals do. They have a hidden perception and delicate sense like plants and animals. Minerals exist potentially in the earth and become actualized at the surface. They are grown as are the animals by the inception of the male sperm in the female womb of the earth and participate in the life of nature which permeates all things."[121]

By the Thirteenth century CE, al-ʿIraqi[122] offered an entirely allegorical treatment while preserving the three groups of earthly things:

This prime matter which is proper for the form of the Elixir is taken from a single tree which grows in the lands of the West. It has two branches, which are too high for whoso seeks to eat the fruit thereof to reach them without labour and trouble; and two other branches, but the fruit of these is drier and more tanned than that of the two preceding. The blossom of one of the two is red [corresponding to gold] and the blossom of the second is between white and black [corresponding to silver]. Then there are two other branches weaker and softer than the four preceding, and the blossom of one of them is black [referring to iron] and the other between white and yellow [probably tin]. And this tree grows on the surface of the ocean [the *materia prima* from which all metals are formed] as plants grow on the surface of the earth. This is the tree of which whosoever eats, man and jinn obey him; it is also the tree of which Adam (peace be upon him!) was forbidden to eat, and when he ate thereof he was transformed from his angelic form to human form. And this tree may be changed into every animal shape.[123]

Arabic alchemy began to reach Europe in the Twelfth century. The *Kitāb Sirr al-khalīqa* and the *Secretum secretorum* (above) extolled sulphur-mercury theory and the philosopher's stone respectively. Robert of Chester translated the *Kitāb al-Kīmyā* attributed to Jābir, and the *Epistle of Maryānus to Prince Khālid ibn Yazīd*, the latter as *Liber de compositione alchemiæ*, in 1144 CE.[124] In *Liber de compositione* Khālid, who has long sought the wisdom of Hermes, travels to the "Mountains of Jerusalem" to meet the "ancient hermit" Maryānus, who reveals detailed if cryptic instructions for producing the philosopher's stone. The *Turba philosophorum* (*Assembly of the philosophers*)[125] likewise strikes alchemical themes, and puts into the mouth of Pythagoras the claim that angels consist only of fire; the sun, moon, and stars of fire and air; terrestrial animals of these plus earth; flying things of fire, air, and water; and vegetables of earth, water, and air. Only man is constituted of all four.[126]

Albertus Magnus[127] and Artephius[128] were exponents of sulphur-mercury theory, while Arnaldo de Villanueva was regarded as an adept. Vincent of Beauvais, who discussed both theoretical and practical aspects of alchemy in his *Speculum maius*, is said to have learned the art.[129] Alchemy was also deeply intertwined with Kabbalah.[130] Despite ecclesiastical bans,[131] interest in alchemy spread. In *Pretiosa margarita novella* (*Precious new pearl*, 1338) Petrus Bonus marshalled arguments for and against the transmutation of metals, including some based on the generation of "minerals, vegetables and animals", before deciding—despite his own lack of success—in its favour. Raymund de Tarrega, a shadowy figure eventually found dead in prison in 1371, described how "to extract the four elements from all things, namely, plants, animals, and metals".[132]

Thus, on the eve of the Italian Renaissance, alchemy—with its admixture of Hermeticism and the occult—was no stranger to Church, university, or the broader society. The hundreds of surviving manuscripts, in vernacular languages as well as Latin, point to the *Secretum secretorum* as one of the most-popular texts of the Middle Ages.[133] Thrice-great Hermes had been reborn as a Platonist and all but baptized into Christianity; the macrocosm was God's handiwork, while the diabolical corners of magic were not discussed in polite company. Animals, vegetables, and minerals ran throughout this entire occult literature, even if minerals were no longer entirely inert and soulless. Intermediates between minerals and plants, and between plants and animals, were mentioned only in the *Liber de secretis naturae*, a book associated with the *Premnon physicon* also known to Albert (Chapter 8).

Three Renaissance humanists: Ficino, Pico, and Agrippa

We have already encountered Marsilio Ficino, translator extraordinaire and head of the Accademia Platonica. He received a traditional education, which in Fifteenth-century Florence meant a scholastic framework, Porphyry's *Isagoge* and the *Organon*.[134] His first philosophical work, *Tractatus de anima editus per Marsilium* (1454/1455), identifies the four faculties of soul as *vegetativa, sensitiva, secundum locum motiva*, and *intellectiva*:[135]

> They are revealed in the four genera of ensouled beings: the vegetative in plants, the sensitive in certain marine shellfish [*quibusdam conchilibus marinis*], that moving-to-another-location in brutes or animals, in which it joins the other two;

150 KINGDOMS, EMPIRES, AND DOMAINS

finally the intellective, in man alone, who with the augmentative, sensory, locomotory and intellective, is said to possess four souls.[136]

These are Aquinas's four grades or modes of life (Chapter 8). Ficino went on to translate the Orphic hymns (1462), the *Corpus Hermeticum* (1463), the works of Plato (1463–1468), and texts by a range of classical authors,[137] none of which mentions *zoophytes*. Ficino returns to these organisms in his *Commentary on Philebus*, situating man within a hierarchy of beings in which *spongiæ, purpuræ, conchæ*, and *ostreæ* are below the brutes, and fungi (*tuber*) are almost identical with inanimate things.[138] In *Theologiæ Platonicæ* he describes a succession of forms ascending from the elements through vapours, stones, metals, plants, animals, and man. Among plants, the *tuber* "is clearly superior to metals because it shows more obvious signs of nutrition and growth, but it does not exceed them by much because it has no order of various parts". Similarly, among animals "the oysters are superior to plants only in that they have a sense of touch, but they stay rooted to the bottom and are nourished more or less like trees". Some men are dull-witted like animals, while others are heroes "next of kin to the divine spirits". In all classes of things "the lowest individuals of the preceding order are linked with and in a way become in turn mingled with the highest of the order that follows".[139]

In *Theologiæ Platonicæ* Ficino argued in detail that a core set of truths runs through all philosophies and religions from Zoroaster, Hermes Trismegistus, and Orpheus to "the divine Plato", culminating in Christianity. Likewise, Giovanni Pico della Mirandola believed in a *prisca theologia* and *philosophia perennis*, famously beginning his *Oration on the dignity of man* with the words of Hermes Trismegistus: "A great wonder, Asclepius, is man!"[140] Pico added Kabbala to this supposed common tradition, not least though his reinterpretation of the six Days of Creation, the *Heptaplus* (1489). In its Second Proem we learn of the three worlds: the divine, with nine orders of angels; the celestial, with nine heavenly spheres; and the elemental, with nine spheres of corruptible forms. Constituting the latter are three types of body devoid of life (the elements, their mixtures—that is, minerals and metals—and atmospheric phenomena); three of vegetable nature (grasses, shrubs, and trees); and "three of sensitive souls, which are either imperfect as in the zoophytes, or perfect but within the limits of irrational phantasy, or what is highest among brutes, capable even of being instructed by men, a mean, as it were, between man and brute, just as the zoophyte is the mean between brute and plant."[141,142] Man is a fourth world, set apart from the hierarchy of being, central to all.[143] Illustrating Pico's three worlds, Nicholas Le Fèvre de La Boderie rearranged the elemental world and integrated numerical and geometric symbols, but preserved zoophyta between plants and animals (Figure 9.2).

From which text or tradition did Pico draw the word *Zoophyton*? It had not yet appeared in the occult, magical, or alchemical literature.[144] By 1489 Pico had learned to read Greek, Hebrew, and Arabic.[145] His extensive library included the *In Aristotelis De anima* by Themistius; the commentaries on *De caelo, Physica*, and *De*

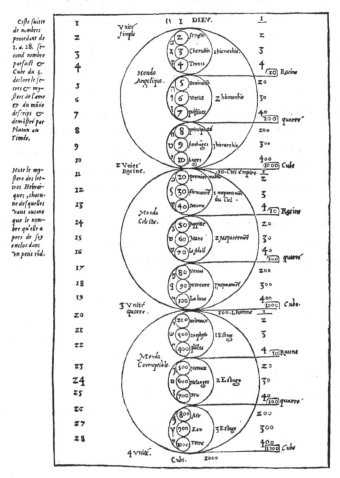

Figure 9.2. Pico's three worlds, as modified by Nicolas Le Fèvre de La Boderie (1579) introducing his brother Guy's translation of Francesco Giorgio's *De harmonia mundi totius cantica tria*, and his own translation of Pico's *Heptaplus*. Original at Niedersächsisch Staats- und Universitätsbibliothek (SUB) Göttingen, 2 TH TH II, 206/56.

anima by Simplicius; those on *De anima* and *Analytica priora* by Philoponus; and the *Logica* (i.e. *Dialectica*) of John of Damascus.[146] His image of zoophytes as a *mean* rather than a genus recalls Dexippus, but is more likely part of the Pythagorean theme (three worlds, three × three spheres). Gemistus Pletho or the *Premnon physicon* are remote possibilities.

Ficino raised clerical eyebrows with his discussion of astrological magic in *On obtaining life from the heavens*, the third of his *Three books of life*.[147] Pico rejected

152 KINGDOMS, EMPIRES, AND DOMAINS

crude astrology as incompatible with the free will of man,[148] instead focusing on magic in the elemental world, so-called natural magic. He wrote that magic is the "apex and summit of all philosophy", presupposes an "exact and absolute understanding of natural things", and "is the practical part of natural science".[149] Opinion differs on precisely what practises Pico meant to include or exclude, but he identifies an understanding of the "powers and activities of natural agents, and their mutual applications and proportions, and their natural strengths, [and] what they can and cannot do through their own power".[150] At minimum, this includes sympathetic medicine. Even so, great exception was taken to his claim that "there is no science that assures us more of the divinity of Christ than magic and Kabbala".[151]

Pico's nephew and self-appointed literary executor Gianfrancesco[152] refers to *zoophyta* in two of his own philosophical works. In *De studio divinæ et humanæ philosophiæ* (1496) he considers how, aided by human philosophy, we may ascend in the contemplation of God. Philosophers may disagree about the basis of the natural hierarchy (whole or mixed forms, rootedness or locomotion, susceptibility to external or internal impulse), but on our journey upward we pass through four types of animated beings: plants, Zoophyta, brutes, and rational animals.[153] In *Examen vanitatis doctrinæ gentium et veritatis discipline Christianæ* (1520) he remarks that

> The more that animals and those things of diverse type approach toward human cognition, the more they are said to participate in knowledge and moving about by a certain reason: and the more they recede toward the plants and shrubs [*plantarum stirpiumque*] they approach a senseless condition, and there taste of its nature, from whence Aristotle fashioned the name Zoophyta, as if to say animal plants, such that they hold a border against both natures. Thus it can be held that superior and inferior grades can be bound into one genus, fused together in equal proportion, and coalesced from two into one.[154]

The text is admirably clear: zoophytes have a single distinct nature, even if they share (attenuated) properties with plants and animals.[155] However, Aristotle did not fashion the name Zoophyta. To my knowledge, this is the first time Zoophyta was so directly misattributed to the Philosopher; it was not to be the last.

Whereas Ficino wrote of astrological magic and Giovanni Pico the natural sort, Cornelius Agrippa explored magic in all three of Pico's worlds. His writings present difficulties for the modern scholar (as they did for his contemporaries), not least because he came to deny the reliability of sense-perception, indeed of Aristotelian epistemology, as the basis of knowledge; the human soul (unlike the body) is divine, and (as sensed by the ancients, particularly Plato) knowledge is in God.[156] In *De triplici ratione cognoscendi Deum* (*On the threefold way of knowing God*, 1529) he identified three grades of spiritual ascent: sense-perception, rational knowledge, then spiritual knowledge.[157] Yet this same Agrippa wrote the standard text on magic, the *De occulta philosophia* (1533)[158]—becoming the model for the Marlowe's Faustus[159] and Goethe's Faust.[160]

NATURE'S MYSTIC BOOK   153

In the first book of *De occulta philosophia* Agrippa discusses the elemental world. Oysters are "fish" under the power of the Moon,[161] while coral is a "stone" under Venus.[162] Animals (oysters) and vegetables (the ebony tree) participate in an occult sharing of virtues with stones. Indeed, some animals can be turned entirely into stones.[163] He then sets out a chain of stepwise correspondence, from God down to stones:

> By these examples you see how by some certain natural, and artificial preparations, we are in a capacity to receive certain celestial gifts from above. For stones, and metals have a correspondency with herbs, herbs with animals, animals with the heavens, the heavens with intelligences, and those with divine properties, and attributes, and with God himself, after whose image, and likeness all things are created.
>
> Now the first image of God is the world; of the world, man; of man, animals; of animals, the zoophyton; of those, truly, plants; of plants, moreover, metals; and of these, stones by their similarities and imaginations are represented. And again in things spiritual, the plant agrees with a brute in vegetation, a brute with a man in sense, man with an angel in understanding, an angel with God in immortality. Divinity is annexed to the mind, the mind to the intellect, the intellect to the intention, the intention to the imagination, the imagination to the senses, the senses at last to things.[164]

In the second book, Agrippa shifts focus to the celestial world. He presents *scalæ* setting out correspondences across six worlds—exemplary, intellectual, celestial, elemental, lesser, and infernal—for the numbers one through ten, and twelve (eleven is "imperfect" and thus has no scale). The scale for *four* identifies the "four perfect kinds of mixed bodies" in the elemental world as animals, plants, metals and stones, while that for *five* identifies the "five kinds of mixed bodies" as animal, plant, metal, stone, and zoophyton.[165]

Paracelsus and the alchemists

Like Agrippa, Paracelsus was a student of the occultist, cryptographer, and Benedictine abbot Johannes Trithemius. Against fierce opposition from the medical professors and apothecaries, Paracelsus sought to overturn Galenic medicine and replace it with an alternative science based on alchemy and natural magic. Some of his insights seem modern: that disease arises externally to the body, affects specific organs, and should be treated with specific chemical agents rather than by rebalancing the humours. At the same time, he believed that occult correspondences link man to the macrocosm, that they can be discovered by signatures, and that for "herbs to be effective, the heavens must be propitious".[166] The aim of alchemy was to make medicines, not gold.[167] From 1533 to 1535 he worked as an itinerant physician, staying at monasteries and treating plague-ridden peasants.[168]

154 KINGDOMS, EMPIRES, AND DOMAINS

Paracelsus accepted that sulphur, mercury, and salt combine to form every bodily substance,[169] and wrote that minerals are animated, grow, and advance in perfection underground.[170] Animals, vegetables, and minerals are "the three chief heads of all earthly things"[171]—although their boundaries could be porous, as under certain conditions vegetables can become perfect minerals and minerals can progress to metals.[172] Ignorant men have supposed the philosopher's stone to be "threefold, and to be hidden in a triple genus, namely, vegetable, animal, and mineral".[173] He treats coral inconsistently, sometimes as a stone,[174] other times as a marine plant.[175] Fungi are imperfect vegetative bodies, as a tumour is to animal flesh, "or the ape to the man".[176] He mentions sponges, although in an uninformative way;[177] he does not use the term *zoophyte*.[178]

Alchemy did not meet with unqualified success in medicine. Paracelsan physicians were seen—not least by the medical fraternity of the day—as "a set of dangerous fanatics, who, in their contempt for the principles of Hippocrates, Galen, and Avicenna, and in their reckless use of powerful remedies, many of them metallic poisons, wrought untold misery and mischief".[179] Nor was its philosophical basis taken seriously in all quarters: the Cambridge Platonist and occultist Henry More wrote of "the rampant and delirious Fancies of that great boaster of Europe *Paracelsus*, whose unbridled Imagination and bold and confident obtrusion of the uncouth and supine inventions upon the world has, I dare say, given occasion to the wildest *Philosophical Enthusiasms* that were ever broached by any either Christian or Heathen."[180]

Enthusiasm nonetheless continued to grow for alchemy as a way to master nature—and coincidentally perhaps, following the example of Nicholas Flamel,[181] to produce untold quantities of the purest gold to endow churches and hospitals. Ancient texts were rediscovered or manufactured, books and manuals written and translated. Much of this literature has been lost or forgotten, but from the Fifteenth century certain books emerged and grew in prestige: among others Thomas Norton's *Ordinall of alchimy* (1477);[182] the anonymous *Rosarium philosophorum* (1550);[183] *Ein kurtz summarischer Tractat von dem grossen Stein der Vralten* attributed to Basil Valentine (1599);[184] *Basilica chymica* (1608) by Oswald Croll;[185] *Novum lumen chymicum* (1614) and *Tractatvs de svlphvre* (1616) probably written by Michał Sędziwój;[186] and *Chymische Hochzeit Christiani Rosencreutz anno 1459* (1616) claimed by Johann Valentin Andreae.[187] These and many others began to be compiled, first in *De alchemia* (1541)[188] and *De alchimia opuscula complura veterum philosophorum* (1550).[189] Later (with the requisite addition of tracts by Hermes Trismegistus, Geber, Albertus Magnus, Ramon Llull, and other supposed adepts) came the six-volume *Theatrum chemicum* (1602–1661),[190] the *Musæum Hermeticum* (1625)[191] later greatly expanded (1678),[192] Elias Ashmole's *Theatrum chemicum Britannicum* (1652),[193] and Jean-Jacques Manget's *Bibliotheca chemica curiosa* (1702).[194]

However much these alchemical tracts differ in philosophical depth, practical detail, and literary merit, they refer relentlessly to macrocosm and microcosm, the four elements (sometimes with a fifth, "quintessence"), sulphur, mercury and salt, and three natural genera: minerals, vegetables, and animals (Figure 9.3). We find occasional hints of a continuum in nature,[195] and imaginative symbolism: the planet Mercury is a mineral, the moon a plant, the sun an animal.[196] So far as I can determine, zoophytes first appear in *Atalanta fugens* (1617), a classically themed emblem-book by the German alchemist, Rosicrucian and amateur composer Michael Maier. In discussing the origin of the Philosopher's stone, he advises that

Figure 9.3. The Fountain of Mercury, from *Rosarium philosophorum* (1550) in the facsimile edition of Telle (1992), page 10. The two-headed snake at top is labelled *a[n]i[m]alis, mineralis, vegetabil[is]*; the inscription around the fountain reads *M[ercurius] mineralis M[ercurius] vegetabilis M[ercurius] a[n]i[m]alis un[um] est*. Republished by permission from John Wiley & Sons Ltd © 1992. Permission conveyed through PLSclear.

156 KINGDOMS, EMPIRES, AND DOMAINS

Philosophers call their Stone a vegetable, because it vegetates, grows, is increased and is multiplied in the image of a plant: this to the ignorant seems wondrous and alien to truth, as they hold certain that stones neither vegetate nor grow in this manner. . . . But therein are they deceived in their judgement. . . . For who would ever have believed a stone to grow under the water, or a plant generated there should become stone, unless experience and the consistent testimony of writers had acknowledged it? . . .

Moreover, in other places the sea yields three medicinal stones produced in part from the vegetable kind, in part from the animal, or rather in part from the secrets of nature; namely pearl, amber, and ambergris. . . . For we have seen some veins of iron and silver growing upon the amber, which could not be done except in the earth. But that flies, gnats, spiders, butterflies, frogs, and serpents should be distinguished in some little pieces [of amber] (we ourselves had 120 beads turned out of amber, [and each] contained some flies, gnats, spiders, and butterflies; and one of the pieces, not without a singular miracle of nature, had nine) happens by the influence and imagination of the heavens. . . . Although some will reply that [ambergris] is the juice or gum of trees (as was said before for amber), nevertheless those who consider it to come forth from veins of the earth judge more plausibly: for trees that bear amber and ambergris have nowhere been seen. . . . We therefore ascribe both kinds of amber to subterranean veins or stones, just as we ascribe pearls to Zoophyta, and coral to the vegetable.[197]

Likewise we find zoophytes in the 1622 *Aurei Velleris sive sacræ philosophiæ* (*Golden Fleece or sacred philosophy*) of Guillaume Mennens, once in a list of substances (earths, stones, minerals, metals, plants, Zoophites, animals, [and] the human body), a second time in a description of the unity of all things: "for the earth meets with the stones, the stones with the plants, the plants with the Zoophites, the Zoophites with the animals, the animals with man, mankind with heaven, heaven with the intellects, the intellects with the supreme workman, God."[198] Eighty years later, a chemical lexicon in the *Bibliotheca chemica curiosa* defines *sponge* as "a Zoophyton, neither animal nor fruit, but having a certain third nature". It also explains that *elementary parts of the world* are of earth (associated with sulphur), water (salt), or air (mercury); those of water are either of imperfect mixture, as aquatic Zoophytes, or perfect, as Fish.[199]

From the early Seventeenth century, the three divisions of nature began to called *kingdoms*—often capitalized—in the alchemical literature. Perhaps the first was *Duodecim tractatus: de lapide philosophorvm* (1604) by Michael Sendivogius,[200] which states that "natura in tria regna divisa est . . . Est regnum minerale, vegetabile, & animale."[201] From 1614 this work appeared under the title *Novum lumen chymicum*. His 1616 *Tractatvs de svlphvre* likewise refers to Nature as divided into Regna Minerale, Vegetabile and Animale.[202] Both treatises were included in the 1678 *Musæum Hermeticum*, as were three others that refer to these as kingdoms: *Tractatus aureus*, *Via veritatis*, and *Vitulus aureus*.[203] Four other tracts in the fourth volume of *Theatrum chemicum* (1613) discuss Kingdoms of minerals, vegetables, or animals,[204] as do sixteen in Manget's *Bibliotheca chemica curiosa* including *De lapide philosophorum differtatio* excerpted from the *Mundo subterraneo* of Athanasius Kircher.[205] Basil Valentine's *Last will and testament* (1671) refers to the three kingdoms both literally and in allegory.[206]

But the hour was drawing late for alchemy. The first book of chemistry, the anonymous *Sceptical chymist*,[207] appeared in 1661 from the pen of Robert Boyle. From its first pages it challenged the ancient ideas of four elements, the three "chymical principles of mixt bodies" (sulphur, mercury, and salt) and much more. He eventually concludes that

> when in the writings of *Paracelsus* I meet with such Phantastick and Un-intelligible Discourses as that Writer often puzzels and tyres his Reader with, father'd upon such excellent Experiments, as though he seldom clearly teaches, I often find he knew; me thinks the Chymists, in their searches after truth, are not unlike the Navigators of *Solomons Tarshish* Fleet, who brought home from their long and tedious Voyages, not only Gold, and Silver, and Ivory, but Apes and Peacocks too; For so the Writings of several (for I say not, all) of your Hermetick Philosophers present us, together with divers Substantial and noble Experiments, Theories, which either like Peacocks feathers make a great shew, but are neither solid nor useful; or else like Apes, if they have some appearance of being rational, are blemish'd with some absurdity or other, that when they are *Attentively* consider'd, makes them appear Ridiculous.[208]

The three natural kingdoms escaped Boyle's scorn,[209] and modern chemistry, like alchemy before it, accepted the three Kingdoms as its default perspective on the world.

Bruno, Fludd, and the nature-mystics

By the later Renaissance, alchemists held no monopoly on the occult. Philosophers of all sorts pondered the new discoveries concerning man (Leonardo da Vinci, Andreas Vesalius), earth (Christopher Columbus), and the heavens (Galileo Galilei, Nicolaus Copernicus) and, as a consequence, lost faith in Aristotelianism. For some, the new knowledge validated scientific method and critical observation. Others followed Ficino and Pico more deeply into the newly available works of Plato, building elaborate syntheses with Pythagorean number-mysticism, Hermeticism, Kabbalah, and magic that reinforced a mystical *prisca theologia* and *philosophia perennis*. These syntheses were often Christian; but Christianity itself was in crisis, as localized rebellions against papal authority (Jan Hus) spread to Florence (Gerolamo Savonarola) and more widely (Martin Luther). A secret if imaginary society, the Rosicrucians, combined Hermeticism, religious reform,

158 KINGDOMS, EMPIRES, AND DOMAINS

and alchemical medicine in aid of the poor. Even conservatives such as Marin Mersenne, founder of the Académie Parisienne, eventually drew back from rigorous Aristotelianism.

A particularly extreme step was taken by the Dominican friar Giordano Bruno who, inspired by the Copernican model, set about to build a new philosophy complete with "full magical Egyptian" Hermeticism and a pagan Kabbalah.[210] Bruno's universe was infinite and pulsating, extending itself like a gastropod (*limax*),[211] and held countless worlds in addition to our own. He wrote that "nature is God in things",[212] and like Agrippa considered the world itself to be living and sensible.[213] In *Lampas triginta statuarum* (*The lantern of thirty statues*, 1587) he set out multiple thirty-step scales, among them a very non-Aristotelian scale of nature that (according to Bruno) represents a consensus view of the philosophers of old. It links the four basic simples (vacuum, shadow, matter, and the atom), their four properties (dryness, water, vapour, and exhalation), three genera of imperfect composition (the seas, snow, winds and such), three of perfect composition (stones and gems; metals; plants, trees, and herbs), and the five genera of animals: *zoophyta*, *bruta*, *imaginativa*, *substantia rationalis*, and *substantia heroica*. Zoophytes exhibit only the sense of touch, placidly contracting and extending themselves in place.[214] Bruno likewise counterposed animals, plants, and minerals[215] in *De monade numero et figura* (1591), a treatise on occult meanings of the numbers from one through ten. At number ten—within an eclectic list that cites Moses, Hermes, Empedocles, and Democritus—we read that "Zoophita are between the animals and plants, which genus you may be able to reduce to either of the two sides; if only as the highest of one order, or as able to hold only the bottommost grade of the other."[216]

Robert Fludd defended the Church (of England), and practised Galenic as well as Paracelsan medicine while, like Bruno, seeking to build a new philosophy. Macrocosm-microcosm was his point of constant reference, although even here Aristotelian, Mosaic, Hermetic, and Paracelsan ideas jostled with one another: his primary elements are darkness, light, and the waters (or the Spirit of the Lord), his secondary elements earth, air, fire, and water, or sulphur, mercury, and salt. Earth is dense water, water is dense air, air is dense fire.[217] Meteorological phenomena are imperfect creatures, while minerals, vegetables, and animals are more perfect.[218] Blood circulates in the human body in imitation of the course of the sun in the heavens, influenced by the Moon and carrying the spiritual impression of the Zodiac.[219]

Fludd accepted the three natural genera and referred to them repeatedly as Kingdoms.[220] Sometimes they were Aristotelian, with superadditive vegetative and animal souls; indeed, he describes experiments to prove that mineral and animal natures are able to coexist.[221] Elsewhere—in a tract directed against Johannes Kepler—he related animal, vegetable, mineral, and angelic bodies to geometrical solids. As Kepler appreciated, this was Hermeticism, not mathematics.[222] Zoophytes make no appearance.

Disaster befell the nature-mystics in the year 1614: Isaac Casaubon,[223] arguably the most learned man in Europe, demonstrated by philological analysis that the *Corpus Hermeticum* was a Christian forgery dating from the Third or Fourth century.[224] Without warning, the ground of ancient authority was cut out from under *prisca theologia* and its thread of occult truth from Hermes through Orpheus, Pythagoras, and Plato to Christianity. Some occultists simply ignored Casaubon; others made do with macrocosm-microcosm, number-mysticism, Kabbalah, and magic in various proportion. The three natural genera—often although not always called *kingdoms*—appeared in works by all manner of authors including Jakob Böhme,[225] Heinrich Nolle,[226] Thomas Vaughan,[227] and (with poetic license) his brother Henry:

> I would I were a stone, or tree,
> Or flowre by pedigree,
> Or some poor high-way herb, or Spring
> To flow, or bird to sing![228]

With our final nature-mystic, Anton Josef Kirchweger,[229] we reach the mid-Eighteenth century—not the Eighteenth century of Buffon, Linnæus, Hume, or Kant, but instead one in which Nature is the universal fire or Anima Mundi, water and earth are female, air and fire are male, water and air are impregnated with the Universal Sperm, and minerals, vegetables, and animals proceed from the Universal Spirit.

> For there is slight difference amongst them all, by which they all have sprouted forth from their sole Material, and the minerals are fixed vegetables, the vegetables [are] fleeting minerals, as the vegetables [are] fixed animals, and the animals fleeting vegetables, and one can easily transmute into another. For the vegetables are eaten as food by man and beast, and made animal by their Archeum, and when man or beast dies, it will be buried in the earth, and therefrom vegetables grow again. If the vegetables receive the mineral vapour, which [passes] fleetingly through the earth [and] rises up into their root, all becomes vegetable [again]. However, if the vegetables putrefy and [assume] a nitreous saline nature, [which] is dissolved into water and is carried through gaps and fissures in the earth, or by a river to the sea and thence to the centre of the earth, it then arises therefrom to the mineral nature.[230]

Liberated from context, this might be read as chemistry, physiology, or even ecology. The context, however, is that of a Golden Chain (attributed variously to Homer, Plato, or Hermes Trismegistus) that connects primordial Chaos with the Perfect consummation, or Five Essences of the Universe (Figure 9.4). It is no surprise that Animals, Vegetables, and Minerals are links in this chain. It is rather more surprising that Animals are closer to Chaos, and Minerals closer to Perfection.

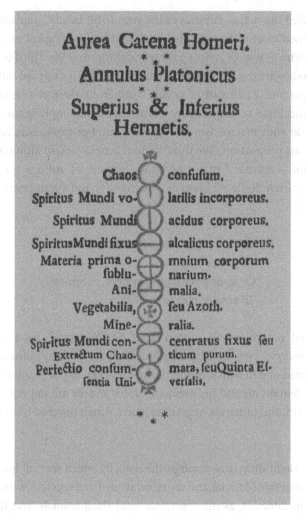

Figure 9.4. Kirchweger, *Aurea catena Homeri* (1723), facing title page. The links are more clearly interlocked in the corresponding figure in the 1757 German edition.

Summary and questions

The authors we have met in this chapter—occultists, alchemists, mystics—were not lonely souls toiling day and night at the writing-desk or bellows. Many maintained high-level patronage as political alliances shifted and war, plague, and religious reformation reshaped Europe. Agrippa was a soldier and diplomat, a participant in the Council of Pisa, and a lecturer at universities in Dôle (Burgundy), Köln, and Pavia; he corresponded with Erasmus, Budé, Melanchthon, and Lefèvre d'Étaples. Others likewise travelled far afield (Pico, Paracelsus, Bruno, Sędziwój, Maier), lectured, served as court physicians, were investigated by the Church,[231] and corresponded with mainstream figures in the Church and civil society.

NATURE'S MYSTIC BOOK   161

Nor was there clear demarcation between occult and mainstream, except for practitioners of black magic. Leonardo da Vinci accepted that the human body is "the world in miniature", and considered the earth itself to possess *anima vegetatiua*.[232] Mathematician Gerolamo Cardano asked "what then is a mine, if not a plant covered with earth and stones?"[233] Natural magic came to mean almost anything from experimental protoscience to nature-mysticism[234]—as illustrated by the *Magia naturalis* (1558) of Giambattista della Porta, a compendium of superstition, unlikely tales, practical advice, and party tricks. Porta demystified optical lenses and distillation even as he acknowledged the web of sympathies and antipathies linking earth and heavens. The colewort shows "deadly hatred, and open enmity" against the vine; tin can be transmuted into silver, fearless dogs generated from tigers, and images of Cupid, Adonis, and Ganymedes in the bedchamber will work their sympathetic magic to ensure one's children are beautiful.[235]

It was Francis Bacon who finally abandoned macrocosm-microcosm theory:

> Neither yet are we so senselesse, as to imagine with *Paracelsus*, and the Alchymists; *That there are to be found in mans Body certaine Correspondences, and Parallels to all the variety of specifique Natures in the world* (as *Starrs, Minerals,* and the rest) as they foolishly fancy and Mythologize; straining, but very impertinently, that embleme of the Ancients, *That man was Microcosmus, an abstract, or modell of the whole world,* to countenance their fabulous, and fictious invention.[236]

With *Novum lumen chymicum*[186] we encountered mineral, plant, and animal *kingdoms.* Was this truly an innovation within this allegory-ridden literature, rife with *faux* tales of ancient caliphs and kings, ever-ready to aggrandize its subject and flatter its would-be patrons among the minor nobility? In Chapters 11 and 12 we shall search more rigorously among mainstream fields that developed at the same time including lexicography, botany and zoology. Before that, however, we have unfinished business in another area out of the mainstream: that of popular myths, superstition, and travellers' tales in Medieval and Renaissance Europe.

10

Allegory, myth, and superstition

> So, slow *Boôtes* underneath him sees,
> In th' ycie *Iles*, those Goslings hatcht of Trees,
> Whose fruitfull leaves, falling into the Water,
> Are turn'd (they say) to living Fowls soon after.
> So, rotten sides of broken Ships do change
> To *Barnacles*; O Transformation strange!
> 'T was first a green Tree, then a gallant Hull;
> Lately a Mushrom; now a flying Gull.
>
> —Josuah Sylvester, *The Divine Weeks*[1]

We now venture in a different direction beyond Church, state, and university into the everyday society of Medieval and Renaissance Europe. This society was hierarchical and unequal, often cruelly so for peasants, serfs, and those displaced by war, disease, or famine. But even these unfortunates, like their cousins in settlements along increasingly prosperous trade routes and in nascent towns and cities, typically had some contact with the broader world: at the village fair, perhaps, or through an itinerant merchant, mendicant friar, or returning crusader. In such circumstances wondrous new tales could take root and spread. In this chapter we explore "intermediate organisms" in allegory, myth, and superstition in Medieval and Renaissance Europe, focusing where possible on popular culture.

Whether we start the clock on this chapter in 500 or 1000 CE, the default worldview was pervaded with myth and superstition. Boëthius drew consolation from the goddess Philosophy, but it was her Wheel of Fortune[2] that spoke to the capriciousness of the times. The far-off heavens enmeshed our world in occult sympathies and antipathies that too often eluded both priest and physician. The Church acknowledged the power of angels, Satan, spirits, demons, visions, and prophecy. The faithful were assured that Lot's wife was turned into a pillar of salt,[3] Aaron's rod became a serpent,[4] the sea parted to reveal dry land,[5] Christ walked on water and rose from the dead, and the Eucharist becomes His very body. The practically minded believed that base metals can be perfected into gold, wheat degenerates into oats, men become wolves, and frogs can rain down from the skies. It was truly a demon-haunted world.[6]

Kingdoms, Empires, and Domains. Mark A. Ragan, Oxford University Press. © Oxford University Press 2023.
DOI: 10.1093/oso/9780197643037.003.0010

This chapter is organized around five steps, each taking us progressively farther from the ordinary. With our first step we come upon events that may be a little difficult to believe, but it is their allegorical meaning, not their actual veracity, that is transparently the main story. Thereafter we encounter beings with exaggerated or oddly arranged features, intended (one imagines) more to arouse wonder or merriment than to inspire reflection on moral issues. In a third step we hear of chimæras typically between beast and man, although in the *borametz* between plant and animal. Our fourth step brings an ongoing transgression of boundaries, notably a two-stage transformation of vegetable matter (usually wood) into a barnacle and thence into a goose. Is this goose thereby a fish, or even a vegetable? A fifth step brings us to wholly incredible incidents that could happen only by miracle or magic. We then conclude with a Sixteenth-century mélange of monsters and marvels, an important and fascinating border-post between ancient scandal and superstition on one hand, and modern medicine and humanism on the other.

Allegory

One need not be a student of Philo or Origen to appreciate that words and images can carry meaning beyond the strictly literal. Just as the carved dragon-heads and intertwined serpents on Norwegian stave churches hint at a hedging of bets on Christianity,[7] even the illiterate could understand that whether or not Saint George had slain an actual dragon, the real story was the triumph of Christianity over paganism.[8] It is unlikely that many thought Hellmouths and other apocalyptic creatures—some composite—to be actual animals, but their depictions in painted stone, stained glass, or illuminated manuscripts were effective as "clear and constant reminders of what awaited the sinner on Judgement Day".[9]

In Chapter 4 we considered the *Physiologus*, a heavily allegorical bestiary intended to guide the reader towards knowledge of the Christian god. Latin translations, differing somewhat from each other in the roster and arrangement of animals (plus a few stones, plants, and men), began to circulate from the late Fourth century CE.[10] Some were later taken into the major European languages as they themselves developed. Illustrations enlivened the stories for readers and reached out to the illiterate. Its allegories were sometimes forced, but with few exceptions (the phœnix, centaur, aspidochelone, and unicorn) the animals themselves were real. An early Eleventh-century edition claiming to be *dicta* of Saint John Chrysostom[11] was the first to focus exclusively on animals,[12] separating them into beasts and birds. The Thirteenth-century *Bestiaire d'amour* of Richard de Fournival, chancellor of Notre-Dame at Amiens, took the tradition in a secular direction, redirecting the allegories to profane love: the viper to warn against envy, the sawfish to rebuke insincere chivalry, the turtledove to celebrate fidelity.[13] Again, some animals were imaginary (the dragon, hydra, phœnix, sawfish, and unicorn) but offered useful moral lessons.

164 KINGDOMS, EMPIRES, AND DOMAINS

Unicorns in particular provided Christian as well as secular allegory, and remain part of popular romantic sensibility to the present day.[14]

Beings with exaggerated features

Exceptionally large or strangely formed beasts, sea creatures, men, and women populate travellers' tales from all ages. For two millennia India was identified as the motherlode of such wonders, thanks to a Fourth-century BCE report by Ktesias of Knidos, court physician for Artaxerxes Mnemon of Persia,[15] and to a Third-century BCE text by Megasthenes, ambassador from Seleucus Nicator to the court of Chandragupta at what is now Patna, on the Ganges.[16] Strabo summarized their reports as fabrications:

> for they are the persons who tell us about the men that sleep in their ears, and the men without mouths, and men without noses; and about men with one eye, men with long legs, men with fingers turned backward; and they revived, also, the Homeric story of the battle between the cranes and the pygmies who, they said, were three spans tall. These men also tell about the ants that mine gold.[17]

But it was Pliny, not Strabo, whose works became known in Europe, and Pliny accepted the reports of Ktesias and Megasthenes wholesale:

> India and parts of Ethiopia especially teem with marvels. The biggest animals grow in India. . . . Indeed the trees are said to be so lofty that it is not possible to shoot an arrow over them. . . . It is known that many of the inhabitants are more than seven feet six inches high . . . and that the sages of their race, whom they call Gymnosophists, stay standing from sunrise to sunset, gazing at the sun with eyes unmoving, and continue all day long standing first on one foot and then on the other in the glowing sand.[18]

Chimæras: the *borametz*

It is not intrinsically unlikely that people in remote lands might be particularly tall, short, swift, or brave, or practise odd customs (although many reports went farther, supposing curious rearrangements of body parts). Other wonders, however, transgressed boundaries between species, more blatantly offending against the natural order. One such story told of Antony and the satyr. According to Saint Jerome, the ancient hermit Antony set out into the desert to find an even older hermit, Paulus, and encountered "a creature of mingled shape, half horse half man, called by the poets Hippo-centaur". Then after walking farther, "he sees a mannikin with hooked snout, horned forehead, and extremities like goats' feet" who

ALLEGORY, MYTH, AND SUPERSTITION   165

revealed himself as "a mortal being and one of those inhabitants of the desert whom the Gentiles deluded by various forms of error worship under the names of Fauns, Satyrs, and Incubi".[19]

Such reports made the Church uncomfortable; earthly beings had been created "after their kind",[20] and its stamp of approval might give license to pagan beliefs. Isidore was a sceptic, choosing his words carefully ("people imagine . . .") and pointing out how confusion or embellishment can get out of hand. He accepted that some monstrosities are one-off portents sent by God, but these do not perpetuate themselves or even survive beyond birth.[21] For Abelard, *chimæra* and *goat-stag* were examples of things that "are located in empty opinion" and "do not give rise to a rational understanding".[22] Aquinas taught that a chimæra, like a syllogism or negation, could be an *ens rationis* but not an *ens reale*,[23] although he later allowed that the ox-men imagined by Empedocles might have arisen long ago by the corruption of some natural principle, even if they could not and did not persist.[24] A century later, theological faculty at the University of Bologna were prohibited from teaching that the word *chimæra* even makes sense.[25] These learned opinions were as effective against popular belief as they were in preventing stonemasons from decorating churches with gargoyles and grotesques.[26]

Thus breathless tales of animal-men come to us from the monk Ratramnus (*Epistle to the Dog-heads*) in the mid-Ninth century; the Vitellus manuscript about 1000 CE; the Monte Cassino illustrated manuscript in the Eleventh century; Foulcher de Chartres in the Twelfth; the Ebstorf and Hereford maps and Jacques de Vitry in the Thirteenth; Jourdain de Séverac, "Sir John Mandeville", and Friar Odoric in the Fourteenth; Thomas of Cantimpré and Conrad von Megenberg in the Fifteenth; Conrad Gessner, Sebastian Münster, Sir Walter Raleigh, André Thevet, and Edward Wotton in the Sixteenth; and Ulisse Aldrovandi, John Bulwer, and Edward Topsell in the Seventeenth.[27] In a chimæra the face, head, or upper torso of a man was typically joined with the lower body of a beast, or the other way around. For some, this mingling of bodies and souls crossed a philosophical or doctrinal red line. An animal-plant chimæra would necessarily transgress even more egregiously, as plants lack the sensitive soul and (according to Saint Basil and Ibn Sīnā) are not even alive. Yet one compound animal-plant being, the *borametz* or vegetable lamb, was everywhere welcomed into the menagerie.

Herodotus wrote that in India, where "every species of birds and of quadrupeds, horses excepted, are much larger than in any other part of the world . . . they possess likewise a kind of plant, which, instead of fruit, produces wool, of a finer and better quality than that of sheep: of this the natives make their cloathes."[28] Later Hellenic and Roman reports confirmed much the same.[29] Theophrastus[30] wrote of trees on the island of Tylos in the Persian Gulf that bear wool in a pod the size of a μῆλον— an apple, although the word could also mean *sheep* or *goat*.[31] The Arabs called this material *al-ḳuṭun*, while an early Hebrew term was *ṣemer gefen*, or *wool of a vine tree*[32]—not unlike the German *Baumwolle*. According to Talmudic tradition, at the time of *Leviticus* a fierce chimæric man-plant known as *jeduah* could be found in

166 KINGDOMS, EMPIRES, AND DOMAINS

the mountains, growing like a gourd or melon with its navel connected by a large stem to a root in the earth. If torn loose by its stem, the being dies.[33] A similar tale may have spread from Syria around the Eighth century CE, and as retold in Chinese annals of that period it is a lamb, not a wild man, that grows attached to the ground by its umbilical cord.[34]

It is scarcely possible to disentangle these stories from one another, nor to determine if they arise from the same natural phenomenon. In any event, two quite similar versions reached Europe in the early Fourteenth century. In 1330–1331 Friar William of Solagna took notes "in unadorned Latin" as fellow Franciscan Odoric recounted his twelve years of travels in the Far East. William later presented the narrative as *The journal of Friar Odoric*. Odoric had heard about a gourd whose fruit encloses a small beast "like unto a young lamb", although had not seen it with his own eyes.[35] A slightly later travelogue attributed to "Sir John Mandeville" likewise gives secondhand reports from a kingdom beyond Cathay, where

> there groweth a manner of fruit, as though it were gourds. And when they be ripe, men cut them atwo, and men find within a little beast, in flesh, in bone, and blood, as though it were a little lamb without wool. And men eat both the fruit and the beast. And that is a great marvel. Of that fruit I have eaten.[36]

The plant became known as the *borametz* (from the Slavonic *баран*, *ram*),[37] or alternatively the *Scythian lamb* or *vegetable lamb*. Illustrations were duly published,[38] and pelts appeared in cabinets of curiosities.[39] In 1549 a European diplomat formerly based in Muscovy, Sigismund, Baron von Herberstein[40] reported third-hand that the lamb grows from a seed like that of a melon, had flesh like that of a crab, and grazed upon whatever grass and herbs it could reach. In translating this passage into English, Lee has one of Sigismund's sources call it a zoophyte. In reality, Sigismund wrote only that it is "animal in the likeness of a plant fixed to the earth".[41] As imagined by Guillaume de Salluste, Sieur Du Bartas, in the verse of Sylvester:

> O wondrous vertue of God onely good!
> The Beast hath root, the Plant hath flesh and bloud:
> The nimble Plant can turn it to and fro;
> The nummed Beast can neither stir nor go;
> The Plant is leaf-less, branch-less, void of fruit;
> The Beast is lust-less, sex-less, fire-less, mute:
> The Plant with Plants his hungry panch doth feed;
> Th' admired Beast is sow'n a slender seed.[42]

Such integration is not only structural and functional, but reaches its very soul(s) and the associated faculties: the lamb is vegetative, senseless, without locomotion; the plant animate, nimble, hungry. There was much reason to doubt that a being so transgressive could exist. Gerolamo Cardano asked how it might be propagated and develop. Referring to Sigismund's report, he argued that a warm-blooded animal can neither grow from a seed nor be sustained during embryonic development by the earth alone, as can a melon.[43] Nonetheless Cardano managed to convince himself that the borametz is probably real.

The question inevitably arose, *is the borametz a zoophyte?* Unlike a sponge or coral, the borametz displays nearly a full complement of animal parts and senses, not to mention (in some variants of the tale) a stalk or stem that enables a sort of motion-in-place. Andreas Libavius pointed out that other zoophytes (the murex snail, for instance) feed on grass, and that terrestrial species of zoophytes, like those of other animals, might be more perfect than their marine counterparts.[44] As for motion-in-place, Fortunio Liceti[45] reminded his readers that sea-nettles—undoubted zoophytes—remain affixed to one spot, but reach out with their tentacles to graze upon passing fish. Even Engelbert Kaempfer, who had failed to uncover evidence of the borametz during his travels in Tartary, called it a zoophyte.[46]

So—*pace* Aristotle—are there terrestrial zoophytes? Cardano speculated that plants "having sensation, and also imperfect flesh, such as that of mollusks and fishes" might perhaps live on land "where the atmosphere was thick and dense"— in Tartary, for instance.[43] Julius Cæsar Scaliger ridiculed Cardano's credulity,[47] but Liceti, a physician and author of a treatise on monstrosities of nature, found the evidence convincing.[45] The borametz would be a new type of zoophyte, but Liceti proposed that even uterine tumours are zoophytes.[48] Nieremberg followed Liceti, plagiarizing his words at length.[49] For Duret[50] the borametz is a zoophyte, for Thomas Vaughan[51] a *plantanimal*. The tale survived into the Eighteenth century[52] before eventually retreating into poetry.[53]

Active transformation: the barnacle-goose-tree

It is one thing for God to have created, long ago, a race of satyrs, which have since perpetuated themselves in kind. It is quite another for animals to arise routinely from inanimate earth, air or water, or from vegetable matter, generation after generation to the present day. One can only suppose that our medieval forebears were numb to the enormity of the idea, as from time immemorial it had been accepted that worms arise from mud, and insects from carcasses.[54] Was not the earth itself imbued with generative force?[55] Yet the instant of transformation was rarely observed: newly formed worms are too small to be seen, fungi appear at night, while metals are perfected underground. Other transformations took place only in remote corners of the world—corners that tended to recede, mirage-like, as the investigative traveller approached. So discovered Aeneas Piccolomini, later Pope Pius II, as he slogged through the snows of Scotland in the late autumn of 1435, seeking the elusive natal home of the barnacle goose—only to be told these trees are to be found yet farther north, in the Orkneys.[56]

168 KINGDOMS, EMPIRES, AND DOMAINS

In Medieval times this goose,[57] known simply as *bernekke*, was common in winter along the western coasts of Scotland and Ireland, but with the approach of spring disappeared to parts unknown.[58] By the late Twelfth century, a belief had arisen that these geese arise from pine or fir wood soaked in the sea. According to Alexander Neckam, such wood exudes a "certain viscous humour, which in the course of time assumes the form of a little bird clothed in feathers, and it is seen to hang by its beak from the wood".[59] Gerald de Barri (Giraldus Cambrensis), who travelled in Ireland in 1185 with Prince John, Lord of Ireland and son of Henry II, returned with reports of birds that are

> at first gummy excrescences from pine-beams floating on the waters, and then enclosed in shells to secure their free growth, they hang by their beaks, like seaweeds attached to the timber. Being in process of time well covered with feathers, they either fall into the water or take their flight in the free air, their nourishment and growth being supplied, while they are bred in this very unaccountable and curious manner, from the juices of the wood in the sea-water. I have often seen with my own eyes more than a thousand minute embryos of birds of this species on the seashore, hanging from one piece of timber, covered with shells, and already formed. No eggs are laid by these birds after copulation, as is the case with birds in general; the hen never sits on eggs in order to hatch them; in no corner of the world are they seen either to pair, or build nests. Hence, in some parts of Ireland, bishops and men of religion make no scruple of eating these birds on fasting days, as not being flesh, because they are not born of flesh.[60]

From the early Thirteenth century, variants of the tale appeared in England, France and Germany in which the birds grow on trees, and when mature drop down like a ripe fruit; or emerge from leaves that have fallen into the sea.[61] Hector Boece added that the geese can be formed from trees that have been cast into the sea, from ships' timber, or via worms that appear in submerged apples or fruits.[62] Albertus Magnus heaped scorn on such tales, pointing out that he and his friends had seen these geese mate, lay eggs, and feed their young;[63] but the myth was well entrenched. Like the borametz, the barnacle-goose-tree seems to have had Eastern antecedents: according to Peter Damian, on the Indian isle of Tylos leafless trees bear fruit from which feathery birds break forth[64]—the same island, and probably the selfsame wool-trees, described by Theophrastus (above). Analogous tales are now known from the Middle East and China;[65] fledgling geese emerge from leaves on an ossuary from ancient Mycenæ (1600–1100 BCE).[66]

In the passage above, Giraldus reported that "bishops and men of religion" consumed barnacle geese on days when the Church forbade the eating of flesh, arguing that by virtue of their origin these geese were actually fish.[67] They might equally have been deemed vegetables, and indeed Vincent of Beauvais hints at this.[68] But (the story goes) Pope Innocent III, in the Fourth Council of the Lateran (1215), issued a decree banning the consumption of goose-flesh on fasting days. Innocent

III was well-educated,[69] and it would be interesting to understand the reasons behind his ruling. Did he doubt that the gummy excrescence was a fish, or that geese truly arose from it? Was the barnacle-goose an unnatural abomination? Or was this for him an opportunity to assert Papal authority while further marginalizing the old pagan ways?

Except on fasting days, the Church did not prohibit the consumption of goose flesh. Geese had become an article of commerce by the time of the Apostle Peter, and by the Ninth century were raised and marketed in huge numbers particularly in Gaul, even if many in England and Wales remained superstitious about eating them.[70] At a banquet on Sunday, 19 May 1342 in Avignon celebrating the coronation of Pope Clement VI, three thousand guests dined on 1195 geese (and much else).[71]

Thomas of Cantimpré mentions Innocent III's injunction in *De natura rerum*, which he wrote between 1230 and 1244 while a student of Albertus Magnus.[72] Vincent of Beauvais, who drew liberally from Thomas's unpublished manuscript, mentions the injunction in his encyclopædic *Speculum maius*, where the next sentence reads "these birds live upon herbs and seeds like geese"[73] as if by explanation. In the mid-Fourteenth century Conrad von Megenberg modelled his *Puch (Buch) der Natur* on Thomas's *De natura rerum*, and likewise noted the Papal injunction.[74] Indeed, the Church continued to engage the question through the Fourteenth and Fifteenth centuries.[75] In 1599 the Jesuit zealot Martin del Rio likened the origin of these geese to that of "flies, mice and the like" and warned of the involvement of magicians calling upon demons;[76] but his remained a minority view.

The papal injunction—sometimes referred to as a *bull*—has long been a fixture of the story, but is never identified by title or precise date. None of the seventy canons of the influential Fourth Council of the Lateran (1215) even remotely concerns barnacles or geese;[77] nor does the matter appear in Innocent III's *Opera*.[78] Innocent rejuvenated a system under which official documents and letters were noted in formal ledgers, or *Registra*. As I write, a project is underway to publish and index those from Innocent III's papacy;[79] many have appeared in other venues over the years.[80] To date, I have found no primary evidence that Innocent III ever engaged this matter. However, his office issued well over five thousand letters,[81] not to mention other types of documents;[82] some have been lost, others were never delivered, and forgeries abound.[83] It is not impossible that the matter of the barnacle-goose-tree was indeed considered at the Fourth Council, which defined Church doctrine on transubstantiation, reinforced papal authority, and sought to hold the clergy to higher moral standards. But to date, the actual words of Innocent III have not come to light. Nor, for that matter, has a supposed *sententia judicatum* from the Sorbonne, ruling that the geese are fish and can lawfully be eaten at Lent.[84]

Then as now, the barnacle-goose myth was the most famous tale of transformation, thanks in part to the perhaps apocryphal intervention by Innocent III. But even in the Sixteenth century, other such tales were afoot. Paracelsus wrote that if a chick is burnt to powder in a glass vessel, and the powder mixed with an extract of horse

170   KINGDOMS, EMPIRES, AND DOMAINS

dung, a new chick will emerge; indeed, men can be produced artificially in the same manner.[85] Porta accepted that blackbirds could arise from rotting sage, eels from rainwater, fish and shellfish from mud.[86] In 1521 naturalist Antonio Pigafetta and his fellow sailors with Ferdinand Magellan found a tree on the island of Cimbubon near Brunei, the leaves of which become animated after falling to the ground, and scamper away when touched.[87] In 1596 Jean Bodin asked whether barnacle-geese might be considered zoophytes.[88] William Turner mentions the myth in *Avivm praecipvarvm qvarvm apvd Plinivm et Aristotelem* (1544),[89] considered the first book of ornithology, and (as we shall see in Chapter 12) it persists in the great herbals and zoologies of the Sixteenth and Seventeenth centuries. Indeed, these miraculous beings continued to turn up in the zoological literature until the early Eighteenth century, in the occasional history or geography book until the 1780s, and on Lenten dinner-tables until the early Twentieth century.[90]

Return from the dead

Fabulists did not draw the line at the spontaneous generation of worms, insects, fish, and frogs. Near-Eastern gods,[91] classical heroes,[92] and associates of Jewish prophets[93] returned from the dead, as did Lazarus[94] and Jesus.[95] Saint George was martyred three times—once chopped to bits, on another occasion buried deep underground, thirdly devoured by fire—but each time was resurrected intact by God.[96] In the Jewish and Christian religions, bodily return from death was the most problematic transformation of all, reserved for God himself or His direct intervention in human affairs. No gummy excrescence, putrefaction, or mud was involved (and in the story of Lazarus was explicitly ruled out): the body was reanimated without passing through an intermediate state. Unless the reanimation was by magic— which for the prophets, Lazarus and Jesus would entail sacrilege of the highest order—such a transformation must be a miracle: which was, of course, the point.

Monsters and marvels

In the late Sixteenth century, a remarkable book appeared from the pen of Ambroise Paré, a provincial barber-surgeon who set up practise in Paris and against the odds became chief surgeon to Charles IX (reigned 1560–1574) and Henri III (1574– 1589). *Monstres et prodiges*[97] was a literary dualizer, rehearsing classical fables and hearsay while drawing on contemporary science and Paré's own experience as a surgeon. His narrative ranged widely over congenital defects and anomalies, urinary-tract stones, tumours, parasitic conditions, human-animal chimæras, beasts real and imaginary, volcanoes, earthquakes, comets, other portentous celestial phenomena, and, delightfully, "other stories not off the subject".[98] Wheat, fish, milk, or blood can rain down from the sky; fruit trees can bear wheat.[99] Paré illustrated two examples of Rondelet's "insect fish, that is to say [fish] which are by nature half way between plants and animals"[100] without commenting on their intermediate nature.

Written in vernacular French, *Monstres et prodiges* was intended for trainee surgeons unschooled in Latin, although not for "idiots and mechanics".[101] Of particular note is its focus on underlying causes, which might be rational (reproductive physiology, injury, disease), divine, moral, or out-and-out demonic. *Monstres* can seem almost modern: overt allegory of the *Physiologus* sort is absent, macrocosm-microcosm serves as literary device[102] as much as explanation, and illustrations are used effectively. His compassion as a surgeon is evident, and he presents hermaphroditism, conjoined twins, and spontaneous sex change as consequences of physiological imbalance during reproduction, not of moral failure or divine retribution. Yet Paré can be moralistic, credulous, and superstitious, and philosophical argument is notable by its absence. Thus, marine monsters, including Rondelet's "insect-fish", are not caused by the fusing and mixing of seed as are other monsters, but merely reflect Nature "disporting herself in her creations".[103] Intermediacy—an issue his book is singularly well-positioned to discuss—is submerged into a broader story of things that happen outside, or contrary to, the usual course of nature.

Following in the same genre was *Histoire admirable des plantes et herbes esmerueillables & miraculeuses en nature: mesmes d'aucunes qui sont vrays Zoophytes, ou Plant'-animales, Plantes & Animaux tout ensemble, pour auroir vie vegetatiue, sensitiue & animale* (1605) by Claude Duret, a jurist and historiographer close to the French royal court. Duret was particularly interested in transformation—in his historiography as well as his botany—and *Histoire admirable* is a veritable compendium of trees that retract their limbs when animals or men approach (the Arbre Vergongneux); goose-trees, both with and without barnacles; other trees whose fruits or leaves drop off and change into birds or fish, or grow legs; trees that are "sensitive and animated like the sponges, sea-nettles and sea-lungs"; and, of course, the borametz. Many of these wonders are depicted in full-page illustrations, some quite surreal. He may have adopted the term *zoophyte* from Guillaume Rondelet,[104] but falsely credits Aristotle ("l'esponge de mer qu'un chacun eçait avoir esté nommée par le Philosophe Zoophyte ou Plant'animal").[105]

Ancients and Moderns

Across Europe, a confrontation played out through the Sixteenth and Seventeenth centuries between (to distil its diversity into single words) Ancients and Moderns. The former argued that we can do no better than to emulate the best among the classical Hellenes and Romans; the latter, that modern learning had already surpassed theirs. Battle lines were drawn across literature and poetry, but the confrontation was broader: authority versus scientific method; Aristotle, Pliny, and Galen versus Descartes, Galileo, and Newton. We have already met some Moderns (Cardano, Paré) and will encounter many others over the next few chapters. The confrontation

172   KINGDOMS, EMPIRES, AND DOMAINS

was particularly acute within the Académie Française, but the two cultures quarrelled in Italy, England, and elsewhere.

Modernism confronted classicism in the visual arts as well. It must remain for a different work than this to follow how intermediate beings or states were depicted in the painting, tapestry, sculpture, and architecture through the late Middle Ages, Renaissance, and Enlightenment. A broad spectrum of intent and meaning is on offer, from the tree-man of Bosch[106] and crouching sphinx of Bronzino[107] to the zoophytes of Blake[108] and the chimæras of Moreau.[109]

Although several of our subsequent chapters explicitly begin with Ancients, from here forward our narrative will emphasize the modern. In the next chapter we examine plants, animals, and intermediate organisms in general intellectual discourse including philology, law, political theory, and literature in Sixteenth- and Seventeenth-century Europe. We begin by following the term *zoophyta* as it spread from Latin into the evolving Castilian, French, and English languages. In Chapter 12 we narrow our focus to the new fields of botany and zoology. Aristotle's "things in the sea" found a continuing if modest home within the Christian-Neoplatonist synthesis and its scholastic reencounter with Aristotelianism; but we have also seen challenges from terrestrial chimæras dramatically different from sponges, oysters, or corals. What would the Moderns do with zoophytes?

Finally, let us appreciate that myths, superstitions, and general bad judgement about intermediate organisms did not suddenly vanish in the year 1700. The first five editions of Linnæus's *Systema naturæ* (1735–1747) included the hydra, unicorn, satyr, borametz, phœnix, *bernekke*, and dragon in a special group called "Paradoxa" or "Animalia paradoxa".[110] The first reports of entomopathogenic fungi used unreformed language including "vegetable fly",[111] and referred to worms that "lived in old trees and changed into a small bush which had stems and leaves and reached a foot high".[112] Other examples await us in the upcoming chapters. In less-technical contexts, writers have long exploited the didactic and dramatic opportunities presented by monsters, chimæras, and transformation. In the Sixteenth century Giambattista Gelli told of a fisherman transformed into an oyster, who nonetheless preferred his new lot over the disrespect and vexations accorded him when he was a man.[113] Lilliputians, Dr Frankenstein's monster, triffids, hobbits, ents, and comic-book superheroes have entertained, inspired, and terrified generations of readers. And Jorge Luis Borges amuses and delights us with some 120 imaginary beings from literature and folklore, many of which are chimæras involving men or women and beasts, serpents, birds, or fish.[114]

11

The return of the zoophytes

> **Zoophyta**. Those things said to have a middle nature between the sensible
> and vegetable, such as oysters, sponges and the like among animals. Perhaps
> (says Budæus) we can say *plant-animates* or *plantanimals*.
> —Gerardus Morrhius, *Lexicon Graecolatinvm*[1]

In this chapter we focus on the term *zoophyta* and its uptake into European languages other than Greek and Latin from about the year 1500. As we have seen, most classical texts that employ the term ζωόφυτα were translated into Latin during the Thirteenth century, and found a receptive audience in the main centres of learning. Scholastics including Grosseteste, Albert, and Aquinas referred occasionally to intermediate organisms, but did not promote them as a separate group and did not use the term *zoophyta*. Gemistus Pletho established a Hellenic bridgehead on the Italian peninsula in 1453, but his ζωοφύτων (*De differentiis*) became *zoophyta* only with Pico della Mirandola (*Heptaplus*, 1489) and Gianfrancesco Pico (*De studio*, 1496; *Examen vanitatis*, 1520). So far, we have encountered *zoophyta* in Agrippa (*De occulta philosophia*, 1533), Bruno (*Lampas*, 1587; *De monade*, 1591) and Maier (*Atalanta fugens*, 1617), and have mentioned its use by Duret (1605), Liceti (1618), Nieremberg (1635), and Thomas Vaughan (1650) without, however, searching more broadly in the humanistic and technical literature that began to flourish in Sixteenth- and early Seventeenth-century Europe. We now initiate this search, starting with an overview of how words became standardized, and spread.

Dictionaries

Johannes Gutenberg set up a printing press in Mainz in 1449 or 1450, and produced his famous 42-line Bible in 1455. Printing soon spread to Bamberg (1459–1460), Strasbourg (1460), Köln (1465), Basel (1468), and Nürnberg (1469), and across the Alps to Subiaco (1465), Rome (1467), and thence to Naples, Milan, and Florence.[2] Johannes Speyer relocated from Mainz to Venice and printed Pliny's *Natvralis historiæ* there in 1469. In 1494 Aldus Manutius, a friend of Giovanni Pico della

174   KINGDOMS, EMPIRES, AND DOMAINS

Mirandola, established a printing house in Venice with the aim of publishing classical works.[3] With superb access to manuscripts, a commitment to typographical excellence, and a monopoly (within the Venetian Republic) on printing in Greek, the Aldine press became a cornerstone of Renaissance humanism. Its five-volume Aristotle (1495–1498)—which began with Porphyry's *Isagoge* and included commentaries, vitæ, and works attributed to Alexander of Aphrodisias, Galen, Philo Judæus, Theophrastus, and others—was particularly notable. Likewise in Florence, Padua, Milan, Paris, Lyon, Basel, Köln, Amsterdam, and elsewhere, classical works were printed alongside those of a religious, medical, or contemporary nature. Classical texts using the word ζωόφυτον were printed in 1480 (Themistius and Simplicius),[4] 1497 (pseudo-Philo),[5] and 1500 (Ammonius), and a Latin translation of Ammonius (*zoophyta*) followed in 1503.[6] The others followed in succeeding decades.[7]

Aldus hired Greeks as assistants—as many as thirty were on his payroll at any given point in time—but more broadly across Europe, Greek-language skills had been lost for nearly a millennium.[8] Latin too had evolved since the time of Pliny, and classical texts presented unfamiliar words and long-lost senses of familiar ones. Dictionaries were needed, first simply to understand classical literature in its own terms, and thereafter to support scholarship into the structure, evolution, and relationships of classical languages and their role in ancient thought, culture, and society. Considine refers to these roles as humanistic lexicography, and philological or antiquarian lexicography, respectively.[9] Dictionaries could also serve to consolidate vernacular languages, or to define specialized terms in fields such as law, mathematics, or botany.

One of the first dictionaries to be printed had originated fifteen centuries earlier. Verrius Flaccus, tutor to the grandsons of Emperor Augustus, compiled a great dictionary of Latin words describing all aspects of Roman life, history, religion, and culture. All forty volumes had long been lost, but in the Second century Sextus Pompeius Festus had produced an abbreviated edition. Less than half of Festus's text had in turn survived, but between 781 and 787 Paulus Diaconus, a scholar in the court of Charlemagne, made an epitome of Festus's edition. A copy of this *Epitome Festi De verborum significatu* was discovered in 1416 and printed in Venice in 1474; it does not mention *zoophyta*.[10]

In 1492 the Spanish humanist Antonio de Nebrija published a Castilian grammar[11]—the first for any European vernacular language—and a Latin-to-Castilian *Lexicon*.[12] Ignoring existing Latin glossaries and scorning medieval sources, he collected terms for his *Lexicon* directly from ancient Latin texts with one exception: words related to plants and animals, which he drew from Theodorus Gaza's 1476 Latin translation of Aristotle's *Historia animalium*.[13] In 1495 he reworked his *Lexicon* into a Castilian-to-Latin *Dictionarium*,[14] again a first for a European vernacular. Neither mentions zoophyta, but the subsequent edition of his *Dictionarium*, published in Seville in 1516, succinctly defines "Zophyton animal ex parte & ex partem planta."[15] This wording remained unchanged in successive editions.[16] Gaza did not use *zoophyta* in his 1476 *Historia animalium* (nor, so far as I can determine, in his other translations of Aristotle and Theophrastus), so Nebrija's precise source remains unknown.

Zoophyta appeared next, quite independently of Nebrija, in the 1530 *Lexicon Graecolatinvm*, the first work to appear from the press of Gerardus Morrhius at the Sorbonne.[17] Morrhius, like Speyer a German, drew on his friend Desiderius Erasmus and other well-known humanists including Guillaume Budé, Lorenzo Valla, Ermolao Barbaro, and Angelo Poliziano for his elegant and expensive *Lexicon*. The definition itself, which in English translation heads this chapter, was drawn closely from Budé's *Annotationes in Pandectas* (1508), which we consider in detail below. Perhaps in the interest of brevity, Morrhius changed Budé's "inter animalia et stirpes" (*between animals and plants*) to "inter animalia" (*among animals*), undermining to some extent Budé's comment that these organisms might be called *plantanimes* or *plantanimalia*. In any event, Morrhius's definition was widely copied (or more often, stolen word for word): more than half of the dictionaries from this period that define *zoophyta*, including editions from Robert Estienne (*e.g.* Paris, 1543; Geneva, 1573), do so using some form of Budé's words.[18]

Whereas the 1530 *Lexicon* supported philological lexicography, two other early reference works were notable examples of humanistic lexicography: the *Dictionarium græco-latinum* by Giovanni Crastone, and the *Dictionarium latinum* of Ambrogio Calepino. The former emerged in 1476 from earlier word lists, and was intended to assist those who wished to write in Greek.[19] The five editions I examined[20] contain entries for ζῴδιον (little animal) and ζωύφιον (animalcule) but not for ζωόφυτον. Indeed, we search in vain for *zoophyta* in dictionaries of the Fifteenth century.[21] Calepino, whose name was to become synonymous with *dictionary*, met with one of the greatest publishing successes of all time, as his *Dictionarium* went through 211 editions from 1502 through 1779.[22] Among the fourteen editions I examined,[23] *zoophyta* appears first in the 1548 Aldine, in a slight variant of the Morrhius/Budé definition. These same editions state that *spongia* is of a third order of nature, neither animal nor vegetable, but of a third sort between the two "which the Greeks call ζωόφυτον".

Words spread along multiple paths—via books, manuscripts, civil and papal decrees, correspondence, even personal contact—and not by dictionaries alone. Erasmus greatly expanded his knowledge of Greek while working at the Aldine press in 1508. Yet dictionaries can reach an unparalleled range of constituencies, from royal libraries and monasteries to merchants and independent scholars. Let us consider ten case studies illustrating the diffusion of *zoophyta* in Sixteenth- and early Seventeenth-century Europe.

Guillaume Budé: Roman law (1508)

For Guillaume Budé, humanism was not mere prose style or an æsthetic, but entailed an organic appreciation of ancient society and cultural dynamics. One

176 KINGDOMS, EMPIRES, AND DOMAINS

path to this appreciation led through lexicography, which Budé elevated to a new level of rigour and applied with notable success to the study of Roman law. Over several centuries a massive if disorganized body of laws—the *Pandects* assembled for Emperor Justinian I[24]—had been the subject of painfully detailed scholastic commentary,[25] but this effort had neither illuminated the dynamic by which these laws had come about in the first place, nor indicated whether or how they might be a model for contemporary Europe. In *Annotationes in XXIV libros Pandectarum* (1508) Budé used his lexicographic skills to rethink Roman law and build a "new relation to the past".[26]

It cannot be said that Budé brought order to his subject: one sympathetic reviewer calls his *Annotationes* "technical, formless, and unreadable".[27] One of his points of discussion begins with terminology on twins and triplets, followed by anecdotes from Livy and Trogus about twins and triplets; then comments on superfecundation, and Aristotle's argument (quoted in the original Greek) that man dualizes (ἐπαμφοτερίζειν) in regard to number of offspring. Budé then reminds us of another sort of dualization, that of the sea-nettle, a thing median between animals and plants: and

> Of what kind they are that have a median nature between the sensible and vegetables, these are called Zoophyta (perhaps we can say plant-animates or plantanimals) such as oysters, sponges and the like, dualizing between animals and plants.

He follows this by generalizing duality, as he had twins to triplets:

> It is for no other reason is it said in the last chapter of *Mark*, "go ye into all the world, and preach the gospel to every creature". For man shares the vegetal soul with plants, the senses with brute animals, mind and reason with the intelligible natures, that is, with the angels. For this reason he [man] is signified by the name "all creation".[28]

Budé does not indicate his source of *zoophyta*, but it was not Aristotle per se. The word occurs in the *Peri khosmos* falsely attributed to Philo Judæus,[29] a work Budé translated, although not until much later (published 1533). By his own report, he had been enthusiastic about classic literature since 1491, so he may have noticed the word in the *Peri khosmos* bound with the Aldine Aristotle (1497), or in Pico's *Heptaplus* or Gianfrancesco's *De studio*. "Oysters and sponges" recalls the commentary on Porphyry's *Isagoge* attributed to Ammonius.[30]

Otto Brunfels: *materia medica* (1534)

Otto Brunfels—ex-monk, state official, educational administrator, and polemicist for religious reform—received his doctorate of medicine at Basel in 1532.

He had long been interested in plants as *materia medica*, and during the final years of his life published a beautifully illustrated *Eicones* of local plants with descriptions in German, earning a place among the German "fathers of botany".[31] According to his 1534 *Onomastikon medicinae*, "ζοώφυτα, id est, plantamina, ab natura inter animal, planta[m]q[ue] ambigua. Philo de mun[do]".[32] The definition is adequate (apart from misspelling ζωόφυτα and *plantanima*) although as we have seen, this *Peri khosmos* (*De mundo*) was not truly from the hand of Philo. Brunfels probably discovered it in Budé's Latin translation, which appeared in Basel in 1533.[33] In the posthumous 1553 *Onomastikon* the attribution was changed to "Philosophus De mun[do]", *i.e.* Aristotle *De mundo*, but this too is false.[33]

François Rabelais: literature in the vernacular (1546)

Budé corresponded with numerous humanists and intellectuals, including (in 1522–1523) François Rabelais,[34] then a monk at Fontenay-le-Compte. Rabelais later left the monastery, became a Benedictine, and by about 1530 had become a secular priest. He then studied medicine (1530–1532) at Montpellier, where he edited the *Aphorisms* of Hippocrates and the *Ars parva* of Galen,[35] and in 1537 lectured on Hippocrates's *De prognosticis* from the original Greek text.[36] Today we enjoy his *Gargantua* and *Pantagruel*[37] as a marvellous example of early secular literature in French. Playful, often vulgar, his language "is his own private brew, in part remnant of older French, in part dialect or patois (even jargon), in part invented for his own artistic purposes, sometimes from Greek roots when available French words did not lie at hand . . . he writes for himself and his private public, which he creates as we read."[38] In Book 3, Pantagruel challenges his companion Panurge with the proposition that the codpiece is the most important piece of armour worn by warriors. Panurge agrees, and continues:

> See how Nature, intending that the plants, trees, herbs and zoophytes created by her should endure and be perpetuated through all succeeding ages—the individuals dying but the species surviving—carefully clad in armour the seeds and buds in which such perpetuation lies; and so with remarkable cunning she furnished and covered them with pods, shucks, teguments, husks, calyxes, shells, spikes, egrets, skins or prickly spines, which form fine, sturdy, natural codpieces for them. Manifest examples are found in peas, beans, chick-peas, nuts, peaches, cotton, bitter-apples, corn, poppies, lemons, chestnuts: in all plants generally, in which we can clearly see that the seeds and buds are covered, protected and armour-clad more than any other part of them.
>
> Not so has Nature provided for the perpetuation of the human race: in the state of innocence, in the primeval Age of Gold, she created Man naked, tender, fragile, with offensive or defensive armour, as an animate being not a plant: an animate being, I say, born for peace not war, an animate being born for the mirific

178　KINGDOMS, EMPIRES, AND DOMAINS

enjoyment of all fruit and vegetable life, an animate being born for the peaceful dominion over all the beasts.[39]

This is the first appearance of *zoophyte* in the French language,[40] and the first in any modern European vernacular (1546). As *Zoophyta* was by that point well-established in Morrhius's *Lexicon* and reached Calepino's *Dictionarium* in 1548, it could be difficult to establish *Pantagruel* as the source of further diffusion. However, we should not discount the role of personal and professional connections in the spread of words and concepts. Rabelais was doubtlessly aware of the term well before *Pantagruel*. He and Budé were guests at an intimate banquet of intellectuals in Paris in 1537.[41] From 1536 to 1568 Université Montpellier was headed by Bishop Guillaume Pellicier, a humanist and natural-history enthusiast whose library contained some 1100 Greek manuscripts.[42] In 1530–1532 Rabelais became a friend of fellow student Guillaume Rondelet, who (after studying Greek in Paris) returned to Montpellier in 1537, becoming Regius Professor of Medicine in 1545. Rondelet wrote extensively on fish and zoophytes (*Vniuersæ aquatilium historiæ pars altera*, 1555); his students included Jean Bauhin, Jacques d'Aléchamps, Charles de l'Écluse (Clusius), Matthias de l'Obel (Lobelius), and Leonhard Rauwolf, most of whom we shall meet in Chapter 12.

Jean Bodin: political theory (1576)

Jean Bodin of Angers—jurist, professor of law, and member of the Parisian *Parlement*[43]—sought to be a moderating influence as social and religious conflict threatened to tear France asunder. A Catholic (and former novice friar), he counselled for humanistic public education and the toleration of Huguenots, and seems to have believed in a *prisca theologia* encompassing Neoplatonism, Judaism, Christianity, and Islam.[44] His knowledge of the classics was broad, even if detractors implied it was superficial. At the same time, he was an unwavering opponent of sorcery and witchcraft, and adjudicated at a trial (1578) in which the accused confessed to scarcely credible activities "even without torture" and was inevitably sentenced to death.[45]

His *Les six livres de la republique* (1576) describes *l'état* as a separate entity, and treats the respective roles of sovereign, subject, and the law.[46] In *Republique* he divides history into three branches,[47] and over his lifetime devoted a treatise to each: the human (*Methodvs, ad facilem historiarvm cogitationem*, 1566), natural (*Vniversæ natvræ theatrvm*, 1596) and divine (*Colloquium heptaplomeres*, unpublished at his death).[48,49] Already in *Methodvs* Bodin argued that just as Aristotle and Theophrastus had described the history of animals and plants, and on that basis Pliny all nature, so too can we apply method to history in a comprehensive way.[50] Beginning with *Republique* and continuing through *Theatrum* and *Colloquium*,

Bodin emphasized one particular feature of nature: its cohesion. Nature—unlike his France—is cohesive because God created intermediates that unite worldly beings in a "perpetuall Harmonicall bond":

> So we see the earth and stones to be as it were ioyned together by clay and chaulke, as in meane betwixt both: and so betwixt the stones and mettals, the Marcasites, the Calamites, and other diuers kinds of minerall stones to grow: So stones and plants also to be ioyned together by diuers kinds of Corall, which are as it were stonie plants, yet hauing in them life, and growing vppon roots: Betwixt plants and liuing creatures, the Zoophytes, or Plantbeasts, which haue feeling and motion, but yet take life by the roots whereby they grow. And againe betwixt the creatures which liue by land onely, and those which liue by water onely, are those which they call *Amphibia*, or creatures liuing by land and water both, as doth the Beuer, the Otter, the Tortoise, and such like: as betwixt the fishes and the fouls are a certaine kind of flying fishes: So betwixt men and beasts, are to bee seene Apes and Munkies; except we shall with *Plato* agree, who placed a woman in the middle betwixt a man and a beast. And so betwixt beasts and angels God hath placed man, who is in part mortall, and in part immortall: binding also this elementarie world, with the heauens or the celestiall world, by the æthereall region. . . . So also a well ordered Commonweale is composed of good and bad, of the rich and of the poore, of wisemen and of fools, of the strong and of the weake, allied by them which are in the meane betwixt both.[51]

Theatrum explicitly counterposes metals, plants, and animate beings as three genera of earthly things[52] and the text is organized as an ascent towards the Platonic Ideas and the angels. Zoophytes (*pectunculi*,[53] sponges, and oysters) appear in a somewhat odd position along this chain, immediately after mistletoe (*viscum*), fungi, and boletes and before bloodless animals, reptiles, and terrestrial insects. Bodin maps all these forms onto branching trees; this is the first time *Zoophytes* appears in this logical structure.[54] *Theatrum* also makes it clear that it is not the progression of forms themselves, but rather of soul, that constitutes the universal bond.[55] The theme of universal harmony continues in *Colloquium*, where zoophytes are again identified as between animate things and plants.[56] This universal linkage is also symbolized by the *pantotheca*, a collection of 1296 artefacts and images that invite reflection upon similarities and progressions beyond the purely natural, and along multiple axes.[57] It is safe to imagine that mollusc shells, and fragments of sponge and coral, were among these 1296 artefacts.

Remarkably, zoophytes appear even in Bodin's alarmist book on witchcraft, *De la démonomanie des sorciers* (1580). He asks whether demons are part of the "harmonic whole" of the universe, and in describing this harmonic whole he returns to the progression among nature with "zoophytes: ou plantbestes" one of the links.[58]

180 KINGDOMS, EMPIRES, AND DOMAINS

Jacopo Zabarella: Aristotelian logic (1606)

Aristotle reached the Renaissance by a long and tortuous path. For centuries his logical works, collectively the *Organon*, followed Porphyry's *Isagoge* as a technical introduction to Plato, while his physical works (notably the *Physica*) were relentlessly "unified" into Neoplatonism. Ibn Rushd sought a more-authentic Aristotle, but added his own ideas (*e.g.* on the unity of the material intellect) which, in turn, evoked commentary from Aquinas and other scholastics. Aristotle's principles of circular motion and perfection of the heavens did not survive the encounter with Galileo's primitive telescope.[59] Yet Aristotle still had much to teach us, not least on the value and application of scientific observation in resolving first principles. The University of Padua, where both Galileo and Zabarella were professors (of mathematics and logic respectively), had been a centre of Averroist teaching since the early Fourteenth century.[60] With the fall of Constantinople in 1453, however, had come an influx into the Latin world of people schooled in the Greek language and adhering to an older, more-conservative Aristotelian philosophical tradition particularly in regard of the divisions of logic—analytics, dialectics, and sophistics rather than the newfangled *prædicabilia* and *prædicamenta*.[61]

Against this background, Zabarella sought to develop logical tools that can distinguish cause from effect, while incorporating observation. Already his teacher and predecessor in the Chair of Logic, Bernardinus Tomitanus, had taught that a combination of demonstration *quia* (induction) and *propter quid*, known as *regressus*, was necessary to establish a philosophy of nature. Zabarella clarified this method further, opening up the first principles themselves as objects of investigation: "no longer are the first principles of natural science taken as indemonstrable and self-evident: they have become hypotheses resting upon the facts they serve to explain."[62] His approach "remained the method and ideal of science" until Locke and Berkeley.[63]

Zabarella could read Greek, and in his first major work, *Opera logica* (1578), repeatedly cited Ammonius, Philoponus, Simplicius, and especially Themistius.[64] In *De methodus*, he offers the zoophyte as an example of something knowable by reference to the thing itself, not from the method of logical division:

> It is not clear enough whether this lives the life of an animal or of a plant, because whether it senses or not is not at all known. Let us take body as the known genus, for certainly Zoophyte is a body. Of course, if we first divide body by means of animate and inanimate, these differentiæ having been set out, it is immediately apparent that Zoophyte is animate, not inanimate. Nevertheless, it is not that the division really does this for us, but because it is known *per se* that Zoophyte takes nourishment and is nourished. But then afterward, when animate body has been accepted, we divide that by means of sentient and insentient. By means of this division it is not apparent that Zoophyte is more sentient than insentient, because where in a given thing neither differentia is known *per se*, division has no power at

THE RETURN OF THE ZOOPHYTES 181

all to make [something] known. At that point, therefore, unless we take refuge in some method that has inferential power, division itself furnishes us no knowledge. We could, however, following sense, take refuge in some accident and use resolutive method—just as if we showed that Zoophyte feels pain when pricked—for from this pain we could gather that sense belongs to Zoophyte, something that division itself can in no way show. This appears to me to be so manifest that it needs no proof.[65]

Zabarella's commentary on Aristotle's *De anima* was published posthumously by his sons in 1605. Others since Aquinas had of course commented on *De anima*, including Jean de Jandun[66] and Tommaso de Vio,[67] but so far as I can determine only Zabarella mentions zoophytes. Following Aristotle, animals are distinguished from other living things because they are sensate, not because they can move; perfect animals are stimulated to move from place to place, whereas "imperfect animals called zoophytes" are likewise sensate but are moved only in place. Thus

> From this passage Aristotle's assertion was clear, that ζωοφύτα may be animals and may have sense, but neither can it be doubted that we owe to [Aristotle himself] in his *Historia animalium* Book 8, chapter 1, that these things [zoophytes] stand in the middle between animals and plants; such is the standard of history and of the more popular manner of speaking, these are referred to only as are found by practical experience, and without the use of reason, and so it is sufficient for the perfect animals to be called animals, which by reason is true, because ζωοφύτα are in the middle between plants and animals: now, on the other hand, speaking in a scholarly way it can be ascertained, that they are animals and have sense, which afterward in the following text it will be said to be the tactile sense.[68]

Zabarella thus finds two readings in Aristotle, one non-technical ("nature proceeds little by little") in which these organisms can be said to be between plants and animals, the other scholarly and according to which they are imperfect animals. Elsewhere in *De anima*, Zabarella discusses their powers of nutrition and sensitivity, and their ability to distinguish *down*, *forward*, and *back* but not *left* or *right*.[69]

Johann Thomas Freig: Ramist natural history (1579)

Zabarella's emphasis on *methodus*, and his contention that two types of logic are at play in Aristotle—one of everyday speech, the other more systematic and rigorous—point to the influence of Ramism. Although today a footnote in the history of logic or educational theory, from the late 1530s the ideas of Petrus Ramus helped shape the Northern Renaissance. Ramus argued that the university curriculum would be more useful if it were less compartmentalized, and set out new approaches that would better integrate the *trivium* of grammar, logic, and rhetoric. One of his logical methods was *distributio*, the comprehensive subdivision of a subject (for example, grammar) on a bifurcating tree.[70] Porphyry's tree of categories was well-established in medieval logic,[71] but Ramus followed all sub-branches systematically to yield an overview that obviated much of the repetition and verbiage that had been typical of scholasticism.

Johann Thomas Freig, known as Freigius, was an enthusiastic proponent of Ramism particularly in natural history and philosophy. Remembered today for having coined the word *psychologia*,[72] Freigius wrote a *Quæstiones physicæ* (1579) in which he applies Ramus's method, including *distributio*, to the physical world. In one diagram (Figure 11.1) he divides corporeal bodies into the inanimate and animate; the animate into spirit and embodied; the latter into vegetative and sentient, that is, Animal; the latter into Ζωόφυτον and true animals; and those into Irrationale

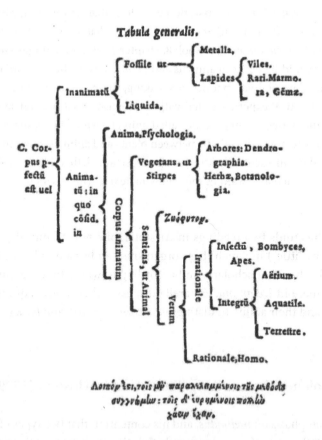

Figure 11.1. Freig, division of corporeal bodies following the method of Ramus (Freig 1579, reprinted 1585), page 8. Note *Zoophyton* as a class of sensate animate bodies, although not "true animals". Bibliothèque interuniversitaire de santé (Paris) *via* Internet Archive.

and Rationale. Thus zoophytes (unlike plants) are sensate, but are not true animals.[73] John of Damascus[74] would recognize these *differentiæ*, but zoophytes are now sentient and therefore animals, even if not "true animals".

Robert Burton: English vernacular (1621)

Freig coined the word *psychologia*, but its most memorable exposition came from the pen of Robert Burton, a scholar at Oxford. Taking as his starting point one human emotion—melancholy—Burton ranges exuberantly through its imaginable varieties, causes, symptoms, and cures, providing layers of anecdotes, examples and quotations, dense marginal references, and gigantic Ramist trees. As *Pantagruel* celebrated Sixteenth-century French vernacular, Burton's fascinatingly impenetrable *Anatomy of melancholy*[75] puts on display a sizeable proportion of early Seventeenth-century English vocabulary. At one point, digressing within a section that notionally deals with the cure of melancholy, Burton lists wondrous places to visit and things to see, including

> Many strange creatures, mineralls, vegetalls, Zoophites were fit to be considered in such an expedition, & amongst the rest that of *Herbastein* of his *Tartar* lambe, *Hector Boethius* goose-bearing tree in the *Orchades*, *Vertomannus* wonderfull palme, that fly in *Hispaniola* that shines like a torch in the night.[76]

In marginal references Burton cites Baron von Herberstein's *Rerum Moscovitarum commentarii* regarding the Tartary lamb, and Boece's *Scotorum historia* for the goose-tree.[77] He does not, however, indicate his source for *Zoophite*.

Juan Eusebio Nieremberg: baroque nature (1635)

We met Nieremberg briefly in Chapter 9, where he argued that every sponge and sea cucumber has its own star; and in Chapter 10, where he cast doubt on the tale of the borametz. These are only two of the many incongruities presented by his *Historia naturæ*.[78] Nieremberg attained the grade *metafísico* at Universidad Alcalá, then took holy orders and from 1625 was professor of physiology at the Jesuit Colegio Imperial in Madrid, teaching not only natural history but also grammar, history, and Holy Scripture.[79] He was best-known at the time for his mystico-devotional works, particularly *De la diferencia entre lo temporal y lo eterno: crisol de desengaños* (1640), in which he presents the sensible world as "the dream of a shadow" and man as "an imperfect being who lives surrounded by temporal things that are vain, ephemeral, and deceitful".[80]

His *Historia naturæ* (1635) is perhaps best characterized by its subtitle: "in which the rarest arcana of Nature, not only things astronomical, & strange animals

184 KINGDOMS, EMPIRES, AND DOMAINS

of the Indes, quadrupeds, birds, fish, reptiles, insects, zoophytes, plants, metals, precious stones, & other minerals, and of the rivers & elemental conditions, but also their medicinal properties, are described; new & most-curious questions are disputed, and many passages of sacred Scripture are eruditely explained". The Indes in question were Spanish America, and about half of the *Historia* reproduces text and illustrations sent back to Spain by Francisco Hernández, who spent seven years in México describing its flora and fauna for Philip II. Yet in the same pages Nieremberg asks whether the condor is perhaps the legendary griffin; proposes that guardian angels had transported American animals to Noah's ark (and back, after the Deluge had subsided); and speculates whether the native peoples of America might be a lost tribe of Israel.[81] Not without cause does Pimentel call this *baroque nature*.[82]

Nieremberg does not simply leap into the wonders of nature, but begins by providing context: natural history has dignity and utility—not only medicinal but also political, ethical, and moral—and offers pleasure. The heavens are animate, perhaps planets and comets too. Sensitive substance extends downward from the heavens in steps or levels (*gradus*), encompassing earthly animals and plants and rendering all nature a simulacrum. Sponges, sea-nettles, sea-lungs, sea-slugs, starfish, sea-urchins, and the like possess only the sense of touch, and cannot move from place to place:

> And so the grades of sentient nature are apportioned according to its own little steps. Some animals have been endowed with many and more-excellent senses, others are satisfied with fewer and blunter [senses]. That is to say, nature rises up by connections, little by little and without leaps, as though it proceeds by an unbroken web, in a leisurely and placid uninterrupted course. There is no gap, no break, no dispersion of forms: these [forms] have, in their turn, been connected, like a ring within a ring. That very golden chain is universal in its embrace. Therefore, just as each sentience has advanced within itself by steps, [with] its little knots [tied] between [these sentient things] themselves, then the next knot connects sentient nature with the spiritual; thus it is required that there be a clasp [*fibulam*] between plants and sentient things. That [thing] will in some way be truly a plant, and truly sentient: just as man is the bond between animals and the Intelligences because he is truly an animal, and truly has understanding. This link of vegetables with the sensate is seen to be the sponge which, having the form of plants, preserves sensation.[83]

For Nieremberg, Aristotle's dualizers not only face both ways, but grasp their neighbors above (animals) and below (plants). The metaphors are mixed: individual types of being are connected ring within ring (*velut anulus anulo*) forming a golden chain (*aurea catena*), yet their junctions are little knots (*noduli*) that connect an unbroken web or fabric (*trama*).[84] Even so, this may be the first description of the living world as a network.[85]

THE RETURN OF THE ZOOPHYTES 185

In this passage Nieremberg does not refer to the organisms that clasp vegetables together with sensate animals as *zoophyta*, but (as mentioned above) the word occurs in the subtitle of *Historia naturæ* and he returns to it on several occasions, asking whether we should consider zoophytes not only as plant-animals but also animal-plants (he never quite answers his own question),[86] and whether the borametz is a zoophyte (above). For their mode of generation he relies on Aristotle, Pliny, fellow Jesuit Girolamo Dandini, and someone we shall meet in the next chapter, Ferrante Imperato.[87] Nieremberg might have discovered the word *zoophyta* in Nebrija or Calepino, but a more-likely source is Guillaume Rondelet, whom he cites frequently in connection with these and other marine organisms.

David Person: rare and excellent matters (1635)

The year 1635 saw the publication of another miscellany, *Varieties: or, a svrveigh of rare and excellent matters, necessary and delectable for all sorts of persons. Wherein the principall Heads of diverse Sciences are illustrated, rare secrets of Naturall things unfoulded, &c.* by David Person, a Scottish gentleman. It too begins by setting our earthly world in broader context, and like the *Historia naturæ* offers a steady stream of rare, albeit not always excellent, questions: can herring fly? It is possible, as the Creator has made amphibia, which inhabit both water and land. How is it that the adamant stone cannot be broken, but falls to pieces in a dish of hot goats' blood? Goats live on rocky cliffs, thus feed on plants that have "rare pearcing and penetrative vertues"; these virtues create an analogy in the blood. The barnacle-goose-tree? Aristotle knew that insects are engendered from clay, and barnacles are insects of the sea.[88] Can there be birds that lack feet and feathers?

> Yea, for as the great Creator hath ordained in nature betwixt himselfe and us men here, Angels, yea good and bad spirits; betwixt sensitive and insensitive Creatures, mid creatures which wee call Zoophyta, and Plantanimalia, as the Fishes *Holuthuna, stella marina, Pulmo marinus*, &c. Even so betwixt fowles and fishes, nature produced middle or meane creatures, by the Greekes called ἀμφίβια, or beasts of two lives; partly living by waters, partly by earth; And of this sort these fowles must be, as betwixt land beasts and fishes, are frogs, and Crocodills; and some others the like.[89]

In a section on numerology, Person holds that "Three of all Numbers should be held in greatest veneration." There are three Persons of the Godhead; three kinds of creatures ordained for the use of man (birds in the air, beasts on earth, fish in the sea); three kinds of living things (intellectual, sensitive, and vegetable) as Men, Beasts, and Plants.[90] He does not attempt to square this with zoophytes being "between sensitive and insensitive Creatures". Indeed, *Varieties* offers neither the erudition nor the focus on natural history of *Historia naturæ*. It is, however, relentless

186　KINGDOMS, EMPIRES, AND DOMAINS

in its message that by the study of nature "we might be brought to the knowledg(e), admiration, and adoration of our great and powerfull God, the Maker of Nature.... There is nothing so meane in Nature, which doth not represent unto us the Image and Power of the Maker; and argue, that none but He could have been their Former."[91] This is a time-honoured motivation, and (as we shall see) was to enjoy a revival in England and northern Europe as the Seventeenth century progressed.

Henry More: the Spirit of Nature (1682)

Our final case study involves the Cambridge theologian Henry More FRS, author of the *Divine dialogues* (1668) and *Manual of metaphysics* (1671). Like his better-known rationalist counterpart René Descartes, More sought to use reason to prove the existence of God. However, More came to accept that some natural phenomena—magnetism and gravity were two examples—lay beyond the remit of Cartesian philosophy. To explain those he invoked an emanative Platonic Spirit of Nature, which he considered to mediate between God and the physical world.[92] Specifically, he backed away from Descartes's position that plants and animals (and by implication, man) are without soul, "more splendid versions of artificial automata".[93] This Spirit was itself hierarchical, responsible for the vegetative functions of plants and the "sensuall Nature" of animals.[94]

More and his disciples wrote against opponents, not least Baruch Spinoza, whose rejection of Cartesian mind-body dualism in favour of body (substance) they considered dangerously far down the slippery slope to atheism. One such disciple was Joseph Glanvill, who however had died before his tract *Lux orientalis* could be published. More published it, together with *Discourse of truth* by his former student George Rust, also recently deceased, appending his own extensive commentary.[95] *Lux orientalis* is an argument for the preexistence of soul. Glanvill presented Divine Goodness as one argument in favour of preexistence, then raised a series of possible objections which, of course, he proceeded to counter. Two of these objections involved the fallibility of human reason in regard of what God should or should not do, and specifically why God did not make man immune from sin and misery. Glanvill replied that if man were impeccable, he would not be man: he could not exercise free will, and there would be no virtue in the world, no patience, faith, or hope.[96] To Glanvill's argument, More added that

> To cavil against Providence for creating a Creature of such a double capacity, seems as unreasonable as to blame her [Nature] for making *Zoophiton's*, or rather *Amphibion's*. And they are both to be permitted to live according to the nature which is given them. For to make a Creature fit for either capacity, and to tye him up to one, is for God to do repugnantly to the Workmanship of his own hands. And how little hurt there is done by experiencing the things of either Element to Souls that are reclaimable.[97]

In this way zoophytes and amphibians are a sort of parallel to man, allowed to choose one or both soul or element, plant or animal, water or land, evil or good. This may be the first figurative use of the term *zoophyte*.[98]

Concluding comments

The same body of classical and medieval texts that offered up *zoophyta* to newly confident European vernacular languages overwhelmingly presented the natural world as one of animals, vegetables, and minerals. Nor did the stirrings of renaissance sweep away myths and legends, occult correspondences, or the vigilance of church and king. Even so, enquiry into nature began to venture beyond logic, analogy, and the exegesis of Aristotle or Moses. Oysters, corals, sponges, and fungi were examined with new eyes, new logical tools, indeed with dissecting trays and, eventually, the microscope. In the next chapter we follow *zoophyta* from about the year 1475, as scientific botany and zoology rose up on each flank.

12

Plants and animals

> The differentia ... is the unity of opposites, of what is determinate and what
> is in itself universal. . . . Observation, which kept them properly apart and
> believed that in them it had something firm and settled, sees principles
> overlapping one another, transitions and confusions developing. So it is
> that observation which clings to passive, unbroken selfsameness of being,
> inevitably sees itself tormented just in its most general determinations—
> *e.g.* of what are the differentiae of an animal or a plant.
>
> —GWF Hegel, Phenomenology of spirit[1]

We now turn from *zoophyta* to the organisms on either side, the plants and animals. Our focus will be on the distinctiveness and delineation of each group—what organisms are included and excluded, and why—and on the arrangement or ordering inside each group.

As we saw in Chapter 1, folk taxonomies often stop short of identifying plants and animals as top-level groups. Moreover, the boundary between them is often blurred or transgressed, for example by chimæric organisms or metempsychosis. Even so, by about two generations before Aristotle, it seems to have been accepted that earthly beings other than man are either plants or animals. Aristotle defined each in relation to the faculties of soul. Soul in plants is vegetative, with only the power of self-nutrition, the end of which is generation; by contrast, animals are defined by the superaddition of animate soul with sensory, appetitive, and locomotive powers. Not all animals partake equally in these additional powers: appetite is always present, but sensation may be reduced to touch alone, and locomotion may be absent altogether.[2]

In principle, the differential presence or absence of these faculties might allow animals to be ordered from the least- to the most-perfect, which in turn should make it easier to locate the precise boundary with plants, perhaps without invoking *zoophyta*. But Aristotle observed that animals share the faculties of soul in various combinations, such that they cannot be ordered in a coherent manner using Platonic dichotomy alone. Consistency might be achieved by prioritizing one or another attribute, for example the number of feet, but we would inevitably end up with groupings that fail to capture true essences, and violate our sense of the natural

Kingdoms, Empires, and Domains. Mark A. Ragan, Oxford University Press. © Oxford University Press 2023.
DOI: 10.1093/oso/9780197643037.003.0012

(our folk-taxonomy). Rather than devise a utilitarian workaround, Aristotle wrote that the "instincts of mankind" led to the recognition of natural groups, each of which "combines a multitude of differentiae, and is not defined by a single one as in a dichotomy. . . . It is impossible then to reach any of the ultimate animal forms by dichotomous division."[3]

The situation with plants is both more- and less-complicated. Theophrastus sought to understand the essential structure of nature, assembling "everything that vegetates" into *plants*. Based on their form (rather than, for instance, the element in which they live) he recognized four types: herbs, undershrubs (bushes), shrubs, and trees. The latter are most-perfect by virtue of their essential parts, not because they are farthest from stones or closest to animals. Perhaps because the vegetative soul is simpler than the animate, Theophrastus did not observe the conflicting distributions of attributes that prevented Aristotle from organizing the animals into a coherent system. This division into herbs, undershrubs, shrubs, and trees held up for nearly two millennia;[4] how a particular ecological, agricultural, or pharmaceutical character mapped onto it was considered a matter of accident.

Herbals (from 1475)

Of course, many treatments of plants or animals bore little pretence of philosophy: even Theophrastus produced a descriptive *De causis plantarum* alongside his philosophical *Historia plantarum*.[5] This utilitarian bent was seen most clearly in herbals, where the aim was to help physicians identify plants for use as *materia medica*. We have discussed the roots of this tradition, and mentioned the herbal writings of Hildegard.[6] In the Thirteenth century Albertus Magnus treated plants philosophically, but nonetheless devoted one book of *De vegetabilibus* (1256)[7] to individual plant types arranged alphabetically within each of the four Theophrastian classes. His colleague Bartholomæus Anglicus likewise arranged plants alphabetically in *De proprietatibus rerum*.[8] In the mid-Fourteenth century, Conrad von Megenberg modelled his *Puch der Natur*[9] on the *De natura rerum* of Albert's student Thomas of Cantimpré. Conrad divided plants into trees, aromatic trees, and herbs, and within each arranged species in more-or-less alphabetical order by Latin name.[10] Printed editions (from 1475) were illustrated using woodcuts, earning *Puch der Natur* pride of place as the first printed illustrated herbal.

Other herbals circulated in Europe during the Middle Ages, notably the Fifth-century *Herbarium* attributed to Apuleius, called Barbarus or Platonicus to distinguish him from the author of the *Metamorphoses*. With text drawn mostly from Dioscorides and Pliny, it necessarily focused on plants of the Mediterranean region.[11] In mid-1480s Mainz, however, two herbal traditions more attuned to the flora of central and northern Europe made the transition to the printed page: the Latin *Herbarius Maguntie* (1484), and the German *Gart der Gesundheit* (1485). Each subsequently appeared in numerous (often bootlegged) editions, and adaptations were made into other European languages.[12] Supplemented with text on animals and coral, the German *Herbarius* was reissued in 1533 by Eucharius Rösslin as *Kreutterbůch von allem Erdtgewàchs*, then from 1557 to 1783 as *Adams Lonicers vollständiges Kràuter-Buch oder Buch der drey Reiche der Natur*.[13] A separate offshoot from the German *Herbarius*, the *Hortus* or *Ortus sanitatis*, appeared in 1491[14] with further material on precious stones, real and mythical animals, and a treatise on medical diagnosis from urine. A Dutch edition was abstracted into English as *The noble lyfe & natures of man, of bestes, serpentys, fowles & fisshes yt be moste knowen* (1491),[15] bearing little evidence of its roots in the herbal tradition.

However much we might wish to recognize distinct utilitarian and philosophical lineages of plant-books, in reality philosophical botany became moribund after Theophrastus. It made a brief reappearance in the mid-Thirteenth century with Albert, then was independently rebuilt atop a utilitarian base by Andrea Cesalpino from 1583 (below). Renaissance botany was born utilitarian, and remained largely so through the Seventeenth century. Even so, from about 1490 a new professionalism began to emerge among authors of plant-books, bringing higher standards of observation, documentation, illustration, and publication.

The rise of scientific botany 1: 1490–1580

The Italian Renaissance brought renewed interest in the classics. A first order of business was often philological: comparing ancient texts, understanding usage, and editing out centuries of scribal errors. In 1492 Ermolao Barbaro exposed five thousand supposed errors in Pliny's *Naturalis historia*,[16] prompting a debate in humanist circles as to whether Pliny had indeed made a terrible hash of Greek terminology about medicinal plants (and much else).[17] For Niccolò Leoniceno, the significance was not just philological: as a physician, he wanted to ensure he was using the correct plant to treat his patients. He amassed an extensive library of Greek and Latin medical texts, and taught his students to collect plants themselves for comparison against the original descriptions. Dioscorides provided the standard: unlike Galen, he had actually described medicinal plants, not merely reported their names.[18]

Leoniceno's students adopted his approach. Euricus Cordus used it in *Botanologicon* (1534) and introduced it to Leonhart Fuchs, who later became professor of medicine at Tübingen and physician to the king of Denmark. Valerius Cordus, son of Euricus, died early but his high-quality plant descriptions were later published by Conrad Gessner.[19] Meanwhile the physician Jean Ruel translated Dioscorides into Latin (*De materia medica*, 1516), easing its path into the medical curriculum at Padua, Bologna, and Montpellier during the 1530s.[20] Gardens of medicinal plants were established at Padua and Pisa in the same decade. Paralleling our narrative from Chapter 11, these physician-botanists expanded their *materia medica* to include local plants, and typically published not only in Latin but in the local vernacular as well. The plant-books of Pierandrea Mattioli and Otto Brunfels set new standards of realism in botanical illustration, the latter employing Hans Weiditz, a disciple of Albrecht Dürer.[21]

This generation of physician-botanists[22] did not attempt to delineate the plants, nor position them between stones and animals. In *De historia stirpium commentarii* (1542) and its German edition, *New Kreüterbůch* (1543), Fuchs presented some 400 local and 100 foreign plants, mostly herbs. A typical entry presents the genus name and synonyms, its species if any, its form, where and when it can be gathered, its taste, and its medicinal use according to Dioscorides, Galen, and/or Pliny. A large, realistic (if slightly idealized) illustration appears on a nearby page.[23] Entries are ordered alphabetically by the Greek name, or in the case of genera unknown to Greek antiquity, positioned after some other genus of similar form.[24] Fungi, lichens, corals, seaweeds, and mythical plants are absent.

The *New Krëutter Bůch* (1539) of Jerome Bock, known as Hieronymus Tragus, appeared first in a German-language edition without illustrations.[25] Without a corresponding illustration to fall back upon, each verbal description had to stand on its own,[26] and those crafted by Bock were original and exceptionally clear. The resulting international attention called for an edition in Latin, for which the publisher commissioned illustrations at his own expense.[27] Bock excluded particularly well-known genera, and ordered the others according to similarity of vegetative form, such as cross-sectional shape of the stem or relative arrangement of the leaves.[28] Herbs, well-represented among *materia medica* and thus in Bock's herbal, are presented in Book 1, then (with numerous exceptions) give way to shrubs and trees in Books 2 and 3 respectively. There is, however, no sense of an overarching gradient of form, and natural (folk-taxonomic) groups are sometimes dispersed or interrupted for extraneous, even poetic, reasons.[29] Fungi and lichens—some thirty species in all—are treated at the beginning of Book 3 just before mosses, mistletoe, heather and myrtle,[30] with the proviso that

> Fungi and truffles are neither plants, nor roots, nor flowers, nor seeds, but nothing more than the superfluous humidity of the earth, trees, rotting wood and other putrefying things. It can therefore be concluded that all fungi and truffles, especially those that are eaten, often grow most frequently in conditions of thundery and rainy weather, as Aquinas the poet affirmed: *And the longed-for thunderbolts will make us a pleasant dinner.*[31]

Other herbalists of this generation likewise ignored fungi, seaweeds and algae. Brunfels limited his *Herbarvm vivæ eicones* (1530) to herbs used as *materia medica*.[32] In *De natura stirpium* (1536) Ruel divided plants into aromatic trees, other woody plants, aromatic medicinal herbs, cereals, legumes, edible herbs, and others; some sections proceed alphabetically by (Latin) genus name, while others are arranged as in Dioscorides. The immensely popular *Commentarii in sex libros Pedacii Discorides* (1544) by Mattioli offered new descriptions of plants (and some

192 KINGDOMS, EMPIRES, AND DOMAINS

animals, including the sea-hare) used as *materia medica*. The truffle is called a root and is presented among herbs.[33]

This generation of naturalists eventually gave way to one whose leading figures, although physicians, viewed the study of plants as more than simply the handmaiden of medicine. They forged links with collectors (who in turn were becoming more sophisticated) and patrons, and were highly attuned to the demands of the book market. Three in particular joined forces in an unprecedented collaboration: Rembert Dodoens, Mathias de L'Obel (Lobelius), and Charles de l'Écluse (Clusius). Their collective success helped shape the new field, even as it gave rise to a profusion of plant names and descriptions.

Dodoens, eldest of the three, was a transitional figure. His *Crüÿdeboeck*, published in Flemish in 1554, was an illustrated *materia medica* in the earlier tradition, even re-using hundreds of woodcuts from Fuchs. It divided plants into six large groups: miscellaneous herbs; aromatic herbs; medicinal herbs; cereals and pulses; edible herbs, roots, and fruits; and trees, shrubs, and bushes. To call these groups unnatural would be an understatement, although smaller-scale groupings by common form of roots, leaves, stems, floral parts, or thorns can sometimes be discerned. Horsetails appear in the first group, ferns, mosses, and marine plants in the third. The *Crüÿdeboeck* was nevertheless attractively produced and well-indexed; French and English editions followed, the former translated by l'Écluse. Eventually Dodoens brought together material from this and smaller-scale publications into his acclaimed folio *Stirpium historiae pemptades sex* (1583).

L'Obel studied medicine under Rondelet, and while in Montpellier joined with fellow student Pierre Pena to write *Stirpivm adversaria nova* (1570–1571).[34] Although not lavishly illustrated, the *Adversaria* broke new ground on several fronts: maritime and marine vascular plants are relatively well-represented; groups of genera, and species or varieties within genera, are indicated by Ramist trees; and most notably, the authors implement a system of classification in which genera are distinguished by the form of the leaf. The *Adversaria* begins with grasses, followed by cereals, and (without making the point explicit) treats mostly plants we now call monocots before moving on to dicots.[35,36] Idiosyncratic associations nonetheless abound including horsetails with asparagus, and ferns with thistles.[37] Bracket fungi are briefly mentioned as a sort of excrescence,[38] while mushrooms and truffles are ignored. The book concludes with shrubs and trees followed by cacti, a sundew, a sponge, the barnacle-goose-tree, and *Litoxyla*.[39]

Christoffel Plantin—who also published Dodoens and l'Écluse—convinced L'Obel to enlarge the *Adversaria*, and in 1576 the larger (and much better-illustrated) edition appeared as *Plantarum seu stirpium historia*. This in turn was translated into Flemish as *Kruydtboeck* (1581).[40] The new material only exacerbated the problem of where to place fungi (between the conifers and palms), mosses, lichens, corals, and seaweeds (after the palms).[41]

The third of these botanist-physicians, l'Écluse, is best known for his descriptions of rare plants, notably of Iberia (1576) and central Europe (1583).[42] Although the nature of his material made it difficult for him to arrange plants systematically, his books start with trees, the most noble among plants, and proceed downward. No plant is less noble than fungi, and his *Rariorum plantarum historia* (1601)—the grand summary of his earlier *rariora*—ends with a separate section, *Fvngorvm in Pannoniis observatorvm brevis historia*, in which fungi are neither championed as a distinct genus of life, nor dismissed as an excrescence or pathology. This is the first (almost) independent monograph of fungi, and a founding text of mycology.[43]

Andrea Cesalpino

The Orto botanico di Pisa was already a hub of Renaissance natural history when Cesalpino became its director in 1555. When re-establishing the Università in 1543, Cosimo I de' Medici had endowed a chair of medical botany, and set aside land for a botanical garden for use in teaching. Under Luca Ghini[44] both the discipline and garden flourished. Ghini became famous for his practically oriented lectures and skill in drying plants, and in due course became court physician. From across Europe scholars, practitioners, and students (including Rondelet, L'Obel, Valerius Cordus, William Turner, and Ulisse Aldrovandi) flocked to Pisa.[45] Cesalpino had studied under Ghini,[46] and like most Italian physicians and academics in those years considered himself an Aristotelian. But he read Aristotle with new eyes, and against the backdrop of a garden—indeed, a world—of plants whose natural order had not yet been properly discerned.

In *Peripateticarum quaestionum libri quinque* (1571) Cesalpino ranges widely over issues including the division of substance, the motion of planets, and the circulation of blood, taking as given the superaddition of vegetative, sensitive, and intellective soul.[47] It was, however, his application of Aristotelian principles to plants that later prompted Linnæus to call him "the first true systematist".[48] In *De plantis libri XVI* (1583) Cesalpino focused on the two powers of vegetative soul: self-nutrition and reproduction. The former is enabled by the root and stem, the latter by the fruit and parts of fructification. As soul is not spatially divisible, in plants it must be located where root and stem meet; indeed, the strength and hardness of this region allows *arbores* (shrubs and trees) to surpass *herbæ* (herbs and bushes).[49] That is, the nutritive power provides the first *differentia* within plants. For its part, the reproductive power is enabled in flowers, fruits, and seeds, and Cesalpino was confident that their multitude and distinctiveness of form would suffice to resolve all genera in which they are perfected. As vegetative soul is nothing more than self-nutrition and reproduction, no further differentiæ can be needed; all else is accident—the shapes of leaves and roots (*contra* Fuchs and L'Obel), colour, odour, habitat, and medicinal properties.[50] Cesalpino had not only identified the essential characters of plants, but had done so in a way that requires a particular subordination, and rules out alternatives: avoiding *doxa*, he had grasped *epistēmē*.[51]

194 KINGDOMS, EMPIRES, AND DOMAINS

The remaining fifteen books of *De plantis* set out his groups of plants and arrange them from most- to least-perfect.[52] As we shall see later, a hierarchical classification on a branching tree does not always yield a smooth gradient of ascent or descent as judged against an external criterion, much less one as subjective as "perfection". For Cesalpino, however, trees are the most perfect: they are grand, have the strongest pith, and transport nutrients the greatest distance, from root to branches.[53] At the other end of the scale, "plants that bear absolutely no seed are surely to be considered the most imperfect; they rise up only from decaying things. For that reason, they need only to obtain nutrition and grow, and cannot produce offspring similar to themselves; [they are], so to speak, between plants and inanimate things, in the manner of Zoophytes between plants and animals. [These are] the genus of fungi; Lenticula palustris, lichens, [and] many marine shrubs."[54]

More than any other naturalist we have met so far, Cesalpino devoted substantial attention to seedless plants. Ferns and horsetails possess root and shoot; others lack even these fundamental organs of nutrition: liverworts, lichens, mosses, seaweeds, corals, and (concluding the final chapter) numerous types of fungi, whose *natura inter plantas maxime peculiaris est*.[55] Certain details are incorrect—notably, corals are not plants, although this would not be settled for another two centuries—but Cesalpino drew all these forms into a single large group within the plants, recognized natural types, arranged them in an order little different from how they are presented in first-year textbooks today, and excluded zoophytes both real and mythical. This was not entirely unprecedented for a work of philosophy: Aristotle had gotten much of this right, if in less detail; Theophrastus, the pseudo-Aristotelian *De plantis*, Albert, Aquinas, and even Paracelsus rather less. It was, however, a major step forward from Brunfels, Fuchs, Bock, and the Flemish herbalists.

The rise of scientific botany 2: 1580–1680

The first major herbal after Cesalpino's *De plantis* was the *Historia generalis plantarvm* of Jacques d'Aléchamps (1586–1587).[56] D'Aléchamps trained as a physician under Rondelet, and in addition to *Historia generalis* translated classical texts in medicine and natural history including the works of Theophrastus. This background is evident in the *Historia generalis*: plant names are linked to Greek words where possible, and classical authors are quoted relentlessly. The work provided the first description of many plants of the region around Lyon, and is profusely illustrated,[57] but is organized in the tradition of l'Écluse's *Stirpium historiae* without mentioning Cesalpino or engaging his argument. Trees are presented first, but subsequent chapters are organized variously by form, ecotype, geography, and medicinal properties. Horsetails appear among the marsh plants; corals, seaweeds, and the barnacle-goose-tree with maritime and marine plants; fungi among "plants with bulbs, and those with succulent and knotty roots"; the borametz among "foreign plants".[58] A "bashful plant" that can perceive the approach of man or animals is compared to the sponge, which "according to Aristotle" is a type of zoophyte.[59]

By the late Sixteenth century thousands of European plants had been described, and every local or regional flora added hundreds more. Neither students nor practitioners could find their way through such detail. In *Methodi herbariæ libri tres* (1592)[60] the Bohemian physician Adam Zaluziansky à Zaluzian argued that botany (*herbaria*) should be treated as a branch of *physica* separate from medicine,[61] and used Ramist trees to organize how it should be taught.[62] Whereas Cesalpino had rejected medicinal virtues as differentiæ because they are mere accidents, Zaluziansky rejected them because they pertain to a different discipline. He allowed various sorts of botanical differentiæ, however, with the result that *Methodi herbariæ* is philosophically incoherent without finding itself notably closer to a natural arrangement. In the lengthy second book Zaluziansky offered a *historia*, beginning with fungi and ending with deciduous trees. Just as sponges, sea-nettles, crabs, and sea-urchins among animals are *informia et indigesta* compared to other animals, so too fungi and "mosses" are *ruda et confusa* in comparison with other plants. These "mosses" include corals, sea-fans, seaweeds, lichens, and "things of a similar nature".[63]

Although practitioners were surely dismayed by the lack of consensus on natural affinities among genera, the explosive growth in number of plants was largely due to the recognition of new species and variants. Fortunately, an existing if underutilized remedy was at hand: nomenclature. In *Phytopinax* (1596) Gaspard (Caspar) Bauhin set out herbaceous plants in a highly structured format that he called a πίναξ, or *register*. Starting with the grasses, he introduced each genus by its one-word name, followed by synonyms and a succinct morphological description. Next come numbered sections, one for each species in that genus. Each begins with the species name, composed of the genus name plus one or a few distinguishing words—not formal *differentiæ* nor medicinal virtues, but all together simply a name, informative for identification and easy to remember.[64] Each section concludes with detailed references and further technical description.[65]

This new approach is on display in Bauhin's *Pinax theatri botanici* (1623), in which more than six thousand species are registered in only 522 unillustrated pages.[66] Again he begins with grasses but now describes all known types of plants, through conifers and palms. In Book 10 he treats in succession ferns and their allies; mosses and lichens; seaweeds and certain marine vascular plants; stony and soft corals, and (under "Zoophyta") sponges and sea-nettles; fungi and truffles; and thistles and gummy shrubs. This Zoophyta is not a genus—its divisions are genera, not species—but represents a more-inclusive if unnamed rank:

Zoophyta. Ζωόφυτα sunt quæ nec animalium, nec plantarum, sed tertiam ex utroq[ue], naturam habent, & ut Aristoteles loquitur, τὰ ἐπαμφοτερίξοτα φυτῷ και ζῷ, quæ ambigunt inter plantam & animal, quæ Plantanimalia dici possunt: talia sunt Spongiæ, Urticæ marinæ,[67]

Although in the *Pinax* zoophytes are not placed at the beginning or end of the plants, they nonetheless constitute a natural group in the context of pre-Darwinian botany.[68] Horsetails are treated in Book 1, following *Sparganium* (bulrushes).[69] Bauhin concludes the *Pinax* with a short section of *Herbæ & arbores admirandæ* including the borametz following Herberstein and Scaliger, the tree with ambulant leaves following Duret, various wondrous trees from the Indies and New World, and the barnacle-goose-tree, which he duly classifies into three species based on whether the geese arise from worms in decaying wood, or from fruit or leaves fallen into the sea.[70]

Like his older brother Gaspard, Jean Bauhin was a botanist, but he died before his major work was complete; a further thirty-seven years transpired before his son-in-law Jean-Henri Cherler could see it through to publication. Unlike the *Pinax* (above), their *Historiae plantarum universalis* (1650–1651) was illustrated, and arranged from trees downward to truffles. Aquatic and maritime vascular plants (Book 38) are separated from seaweeds, corals, and sponges (Book 39), the latter two placed together on the authority of Ferrante Imperato, whom we meet below. Book 39 concludes with the barnacle-goose-tree, unabashedly called *Arbores conchiferæ vel anatiferæ falsò dictæ.*[71] The concluding Book 40 treats eighty-three genera of *excrementa terræ*: fungi and truffles.

Other botanical books followed over the succeeding decades including new and reissued herbals, natural histories of exotic lands, and experimental works on plant anatomy and reproduction.[72] We conclude this section, however, with a remarkable work from Joachim Jung. Better known today as a mathematician and logician, Jung was also a physician and professor of natural science. He published little in his lifetime,[73] but his students later edited his treatises and saw to their publication; several of these were brought together as *Opvscvla botanico-physica* (1747). His *Isagoge phytoscopica*, edited in 1678 by Johann Vaget, is a comprehensive outline of theoretical botany. At the highest level it is organized as two blocks, one (chapters 1–12) dealing with augmentative powers and the associated morphological structures, the other (chapters 13–28) with generation. Concepts and terminology are presented in aphorisms. The first chapter defines plants and sets out their distinctive characters; organized as six points, it bears translation in full:

1. A plant, φυλὸν, is a body living but non-sentient; or a body affixed to a certain spot, or to a certain substratum, whence it can gain self-nutrition, grow in size, and finally propagate [itself].

2. It is said to gain self-nutrition when it transforms the nourishment that it attracts into the substance of its parts, to make up for that which has been dissipated by its natural heat and internal fires.

3. A plant is said to grow in size, or be augmented, when more substance is restored to it than [has been] dissipated, and so more is restored not to its secondary parts alone, nor to those sufficiently mature, [but] in all dimensions; and also, it generates new parts in addition.
4. Thus the growth of plants is distinguished from the augmentation of animals. Firstly, not all their parts are augmented simultaneously. For the leaf of a tree, when it has expanded enough, does not grow larger; and likewise, the young shoot or twig, when extended sufficiently, is not extended farther in length. The new parts generated are: new leaves, new shoots, stalks, flowers.
5. A plant is said to propagate [itself], when it brings forth another of its own kind. The word *propagandi* [something propagated] is thereby used in this broad sense.
6. Another thing is said in particular of a plant: other rudiments, perhaps, of plants, or of failed conditions of generation, [are] *abortus*. The view that such are fungi, truffles, mosses, duckweeds and the like, we cast into the realm of speculation.[74]

In *Doxoscopiae physicæ minores*[75] Jung aphorizes on diverse topics relevant to scientific botany, from the formation of genus names to plant morphology and physiology. He finds that although corals, "other" mosses, fungi, and the like show a certain inclination towards life (*quibusdam conatum tantum vitæ habere*), they are not alive.[76] He notes the presence of seeds (including what today we call spores) in supposedly seedless plants including fungi, and divides plants into trees and herbs, thereby collapsing shrubs into the trees.[77]

Medieval and early Renaissance animal-books

Many Medieval animal-books were based on the *Physiologus*.[78] From about the Twelfth century the *Physiologus* itself was greatly expanded with new material from Isidore, Solinus, the *Hexaemeron* of Ambrose, Hrabanus Maurus, and/or the *Pantheologus* of Peter of Cornwall. Some versions gained sections on trees and on the so-called Ages of Man, then a popular theme.[79] All major versions were translated into dialects of French between about 1120 and 1225; one of these translations, *Li livre des créatures* prepared by the Anglo-Norman poet Philippe de Thaun for Adeliza de Louvaine on the occasion of her marriage to Henry I of England in 1121,[80] is among the oldest extant examples of the Anglo-Norman language.[81] Other Thirteenth-century animal-books include the *Bestiaire* of Gervaise of Normandy, the *Bestiaire d'amour* of Richard de Fournival,[82] *Li livres dou trésor* by Brunetto Latini, and the Waldensian *De las propriotas de las animançes*. The latter two group animals into birds, beasts, fish, and serpents.[83] *De proprietatibus rerum* of Bartholomaeus Anglicus, derived in part from the *Physiologus*, was translated

198   KINGDOMS, EMPIRES, AND DOMAINS

into English by John Trevisa in 1397.[84] The anonymous Fifteenth-century *Libellus de natura animalium*, first printed around 1508,[85] states that animals are divided into celestial bodies, beasts of burden, reptiles, and wild beasts[86] but in addition describes birds, insects, the salamander (illustrated sitting in a fire, looking rather like a hyena), unicorn, griffin, siren, whale (with a boat stranded on its back), the *serra*, dragons, the hydra, and various serpents.

Animal-books of a practical nature from those years include the *Aviarum* by Hugo de Folieto (about 1132–1152),[87] Latin translations of Arabic and Persian texts on falconry, and the famed *De arte venandi cum avibus* by Frederick II of Hohenstaufen (about 1244–1250).[88] The latter can almost be considered philosophical, as Frederick—patron of Michael Scot—squandered no opportunity to correct Aristotle, whom he (no doubt rightly) accused of being ignorant in the art of falconry. Giordano Rufo wrote *Medicina equorum* (1250), the first book on hippiatric medicine, at the behest of Frederick II.[89] The *Ruralium commodorum* of Pietro de' Crescenzi (1304–1309), an update of Columella's *De re rustica*, discusses the keeping of domestic animals and bees, game hunting, fishing, and falconry.[90] A text on sport-fishing in England seems to predate the year 1406.[91] Illustrated manuscripts show the collection of butterflies.[92] One searches in vain for contemporaneous botanical counterparts of these and similar works. On the other hand, early zoology enjoyed no systemic driving force of the weight and worthiness with which medicine drove early botany.

Following the advent of the printing press, Albert's *De animalibus* (1479) joined Pliny's *Natvralis historiæ* (1469) and Latin translations of Aristotle's books on animals (1476, 1494–1498), Oppian's *Halieutica* (1478), and Aelian's *Historia animalium* (1533, 1535) in printed copy. But no zoological counterpart of Leoniceno came forward, and scientific zoology got off to a comparatively slow start.

The rise of scientific zoology 1: 1520–1550

The earliest Renaissance books of zoology simply drew together classical Greek and Roman descriptions of fish, birds, and beasts. In *De Romanis piscibus libellus* (1524) Paolo Giovio listed some 233 species of "fish" known to Varro, Columella, Pliny, and other classical writers. The final three chapters treat the octopus, cuttlefish, and squid; crustaceans; and oysters and other shellfish, including the *urtica* (sea-anemone). No mention is made of sponges or zoophytes. Although better known for his herbal, William Turner wrote a small treatise on the birds known to Pliny and Aristotle (1544);[93] it is considered the first book of ornithology. Michael Herr's German-language work on quadrupeds (1546), notable for its woodcuts and extraordinary Gothic typefaces, arranges fifty-eight animals in no particular order, for each describing its physical features, habitats, reproduction, means of capture, and use in medicine.[94] Finally, *De animantibus subterraneis liber* (1549) by Georgius Agricola—better known for his *De re metallica*, the standard work on mining and metallurgy—describes animals that are generated, live, or hibernate underground including not only ants, mice, and rabbits but also demons and *cobalos*, a mischievous sort of gnome.[95]

The rise of scientific zoology 2: the momentous 1550s

During the 1550s six works appeared that, despite their differences, collectively mark the beginning of scientific zoology. One was *De differentiis animalivm* (1552) by Edward Wotton, an Oxford graduate who trained in medicine at Padua before returning to England, where he became president of the College of Physicians.[96] *De differentiis* is an annotated classification of animals based on Aristotle's *Historia animalium*. In its Preface, Wotton acknowledges that Ruel and Agricola have already described plants and minerals, so he will restrict his scope to the animals, taking Aristotle as his guide but moving forward gradually using the animals themselves as a starting point.[97] He proceeds systematically from the general—Aristotle's major groups—to the specific, where possible arranging the animals according to the number and features of their feet.[98] His descriptions cite the ancients at every turn, notably Pliny but 214 others as well, from Aelian to Zoroaster.[99]

De differentiis bears closer examination. It is set out in ten books, the first two offering general considerations on animals—their parts, physiology, reproduction, habitats, diets, and utility for mankind. An *animal* is an ensouled body endowed with sense-perception.[100] Next Wotton turns to animals with blood, describing the features they share in common (Book 3), and those that distinguish man; viviparous quadrupeds; oviparous quadrupeds and serpents; birds; and aquatic animals (Books 4 through 8). Book 9 treats insects and other bloodless animals, and Book 10 "those distinguished as bloodless aquatic [animals], that is to say *mollia*, crustaceans, shellfish, and those things called zoophytes, all of which some still refer to as fish".[101] Wotton reaches zoophytes at Chapters 248–251, where his descriptions scarcely take us beyond Aristotle. He reminds us that these organisms are of *ancipiti natura*, neither perfect animal nor perfect plant,[102] and admits ascidians, holothurians, sea-stars, sea-lungs, sea-anemones, sea-slugs, and sponges—the latter, he says, are only slightly distinct from plants. Fungi, on the other hand, are to be grouped with plants. In the final chapter (Chapter 252) he mentions some less-than-worthy reports of animals perhaps rare, sui generis, misinterpreted, or out-and-out imaginary, but steers clear of mythological creatures.[103] He affords Zoophyta the same (unnamed) rank as the Aristotelian Mollia, Crustata, and Testata,[104] but stops short of attributing Zoophyta to Aristotle.[105] Fungi are plants, zoophytes are animals, and no third genus inhabits the miniscule gap between sponges and plants. Aristotle would, we may suspect, find that satisfactory.

Our second work of scientific zoology came from the pen of the physician-naturalist Pierre Belon, a student of Valerius Cordus. Belon wrote on Egyptian antiquities, conifers, and *estranges poissons marins* (including dolphins and the hippopotamus), but became renowned for the critical eye on display in *De aquatilibus* (1553). Like Wotton, he divides aquatic animals into the sanguinous (see below) and exsanguinous, and the latter into *mollia*, crustaceans, testaceans, and zoophytes:[106]

> Zoophyta, which might be called plant-animals, Aristotle indeed reports to differ very little from plants. For to such an extent is their nature two-faced, that it is neither possible to declare that the animate is held perfectly in them, nor however for them to be adjudged plants. As a matter of fact they meet together with [*conveniunt*] plants and fungi, except that they do not live attached [to something]. Truly in some way they are endowed with flesh, [and] may be seen to have one or another of the senses, this too they have in common with the animals. Pliny preferred to refer them neither to the animals nor the genus of plants, but to a certain third nature partaking of both. In this they absolutely have a certain life, but without any manifest sensation, in the same way together with the plants, and Holoturia [*sic*], *aures*, and many [things] of that kind. Others they truly call only attached, such as Tethya, and those they call Armorici Pollicepedes.[107]

Belon included sponges as the last-mentioned genus among zoophytes.[108]

Arguably more important for marine zoology was the *Libri de piscibus marinis* (1554–1555) of Guillaume Rondelet. We met Rondelet earlier in this chapter, as mentor to a generation of botanists. He was likewise a central figure of mid-Sixteenth century zoology, not least for having convinced Ulisse Aldrovandi (below) to take up the study of animals. The first part of *Libri de piscibus* describes bony fish, sharks, rays, sea turtles, dolphins, and whales in sixteen books, followed by Molles (squid, octopi, sea-hares, sea-anemones, and jellyfish) and Crustacea (crabs, lobsters, sea-urchins). The second part (1555) includes two books on Testacea (limpets, barnacles, chitons, abalone, and bivalves; *Murex*, conches, turban-shells and the like), and a book of "insects and zoophytes".[109] Insects are distinguished by deep indentations in their body; they have neither bones nor flesh, but consist of a sort of intermediate substance. With these he admixes

> the Zoophytes, which are themselves neither of the animals nor of the shrubs, but have a third nature from both, and which it is impossible to name in Latin unless we call them plant-animates, or plant-animals. Some of these Aristotle enumerated among the testaceans, such as *Tethya* and *Stellas*, which on account of the hard coat with which they are protected [are called] testaceans, [and] because they are intermediate between the natures of plants and animals can be explained as Zoophytes. Hereafter we join the insects and zoophytes together in [one] book, because it is the zoophytes that resemble the nature of the insects, and [conversely] the insects the nature of the zoophytes. From these we arrive all the way to those things that are not in any straightforward manner animate, and which altogether

resemble plants, and as well are deficient in every nature of the animals, of which type are the sponges.[110]

Rondelet's "insects and zoophytes" include various worms, a leech, sea lice, the sea-horse, sea-stars, sea cucumbers, sea-pens, sea-lungs, and sponges.

Our fourth book of scientific zoology is the sprawling *Historiæ animalium* (1551–1558) of Conrad Gessner.[111] Boldly appropriating the Aristotelian title, Gessner set out to compile everything that was known about the form, habitat, behaviour, diet, character, and culinary and medicinal uses of animals. A physician and professor of Greek, he drew from classical, medieval and contemporary literature alike,[112] and gave over vast tracts of text to etymology, animal names in various languages, folk-tales, proverbs, symbolism, and metaphor. Illustrations were sourced from correspondents or copied from other publications; an edition with coloured prints was offered at a higher price.[113]

Gessner divides animals into viviparous quadrupeds, oviparous quadrupeds (amphibians), birds, fish and aquatic animals, serpents, and insects.[114] This suggests a six-volume work, although he lived to see only the first four published. Within each, genera are arranged alphabetically.[115] Amongst the mass of information it can be difficult to discern Gessner's own views, but the volume on fish and aquatic animals draws significantly on Rondelet and Belon, who are credited on the title page.[116] Sponges are zoophytes, insentient or barely sentient; but (following Belon and Rondelet) Gessner admits them as animals. In a Ramist tree of doubtful scientific validity he distinguishes three types of sponge, two exhibiting a full range of senses, one (aplysiæ) with only limited sense-perception.[117] Sea-slugs and sea-cucumbers are zoophytes too, as are sea-pens, sea-lungs, sea-stars, and sea-urchins, although some forms may simply be *purgamenta*, cast-off material such as egg cases.

Gessner cuts more to the chase in his illustrated *Nomenclator* (1560), arranging marine animals into seventeen Orders from bony fish through sharks and rays; cetaceans (including sea turtles and various imaginary monsters); gastropods, sea-urchins, and jellyfish; crustaceans; shelled animals, sea-stars and holothurians; marine insects, including the sea-horse, sea lice, and various worms; and finally zoophytes, for which the brief introductory section reads as follows:

Order XVII: On marine Zoophytes, those things which do not have a hard protective coat as do the Testaceans (because we make those last among the Testaceae), but [have] a covering sui generis.

Holothurians and Tethya, which Rondelet assigns to the Zoophytes, I enumerate in the final position among the Testaceae (where I give the reason for having done this).

It is the nature of sponges that they perceive other things poorly: in our account of them we show that they do not have sense-perception. There are truly certain steps of nature, just as elsewhere, wondrously always approaching *in the middle and dualizing*, in this way crossing over from shrub to animal. After inanimate

202 KINGDOMS, EMPIRES, AND DOMAINS

bodies, it happens that a certain middle follows, (as if those things that arise between stones and metals perhaps have a power): a third [group] of living things, as the plants. At the terminus of the plants, the beginning of the Zoophytes, are the sponges, at first said simply: thereafter the aplysiae: [and] after those, the Pulmones, Holothuria, Tethya, and many Zoophytes in succession, each a little closer to perfection, all the way to the sea-shells, which surpass the snails, and so forth it ascends all the way up to mankind.

We call Zoophytes (says Rondelet) those things have the nature not of the animals nor of the shrubs, but of a third genus of them both: which we cannot name in Latin, unless we call [them] Plant-animates, or Plant-animals. Aristotle enumerated some of these among the testaceae, as Tethya and Sea-stars: those on account of their hard protective coat as testaceans: those truly facing both ways between plants and animals can be set forth as Zoophytes. We unite them as a genus [that is] a balance of both: certain of those may be as Zoophyta to the Insects: and certain Insects, which approach the nature of Zoophytes. This is how it is.[118]

Our fifth work initiating scientific zoology is the *Aquatilium animalium historia* (1554–1558) of Hippolito Salviani, later papal physician.[119] The book is presented in two sections. First comes a table, running over 112 pages, that pulls together bibliographic information on aquatic animals: not only fish but also the beaver, crab, crocodile, frog, hippopotamus, otter, shellfish, sponge, and various sea monsters. Ten parallel columns display for each genus its name in Latin, Greek, and vernacular languages; "attributes" including its form, habitat, and diet; and detailed references to works by Aristotle, Oppian, Pliny, Athanæus, Aelian, and other authors. Thereafter follows a *historia* for just under one hundred fish including sharks and rays, plus the octopus, cuttlefish, and squid. For each, Salviani gives the name, description, location, reproduction, parts, behaviour, taste, nutritional value, mode of preparation, and use in medicine, sometimes citing classical authors at length. Each entry is accompanied by at least one full-page illustration printed made from a copper engraving, yielding an image much superior to those in Belon, Rondelet, and Gessner, which were printed from woodcuts.

Let us take stock of these works so far. From Wotton to Gessner the number of recognized animal genera increased dramatically, not least among the bloodless animals, which began to be observed firsthand, dissected, grouped more naturally into genera and species, and carefully illustrated. Rondelet in particular described many of these genera for the first time. Free-living life cycle stages and so-called *purgamenta* began to be recognized as such. Zoophyta emerged within the animals as a formal group (for Gessner, an Order) and gained focus as nudibranchs, crustaceans, and echinoderms were returned to their original Aristotelian group.[120] The Aristotelian faculties of soul were still used to relegate sponges to the bottommost grade among animals, but were increasingly joined by new characters based on internal or external form. Gessner's common group of zoophytes and insects was, however, a retrograde step, as it obscures useful *differentiæ*.

Yet this zoology was not science as we know it today. Belon wrote on aquatic animals in general (*aquatilis*) and as such admixed fish, dolphins, seals, and whales (shown threatening a ship and its hapless sailors) with the beaver, chamæleon, crocodile, hippopotamus, various lizards, otter, and turtles. Rondelet excluded quadrupeds (except tortoises), but even so treated cetaceans as *pisces* against Aristotelian precedent.[121] Gessner placed bats among the birds,[122] and recounted (albeit more as historiography than as fact) the story of the barnacle-goose.[123] Belon, Rondelet, and Gessner described and illustrated imaginary animals including Neptune's horse and the infamous monk- and bishop-fish.[124] Belon remarked on the "coldness" of the torpedo ray, and Rondelet added that this was not due to temperature alone but also to some occult property.[125] Gessner in particular sought out animal proverbs, symbolism, and emblematic material.

I promised six works that collectively introduce scientific zoology, but have so far discussed only five. The sixth—Belon's *L'Histoire de la nature des oyseaux* (1555)—does not mention sponges or zoophytes, and refers to plants only in passing; but as a foundational work of comparative anatomy,[126] it introduces concepts that in the Nineteenth century would reshape our thinking about the animal kingdom. Yet (as we shall see in Chapter 17) these too leave zoophytes, which offer relatively few anatomical features for comparison, on the lowest steps of the animals.

The rise of scientific zoology 3: the encyclopædists 1560–1660

Like Gessner, Ulisse Aldrovandi undertook a comprehensive multi-volume work on animals, and like Gessner did not live to see it to completion. When the final part appeared in 1645 his zoological encyclopædia ran to eleven volumes: of birds, insects, other bloodless animals (including *mollia*, crustaceans, testaceans, and zoophytes), fish, quadrupeds with non-cloven or cloven hoofs, serpents and dragons, viviparous and oviparous quadrupeds, and monstrosities.[127] By and large these groupings improve on those of Gessner, as cetaceans are distinguished from fish, and insects from molluscs. Aldrovandi even depicts skeletal anatomy where available. Yet much old-style natural history remains: he mixes an even richer broth of fable, allegory, adages, sympathies, antipathies, omens, physiognomy, and miracles than did Gessner,[128] while dragons, sea-serpents, a saw-headed cetacean, and even the barnacle-goose-tree spill over from his monster-book into the other volumes.[129] For its part, his volume on monsters easily surpasses in weirdness those of his near-contemporaries Paré and Duret (Chapter 10); we find there not only Belon's monk-fish but all manner of mythical men, chimæras, fœtuses, animals with supernumerary body parts, monstrous plants, and portentous atmospheric phenomena.[130]

Aldrovandi further modified Gessner's classification of exsanguinous animals, for example moving sea-anemones from mollia to testaceae. Less justifiably, he divided

204 KINGDOMS, EMPIRES, AND DOMAINS

insects into terrestrial and aquatic; aquatic insects into those with and without feet; and placed into the latter various worms, leeches, the seahorse, and starfish.[131] His zoophytes include sea-anemones, jellyfish, holothuria, tunicates, a "marine fungus", the sea-pen, and the "sea-hand" (possibly a bryozoan).[132] Without offering much justification, he banished sponges from the animals altogether.[133]

At about the same time, English cleric Edward Topsell recast and paraphrased Gessner's volume on viviparous quadrupeds as *The historie of foure-footed beastes* (1607), adding bits of text from other sources and, not reassuringly, selecting a woodcut of the gorgon—a "strange Lybian Beast"—to grace the title page.[134] Next came *The historie of serpents* (1608), likewise based on Gessner but appropriating chapters on bees, spiders, and earthworms from other authors.[135] After Topsell's death these were bound with a translation of *Insectorum sive minimorum animalium theatrum* begun by Thomas Mouffet, which itself arose from a study of Gessner's insect collection[136] and was completed after Mouffet's death by Théodore Turquet de Mayerne. By this circuitous route arose the first (reasonably) comprehensive zoology in the English language (1658), its 140-word title often shortened simply to *Topsell*.[137]

The volumes on quadrupeds and serpents are set out alphabetically, albeit with exceptions: *toad* follows *frog*, *ram* follows *sheep*. Less justifiably, his "genera" sometimes include dissimilar animals that happen to share a name (the so-called seahorse, *i.e.* hippopotamus, as a species of horse) or a supposed diet (the mythical manticore as a species of hyena).[138] Insects are divided into those with wings (Book 1) and those without (Book 2); bees are discussed first because of all insects, they "are the principal and are chiefly to be admired, being the only creature of that kinde, framed for the nourishment of Man".[139] A Ramist tree promises to guide the presentation of insects without wings,[140] but is not rigorously followed. Birds, and bloodless animals other than insects, are not treated.

Gessner, Aldrovandi, and Topsell are collectively called the *encyclopædists*, together with a fourth author famed in his time but lesser-known today: John Jonston, a scholar and physician of Scottish descent who was based in the Polish-Lithuanian Commonwealth. Jonston wrote on many topics including minerals and trees, but is remembered mostly for his five-part zoological encyclopædia *Historiæ naturalis* (1649–1657).[141] Jonston divides animals into quadrupeds; birds; fish; insects, serpents and dragons; and *exangiuibus aquaticis*. He pared away most of the etymology, poetry, symbolism, epithets, culinary and medicinal utility—and all the anatomy—of earlier authors, and furnished the work with grand illustrations which, although at root recycled from Gessner, Aldrovandi, Rondelet, and Georg Marcgrave's *Historia natvralis Brasiliae* (1648),[142] were rendered skilfully in copperplate by Matthäus Merian the Younger and made available in colour. Although not a critical success, *Historiæ naturalis* was well-received by the broader public.[143]

A notable feature of *Historiæ naturalis* is its multi-ranked hierarchical structure. The volume on birds, for instance, is successively divided into books (*e.g.* on terrestrial carnivorous birds) and these into titles (*e.g.* parrots), then chapters, articles, and points, within which names of individual species are set in capital letters. Each *liber, titulus, caput, articulus*, and *punctus* is numbered. This is not a Linnæan hierarchy: the first chapter within each title has a different role, *genus* and *species* are relative (as in Plato's logical division of the arts), and for less-speciose groups the hierarchy can be shallow. Nor is each subdivision argued on the basis of explicit differentiæ. Nonetheless it is remarkable to find a seven-rank hierarchy a century before Linnæus, and alongside illustrations of unicorns, seven-headed dragons, and mermaids.

Jonston divides bloodless aquatic animals into the familiar four groups: *mollia* (the octopus, cuttlefish, and squid), crustaceans (including sea-turtles and starfish), testaceans (including sea-urchins), and *Zoophytis sive Plant-animalibus*, the latter largely following Aldrovandi and Rondelet. His final chapter lumps together eight species—holothuria, cnidaria, a tunicate, Aldrovandi's marine fungus and sea-hand among them—with minimal commentary.[144] Following Aldrovandi, Jonston excludes sponges from the animals.

The rise of scientific zoology 4: curiosities and specialization

There was more to Seventeenth-century zoology than the weighty tomes of the encyclopædists. Explorers, diplomats, and missionaries returned with stories, drawings, and specimens. Ferrante Imperato in Naples,[145] Ole Worm in Copenhagen,[146] Johann Jakob Spener in Berlin,[147] and others assembled large collections of natural objects, known as curiosity cabinets or *museæ*, which attracted visitors and were often catalogued in illustrated volumes that, in turn, sometimes pushed the boundaries of mineralogical, botanical, and/or zoological classification.[148]

Imperato was a wealthy apothecary who corresponded with Porta, Clusius, and Aldrovandi. His cabinet, which attracted visitors from across Europe, was notable for its emphasis on marine organisms including corals and seaweeds, many of which were described and illustrated in *Dell'historia natvrale* (1599).[145] Imperato thought corals to be plants, even if some are articulated like the bones of animals;[149] he called some *Coralli*, while grouping others together as *Pori*.[150] Fungi are vegetables "of a nature close to plants", sponges have a nature close to fungi (albeit with sensitivity otherwise associated with the animal soul), and alcyonia (soft corals) are not unlike sponges.[151] He was one of the first to illustrate marine algae, which he considered to be plants as well.[152]

The cabinet of Basilius Besler, a Nürnberg apothecary, botanist, and (from 1598) curator of the Hortus Eystettensis for prince-bishop Johann Konrad von Gemmingen,[153] served as the basis for an elegant if thin volume of thirty-four plates prepared for his nephew Michel Rupert Besler. In its title, *Gazophylacium rerum naturalium de regno vegetabili, animali et minerali* (1642), we find the first prominent presentation, beyond the alchemical literature, of three Kingdoms of nature.

206 KINGDOMS, EMPIRES, AND DOMAINS

To the extent Besler's cabinet is fairly represented by *Gazophylacium* (and by his own earlier *Facicvlvs rariorvm*), however, its placement of corals, fungi, and fossilized wood and shells fell short of contemporary standards.[154]

Walter Charleton's *Onomasticon zoicon* (1668) was based on the royal menagerie in St James Park, London. Charleton emphasized nomenclature ("if one doesn't know the names, one loses knowledge of the things"),[155] and arranged the *Onomasticon* in a hierarchical manner. Animals he divided into *Rationale* and *Irrationale*; the latter into quadrupeds, birds, reptiles, and fish; fish into those with blood, and those without; and the *Exsanguia* into *Mollia seu mollusca*, Crustacea, and Testacea.[156] He arranged genera (in capital letters) under Classes, and species (in italics) under genera. Charleton set out twenty-one classes under *Pisces* including Piscium Cetaceorum, Piscium Mollium, Crustatorum, Testaceorum Turbinatorum, Univalvium, and Bivalvium (including sea-urchins), following which we find a separate Zoophytorum sive Plant-animantium.[157] His brief treatment of zoophytes largely followed Aldrovandi and Jonston.[158]

The Seventeenth century also brought interest in individual groups of animals, not only quadrupeds and birds but "difficult" groups as well.[159] Thomas Willis described the internal anatomy of the lobster, earthworm, and oyster,[160] as did Martin Lister of spiders, molluscs, and cuttlefish.[161] Insects in particular came into their own. Antony van Leeuwenhoek described their microscopic anatomy,[162] and Jan Goedart[163] and Jan Swammerdam their development. In *Historia insectorum generalis* (1669) Swammerdam identified insect eggs, and developed a "solid and unified hypothesis" that insects pass through intermediate stages before emerging as adults. Owing to his untimely death, many of his observations were published only much later, in the bilingual *Bybel der nature / Biblia naturae* (1737),[164] whence they helped seal the demise of the ancient idea that insects arise spontaneously from mud or decaying carcasses—an idea already much weakened by the experiments of Francesco Redi on flies.[165] Joachim Jung does not appear to have left us a summary of theoretical zoology, as he did for botany (above).[166] His notes on insects, later brought together by Vaget as *Historia vermium*, discuss their "polymorphism".[167]

Well before the end of the Seventeenth century, then, membership in the Animals had largely solidified around the genera identified by the encyclopædists. Even the order of presentation, from quadrupeds "downward", became standard. In Chapter 9 we remarked that in the alchemical literature Animals, Vegetables, and Minerals began to be called Kingdoms from the first decade or two of the Seventeenth century. The herbalists and zoological encyclopædists above did not follow their lead, although in 1634 Mayerne (in his Epistle Dedicatory to Mouffet's *Theatre of Insects*) referred to "the three Kingdoms of the universal spirit, (the Vegetable, Animal, and Mineral)".[168] So far as I can ascertain, the earliest such usage outside the alchemical literature was by Johann Heinrich Alsted in 1615: "suntque mineralia, vegetabilia, & animalia. Hæc enim sunt tria veluti regna, in quæ diuiduntur omnia elementa perfectè mixta."[169] In a 1639 letter to Marin Mersenne, Descartes wrote "Quemadmodum Deus est unus et creavit Naturam unam, simplicem, continuam, ubique sibi cohærentem & respondentem, paucissimis, constantem principiis elementisque, ex quibus infinitas propemodum res, sed in tria regna, Min(erale), Veg(etale) & Animale, certo inter se ordine gradibusque distincta perduxit."[170] Apart from a cameo appearance in the younger Besler's *Gazophylacium*, this usage became commonplace only after mid-century.

Zoophyta: a fourth division of nature?

We met Andreas Libavius briefly in Chapter 10 in connection with the borametz. He was attracted to controversy, and in 1599—the year of Imperato's *Dell'historia naturale*—presented a lecture on zoophytes.[171] Libavius begins by making a claim— that zoophytes are neither entirely animal nor entirely plant, but of an essence of both natures, "here somewhat more, there somewhat less"—only to attack it:

> it is illogical and unphilosophical that something might be of two natures; such a being would require its own law of physical matter; we might as well believe in centaurs. We might better refer to plantanimals as riven in two as the hoof of a deer, sloughed off each year and regrown. But then, zoophytes participate in life: did not Pliny argue that sponges are two-faced? What is sensate is animal, what is non-sensate is not animal; imperfection reflects steps in the works of the Creator. But then, why not zoanthropons and stœchiomicta? Aristotle did not use the word *zoophyte*. Plants are fixed to the earth but zoophytes are capable of motion, and of sensation and appetite too. Eggs and worms and balls of slime might better be called zoophytes than those things that actually are so named.

And so Libavius proceeds, flinging claims back and forth on behalf of Aristotle and Pliny. *Natura inquit non facit saltum, sed continuat substantiarum progressus.* The small steps from the inanimate to the animate include the lithodendron, truffles, and fungi. Nature makes new genera in the sea. Sponges have blood. Aristotle acknowledged that sponges resemble plants. Pliny called for a third genus between animals and plants. The fœtus is affixed to its substratum, but is not a plant. Sea-urchins and insects lack true flesh but remain animals, just as the *arbor pudica* [mimosa] and sun-seekers remain plants.

Libavius enumerated what Aristotle and Pliny considered as dualizers, adding a few observations from Scaliger and others[172] while displaying no awareness of Belon, Gessner, Rondelet, or Salviani. It would be absurd for the entire genus of zoophytes to dualize between plants and animals: by weight of evidence, they are animals.[173] Are there zoophytes on dry land? (Apparently yes). Are eggs, worms, balls of mucus, abortive fœtuses, and monsters zoophytes too? (The philosophers say yes on the first four; the latter are not ordinary genera at all). Finally, he returned to the interface between the inanimate and plants, where we find corals, pearls, gemstones, and crystals that grow in the earth.

208 KINGDOMS, EMPIRES, AND DOMAINS

Such was the dialectic by which Libavius delineated and subdivided Zoophyta. We should not speak of them as a peculiar genus, he wrote, nor as composed of a separate class of substances, but as diverse things with a common disposition [*unam quandam affectionem communem*]. By approaching both plants and animals, they merit the name Zoophyta. Their genus is duplex, both terrestrial and aquatic. The aquatic is differentiated into the testate, fleshy, and spongy, the terrestrial into the testate and fleshy: the spongy are absent, or have degenerated into the non-living, to which they are proximate.[174] Libavius did not (to my reading) explicitly claim Zoophyta as a fourth grade of organized things, but the implication is there for the taking.

Plants and animals in 1680

When Johannes de Spira printed Pliny's *Historiæ naturalis* in 1469, plants and animals were defined by Aristotle's types of soul, and identified by traditional names scattered throughout ancient texts, herbals, and the *Physiologus*. By 1680 plants and animals had been delineated, ordered, described, and professionally illustrated. Ancient references had been cross-indexed and attached to modern names, some binomial. Classifications—some hierarchical—relied more on vegetative or reproductive features than on ecology or medicinal virtues. Comprehensive botanies and zoologies were popular, as were texts focused more narrowly (for example) on the plants of a particular region, or on birds or fish. Fungi were typically grouped with plants, whether as living beings in their own right or a sort of excrecence. Octopi, jellyfish, oysters, and sea-urchins were animals, while doubt remained about sponges—although their supposed sense-perception often tipped the balance in favour of animality. The most-egregious mythical beings had been struck off the rolls, while those remaining were typically relegated to a carefully worded final chapter, or to a separate book of monsters and marvels. Natural history became internationally collaborative and increasingly professional, linking collectors, patrons, publishers, and illustrators.

As we have seen, animals, plants, and minerals began to be called *Kingdoms* only in the Seventeenth century, first by alchemists (Sendivogius *et al.*), generalists (Alsted), and physician-botanists (the Beslers), not by front-line naturalists.[175] Nicolas Lemery, an early pharmaceutical chemist, divided medicaments into those derived from minerals, from vegetables, and from animals,[176] but did not refer to these groups as kingdoms.[177] Emanuel König's *Regnum minerale*, *Regnum vegetabile* and *Regnum animale*[178] fall largely into this tradition: König was professor of theoretical medicine at Basel. His book on animals can (generously) be seen as scientific zoology, and includes a brief section on classification based on Ramist trees;[179] but, like his other books, emphasizes dietary and pharmaceutical applications.

Loose ends certainly remained. Through the Sixteenth and Seventeenth centuries (and in fairness, well into the Nineteenth), zoologists struggled to identify natural groups among the bloodless animals. Some worms turned out to be immature insects, other marine genera, egg-masses, or egg-cases. While accepting fungi as plants, botanists were oddly incurious about seaweeds. The old belief that corals are marine shrubs that can turn to stone cried out for critical analysis. Zoophyta, reduced to sponges plus a dozen nondescript genera that few naturalists had ever seen, seemed collectively unworthy of a separate Kingdom; yet the precise boundary between plants and animals remained vague. As we shall see in Chapter 14, much of this would soon change.

Our botanical and zoological narratives in this chapter run to (about) 1680 for several reasons. In the mid-Seventeenth century Robert Hooke devised the compound microscope and described, for the first time, the fine-scale structure of diverse natural objects including moulds, sponges, and insects.[180] Soon thereafter Leeuwenhoek, a Dutch linen-draper, learned to make single-lens microscopes that afforded considerably greater resolution,[181] and during the second half of the 1670s discovered the world of tiny organisms; he was elected to the Royal Society (London) in 1680. In Chapter 13 we accompany the microscopists as they describe new organisms and attempt to place them within the newly minted Plant and Animal Kingdoms. Soon thereafter John Ray abandoned single differentiæ (*Methodus plantarum nova*, 1682) and introduced a biological definition of species (*Historia plantarum*, 1686–1704) in classifying the plants. We return to Ray and the new taxonomy in Chapter 15.

The following three chapters should be read as a unit, as their respective themes—the microscopic world, continuity in nature, and classification—are closely intertwined. In his biography of Buffon, Jacques Roger distinguishes *observers* and *classifiers*; painted with that particular broad brush, our next two chapters bring stories of observation, Chapter 15 of classification. Roger remarks that in the first half of the Eighteenth century, "the two groups of scholars more or less ignored each other".[182] Some *observers* (Leeuwenhoek, Needham) are true to form, but we shall find the dichotomy honoured more often in the breach. Buffon famously attacked Linnæan taxonomy,[183] but himself was perhaps less an observer than one who assembled facts in search of deeper explanation.[184] Failing eyesight forced Charles Bonnet (Chapter 14) to abandon microscopy (and reading and writing) at age twenty-seven,[185] yet his Chain of Being was the very antithesis of classification. Linnæus travelled with a microscope,[186] corresponded with microscopists (Baker, Münchhausen), and studied the "invisible world" of pathogenic fungi.[187] Even so, these entanglements cannot channel Eighteenth-century natural history into a single historical narrative. We begin Chapter 13 in Delft.

13

The most wretched creatures

You have made your way from worm to man, and much in you is still worm.
—Friedrich Nietzsche, *Also sprach Zarathustra*[1]

Multiple worlds

In August 1674, when Antony van Leeuwenhoek first glimpsed what today we call unicellular organisms,[2] nearly a decade had passed since Robert Hooke's discovery that insects, sponges, moulds, and many other commonplace objects were intricately structured at microscopic scale, with complex features arranged in striking geometric patterns. This new world of extraordinarily tiny animals swimming merrily about in swamp water or vegetable infusions initially evinced consternation and disbelief.[3] But neither the reality (soon confirmed) nor the immense number and wondrous diversity of these little beings precipitated an intellectual crisis: quite the opposite, as microscopists and philosophers alike acknowledged the wisdom of God manifest in his works of Creation. But there was more to this lack of surprise: microscopic beings were not entirely unanticipated, and indeed might resolve certain longstanding problems about living nature including the mechanisms of generation and infectious disease. More generally, the discovery of new lands beyond the seas, the invention of the telescope, and new ideas about the universe had already given rise to scientific and popular literature on "multiple worlds"—even if these had not been imagined to exist under, or even within, our very noses.

As we have seen, Plato considered the cosmos itself a living being, endowed with soul and intelligence.[4] The subsequent reconciliation of Neoplatonism and Christianity gave license to hierarchies of "intelligences" (seraphim, cherubim, thrones, dominations, virtues, powers, principalities, archangels, angels)[5] and the association of human souls with planets and stars. Our world was likewise animate, possessed of vital force. So long as the cosmos was acknowledged to have had a beginning in time and heavenly beings did not become objects of worship, the Church could give its benediction. But Plato had gone farther, arguing that for the cosmos to be perfect it must contain all possible kinds of beings. Thus Epicurus: "there are infinite worlds both like and unlike this world of ours".[6] Aristotle disagreed with Plato on this point, asserting that "it is not necessary for everything potential to be actual" and "that which has a potency need not exercise it".[7] He argued reductio

Kingdoms, Empires, and Domains. Mark A. Ragan, Oxford University Press. © Oxford University Press 2023.
DOI: 10.1093/oso/9780197643037.003.0013

ad absurdum against a plurality of worlds: an earth cannot by nature tend towards different places, nor can there be a void in the heavens.[8] Simplicius and Ibn Rushd agreed.[9]

Even so, other ideas about multiple worlds were afoot. Anaximander seems to have accepted that multiple worlds could exist in temporal succession; some Pythagoreans believed in simultaneously inhabited heavenly bodies; the atomists Leucippus and Democritus allowed both spatial and temporal multiplicity.[10] Origen wrote that "not then for the first time did God begin to work when He made this visible world; but as, after its destruction, there will be another world, so also we believe that others existed before the present came into being. And both of these positions will be confirmed by the authority of holy Scripture."[11] Augustine, on the other hand, argued that belief in a plurality of worlds, whether spatial or temporal, conflicted with the *Genesis* account of Creation.[12] Albert followed Aristotle and Ibn Rushd,[13] while Aquinas emphasized that the uniqueness of our present world was central to its perfection[14] but acknowledged that God could do "whatever he pleased short of a logical contradiction".[15]

With the so-called Condemnation of 1277 Étienne Tempier, Bishop of Paris, sought to put the philosophers on a short leash. Among the ideas to be condemned was the proposition that God cannot create multiple worlds.[16] Ironically, this stimulated new thinking about the physical world,[17] and opened hitherto risky lines of argument that soon resulted in claims that multiple worlds actually exist and may be inhabited. William Vorilong speculated on whether humans on other worlds had experienced the Fall and if so, whether Christ had redeemed them.[18] Nicholas of Cusa—not yet a Cardinal—discussed "in ignorance" whether beings might inhabit the sun, moon, and stars. He supposed that the inhabitants of the sun are more "solar, bright, illustrious, and intellectual" and "spiritlike" than those on the moon, whereas we on earth are "material and solidified".[19] Somewhat later, Giordano Bruno identified these notional other worlds with the stars and Copernican planets,[20] a step too far for the Church at that moment. Two generations before Leeuwenhoek, Galileo observed that the moon was not a perfect translucent sphere but rather a world acceptably alike our own, with unevenly spaced mountains and depressions.

Many of these ideas came together in the influential *Entretiens sur la pluralité des mondes* (1686) by Bernard de Fontenelle.[21] In it a philosopher (Fontenelle himself) describes to his hostess, the Marquise, the worlds newly revealed by Galileo and Leeuwenhoek. The planets are likely home to humans like ourselves, he assures her, although the farther we venture from earth the more different they will be. Here on earth, there are as many types of microscopic animals as visible ones: in the second edition of *Entretiens* (1687) Fontenelle speaks of liquids *remplis de petits Poissons ou de petits Serpens que l'on n'auroit jamais soupçonnez d'y habiter*.[22] Christiaan Huygens was less cautious in *Kosmotheoros* (1698), envisioning that other planets are home to advanced humanoid civilizations and diverse animals and plants.[23] Together the two works were instrumental in convincing the educated public that

212 KINGDOMS, EMPIRES, AND DOMAINS

multiple worlds—cosmological and microscopic—are consistent with modern philosophy and science, and indeed necessary for nature to be complete and perfect.[24]

Invisible airborne seeds

Twenty-one centuries ago, Marcus Terentius Varro advised that farmhouses should be situated far from swampy ground "because certain minute animals, invisible to the eye, breed there, and, borne by the air, reach the inside of the body by way of the mouth and nose, and cause diseases".[25] Columella cautioned that swamps breed stinging insects which carry pestilence,[26] while Virgil described the spread of disease among cattle.[27] Lucretius concluded *De rerum natura* with a consideration of the plague at Athens in 430 BCE: "just as there are seeds [*semina*] of things helpful to our life, so, for sure, others fly about that cause disease and death."[28] As for airborne seeds helpful to life, we saw in Chapter 1 that, according to the *Iliad*, the west wind Zephyrus fathers wind-swift mares; Aristotle, Varro, Columella, Virgil, Pliny, Solinus, and Augustine[29] repeated variants of the story. According to Philo of Alexandria,

> [Angels and dæmons] are souls that fly in the air. And let no one assume that what is said here is a myth. For the universe must be animated through and through and each of its primary and elementary parts encompasses the life forms that are akin and suited to it. . . . The earth contains the land creatures, and the sea and the rivers those that are aquatic, fire the fire-engendered . . . and heaven the stars. . . . The air, too, must therefore be filled with living things, though they are invisible to us, since even the air itself is not visible to sense. We ought not conclude, however, inasmuch as our vision is incapable of forming images of souls, that there are consequently no souls in the air, but they must be apprehended by the mind, that like may be discerned in conformity with its similars. Moreover, do not all land and water creatures live by air and breath? And is it not true that when the air is polluted pestilential conditions usually arise, suggesting that it is the principle of animation for all and each?[30]

These notions survived remarkably intact into the Renaissance. Girolamo Fracastoro, who had studied alongside Copernicus at Padua, wrote that contagions are mediated by imperceptibly small seeds (*seminaria*) that can be spread in three ways: by direct contact, on *fomites* (carriers such as clothing), or through the air. Once inside the body of a new host they multiply after their own kind, bringing about disease.[31] William Harvey argued that not only contagious diseases, but also insects and some other animals, are generated from "elements and seeds so small as to be invisible, (like atoms flying in the air,) scattered and dispersed here and there by the winds".[32] Not only horses but also partridges, tigers,[33] and chambermaids[34] might be impregnated by the wind.

THE MOST WRETCHED CREATURES 213

Leibniz and monads

Over the four decades from 1676 to his death in 1716, Gottfried Leibniz set out—in an immense number of letters, articles, and pamphlets—arguably the most-ambitious philosophical system since the Neoplatonic synthesis. A rationalist, he pointed to

> two famous labyrinths, in which our reason often goes astray: the one relates to the great question of *liberty* and *necessity*, especially in regard to the production and origin of *evil*; the other consists in the discussion of *continuity* and of the *indivisible points* which appear to be its elements, and this question involves the consideration of the *infinite*.[35]

The elements of the latter labyrinth—continuity and indivisible points—bear directly on our question. Leibniz approached the problem in two ways: one mathematical, the other idealistic and physical. His calculus describes sequences of infinitely close points on graphs, while his latter approach is based on infinitely small particles, or *monads*. He regarded "souls, or rather Monads, as *atoms of substance*, since, in my opinion, there are no *atoms of matter* in nature and the smallest portion of matter has still parts." These monads are "not parts of bodies, but presuppositions of them";[36] they are indivisible, self-moving, spiritual yet real, mind yet body, intensive rather than (like Cartesian corpuscles) extensive. In a sense each contains, or has its own perspective upon, the entire universe:[37]

> Whence it appears that in the smallest particle of matter there is a world of creatures, living beings, animals, entelechies, souls. Each portion of matter may be conceived as like a garden full of plants and like a pond full of fishes. But each branch of every plant, each member of every animal, each drop of its liquid parts is also some such garden or pond. And though the earth and the air which are between the plants of the garden, or the water which is between the fish of the pond, be neither plant nor fish; yet they also contain plants and fishes, but mostly so minute as to be imperceptible to us.[38]

As monads are infinite in number and no two are identical, it is possible to think of them as constituting a series, each differing from the next by an infinitesimally small degree. For Leibniz it was then a small step to a Chain of Being:

> I think, then, that I have good reasons for believing that all the different classes of beings, the totality of which forms the universe, are, in the ideas of God, who knows distinctly their essential gradations, merely like so many ordinates of one and the same curve, the relations of which do not allow of others being put between any two of them, because that would indicate disorder and imperfection. Accordingly men are linked with animals, these with plants, and these again with

214 KINGDOMS, EMPIRES, AND DOMAINS

fossils, which in their turn are connected with those bodies which sense and imagination represent to us as completely dead and *informes*. But the law of continuity requires that, *when the essential determinations of any being approximate to those of another, all the properties of the former must gradually approximate to those of the latter*. Therefore all the orders of natural beings must necessarily form only one chain, in which the different classes, like so many links, are so closely connected with one another that it is impossible for sense or imagination to determine exactly the point where any one of them begins or ends; all the species which border upon or which occupy, so to speak, *régions d'inflexion et de rebroussement* being necessarily ambiguous and endowed with characteristics which may equally be ascribed to neighbouring species. Thus, for instance, the existence of zoophytes, or, as Budæus calls then, *Plant-animals*, does not imply monstrosity, but it is indeed agreeable to the order of nature that they should exist. And so strongly do I hold to the principle of continuity that not only should I not be astonished to learn that there had been found beings which, as regards several properties—for instance, those of feeding or multiplying themselves—might pass for vegetables as well as for animals, and which upset the common rules, founded upon the supposition of a complete and absolute separation of the different orders of beings which together fill the universe: I say, I should be so little astonished at it that I am even convinced that there must be such beings, and that natural history will perhaps some day come to know them, when it has further studied that infinity of living beings whose smallness conceals them from ordinary observation, and which lie hid in the bowels of the earth and in the depths of the waters.[39]

For Leibniz, bodies are not *composed of* monads (as monads are not extensive), but rather are phenomena *arising from* them. Monads, not matter, are the primary reality. Each living body reflects an aggregate or plenum of monads, among which a dominant monad best represents the others.[40] The dominant monads of plants are *perceptive*, each representing (from its own viewpoint) all creation only in an "obscure and confused" manner. Those of brute animals are *apperceptive*; they mirror creation clearly with attention and memory, and merit the name *souls*. Monads which by an act of self-reflection know themselves—that is, are imbued of reason—we may call *spirits*.[41] Nor are monads cells, or little animals. Living bodies arise from pre-existing seeds,[42] with birth and development reflecting the unwrapping (*evolutio*) of a pre-existing individual, very much as "when worms become flies and caterpillars become butterflies",[43] while death is a wrapping up (*involutio*) of individuality, "a kind of pupation".[44] Leibniz followed the reports of the microscopists, and later corresponded with Leeuwenhoek regarding spermatozoa and preformation.[45]

With *Monadology*, Leibniz broke not only with Descartes on body, but with Aristotle and Aquinas on form. Needham points to Leibniz's monads as "the first appearance of organisms upon the stage" of Western theory.[46] In a less-complimentary vein, Kant called Leibniz's universe "a kind of enchanted world".[47] Yet such theories were, so to speak, in the air: Fontenelle had asked

What is an animal formed of? Of an infinity of corpuscles which were scattered in the grass he has eaten, in the water he has drunk, in the air he has breathed; he is a composite whose parts were brought together from a thousand different places in this world, these atoms circulate constantly, forming now a plant, now an animal, and after having formed one it is no less possible for them to form another.[48]

In the next chapter we return to Leibniz's continuum in nature and its consequences for intermediate genera.

Leeuwenhoek and Joblot: little animals observed

To observe tiny objects, a magnifying apparatus of sufficient power and resolution is needed. Optical lenses were known in antiquity:[49] Aristophanes refers to a glass or crystal useful to kindle fires,[50] and Seneca used a water-filled glass globe to assist in reading.[51] Lenses and spectacles were available in medieval Europe,[52] although craftsmen and guilds kept lenscraft a closely guarded secret; some details were published in 1569[53] and became widely accessible through Porta's *Magia naturalis* (1597).[54] In the first decade of the Seventeenth century Galileo constructed a microscope and used it to examine insects; within twenty years lens-makers were selling compound microscopes in the Netherlands, Paris, and London.[55] Both the earliest-known engraving of objects (bees, and a weevil) seen through a microscope,[56] and the word *microscopio*, date to 1625,[57] while the first book of micrographia appeared in 1656.[58]

By and large, microscopy was welcomed as a technology that might contribute to established areas of enquiry: the functional anatomy of plants, animals, and humans; the circulation of blood; reproduction and generation; the structure and propagation of insects, worms, and fungi. Although the existence of microscopic beings had been anticipated to some degree (above), none had been seen until Leeuwenhoek turned his bead-lens microscope onto bog-water in the late summer of 1674.[59] Over the next several years he recorded observations on natural waters, vinegar, infusions, saliva, and the like, describing a host of tiny beings now known as bacteria, ciliates, euglenoids, flagellates, foraminifera, colonial and filamentous green algae, rotifers, nematodes, and the freshwater polyp *Hydra*,[60] as well as spermatozoa. Lacking technical terminology, he called them all (in Dutch) *(little) creatures* or *(little) animals*, referred to them by shape (globules, ovals, filaments), and/or drew analogies (serpent-like motion). He observed (but misinterpreted) their reproduction,[61] described subcellular features and, assuming their bodies to be scaled-down versions of those of macroscopic animals, computed the size of their notional muscles.[62] Leeuwenhoek is not known to have suggested that the little animals caused contagion or disease, although others quickly made the connection.[63] His verdict was clear: "these little animals are the most wretched creatures I have ever seen".[64]

216 KINGDOMS, EMPIRES, AND DOMAINS

Further reports by Leeuwenhoek and others soon followed, many appearing in the Royal Society's journal *Philosophical Transactions*.[65] The animality of these little beings seems not to have been questioned: many were actively motile, sported little "legs" (various extracellular appendages including flagella), exhibited sensitivity,[66] and could be killed or immobilized by poisons.[67] Indeed, some were fully as perfect as macroscopic animals.[68] On the other hand, Leeuwenhoek seemed reluctant to label motile cells of the colonial green alga *Volvox* animalcules, comparing them instead with seeds of plants.[69]

The first monograph of little animals, *Descriptions et usages de plusieurs nouveaux microscopes*, appeared in 1718 from the pen of Louis Joblot, professor of mathematics at the Académie Royal de Peinture et Sculpture in Paris.[70] Initiating a tradition that would persist into the Twentieth century, Joblot began his book with a technical description of microscopes and the practise of microscopy, then presented a series of illustrated biological studies.[71] The book is also noteworthy for its accessible style, imaginative names for the little animals (including *slipper* for the ciliate *Paramecium*[72]) and—fifty years before Spallanzani—experiments with hay infusions that speak strongly against spontaneous generation.[73] With Francesco Redi's experiments in mind,[74] George Adams, whose *Micrographia illustrata* (1746) drew on Joblot's *Descriptions*, concluded that the little animals are

> produced from the Spawn of some invisible volatile Parents, and generated like Gnats and several other Sorts of Flies, which are bred and undergo several Changes in the Water before they take Wing; that some of them originally may be Water Insects, or really Fish, small enough to be raised in Spawn with the Vapours, and to fall down again in rain, and to grow and breed in Water that is kept.[75]

The idea that the little animals in waters and infusions routinely spread through the air as imperceptibly tiny seeds or "spawn" met with resistance. Redi had demonstrated that flies do not arise de novo, but thanks to Hooke's *Micrographia* the day had long passed when flies were scorned as among the least-perfect animals.

Buffon, Needham, and Spallanzani: spontaneous generation

In 1739 Georges-Louis Leclerc, later ennobled as Comte du Buffon, *seigneur* of Montbard in Burgundy and a member of the Academy of Sciences,[76] became *intendant* (director) of the Royal Botanical Garden in Paris, and thereby also of the neglected royal Cabinet of Natural History.[77] Buffon developed contacts within the royal court, modernized the buildings, and expanded the collections, but spent much of his time in Montbard preparing the work for which he would become known across Europe: *Histoire naturelle, générale et particulière*. Although Buffon intended it to illuminate all of nature including *des Insectes de la mer* and *des Animaux microscopiques*, in his lifetime the series progressed little beyond

THE MOST WRETCHED CREATURES 217

minerals, quadrupeds, and birds.[78] Even so, thanks to its luminous (or according to his many detractors, pompous) style and its focus on nature as a "vast spectacle", the *Histoire naturelle* captured the *grande bourgeois* imagination across Europe. Simultaneously it enraged many natural historians for ignoring their hard-won hierarchical classifications, and raised suspicion among conservative elements in the Church for implying that nature may not require the ongoing hand of God.[79]

Buffon's *Histoire naturelle* was popular *en particulière*, not least for presenting individual animals in a way that "enchants and magnifies [*agrandit*], so to speak, the reader".[80] As natural history *en générale* its success was less obvious, as few of Buffon's theories were taken seriously in his lifetime.[81] Informed by books and *pensées* rather than through specialist training or observation, Buffon put forward ideas on the formation of earth and the planets, the properties of matter (including living matter), and the interrelationship of all living things.[82] Here we are particularly interested in his theory linking form, growth, and reproduction, set out in the first five chapters (completed 6 February 1746) of the second volume, *Histoire générale des animaux*.[83] According to Buffon, organisms are assemblies of living particles. The form of an animal or plant is determined by an "inner mould".[84] Particles—which themselves have, so to speak, stepped over the line from the mineral to the living[85]—are attracted to a body by an internal force not unlike gravity.[86] Just as rock salt always expands in cubical form, so too can we consider

> that there really exists in nature [an infinity] of small organized beings, alike, in every respect, to the large organized bodies seen in the world; that these small organized beings are composed of living organic particles, which are common to animals and vegetables, and are their primitive and [incorruptible] particles; that the assemblage of these particles forms an animal or plant, and consequently that reproduction, or generation, is only a change of form made by the addition of these resembling parts alone, and that death or dissolution is nothing more than a separation of the same particles.[87]

When a body has expanded to its full size, further particles assimilated from outside collect in certain of its parts and—directed by the same force that governed its growth—serve in reproduction. The simpler the body, the greater the number of its parts that can collect these additional particles and thereby give rise to a new body.[88]

Meanwhile John Turberville Needham, an English Catholic priest and experienced microscopist, had formed the idea that spermatozoa are not themselves living entities, but rather are pneumatic "machines" that carry and can expel tiny living globules.[89] Buffon and Needham joined forces in March 1748 to test their complementary hypotheses by microscopic examination of vegetable infusions. Using a single-lens microscope equipped with a mirror to collect and focus light onto the specimen,[90] they observed that seed infusions soon "swarm'd with Clouds of moving Atoms, so small, and so prodigiously active; that tho' we made use of

218   KINGDOMS, EMPIRES, AND DOMAINS

a Magnifier of not much above half a Line focal Distance, yet I am persuaded nothing but their vast Multitude render'd them visible".[91] Buffon described these experiments in *Histoire générale des animaux*. Later Needham boiled mutton gravy for ten minutes in the expectation that the heat would kill any pre-existing eggs or seeds, then closed it inside a glass vial using a cork: vast numbers of little animals promptly appeared.[92] Since the atmosphere is not noticeably thick with them, in infusions familiar types succeed one another without delay or seasonality, and some are too large to become airborne or unfit to travel over dry land, therefore (he reasoned) infusion-animalcules must arise in situ and de novo.[93]

Working on his own, Needham then examined infusions of crushed wheat-seeds. From the gelatinous mass in his flasks, he again soon observed "Clouds of moving Atoms". Then after two weeks during which little happened,

> From these [particles] uniting into one Mass sprung Filaments, Zoophytes all, and swelling from a Force lodged within each Fibre. These were in various States, just as this Force had happen'd to diversify them. . . . These Filaments were all Zoophytes, so teeming with Life . . . nothing can more perfectly than these wheaten Filaments, represent in Miniature Corals, Coralloids, and other Sea Plants, which have long been observ'd to be teeming also with Life, and have been suppos'd to be the Work of Animals. . . . Are not therefore all these in the same Class, and is not their Origin similar?[94]

And further:

> It seems plain therefore, that there is a vegetative Force in every microscopical Point of Matter, and every visible Filament of which the whole animal or vegetable Texture consists: And probably this Force extends much farther; for not only in all my Observations, the whole Substance, after a certain Separation of Salts and volatile Parts, divided into Filaments, and vegetated into numberless Zoophytes, which yielded all the several Species of common microscopical Animals; but these very Animals also, after a certain time, subsided to the Bottom, became motionless, resolv'd again into a gelatinous filamentous Substance, and gave Zoophytes and Animals of a lesser Species.
>
> This is not only true of all the common microscopical Animalcules, but of the spermatic also; which, after losing their Motion, and sinking to the Bottom, again resolved into Filaments, and again gave lesser Animals. Thus the Process went on through all visible Degrees, till I could not any longer pursue them with my Glasses: And thus evidently the spermatic are to be class'd with the common microscopical Animals.
>
> Hence it is probable, that every animal or vegetable Substance advances as fast as it can in its Resolution to return by a slow Descent to one common Principle, the Source of all, a kind of universal *Semen*; whence its Atoms may return again, and ascend to a new Life.[95]

These are beautiful words, but Needham had designed his experiments poorly. When Lazzaro Spallanzani extended the boiling time to one hour, sealed each flask by fusing its neck, and instituted adequate controls, no animalcules appeared.[96] Needham countered that Spallanzani had damaged the vital force, or the air inside the vessels; Spallanzani replied with ever-more-meticulous experiments, over one hundred in all, all consistent with his contention that the little animals arise from eggs or seeds.[97] Had the story ended there, Needham's theory might have faded into obscurity; but as we shall see later, new observations—accompanied by new misinterpretations—would ensure for infusion-animals an ongoing role in the convoluted debate over the biology of sexual reproduction.[98]

Let us examine more closely Needham's experiment with crushed wheat-seeds. After some time in water, salts and volatile material evaporated, and the remaining (plant) material "became softer, more divided, and more attenuated" and felt gelatinous.[99] Microscopic examination revealed filaments (fungi, but interpreted by Needham as zoophytes), which subsequently liberated little beings that exhibit self-movement (animals). Needham concluded that plants can give rise to animals, these again to plants, and so on in an alternating series.[100] He appreciated that this "strange Vicissitude"[100] might call into question the nature of the little animals: the "Animalcules, if they may be call'd indifferently by that Name, manifestly constitute a Class apart."[101] Elsewhere he refers collectively to the filaments and animalcules as "a new Class of Beings".[102] In 1748 Needham did not venture to name or delineate this class, but this is the first time that the little animals had been set apart from macro-animals, or grouped with zoophytes.[103] Thirteen years later he explicitly positioned this class between plants and animals,[104] allowing these organisms irritability but not sensitivity.[105] Later yet (1774) Needham identified four "ideas" within the natural world: the *minéro-végétal* (minerals), *végéto-végétale* (plants and trees), *végéto-vital* (zoophytes), and *sensitif* (animals including man, who is further set apart by his intellectual soul). He assigned vegeto-vitality to "polyps, and microscopic animals in general, and above all the so-called spermatic animals, to starfish, to several other marine organised substances in the lower classes of animation, and in general to any organised species which multiples by division."[106] Led by beings unknown to Belon and Rondelet, Zoophytes had regained their place as a distinct group of intermediate organisms.

A class of their own?

Needham was not alone in treating the little animals as a single (if diverse) group of beings. In the third volume of his *General natural history* (1748–1752) John Hill, an English physician and botanist,[107] recognized two classes of "lesser animals": Animalcules and Insects.[108] The former, defined purely by size, encompass nematodes (genus Enchelides), rotifers (Brachionus), spermatozoa (Macrocercus), ciliates, and other microscopic beings. Despite their immense number, he

220 KINGDOMS, EMPIRES, AND DOMAINS

writes, Animalcules are of extremely few species. He arranged them into three classes: Gymnia (without tails or visible limbs: four genera), Cercaria (with tails but no visible limbs: two genera), and Arthronia (with visible limbs: two genera).[108] He then subdivided each genus into species by shape, proportion, or the presence of a mouth. In no fewer than seven of these genera[109] he found evidence of an intestine. As for Zoophyta, Hill declared that "there are no such creatures"; the term is "not founded on any thing in Nature", and many alleged zoophytes are "true and perfect animals possessed of that great Characteristick Quality Locomotion".[110] Sponges, corals, and fungi, on the other hand, are plants.[111]

The early decades of the Eighteenth century witnessed an enthusiasm for microscope-building, especially in England.[112] From the 1730s steps were taken to reduce the chromatic and spherical aberration to which multiple-lens instruments were notoriously susceptible. Henry Baker's *The microscope made easy* (1742) offered practical tips for preparing, observing, and measuring microscopic objects; making infusions; and studying simple physiological processes in vivo.[113] For some of a technological bent, microscopy began to appeal as a leisure activity: physicians, clergymen, and others built or purchased microscopes and turned them upon crystals, flowers, insects, fungi, or infusion-animals.[114] Baker's *Employment for the microscope* (1753) contains the first description of the phosphorescent dinoflagellate *Noctiluca*,[115] and of a shape-shifting animalcule he called the Proteus[116] after the sea-god of the same name.[117] Baker refers to all these little beings as animals, animalcules, or insects.[118]

Miniature-painter August Rösel von Rosenhof studied insects for a decade before releasing his exquisitely illustrated four-volume *Insecten-Belüstigüng*, or *Amusements with insects* (1746–1761).[119] Unlike Baker, Rösel arranged genera systematically and positioned Insects as a step in the Great Chain of Being.[120] Rösel listed four features that, in his judgement, distinguish insects from other beings: they lack bones; they are equipped with a straw [*Saugrüssel*] or spike [*Stachel*], or with a mouth that opens transversely or left-to-right, not up-and-down; their eyes (if any) are large and uncovered; and they breathe through small pores in the sides of their frame. Any being distinguished by these criteria—including crayfish, *Hydra*, spiders, and scorpions—was, for Rösel, an insect.

In the third volume of *Insecten-Belüstigüng* (1755) Rösel took a substantial digression to describe small aquatic organisms including *Hydra* (to which we return in the next chapter), bryozoans, and vorticellids. Admittedly, by his own criteria *Hydra* might not qualify as an insect; it lacks bones and its mouth opens transversely, but experts had not yet determined if it has eyes or breathes air. However, when one is cut in half the pieces regenerate, as a tree regenerates from a branch. For Rösel, all this was evidence enough to situate *Hydra* between insects and plants on the scale of life.[121] At the end of his digression Rösel thanks his readers for their indulgence, as he has "mixed these creatures in with the insects". But before returning to insects in (God willing) the fourth volume of his work, he will describe two further aquatic organisms: the so-called globe-animal,[122] and the lesser Proteus.[123] Leeuwenhoek[69] and Baker[124] had earlier seen the globe-animal, today known as the colonial green alga *Volvox*. But Rösel's Proteus was much smaller than that described by Baker (hence *der kleine*) and indeed was a new little animal, today *Amoeba proteus*. Although we find *Hydra*, *Volvox*, and *Amoeba* in *Insecten-Belüstigüng*, Rösel did not consider them insects.[125]

Two micrographia more-or-less contemporary with Rösel's *Insecten-Belüstigüng* have been claimed to treat Infusoria as a separate class of beings. Martin Ledermüller, a notary and amateur microscopist, published two well-received tracts on spermatozoa[126] before embarking upon *Mikroskopische Gemüths- und Augen-Ergötzung* (*Microscopic delight for mind and eye*, 1760–1763), a work of popularization illustrated with hand-coloured plates.[127] In it he treats the usual subjects (crystals, flowers, fibres, feathers, insects, *Hydra*) in seemingly random order, and devotes two sections to infusion-animals.[128] Ledermüller states that the "worms" in hay infusions "belong in the class of the infusion animalcules", but there is no reason to think that he intended this as a formal systematic designation; indeed, he considered them larvæ of field insects,[129] a view maintained in the French edition.[130]

Heinrich Wrisberg released *Observationvm de animalcvlis infusoriis satvra*[131] in the year (1765) he became medical director of the clinic for gynæcology and midwifery at the University of Göttingen.[132] Like Buffon and Needham, he observed infusions made from decaying animal or plant substances, typically over three or four weeks.[133] He describes the formation of small globules or bubbles, after which little corpuscles congeal to form one or another kind of infusorial animalcule, and later various sorts of polyps. On about seventy occasions through the book Wrisberg uses a variant of *animalcula infusoria* or *infusoria animalcula*,[134] but he never refers to the little animals simply as *infusoria*, much less as constituting a class Infusoria.[135]

Summary: one hundred years of little animals

This chapter has focused on the discovery and description of microscopic forms of life from 1674 (Leeuwenhoek) to 1765 (Wrisberg). During this period, microscopes evolved from simple hand-held wooden devices with a single homemade bead-lens, to showy commercial multi-lens instruments with condensers, mirrors, and accurate focusing mechanisms. Needham (and perhaps Baker) also had access to a single-lens microscope with superior illumination and optics. Further improvements were introduced through the latter part of the Eighteenth century, and the discovery of little animals continued.[136]

In parallel, the Seventeenth and Eighteenth centuries brought new ideas about the biology of reproduction and generation, particularly in animals. Indeed, the two stories are deeply intertwined: infusion-animalcules and spermatozoa were studied using the same microscopes, described in the same scientific papers, illustrated in

222 KINGDOMS, EMPIRES, AND DOMAINS

the same books, classified as sister genera (Hill), and viewed similarly (as transitory assemblages of living globules) in the Buffon-Needham hypothesis. As a potential material basis for human reproduction, spermatozoa attracted close attention from philosophers and theologians. The little animals in pond water and vegetable infusions were not without philosophical interest (Fontenelle), but were as yet little-understood. Much revolved on whether they represent a level of organization in the assembly of plant and animal bodies from living corpuscles (Buffon, Wrisberg), are developmental stages of microscopic (Adams) or macroscopic (Ledermüller) insects, or are self-contained life forms which perpetuate themselves after their own kind (Baker).

Of these, the corpuscular hypothesis presented the most-fundamental challenge to the established view of the natural world. It was not inconceivable that natural bodies are particulate in one sense or another (Leucippus, Descartes, Leibniz). Vegetative propagation in plants, and growth of the animal embryo from an imperceptibly small point, called out for a unified mechanistic explanation. Needham believed that the "Clouds of moving Atoms" he observed through his single-lens microscope were in fact the living globules required by this hypothesis. Not until 1827 would Robert Brown discover that in a fluid, suspended particles of this size move randomly due to physical processes unrelated to vitality.[137]

Other early microscopists who sought living corpuscles, globules, or atoms necessarily worked at (or more likely, beyond) the limit of optical resolution of their instruments, and typically had to contend with inferior illumination, and residual spherical and chromatic aberration.[138] The results were predictably spotty. Even positive results had to be treated with caution not only for technical reasons, but because observers sometimes see what they expect to see.[139] In the 1740s, expectations were shaped by respected authorities (Buffon), philosophers (Descartes, Leibniz), and/or institutions (the Royal Society), all favourably disposed to little particles.

We are not yet done with this story. Two more major threads, with roots in other historical and philosophical traditions, remain to be woven in: the idea of a gradient or continuum in nature; and a unitary classification embracing plants, animals, and zoophytes. We consider these in the next two chapters.

14

Continuity in the living world

> Man is not moulded from a costlier clay; nature has used but one dough, and has merely varied the leaven.
>
> —Julian Offray de La Mettrie, *Man a machine*[1]

The idea that all things in the universe are connected has featured in many cultures over at least four millennia. It is intrinsic to religious or philosophical systems in which the heavens and earth are considered to have come into being by emanation from or within a First Principle or Being. Although (as we have seen) emanationist cosmologies may differ in important ways—whether the universe is eternal or was created, God transcendent or immanent, the void existent or impossible—they all require that material objects including animals, vegetables, and minerals (or the corresponding ideas) form a progression or series such that individual types are in some sense adjacent. In some versions, souls can migrate upward and downward within this series and may eventually journey back to God, or to a heavenly or spiritual realm.

As formulated by Arthur Lovejoy, the combination of Aristotle's linear gradation of soul with Plato's assumptions of plenitude and continuity of form yielded the Great Chain of Being, an immense (in principle, infinite) "number of links ranging in hierarchical order from the meagerest kind of existents, which barely escape non-existence, through 'every possible' grade up to the *ens perfectissimum* . . . every one of them differing from that immediately above and that immediately below it by the 'least possible' degree of difference". To be sure, "from the Platonic principle of plenitude the principle of continuity could be directly deduced".[2] But in the first instance, it is examples of continuity that we observe in nature.

Connectivity or continuity in nature has been described using the metaphor of a cord, chain, ladder, or tree.[3] In principle, these images might emphasize different aspects of connectivity. A cord is of one piece from end to end, although at finer scale it consists of multiple (perhaps shorter) strands woven together. A chain is a series of discrete links that serially overlap and interlock with one another, whereas a ladder (*scala*) has separate rungs or steps, and stiles or rails that might respectively symbolize upward progress and downward descent, or emanation and return.[4]

Kingdoms, Empires, and Domains. Mark A. Ragan, Oxford University Press. © Oxford University Press 2023.
DOI: 10.1093/oso/9780197643037.003.0014

224 KINGDOMS, EMPIRES, AND DOMAINS

A tree is organic, rooted in the earth, and reaches up unidirectionally towards the heavens. Let us consider each in slightly more detail.

Amidst the tenth year of the Trojan War, Zeus assembled the immortals on highest peak of Olympus to put a stop to their meddling on behalf of one or another of the warring parties. He reminded them of his supremacy:

> Come, you gods, make this endeavour, that you all may learn this.
> Let down out of the sky a cord of gold [σειρήνχρυσείην]; lay hold of it
> all you who are gods and all who are goddesses, yet not
> even so can you drag down Zeus from the sky to the ground, not
> Zeus the high lord of counsel, though you try until you grow weary.
> Yet whenever I might strongly be minded to pull you,
> I could drag you up, earth and all and sea and all with you,
> then fetch the golden rope about the horn of Olympos
> and make it fast, so that all once more should dangle in mid air.
> So much stronger am I than the gods, and stronger than mortals.[5]

As antecedent to the Great Chain of Being, this is all very curious. This golden cord had not been in place from time immemorial; nor, it seems, did Zeus follow through on his threat to have it let down from the sky. Although in principle the cord was bi-directional, Zeus boasted that he could use it to drag the lesser immortals, the earth and sea too, up to Mount Olympus or even suspend them in mid-air, but they could not pull him down to earth. Explicitly citing this passage as a metaphor of order in the cosmos, Plato[6] and Aristotle[7] likewise use the word σειρᾱ, which to them meant *cord* or *rope* but later came to mean *line*, *thread*, or *chain* as well. The metaphor persisted through middle and Neoplatonism[8] to Macrobius:

> Accordingly, since Mind emanates from the Supreme God and Soul from Mind, and Mind, indeed, forms and suffuses all below with life, and since this is the one splendor lighting up everything and visible in all, like a countenance reflected in many mirrors arranged in a row, and since all follow on in continuous succession, degenerating step by step in their downward course, the close observer will find that from the Supreme God even to the bottommost dregs of the universe there is one tie, binding at every link and never broken. This is the golden chain of Homer [*Homeri cathena aurea*] which, he tells us, God ordered to hang down from the sky to the earth.[9]

Through Macrobius the Golden Chain passed into European thought,[10] where it eventually became encumbered with alchemical and occult symbolism. Even so, it continued to offer the image of adjacent genera including animals, vegetables, and minerals.[11]

Whereas the cord and chain are of Hellenic origin, the ladder metaphor was Hebraic,[12] or Middle Eastern more generically. According to *Genesis*, in a dream

Jacob, a grandson of Moses, beheld a ladder or staircase joining heaven and earth, on which angels were ascending and descending.[13] Jacob's ladder was appropriated by Greek and Latin Church Fathers[14] as an allegory of the journey of the soul to God, or of the challenges that a Christian must face in life.[15] In the Egyptian *Book of the dead*, the soul ascends on a ladder to the next world.[16] Origen quoted Celsus at some length on the Persian mysteries:

> These truths are obscurely represented by the teaching of the Persians and by the mystery of Mithras which is of Persian origin. For in the latter there is a symbol of the two orbits in heaven, the one being that of the fixed stars and the other that assigned to the planets, and of the soul's passage through these. The symbol is this. There is a ladder with seven gates and at its top an eighth gate. The first of the gates is of lead, the second of tin, the third of bronze, the fourth of iron, the fifth of an alloy, the sixth of silver, and the seventh of gold. They associate the first with Kronos (Saturn), taking lead to refer to the slowness of the star; the second with Aphrodite (Venus), comparing her with the brightness and softness of tin; the third with Zeus (Jupiter), as the gate that has a bronze base and which is firm; the fourth with Hermes (Mercury), for both iron and Hermes are reliable for all works and make money and are hard-working; the fifth with Ares (Mars), the gate which as a result of the mixture is uneven and varied in quality; the sixth with the Moon as the silver gate; and the seventh with the Sun as the golden gate, these metals resembling their colours.[17]

A fourth metaphor of continuity between heaven and earth was more common throughout the Middle East: that of the tree.[18] A tree reaches up from earth (indeed, from within the earth) towards heaven, typically unfolding multiple branches that offer certain metaphorical uses ("holding up the sky") but may be inappropriate for others. Unlike a cord, chain, or ladder, a tree can sprout from a seed or cutting, grow, and die. A tree can also be felled and fashioned into wooden artefacts which themselves can serve as metaphors: Enkidu and Gilgamesh cut down the tallest cedar to provide timber for the door to the Temple of Enlil,[19] and the Holy Cross became a central symbol of Christianity. In commenting on the creation of heaven and earth in *Genesis*, the book *Bahir* adds text in which God says he has planted and rooted a tree that is the origin of souls.[20] In the *Midrash Konen*, the image is inverted such that the cosmic tree extends downward, with its roots above in the celestial Garden of Eden and its tip in the terrestrial paradise; on it the souls of the righteous ascend and descend as on a ladder.[21]

We would be unwise to overinterpret differences among the cord, chain, ladder, and tree as representations of continuity in nature. Just as souls ascend and descend on the cosmic tree "as on a ladder", later authors sometimes used these metaphors interchangeably. The tree and ladder are equated in a Good Friday homily once attributed to Saint John Chrysostom.[22] In the late Fifteenth and Sixteenth centuries, as Pico della Mirandola and others were drawing Kabbalah into the

226 KINGDOMS, EMPIRES, AND DOMAINS

supposed common Hermetic-Neoplatonic tradition, Johann Reuchlin claimed that the Homeric golden chain and Jacob's ladder were equivalent.[23] According to Giambattista della Porta,

> Seeing then that the Spirit cometh from God, and from the Spirit cometh the soul, and the soul doth animate and quicken all other things in their order, that Plants and bruit beasts to agree in vegetation or growing, bruit beasts with Man in sense, and Man with the Divine creatures in understanding, so that the superior power cometh down even from the very first cause to these inferiours, deriving her force into them, like as it were a cord platted together, and stretched along from heaven to earth, in such sort as if either end of this cord be touched, it will wag the whole; therefore we may rightly call this knitting together of things, a chain, or link and rings, for it agrees fitly with the rings of *Plato*, and with *Homers* golden chain, which he being the first author of all divine inventions, hath signified to the wise under the shadow of a fable, wherein he feigneth, that all the gods and goddesses have made a golden chain, which they hanged above in heaven, and it reacheth down to the very earth.[24]

Later, John Milton employed the chain in Book 2 of *Paradise lost* (1668) as metaphor for the linkage between heaven and earth, but Jacob's stairway in Book 3.[25]

From Boëthius onward, nearly everyone who took a philosophical or scientific approach to the natural world acknowledged animals, vegetables, and minerals as the main (or only) genera of earthly things, and (if the question arose) ranked these genera as had Aristotle, with animals closest to man and minerals the most distant. As we have seen over recent chapters, this was true for Eriugena, Marius, Bernard Silvestris, Grosseteste, Albert, Aquinas, and Bonaventure; Ficino, both Picos, Agrippa, Paracelsus, Bruno, and Kirchweger; Cardano, Scaliger, Liceti, and Paré; Budé, Bodin, Zabarella, Nieremberg, and Freig; Cesalpino, Wotton, Belon, Rondelet, Gessner, Aldrovandi, Jonston, Imperato, and Libavius; Leibniz, Hill, Rösel, Needham, Buffon, and Spallanzani. To these we could add Jean de Meun,[26] Llull,[27] Sebonde,[28] Fortescue,[29] de Bouelles,[30] Francis Bacon,[31] Mersenne,[32] Bourguet,[33] and any number of others. Most accepted Aristotelian continuity and (at least in principle) a Platonic plenitude, if only in the sense that God could create whatever He wished. Whether animal, plant, or zoophyte, sponges and the like were the poster children of continuity, and Nieremberg explicitly assigned them to the point in the golden chain where the vegetable and animal links clasp one another.[34] But while these organisms exemplified plenitude, not even Cardano or Libavius invoked plenitude to argue that zoophytes must exist.

The Great Chain under attack

As we have seen, by 1440 Nicholas of Cusa was speculating that the sun, moon, and stars might be inhabited. A century later Copernicus published his astronomical model in which the earth was no longer the centre of the cosmos.[35] Indeed, there may exist multiple worlds, multiple humanities. In *De l'infinito universo et mondi* (1584) Giordano Bruno has Burchio pose the question of where we should now find "that beautiful order, that lovely scale of nature"; Fracastoro answers "In the realm of dreams, fantasies, chimeras, delusions".[36] Scholarly opinion differs on whether these developments fatally compromised the Great Chain of Being,[37] but at minimum they fired a first shot across its bow.

A new crisis befell the Chain in the 1640s, as Descartes resolved reality into God, mind, and matter. Organized bodies were now qualitatively different from brute matter, as was man from beast, and God from man. Animals became automata, sensate and impulsive but without mind or soul.[38] To be sure, Descartes was in no hurry to dismember the earthly portion of the Great Chain: recall (Chapter 12) that he considered nature "one, simple, continuous, everywhere self-coherent and respondent by the fewest constant principles and elements, from which [arise] the near-infinity of things, but in three kingdoms, Mineral, Vegetable and Animal, certainly among themselves [extended by God] in order and distinct grades" just as God himself is unitary, simple, continuous, and uninterrupted.[39] But if every natural thing is governed by the same few principles, surely all animals share analogous structures for taking in and digesting food, circulating the blood, and reproducing in kind. Little scope remained for the intermediate states that must occur in zoophytes.[40]

Over time the Cartesian system was softened; animals regained a lesser sort of soul[41] and, as we have seen, by folding reality into monads Leibniz returned life to the cosmos. In the decades before Needham, Christian Wolff in Germany sought to reconcile Cartesian matter with the *Monadology*, while in France Pierre-Louis Maupertuis and Voltaire popularized the ideas of Newton against some resistance.[42] We have already considered how the discovery of the microscopical world affected ideas of continuity and plenitude. In this chapter we consider two further challenges to the Great Chain: the nature of corals, and of the freshwater polyp *Hydra*.

Richard Bradley: *A philosophical account*

Copernicus and Descartes had undermined the Great Chain, but as the Eighteenth century dawned no one, it seems, took much notice. According to Lovejoy, "there has been no period in which writers of all sorts—men of science and philosophers, poets and popular essayists, deists and orthodox divines—talked so much about the Chain of Being, or accepted more implicitly the general scheme of ideas connected with it."[43] Even so, there was no *Book of the Great Chain*.[44] This lacuna was noticed by Joseph Addison, an essayist and poet who himself popularized the Great Chain in his newspaper *The Spectator*.[45] In the 19 July 1711 issue, Addison called on "our Royal Society [to] compile a body of Natural History, the best that could be gathered together from Books and Observations" although it would necessarily be

228 KINGDOMS, EMPIRES, AND DOMAINS

incomplete, as many beings are hidden by deserts and oceans; and "Besides that there are infinitely more Species of Creatures which are not to be seen without, nor indeed with the help of the finest Glasses."[46] Inspired by Addison (who died in 1719), Richard Bradley wrote *A philosophical account of the work of nature. Endeavouring to set forth the several Gradations Remarkable in the Mineral, Vegetable, and Animal parts of the Creation. Tending to the Composition of a Scale of Life* (1721).

Bradley was not the obvious person to rise to the call. A horticulturalist without a university education, he had become known through his books on husbandry, gardening, and the propagation of succulent plants. On that slender foundation he was elected to fellowship in the Royal Society, and four years later was appointed the first professor of botany at the University of Cambridge, where his perceived unsuitability—he lacked Greek, Latin, and the promised wherewithal to endow a botanic garden—excited scandal.[47] These circumstances make his Great Chain all the more interesting, as he did not simply draw on the usual authorities. Bradley begins with earths and minerals, which grow (albeit very slowly) in their strata; thus

> we need not scruple to allow them *Life* too, however slow it be; these indeed have no *Local Motion* no more than Plants; but Animals that have *Local Motion* are yet analogous to Plants in *Generation* and *Circulation* of Juices thro' their Bodies, and have *Sensation* more than Plants: Plants then want *Local Motion* and *Sensation* to be equal to Animals; but I suppose have only the Powers of *Visible Growth* more than Minerals, and of being transplanted from place to place, and yet retain the Power of *Growth*; but where must we transplant the Earth to make it grow, or improve it?[48]

In the second chapter Bradley treats "the *Coralline, Truffle, Fungus, Sponge,* and such Bodies which possess the first Degree of Vegetative Life, and are seemingly the Passage between Minerals and perfect Plants". No vegetable grows as slowly as the minerals, unless we include coral among the vegetables, as Bradley was disposed to do on account of its branched figure, diverse species, and analogies to his beloved succulents.[49] The sponge is "a Subject leading to Vegetation, and is what I believe is allow'd by all to be a Plant, tho' it is indeed seemingly imperfect . . . but its Vessels are so nicely woven into one another, that every part is equally supply'd with Juices." Sponges can be driven about by the sea, as are some aquatic plants. Next "but little more perfect, seemingly, either in Figure or Parts" are the truffle, puff-balls, and fungi of increasing structural complexity.[50]

From fungi, Bradley proceeds to plants that (in his opinion) lack leaves or flowers, or bear inconspicuous leaves. He abandons growth rate as a measure of rank, as some imperfect plants (and animals) grow very quickly indeed.[51] He inserts an essay by Antoine de Jussieu on analogies between plants and animals,[52] then returns to gradations in nature with "those *Plants* that are very visibly endued with all the Parts required in *Vegetables,* viz. *Roots, Trunks, Bark, Pith, Branches, Leaves, Flowers* and *Fruits*; and they are of three Kinds, *Herbs, Shrubs,* and *Trees.*"[53]

CONTINUITY IN THE LIVING WORLD   229

Of these, the tree is "most lofty in its Growth, and has its Parts more robust, firm and lasting". Various "Vegetable Bodies" depend on trees: stamens ("for the most part like so many *Fungi* taking Root in the Foot-stalk of the *Flower*"), petals, mistletoe ("at their Extremities like the Mouths of *Leeches*"), mosses, leaves, twigs and branches ("really so many *Plants* growing upon one another . . . *Buds* explained").[54]

Next come "*Immoveable Shell-Fish*" including the *Oyster*, *Muscle*, *Cockle*, *Barnicle*, *Scallop*, and *Pectunculæ*, the shells of which "have a kind of Vegetative Growth". These are fixed in place like plants, but "have such a Share of Animal Life as to afford them the Power of Sensation".[55] Thereafter come "such *Shell-Fish* as have *Local Motion*, such as *Lobsters*, *Crabs*, *Star-Fish*, &c.", then "such other *Shell-Fish* as move from place to place, by means of an undulating Motion of their fleshy Parts out of the Shell, after the manner of *Snails*", and "such *Fish* as are Inhabitants of the Salt and Fresh Waters, that are framed for swimming only" including cetaceans.[56]

From these Bradley moves to "*Serpents*, the *Crocodile*, *Lizard*, *Camelion*, and others of the Scaley Tribe, which are Amphibious, and Inhabitants of the Land; [and] *Flying Lizards*, &c. which seem to be the immediate Passage between the Fish and Bird kind." Serpents and snakes resemble fish in their scales, eel-like motion and "Degree of Life".[57] Thence it is but a short step to *Birds* and *Fowls*, the *Batt*, *Flying Squirrels*, and the like, which seem to be the Passage between Fowls and Four-footed Beasts. Birds are oviparous like serpents and lizards, have "four Branches to their Bodies" like lizards, and of course fly. Hibernation is a further analogy.[58] Bats and the flying squirrel, on the other hand, are viviparous and suckle their young "like Quadrupeds".[59]

As for the quadrupeds, Bradley ranges without apparent plan over their types, size, motion, "Strength and Spirit", hair, horns, diet, Use and Service to Mankind, and voice.[60] He proceeds from horses to oxen, sheep, dogs, apes, and monkeys,[61] then enters into a rambling discussion of reproduction and generation, which in some quadrupeds "is nearly agreeing with that of Mankind".[62] He pursues all manner of analogies between insect- or animal-eggs and the grains or seeds of plants, finding uniformity "relating to the Generation of all living Bodies".[63] Dismissing the distinction between "*Perfect* and *Imperfect Animals*" as an "antient Error",[64] Bradley finds great "Art and Contrivance" in the structure of insects, analogizes the metamorphoses of insects and frogs ("a *Frog* is a *Fish* in its Beginning"), and praises the "extraordinary Perfection, and even more than Man himself can boast of, that Gift of Power in tasting Life successively in different States and in d[i]fferent Elements".[65] He concludes his account of generation by observing that

> the *Egg* of the *Female*, before Impregnation, seems to possess a Degree of Growth or Life, somewhat like that in *Minerals*; when the same *Egg* is impregnated, it then possesses a kind of *Vegetative* Growth; and takes upon it the *Animal* Life and Growth as soon as it quickens, at which time it only begins to enjoy the Power of Sensation.[66]

230   KINGDOMS, EMPIRES, AND DOMAINS

After a paragraph on amphibious quadrupeds (the *Hypopotamus*, *Otter*, *Beaver*, and *Seal*) Bradley is "led naturally to treat of such Creatures with four *Legs* as are partly *Animal*, partly *Insectal*, such as *Frogs* and *Toads*". But before leaving the quadrupeds:

> I suppose it may be wonder'd at, that hitherto I have not mention'd Mankind, who is so remarkable a Creature, and Lord of all the rest; I confess, was I to have placed him where the Parts of his Body would most agree with those of the created Bodies mention'd in this Treatise, I must have set him in the middle of this Chapter; but I suppose my Reader will excuse me, if I shew him so much regard, that I rather speak of him in the summing up of my Scale, than let him be encompass'd with wild Beasts.[67]

Like quadrupeds, frogs have four legs and a fleshy body, and hibernate in winter; but their metamorphosis places them closer to insects, which come next in his scale of nature: first "irregular insects" (snails, earthworms, centipedes, millipedes, spiders, and insects without wings), then butterflies, bees, wasps, beetles, locusts, mayflies, flies, and the "walking leaf".[68] After these come tiny insects: cheese-mites, vinegar-eels, and the microscopic animalcules of paste and pepper-water. He supposes that infusion-animals hatch from eggs deposited by equally microscopic *Mother Insects* in the air, and concludes with "*Animalcula in Semine Masculino*" without connecting these to infusion-animals.[69]

Like numbers in a series, Bradley concludes, all created bodies are interdependent: "if any one was wanting, all the rest must consequently be out of Order". Indeed, "the Laws or Rules of Mathematicks, as they now are, could not be just, if Nature's Laws were different from what we now observe them to be".[70] And like numbers or musical tones, nature is infinitely divisible, whether in the vegetation of plants or the life and growth of animals: thus his "*Scale of Life*, or a Chain of created Beings".[71] In truth, however, Bradley offers no such continuous scale. His is a cord woven of local affinities, few of which reach beyond the immediate neighbourhood, and none of which resonates from one end of the Scale to the other. We find in *A philosophical account* no single continuum of form or essence, no superadditive soul, no Pythagorean harmony. Some important beings (*Mimosa*, jellyfish, cephalopods, *Hydra*) are inexplicably absent, and Bradley lapses far too often into disquisitions on mushroom propagation, the fecundity of elm trees, or practical tips on fishponds and vineyards.

Even so, let us not conclude in haste that Bradley made a poor fist of the hand he was dealt. His Scale encompasses the three Kingdoms at some granularity. Internal structures (or their absence) help him position corals, fungi, and sponges between minerals and plants, while he arranges molluscs and crustaceans stepwise between plants and animals based on motility. And given that insects are so intricately structured, and their lives span multiple stages and environments, why should they not be considered more perfect or praiseworthy than quadrupeds? To be sure, this—or more precisely, the concomitant relegation of Mankind to a position amidst the beasts—became a lightning-rod for criticism.[72] But man had long been a middle link in the Chain, albeit more accustomed to looking upward to the celestial hierarchy than to frogs and insects. The Church, and Enlightenment philosophy, bade man to humility.[73] Copernicus and Descartes had removed man from the centre of the universe: and someone had, after all, taken notice.

Corals: an ancient enigma resolved

Mankind has long been interested in coral as ornament or medicine, but for many centuries there was little consensus on its precise nature. Most early authors followed Theophrastus in believing corals to be stones, or a plant that can turn to stone; others thought them intermediate between the two, while al-Bīrūnī alone considered them animals.[74] By the Renaissance, opinion had largely swung towards the opinion that corals are plants, and monographs by Joseph Pitton de Tournefort in 1694 and 1700 helped ensure that this remained the default view in France and beyond.[75] Another possibility was that a vegetable "bark" (*écorce*) surrounds and creates a stony core.[76]

By the Seventeenth century the coral trade was flourishing, from Marseilles in the west through the Middle East to India and beyond. As described by Luigi Ferdinando Marsigli, large pieces might be fashioned into handles for swords, knives, or canes, or into statuary, while smaller pieces were used to make jewellery, rosary beads, or buttons.[77] For medicinal purposes corals were powdered and mixed into concoctions, or used as adsorbents.[78] Alchemists had long been interested in corals as well. In the 1660s Robert Boyle collected coral at Marseilles for experiments on the liberation of gas (later identified as carbon dioxide) under the action of an acid.[79] In 1706 Count Marsigli, whose military career had come to an unhappy end, set up a laboratory for marine studies at Cassis near Montpellier[80] and soon thereafter went to sea with the *corailleurs*, to determine with his own eyes whether coral be mineral (as he believed) or vegetable.[81] Chemical studies he and others carried out for the Société royale de Montpellier pointed unexpectedly to an animal nature, but were deemed inconclusive, and Marsigli persisted in believing coral to be mineral.[82]

Marsigli returned to sea with the *corailleurs* in December 1707, and this time brought branches of coral back to shore in glass vessels filled with seawater; the next day he was amazed to find the branches covered with white flowers. Resolved to examine the flowers under his microscope, he removed a branch from the water, only to see the flowers immediately retract. He then re-immersed the branch in the seawater, and within ninety minutes the flowers again emerged.[83] In a different telling of the event, François Xavier Bon[84] observed movement in the incipient flowers, and concluded that they were parasitic insects, analogous to those that feed on the sap of terrestrial plants; but Marsigli would hear nothing of it, and Bon's observations went unpublished.[85]

232 KINGDOMS, EMPIRES, AND DOMAINS

While in Marseille, Marsigli and Bon resided with Charles Peyssonel, head physician at the Hôtel-Dieu, and Marsigli showed the coral-flowers to Peyssonel's son Jean-André. In 1723 Jean-André Peyssonel—now a physician and naturalist in his own right—went to sea with the *corailleurs*, confirmed Marsigli's observations, and sent a report on coral-flowers to the Abbé Jean-Paul Bignon, president of the Académie des Sciences. But in 1725 the younger Peyssonel discovered that the coral-flowers retract if touched or exposed to acid. He concluded that the flower is *une petite ortie ou poulpe*, and the coral-body a hive or nest built up from their secretions.[86] The obligatory letter to Bignon was passed to the leading entomologist of the day, René-Antoine de Réaumur, who arranged for it to be read at the Académie in June 1726 without revealing the author's identity. Such a misguided report could not be published, but Réaumur replied graciously in print, referring to the author as *le Phisicien* and offering an alternative explanation surprisingly like that of Bon: the supposed coral-flowers are metamorphoses of actual insects that infest the vegetable bark.[87]

Soon thereafter Peyssonel was posted to Guadeloupe, where he quietly continued his investigations. Finally in 1751 he dispatched a 400-page treatise to the Royal Society in which he referred to the coral-insect as "a little *urtica, purpura,* or *polype*".[88] Summarizing the treatise for *Philosophical Transactions*, William Watson wrote that although "some will still consider these marine productions as plants, they are truly zoophytes, formed by the labour of the animals, which inhabit them, and to which they are the stay and support".[89] In the meantime, the French scientific establishment had come to much the same conclusion. Elegant microanatomical studies on the colonial bryozoan *Lophopus* by Abraham Trembley,[90] and by Bernard de Jussieu and Jean-Étienne Guettard on other species likewise thought to be marine plants,[91] had convinced Réaumur that Peyssonel had been right in 1726. In a lengthy preface to the final volume of *Memoires pour server a l'histoire des insectes* (1742), Réaumur set out the new consensus: coral-polyps are animals, and are to the coral-body what wasps are to a wasp-nest (*guêpier*). He named Peyssonel as the once-anonymous Phisicien, and predicted that many more such *polypiers* would be discovered.[92]

Vitaliano Donati explicitly embraced the new consensus,[93] but stopped short of admitting corals into the animals.[94] He championed the idea of a Great Chain, and speculated that mosses and fungi link terrestrial plants with the insects.[95] As for marine beings, he recognized three levels between plants and animals: *Polipari, Piante-animale,* and *Animale-piante.* Polyparies (corals, madrepores, and millepores) resemble plants in overall aspect, while their insect-like polyps serve for reproduction. In *Della storia* at least, the other two stages are represented by sponges.[96] Donati's Chain was unusual in other ways too:

When I observe the productions of Nature, I do not see one single and simple progression, or chain of beings, but rather I find a great number of uniform, perpetual and constant progressions.

And again:

> In each of these orders, or Classes, nature forms its series and presents its almost imperceptible passages from link to link in its chains. In addition, the links of the chain are joined in such a way with the links of another chain, that the natural progressions should have to be compared more to a net [*rete*] than to a chain, that net being, so to speak, woven of various threads which show, between them, changing communications, connections, and unions.[97]

Donati singled out two of these "beautiful" links: the earthworm, and Trembley's *Hydra*. In terrestrial plants we find uniform progressions among the different sorts of flowers, the parts of these flowers, among seeds, fruits, and parts used in propagation. Nature moreover extends the series of plants and animals into the sea, where perhaps the passages between links proceed with particular ease.[98] Does Donati envision (at least) two terrestrial chains, one of animals and another of plants, each extending into the sea—or separate terrestrial and aquatic chains? His comment about mosses and fungi implies the latter. In either case their links are further woven into a net—the first since Nieremberg.[99]

It is tempting to conclude that by the mid-1740s our protagonists had resolved the enigma of coral. Their conclusion—that animate polyps form corals—has stood the test of time, even if the analogy between *polypier* and *guêpier* proved overly facile.[100] Yet a surprising number of their contemporaries continued to view corals as zoophytes, plants, or even stones.[101] It did not help that beings such as *Corallina*, which were eventually recognized as algae, were too often lumped together with true corals, madrepores, and millepores. In the next chapter we investigate how corals, coralline algae, sponges, and similar beings were assimilated into the all-inclusive and hugely influential taxonomic systems of the later Eighteenth century, notably that of Linnæus. But first we must consider the most troublesome being of all: the so-called freshwater polyp *Hydra*.

Hydra: a new enigma

Hydra—the fearsome multi-headed serpent of Greek mythology—lived in a lake near Lerna, on the east coast of the Peloponnesus. The terrorized locals had been unable to kill it because when one of its heads was severed, several would grow in its place. Hydra was finally slain by Heracles and his nephew Iolaus, although details differ among versions of the myth. A much smaller multi-armed aquatic creature with unexpected powers of regeneration would be at the centre of perhaps the most-remarkable episode in Eighteenth-century natural history.

The small green being with multiple arms was first described by Leeuwenhoek in a letter written on Christmas Day 1702 and read to the Royal Society in early 1703.[102] He called the otherwise unnamed organism a *Dierke* (Animalculum),

234 KINGDOMS, EMPIRES, AND DOMAINS

described its "tentacles" (which could be extended and retracted) and "abdomen", and observed its generation from a "mother". Soon thereafter an anonymous "Gentleman in the Country" reported independent observations of "so odd an Animal" with "arms" that issue from around "a small knob, which I take to be the Head".[103] Neither Leeuwenhoek nor his English counterpart revealed the slightest doubt that it was an animal.

Over the next few decades, the little animals (or their close relatives) were occasionally noted by amateur microscopists and scholars including Bernard de Jussieu, none of whom is on record as having found it particularly remarkable. This was soon to change. In 1739 a young Genevan, Abraham Trembley, arrived at Sorgvliet, outside The Hague, to begin duties as tutor to the sons of Count Willem Bentinck.[104] Trembley's *Mémoires, pour servir à histoire d'un genre de polypes d'eau douce* (1744) record the evolution of his thinking about the organism, which he first observed in June 1740 and on the suggestion of Réaumur called *the freshwater polyp*.[105] At first unaware of the two earlier reports, Trembley wrote that a casual observer could "scarcely avoid taking the freshwater polyp for a plant".[106] But the contractility of its arms, and its ability to move from place to place like an inchworm, implied that it was an animal.[107] It was also attracted "like a moth" to light.[108] Trembley noticed that some polyps have six arms, others eight. Animals have a fixed number of arms or legs, whereas plants of a given genus can exhibit different numbers of branches or roots.[109] Although by that point Trembley was "much more inclined to think of them as animals", he decided to test whether, like a plant, the polyp could be propagated by cuttings. He sectioned a polyp transversely, and was amazed to find that each part regenerated the missing portion.[110] He also observed a polyp "that was beginning to produce a little one ... very closely akin to the way plants multiply when they give off *shoots*".[111] Trembley saw polyps swallow, digest, and be nourished by worms—observations that "certainly provided convincing evidence that they were animals".[112] He went on to describe their arms, legs, mouth, stomach, and skin,[113] but found no structures corresponding to eyes.[114] By now aware of the earlier reports, Trembley carefully observed the development and parturition of polyps, interpreting the process as vegetative, like a plant giving off shoots.[115]

News of the discovery spread quickly, well before Trembley's *Mémoires* appeared in print. Trembley notified Réaumur by letter in December 1740, and subsequently sent live specimens. Réaumur—sure that it was an animal—confirmed Trembley's results, read letters from Trembley to the Académie, and even brought living polyps for by inspection by his fellow *savants*, presumably including Buffon. Buffon then informed the Royal Society.[116] Réaumur also described the polyp in the aforementioned preface to the sixth volume of his *Memoires*.[117] The news elicited the predictable spectrum of reaction, from enthusiasm to disbelief and even ridicule.

Why did the freshwater polyp arouse such passions? One reason lay in the nature of its challenge to the Great Chain of Being. Unlike a sponge or Needham's water moulds, the freshwater polyp displayed full sensory powers: it could detect and move towards light; capture, internalize, and digest food; and eliminate waste. Yet it could regenerate missing parts, and—in Trembley's interpretation—propagate by shoots. Réaumur and his colleagues soon showed that many simple animals—earthworms, millipedes, starfish, certain stages of insects—could regenerate severed parts.[118] Propagation by shoots was another matter, but for the freshwater polyp this depended on Trembley's interpretation.

At the end of *Mémoires*, Trembley reflected on whether the freshwater polyp is best considered an animal, plant, or zoophyte. We must bear in mind, he wrote, that many plants and animals remain unknown to us, and doubtlessly possess traits we do not currently associate with the respective group.[119] Further, the traits we consider to characterize all plants or animals have been extrapolated from just a few genera—oak trees, perhaps, or tigers; but "the more one approaches the general idea of an animal and that of a plant, the less one will find differences between them."[120] Trembley pointed to Herman Boerhaave, who after great effort could identify only a single "general and essential difference": that plants draw in nourishment through external roots, whereas animals are nourished via internal vessels.[121] Nor (Trembley continues) did Boerhaave assert that this distinction applies without exception. That was just as well: polyps turned inside-out fare perfectly well.[122] The best conclusion Trembley could salvage was that freshwater polyps—and for that matter, all beings known as zoophytes—should be considered "animals which show more noteworthy similarities to plants than do other animals".[123]

Savants of all stripes took note. For Buffon, the ability of the freshwater hydra to regenerate completely from any portion—basal, central or apical—supported his theory that organisms are assemblies of tiny living particles.[124] Moreover, as *la marche de la nature* is uniform and imperceptible,

> there are beings which are neither animals, nor plants, nor minerals, and which one would try in vain to relate to one or the other . . . [like Trembley's polyp,] the moving bodies which are found in seminal liquors, in the infused flesh of animals and in the seeds and other infused parts of plants, are of this species; we cannot say that they are animals, we cannot say that they are plants, and surely we will say even less that they are minerals.
>
> We can therefore assure, without fear of advancing too far, that the great division of nature's production into animals, plants and minerals does not contain all material beings; there are, as we have just seen, organised bodies which are not included in this division. We have said that the progress of nature is made by nuanced and often imperceptible steps; and that it passes, by insensible nuances, from animal to vegetable; but from the vegetable to the mineral the passage is abrupt, and this law of going only by nuanced degrees seems to be contravened. This made me suspect that by examining nature closely, one would come to discover intermediate beings, organised bodies which, without having, for example, the power to reproduce like animals and plants, would nevertheless have a species of life and movement; other beings which, without being animals or plants, could well enter

236 KINGDOMS, EMPIRES, AND DOMAINS

into the constitution of one or the other; and finally of other beings which would be only the first assembly of organic molecules of which I have spoken in the previous chapters.[125]

Not everyone accepted that the polyp demonstrated continuity between plants and animals. Rousseau wrote that the nature of the polyp's generation was a major scientific and philosophical problem.[126] Voltaire considered it a plant,[127] denied the existence of a Chain,[128] and in *Micromégas* had wicked fun with concepts of size and scale (Maupertuis and his fellow-philosophers as microscopic insects or thinking atoms), soul, reason, and mankind as the crown of creation.[129] For those convinced by Trembley, however, the polyp was evidence for the Chain, and for fundamental analogies between animals (including man) and plants.[130] Buffon and Bonnet saw the polyp as the very point of passage from plant to animal.[131] Julien La Mettrie placed it among the animal-plants—beings that "begin to show life"—and predicted that further animal-like properties will be discovered among them.[132] Jean-Baptiste Robinet took a more-extreme position, arguing that all beings are variants on a single idea or prototype: as no quality can be essential to any group, there can be no intermediate types. Everything—heavenly bodies, man, animals, plants, stones, fire, water, air, earth, brute matter—is organic, living, and ensouled. *Tout est animé: tout est animal.*[133] Denis Diderot speculated about "human polyps" on Jupiter and Saturn that might one day give rise to beings more advanced than ourselves.[134]

Other, more-dangerous implications lay in wait. If soul inheres in matter, by chopping up a polyp do we subdivide a soul, or perhaps create new souls? Does its original soul remain intact in one segment, while fresh souls rush into the others? Might this redistribution of souls—brought about not in the natural course of generation, but on the whim of man—call down the hand of God?[135] And is it not but a small step from the inherence of soul in matter to full-blown materialism, or even atheism?[136] Bonnet intended his theory of palingenesis as an alternative to atheistic materialism: all living things pre-exist as an infinitely recursive series of germs, each a miniature of the adult. These are embodied in the ovum, and what we see as (re)generation is simply a mechanical unfolding into predetermined form.[137] Trembley's polyp can regenerate as it does because these eggs are "scattered throughout its body".[138]

Charles Bonnet: the canonical Great Chain of Being

Earlier in this chapter we considered Bradley's Scale as a special case of natural-historical thought *circa* 1721. We now revisit the Great Chain at the apex of its influence, as set out by Charles Bonnet in *Traité d'insectologie* (written 1743, printed 1744, published 1745) and, with a few modifications, further in *Contemplation de la nature* (most of which was written 1747–1752; published 1764). Bonnet then added voluminous footnotes (themselves footnoted) to *Contemplation* during 1779–1782 for its inclusion in the eighteen-volume *Œuvres d'histoire naturelle et de philosophie de Charles Bonnet* (1779–1783).[139] Like Bradley, Bonnet preferred the metaphor of a scale (ladder). The version he presents in these works may be considered its canonical late-Eighteenth-century form.[140]

From the age of seventeen until his death, Bonnet was intimately involved with the discoveries and controversies discussed in this and the previous chapter. He was a relative of Trembley,[141] an ally of Needham,[142] and a corresponding member of the Académie des Sciences (1740) and the Royal Society (1743). He had become interested in insects upon reading Réaumur's *Memoires*,[143] and in 1737 initiated a correspondence with Réaumur that lasted until the latter's death in 1757.[144] Bonnet discovered parthenogenesis in aphids[145] and, following Trembley's observations on the freshwater polyp, demonstrated regeneration and grafting in earthworms.[146] He advised Spallanzani in his experimental studies against the Buffon-Needham "living globule" hypothesis;[147] Spallanzani later reciprocated by translating *Contemplation* into Italian.[148]

Unlike Bonnet's later, more-philosophical works, *Traité d'insectologie* describes his experiments on aphids and other insects. In the *Préface* to *Insectologie* he set his work in the context of earlier work by Redi, Malpighi, Swammerdam, Leeuwenhoek, Vallisnieri, and Réaumur on insects, and by Trembley and Lyonnet on the freshwater polyp. He then turned to the grandest context of all:

> Indeed, if we go through the main productions of Nature, we will easily believe that between those of different classes, and even between those of different genera, there are some that seem to hold the middle and, in this way, form points of passage or *liaison*. This is especially evident in the Polyps. The admirable properties which they have in common with the Plants, namely multiplication *by cuttings* and *by shoots*, indicate sufficiently that they are the link that unites the vegetable kingdom with the animal. This reflection gave me the perhaps rash thought to draw up a Scale of natural Beings [*Échelle des Êtres naturels*], which can be found at the end of this Préface. I produce it only as a trial, but enough of one to make us comprehend the greatest ideas of the system of the world, and of the Infinite Wisdom that has formed it and combined its different parts. Let us pay attention to this beautiful spectacle. We see this innumerable multitude of organised and unorganised bodies placed one above the other according to the degree of perfection or excellence that is in each. If this order does not appear everywhere equally continuous, it is because our understanding is still very limited; the more scales or degrees we discover, the more it will be augmented. When nothing more remains to be discovered, our understanding will have attained its greatest perfection. But can we hope for this here on earth? Apparently, only the celestial Intelligences can enjoy this advantage . . . . And if, as I think, all these Scales, of which the number is nearly infinite, [together] form a single one which brings together all possible

238 KINGDOMS, EMPIRES, AND DOMAINS

orders of perfection, it must be admitted that one could not conceive of anything more grand or elevated.[149]

In *Contemplation*, Bonnet gives a sense of the immensity of the full Scale by supposing that each step corresponds to a known species.[150] Yet "this order does not appear everywhere equally continuous", as we observe countless local continuities. A single Great Chain spans Creation, but we mortals may never glimpse it in full. This notion is strikingly reminiscent of Nieremberg, Donati, and Bradley (above).

As promised, Bonnet's Scale spans Nature, from the "most subtle matter" to man. The order he presents in *Contemplation* differs somewhat from that in the figure in *Insectologie*, particularly amongst the unorganized bodies: the elements, earths, sulphurs, metals, salts, crystals, stones, precious stones, and amianthus. In *Insectologie* the next steps upward are lithophytes; corals and coralloides; truffles; mushrooms and agarics; mosses; lichens; then plants. In *Contemplation* he passes quickly over these beings, adds a few others, and refers the reader to Donati's *Della storia* (above), commenting favourably on Donati's network metaphor.[151] Bonnet divides organized bodies into two general classes, Vegetables and Animals, although

> One cannot discern precisely where the Vegetable ends, and where the Animal begins. This is a consequence of the gradation that the Author of Nature has observed in his Works. Neither the greater or lesser simplicity of organisation; nor the manner of their birth, nutrition, growth and multiplication; nor the locomotive faculty furnishes sufficient characters to distinguish between these two orders of Beings.[152]

In fact (he continues), there are some animals whose entire structure is as simple as that of plants. Seed and germ are analogous to egg and embryo; both internalize matter; some plants are sexual, while many animals multiply by cuttings and shoots; some animals pass their entire lives fixed in place. Bonnet explores these and other analogies between plants and animals at length.[153] If one character is unique to animals it is the presence of nerves, but even here there are exceptions: "sensitive" plants[154] and polyps. The latter are true animals but seem not to have nerves per se, although Bonnet is confident that they have analogous organs that enable sensation.[155] From polyps or mussels there is but a *bien petite* distance to plants.[156] As for zoophytes or animal-plants,

> I beg your pardon for this barbaric expression, which is not even philosophical. I wish to render with a single word those properties so remarkable, common among diverse Insects, and which seem to bring them closer to Plants. Animals which multiply like them, by cuttings and shoots, animals which can be grafted, are true *Zoophytes* or *Animal-plants*. I know well that they are at the bottom [*au*

fond] of the pure Animals, but they have more affinity with plants than with those Animals more generally known; and it this sort of affinity that the word *Zoophytes* must awaken in the mind.[157]

And in *Considérations sur les corps organisés*, referring to Needham's water moulds:

> Those filaments that Mr Needham transforms *into perfect Zoophytes*, are they really such? Or rather, do we have proof that real *Zoophytes* exist; I mean, beings that are at the same time, and in the proper sense of those two terms, Vegetables and Animals? To adjudicate that question, one would have to know the *character* that differentiates the Animal from the Plant; and those who have meditated most on that subject, confess in good faith their ignorance. When one abstracts from the Animal everything that it has in *common* with the Plant, one is surprised to see that no character remains that could be considered as *distinctive* .... An *animal-plant* would be, properly speaking, neither Animal, nor Plant; it would form a class apart, a new shading [*nuance*], a new rung [*échellon*] in the scale of Nature.[158]

Even so, Bonnet accepts the term as a matter of convenience, including among zoophytes not only Trembley's *Hydra* but also various worms, millipedes, and microscopic beings he calls cluster-, bell-, bulb-, funnel-, and net-polyps.[159] At least some of the latter microscopic polyps are presumably organisms today known as protozoa, including the ciliates *Vorticella* and *Stentor*.[160] Bonnet's zoophytes exclude corals (above) and sponges.[161] Bonnet then proceeds to insects, other animals, man, and (in what seems a throwback to earlier days) the Celestial Hierarchies: Angels, Archangels, Seraphim, Cherubim, Thrones, Virtues, Principalities, Dominions, Powers, in whose centre shines the Sun of Justice.[162] *Contemplation* offers vignettes of how animals resemble each other in structure, nutrition, sensation, locomotion, social relations, language, and the like—but unlike in *Insectologie*, we find no unitary progression. Bonnet instead muses that the Scale of Nature may ramify: insects and shellfish as "two lateral and parallel branches of the great Trunk", the frog and lizard an offshoot from insects, the crayfish and crab a branch off shellfish.[163]

Let us summarize with a few thoughts on Bonnet's scale of nature. First, we observe it only in portions. If our understanding were complete, these portions would likely join together in a Scale of surpassing grandeur; but for now, the step from unorganized to living beings looms large, while some other regions seem to run in parallel, or even to branch. Second, animals and plants grade thoroughly into each other. No character distinguishes them, no zoophytic borderland lies between. Bonnet's later *Tableau des considérations sur les corps organisés* downplays Vegetables, Animals, and Man, instead contrasting living beings with the non-living.[164] Third, whatever discontinuities beset our earthly Scale, innumerable other worlds exist: some with imperfect beings, others with beings far superior to those here on earth. On these superior worlds, stones may be organized, plants sentient,

240   KINGDOMS, EMPIRES, AND DOMAINS

animals able to reason, humans angelic.[165] Finally, Bonnet's Scale is dynamic, in the sense that all species are destined for greater perfection:

> I have stopped elsewhere to consider this marvellous gradation that reigns among all living Beings from the Lichen and the Polyp, to the Cedar and Man. The Metaphysician can find in the Law of Continuity the reason for this progression; the Naturalist confines himself to establishing it on facts. Each Species has its own characters that distinguish it from all others. The *ensemble* of these characters constitutes the nominal Essence of the Species. The Naturalist searches for these characters; he studies them, describes them and composes these learned Nomenclatures, known under the names *Botany* and *Zoology*. It is in striving to arrange all these organic Productions into Classes, Genera and Species that the Naturalist becomes aware that these Divisions of Nature are not cut like those of Art; he observes that between two Classes or adjacent genera, there are some species in common [*mitoyennes*] which appear not to belong more to one than to the other, and which disturb his methodical Distributions to a greater or lesser extent.
>
> The same progression that we discover today among the different Orders of organised Beings will be observed, without doubt, in the future State of our Globe: but it will follow other proportions that will be determined by the degree of perfectibility of each Species. Man, by then having been transported to another [place of] sojourn more suited to the eminence of his Faculties, will leave to the Ape or the Elephant that premier place he once occupied among the Animals of our Planet. Upon this universal restitution of the Animals it will be possible to find a Newton and Leibniz among the Apes and Elephants, a Perrault and Vauban among the Beavers, and so forth.
>
> The lowest species, such as Oysters, Polyps and so on, will be to the highest species of this new Hierarchy as the Birds and Quadrupeds are to Man in the present Hierarchy.
>
> Perhaps there will be a continuous and more-or-less slow progress of all the Species toward a higher perfection, ensuring that all the degrees of the scale will be continually variable in a fixed and constant relation: I mean to say, that the mutability of each degree will always have its reason in the degree which immediately precedes it.[166]

All species must progress in lockstep, lest a rift appear in the Scale. Even "Brute Matter has, as its ultimate goal, organic Matter".[167]

The Great Chain after 1780

Continuity in nature remained an important point of reference in late Eighteenth-century thought. Soame Jenyns[168] and Saint-Pierre[169] held it up as evidence for Divine wisdom, power, and beneficence; Johann Gottfried Herder found therein

God's order in the universe.[170] Alexander Hunter[171] and Robert Peirson[172] structured essays on rural economy around the Chain; Benjamin Franklin[173] worked it into a short story. Indeed, even today the Chain lives on in literary and poetic settings. In scientific contexts we find continuity resplendent in Richard Watson's *Chemical essays*,[174] and in Charles White's *Account of the regular gradation in man, and in different animals and vegetables*.[175] Thienemann presents a detailed if wildly idiosyncratic Chain from 1780 that combines salt-sulphur chemistry, Bonnet's scale, Linnæan systematics, and religious concepts.[176]

We end this chapter with *Flore française* (1778), written by Jean-Baptiste Lamarck at the end of his ten-year apprenticeship under Bernard de Jussieu. Its *Discours préliminaire* includes an essay on *l'Ordre naturel* and a genus-level Chain of Being, although these may owe more to the Abbé René-Just Haüy than to Lamarck himself.[177] Writing in 1800, however, Lamarck made it clear that

> In speaking of this even gradation in the complexity of organization I do not mean at all to suggest the existence of a linear series with regular intervals between the species and genera: such a series does not exist; but I speak of a series in which the principal masses, such as the great families, are almost regularly spaced; a series which quite certainly exists among both animals and plants, but which when considered in terms of genera or more particularly of species, forms in many places lateral ramifications of which the terminal ends provide truly isolated points.
>
> If there exists a graded series among living things, at least of the principal groups, in respect of the complexity or simplicity of their organization, it is clear that in a truly natural classification either of animals or plants, one must necessarily place at the two ends of the scale the forms which are most dissimilar, the furthest apart in affinities, and which consequently form the extreme terms presented by animal or plant organization.
>
> Any classification which departs from this principle seems to me to be faulty because it cannot conform to the order of nature.[178]

As we shall see in Chapter 17, Lamarck eventually abandoned much of this position too; but it captures important elements of Bonnet's position including the recognition of higher-level order in nature;[179] acknowledgement that individual genera and species may not fit neatly into this order; and that some forms deviate from the main order, like branches from the trunk of a tree. By definition, continuity erodes discrete taxa and, pressed far enough, might erase them entirely, coalescing everything into organized nature versus unorganized, living versus non-living. In non-technical contexts it may prove convenient to refer to plants and animals, indeed to fish, birds, quadrupeds, and the like; but (as Aristotle concluded long ago) for some genera we will be at a loss to determine whether they should be included in or excluded from a particular group. From this perspective, Bonnet was perfectly correct to refuse to countenance Zoophyta; but a slippery slope leads towards mere poetry.

242 KINGDOMS, EMPIRES, AND DOMAINS

In Chapter 15 we shall examine the countervailing tendency, towards a comprehensive unitary classification of (at least) the living world. As developed in the late Seventeenth and Eighteenth centuries, no genus should be excluded or ignored: a place should be found for every being including infusorial animalcules, freshwater polyps, and zoophytes. And as this was a period of exploration and empire-building, new types of beings came to the attention of European scientists.

15

Classifying God's handiwork

> O Lord, how manifold are your works!
> In wisdom you have made them all;
> the earth is full of your creatures.
>
> —*Psalm 104:24*[1]

Taxonomy—the recognition of groups of similar beings—may be "everywhere as old as language".[2] Arranging these groups in a comprehensive hierarchical system according to general philosophical or scientific principles is, however, more recent. In earlier chapters we witnessed important steps in this direction by Aristotle and Theophrastus, Wotton and Cesalpino, Aldrovandi and Jonston. Most of these efforts were limited to plants, or to animals; included mythological and imaginary beings; gave credence to folk tales, such as the supposed ability of sponges to sense the approach of danger; and used whatever data happened to be at hand—form, behaviour, medicinal properties, or habitat. As we have seen, infusorial animalcules and the freshwater hydra were unknown before the late Seventeenth century. On a positive note, important practises made their (sporadic) appearance through the late Renaissance: classical quotations began to be omitted, multiple taxonomic levels (order, class and genus) recognized, and genus names simplified.

All this nonetheless leaves us a few steps shy of the comprehensive systems we associate with Eighteenth-century taxonomy. These steps were taken by taxonomists active in the late Seventeenth and very early Eighteenth centuries, notably Pierre Magnol, Joseph Pitton de Tournefort, and John Ray: we meet each of these below.

As we saw in the previous chapter, the Eighteenth century also marked the heyday of the Great Chain of Being. A multilevel hierarchical classification might, or might not, be consistent with a particular notion of continuity in nature. In any event, taxonomy and continuity represent different perspectives, address different aims, and answer to different criteria. If natural beings grade imperceptibly into one another, it might be impossible, or unnaturally arbitrary, to recognize taxa at all. In 1689 John Locke raised the philosophical stakes by arguing that groups in nature are mere mental constructs, devoid of independent reality, and with boundaries "as men, and not as nature makes them".[3] Locke, who accepted a close continuity across

Kingdoms, Empires, and Domains. Mark A. Ragan, Oxford University Press. © Oxford University Press 2023.
DOI: 10.1093/oso/9780197643037.003.0015

244 KINGDOMS, EMPIRES, AND DOMAINS

nature and was familiar with botanical taxonomy—he had compiled a herbal of English plants, based on 1600 specimens[4]—argued that

> the animal and vegetable kingdoms are so nearly joined, that if you will take the lowest of one, and the highest of the other, there will scarce be perceived any great difference between them; and so on, till we come to the lowest and the most inorganical parts of matter, we shall find every where, that the several species are linked together, and differ but in almost insensible degrees.[5]

For Bradley and Bonnet, the Chain was evidenced only in parts; nature might actually have made a leap from the unorganized (minerals) to the living, while other regions of the Chain seem to run in parallel, or veer off into side branches. Hierarchical taxonomy could accommodate such imperfections, but the absence of a single unifying character might nonetheless pose a greater danger for taxonomy than for the Great Chain.

These years brought new opportunities—and potential problems—for both continuity and classification. Explorers, sea captains, and conquistadors sent strange new plants and animals to European capitals from Southeast Asia, coastal Africa, and the Americas. In 1623 Gaspard Bauhin had recognized more than six thousand species of plants,[6] and this number had continued to climb inexorably ever since. The same was happening with animals and minerals too, if less dramatically. Such diversity cried out for a classificatory method that was universally applicable, generated a multilevel hierarchy, and could be applied quickly at scale.

Finally, if a classification is to be comprehensive, decisions must be made about so-called intermediate organisms. Proponents of continuity might get away with poetic words about sponges (or corals, hydra, or sensitive plants) being somehow part plant and part animal, or ignore these beings entirely. A unitary hierarchical classification leaves no such place to hide.

Magnol and Tournefort

As in the Renaissance, botanists were at the forefront of new developments in classification. Pierre Magnol contributed the alphabetically arranged *Botanicum Monspeliense* (1676) and, posthumously, *Novus caracter plantarum* (1720). In the latter he divided plants into three groups: those with the calyx external, internal, or both external and internal.[7] Those with an external calyx he classified into eleven sections, including marine and aquatic plants (stony and soft corals, corallines, sponges, fuci, algae), ferns, mosses, and fungi. More importantly, taking up the challenge posed by Locke, he sought to clarify the properties of natural taxa:

> Speaking of animals and plants offers me the occasion to reduce the plants to certain families (I call these families, so that they may be compared to families of

men), but if the features of families are marked only by fructification, it is impossible to see the various parts of the plant among which the chief distinctions and characters are to be discovered, that is to say the roots, stalks, leaves, flowers and seeds; there is a certain similitude and affinity in many plants which does not consist in the parts taken separately, but in the total composition, which strikes the sense but which cannot be expressed in words.[8]

Magnol wrote these words more than a century and half before Mendel or Darwin, but "families of men" more than hints at genealogical relationship. Joseph Pitton de Tournefort focused instead on the genus, offering formal descriptions for 698 genera of plants.[9] In *Élémens de botanique, ou methode pour connoître les plantes* (1694) he argued for Cesalpino's fructification-based approach:

> To make a Botanical system, it is not sufficient to know the characters of the genera of plants; it is necessary to arrange these kinds in certain classes, and then arrange these classes in a simple and natural order. One could not, says Cesalpino, remove the confusion of this science, unless one arranges, so to speak, the plants as the soldiers of an army. According to this author's thought, the genera of plants are like so many companies, and the classes of plants can be compared to regiments.
>
> To establish the classes of the plants is precisely to discover what several genera of plants have in common, which distinguishes them essentially from all other genera, and to enclose these same genera in certain orders which we have called the classes of plants. All the researches which I have made in examining the different ways in which these classes can be composed, have convinced me that their composition was not arbitrary; but that it was a continuation of the course of action which one took in the establishment of the genera, and that it was absolutely necessary to have regard only to the true relations between them; that is to say, to the relationship extracted from the structure of their essential parts, these being the flower and the fruit.[10]

To be sure, this system cannot be extended to plants that lack flowers and fruits, for which one must use other characters including the means of propagation, *habitus*, and external appearance.[11] Tournefort's approach and resulting classification were influential on the Continent, particularly given his position as Demonstrator of Plants at the Jardin du Roi in Paris. Linnæus too thought highly of his work.

Ray and natural theology

John Ray, an older contemporary of Magnol and Tournefort, was exposed to the same philosophical currents particularly during his travels (1663–1666) on the Continent with Francis Willughby. In his early writings he arranged plants alphabetically[12] or based on their leaves, flowers, seed-vessels, fruit, seeds, even gums and

246 KINGDOMS, EMPIRES, AND DOMAINS

resins.[13] By 1673, however, he had significantly sharpened his analysis, identifying the *species* as the basis of classification and implying that it is defined by genealogy.[14] Ray made this explicit in *Historia plantarum* (1686):

> In order that an inventory of plants may be begun and a classification of them correctly established, we must try to discover criteria of some sort for distinguishing what are called "species". After long and considerable investigation, no surer criterion for determining species has occurred to me than the distinguishing features that perpetuate themselves in propagation from seed. Thus, no matter what variations occur in the individuals or the species, if they spring from the seed of one and the same plant, they are accidental variations and not such as to distinguish a species.... Animals likewise that differ specifically preserve their distinct species permanently; one species never springs from the seed of another nor vice versa.[15]

Even so, the question remained as to how species might be identified and delineated in practise. Ray praised Cesalpino and his emphasis on the number, location, and form of the parts of fructification,[16] but drew liberally on other characters.[17] He was hardly the first to do so,[18] but he annoyed many of his contemporaries by elevating the practise to a principle:

> In summary, the best [classificatory] Method for plants is that in which all genera, the highest as much as the subordinate and lowest, have the more attributes in common or agree in the more parts or accidents. I suppose that Nature does not admit such a classification. It is approximated by that [Method] in which all plants which have the more attributes in common are referred to their genera; and the remainder, which offer up no conformity or similitude in [their] several parts or accidents, are divided into genera by more-general customary characters [*notis generalioribus*] such as conformity in some one conspicuous part, as evidently was suitably [*commodissimum*] done according to the structure of the flower, the regularity or number of petals, or simplicity of the fruit or its division into sections.[19]

It did not help that his resulting classification differed from that of Tournefort. Ray accepted a continuum of species in nature:

> Truly Nature (as the saying goes) *makes no leaps, and passes from extreme to extreme only through a mean*, between things higher and lower there are ordained species intermediate and of ambiguous status, as if they link one product [of nature] with another and participate in both classes, such that one cannot know to which it belongs entirely: as for example the so-called *Zoophyta* between *Plants* and *Animals*.[20]

Ray took holy orders in 1660, and as a Fellow of Trinity College read sermons at Cambridge. From these sermons emerged his influential *The wisdom of God manifested in the works of the Creation*, published in seven editions from 1691 to 1717, in which he referred to creatures as forming a series of perfection.[21] Within the plants too, his Method proceeds gradually from the less perfect to the more perfect, or from the lowest orders to the highest.[22] Among animals, man is sentient, brute animals merely capable of sensation, while "*worms, the oyster, the sponge* and that meagre [*exilia*] animal detected recently by the microscope, truly these are imperfect, as no one freely concedes to them perception and memory, let alone immortality."[23] Algae, fuci, sponges, stony and soft corals, fungi, mosses, and lichens are, by contrast, Plantæ imperfectæ.[24]

Ray began *Historia plantarum* with Jung's definition of plants,[25] and *Synopsis methodica animalium quadrupedem* with Aristotle's definition of animals.[26] He refers in passing to "the so-called Zoophyta" (above) and "animals . . . which are imperfect or of a two-faced [*ancipitis*] nature, which we usually call Zoophytes",[27] but did not include Zoophyta in his classification.[28] Ray considered it unreasonable that "Plants being of a lower Form or Order of Being, should produce Animals".[29] Nature may not leap from plants to animals, but Ray found it possible to draw a sharp line between the two.

Ray's *Wisdom of God* restated the ancient theme that the wondrous complexity of natural things, their fitness for purpose and place in the natural order can be taken as evidence of design by a wise and powerful God. Strands of this argument can readily be found in Socrates,[30] the Stoics, al-Ghāzalī,[31] Aquinas,[32] Sebonde,[33] and Newton,[34] and came to be called *physico-theology* or *natural theology*. The Church much preferred the evidence of scripture or religious experience, and from time to time worried that argument from design comes rather too close to equating God with nature. Natural theology nonetheless found a particular resonance in late Seventeenth- and Eighteenth-century England, where a village parson might be an avid microscopist and correspond with the Royal Society.[35] In microscopic animalcules, the clergyman William Derham found

a whole and compleat Body, as exquisitely formed, and (as far as our Scrutiny can possibly reach) as neatly adorn'd as the largest Animal. Let us consider that there we have Eyes, a Brain, a Mouth, a Stomach, Entrails, and every other Part of an Animal-Body, as well as Legs and Feet; and that all those Parts have each of them their necessary *Apparatus* of Nerves, of various Muscles, and every other Part that other Insects have; and that all is covered and guarded with a well-made Tegument, beset with Bristles, adorn'd with neat Imbrications, and many other Fineries. And lastly, let us consider in how little Compass all Art and Curiosity may lie, even in a Body many Times less than a small Grain of Sand; so that the least Drop of Water can contain many of them, and afford them also sufficient Room to dance and frisk about.[36]

Cotton Mather brought Derham's words to New England,[37] where Jonathan Edwards regarded God's lesser creatures rather less positively, comparing sinners

248 KINGDOMS, EMPIRES, AND DOMAINS

with a "spider, or some loathsome insect" held over the flames of Hell by an angry God.[38] Somewhat later John Wesley collected descriptions of a wide range of individual marvels—microscopic animalcules, corals and corallines, freshwater polyps, tapeworms ("indeed a Zoophyton"), even the barnacle-goose. Apparently in reference to Needham's observations, Wesley accepted that plants can be "translated" into animals, and vice versa.[39] Moreover,

> As Barnacles seem to be a Medium between Birds and Fishes, altho' they more properly belong to the former, so is a Polypus, (altho' it is doubtless an Animal) between Animals and Plants.[40]

These and other works of natural theology were reprinted in multiple editions through the Eighteenth century. The theme continued into the Nineteenth with William Paley's *Natural theology* (1802), and culminated in the by-then anachronistic Bridgewater Treatises (1833–1840). We shall return to this tradition again in Chapter 20.

Linnæus

The name of Carl Linnæus is today joined with *classification*, and not without cause: modern nomenclature of vascular plants begins with his *Species plantarum* (1753), and zoological nomenclature with his *Systema naturæ* (1758). But Linnæus offers more than a unified hierarchy of organisms named by genus and species. He developed a philosophy of classification, and advised on its practise. He underpinned his classification with a theory of form that would not have been out of place in the Sixteenth century.[41] Through 171 student dissertations, collected as the *Amœnitates academicæ*, he explored diverse topics in mineralogy, botany, zoology, ecology, anthropology, and medicine.[42] In Linnæus we find keen observation, but not experiment; Aristotle and Cesalpino, but not Spallanzani.

Like Ray and others before him, Linnæus accepted that "nature makes no leaps", even if we can as yet see the natural system only in fragments.[43] Nature offers "testimony to the glory of God and the moral order of the world", although his was the stern, retributive Old-Testament God of Jonathan Edwards.[44] Later in life Linnæus sought order in numerology, citing Ovid and Virgil.[45] It is easy to dismiss his views as old-fashioned, but the idea that living nature was arranged geometrically (based, for instance, on the number five) came back into fashion in the 1830s,[46] while Goethe expressed intellectual indebtedness to Linnæus.[47] More broadly, some find in his theory of form a prefiguration of *Naturphilosophie*.[48]

Linnæus was a prolific author and lecturer, wrote four autobiographies, and corresponded with scores of colleagues.[49] Here we take three approaches to his classifications of plants, animals, and intermediate beings, focusing first on his System of Nature, then on his underlying theory, and finally on his third kingdom of living beings—*Regnum neutrum* or *Regnum chaoticum*.

Systema naturæ

Not since Albert (or, more problematically, König) had anyone tried to systematize all the productions of nature. For three years after receiving his medical degree (for a thesis on malaria) Linnæus remained in the Netherlands, where he was mentored by Herman Boerhaave, Jan Frederik Gronovius, and Johannes Burman, and served as personal physician to George Clifford III, a director of the Dutch East India Company.[50] In these years Linnæus wrote nine books: six classifications and floras, and three of botanical principles and practise.[51] The most ambitious was *Systema naturæ, sive regna tria naturæ systematice proposita per classes, ordines, genera, & species* (1735). Its twelve printed pages set out Regnum vegetabile, Regnum lapideum, and Regnum animale:[52]

> *Lapides* grow. *Vegetabilia* grow and live. *Animalia* grow, live and are sensate. Thus are the limits among these Kingdoms constituted.[53]

This formulation points to a Chain of Being, and indeed Linnæus intended *Systema naturæ* to present nature as "an orderly chain".[54] He divided each kingdom into classes and orders. Plants he formed into twenty-four classes[55] based on the parts of fructification, concluding with Cryptogamia,[56] among which sexual reproduction is "hidden" and flowers "concealed". In 1735 these included ferns, mosses, algae, fungi, and lithophytes (sponges, stony corals, and soft corals).[57] Animals he formed into six classes ending with Vermes. As for the unicorn, satyr, borametz, barnacle-goose-tree, and the like, he identified some with real (non-paradoxical) beasts, and swept the others into the dustbin of history.[58]

Systema naturæ* soon became famous, and over the next six decades appeared in more than a dozen further editions[59] among which the Tenth (1758) is of particular note, as it has since been made the starting point for modern zoological nomenclature. Taxa were added, moved, or dropped as knowledge improved, not least among Vermes, within which he recognized three (1735–1747), four (1748–1756), or five (1758–1766) orders (Table 15.1).

In the first nine editions, Reptilia included the tapeworm, earthworms, leeches, and slugs, while Zoophyta held marine invertebrates including sea-slugs, sea-urchins, starfish, jellyfish, squid, cuttlefish, and tunicates. From the Sixth edition Linnæus moved Lithophyta from Cryptogamia to Vermes,[60] and added the freshwater polyp (as *Hydra*) to Zoophyta. Genus Spongia, a member of Cryptogamia Lithophyta from the First through Fifth editions, was omitted from the Sixth through Eighth, re-established in Cryptogamia Algæ in the Ninth through Eleventh, then

250 KINGDOMS, EMPIRES, AND DOMAINS

Table 15.1. Major editions of *Systema naturæ*. (L) indicates written by Linnæus himself.

Edition	City	Year(s)	Orders within Vermes
1 (L)	Leiden	1735	Reptilia, Testacea, Zoophyta
2 (L)	Stockholm	1740	Reptilia, Zoophyta, Testacea
3	Halle	1740	Reptilia, Testacea, Zoophyta
4	Paris	1744	Reptilia, Zoophyta, Testacea
5	Halle	1747	Reptilia, Zoophyta, Testacea
6 (L)	Stockholm	1748	Reptilia, Zoophyta, Testacea, Lithophyta
7	Leipzig	1748	Reptilia, Zoophyta, Testacea, Lithophyta
8	Stockholm	1753	(not applicable)
9	Leiden	1756	Reptilia, Zoophyta, Testacea, Lithophyta
10 (L)	Stockholm	1758–59	Intestina, Mollusca, Testacea, Lithophyta, Zoophyta
11	Halle	1760–70	Intestina, Mollusca, Testacea, Lithophyta, Zoophyta
12 (L)	Stockholm	1766–68	Intestina, Mollusca, Testacea, Lithophyta, Zoophyta
12a ("13")	Vienna	1767–70	Intestina, Mollusca, Testacea, Lithophyta, Zoophyta
13 Gmelin	Leipzig	1788–93	Intestina, Mollusca, Testacea, Zoophyta, Infusoria

relocated to Vermes Zoophyta in the Twelfth. From the Tenth edition he moved Reptilia (although not its constituent genera) from Vermes to Amphibia; created Intestina and Mollusca within Vermes; moved sea-slugs, sea-urchins, jellyfish, star-fish, squid, and the like to Mollusca; and added *Volvox globator* (a colonial green alga) and *Volvox Chaos* (an amœba) to Zoophyta. In the Twelfth edition—the last from his own hand—he accepted more microscopic animalcules into *Volvox, Furia,* and *Chaos,*[61] and mentioned six "living molecules" which merit further study:

α. The *contagion* of eruptive fevers?
β. The *cause* of paroxysmal fevers?
γ. The *moist virus* of syphilis?
δ. *Leeuwenhoek's* spermatic animalcules?
ε. The æry mist *floating in the month of blossoming* [*i.e.* May]?
ζ. *Münchhausen's septic agent* of fermentation and putrefaction?[62]

We return to these "living molecules" below, but first let us look more closely at how Linnæus described Zoophyta in the Tenth edition:

ZOOPHYTA are composite animals, constituted where the Animals and Vegetables meet. Most *take root* and *grow up* into *stems*, multiplying life in their branches and deciduous *buds*, and in the transformation of their animated blossoms, which are endowed with spontaneous motion, into *seed-bearing capsules*. And if Plants resemble Zoophytes, but are destitute of sensation and loco-motion, so too are Zoophytes as true plants, but furnished with a system of nerves, sensation and organs of motion: *indeed they possess a sense-perception which has the nature of neither the animal nor the vegetable, but is of a certain third nature which draws upon them both: Pliny.*[63]

In the Twelfth edition, Linnæus added "*and such that Plants by natural meta-morphosis are transformed into Animals*".[64] *Composite animals, natural metamor-phosis*: to understand these terms we must consider the theory Linnæus developed in parallel with his works of systematics.

Linnæus the theoretician

Beyond the classifications lie other Linnæan writings: unfamiliar, speculative, sometimes quite fanciful, often highly relevant to our question. Here we find a body of theory—if such is the word, as little of it is testable except by the rules of logic. Lindroth labels it "a kind of religion, a system of botanical dogmas . . . the work of a scholastic legislator, not of an empirical natural historian humbly searching for truth."[65] If so, the aphorisms of *Philosophia botanica* (1751) are its Commandments, the *Amœnitates* its books of the Prophets. Our point of departure is his certainty that genera and species actually exist in nature.[66] Some species were created in the Beginning and remain unchanged, while others have arisen since by hybridization.[67]

Plants are composed of two substances, medulla (marrow) and cortex (bark).[68] Medulla, the motherly substance, bears life and seeks to produce new plant material. This happens asexually through buds. When medulla breaks free of the restraint imposed by the fatherly protective cortex, the plant transforms (metamorphoses) into a (sexual) blossom. This blossom—not the root, stalk or leaf, all of which fall away—is the plant itself, naked (so to speak) and thus the proper object of classifica-tion. The cortex gives rise to stamens and pollen, the marrow to pistils and ovaries, and in due course the marrow is gathered into seeds.[69] This normally happens within a species,[70] but hybrids arise when medulla is contributed by one species, cortex by another. As any one species might be able to hybridize with several others, plants display affinities not along a single linear series but rather on all sides, as ter-ritories on a geographical map.[71] Animals too have medulla and cortex: the spinal cord and nerves, which carry sensation (a defining feature of animals), are threads of medulla. As in plants, medulla is usually constrained by cortex, for example the vertebral column or, in insects, the pupal encasement.[72] In Vermes, however, medulla is scarcely constrained at all; and animalcules of the final species, *Volvox Chaos*, are "naked medulla".[73]

The various parts of a plant[74] are fundamentally similar, and to some extent in-terconvertible: buds metamorphose into branches, leaves into petals. So too a

252 KINGDOMS, EMPIRES, AND DOMAINS

caterpillar pupates, and the imago breaks through the pupal cortex to emerge as a perfect adult, likewise the proper object of classification.[75] Viewed in another way, the parts of a specific inflorescence—bracts, corolla, stamens, and pistil—are manifestations of what would have become (through a process of metamorphosis) leaves in a succession of later years, had the plant not flowered.[76] Linnæus called this prefiguration or anticipation *prolepsis*.[77] Combined with metamorphosis, it explained other phenomena too, including parthenogenesis in aphids (Bonnet), and the nested generations of *Volvox globator* (Leeuwenhoek).

In plants, the medulla pushes through the cortex to form buds (hence branches) and flowers. By contrast, in most animals it remains constrained, and animals remain simple. But as we have just seen, in Vermes the medulla is scarcely constrained, so like plants they are often *composite*. That is, multiple individuals are joined together by attachment to a common base (corals), on a branched stalk (some freshwater polyps), or as articulated sections, each in effect an independent animal (the tapeworm *Tænia*).[78]

This brings us to the interface between animals and plants. From the Sixth through Ninth editions of *Systema naturæ*, Linnæus had positioned Lithophyta (corals) as the final order within Vermes, but in the Tenth he reorganized Vermes, moving sea-slugs, jellyfish, and the like to Mollusca, transferring *Tænia* from Reptilia to Zoophyta, then slotting his reformed Zoophyta into last place instead, implicitly adjacent to plants. Why did he consider these two orders the nearest to plants, and on what basis did he place zoophytes at the very boundary?

Lithophytes are "composite molluscan animals, sprouting out from a stony underlying coral in which they are inserted and which they build".[79] In 1745 (*Corallia Baltica*) Linnæus could point to reasons (and precedent) to place corals among the fossils (*i.e.* stones), the plants, or the animals, but the work of Peyssonel and Bernard de Jussieu soon convinced him of the latter.[80] Even so, in the Tenth edition he wrote that the limits of all three Kingdoms of Nature "run together in the Lithophytes".[81] By contrast, zoophytes are "composite flowering animals; *stirps vegetans*".[82] Linnæus knew well that Renaissance botanists used *stirps* to mean *stem* or *shoot*; and *vegetans* means *enlivened*, but this could be said without invoking vegetables.[83] For Linnæus, the inflorescence (hence the medulla) of zoophytes is animal, the stem or stalk (cortex) vegetable. He made this clear in a letter to John Ellis:

> *Zoophyta* are constructed very differently, living by a mere vegetable life, and are increased every year under their bark, like trees, as appears from the annual rings in a section of the trunk of a *Gorgonia*. They are therefore vegetables, with flowers like small animals, which you have most beautifully delineated. All submarine plants are nourished by pores, not by roots, as we learn from *Fuci*. As Zoophytes are, many of them, covered with a stony coat, the Creator has been pleased that they should receive nourishment by their naked flowers. He has therefore furnished each with a pore, which we call a mouth. All living beings enjoy some motion. The Zoophytes mostly live in the perfectly undisturbed abyss of the ocean.

They cannot therefore partake of that motion, which trees and herbs receive from the agitation of the air. Hence the Creator has granted them a nervous system, that they may spontaneously move at pleasure. Their lower part becomes hardened and dead, like the solid wood of a tree. The surface, under the bark, is every year furnished with a new living layer, as in the vegetable kingdom. Thus they grow and increase; and may even be truly called vegetables, as having flowers, producing capsules, &c. Yet as they are endowed with sensation, and voluntary motion, they must be called, as they are, animals; for animals differ from plants merely in having a sentient nervous system, with voluntary motion; nor are there any other limits between the two.[84]

Thus, it seems, both corals (in which all three Kingdoms run together) and zoophytes (which metamorphose, as it were, from plant to animal) have some claim on the boundary between plants and animals. But lithophytes[85] are sensitive, whereas in zoophytes only the flower is animate. Like plants, zoophytes (but not corals) metamorphose, revealing a flower. *Volvox globator* shows prolepsis.[86] *Volvox Chaos* is a mere point, unconstrained of form, pure marrow. As summarized by Stevens and Cullen:

> Linnaeus toyed with the idea that the relationship between the plant and animal kingdoms was similar to the metamorphosis that occurred in the development of plants and insects, that of the plant in particular occurring during prolepsis and the shedding of its covering. . . . Zoophytes also showed this metamorphosis, but becoming more like animals in the process; hence, plants could possibly metamorphose into animals—witness Münchhausen's findings.[87]

Linnæus, Münchhausen, and the invisible world

We have seen above how Linnæus classified many of the problematic organisms, including corals (as Lithophyta), the freshwater polyp, *Taenia*, Leeuwenhoek's *Volvox*, and Rösel's little Proteus (as Zoophytes). Two problematic groups remained: fungi, and most of the infusorial animals.

Giambattista della Porta is usually credited with discovering that fungi produce "seeds" (in reality, spores);[88] this makes them organisms (not excrescences), indeed plants. Pier Antonio Micheli found seeds (and flowers!) in a broad range of fungi, and demonstrated that the presumed seeds give rise to new fungi.[89] A few years later Albrecht von Haller reported, in a letter to Linnæus, that many more fungi probably produce seeds.[90] Others were unconvinced.[91] Such was the situation when Freiherr Otto von Münchhausen first wrote to Linnæus on 18 April 1751.[92] Münchhausen, then *Landdroste* for Steyerberg in Lower Saxony, sought to understand the diseases that periodically damaged grain crops and fruit trees.[93] Examining smut (ustilago) with his microscope, he found a black powder that "consists of nothing but small,

254 KINGDOMS, EMPIRES, AND DOMAINS

transparent globules with small black spots inside, which are the eggs of infinitely minute insects, or rather of little worms".[94] More generally,

> When fungi become old, and especially Lycoperda, and all moulds, they scatter about themselves a blackish dust; if we observe this under a good magnifying-glass, we find semi-transparent little globules, filled inside with black points, and of a substance not dissimilar to that of the previously described polyps. I have placed this dust into water and allowed it to stand in a warmish condition, and the globules gradually swelled up, and changed into motile ball-like animals. These animalcules (at least I shall call them that, on account of their resemblance) run about in the water; when one gives them further attention over some days, it is noticed that they come together in clumps of a hard web, and from these further arise either moulds or fungi: where fungi grow, at first one sees white veins which, to be sure, are usually regarded as roots, but in fact are fact are nothing else but tubes in which the polyps move back and forth, [and] soon form a large structure...
>
> ... it seems to me without doubt that all corals, fungi, moulds, lichens, smut in grain, yes indeed almost every fermentation ... might originate from polyp-like creatures, which I cannot recognise as complete [*völlige*] animals.[95]

Linnæus confirmed these observations himself.[96] Like Needham (but without his little globules), Münchhausen and Linnæus agreed that rusts and smuts give rise to animalcules, which then create a web of filaments that regenerate the original rust or smut. As fungi are plants and the motile animalcules are animals, the two kingdoms are connected by reciprocal metamorphoses. Ellis disputed this interpretation, suspecting (correctly) that the motile animalcules, which arise in many infusions, are not involved in the generation of fungi.[97] In the Twelfth edition of *Systema naturæ* Linnæus elevated *Chaos* to a genus in its own right, containing not only infusorial animalcules and amœbæ, but also fungal seeds that metamorphose from vegetables to animals and vice versa. But he came to suspect that Ellis might be right, and queried whether the *Chaos Fungorum* of *Mucor* was "sufficiently distinct" from the infusion-animalcules of the microscopists.[98]

Münchhausen's conclusion that fermentation is caused by living organisms is notable, as the matter remained open until the work of Louis Pasteur nine decades later.[99] Linnæus was even more interested in the role of microorganisms in human disease, and in drawing up genus *Chaos* for the Twelfth edition (above) he grouped Münchhausen's "*septic agent* of fermentation and putrefaction" alongside human fevers.[100] The first disease organisms to be formally classified were thus species of the zoophyte *Chaos*.

Our story is not quite complete. As early as 1754, Münchhausen suggested to Linnæus that fungi, lichens, corals, and hydroids might be placed in a "Regnum quasi intermedium inter Regnum animale et vegetabile".[101] Münchhausen returned to the idea in *Der Hausvater* (1766), devoting a section to "Regnum neutrum. On doubtful productions of Nature". Animals have a circumscribed body with limbs; they move willingly from place to place, perform deeds voluntarily, digest food, and reproduce sexually. Plants germinate from seed, are fixed to the earth with a root, and produce a flower and seeds. But there exists a third kind of body that does not share fully in the characters of either kingdom; for these he set up a new *Mittelreich* or *Regnum neutrum* with three divisions: Hydræ (freshwater polyps), Lithophyta (corals), and Fungi. These organisms, along with all other ferments, "offer the median between the plants, animals and minerals".[102]

Linnæus held Münchhausen in high regard,[103] but one might expect him to reject such a thin argument. Instead, Linnæus set a student thesis[104] on the "invisible world", not only recent discoveries but "things more obscure, to be committed to the diligence of men in the future".[105] The result, *Mundus invisibilis*, is an odd and unsatisfying document. The microscope (he argues) has unlocked new rooms in the palace of nature, revealing even mosses and flies as wonderful works of the Creator. Réaumur, Bernard de Jussieu, Trembley, and others studied the freshwater polyp, and Peyssonel and Ellis corals, although it remained for Linnæus himself to bring these formally into Zoophyta. Münchhausen has observed that seeds of ustilago, mucor, mushrooms, and other fungi germinate into little worms, which then weave a web that grows again into the original fungus. In the same way, exanthematic and contagious fevers may contain animalcules; analogies may exist between seminal animalcules and pollen; and the little ethereal clouds that Réaumur saw hovering in the air in the month of flowering may likewise consist of animalcules.[106]

Turning the tables on his query whether mucor-seeds are sufficiently distinct from infusion-animalcules, Linnæus asks whether the animalcules of infusions, and those of pepper-water, arise from seeds of mucor, which seem to be ubiquitous in the environment. In reply, he offers only that "these animals will truly one day bring the investigators of nature to the origin of the animate in animals, since it seems as if the first animated molecules are perceived here."[107] Even the medulla of plants might conceivably become animate, although it remains bereft of nerves and cannot move voluntarily. And

> the Order Fungi now seems to be so distinct from the order of plants and the order of Zoophytes, that they are indeed at least on the border between the animal and plant kingdom, in such a way that no one could until now capture the whole sweep of nature [*geodætes naturæ*] to see to which of these kingdoms this province should be rightly attributed. . . . In a word: just as Zoophyta flourish from a plant into small little animals, so according to an opposite law of nature, if this origin of Fungi could be confirmed, Fungi grow from Animalcules into plants.[108]

A few years later Linnæus went even farther: is not man himself a fungus, propagated by the spermatic animalcule?[109] The spermatic worm is "a bit of

256 KINGDOMS, EMPIRES, AND DOMAINS

released, expansive and proteus-like medulla. . . . The medulla is found everywhere in nature and is the carrier of the growth force and life force, attired in innumerable metamorphoses it transforms itself into man also. Linnæus suspects too that the medulla deep down is driven by a cosmic force, an everywhere flowing electricity."[110]

Mundus concludes with fourteen scholia,[111] eight of which speak to genus *Chaos* as set out in the Twelfth edition. The final scholium asks "Are those very small animalcules, called *chaotica*, merely medullary, and almost without an organic body?"[112] This is why the possible new kingdom might be called Regnum Chaoticum: its members (like those of genus *Chaos*) are mostly or entirely medulla, scarcely constrained by cortex and thus plastic in form.[113]

Little came of the Neutral or Chaotic Kingdom. Neither Linnæus himself, nor Gmelin later, incorporated it into the *Systema naturæ*. Corals are not a middle ground, in any meaningful sense, between animals (or plants) and the Mineral Kingdom. Reciprocal metamorphosis between plants and animals was a bridge too far for many natural historians;[114] animalcules and fungi were not time-honoured neighbours in the Great Chain of Being. In an age of specialization, infusion-animals, freshwater polyps, and fungi were increasingly studied by distinct research communities. As we shall see below and in subsequent chapters, other Middle or Neutral Kingdoms came and went, albeit with different membership and theoretical justifications. Even so, the idea that plants can transform into animals would persist into, or reappear in, the Nineteenth century.

Linnæus: reception

Linnæan classification remains a point of reference even today. He formally brought the three kingdoms into a single Imperium Naturæ.[115] He swept away mythological beings, implemented binomial nomenclature, arranged the plants by their organs of fructification, consolidated Zoophyta within the animals, found places for all the problematic beings, and classified agents of plant and human disease. His bottom-up approach to diversity avoided the pitfalls of top-down logical division, and helped classification support continuity. He moreover developed a body of theory—or, less charitably, scholastic dogma—that justified his sexual system and guided his decisions about zoophytes, fungi, and the boundary between animals and plants. His theory of form inspired later thinking about plant morphogenesis.

Not all of these efforts were universally applauded. The editors of *Encyclopædia Britannica*, for example, excoriated his claim (in *Sponsalia plantarum*) that plants, like animals, spontaneously propel humours and are therefore alive:

> Strange, that a man of Linnæus's capacity, or indeed of any capacity at all, should seriously employ an argument pregnant with every degree of absurdity![116]

They also objected to his ideas about hybrids, cryptogams, and plant sexuality in general (the delicate parts of which they quoted only in Latin):

> There is not any science which has so little connection with theory as botany. . . . A man would not naturally expect to meet with disgusting strokes of obscenity in a system of botany. But it is a certain fact, that obscenity is the very basis of the Linnæan system. The names of his classes, orders, &c. convey often the vilest and most unnatural ideas. . . . Men or philosophers can smile at the nonsense and absurdity of such obscene gibberish; but it is easy to guess what effects it may have upon the young and thoughtless.[117]

In 1759 the Church suppressed the books of Linnæus for grouping man with the Primates; the ban was rescinded fifteen years later by Pope Clement XIV.[118] But objections to Linnæan classification ran deeper than prudery, or its affront to man's self-importance. Botanists disagreed with the use of stamens and pistils as characters;[119] historians have excoriated his "scholastic sophistry";[120] others disparaged his dogmatism[121] and heavy hand of authority.[122] Meanwhile (as we have seen), Buffon led natural history in France away from formal systematics, presenting nature instead as *un vaste spectacle*. This is not to say that Linnæus was without influence. As noted above, Gmelin produced an expanded Thirteenth edition after Linnæus's death, and all manner of "reformed" Linnæan classifications sprang up.[123] Erasmus Darwin and colleagues (including Samuel Johnson) translated the *Systema vegetabilium* into English,[124] and Darwin drew on Linnæan botany for his poem "The loves of the plants" (1791).

What, then, are fungi?

Linnæus left fungi in an untenable position: the fruiting body classified under Cryptogamia Fungi within Plantae, and connected by reciprocal metamorphosis with a supposed motile animalcule stage under Vermes Zoophyta within Animalia. This situation was susceptible to attack on multiple fronts. Against the classical view that fungi are plants, it was argued that fungi lack roots, flowers, and seeds, the supposed seeds being eggs of insects that form, inhabit, or prey upon the fungus.[125] If without seed, fungi may not be organisms at all, but rather "putrescent fermentations"[126] or "vegetable crystals".[127] Others argued that since fungi have a motile (animalcule) stage, they must be animals.[128] In response, Ellis pointed out (above) that motile animalcules appear in any decaying matter.[129] Analogies were liberally drawn with corals and/or zoophytes[130] which, of course, were equally misunderstood during much of this period. Even the long-discredited idea that fungi are minerals resurfaced.[131] Each of these positions was disputed in turn.[132]

258　KINGDOMS, EMPIRES, AND DOMAINS

It is not difficult to see how Münchhausen's separate kingdom might appeal, particularly to those who could not accept a perpetual (daily) transgression of the boundary between plants and animals. Already in 1768, Otto Friedrich Müller concluded that fungi cannot be accommodated in any of the three standard kingdoms.[133] Natalis Joseph Necker referred to the Linnæan metamorphosis as "poetic delirium", comparing it with tales from Ovid and the apocryphal borametz.[134] Zoophyta and Lithophyta are new steps on the ladder of nature, he wrote, and fungi might be yet another.[135] Later, in *Traité sur la mycitologie* (1783), Necker argued that fungi are not animals, as they do not arise from animalcula, and share none of the *caractères essentiels* nor faculties proper to animality. But neither are they plants, as they lack sexual reproduction and the organs characteristic of plants, and do not regenerate themselves annually. They differ from minerals by their mode of growth (by intussusception, not juxtaposition) and taking in of nourishment. Thus

> Seeing that the fungi do not belong to any of the three kingdoms, as I believe I have sufficiently proven: it is a necessity to establish a proper and particular one for them. I propose the name (Regnum mesymale) *Regne mésymale*. This new kingdom, once [it is] received and adopted by naturalists, adjoins immediately to that of the plants; then the fungi would form the third kingdom, and the minerals the fourth.
>
> The term mésymale, derived from ΜΕΣΟΣ *or, ò medius s[eu] neuter*, middle or neuter, that is to say, a kingdom midway or intermediary containing the productions of the plants.[136]

This new kingdom would adjoin the algae and lichens.[137] A similar conclusion was reached by Pierre Villemet:

> What can be concluded from this collection of such disparate objects, of this astonishing variety, which proves to us that if mushrooms have some properties of vegetables, they also seem, in certain respects, to hold to the animal kingdom, and perhaps even to claim, if not a place, at least a presence in the third. One might quickly create a new class for these beings. I offer to call it *pseudo-zoo-litho-phytes*; of course, it deserves to be subdivided into distinct species. The common mushroom, for example, would be a *pseudo-phyto-membrano cellulaire*, spongy, suberous [cork-like], with a pedicle, which carries a convex capital above, concave, *feuilleté* [layered like a puff-pastry] and fistulous [hollow-stemmed] below.[138]

Neither Kingdom Mésymale nor Class Pseudozoolithophyta gained traction, and a few years later Christiaan Persoon cut through the analogy and muddy thinking to argue that fungi are true plants because they exhibit differentiation, development, irritability, secretion, and reproduction.[139] Even so, arguments about where fungi might fit into the system of nature rumbled on well into the Nineteenth century, even as motile reproductive stages were increasingly recognized in other Linnæan plants.[140] As late as 1820, Nees von Esenbeck recognized Fungi as a separate kingdom of living nature.[141]

Adanson, Scopoli, and de Jussieu

The classification of nature did not end with Linnæus. Beyond the host of "improved" but fundamentally Linnæan systems, at least three noteworthy classifications[142] ignored key tenets of the *Philosophia botanica*.

First to appear was *Familles des plantes* (1763–1764) by Michel Adanson, a student of Bernard de Jussieu and Réaumur who spent five years studying the natural history and languages of Sénégal.[143] Finding the Linnæan system inadequate for such exotic flora and fauna, Adanson developed an alternative approach that, like that of Ray, recognized groups that share the greatest number of features in common, not only those of fructification but also of roots, branches, leaves, seeds and all others that collectively produce the "true *physique* of the plants".[144] There may be continuity in God's eyes, he wrote, but we see groups, and that is sufficient.[145] Gaps are widest between kingdoms, less so between classes, and slightest between varieties. Nor (*contra* Buffon) are species necessarily delineated by genealogy, as minerals do not reproduce, and even some plants and animals are generated asexually.[146] His fifty-eight families begin with *Byssi* (diverse cryptogams including *Aspergillus*, *Botrytis*, *Conferva*, and *Tremella*), then *Fungi* in seven sections (united for the first time with lichens), *Fuci*, *Hepaticæ*, and *Filices*, and conclude (surprisingly) with *Musci*.[147] His refusal to use binomial nomenclature, and his often-difficult relationships with the Parisian scientific establishment, ensured for *Familles* a modest reception.[148]

Giovanni Antonio Scopoli intended his *Introductio ad historiam naturalem* (1777) to be a *Systema naturæ* encompassing all three Kingdoms, but better aligned with the Chain of Being. Already in *Flora Carniolica* (1760) he had abandoned a rigorous Linnæan structure, presenting the cryptogams (fungi, algae, mosses, and ferns) first, adding a Class Inundatæ for aquatic plants including *Chara*, and utilizing a range of characters, not only those associated with fructification.[149] The *Introductio* deviates farther. It begins with Regnum Lapideum, followed by Regnum Vegetabile ordered from the tribes supposedly closest to minerals (Linnæan cryptogams) to those most like animals (also Linnæan cryptogams):

Tribe 1: Michelii or *Incompletæ*: lichens and seaweeds
Tribe 2: Plumerii or *Obsoletae*: ferns
(…)
Tribe 35: Dillenii or *Muscoideæ*: lycopods, mosses, and liverworts
Tribe 36: Battaræ or *Fungadeæ*: fungi, (more) lichens, and *Tremella*.

Plants and animals are united by a common plan based on cortex and medulla; the latter is responsible for movement, sensation, and generation.[150] *Introductio* concludes with Regnum Animale ordered from Müllerii or *Infusoria* (including rotifers and *Hydra*) to Kleinii or *Mammalia*. Neither Lithophyta nor Zoophyta is retained. The final genus is *Homo*, created in the image of God. In the words of Frans Stafleu, Scopoli "was a searching soul but did not find the answer".[151]

Philosophers have long considered that nature is complete or perfect. As such, it must be ordered according to some principle, whether mathematical harmony or the mind of God. Aristotle taught that natural groups exist and can be recognized on the basis of multiple characters.[152] Cesalpino and Linnæus sought to approximate the natural system using features of fructification.[153] While this approach offered certain advantages (notably its simplicity), it sometimes grouped otherwise dissimilar plants, dispersed well-accepted genera, and offered no guidance where the number of stamens is variable, or flowers are altogether absent. And whatever the (doubtful) merits of cortico-medullary theory, the sexual system was arbitrary, and even Linnæus acknowledged that it let us glimpse the natural order only in fragments.[154]

Antoine-Laurent de Jussieu sought to grasp the natural system more directly. In *Genera plantarum* (1789) he argued that classification involves two separate activities: the recognition of natural groups, and the arrangement (*ordinatio*) of these groups to reflect continuity in nature. Species are stable and perpetual, "successively reborn by continued generation"; the members of a species are uniform in all important respects. Species which share many characters are then grouped into genera; characters important for the delineation of one genus may be unimportant to another. Orders and classes are recognized on the basis of even fewer characters, often found among the organs of fructification.[155] The three kingdoms should be replaced with only two, the Organic and Inorganic.[156] Continuity is not strictly linear, and might instead be thought of as a geographical map with adjoining territories;[157] or better, as bundles of sticks.[158]

Applying this system to the plants, Jussieu recognized fifteen classes, one hundred orders, and a large number of Plantæ incertæ sedis (which he grouped by number and arrangement of petals). The first class, Acotyledones, is essentially the Linnæan Cryptogamia in reverse order (Fungi, Algæ, Hepaticæ, Musci, Filices) plus aquatic plants (Naïades), expanded from Scopoli's Inundatæ. Fungi are analogous in part to the animal zoophytes, and are (as it were) intermediate between these and certain algae, and very different from other plants in structure, flowering and habitat.[159] Jussieu's *Genera plantarum* implemented binomial nomenclature, and was published with accolades from the Académie royale des Sciences and the Société royale de la médecine, and under their privilege. It too suffered criticism for perceived errors in forming natural groups, but together with Adanson's *Familles* and Scopoli's *Introductio* made a powerful non-Linnæan case against the need for a kingdom of beings intermediate between plants and animals.

Zoophyta as animals

Linnæus incorporated Zoophyta into Animalia in 1735 (above), and there it remained in all subsequent editions of *Systema naturæ*. Its membership evolved—Hydra was added in 1748, *Spongia* and *Volvox* in 1766—and genus descriptions were modified as new species were reported. Yet knowledge of these organisms remained less than satisfactory. Apart from locomotion, irritability, or feeding behaviour, there was not even a good general-purpose test to distinguish animal from plant.[160] One consequence was that coralline algae were often grouped alongside corals as Lithophyta or Zoophyta. Peter Simon Pallas, a young medical doctor from Berlin then based in The Hague, cast doubt on the classification of several Linnæan species (particularly Vermes) in *Miscellanea zoologica* (1766); and in *Elenchus Zoophytorum* (1766) ventured a major rearrangement of Zoophyta, incorporating the lithophytes while removing *Tænia* (to the intestinal worms), *Volvox* (infusoria), and *Corallina* (plants).[161] Zoophyta represented the terminus of the animals:

> In the Zoophyta, as we will rightly be calling them, we become acquainted with vegetable nature so mixed with the animal, that it is truly two-faced and everywhere doubtful; we very clearly read *progress* from vegetative to fragments of the innate animal; we uncover an organic *fabric*, in the more noble animals gradually more complicated, in the most imperfect or, as is rightly said, in the *simplest manner*, of plants (as may be said) bordering on the more imperfect; we are astounded at the beginnings of animals, thence from living molecules, through the other animate classes, up to the culmination of perfection in human, gradations as well as modifications; from this newly shines forth the light of a theory of generation and strong psychology; the universal Nature has begun to be revealed, and we get to know of *living bodies, the* EMPIRE *of the organics, the insensible material* of our globe as the basis to be built upon, from a twofold animate as well as vegetative stock, joined together by the highest affinity, and consisting of lands, just as fields uncultivated by farmers [*inculto agro coloni*], up against one another.[162]

This is an image not only of Linnæan territories on a geographical map, but of virgin lands adjacent to one another, awaiting colonization or cultivation. Pallas recognized fifteen genera of zoophytes,[163] of which eleven are "Animal vegetans" and a twelfth (*Madrepora*) "Animal modo simplex, modo vegetans". *Brachionus* is "Animal minutissimum", *Tubipora* "Animal compositum, anomalum", and *Spongia* "Animal ambiguum, crescens, torpidissimum". *Tænia*, *Volvox*, and *Corallina* were set apart as "Genera ambigua".

Regarding continuity in nature, Pallas agreed with Bonnet that the three kingdoms—Animals, Vegetables, and Minerals—are utterly arbitrary and imaginary. Within the Empire of Organic Bodies lie two kingdoms, the farthest reaches of the Animal kingdom being a region of incertitude and transition

262 KINGDOMS, EMPIRES, AND DOMAINS

to the Vegetable.[164] Zoophytes share analogies with plants as well as animals. *Volvox* is the simplest animal, *Lycoperdon* the most-imperfect vegetable.[165] Animal life passes over from sponges to fungi or Byssi,[166] with no gap between them.[167] Indeed, *Natura non facit saltum.*[166] Species are bound closely by affinities into genera, these into orders, orders into classes, and classes themselves are interwoven [*inter se contexuisse*]:

> From this, various authors desire a certain pleasing *Scale* in *Nature,* of such excellence as will never be found, such as *Bradley* and *Bonnet* wish for. The gradation can be expressed no less well, indeed very much better, as various affinities in polyhedral figures, [with] genera of organic bodies distributed close by in numerous small spaces by turns. And as *Donati* has already judiciously observed, the works of Nature are not connected in series in a Scale, but cohere in a Net. On the other hand, the whole System of organic Bodies may be well represented by the likeness of a Tree that immediately from the root would extend from the simplest Plants and Animals a duplicated, variously contiguous trunk, Animals and Vegetables; the former advancing from Mollusca to Pisces, with a great lateral branch of Insecta sent out between these, [and] thence on to Amphibia; and at the farthest tip it would sustain the Quadrupeds, but [with the] Aves thrust out below the Quadrupeds as an equally great lateral branch. . . . The trunk is fully composed of the most-principal genera joined as neighbours, and everywhere send out genera as twigs, yet those lateral branches cannot connect among themselves.[168]

By the 1760s the Great Chain was a shadow of its former self: severed into organic and inorganic kingdoms, visible only in parts, suspected of parallel regions and the occasional branch. Pallas then dismembered it and rearranged the organic portions into a tree divided at ground level into contiguous plant and animal trunks, each representing principal genera. From the animal trunk extend great lateral branches of insects and birds; at finer scale, genera are everywhere thrust out like twigs. The 1787 German translation of *Elenchus* reads "The trunk, composed of the principal series of genera that are related and stand compact to one another."[169] Unfortunately, Pallas did not provide an image. Later scholars have found in this passage a first distinction between *analogy* (superficial and idealized relatedness) and *homology* (structure and mode of generation),[170] and "the first known scheme to express the sequential development of animal organisms in terms of a family tree".[171] At minimum, it is the first explicit use of the tree metaphor to describe relationships among modern classes of animals.

A decade later, the Danish naturalist Otto Friedrich Müller recognized five orders within Vermes: Infusoria, Mollusca, Testacea, Helminthica, and Cellularia (inhabitants of cells: the Linnæan lithophytes and remaining zoophytes).[172] His Mollusca included *Hydra*, cephalopods, comb jellies, sea cucumbers, and jellyfish; his Testacea sea cucumbers; and his Cellularia corals, sponges, and the fungus *Clavaria.* The non-marine infusoria, helminths, and testacea he described further in *Vermium terrestrium et fluviatilium* (1773–1774), and this in turn served as the basis for *Animalcula infusoria fluviatilia et marina* (1786), considered the first comprehensive taxonomic treatment (379 species in seventeen genera) of the organisms now known as protozoa.[173] Blainville[174] and Bory[175] credit Müller with having introduced the taxon *Infusoria*, subsequently adopted by Gmelin in the Thirteenth edition of *Systema naturæ*.

Johann Friedrich Blumenbach (*Handbuch der Naturgeschichte*, 1779) retained the Würmern in five orders: Mollusca, Testacea, Cartilaginea (sea-urchins, starfish, sea-palms), Corallia, and Zoophyta (*Pennatula*, *Hydra*, *Vorticella*, *Volvox*, and *Chaos*, the latter including vinegar-worms and spermatia). August Batsch (*Kenntniss der Geschichte der Thiere und Mineralien*, 1788–1789) recognized eleven families of Würmern including Blumenthiere (polyps and corals), Sonnenthiere (vorticellids), and Infusionsthiere. He equated these latter three with Zoophyta or Pflanzenthiere; again, they take us to the very terminus of the animals—although for Blumenbach the animals adjoin the minerals.[176]

Finally, we may note *Die Pflanzenthiere* begun by Eugen Esper, professor of natural history at Erlangen.[177] He wrote that systematic classification has yielded no surprises; but the wondrous species in one *Abtheilung* of nature—the Phytozoa, or Pflanzenthiere—make us ask whether the accepted kingdoms of nature are the most appropriate. With these beings, the Lithophyta and the Zoophyta of Linnæus,[178] the boundaries of the three kingdoms are nearly inseparable; it is here we catch sight of the true order of steps in the scale of creatures.[179] Indeed, both the plant and animal kingdoms are united in single genera.[180] Esper seems to imply that the Chain may not ultimately pass through a single transitional genus or point of contact.[181]

Thus, by the final decades of the Eighteenth century, Zoophyta was securely ensconced among the animals, even as its precise membership remained elusive. In Germany, Russia, Scandinavia, and Great Britain, *Zoophyta* usually included hard and soft corals, bryozoans, hydroids, vorticellids, sea-pens, and *Hydra*; often corallines and tapeworms; and sometimes amœbae and infusoria. Where the Linnæan writ ran less strong (notably France), *Zoophyta* might also include jellyfish, sea-urchins, and starfish. As we shall see in Chapter 17, in the second decade of the new century Georges Cuvier fundamentally reimagined relationships among animals, and *zoophyte* gradually became an informal term rather like *beast*, *brute*, or *herb*.

Summary

Classification and order were hallmarks of Eighteenth-century thought, not only in natural history.[182] This may have been particularly true in Sweden, where (it has been argued) social and bureaucratic organization and discipline had made the country powerful.[183] Even so, desire for order in nature reached across much of Europe. Johan Wallerius sought to establish a Kingdom of Water, with classes,

264   KINGDOMS, EMPIRES, AND DOMAINS

orders, genera, and species;[184] Johann Denso added a Kingdom of Fire,[185] Johann Becher a Kingdom of Air.[186] Johann Tietz grouped æther, air, and water with stones, salts, sulphur, and metals into a *Materialienreich*,[187] while Georg Borowski arranged air, fire, water, and earth into an *Elementarreich*.[188] Luke Howard classified clouds, and described the metamorphosis of one cloud type into another.[189] Minerals,[190] crystals,[191] chemical substances,[192] medicinal compounds,[193] diseases,[194] even botanists and botanical writers[195] were duly classified. Johann Beseke then classified these classifications.[196]

For the Eighteenth-century savant, it did not suffice to situate each fungus, insect, or worm into an appropriate taxon. It was also important that the classificatory hierarchy be well-constructed, the characters stable, the names brief but informative. In particular, naturalists should diligently search out nature's own system.[197] At the beginning of the century this was assumed to mean a single linear scale; but by the 1760s naturalists were no longer so certain, and alternatives were being mooted. Intermediate kingdoms or other high-level taxa—neutral, chaotic, *mésymale*, pseudozoolithophytic—were proposed within the familiar linear chain. It was less clear that they were compatible with the parallel, branched, or treelike alternatives.

We passed without comment over a puzzling comment by Pallas (above):

> The gradation [in nature] can be expressed no less well, indeed very much better [than as the Chain of Bradley or Bonnet], as various affinities in polyhedric figures, [with] genera of organic bodies distributed close by in numerous small spaces by turns.[168]

What did Pallas mean by *figuræ polyedræ*? Neither a linear scale nor a tree is a polyhedron.[198] The networks of Nieremberg and Donati were presumably rectangular grids, with natural series of genera (the vertical lines) connected by affinities (horizontal lines), or vice versa. We know the natural world only imperfectly, so the fabric is incomplete and loosely woven; but genera form the threads, not the "small spaces". Although (thanks to Paul Giseke) we think of the territories on Linnæus's geographic map as rough circles, the map did not appear until 1789,[199] whereas Pallas mentioned *figuræ polyedræ* in 1766. Did Pallas envision the Linnæan territories as polygons? In any case, the early Nineteenth century did bring geometric representations of the natural world, some quite remarkable. These we explore in the next chapter.

16

Beyond the end of the Chain

> One will never succeed in arranging the plants in their natural order, in such a way as to mark all the relationships that exist between families, if one arranges them along one or even along several lines. Any plane shape, however complex a form one grants to it, will represent the natural order only imperfectly.
>
> —Augustin Augier, *Observations sur l'ordre naturel des végétaux*[1]

As we have seen, the Great Chain of Being did not survive the Eighteenth century intact. Continuity remained an important point of reference in social and religious contexts, but by the 1790s it was old-fashioned to arrange natural beings in a simple linear scale or chain. Different naturalists saw the problem differently. Many accepted that some genera form local series without cohering into a single scale, while other genera run in parallel, or appear as side-branches. Félix Vicq d'Azyr[2] and Louis-Jean-Marie Daubenton[3] found that different types of characters imply different series and groups of animals. A better case could be made for a chain of classes (not genera), but classes are artificial and defined by only a few characters. Johann Blumenbach regarded the search for a Chain as a harmless intellectual game:

> I am indeed very much opposed to the opinions of those, who, especially of late, have amused their ingenuity so much with what they call the continuity or gradation of nature; and have sought for a proof of the wisdom of the Creator, and the perfection of the creation in the idea, as they say, that nature takes no leaps, and that the natural productions of the three kingdoms of nature, as far as regards their external conformation, follow one upon another like the steps in a scale, or like points and joinings in a chain. But those who examine the matter without prejudice, and seriously, see clearly that even in the animal kingdom there are whole classes on the one hand, as that of birds, or genera, as that of cuttle-fish, which can only be joined on to the neighbouring divisions in those kinds of plans of the gradation of natural productions but indifferently and by a kind of violence....
>
> And in this kind of systems, so far from their being filled up, there are large gaps where the natural kingdoms are very plainly separated one from another.

266 KINGDOMS, EMPIRES, AND DOMAINS

There are other things of this kind; and so although after due consideration of these things, I cannot altogether recognize so much weight and importance in this doctrine of the gradation of nature, as is commonly ascribed to it by the physico-theologians, still I will allow this to belong to both these metaphorical and allegorical amusements, that they do not throw any obstacle in facilitating the method of the study of natural history.[4]

Antoine-Nicolas Duchesne went further, drawing up a "Table of Pythagoras"[5] in which the main groups of beings (mammals, birds, reptiles, fish, insects, worms, vegetables, minerals) are assigned rows, and their distinguishing characters columns. The Chain lies along the diagonal, but Duchesne could annotate almost every square: features supposedly distinctive to one group are in fact widespread. He concluded that the chain is a "seductive illusion", and belief in so-called intermediate beings "ridiculous".[6] Taking aim at Charles Bonnet, Duchesne asserted that naturalists laugh at the Chain of Being; only "*littérateurs* who think themselves metaphysicians" continue to traverse his ladder. With this analysis Duchesne might have argued for a network of beings, or for no Plan at all. Yet many of the characters in his table were (in the words of a previous generation) accidents—size, colour, vocalization—and his Table sank without a trace.[7]

What, then, is the shape of the Plan of Nature? In this chapter we explore a range of proposals including a map, network, tree, spiral, circles, and polygons symmetrical or irregular. Most have little in common except their nonlinearity. Some were motivated by a Pythagorean notion that number or geometry underlies nature. Many were put forward between about 1770 and the early 1830s, although the map (Linnæus), network (Nieremberg, Donati), and tree (Pallas) were introduced earlier. I do not include Georges Cuvier's arrangement of animals into four *embranchemens*, as that was driven by comparative anatomy, not by a supposed plan of nature.

Nature as a map

On 13 February 1750, Linnæus wrote to Johann Gesner that

The plants themselves have been disposed by the great Creator according to their affinities, in such a manner as if they had been written on a chart or geographical map in which the boundaries between closely adjacent regions are very difficult to distinguish. They acknowledge a dual affinity which no mortal could easily overthrow. A natural system is certainly the chief requirement but, while we hold to it, it is the most difficult of all to bring to completion, since the boundaries [of the genera] are scarcely defined, as is clear today from the natural orders.[8]

The map soon became a tenet of botanical theory.[9] There is no evidence that Linnæus himself ever drew out such a map; his student Paul Giseke made a rough sketch in 1766 based on a lecture by the master, then redrafted it for publication in 1789, labelling it *tabula genealogico-geographica affinitatum plantarum*.[10] This map (Figure 16.1) shows fifty-six orders of plants, each represented as a wavy circle with an area corresponding, more or less, to the number of its genera.[11] The vertical and horizontal axes impart scale, but otherwise bear no particular meaning (but see below). Orders sharing the greatest affinity are placed adjacent or proximate to one another. Charles-Louis l'Héritier de Brutelle imagined a similar but spherical map encompassing the classes, orders, and genera of all three kingdoms.[12]

Affinity is a slippery concept, but basically refers to closeness (in the scheme of nature) inferred from similarities in form, structure, or physiology. From *Philosophia botanica* we learn that affinity can exist between characters (especially essential characters) or taxa,[13] and can be quantitative, varying in proportion with the number of parts of fructification.[14] Later authors including Antoine-Laurent de Jussieu drew an explicit analogy with chemistry, where affinities "determine how chemical substances attract each other and form compounds of fixed stoichiometry".[15] Similar regularities might occur among living nature, and help us understand its plan. Affinity (similarity within a series) was sometimes distinguished from *analogy* (similarity between parallel series).[16] Elias Fries sought to reinterpret

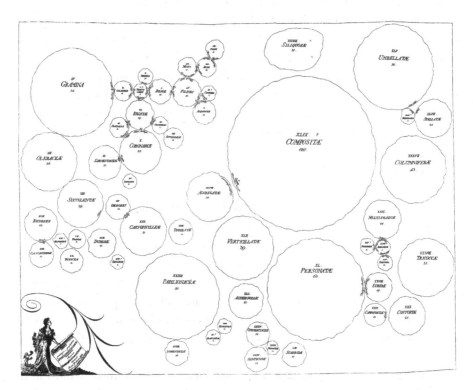

Figure 16.1. Genealogical-geographical table of the affinities of plants, after the Linnæan natural orders, drafted by Giseke (1789) and appearing in *Prælectiones* (1792), facing page 623.

268 KINGDOMS, EMPIRES, AND DOMAINS

the horizontal axis of Giseke's map as affinity, and its vertical axis as analogy.[17] Alas, none of this greatly clarifies what Linnæus meant by *dual affinity*.[18]

The word *genealogical* in Giseke's title of the Linnæan map is more puzzling. Linnæus believed that species were created by God, and have been maintained ever since by shared ancestry of cortex and medulla; but given our imperfect knowledge, higher taxa are often artificial. If we could somehow constitute genera, families, and orders according to the natural order, they too would become more genealogical— although not in the sense of direct ancestry, but instead like cousins.[19] But there is no sense in Linnæus (or Giseke) of genealogy on a branching tree.[20]

Although never as ubiquitous as the Chain or Scale, two- or three-dimensional maps of nature remained in use until at least 1871.[21] The format allows for the inclusion of intermediate kingdoms, but none was recognized by Linnæus, Giseke, or l'Héritier.

Nature as a network

As long ago as 1635 Nieremberg fleetingly compared nature to a network, perhaps more as a rhetorical flourish than a careful statement based on observation and analysis.[22] In 1750 Donati revived the metaphor to describe continuities among terrestrial and aquatic organisms. On land, mosses and mushrooms probably link plants to insects;[23] in the kingdom of the waters, however, we find animal-plants such as *Tethya* that are sensitive (although lacking a head or eyes) and drift from place to place, yet resemble plants in the simplicity of their structure and mechanism.[24] Earthworms (which live in humid soil) and the freshwater polyp are further links.[25] Donati did not provide a full network of nature, but he was widely credited, through the second half of the Eighteenth century, with having introduced nature-as-network.[26]

The network metaphor was subsequently adopted by Giuseppe Olivi:

Meditation upon the objects that inhabit the sea [allows us to] recognise as evidence, that almost every order is joined with various other orders, and that every species, each coming close to other similar ones through many analogous characters, then moves away from the same [species] in other essential qualities, such that it approaches objects of different orders. Such reflections would lead me to adopt the other imagined opinion, and, to tell the truth, imperfectly sketched by our Donati, that is to consider the progression of organic bodies as a linear series, which in approaching and moving away [from others], form almost a network.[27]

Other descriptions were less abstract. Jacques-Henri Bernardin de Saint-Pierre depicted a network of mollusc shells based on their sphericity.[28] Lorenz Oken sought to explain the reproductive modes and potentials of animals, plants, polyps, and corals using triangular networks.[29] Johann Rüling presented a large *Tabvla phytographica vniversalis affinitates ordinvm natvralivm plantarvm* in which orders are represented simply by name, not as a territory. Local series branch, run in parallel, and/or are connected by horizontal lines representing affinities. Nor are the series strictly Linnæan: an unbranched series, for example, connects fungi (Mucorales) through algae, lichens, mosses, ferns, and palms to lilies.[30] Johann Hermann offered an immense *Tabula affinitatum animalium*[31] partitioned into the six Linnæan classes. Affinities are abundant and run both horizontally and vertically, including between animal taxa and their supposed counterparts among the Vegetables and Minerals. Hermann provides more detail throughout the text: the rays (*Raja*), for example, share affinities with mammals, amphibia, insects, birds, fish, and worms.[32] The *Tabula affinitatum regni vegetabilis* of August Batsch is a tangle of horizontal, vertical, and diagonal affinities.[33]

Gottfried Treviranus distinguished affinities between characters from those connecting organisms:

> If one now asks whether the epitome of these forms, as Bradley and Bonnet would have it, forms a ladder, or, as Donati and Olivi claimed, a net, then each of these questions can be answered in the affirmative, and each in the negative, depending on which point of view one proceeds from. Those forms make a ladder as soon as one takes into consideration only the individual parts of their organization; they make a network, not a ladder, if one brings their entire organization into contention.[34]

As this is found in the first book to use the word *Biologie* in a modern sense, "it can be said that nature-as-network was present at the dawn of modern biology".[35] Not long afterward, botanist Robert Brown found a network preferable to a linear chain.[36]

Networks could represent relationships other than affinity. In *Histoire naturelle*, Buffon depicted a network of canine breeds—"if you will, a sort of genealogical tree . . . oriented like the geographical maps" (Figure 16.2). The solid lines represent genetic continuity during translocation of the breed to a different country: the sheepdog, for instance, to Lapland (where it degenerated) and to Iceland, Russia, and Siberia (where its form has been maintained, even perfected). The dashed lines identify interbreeding, the axes geographical latitude and longitude.[37]

Nature as a polygon or Easter egg

Maps and networks focus on individual taxa and their relationships: their story lies in the details, not in an overall emergent pattern. By contrast, from about 1812 nature began to be depicted as one or another geometric figure, often with a plane of symmetry. This trend began in northern Europe[38] and involved some of the leading naturalists of the day.

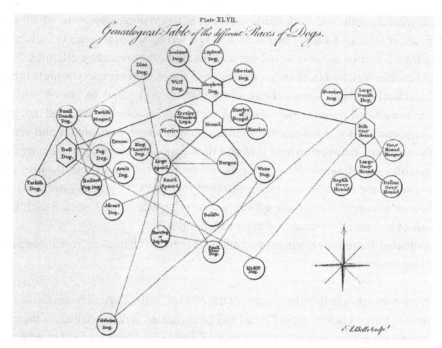

Figure 16.2. Genealogical table of the different races of dogs, from the Third edition (1791) of Smellie's translation of Buffon's *Histoire naturelle*, volume 4 (1791), Plate 47 preceding page 49. The French original first appeared in 1755. Wellcome Library, Creative Commons Public Domain Mark 1.0.

In 1812 Karl Rudolphi, a specialist on parasitic worms, reclassified animals based on features of the nervous system. His system was to be a modified linear scale, as "it has long been understood that the animals cannot be reduced to a ladder".[39] He divided animals into Phaneroneura (those with a dorsal spinal cord, plus a system of ganglia comprising the sympathetic nervous system) and Cryptoneura (in which the nervous substance is indistinct), and subdivided the former into Diploneura (mammals, birds, amphibians, fish) and Haploneura. Within Haploneura he recognized Myeloneura (with a ventral spinal cord: crustaceans, insects, annelids) and Ganglioneura (with a ganglial system but no spinal cord: molluscs, rays). Neither Myeloneura nor Ganglioneura has primacy over the other: they proceed side by side, each meeting Cryptoneura (filamentous intestinal worms, jellyfish, radiata, and "other zoophytes") in the inevitable ambiguous genera.[40] Rudolphi depicted this arrangement in a small, somewhat redundant diamond-shaped figure[41] without saying precisely what the lines represent.[42]

Five years later Georg August Goldfuss,[43] then secretary of the Imperial Leopoldinian-Carolinian Society of Naturalists in Erlangen, dedicated the remarkable *Ueber die Entwicklungsstufen des Thiere* (1817) to his older colleague Christian Nees von Esenbeck.[44] Subtitled *Omne vivum ex ovo*,[45] the tract is remembered mostly for its depiction of the animal kingdom as a *Thierey* (animal-egg), indeed an Easter egg (Figure 16.3):

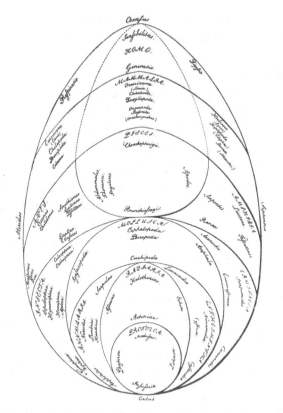

Figure 16.3. Goldfuss, animal-globe or Easter egg, from *Ueber die Entwicklungsstufen des Thieres. Omne vivum ex ovo* (1817).

I send you, dear friend! here an animalistic Globe [*Weltkugel*]. It has its north and south poles, and its eastern and western halves;[46] but it is quite strongly pressed inward at the poles. In this way it has admittedly lost the form of an apple, but instead rather resembles an egg, in truth an Easter egg [*Osterey*], namely an egg of the Resurrection, of east to west, to evening, to the downfall of advancing life. At the same time, it may be the materialization (from the World-egg) of the oft-disputed saying: *omne vivum ex ovo*.

On the eastern half of this animal-globe, animal life sprouts up [into sight] from its organizable primal matter, and strives to become individual on three sides. If it did not grow in the east, but instead in the North Pole, it would take root there, rise straight to the south, spread out east and west like the surfaces of a leaf, and contract again to potential unity in the south, and therefore become Plant. But the animal-egg, whether or not it arises from the water like all life, is not tied down and linear like the plant, defeated, but keeps its mobility, changeability, and thus becomes free and something in itself. To be sure, the animal also has a root which strives to push forward stem, leaf and flower. Just as this root continually dissolves into points, spheres and lines, so also the animal stems, leaves and flowers strive restlessly toward separation and independence.[47]

272 KINGDOMS, EMPIRES, AND DOMAINS

The tract continues in overwrought comparison of animal, vegetable, and earth, painful to read even by the standards of *Naturphilosophie*:

> Man is the flower of the animal kingdom. He propagates himself, digests, breathes, and thinks. Through the act of procreation, he stands in connection with the universe [*All*] and becomes one with it; through the process of digestion, he ensures the preservation of his individuality, and thus separates [himself] egoistically from the Universe; through the act of respiration, he is reflected onto the external world, and by the brain he reflects upon himself. What his brain is for his own body, is the whole man for all Animality. By representing the highest synthesis of brain, respiratory, digestive and sexual organs, the individual animal classes must themselves behave in the same way as those organs in man; each class must be formed excellently respective to one of these organs, and must therefore be placed as part in relation to the organic whole of Animality.[48]
>
> The middle rank of the mammals, the fishes, molluscs, radiates and protozoa are the root by which the brain-flower [*Gehirnblüthe*] of man becomes one with the Universe; at the northern rank of the mammals, the amphibians, crustaceans and intestinal worms the digestive organs are predominant, corresponding to the stem [*Stamm*] of plants; the carnivorous mammals, birds, insects and annelids, stand[ing] opposite these in the south, are respiratory animals, the leaves of the animal-stem.[49]
>
> The animals of the first and lowest class are the *Protozoa*, seed-animals of the primæval animal. Determined directionless, seemingly arbitrary movement, mostly in turning on its own axis and progressing in turns like moons around a planet, is the first evident act of animal life.[50]
>
> The brain-animal of the mammals, the flower of the animal kingdom, the animal of animals, is man. In him all animal organs are united in the utmost perfection and harmony.[51]

After a reply by Nees von Esenbeck—including a poem invoking lotus buds and the radiance of Krishna—Goldfuss has the final word:

> So here you have it, dearest friend! my animal-egg. Take it as a merry joke, or as serious. It is the same to me.[52]

It may indeed be difficult to take the figure or tract seriously, but Goldfuss clearly viewed even "the simplest infusorial forms" as animals, indeed the source of all animals.[53] By depicting taxa as sharing regions of the egg (not, as in Hermann or Batsch, connected by lines of affinity) he conveys inclusion and unity—themes central to *Naturphilosophie*, and to German Romanticism more broadly.[54] We also note the new term *Protozoa* (encompassing Phytozoa, Medusae, Lithozoa, and Infusoria), here introduced as the "seed-animals of the primæval animal". Goldfuss formalized this taxon in 1820.[55]

Merry joke or not, the egg diagram briefly caught the fancy of the young Karl Ernst von Baer, who reimagined it as a series of four stages of increasing morphological complexity and scope of habitat (Figure 16.4).[56] A central axis connects infusoria through eleven intermediate taxa to man, while eight other taxa appear to the side.[57] Soon thereafter von Baer forswore such "regularly schematized representations".[58]

In October 1821 Karl Eduard Eichwald submitted *De regni animalis limitibus* as his habilitation thesis at Dorpat (Tartu) University.[59] Prefaced with Aristotle's "Nature proceeds little by little from things lifeless to animal life",[60] it identifies supposed transitions (not just affinities) among classes and orders of animals. And not only within the animals: certain organisms rejoice in being both animal and plant.[61] Some naturalists call them animals, others plants, but in truth the freshwater polyps and infusoria are neither. Nor, for that matter, do they belong to an intermediate kingdom; instead, they dissolve the boundary between plants and animals.[62] Insofar as infusoria are animals they are the simplest and least perfect, the primordium of animal organization, and can arise by spontaneous generation.[63] Eichwald's *Tabula transituum animalium* (Figure 16.5)[64] identifies these transitions and illustrates his concept of grades, but its left-right symmetry is a matter of artistic license.

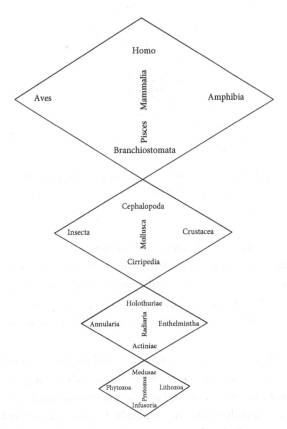

Figure 16.4. Von Baer, schematic representation of animals, from an early notebook [1819]. Redrawn after Raikov (1951), pp. 91 and 95.

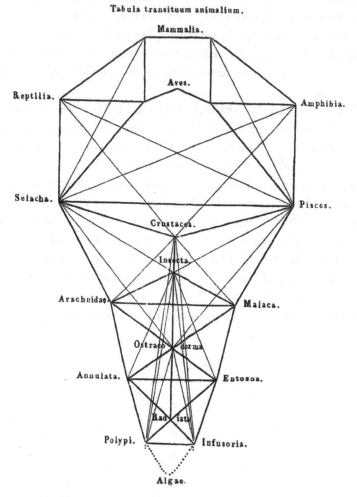

Figure 16.5. Eichwald, table of transitions among animals, from *De regni animalis limitibus atque evolutionis gradibus* (1821).

A few years later, in Paris, botanist Adrien de Jussieu turned his attention to Rutaceæ, a group of aromatic shrubs earlier recognized by his father Antoine as an order[65] but now in need of revision following discoveries in Australia and South America. The constituent genera could not be organized into or around a single linear scale, so Adrien drew up a Linnæan map of affinities. However,

> this ingenious idea of Linnæus, which M. de Candolle has developed with details no less ingenious, the idea of comparing to a geographical map, plants arranged according to their natural relations, I still believe insufficient. I believe we cannot express on one surface these affinities, multiplied and intersecting in all directions; and I would rather attempt to compare in this respect the groups of organized beings with those systems of bodies scattered everywhere in space, where many

are held at different distances around a common centre and can themselves become secondary centres; while each of the systems touches at the same time with its vast and innumerable circumferences a host of neighbouring systems, and that several other bodies, floating indecisively amongst them, escape for a long time in their eccentric race to the inadequate observation which pretends to assign to them laws and a certain place.[66]

Such was his intent for his *Carte des genres disposés suivant leurs affinités mutuelles* (Figure 16.6),[67] an ambitious hybrid of star-map and asymmetric polygon. Internally, double lines dissect Rutaceæ into five main subgroups of undefined taxonomic rank.[68] Genera (shown as dots of different size according to number of species) are connected by lines of affinity, many radiating out from *Zanthoxylum*. A few genera outside Rutaceæ, but linked into the group by lines of affinity, are shown for good measure. The nine external sides are labelled, those around Diosmeæ by geographical distribution, the others indicating taxa with which those Rutaceæ had

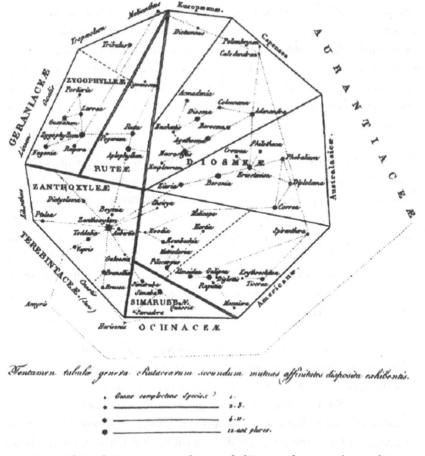

Figure 16.6. Adrien de Jussieu, map of genera [of Rutaceæ] arranged according to their mutual affinities (1825), Plate 29, lower panel.

276 KINGDOMS, EMPIRES, AND DOMAINS

theretofore been associated. Adrien de Jussieu does not explicitly claim that the figure captures much about the plan of nature, and it is telling that he moved on (below) to explore other representations.

Nature as a branched tree

Trees have long served as a metaphor of continuity in nature, although until the mid-Eighteenth Century, emphasis fell on the trunk, connecting earth to heaven. With Charles Bonnet and Peter Pallas, the branching structure became important too. Bonnet suggested that insects, shellfish and certain other groups might branch off the Scala Naturæ,[69] while Pallas represented the system of organic bodies as a tree.[70] Linearity could survive a few anomalous side branches;[71] but for Pallas, parallel trunks and "great lateral branches" were the main story. Although Pallas explicitly contrasted his tree with Donati's network and denied major lateral relationships, Rudolphi managed to find a network in his words:

> [Pallas] irrefutably showed that the organic kingdom does not permit a strict separation among its creatures, that the animals and plants grow together with the zoophytes, that a ladder of nature is not to be thought of, but that the natural bodies rather together form a network. One could best think of the system of organic bodies as a tree, which stands on the inorganic kingdom as upon the ground, [and] from the root divides itself into two trunks which now and again approach one another.[72]

Rudolphi had no access to the image in Pallas's mind; nor do we. Curiously, in the same year that Pallas suggested the tree metaphor, the nineteen-year-old Antoine-Nicolas Duchesne depicted a tree of relationships among strawberries.[73] As in Buffon's network of dog breeds, the lines in Duchesne's *Généalogie des fraisiers* (Figure 16.7) depict genetic continuity under human intervention (transplantation and breeding). Pallas became famous and his tree metaphor was widely known,[74] but thirty-five years passed before another tree was published. Augustin Augier, a teacher in southeastern France, set out to arrange the families of plants in continuous series, but met with great difficulties:

> I was then convinced that plants formed different series united by their base, observing between them a gradation like those of the branches of a tree: I then worked to make different series, and to establish their gradations . . .
> The order that I established among plants is found equally in the three kingdoms of nature, and that seems to me a favorable precedent for it to be regarded as natural. The three kingdoms form three major series which begin with the least perfect beings and end with the most perfect. Under the similarity of their organisations, they are themselves made up of several series or smaller

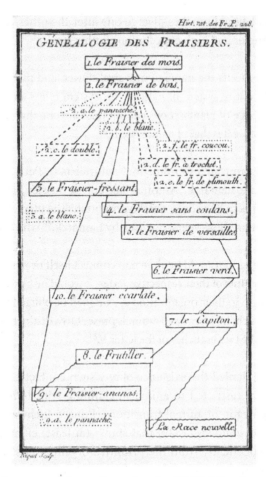

Figure 16.7. Duchesne, genealogy of strawberries (1766) following page 228, as engraved by Niquet. Based on image provided by Société Nationale d'Horticulture de France.

families which are united by beings which, although appearing to take the nature of two or several families, properly belong neither to one nor the other, and by this form transitions; this makes it difficult to find decisive characters. Zoophytes unite the three kingdoms; mammals are united to fish by the whales, and birds to quadrupeds by bats, etc.[75]

The resulting tree is largely symmetrical, and intentionally so:[76]

The different classes of beings do not form a continuous series that one could arrange along a single line. Had that been the case, the source of nature would have been known since a long time. The different classes and families form very distinct groups, arranged in a very symmetrical and admirable order with respect to one another; they have only points of contact between each other.[77]

278 KINGDOMS, EMPIRES, AND DOMAINS

Even so, the main lateral branches diverge one after the other, not symmetrically, from the trunk. Augier had high expectations for his system:

> This shape justly merits the name of universal flower, and the unfolding of this flower forms the entire system of nature. This system offers an easy method for acquiring knowledge of plants, more natural than anyone that the human spirit could invent, because it is that of nature itself.[78]

His *Essai* was communicated to influential naturalists in Paris including Georges Cuvier, Jean-Baptiste Lamarck, Antoine-Laurent de Jussieu, and Étienne-Pierre Ventenat,[79] but seems to have been quickly forgotten. The plan of nature was under active consideration in those years, not least by Lamarck,[80] who later wrote that

> if the vegetable kingdom could be shown to connect itself or pass into the animal kingdom by any points of their respective series, it would be by those alone which are the most simple in their organization . . . we shall be obliged to admit that instead of forming a chain, plants and animals present two distinct branches, united at their base like the two branches of the letter V.[81]

Lamarck, however, denied the existence of any such contact-points, and his own skeletal trees[82] were restricted to animals beginning with *Infusoires*. Moving on from his earlier polygon of animal transitions, Eduard Eichwald offered an idiosyncratic *arbor vitæ animalis* (1829) that shows (at least) eight main lines of animals arising independently from the primitive organic mass (Figure 16.8). The accompanying text explains:

> *Chaos* because *animal*, as globules of primitive mucus of varied forms, but arranged in order and aggregated, abundant in the stagnant water, continually generated all true marine animals; therefore in the conditions that soon followed, [it is] as if the lowest branch of the tree, rooted in the primitive mucus, sprouted forth the lowest genera of *phytozoans*, immediately in succession diversified from this chaotic mass, perhaps with organs both internal and external, spontaneously motile, following a path. In the same way, on the other hand, those of the *phytozoans* among the throng of these smallest beings, standing in the stagnant water, spontaneously produced the vibratory and rotatory polyps, thus in marine water, in the same way the organic chaotic heavy mass, as if the second branch of the tree of animal life, the lowest *cyclozoa*, medusae, of a circular bodily form, conspicuous; of which a higher branch was indeed grafted onto the tree of life . . . the greater elevation of this branch can of course be explained, if it is allowed that the lower can itself send forward a new sprout at the place that is highest of this antecedent class, a series sprouting forth in the chaotic mass, from the root of the common tree.[83]

Figure 16.8. Eichwald, tree of animal life, from *Zoologia specialis*, volume 1 (1829), facing page 41.

In Eichwald's depiction the animal lineages are closely appressed, with genera thrust out everywhere like twigs;[84] but this is not quite the tree of Pallas, as the second main trunk is absent, and with it the close contiguities between animals and plants.[85] As we saw above, Rudolphi (1812) claimed that Eichwald's tree stood upon the inorganic kingdom; Eichwald himself wrote instead that "that chaotic mass substance to which we refer, organic [and] ambiguous, contains distinct in itself neither the animal nor the vegetable, [but] has the strength to transform truly into either, as the opportunity is given it."[86]

In his influential *Cours élémentaire . . . botanique*, Adrien de Jussieu accepted that natural relationships were better represented as a tree than as a chain, geographical map, or network, in part because a tree "does not exclude the idea of a general series". As in Augier's tree (only more so), families can "all converge toward the trunk and leave it one after the other in a single line unrolled from the bottom up".[87]

Nature as a spiral

Pavel Horaninov, a medical doctor, was professor of botany and pharmacology at the Petersburg Medical-Chirurgical Academy when Eichwald joined the faculty to teach zoology and mineralogy.[88] The two held similarly transformist views. In *Primae lineae systematis naturae* (1834) Horaninov divided nature into Regnum Amorpho-anorganicum and Regnum Organicum, but thereafter divisions are fourfold, or in multiples of four. He recognized four classes of amorpho-inorganics, and four main divisions of organics (Vegetabilia, Phytozoa, Animalia, and Hominem). Vegetabilia contain four "circles" with twenty classes in all, Phytozoa four classes (Algae, Fungi, Polyparii, Acalephi), and Animalia twelve. At finer scale Algae and Fungi display four orders each, Acalephae four sections with twelve orders.

Such Pythagoreanism merits a diagram, and Horaninov does not disappoint, depicting nature as a grand spiral (Figure 16.9).[89] Outermost we find thirteen

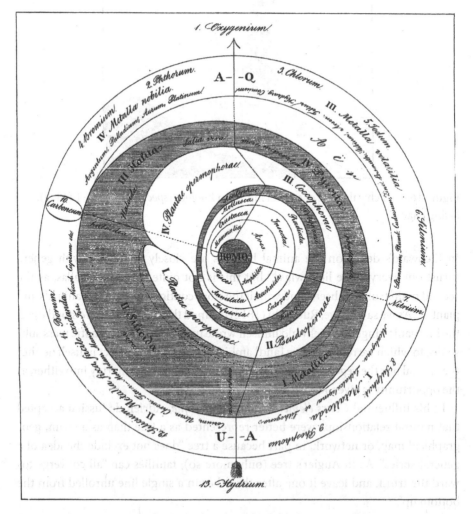

Figure 16.9. Horaninov, schematic diagram of nature, from *Primae lineae systematis naturæ* (1834).

chemical elements,[90] hydrogen and oxygen forming the principal polarity, nitrogen and carbon a secondary axis. Hydrogen ("absolute positive") and oxygen ("absolute negative") combine to produce water, and the letters A Q U A remind us that water is the *fluidum primordiale*, the *menstruum materiale commune omnium*.[91] Next come twenty-four more elements, then Air, followed by the four classes of amorpho-inorganics likewise in clockwise orientation. A dotted line leads us to the beginning of the vegetable series, and we follow the *prima linea* counterclockwise through phytozoa and animals to Man at the centre. Oddly perhaps, one of his four classes of amorpho-inorganics, Æther or Ignis, is nowhere to be found; nor does he make use of the many contiguities on offer. Ignoring the substantial body of opinion (reaching back to *Philosophia botanica*) that the least-perfect animals adjoin the least-perfect plants, Horaninov places infusoria immediately after complex flowering plants.

Horaninov's *Tetractys naturae seu systema quadrimembre omnium naturalium* (1843) was a direct extension of his *Primae lineae*, albeit with the nature-mysticism toned down and plants accorded more emphasis. Unfortunately (for it would have been most interesting) he does not actually depict nature as a tetractys, but his focus on the number four is relentless. We again find four kingdoms; comprising the (renamed) Amphorganic Kingdom are Fungi in four orders, Algae in eight, Polyparii in four (each with four families), Acalephae in four as well. In *Characteres essentiales* (1847) he retained Regnum Amphorganicum and its four classes, albeit with substantial revisions.[92]

Nature as a circle

In 1808 Gotthelf Fischer von Waldheim, professor of natural history at Moscow University, depicted plants, animals and polyps as on two *cercles de mouvement* which touch, or perhaps intersect, one another at two points (Figure 16.10). Natural bodies, he wrote, are either organized or brute. In the left-hand cycle, plants assimilate brute matter and transform it into organized matter; on the right, polyps recycle organized matter to the earth. At the upper point (labelled 1), mould is produced by the corruption of animal matter; at the lower (2), infusorial animals arise from vegetable matter.[93] Even bearing in mind that this is not a Venn diagram,[94] complete interpretation is elusive, given that Fischer von Waldheim classified polyps among the animals.[95]

From about 1816, botanists in France began to represent individual orders or families as a circle, with the constituent taxa spaced at intervals like numbers on the face of an analog clock. In some cases, the taxa form a series which comes full circle and rejoins itself; in others, no series was recognized. Some circles were simple, divided into sections like a pie.[96] Others sported concentric rings of annotation, colour-coded by character.[97] Augustin de Candolle depicted family Melastomeæ and sub-order Chariantheæ as two intersecting circles.[98] In another diagram, he shows lines of affinity radiating from within a circle (family Crassulaceæ) to nearby

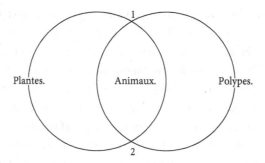

Figure 16.10. Fischer von Waldheim, organized nature as two *cercles de mouvement*. From *Tableaux synoptiques de zoognosie* (1808), page 181. Smithsonian Libraries *via* Biodiversity Heritage Library.

families.[99] Alexandre Henri de Cassini emphatically rejected geographical maps and networks as representations of nature; the linear series "is, and always will be, despite its imperfections, the best and most natural of all dispositions",[100] but these imperfections could be attenuated, and affinities best shown, by depicting nature—in this case, tribes of family Synanthérées—as a bracelet of nineteen circles.[101] To my knowledge, these circles were never used to depict the entire plant kingdom, much less all of nature.

A somewhat mysterious diagram with radial symmetry comes to us from Eichwald.[102] It is not explained (or even mentioned) in the accompanying article, but from context may be taken to depict connections (transitions?) among animal types based on functional features.[103] The south-to-north axis, from Phytozoa to Therozoa through Sporozoa, is one of increasing perfection.

Quinarian nature

Although William Sharp Macleay had not been specifically trained in natural history, he was familiar with current biological thought through his father Alexander, secretary of the Linnean Society in London from 1798 to 1825. William served as an attaché to the British embassy in Paris from 1815–1818, and came to know leading French naturalists including Georges Cuvier and Étienne Geoffroy Saint-Hilaire[104] at a time when the development, unity, and transformation of animal form were under intense discussion.[105] Upon returning to Cambridge, Macleay developed a philosophy of science in which hypotheses are put forward in the absence of evidence, then tested rigorously as evidence becomes available.[106] Taking the position that nature is continuous yet non-linear,[107] he drew the animal kingdom as a multi-level system of nested circles (see below). At the top level, an animal and a vegetable circle touch each other at a single point (Figure 16.11).[108]

Macleay recognized five classes within Animalia (see below), and positioned them at equal intervals around the animal-circle. He knew of only two classes

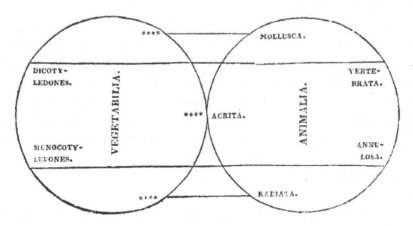

Figure 16.11. Macleay, general view of organized matter as two circles, indicating analogies between known or supposed classes of vegetables and animals. *Horæ entomologicæ* (1819–1821), page 212. Mann Library, Cornell University *via* Biodiversity Heritage Library.

within Vegetabilia—Monocotyledones and Dicotyledones—but supposed that three more would eventually be formed among the algae and fungi, and marked their places with sets of asterisks. This allowed Macleay to draw parallel lines of analogy between the classes (or asterisks) at corresponding mirror-image positions. The dicots, for example, are "the *Vertebrata* of the vegetable kingdom, their hard or osseous parts being as it were in the middle, and thus affording the most perfect and intricate plan of vegetable construction". Monocots are annulose and often articulated; many fungi are radiate, while molluscs may eventually find counterparts among the fuci or liverworts.[109] These analogies he distinguished from affinities, which determine the arrangement of taxa within a circle. Affinities are many and reinforce one another; analogies stand out "against a backdrop of overall dissimilarity".[110]

In those years, the classification of animals remained unsettled: Linnæus (1766) had recognized six classes; Lamarck eleven (1801) or fourteen (1809, 1812); Treviranus (1802) eight; Oken five (1804), seven (1811), or eight (1817); and Cuvier four *embranchemens* with sixteen (1812) or nineteen (1817) classes.[111] Macleay removed the infusoria, intestinal worms, and three orders of *polype*s from Cuvier's Radiata—not unreasonably, as they are scarcely radiate—and made from them a new class, Acrita.[112] Guided by his interpretation of affinities, Macleay ordered his five animal classes Acrita—Mollusca—Vertebrata—Annulosa—Radiata (and back again to Acrita). This was not a standard "monad to man" series: perfection is

284   KINGDOMS, EMPIRES, AND DOMAINS

approached in two groups, Vertebrata and Mollusca.[113] The Acrita do not even present a unitary type; even so,

> Nature, so far from forgetting order, has, at the commencement of her work, in these imperfect animals given us a sketch of the five different forms which she intended afterwards to adopt for the whole animal kingdom. In the soft mucous sluggish *Intestina* she has given the outline of the *Mollusca*. In the fleshy living mass which surrounds the bony and hollow axis of the *Polypi natantes*, she has sketched a vertebrated animal. In the crustaceous covering of the living mass, and the structure more or less articulated in the *Polypi vaginati*, we trace the form of the *Annulosa*; while the radiated form of the *Rotifera* and the simple structure of the *Polypi rudes* may in general remind us of the radiata.[114]

At each tangent he placed a so-called osculant taxon,

> smaller links of the great chain [which] appear to have no very distinct type of peculiar construction. They are all very imperfect beings, and seem in general to be compounded of properties which more peculiarly belong to the two great divisions which they link together; or, if their structure may be referred to any one type, it is undoubtedly to that of the circle of *Acrita*.[115]

Each osculant taxon displays affinities with the adjoining groups, ensuring a path of fine-scale continuity through the entire figure (Figure 16.12). Tunicates, for example, are polyps "by means of which we may leave the Acrita and proceed to explore our way into a more complicated region of organization".[116] Moreover, each circle opens to reveal another circle of five circles, and so on through successive taxonomic ranks down to genera and species.[117] Macleay found osculant groups at every level of analysis within the animals, although not between Animalia and Vegetabilia. At that singular position, an osculant might not be a Third Kingdom (because osculants are of lower taxonomic rank than the circles they connect), but it would be an intermediate or transitional taxon of some sort. Macleay does not open this door.

Five circles per level, five members per circle: and where the count falls short, it is due to "our ignorance of the productions of Nature".[118] He denied being a number-mystic; the sets of five, the symmetry, are (he assures us) purely empirical.[119] Perhaps animal form is fundamentally mathematical.[120] Macleay was not the first to entertain this thought: his contemporary Julien-Joseph Virey found the number *five* to underlie the form of radiates, zoophytes, and many plants, as did *three* for monocots, and *four* for certain other beings.[121] Linnæus described twelve stages of human life, strength, appetites, passions, judgement, and other facilities.[122] As for classification, Elias Fries recognized four taxa at each rank of the fungal hierarchy,[123] as did Horaninov more broadly (above). William Kirby and William Spence preferred *seven*, pointing to support from both nature and Scripture.[124] According

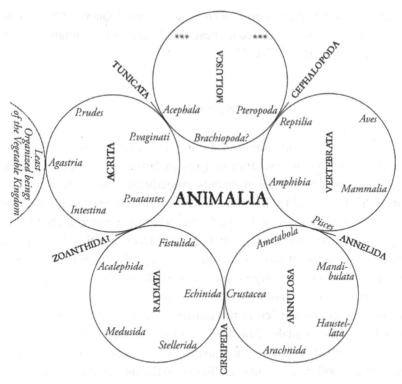

Figure 16.12. Macleay, quinarian arrangement of the animal kingdom. *Horæ entomologicæ* (1819–1821), page 318 as redrawn by Michal De-Medonsa. Republished by permission from Springer Nature Customer Service Centre GmbH: Springer Nature, *Journal of the History of Biology*, A Novick, On the origins of the quinarian system of classification, © 2015.

to Thomas Browne nature is ordered in fives, as in a quincunx.[125] Reaching back farther, nearly any small integer one wants can be drawn from Ficino, Agrippa, or Bruno.[126]

The quinarian system enjoyed some success, mostly in Great Britain, in no small measure due to its popularization by William Swainson.[127] From 1841 it fell out of favour, as Hugh Strickland argued that nature was patently irregular and discontinuous, and that the quinarians had manufactured regularities and evaded genuine tests.[128] By then Macleay had emigrated to Australia, and Swainson to New Zealand.[129] Even so, a sympathetic treatment appeared in the *Vestiges of the natural history of Creation* (1844).[130] Hugh Miller arranged fossil plants in a pentagon in his popular *Testimony of the rocks* (1857),[131] as did Andrew Pritchard the foraminifera in *History of infusoria* (1861), citing Macleay.[132] On the Continent, Johann Jakob Kaup held that at each taxonomic rank, one of the five anatomical systems, one of the five sensory organs, and one of the five bodily regions of animals attains primary development. Situating the birds within this system, he depicted Familie Corvidae (crows) as a set of five pentagons (1854).[133,134] Across the Atlantic a young Edward

286 KINGDOMS, EMPIRES, AND DOMAINS

Drinker Cope sketched a quinarian arrangement of amphibians (1857),[135] while in Toronto William Hincks classified molluscs, birds, and other animals according to quinarian principles until 1870.[136]

Summary

As the Eighteenth century drew to a close, the *Scala naturæ* was in trouble. Naturalists were abandoning strict linearity in favour of maps, networks, trees, circles, polygons, or other arrangements. Nonetheless it was almost universally assumed that there *is* a Plan in nature, doubtlessly one of great elegance (befitting the Creator) and hence probably based on number or symmetry. The plan might be linear in some general sense,[137] but it had to be more than that. Linearity can be found along the east-west axis of Goldfuss's Easter egg, the central axis of von Baer's diamonds and Eichwald's polygon, and the *prima linea* of Horaninov's vortex; but these are not the simple series of Bradley or Bonnet.

Figures can also be implicit in the words of a text: the Linnæan map became a metaphor of the vegetable kingdom long before it was rendered graphically. Linnæus explicitly signalled his intent to propose a new arrangement of nature; he chose his words with care, and likened his map to the familiar image of a chart of geographical territories. Alas, few instances are so clear-cut. Pallas used the metaphor of a tree, but we depict this tree at our peril: precisely how do Mollusca advance to Pisces? Given that the animal and vegetable trunks remain variously contiguous, is his tree really (in part) a network?[138]

In this chapter, I have presented alternative arrangements of nature—maps, trees, and the like—under seven headings. It does not follow that seven distinct metaphors vied for primacy. What, for example, distinguishes a network from a map? It is not whether the territorial boundaries abut each other (continental Europe) or not (island nations). We might surmise that Giseke preferred not to clutter his map with lines of affinity that were implicit (and in some cases, explicit) in his text.[139] Had he decided otherwise, we might be discussing the Linnæan network. Strickland later drew a large map of Order Insessoria (perching birds) in which tribes are large "islands" yet contain genera linked, within and among islands, in a network of affinities.[140]

So maps can imply or contain networks. A tree is a special case of a network (one that does not contain a cycle), so maps can contain trees. Quinarian diagrams, it seems, can contain trees too: Macleay went to some length to show that Lamarck's trees can be redrawn as his own animal-circle "with scarcely any alteration".[141] For his part, Lamarck would have rejected the cycle metaphor on the grounds that it undermines his argument, for example that annelids arose independently from worms.[142] Giulio Barsanti calls Robert Morison's curious diagrams *maps*;[143] to my eye, they are trees very much within the tradition of Medieval *arbores*.[144]

To complicate matters, any hierarchy can be reformatted as a tree.[145] Hierarchical classifications predate natural history, yet for millennia the *Scala naturæ* was the sole metaphor for heaven and earth. No tree metaphor lay underneath ancient folk taxonomies; nor is one somehow implicit in Aristotle, Ibn Sīnā, Albert, Wotton, Gessner, Ray, or Linnæus.[146] Hierarchical classifications sit comfortably alongside the postlinear diagrams of Eichwald (1821) and Horaninov (1834), which are not trees either. On the other hand, a tree metaphor *is* implicit in von Baer's Darstelling des Fortschrittes der Entwickelung, the hierarchy of embryological transformations that (by his analysis) gave rise to present-day animals.[147] We know this because his contemporaneous sketches of (botanical) trees, densely annotated with the very same technical terms as in his Darstelling, have been preserved.[148]

Further, there may be little to distinguish circles, polygons with radial symmetry, and some quinarian diagrams. Botanist John Lindley displayed the "exogens" (dicots) as a five-pointed star with ten orders in each arm.[149] Arrangement within each arm is by affinity, while lines of analogy connect orders at corresponding positions in neighbouring arms.[150] Lindley does not suggest that each arm could be opened to reveal another five-pointed star at a lower taxonomic rank; nor does he mention Macleay, Strickland, or quinarianism. The figure could easily be reshaped into a pie, and sliced into five servings.

Of the metaphors presented in this chapter, the map, network, quinarian circle, polygon, and spiral were engaged at kingdom level, although only Horaninov recognized a kingdom intermediate between plants and animals. In the *Dictionnaire classique d'histoire naturelle* (1822), Isidor[e] Bourdon described the animal and plant kingdoms as adjacent pyramids, "intimately united at their base, extremely divergent at their summit".[151] Circles (and polygons with rotational symmetry) are less-obvious choices for intermediate taxa because every element is, in a sense, intermediate. But the infrequency of Third Kingdoms of Life in this episode of natural history was a consequence of how naturalists were thinking about nature, not of intrinsic limitations imposed by the metaphors themselves.

The first half of the Nineteenth century was formative for modern biology, but these years played out differently in France, Germany, and Great Britain. In France, science enjoyed high social and political status; scientists were widely employed in higher education and the civil service, welcomed into the political elite, and became advisors on matters of state policy. The leading institutions, including the Muséum d'histoire naturelle, were lavishly supported.[152] By contrast, in Germany (and central Europe more generally) the state was autocratic, society hierarchical and conservative. This was mirrored in the seats of scientific authority, the universities, where lectures and theses were usually delivered in Latin,[153] and "each sanctioned body of knowledge was the province of a single full professor".[154] Yet, from about 1805 to 1825, a strain of romanticism swept across German science: speculative, poetic, sometimes bordering on the mystical, *Naturphilosophie* was conducive to ideas of unity, potential, and transformation.[155] Great Britain was different

288   KINGDOMS, EMPIRES, AND DOMAINS

again: there were, of course, seats of scientific authority (the Royal Society), but universities discouraged the study of natural history, and professors were paid poorly. Many renowned British naturalists were physicians, clergymen, landed gentry, or workingmen, and a vigorous popular science culture developed.[156]

The next four chapters take us inside biological science as it developed in these three countries. In Chapter 17, French biologists move beyond *histoire naturelle*, creating new sciences of *anatomie* and *morphologie*. In Chapter 18, German biologists explore the fine structure of infusion-animalcules (with decidedly mixed results), and develop modern cell theory. In Chapter 19 we remain on the Continent, as botanists and zoologists alike struggle to come to terms with the motile reproductive cells of algae and fungi—and motile plants that are complete organisms in their own right. In Chapter 20, English, Scottish, and American naturalists recognize Third Kingdoms, and debate evolution.

17

From *histoire naturelle* to *anatomie* and *morphologie*

> One might say that the plant is a rough draft, a framework of the animal, and that to form the latter, it sufficed only to cover this framework with a system of exterior organs specific to establishing relations [with its environment].
> —Xavier Bichat, *Recherches physiologiques sur la vie et la mort*[1]

Had Tournefort or Ray stepped a hundred years forward in time—to, let's say, the year 1795—he would scarcely have recognized contemporary natural history. Nature was no longer a static linear Chain. A chasm had opened between minerals and living beings.[2] Corals and hydra, perhaps sponges too, were now animals. Intrinsic life processes—physiology, irritability, sensation, reproduction—had supplanted vegetative and animate soul. There were hints of interplay among genealogy, environment, variation, and perfection of form. Natural history was giving way to biology.

It would have mattered whether our time-travellers materialized in Paris, Berlin, or London, as distinct traditions were developing in natural history. In France—our focus in this chapter—the tradition built on a long history of publications in the vernacular including Descartes's *Discours de la méthode* (1637), Fontenelle's *Entretiens sur la pluralité des mondes* (1686), and Tournefort's *Élémens de botanique* (1694). The French professional classes preferred nature-as-spectacle as offered by Pluche (*Spectacle de la nature*, 1732–1751) and Buffon (*Histoire naturelle, générale et particulière*, 1749–1788) to dry systematics; even binomial nomenclature came late to France.[3]

Natural history *à la française* was further distinguished by the prominence of ambitious encyclopædias, most notably the *Encyclopédie, ou dictionnaire raisonné des sciences, des arts et des métiers* (1751–1772) edited by Denis Diderot and Jean d'Alembert. Modelled on the *Cyclopædia* (1728) of Ephraim Chambers,[4] the *Encyclopédie* provided authoritative treatment of all areas of knowledge including natural history and its application in (for example) agriculture and medicine. The *Encyclopédie* can be seen as growing out of the pan-European Republic of Letters;

Kingdoms, Empires, and Domains. Mark A. Ragan, Oxford University Press. © Oxford University Press 2023.
DOI: 10.1093/oso/9780197643037.003.0017

290 KINGDOMS, EMPIRES, AND DOMAINS

its editors and key contributors sought to promote Enlightenment ideas and ideals, which of course were not exclusively French. Nonetheless almost all contributors were French by birth or domicile. The *Encyclopédie méthodique par ordre des matières* (1782–1832) was a reorganization of the *Encyclopédie* by subject, albeit in much greater depth.[5] The *Nouveau dictionnaire d'histoire naturelle* (1803–1804) and its *Nouvelle édition* (1816–1819), *Dictionnaire des sciences naturelles* (1816–1830), and *Dictionnaire classique d'histoire naturelle* (1822–1831) focused specifically on natural history.

French science was also impacted by the Revolution (1789–1799). Royal patronage came to an end, as did the role of the Church in education and the universities. The Jacobins dissolved the Académie royale des sciences (1793), and sent Lavoisier to the guillotine (1794).[6] From 1795, however, the Directorate took a different course, reestablishing scientific institutions and creating others. The Muséum d'histoire naturelle in particular emerged with a remarkable cohort of professors including Louis-Jean-Marie Daubenton, Étienne Geoffroy Saint-Hilaire, Antoine-Laurent de Jussieu, Jean-Baptiste Lamarck, and (from 1802) Georges Cuvier. French scientists were at the forefront of biological thought in the early decades of the Nineteenth century, not least concerning the structure, development, and relationships of animals. Ideas of an intermediate kingdom of nature were pursued by Jean-Baptiste Bory de Saint-Vincent. We meet all of these in this chapter.

Denis Diderot and Jean Le Rond d'Alembert

Diderot and d'Alembert intended their *Encyclopédie* to classify, describe, and interrelate all knowledge, including the manual arts and crafts, and thereby advance the ideas of the Enlightenment.[7] To this end they assembled a remarkable cast of contributors including the Baron d'Holbach, Jean-Jacques Rousseau, and Voltaire. Many of the key entries on natural history came from the pen of Daubenton, a member of the Académie des sciences and, until the mid-1760s, Buffon's principal collaborator on *Histoire naturelle*.[8] Daubenton's entries were up to date, but broke no new conceptual ground. Nature was divided into three principal sections (minerals, plants, animals)[9] occasionally called *règnes*.[10] His entry on *Animal* (1751) drew heavily on the first volumes of *Histoire naturelle*,[11] while those on plants drew on Tournefort, and those on coral and zoophytes on Marsigli, Peyssonel, and Donati.[12] Zoophytes were "animals whose nature seems to be as much related to that of plants, as to that of animals".[13] There was no entry for *Chaîne des êtres*, *Échelle des êtres*, or *Scala naturæ*. Daubenton remained a central figure in French natural history until nearly the end of the century (see below).

Diderot's own view of nature was less constrained. In *De l'interpretation de la nature* (1753) he spoke of the animal kingdom (and implicitly, the plant kingdom) as displaying all possible variations on a prototype. As metamorphoses of the animal prototype approach, by insensible degrees, those of the plant prototype, we find "uncertain, ambiguous beings" that have been stripped of most of the "forms, qualities, and functions" of their own kingdom, and clothed in those of the other.[14] How, he asks, can this fail to lead us to believe in a prototype common to all beings? Later, in *Entretien entre d'Alembert et Diderot* (1769), he had d'Alembert champion "this passage from marble into humus, from humus to the vegetable kingdom, from the vegetable to the animal kingdom, to flesh".[15] In *Éléments de physiologie* (1778) he imagined a dynamic chain of being with soft or easily traversed internal boundaries, including between plant and animal. It is not difficult to hear echoes of Needham and Bonnet:

> One must classify beings from the inert molecule, if there is one, to the living molecule, to the animal-plant, to the microscopic animal, to the animal, to man.
>
> The chain of beings is not interrupted by diversity of forms. Form is often but a deceptive mask, and the link that appears to be missing perhaps resides in a known being which the progress of comparative anatomy has not yet been able to assign its true place . . .
>
> Vegetation, life or sensibility, and animalization are three successive operations. The vegetable kingdom may well be, and have been, the primal source of the animal kingdom, and have taken its origin in the mineral kingdom; and the latter could well have emanated from heterogeneous universal matter . . .
>
> What is a plant? What is an animal? a coordination of infinitely active molecules, an *enchaînement* of small forces, all of which combine [only] to separate: it is therefore unsurprising that these beings pass so quickly.
>
> The difference between the vegetable kingdom and the animal kingdom: mobility in animal principles; fixity in plant principles.
>
> These are two effects of *nisus* [potential energy] preserved or destroyed. The gelatinous substance of one and the other shows a middle state between animal and plant . . .
>
> Upon analysis a mushroom yields a volatile alkali, a characteristic sign of the animal kingdom; also the mushroom seed is living: it oscillates in water, moves about, restless, avoids obstacles, and seems to swing [*balancer*] between the animal kingdom and the vegetable kingdom before settling on the latter . . .
>
> Zoophytes have only the sensation [characteristic of] life. Freshwater polyps have sensation, life [itself] and digestion, they are animal-plants . . .
>
> Adanson was the first who perceived a singular movement in an aquatic plant called *tremella*; he denied life and sensation to this plant, and therefore animality. Fontana makes [it] the transition from the vegetal kingdom to the animal kingdom, it is, according to him, simultaneously a true plant [and] a true animal.[16]

Louis-Jean-Marie Daubenton

In 1778 the Parisian bookseller Charles-Joseph Panckoucke began to plan a new sort of encyclopædia, the *Encyclopédie méthodique*, to be set out in twenty-six

292 KINGDOMS, EMPIRES, AND DOMAINS

subject areas. *Histoire naturelle des animaux* was placed under the editorship of Daubenton,[17] who delivered the *Tome premier* in 1782. Panckoucke had specified that each area be introduced with a *Discours préliminaire*, and Daubenton provided no fewer than four introductory essays.[18] In the first, *Introduction a l'histoire naturelle*, he admits the attraction of a *scala naturæ*, but either the Author of Nature has not arranged beings in that way, or we humans are too dim to apprehend it. This is why naturalists have resorted to arbitrary characters. In reality, the productions of nature succeed each other along *plusieurs lignes obliques*, by which he means that different characters offer conflicting information.[19] In the second introduction, *Les trois regnes de la nature*, Daubenton again takes aim at the Great Chain. No connection is possible between the mineral kingdom and either the plant or animal kingdom because no body can pass from an unorganized to an organized state, participate simultaneously in both, or be an intermediate. If nature were linear, intermediates should be sought between plants with the most organs (trees) and animals with the fewest (worms); yet naturalists search instead among the simplest organisms of each kingdom. Linnæus, Pallas, and Ellis tell us that lithophytes, corals, and polyps are animals, not intermediate beings, and they are right.[20]

In 1793 Daubenton, at seventy-seven the elder statesman of *histoire naturelle*, was appointed professor of natural history at the Muséum, and was selected to lecture on diverse topics in natural history at the new Écoles normales. In these lectures, delivered in 1795, he returned to the question of an intermediate kingdom,[21] at first taking a more openminded position:

> Here we have organized bodies; they must be either animal or vegetable; but we have never clearly distinguished and characterised well *animality*, *vegetality*, if I may be permitted to use these expressions. There are many animals, such as the ones you mentioned, the *polyps*, which, perhaps, occupy a place between the animal and the vegetable. They seem to offer us a new division in natural history, and might perhaps shed new light on beings intermediate between plants and animals.[22]

And again:

> Among animals and plants there are differences large enough to be distinguished into two sections. I go even further; I ask if the polyps, the acetabulum, the infusionary animals, etc. do not have an organization sufficiently different from that of most animals to have another name? I ask if the confervas, mushrooms, moulds, lichens, etc., are true plants? I might report here many other observations which tend to prove that there is a very large quantity of organized beings which are neither true plants, nor true animals; but I have said enough to draw the attention of naturalists to this matter, which is one of the most important and most difficult in natural history. It is only by delving into this more deeply by observations and meditations that we can clearly distinguish the true plants and true animals,

[and] the other organized beings which differ enough to have another name and another rank in the methodical division of the productions of nature.[23]

In a later lecture he stuck closer to his 1782 position, but concluded

Let us consider nature without ruling out any system of continuity or interruption in the order of its productions. We will see it as it is, and judge it the better when we have acquired more knowledge.[24]

Jean-Baptiste Lamarck

Few figures in science attract such a disparate range of commentary as Jean-Baptiste Lamarck, the "mythical precursor" of evolutionary biology.[25] After an injury ended his military career, he studied medicine, then (under Bernard de Jussieu) botany. We have already mentioned his first major work, the *Flore françoise*,[26] and its static linear chain of plant genera. Respected by his Parisian colleagues and promoted by Buffon, Lamarck became a royal botanist (1781), then keeper of the herbarium of the Jardin du Roi (1788).[27] At the height of the Revolution (1793) he was appointed the first professor of "insects, worms, and microscopic animals" at the Muséum.[28]

His contributions to science over forty years cannot be summarized in a few words. He held anachronistic views on chemistry (the four elements) and geology (uniformitarian), but developed a theory of evolution that involved environment, mechanism, and genealogy. He was one of the first to use the term *biologie* in a modern sense.[29] There is little evidence that he ever dissected a worm or mollusc, or observed its life cycle in nature, but he was arguably the first[30] to appreciate the fundamental dichotomy between animals with and without vertebrae.[31] All this suggests that we can understand Lamarck as a transitional figure linking Eighteenth-century natural history with Nineteenth-century biology. Here we are less interested in his contributions to invertebrate zoology or evolutionary theory, than in his (not unrelated) views on intermediate kingdoms of life.

Lamarck began the *Flore françoise* by classifying beings into inorganic and organic, and the latter into plants and animals, yielding three kingdoms of nature.[32] Following Buffon, he warned that classification is "artificial and arbitrary" and often gets in the way of appreciating *la marche de la Nature*.[33] The natural order among plants is a linear scale from fungi to grasses, which he illustrated at the rank of genera.[34] In *Flore françoise* there is no sense of temporal progression within this scale; nor does he situate it within a comprehensive *scala naturæ*. Lamarck went on to produce other significant works of practical botany, oriented towards description and identification.[35]

With the revolutionary reorganization of the Muséum, Lamarck hoped to be named professor of botany; instead, he was assigned the organisms Linnæus had

294 KINGDOMS, EMPIRES, AND DOMAINS

called insects and worms.[36] This was not entirely without cause, as Lamarck had assembled a substantial collection of mollusc shells. The appointment brought opportunity, as the Muséum grew rapidly with the appropriation of private collections and the plunder of museums and cabinets across Europe by French armies.[37] Lamarck recognized among the animals a progression from simple to complex, as he had earlier among the plants. But

> In speaking of this even gradation in the complexity of organization I do not mean at all to suggest the existence of a linear series with regular intervals between the species and genera: such a series does not exist; but I speak of a series in which the principal masses, such as the great families, are almost regularly spaced; a series which quite certainly exists among both animals and plants, but which when considered in terms of genera or more particularly of species, forms in many places lateral ramifications of which the terminal ends provide truly isolated points.[38]

Lamarck explicitly denied that a single chain runs through both plants and animals,[39] and eventually disowned the idea even within plants or animals (below). Nor did he accept that natural relationships among plants or animals could be seen as forming a network, or points on a geographical map.[40]

Système des animaux sans vertèbres (1801)

In *Flore françoise* Lamarck had drawn a conventional boundary between plants and animals: both live, grow, and develop, but animals are, in addition, sensible, show spontaneous movement, and (in the special case of man) enjoy the light of reason.[41] As professor of "microscopic animals" he drew the same boundary in *Système des animaux sans vertèbres* (1801), adding only that plants lack organs of digestion.[42] Animals themselves, however, fall into two great branches or series: vertebrates and invertebrates.[43] A progressive diminution of organization and faculties can be distinguished within each series, such that by the seventh and final class of invertebrates, the Polyps, the organs of respiration, circulation, and sensation cannot be detected and "appear not even to exist".[44] These beings represent "a sort of rough draft of animality":[45]

> Although the Polyps are the least-known of all animals, they are without doubt those whose organization is the simplest, and which consequently have the fewest faculties. We find in them no organ of sensation, nor of respiration, nor any intended for the circulation of fluids. All their viscera are reduced to a simple alimentary canal which, like a more-or-less elongated sac, has only one opening, at the same time both mouth and anus; and this alimentary canal is apparently surrounded by absorbent globules that contain fluids maintained in a sort of motion by suction and transpiration.

The animalcules which are at the end of the last order of polyps are no more than animated points, mere gelatinous corpuscles, of a simple form, and contractile in almost every direction.[46]

In *Système* Lamarck recognized three orders of polyps: Polypes à rayons, Polypes rotifères, and Polypes amorphes ou microscopiques. The first of these, so named because their tentacles are arranged as spokes, he further divided into Polypes nus (including *Actinia*, *Hydra*, and *Pedicellaria*) and Polypes coralligènes (millepores, corallines, corals, and sponges). His Polypes rotifères include *Vorticella*, while his Polypes amorphes ou microscopiques include infusoria, *Vibrio*, *Proteus*, *Volvox*, *Monas*,[47] and spermatic animalcules.[48] He considered it a "striking analogy" that amorphous polyps and fungi appear, possibly by agglomeration of gelatinous molecules, when animal or vegetable matter begins to putrefy.[49] Lamarck was non-committal about reports that microscopic animalcules multiply via eggs, but it is *véritablement admirable* that they can multiply by longitudinal or transverse fission.[50] As the internal structure of the invertebrates became progressively better-known (not least through the efforts of Georges Cuvier: below), Lamarck revised his classification of invertebrates, and in 1807 removed the infusoria from the polyps.[51]

Philosophie zoologique (1809)

In *Philosophie zoologique* (1809) Lamarck set out a theory of evolution according to which the simplest beings (moulds, fungi, infusorial animalcules) arose spontaneously through the action of material forces (caloric and electrical fluids).[52] These forces then caused beings of greater complexity to arise in temporal succession: polyps from infusoria, radiaria from polyps, and so on all the way to mammals. Animals advance or regress along this main axis through the use or disuse of structures or organs, and lineages might further diversify through the inheritance of acquired characters.[53] In the main text—that is, until late June 1809—Lamarck recognized ten classes of invertebrates, and four of vertebrates, along this axis.[54] Just prior to publication, however, he decided that not only infusoria but also worms had arisen spontaneously.[55] In a hastily written section called Additions he proposed that animals must therefore have evolved along two paths: an all-aquatic linear series from infusoria to polyps and radiaria, and a branched path from worms to mammals. To be sure, an uncomfortably large gap separates molluscs from fish, but he was confident that it would eventually be bridged.[56] His famous tree of animal origins depicts these two paths, not as an aid to classification but as a depiction of "the actual order followed by nature in the production of animals".[57]

In his Muséum lectures of the same year, Lamarck abandoned his earlier position that all animals are sensate and move voluntarily. He now taught that volition is largely restricted to birds and mammals, while the lower invertebrates are devoid of feeling (*apathique*).[58] Infusoria are the only creatures which do not have to carry on any digestion when feeding, and which in fact only feed by absorption through the pores of their skin and by an internal imbibition.

In this they resemble plants, which live entirely by absorption, carry on no digestion and in which the organic movements are only achieved by external stimuli; but the infusorians are irritable and contractile and perform sudden movements which they can repeat several times running; this it is that indicates their animal nature and distinguishes them essentially from plants.[59]

Whatever their analogies with plants, infusoria were progenitors of animals and thus secure in the Animal Kingdom.

Histoire naturelle des animaux sans vertèbres (1815–1822)

Lamarck expected that that his seven-volume *Histoire naturelle des animaux sans vertèbres* (1815–1822) would be received as a second edition of *Système des animaux sans vertèbres*—that is, as a systematic treatise. He further hoped that his descriptions of the classes, families, genera, and principal species of invertebrates would be seen to support the theory he had set out in *Philosophie zoologique*,[60] for example by revealing series of transitional forms. More importantly for our question here, in *Histoire naturelle* Lamarck presented his mature (if not quite final) views on the distinction between animals and plants, and the relation of infusoria to other animals.

As in *Système*, Lamarck insisted that no single chain connects all beings. An "immense distance" separates organic from inorganic bodies. Plants and animals form distinct branches, approaching each other only at their base, and even there only by virtue of common simplicity. Specifically, "plants do not grade into [*ne se nuancent*] animals at any point in their series".[61] There are no animal-plants, no plant-animals, and we must avoid terms such as *zoophyte* that express a false idea.[62]

As before, infusoria are the simplest and least-perfect animals. *Monas*, *Volvox*, and others lack a mouth, alimentary sac, discrete organs, voluntary movement, and sensation; were it not for the nearly uninterrupted series leading to polyps and radiaria, we might doubt that *Monas et al.* are animals at all.[63] At the same time, this tells us that animals cannot be defined by voluntary movement or sensation.[64] Lamarck now recognized three primary divisions among the animals: *Animaux apathiques* (infusoria, polypes, radiaria, worms), *Animaux sensibles* (insects, spiders, crustacea, annelids, barnacles, molluscs), and *Animaux intelligens* (fish, reptiles, birds, mammals). The *Animaux apathiques* and *Animaux sensibles* together constitute the *Animaux sans vertèbres*.[65] Running through these fourteen classes he discerned "a general series that indicates, more-or-less as a whole, the *marche* that nature has followed in imparting existence to the difference races of these beings".[66]

No sooner had Lamarck written these words, but (as in 1809) he changed his mind—or at least, his emphasis.[67] Animals have in fact arisen in two series, one from infusoria, the other from worms—although in *Philosophie zoologie* he had gotten some of the details wrong. Citing new work by Savigny, Lesueur, and Desmarest[68] he removed ascidia, and bivalves and their relatives, from the molluscs, revealing a path from radiaria through the two new classes (Ascidiens, Acéphales) to the revised molluscs.[69] Cephalopods might be removed as well, and be placed at the termination of this series.[70] Thus infusoria have given rise to animals that are not articulated and lack an organized nervous system, while articulates, including those with a medullary cord and ganglia, have arisen from worms.[71] The Animaux sensibles, like the Animaux apathiques, are biphyletic.

Loose ends remained. Lamarck suspected that annelids were incorrectly placed; gaps remained between insects and spiders or crustaceans, and between crustaceans and barnacles, while the origin of Animaux intelligens (vertebrates) remained obscure.[72] Other animal series might exist, as yet unrecognized.[73] But there was much to like about his new plan of animal relationships: free-living worms must have arisen (perhaps from the intestinal variety) well after the origin of infusoria,[74] articulates later yet. Lamarck wrote confidently of the *order of formation*.[75]

Système analytique des connaissances positives de l'homme (1820)

In 1820 Lamarck, blind and largely estranged from his erstwhile colleagues, dictated a final monograph befitting an Eighteenth-century *savant*: an overview of human intelligence, reason, and imagination.[76] He began by establishing foundational concepts including matter, nature, and sensation, and took the opportunity to revisit his earlier definitions of life, plants and animals, and his system of animal origins. Ten general features distinguish living from inorganic bodies.[77] Plants— living non-irritable bodies—exhibit nine essential characters.[78] The first plants (filamentous fungi, other fungi, and perhaps lichens) arose from inorganic matter on one or a few occasions; then cryptogams, monocots, and "polycots" followed in succession along a highly branched path.[79] Animals—living irritable bodies—likewise exhibit nine essential characters.[80] The first animals arose from non-living matter at the same time as did plants, but altogether independently of them. No chain unites plants and animals, and no gradation exists between them, even at their simplest.[81] Up to this point, Lamarck has scarcely taken us beyond the *Histoire naturelle des animaux sans vertèbres* (above).

Without warning or explanation, Lamarck then revised his earlier tree of animal relations. As before, a *Monas*-like infusorian was the first animal,[82] and an almost unbroken succession of intermediates leads thence to polyps; but thereafter three (not two) branches emerge. As before, one leads to radiaria, another to tunicates, bivalves, and molluscs; but now a third branch leads to worms and articulates.[83] He did not belabour the point, but if worms have not arisen directly from inorganic

298 KINGDOMS, EMPIRES, AND DOMAINS

matter, then all animals presumably share a common origin.[84] Man—the subject of this work—is "the most eminent term of this great series of productions".[85] In an echo of Augustine, Lamarck reminds us that an "intelligent and boundless power" could have "established an order of things" rather than just creating beings "immediately and without intermediary".[86]

Georges Cuvier

Jean-Léopold-Nicolas-Frédéric Cuvier, known from an early age as Georges,[87] arrived in Paris in 1795 and quickly secured an appointment as assistant to Jean-Claude Mertrud, the professor of animal anatomy at the Muséum d'histoire naturelle. He succeeded Daubenton as professor of natural history at the Collège de France in 1799, and Mertrud in 1802.[88] Since at least 1790 Cuvier had rejected the idea that a single chain joins all beings, and following Buffon he viewed taxa above the rank of species as mere abstractions.[89] Inspired by Jussieu's *Genera plantarum*, however, and based on his own initial investigations, he began to reconsider his views on higher taxa.[90]

By April 1795 Cuvier had concluded that it should be possible to discern the natural order of animals, including supraspecific taxa, if we understood characters well enough to use only those that best reflect affinity. The path to this understanding lay through comparative anatomy. The Creator has designed each animal for its intended environment and way of life, in particular by arranging its body parts—bones, organs, blood vessels—so they work harmoniously together to deliver sensation, respiration, and other physiological functions. Some systems (the nerve cord, the heart) are so life-critical, and their major parts so perfectly integrated, that subsidiary organs (ganglia, gills) must adjust to them, perhaps differently in different orders or families. The most-critical systems offer characters that reflect affinity among broad groups of animals—that is, taxa of the highest rank—whereas subordinate organs provide characters that are more narrowly informative, for example within a single order or family. One distinguishes critical from subordinate organs, and thereby primary from secondary characters, through "reasoning and experience".[91] The very next month, Cuvier arranged six non-abstract classes of "white-blooded animals" in a series of decreasing "organic perfection" using features of their circulatory and nervous systems.[92] In a fully fledged *méthode naturelle* one would use conserved features of the most-important organs to delineate the highest ranks, and characters of progressively subordinate organs to form the lower ranks.[93]

Under Cuvier, the *chaire d'Anatomie des animaux* at the Muséum was renamed *Anatomie comparée*, reflecting not only his new *méthode* but also his hands-on expertise in dissection—a skill not widely shared by his professional colleagues. His *Leçons d'anatomie comparée* (1800–1805) further emphasized the importance of knowing the internal parts of animals. By 1812 Cuvier had become convinced that he had identified the most important organ system of all:

> Reflecting then upon the principal organs which have determined this resemblance between the animals of each form, I promptly found a satisfactory reason for this resemblance. The nervous system is the same in each form; now, the nervous system is basically the whole animal; the other systems are there only to serve or to maintain it; it is therefore not surprising that it is according to [the nervous system] that [the other systems] adjust themselves.[94]

On this basis Cuvier identified four principal forms or plans among the animals, and described four *provinces* or *embranchemens*:[95] vertebrate or skeletal animals (mammals, birds, reptiles, fish), molluscs (cephalopods, gastropods, pteropods, acephales), articulates (annelids, crustacea, spiders, insects), and zoophytes (echinoderms, intestinal worms, polyps, infusoria).[96] These *embranchemens* do not form a series, but instead are coequal.[97] That is, vertebrates are taxonomically equivalent to molluscs (or articulates, or zoophytes), not to invertebrates as a whole. Likewise, the sixteen classes are coequal among themselves: infusoria are not prior or inferior to polyps or worms. Surprisingly perhaps, Cuvier retained the name *zoophytes* for the fourth branch:

> In our fourth and last *embranchement*, regularity is based on a new plan that recalls ordinary forms in plants. It is what has caused various naturalists to call some of them zoophytes, and has induced me to generalize this name to all the animals of this *embranchement*. They could also be called radiated animals, because their organs, both animal and vital, are almost always arranged around a centre like the rays of a circle. By this word everyone remembers the starfish, sea urchins, jelly-fish, actinias, and innumerable polyps, either naked or corallagenous; but it is still right to include animals for which the radiate arrangement is less apparent, but nonetheless real, such as sea cucumbers, sipunculids, and most intestinal worms; regarding which I must remark that the parts arranged on two lines must also be considered as radiate when they are organs found unpaired in the three great [bilaterally] symmetrical branches.[98]

These four *embranchemens* later formed the backbone of *Le règne animal* (1817).[99] Cuvier begins with a succinct historical preface, and a theoretical introduction that restated his *méthode naturelle* (above), then steps systematically through the animal kingdom, explaining the characteristic anatomy and physiology of each taxon, introducing genera and representative species, and commenting on (for example) diet, habit, and reproduction. Zoophytes are now divided into five classes: echinoderms, intestinal worms, acalephs (sea-anemones, jellyfish, comb jellies), polyps (vorticellids, corals, corallines, madrepores, sponges), and infusoria

300 KINGDOMS, EMPIRES, AND DOMAINS

(divided further into rotifers and "homogeneous infusoria").[100] Infusoria may not be obviously radiate, but

> It is customary to place at the end of the animal kingdom [these] beings so small, that they escape being seen with the naked eye, and have been distinguished only since the microscope revealed to us this (so to speak) new world. Most of them present a gelatinous body, of the most extreme simplicity, and these must indeed find their place here; but some [naturalists] have left among the infusoria animals seemingly much more complicated, and which resemble [the others] only in their small size, and in the habitats in which they are usually found.[101]

Life is possible only if all bodily systems—neurological, circulatory, respiratory and the others—interact in harmony. Cuvier believed that these harmonies had been established once-and-for-all-time at Creation. To be sure, subsequent cataclysms have reshaped shorelines, toppled mountains, and extinguished species; animals and plants have invaded and repopulated affected regions.[102] But (*pace* Augustine) no divine plan of ongoing creation or evolution has unfolded over time. Complex organ systems could not have been picked apart and rebuilt, time and again, along series of evolutionary intermediates, as Lamarck would have us believe. Cuvier dismembered not only the *scala naturæ*, but continuity more generally. In its place we are left with four animal forms or plans, static and timeless.

Étienne Geoffroy Saint-Hilaire

Étienne Geoffroy Saint-Hilaire joined the Muséum in 1793 initially in a modest role, but following the revolutionary reorganization emerged as professor of quadrupeds, birds, reptiles, and fish.[103] He was appointed professor of zoology at the Faculté des sciences in 1808, and held both appointments until his death in 1844. From early in his zoological career Geoffroy denied the existence of a *scala naturæ*, and held that a common organizational plan can be discerned within each class of animals.[104]

> The forms in each class of animals, however varied, all result in the end from organs common to all. Nature refuses to employ new ones. Thus, all the most essential differences which affect each family within the same class come only from another arrangement, from another complication, in short, from a modification of these same organs.[105]

This summary set off no alarms: Buffon had, after all, written that a common plan extends from man to zoophytes and plants.[106] Geoffroy and Cuvier became close friends, wrote articles together (1795–1798), and planned a major collaborative work on mammals.[107] But their research programs soon diverged. In the hands of

Cuvier, *anatomie comparée* was empirical and diagnostic; Geoffroy sought to make it philosophical and unifying. To this end he disregarded both form[108] and function, focusing instead on the number, relative positions, symmetry, and mutual compensation of body parts—that is, on organizational plan. Guided by a *principle of connections*, he matched bones and organs with their analogues in other animals, at first within the vertebrate *embranchement*,[109] then—much more controversially—between vertebrates and articulates, arguing (for example) that the insect exoskeleton is analogous to the vertebral column. A common plan thus unifies these two *embranchemens*. Cuvier had severed the *embranchemens* from one another, but Geoffroy joined them back together, invoking a familiar image:

> at its first point of subdivision the zoological tree will be divided into two main trunks, the *vertebrates* and the *invertebrates*. The first trunk will remain composed of its two original *embranchemens*: one, containing the animals of the higher classes, or the *higher vertebrates*, and the other, the articulated animals, which one might perhaps prefer to designate by the name *dermo-vertebrates*, because in these animals the vertebrae adhere to the dermis.[110]

In 1829 two naturalists about whom little else is known, Meyranx and Laurencet, sent a manuscript to the Académie des sciences arguing that molluscs (in this case a cuttlefish) were, so to speak, rearranged vertebrates. Geoffroy seized on the paper as evidence that the vertebrate plan extends to the molluscan embranchment, thereby setting off the famous debate with Cuvier that played out before the Académie (and the broader public) in early 1830.[111] Zoophytes were scarcely mentioned, but in his subsequent book on the debate, *Principes de philosophie zoologique* (1830), Geoffroy refers inclusively to the animal plan extending to *les autres embranchemens*.[112] Earlier, he surely had zoophytes in mind when he argued that elements of the vertebrate plan can be discerned among invertebrates if connections are drawn with embryological stages rather than the adult vertebrate.[113]

It was not difficult to see where *anatomie philosophique* might lead. Structure, or at least specific structures, would be decoupled from function. Bones and organs per se would tell us little about the conditions of existence. The four anatomical plans, distinct and timeless, would be revealed as illusory. The door would again open to transformation and temporal change. It would be necessary to rethink the primacy of the nervous system, indeed the principle of character subordination, in mapping the natural order. Fanciful analogies might be drawn between the skeleton of animals and lignified tissues of plants.[114] From the perspective of Cuvier, idealism would have defeated careful descriptive science.

Toby Appel has reminded us that "both Cuvier and Geoffroy defended extreme positions" and, in that sense, both lost the debate. The French zoologists who followed took more-moderate positions, drawing from both.[115] Indeed, researchers beyond the walls of the Muséum made important contributions to our understanding of corals, polyps, or infusoria without becoming entangled in

302 KINGDOMS, EMPIRES, AND DOMAINS

the controversy. We have already noted the opinions of Adanson (1763), Bonnet (1764), and Augier (1801) on a third kingdom,[116] and will encounter others in upcoming chapters, notably those of Jean Lamouroux, François-Benjamin Gaillon, and Félix Dujardin. Here we survey seven other naturalists who wrote in French between 1786 and 1856, to determine how they engage our question of a kingdom intermediate between plants and animals.

Félix Vicq-d'Azyr: *le règne vivant*

We begin with the anatomist Félix Vicq-d'Azyr, a member of the Académie des sciences (1774) and Académie française (1788), perpetual secretary of the Société royale de médecine, and physician to Marie Antoinette. In *Traité d'anatomie et de physiologie* (1786) he wrote:

> Of the three kingdoms that embrace the whole of Nature, two are so mixed together [*confondent*] that it is nearly impossible to establish their limits. The great differences that we observe between the [farthest] ends of their chains disappear as we approach the point that unites them: mushrooms, vesicular and articulated plants, corallines, and those vegetations in which a family of animals works in common, and which, solidly attached by their base, can move only in their branches; all these substances seem to hold the *milieu* between animals and plants, or at least leave little interval between these two orders. It is not the same with minerals: governed immediately by the known laws of mechanics and elective attractions, receiving no increase and acting only on their surfaces, they form a great circumscribed system in all its points. This is not equivocal in any reports.
>
> To this great class one can thus oppose another in which the masses animated by particular and spontaneous movements reproduce by germs, in which the elements move ceaselessly, collide, and combine in a thousand ways, and whose parts, after having increased by an interior force, finally decay and return to their first kingdom, to which death seems to render [again] what life has taken from it.
>
> These effects are common to plants and animals; in some as in the others, humours circulate, juices are separated, the air is attracted and flows in particular vessels; the sexes are distinct and fecund, and all experience the development which gives them each year a layer or new productions.
>
> There are then only two kingdoms in Nature, of which one enjoys, and the other is deprived of, life.[117]

In 1792 Vicq-d'Azyr attached a name, if informally, to the former: *règne vivant*.[118] He found analogies between molluscs and zoophytes (on one hand) and algae and fungi (on the other),[119] but the presence of a digestive channel and a nervous system clearly distinguish animals from plants.[120] Fungi are plants, zoophytes are animals; all are members of a common *règne vivant*. The idea was scarcely new. For

Aristotle and generations of philosophers thereafter, beings above some point on the *scala naturæ*—plants, animals, man, indeed heavenly bodies—were ensouled, alive.[121] Bonnet explicitly contrasted the living with the non-living;[122] Antoine-Laurent de Jussieu promoted a Kingdom of Organic Bodies, Pallas an Empire.[123] No doubt mindful of recent events on Place de la Révolution, Daubenton advised that the word *kingdom* was "improper" and "unintelligible" in natural history.[124] Vicq-d'Azyr kept his *règnes*, including *règne vivant*, in lower case.

Jean Guillaume Bruguière: a new arrangement of Vermes

The section of *Encyclopédie méthodique* dealing with worms—that is, following Linnæus, the intestinal worms, molluscs, testacea, lithophytes, and zoophytes—was assigned to Jean Guillaume Bruguière, a specialist on mollusc shells. In a bold move for a naturalist lacking a secure position, he subdivided the class in a new way: into infusoria, intestinal worms, molluscs, echinoderms, testacea, and zoophytes, in that order.[125] Bruguière completed only a first volume, after which *Histoire naturelle des vers* was taken over by Lamarck, and brought to completion by Gérard Paul Deshayes.[126] The volumes are noteworthy for their lack of *scala naturæ* imagery. These beings are "endowed with organs as perfect as those of the largest animals, since they reproduce as well [*de même*], and hold in nature a rank so unequivocal, if less suspected".[127] Infusoria are no longer chaotic, nor (implicitly) the terminus of the animal kingdom.

Julien-Joseph Virey: *evolution* along parallel chains

Writing in the *Nouveau dictionnaire d'histoire naturelle*, Julien-Joseph Virey, a physician and writer on anthropology, perceived an "admirable chain of organisation among the animals and the plants":

> All beings tend to their vital perfection; thus, each individual receives a greater development of faculties as it advances in age. Likewise, the most imperfect beings aspire to a more-perfect nature; this is why species constantly climb up in the chain of organized bodies by a kind of vital gravitation. For example, the polyp tends to the nature of the worm, the latter tends to the organization of the insect, the insect aspires to the conformation of the mollusc, this tends to become fish, and so on until the man. . . . In plants we observe the same gravitation, because nature always aspires to the perfection of its works.
>
> Animals tend to man, the plants aspire to animality; minerals seek to approach the plant.[128]
>
> Nature, then, needed to vary only a few of the diverse generations of the same plant, the same animal, to create a multitude of neighbouring beings, which we

304 KINGDOMS, EMPIRES, AND DOMAINS

call *species*. The most remarkable variations are for us genera, families, classes, and all this scaffolding of methods invented by the human mind to facilitate its knowledge of objects, but which are by no means the work of nature. With a single bird, nature was able to create by successive modifications all the other birds. A single grass could be transformed by divine power into all possible grasses, in the course of time and [under] the influence of circumstances. We will say the same for all the breeds of plants and animals that populate our world. *Nature* has produced at first only an animal, a very simple vegetable, which has varied infinitely, and complicated by nuances to the most-perfect species.[129]

Virey seems to imagine this "chain of organized bodies" to form a single linear scale. He returned to the theme in *Philosophie de l'histoire naturelle* (1835), but now with two linear scales:

> In the *marche ascendante* of nature, the simplest animals do not correspond at all to the most-complete plants, but each kingdom follows a parallel development, such that the least-developed plant (such as a conferva) holds the same rank in the phytological scale as the infusorial animalcule in the zoological series. These two kingdoms are born almost simultaneously in stagnant waters. As plant and animal organisms are deployed [*déploient*] into more-perfect species, they become increasingly distinguished by very distant differential characters. Thus a large tree can never be confused with an animal elevated on the zoological scale, while at the base of the two kingdoms, one sees an approach of the races of plants and animals [to each other], such that one might believe in the existence of true zoophytes, transformable into one another, such as the zoocarps, psychodinea, etc., of some naturalists.[130]

At this common base we find

> inferior or protogenic races, [with] pulpy or cellular tissue, their forms either radiate or amorphous, [and] living together in humid or aquatic places. These plants and these animals approach [each other] so closely, in this intimate fraternity, that many naturalists can scarcely bring themselves to set a boundary separating them within this *règne chaotique*, if indeed it exists, and if there is not in several confervas and infusoria, as in the zoophytes, a mixture of vegetation and animalization.[131]

The image he paints here is of parallel scales arising from a common pool of reciprocally transformable life forms that may (or many not) be called a chaotic kingdom—the designation recalling Linnæus. Virey refers to the *marche ascendante* as *evolution*,[132] and extends a *Loi d'ascension chez les animaux et les végétaux* to cover not only the progressive elaboration of species, but also the development of an individual from embryo to reproductive maturity.[133] Confusingly, though, he concludes *Philosophie de l'histoire naturelle* by presenting incompatible models and metaphors of nature—perhaps as a belated (and poorly structured) literature review, perhaps not.[134]

Pierre-Jean-François Turpin: *végéto-animaux*

Pierre Turpin, a soldier and amateur botanist, is best known for his precise, elegant illustrations of plants and animals.[135] He developed a detailed theory of plant and animal organization based on stepwise assembly from cells to single filaments, parallel filaments, homogeneous aggregations, then simple and branched tubules.[136] He later complicated his theory, adding three stages of aggregation of *globulines*, tiny vesicles or granules of different colours that constitute plant cells.[137]

In a remarkable image (our Frontispiece), Turpin illustrated Règne Organique as two linear series of beings, respectively Branche Vègètale and Branche Animale, arising from a common base labelled VÉGÉTO-ANIMAUX.[138] The plant series culminates in the *renoncule* (buttercup), the animal series in Man. He divides each line into two main sections, plants into "simple or axiferous" and "compound or appendicular", animals into invertebrates and vertebrates. Végéto-animaux are "mixed beings, of first formation, organized spontaneously of matter on dissolution, and giving birth, *par gradation*, to the Vegetal and Animal Branches".[139] He asserts that this figure is a geometric reduction of another, on which he has been working for a long time, and on which many beings at all points on the chain will be represented from nature.[140] What are these végéto-animaux? Turpin was responsible for the iconography, particularly of plants, in the second edition of *Dictionnaire des sciences naturelles* (1816–1830).[141] Introducing the first volume of plates of the Vegetables, he identifies 22 species in seven genera of

> organized mixed beings provided with a remarkable movement of locomotion. These mixed beings, which could just as easily start the vegetable kingdom as the animal kingdom, were placed at the head of the former, but with a distinct series of numbers.[142]

The genera—*Bacillaria, Navicula, Lunulina, Stylaria, Echinella, Palmetina,* and *Surirella*—are all diatoms.[143] Curiously, entries in the *Dictionnaire* itself identify three of these as plants,[144] while Turpin himself assigns *Surirella* to "Botanique zoologique? microscopique" and writes:

> After examining this production, as I have just described it, one naturally wonders, is it vegetable or animal? By first considering only the brittle and calcareous nature of the valves, we would decide in favour of animality; but then when we assure ourselves that this being, at whatever age we observe it, is perfectly inert and simply vegetative; that, on the other hand, its reproduction is similar to that of a

306   KINGDOMS, EMPIRES, AND DOMAINS

confervoid plant, that is to say, reduced to seminules or green reproductive bodies, we remain in indecision, waiting for new observations to come to enlighten us in this regard.[145]

Henri Marie Ducrotay de Blainville: infusoria as an appendage

Henri de Blainville, who briefly enjoyed the patronage of Cuvier at the beginning of his career as a naturalist,[146] argued that external features are much more useful than internal anatomy in zoological classification.[147] Animals obviously form a series,[148] and he set about to join up Cuvier's *embranchemens*, arranging animals into three subkingdoms and twenty-five classes that reflect this series.[149] In the same article, he offered a second classification based on "form and structure": Empires of Organized and Unorganized beings, the former divided into Animal and Plant kingdoms, and animals into True and Doubtful subkingdoms. Blainville annoyed his colleagues by creating, dissolving, merging, renaming, and changing the ranks of inverte-brate taxa; some he claimed to be intermediates, for example Malacentomozoa ("articulo-molluscs") between arthropods and molluscs.[150] Even so, he accepted that the animal series is not smoothly continuous. Blainville recognized a Doubtful subkingdom among plants as well, but kept sponges and infusoria at the base of his animal scale as Hètèromorphes (1816), Douteux (1816)[150] or Amorphozoaires (1822), the latter including corals as well.[151] He defined plants and animals using a curious combination of characters:

A plant is an organized being (that is to say cellulosic, inhaling and exhaling, ca-pable of nourishing and reproducing itself), strongly carbonated, most-often complex, without an intestinal canal, without visible contractile fibres, without ob-vious excitatory fibres, and consequently not digesting, not moving, not feeling its relationships with external bodies, although sometimes it seems to us that by slow and successive changes it is producing toward a determined goal.

An animal, on the contrary, is an organized being, strongly nitrogenous, most-often simple, constantly provided with a more-or-less complete intestinal canal, with contractile and excitatory fibres, nearly always visible, consequently digesting, and sensing more-or-less its relations with external bodies, and demonstrated to us by the sudden movements that we see it execute toward an obvious goal.

But as this definition does not include all the organized beings which one often ranks, without really knowing why, among animals, we are forced to make those [beings] that could not enter the first section of the organic empire only because they have no traces of the intestinal canal, a kind of appendage of the animal kingdom. This is how we place the sponges, the false alcyonia, the *moleculaires* and even the corallines, although it is impossible to apply the definition we have just given of the animal.

It is perhaps for the same reason that phytologists are obliged to include among plants the mushrooms and lichens, although, without doubt, they are much less distant from the definition of the vegetable, than the organized bodies of which we have just spoken are from the animal; but which however depart from it in several important respects.

We therefore accept, under the name of animal, a certain combination of organs producing certain forces, *inter alia* a digestive force and a locomotive force, affecting a determined form, and acting on the external circumstances in a manner equally determined.[152]

Attaching corals, sponges, and infusoria as "a kind of appendage" to the animal kingdom "without really knowing why" surely reflects a failure of analysis. Later, as successor to Lamarck (1830) then Cuvier (1832) at the Muséum d'histoire naturelle,[153] he returned the corals to Actinozoa,[154] and referred to Amorphozoa (that is, sponges and infusoria) as including "the beings most obviously neighbours of the parenchymatous plants".[155] Blainville advised that oscillatoria, bacillaria, and some other unicells are actually plants,[156] but avoided any idea that the two types grade into one another, or are interconvertible.

Henri Milne-Edwards: embryology and classification

Henri Milne-Edwards, a student of Cuvier and successor to Geoffroy Saint-Hilaire in the chair of zoology at the Muséum, is credited with applying embryological characters to animal systematics. Karl Ernst von Baer,[157] Martin Barry,[158] and Geoffroy had pointed to parallels between the developmental series and natural classification, but Milne-Edwards brought their strands of argument together into a unified theory. During development, form progresses from the type common to the broader taxon (in early stages), to characters particular to the genus and species (in the adult). As this takes place in all but the simplest animals, we have the image of a hierarchically branching tree.[159] Milne-Edwards supposed that the developmental series of a mammal, for example, might pass through a stage resembling the embryo (but not the adult) of a radiate or mollusc, "or perhaps even the permanent state of some lower zoophytes, such as the amœbæ".[160] Although infusoria (including amœbæ) do not themselves exhibit developmental series, they are confirmed as animals by, so to speak, the back door.

Jean-Baptiste Bory de Saint-Vincent: Règne Psychodiaire

Inspired, perhaps, by his uncle Bernard-Germain de Lacépède, who had collaborated with Buffon and was the first professor of Reptiles and Fishes at the Muséum, Jean-Baptiste Bory de Saint-Vincent showed interest in natural history from an early age. He made his career first, however, in Napoleon's *Grande Armée* (1805–1814), and thereafter as an editor, *voyageur naturaliste*, and politician. Bory was on friendly terms with naturalists in France (Cuvier, Lamarck, Latreille),

308 KINGDOMS, EMPIRES, AND DOMAINS

Germany (Humboldt, Jacquin, Willdenow) and elsewhere,[161] but developed decidedly idiosyncratic views on the structure and plan of nature.

Bory held that matter is perceptible in six states: mucous, vesicular, active (*agissant*), vegetative, crystalline, and earthy. These combine in obscure ways to produce natural bodies, and are released again when plants and animals decompose.[162] Natural bodies are of two types, inorganic (eternal) and organized (perishable). The inorganics form two kingdoms, Règne Ethéré and Règne Minérale, and organized bodies three, Règne Végétale, Règne Psychodiaire, and Règne Animal. Plants are insensible, unconscious of their existence or of time, and unable to move from place to place. Psychodiaires (*arthrodiées*, sponges, most polyps) vegetate and live successively: mature individuals are *apathique*, but their propagules are animated and seek out a safe place to live. Animals (radiates, molluscs, articulates, vertebrates) vegetate and live simultaneously: each individual is sensible, conscious of its existence, and endowed with the faculty of locomotion.[163]

According to Bory, droplets of mucous matter were the first bodies produced by nature. He assigned these bodies to genus *Chaos*, but this is not the Linnæan grab-bag of simple forms; rather, it is the first genus in Chaodinées, a family characterized by "a particular mucosity manifested only by touch, the transparency of which prevents the appreciation of its form and nature, and in which the microscope does not help to distinguish any organization".[164] Bory was unsure how to classify *Chaos*, first calling it the simplest genus of plants,[165] then perhaps an animal *and* a cryptogam,[166] and finally an intermediary, neither plant nor animal.[167] In any case, these droplets can be penetrated by other forms of matter—vesicular, active, and/or vegetative—yielding simple plants or animals.[168] From these, in turn, arise beings of greater complexity and (in the case of animals) sensory powers.[169] He recognized three orders and sixteen genera within Chaodinées, and identified some as the likely source of particular plant genera.[170]

The first organisms to have crossed over into animality were those Bory called *microscopiques*—more or less the infusoria as understood by Blainville.[171] In modern terms these include free-living ciliates, certain dinoflagellates (*Tripos*), euglenozoa (*Virgulina*), unicellular and colonial green algae, a few bacteria, amœbæ, rotifers, various nematodes (*Vibrio*) and trematodes (*Histrionella*), and spermatozoa.[172] He assigned gelatinous cyanobacteria (*Nostoc*) and algae (*Palmella*), charophytes, fungi and mushrooms, various seaweeds, lichens, liverworts, mosses, horsetails, ferns, and lycopods to the plants.[173]

There remained, for Bory, a third kingdom: the Psychodiaire, or two-souled beings. They are vegetative, and in addition possess a "first degree of animality, but not that complete animality which results from intellect added to simple instinct".[174] His three classes do not map neatly onto modern taxa, but encompass diatoms, filamentous cyanobacteria (*Oscillatoria*) and algae (*Spirogyra, Zygnema*), coralline algae (*Amphiroa*), testate amœbæ, stalked ciliates (*Vorticella, Zoothamnion*), sponges, sea-fans, sea-pens, tunicates, bryozoans, and corals.[175] Of all beings Psychodiaire were created first, appearing as mucous, vesicular, and active matter in the primordial ocean. The subsequent addition of other forms of matter yielded articulated, flexible, and tubular polyps.[176] Many families of Psychodiaire grade "by imperceptible nuances" into plants or animals, thereby forming an immense network.[177] Bory is by turns cautious, ambiguous, and self-contradictory in describing how the three kingdoms are related to one another, but he clearly rejects a linear scale in which Psychodiaire are above plants but below animals.

Summary: France

In this chapter the *histoire naturelle* of Pluche and Buffon gave way to *biologie*, with an early-modern look and feel that extended to institutional and professional contexts. The Revolution drove the Church to the sidelines, although some individual savants remained devout.[178] The Chain of Being was largely cast aside, and with it superadditive souls, fine-scale continuity, occult affinities, and plenitude. Lamarck was not the first to embrace transformationism, but he built a modern *système* on it. Cuvier argued that function determined form, Geoffroy the reverse.[179] Entire fields of enquiry—some old, many new—muscled in on each other: anatomy, morphology, physiology, chemistry, embryology, palæontology, evolution, classification.

This was not purely a period of *pensée* and *philosophie*. Cuvier was acclaimed as a hands-on anatomist, as was Milne-Edwards later. Illustrations in the primary literature of the day—monographs, encyclopædia entries, journal articles—were designed to distinguish taxa and illuminate specific matters under discussion, not to present nature as *une grand spectacle*. Taxonomic diagnoses provided by Lamarck, Cuvier, Blainville, Bory, and others remain valuable today. Milne-Edwards pioneered underwater collection using a rudimentary diving bell.[180]

Regardless of whether the main animal taxa were thought to form a single series (Virey, Blainville), parallel or diverging series (Virey, Turpin, Geoffroy), a branching tree (Lamarck, Milne-Edwards) or independent embranchments (Cuvier), intermediate kingdoms gained little traction. Likely candidates for membership—infusoria, sponges, corals—were instead drawn into classes of more-complex beings on the basis of shared characters real or imagined (nervous or digestive systems, motility), as primordial forms, or as analogues of early embryological stages. The exception was Psychodiaire, Bory's kingdom of apathic beings that reproduce by animated propagules. Despite its ambiguities and loose ends, Kingdom Psychodiaire was not a simple borderland. Bory described its independent generation from defined states of matter, a temporal succession of plant and animal stages, and a network of evolutionary transformations that links these beings to plants and to animals. His proposal was not without context, and in Chapter 19 we shall explore the disorienting world of diatoms, desmids, zoospores, and alternating generations.

We have presented French science of this period as largely self-contained. This is not entirely incorrect, as revolution and wars complicated scientific exchange across Europe. Buffon, the encyclopædias, and the Écoles normales had substantially

310   KINGDOMS, EMPIRES, AND DOMAINS

greater impact within France than externally. Centralization of biological research in Paris, while not absolute, created a critical mass unmatched at the time. The idealism that ran through the works of Geoffroy was distinct from German *Naturphilosophie* or British nature-theology. Conversely, experimental microbiology and cell theory came late to France. But other developments transcended national borders, not least the integration of animal embryology with classification. Our story now moves to Germany, where we find a very different social, institutional, and scientific landscape.

18

Naturphilosophie, polygastric animalcules, and cells

> Life is not a property or product of animal matter, but rather, conversely, matter is a product of life. Organism is not a property of certain natural things, but rather, conversely, the certain natural things are, as much, [the] many restrictions or individual perspectives [upon] the universal organism. . . . Things are likewise not principles of organism, but rather, conversely, the organism is the principle of things.
>
> —Friedrich Schelling, *Über den Ursprung*[1]

In the previous chapter our time-travellers visited Paris, home to the leading French scientific institutions and centre stage of the broader Enlightenment. As they now arrive in the German lands,[2] they find instead a multitude of states large and small, decentralized, diverse, with multiple centres of learning. Jena was famously free-thinking; Würzburg renowned for its medical faculty; the Königliche Akademie der Wissenschaften in Berlin scientifically progressive. A dynamic landscape of political alliances, economic fortune, and war facilitated the intermittent capture of individual university faculties by theologians, cultural nationalists, or partisans of one or another philosophical system.[3]

The Enlightenment took a different form in Germany as well. To be sure, Frederick II "the Great" (in power from 1740 to 1786) championed universal values, reformed the Prussian state and society, and recruited savants (including Euler, Lagrange, d'Alembert, Maupertuis, Voltaire, de Condillac, and La Mettrie) to his court and the Akademie.[4] Enlightenment ideals were apparent in music, theatre, literature, and the rebirth of classicism, although these sometimes led in a nationalistic direction.[5] But more broadly, Germany "had never fallen beneath the spell of Newton and Locke". Well into the Eighteenth century it was still the land of Paracelsus, Böhme, and the Rosicrucians, of alchemy and mysticism.[6] German society was conservative, northern Lutherans (at least) as much as southern Catholics. In 1723 Christian Wolff, rector and professor of philosophy at the University of Halle, was banished from Prussia for impiety and teaching determinism;[7] seventy-six years

Kingdoms, Empires, and Domains. Mark A. Ragan, Oxford University Press. © Oxford University Press 2023.
DOI: 10.1093/oso/9780197643037.003.0018

312 KINGDOMS, EMPIRES, AND DOMAINS

later philosopher Johann Fichte was dismissed from Jena for supposed atheism.[8] Latin held on longer in Germany, not least in the universities, than elsewhere in Europe.[9]

Johann Gottfried Herder

From this unpromising soil interesting shoots appeared. Johann Herder, a Lutheran clergyman, literary critic, and historian, is difficult to categorize: a social conservative who welcomed the French Revolution; a student of Kant who became a critic of transcendental idealism; the "father of cultural nationalism" who placed humankind above fatherland.[10] He began *Ideen zur Philosophie der Geschichte der Menschheit* by treating plants, animals, and man at considerable length, arranging them explicitly in a Chain and discussing their form, physiology, and various powers:

> From stone to crystal, from crystal to the metals, from these to the *Pflanzenschöpfung*, from the plants to animal, from these to Man we see the Form of Organization ascend, [and] with it also the powers and impulses of the creature become more heterogeneous and finally of all, are united in the form of Man, to the extent they can grasp it.[11]

All English translations I have seen render *Pflanzenschöpfung* as *plants* or *the plant world*, and this is surely what it means in its single other occurrence in *Ideen*.[12] Elsewhere in *Ideen*, Herder refers to "*Pflanzengeschöpfen*, snails and insects" as the "lowest types of animal",[13] with *Pflanzengeschöpfen* seemingly a synonym of *zoophyte*. He identifies "zoophytes and insects" as the animals that "come closest to plants", and speaks of the transition (*Ubergang*) from plants to zoophytes.[14] He also refers to the "Reich der Pflanzen und Pflanzenthiere":[15] consistency was not Herder's strong point. Within the Chain we see "a prevailing resemblance of *Hauptform*",[16] but he does not develop a theory of archetype.

Johann Wolfgang von Goethe

Johann Wolfgang von Goethe studied law first at Leipzig, then in Strasbourg, where he found it more enjoyable to attend lectures on physics, chemistry, and anatomy, pursue friendships with men of science and medicine,[17] and try his hand at poetry and drama. In Strasbourg he also met Herder, who introduced him to Shakespeare and to epic, folk, and nature poetry. His sentimental novel *Die Leiden des jungen Werthers*[18] brought acclaim, notoriety, and an invitation to the court of Karl August, Duke of Saxe-Weimar and Saxe-Eisenach.[19] Goethe went on to serve Karl August in various roles, including as Privy Councillor (*Geheimrat*), for more than half a century. These duties were rarely burdensome, and Goethe assembled a mineral-cabinet, took lessons in anatomical drawing, studied the bone structure of animals and man, and read Linnæus.[20] He committed his thoughts to notebooks, letters, and essays, some of which appeared in print only decades later, or not at all.[21]

Under the tutelage of Justus Loder, physician to Karl August, Goethe discovered that man, like the quadrupeds, possesses an intermaxillary bone.[22] Blumenbach and others had denied this, and on that slender ground held that man could be distinguished from the animals not only by his mental facilities, but even osteologically. For Goethe, his discovery meant that man stands atop the animals not in anatomical detail, but in the perfection with which structure "furnishes the scaffolding for the development of the spirit".[23]

Turning his attention to plants, Goethe asked whether the immense diversity of plant form can likewise be subsumed into a common plan. Linnæus taught that branches arise from buds, petals from leaves; and more controversially, that the parts of an inflorescence are manifestations of what would have become leaves in future years, had the plant not flowered.[24] Goethe rejected the latter idea, but came to see most parts of flowering plants—cotyledon, foliage-leaf, nectary, sepal, petal, stamen, pistil, fruit, seed coat—as transformations of a single conceptual structure, the symbolic or transcendental leaf.[25] In 1786 he then made a further intuitive leap to the idea of the *Urpflanze*.[26]

What, precisely, did Goethe mean by *Urpflanze*? *Ur-* can connote *primæval* (belonging to an early age) or genealogically *ancestral*, and Goethe surely reflected on these meanings: in 1784 his muse Charlotte von Stein had written to poet Karl Ludwig von Knebel that

> Herder's new work [*Ideen* 1] makes it probable that we were first plants and animals; what Nature will now stamp out of us, remains for us quite unknown. Goethe is now brooding over these things, and anything that has passed through his imagination becomes extremely interesting. Such is how, through him, I [now] find those repulsive [*gehässigen*] bones and the bleak Stone Kingdom.[27]

However, the *Urpflanze* that emerged from Goethe's imagination was neither relict nor ancestor, but an archetype—a symbol or model[28] that subsumes all possible plant types, and from which they all could be intuited:

> The *Urpflanze* will be the most wondrous creature in the world, one for which Nature herself will envy me. With this model and the key thereto, one can then endlessly invent plants that must be consistent, that is, even if they do not exist, they *could* exist and be not just artistic or poetic shadows or semblances, but have an inner truth and necessity.[29]

Symbol, relict, and ancestor can be difficult to disentangle, and Goethe mixed-and-matched these meanings in different ways. In studying the vertebrate skeleton, he

314   KINGDOMS, EMPIRES, AND DOMAINS

had hoped to find "the *Urthier* . . . that is, the concept, the idea of the animal".[30] He propagated houseplants and was partial to *Bryophyllum*, a curious plant that "manifests the triumph of metamorphosis";[31] but the *Urpflanze*, like the *Urthier*, was an Idea. On the other hand, the third kingdom of nature had an actual *Urgestein*: granite. Goethe was a Neptunist: that is, he believed that most rock types had originated in the primæval ocean. Granite, the first type to have emerged as the ocean receded, was supposedly the co-crystallization product of quartz, feldspar, and mica in perfect proportions.[32] Goethe theorized that porphyries (rocks with particles of different sizes and distinct composition) arose from granite as its compositional balance became upset, for example by weathering. In this way granite contains within itself all other rock types, and they arise from it by metamorphosis.[33] For the poet, the very foundations of Earth emerge from chaos, diversity from unity. Nor were these his only *Ur*-models. Moses had received the Commandments graven into *Urstein*.[34] The magnet was an *Urphänomen*. Elsewhere we find the *Urbild, Urform, Urding, Urwelt, Urgebürge, Urfelsen, Urmetall, Urfarbe, Urwesen, Urglieder, Urgefühl*, and *Urworte*: some primæval, others transcendental and/or poetic. Indeed, he found it impossible to get his mind fully around *type* in nature.[35] Such are the challenges of idealism.

We cannot deny the role of archetype in his construction of *Morphologie*, a new "doctrine of forms" with mechanical, comparative, temporal, and mental dimensions.[36] To the modern eye, however, Goethe made little further use of archetypes. Unlike Schelling (below), he did not claim that all organic beings share a common archetype. He did not erect a hierarchy of archetypes, or base a system of classification on them.[37] He did not elaborate a course of biological evolution, or consider whether living nature was, as a consequence, linear or branched. He did not use archetypes to delineate the plant or animal kingdom; nor did he show concern that *vertebra* (the organizing principle of his *Urthier*) excludes polyps and worms,[38] or that *leaf* excludes algae. As a consequence, he did not come face to face with the issue of an intermediate kingdom—with one exception: animalcules.

In the spring of 1786 Goethe prepared vegetable infusions and made precise drawings of the infusorial animalcules that appeared.[39] Microscopic beings emerged regardless of whether the infusion was kept in darkness or light, but he interpreted the former as animals, the latter as plants, presumably on the basis of their colouration. Unlike Needham, Goethe did not imagine that decaying vegetable matter was itself transformed to animalcules; rather, a common seed produced animals in the dark, plants in the light. Darkness and light are polar opposites; so too must be animal and plant:

> Regarded in their most imperfect constitution, plants and animals are hardly to be distinguished. A living point, rigid, flexible [*beweglich*] or semi-flexible, is what is barely perceptible to our senses. We dare not decide from these first beginnings, whether there are enough features and analogies for it to be demarcated to either

side, to be led across through light to Plant, [or] through darkness to Animal. This much, however, we can say, that from a relationship [*Verwandtschaft*] in which plant can scarcely be distinguished from animal, creatures gradually come forth, perfecting themselves toward two opposite sides, such that the plant is finally glorified into the tree, enduring and rigid, the animal into Man, of the highest flexibility and freedom.[40]

Immanuel Kant: transcendental idealism

In a formulation often attributed to Hegel, Kant's reconciliation of rationalism (Descartes, Spinoza, Leibniz) with empiricism (Locke, Berkeley, Hume) marks the beginning of modern philosophy.[41] While the French Enlightenment celebrated *l'esprit humain*, German philosophers from Kant onward developed ponderous self-contained systems that set out relationships among the external world, the senses, mind, will, self-consciousness, necessity, and freedom. Many dealt at length with nature as worthy of explanation in its own right, and/or as the arena in which human experience plays out. Here we limit our consideration of these systems to their treatment of the organic world, specifically the interface between plants and animals.

In *Critique of pure reason* (1781), Immanuel Kant argued that things (*noumena*) exist in the world, quite apart from human experience. So too do transcendental ideas, for example laws governing the cosmos. We interact not with *noumena*, but rather with *phenomena*, in part by imposing upon them forms of thought (categories) that order our sensory impressions.[42] Plants and animals require a different sort of explanation than (for example) minerals, as they perpetuate themselves in established forms: that is, they are natural purposes, both cause and effect, and must be understood teleologically, as if their parts were designed to produce the whole organism.[43] Yet like other species, plants and animals are necessarily fixed within clear and stable boundaries; they cannot overlap, merge, or change into one another, as that would undermine the very foundation of reason. When Herder suggested that plants and animals may have arisen, through a succession of increasingly well-adapted forms, from the maternal womb of Earth, Kant rejected the thought as "so monstrous that reason recoils".[44] He cautiously allowed, however, that it is "a legitimate and excellent regulative principle of reason" to use the *scala naturæ* as a "maxim" in seeking out order in nature.[45] At times he seemed to toy with the idea himself:

This analogy of forms, which with all their differences seem to have been produced according to a common original type, strengthens our suspicions of an actual relationship between them in their production from a common parent, through the gradual approximation of one animal-genus to another—from those in which the principle of purpose seems to be best authenticated, *i.e.* from man, down to the

316   KINGDOMS, EMPIRES, AND DOMAINS

polyp, and again from this down to mosses and lichens, and finally to the lowest
stage of nature noticeable by us, *viz*. to crude matter.[46]

Although Kant stopped short of accepting that organisms can ascend on a scale,
he allowed that reason might lead one to suspect this. Again, from a lecture on
metaphysics:

> Admittedly, I find a transition from the mineral kingdom into the plant kingdom,
> which is already a beginning of life; [and] further from the plant kingdom into the
> animal kingdom, where there is also a small degree of life; [but] the highest Life is
> however the freedom that I find among Man.[47]

He quickly pointed out that the biological series, unlike a mathematical continuum,
is neither limitless nor continuous. Later yet, Kant was less generous:

> This is the so-called continuum of forms (*continuum formarum*), according to the
> analogy of the physical continuum (*continui physici*), where the minerals com-
> mence the order, through the mosses, lichens, plants, zoophytes through the
> animal kingdom until human beings. This is nothing more than a dream whose
> groundlessness Blumenbach has shown.[48]

Although no fan of a *scala naturæ*, Kant promoted another old idea onto the
modern agenda: that of a formative force intrinsic to nature. As we have seen, many
early cultures held that the cosmos is ensouled or animate. Ideas of a vital or formative force persisted in various forms: the *Archeus* (Paracelsus, van Helmont), *anima
sensitiva* (Harvey, van Helmont, Stahl), *moule intérieur* (Buffon), medullary forces
(Linnæus), *Lebenskraft* (Medicus), *Nutritionskraft* (Born),[49] *vis essentialis* (Caspar
Wolff),[50] and Blumenbach's *nisus formativus*[51] or *Bildungstrieb*.[52] Kant appropriated
the latter as the mechanical component of his teleology, manifest in processes including generation and embryological development.[53] The term was then taken up
by Fichte (*System of ethics*, 1798), Schelling (*On the world soul*, 1798), and others for
use in their increasingly radical idealistic systems.

   We need not follow idealism step by step down the yellow brick road to Hegel,
for whom even *noumena* are only Idea. His *Philosophy of nature* deals mostly with
"inorganic physics" (more precisely, the corresponding mental representations),
and under "organic physics" (particularly *body*) treats geological nature, vegetable
nature, and animal organism in that order, without touching on transitional or intermediate forms.[54]

German Romanticism

Idealistic philosophy became deeply intertwined with the Romantic movement in German culture.[55] In *Critique of judgement* (1790), Kant contended that

teleological and æsthetic judgement can be alike, in that the categories we impose are not pre-existent, but instead must be worked out afresh by each observer: that is, they are subjective, and as such "inseparably human".[56] In the hands Schelling, Fichte, and Hegel, first *phenomena*, then *noumena* too were subsumed into human self-consciousness. According to Novalis—whose brief, emotionally fraught life was held to be quintessentially Romantic—the universe is not physical at all, but a manifestation of human spirit.[57] But Romanticism was far more than a current in academic philosophy: it celebrated the æsthetic intuition in regard of art, literature, and (not least) nature. Idealism was intellectual, Romanticism sensual. At the personal level—and Romanticism was nothing if not individual—Romantic sensibility involved spontaneity, passion, sometimes tragedy. Interest ran high in symbolism, genius, child prodigies, heroism, folklore, and supposed national characteristics. Some Romantics flouted social convention—which was not difficult, given the conservatism of German society.

The Romantic involvement with nature was particularly important. If nature is inextricably bound up in self-consciousness, experiencing it at first hand becomes a path to understanding, indeed creating, one's self.[58] The Romantic might roam through Italy (Goethe, Haeckel), South America (Humboldt), Sri Lanka, or Malaysia (Haeckel); paint moody landscapes (Carus); or take a contemplative stroll in a meadow. Peter Schlemihl, the noble if flawed hero of the eponymous short story by botanist and explorer Adelbert von Chamisso, ventured to many lands but, unable to travel on to New Holland (Australia) and thereby cap off his self-creation, gazed southward from a Lombok shore and wept:

> New Holland, that extraordinary country, so essentially necessary to understanding the philosophy of the earth, and its sun-embroidered dress, the vegetable and animal world; and the South Sea with its Zoophyte islands, were interdicted to me . . . such is the reward for all the labours of man![59]

Naturphilosophie

Idealism, Romanticism, and the study of nature (as more than an aid to personal *Bildung*) came together in the late 1790s as *Naturphilosophie*. Kant had left an epistemological gap between *noumenon* and *phenomenon*, between deterministic physical nature and ethical self-determination. Fichte sought to bridge this gap by making noumenal nature the object of self-consciousness.[60] Schelling offered a different solution: nature is an expression of Spirit, a never-ceasing productivity that necessarily brings forth products (organisms) in which form and matter are united in particular arrangements (archetypes). These arrangements ensure that effect[61] is not dissipated, but instead is captured as physiology.[62] These archetypes reflect nature, and constitute a hierarchy from archetype of organism downward through that of animal (or plant), of quadruped, and so forth.[63] Yet just as Spirit is unitary, so too the products of nature "are to be seen as only *one* organism inhibited at various stages of development".[64]

318 KINGDOMS, EMPIRES, AND DOMAINS

Schelling has more to say about these stages. Moving downward from animals—Schelling does not explicitly mention a *scala naturæ*—sensibility (in higher animals) gives way to mere irritability (in lower animals and a few plants), which in turn dissipates (in plants), leaving only the force of reproduction.[65] Zoophytes, like most plants, lack both sensibility and irritability, but Schelling does not hesitate to consider them animals.[66]

If one is prepared to accept nature-as-idea, none of this so far is particularly objectionable. Schelling intended his *Naturphilosophie* to complement Fichte's *Wissenschaftslehre* by subsuming ego into unitary, animated nature.[67] Even today, thoughtful people may find value in regarding the Earth, including humankind, as a single holistic system.[68] Cuvier, Saint-Hilaire, and Richard Owen studied the doctrine of archetype.[69] Schelling tried hard (often too hard) to incorporate the discoveries of empirical science (galvanic forces, oxygen) into his system, and conducted experiments to confirm specific points.[70] On the surface, all this is more typical of science as we recognize it today, than of rampant metaphysics. But Fichte and Schelling were not justifying Baconian method: their science of nature was to be deduced from first principle (*Grundsatz*).

Kant had argued that forces of attraction and repulsion undergird the physical world, and illustrated their synthesis by reference to the poles of a magnet.[71] On this foundation Schelling and Oken built a baroque edifice of supposed polar opposites, analogies, parallels, and empathies.[72] In *Darstellung meines Systems*, Schelling set plants and animals as polar opposites[73] and entangled them with other polarities: animals are positive, nitrogen, south, male, warm, earth, and iron; plants negative, carbon, north, female, cold, sun, and water. Such are the origins of the four cardinal directions in the Easter egg diagram of Goldfuss,[74] and the carbon-nitrogen and oxygen-hydrogen axes in the spiral diagram of Horaninov.[75] Fundamental polarities need not exclude a third kingdom: recall the Phytozoa of Horaninov, intermediate between plants and animals. Schelling, however, chose to emphasize instead the "perfect relation of identity" between plants and animals.[76]

Lorenz Oken

Oken, a friend and disciple of Schelling,[77] was for decades the most-visible proponent of *Naturphilosophie*. He was appointed a professor of medical science at Jena (1807), where Goethe was rector; after disputes with various civil authorities he accepted an invitation to Zürich, where he finished his career as professor of natural history.[78] His journal *Isis* (1817–1848), and the Gesellschaft Deutscher Naturforscher und Ärtze, which he founded in 1822, helped to professionalize German science, rebuild international linkages after the Napoleonic wars, and promote science beyond the universities.[79] Level-headed contemporaries praised his clear-sighted observational research and effectiveness as a teacher.[80] However, these virtues are not always obvious in his written work. Perhaps he sought to develop the pure *Wissenschaft* that Karl Friedrich Burdach had contrasted with *Erfahrungswisse nschaft*.[81] In any case, it is next to impossible to make sense of many of his writings, as he heaps analogy upon abstraction, metaphor upon symbol, contradicting himself—or so it seems—time and again. Not a few observers have suspected Oken of having his readers on, "since there is no other explanation to be offered for what appears to be meaningless gibberish".[82] With these important caveats, let us try to pin him down on three issues pertinent to our question.

The kingdoms of nature

For Oken, the galvanic polarity[83] is the principle of life.[84] In *Outline of the system of biology* (1805) Oken described the coral stem as a "living galvanic column", its flowers "not animals at all, nor even plants, but only the representative of the liquid in the galvanic column". Combining the galvanic positivity of Coral with its polar opposite, the negativity of Plant, yields a "synthetic organism", Animal.[85] Oken connected infusoria with (coral) polyps:

> The granular [*körnichte*] mass of polyps, or as we can call it with the consent of natural scientists, the infusorial mass, is the *Ur* of sexual union [*Begattung*], it necessarily precedes all else, it is itself the material in all organic origin [*Entstehung*], which is metamorphosed into the organic forms [*Gebilde*], which gathers the inorganic masses of earth, air, water, metals, sulphur, and salts and forces them to form the organic trunk [*Stamm*], which then inhabits them as the pure Living, although according to the order in which they are assembled, sometimes as Coral, sometimes as Plant, and finally if it was able to work in the middle, as Animal.[86]

Accordingly, in *Die Zeugung* (1805) he divided the organic world into infusoria, plants, and animals,[87] and referred to the Kingdom of Corals.[88] On the other hand, in his seven-volume *Allgemeine Naturgeschichte* (1833–1841) Oken recognized the three standard kingdoms (mineral, plant, and animal), placing fungi among the plants, and infusoria and polyps with the animals.[89] In *Lehrbuch der Naturphilosophie* he asserted that "there can be only three kingdoms of nature".[90]

The plant-animal interface

Oken described the plant-animal interface in different ways. We have already mentioned one, involving the galvanic dipole of polyp and plant. Another can be found in the Foreword to his *Lehrbuch* (1843):

320 KINGDOMS, EMPIRES, AND DOMAINS

> The Mineral, Vegetable, and Animal classes are not to be arbitrarily arranged in accordance with single or isolated characters, but [are] to be based upon the cardinal organs or anatomical systems, from which a firmly established number of classes must of necessity result; moreover, each of these classes commences or takes its starting-point from below, and consequently all of them pass parallel to each other.[91]

This seems to rule out a linear scale in nature, but might be compatible with separate infusorial origins of plants and animals (below). In the same work, however, he explains the natural world differently again:

> Metal and sulphur have, in the Geogeny, announced themselves as the precursors, or harbingers, of the vegetable world. In this respect, also, can the vegetable kingdom be regarded as the mineral kingdom, that, having continued to grow, has become alive. The ore, which becomes organic, becomes carbon or plant.[92]
>
> ... The vegetable blossom loses its definition as plant, so soon as it has acquired self-substantial life.... The self-moveable or automatic blossom has consequently passed over into a new kingdom, or into one whose very definition is the self-substantial motion ... such a blossom is an Animal. An animal is a blossom without stem or a flower, which of itself produces its stem, this being the reverse of what takes place in the plant.[93]
>
> A plant is an animal retarded by the darkness, the animal is a plant blooming directly through the light, and devoid of root.... The animal is a whole solar system, the plant only a planet. The animal is therefore a whole universe, the plant only its half; the former is microcosm, the latter microplanet.[94]
>
> ... Bone, flesh and nerve are the highest organs of the animal; the viscera, which mostly consist of cellular tissue, will indicate the Vegetative in the animal.... What is not bone, flesh or nerve, is not animal, but vegetable.[95]

These aphorisms speak of evolution from mineral to plant, and from plant directly to animal, without a role for polyp. Oken calls many systems in the animal body *vegetative* or *vegetable*, but does not claim that animals "retain" them. His claim is stronger: that animals *are* plants, albeit with additional systems: "the heart is the animal in the plant".[96]

The infusorial origin of plants and animals

Oken held that just as plant and animal matter decomposes into infusoria (recall Needham and Buffon), so too conversely

> all higher animals must arise from [infusoria], as their constituent animals. For this reason we call them [infusoria] *Urthiere*, of which I claim (admittedly without

being able to give the reasons here) that they originated in the Creation just as generally and irreversibly as did Earth, Air and Water; that they, like these elements in their sphere, are elements in the organic world, and comprise not only the *Urstoff* of animals, but also of plants . . . ; and for which the transition [*Uebergang*] of infusoria into tremellas, and this into that, may nevertheless be evidence: in this broader sense they can be called the *Urstoffe* of the Organics.[97]

Despite having originated in the Creation, these infusoria must (it seems) compose themselves anew from the inorganic:

the first transition from the inorganic into the organic is the transformation into a thermic vesicle, which in my theory of generation I called Infusorium, which . . . in water is determined to animal, but in the air to plant. Animal and plant are nothing more than a multiply branched or reiterated vesicle.[98]

An Infusorium coagulates [*zusammengerinnt*] from slime [*Schleim*] . . . a drop of slime is already an Infusorium.[99]

The cellular tissue of plants is a "multiplication" of the primary vesicle;[100] and likewise, the "animal body is nothing else than a compound fabric of Monads".[101] Several decades later, Oken offered a retrospective on his theory:

I first advanced the doctrine, that all organic beings originate from and consist of *vesicles* or *cells*, in my book [*Die Zeugung*] . . . These vesicles, when singly detached and regarded in their original process of production, are the infusorial mass, or the protoplasma [*Ur-Schleim*] from whence all larger organisms fashion themselves or are evolved. Their production is therefore nothing else than a regular agglomeration of Infusoria; not of course species already elaborated or perfect, but of mucous vesicles or points in general, which first form themselves by their union or combination into particular species.[102]

Insofar as infusoria are organisms, Oken classified them,[103] along with polyps, corals, and acalephs,[104] as Oozoa (egg-animals), the first [division] of the animal kingdom. Oozoa

represent those products of nature which are prior or antecedent to the animal world; namely, first of all plants, and further still the inorganic kingdom also, or the earth, since they have originated in the water and can be as well developed from the stones as the Lichens. There are therefore Lithozoa or Stone-animals, and Phytozoa or Plant-animals, among the Oozoa.[105]

Thus, infusoria are, in several senses of the word, *Urthiere*. They are also coagulated slime; the constituent vesicles (cells) of plant and animal tissues, or their immediate precursors; animals in their own right; and representatives of the primæval ocean.

322 KINGDOMS, EMPIRES, AND DOMAINS

By the 1830s most scientists in Germany (and elsewhere) would have used distinct terms to distinguish concepts as diverse as these. *Naturphilosophie* trafficked in analogy and symbolism, of course, but Oken would probably have defended this muddy terminology as emphasizing the unity of Idea in nature.

Over time *Naturphilosophie* lost influence in German science. The reasons were multiple: generational change, pushback against Hegelian idealism (not least against the revolutionary politics of a Hegelian left in the decade before 1848), greater exposure to ideas from France and Britain, the rise and institutionalization of specialized disciplines within life science, and attacks by respected authorities— Justus Liebig famously comparing *Naturphilosophie* with the Black Death, and its proponents with madmen and mass murderers.[106] One of its final statements came in *Natur und Idee* by Gustav Carus (1861), to which we return at the end of this chapter. Although no natural scientist working in central or northern Europe in the Nineteenth century was untouched by transcendental idealism, Romanticism, and/ or *Naturphilosophie*, some embraced its methodology much more than others. We now consider four such scientists, inquiring how they engaged the issue of intermediate kingdoms of life.

Gottfried Reinhold Treviranus

Treviranus trained in mathematics and physiology at Universität Göttingen (where he studied natural history under Blumenbach), receiving his medical degree in 1796. He then set up a medical practise in his home city, Bremen. His major work, the six-volume *Biologie, oder Philosophie der lebenden Natur für Naturforscher und Aerzte* (1802–1822), avoids the substance (if such be the word) and language of *Naturphilosophie*. Instead, he offers empirically based presentations of the anatomy, physiology, chemistry, distribution, generation, nutrition, and economy of living organisms that individually would not have been out of place in a French textbook of the period. *Biologie* is not primarily a work of classification, but Treviranus situated *Reiche* of plants, zoophytes, and animals across "two great divisions of living nature": nitrogen predominates in animals and Thierpflanzen, carbon in Pflanzenthiere and plants.[107] His Thierpflanzen include sea cucumbers, sea urchins, starfish, jellyfish, hydroids, sea-feathers, sea-pens, soft and stony corals, and infusoria; his Pflanzenthiere encompass fungi, confervas, seaweeds, lichens, liverworts, mosses, ferns, horsetails, and duckweeds.[108] After so many centuries, this remarkably broad assemblage was *das Reich der Zoophyten*—the first zoophyte Kingdom.[109]

Alexander von Humboldt

After desultory studies in economics, Alexander von Humboldt enrolled at Universität Göttingen (where Blumenbach was professor of natural history), and later completed his studies at the Bergakademie Freiburg (1792), where he studied geology under Werner. He became an inspector of mines for the Prussian government, but found time to attend anatomy seminars given by Loder at Jena, sometimes attending alongside Goethe. Humboldt too became fascinated with botany, and his *Floræ Fribergensis*—or more likely the appended *Aphorismi ex doctrina, physiologiæ chemicæ plantarum*[110]—led to a collaboration with Goethe on the electrical activation of muscle fibres.[111] In *Floræ Fribergensis* Humboldt described algae (including lichens) and fungi he found growing in mines. The *Aphorismi* arose from his experiments in plant physiology, including studies of the response of plants to various gases, chemicals (including opium), and stimuli (heat, light, electricity). He argued that the vital force[112] is present in plants as well as animals, identified the structures in which it is most in evidence, and speculated on its chemical and physiological basis.[113] In a lengthy footnote he decried how the distinction between plants and animals had been driven by attention to "the extreme limits of nature"—trees and man.[114] He commented on Trembley's hydra, and on polyps and zoophytes, but found no cause to recognize a third or intermediate kingdom.

From 1799–1804 Humboldt, together with botanist Aimé Bonpland, explored remote regions of South America. Their systematic data on topography, temperature, river basins, and the distribution of plants initiated the disciplines of physical geography and plant biogeography. Humboldt's studies on volcanoes, and on the igneous origin of rocks formerly thought to be sedimentary, did much to disprove the Neptunist geology of Werner and Goethe. Although the expedition and subsequent books were considered an epitome of German Romanticism—his moving reflections on nature fill many pages—the æsthetic did not blind him to the terrible hardships endured by indigenous peoples in these lands. On his return to Europe, Humboldt based himself in Paris, where he frequented Cuvier's *salon* and attended his series of lectures during the debate with Geoffroy. Humboldt was later recalled to Berlin, where he presented the lectures (1827–1828) that eventually became his acclaimed *Kosmos* (published 1845–1862), and chaired the 1828 meeting of Oken's Gesellschaft der Naturforscher und Ärtze, the first to separate presentations by discipline. In 1829 Humboldt led a scientific expedition across the Russian empire, accompanied by Christian Gottfried Ehrenberg, whom we meet below. In his later years he recanted his early view on the vital force, admitting the "difficulty of satisfactorily referring the vital phenomena of organic life to physical and chemical laws".[115]

Humboldt's many contributions to biology do not extend to classification, which he considered an "obscure domain" irrelevant to understanding the earth as a physical system.[116] Even so, to the end of his life he remained *au fait* with developments in plant and animal science, pointing to cyclosis (cytoplasmic motion or "streaming") as refuting a simple delineation between plants and animals based on movement,[117] and praising the discovery by Ehrenberg, Ross, and others that the ocean everywhere teems with microscopic beings (which he assumed to be animals).[118] Notably, he did not assume that green colouration marks an infusorian

324   KINGDOMS, EMPIRES, AND DOMAINS

as a plant.[119] And even with material as technical as oceanic microbiology, his summary in *Kosmos* centred on the "strong and beneficial influence exercised on the feelings of mankind by the consideration of the diffusion of life throughout the realms of nature".[120]

Karl Ernst von Baer

We briefly met Karl Ernst von Baer in Chapter 16, in regard of his early flirtation with geometric representations of the animal kingdom. Von Baer was exposed to *Naturphilosophie* first as a medical student at Dorpat (Tartu), where he studied under Burdach, then more formally at Würzburg; despite initial curiosity, he found *Naturphilosophie* "as empty as it was laboured".[121] He studied botany and zoology—his autobiography mentions a fascination with the shapes of infusoria—built a herbarium, and learned about fungi and algae from Nees von Esenbeck. He later won renown for discovering the notochord, describing the blastula, and for his eponymous laws of embryology:

1. That the more general characters of a large group appear earlier in their embryos than the more special characters.
2. From the most general forms the less general are developed, and so on, until finally the most special arises.
3. Every embryo of a given animal form, instead of passing through the other forms, rather becomes separated from them.
4. Fundamentally, therefore, the embryo of a higher form never resembles any other form, but only its embryo.[122]

These laws went to the heart of several issues in contemporary biology. The first two favour an epigenetic model of animal generation over preformation.[123] His third and fourth are what one might expect if animals have evolved in a tree-like manner. Whether von Baer accepted evolution, however, is a matter of perennial dispute.[124] He denied that genera form a linear series, and that species can transmutate, although he did accept that species might adapt, in limited measure, to local conditions. Nor could he agree that the course of individual embryonic development had much to teach us about relationships within the animal kingdom.[125] Instead, he saw in an embryological series the unfolding of a morphological model, the *Haupttypus*, through a hierarchy of submodels corresponding to classes, families, and so on.[126] It would remain for others—notably Serres, Meckel, Owen, and Haeckel—to explore how ontogeny might recapitulate phylogeny. As for the classification arising from von Baer's approach, its first division was into four so-called provinces (articulates, radiates, molluscs, vertebrates): that is, the four *embranchemens* of Cuvier, although he criticized Cuvier for conflating type and grade, with the consequence that "all the animals of low organization are thrown among the Radiata, although very many of them are by no means radiate in their structure".[127] His classification did not rely on single characters, so it had not been led astray by "so-called affinities".[128]

Where, then, did microscopic beings fit into von Baer's classification? His studies on invertebrates revealed their diversity of type,[129] and the situation deteriorated further as one progressed into the infusoria. OF Müller had relied on characters "little suited to yield natural groups",[130] and while Lamarck had removed some animalcules with a mouth-opening and digestive cavity, the remaining beings were still not of a common type.[131] The diatoms (Bacillariæ) appear not to be animals at all, but rather plants, although von Baer did not explain precisely why.[132] We can, however, recognize that some animalcules are thread-like, others spherical, or circular, or elongated and flat,[133] and on this basis we can recognize them as prototypes [*Vorbilder*] of animals of higher grades: vibrios of roundworms, nematodes, and insects; paramecia of trematodes; rotifers of molluscs; a spherical animalcule of ascidians, medusæ, and sea-urchins.[134] Protozoa[135] are not a natural group at all, but variously the lowest grade of articulates, radiates, or molluscs.

Christian Gottfried Ehrenberg

After a brief and undistinguished career in theology, Christian Gottfried Ehrenberg studied medicine at Berlin, where he was exposed to Romanticism and *Naturphilosophie*: his dissertation concludes with a verse from Schiller, and he defended theses on astrology, Oken's *mathesis*, and whether the agent of eruptive fevers [*exanthemata*] might be classified among the plants.[136] Under the guidance of Heinrich Link, professor of botany and director of the botanic garden, Ehrenberg demonstrated that fungi arise not from worms, infusoria, or inorganic material, but from spores.[137] He classified fungi from Chamisso's voyage around the world,[138] then on Humboldt's recommendation joined an archæological expedition to North Africa (1820–1825) which he was fortunate to survive.[139] Later he accompanied Humboldt across the Russian empire (above) on an expedition which, owing to the Tsar's sponsorship and Humboldt's fame, had a "carnival-like atmosphere".[140]

Applying his skills in microscopy, over the course of his career Ehrenberg described thousands of living and fossil diatoms, radiolarians, ciliates, and other microscopic beings.[141] He demonstrated that microscopic life is widespread and abundant on our planet including in the ocean, polar regions, and atmosphere, and that microbes constituted or had caused the deposition of various rocks, sediments, and mineral deposits. He studied the development and feeding behaviour of coral polyps, and showed that some infusoria are parasites, including on other microbes. Even as a professor, then Rector of Universität Berlin, he continued his research: Charles Darwin sent him samples collected during the voyage of the *Beagle*.

By the 1820s experimental science was finding its way into zoology, and Ehrenberg showed that some infusoria could ingest food coloured with indigo,

326 KINGDOMS, EMPIRES, AND DOMAINS

carmine or *Saftgrün*. Inside the organism the food was sequestered in vesicles he not unreasonably called stomachs (*Mägen*). We know today that food vesicles, like other types of vacuoles, do not simply float around in the cytoplasm, but are actively moved about, merge with other membranous structures, and are eventually recycled to the surface. Ehrenberg could not possibly have observed all these details with the modest microscope he insisted on using.[142] Even so, he drew beautifully detailed diagrams of complete alimentary systems in organisms such as the diatom *Navicula* and the ciliate *Stentor*.[143] On this basis, Ehrenberg grouped these organisms in a class he named Magenthiere in German, Les Polygastriques in French, and Polygastrica in Latin.[144] They are

> animals without a spinal column and free of a pulse, with their alimentary canal divided into numerous vesicular stomachs, with (due to bud-formation or self-division) an indefinite body shape, with both sexes united in one, moved by (oft-times whirling) pseudopodia and without true articulated feet.[145]

Within the group he recognized twenty-two families, twelve Anentera (stomachless) and ten Enterodela (stomach-bearing). According to Oken, a digestive system was the first form of animality:

> The intestinal system is the first form of body . . . the protoplasma is a hollow globule. The intestinal system is therefore nothing else than the original cystic form. Thus, there are *Cystic animals* like the Infusoria.[146]

Nor is this the only hint of *Naturphilosophie* in Ehrenberg: he describes Magenthiere as worms with multiple stomachs, snails without a heart, insects without articulation, fish without a spinal cord.[147] Oken was hardly the first to find an alimentary system in animalcules, or to use its supposed presence to distinguish animals from plants: in 1752 John Hill had found an intestine in *Cyclidium* and *Paramecium*,[148] and Vicq d'Azyr (1792), Lamarck (1801), and Blainville (1822) considered it a criterion of animality.[149] Audouin and Milne-Edwards had used details of the digestive tract to reclassify polyps,[150] as had Ehrenberg himself in separating bryozoa from corals.[151]

But Ehrenberg went much farther, claiming (even in the title of his monograph) that infusoria are complete or perfect [*vollkommene*] organisms, possessed not only of a mouth and alimentary apparatus, but also appendages of touch, food-capture and locomotion, reproductive organs, a visual apparatus (today called the *stigma* or *eyespot*), associated muscles and nerves,[152] and "full clear mental capacities, like other animals".[153] He was not the first to suggest this either, although we must look to William Derham (1713) for precedent.[154] For Ehrenberg, infusoria were neither prototypes, nor beings at the margin of animality, tenuously joined to worms and insects by exterior form, type, idea, or lack of a palatable alternative: they were complete organisms, indeed complete animals: very small ones to be sure, and no less wonderful for that.

Ehrenberg appreciated that with "a sharper definition of animal in general . . . all plants and minerals [can be] separated sharply and strictly by their lack of animal-organic systems".[155] Thus he provisionally considered *Oscillatoria* and *Spirogyra* to be plants, not infusoria, because they lack an oral aperture, reproduce and grow only by budding, are rigid like plants, are fertilized as in some fungi, exhibit intracellular crystals, and apparently lack voluntary movement.[156] Today we consider *Oscillatoria* a cyanobacterium, and *Spirogyra* a green alga. But Ehrenberg's venture with "complete organisms" ended badly. In 1835 Félix Dujardin, a chemist and applied scientist from Tours, described in foraminifera and amœbæ a "glutinous, diaphanous material" which fills the body and extrudes to form pseudopodia. This basal contractile material or *sarcode* has the property of giving rise to *vacuoles adventives*, and it is these that Ehrenberg had mistaken for stomachs and other organs.[157] Dujardin found sarcode abundant in worms and insect larvæ, but did not claim it to be present in higher animals— although he had chosen the name to indicate that it "forms the passage to flesh [*chair*] proper, or is destined to become flesh itself".[158] Gustaf Focke, Thomas Rymer Jones, and Edward Forbes joined Dujardin in opposing Ehrenberg's interpretation.[159]

With stomachs and sarcode, the world of infusoria intersected a major development in early Nineteenth-century biology—cell theory. The encounter would prove momentous for both, and for third kingdoms of life.

Cell theory

What we know today as *cells* were first observed in plant tissues by Hooke (1665), Malpighi (1671), and Grew (1672), and in blood by Swammerdam (early 1660s). In *Theoria generationis* (1759) Caspar Friedrich Wolff described the cell wall and cell contents in developing plant tissues.[160] As we have seen, Needham and others considered that plant and animal tissues are interconvertible with small bodies akin to infusoria, and some investigators identified these bodies with cells. Oken staked personal claim to having initiated cell theory (in *Die Zeugung*, 1805), although Gottfried Treviranus wrote in the same year that

> The first beginning of all organization of the living being is an aggregation of vesicles [*Bläschen*], which have no connection with one another. From these arise all living bodies, just as they are all dissolved into them again.[161]

A more-comprehensive cell theory came together in 1838–1839—that is, during the dispute between Ehrenberg and Dujardin—in publications by Matthias Schleiden and Theodor Schwann. The former, a professor of botany at Jena, held that all plants are made of cells.[162] He was highly critical of idealistic morphology as introduced by Goethe,[163] and regarded cells as structural and physiological units. Schwann had trained in medicine at Würzburg and Berlin, and in 1839 became professor of anatomy at Louvain. He defined *Zellentheorie* as a *Bildungsprinzip* applicable in all organic tissues, animals as well as plants.[164] Schleiden and Schwann understood cell formation somewhat differently, but both focused on the organizational role of the nucleus, and on the delimiting wall or membrane. Botanists Hugo Mohl and Carl Nägeli shifted the emphasis to the amorphous contents—Dujardin's *sarcode*—which Mohl called *Protoplasma*.[165] These concepts were subsequently unified by Max Schultze, based on his observations of muscle cells and rhizopod amœbæ: "a cell is a little clump of protoplasm, in whose interior lies a nucleus".[166] For Schultze, the concept *cell*—as the unit of physiology, indeed of life, not simply of structure—applied beyond plants and animals: diatoms, for instance, are organized as cells ("neither more nor less") although they cannot be assigned to either kingdom.[167] The tenets of cell- and protoplasm-theory have been summarized;[168] here we focus on implications for the plant-animal interface, and for a possible third kingdom.

During the 1840s consensus slowly emerged that, to at least a first approximation, plants and animals are built of structurally equivalent units (cells). By 1850, Ferdinand Cohn was able to emphasize that both plant and animal cells contain the same (or a highly analogous) substance, the protoplasm; cells in plants differed from those in animals only by the presence of a cellulosic membrane.[169] Rudolf Virchow—famous for the aphorism *omnis cellula e cellula* (all cells arise from cells)[170]—considered it "very difficult to decide whether there are rigid differences between plant and animal cells".[171] Schwann held that

> The most intimate connexion of the two kingdoms of organic nature [can be proven] from the similarity in the laws of development in the elementary parts of animals and plants.[172]

There was, however, no immediate call for plants and animals to be joined in a common kingdom to the exclusion of non-cellular beings. Many early cell theorists held that each cell of a multicellular organism "corresponds in certain respects to the whole body" of simple beings such as infusoria, and that multicellularity arose when they continued to adhere to each other after division.[173] If plants and animals are little more than self-perpetuating alliances of sticky infusoria, recognizing the species not caught up in multicellular beings as a separate kingdom might break apart natural groups. These issues can be dissected in various ways, but at their core we find two questions that almost mirror one another:

1. Should infusoria, amœbæ, algae, and such beings be considered (uni)cellular?
2. Do individual cells of multicellular plants and animals correspond to the whole body of simple organisms such as infusoria?

Today we would frame these questions using concepts from genetics, evolution, and developmental biology, and call on data from molecular biology, biophysics, and advanced imaging. These options were not open to our counterparts in 1840, of course; but even so, a lively and thoughtful debate ensued, with consequences that continue to resonate today.

As we have seen, Trembley observed diatom populations to grow through self-division,[174] and as did OF Müller with desmids.[175] In the late 1830s explicit parallels were drawn between the self-division of protozoa and the cleavage of fertilized eggs in animals, yielding so-called *blastomeres*. These, in turn, were recognized as cells by Martin Barry,[176] Karl Reichert,[177] and Carl Bergmann,[178] completing the analogy. According to Albert von Kölliker, "blastomeres always multiply by division, like infusoria".[179] Kölliker and his colleague Karl von Siebold used their *Zeitschrift für wissenschaftliche Zoologie* to promote the idea that protozoa (and their counterparts in the plant kingdom) are true cells, and have much to teach us about cellular function.[180] In 1848 Siebold chose his words carefully in diagnosing Protozoa as a phylum within the invertebrate animals:

> First principal group. Protozoa. Animals in which the different organ systems are not clearly distinguished, and whose irregular form and simple organization can be reduced to a single cell.[181]

Siebold further pointed out that nearly all infusoria and rhizopods possess a nucleus that divides in coordination with fission of the organism; these nuclei make infusoria resemble cells (of multicellular organisms).[182] Siebold grouped the infusoria (ciliates) and rhizopods together as Protozoa, excluding vibrios, most unicellular algae, diatoms, desmids, and sponges. This delineation had the virtue of integrating protozoa into the animals on the basis of anatomy and physiology, rather than by idealized form, purposeful movement, or archetype. Less happily, it separated protozoa from similar beings in the plant kingdom.

The idea (however carefully worded) that protozoa are free-living cells did not, however, meet universal approval. Schwann's mentor, the renowned physiologist Johannes Müller, was reluctant to accept that any organism can remain in a single-celled condition through its entire life history.[183] Ehrenberg thought that relatively few animalcules multiply by self-division (*Theilung*).[184] Richard Owen, a leading partisan of archetype, accepted Ehrenberg's belief that infusoria are complete organisms—in his 1843 *Lectures* he even sketched a ciliate with a succession of stomachs along a well-defined alimentary tract[185]—and argued that

> No mere organic cell, destined for ulterior changes in living organisation, has a mouth armed with teeth, or provided with long tentacula; I will not lay stress on the alimentary canal and appended stomachs, which many still regard as "sub judice"; but the endowment of distinct organs of generation, for propagating their

330 KINGDOMS, EMPIRES, AND DOMAINS

kind by fertile ova, raises the Polygastric Infusoria much above the mere organic cell.[186]

Three other objections were raised against the unicellular hypothesis. The first came from Maximilian Perty, a *Naturphilosoph*[187] and professor of natural history, zoology, and comparative anatomy at Bern. Perty claimed to have observed many simple beings including infusoria, amœbæ, and rotifers to reproduce by the release of *Blastien*, a "certain class of vesicles and corpuscles"—basically, immature daughter cells—that are normally present within the body.[188] Contemporaries viewed the idea as a halfway house between protozoa as single cells (Siebold) and complete animals (Ehrenberg). In *Zur Kenntniss kleinster Lebensformen* (1852) Perty reclassified the protozoa, making Infusoria and Rhizopoda sister classes within subkingdom Archezoa or Urthiere.

A second difficulty was that many infusoria, notably ciliates and opalinids, display multiple nuclei.[189] Before it was known how cells divide or protozoa multiply, it was possible to interpret the presence of multiple nuclei in different ways, including as evidence that the ciliate organism is a fusion of multiple cells, or represents a special grade of organization.[190] As we shall see in Chapter 21, the idea that ciliates (but not, for instance, flagellates) are something more than simple unicells remained a serious alternative to Siebold's position well into the 1860s.

A third approach was taken by Thomas Henry Huxley. As a medical apprentice he had accepted cell theory, but in an influential review[191] turned against it, citing a variety of reasons. Marsha Richmond has argued that Huxley's objections were ultimately theoretical (even metaphysical), including a concern that the theory invested vital powers in discrete structures, notably the nucleus.[192] Drawing on his own observations, Huxley dismissed as superficial Schwann's "great principle" of the identity of plant and animal cells. Nor, he wrote, could evidence be adduced of any correspondence between protozoa (or their vegetable counterparts) and the cells of plants or animals. Individual protozoa are neither multicellular nor unicellular.[193] His arguments and so-called epigenetic theory of the cell profoundly influenced generations of zoologists including Adam Sedgwick,[194] Charles Whitman,[195] Sydney Vines,[196] and Clifford Dobell.[197] Much later, Dobell looked back on cell theory as having had a "paralysing effect" on the study of acellular beings: so long as they "are 'primitive unicellular organisms', so long will their biological significance remain unrecognised".[198] As for our second question above,

> An absolutely fundamental point which must be recognized at the outset of our analysis is this: one whole protist individual is a complete individual in exactly the same sense that one whole metazoan individual is a complete individual. Amoeba is an entire organism in just the same sense that man is an entire organism . . . a protist is no more homologous with one cell in a metazoon than it is homologous with one organ (*e.g.* the brain or liver) of the latter. Only the cytologist blinded by what he sees through the microscope could ever believe in such a preposterous

proposition. . . . To the man who has not been led astray by the cell theory, this proposition is self-evident.[199]

Cell- and protoplasm-theory, among the great unifying ideas of Nineteenth-century biology, did much to divide biologists. During the 1840s and 1850s, however, biologists on the Continent were united in finding no justification in cell theory to recognize unicellular or acellular beings as a separate kingdom.[200] Unexpectedly, though, a third kingdom emerged from the idealistic tradition.

The last *Naturphilosoph*: Carl Gustav Carus

Carl Gustav Carus,[201] an idealist and Romantic, was an obstetrician, zoologist, classicist, acclaimed landscape painter,[202] friend of Goethe, physician to King Frederick Augustus II of Saxony, and President (1862–1869) of the Leopoldina. Like Oken and Goethe, Carus studied vertebrae and skulls early in his academic career.[203] In the first edition of *Lehrbuch der Zootomie* (1818) he arranged the animal kingdom in seven classes according to the perfection of the nervous system including the vertebral column, using standard names for the taxa.[204] Plants and animals are the only organic kingdoms, and he described their relationship in an almost Aristotelian manner:

> When we remark in the Life of Animals the recurrence of all the functions already existing in Plants, such as Nutrition, Growth, Respiration, Secretion, and Reproduction, with the superaddition of a higher gradation of Life, consisting in the exercise of the Nervous, Muscular, and Sensorial Systems, we must be convinced that the Unity of the Life of Animals consists in the combination and mutual dependence of two distinct Spheres, which we shall henceforward name *vegetative* and *animal*.[205]

The classes of animals "do not necessarily constitute a perfect series", however, as the seven organ systems are "mutually combined and intermixed".[206] By contrast, the second edition of his *Lehrbuch* (1834) had a more nature-philosophical flavour: taxon names remind us of those introduced by Oken, and Carus depicted the classes as concentric circles with Man in the centre.[207] Thereafter his written work focused on medical topics including physiognomy, psychology, and the unconscious, and on art and æsthetics, occasionally straying into mysticism.[208] It is a surprise, then, to find in his *Natur und Idee* (1861) three *Reiche* of epitelluric (earthly) organisms: *Protorganismen*, plants, and animals. His nine-point argument is convoluted and wordy, but can be summarized as follows:

> Anyone who thinks clearly about the contrast between the two widespread earthly kingdoms, plants and animals, will become convinced that they must have

332 KINGDOMS, EMPIRES, AND DOMAINS

developed from a prior Indifference [*vorausgehenden Indifferenz*]. The genera of *Protorganismen* cannot be placed among plants or animals, but instead must be united in their own kingdom. Most scientists have remained blind to this, however, because deep philosophy is alien to them.

1. Because the classification of earthly individuals must follow the stages and developmental phases of the highest organism, Man, it follows that the prior Indifference must be represented in the first division of the human organism, that is, in the Urzelle of the as yet unfertilized human egg. No Protorganismus can rise above the concept of a genderless, multiply crystallizing cell that multiplies by self-division. The primary sign of a Protorganismus is thus the complete absence of sexual organs.

2. Like the first indifferent human egg, Protorganismen are microscopically small. They arise and remain in water.

3. The bond between Protorganismen and the primordial fluidity of Earth, i.e. water, imparts a plasticity to Protorganismen such that they exist in an inexhaustible variety of geometric shapes (balls, lines, surfaces, linear bodies), as we see in the crystallization of water into snowflakes.

4. Protorganismen are sometimes classified as genera of infusoria, sometimes as genera of algae: Protococcus, Monas, Gonium, Achlya, Oscillaria, Spirulina, Gallionella, Navicula, Bacillaria, Desmidium, Sphaerastrum, Volvox, Xanthidium, Eunotia, Echinella, Evastrum, Ursinella, and many others. Just as the planets crystallized from the elemental plastic, so too these beings, like the Earth itself, contain silica.

5. Through their great numbers, these tiny beings have contributed to the volume of the planet, for example in river deltas, shell beds and deposits, and certain rock types.

6. Philosophically, it is very remarkable that individual Protorganismen self-divide to yield several or many. It will be highly significant if plants and animals prove to enlarge themselves in the same way, that is, through the ongoing self-division of their individual units.

7. Protorganismen lack not only sexuality but also other internal organic reproductive systems, and thus can lose their individuality only by self-division. Given their great variety, it is difficult to know where to draw the line between Protorganismen and animals. Protorganismen are the simplest beginnings of a single cell. The concept "egg" can be applied to their reproduction by Blastien; and we must add the concept "fertilization" to allow the separated part to develop as a whole.

8. Protorganismen are small because their internal structure is simple. When a Protorganismus self-divides, the new individuals remain at the first stage of Bildung.

9. The concepts species and individual are more difficult among the Protorganismen than elsewhere among organic beings. A certain arbitrariness must remain in their classification, as it is difficult to define a type. It may be

best simply to distinguish groups and tribes [Sippschaften] that share a general design, for which reference to geometrical shapes would be decisive: the point (Monas, Volvox, Ehrenberg's polygastric infusoria), line (Oscillaria, Vaucheria, Vibrio, Gaillonella, diatoms, bacilliaria, Hydronema, Achlya), and surfaces (Amoeba, Gonium, Evastrum, Desmidium, Micrasterium, Polythamnium).[209]

The point-species, surface-species, and certain of the line-species can undergo a transition to certain types of animals, whereas other line-species metamorphose to plants.[210] In other words, *Protorganismen* are prior to plants and to animals, not intermediate between them. Before moving on to the other two epitelluric kingdoms, Carus reflects on how individual organisms are compelled by their relation to the macrocosm to assert life by transformation and metamorphosis; and how in death, even as the individual is increasingly subordinated to the whole, great geological formations can be built up. As for those not philosophically inclined,

Anyone who can see such a pure encounter of true speculation with observation, without being imbued with the sublime wisdom and beauty of eternal Becoming [*des ewig Werdenden*], for him there is much, including this book, that is not written.[211]

Summary: Germany

This chapter has focused on the understanding of plants, animals, and microscopic beings in Germany from the 1770s until the 1850s. This was the period of idealism in philosophy, Romanticism in æsthetics, and *Naturphilosophie*. We might have begun our narrative earlier, with (for example) Georg Stahl or Albrecht von Haller,[212] or mentioned others along the way (Johann Ritter, Carl Willdenow, Carl Kielmeyer); but they did not introduce new ways of thinking about the plant-animal interface, or delineate third kingdoms of life.

Historians have treated this period unkindly: Romanticism was awash with "glittering, boneless generalities",[213] the philosophers led us into endless halls of mirrors, and *Naturphilosophie* proved to be a house of cards. Yet Kant made useful contributions in earth science and cosmology;[214] Schelling conducted experiments to refine his ideas in *Naturphilosophie*. Oken was the first in Germany to offer a replacement for the Linnæan system,[215] and was instrumental in professionalizing, internationalizing, and popularizing science. Humboldt gave us physical biogeography, Goethe a sense of *natura naturans*. To different degrees morphology, archetype, and protoplasm moved beyond their idealistic and Romantic origins.

Throughout this chapter (and others) I have referred to *microscopic* or *simple beings* where presumed animal and vegetable forms were collectively under discussion, or where context required a neutral term. Biologists of the period were sometimes less scrupulous: *infusoria* did not necessarily arise from infusions,

334 KINGDOMS, EMPIRES, AND DOMAINS

while *animalcula* and *protozoa* might include simple plants. In the mid-1820s Bory took an important step, separating *microscopiques* from *psychodiaire* and placing the latter into a purpose-built third kingdom; but *psychodiaire* included sponges, sea-fans, and corals as well as diatoms, algae, amœbæ, and ciliates.[216] As we saw above, in 1848 Siebold established phylum Protozoa for unicellular beings he considered invertebrate animals. Carus's *Protorganismen* were neither plants nor animals, although some could metamorphose to plants, others to animals. Other third kingdoms would appear in the 1860s. In the meantime, there was work to be done: the world of (mostly) microscopic beings was becoming more confusing in ways that undermined the old plant-animal boundary.

19

Green matter, zoospores, and diatoms

There are more things in heaven and earth, Horatio,
Than are dream't of in your philosophie.
—William Shakespeare, *The tragedie of Hamlet, Prince of Denmarke*[1]

From the previous two chapters, one might be forgiven for thinking that microscopic beings were primarily a problem for zoologists. Lamarck, Cuvier, Blainville, Milne-Edwards, von Baer, Ehrenberg, Oken, and Siebold classified infusoria within the animals. Bory, Goethe, and Carus acknowledged that certain microscopic beings might transform into plants, but did not pursue the matter in detail. But as we shall see in this chapter, the gravest risks to the old animal-plant dichotomy—and the most-innovative interpretations—were springing up at the interface between infusoria and algae. Before we venture into this notoriously complicated area, let us take stock of the knowledge of microscopic beings and algae *circa* 1780.

Simple animals, simple plants

Linnæus speculated about an intermediate kingdom, but his authoritative *Systema naturæ* decreed that all beings are either animals or plants.[2] Infusoria, including *Bacillaria* and *Volvox*,[3] were Vermes; algae and fungi were Cryptogamia, plants with hidden sexuality.

Linnæus was not himself a microscopist, and the later editions of *Systema naturæ* largely adopted the genera set out in OF Müller's well-regarded *Animalcula infusoria*.[4] These genera were based on form: and rather gross form at that, as Müller's microscopes afforded only modest resolution.[5] With only seventeen genera available to accommodate 378 species, some genera were unwieldy, notably *Trichoda* with eighty-nine species, and *Vorticella* with seventy-five. The same was true among the algae: by the Thirteenth edition of *Systema naturæ*, genus *Fucus* held 141 species, *Ulva* thirty, *Conferva* fifty-nine, and *Byssus* twenty-three.[6] By about 1800 it was abundantly clear that both Müller's infusoria and the Linnæan cryptogams were in dire need of taxonomic revision.[7]

Kingdoms, Empires, and Domains. Mark A. Ragan, Oxford University Press. © Oxford University Press 2023.
DOI: 10.1093/oso/9780197643037.003.0019

336 KINGDOMS, EMPIRES, AND DOMAINS

Of the two, infusoria presented the lesser challenge. The path from Müller to the classifications of Siebold,[8] Perty,[9] and Stein[10] was not without errors and blind alleys, but (as we have seen) Lamarck, Bory, Ehrenberg, and others distinguished protozoa from worms, polyps, rotifers, and spermatozoa. Genera and species were delineated on the basis of stable characters, and various higher taxa were established. Protozoa were retained among the animals whether as primordial forms, analogues of early embryological stages, or simply because they were self-motile. Few authorities were perturbed that some are pigmented, or form small linear (*Bacillaria*) or globular (*Volvox*) colonies. Infusoria and amœbæ might be limiting cases of animality, but with few exceptions they did not fundamentally challenge the animal-plant dichotomy.

The same cannot be said of algae. From the start, there was dissatisfaction (especially in France) with the name Cryptogamia. Most of the ancients believed that ferns, mosses, algae, and fungi are flowerless. Tournefort (1694) wrote that flowers and seeds are "ordinarily unknown" among ferns, mosses, and algae.[11] Adanson (1763) considered that, in general, members of his heterogeneous family Bissus (simple filamentous organisms), as well as mushrooms and seaweeds, are asexual and lack sex organs.[12] Decades passed, but no cryptic flowers were uncovered. In 1819 AP de Candolle held algae, fungi, and lichens to be asexual,[13] while Achille Richard concluded that all the Linnæan cryptogams are devoid of sex organs.[14] Across the Channel, William Hooker[15] and John Lindley[16] followed Richard. Alternative names for Cryptogamia that did not imply a hidden sexuality were put forward including Cellulares, Acotylédonés aphylles, Agames, Inembryonées, Thallogens, and Thallophyta.[17]

In the absence of floral characters, by what criteria might algae be classified? Chemical analysis could scarcely distinguish plants from animals. Jean Lamouroux divided Thalassiophyta into six orders distinguished in part by colour: Fucacées (olive-greenish, darkening in air), Floridées (purple-reddish, becoming shiny in air), Dictyotées (greenish, not darkening in air), Ulvacées (green, becoming yellow or white in air), Alcyonidiées (earthy tawny-olive, darkening in air), and Spongodiées (green, fading in air).[18] This approach had the further merit of aiding rapid identification in the field. An Aristotelian would, of course, object that colour is an accident; and there are, so to speak, grey areas between (for instance) olive-greenish and greenish. Indeed, Fucacées and Dictyotées have since been subsumed into Phaeophyceae (brown algae). Even so, it is remarkable how well Lamouroux's algal orders have fared to the present day.[19]

How, then, do algae reproduce?

Botanists and marine biologists had many reasons to study algae, not least to determine how they are generated in the absence of structures that could credibly be interpreted as flowers, or (for smaller forms) even cortex and medulla.[20] Several options presented themselves. Some might propagate by simple fragmentation. Tournefort had followed Cesalpino and Jung in suspecting that the powder on the underside of fern-leaves is seed, and supposed the same for mosses.[21] John Lindsay regenerated *Polypodium* ferns from such powder,[22] while John Stackhouse regenerated juveniles of *Fucus* using "seeds" he collected from "pericarps" of three species.[23] Seedlike but asexual particles might furthermore explain a variety of old and new observations in mosses, ferns, and fungi, and a consensus developed that flowerless plants can reproduce asexually via tiny seedlike bodies. These became known as *spores*.[24]

More controversial was *generatio æquivoca*, or spontaneous generation. Despite some evidence to the contrary,[25] belief persisted well into the Nineteenth century that algae, fungi, and sponges could simply spring up in water, just as infusoria, worms, and certain other animals could arise in mud or in the body of a host animal.[26] Old ideas of progression from mineral to plant had not entirely disappeared, and were broadly compatible with Nineteenth-century transformism and *Lebenskraft*. Lamarck, for example, imagined that infusoria and worms arise spontaneously, while Kant, Treviranus, Oken, Eichwald, Goethe, and Carus thought the same for living points or globules, infusoria, or *Protorganismen*.[27] Nor could sceptics explain how worms appeared inside living animals. Insensible airborne seeds might give birth to microscopic beings in natural waters and infusions, but the "internalist" school of parasitologists (including Bloch, Goeze, Bremser, and Rudolphi) rejected a similar explanation for intestinal worms, arguing that the host gut offered an impenetrable defence from the environment.[28] As we will soon see, spontaneous generation was invoked to explain another curious phenomenon.

The early decades of *phycology* (the study of algae) form a particularly underappreciated chapter in biological science. There was no overarching clash of philosophical systems, no new marble temple on the royal parade, no Goethe or Cuvier. The primary literature is notoriously opaque, as the front-line researchers struggled to describe structures and phenomena for which no technical terms existed. Some species were easier to find (growing in the local ditch), identify, and work with than others. Too many experiments were poorly designed and/or misinterpreted. It is telling that, two hundred years on, no modern historical synthesis has been published.[29] Indeed, some basic biology remains unclear to the present day.[30] Fortunately, our aim here is not to survey reproduction in the algae, but instead to examine how researchers relaxed kingdom boundaries, or called upon an intermediate kingdom, to explain their observations. We do so through a series of illustrative case studies: the "green matter" debate initiated in the 1770s; the recognition of motile "zoospores" from about 1803–1845; experiments on algal metamorphosis from the 1810s through the 1840s; and the diatom dilemma of the 1840s.

338   KINGDOMS, EMPIRES, AND DOMAINS

Case study 1: Priestley's green matter

In 1771 Joseph Priestley, an English dissident theologian and political activist, was conducting experiments to understand the composition of air. In one series of experiments, he enclosed air in a glass vessel, and vitiated it with a burning candle or the respiration of a mouse. He then showed that it could be refreshed (such that it would again support a candle-flame or mouse) by adding a sprig of mint (or another plant) and exposing the vessel to sunlight. If the sprig were immersed in water and exposed to sunlight, he might observe bubbles forming on its leaves and stalk; these turned out to be "dephlogisticated air" (oxygen).[31] In 1778 he recorded that the inside surfaces of the vessels sometimes became coated with green matter, which in sunlight likewise released dephlogisticated air. Given that the vessels were closed, he initially concluded

> That the external air, or animalcules in it, have nothing to do in the formation of this green matter. . . . The production of this green matter in close[d] vessels seems to prove that it can neither be of an animal or vegetable nature, but a thing *sui generis*, and which ought, therefore, to be characterized by some peculiar name; and all the observations that I have made upon it with the microscope agree with this supposition.[32]

Moreover, if this green sediment releases oxygen, then oxygen release is not a sole property of plants. From here Priestley lost his way, for a time, amidst what he saw as conflicting data. Eventually his friends brought him to the "fullest conviction" that his "green matter" was, after all, a plant,[33] indeed probably a *conferva*.[34] He later rejected any thought that it might have arisen spontaneously.[35]

Other savants soon set up similar experiments, mostly to study the composition of air; but some took a closer look at the green matter itself, and at microscopic beings in the surrounding water. Jan Ingen-Housz, personal physician to Empress Maria Theresa of Austria, described the green matter as a filmy substance that incorporates numerous irregularly shaped saline or stony crystals. He observed many small beings, of green colour, swimming freely about; these he called "true insects" on account of their ability to locomote. Other animalcules, of identical appearance, seemed trapped in the filmy crust.[36] Recall that from 1742, when Réaumur accepted Peyssonel's argument that coral-polyps are animals, a consensus emerged that coral-polyps are to the macroscopic coral-body (*polypier*) as wasps are to a wasp-nest (*guêpier*). Some savants speculated that sponges and seaweeds might likewise be a sort of polypary.[37] Ingen-Housz was read as implying that green matter houses small insects (which can otherwise swim freely), much as a polypary houses coral-polyps.

Between about 1782 and 1784 Ingen-Housz became convinced, perhaps correctly, that the green matter in his experiments was of a different genus than that studied by Priestley. The green matter of Jean Senebier (below), and that of Felice

GREEN MATTER, ZOOSPORES, AND DIATOMS   339

Fontana,[38] might be different again. Depressingly many pages of their reports were given over to inconclusive speculation about whether one or the other might be a species of *Conferva*, or perhaps of *Tremella*. In any event, Ingen-Housz now reported *Wasserfaden* ("water threads") in his green matter: near-colourless tubes that house round green "insects" which over several days become increasingly lively, and eventually break free and swim away.[39] In due course the insects settle, and give rise to new tubes which elongate like plants. Ingen-Housz's explanation was nuanced:

> After all I have said about *tremella* and *conferva rivularis*, one might doubt whether these two beings deserve a place among the plants. Truly they grow, and spread like plants: the *conferva* even grows branches exactly like them; but freshwater polyps grow branches in this same manner; and meanwhile, after Mr Trembley examined more carefully the economy of these beings, they have generally been placed in the animal kingdom, and have even been accepted as true animals. Would there not be some probability that this tremella and the *conferva rivularis* are intermediate beings between the animals and the plants, similar, at least in some respects, to several of these bodies which are called *zoophites*, the *animal-plants*, and that the insects, which are the first rudiments of the green matter of Mr Priestley, compose the tubes of the *conferva*, and withdraw into their hollows more or less as do those insects which compose the greater part of the corallines, and several others, so-called marine plants, according to the observations of Mr Bernard de Jussieu, and especially of Mr Ellis?[40]

Ingen-Housz did not doubt that the motile insects are animate; and although an insect can give rise to a form that elongates, branches and spreads like a plant, this green matter was at first "a purely animal substance", and

> in all the metamorphoses which it [subsequently] undergoes, it seems at least probable that it does not completely abandon its primitive nature, even though it finally takes on the figure of a being (a tremella) which so far has been ranked among plants.[41]

Thus Ingen-Housz denied that the green matter is fully a plant; it might be an intermediate being, a sort of zoophyte perhaps, but in the sense that a coral is a polypary, with animate polyps. Ironically, in reaching this conclusion Ingen-Housz, like Priestley, came to deny that only plants can release oxygen.[42] Justin Girod-Chantrans, who championed the idea that many kinds of algae are polyparies, predicted that

> the moment is not far off when these productions of waters, without any kind of apparent fructification, which have so far been ranked among the plants of the lowest order, will again climb the ladder of beings to take their systematic place at the

340 KINGDOMS, EMPIRES, AND DOMAINS

head of the vegetables, immediately below the marine and freshwater polyps. This is not to say that all the *conferves* of the botanists must add to the numerous genera of polyparies: perhaps several will remain, with reason, among the cryptogams; but we can hardly believe that [three species of *Conferva*] can maintain their rank among the plants.[43]

Another savant who considered the green matter at length was Jean Senebier, chief librarian of the Republic of Geneva and editor of Spallanzani's journals. He argued by analogy: the animalcules seen swimming near the green matter are more like cattle grazing upon a field of grass, than like bees swarming around their hive. Different animalcules appear successively in the water, just as the cattle might be replaced by a flock of turkeys. No analogy holds between green matter and corals: filaments of the former are flexible, whereas the coral polypary is a "stony case". Senebier concluded that the green matter is a true vegetable analogous to *Ulva intestinalis* or *Noctoc*, and is unrelated to the swimming animalcules. Nor is the liberation of oxygen by plants analogous to respiration by animals. Self-locomotion was not his only criterion of animality, but he emphasized it time and again.[44]

These worthy opinions did not, however, settle the matter. Benjamin Thompson considered the green matter to be "evidently of an animal nature, being nothing more than the assemblage of an infinite number of very small, active, oval-formed animalcules, without any thing resembling *tremella*".[45] Franz Gruithuisen maintained that Priestley's green matter consists "merely of green infusoria, [and] bears in itself a vegetable nature, and is a true intermediate [*Mittelding*] between an infusion-animal and a plant". Gruithuisen also claimed that adding distilled water to most minerals, rocks, and especially granite produces infusoria, and that "when a new cryptogamic plant arises, a type of infusorian has provided the seed every time".[46]

Our first case study concludes with few definitive outcomes. As phlogiston theory was abandoned, the interplay of atmospheric gases with plants and animals became less of a mystery. No consensus was reached on the precise identity of green matter, or on the extent to which it might offer broader truths for plant or animal biology. Ingen-Housz, Girod-Chantrans[47] and others observed the release of motile bodies from attached filaments, but only Ingen-Housz connected the phenomenon to propagation.[48] There was, however, general agreement that the motile beings must be animals, and the sessile forms plants or polyparies. Ingen-Housz refused to rule out a spontaneous origin for the green matter,[49] and described its various transformations (including the production of motile bodies) as *metamorphosis*.[50] Ingen-Housz died in 1799, Priestley in 1804, Fontana in 1805, Senebier in 1809, and the "green matter" episode passed into history as an odd and inconclusive chapter in early biological science.[51] Fortunately, the baton had already been taken up by a new generation.

Case study 2: zoospores

We should not underestimate the challenges faced by this next generation. Even in the best-studied genera growth, cell division, gametes, and fertilization were very poorly understood, and would remain controversial until the 1850s.[52] From seemingly well-designed experiments, Spallanzani argued that neither spermatozoa nor pollen had a role in fecundation. In France and Italy, belief was widespread that eggs and seeds embody within themselves, preformed, all the parts necessary to produce a new being; seminal fluid and pollen merely stimulate development, perhaps via a chemical substance or vapour. Spores differ from seeds by containing only a liquid (*i.e.*, not preformed parts), and presumably receive their developmental prompt from soil or water, without the involvement of a sexual act. German biologists preferred a less-mechanistic explanation: polar forces, described as fluids or chemical elements, triggered epigenesis (development from a homogeneous, undifferentiated state) with, again, no material role for spermatozoa.[53] Plants were either sexual, with flowers and seeds, or asexual, reproducing through spores.

A fresh view was offered by Jean-Pierre Vaucher, professor of the history of theology at Université de Genève. An avid botanist, he sought to understand the processes that underlie plant growth and reproduction.[54] His early work focused on the Linnæan genus *Conferva* (filamentous algae). Unlike Priestley or Ingen-Housz, Vaucher collected confervas from natural waters at different times of the year, described and illustrated them carefully, and (appreciating the diversity within the group) classified them into six families defined by mode of reproduction.[55] His default position was that all these modes are sexual, even when he could not say precisely what structure might serve as a flower, ovary, or stamen. In one family, Conjuguées,[56] fertilization incorporates an animal-like step in which "individuals approach and meet",[57] and he depicted the transfer of material from one strand to the other.[58] But he strongly denied reports (by *e.g.* Girod-Chantrans) that motile animalcules are involved in the reproduction of confervas.[59] He took important steps into a difficult area—for example, describing sexual isomorphism in *Spirogyra*[60]—although it eventually became clear that he had misinterpreted many reproductive structures.[61]

Vaucher also demonstrated the value of working with well-characterized species. He pointed out species amenable to study in the home laboratory, described where and when they could be collected, and opened minds to a potential variety of reproductive structures and modes. Among the biologists who took up the challenge, Johann Trentepohl and Christian Nees von Esenbeck observed the release of motile bodies by conferva. Trentepohl saw them escape from dark structures at the tip of the filament, swim away, and after a few hours settle on the glass and produce a new conferva. He asked—but could not answer—how a plant can change so rapidly into an animal.[62] Nees von Esenbeck called the motile bodies *infusoria*.[63] But

342   KINGDOMS, EMPIRES, AND DOMAINS

microscopes capable of resolving the structure of spores, gametes, and animalcules, and the morphological detail of sporulation or fertilization, did not become available for several decades.[64]

An important step was taken by Franz Unger, then a physician in Lower Austria.[65] On 5 March 1826, Unger collected a conferva[66] near Vienna, took it home in a flask, and set it in a window. Over the course of the next twelve days he observed a remarkable succession of events: new filaments appeared, then small spheres (*Kügelchen*)—some *unbelebten*, others identical in appearance but *belebten*, hurrying about in the water like little animals.[67] The next day the filaments and infusoria were covered with air-bubbles, which Unger supposed (but did not prove) to be oxygen.[68] Using a "good magnification"[69] he observed the release of motile, ovoid spheres from clublike structures on the filaments.[70] The spheres eventually settled, changed shape and colour, and pushed out small *Fortsätzen* that anchored a new conferva. An entire cycle took twelve days.

Unger was justifiably amazed. Before his eyes, an undoubted plant—Vaucher had called *Ectosperma* "as perfect in its species as [plants] of a higher rank"[71]—had produced self-motile (hence animal) bodies. These settled down and gave rise to a plant, which in turn developed *infusoriellen Frucht*.[72] One after another, respected authorities weighed in with explanations that did not involve transgression of the animal-plant boundary. Oken, never one to shy away from metamorphosis, suspected a physical phenomenon related to neutral buoyancy.[73] Ehrenberg denied that motility meant animality, and likewise proposed a physical basis.[74] Julius von Flotow, a retired military officer and specialist in lichens, offered several opinions, including that some infusoria are animal, others vegetable: motility does not require stepping across the boundary.[75]

The 1830s brought small steps towards better understanding such life cycles, including spore germination[76] and cell division,[77] although more effort was directed to ferns and mosses than to algae.[78] But as we saw in Chapter 18, this was also the decade of cell theory. Matthias Schleiden brought the new cell-perspective to bear on plant reproduction with his "pollen theory", which equated pollen grains with spores.[79] He held that fluid in the pollen tube gives rise to cells, which are then implanted into the embryo sac:[80] that is, sexual reproduction (in flowering plants) is just a form of asexual reproduction via spores. Schleiden did not entirely resist the urge to draw analogies with reproduction in animals, but Stephan Endlicher carried the analogies much farther: the sporangium (in cryptogams) and anthers (flowering plants) with the (animal) ovary; spores and pollen with the egg.[81] Pollen theory became popular in Germany, if less so in France. It did not stand the test of time,[82] but had the effect of adding cellularity and reproduction to the list of features that bind the supposedly asexual plants into the plant kingdom.

Against this background, in 1842 Unger, now armed with a Plössl microscope,[83] returned to *Ectosperma*, by then renamed *Vaucheria*. He confirmed his earlier observations, adding finer detail and superior illustrations. Notably, he observed that the swimming motion of the spores is caused by fine whiplike threads on their surface.[84] He called the motile bodies *infusoria* and "animal embryos which cannot rise above this stage of life, and after a short time again take up the vegetable nature from which they originated".[85] Nor was he reluctant to paint a larger picture:

As Oken has already so suitably pronounced: the plant-world is the womb of the animal-world . . . just an indentation of the blastula, and the animal is ready, its existence secured forever. . . . The animal in its individual continuance, guided by the need for nutrition, had to cling to the ever-rejuvenating maternal breast of the plant world. Plants and animals are closer to each other than one usually assumes, and I see, at least, nothing contradictory in considering the animal world as the second birth of creative omnipotence, which the plant world had to precede.[86]

Unger's new report, structured as eighteen letters, bore the title *Die Pflanze im Momente der Thierwerdung* (*The plant in the moment of becoming animal*). Not for the first time, Unger's colleagues were dismayed. Endlicher, to whom Unger had dedicated *Thierwerdung*, reminded him that motion does not prove animality. Lamenting the direction Unger's speculation had taken, Endlicher offered his services as editor: he would delete sections, and change the title.[87] Schleiden came directly to the point:

Only a science crazy with fantastical mysticism, and far removed from a clear, self-intelligible Natural Philosophy, could entertain the dreamy notion that creatures may be at one time animals and at another plants. Were this possible, it would necessarily much more readily happen that a being should be now a fish and now a bird; or at one time a conferva, at another a rose; and then what would all our natural science be but folly! This perplexity of ideas . . . has latterly been carried to great length by Unger (*Die Pflanze im Momente der Thierwerdung*) and Kützing (*Phycologia generalis*). It can only be regretted that such able inquirers should be so entirely without any philosophical insight.[88]

Philosophy aside, however, this time Unger's observations were confirmed by Gustave Thuret, a retired diplomat and well-regarded amateur botanist.[89] Working at first with the large brown seaweed *Fucus*, and later with *Vaucheria* and other filamentous algae, Thuret described not only (asexual) zoospores but, in addition, two types of spore-like bodies, one larger, the other smaller and motile.[90] He went on to show that the smaller bodies can induce the larger ones to germinate: algae (some of them, at least) are sexual,[91] indeed both sexual *and* asexual. He classified all algae with motile spores into Algæ Zooporeæ with two sections, Chlorosporeæ and Phæosporeæ.[92] He acknowledged that some zoospores resemble green infusoria such as *Chlamydomonas*, *Euglena*, *Volvox*, and *Protococcus*, but he had never observed the latter to metamorphose into filaments, much less into a moss or lichen.

344 KINGDOMS, EMPIRES, AND DOMAINS

But while that allows us to distinguish zoospores from infusoria, it does not help us distinguish plants from animals:

> While believing that we should not confound what is distinct or reunite what nature has separated, I am nonetheless disposed to recognize that the extreme analogy of animals and lower plants does not permit us to trace a precise line of demarcation between the two branches of the organic kingdom. The presence of vibratory flagella on the reproductive bodies of algae is just one more argument in favour of this unity, which is confirmed every day by new observations.... As one descends the scale of beings, the distinctive characters of the animals and plants tend to be erased, and one finally arrives at these ambiguous productions which the observer hesitates to classify on one side rather than the other. The group of green-coloured Infusoria, of which I have just spoken, offers an example; for, regardless of which kingdom into which we relate it, it will always have the closest connection with the neighbouring kingdom.[93]

With this we close our second case study. Algae were found to produce, and in turn to develop from, motile bodies that could be (and were) mistaken for infusoria. It eventually became apparent that the life history of many (but not all) algae involves a motile spore or gamete. Algae did release motile spores, not as a polypary (*i.e.* a colonial animal), but in the normal course of vegetable generation. Early reports that algae reproduce sexually were confirmed: musty old Linnæus had been right, and Schleiden's pollen theory was rendered superfluous. Even more surprisingly, individual algae can reproduce sexually *and* asexually. Motile spores and gametes were discovered in ferns, mosses, and fungi as well,[94] and the idea, falsely blamed on Aristotle, that only animals are capable of self-motion lost further ground. Nonetheless Unger interpreted the motile spores of *Vaucheria* as animal, their production as *Thierwerdung*. As our next case study shows, he was not alone.

Case study 3: metamorphosis

Our third case study reprises a theme from earlier chapters: whether a being can take on the form of another. Metamorphosis enjoyed a fine classical heritage, but over the years had lost ground to the idea that genera are fixed, and species reproduce in kind. We pick up the theme in the second decade of the Nineteenth century with Carl Adolph Agardh, professor of botany and practical economy at the University of Lund.[95] Like Linnæus, Agardh believed that floral characteristics should be the basis of plant classification,[96] but he accepted that algae, fungi, and lichens are asexual and lack flowers.[97] Especially at higher taxonomic ranks his system differed somewhat from that of Linnæus, and his *scalæ* of genera based on organizational patterns[98] add an idiosyncratic touch. Agardh held that Order Nostochinæ "begins at the limit of the animal kingdom", and genus *Protococcus* "vacillates in position between the infusorial animalcules and algae".[99] On the whole, however, his descriptive and systematic work rests solidly within the established tradition, and is remembered today for its Class *Algæ* with membership not too dissimilar from what can be found in modern textbooks of phycology.[100]

It comes as a shock, then, to read *Dissertatio de metamorphosi algarum*, a thesis written by Agardh and defended by his student Joachim Åkerman in 1820.[101] By then the excitement over Priestley's green matter had largely abated, and Vaucher, Trentepohl, and Nees von Esenbeck had described motile spores in various algae. The *Dissertatio* relates that on 1 September 1815, Agardh placed a filamentous green alga (*Zygnema*) into a vase of purified water; green globules soon appeared, some moving actively through the water, others immobile on the submerged surfaces. The latter soon produced green but membranous algae, seemingly *Ulva*, a genus of distinct form. Agardh is clear:

> Accordingly, the cycle of its metamorphosis was, *Zygnema, Animalcula Infusoria, Ulva*. From vegetable to animal, from animal to vegetable.[102]

But there is much more. In an earlier experiment (1811), the infusorian *Enchelis* became quiescent ("died"), then metamorphosed into filamentous algae similar to *Oscillatoria*. A gelatinous *Oscillatoria* produced self-motile filaments similar to a *Nostoc*. A fungus became a conferva, an alga became a lichen; algae metamorphosed from one order or genus to another.[103] Moreover, some algae constitute parts of other algae: *Vaucheria* is a nomad, roaming lawlessly across taxonomic boundaries; *Codium* is civil, lawlike, and finitely circumscribed. Plants of all other natures and genera are made up of algae: Agardh assures us that the algal filaments can be dissected out, and grown separately in water. We could expect nothing else, given that plants require only water and dampness to spring into being, and these are the native land of algae alone.[104] He summarizes:

> Thus nature progresses little by little. At first, [nature that is] weak and generally uncertain does not accurately distinguish between organisms as animals and plants; later it defines them better and better, so that at last the forms are permanently set in place; but nature so loved her first-born Algal daughters that in the more-perfect plants she repeats them again and again. So, just as innumerable spirits inhabit the higher world, now too Algae live in all vegetable nature, which would not be able to be the case, unless this nature were composed of them, as if by monads.[105]

It is difficult to know what to make of this. Lichens do, in fact, have algal and fungal components. Algae (notably *Ulva*) and fungi can be remarkably pleomorphic, their form varying according to the conditions of growth.[106] Anton Lichtenstein had earlier interpreted fungal genera as a series of transformations.[107] French botanists of the time described plants as consisting of *vaisseaux* (tubes).[108] As for

346 KINGDOMS, EMPIRES, AND DOMAINS

algae constituting all plants, Agardh might have drawn, in his own way, on Goethe, Linnæus, Münchhausen, Needham, or Leibniz. The previous year, prize-winning botanist Heinrich Link had claimed that mosses and perhaps grasses arise from confervas,[109] while Christian Friedrich Hornschuch believed that an archetypal vegetable infusorium (*Monas lens*) gives rise to Priestley-like green matter from which algae, lichens, and mosses then develop.[110] Franz Meyen independently confirmed that *Protococcus* undergoes metamorphosis, first to a filament (which he named *Priestleya*), then to *Ulva*.[111] On the other hand, Agardh must have appreciated that rampant metamorphosis could only subvert order in nature.[112]

As it happened, developments in nearby countries began to cast a new light on the generation of marine animals. In 1819, Adelbert von Chamisso (whom we met briefly in Chapter 18) showed that individuals of the tunicate *Salpa* are singular and free-living in one generation, but bound together in chains in the next; he called this *Generationswechsel*, drawing a direct analogy with caterpillars and butterflies.[113] A student in Edinburgh, Charles Darwin, found that a free-swimming marine invertebrate is the larval stage of the colonial bryozoan *Flustra*.[114] Norwegian theologian and marine biologist Michael Sars and his colleagues made similar discoveries for various molluscs, starfish, and jellyfish.[115] The Danish zoologist Johannes Steenstrup believed that generations alternate in all "lower classes" of animals.[116] Apart from Chamisso, none of these investigators called these cyclical changes of form *metamorphosis*.

Benjamin Gaillon

Agardh's speculations might have been quietly forgotten, were it not for the observations of two amateur naturalists. Around 1816 Benjamin Gaillon, a collector of customs duties in Dieppe, began to examine local marine algae with his microscope. In several species he found fructification-bodies and spores, which he described as asexual.[117] He observed a membranous *Ulva* produce spores which, in turn, gave rise to filaments that could be assigned to a different species—a change he did not hesitate to call metamorphosis.[118] When he disrupted a filament of *Conferva comoïdes* using a pin, corpuscules he called seeds (*la graine ou les seminules*) were liberated; they moved "gravely and slowly" at first, but soon became "endowed with a movement sudden, iterative, measured and voluntary".[119] He identified the corpuscles as diatoms.[120] Observing these (and presumably other) corpuscles over a year, he saw them line up in single file, secrete a mucus sheath, and again become a filament; branches appeared in the same manner, such that "the same species seen in different states would be taken for different species".[121] For this, Bory de Saint-Vincent dubbed Gaillon the "Ovid of Algology".[122]

In 1820, Gaillon determined that the green colouration that appeared every summer in local oysters was caused by an accumulation of an animalcule he called

Vibrio ostrearius.[123] Bory recognized the animalcule as a diatom and renamed it *Navicula ostraria.* As such, it joined his kingdom Psychodiaire.[124]

We hear no further from Gaillon until 1832–1834, when he generalized his earlier observations on *Conferva comoïdes,* by then renamed *Girodella comoïdes.*[125] He referred to the motile corpuscles as animalcules, and (without presenting evidence) claimed that many freshwater and marine beings are likewise constituted of "internal corpuscles endowed with animation and, at a certain time of their existence, with the locomotive faculty". These corpuscles, or *zoadules,* ensure the reproduction of the filament, membrane, or envelope (*nemate*) that others have mistaken for a plant. Some zoadules are globular like *Monas,* while others are elongate and pointed like *Navicula.* From each sort two series arise, collectively encompassing a wide range of membranous algae and water moulds.[126] Gaillon called this group Némazoaires,[127] and considered that they represent the

> point of junction, transition and passage from the animal kingdom to the vegetable kingdom, since in some cases they affect the colour and immobility of plants, while in others they are endowed with activity of the most agile animals.[128]

Friedrich Traugott Kützing

The second, ultimately more-influential series of observations and experiments was initiated in 1829 by Friedrich Traugott Kützing, then an assistant pharmacist in Schleusingen (Thuringia).[129] Aware of Hornschuch's claim (above) that mosses develop from green monads, over several years Kützing examined growths on trees and walls near his apartment. His notes were a jumble: he saw filaments develop from green globules he identified as *Protococcus.* Other filaments came together suddenly "as if by a stroke of magic", yielding the moss *Bryum.*[130] Infusoria in a glass of water died, then *Protococcus* appeared; a simple *Conferva* was perfected into an *Inoderma.* Kützing wove these scattered observations into a single narrative: depending on environmental conditions, a single primary form (for example, *Protococcus*) might develop into different higher genera; conversely, a given higher form might arise from different basic forms.[131] Hornschuch, and for that matter Agardh, Ingen-Housz, and Priestley, had been far too conservative.

In 1835–1836 the Royal Holland Society of Sciences offered a gold medal and 150 guilders for a work judged to prove "beyond all doubt" whether or not lower plants can pass from one kind into another.[132] Kützing later reflected that "this prize-question was made just for me".[133] His approach remained as before: from time to time after a rainfall, he scraped greenish growths off "walls, stone pavements, earth, pieces of wood, boards, stems of dead plants, rotten wood, roof tiles, tree-bark and the like", and observed them under a microscope.[134] At first he might find only *Protococcus,* but more-differentiated forms soon appeared that he could not

348 KINGDOMS, EMPIRES, AND DOMAINS

confidently assign as an alga, fungus, lichen, or moss. These forms were neither chaotic (Bory), nor frivolities of nature (Meyen), but revealed a law of nature: that lower forms not only develop into higher forms, but become bound into them, such that the higher form is dissolved and disintegrated into the lower.[135]

Nor is this restricted to algae: all cellular cryptogams can be generated from the selfsame organic matter by *Urbildung* (*generatio originaria*), *Fortbildung* ("morphosis") and *Umbildung* (metamorphosis). The lower plants display all five stages of plant development: the *spherical* (the original form), *nematic* (extending the sphere in one dimension: filament, fibre, and tube), *phylloid* (adding breadth: the leaf), *stelechoid* (extending further in all dimensions: the stem), and *sorenmatic* (comprehending all the above in preparation for "higher individual unity").[136] There are, in addition, secondary forms and combinations: thanks to the greenish growths, Kützing could describe and illustrate them all.[137]

Kützing concluded his prize-essay with the claim that, as a consequence of his work, "the concept of species, genus, family, and class among the cryptogamic cellular plants must be *schwankend*"—*unsteady* or *fluctuating*. This is particularly true for lichens, sponges, and algae, which produce forms so *schwankend* that it is often impossible to assign them to a class.[138] He then applied these ideas in his *Phycologia generalis* (1843):

It should not remain unmentioned, that among the seaweeds [*Tangen*], especially in the lower groups, species cannot be accepted in the sense that one is used to accepting them among the phanerogamic plants. The species of the lower seaweeds are, strictly speaking, only forms, either of developmental stages or of developmental series, and accordingly there are only two paths that could be followed for their systematic treatment. Up to now, one has followed only the path by which the different forms are presented according to their developmental stages. Our knowledge of these forms has not yet progressed so far, that the other path to their systematics could be taken. Therefore, in this work I have arranged the forms according to their stages of development.[139]

It remained only for Kützing to attack the animal-plant boundary. Already in *Phycologia generalis* he had commented that

As the plant kingdom begins its great form-series [*Formenreihe*] amongst the lower structures [*Gebilde*] of the algae, as does the animal kingdom amongst the infusoria, so we cannot be surprised, that the two touch each one another in their points of origin. In many instances it is impossible to determine the exact boundary between the two kingdoms.[140]

Schleiden had by this point labelled such claims of metamorphosis "mystical dreams" and "unscientific fantasy-games".[141] Kützing replied in a tract subtly titled

GREEN MATTER, ZOOSPORES, AND DIATOMS 349

Sophists and dialecticians, the most-dangerous enemies of scientific botany (1844), accusing Schleiden of untruth, intellectual cowardice, and much else.[142] Kützing denied that he (Kützing) had sought to undermine the absolute concept of species; he had merely pointed out that certain species are less well-delineated than is usually assumed.[143] As for the animal-plant boundary, Ehrenberg had claimed all infusoria as animals, yet he (Kützing) had shown that *Microglena monadina* develops into *Ulothrix zonata*. There are but three possible explanations: *Ulothrix*, hence all plants, are only the wombs of infusoria; *Microglena*, along with all other infusoria, spirilla, vibrios and the like, are plants; or *Microglena* is an animal and *Ulothrix* a plant, and

> there are either moments when one and the same substance is alternately animal or plant, or one must assume that in these bodies the plant-element is united with the animal-element and, depending on which one develops before the other, the body joins more to a plant- or animal-structure.[144]

Kützing opted to explain *Microglena* and *Ulothrix* as animal united with plant. But the two kingdoms touch most closely in their simplest forms, and he addressed these forms in *Über die Verwandlung der Infusorien in niedere Algenformen* (1844). *Verwandlung* begins with a lengthy (and somewhat tendentious) literature review: since OF Müller, investigators had struggled to decide whether microscopic beings are plant or animal. Ehrenberg, the leading authority on infusoria, considered the red-snow organism[145] a plant, while the botanist Meyen thought it an animal. Faced with plain evidence, these savants clung to the dogma that plants cannot change into animals, nor animals into plants, and found themselves in blind alleys, peddling claims about infusorial gonads (which turned out to be oil-droplets), digestive tracts (an illusion), or the eyespot (which has nothing to do with vision).[146]

Kützing's approach in *Verwandlung* was much as before. On six occasions from June to August 1844, he collected samples from green growths in the pond in front of his house in Nordhausen. The first samples showed mostly the infusorian *Chlamydomonas pulvisculus*, a few of which had already begun to produce *Conferva*-like filaments. As the weeks passed, he recorded a greater diversity of forms: some simple but of different sizes, shapes, or colours, others thread-like, bead-like, comb-like, or gelatinous. Some of the thread-like forms gave rise to the alga *Stigeoclonium stellare*, while gelatinous forms produced *Palmella botryoides* and *Gloeocapsa*.[147] The transformation of animals into plants, and vice versa, had been "undoubtedly proven" yet again.[148]

Kützing had earned credibility as a scientist. He discovered that the frustule (shell) of diatoms is made of silica, and that fermentation is caused by yeast.[149] He was the first to classify diatoms and desmids in *Hauptgruppen* of equal taxonomic rank.[150] He described a large number of genera, and improved the delineation of many others.[151] By all reports he was a well-respected, civically minded *Bürger*.[152]

350 KINGDOMS, EMPIRES, AND DOMAINS

Yet beings do not change genus from one week to the next. Where, precisely, did his research run off the rails? Few experiments benefit from blatant disregard of aseptic technique; but it is one thing to observe a succession of forms in a flask or pond, quite another to convince oneself that they are connected in a real-time developmental series. It is clear from his *Umwandlung*, *Sophisten*, and *Verwandlung* that Kützing believed that any being that looked like a *Protococcus* was, in fact, a *Protococcus* and not (for example) an alga, moss, or fern that, as part of its normal life cycle, passes through a *Protococcus*-like stage. He was ever-ready to assign such forms, however transient, to an existing genus or species; and where no suitable taxon was available, he created one: nearly forty in his 1833 report alone. His claim that metamorphosis renders taxa *schwankend* is not without irony.

But yet: it is clear from the *Schlusswort* to his *Verwandlung*[153] that Kützing was alert to much of this. Schleiden accused him of *Naturphilosophie*, but Kützing hearkened back to an ancient distinction in Western philosophy: we can consider an object *ein Fertiges*, or *ein Werdendes*.[154] The former is analytic, based in a moment in time, and directs us to Linnæus, to systematics; the latter is synthetic, considers the sequence of phenomena over the lifetime of an object, and points us to Goethe and physiology. Systematics is based on definition, on the sharpest boundaries possible, but is ultimately empirical, hence arbitrary. Developmental history takes us closer to the Idea of an object; yet developmental history cannot be defined, only exposed, and to do this clearly and intelligibly we need taxa with clear boundaries. The more strictly we adhere to the phenomenon itself, the less we risk being drawn into abstruse metamorphosis-doctrine.[155] What, he asks, truly separates animals from plants? It is arbitrary movement, and the presence of organs that enable it: and not locomotion alone, but more generally the "voluntary extension and contraction, arching up and flexing, opening and closing certain parts".[156] These, he asserted, are the plain facts; it is sophistry to call this mere interpretation.[157]

Case study 4: diatoms and desmids

Of all the beings known to early Nineteenth-century science, diatoms presented the greatest challenge to a clear delineation of the kingdoms of organic nature. Leeuwenhoek observed solitary diatoms in late 1702,[158] while the anonymous Gentleman in the Country pictured colonial diatoms in 1703.[159] For many years thereafter, solitary diatoms were generally regarded as infusoria, and colonial forms algae of genus *Conferva*. Solitary diatoms can be seen to glide slowly across a surface; once microscopes of sufficient resolution became available, this was found to be brought about by a slit-like structure, the *raphe*. Predictably, their motility was taken as evidence that diatoms are animals.

The diatom debate played out over the quarter century from 1824 until the late 1840s. For much of this period, desmids—microscopic beings now known to be related to *Chara* and land plants—were classified within or alongside the diatoms.[160] Early microscopists might be forgiven this error, as desmid cell walls are often ornamented (although not silicaceous), and colonial forms are known. Abbot Bonaventura Corti, successor to Lazzaro Spallanzani as professor of physics at Università di Reggio Emilia, was the first to describe a desmid (*Closterium*); he commented on its slow but spontaneous locomotion, and called it a *piantanimal*.[161] The following year *Closterium* was independently discovered by Johann Eichhorn, pastor of St Catherine's in Gdansk; he remarked on its slow, infrequent movement, and supposed it to be an animal.[162]

During the 1840s almost every biologist interested in microscopic beings ventured an opinion on whether diatoms are plants, animals, both, or neither, usually focusing first on their motility. Those who denied that self-motility and locomotion make algal zoospores animals, could scarcely argue that these same properties, expressed in attenuated fashion, make diatoms animals. Ehrenberg analogized the raphe with a snail's foot,[163] but Kützing could find no foot-like organ.[164] John Ralfs, author of *The British Desmidieæ* (1848), pointed out that desmids are no more motile than many algae, indeed less so than many algal zoospores, including their own.[165]

Beyond motility, every character that might tip the balance towards animal or plant was brought to bear on the question. Given their specialized forms, it was difficult (*pace* Gaillon) to interpret diatoms or desmids as the base of a metamorphic series leading to any type of higher organism. Ehrenberg claimed that diatoms (and perhaps desmids) have multiple vesicular stomachs, albeit no intestinal canal;[166] as we have seen, Dujardin and Meyen soon disproved his polygastric theory.[167] Other features supposed to favour animality included overall form (the desmid *Staurastrum* as "strangely insect-like"), the mineralized shell (*cf.* oysters), ornamentation of the cell wall, reproductive structures or openings, transverse cell division, and the presence of nitrogenous compounds.[168] Characters put forward as favouring a vegetable nature included conferva-like colonies, reproduction by conjugation (in desmids), production and release of zoospores, structural rigidity, resistance to desiccation, their intracellular structure, cytoplasmic streaming, the presence of chlorophyll and starch (in desmids), the absence of nitrogenous compounds,[169] and sunlight-driven production of oxygen by the diatom *Navicula*[170] and desmids.[171] Each and every claim was met by a counterclaim: *all* cells divide by transverse fission; green monads (which Ehrenberg insisted are animals) contain chlorophyll and liberate oxygen; the starch in desmids might come from plant material in their stomachs. By recognizing diatoms and desmids as separate taxa, Kützing bought some flexibility.[172] Even so, it proved impossible to untangle this web of legitimate points, factual errors, over-interpretation, misleading analogies, and false choices. Unsurprisingly, no consensus emerged.

Already in 1819, Heinrich Link had regarded diatoms as intermediates between plants and animals.[173] In 1844, however, Kützing had a different solution: like *Microglena* and *Ulothrix* (above), a being can be both plant *and* animal, until eventually one element becomes predominant. In *Die kieselschaligen Bacillarien*, Kützing makes much the same argument for diatoms, although his words are slightly ambiguous, and he seems to appreciate that the case for a vegetable nature is stronger than that for an animal nature.[174] But he does not stop there:

> we therefore would have to accept that there are three substances in diatoms, namely:
>
> 1) *one chemical [and] inorganic*, the silica that forms the shell;
>
> 2) *one organic, animated as vegetable*, from which emerges in part the coloured gonimic substance [inside the cells], in part the mucilaginous and gelatinous structures that represent the formless common covering of some Naviculæ, the tubules of the Schizonemæ and the stalks of the attached forms (Achnanthes, Gomphonema etc.);
>
> 3) *one organic, animated as animal*, that would be used to form the organs of movement. (...)
>
> In their compound forms, all diatoms appear as decided plant-organisations; only the simplest forms show phenomena which are reminiscent of those of the infusoria; in this, however, they resemble only all lower plant-organisations in which the same occurs.[175]

Kützing stops just short of claiming that diatoms are animal, vegetable, *and* mineral.[176]

Summary

In this chapter we have explored the interface between infusoria and algae, focusing on the period from the 1770s through the 1840s. Knowledge of both groups increased dramatically during this period: from seventeen (OF Müller) to 382 genera of infusoria,[177] and from four (Linnæus) to 596 genera of algae.[178] The scope of infusoria narrowed as spermatic animalcules, worms, and polyps were excluded, while rotifers, green monads, and diatoms fell into contention. By contrast, algae expanded in scope: lichens, mosses, and fungi were banished, but red algae were reclaimed from the corals, as were pigmented unicells from the infusoria. Carl Nägeli, then a Privatdozent at Universität Zürich, defined algae as plants with

> cell contents consisting partly of starch granules and pigmented vesicles; no spontaneous generation; reproduction asexual, by spores [*Keimzellen*].[179]

Nägeli considered this to distinguish algae from fungi, which might arise spontaneously,[180] without requiring algae to be aquatic.[181] As we have seen, however, there had long been circumstantial evidence that some algae reproduce sexually. Nägeli remained unconvinced until 1855, when Nathaniel Pringsheim settled the matter by observing spermatozoa to penetrate into the *Vaucheria* egg.[182] On the other hand, Nägeli took the lead in classifying pigmented infusoria as algae, notably in his *Gattungen einzelliger Algen* (1849), against the opposition of Schleiden, Cohn,[183] and others. In due course green, brownish, and red unicells were grouped respectively with the green, brown, and red macrophytes of Lamouroux, yielding the core of modern phycological systematics.[184]

Our journey in this chapter, from Linnæus and OF Müller through Priestley, Ingen-Housz, Vaucher, Agardh, Gaillon, Ehrenberg, and Unger, to Kützing, Nägeli, and Pringsheim has not been a linear one. Even at mid-century, no consensus had emerged on whether fungi can arise de novo, whether self-motility implies animality, or whether diatoms are animals, plants, or perhaps both at once. Green matter, zoospores, apparent metamorphoses, and diatoms forced savants to rethink their beliefs about animals and vegetables. The process was often far from pretty, but a deeper understanding emerged of cellular structure, physiology, and generation: and not only among algae, as alternating sexual and asexual stages were recognized among mosses, ferns, and eventually seed-bearing plants,[185] thereby returning a measure of unity to the plant kingdom. As for our question, during this period there developed widespread, although not universal, appreciation that:

- The plant kingdom does not begin with *Conferva*-like filaments, but includes simple (unicellular) beings including *Protococcus*. This understanding was reached outside the framework of metamorphosis theory or *Naturphilosophie*.
- Infusoria (as understood by OF Müller and Ehrenberg) do not constitute a natural group, as they include plants as well as animals.
- It is not a contradiction in terms to speak of unicellular animals, or of unicellular plants. Even so, many biologists trod cautiously on this point.[186]
- Plants and animals do not differ fundamentally in the structure or division of their cells (or protoplasm). Were this otherwise, Kützing could not have claimed diatoms as both animal and plant. His theory was opposed on various grounds, but not as cell biology.
- Self-motility and locomotion are not exclusive to animals.
- Animals do not produce chlorophyll[187] or starch,[188] or release oxygen (*contra* Ehrenberg).

Our inventory of possibilities for the animal-plant interface now includes:

- A boundary (perhaps sharply defined, perhaps less so), such that beings are, and remain, either plant or animal;
- A self-contained intermediate taxon perhaps at the rank of kingdom, *e.g.* Zoophyta;

354 KINGDOMS, EMPIRES, AND DOMAINS

- Stepwise ascent, over time, from plant to animal, passing from one stable genus to another as up a ladder of being (Bonnet, Lamarck). The genera (forms) remain plant, zoophyte, or animal, but the beings ascend;
- An uncommitted or indeterminate state from which organisms can develop, becoming either plant or animal. This passage might be conceptual (Turpin), take place over time (Bory, Carus), or occur at an early point in the life of each individual (Goethe). Schultze did not indicate whether his *Urorganismen* might develop further.
- A state in which individuals are simultaneously both plant and animal, and either remain so (Eichwald), or commit to one or the other during the individual's lifetime (Kützing); and
- A temporal alternation or cycle, such that an organism is successively plant, then animal, then plant again (etc.). The plant-animal boundary is crossed with the passage to each successive state (Needham, Münchhausen, Unger, Gaillon).

Simultaneous or alternating animality and vegetality did not prove to be popular options, although in 1910 Henri Bergson imagined that "the first living organisms oscillated between the vegetable and animal form, participating in both at once", and that "the characteristic tendencies of the evolution of the two kingdoms, although divergent, coexist even now, both in the plant and in the animal. The proportion alone differs."[189]

We have one more stop before returning to pick up our story mid-Century. Let us now follow our time-travellers across the Channel, to Britain.

20

Temples of Nature

So, view'd through crystal spheres in drops saline,
Quick-shooting salts in chemic forms combine;
Or Mucor-stems, a vegetative tribe,
Spread their fine roots, the tremulous wave imbibe,
Next to our wondering eyes the focus brings
Self-moving lines, and animated rings;
First Monas moves, an unconnected point,
Plays round the drop without a limb or joint;
Then Vibrio waves, with capillary eels,
And Vorticella whirls her living wheels;
While insect Proteus sports with a changeful form
Through the bright tide, a globe, a cube, a worm.
Last o'er the field the Mite enormous swims,
Swells his red heart, and writhes his giant limbs.

—Erasmus Darwin, *The temple of nature*[1]

Our time-travellers' final stop, Britain[2] *circa* 1770, differed in important ways from France and Germany. London was the most-populous city in Europe,[3] but unlike Paris was not a major centre of learning, and did not even boast a university until 1826.[4] London did host the learned societies,[5] but the Royal Society had become a gentleman's club,[6] and royal patronage was desultory at best. The towering scientific figures of the past—Francis Bacon, William Harvey, John Ray, Isaac Newton—remained a source of pride, but no British Buffon or Goethe had taken their place. The ancient universities at Oxford and Cambridge were little more than branches of the Church of England, while the Test and Corporation Acts[7] limited the political, commercial, and educational rights of Catholics and nonconformists. As we shall see, Methodists, Quakers, and Unitarians were prominent in scientific and industrial innovation, often in regional cities such as Birmingham.

The situation differed somewhat in Scotland. The University of Edinburgh was a hotbed of the Scottish Enlightenment, but natural history took a back seat to mathematics, physics, chemistry, and geology. William Smellie, the university's printer,

Kingdoms, Empires, and Domains. Mark A. Ragan, Oxford University Press. © Oxford University Press 2023.
DOI: 10.1093/oso/9780197643037.003.0020

edited the *Encyclopædia Britannica* (first edition, 1768–1771), competed unsuccessfully for the professorship of natural history (1779), and translated Buffon's *Histoire naturelle* (1780–1785).[8] In *The philosophy of natural history* (1790) Smellie denied that plants are endowed with life, and took issue with Linnæus regarding their sexuality.[9] He held that "the whole universe is linked together by a gradual and almost imperceptible chain of existences both animated and inanimated" and considered this to be evidence for the unity of God and his design in nature,[10] but maintained the division of natural bodies into mineral, vegetable, and animal kingdoms.[11]

Agriculture remained the basis of the economy, but times were changing. As the Eighteenth century progressed, canals were dug, mills constructed, and steam engines put to work. The gradual advance of literacy and economic prosperity went hand in hand with a broad popular interest in science. Practically minded persons—landowners, industrialists, shopkeepers, artisans—joined scientific clubs and societies, flocked to lectures, gazed down microscopes, and tried their hand at experiments. Chemistry, mineralogy, and geology were ascendant, while natural history was valued as much for æsthetic reasons as for its connection with the rural economy. The emphasis on applied arts and sciences is apparent in (for instance) the essays of the Reverend John Walker, the successful candidate in 1779 for the professorship of natural history at the University of Edinburgh.[12] Meanwhile, improvements in optics and microscopy were exploited by John Turberville Needham and Henry Baker.[13]

The parson-naturalist had been an established fixture of British science since William Turner in the Sixteenth century, John Ray in the Seventeenth, and William Derham in the early Eighteenth.[14] Ministers, priests, and the occasional bishop contributed to natural history on the Continent as well,[15] but the tradition flourished in Britain through the Eighteenth century and continued strongly into the Nineteenth,[16] even as French and German science was specializing and becoming professionalized. To be sure, even in Britain some of these clerics were based at universities, or enjoyed fellowship of the Royal Society; but others tended quietly to their rural flock, surrounded by nature. The exemplar was Gilbert White. After studying at Oxford (and serving as Dean of Oriel College), he returned to rural Hampshire as a curate, and eventually inherited the grand home in Selborne where he had been born.[17] His *Natural history of Selborne* (1789) became immensely popular for its simplicity, close observation of nature and its seasons, and an idealized harmony between mankind and nature.[18] Even at this remove from formal systematics, White contrasted animals with vegetables, with no hint of an intermediate taxon.[19]

Britain: three Linnæan kingdoms

For John Ray, nature had been designed and created by an omniscient God, and was therefore orderly and fit for purpose. Ray's writings were held in high esteem in Britain, not least his *Historia plantarum* (1686–1704), from which his *Synopsis methodica stirpium Britannicarum* (1690) emerged as the first taxonomically arranged British flora.[20] Ray had relatively little to say about the plant-animal interface. As we have seen, in 1682 he allowed that it may not be possible to determine whether zoophytes are entirely plants or animals; and in 1693 he called worms, oysters, sponges, and microscopic beings imperfect animals, and referred to "the so-called Zoophyta" as "animals . . . which are imperfect or of a two-faced nature".[21] These were not the clearest of statements, but do not call for an intermediate kingdom.

The subsequent generation of British naturalists found no reason to reconsider. In his three-volume *General natural history* (1748–1752), subtitled *New and accurate description of the animals, vegetables and minerals, of the different parts of the world*, John Hill wrote that "there are no such creatures" of a middle nature between animals and vegetables.[22] His contemporary Henry Baker considered microscopic beings to be animals,[23] and speculated that corals or corallines might variously be referred to the animals, plants, or minerals.[24] John Ellis, who corresponded with Linnæus, asserted that marine productions must be either plants or animals, and decided that corals are the latter.[25]

With the appearance of the Tenth edition of *Systema naturæ* (1758–1759), British natural historians finally warmed to the Linnæan system: perhaps its pragmatism was suited to British practicality.[26] James Lee's *Introduction to botany*, an adaptation of *Philosophia botanica*, appeared in nine editions from 1760 to 1811.[27] From 1762 John Hope, professor of medicine and botany at the University of Edinburgh,[28] promoted Linnæan classification. London apothecary William Hudson adopted the Linnæan classes and binomial nomenclature for his *Flora Anglica* (1762), the first since Ray. William Withering (*A botanical arrangement of all the vegetables naturally growing in Great Britain*, 1776) removed the sexual references in the names of the Linnæan classes and orders due to

> an apprehension that Botany in an English dress would become a favourite amusement with the Ladies, many of whom are very considerable proficients in the study, in spite of every difficulty.[29]

A stream of Linnæan botanies followed: from John Lightfoot (*Flora Scotia*, 1777), Robert Waring Darwin (*Principia botanica*, 1787),[30] Thomas Martyn (*Thirty-eight plates . . . intended to illustrate Linnæus's System of vegetables*, 1799),[31] John Hull (*British flora*, 1799; *Elements of botany*, 1800), Robert Thornton (*New illustration of the sexual system of Carolus von Linnaeus*, 1807),[32] Thomas Green (*Universal herbal*, 1816),[33] James Millar (*Guide to botany*, 1818), and Robert Kaye Greville (*Flora Edinensis*, 1824). In North America, Henry Muhlenberg's *Catalogue of the hitherto known native and naturalized plants of North America* (1813) followed Linnæus. James Jenkinson brought excerpts from *Genera plantarum* and *Species plantarum* into English as *A generic and specific description of British plants* (1775); Hugh Rose

358   KINGDOMS, EMPIRES, AND DOMAINS

translated *Philosophia botanica* as *Elements of botany* (1775);[34] and "A Botanical Society, at Lichfield"[35] organized a translation of *Systema vegetabilium* (1783).

We do not find a corresponding body of Linnæan zoologies, perhaps because Buffon was proving popular in translation (as would Cuvier later).[36] Thomas Pattinson Yeats's *Institutions of entomology* (1782) offered extracts from *Systema naturæ*; Robert Kerr's *Animal kingdom* (1792) translated its zoological volumes;[37] and William Turton's *General system of nature* (1806) updated the Gmelin *Systema naturae*. Further competition came from Edinburgh bookseller Peter Hill, a friend of Smellie and Robert Burns, whose anonymous *New system of the natural history* (1791–1792) owed more to Buffon than to Linnæus; it broke no new ground in distinguishing animals from vegetables.[38]

Knowledge of foreign flora and fauna increased dramatically from the mid-1760s. A few of Linnæus's students brought samples back from North America, Africa, or the Far East,[39] but their efforts were greatly overshadowed by Joseph Banks, a sometime student at Oxford and heir to an estate in Lincolnshire. Banks collected plants and animals in Newfoundland and Labrador (1766), assigning them Linnæan binomials. He then sailed with James Cook on the HMS *Endeavour* (1768–1771), returning with tens of thousands of specimens.[40] Later, as president of the Royal Society (1778–1820) and advisor to George III, Banks commissioned expeditions to many parts of the globe, greatly enriching the collections of Kew Gardens and introducing some 7000 species of exotic plants into England—all of which needed to be named and classified.

Fortunately, help was at hand. Upon Linnæus's death in 1778 his library, papers, and collections passed first to Carl, his son and successor in the chair of practical medicine at Uppsala; then to Sara Lisa, wife of the elder Linnæus.[41] To fund dowries for her four daughters, Sara Lisa sold the lot to James Edward Smith, a wealthy twenty-four-year-old medical student from the University of Edinburgh. In September 1784 Smith had the Linnæan heritage shipped to London, where in 1788 he founded the Linnean Society. The latter society, in turn, purchased the collections from Smith's widow Pleasance in 1829.[42] Smith presented translations of Linnæus's *Disquisitio de sexu plantarum* (1786), *Lachensis Lapponica* (1811), and selected letters (1821), and applied Linnæan concepts and classification in his own written work including *Flora Britannica* (1800–1804).[43]

Smith had little occasion to comment on the plant-animal interface, although in *Introduction to physiological and systematical botany* (1807) he held that plants are living beings that may move spontaneously if involuntarily, while "those half-animated beings called Corals and Corallines" are "fixed, as immoveably as any plants, to the bottom of the sea, while indeed many living vegetables swim around them, unattached to the soil, and nourished by the water in which they float".[44] Nonetheless he helped consolidate in Britain a three-kingdom view of nature through not only the Linnæan collections, but also the *Encyclopædia Britannica*. As we have seen, its first editor, William Smellie, was hostile to the idea that plants are alive and reproduce sexually.[45] By contrast, the editors of the Second (1778) and

Third (1797) editions held that the matter could not yet be decided.[46] The Third through Sixth (1823) editions presented the case for plant sexuality by quoting at length from Smith's translation of *De sexu plantarum*, while the Fourth (1810) through Sixth editions glorified Linnæus as the equal of Bacon, Newton, and Locke, and set out a Linnæan classification of plants in immense detail.[47] In the important 1824 Supplement, Smith himself described botany as a philosophical as well as practical pursuit, explained how the Linnæan collections supported scholarship, and pointed to similarities between the systems of Linnæus and Jussieu.[48]

If the article in *Encyclopædia Britannica* marked 1810 as the apogee of Linnæan systematics in Britain, the appearance of Robert Brown's *Prodromus floræ Novae Hollandiæ et Insulæ Van-Diemen* in the same year demonstrated that the so-called natural system of Jussieu could accommodate the distinctive floras of Australia and Tasmania.[49] It was both appropriate and ironic that it fell to Smith, in the 1824 Supplement, to remind British readers of Jussieu's system. John Loudon's *Encyclopædia of plants* (1829) described both systems, and subsequent editions of *Encyclopædia Britannica* were even-handed.[50]

Eighteenth-century British naturalists sometimes referred to a kingdom of fossils, rather than of minerals. In contemporary English, a fossil(e) was *that which is or may be digged out of the Ground*[51] including earths, crystals, stones, gems, salts, and minerals, while

> The extraneous Fossils; which are bodies of the animal or vegetable kingdoms accidentally bury'd in the earth, belong properly to the histories of plants and animals.[52]

We find such a Fossil(e) Kingdom in Benjamin Martin's *Bibliotheca technologica* (1737), John Berkenhout's *Outlines of the natural history of Great Britain and Ireland* (1767–1772), Richard Pulteney's *General view of the writings of Linnæus* (1781), the Second and Third editions of *Encyclopædia Britannica*,[53] and Smith's *Introduction to physiological and systematical botany* (1807) among others, and in a host of popular books, journals, and magazines well into the 1840s. By contrast, John Whitehurst (*Inquiry into the original state and formation of the earth*, 1778), James Hutton (*Theory of the earth*, 1788), Pulteney (*General view*, Second edition, 1805), James Parkinson (*Organic remains*, 1808–1811), and the *Encyclopædia Britannica* (from the Fourth edition, 1810) refer instead to the Mineral Kingdom.[54]

Erasmus Darwin

Today as in his lifetime, Erasmus Darwin defies ready summary. His fame as a physician extended beyond his native Midlands to the royal court and Continent. He designed, and sometimes built, a remarkable range of mechanical devices of varying practicality. He favoured social reform, promoted education for young

360 KINGDOMS, EMPIRES, AND DOMAINS

women, opposed slavery, drew hope from the French Revolution, and was a relentless critic of religion. He was widely if fleetingly acclaimed as Britain's leading poet. As for his science, the earth was many millions of years old; life had arisen in the sea; plants and animals had evolved from a few original types; the strong devour the weak, while the superfluous are swept away; embryonic forms analogize earlier stages of life; and nerve fibres conduct electricity. He held that animality possesses "the faculty of continuing to improve by its own inherent activity"—that is, without divine intervention. He read Buffon, translated Linnæus, befriended Priestley, and antagonized Withering. According to biographer Desmond King-Hele, he "took all knowledge as his province".[55]

Darwin set out his natural history in four major works: *The botanic garden*, made up of *The economy of vegetation* (1791) and *The loves of the plants* (1789); *Zoonomia; or, the laws of organic life* (1794–1796); *Phytologia; or the philosophy of agriculture and gardening* (1800); and *The temple of nature; or, the origin of society* (1803).[56] The first and last are poems, albeit with extensive "philosophical notes"; the other two, no less imaginative, are set in prose. He refers to *kingdoms* or *worlds* of animals and vegetables; only in *Botanic garden* does he acknowledge a mineral kingdom.[57] Animals differ from plants in a number of respects, notably their locomotion and mode of nutrition,[58,59] but beings are linked by analogies so robust that Darwin refers to muscles, veins, arteries, and a pulmonary system in plants, indeed nerves, a brain, sensation, and volition; and he imagines that vegetables receive "a degree of pleasurable sensation" from the activity of their vascular systems, and from the reproduction of their species.[60]

Darwin did not explicitly arrange organisms in a scale,[61] although vegetables are inferior or imperfect animals,[62] and fungi "make a kind of isthmus connecting the two mighty kingdoms of animal and of vegetable nature".[58] Fungi "approach to animal nature" because they are "generated without seed by their roots only, and without light",[63] and indeed "appear to be animals without locomotion";[64] but he does not advocate their formal reclassification. For that matter,

> anthers and stigmas [of flowers] are real animals, attached indeed to their parent tree like polypi or coral insects, but capable of spontaneous motion . . . they are affected with the passion of love, and furnished with powers of reproducing their species, and are fed with honey like the moths and butterflies, which plunder their nectaries.[65]

Darwin accepted that various present-day beings—microscopic animals and vegetables, Priestley's green matter, certain fungi[66]—arise spontaneously, and devout Christians raised their eyebrows when he proposed a similar, all-natural origin for life on earth:

> Hence without parent by spontaneous birth
> Rise the first specks of animated earth;
> From Nature's womb the plant or insect swims,
> And buds or breathes, with microscopic limbs.[67]

In another passage, life began as a single living filament:

> Or, as the earth and ocean were probably peopled with vegetable productions long before the existence of animals; and many families of these animals long before other families of them, shall we conjecture, that one and the same kind of living filaments is and has been the cause of all organic life?[68]

Here as elsewhere, precision comes off second best to Darwin's poetic enthusiasm;[69] but he appears to entertain the possibility that plants and animals arose separately, perhaps on multiple occasions. No intermediate kingdom was needed, as either common source or transition.

Darwin's achievements were many and his scientific vision remarkable, but he was not a British Goethe. *Botanic garden*, with its French chemistry[70] and pæans to the French and American Revolutions,[71] was read with different eyes after Louis XVI and Lavoisier went to the guillotine and a French revolutionary army marched into Holland. William Pitt's government encouraged the reporting of spies and radicals; riots in Birmingham targeted Priestley (who escaped to America) and other non-conformist intellectuals, and for a time it seemed that Darwin might be tried for sedition.[72] Popular taste in poetry changed too: the Arcadian settings, flowery couplets, and didactic intent of *Loves of the plants* gave way to romantic forms inspired by *Werther*, beginning with the 1798 *Lyrical ballads* of Wordsworth and Coleridge.[73] Darwin's *Loves of the plants* was parodied by George Canning and John Hookham Frere as *Loves of the triangles*,[74] while his science (notably his use of analogy) was attacked broadly by Thomas Brown,[75] and more narrowly by John Mason Good.[76]

Natural theology

As noted in Chapter 15, the idea that "the wondrous complexity of natural things, their fitness for purpose and place in the natural order can be taken as evidence of design by a wise and powerful God" can be traced back to Socrates, and was embraced to a greater or lesser extent by many, although not all, schools of thought through the Middle Ages, Renaissance, and early modern eras. The corresponding body of teachings, known variously as physico-theology or natural theology,[77] took some root on the Continent,[78] but rose to particular prominence in Britain from about 1690 to 1844. This tradition interests us here because it offered a perspective independent of the great systems of biological classification, and invited the contemplation of simpler beings:

> Consider the lilies how they grow: they toil not, they spin not; and yet I say unto you, that Solomon in all his glory was not arrayed like one of these.[79]

Both Church and Academy were, however, suspicious of full-blown natural theology. The Church had long benefitted from the tenet, dating to William

362 KINGDOMS, EMPIRES, AND DOMAINS

of Ockham, that Scripture does not admit proof by reason; correspondingly, philosophers and naturalists thanked Francis Bacon for separating theology and science.[80] Robert Boyle held that miracles have no place in the interpretation of natural phenomena,[81] but alongside *The sceptical chymist* (1661) wrote *The excellency of theology compared with natural philosophy* (1665), in which he drew precisely the comparison above: we see "how admirably every animal is furnished with parts requisite to his respective nature", and

> that sure God, who has with such admirable artifice framed silk-worms, butterflies, and other meaner insects, and with such wonderful providence taken care, that the nobler animals should as little want any of all the things requisite to the compleating of their respective natures; and who, when he pleases, can furnish some things with qualifications quite differing from those, which the knowledge of his other works could have made us imagine, (as is evident in the load stone, and in quick-silver among minerals, and the sensitive plant among vegetables, the camelion among animals, &c).[82]

Across his written work, Boyle referred repeatedly to the animal, vegetable, and mineral kingdoms.[83] He endowed a series of eponymous Lectures "to demonstrate from nature the wisdom and goodness of God".[84] Lecturers have included Richard Bentley (1692, 1694), Josiah Woodward (1710), William Derham (1711–1712), and Alfred Barry (1876).[85] We have already noted that Ray, Derham, and John Wesley in England, and Cotton Mather and Jonathan Edwards in the American colonies, found no cause to invoke a third or intermediate kingdom of beings.

William Paley was already well-known for his *Principles of moral and political philosophy* (1785) and advocacy against slavery, when his *Natural theology* (1802) greatly popularized the analogy[86] that just as a watch implies a skilful and intentioned watchmaker, so too nature is evidence of a supremely intelligent Designer. Paley drew examples from diverse sources,[87] but had nothing to say about microscopic animalcules or green matter, and cast scorn on Buffon's organic molecules and internal moulds.[88] He referred often to animals and vegetables, and seemed to consider them as the only types of organized bodies.[89]

Natural theology took a further step towards Buffonian natural history through the work of William Swainson. At first attached to the customs and military services, Swainson travelled to Malta, Sicily, and coastal Brazil, but retired at age twenty-six due to ill health. A talented illustrator, he was one of the first to adopt lithography (1820) in producing works of natural history. He then wrote and illustrated the natural-history volumes for Dionysius Lardner's *Cabinet cyclopædia*, a 133-volume series (1829–1846) marketed as self-improvement for the middle class.[90] Swainson believed that design, hence evidence for a perfect Designer, can be found in "every branch of the animal kingdom".[91] He sought (unsuccessfully, in the opinion of his contemporaries) to fit animals into five quinarian circles, and wrote that

It has long been customary, not only in science, but in ordinary parlance, to designate the three great divisions of ponderable matter as the *animal*, the *vegetable*, and the *mineral* kingdoms of nature; and, although it is not yet ascertained in what precise manner the vegetable, or, perhaps, also, the mineral kingdom describe their own circles, yet it is sufficient for our present purpose that the animal kingdom forms a circular group, comprehending all beings which usually pass under that name.[92]

Natural theology, at least in Britain, culminated in the Bridgewater Treatises. In 1825 Francis Henry Egerton, the Eighth Earl of Bridgewater, directed the trustees of his estate to provide the President of the Royal Society with funds to commission one or more treatises

On the Power, Wisdom, and Goodness of God, as manifested in the Creation; illustrating such work by all reasonable arguments, as for instance the variety and formation of God's creatures in the animal, vegetable, and mineral kingdoms; the effect of digestion, and thereby of conversion; the construction of the hand of man, and an infinite variety of other arguments; as also by discoveries ancient and modern, in arts, sciences, and the whole extent of literature.[93]

Upon the Earl's death in 1829, the President, assisted by the Archbishop of Canterbury and the Bishop of London,[94] appointed eight authors. Given the terms of commission, it is unsurprising that most of the resulting treatises affirm the three kingdoms. That by parson-naturalist William Kirby was the most explicit in regard of the interface between plants and animals, rejecting the idea that organized bodies might "partake of two natures, that are either animal at one period of their existence and vegetable at another, or else are partly animal and partly vegetable". Kirby dismissed the "opinions" of Agardh and Unger, supposing instead that the oscillatory motion of filamentous algae is due to Brownian motion, and that

the motions of these seeds or germes, may be merely mechanical, and may be necessary to enable them properly to fix themselves, somewhat analogous to those mechanical contrivances by which the seeds of numerous plants, as those of the dandelion and cranesbill, are transported to a distance and enabled to enter the soil and fix themselves in it.

That any creature should begin life as an animal and end it as a plant seems to contradict the general analogy of creation, and requires much stronger proofs than appear to have been adduced in the present case, before it can be admitted.[95]

Curiously, the Treatise by John Kidd, Regius professor of medicine at Oxford, sets out nature in *four* "kingdoms or divisions": the atmospherical, mineral, vegetable, and animal. In some passages he explicitly refers to an atmospherical kingdom, but elsewhere reserves *kingdom* for the standard three, and calls the atmosphere

364   KINGDOMS, EMPIRES, AND DOMAINS

a *department*.[96] In any case, he asserts that even in monsters and *lusus naturæ*, all component parts of an individual are necessarily of the same species, so there can never be confusion between kingdoms.[97]

By the 1830s, biologists were engaged with cell theory, and with intricacies of the generation, development, and physiology of plants and animals. Explanations were being drawn, with varying degrees of success, from chemistry, hydraulics, heat exchange, even galvanism. Despite the Bridgewater Treatises, natural theology was increasingly seen as an anachronism, until in 1844 its weakened grip was finally broken by the anonymous *Vestiges of the natural history of Creation*. We return to the *Vestiges* below, but first must introduce the single most-important transitional figure of the period.

Richard Owen

Britain may not have brought forth a Buffon or Goethe, but it produced a Cuvier in the person of Richard Owen. Trained in medicine, Owen instead developed a career as conservator of museum collections, first at the Royal College of Surgeons (from 1827), then at the British Museum. He is remembered today for his work on the comparative anatomy of living and fossil vertebrates,[98] and for establishing the Natural History Museum[99] in a grand new building; but several of his early publications dealt with invertebrates, and he was the first president of the Microscopical Society of London (1839).

Until at least 1842, Owen viewed each type of animal as had Cuvier: as designed by the Creator for its environment and way of life, with organs arranged in integrated systems that deliver physiological functions including circulation, respiration, and sensation. This view was eminently compatible with natural theology, a fact not lost on William Buckland, an eminent theologian, geologist, and author of one of the Bridgewater Treatises: it was Buckland, William Whewell (also a Bridgewater author), and their Oxford and Cambridge colleagues who anointed Owen the British Cuvier.[100] This organ-centric approach is on display in Owen's *Remarks on the entozoa* (1835).[101]

By the early decades of the Nineteenth century taxonomists had abandoned a simple Chain of Being, but could not agree on a natural arrangement of the simplest animals. Worms were particularly troublesome, owing to their diversity of form yet paucity of morphological features that might distinguish natural groups. Rudolphi set up Entozoa with five orders,[102] but later excluded Order Nematoidea on the grounds that its members possess a nervous system, albeit without ganglia.[103] Rudolphi made the nematoids a family within the annelids (where they fit uncomfortably at best), and considered the remaining Entozoa to be part of Cuvier's Zoophytes ou Animaux Rayonnés,[104] commenting that the latter group was a "regnum chaoticum".[105] As we have seen,[106] Macleay separated Radiata into those with a radial body plan (echinoderms and acalephs), and those without (infusoria, polyps, and nematoids: collectively Acrita). However, the discovery of ova in *Taenia*, and (according to Ehrenberg) of alimentary organs in infusoria, rendered Macleay's Acrita heterogeneous too.

Surveying the situation, Owen concluded that "it is only with respect to the nervous system that we can attribute a community of structure to a primary division of the animal kingdom".[107] He rearranged Cuvier's Zoophytes[108] into two natural groups: polygastric infusoria, polyps, parenchymous entozoa, and acalephs, for which he retained the term Acrita; and echinoderms, cœlomate entozoa, epizoa, and rotifers, which he called Nematoneura. The former show only "obscure traces" of nerve filaments, or their nervous system is diffuse or absent altogether, whereas in the latter "nervous filaments are always distinctly traceable".[109] To be sure, his newly reformulated Acrita remained something of a grab-bag, but this was because

> They are analogous to the ova or germs of the higher classes . . . as the changes of the embryo succeed each other with a rapidity proportionate to the proximity of the ovum to the commencement of its development, so also we find that in each class of Acrita there are genera which advance into close approximation with some one or other of the classes belonging to the higher divisions of the animal kingdom. It results, therefore, from this tendency to ascend in the scale of organization that there is greater difficulty in assigning constant or general organic characters to the Acrita than to any of the higher divisions of animals. Even in the nervous system, we find as we are led step by step from the hydra to the actinia in the class Polypi, that the nervous globules begin to manifest the filamentary arrangement about the oral orifice in the last named genus . . .
>
> For the most part all the different systems seem blended together, and the homogeneous granular parenchyma possess many functions in common.
>
> Where a distinct organ is eliminated it is often repeated indefinitely in the same individual. Thus in the polypi the nutritious tubes of one individual are generally supplied by numerous mouths, and it has, consequently, the semblance of a composite animal; the polygastrica derive their name from an analogous multiplication of the digestive organ itself. . . .
>
> The formative energies of the Acrita being thus expended on a few simple operations, and not concentrated on the perfect development of any single organ, it is not surprising that the different classes should exhibit the greatest diversity of external figure. But it has been well observed that Nature, so far from forgetting order, has, at the commencement of her work, in these imperfect animals given us a sketch of the different forms which she intended afterward to adopt for the whole animal kingdom. Thus in the soft, sluggish sterelmintha[110] we have the outline of the mollusca; in the fleshy living mass which surrounds the earthy hollow axis of the polypi natantes, she has sketched a vertebrated animal; and in the crustaceous covering of the living mass, and the structure more or less articulated of the polypi vaginati we trace the form of the annulose or articulate classes.[111]

366 KINGDOMS, EMPIRES, AND DOMAINS

Owen was not the first to find such analogies: as we have seen, von Baer considered that microscopic animalcules offered prototypes of animals of higher grades,[112] and Geoffroy found elements of the vertebrate organizational plan in zoophytes.[113] From 1843, Owen replaced his organ-centric (functional) approach with one based on taxonomy (form) and, like Blainville, Milne-Edwards and others, forged a middle way between Cuvier and Geoffroy.[114] Dov Ospovat found in this the basis of Owen's famous distinction between analogue ("a part or organ in one animal which has the same function as another part or organ in a different animal") and homologue ("the same organ in different animals under every variety of form and function").[115] In his 1843 Hunterian lectures, Owen described in substantial detail how

> every animal in the course of its development typifies or represents some of the permanent forms of animals inferior to itself; but it does not represent all the inferior forms, nor acquire the organization of any of the forms which it transitorily represents....
>
> There is only one animal form which is either permanently or transitorily represented throughout the animal kingdom: it is that of the infusorial Monad ... which is to be regarded as the fundamental or primary form.[116]

Unsurprisingly, Owen considered monads "true animals". More surprisingly perhaps, his 1843 lectures maintained the full Ehrenbergian view of infusoria as complete organisms, with not only stomachs but a mouth, teeth, intestines, muscles, nerves, and generative organs.[117] He cited Dujardin once, on entozoa; sarcode theory not at all. Not every microscopic being was necessarily an animal—Owen called *Protococcus* "the lowest form of cryptogamic plants"[118]—but

> An animal differs from a plant in having a stomach and a mouth, it is thereby qualified to exert its most conspicuous animal property, that of locomotion.[119]

Eventually Owen had to accept that Ehrenberg had been mistaken. For his 1852 Hunterian lectures, Owen opened a door to organisms other than plants and animals. Animals share certain specialized characters, plants share others; but

> there are very numerous living beings, especially those that retain the form of nucleated cells, which manifest the common organic characters, but without the distinctive superadditions of either kingdom. Such organisms are the *Diatomaceæ, Desmidiæ, Protococci, Volvocinæ, Vibriones, Astasiææ, Thalassicolæ,* and *Spongiæ*; all of which retain the character of the organised fundamental nucleated cell, with comparatively little change or superaddition.[120]

In 1858 he identified these superadditions, modified the roster of organisms that lack them, and gave these organisms a collective name:

> When the organism can also move, receive the nutritive matter by a mouth into a stomach, inhale oxygen and exhale carbonic acid, develop tissues the proximate principles of which are quaternary compounds of carbon, hydrogen, oxygen, and nitrogen, it is called an "animal". When the organism is rooted, has no mouth or stomach, exhales oxygen, has tissues composed of "cellulose" or of binary or ternary compounds, it is called a "plant". But the two divisions of organisms called "plants" and "animals" are specialized members of the great natural group of living things; and there are numerous organisms, mostly of minute size and retaining the form of nucleated cells, which manifest the common organic characters, but without the distinctive superadditions of true plants or animals. Such organisms are called "Protozoa," and include the sponges or *Amorphozoa*, the *Foraminifera* or Rhizopods, the *Polycystineæ*, the *Diatomaceæ*, *Desmidiæ*, and most of the so-called *Polygastria* of Ehrenberg, or infusorial animalcules of older authors.[121]

He repeated this passage almost word for word in *Palæontology* (1860), adding *Gregarinæ* to the list of protozoa.[122] Without explanation, however, in *Palæontology* he made Protozoa a Kingdom alongside Animalia:[123]

KINGDOM PROTOZOA
  Class Amorphozoa
   Rhizopoda
  Sub-Class Polycystineæ
  Class Infusoria
KINGDOM ANIMALIA
 Sub-kingdom Invertebrata
  PROVINCE RADIATA
  Class Hydrozoa
 (. . .)

As protozoa lack the "distinctive superadditions of true plants or animals", their kingdom is separate from, and in a sense prior to, those of the animals and vegetables—not unlike the *Protorganismen* introduced by Carus the following year.[124] Owen accepted that organisms evolve—he referred to "the axiom of *the continuous operation of the ordained becoming of living things*"[125]—and in 1868 set out a theory of evolution.[126] Notably, he found in the above analogies "many roots from which the higher grades have ramified". He denied that animals have unfolded in unilinear sequence,[127] but did not engage the tree metaphor. Increasingly drawn to other issues and occupied with the new Natural History Museum, Owen did not return to this question.[128]

368 KINGDOMS, EMPIRES, AND DOMAINS

Vestiges of the natural history of Creation

The anonymous *Vestiges* created an immediate sensation upon its appearance in October 1844. Written for the non-specialist, *Vestiges* joined the physical, biological, and social sciences into a narrative of progressive change governed by natural law. Ideas of temporal change were scarcely new within astronomy, geology, botany, zoology, or anthropology: even some Bridgewater authors thought progress to be part of the divine plan.[129] French *philosophes*, German nature-philosophers, and Erasmus Darwin had often enough presented Man as the end-product of swirling galaxies and chemical forces. But the author crossed a line by treating the ultimate mystery—the origin of life on earth, and the subsequent appearance of plants, animals, and Man, even passions, intellect, and morality—as subject to natural law, and thus within the remit of scientific inquiry. The death knell had rung on natural theology as an explanation of living nature.

Vestiges set off an avalanche of criticism, which only increased its notoriety (and sales).[130] Practicing scientists drew attention to its factual errors and amateurism, while clerics excoriated its materialism. The most egregious errors were corrected over subsequent editions, although the odd idea that insects and fungi could be generated de novo by the use of electricity[131] was retained. Importantly, *Vestiges* argued that ordinary reproduction led to the

> gradual evolution of higher from lower, of complicated from simple, of special from general, all in unvarying order, and therefore all natural, although all of divine ordination.[132]

The author—posthumously identified as the Edinburgh publisher Robert Chambers[133]—drew an analogy between embryological development and the serial appearance of species in geological strata. He argued that life arose independently on different continents; the same forms arose where conditions allowed, while minor local variations in these conditions resulted in taxa unique to each region. Nor did he contend that species continue to evolve: we see only vestiges of earlier evolution.[134] He described organic nature as branching into parallel series:

> The nucleated vesicle, the fundamental form of all organization, we must regard as the meeting-point between the inorganic and the organic—the end of the mineral and beginning of the vegetable and animal kingdoms, which thence start in different directions, but in perfect parallelism and analogy. We have already seen that this nucleated vesicle is itself a type of mature and independent being in the infusory animalcules, as well as the starting point of the foetal progress of every higher individual in creation, both animal and vegetable.[135]

An "obvious gradation" exists in each kingdom, "from the simple lichen and animalcule respectively up to the highest order of dicotyledonous trees and the mammalia", although it may resolve into "branching or double lines at some places; or the whole may be in a circle composed of minor circles".[136] It is "impossible to say where vegetable ends and animal begins",[137] but infusoria and sponges are animals, fungi and lichens are plants;[138] no intermediate taxon is required.

Although much in *Vestiges* was controversial, the metaphor of nature as diverging tree-trunks was not. Nor did Chambers have to reach back to Pallas[139] for this image: the previous year, Edward Forbes, curator of the museum of the Geological Society of London and professor of botany at King's College London, had written

> The two great kingdoms of organized nature seem to spring, as it were, from one root, and to branch out into correspondent trunks, which, even at their most distant ramifications, exhibit mutual analogies. The lowest forms of each so closely approximate as to furnish subjects for continual discussion; and the number of species, the position of which, whether in the animal or vegetable kingdom, is as yet disputed or undetermined, proves the close alliance of the animal and vegetable natures toward the point of union. Creatures which Ehrenberg figures as animalcules, Meyen describes as plants; and naturalists have not yet ceased debating on the nature of sponges and corallinæ. As we ascend in each great series, the animal or vegetable nature of their respective members becomes more and more decided and unquestionable, while there is still retained a close resemblance of external form. . . . Ascending higher we find all resemblance of form disappear, but still there is a true analogy.[140]

Charles Darwin

In the previous chapter we briefly encountered Charles Darwin as a medical student in Edinburgh, investigating the life cycle of the bryozoan *Flustra*. What unfolded thereafter is well-known: his voyage on HMS *Beagle* (1831–1836); study of Lyell's *Principles of geology*, finches, and coral reefs; mixed feelings about the *Vestiges*; his notebooks, monographs on barnacles, and correspondence with Alfred Russel Wallace; the *Origin of species* (1859), *Descent of man* (1871), and other substantial works.[141] In *Origin*, Darwin argued that over time, species arise as a consequence of heritable variation combined with natural selection, as represented by a tree diagram.[142] Darwin did not attempt a panoramic "nebula to Man" story and had little to say about the origin of life, although he suspected that vegetables and animals arose from a common primordial form,[143] and supposed that

> The parent monad form might perfectly well survive unaltered and fitted for its simple conditions, whilst the offspring of this very monad might become fitted for more complex conditions. The one primordial prototype of all living and extinct creatures may, it is possible, be now alive![144]

370 KINGDOMS, EMPIRES, AND DOMAINS

Cautious as always, Darwin finally addressed these issues in concluding the Third edition of *Origin* (1861). He stated his belief that "animals have descended from at most only four or five progenitors, and plants from an equal or lesser number", and continued:

> Analogy would lead me one step farther, namely, to the belief that all animals and plants have descended from some one prototype. But analogy may be a deceitful guide. Nevertheless all living things have much in common,—in their chemical composition, their cellular structure, their laws of growth, and their liability to injurious influences. . . . In all organic beings the union of a male and female elemental cell seems occasionally to be necessary for the production of a new being. In all, as far as is at present known, the germinal vesicle is the same. So that every individual organic being starts from a common origin. If we look even to the two main divisions—namely, to the animal and vegetable kingdoms—certain low forms are so far intermediate in character that naturalists have disputed to which kingdom they should be referred, and, as Professor Asa Gray has remarked, "the spores and other reproductive bodies of many of the lower algæ may claim to have first a characteristically animal, and then an unequivocally vegetable existence." Therefore, on the principle of natural selection with divergence of character, it does not seem incredible that, from some such low and intermediate form, both animals and plants may have been developed; and, if we admit this, we must admit that all the organic beings which have ever lived on this earth may have descended from some one primordial form. But this inference is chiefly grounded on analogy, and it is immaterial whether or not it be accepted.[145]

Darwin seems to know the ideas of Unger and Kützing only through his friend in America, Harvard professor Asa Gray.[146] Darwin's "low and intermediate form"— intermediate, that is, between plants and animals—catches the eye, but he does not make a kingdom of it,[147] or even care whether the obvious inference (that all living beings share a common origin) is accepted.

John Hogg

John Hogg, a lawyer and amateur antiquary, wrote a natural history of his native County Durham,[148] and published unremarkable items on plants, fish, and birds. In an initial venture into zoophytes, he supposed the bryozoan *Flustra arenosa* to be the *nidus* of the moon snail *Natica glaucina*.[149] Later, discussing whether sponges are animals or plants, he labelled "monstrous" the idea that "certain bodies belonging to several Cryptogamous plants are at first animals, at that after a time they change into true vegetables".[150] Simple chemical tests cannot prove that sponges are animals; furthermore,

They have no tentacles, no cilia, no mouth, no œsophagus, no stomach or gastric sac, no gizzard, no alimentary canal, no intestine, no anus, no ovaria, no ova, no muscles or muscular fibres, no nerves or ganglia, no irritability or powers of contraction and dilatation, no palpitation, and no sensation whatsoever.[151]

Eight years later, now as president of his local Naturalists' club, Hogg proposed that sponges be classified in a new order between algae and fungi, or (even better) within the fungi.[152] As for infusoria: some are animals, other vegetables,

Although, strictly speaking, in Nature there *may be no* actual distinction between these two kingdoms; and that life, in the lowest animal and that in the simplest plant, *may* be the *same*, both beings having the same properties of existence, in their receiving nourishment, in their power of increasing in size, in their propagation, as well as in their being subject to the same penalty of life—namely, *death*—still the Naturalist must endeavour to draw a line of demarcation between these two great provinces, for the sake of the arrangement and classification of the infinitely numerous living beings, or organisms, existing in the world. And, for this purpose, the clearest and most certain distinction between an animal and a vegetable seems to be the presence of a stomach, or a stomachic sac, and of a muscular apparatus in the former, and entire absence of them in the latter.[153]

A bolder step was to come. On 28 June 1860—two days before the famous confrontation between Thomas Henry Huxley and Bishop Samuel Wilberforce—Edwin Lankester read a paper by Hogg before the British Association for the Advancement of Science. In it, Hogg reminded his audience how every attempt to clarify the animal-vegetable interface had come to naught.[154] Sponges, diatoms, desmids, and infusoria were particularly problematic. Part of the problem was that Linnæus had been "much too concise" in his definitions, which could usefully be relaxed as follows:

Minerals are bodies, hard, aggregative, simple or component, having bulk, weight, and often regular form; but inorganic, inanimate, indestructible by death, insentient, and illocomotive.
Vegetables are beings, organic, living, nourishable, stomachless, generative, destructible by death, possessing some sensibility; sometimes motive, and sometimes locomotive in their young or seed state; but inanimate, insentient, immuscular, nerveless, and mostly fixed by their roots.
Animals are beings, organic, living, nourishable, having a stomach, generative, destructible by death, motive, animate, sentient, muscular, nervous, and mostly spontaneously locomotive, but sometimes fixed by their bases.[155]

But (Hogg continued) naturalists could not even agree on the number of kingdoms: some recognized two, others three. Owen himself—who was

372 KINGDOMS, EMPIRES, AND DOMAINS

probably in the audience[156]—had called the entire problem "insuperable".[157] As a consequence,

> it appears to many desirable to place those creatures, or organic beings, whose nature is so doubtful in a fourth kingdom. And although I at present do not feel quite convinced of the immediate necessity of doing so, or that it will ever remain—notwithstanding the progress which we hope will continue to be made in physical science—impossible for man to determine whether a certain minute organism be an animal or a plant, I here suggest a *fourth* or an additional kingdom, under the title of the *Primigenal* kingdom,
>
> REGNUM PRIMIGENUM,
> continens PROTOCTISTA, *i.e.*,
> PROTOPHYTA ET PROTOZOA.—
>
> This *Primigenal* kingdom would comprise all the lower creatures, or the primary organic beings,—"Protoctista,"—from πρῶτος, *first,* and χτιστά, *created beings*;—both *Protophyta*, or those considered now by many as, lower or primary beings having more the nature of plants; and *Protozoa*, or such as are esteemed as lower or primary beings, having rather the nature of animals. And to those formless or amorphous beings, whether partaking more of a vegetable or of an animal nature, I give the name of *Amorphoctista*—ἀμὸρφοχτιστά—instead of *Amorphozoa*, originally bestowed on them by the French writer M. de Blainville. (...)
> The *Primigenal* kingdom might be placed either the fourth and last, or between the vegetable and animal kingdoms.[158]

There is quite a bit to unpack here. Owen had placed sponges in Amorphozoa, the first class in his Kingdom Protozoa.[159] Hogg agreed that sponges are amorphous, but (believing them to be plants) disputed the -zoa.[160] Why (we might ask) did he not simply rename the taxon Amorphophyta, and move it to the vegetable kingdom? Doing so would admix formed and formless beings, but Hogg had not explicitly mentioned form in defining vegetable and animal (above). Perhaps he was hedging his bets on the nature of sponges: if, in due course, they were determined to be plants (or animals), they could be reclassified accordingly, and the (slightly diminished) Primigenum situated nearby. In the meantime diatoms, desmids, and infusoria would have a taxonomic home.

Hogg does not tell us whether Amorphoctista is part of Primigenum, although the extended abstract included it.[156] Nor does he indicate its rank: is it coequal with Protoctista,[161] or perhaps a subdivision (subkingdom?) alongside Protophyta and Protozoa? For that matter, is Protoctista even a taxon? Unlike Owen in *Palæontology*, Hogg offers few details.[162]

Finally, let us return to Hogg's suggestion that Primigenum might be placed "fourth and last" among kingdoms. He provided a coloured plate (Figure 20.1) with this explanation:

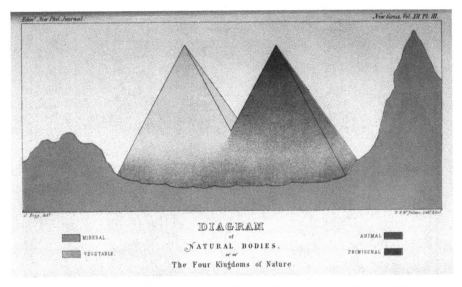

Figure 20.1. Hogg, diagram of natural bodies, or of The Four Kingdoms of Nature. *Edinburgh New Philosophical Journal, New Series* 12 (July–October 1860), Plate 3.

the vegetable and animal kingdoms have been well compared to two lofty pyramids, which diverge from each other as they ascend, but are placed on, or united in, a common base;[163] this base, then, might fairly represent the *Primigenal* kingdom, which includes the lower creatures or organisms of both the former, but which are of a doubtful nature, and can in some instances only be considered as having become blended or mingled together.[164]

If there is a last place on offer in the diagram, it can be only that of the Mineral kingdom. No philosopher or natural historian, whether nebulist or biblical literalist, would consign any living being, formed or formless, to a position below the minerals. If "fourth and last" is not a mental lapse, what are the other three kingdoms of living beings? In the wake of the *Origin of species*, this was a charged issue. Since it first appeared in 1840, Edward Hitchcock's popular *Elementary geology* had featured a chart of fossil and living plants and animals as branched series culminating in Palms and Man respectively, against a geological timescale (Figure 20.2); the chart disappeared from the thirty-second (1860) and subsequent editions.[165] In 1857 Owen, invoking Linnæus and Cuvier, had accorded man a bespoke subclass (Archencephala) within Mammalia,[166] but in 1859 (to a broader audience) emphasized man's special moral and religious status,[167] and in 1860 was coy about the matter.[168] By "fourth and last", was Hogg hinting at a kingdom of Man?

Thomas B. Wilson and John Cassin

We do not know what prompted Thomas Wilson, a wealthy benefactor of the Academy of Natural Sciences of Philadelphia, and John Cassin, a curator at

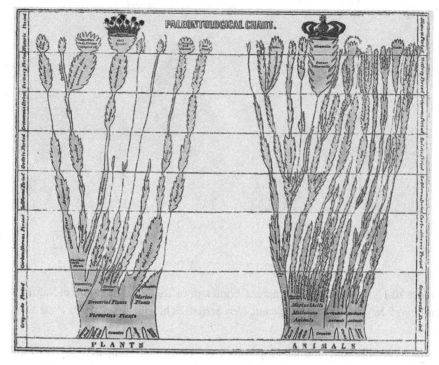

Figure 20.2. Hitchcock, palæontological chart showing plants and animals as branched series culminating, in the Historic Period, in crowned groups, respectively Palms (tiara with palm trees) and Man. *Elementary geology* (editions 1840–1859 only), frontispiece.

the Academy and a highly regarded ornithologist, to turn their attention to the kingdoms of nature,[169] although it may be no coincidence that Asa Gray (above) was an active member of the Academy, and curator Joseph Leidy had published on entozoa and entophyta.[170] In 1863, amidst the Civil War, Wilson and Cassin[171] returned to issues raised by Owen and Hogg. Although once it had been thought that living beings formed a linear chain of being, now most authorities accept that

> from a point of the first manifestation of life, its progress of evolution or development is into two series or great classes of existences,—animal and vegetable... In our opinion it may be demonstrable, that the first assumption of life manifests itself in objects constituting a primary great class or kingdom of more simple organization than either the animal or vegetable kingdom, and possessing also an equally characteristic specialization in its structure and functions.[172]

Owen (above) had recognized a group (later, a kingdom) of beings, protozoa, based on their lack of the distinctive superadditions that define animals and vegetables. By contrast, Wilson and Cassin undertake to demarcate a third kingdom using superadditions as distinctive as those cited by Owen.[173] They argue as follows: specialization is manifest in organ systems that serve reproduction, nutrition, and sensation. Although these functions are common to all organized beings, organs of nutrition become specialized and attain "prominence, or dominant prevalence" in vegetables, as do organs of sensation in animals,[174] while

> The Reproductive function ... beginning with mere cellular conjugation, becomes specialized first in a great group of organized beings of more simple structure than either Vegetables or Animals, which we regard as eminently and demonstrably a primary division or kingdom, and apply to it the name *Primalia*. In this kingdom organs of Reproduction are temporarily formed, and no other.[175]

In plants the reproductive organs are temporary but "of greatly increased importance", while in animals they are more-highly organized yet, and in the "higher sub-kingdoms [of animals] attain permanency of structure". Wilson and Cassin condense this after the style of Linnæus:

> ANIMALIA, corpora organisita, generantia, spirantia et sentientia.
> VEGETABILIA, corpora organisita, generantia, spirantia, *non sentientia*,
> PRIMALIA, corpora organisita, generantia, *non spirantia, nec sentientia*.[176]

They assert that Primalia contains all organisms known to be non-vascular and lacking organs of respiration and circulation: algæ, lichens, fungi, sponges, and conjugates, recognized as subkingdoms.[177] They regard Primalia "as containing the whole of the Kingdom *Protozoa* of Professor Owen",[178] but do not assign Owen's protozoa to specific subkingdoms.[179]

Their work represents an important, if imperfect, step forward in classification. Important, because they enunciate the principle that all taxa, even the most primary, should be based on unique sets of superadded characters.[180] Imperfect, in that distinctiveness need not be all or nothing, but can be (indeed, is) a matter of prominence, prevalence, permanence through the life cycle, or order of appearance.[181] Nor is Primalia a natural group, in any sense of the word. But unlike Primigenum and most third kingdoms before it, Primalia was neither an undifferentiated base nor a borderland, but a coequal third kingdom. It is an open question whether Wilson and Cassin viewed organisms as evolving over time, but Asa Gray and Charles Darwin were members of the Academy of Natural Sciences,[182] and Wilson and Cassin were surely familiar with the concept through Leidy, not to mention Owen's encyclopædia article and book.

Popular natural histories in Victorian Britain

The Victorian age[183] brought an immense number of popular works on natural history: earnest books of self-improvement for workingmen or the middle class;

376   KINGDOMS, EMPIRES, AND DOMAINS

pocket guides to the identification of ferns, butterflies, or sea-shells; weighty volumes of microscopic technique; and for the drawing-room, gloriously illustrated series of birds or mammals, set against exotic landscapes. Anyone could be an author: academics, journalists, landed gentry, stonecutters, vicars, women. The works themselves ranged from semi-technical to devotional, but typically shared certain features: a romantic idealization of nature, presentation of nature as the handiwork of an all-powerful Creator, and broad appeal across social classes.[184] They complemented and reinforced currents in British society including the continued expansion of literacy and readership; the popularity of museums, zoos, and public gardens; the rise of domestic tourism and seaside holidays; and leisure activities such as keeping a home aquarium or breeding pigeons.

Towards the technical end of the spectrum we find works such as Charles Darwin's *Formation of vegetable mould through the action of worms* (1881), and fat volumes on microscopy by William Benjamin Carpenter[185] and Jabez Hogg.[186] Books on microscopy or infusoria usually mention that opinions may differ on diatoms and desmids,[187] but rarely dispute the canonical kingdoms. An interesting exception is *Minerals and metals* (1835), authored anonymously for the Society for Promoting Christian Knowledge, which asserts that five kingdoms are required:

> First, The animal; the second contains those bodies which partake of the nature of the first, and of the following, such as polypes, sponges, madrepores, &c.; third, The vegetable; fourth, The mineral; and the fifth embraces the four imponderable agents, heat, light, electricity, and magnetism.[188]

Scottish stonemason Hugh Miller, an evangelical Christian and self-educated geologist whose highly entertaining popular works included serious geology, wrote that

> I reckon among my readers a class of non-geologists, who think my geological chapters would be less dull if I left out the geology.[189]

Indeed, popular natural histories of the era were usually written for a broad non-technical audience. To assess how the kingdoms of nature were presented in mainstream Victorian natural histories, I examined forty-four popular works published between 1844 and 1884 that, given their subject areas, might treat kingdoms and/or the animal-vegetable interface: books on aquatic or marine life, the seashore, algae, zoophytes, microscopic beings, and geological strata.[190] Ten of the thirty-four authors were, or would become, fellows of the Royal Society; seven held a religious title; six were women. Of the forty-four works, fully forty-two accept the canonical kingdoms of nature.[191] One also offers a non-standard kingdom, variously identified as the *Human* or *rational* kingdom.[192] We may grant three others poetic license for the *kingdom of nature*, the *kingdom of heaven*, the *fire kingdoms*, and the *star kingdoms*;[193] the *marine animal kingdom*;[194] and the *Sponge-kingdom*.[195] None takes up Protozoa or any of the other mid-Nineteenth century third kingdoms, although fifteen of the works allow that the animal-vegetable boundary might be uncertain (more often because of sponges than diatoms or desmids), and two explicitly mention the ideas of Kützing or Unger.[196]

From such works the animal, vegetable, and mineral kingdoms marched forth, with only the slightest nuances, into the broader English-language popular culture: fiction, poetry, a game for children, and (from 1952–1959) a television quiz show on the BBC. And (try as we may), who can forget

> I'm very good at integral and differential calculus,
> I know the scientific names of beings animalculous;
> In short, in matters vegetable, animal, and mineral,
> I am the very model of a modern Major-General.[197]

The game was not over, but the three canonical kingdoms held most of the cards.

Summary: Britain

In this chapter we have followed several currents in Eighteenth- and Nineteenth-century British natural history—Linnæan classification, natural theology, the influence of Owen and two Darwins. Although each current had its own origins and dynamic, they were linked by a dense web of interconnections, some apparent, others behind the scenes. Natural theology continued in its softer Gilbert White version well into the Victorian era, one rivulet flowing into the mainstream sense of wonder before nature and respect for its Creator. Among the Bridgewater authors, Buckland had already broken from *Genesis* literalism, and later reinterpreted the Kirkdale hyena bones as evidence of glaciation not Deluge;[198] and Darwin quoted Whewell, on general laws in nature, at the front of *Origin of species*. To be sure, Owen and Darwin had little time for each other; but scientifically they were allies against Bishop Wilberforce, as Wilberforce and Huxley had been against the *Vestiges*.

The victory in Britain of the canonical kingdoms might be written off as ignorance of, or aversion to, ideas from France or Germany; but the third kingdoms we met in this chapter were home-grown.[199] If the British were indeed a practical folk, why did they not appreciate that a third kingdom of organisms would (as explicitly argued by Hogg) make for a cleaner delineation of Animals and Vegetables? If distinctive superadditions were needed to ensure that the third kingdom was not simply a dumping-ground, Wilson and Cassin offered a way forward. We may suspect that third kingdoms made little headway in Britain because the status quo drew support from so many directions: Ray, Linnæus, Paley, White, Smith, the Royal Society, microscopists, Bridgewater authors, countless popular natural histories, even Owen (before 1852), Darwin, and Huxley.

378 KINGDOMS, EMPIRES, AND DOMAINS

By 1864, six main intermediate kingdoms were on offer: Psychodiare (Bory), Acrita (Macleay), Protozoa (Owen), Primigenum (Hogg), Primalia (Wilson and Cassin), and *Protorganismen* (Carus).[200] Each offered different philosophical or theoretical foundations, and a somewhat different membership; none was simply a borderland between animals and plants, as Zoophyta had been. In Chapter 21 we reach the most influential third kingdom of all.

21

Ernst Haeckel and Protista

The last Olympian figure of the science of evolution in the nineteenth century is that of the venerable, resolute and optimistic Ernst Haeckel. Like a Titan who has laid aside his arms and implements and composed himself to rest after a laborious and stormy life, he sits throned upon a mountain of masterly accomplishment and looks with a smile upon the new days in which the harvests of his thought are growing. This silver-haired, blue-eyed Luther of Science, gifted with the spirit of perennial youth, is the last of a small but mighty band. Darwin, Huxley and Spencer were the other great and contemporaneous lights of this system of suns. But as the most modern and the most active of them all, he may be said to embody in himself and his labors the latest expression of the theory of evolution.

—Herman Scheffauer, in *North American Review*[1]

Haeckel is a fool. That will be apparent one day.

—Rudolf Virchow, as quoted by Carl Ludwig Schleich[2]

Born in Potsdam in 1834, Ernst Haeckel grew up in Merseburg (Saxony). His father was a privy councillor to the Prussian court, and young Ernst immersed himself in the natural science and travels of Goethe, Humboldt, and Darwin.[3] On his father's advice he enrolled in medicine at Würzburg (1852), where he studied general anatomy under Albert von Kölliker and Rudolf Virchow, and took courses from Carl Gegenbaur, Franz Leydig, and August Schenk.[4] During 1854–1855 he studied in Berlin under Johannes Müller and Alexander Braun before returning to Würzburg. Following the example of an earlier generation of scholars, in summer 1856 Kölliker and Haeckel decamped to Nice to study marine organisms; there they met Müller, who was researching radiolaria. Back in Würzburg, Haeckel submitted his medical dissertation on the tissues of crayfish (1857). He then made his own Italienische Reise during 1859 to study art, architecture, and marine invertebrates. Towards the end of the year, he decided to focus his research on radiolaria.[5] Haeckel returned to Germany in early 1860, and after an interlude in and around Berlin he accepted Gegenbaur's invitation to Jena.

Kingdoms, Empires, and Domains. Mark A. Ragan, Oxford University Press. © Oxford University Press 2023.
DOI: 10.1093/oso/9780197643037.003.0021

380 KINGDOMS, EMPIRES, AND DOMAINS

Haeckel's professors, lecturers, and mentors were among the most renowned and progressive in the German lands at the time. It is informative to survey their views on plants, animals, and a possible intermediate taxon, as they surely contributed to Haeckel's default positions.

Johannes Müller was the leading physiologist of his day. In *Handbuch der Physiologie des Menschen* (1834–1840) he distinguished irritability from sensibility, plants from animals, and held sensation and voluntary motion to be "the more remarkable [of] animal properties". It was "still a matter of doubt whether certain simple organised beings, such as the sponges and several so called alcyonia, are animal or vegetable" (although he went on to treat alcyonia as animals), and he mentioned Bory's intermediate beings in a noncommittal manner. Müller accepted Ehrenberg's view of infusoria as "perfect animals", but remained agnostic about diatoms.[6] Although not entirely free of Romanticism including a *Lebenskraft*, his *Handbuch* abandoned the nature-philosophy of Oken and Schelling, while his scientific expeditions to the seashore, use of plankton nets, and studies on echinoderms, holothuria, and radiolaria set an example for successive generations of German zoologists. His untimely death in 1858 deeply affected his students and colleagues, including Haeckel.

According to Kölliker's influential *Handbuch der Gewebelehre des Menschen* (1852), plants and animals arise from cells, and their bodies are composed of cells; the simplest plants and animals are probably unicellular, even if this had not yet been demonstrated for some protozoa, notably the rhizopods.[7] By contrast, Virchow cautioned against the idea that plant and animal cells are fundamentally the same,[8] but held that *omnis cellula e cellula*, and denied that infusoria, fungi, algae, or intestinal worms arise spontaneously.[9] Leydig, who worked with Kölliker, reminded his readers of the earlier cell theories of Bichat and Oken, and likewise described organisms as cellular.[10] None of these zoologists was greatly interested in the precise line between plants and animals.

In *Verjüngung in der Natur* (1850), Braun set out a comprehensive theory of the cellular basis of plant life.[11] He recognized several grades of unicellular plants—those for which "the whole cycle of life is completely shut up in one cell" versus those with a life cycle that involves independent reproductive cells—and concluded that it is impossible to draw a sharp line between unicellular and multicellular plants. Instead, he positioned various algae, including diatoms and desmids, along this gradient.[12] In describing spore formation in the green alga *Chlamydomonas*, he stated that if one were "ignorant of the rest of its history, one would be led, by the form and mode of division of the cells, to regard these crusts as belonging to a *Pleurococcus*",[13] a remark that can be taken against the transformationism of Kützing. It is not difficult to detect a streak of Romanticism in Braun—for example his interest in individuality and symmetry—from which the historian Eric Nordenskiöld draws a direct line to Haeckel.[14]

As microscopes improved, morphologically simple parasites were discovered in tissues of plants and animals, not least marine and freshwater invertebrates: Kölliker, for example, described a range of gregarines and argued that they are unicellular animals.[15] In *Verjüngung*, Braun returned to his earlier discovery of *Chytridium olla*, a microscopic being that reproduces via motile zoospores and is parasitic on freshwater algae; he remained unsure whether it should be considered an alga or a fungus.[16] Siebold, Pringsheim, and others had reported parasites much like it on a range of plants and animals, and in 1855 Braun concluded (in part by analogy with the water moulds *Achlya* and *Saprolegnia*) that *Chytridium* was probably a plant.[17] In the late 1850s Schenk described another chytrid, *Rhizidium intestinum*, classifying it among the plants on the basis of similarity of reproductive structures, and concluding that "an absolute border cannot be drawn at the moment between the plant and animal kingdoms".[18]

We do not know what precise combination of these influences led Haeckel to marine invertebrates: the Romantic idea that the simplest forms offer the most-fundamental insights; the gothic symmetry of radiolaria; the inspiring but tragic example of Müller; or simply that marine biology was not medicine, which he had increasingly come to dislike. Among the reasons, however, was surely the example of his slightly older friend and colleague Carl Gegenbaur. Like Haeckel, Gegenbaur had abandoned medical studies to pursue animal morphology; travelled to the coast with Müller and Kölliker; and studied siphonophores and medusæ in Messina.[19] In 1858 Gegenbaur acceded to the chair of anatomy and physiology at Jena on the condition that the two fields be separated, leaving him responsible only for anatomy, which he proceeded to make the sole basis of his zoology. In *Grundzüge der vergleichenden Anatomie* (1859), he argued that

> Just as the virtual meaning of the egg as a certain animal is not taken from the properties of the egg, nor from its form and composition, nor yet from the life phenomena that otherwise allow it to be perceived as an egg, but rather from the developmental life that is hidden in it, which gradually allows a certain animal form to be revealed: so too the meaning of those beings as animals or otherwise as plants does not have to be derived from their specific condition, but must arise from the phenomena that appear in their overall development, and from the forms that already bear a distinct plant- or animal-nature that stand in close connection to those dubious organisms.[20]

From amœbæ to higher animals, we find "an uninterrupted developmental series of forms" along which somatic cells surrender their independence to tissues and organisms. By contrast, in the plant kingdom, somatic cells largely retain their individuality.[21] Gegenbaur took a further step in his entrance address to the medical faculty at Jena (1860), arguing that as we work our way downward from higher animals, we eventually reach beings whose cells exhibit a bare minimum of interdependence, or are internally differentiated: infusoria, gregarines, and certain rhizopods.[22] This is our animal series, beyond which all remaining forms (that is, the truly unicellular) can be considered plants.[23] Pointing to exceptions on both

382 KINGDOMS, EMPIRES, AND DOMAINS

sides, Carl Claus criticized Gegenbaur's line as "one-sided and artificial".[24] Claus systematically dismissed a long list of potential morphological, physiological, and chemical distinctions, concluding that "there is no fixed and absolute boundary between animals and plants". The simplest creatures are not animals or plants in the usual sense, but merely bear a greater analogy to one or the other.[25]

Die Radiolarien (1862)

In his doctoral dissertation on crayfish (1857) Haeckel drew the standard analogies between plant and animal tissues.[26] Radiolaria, however, forced him to confront the animal-plant boundary among unicellular organisms. In his gloriously illustrated *Die Radiolarien* (1862) dedicated to the memory of Müller,[27] Haeckel posed the question *Thiere oder Pflanzen?* and, like others before him (above), dismissed potential distinctions one by one: the "animal functions" of sensitivity and voluntary movement, the "vegetative phenomena" of nutrition[28] and reproduction. With physiological criteria inadequate, he concluded, we must follow Gegenbaur in relying on morphological characters and fine structure:[29]

> Gegenbaur conceives of the entire kingdom of organized natural bodies as a continuously connected series [*eine continuirlich zusammenhängende Reihe*], within which a stepwise differentiation and development of organization takes place from a common midpoint outward in two opposite directions. The essential and characteristic properties that distinguish the animals and plants at the two endpoints of this connected chain become obliterated as each approaches ever-nearer the midpoint, such that the deepest organisms in each kingdom do not, in general, appear susceptible of a deep and absolute distinction, as has so far always futilely been sought for them. If one nonetheless finds it necessary for there to be a boundary-post between the two, one must seek it in their elementary histological structure.[30]

Haeckel then restated Gegenbaur's *trefflich* description of plant cells remaining autonomous in their stiff cellulosic membrane, whereas animal cells fuse into differentiated tissues; and immediately set about applying it to rhizopods. Some amœbæ, to be sure, are "neglected developmental states of lower plants and animals", but the bodies of true rhizopods without contractile vacuoles (including the radiolaria) are without exception either wholly or in part a complex of fused cells, and as such certainly belong in the animal kingdom. It is of greatest significance, he wrote, that the central sarcode is fused (fully in acyttaria, partially in radiolaria): indeed, this is why these rhizopods must be considered animals.[31] By the same consideration, many of Ehrenberg's polygastrica, as well as diatoms, desmids, volvocinea, and various monads, are plants. Given the current state of knowledge, Haeckel deferred any decision regarding sponges, gregarines, and myxomycetes.[32]

ERNST HAECKEL AND PROTISTA   383

In *Radiolarien* we also find Haeckel's initial reaction to Darwin's theory of descent.[33] Haeckel had indeed found variability within genera, and numerous transitional forms that collectively trace a "fairly uninterrupted chain of related links". He was disappointed that Darwin had not dealt with the origin of the primordial organism, but admired this "first serious [*ernstlich*] scientific attempt to unify all phenomena of organic nature", in particular to "replace incomprehensible miracle with comprehensible natural law". Haeckel surely identified with Darwin's call for "young and aspiring naturalists" to step forward to test his theory, and obliged by offering a polyphyletic *Verwandtschaftstabelle* (relationship-table) of families, subfamilies, and genera of radiolaria. On this basis Haeckel inferred that the "primal radiolarium" was a spherical lattice-ball with a central capsule and centrifugal spines, much like *Heliosphaera actinota*.[34]

In his opening plenary at the 1863 congress of the Gesellschaft Deutscher Naturforscher und Ärtze in Stettin, Haeckel praised Darwin's epochal theory of descent—he predicted it would prove as significant for organic nature as had Newton's law of gravity for the inorganic—but again noted that Darwin had not provided any clue on the origin of the first cell or cells:

> Was it a simple cell, such as those many that still exist as independent beings on the dubious border of the animal and plant kingdoms, or one that represents the eggs of all organisms at all times? Or was it in an even earlier time just a simple living little clump of slime capable of nutrition, reproduction and development, a moneran [*Moner*], like certain amœboid organisms that appear not even to have reached the organizational height of a cell?[35]

In early 1864 Haeckel's beloved wife Anna died suddenly.[36] After a few months, he began pouring his ideas into a sprawling 1228-page work, *Generelle Morphologie*.[37]

Generelle Morphologie (1866)

> I have ventured, in 1866, in my *General Morphology*, to make a first attempt to arrive at the natural system of organisms with the help of the biogenetic principles of the theory of descent, and . . . to establish phylogenesis as the basis of the natural system.[38]

Darwin may have been overly cautious in *Origin of species*, but Haeckel would not make the same error: his *Generelle Morphologie* left few stones unturned, from the atomic basis of matter to the unity of God and Nature. The work defies brief description, but let us try to capture its main arguments in regard of a third kingdom of organisms.

We begin with its title: *General morphology of organisms. General elements of the science of organic forms, mechanically grounded on descent-theory as reformed*

384 KINGDOMS, EMPIRES, AND DOMAINS

by Charles Darwin. For Haeckel, "morphology or form-theory of organisms is the entire science of the internal and external form-relationships of living natural bodies, animals and plants, in the broadest sense of the word." The aim of organic morphology is the knowledge and explanation of these form-relationships, that is, tracing their appearance back to certain natural laws.[39] General (as opposed to special) morphology focuses on the "uppermost and most-general laws of organic form [*Formbildung*] which apply to the whole of organic nature, both anatomical and genetic laws".[40]

Descent-theory merits specific mention in the subtitle, and Haeckel dedicated the second volume of *Generelle Morphologie* to its founders: Goethe, who infused morphology with ideas of archetype and transformation; Lamarck, who put descent theory on a scientific basis; and Darwin, whose *Origin of species* made it part of the "living commons of biological science".[41] Goethe was a constant touchstone: quotations from his *Metamorphose der Pflanze*, *Osteologie*, and other works introduce every book and chapter of *Generelle Morphologie*. For Goethe, antagonism between the endless diversity and undeniable similarity of organic forms pointed to a sacred riddle or secret law; Haeckel found its "happy solution-word" in *Stammverwandtschaft*, genealogical stem-relationship.[42]

Haeckel divided morphology into *anatomy* and *morphogeny*. The former, the science of accomplished [*vollendeten*] form, has as its two branches *tectology*, the study of structures, and *promorphology*, the study of *Grundform*. Morphogeny or developmental history is the science of developing [*werdenden*] form; its two branches are *ontogeny* (or embryology) and *phylogeny* (or palæontology). Ontogeny is the developmental history of organic individuals (*bionts*), phylogeny the developmental history of organic *Stämme*.[43] These distinctions unfold in the organization of *Generelle Morphologie*: volume 1 is *General anatomy of organisms*, volume 2 *General developmental history* [*Entwickelungsgeschichte*] *of organisms*.[44] Anatomy, morphogeny, ontogeny, and phylogeny are presented in Books 3–6 respectively.

Haeckel began volume 1 by setting the scientific and philosophical groundwork, then argued that living beings must have arisen from inorganic nature—a process he called self-generation or *autogony*,[45] presenting it as an extension to the Kant-Laplace theory of earth formation.[46] There is not so great a difference, he wrote, between the most-perfect inorganic individuals (crystals) and the least-perfect organic individuals (monera), and we can think of autogony as a sort of organic crystallization in which an initial atom-group (perhaps a protein) attracts others, grows, and becomes a homogeneous individual. In the balance between growth and dissolution we find the beginnings of nutrition and reproduction. Many generations of amœboid beings populated the *Urmeer*; the nucleus appeared later, and multicellular organisms arose by incomplete separation of units after division.[47] Autogony presumably happened at different points over the earth's surface, and a struggle for existence ensued;[48] indeed, beings may still be developing by autogony.[49]

Autogony yielded monerans, mere lumps of homogeneous plasma at the boundary between the animate and inanimate.[50] The simplest organisms today are monera, and all types of organisms have descended from one or a few monera. He called each main group a *Stamm* or *phylon*; each is a genealogical unit, with a specific moneran as its stem-form. Haeckel supposed that the vertebrates, cœlenterates, diatoms, and so forth had their own stem-forms, but he could not exclude that they all may have arisen from a common form,[51] or are polyphyletic.[52] Given that he is refounding morphology on Darwinian descent theory, the natural system of organisms is their *Stammbaum*.[53] We can come to know this *Stammbaum* by comparing the palæontological (phyletic), embryological (of individual bionts), and systematic development of organisms.[54] At the end of the second volume, Haeckel provided a *Stammbaum* for all organisms, and separate *Stammbäume* for plants, the five main *Stämme* of animals, and mammals including man.

Individuality had long been a theme of German Romanticism, and Haeckel does not disappoint, recognizing morphological, physiological, and genealogical individuals.[55] Morphological individuals—the sort discussed most in *Generelle Morphologie*—can exhibit six orders of *Ausbildung*: the (anucleate) cytode and (nucleate) cell; organ; antimer; metamer; person; and cormus (colony).[56] Monera are "nothing but individual isolated living cytodes", "organisms without organs".[57] With this formulation, Haeckel sought to replace cell theory with plastid theory.[58]

Generelle Morphologie is not primarily a work of systematics, the details of which are necessarily "special" rather than "general"; but as the genealogical foundation of the natural system is valuable in understanding the history of development, Haeckel allowed that it would "not be superfluous" to attempt a brief overview of the natural system.[59] From a very early time, even *die einfache Naturmensch* distinguished plants from animals,[60] but Aristotle encountered "insurmountable difficulties" in drawing a clear distinction, and the thoughtless and dogmatic disciples of Linnæus fell into ever-deeper confusion.[61] Perhaps plants and animals arose from separate monera, or from a common moneran; more likely, he asserts, present-day organisms arose from a smallish number of monera, although more than two.[62] This assumption opens the door to a better resolution:

> All disputes over this difficult point will be resolved once we assume that the numerous organisms, which cannot be classified into either the animal or the plant kingdom without apparent compulsion, belong to several independent *Stämme* of living beings, which have developed independent of the *Stämme* of the animal and plant kingdoms. In the known facts, we find absolutely no need to assume that all organism-*Stämme* must be either animals or plants. All the more, we must consider the previously accepted exclusive binary division into animal and plant kingdoms not to be well-founded in this regard. It has already been pointed out from various sides that it would be a great advantage for both zoology and botany if the many dubious creatures, which are neither true animals nor true plants, were to be united in a special middle kingdom or kingdom of original beings; as far as we know, no one has yet attempted to determine the content and scope of such a new kingdom, and to ground and justify its delimitation scientifically. Here we

386 KINGDOMS, EMPIRES, AND DOMAINS

venture to attempt this on the basis of the above deductions, and propose to bring together all those independent organism-*Stämme* which with complete certainty and without contradiction can be attributed to neither the animal nor the plant kingdom, under the collective name *Protisten*, firstlings [*Erstlinge*] or primæval beings [*Urwesen*].[63]

With the animal and plant kingdoms swept clean of problematic beings, Haeckel set out his system: five phyla of animals (vertebrates, molluscs, articulates, echinoderms, cœlenterates), eight of protists (sponges, noctilucids, rhizopods, protoplasta, monera, flagellates, diatoms, myxomycetes), and four of plants (phycophytes, characeæ, nematophytes, cormophytes).[64] It is quite possible, he held, that each of these phyla originated from a different moneran.[65] He made Infusoria a subphylum of Articulata (alongside arthropods and worms), thereby (like Gegenbaur) dividing the protozoa between animals and protists. Likewise, he distributed algae among phycophytes, characeæ, and flagellates.[66]

Haeckel acknowledged that the contemporary state of knowledge did not allow him to say a great deal about the protist kingdom, but—quoting Goethe to the effect that "a bad hypothesis is better than none"—he identified eight *Stämme* of protists (above). These *Stämme* share no common *Bauplan* or features apart from their simplicity of form—some are amœboid, others spherical or cylindrical—and the absence or imperfection of organs. Little was known of their chemistry or metabolism: some are oxidative like plants, others reductive like animals, while many others are intermediate. Some protists, he accepted, may yet prove to be simple plants or simple animals, and more protists surely remained to be discovered. If the protistan kingdom were disallowed, he would assign the diatoms and myxomycetes to the plants; the rhizopods, noctilucæ, and sponges to the animals. Flagellates, protoplasts, and monera remain doubtful in any case.[67] None of this argued against collecting them in a common kingdom, as Haeckel did not require kingdoms to be monophyletic.[68]

Haeckel was hardly the first to apportion microscopic organisms between different kingdoms: as we have seen, Gegenbaur had done so on the basis of supposed multicellularity;[69] and before him Bory, Flotow, Kölliker, Siebold, and others, on various grounds. But as Haeckel was now arguing from descent theory, his precise placement of the border, and reasons for doing so, bear close examination. The primary question here must concern why he assigned the infusoria to the articulates. As for Articulata itself,

We retain the *Stamm* of the articulates, [after] accounting for minor modifications, in almost the same scope in which von Baer and Cuvier established it. It encompasses the two powerful subphyla of worms (Vermes) and *Gliederfüsser* (Arthropoda), which are now almost universally quoted as two separate types or subkingdoms, and as such would correspond to two independent phyla. In our view, however, the arthropods, which differ only in the greater differentiation (heteronomy) of the metamers (trunk segments) and the arrangement [*Gliederung*] of the extremities located on them, represent only a more highly developed branch of the worm strain. The overall organisation of both subphyla otherwise agrees so completely that we cannot separate them.

As a third subphylum of the articulate strain, we join to the unitary worms and arthropods the class of true infusoria, which alone of all segments of the dissolved protozoan-circle can be held to be true animals. We regard the infusoria as the surviving remnants of the old common stem-form of the articulates, and indeed it seems that from these the flatworms or turbellaria were the next to develop, [and] from whose differentiation the other worms then emerged. (...)

The kin-relationships of the articulate-stem, and especially the subphylum of the worms, are the most complex of all animal *Stämme*, apart from the possible genealogical connection with the other animal *Stämme*. Comparative anatomy gives us the most information. In contrast, we know the ontogeny of most articulates only very incompletely; and palæontology offers numerous remains only of the hard-shelled and water-dwelling crustaceans, [while] of the remaining, mostly land-based arthropods only relatively very few and insignificant, [and] of the enormous mass of worms and infusoria, due to their soft and destructible body structure almost no remains of any note, or recognizable. For these reasons, here we can sketch only the most-general and volatile outlines of the *Stammbaum* of the articulates, and must leave the best part of this interesting and difficult task to the better-informed future. Forms immediately related to the autogonous monera, which must have provided the origin of the infusoria, are perhaps still to be found today among the monera of the protist kingdom, as perhaps are their further developmental stages among the protoplasts and flagellates, which already seem to lead over to the infusoria.[70]

Haeckel then describes Infusoria, the first subphylum of Articulata:

The sub-stem of infusoria, as we consider them here as the starting group of the articulate-stem, embraces only the two classes of ciliates (cilia-infusoria) and suctoria (acinetes). These two classes, which are so far apart in their metaphase, appear however from their ontogenesis to have the closest connection. According to more-recent investigations, both the ciliates and the acinetes seem to reproduce through acinete-like larvæ or so-called "swarm-sprouts" which cannot be distinguished between the two classes. Like the fixed acinetes they bear suction-tubes,[71] and swim about by means of a coat of cilia as do the ciliates. These common juvenile forms seem to us to prove an indubitable phylogenetic (but not ontogenetic! but however) connection. We consider infusoria that persist through their life in the form of ciliated acinete-like larvæ to be the ancient progenitors of all infusoria (and thereby of all articulates), and assume that both the acinete and ciliate classes developed from them as two divergent branches: the acinetes lost the cilia of the larvæ by adaptation to a fixed way of life, and retained the suction

388 KINGDOMS, EMPIRES, AND DOMAINS

tubes; conversely, the ciliates retained the cilia and lost the suction tubes. As for the deeper origin of the group, the ancient ancestors that built the bridge between the autogonous moneran form of the articulate-stem (and thereby perhaps of the entire animal kingdom) and the ciliated acinete-larvæ are perhaps still to be found today among protists of the flagellate- or the protoplast-stem.[72]

In this way Haeckel connected ciliates and suctoria through their common juvenile form; under the circumstances, we might consider this ontogeny. As for the place of infusoria among the annelids (hence animals), he says in regard of turbellaria (flatworms):

> Of all worms, these undoubtedly stand closest to the infusoria, and are themselves connected to the ciliates through some transition forms so dubious, that their descent therefrom cannot be denied.[73]

The lack of certitude is hardly reassuring, but Haeckel sought support for this notion from Rotatoria (rotifers), which he recognized as the third group of worms:

> In our opinion, the class of the wheeled animals is a very old remnant of the branch of the articulate-stem from which the crustaceans, and hence further the arthropods, have developed. On the one hand, the rotifers, through their deepest forms, are so intimate with the turbellaria (Rhabdocœlæ) and even with the infusoria, [while] on the other hand, through their highest forms, they are bound so close to the crustaceans (Entomostraca) and therefore with the arthropod subphylum, that we must consider them an intermediate form between the scolecids and the arthropods, that is, as ancient and very-little-changed direct descendants of those worms from which the arthropods have developed.[74]

Haeckel did not mobilize other opinion in support of these connections, but he surely knew that Oscar Schmidt had argued that the fine structure of ciliates indicates they are not unicellular organisms, but are instead connected with turbellaria.[75] Many marine and freshwater invertebrates pass through a free-swimming ciliated larval stage; Johannes Müller had described ciliated larvæ of turbellaria, and pointed out their general similarity to infusoria.[76] Haeckel sought to trace the animal lineage farther back, supposing that ciliates share an ancestor with flagellates, which in turn shares an ancestor with (nucleate) gymnamœbæ, which finally share an ancestor with (homogeneous, anucleate, autogonous) monera.[77]

New classes of Protista

After *Generelle Morphologie*, Haeckel resumed his studies on lower organisms. Whether his primary interest lay in primitive life-forms, invertebrate embryology, or phylogenetic systematics is beside the point: all were interrelated, and could help him uncover the laws and system of nature. Nor is it particularly useful to reconstruct his studies in temporal order, as he pursued these interests in parallel, and on occasion presented results before publishing the supporting evidence.[78] As his understanding evolved and new organisms were discovered, he adjusted the membership of Protista. Some of these refinements proved successful, whereas others led him into a maze of conflict and inconsistency. We begin with the discovery of new life-forms.

Between 1863 and 1867 Haeckel described five new species of monera, the simplest organisms alive today: the "absolutely structureless and homogeneous" *Protamœba primitiva*, and the slightly structured *Myxastrum radians*, *Myxodinium sociale*, *Protomyxa aurantiaca*,[79] and *Protogenes primordialis*.[80] Lev Cienkowski, professor of botany at Imperial Novorossiysk University in Odessa, added *Monas amyli* and three species of *Vampyrella*.[81] As Haeckel believed that autogony had occurred at multiple locations around the globe (and was probably still occurring today), it was reassuring to find a diversity of monera.

Of greater interest was a substance recovered from the floor of the North Atlantic in 1857 by the cable-laying ship HMS *Cyclops*. Samples preserved in alcohol made their way to TH Huxley, who reported that they contained immense numbers of foraminifera, plus a smaller number of rounded, acid-soluble structures which he named coccoliths.[82] George Wallich interpreted the coccoliths as having fallen off spherical bodies (coccospheres) which he supposed to be larvæ of foraminifera.[83] A decade later Huxley—now in possession of an advanced Ross microscope—returned to the samples, describing three components: two types of coccoliths, the corresponding coccospheres, and lumps of "transparent gelatinous matter" that he described as a

> deep-sea "Urschleim," which must, I think, be regarded as a new form of those simple animated beings which have recently been so well described by Haeckel in his "Monographie der Moneren." I proposed to confer upon this new "Moner" the generic name of *Bathybius*, and to call it after the eminent Professor of Zoology in the University of Jena, *B. Haeckelii*.[84]

Haeckel was amazed that Oken's *Urschleim*—the primordial source of all organisms—might actually have been discovered, and his initial caution soon gave way to missionary zeal, as he proclaimed *Bathybius* a product of autogony.[85] Samples without coccoliths, dredged up near Greenland in 1872, were duly proclaimed an even more-primitive moneran, *Protobathybius*. But in 1875, to the great surprise of all concerned, experiments conducted aboard HMS *Challenger* in the western Pacific Ocean indicated that this *Urschleim* was no organism, nor even organic, but instead simple clumps of calcium sulphate gel that had precipitated when preservative (alcohol) was added to seawater onboard the ship. Huxley apologized gracefully, while Haeckel fought on, arguing that *Bathybius* might live only in the North Atlantic.[86]

390 KINGDOMS, EMPIRES, AND DOMAINS

Other newly discovered organisms were real, and required new classes within Protista. For *Labyrinthula*, a curious net-like microorganism discovered by Cienkowski on seaweeds in Odessa harbour,[87] Haeckel set up class Labyrinthulea.[88] For *Magosphæra planula*, a spherical colony of ciliated cells with unicellular free-swimming and amoeboid juvenile stages, he created class Catallacta;[89] the organism (if such it was) has not been observed since.[90] Much later he added another idiosyncratic type of protist, the calcocytes—today known as coccolithophorids—with nary a mention of *Bathybius*.[91]

Sponges and gastraea theory

The first organisms to lose their status as protists were the sponges. As we have seen, sponges had been recognized as problematic since Aristotle. At first Linnæus considered them plants, but in 1767, following Ellis and Pallas, he moved them to the animals. Robert Grant's discovery that sponge larvæ bear cilia convinced many that sponges are in fact animals.[92] Even so, it remained unclear precisely where they fit into the animal kingdom. The amoeboid nature of isolated sponge cells,[93] the presence of cells with flagella lining the internal canals, and (in some sponges) the occurrence of silicaceous spicules[94] all suggested a connection with protozoa. In 1868 Haeckel transferred the sponges from Protista to Cœlenterata, *i.e.* to the animals.[95] His detailed argument appeared in 1870, and relied on a new argument: that the sponge body, like that of cœlenterates (but unlike that of protozoa), consists of two hulls (*Bildungshäuten*) or germ-layers (*Keimblättern*), *entoderm* and *ectoderm*, that develop from the invaginated egg.[96] Within Cœlenterata, the sponges might be grouped with the corals, or perhaps be recognized as a sister-group to the other cœlenterates.[97]

According to Haeckel, the sponges provide a splendid demonstration of Darwin's theory of the origin of species.[98] Their morphology varies even within a species, and he could easily discern evolutionary trajectories forward from the presumed ancestral type.[99] In the calcareous sponge *Olynthus*[100]—a simple sack terminating in a single pore—Haeckel saw the ancestral sponge-type, the infolded larva, which he named the *gastrula* or little stomach.[101] Against the background of Oken's teaching that the digestive system is the first form of animality,[102] this was not a value-neutral term. Another calcareous sponge, *Guancha blanca*, seemed to present multiple transformations from this ancestral sponge-form.[103] The natural history of sponges—their family tree—was an elaboration of forms derived from this ancestor by the addition of more pores, ramification of the internal canal-system, and the appearance of spicules.[104] Nor was it difficult to draw analogies (probable homologies) with the structure of corals. He soon formalized much of this with his gastraea theory, according to which

From this identity of the gastrula among representatives of the most different animal-*Stämme*, from the sponges to the vertebrates, I conclude, according to the biogenetic law, that the animal phyla have descended in common from one unique unknown *Stamm*-form, which in essence was of the same structure as the gastrula: Gastraea.[105]

Haeckel developed gastraea theory further in 1874, explicitly dividing the animal kingdom into protozoa and metazoa:

> The real purport of this Gastraea-theory depends on the conception of a true homology of the primordial rudiment of the intestine [*Darmanlage*], and of the two primary germ-lamellæ in all animals except the Protozoa, and may be briefly summed up in the following words:—The entire animal kingdom divides into two chief divisions: the older, lower group of the Protozoa (Urthiere), and the younger, higher group of the Metazoa (Darmthiere). The main group of the Protozoa or Urthiere (animal Monera and Amœba, Gregarina, Acineta, Infusoria) always increases only by the development of the animal individuality of the first or second order (Plastide or Idorgan); the Protozoa never form germ-lamellæ, never possess a true intestinal canal, and, especially, never develop a differentiated tissue; they are probably of polyphyletic origin, and branch off from many different primevally generated Monera. The main group of Metazoa, or Darmthiere (the six races of Zoophyta, Vermes, Mollusca, Echinodermata, Arthropoda, and Vertebrata) is, on the contrary, probably of monophyletic origin, and arises from a single common root form, the Gastraea, which has sprung from a Protozoan form; it always multiples by developing the animal individuality of the third or fourth order (Person or Cormus); the Metazoa always form two primary germ-lamellæ, always possess a true intestinal canal (a few retrograded forms only excepted), and always develop differentiated tissues; these tissues always arise from the two primary germ-lamellæ only which have been transferred as an inheritance of the Gastraea of all the Metazoa, from the simplest sponge up to the man.[106]

By accepting protozoa as unicellular animals, Haeckel acceded to Siebold's position from 1848, with about the same membership.[107] The protozoan body might consist of a simple cytode (monera, monothalamia), an aggregate of cytodes (polythalamia), a simple cell (amœbæ, unicellular gregarines, infusoria), an aggregate of identical simple cells (polycellular gregarines, synamœbæ), or even show slight differentiation; but it never forms germ-lamellæ, and encloses no true intestinal cavity.[108] Ciliates and articulates had parted ways forever.

With sponges removed to the animal kingdom, the remaining protists were (so far as was then known) entirely asexual, except for *Volvox* and its relatives within Flagellata.[109] Haeckel pointed out that by transferring fungi and phycochromaceæ[110] from the plant kingdom to the protists, and conversely making *Volvox* a plant, exclusive asexuality would become a unique feature of protists.[111] Fungi are not obvious candidates for membership in Protista, not least because the most-advanced forms (mushrooms) differentiate tissues. Haeckel included fungi and phycochromaceæ in

392 KINGDOMS, EMPIRES, AND DOMAINS

his 1868 list of protists[112] but in 1870, acknowledging that fungi are sexual,[113] did not press the point.[114] He treated fungi as protists again in 1878.[115]

Monera, protozoa, and protophyta

In *Generelle Morphologie* and his subsequent Monographie der Monera (1868), Haeckel put into play four concepts highly relevant to a third kingdom: protists, protozoa, protophytes, and monera.[116] Protists are beings that, at the current state of knowledge, cannot be assigned as animals or plants. They share no common *Bauplan* or features, apart from simplicity of form. Protozoa are animals that do not form germ-layers or a rudimentary intestine, do not differentiate tissues, and have not descended from a gastraea-ancestor. They are "the phylogenetic root-form of the animal kingdom", as protophytes are "the phylogenetic root-form of the plant kingdom".[117] As for monera, they are

> such peculiar organisms that they can be classed with equal propriety, or rather with equal arbitrariness, as primitive animals or as primitive plants. They may just as well be regarded as the first beginnings of animal as of vegetable organization. But as no one mark of distinction inclines them more to one side than to the other[,] it seems most correct at present to class them as intermediate between true animals and true plants; and to assign them with the Rhizopoda, Amœbæ, Diatomaceæ, Flagellata, &c., to that ill-defined kingdom between and animal and vegetable kingdoms which I have called the kingdom of primitive forms, or Protista.
>
> The Monera are indeed Protista. They are neither animals nor plants. They are organisms of the most primitive kind: among which the distinction between animals and plants does not yet exist. But the term "organism" itself seems scarcely applicable to these simplest forms of life; for in the whole conception of the "organism" is especially implied the construction of the whole from dissimilar parts,— from organs or limbs. At least, two separate parts must be united to complete the description of a body as an organism in this original sense ... [but] the Monera are strictly "organisms without organs". Only in a physiological sense can we still call them organisms; as individual portions of organic matter, which fulfil the essential life-functions of all organisms, nourishment, growth, and reproduction.[118]

Haeckel supposed that monera have arisen on multiple occasions; and although many subsequently died out, those alive today are (almost certainly) polyphyletic.[119] Animals, plants, and protists evolved from monera. The different *Stämme* of protists, so dissimilar from one another, have surely arisen from different, unrelated monera.[120] As for animals and plants, Haeckel did not pretend to know whether each great kingdom is monophyletic, or arose from a few moneran ancestors; not unreasonably, he offered *Stammbäume* depicting each case.[121] In any event, the stem-forms of all three kingdoms were monera. How, then, should we classify living monera?

> It will perhaps seem strange that I should here again begin with the remarkable *Monera* as the first class of the Protista kingdom, as I of course look upon them as the most ancient primary forms of all organisms without exception. Still, what are we otherwise to do with the *still living Monera*? . . . it would evidently be just as arbitrary and unreasonable to assign them to the animal as it would be to assign them to the vegetable kingdom. In any case we shall for the present be acting more cautiously and critically if we comprise the still living Monera—whose number and distribution is probably very great—as a special and independent class, contrasting them with the other classes of the kingdom Protista, as well as with the animal kingdom.[122]

Trees can be read in two ways: forward in time, from the base to the stems and leaves; or the reverse. The former emphasizes the branching process (cladogenesis), and invites us to explore diversity; for the latter, however, we must choose a starting point. Haeckel knew that his audience—particularly of his popular *Natürliche Schöpfungsgeschichte* and the complementary *Anthropogenie*[123]—were particularly interested in the evolution of man. For them he described the journey backward—in phylogenetic and embryological development, for these were cause and effect—from man through ever-deeper stages: to the cell-colony, the single cell, the cytode. After *Gastraea-theorie* (1873) he treated metazoa as monophyletic, and thanks to discoveries such as *Magosphæra* he could reconstruct a detailed ancestral series from the first metazoan back to monera.[124] Not unreasonably, he wanted to call these forms animals—specifically, protozoa.

Haeckel attempted this in two ways. First, he began to distinguish "animal", "plant", and "neutral" monera, in a low-key manner in the first edition of *Natürliche Schöpfungsgeschichte* (1868),[125] then more explicitly in the Second (1870) through Seventh (1879) editions.[126] In parallel, he distinguished protozoa (amœbæ, gregarines, acinetes, ciliates) and unidentified protophytes from "neutral protists" (monera, flagellates, catallacts, labyrinthulids, myxomycetes, rhizopods).[127] Protoplasts (amœbæ and gregarines) became protozoa, while (confusingly) he presented monera as neutral. He also considered moving noctilucæ to the animals.[128]

In addition, in 1873 Haeckel designed *Stammbäume* in which monera, amœbæ, and gregarines are labelled Protozoa. One variant identifies Protista ("neutral monera", catallacta, flagellates, noctilucæ, myxomycetes, rhizopods) as Protozoa (Figure 21.1).[129] With the concepts of protista, protozoa, protophyta, and monera now hopelessly entangled, and his earlier dictum that "monera are indeed protista, neither animals nor plants" in tatters, Haeckel admitted that

> Of course, it is a question of minor importance—and personally I do not care at all whether the two main divisions of the animal kingdom are compared as

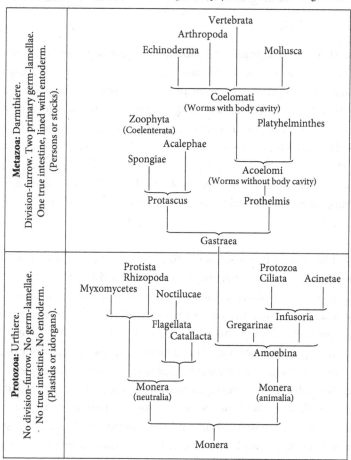

Figure 21.1. Haeckel, phylogenetic table of stem-relationships of the phyla of the Animal Kingdom. Redrawn after *Jenaische Zeitschrift für Medizin und Naturwissenschaften* 7 (1873), page 560. Translation by MAR.

Protozoa and *Metazoa*, as I have done in *Gastraea-theory*, or as *Protista* and *Animalia*, as I proposed ten years ago in *General Morphology*. The deep divide which separates the two main groups, and which is clearly determined by the many important organizational differences emphasised, remains the same in both cases. However, in the meantime (in several editions of *Natürliche Schöpfungsgeschichte* and the *Anthropogenie*) I have attempted to separate protists and protozoa, and to preserve alongside the completely indifferent and neutral protist-kingdom (rhizopods, myxomycetes, flagellates, etc.) a division of protozoa which contained the oldest phylogenetic stages of development of the animal kingdom, from the Monera to the Gastraea. But I now confess that I find this attempt to have failed and, in practice, to be unrealizable. Of course, in theory one can hold protists and protozoa to be phylogenetically distinct; on one side one can regard the genuine protists (or "primordial organisms") as indifferent and completely neutral organisms of the lowest order, related neither to real animals (metazoa) nor to real plants, and which are most probably polyphyletic, of completely independent origin (especially the richly shaped group of Rhizopods, Acyttaria, Radiolaria etc.); [and] on the other side, protozoa (or real "primæval animals") can be regarded as the simplest organisms that form the root of the metazoan *Stammbaum* (—"from monera to gastraea"—) (monera, amœbæ, synamœbæ, planæada). But just as this phylogenetic separation of protozoa and protists is justified in theory, it seems so worthless in practice. Because we lack—and probably will always lack—all the clues to establish with certainty a sharp boundary between those two groups, although both are likely to have different polyphyletic origins and originate from different, autogonously appearing monera. The indifferent and neutral character of those lowest life forms, which are mostly unicellular, gives no hope that that important difference in origin will ever be discovered. Such completely indifferent unicellular organisms, such as the amœbæ and euglenæ, and such characterless cell-aggregates, such as the catallacta and volvocines, can just as well be protozoa as neutral protista, or for that matter protophytes. For these reasons, it will be the most expedient—at least for the time being—to allow the boundary between protozoa and protists to drop, and to contrast the two main groups of the animal kingdom *either* as *Protista* and *Animalia, or* as *Protozoa* and *Metazoa*.[130]

Das Protistenreich (1878)

Haeckel chose the former path for his non-technical booklet *Das Protistenreich*. Setting as border-posts gastrula formation on the animal side, and thallus formation on the plant side, he declared all organisms in-between to be protists: amœbæ, gregarines, flagellates, volvocines, noctilucæ, infusoria (ciliates and acinetes), rhizopods, myxomycetes, fungi, diatoms, labyrinthulæ, catallacts, and monera (including bacteria). The word *protozoa* appears only twice.[131] Haeckel had a gift for the memorable phrase, and he does not disappoint: plants are cell-republics, animals cell-monarchies; individual cell-citizens surrender part of their personal freedom to the law of the state, and share in the work of life. By contrast to these well-organized *Culturstaate*, multicellular protists are as raw hordes of primitive peoples, while most protists are cell-hermits, spending their entire life as simple isolated cells.[132]

Protistenreich reprised many of his ideas: monera arose on multiple occasions by autogony, struggled for existence, and were driven to greater specialization and perfection; all organisms have evolved from monera; protists with a plant-like metabolism preceded those with an animal-like metabolism. A few protists approach a simple form of sexual generation,[133] but only plants and animals are sexual. As for the gastrula, the outer germ-lamella (from which organs of sensation arise) is the *animal* layer; the inner germ-lamella, which encloses the nourishing cavity, is the

396   KINGDOMS, EMPIRES, AND DOMAINS

vegetative layer.[134] The gastrula—"the true animal in its simplest form"—separates animals from protists (not from protozoa).[135]

Also in *Protistenreich* we encounter another of Haeckel's *idées fixes*: *Seele*, usually translated *soul* but (per *Generelle Morphologie*) "a sum of different, highly differentiated functions of the central nervous system, among which the will and sensation are the most important".[136] Haeckel, ever-vigilant against any duality of mind and matter, first described *Seelen* in man and the higher vertebrates as having evolved from simpler reflex movements common to the lower animals, plants, and protists.[137] But nature is one, and *Seelen* are to be found in cells, cytodes, plastidules, inorganic crystals, molecules, atoms, even electricity.[138] Ciliates, he wrote in *Protistenreich*, are true unicells that exhibit a high degree of *Empfindlichkeit* (sensitivity) and *Willens-Energie*, and as such provide the highest demonstration of the cell-soul.[139]

Protists and Histones

By the late 1880s Haeckel had concluded that the "boundary question" could not be decided by empirical science: only a "philosophical and deductive path of logical definition" would bring resolution.[140] He provided these definitions in 1889, contrasting beings that form tissues (animals and plants, collectively *histones*) with those that do not (protists).[141] On successive pages of the Eighth edition of *Natürliche Schöpfungsgeschichte*[142] he offered alternative classifications of the organic world:

> System of the organic world on a morphological basis
>> First organic kingdom: the unicellular beings: Protista
>>> Subkingdom Urpflanzen (Protophyta)
>>> Subkingdom Urthiere (Protozoa)
>> Second organic kingdom: the multicellular beings: Histones
>>> Subkingdom Gewebpflanzen (Metaphyta)
>>> Subkingdom Gewebthiere (Metazoa)

> System of the organic world on a physiological basis
>> First organic kingdom: Pflanzen (Plantae)
>>> Subkingdom Urpflanzen (Protophyta)
>>> Subkingdom Gewebpflanzen (Metaphyta)
>> Second organic kingdom: Thiere (Animalia)
>>> Subkingdom Urthiere (Protozoa)
>>> Subkingdom Gewebthiere (Metazoa)

If we find morphological characters compelling, we can divide organisms into protists and histones. If we prefer a physiological perspective, we can maintain a traditional division into plants and animals. Remarkably, Haeckel drew a boundary-line between plants and animals from top to bottom (Figure 21.2),[143] even as he reminded us that most unicellular organisms are neutral protists.[144] All protists are either protophytes (*protista vegetalia*) or protozoa (*protista animalia*).[145] Protophyta embraces *phytarchs* (plant monera), diatoms, desmids, and simple green algae,[146] while *zoarchs* (animal monera including bacteria), lobose amœbæ, gregarines, flagellates including catallacts, ciliates, acinetes, and rhizopods comprise Protozoa.[147] Phytomonera (phytarchs) and zoomonera (zoarchs) can be distinguished only by metabolism:[148] the former build plasma (like plants), the latter consume plasma (like animals).[149] In this way, even his morphologically based classification depends in part on physiology. Protophytes are plants *and* protists; protozoa are animals *and* protists.

Haeckel retained this dual arrangement in *Plankton-Studien* (1890) and in the Fourth edition of *Anthropogenie* (1891), although in the latter he emphasized the animality of protozoa, and added further ambiguity by moving the "permanent blastosphere" *Magosphæra* into the cœlenterates[150] while retaining it as a protozoan[151] and protist.[152]

Four kingdoms of life

Systematische Phylogenie (1894) was Haeckel's final major scientific work. Robert Richards describes it standing "like a snow-capped volcano, only vaguely reminiscent of the fire-belching *Generelle Morphologie* that threatened the orthodox three decades earlier",[153] but this ignores its innovative—or, less charitably, mad—kingdom-level classification. Returning to his morphologically based system (above), Haeckel wrote that

> while the double kingdom of the Histones is generally divided into the two large groups of the animal and plant kingdoms, the corresponding two-way division of the protist kingdom runs into significant difficulties. To be sure, in systematic practice today, almost half of the protist kingdom (with vegetal metabolism) is placed in the plant kingdom, the other hand (with an animal form of nutrition) in the animal kingdom. . . . Although this two-way division practically corresponds to the old custom and to the usual division of labour between botany and zoology, and is likely to be retained in practice for a long time, it is, at its base, phylogenetically not feasible.[154]

Even so,

> there is no objection against this *artificial* bisection of the protist kingdom, as long as one remembers that it has no phylogenetic meaning. In the present imperfect state of our *protistology*, it is even indispensable; for we must first and foremost

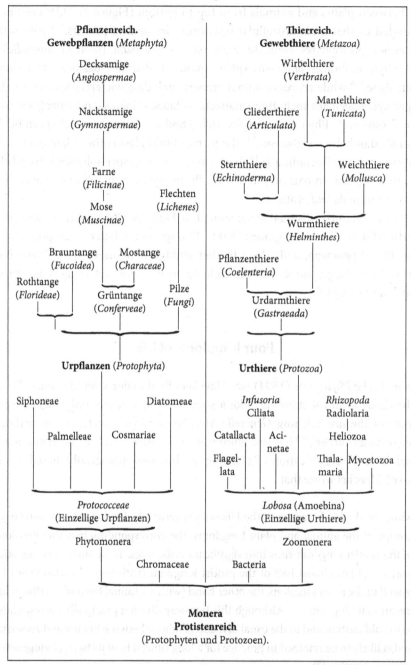

Figure 21.2. Haeckel, *Stammbaum* of the organic world, with vertical line demarcating Plant- and Animal-kingdoms. Redrawn after *Naturliche Schöpfungsgeschichte*, Eighth edition (1889), page 454. Marine Biological Laboratory, Woods Hole Oceanographic Institution Library via Biodiversity Heritage Library.

define the individual larger and smaller protist groups sharply and clearly, before we can think of trying to disentangle their most-complex and difficult phylogenetic relationships.[155]

In *Generelle Morphologie*, Haeckel offered his protist kingdom as a stopgap solution to a vexing practical problem; in *Systematische Phylogenie* he presented its binary division—on a non-phylogenetic basis, if to a phylogenetic end—in much the same way. Morphology, anatomy, and ontology have been of no use; only metabolism and diet offer a way forward.[156] It seems that the promised bisection will, however, be incomplete:

> If we carry out this *artificial* (but not therefore logical) divorce, three different main groups of the protist kingdom emerge for our phylogenetic systematics of the protists, namely 1- asemic protist-*Stämme*, 2- typical protophyte-*Stämme*, [and] 3- typical protozoa-*Stämme*.[157]

"Typical protophytes" are protists "in which the vegetal character of the unicellular organism emerges in the union of [certain] morphological and physiological features", which Haeckel enumerated. Likewise, "typical protozoa" are identified by the union of (different) morphological and physiological features.[158] In contrast, *protista asemica* are those

> in which the simple organism of the plastid does not yet show a pronounced relationship to the typical *protozoa* and *protophytes*: either the formation of the plastid is still completely indifferent (at the lowest stage); or morphological and physiological features are so admixed, that neither its vegetal nor its animal character steps forward clearly and unambiguously.[159]

These *protista asemica* embrace archebionts (anucleate protists), mastogophores (phytomonads, zoomonads, catallacts, dinoflagellates, noctilucæ, volvocinæ), and "fungilli" (chytrids, gregarines, phycomycetes, zygomycetes).[160] The inclusion of gregarines and volvocinæ is surprising, but the tension between morphology and physiology has been broadly constructive. If, however, we expect Haeckel to position the asemic protists neutrally astride the plant-animals boundary, we would be quite mistaken: as in his 1889 *Natürliche Schöpfungsgeschichte*, he extended the plant-animal boundary through the asemic protists too.[161] The corresponding *Stammbäume* are not always easy to interpret.[162]

His "philosophical and deductive path of logical definition" (1889–1894) led to four main groups of organisms: protophytes (*protista vegetalia*, *protista plasmodoma*), protozoa (*protista animalia*, *protista plasmophaga*), metaphytes (*histones plasmodomi*), and metazoa (*histones plasmophagi*). Utterly in-character, Haeckel then declared each a kingdom of the organic world. In parallel (it seems) with the kingdoms of plants and animals, and/or those of protists and histones, he

400 KINGDOMS, EMPIRES, AND DOMAINS

welcomes the kingdoms of Protophyta (*Urpflanzen*), Metaphytes (*Gewebpflanzen*), Protozoa (*Urthiere*), and Metazoa (*Gewebthiere*).[163] This four-kingdom system—in overlapping parts morphological, physiological, non-phylogenetic, monophyletic, and polyphyletic[164]—stands as a remarkable end-point for his equally remarkable journey from 1866, when he set out to reinterpret the organic world on the basis of Darwinian descent theory.

After *Systematische Phylogenie*, Haeckel turned his attention more fully to other issues: secular education, monism, and the place of man in the universe. His observational and theoretical work discussed above, however, initiated a protistological tradition, in which morphologically simple organisms are of interest in their own right, not (or not only) as "lower plants" or "lower animals". This viewpoint competed to varying extents with established (although evolving) traditions in protozoology and phycology, and with the emergence of bacteriology as a new field. We conclude this chapter with an overview of these traditions from mid-century to about 1910, focusing on their receptivity to a third kingdom of life.

The protozoological tradition

The implications of Siebold's recognition (1848) of rhizopods and infusoria as animals "whose irregular form and simple organisation can be reduced to a single cell"[165] played out in complex ways over the next several decades. If Siebold were right, the nucleus (or nuclei) of protozoa should be structurally and functionally equivalent to the nucleus of an animal cell. Yet observations by Johannes Müller[166] and Édouard-Gérard Balbiani[167] suggested instead that the ciliate nucleus was a sort of gonad,[168] an idea more in line with Ehrenberg's *vollkommende Organismen*. Friedrich von Stein took a similar view for ciliates and flagellates.[169] Nor did conjugation in ciliates have an obvious counterpart among somatic cells in animals.

A complete reconceptualization was needed: of fertilization in animal cells, conjugation in protozoa, cell and nuclear division in both. This was provided by Otto Bütschli, an outsider to the relevant academic camps,[170] in his *Studien über die ersten Entwicklungsvorgänge der Eizelle, die Zelltheilung und die Conjugation der Infusorien* (1876).[171] Many of the specific matters he discussed are now of historical interest only; but taken together, he succeeded in unifying disparate observations around a model in which protozoa are indeed unicellular, albeit not in the sense of a single metazoan somatic cell. Rather

> If we retain the comparison between the conjugating infusorian and fertilisation (and I believe I have shown that justification for this is not lacking), the possibility also follows from this, to compare the further behaviour of the conjugation-products, *i.e.* in one case that of the rejuvenated infusorian, in the other that of the fertilized egg-cell. In both cases we see that there arises an energetic increase through division, which in the latter case leads to the formation of a multicellular

organism, in the former, on the other hand, to the rise of a series of generations, of which one may suppose that it has reached its conclusion when two animals first prepare again to conjugate. Morphologically, one would therefore have to compare the sum of all the single individuals of these generations with a multicellular higher organism that has emerged from the egg-cell and is itself again preparing for egg formation. Let us think of those individuals that have arisen from a conjugated infusorian, for example of a *Vorticella* with bud-shaped conjugation, all united into a colony (*e.g.* similar to the flagellate-colonies and to Haeckel's catallacts), and let us allow the ability of conjugation gradually localizing itself, through the onset of the division of labour, to certain cells of the colony (perhaps originally only those that are most advanced in division), which cells would then become the progenitors of new colonies; thusly we obtain the picture of a simplest animal organism with sexual reproduction.[172]

In *Studien* he also recognized that the structure now known as the ciliate micronucleus is in fact the primary nucleus.[173] His demonstration of nuclear continuity between generations, and characterization of the nucleus as a physiological (not structural, organizational, or fecundative) structure, were foundational for the theory of heredity.[174] His monograph on protozoa (1880–1889) in Bronn's *Klassen und Ordnungen der Thier-Reichs* helped establish protozoology as a discipline.[175] In the introduction to this monograph, Bütschli set out his own view of the kingdoms of nature.

Bütschli agreed with Haeckel "and most biologists" that "the question of the border between the two kingdoms and the place of the unicellular beings can be solved only by genealogical means".[176] This requires a *Stammbaum*. In Figure 21.3,[177] a horizontal line demarcates multicellular from unicellular organisms. Above the line we find the standard kingdoms of animals and plants, each potentially diphyletic.[178] Below the line are twenty lineages of "unicellular organisms (Protista)". True fungi do not appear, for which Bütschli begs a lack of personal expertise.[179] After a few further *caveats*, he declares these organisms a middle kingdom much like Haeckel's Protista, albeit with certain modifications.[180] He would have preferred to include the protococcoidea (unicellular algae), but they are better grouped with the higher plants.[181] Unlike Haeckel, he did not carve out a special (moneran) status for schizophytes and bacteria, as he suspected that they have a very simple nucleus.[182] In his view flagellates, not protamœbæ, were the main fount of diversity. Not all of these unicellular organisms found a place in his monograph, but his assignment had been to present "the so-called protozoa" as generally understood, not to engage in reform.[183] The series was, after all, *Klassen und Ordnungen der Thier-Reichs*.

For someone now considered a founder of modern protozoology, Bütschli was remarkably accepting of Haeckel's Protista. The same cannot be said of William Saville Kent, who had trained under Huxley and largely adopted his position against cell theory.[184] Saville Kent raised two objections: that Protista "has no real

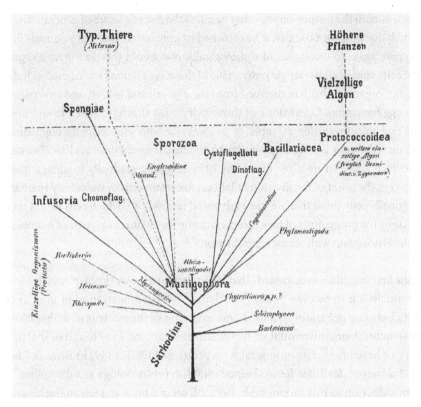

Figure 21.3. Bütschli, *Stammbaum* of organisms, with horizontal line separating multicellular from unicellular organisms. In *Bronn's Klassen und Ordnungen des Their-Reichs, Band 1. Protozoa*, page xii (1888). Marine Biological Laboratory, Woods Hole Oceanographic Institution Library via Biodiversity Heritage Library.

existence", as all protists can "with tolerable, if not absolute certainty" be referred to the plant (diatoms) or animal kingdoms (all the others); and that the addition of an intermediate kingdom means that we must draw two demarcation lines, not just one. He also disputed the existence of monera, as nuclei may be observable only in mature or reproductive phases.[185] Saville Kent returned Spongia to the protozoa because some of their cells resemble choanoflagellates.[186] He recognized twenty-one orders of protozoa and arranged them in two ways: by type of appendage (into classes), and by "the nature of the oral apparatus or systems subordinated to the function of food-ingestion" (into sections).[187] In a remarkable diagram he depicted "relationships, and presumed phylogeny, or lines of evolution" among these sections, classes, and orders.[188] These lines of evolution show the protozoa evolving from protamœbæ (per Haeckel rather than Bütschli), but he offered no indication how protozoa are related to other animals, or to plants.

The years 1882 to 1912 brought a succession of weighty treatises on protozoa in English, French, and German, consolidating the protozoological tradition. Some of these authors drew upon Haeckel, accepting his monera,[189] gastraea theory,[189,190] and/or histones,[191] even referring to *Protista* on occasion.[192,193] Others denied that organisms could be anucleate,[194] objected that Kingdom Protista was so large and diverse as to make its study difficult,[195] and/or held that many so-called protists were perfectly good plants or animals.[191,195] Minchin concluded that Protista "can only be regarded as a convenient makeshift or compromise, rather than as a solution of a difficult problem—that, namely, of giving a natural classification of the most primitive forms of life".[196] All, however, presented protozoa as simple animals. Ambiguous forms, anomalies, and exceptions were inevitable, but were not allowed to distract from this main message.

The phycological tradition

Algae have been considered plants since Theophrastus described marine algae in *Historia plantarum*.[197] Linnæus placed algae second-last among Cryptogamia, followed only by fungi.[198] As we saw in Chapter 19, Jean Lamouroux divided thalassiophytes (algae, soft corals, and sponges) into orders according to colour.[199] Building on these distinctions, Carl Agardh[200] and William Henry Harvey[201] established high-level taxa of green, brown, and red algae (plus diatoms)—groups recognizable today as the core of modern phycology. In the absence of fructification, genera of algae were distinguished by form, so it was perhaps an obvious step to arrange algal genera into form-series within each colour group: for Harvey, individual *cellules* cohere end to end, forming filaments; simple or branched filaments then join together, yielding fronds.[202] In this way parallel form-series, from simple filaments up to complex fronds, came to be recognized for the green, brown, and red algae.[203] From green algae, the series continued to the higher plants.

As we have seen, extension of form-series in the other direction—that is, to pigmented flagellates and monads—has a fraught backstory (Priestley, Agardh, Kützing). But in 1847, against a background of the new cell theory and greater understanding of life-histories and reproduction, botanist Carl Nägeli defined algae as plants that reproduce asexually by spores,[204] implicitly welcoming pigmented unicells. The welcome soon became explicit in *Gattungen einzelliger Algen* (1849). Ferdinand Cohn quibbled with Nägeli's definition of unicellularity,[205] but the stage seemed to be set for phycologists to stake a formal claim to (at least) the pigmented flagellates and monads.

What actually transpired, however, was somewhat more complicated. Studies on aquatic fungi by Nathanael Pringsheim[206] and others revealed reproductive structures similar to those of filamentous algae, consistent with the idea that these fungi had arisen from algae on multiple occasions by loss of chlorophyll. Cohn[207] proposed a Class Thallophytæ in which filamentous fungi were admixed with green, brown, and red algae, while Julius Sachs[208] arranged fungi and algae into parallel form-series, again collectively as thallophytes. Some of the groupings implied by this parallelism, for example of diatoms with conjugate algae, were highly

404 KINGDOMS, EMPIRES, AND DOMAINS

imaginative. Within a few years Anton de Bary[209] called Thallophyta an unnatural taxon, while Alfred Bennett disassembled it to recover blue-green, green, brown, and red algae.[210] The idea of a common Thallophyta persisted in some quarters, but from the 1880s was increasingly abandoned by those seeing themselves as phycologists or as mycologists. Nordal Wille made a particularly influential case for form-series among Chlorophyceæ,[211] and Adolf Pascher among brown and golden-coloured algae,[212] but form-series ascending from unicells can be found in many phycological articles of this period.[213]

In *Mechanisch-physiologische Theorie der Abstammungslehre* (*Mechanical-physiological theory of descent-theory*, 1884), Nägeli declared *phylogenetic union* a general law of organic nature, and pointed to the emergence of higher plants from unicells via a stepwise form-series as an example.[214] Nägeli and Haeckel disagreed on certain biological issues (including the role of natural selection) but agreed on many more, and their scientific interactions were conducted with mutual respect. In *Mechanisch-physiologische Theorie*, Nägeli argued that organisms have arisen from inorganic nature on multiple occasions. Initially, protein molecules arrange themselves into micelles, which combine into a plasma. This plasma grows by taking up organic and inorganic substances, produces new micelles, and differentiates into regions of different physicochemical properties: that is, exhibits life. Driven by a slow mechanical *autonome Vervollkommnung* (autonomous perfecting-principle) and adaptation to the environment, early versions of cell division, reproduction, and tissue formation came under the control of heritable *Anlagen* in the solid region of the plasma (*Idioplasma*), setting the stage for the subsequent emergence of modern life-forms, and a general parallel between ontogenesis and phylogenesis. Nägeli refused, however, to outline a phylogeny of even the plant kingdom, pointing to multiple independent origins and frequent extinctions.[215]

Nägeli recognized two stages in the process described above: the abiogenesis of proteins and their organization into micelles; and the development of this primordial plasma-mass up to the simplest organisms known to us. For the (living) beings of this second intermediate stage he chose the name *Probein*, as "they precede the beginnings of life known from experience".[216] On at least thirteen occasions in *Mechanisch-physiologische Theorie* he referred to the probial kingdom,[217] but never to "three kingdoms of life",[218] presumably because he regarded the former as having preceded the plant and animal kingdoms.[219] Nägeli's *Probienreich* may be the only kingdom to have gone extinct.

Phycologists of this period did not hesitate to classify algae within the plant kingdom, but some employed concepts and terminology from Haeckel including *protists* and *the protist kingdom*.[220] Georg Klebs called "all departments of the thallophytes and protozoa" *Protobionten*,[221] although it is unclear whether he intended this as a formal taxon. By the 1890s bacteria had come into prominence in their own right (below), and Klebs sought to understand relationships among bacteria, phycochromaceæ (now cyanobacteria), flagellates, and the various types of multicellular beings (brown algae, red algae, higher plants, metazoa). In

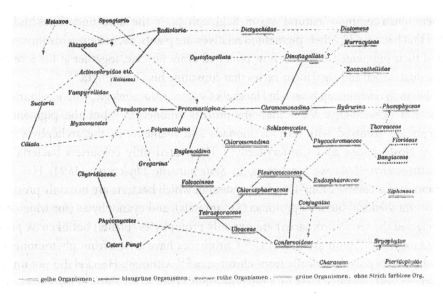

Figure 21.4. Klebs, relationships among the lower organisms. *Zeitschrift für wissenschaftliche Zoologie* 55 (1893), page 428. Smithsonian Libraries via Biodiversity Heritage Library.

Flagellatenstudien (1893)[222] he summarized how animals (at the upper left) and plants (lower right) might be related through a network of monads, flagellates, and filamentous forms (Figure 21.4). The lines represent kinship relations of greater or lesser certainty.[223] It is not entirely clear where the starting point(s) may be: many divisions, he asserts, have originated from amœboid forms.[224] Alternatively, bacteria might be basal.[225] In place of a simple form-series of green algae at the base of the higher plants, Klebs offered a polyphyletic origin within this network.[226]

The bacteriological tradition

Organisms now known to be bacteria were observed by Leeuwenhoek, who called them *little animals*.[227] In his books on worms (1773) and infusoria (1786), OF Müller placed bacteria and other tiny beings into the genera *Monas* and *Vibrio*.[228] Ehrenberg introduced *Bacterium* as a genus name,[229] and in *Infusionsthierchen* (1838) recognized family Vibrionia containing the genera *Bacterium*, *Vibrio*, *Spirillum*, *Spirochæte*, and *Spirodiscus*.[230] Dujardin (1841) agreed that bacteria are infusoria.[231] Nägeli excluded bacteria from his system of algae, but accepted more than a dozen genera of microscopic beings that resemble bacteria but are blue-green (cyan) in colour. Most of these he placed in order Chroococcaceæ.[232] He later named the colourless bacteria Schizomycetes, considering them allied with moulds and yeasts.[233]

Thereafter Cohn, recognizing that the close morphological similarity between Schizomycetes and Chroococcaceæ implied a relationship [*Verwandtschaft*], united

406 KINGDOMS, EMPIRES, AND DOMAINS

them into a common "natural" taxon, Schizophyta.[234] The name underlines his belief that bacteria, like their pigmented relatives, are plants. Sachs kept schizomycetes and their pigmented relatives, now renamed cyanophyceæ, together at the foot of the dual fungal and algal form-series that constitute his Thallophyta.[235]

No such relationship is seen in Haeckel's *Generelle Morphologie*, where colourless bacteria (Ehrenberg's Vibrionia) are protists (monera),[236] but the pigmented forms are admixed with various monads and filaments in the archephyte orders Nostochaceæ and Confervaceæ.[237] He treated only colourless bacteria in *Protistenreich*.[238] As we have seen, in *Systematische Phylogenie* (1894) Haeckel elaborated a complex four-kingdom system in which bacteria are not only protists (*protista asemica*) but also protozoa (*i.e.* animals), and cyanophytes (for which he preferred the term chromacea) are not only protophytes (plants) but likewise *protista asemica*.[239] Both are monera.[240] Chromacea have arisen from phytomonera, and bacteria polyphyletically from chromacea.[241] Although Haeckel did not unite bacteria and cyanobacteria in a purpose-built taxon like Cohn's Schizophyta, he was clear that both lack nuclei (and are therefore not cells, but rather cytodes), pigment bodies, and sexual reproduction.[242]

The most-compelling motivation for studying bacteria, however, was undoubtedly their causative roles in human disease. In Chapter 13 we saw that ancient ideas of "invisible airborne seeds" persisted into the Renaissance in explanation of contagious disease. Linnæus later speculated on "living molecules" as the cause of fevers and putrefaction.[243] Rapid population growth in cities such as London during the Industrial Revolution brought outbreaks of cholera and typhoid. Would-be reformers blamed microscopic infectious particles or "bad air", but lack of scientific evidence allowed others to contend that

> The belief in contagion, like the belief in astrology and witchcraft, seems destined to die out; and as we have got rid of all regulations for consulting the stars or attending to omens before we begin any undertakings, and of all the laws against feeding evil spirits and punishing witches, so we shall no doubt in time get rid of the quarantine regulations that were established from the old belief in contagion.[244]

Studies on anthrax by Casimir Davaine, Pierre Rayer,[245] and others from mid-century culminated in the work of Robert Koch, who developed methods to culture the anthrax bacillus in his home laboratory, identified its spores, and demonstrated their infectivity: that is, he mapped the life history of the bacterium.[246] His report—in Cohn's *Beiträge zur Biologie der Pflanzen*—initiated a thirty-year "golden age of bacteriology" in which twenty main infectious diseases of human were linked to bacterial pathogens.[247] Numerous books on bacteriology appeared, most of which mentioned in passing that bacteria are considered plants.[248] To be sure, attention was focused on bacteria as agents of disease; but bacteriology was further dissociated from natural history by the well-known hostility of Koch and Louis Pasteur to the theory of evolution, on the grounds that evolution is an extension of spontaneous generation.[249] In any event, perhaps it was best for bacteria to be plants:

> The mystic words "microbes" and "bacteria" have been hurled at the popular head with so much emphasis and so little explanation that it would not be surprising to find many people living under the misapprehension that they are minute "fiery serpents," which are always on the look-out for victims, and crawl about them day and night. Not a few people feel comforted by the knowledge that microbes, harmless or harmful, belong to the vegetal rather than to the animal kingdom. Such knowledge takes away the element of repulsiveness arising from the notion of microbes being internal animal parasites or entozoa.[250]

The protistological tradition

We conclude this chapter with the protistological tradition, in which morphologically simple organisms are considered in their own right, not as simple plants or animals. This may be as strong a characterization as the evidence will bear, as Kingdom Protista itself met with only limited acceptance. It cannot have helped that Haeckel himself redrew its boundaries, carved out Monera, then (in the Eighth edition of *Natürliche Schöpfungsgeschichte*, and in *Systematische Phylogenie*) introduced a matrix classification in which some organisms are protists and plants, others are protists and animals. Inexplicably, he reprised his *Generelle Morphologie* in *Prinzipien der generellen Morphologie der Organismen* (1906), as if *Systematische Phylogenie* had never been written. Kingdom Protista was a moving target.

To be sure, Protista won some followers. During 1879–1882 (that is, between *Protistenreich* and *Systematische Phylogenie*), zoologist Entz Géza wrote a critical history of protistology, although it did not see print until 1888.[251] Later, his introduction to *Az élők világa növény- és állatország* (*The world of living plants and animals*, 1907) began in the spirit of *Natürliche Schöpfungsgeschichte* or *Systematische Phylogenie*:

> Organisms familiar to us are divided into two major groups, kingdoms, that is, the animal and plant kingdoms. This roster is sufficient for the living beings we encounter day to day; but, as we will have the opportunity to explain below, at the lowest stage of life we find innumerable living beings of which it is impossible to tell whether they are animals or plants. These organisms of the lowest order, which connect the two great kingdoms, are called protists (*Protista*), and among them those that feed like animals are called animal protists (*Protozoa*), and those that have a diet like plants, plant protists (*Protophyta*).[252]

Entz went on to emphasize that protists form a transitional group, like a third kingdom, that ties the plant and animal kingdoms closely together. He illustrated

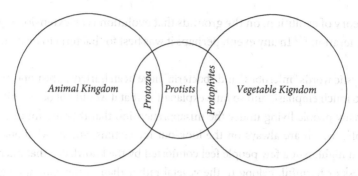

Figure 21.5. Entz, relationships in the organic world, showing the connection between the animal and plant kingdoms. From *Az élők világa növény- és állatország* (1907), page 50. Labels translated by MAR.

this using a Venn diagram (Figure 21.5) in which some (seemingly many) protists are neither animals nor plants.[253] This is more in the spirit of *Generelle Morphologie* or *Protistenreich* than of *Natürliche Schöpfungsgeschichte* and *Systematische Phylogenie*.

Protista shared the limelight in an 1890 publication by cell biologist Emma Leclercq, the first female graduate of Université libre de Bruxelles. Drawing on Bütschli as well as Haeckel, she sketched the polyphyletic evolution of flagellates from "protists of the first order" including protamœbæ and bacteria. For Leclercq, Protista had the membership set out by Haeckel in *Protistenreich*, although she set aside catallacts and labyrinthuloids on the grounds that they are little-known, and fungi because they are multicellular. Delightfully, she described her phylogenetic diagram as "not a tree, [but] rather a thallus, a prothallus!"[254]

The mixed fortunes of Kingdom Protista were on display in the biological volumes of *Die Kultur der Gegenwart*, a major reference series (initiated in 1905) that sought to make the case for German cultural leadership in Europe. In his essay, botanist Eduard Strasburger referred to *Protistenreich* once; Heinrich Poll and Wilhelm Roux mentioned protists in passing. Max Hartmann used the term *Protisten* throughout his essay "Mikrobiologie. Allgemeine Biologie der Protisten" but spoke of a *Protistenreich* only once, in direct reference to Haeckel (followed by a comment that its boundary with plants should be redrawn). Richard Hertwig, a former student of Haeckel, failed to mention Protista or protists in his thirty-eight-page essay on unicellular organisms.[255] In his earlier *Lehrbuch der Zoologie* (1893), Hertwig claimed that Protista had met with little approval.[256] Historian Charles Singer went farther, calling Protista "an untenable concept".[257]

There is, however, another way to think of the protistological tradition: as a new openness to the comparative study of morphologically simple organisms across traditional boundaries. Protists are fascinating and diverse; exemplars of fundamental biological processes (sexual reproduction, complex life histories); parasites or pathogens. This was the vision Fritz Schaudinn sold to publisher Gustav Fischer, and proclaimed to the scientific community in launching *Archiv für Protistenkunde*.

The *Archiv* would provide focus, attracting manuscripts that would otherwise be dispersed across journals of zoology, botany, human physiology and anatomy, pathology, and hygiene.[258] It was the first specialist journal of "lower" organisms.[259]

Richard Hertwig led off the first volume of *Archiv* (1902) with a review of the contribution of protozoa to cell theory.[260] Bütschli and Schaudinn reported studies on bacteria including cyanobacteria.[261] Franz Doflein set out a system of protozoa that incorporated pigmented forms, but not bacteria.[262] Other articles treated ciliates, coccolithophorids, diatoms, foraminifera, gregarines, suctorians, trichomonads, and trypanosomes. Two articles were in English, one in French; a review of recent textbooks of protozoology covered works in English, French, German, and Italian. The word *protist* appears only five times over its 498 pages, *Protistenreich* (or its equivalent) not at all.[263]

Untethered from a formal Kingdom Protista, the term *protist* became a convenient catch-all for unicellular (or acellular) beings in general. After Schaudinn died (of an amœbic infection) at age thirty-four, editorship of the *Archiv* passed to Hartmann and Stanislaus von Prowazek.[264] Under their guidance the *Archiv* continued to accommodate diverse viewpoints, including a notable article in which Clifford Dobell[265] railed against cell theory in general and unicellularity in particular, and argued for a Protista that does not imply that protists are primitive or primordial.[266] Another quarter century would pass before Protista made its formal return (Chapter 22), but *protist* had entered both the language, and the scientific consciousness.[267]

Summary: Haeckel and Protista

Haeckel called himself "wholly a child of the Nineteenth century".[268] This is surely true for better and for worse, and lengthy dockets have been tabled on each side. For the biologist, however, the Nineteenth century brought Darwin's theory of evolution, and Haeckel became its untiring proponent. From *Radiolarien* onward, descent theory ran through his published work. The tree in *Origin of species* was abstract and sparse; Haeckel gave us instead a sturdy oak with a common moneran trunk,[269] three great kingdom-level branches, and lush foliage.

The Protista of *Generelle Morphologie* was a "special middle kingdom" in a phylogenetic and ontogenic sense. In principle, whatever borders it shared with animals and plants arose because one cannot easily determine which amœbæ or flagellates gave rise to the other kingdoms, and which did not. These disputed areas did not enclose the protistan territory in the way that the plant and animal borders once delimited Zoophyta in the Great Chain. But even this level of ambiguity proved unworkable in practise, and in 1889–1894 Haeckel granted protists dual nationality, requiring Protista to share sovereignty. It is a stretch to consider the Protista of *Systematische Phylogenie* a middle kingdom, or (perhaps) a kingdom at all; but with the tension between continuity and boundary-lines as old as Parmenides and Zeno, Haeckel's four kingdoms were an innovative if imperfect solution.

410   KINGDOMS, EMPIRES, AND DOMAINS

The scientific traditions in regard of protozoa, algae, bacteria, and protists all rolled forward into the Twentieth century. As we shall see in our final chapters, the biological world, like Europe, was soon to be divided in new, more-fundamental ways. With these divisions—and the resulting third (and fourth, and fifth . . .) kingdoms—came new connections too, not anticipated by Haeckel or most of his contemporaries.

22

Beyond three kingdoms

> There is no virtue in the hoary three [kingdoms]. In the glorious future that
> is before her, Science may discover ten; or she may reduce them all to one.
> Let her do as Nature bids her, and she will do truthfully and well.
> —William Henry Dallinger, in *Proceedings of the Liverpool Literary and
> Philosophical Society*[1]

The Twentieth century became a fractious period in biological science, as in
many other areas of human activity. Darwin's *Origin of species* and *Descent of man*
endured as bottomless wells of controversy and polemic.[2] Haeckel proved no less
polarizing. Rival camps debated the mechanism of heredity, ontogenetic recapitu-
lation, theories of the cell, and whether microbes are mono- or pleomorphic.[3] Some
microbiologists aspired to put their discipline on a phylogenetic footing, while
others scorned this as unrealistic and unnecessary.[4] Academics and popularizers
carried an often-distorted banner of Darwin or Haeckel into anthropology, linguistics, and psychology, not infrequently ending up in the tar-pits of eugenics, racism,
or nationalism.[5]

Other fissures opened over the foundations of biological classification, with
opinions colliding and combining in ways that defy simple summary. Genera and/
or species had long been accepted as real, whether as objects of divine creation or
genealogical descent. By contrast, higher taxa—families, orders, classes, phyla, and
kingdoms—were typically held to be mental constructs that nonetheless might, in
favourable cases, capture or approximate actual groupings in Nature. As we have
seen, some authorities sought these groupings in patterns of mechanical or developmental constraint; others imagined a hierarchy of genetic relationships, as
in the greater and lesser branches of a tree. Yet in the absence of a direct window
into the mechanistic basis of heredity, systematists of all camps could but draw on
a common set of observable features describing external form, internal parts, life
histories, embryological series, and cellular structure. Some systematists arranged
taxa along form-series; others assigned them to branches of a tree. Haeckel did
both, sometimes in a single diagram.

We concluded Chapter 21 with four perspectives on simple organisms. In practise these often graded into one another, such that a protozoologist might describe

Kingdoms, Empires, and Domains. Mark A. Ragan, Oxford University Press. © Oxford University Press 2023.
DOI: 10.1093/oso/9780197643037.003.0022

412   KINGDOMS, EMPIRES, AND DOMAINS

studies on unicellular algae in the pages of *Archiv für Protistenkunde*. These perspectives continued to inform views on biological kingdoms well into the new century, but were joined by three developments of such scope and consequence that all subsequent classifications bear their imprint. We examine these in our final two chapters. In this chapter we encounter new high-level taxa that outrank and subsume kingdoms, and a consequential new theory of the origin of cells. In the final chapter we enter the brave (but disorienting) new world of molecular phylogenetics and phylogenomics. Individually and in combination, these developments continue to revolutionize the way that biologists view the living world, including at the highest taxonomic levels.

Kingdoms and superkingdoms

So far as is known, living things were first divided into man, animals, and vegetables at the time of Philolaus, two generations before Aristotle.[6] Later, Porphyry classified animate corporeal substance into the sentient (animal) and non-sentient (vegetable); Ammonius Hermiæ, John Philoponus, and John of Damascus added zoophytes.[7] From the early 1600s, the most-inclusive taxa of natural beings began to be called *kingdoms*.[8] Third kingdoms came and went, sharing their fleeting moment with Animals and Vegetables. From time to time, all current kingdoms of organic beings were swept into a single more-inclusive supertaxon: the Imperium Naturæ of Linnæus,[9] the Organicorum imperium of Pallas,[10] the Regnum Organicum of Jussieu,[11] the *règne vivant* of Vicq-d'Azyr.[12] Until now we have not, however, encountered an organic supertaxon that excluded plants and animals.[13] Plants and Animals have always taken, or shared, centre stage.

The first supertaxon to exclude plants and animals emerged in the Twentieth century. To understand how it came about, we must return to the story of bacteria. In the late 1840s, Nägeli removed the colourless bacteria from algae, but retained similar beings of a blue-green colour; he later named the former Schizomycetes, as he thought them to be allied with fungi.[14] Over the subsequent decades and well into the new century—that is, paralleling the birth and consolidation of bacteriology as a discipline—colourless bacteria were classified in almost every conceivable way: as animals, plants, fungi, or protists.[15] By contrast, blue-green algae became ensconced at the base of the green algal (hence plant) form-series, or on a basal branch of the corresponding tree. Cohn joined bacteria and blue-green algae into a single taxon, Schizophyta,[16] although this was not widely adopted.

Consensus was particularly elusive on the question of a bacterial nucleus. Schleiden, Schwann, Remak, and others insisted that cells are organized by the nucleus (or nucleolus).[17] Haeckel asserted that bacteria and blue-green algae lack a true nucleus, and are thus cytodes, not cells.[18] New staining methods, however, revealed zones inside bacteria. Some authorities concluded that "the existence of a formed nucleus even in the simpler bacteria (Eubacteriales) has been demonstrated beyond doubt";[19] others were more circumspect.[20] In a 1941 review, Isaac Lewis discussed eight hypotheses on the bacterial nucleus:

1. Bacteria do not possess a nucleus, or its equivalent;
2. The bacterial cell is differentiated into a chromatin-containing central body, and a peripheral cytoplasm;
3. The bacterial body *is* a nucleus devoid of cytoplasm: a naked nucleus or nuclear cell.
4. The nucleus consists of chromatin bodies scattered throughout the cytoplasm;
5. The nucleus is polymorphic, depending on the stage of the bacterial growth cycle;
6. The nuclear substance is diffuse, finely dispersed through the cytoplasm;
7. The protoplast contains one or more true vesicular nuclei; and
8. The bacterial nucleus is an "invisible gene string" analogous to a single chromosome.

Unsurprisingly, the issue became entangled in the battle over cell theory; as late as 1955, one eminent cytologist asked "whether it is legitimate to speak of a nucleus in the Cyanophyceae and bacteria".[21] Other biologists maintained that bacteria and/or blue-green algae are polyphyletic, and thus should not be joined into a single taxon at any rank.[22] Even those who accepted a single origin for bacteria, or were unperturbed by polyphyletic taxa, might hesitate to base a taxon—much less a kingdom—on the absence of a single character.[23] Despite these obstacles, taxa defined or characterized by the absence of a nucleus began to appear. Many were of middling rank, but at kingdom level we find *Mychota*, *Procaryotes*, variants of *Akaryonta*, and a reformed *Monera*. Two of these became superkingdoms that excluded plants and animals. Let us consider each in turn.

Mychota

By the mid-1910s, the dispute over microbial pleomorphism—the supposed variation of exterior and interior form, physiology, and pathogenicity according life-history stage and environment—was in full tilt. In 1925 one of the protagonists, Günther Enderlein, set out a theory of pleomorphism notable for its idiosyncrasy, detail, and near-impenetrable terminology.[24] He held that cells arise when a *Mych* (a primitive nucleus or *Urkern*) brings life to an *Urzell*, attracts protoplasm (forming a *Mychit*), and combines with other such bodies.[25] Bacteria are cellular, have a nuclear apparatus and cytoplasm, undergo meiotic division, and reproduce sexually.[26] Microbes moreover contain tiny transmissible particles of plant origin (*protits*) which, stimulated by a nocuous environment, can develop into a pathogenic virus, bacterium, or fungus.[27] From *Mych* came *Mychota*, his term for the *Stamm*

414 KINGDOMS, EMPIRES, AND DOMAINS

of colourless bacteria.[28] Herbert Copeland later established Kingdom Mychota for "the organisms without nuclei, bacteria and blue-green algae".[29]

Procaryotes

A second proposal from 1925 was as minimal as Enderlein's was complex. As a student at the Sorbonne, Édouard Chatton discovered an odd amœba in the digestive tract of the crustacean *Daphnia*, and named it *Pansporella perplexa*.[30] In 1923 Chatton, by then a professor of biology at Université de Strasbourg, undertook to classify it properly. He described its vegetative forms, encystment, sporulation, and germination, emphasizing the appearance and integrity of the nucleus at each stage of its life history.[31] He then systematically compared these details with those of other protozoa, and of protists more broadly, concluding that a new family (Sporamœbidæ) was needed.[32] Chatton took the opportunity to propose a new classification of protists, offering a hierarchical table and a phylogenetic network. In each we find two new high-level taxa of unspecified rank: *Procaryotes* and *Eucaryotes*. He offered no explanation, but the terms imply pre- and true nuclearity respectively.[33] Bacteria, spirochætes, and blue-green algae are procaryotes; as for eucaryotes, his table lists only protozoa (ciliates, flagellates, rhizopods, sporozoa) and a few immediate relatives.[34] In his phylogenetic network we also find radiolaria, heliozoa, brown and red algae, sponges, fungi, metazoa, and archegoniates, although not angiosperms.[35] Some of these, to be sure, are protists; but how many are eucaryotes?[36] Chatton does not say.

By some accounts,[37] Chatton's distinction lay forgotten until its "rediscovery" in the 1960s. That is not entirely factual. André Lwoff, a student and collaborator of Chatton (and eventual Nobel laureate), began his *Recherches biochimiques sur la nutrition des protozoaires* (1932) as follows:

> With E. Chatton (1926) we divide the Protists into:
>
> 1° *Procaryotic Protists*, without a defined nucleus and individualized mitochondria: Bacteria and affiliated forms.
>
> 2° *Eucaryotic Protists*, equipped with a nucleus and mitochrondria. These are the Protozoa in the broadest sense of the term. To facilitate the exposition that follows, we arbitrarily divide the eucaryotic protists, or *Eucaryotes*, into two groups:
>
> *a*) The *Protozoa sensu stricto*, which comprise the Eucaryotes lacking chlorophyll.
>
> *b*) The *Chlorophytes*, which contain groups of Eucaryotes that possess chlorophyll and one or more plastids.[38]

In 1936 Bert Knight brought this passage into English, preserving the ambiguity of whether there exist Eucaryotes other than protozoa:

Lwoff divides the protista into: *Procaryotic Protista*, without nucleus and individual mitochrondria: bacteria and related forms. *Eucaryotic Protista*, having a nucleus and mitochondria. These Eucaryotes (protozoa in the widest sense) are further subdivided by Lwoff.[39]

Lwoff himself later reprised the passage, altering it subtly:

> With E. Chatton (1926), we divide the Protists into Procaryotes and Eucaryotes. According to Chatton's definition, the prokaryotic Protists lack a nucleus and individualized mitochondria: they include bacteria and related forms. The eukaryotic Protists possess a nucleus and individualized mitochondria.[40]

Meanwhile, Chatton expanded (or perhaps clarified) the scope of Eucaryotes. In Twentieth-century France, as elsewhere in Europe even today, a candidate for an advanced degree or academic appointment was expected to submit a *curriculum vitæ* and compendium of previous work. In 1918, Chatton had submitted *Titres et travaux scientifiques 1906–1918*[41] towards a doctoral degree and appointment as a *maître de conférences* (lecturer) at Strasbourg. Two decades later, as an applicant for appointment as professor at the Sorbonne and director of the marine biological stations at Banyuls-sur-Mer and Villefranche-sur-Mer, he drew up a weighty *Titres et travaux scientifiques 1906–1937*[42]. In its introductory section, written in 1937, we read that

> Protistologists agree today to consider autotrophic Flagellates, as the most primitive of the Protozoa with a true nucleus, the Eucaryotes (an assemblage that also embraces the Vegetables and Metazoans), because they are the only ones able to achieve the total synthesis of their protoplasm from a mineral medium . . .
>
> However, at the base of the vegetable kingdom, there is perfect evolutive continuity between the three large groups of autotrophic Flagellates . . . and the major groups of Algae.
>
> Although the continuity is much less evident from Flagellates to Metazoans . . . the animal microgamete also represents the survival, somewhat modified, of the ancestral flagellar state.[43]

With these words, Chatton unambiguously cleaved the living world along a fault-line—the presence or absence of a true (membrane-bounded) nucleus. Animals, vegetables, and protists[44] became subsets of a top-level assemblage—Eucaryotes—that itself did not encompass all earthly living things. He did not formalize Procaryotes and Eucaryotes as taxa, identify Animals and Plants as kingdoms, or remark on the momentous step he was taking.[45]

416 KINGDOMS, EMPIRES, AND DOMAINS

Akaryonta or Anucleobionta

While *Mychota* granted bacteria a primitive (if surprisingly capable) nucleus, and *Procaryotes* indicated a prenuclear status, other taxon names of the period reflected a harder line. In 1930, the Czech botanist František Novák distinguished three groups (*skupiny*) of organisms: Aphanobionta (viruses), Akaryonta, and Karyonta. The first two consist of one kingdom (*říše*) each, likewise named Aphanobionta and Akaryonta. Kingdom Akaryonta has two divisions (*odděleni*), Schizomycetes (bacteria) and Cyanophyta (blue-green algae). Karyonta consists of two kingdoms, Karyophyta (flagellates, dinoflagellates, diatoms, algae, green plants) and Karyozoa (animals).[46] Novák considered that akaryonts have no phylogenetic relationship to other forms of life, including karyonts.[47]

In 1946, Zoologist Jürgen Harms recognized Anucleobionten (bacteria and blue-green algae) and Nucleobionten (algae, protists, multicellular plants, and multicellular animals).[48] Werner Rothmaler formalized the former as kingdom Aucleobionta (1948), and partitioned the latter among kingdoms Protobionta (protists), Cormobionta (plants with stomata), and Gastrobionta (metazoa);[49] in 1951 he changed Anucleobionta to Akaryobionta.[50]

Monera

For Haeckel, monera were "organisms without organs".[51] In his hands, *Stamm Moneres* grew to include colourless bacteria, chromacea (blue-green algae), various supposedly anucleate rhizopods, and *Bathybius*. At first monera were protists, *i.e.* neither animals nor plants.[52] By *Systematische Phylogenie*, bacteria were protists *and* protozoa (animals), blue-green algae protists *and* protophytes (plants).[53] Haeckel eventually accepted that most "rhizomonera" do in fact have a nucleus, but drew the line at *Protamœba primitiva* and *Pelomyxa pallida*, and supposed that other amœbæ might yet be proven anucleate.[54] Against this backdrop, and with debate still swirling around the question of nuclei in bacteria, it required courage for Herbert Copeland (1938) to reconstitute Monera "as a distinct kingdom and distinguished by lack of nuclei".[55] He limited its membership to autotrophic bacteria (including blue-green algae), "ordinary bacteria", and spirochætes. Cornelis van Niel, a scion of the influential Delft school of microbiology, and his student Roger Stanier soon followed.[56]

Each of these terms—Mychota, Procaryotes, Akaryobionta, Monera—had its proponents, but from 1939 began to be recognized as more or less equivalent.[57] The following year a new technology, electron microscopy, was first brought to bear on cell structure.[58] Over time its resolution improved, and practitioners learned to avoid or recognize artefacts; but no membrane-bounded nuclei, or mitochondria or plastids, turned up in bacteria.[59] By contrast, nuclei were readily confirmed in other organisms. In an influential article in 1962, Stanier and van Niel referred not only to cellular organization, but also to organisms, as *procaryotic* or *eucaryotic*.[60] The following year, in the second edition of their influential textbook *The microbial world*, Stanier and colleagues stated that

> this basic divergence in cellular structure, *which separates the bacteria and blue-green algae from all other cellular organisms*, probably represents the greatest single evolutionary discontinuity to be found in the present-day living world.[61]

They did not, however, use this as the basis of a formal classification, instead sorting their three kingdoms according to the presence (Plants, Animals) or absence (Protists) of differentiated cells and tissues; bacteria were "lower protists".[62] Stanier later took the lead in referring to blue-green algae as *cyanobacteria*, further emphasizing the discontinuity.[63]

None of this resolved the question of a "third kingdom". To be sure, with bacteria and blue-green algae safely tucked away into a (super)kingdom of their own, other taxa might be delineated more precisely: green algae, for example, no longer encumbered by the blue-greens, could be drawn fully into Plantae, simplifying Protista in the process.[64] A four-kingdom classification by Copeland,[55] and a five-kingdom variant by Whittaker,[65] became prominent; but others were proposed, offering from one to nineteen kingdoms. We conclude this section by surveying these four- and five-kingdom schemes, and giving a sense of some alternatives up to 1975.

Four kingdoms (Copeland, 1938–1956)

In fashioning his 1938 system, Copeland sought to balance evolutionary relationship against convenience.[66] For the beings least changed since the origin of life, and lacking nuclei, he accepted Haeckel's term Monera.[67] He delineated Plantae[64] and Animalia with little ambiguity, leaving Protista (including red algae, brown algae, diatoms, and fungi) as "lacking the combinations of characters to be listed as characteristic of plants and animals".[68] In 1947 he replaced Monera with Mychota (Enderlein), and Protista with Protoctista (Hogg), citing technicalities related to the rules of nomenclature.[69] Copeland updated his four-kingdom classification in 1956, paying close attention to nomenclature and priority.[70] Fred Barkley,[71] Werner Rothmaler[72] and others adopted his four-kingdom scheme, but retained green algae as protists.

Five kingdoms (Whittaker, 1969)

An ecologist by training, Robert Whittaker objected to the technical direction of systematics taken by Copeland. He argued that kingdoms should reflect the major functional groups of organisms in living communities, as defined by their role in food webs: as *producers* (plants and a few others), *consumers* (animals), or "*reducers, transformers, or decomposers*" (fungi and bacteria). Each of these roles implicates a mode of nutrition (photosynthesis, ingestion, or absorption) made possible by specialized biochemical, cytological, and other features.[73] He sought to capture these "major directions of evolution"[74] in a schematic diagram (Figure 22.1) in which multiple lines of ancestry link groups in Protista to phyla of plants, animals, and fungi.[75] Whittaker allowed that the living world might be divided into one, two, three, four, or more kingdoms, but for the (ecological) reasons mentioned above he preferred a four-kingdom system (1957 and 1959). A decade later, accepting the profound cytological and biochemical gulf between prokaryotes and eukaryotes, he split off Monera as a fifth kingdom (Figure 22.2).[65]

Like Whittaker, Lynn Margulis brought fresh perspectives to biological classification. In 1971 she modified his five-kingdom system to reclaim red algae, brown algae, chytrids, oömycetes, sponges, and certain other taxa into Protista, making Plantae, Fungi, and Animalia monophyletic (Figure 22.3).[76] She later fit his five kingdoms into the Prokaryota-Eukaryota framework.[77] With or without her modifications, the five-kingdom system was widely hailed as the culmination

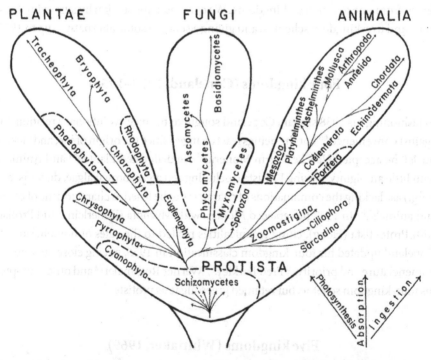

Figure 22.1. Four kingdoms of organisms, with simplified evolutionary relations. Republished by permission of University of Chicago Press—Journals, from On the broad classification of organisms. RH Whittaker, *Quarterly Review of Biology* 34(3), 1959 at page 217. Permission conveyed through Copyright Clearance Center, Inc.

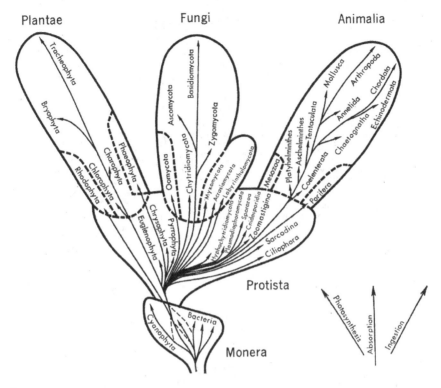

Figure 22.2. Five-kingdom system based on three levels of organization: prokaryotic (kingdom Monera), eukaryotic unicellular (kingdom Protista), and eukaryotic multicellular and multinucleate. Republished by permission of American Association for the Advancement of Science, from New concepts of kingdoms of organisms. RH Whittaker, *Science* 163(3863), 1969 at page 157. Permission conveyed through Copyright Clearance Center, Inc.

of neoclassical (pre-ultrastructural, pre-molecular) systematics at kingdom scale. Yet Protista remained heterogeneous, defined only by negative ("absence of") characters.

Other high-level proposals to 1975

Innumerable alternatives to the Copeland and Whittaker-Margulis classifications were brought forward. Some did little more than demote kingdoms to a lesser rank, or promote phyla to kingdoms.[78] Others offered more substance, for better or for worse. Claiming that bacteria and blue-green algae have "rudimentary sorts of nuclei", Raymond Moore recognized three kingdoms (Protista, Plantae, Animalia), the former with subkingdoms Monera and Protoctista.[79] Edward Dodson,[80] Charles Jeffrey,[81] and Gordon Leedale[82] subsumed protists into the higher kingdoms. Peter Edwards dispersed algae, plants, and fungi into seven kingdoms.[83]

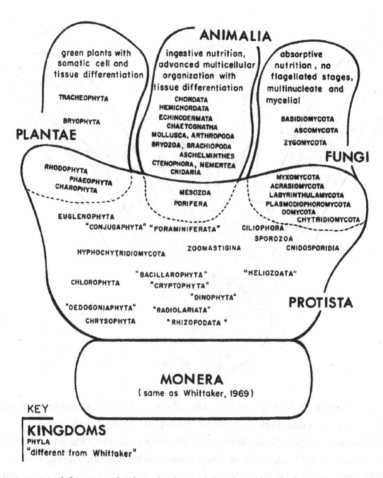

Figure 22.3. Modification of Whittaker's 1969 five-kingdom scheme based on protist phylogeny. Republished by permission of John Wiley and Sons, from Whittaker's five kingdoms of organisms: minor revisions suggested by consideration of the origin of mitosis. L Margulis, *Evolution* 25(1), 1971 at page 244. Permission conveyed through Copyright Clearance Center, Inc.

Ellsworth Dougherty and Mary Belle Allen separated protists into three "major levels of structural organisation": those of *monera* (bacteria and blue-green alga), *mesoprotists* (red algae, which are nucleate but lack flagella), and *metaprotists* (all other organisms except metaphytes and metazoa).[84] For John Dodge, permanently condensed chromosomes and the (supposed) absence of nuclear proteins made dinoflagellates Mesocaryota, above Procaryota but below Eucaryota.[85] The award for thinking farthest outside the box goes, however, to Lawrence Dillon. Taking the nucleus to be a sort of endosome, he arranged cytological states into a Y-shaped series from which bacteria, amoebæ, and algae emerge as side branches; one arm of the Y terminates in Eumetazoa, with brown algae as its sister-group.[86] Given that organisms he considered plants (including bacteria and many protists) "make up the beginning and end points as well as the majority of the branches, it appears most

BEYOND THREE KINGDOMS   421

logical to recognize this single kingdom as the plant kingdom, or superkingdom".[87]
All flesh was grass.

The rise of cellular ultrastructure

By the mid-1970s, transmission electron microscopy was taking hold in biological research. Some protists proved to be difficult subjects, but improved methods brought a flood of data on the fine structure of mitosis, flagella, organelles, membranes, and much else. A new scientific society formed in 1975, the Society for Evolutionary Protistology (later internationalized to ISEP),[88] soon became a venue for sharing these data and debating their phylogenetic implications. Not surprisingly, the next generation of kingdom-level proposals built heavily on characters arising from cellular ultrastructure. Among these we find a succession of classifications from Tom Cavalier-Smith, offering from six to ten kingdoms.[89] The names and membership of his taxa could present a moving target, but at root he focused on the stories told by ultrastructure: whether a plastid (for instance) is surrounded by two membranes, or three, speaks to very different evolutionary histories. To understand why, we must introduce the second remarkable insight of the Twentieth century: that the eukaryotic cell had been assembled not gradually, but in a series of discrete steps.

Eukaryogenesis 1: *Natura facit saltum*

How did the first cell arise? Haeckel had posed the question at Stettin,[90] but over the next quarter century the matter attracted little interest. Before Schleiden and Schwann, it was widely held that tissues of plants and animals are composed of tiny globules.[91] Objects answering to that description could be seen under the microscope, although in retrospect it is hard to know what was observed in each instance—cells, subcellular bodies, or optical artefacts. According to Friedrich Arnold (1836) and Karl Baumgärtner (1842), globules can arrange themselves into a filament or tubule, or close up into a hollow sphere.[92] As the 1850s progressed, the cell came to be seen as a physiological (not merely structural) unit;[93] protoplasm was no longer a passive matrix, but rather a complex system of interacting parts making up *ein lebendig Bau*.[94] In 1861 Ernst Brücke—another former student of Johannes Müller—memorably called the cell an *elementary organism*,[95] and the concept was embraced by Gegenbaur, Haeckel, Bütschli, Reinke, Hertwig, Hartmann, and many others. Brücke, moreover, speculated that the cell might not be elementary:

> I call cells elementary organisms, as we call the bodies which so far have not been broken down chemically, elements. As little as their indivisibility has been proven, just as little can we deny the possibility that cells themselves are perhaps made up

422 KINGDOMS, EMPIRES, AND DOMAINS

of other, still-smaller organisms, which stand in a similar relationship to them as do cells to the whole organism; but as yet we have no reason to accept this.[96]

Societies, organisms, cells; might the latter, in turn, be consortia of even-smaller units? As optical lenses were further refined and cytochemical methods became less destructive, the question became increasingly accessible. To be sure, some investigators were less constrained than others by the paucity of hard data. Antoine Béchamp held that cells are produced by, and contain, *microzymas*—tiny living bodies that can transmute into bacteria and multicellular organisms, or (upon stimulus from the environment) devolve into "morphological manifestations of the morbid state".[97] Reality proved rather more prosaic, as through the 1880s a host of vacuoles, starch- and protein-bodies, fat- and oil-droplets, polyphosphate granules, and the like were characterized in cytoplasm. Yet the suspicion remained that larger, more-subtle bodies were present as well. Walther Flemming described *fila*,[98] while Richard Altmann lumped diverse granules and fibrils together as *bioblasts*, "the morphological unit of living matter".[99] Granular bioblasts, he wrote, are like microorganisms:

> it seems as if the old doctrine of the elementary granules is justified. Cells are not elementary organisms, but colonies thereof, with peculiar laws of colonization; cells do not, however, arise through the assembly of the little globules, but they arose therefrom in those historical periods, which are just as peculiar to the microscopic elements as to the coarse forms of living beings; the elementary granules of the cells, which to this day have their analogous representatives in the microorganisms and which have existed in the cells since those periods, are no longer capable of becoming independent living beings.[100]

It is impossible to know precisely what Altmann had observed. Carl Benda, who described and named mitochondria in 1898, insisted that mitochondria are not bioblasts.[101] But mitochondria too resembled bacteria. The "golden age of bacteriology" was well underway,[102] and unsurprisingly some bacteriologists tried to culture the bacteria-like components of cells. With contamination by actual bacteria an ever-present risk, the results were predictably contentious.[103] Paul Portier and others, however, reported success in culturing mitochondria outside the cell, where they remained metabolically active and, in some cases, gave rise to "true microorganisms".[104] Portier—who worked at Institut océanographique de Monaco, not far from Chatton—supposed that mitochondria had originated when cells took up bacteria as food; some became domesticated and settled down into long-term residence. Even today, he thought, bacteria taken up as food sometimes fuse with existing mitochondria, rejuvenating the existing stock.[105]

Nor were mitochondrial the only such intruders. In 1880 botanist Johannes Reinke reported that chlorophyll granules have a "highly individualized and independent existence", and can remain alive (and multiply by division) even after the mother cell has died and become overrun with fungal filaments.[106] Three years later, Friedrich Schmitz found that plastids in brown, red, and green algae arise from pre-existing plastids.[107] Andreas Schimper, a former student of Anton de Bary, understood what continuity might imply:

> Should it be definitively confirmed that the plastids are not newly formed in the egg cells, their relationship to the organism that contains them would be somewhat reminiscent of a symbiosis. It is possible that the green plants really owe their origin to a union of a colourless organism with one that is uniformly tinted [*tingierten*] by chlorophyll.[108]

By 1885 Schimper had confirmed that plastids in higher plants too are continuous. When a plant becomes green (for example, upon exposure to light), chloroplasts do not spring up de novo; instead, pre-existing colourless plastids (leucoplasts) are converted to green ones (chloroplasts), seemingly without involvement of the nucleus. Thus

> the chromatophores appear from the beginning as completely independent plasmatic bodies; they behave, in terms of their reproduction and their chemistry, much more like proper [*eigene*] organisms than as parts of the plasma-body; they show no relation to the cytoplasm or the cell nucleus, and retain their most important peculiarities in spite of various metamorphoses throughout the plant world.[109]

Analogous associations were turning up elsewhere in nature: bacteria in root nodules of legumes, algae in protozoa and marine invertebrates.[110] Lichens offered a particularly compelling example. In the late 1860s Simon Schwendener, who had trained under Alphonse de Candolle and Carl Nägeli, reported that lichens are dual beings, made up of a fungus living parasitically off an alga.[111] Others confirmed this duality, but disputed the nature of the relationship; it is in this context that Anton de Bary (1878) introduced the term *symbiosis* for such an association, without committing to which component might specifically benefit.[112] Work began to try to pick the lichen symbiosis apart. Andrei Famintsyn and Josif Baranetzky extracted (algal) gonidia from three species of lichens and found that they could reproduce, producing zoospores typical of the green alga *Cystococcus*.[113] Famintsyn tried to culture chloroplasts as well, albeit without success.[114] He viewed the relationship in lichens as of mutual benefit, and as "the first irrefutable factual proof of [Charles Darwin's] theory of the evolution of organisms".[115] He supposed that similar cohabitations might be common in the living world, and identified plant and animal cells as probable examples.[116]

It remained for Konstantin Merezhkowsky[117] to weave the various strands into a unitary theory he called *symbiogenesis*,[118] set out in articles in 1905 (fundamental claims) and 1910 (theory, geological context, and consequences for biological

424   KINGDOMS, EMPIRES, AND DOMAINS

systematics). He then summarized his ideas, and responded to critics, in an entertaining article written in 1918 but published in 1920.[119] Liya Khakhina[120] and Jan Sapp[121] consider his claims and theory at some length; I briefly restate them here because they underpin his recognition of a new kingdom of organisms.

For Merezhkowsky, chromatophores are not organs[122] of the plant cell, but self-perpetuating symbionts that originated as free-living bacteria, and even now remain beyond the control of the nucleus.[123] They are morphologically and physiologically similar to blue-green algae; indeed, the latter can be considered free-living chromatophores.[124] Chromatophores are, moreover, indistinguishable from the (unquestioned) algal symbionts in amœbæ, *Hydra*, and the like.[125] He distinguished two kinds of protoplasm: *mycoplasm* (robust, synthetically active, structurally complex, the bearer of heredity) and *amœboplasm* (less robust, capable of amœboid movement).[126] These recall Haeckel's *plasmodomi* and *plasmophagi*, although Merezhkowsky did not draw the connection.[127] Mycoplasm has given rise to chromatin granules, bacteria, blue-green algae, and fungi,[128] and amœboplasm to the cytoplasmic matrix of plant and animal cells.

In an earlier epoch, monera (amœboplasm) crawled about on the ocean floor, devouring bacteria (mycoplasm). Some bacteria, resistant to digestion inside the moneran, banded together and surrounded themselves with a membrane, becoming the nuclear chromatin.[129] After a period during which free-living bacteria diversified (yielding, inter alia, blue-green algae of various colours), and the newly (proto)nucleate monera diversified into lineages of amœbæ then flagellates, coloured bacteria invaded (some) flagellates, forming symbioses: hence the types of unicellular and multicellular algae. Higher plants arose from green algae, while a few lineages secondarily lost their plastids and became phycomycetes (which thus are not fungi). Lines of amœbæ and flagellates that did not take up coloured bacteria developed into animals, while fungi evolved from bacteria (Figure 22.4).

This stepwise scenario gifted Merezhkowsky a number of all-or-nothing distinctions that had not been available to naturalists before him. Kingdoms could be established, and delineated cleanly, based on the number of symbioses and/or type of partner. He recognized a *Mykoidenreich* for entities of pure mycoplasm, unaffected by symbiosis: bacteria, blue-green algae, fungi, chromatophores, and nuclear chromatin bodies—the first kingdom to claim the very defining parts (so to speak) of organisms in the other kingdoms. A single symbiotic step (resulting in the nucleus) gave Animals, while Plants are defined by a second symbiosis.[130] Kingdom Protista was no more.[131] His animal kingdom is polyphyletic;[132] his plant kingdom even more so.[133] Lichens, which result from a third symbiosis, arose on perhaps fifteen to twenty occasions.[134]

We cannot leave Merezhkowsky without commenting on the evocative language and rich visual imagery of his texts. The plant cell is nothing else, he wrote, but an animal cell with the blue-green algae that have invaded it.[135] We recognize the algal symbionts in turbellarian worms as recent arrivals because they "haven't yet

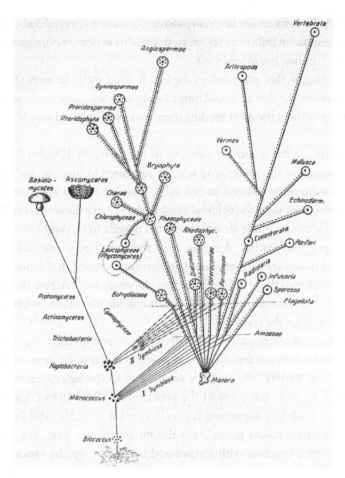

Figure 22.4. Mereschkowsky, representation of the organic world as two *Stämme* arising from independent roots. Thin lines indicate mycoplasm, heavy lines amœboplasm, and broken lines cyanophyceae or chromatophores. *Biologisches Centralblatt* 30 (1910), page 366. Marine Biological Laboratory, Woods Hole Oceanographic Institution Library *via* Biodiversity Heritage Library.

had time to take off their overcoats".[136] Certain amœbæ that host symbiotic algae are "plant-animals . . . plants in the process of formation, animals in the process of transforming into plants".[137] Is a diatom plant, or animal? *Mon Dieu*, the answer is simple: morphologically a plant, physiologically an animal.[138] Above all, we have his idyll of the lion and palm tree:

> Let us imagine a palm tree growing quietly on the bank of a spring and a lion lying hidden in the bushes next to it, all his muscles tensed, with blood-lust in his eyes, ready to leap onto an antelope and clamp his jaws onto its throat [*sie zu erwürgen*]. Only symbiosis-theory allows one to penetrate into the deepest secret of this image and to divine and understand the fundamental reason that two phenomena so enormously different, like a palm tree and a lion, could have been brought

426 KINGDOMS, EMPIRES, AND DOMAINS

forth. The palm behaves so calmly, so passively, because it is a symbiosis, because it contains a myriad of little workers, green slaves (chromatophores), who work for it and feed it. The lion has to feed itself.

Let us imagine that every cell in the lion is filled with chromatophores, and I have no doubt but that he would immediately lie down calmly next to the palm tree, feeling full or at the most needing some water with mineral salts.[139]

Symbiogenesis offered a radical solution to the question Haeckel had posed at Stettin; but much of the underlying science remained in dispute, and autogenous alternatives were never set out in detail.[140] Theodor Boveri speculated that nucleus and cytoplasm were the original symbionts.[141] Boris Kozo-Polyansky (whose *natura facit saltum* we quote above) suspected flagella to be symbionts.[142] Auguste Lumière ripped into Portier, deriding both the supposed universality and likely benefits of *les symbiotes*.[143] Alexandre Guilliermond denied that mitochondria are bacteria, but believed they could change into chloroplasts.[144] Across the ocean, Ivan Wallin claimed to have cultured mitochondria outside the cell, but when challenged could not reproduce the result.[145]

In his influential textbook *The cell in development and heredity*, Edmund Wilson concluded that symbiosis theory "may appear too fantastic for present mention in polite biological society"[146]—perhaps an allusion to the heterodox scientific and philosophical ideas of not a few of the protagonists.[147] In 1946 Edgar Altenburg continued the tradition, supposing that cells contain invisible symbionts (*viroids*) that descend from "naked genes" in a distant precellular past. For these hypothetical ancestors, together with viruses and bacteriophage, he created Kingdom Archetista.[148]

Eukaryogenesis 2: science may discover ten

Cells were first described in the 1660s, but it was in the 1960s that cell biology came into its own.[149] The application of electron microscopy clarified many fundamental issues, including the organization of prokaryotic and eukaryotic cells. With very few exceptions, the latter have membrane-bounded nuclei; mitochondria; and structures (involved in mitosis and flagellar motility) that exhibit a characteristic "9 + 2" structure in cross-section.[150] To be sure, a few questions remained (somewhat) open at first—notably whether mitochondria do, or do not, occur in bacteria, blue-green algae, and red algae.[151] Other distinctions were clear-cut, including many concerning the most-fundamental issue of all: heredity.

The laws of inheritance, now credited to Gregor Mendel, were rediscovered at the very beginning of the Twentieth century. It soon became clear that "the phenomena of Mendelian heredity generally result from combinations, segregations and recombinations of the [nuclear] chromosomes in successive generations".[152] We return to chromosomes and heredity in our final chapter.

A few characters, however—notably patterns of leaf colouration—did not obey Mendel's laws. Suspicion immediately fell upon chromatophores. Erwin Baur (1909) demonstrated that plastids themselves carry mutable hereditary factors;[153] Otto Renner (1934) distinguished the nuclear *Genom* from the plastid-borne *Plastom*.[154] By the mid-1950s there was considerable evidence that plastids contain DNA and RNA.[155] Any remaining doubt was removed in 1962, when cell biologist Hans Ris, working with botanist Walter Plaut, reported that fibrils visible (under the electron microscope) in the chloroplast of *Chlamydomonas* disappear upon treatment with an enzyme that breaks down DNA.[156] They noted the striking ultrastructural similarities between the *Chlamydomonas* chloroplast and blue-green algae, including the presence of ribosome-like particles and fibrillar DNA,[157] and concluded that

> this similarity in organization is not fortuitous but shows some historical relationship and lends support to the old hypothesis of Famintzin (1907) and Mereschkowski (1905) that chloroplasts originate from endosymbiotic blue-green algae. . . . With the demonstration of ultrastructural similarity of a cell organelle and free living organisms, endosymbiosis must again be considered seriously as a possible evolutionary step in the origin of complex cell systems.[158]

The following year Sylvan and Margit Nass reported DNA in chick mitochondria, and acknowledged the "great deal of modern biochemical and ultrastructural evidence that may be interpreted to suggest a phylogenetic relationship between blue-green algae and chloroplasts[,] and bacteria and mitochondria".[159] Inspired by such reports, Lynn Sagan (later Margulis), a former student of Ris and Plaut, set out to understand cytoplasmic heredity.[160] Her review of the classical light-microscopic literature, written in 1964–1965 but published (after some fifteen editorial rejections) only in 1967,[161] set out the following scenario for the evolution of mitosing cells:

1. Prokaryotes evolved under a reducing atmosphere and diversified into anaerobic fermenters, chemo- and photoautotrophs, motile heterotrophs and, later, aerobic heterotrophs and respirers.
2. A heterotrophic anaerobe ingested an aerobe (the proto-mitochondrion). The endosymbiosis became obligate (*i.e.* the symbiont became established as mitochondria), resulting in "the evolution of the first aerobic amitotic amœboid organisms".[162]
3. Some of these organisms "ingested certain motile prokaryotes" which, in turn, became symbiotic in the host, forming primitive amœboflagellates. One of these became the ancestral eukaryote. "In these heterotrophic amœboflagellates classical mitosis evolved" over millions of years, presumably under an oxidizing atmosphere.[163]
4. Mitosis evolved stepwise. Six steps can be recognised based on the intracellular location(s) and function(s) of the 9 + 2 homologue. These steps can be

428　KINGDOMS, EMPIRES, AND DOMAINS

arranged into a plausible sequence by reference to nuclear division in present-day lower eukaryotes.[164]

5. Multiple lineages diverged as each stage progressed, resulting in present-day protozoa, fungi, animals, and other eukaryotes. In some lineages, plastids arose via a further symbiosis involving a photosynthetic prokaryote homologous to present-day blue-green algae. Twenty such endosymbiotic events can be identified, yielding red, green, yellow, and brown plastids.[165] In some cases, plastids were secondarily lost.

The article introduced symbiosis theory to a new generation of biologists (including the present author), and placed the origin of cells into the context of earth systems evolution. The scenario itself, however, is problematic. The 9 + 2 structures neither contain DNA, nor replicate by division.[166] Present-day motile bacteria (*e.g.* spirochætes) exhibit no 9 + 2 profile, encode no homologues of flagellar motility proteins, and swim not by undulation, but instead via the action of numerous tiny "bacterial flagella". Motile bacteria are not plausible precursors of flagella in eukaryotes.[167]

The scenario is taxonomically problematic as well. Margulis believed that it brought order to natural groups that Copeland had left isolated, and broke up taxa wrongly recognized on the basis of plastid colour.[168] It wreaked havoc, however, upon other taxa including green algae, to which it assigned six origins spread over four stages of the evolution of mitosis.[169] More than sixty primary lineages branch off the main trunk, many leading to a single genus or species. Yet (as mentioned above) a few years later she endorsed Whittaker's classification; lineages had to be bundled up, phyla reassembled and assigned to kingdoms.[170] How trade-offs were made—between mitosis and plastid colour, for instance—was not explained.

Margulis also sought to dissect the very first steps of eukaryogenesis. Using present-day organisms and communities as examples, she developed a further scenario in which a consortium of prokaryotes with complementary metabolic and energetic capabilities became stabilized in the form of a cell. Two of these symbionts contributed genes to a proto-nucleus, which at first remained tethered to a 9 + 2 structure that conferred motility. This assembly, the *karyomastigont organellar system*,[171] eventually (on five separate occasions) dissociated into 9 + 2 basal bodies on one hand, and a freestanding nucleus on the other, in the process giving rise to the cytoskeleton.[172] At some point after the symbiosis was consolidated but before the nucleus was liberated, the newly formed cell stepped across the prokaryote-eukaryote border: this "swimming chimera" was "the first protist". In this scenario, the most-recent common ancestor of modern eukaryotes came a little later,[173] and the proto-mitochondrial symbiosis later yet.[174] Modern protist lineages that lack mitochondria, she believed, arose at this pre-mitochondrial stage.[175]

Margulis was the first to infer that plastids and mitochondria arose in separate symbiotic events.[176] Like Whittaker, she drew evolutionary meaning from biological communities whether in thermal muds, termite guts, or a local bog.[177] She relied heavily on the light-microscopic literature,[178] even as ultrastructure and molecular phylogenetics were proving their worth. As a consequence, she made only superficial use of plastids as phylogenetic characters, and had to settle for arbitrary taxonomic boundaries among green algae and plants.[179] She repeatedly affirmed the five kingdoms, even when faced with a deep molecular dichotomy among prokaryotes (Chapter 23). One biologist who made very different decisions was Tom Cavalier-Smith.

At first, Cavalier-Smith did not accept symbiotic origins.[180] He derived the eukaryotic cell from a blue-green alga that, having lost its wall, became capable of endocytosis (and "cytosis"—controlled membrane fusion—more broadly). This, in turn, provided both selective pressure and physical mechanism for the endogenous formation of membrane-bounded compartments, including the nucleus and organelles.[181] Membranes had received little attention in the light-microscopic literature, but electron microscopy revealed that they share a common basic structure.[182] Cavalier-Smith built on this commonality, even after he accepted (*circa* 1980) that plastids had arisen symbiotically.

Electron microscopy reveals that plastids of red algae, green algae, and plants are enclosed in two membranes.[183] The same is true of free-living blue-green algae and other Gram-negative bacteria. A blue-green alga swallowed up (per Merezhkowsky) by a moneran would find itself inside three membranes: its own two, plus the phagosome membrane of the host. To avoid becoming lunch, it had to escape from the latter. This is why primary plastids have only two membranes and are situated in the cytosol, not in the host's intracellular membrane system.

Electron microscopy also shows that plastids in other photosynthetic organisms are surrounded by three or four membranes.[184] In those with yellowish, golden, or brown plastids (cryptomonads, diatoms, and *Vaucheria*, for example) the outermost periplastic membrane (PRER) bears ribosomes on its cytoplasmic face, and is usually confluent with the outer membrane of the nuclear envelope.[185] These organisms are (bi)flagellate at some stage in their life history, and at least one flagellum is decorated with hairs (*mastigonemes*) of a characteristic tubular construction.[186] The PRER mediates the transport of metabolites between plastid and cytoplasm, and is the site of mastigoneme assembly. Cavalier-Smith argued that this complex arrangement is so unlikely to have arisen more than once, and its loss would be so disadvantageous, that its presence defines a kingdom, which he named Chromista.[187]

If the two inner membranes arise (as before) from the proto-plastid, and the outer one from the host phagosome, what of the third? It must derive from the *plasma* membrane of the symbiont, which was therefore a *eukaryotic* alga. This might seem improbable, but in two groups of protists we find the proverbial smoking gun: a *nucleomorph*, a reduced but recognizably eukaryotic relict nucleus, between the second and third membranes—that is, in the compartment that was once the symbiont's cytoplasm.[188] From pigmentation and molecular data, we can infer that

430 KINGDOMS, EMPIRES, AND DOMAINS

a red alga gave rise to the nucleomorph and plastid of cryptomonads, and a green alga to those of *Chlorarachnion*.[189] Some dinoflagellates retain an analogous relict nucleus from a more-recent green algal symbiont.[190] The *Euglena* plastid, with three (rather than four) surrounding membranes, must have lost the symbiont plasma membrane, and therewith any chance of maintaining a nucleolus.[191]

Electron microscopy moreover reveals that the malarial parasite *Plasmodium* contains a plastid-like but unpigmented structure, the *apicoplast*. Surrounded by four membranes, it retains a reduced but active genome: evidence of a former symbiosis with a eukaryote, probably a red alga.[192]

Cavalier-Smith put these endosymbioses to use in his classifications. At the highest level he accepted two superkingdoms, Prokaryota and Eukaryota.[193] Within the latter he followed Whittaker in recognizing Plantae, Fungi, and Animalia, although with sharper definition at the boundaries. He then divided the remaining eukaryotes into Chromista, which exhibit the set of characters described above (and thus descend from a secondary symbiosis); and Protozoa, which do not. Plants descend from a primary endosymbiosis with a blue-green alga, and therefore include red and green algae.[194] Fungi differ from protozoa by the presence of a chitinous wall around the vegetative cells. Animals are "ancestrally phagotrophic multicells with collagenous connective tissue between two dissimilar epithelia", and include choanoflagellates and sponges.[195]

Cavalier-Smith recognized from six to ten kingdoms as he sought a balance among explanatory power, major evolutionary trends (as he saw them), and simplicity. Against a background of burgeoning data, his thinking evolved on certain potentially unifying themes among eukaryotes, including whether the (inferred) number of ancestral 9 + 2 basal bodies is suitable as a high-level taxonomic character (the *unikont/bikont criterion*),[196] and whether present-day eukaryotes are ancestrally amitochondriate (the *Archezoa hypothesis*).[197] He carved out kingdoms of chytrids, cryptomonads, euglenoids, and red algae, then subsumed them into others.[198] Especially within Protozoa, subkingdoms and phyla rose and fell as he refined scenarios for the evolution of cells. He embraced paraphyletic taxa, and held that systematics is as much an art as a science.[199] Nor did he hesitate to value some types or combinations of data over others, or to second-guess trees inferred from molecular sequences. His classifications are grounded in the diversity of eukaryotic cell biology, and he maintained—rightly, in my opinion—that five kingdoms are simply too few to convey this diversity.[200]

Summary

By the 1980s it was widely agreed that the evolutionary discontinuity separating prokaryotes and eukaryotes is the most profound in the living world. No mesokaryotic cell type stands in the middle. Even so, the chasm is spanned by bridges of greater or lesser antiquity. Bacteria contributed mitochondria and plastids to the eukaryotic lineage. If different members of an ancestral consortium contributed DNA to the proto-nucleus, multiple ancestry runs even deeper.[201] Secondary and tertiary endosymbioses mix eukaryote with eukaryote, compounding all the above. Famintsyn and Merezhkowsky used lichens as a case in point, but (by convention) taxonomists ignore their dual nature and classify lichens into the genus to which the fungal partner belongs.[202] By contrast, two modern kingdoms (Plantae and Chromista) are *defined* by symbioses.[203]

With the third of our Twentieth-century developments—molecular sequences— we arrive at the very basis of heredity. The consequences for our view of the living world at kingdom level could scarcely have been foreseen.

23

Genes, genomes, and domains

> Here the last word is with the experimental geneticist; and one cannot deny
> that to-day the *infima species*, the ultimate element, of biological classifica-
> tion is the gene.
>
> —Francis Arthur Bather, President's address to the Geological Society of
> London (1927)[1]

Introduction: the molecular basis of heredity

Genetics emerged as a scientific discipline in the first decades of the new century. Botanists and zoologists investigated the principles of heredity, rediscovering the work of Gregor Mendel.[2] William Bateson introduced the word *genetics*,[3] while Wilhelm Johannsen called the unit of hereditary information a *gene*,[4] and referred to *genotype* and *phenotype*.[5] Debate raged over whether the saltatory mutations often observed in phenotype could be caused by gradual changes in individual genes.

Quite what these genes might be, chemically and physically, proved elusive. Friedrich Miescher had found that nuclei contain nucleic acid and protein,[6] and Albrecht Kossel determined that a basic protein, *histone*, forms a salt-like complex with nucleic acid.[7] Thomas Hunt Morgan and Alfred Sturtevant established that genes are borne on chromosomes, but were incurious as to their chemical nature.[8] By about 1930 it had become apparent that nucleic acid is a high-molecular-weight polymer of deoxyribose to which four types of subunit (two purine and two py-rimidine) are attached.[9] The subunits might be stacked one after another "like pennies in a bank clerk's roll".[10] Problematically though, DNA extracted from diverse sources seemed too uniform in composition to account for the diversity of living forms.[11] Perhaps DNA is just a scaffold, and proteins are responsible for heredity. Franz Hofmeister and Emil Fischer had independently shown that (some) proteins are linear polymers of α-amino acids,[12] although belief persisted that other proteins might be colloids, or form two- or three-dimensional networks.[13]

Help came from an unexpected direction. In 1928 Frederick Griffith reported that a rough avirulent form of the pneumococcus bacterium could be transformed into a smooth virulent form by a non-living but heat-sensitive substance.[14] Oswald Avery and colleagues identified the transforming material as DNA,[15] while Theodosius Dobzhansky drew the analogy between bacterial transformation and

---

Kingdoms, Empires, and Domains. Mark A. Ragan, Oxford University Press. © Oxford University Press 2023.
DOI: 10.1093/oso/9780197643037.003.0023

GENES, GENOMES, AND DOMAINS   433

genetic mutation.[16] George Beadle and Edward Tatum used induced mutations, and genetics, to demonstrate that one gene specifies one protein.[17] Once James Watson and Francis Crick showed DNA to be topologically linear,[18] it became overwhelmingly likely that genetic information is stored in DNA, copied, read out, and translated into protein in a fundamentally linear manner. The biochemical details—the triplet genetic code, messenger and transfer RNAs, translation on ribosomes—followed.[19]

The new molecular biology could not immediately be brought to bear on biological classification. Nucleic acids were particularly difficult to characterize at the required level of detail. Proteins proved somewhat easier to work with, and chemists began to describe families of proteins within which meaningful comparison can be made across taxa. Protein structure was dissected into levels: primary (the linear order of amino acids), secondary (local folds and coils), tertiary (overall three-dimensional shape), even quaternary (assembly of similar or dissimilar proteins into a multimer). To a first approximation, the primary structure (*sequence*) of a protein is specified by that of the corresponding gene, whereas its higher-order (*folded*) structure arises spontaneously through subsequent physical, electrostatic, and ionic interactions, and (together with subcellular localization) is the basis of function. At first, it was immensely difficult to determine the sequence of a protein.[20] Even today, it may be next to impossible to isolate and sequence a protein if it is membrane-bound, transiently expressed, and/or present in only a few copies per cell.

Molecular phylogenetics before sequences

The first proteins sequenced were small, water-soluble, abundant, and readily purified. As with many complex problems, sequencing was approached by breaking the protein into smaller pieces, which were then characterized and computationally reassembled. Already in the 1940s it had been observed that the same protein in different species (say, insulin in cattle and humans) yields similar oligopeptides upon hydrolysis. From about 1954, František Šorm and colleagues used patterns of di-, tri-, and tetrapeptides to compare proteins, and interpreted similarities as the result of biological evolution.[21] From the sketchy data available in 1955, Sidney Fox and Paul Homeyer concluded that "protein synthesis has not, in the main, yet become sufficiently diverse through molecular evolution to yield substantially unrelated proteins".[22]

The first proteins sequenced from multiple genera—insulin, fibrinopeptides, globins, pituitary hormones—occur only in animals (or even mammals), and thus could not be informative on kingdoms.[23] These early comparisons did, however, confirm that sequences are "documents of evolutionary history"[24]—history that could be reconstructed by computationally based analysis. In 1963, Emanuel Margoliash compared sequences of mitochondrial cytochrome *c* in organisms representing two kingdoms (six animals plus *Saccharomyces*), and used a simple statistical model to estimate how long ago they had diverged from one another.[25] By 1968, cytochrome *c* sequences were known from twenty-six animals, three fungi, and a plant (Figure 23.1).[26] According to Margoliash and colleagues,

> the phylogenetic relations depicted fit remarkably well classical zoological concepts of the evolutionary relations of the species listed. One is therefore forced to conclude that, in some cases at least, the molecular evolutionary transformations of a single gene yield a statistically valid sample of the evolutionary changes of the species as a whole.[27]

According to this analysis, wheat cytochrome *c* is about equidistant from the ancestral sequence inferred for animals, and that inferred for fungi. Much about

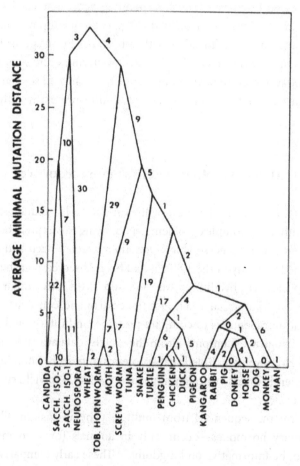

Figure 23.1. Margoliash, Fitch, and Dickerson, statistical phylogenetic tree based on minimal mutation distances between cytochromes *c* of selected fungi, plants, and animals. *Brookhaven Symposia in Biology* 21 (1968), page 274. Hathi Trust.

the tree is data-dependent: had the authors included a cytochrome *c* from (say) a jellyfish or worm, they would have inferred an older ancestral animal sequence.[28] Sequences of cytochromes *c* soon became available from protists[29] and a bacterium, giving the first five-kingdom protein tree.[30] The statistical midpoint, sixteen units along the *Rhodospirillum* cytochrome c_2 branch, was taken to be the starting point from which evolution had run forward, culminating in the branch tips (present-day sequences). Read this way, the tree confirms the prokaryote-eukaryote dichotomy, while fungi and animals are sister groups.[31] Despite its limitations, this was a promising start for protein-sequence phylogenetics.

Even so, the cytochrome tree pointed to difficulties ahead. In eukaryotes, cytochrome *c* is located in the mitochondrion: might it be a marker of mitochondrial (not nuclear) evolution? Its gene is nuclear,[32] but might have been transferred there from the proto-mitochondrial symbiont. Midpoint rooting is statistics, not biology. Nor was useful detail forthcoming for the bacterial branch: *c*-type cytochromes occur in many bacteria including cyanobacteria and their plastid descendants, but (except in *Rhodospirillum* and its relatives) are not of the mitochondrial subtype.[33] Adding these other types to the analysis would give a tree of cytochromes, but not of the corresponding organisms. In any case, bacterial cytochromes may be too divergent in structure or function, too information-poor, and/or at their limit of evolutionary change;[34] their genes were suspected to have been transferred laterally among species.[35] We return below to many of these points.

At a practical level, the ultimate barrier to the use of protein sequences in large-scale classification is simply that so few proteins are represented in all kingdoms of life. A better molecular marker was needed: one that is universal, orthologous, of consistent function, insusceptible to lateral transfer, and of a size, complexity, and dynamic range that makes it informative on phylogenetic diversification at all phyletic depths and taxonomic ranks. Ideally, the molecule should get to the heart of the condition we call *life*. This was a tall order, but—in the view of Carl Woese—a candidate was already on offer. It was not, however, a protein.

The ribosomal RNA Tree of Life

> A basic goal of biology is to account for the evolution of the cell. Emergence of the translation apparatus is the single most important event in this evolution, for capacity to translate is what defines genotype and phenotype.
> —Carl Woese[36]

The translational apparatus—the ribosome, attendant enzymes, and small RNAs—was in every respect worthy of study. Albert Claude recovered RNA-rich particles from animal-cell lysates in the late 1930s; they were linked with protein synthesis in the 1950s, and named *ribosomes* in 1958.[37] A ribosome consists of two subunits of different size, composition, and role in translation; each subunit contains, inter

436 KINGDOMS, EMPIRES, AND DOMAINS

alia, one large rRNA (the *r* indicates *ribosomal*)[38] and several dozen characteristic proteins. Fred Sanger, whose team had been the first to sequence a protein (insulin), undertook to sequence RNA using a similar approach. But when Woese took up a faculty position at the University of Illinois in 1964, no RNA molecule had been fully sequenced.[39] Small RNAs eventually yielded,[40] but the much larger rRNAs were beyond the technology of the day. So as Šorm had compared proteins by their constituent oligopeptides (above), from 1971 Woese (later joined by colleagues in Canada and Germany) digested ssu-rRNAs with nuclease, and drew up lists (*catalogues*) of their oligonucleotides. By 1985, when the approach was superseded by direct sequencing, catalogues were available for ssu-rRNAs of more than 400 organisms.[41]

These catalogues were spectacularly rich in information. Some oligonucleotides are highly conserved across taxa, or vary in a tightly coordinated manner: Woese reasoned that these must be involved in the oldest, most-fundamental steps of translation, and have the most to tell us about the deepest features in the phylogenetic tree. By contrast, other oligonucleotides are much less constrained across taxa, and might characterize a genus, species, or strain. Even the latter, though, are insulated from the vagaries of physiology in a way that (for instance) metabolic enzymes, and the corresponding genes, could never be. We have in rRNA a general-purpose chronometer, informative for classification in all groups of organisms, and at all taxonomic ranks.[42]

The first several dozen[43] catalogues fell into two broad types, corresponding to the ssu-rRNAs of bacteria and plastids on one hand, and the somewhat larger ssu-rRNAs of the eukaryotic nucleocytoplasm on the other. But in 1977, Woese and George Fox reported that ssu-rRNAs of methanobacteria yield a third type of catalogue, distinct in equal measure from the bacterial and eukaryotic patterns.[44] Methanogens lack a nucleus and have a wall, but had received little attention from biochemists, and none (to that point) from molecular biologists. On the basis of rRNA catalogues from four strains of methanogens representing two genera, Woese and Fox proposed a new "urkingdom" of life, the *archaebacteria*.[45] Archaebacteria—later renamed archaea, to avoid any implication that they are bacteria[46]—turn out to be rather common. Some occupy extreme environments (hypersaline lakes, hot springs, deep-sea thermal vents); many others live in unremarkable habitats.[47] Chatton's dichotomy suddenly had unanticipated company.

The gene encoding ssu-rRNA in *Escherichia coli* was sequenced in 1978.[48] Many more followed, and (as expected from the catalogues) showed taxon-specific variations on an obviously orthologous theme. Within a decade, Woese could present a "universal" tree of rRNA sequences from bacteria, archaea, and eukaryotes.[49] This tree, variously redrawn (see below), became an icon of molecular biology, indeed of modern visual culture—the *Tree of Life*.[50] Woese disliked the name: it was a tree of ssu-rRNAs. Yet *Tree of Life* holds an element of truth, as it is through translation that phenotype emerges from genotype, and this is a central part of what cells, and

GENES, GENOMES, AND DOMAINS   437

life, are about. rRNAs, moreover, are faithful markers of lineages. Ribosomes are intricate machines, under relentless selection for accuracy and efficiency; it beggars belief that the cell (and organism) could remain competitive, or even viable, with a foreign rRNA at the heart of its ribosomes.[51]

Unlike cytochrome *c*, ssu-rRNA is encoded in every entity capable of translation: archaea, bacteria, and (in eukaryotes) the nucleocytoplasm, relict nuclei, plastids, and mitochondria. At long last, microbiology would be put on a phylogenetic basis, and all biology united in a common tree. Detailed rRNA trees confirmed the cyanobacterial ancestry of primary plastids,[52] the α-proteobacterial ancestry of mitochondria,[53] and the complex evolutionary histories of chromists. The rRNAs of (supposedly mesokaryotic) dinoflagellates fall solidly among those of eukaryotes. With bacteria and archaea on separate primary branches, biologists were urged to abandon *prokaryote*[54]—or if they must, retain it as an informal term for anucleate (but otherwise quite dissimilar) beings. By contrast, Eukaryota was seen to have a distinct reality, based in the nucleocytoplasmic lineage.

In anticipation of iconoclasm to come (below), let us be precise about ssu-rRNA and the iconic tree. Woese found the third type of ssu-rRNA in the catalogues; no rRNA had been sequenced when his visage graced the front page of *The New York Times*, over the headline Scientists Discover a Form of Life That Predates Higher Organisms.[55] To be sure, the sequences and tree added useful detail. For the latter he handcrafted sequence alignments of deep structural sophistication, but employed a simple statistical method for tree inference per se. The tree made biological sense: in it we find subtrees of metazoa, plants, fungi, and so forth. The (literal) headline act, though, was the third kingdom (later *domain*).[56] Three primary branches, three biological domains: but for Woese, domains were not statistical constructs, branches on a computed tree. He instead defined a domain by its *signature*—the "positions in the molecule that have a highly conserved or invariant composition in one [domain], but a different (highly conserved) composition in one or both of the others". These became fixed early in the evolution of each major lineage, and form or stabilize lineage-characteristic features of the rRNA folded structure.[57]

The first universal rRNA tree was unrooted (Figure 23.2). Woese toyed with midpoint rooting,[58] but did not much trust the branch lengths: the rRNA chronometer is structural, not statistical.[59] We might nonetheless locate the root by some other stratagem. Miyata Takashi and colleagues had used ancestral duplications in other gene families to position the root between bacteria (on one hand) and archaea and eukaryotes (on the other);[60] Woese adopted this root for his rRNA tree (Figure 23.3).[61] With this fateful step, Prokaryota lost any remaining claim to phylogenetic coherence.

Further adventures were to befall the Tree of Life (below); but let us return, for now, to the broader story of molecular phylogenetics in the late 1980s and into the 1990s.

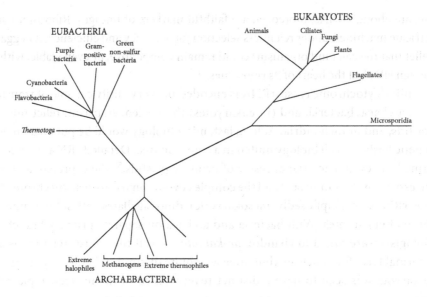

Figure 23.2. Universal phylogenetic tree determined from rRNA sequence comparison. Republished by permission of American Society for Microbiology, from Bacterial evolution. CR Woese, *Microbiological Reviews* 51(2), 1987 at page 231. Permission conveyed through Copyright Clearance Center, Inc.

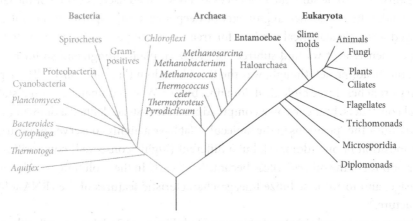

Figure 23.3. Universal phylogenetic tree in rooted form, showing the three domains, after Woese, Kandler, and Wheelis, *Proceedings of the National Academy of Sciences USA* 87 (1990), at page 4578. Redrawn by Mauricio Lucioni (2013) with addition of further branches, at Wikimedia Commons, "Phylogenetic Tree, Woese 1990", Creative Commons Attribution-Share Alike 3.0 Unported license.

The molecular consensus erodes

As the 1980s progressed, biologists increasingly turned their attention to genes that encode proteins. Some proved unsuitable as phylogenetic markers. Many others yielded useful trees, even if only within a kingdom or phylum. To a first approximation, these trees were broadly topologically consistent among themselves, and with the ssu-rRNA tree. Instances of disagreement—often confined to a branch or region, rather than being systemic—could be ascribed to the limitations of a particular gene as a chronometer, or to biased taxonomic sampling, cryptic paralogy, imperfect alignment, or inadequate methods of inference—legitimate concerns, to be sure. It was widely assumed that as more organisms were sampled and inference methods improved, the disagreement would melt away.

Yet disagreement persisted, and by the mid-1990s risked becoming the elephant in the room. Careful analyses of twenty-four protein datasets by Brian Golding and Radhey Gupta,[62] and of sixty-six protein datasets by Jim Brown and Ford Doolittle,[63] made it clear that topological inconsistency was real and, in bacteria and archaea, affected regions of trees inferred from many genes involved in biosynthesis, metabolism, or energetics. Brown and Doolittle concluded that "the evolution of genomes, or parts thereof, is sometimes decoupled from that of the host organism. Lateral gene transfer is one example of a decoupled process."[64]

Their concern was not hypothetical. By then it was well-known that DNA can be carried on viruses, plasmids, and other mobile elements from one unrelated organism to another. Some organisms take up DNA directly from their environment—presumably as food, but with some ending up in the genome. In a laboratory context, these processes form the basis of molecular biotechnology: the reality of their existence was not in doubt, but it was a surprise to encounter them so often in natural populations. Bacteria deploy defensive systems, but the spread of antibiotic resistance in livestock yards and hospitals is daily testament that these defences are not impermeable. Mapped onto a phylogeny of organisms, genes are seen to have moved *laterally* or *horizontally* from one branch of the tree to another. The unsettling conclusion was that in many cases, sequence trees disagree because different regions of the genome actually have different evolutionary histories.

Thinking laterally about genomes

Throughout history, technological innovation has opened new vistas. The microscope gave us entry to a hidden world of microorganisms, zoospores, cells, and (as optical resolution and associated technologies improved) organelles and chromosomes. Electron microscopy revealed supernumerary membranes and relict nuclei, explained by secondary symbioses. Molecular biology gave us sequence trees, closing certain questions (the bacterial origins of organelles) while opening others (why protein trees disagree). Genomics would prove even more revolutionary.[65]

Even in the 1990s, molecular phylogenetics was slow work. It is not easy to clone genes from unfamiliar organisms. Genes may have been lost, replaced (as with bacterial cytochromes) by paralogues, or have evolved to unrecognizability. Many organisms are difficult to obtain and work with; some are dangerous pathogens.

440 KINGDOMS, EMPIRES, AND DOMAINS

Comparative biologists quickly discovered that DNA laboratory protocols had been optimized for mammalian cells and *Escherichia coli*. Would it not make sense for a consortium of research groups, each with specialized expertise in a particular organism or taxon, to sequence *all* the genes in, say, twenty or fifty organisms distributed across nature, then share the data amongst themselves and more widely?

Comparative biologists were not alone in thinking along these lines. Many in the biomedical community predicted that a complete sequence of the human genome would bring new understanding, diagnosis, and treatment of disease. Geneticists hoped for the mouse, fruit fly, worm, and yeast genomes; agricultural scientists for cattle, wheat, and rice. Genome sequences were needed too for agents of infectious or parasitic disease. Some of these (including human) were too large and complex to tackle at first, but strategies, methods, and infrastructure could be developed by sequencing smaller, more-tractable genomes. Many interests were aligning. Fortunately, a few funding agencies reached out to evolutionary biologists for advice on "what to sequence next".[66] Visionaries dared to dream that someday, the sequence of every gene in every organism of interest would be freely available online.[67]

The first bacterial genome sequence was reported in 1995.[68] Within a decade, genomes had been sequenced from more than one hundred genera of bacteria and archaea, and phylogenomic analyses were mounted to map evolutionary relationships and identify instances of lateral genetic transfer (LGT). The results were striking: taken altogether, the data largely confirmed the ssu-rRNA tree. Yet all microbial genomes, and almost all of their gene families, show evidence of LGT. In some bacterial genomes, fifteen percent or more of genes display evidence of lateral ancestry. To a first approximation, genes of replication, transcription, and translation have been less affected than those encoding short-term selectable functions including nutrient uptake, biosynthesis, and metabolism.[69] Nuclear genes of eukaryotes are much less susceptible to LGT, although not immune.[70] LGT has been more frequent (or successful) among closely related organisms, but instances abound of transfer across classes, phyla, and (to a lesser extent) domains.

So as feared, different regions of microbial genomes can have different evolutionary histories. The subversion, however, runs deeper. Genes are not privileged units of lateral transfer.[71] Lateral origins are indicated for "non-coding DNA, portions of genes, intact genes, multi-gene clusters, operons, plasmids, transposable elements and pathogenicity islands".[72] Different regions of *a given gene* can, and often do, have multiple evolutionary histories. This can undermine gene-centric analyses, whether in a tree-based or network-based paradigm.

Genomes and pan-genomes

Equally unexpected was the fluidity of genome size and gene content over evolutionary (and even historic) time. The matter had not been explicitly framed before the era of genomics,[73] but had it been, the observed extent of variation could scarcely have been anticipated. The best-studied case is, unavoidably, *Escherichia coli*. By late 2020, more than 140,000 *E. coli* genomes had been sequenced, and over 10,000 assembled and annotated to high quality.[74] Of these, the smallest contains 3700 genes; the largest, 5946 genes. About half of the genes in an average *E. coli* have counterparts in 99 percent or more of investigated strains, but fewer than 300 genes are represented in absolutely every *E. coli* genome. Many isolates contain genes not previously recorded in the species. As a consequence, the set of genes present in at least one *E. coli* genome (its *species pan-genome*) is gigantic—more than 125,000 in one recent survey—and shows no sign of levelling off as further isolates are sequenced.[75] This is referred to as having an *open* pan-genome. For their part, the pangenes can be shared, to differing degrees, across genera and more broadly.[76]

Not all genomes operate like that of *E. coli*. From limited data, it appears that bacteria that live in biodiverse or mixed communities can have open pan-genomes, whereas an intracellular or solitary lifestyle correlates with a *closed* pan-genome.[77] It is too early to generalize about archaeal genomes. Molecular-sequence trees indicate that archaeal genes can be moved about. Archaea host a rich diversity of viruses—potential vectors of genetic material—some with counterparts among bacteria and/or eukaryotes. Most types of archaeal viruses, however, are taxon-specific and exchange few genes among themselves. Many archaea, and nearly all archaeal hyperthermophiles, have defence mechanisms, similar to those in bacteria, against incoming DNA.[78]

By contrast, genomes of plants and metazoa behave very differently (setting aside the historical relocation of genes from organelles to the nucleus). Chromosomal regions are regularly duplicated, moved about, inverted, and deleted; gene families expand and contract. But genes are rarely lost altogether; nor are they imported from outside the genus. The genome of rice (*Oryza sativa*) is more variable than many, but its pan-genome is (proportionally) two orders of magnitude smaller than that of *E. coli*.[79]

Genomes from the environment

Microbiologists have long bemoaned the recalcitrance of bacteria to laboratory culture.[80] Relatively few bacterial phyla are represented in culture; the situation is worse with archaea. Woese had noted that certain oligonucleotides appear in rRNA catalogues of all taxa. In the 1980s, molecular biologists began to use these universally conserved regions to amplify rRNA genes directly from bulk DNA recovered from environmental samples—soil, ocean water, microbial biofilms, and the like. The amplified rRNA genes were then sequenced and located in the rRNA tree, opening a direct window into environmental biodiversity.[81]

With advances in sequencing technology a genome-scale counterpart became possible, known as *metagenomics*. Bulk DNA is sequenced in depth, and the

442 KINGDOMS, EMPIRES, AND DOMAINS

resulting reads are matched against online databases to determine what taxa, genes, and functions are present. Computational methods can be further employed to distinguish and (in favourable cases) assemble microbial genome sequences from metagenomic data. The success of *genome-resolved metagenomics* depends on a host of factors including the number, size, and distinctiveness of genomes in the sample, and the quality and depth of sequence data. Opportunity abounds for error, including the mistaken inclusion of genes into, or their exclusion from, the draft genomes.[82]

Environmental rRNA sequencing and genome-resolved metagenomics have been successfully applied, sometimes in parallel, to discover new microorganisms—very few of which are subsequently cultured, or even observed under the microscope. The bacterial and archaeal domains turn out to be greatly more diverse than earlier appreciated, with broad new branches often deep in the rRNA tree. Even so, the three Woeseian domains—Bacteria, Archaea, and Eukarya—have survived intact in the rRNA tree.[83] Some of the new rRNAs differ in "universally conserved" regions too, implying that even greater biodiversity may lie undiscovered.[84]

In 1984, Jim Lake and colleagues used electron microscopy to recognize *four* structural types of ribosomes: one each from bacteria, the eukaryotic nucleocytoplasm, and the two recognized major subdivisions of archaea. The type recovered from crenarchaeotes (including *Sulfolobus* and *Thermoproteus*) most closely resembles the eukaryotic type, and on that basis Lake *et al.* recognized Eocyta as the sister kingdom to eukaryotes.[85] Several subsequent small-scale analyses of protein-coding genes supported the *eocyte tree* against the three-domain alternative, as did a 2008 study by Martin Embley and colleagues using a 53-gene set and sophisticated tree-inference methods.[86]

Metagenomics was soon brought to bear on the question. Almost immediately, a third phylum was proposed within Archaea[87] to accommodate the sponge symbiont *Cenarchaeum* and several relatives attested only by their ssu-rRNA sequences.[88] Genome-resolved metagenomic analysis of aquatic sediments revealed a panoply of new archaeal phyla, including some claimed to "bridge the gap between prokaryotes and eukaryotes".[89] The various phyla of archaea, old and new, have been collected into four *superphyla*—Asgard, DPANN, Euryarchaeota, and TACK. In multi-protein trees the Asgard metagenomes appear as a specific sister-group to eukaryotes, and assemble to include genes that specify components of the cytoskeleton, protein-degrading machinery, and membrane-trafficking system[90]—*eukaryotic signature proteins*, involved in characteristic processes and conserved across most eukaryotic lineages.[91]

Yet we must tread cautiously. The protein sequences on which these analyses are based retain little phylogenetic signal at this phyletic depth.[92] Reanalysing the Asgard data, Bill Martin and colleagues could find no evidence that the genes encoding these proteins even arise from a common biological source: the supposed metagenomes may be "unnatural constructs, genome-like patchworks of genes" joined artefactually by the bioinformatics.[93] If so, some or all of their supposed

GENES, GENOMES, AND DOMAINS   443

eukaryotic genes may not be a physical part of these (or any) archaeal genomes.[94] Resolution of the matter is likely to require the analysis of actual organisms.[95]

Retrospective: the domains of life

Given all we currently know about organisms, genes, and genomes, how should we represent the relatedness and wondrous diversity of life? How many biological domains should we recognize? The question has technical aspects, which we address first; but it also reaches deeply into how we conceptualize the biological world.

Over twenty-six centuries, naturalists have sought order in the natural world, using the tools at their disposal: philosophy, logic, analogy, or similarities of form, structure, or development. From species upward, taxa at each rank were aggregated by shared similarity into more-inclusive taxa at the next-higher rank, yielding a taxonomic hierarchy. After Lamarck, Darwin, and especially Haeckel, classification became entwined with genealogy, specifically treelike descent with modification. There is no magic key for turning a tree (much less a network) into a classification. A tree depicts a continuous process; its branches may be asymmetric, irregular, variously sparse or dense. Hierarchical classifications are discrete, with a smallish number of standardized categorical ranks. Nor does phenotypic similarity necessarily track genealogy.

From the early Twentieth century, various criteria and practises were developed to coordinate classification—that is, hierarchical classification based on shared similarity—with genealogy. Criteria based on branch points and subtrees can be objective and precise; decisions involving similarity and difference, or how best to balance the size and diversity of taxa at a given rank, draw instead on established practise, expert judgement, and expediency.[96] Three schools of classification emerged,[97] of which two are relevant here: *evolutionary systematics*, in which both similarity and genealogy are used to delineate taxa; and *phylogenetic systematics* or *cladistics*, in which taxa are based on genealogy alone.[98]

Until now I have skated lightly over technical issues, but we can no longer avoid *monophyly*, *paraphyly*, and *polyphyly*. There is a rich tradition of defining these terms in subtly (or not so subtly) incompatible ways.[99] A minimal definition holds a group monophyletic if all its members stem from a common ancestor.[100] Both schools agree that taxa must, at minimum, be monophyletic in this sense. A group that stems from a common ancestor but does not include all its descendants is *paraphyletic*. Biologists are divided over the validity of paraphyletic groups in systematics. According to evolutionary systematists, a species that evolves beyond the similarities that define a taxon should be placed into a new purpose-built taxon; this renders the former taxon paraphyletic, and that is quite alright.[97] Cladists may not agree among themselves on every theoretical and practical issue,[101] but are united in rejecting paraphyletic groups as formal taxa. Finally, *polyphyletic* means

444 KINGDOMS, EMPIRES, AND DOMAINS

having more than one line of ancestry. No modern school of systematics accepts polyphyletic taxa.

The polyphyletic assemblage most often claimed as a taxon is one of viruses. As we saw in the previous chapter, Novák established Aphanobionta for unseen bionts, while Barkley, Rothmaler, and Jeffrey each set up a kingdom for viruses.[102] More typically, viruses are excluded from such classifications on the grounds that they are not living entities. It is certainly hard to accept that viroids—at smallest, a single 246-nucleotide strand of RNA[103]—are alive; yet the largest viruses encode more than 1000 proteins, including components of the translational apparatus.[104] More-compelling arguments against a domain of viruses include their polyphyly, the volatility of viral genomes through time, and the concomitant absence of ancestral lineages.[105] Acytota, a proposed domain encompassing viruses, bacteriophage, viroids, ribozymes, transposons, and various low-complexity elements of eukaryotic genomes[106] is breathtakingly less justifiable. Here we focus on the domains of life.

The **bacterial domain** arose from a common ancestor. We infer this from the structure and sequences of core translational molecules, not from bacterial morphology (simple), ultrastructure (missing in action), or physiology (continually reshaped by LGT). From the genetic perspective, the domain is a vast, uneven network: vertical transmission predominates in some regions, lateral signal in others. DNA flows (on viruses, plasmids, and the like) across the domain, with some taxa connected by floodways, others by narrow rivulets or dry riverbeds. LGT has not, however, reduced bacteria to physiological uniformity: distinctive taxa are everywhere. Bacteria carve out and occupy specialized niches today, as they have for billions of years.[107] The plastid genomes (and photosynthetic genes more broadly) of green plants and red algae tell us that oxygenic photosynthesis has gone hand in hand with the cyanobacterial lineage of rRNA for a billion years. Other physiological types (gut microbiota, for example) have been assembled more recently. A broad basal radiation, known only from metagenomic data, seems to be composed mostly of obligate fermenters.[108]

We do not yet have an adequate overview of the **archaeal domain**. Morphology is simple (or entirely unknown), but molecular data point to a single evolutionary origin. Their transcriptional and translational systems are strikingly dissimilar from those of bacteria, instead resembling their eukaryotic counterparts in sequence and organization. The domain was initially subdivided into two kingdoms[109] or phyla,[110] but as we have seen, many candidate phyla have subsequently been put forward on the basis of metagenomic data. The diversity of archaeal viruses, and the presence of defensive mechanisms against incoming DNA, imply that LGT is frequent. It remains to be seen whether LGT is as pervasive among archaea as among bacteria. Nonetheless, we are unlikely to be far off the mark in characterizing the archaeal domain, like its bacterial counterpart, as a vast, uneven genetic network. Modest gene flow between the two domains does not, at this point, call into question the coherence or distinctiveness of either.

GENES, GENOMES, AND DOMAINS 445

One group of organisms remains, unique in many respects. Their cells exhibit a membrane-bounded nucleus, an endomembrane system, a protein-based cytoskeleton, mitochondria, and the 9 + 2 structures described in Chapter 22. In the nucleus we find linear chromosomes that end in telomeres.[111] Introns interrupt many nuclear genes, necessitating downstream measures (splicing) carried out on ribonucleoprotein machines (spliceosomes) larger and compositionally more dynamic than ribosomes.[112] The nuclear-encoded rRNAs, and many proteins,[113] indicate a single evolutionary origin. Some members can attain great size and complexity; individuals of one species presume to ask questions such as these. There is much to recommend **eukaryotes** as a **third domain of life**.

Much about eukaryotes is relatively uncontroversial, so let us begin there. They have arisen from a common ancestor,[114] and (broadly speaking) have diversified in a treelike manner. Morphology, ultrastructure, and molecular sequences are largely concordant along the animal, plant, and fungal branches. It has taken longer to work out relationships among the protists, as potentially unifying concepts—archezoa,[115] unikonts versus bikonts,[116] algae versus protozoa—had to be abandoned. Most eukaryotes are now collected into five[117] or six[118] informal but probably monophyletic *supergroups* of an unspecified rank more-inclusive than kingdoms. Internally, each supergroup reflects a historical dynamic involving treelike descent with modification; physiological specialization; primary, secondary, and higher-order symbioses; and secondary loss of genes and structural features. Few data speak to the branching order among supergroups, so they are usually shown as arising, more or less simultaneously, from a phylogenetic "big bang".[119] None, however, is ancestrally amitochondriate. Little ongoing LGT directly from bacteria or archaea, or from one eukaryote to another, is in evidence.[120]

In universal trees, the origins of **mitochondria and plastids** appear as crosslinks: the former from α-proteobacteria to the eukaryotic lineage basal to the most-recent eukaryotic common ancestor (*i.e.*, just below the big bang), the latter from cyanobacteria to plants and red algae. In reality, these events cast a wider network on both ends. In a wide-ranging study Chuan Ku, Bill Martin, and colleagues[121] identified 1060 proteins that almost certainly entered the eukaryotic lineage with a proto-plastid; of these, 406 have since spread to other phyla of eukaryotes via secondary endosymbioses. As expected, almost all of the 1060 have counterparts[122] in many or most cyanobacteria. Many also have counterparts in diverse bacterial phyla: the proto-plastid symbiont brought not only its core genome, but a diverse set of pangenes.[123] A further 1397 proteins, more broadly distributed among eukaryotes, probably arrived via the proto-mitochondrion, although specific α-proteobacterial signal is weak[124]—or, viewed another way, the likely pan-genomic signal is strong. We return to this dataset in a moment.

Eukaryotes seem predisposed to endosymbiosis, and to intracellular associations more broadly. In addition to the primary and secondary symbionts above, eukaryotes host intracellular archaea (methanogens in rumen ciliates), bacteria (*Buchnera* and *Wolbachia* in insects; *Rhizobium* in legumes), and protists (*Chlorella*

446   KINGDOMS, EMPIRES, AND DOMAINS

in *Hydra*; *Symbiodinium* in marine invertebrates). Pathogenic bacteria (*Brucella*, *Listeria*, *Mycobacterium*, *Salmonella*, *Streptococcus*) can become established in animal cells. In these cases, few if any genes are exchanged; the host and guest retain their original taxonomic identities, and we do not need to add a cross-link to the tree.[125]

We know nothing of the physical process by which two dissimilar organisms, soon to merit the labels *host* and *proto-mitochondrion*, initiated their relationship. Following Portier, Wallin, and Margulis[126]—and emboldened by the broad, presumably ancestral predisposition of modern eukaryotes to endosymbiosis—we default to images of invasion or engulfment.[127] Others, focusing on the physiological advantages such an association would unlock, speak more generically of fusion.[128] In any event, the data show an episodic influx of bacterial genes into the eukaryotic lineage—evidence of a discrete, sustained physical relationship.[129]

What, then, was the **host organism** in which the proto-mitochondrion became established? Two alternatives have found support: that it was the (proto-)eukaryotic nucleocytoplasm as a deep lineage in its own right;[130] or that it was an archaeon.[131] In the three-kingdom rRNA tree, the eukaryotes branch off near the archaeal common ancestor; in the eocyte tree, they arise from within archaea.[132] Either way, the two are likely to share genes ancestrally, and perhaps also through LGT. In the late 1990s, it emerged that many *informational* proteins (those involved in replication, transcription, and translation) of eukaryotes have immediate relatives among archaea, whereas *operational* proteins (those involved in biosynthesis, metabolism, nutrient uptake, and the like) often show a specific relationship between eukaryotes and bacteria.[133] The final 128 proteins in the Ku *et al.* dataset (above) occur in almost all eukaryotes and archaea, have few bacterial counterparts, and are predominantly informational.[134] The host was an archaeon, or something quite similar. But which?

Many cellular features of modern eukaryotes, and the corresponding signature proteins, must have arisen in this lineage. The ancestor could engulf and internalize prey (or stabilize a fusion event); extract energy from membrane-bounded compartments; send and receive intracellular signals; localize proteins; and operate its cell cycle.[91] It would have been an attractive and capable host for an aspiring mitochondrion. Until recently, few homologues of eukaryotic signature proteins were known among archaea, but many have now turned up, particularly in the Asgard metagenomes.[134] Confirmation of these results (for example, by single-cell genomics) would—in the absence of further twists and turns in our story—identify the eukaryotic nucleocytoplasm as a special type of archaeon.[135]

Are we archaea?

If future work confirms that the eukaryotic nucleocytoplasmic lineage arose within Archaea, must we abandon Eukaryota as a separate domain? We might find a more-inclusive name, for example *Arkarya*:[136] but do archaea and eukaryotes belong in the same domain, whatever its name, to the exclusion of bacteria?

For some, the answer is obvious. Ernst Mayr maintained that the "world of highly evolved eukaryotes is simply an entirely different world from the world of the two kinds of bacteria, the Prokaryotes". The former is macroscopic, and involves grand phenotypic diversity, sexual reproduction, and species in their tens of millions. Prokaryotes, he argued, offer none of this.[137] To be sure, he greatly underestimated the diversity of bacteria and archaea.[138] But as an evolutionary systematist, he argued that eukaryotes have evolved far beyond any reasonable description of prokaryotes or archaea, and must be classified accordingly. A taxon that accommodates deepsea hyperthermophiles and elephants explains little; it does not help us draw generalizations, make testable predictions, or plan a program of research.

As it happened, archaea began to attract the attention of evolutionary biologists during a period—the 1980s, more or less—when cladism was becoming prominent beyond the systematics community per se. Molecular evolutionary biologists were attracted to the cladists' precise concepts about trees, nodes, and branches; microbiologists could ignore conventions established in the classification of birds and fossil horses. It was perhaps inevitable that the evolutionary and phylogenetic schools would clash over the question just posed: are eukaryotes archaea?

A preliminary round played out between Mayr and Woese.[139] Mayr objected to the division of prokaryotes into two high-level taxa (he called them *empires* rather than *domains*), arguing that archaea are phenotypically indistinguishable from bacteria. He supposed that Woese had made Archaea a domain because Hennig required phylogenetic sister groups to hold the same categorical rank,[140] and for that he labelled Woese an unreconstructed cladist.[141] For Woese, classification was a second-order issue. In his view, Mayr was advocating the selfsame complacent, subjective, parochial birds-and-insects taxonomy that had for so long failed microbiology.[142] With molecules (particularly rRNA) we can probe the origin of the cell, indeed of life on earth:

> the disagreement between Dr Mayr and myself is not actually about classification. It concerns the nature of Biology itself. Dr Mayr's biology reflects the last billion years of evolution; mine, the first three billion. His biology is centered on multicellular organisms and their evolutions; mine on the universal ancestor and its immediate descendants. His is the biology of visual experience, of direct observation. Mine cannot be directly seen or touched; it is the biology of molecules, of genes and their inferred histories. Evolution for Dr Mayr is an "affair of the phenotypes". For me, evolution is primarily the evolutionary *process*, not its outcomes. The science of biology is very different from these two perspectives, and its future even more so.[143]

Mayr posed the question: *two empires, or three?* Woese was not much inclined to further battle, but the distinctiveness of archaea—revealed by ssu-rRNAs, reinforced by informational and other molecules, and cladified in the rooted tree—made it easy for molecular evolutionary biologists to side with Woese against Empire Prokaryota. Cavalier-Smith[144] and Gupta[145] dissented on cell-biological grounds, but Woese's "three domains" had become paradigmatic as the Tree of Life, and was widely seen to depict a fundamental truth about biodiversity, particularly that of microbes. But the three-domain formulation was never tenable from a cladistic perspective,[146] and the Asgard metagenomes further tipped the balance. Ford Doolittle summarized the new two-domain position concisely:

> if "domain" is taken as the deepest division in any tree, and all trees are taken to be bifurcating, there can really only be two domains . . . Bacteria on one side and a clade comprising Archaea and Eukarya on the other.[147]

He allowed that eukaryotes "still seem very different from their archaeal ancestors", and a case might be made to retain domain Eukarya for that reason alone—although with the Asgard data "these differences are now starting to erode".[148] How, then, do we find our way forward? Three paths appear open:

- We might accept the cladistic argument at face value, and embrace eukaryotes as archaea (or arkarya); and nuclei, endomembranes, and the cytoskeleton as autapomorphies.[149] Archaea (Arkarya) remains holophyletic even in conservative interpretation, and there are two domains of life: Bacteria, and everything else.
- We might argue that the symbiosis (or fusion) that created the eukaryotic cell caused such profound genomic disruption, and created a phenotype so far beyond the reach of standard anagenesis, that it must be treated as a special case.[150] Our claim might be broad (that established notions of monophyly, paraphyly, and polyphyly apply only to whole-organism biology), or limited to eukaryogenesis. It remains to be seen how this might be argued in detail, and whether Archaea and/or Eukarya could emerge as valid taxa at domain (or any) rank.[151]
- In any case, we should continue to search for data types that are informative at the very base of the tree, where molecular (including ssu-rRNA) sequences provide weak phylogenetic signal at best. Elements of protein folded structure (confusingly, also called *domains*), for example, are highly conservative, and seem to support a distinct identity for eukaryotes.[152]

Last words on kingdoms, empires, and domains

And so we conclude our journey from the African plain. Man is an animal; animals, together with fungi—those erstwhile least-noble plants—together constitute one of perhaps six major groupings within eukaryotes. Eukaryotes may, in part, be but a late offshoot within Archaea. Our physiology and energetics are of bacterial origin.

Along the way, we have seen the world of organisms depicted as a chain, net, map, and tree. Kingdoms (domains) are recognized differently in each. Given a chain, we decide how many and precisely which links to cut, and how to apportion the fragments of each severed link. With networks and maps, we make many such decisions. With a tree, by contrast, kingdoms (domains) should be the deepest trunks. From the 1860s until about 1990 the tree metaphor was predominant, even if some organisms—notably protists and fungi, but others too—were difficult to place precisely. With LGT, the bacterial and archaeal domains became networks, although with a core treelike signal. Eukaryotes arose in an inter-domain symbiosis but thereafter diverged in a treelike manner, albeit with some topological complication due to secondary and tertiary endosymbioses. Kingdoms proliferated, and were gathered together into supergroups.

As I write in the year 2021, the shape of nature is as contested as at any point since the early Nineteenth century. Does the living world resemble a tree, or a network? Can it meaningfully be both? There are trees in networks, of course, but it would be tendentious to speak of *Darwin's great network*. The situation has attracted biologists and philosophers.[153] Ford Doolittle and Eric Bapteste contend that

> Evolutionists have long acknowledged a diversity of population-level diversification mechanisms (selection, drift, convergence, and parallelism) and (with reservations) clade-level mechanisms that extend beyond the selectionist and gradualist framework mapped out by Darwin. At the genome level, vertical descent and LGT, gene creation, duplication and loss, in all combinations with population-level processes, expand the evolutionary repertoire. A multifaceted process pluralism is now the common view.[154]

They conclude that different "patterns of true relationships among taxa" can be appropriate representations of these multifaceted processes for different groups (taxa), ranks, or purposes, a stance they call *pattern pluralism*.[154] Hierarchical classification may be appropriate for taxa that have diversified in a treelike manner, but inappropriate for those shaped significantly by LGT or endosymbiosis.[155] The supergroups of eukaryotes organize taxa that (thanks to secondary and higher-order symbioses) can be related in complex ways not easily captured in a hierarchical classification. These supergroups may serve as exemplars for biological domains, the top-level categories of earthly beings in their wondrous diversity of pattern and process.

APPENDIX

Victorian popular natural histories

Methodology: I selected these 44 works from an initial list of 86 Victorian-era popular works in natural history, using the following criteria: published in Britain between 1844 (*Vestiges*) and 1885; author British; originally written in English; first or early edition; topic area conducive to discussion of kingdoms of nature and/or a plants-animal interface; maximum two works per author; not primarily academic, technical, or devotional; not a voyage- or expedition-book; not an argument for or against a specific issue *e.g.* evolution or cell theory. I considered the three volumes from the Society for the Promotion of Christian Knowledge as a single work (the series ran to five volumes, but the first two volumes are farther afield from our question). Where a work is in two or more volumes, all volumes were examined.

Arabella Buckley, *Life and her children: glimpses of animal life from the amoeba to the insects.* London (1879)

Robert Michael Ballantyne, *The ocean and its wonders.* London (1874)

Charles Robert Bree, *Popular illustrations of the lower forms of life.* London (1868)

Anges Catlow, *Drops of water.* London (1851)

John Cocks, *The sea-weed collector's guide . . . .* London (1853)

William Sweetland Dallas, *A natural history of the animal kingdom.* London [1860?]

Peter Martin Duncan (ed.), *Cassell's natural history.* Six volumes. London (1876–1882)

Peter Martin Duncan, *The sea-shore. Natural history rambles.* London (1879)

Rev. Robert William Fraser, *Ebb and flow.* Edinburgh (1860)

Rev. Robert William Fraser, *The seaside naturalist.* London [1868]

Mrs Alexander [Margaret] Gatty, *British sea-weeds.* London (1863)

Philip Henry Gosse, *Evenings at the microscope.* New edition. London (1874)

Philip Henry Gosse, *Life in its lower, intermediate, and higher forms.* London (1857)

Samuel Octavus Gray, *British sea-weeds.* London (1867)

John Harper, *Glimpses of ocean life; or, rock-pools . . . .* London (1860) [examined 1861 edition]

[Abbot] Georg Hartwig, *The harmonies of nature or the unity of creation.* London (1866)

[Abbot] Georg Hartwig, *The sea and its living wonders.* London (1860)

William Henry Harvey, *The sea-side book.* London (1849)

Rev. Samuel Haughton, *The three kingdoms of nature briefly described.* London [1869]

Thomas Rymer Jones, *The animal creation.* London (1865)*

Thomas Rymer Jones, *The aquarian naturalist.* London (1858)*

Rev. Charles Kingsley, *Glaucus, or the wonders of the shore.* London (1855)

Rev. David Landsborough, *A popular history of British sea-weeds.* Second edition. London (1851)

Rev. David Landsborough, *A popular history of British zoophytes, or corallines.* London (1852)

Louisa Lane Clarke, *The common seaweeds of the British coast.* London [1865?]

Edwin Lankester, *Half-hours with the microscope.* London (1859)

Rev. Hugh Macmillan, *First forms of vegetation.* London (1874)**

Gideon Algernon Mantell, *The invisible world revealed by the microscope; or, thoughts on animalcules.* London (1846) ***

Hugh Miller, *The cruise of the Betsey.* London (1858)

Hugh Miller, *The old red sandstone.* Edinburgh (1841)

St. George Mivart, *Lessons from nature.* London (1876)

David Page, *The past and present life of the globe.* Edinburgh & London (1861)

John Phillips, *Life on the earth.* Cambridge (1860)

452 APPENDIX

Thomas Lamb Phipson, *The utilization of minute life*. London (1864)
Andrew Pritchard, *Microscopic illustrations of living objects* .... New edition. London (1847)
Andrew Pritchard, *Notes on natural history*.... London (1844)
John Thomas Queckett, *A practical treatise on the use of the microscope* .... London (1848)
Henry James Slack, *Marvels of pond-life*. London (1861)
Society for Promoting Christian Knowledge [attributed to Sarah Windsor Tomlinson]
   The animal kingdom. London (1848)
   The vegetable kingdom. London (1849)
   The mineral kingdom. London (1856)
Mary Somerville, *On molecular and microscopic science*. London (1869)
John Ellor Taylor, *Half-hours at the sea-side*. London (1872)
John Ellor Taylor, *The sagacity & morality of plants*. London (1884)
Rev. John George Wood, *Common objects of the sea shore*. London (1857)
Rev. John George Wood, *Common objects of the microscope*. London (1861)

* In listing Thomas Rymer Jones under "Jones" I follow the *Catalogues of the British Museum (Natural History)*
** This is, in effect, the second edition of his *Footsteps from the page of nature* (1861)
*** A second, very similar edition appeared as *The invisible world revealed by the microscope* (1850)

Acronyms

In the Endnotes, *PG* refers to a work in the 161-volume series *Patrologia cursus completus. Series Græca* edited by Jacques-Paul Migne (Paris, 1857–1866). *PL* refers to a work in the 221-volume *Patrologiæ cursus completus. Series Latina* likewise edited by JP Migne (Paris, 1841–1855) and variously reprinted. Works in *PG* and *PL* are listed in the References under the original Greek or Latin author, *e.g.* Gregory of Nyssa or Hrabanus Maurus.

Likewise in the Endnotes, *SCC* refers to the series *Science and civilization in China* by Joseph Needham and colleagues (Cambridge UP, 1954 ff), listed in the References at Needham J.

In the References, *PNAS* refers to the *Proceedings of the National Academy of Sciences of the United States of America*. *UP* means *University Press*.

Notes

Chapter 1

1. For the purposes of this book, we can take *plant* and *vegetable* as synonymous. Although today English speakers tend to juxtapose *animal* with *plant*, until the late Nineteenth century *animal* was more typically contrasted with *vegetable*. I maintain the term *vegetable* here in recognition of this traditional usage.
2. Beseke (1797:87–119), especially pp. 92–93, and note 3 at page 92.
3. König, *Regnvm animale* (1682); also *Regnum vegetabile* (1688) and *Regnvm minerale* (1686). I examined only the 1696 edition of *Regnum vegetabile*. As we shall see in Chapter 9, König was not the first to call these three groups *kingdoms*.
4. Isidore Geoffroy Saint-Hilaire, *Histoire naturelle* (1854–1862), particularly volumes 1:1–164 (1854) and 2:1–166 (1859).
5. Gibson (2015), particularly pp. 1–36. A history of the three kingdoms by Isidore Geoffroy Saint-Hilaire (1859) is detailed and useful, if not always accurate.
6. Henshilwood *et al.* (2002). Material for making paints, discovered later at the same site, date to at least 100,000 YBP. For consistency I convert years before present (YBP) or years ago (YA) to BCE by subtracting 2000.
7. Braudel (1998, 2001). Sulawesi cave art: Aubert *et al.* (2019) and Brumm *et al.* (2021). Stone tools: note 1 (by Jean Guilane) to Braudel, Chapter 2; flint-tipped spears: Braudel (2001), page 29. *Der Löwenmensch*, now in Museum Ulm; spear thrower 14000–13000 BCE from rock shelter in Montastruc, France: J Cook (2013). The Venus of Hohle Fels (also known as the Venus of Schelklingen), from mammoth ivory: Conrad (2009). The Venus of Dolní Věstonice, from fired clay: Vandiver *et al.* (1989). Caves: Bradshaw Foundation, and Encyclopedia of Art Education. Australian animals: BJ King, National Public Radio (NPR). See also Davidson (2017).
8. The Bradshaw Foundation website is a key resource; also Lawson (2012). Groupings: Lewin (1979); Tauxe (2007); and d'Huy (2011).
9. Human in bison hunt: Lascaux, lower level at back of the Chamber of Engravings. The "sorcerer" or "horned God" at Trois Frères is sketched by Alberch (1989), plate 2; another is at Gabillou Cave (Dordogne, France). Seal-man at Cosquer: Bradshaw Foundation. Interpretation of the Trois Frères image as a horned human has been controversial. The chimæric animal at Lascaux is in the Bull's Chamber: paleolithic-neolithic.com. Animal-headed human figures appear in Australian rock art slightly later, although dating is imprecise (Taçon *et al.* 2020 and references therein).
10. Mateu (2002).
11. Selva Pascuala mural: Akers *et al.* (2011). Rock paintings (9000–7000 BCE) in what is now the central Saharan Desert show a man (god?) with mushrooms all over his body: Samorini (1992).
12. J Krause *et al.* (2007); Kuhlwilm *et al.* (2016); Mozzi *et al.* (2016); Atkinson *et al.* (2018).
13. EL Greene (1983) 1:192.
14. *The book* (or *instruction*) *of Ptah-Hotep* and *The book* (or *instruction*) *of Ka'quemna* are known via the Prisse papyrus, probably XII Dynasty (2758–2714 BCE) although possibly XI Dynasty.

456   NOTES

Gunn (1906) gives somewhat different dates than does Myer (1900); those shown are from Gunn. The stories, however, are much older: *Kaʾquemna* is probably III Dynasty (*ca* 3998–3969 BCE), and *Ptah-Hotep* V Dynasty (3580–3536 BCE). Other records include *The book (or instruction) of Amenemhěet*, XII Dynasty (2778–2748 BCE); the Stele of Mentuhotep (*ca* 2748 BCE); the rock tomb at Beni-Hasân (XII Dynasty, 2758–2714 BCE); *Introduction to the maxims of the scribe Ani* (Papyrus Bulak no. 4, XI or XII Dynasty); the *Papyrus of sayings* now at Leiden (itself *ca* 1300 BCE, although the sayings themselves are doubtlessly earlier, although not before XII Dynasty); and an inscription at El-Kab (XIX Dynasty, *ca* 1730 BCE). *Melon* might alternatively be translated *vegetable*. See Myer (1900) and Gunn (1906).

15. Egyptian *Book of the dead*, Budge translation (1895). Current versions date to about the VI Dynasty, by which time scribes were unable to understand the earlier (II Dynasty) texts. The "most moderate estimate" of the age of *Book of the dead* is 3300–3166 BCE (Budge, page xii).

16. RC Thompson (1936), page xvi.

17. *Beasts of the field*, Budge (1895:251); *beasts, passim*. Budge's mention of *plants* describing the vignette to Plate VII (page 277) does not appear to be a direct translation.

18. Thompson, 1936 (above) and 1949. Six meanings or combinations of *šammu* are found: *herb of the field*; *plant* (all plants); *plant* in relation to its place of origin, *e.g.* garden, mountains, sea, etc.; incorporated into the name of a specific plant; *drug* (vegetable or mineral); and in terms referring to colours of specific plants or drugs (Thompson 1949, especially pp. viii–ix, 3, and 8–9). There is evidence of a tablet beginning simply with Ú: (*šammu*), but unfortunately it has been lost.

19. Notably the *Har-ra*, or *Hubullu*, of Ashur from the Ninth century BCE, a bilingual Sumero-Akkadian lexicon in cuneiform letters: Bodenheimer (1958:151). Plates XI to XV present a systematic enumeration of wild and domestic animals of land, water, and air. Plate XIV names 409 animals, mostly wild and terrestrial, including mythical, symbolic, and real serpents as well as worms, caterpillars, larvæ, flies, wasps, bees, water-strider, spiders, starfish, dragonfly, ants, scorpions, gecko, chameleon, frog, and toads. A German translation of Plate XIV is given by Landsberger (1934). These lists, in Akkadian script, were apparently based on earlier lists in Sumerian, which by the time of Hammurabi was a dead language (Landsberger 1934:45). Prefixes may denote groupings corresponding to modern taxonomic groups, or size, colour, fierceness, or typicality.

20. Sarton, *Ancient science* (1970:83).

21. The most-complete recension, on which these passages are based, dates from the library of King Assurbanipal (Seventh century BCE), a great antiquarian and last king of Assyria. The epic itself appears to have existed in written form in the first centuries of the Second millennium BCE, and doubtlessly dates to many centuries earlier: Sandars, *The epic of Gilgamesh* (1960:7–8).

22. Quotations at Sandars (1960), pp. 81, 82, 94, 105, and 106.

23. Museum number IM19606. See JB Pritchard (1969), vol. 3, figure 502 at page 171, and text at page 308. The top panel shows not only the fertility-goddess Inanna (or her priestess) but also men, a gazelle and lion, and baskets of fruit or vegetables. Under the stalks of grain is a wavy line that may depict water. This vase is usually dated to the Jemdet Nasr Period (*ca* 3000 BCE), although Basmachi (1947:193–201 in Arabic section) and Kleiner & Mamiya (2006) 1:20–21 date it a few hundred years earlier.

24. The correspondence depends in part on methodological assumptions. See RC Thompson (1924), pp. xix–xx; Mayr, Linsley & Usinger (1953); Whorf (1956); Eliade (1978:35–37); B Berlin *et al.* (1973:216); Atran (1985); and Atran (1990:17–46).

25. Berlin *et al.* (1973, 1974); Hunn (1977:42–44).

26. Hunn (1977:134–135).

27. Berlin *et al.* (1974:30).

NOTES 457

28. Aztec (Nahautl): del Paso y Troncoso (1886), Tomo 3; Algonquin: Lahontan (1703); Cree: Harmon (1820); Yoruba: Bowen (1858).

29. J Dawson (1881).

30. Rink (1887).

31. Atran (1985:152–153).

32. Durkheim & Mauss (1901–1902); R Needham (1963), pp. xi–xii.

33. Greene (1983) 1:177 and 1:187.

34. *Genesis* 1:20–21, 24–25; 2:7, 19; Eliade (1978:40–43 and note G, page 208); and Frazer (1935:4–15). Rivers, springs, caves, and caverns were gynæcological symbols (the Earth Mother) in mythologies from Mesopotamia, Babylon, Sumer, Egypt, Greece, and else-where. Men and gods were born from earth, dust, clay, or stones in creation myths of the Egyptians, Babylonians, Incas, Mayans, Greeks, Semites, Caucasians, Polynesians, Maori, and tribes in Africa, Australia, India, and the Americas. Ores and precious stones were sometimes considered to gestate inside the Earth Mother. Ovid (Helvetius edition, 1817) not only considers men to be formed from clay (*Metamorphoses* 1:76–88), but also reports that half-clay, half-living objects can be found in the earth (1:416–437). The Egyptian account involved not only clay, but also the potter's wheel (Budge 1895, page xcvi). Pottery was developed during the Neolithic: for a brief overview of the history of pottery, see Braudel (2001); and Childe (1963:65–66). Pottery appeared around 6000 BCE (Braudel, 2001:41), but the potter's wheel was not in common use in Egypt until 2600 BCE (2001:55).

35. Frazer (1935:6) cites *adamah*, or *ground*, as the feminine form of *man*, *adam*, perhaps re-flecting an Earth Mother; see also Maimonides (*Guide* 1:xiv; Friedländer edition 1969:25). *Genesis* has man being made in the image of God (1:27) but formed from the "dust of the ground" (2:7). In the Torah (*Genesis* 1:2), "The earth was chaos and void, and darkness on the face of the deep"; according to Jewish commentaries, "deep" (*tehom*) refers to the mud and clay on the bottom of the sea: Kaplan (1997) page 75 and note 208 at page 360.

36. Sandars, *Gilgamesh* (1960:62–63).

37. Qumran scrolls "Community Rule" (1QS) and "Hymns" (1QH).

38. The Book of *Enoch the Prophet*, chapters 74–79 (Section 17 of the Paris manuscript). Cf *Genesis* 2:19. Men and women form from shapeless lumps that bubble up from the earth in Empedocles, fragment 62 (MR Wright 1981/1995).

39. Bozic & Marshall (1972), pp. 17 ff, 29 ff, and 79 ff.

40. Frazer (1935:16–17).

41. Tribes of ancient Arabia bore animal names (lion, shark, . . .), but authorities are divided as to whether this was totemistic: see WR Smith (1885); Pellat *et al.*, *Hayawān* (1986:305). At page 314, Pellat notes that many Turkish tribes had animal names, which seems to indicate the existence an animal cult there in pre-Islamic times.

42. Frazer (1904), and (1935:21, 31–33); Durkheim & Mauss (1901–1902), pp. 6–7, 21–22, and 35 ff. Totemistic systems can be complex, with sub- and pseudo-totems. Frazer (1904:475) cites instances in which both animal and plant sub-totems are found within a totemistic group. Phatries (intermediate levels of social organization between tribe and clan) and die-tary restrictions may reflect past totems.

43. PG Kuntz & ML Kuntz (1987:319–334). The tree symbol occurs in diverse cultures and traditions from the Middle East, India, central Europe, Scandinavia, and elsewhere. For more on tree-symbolism, see Frazer (1890).

44. Lum (1944); TH White (1954), especially pp. 255–258. For the Babylonian water-god, reproduced from Lum, see White (1954:252). Other, less-familiar animal deities and monsters from Egypt are described in the *Book of the dead*.

458   NOTES

45. The images at Lascaux include only a few representations of humans, and at most two of plants, both doubtful (Windels 1950:55). For the so-called two-horned unicorn at Lascaux, see Windels (1950) page 52, image 2.

46. The man-scorpion was one of the monsters that, according to the *Enuma elish* (the Babylonian poem of creation), was created by chaos at the beginning of the world (Sandars 1971:37).

47. For the *Physiologus*, see Ley (1967) volume 1, second (unnumbered) page. We return to the *Physiologus* in Chapter 4. Ovid: *Metamorphoses* 15, lines 393–410; Elsee (1908:16); and McCulloch (1960:158). The Phœnix symbolizes Jesus, who like the Phœnix came back to life; see *e.g.* the *Hexaemeron* of St Ambrose (Day 5, chapter 23, verse 79).

48. Rudolph (1949).

49. Ovid, *Metamorphoses* 12, lines 209–536; sirens, verse 552; Topsell (1658) 1:14–15; Herodotus, *Histories* 3.116; Pliny, *Naturalis historia* 7.9–7.32; Isidore, *Etymologies* 12.2.17; C Zirkle (1941). The manticore apparently originated from a description by Ktesias (*Indica*), perhaps *via* the *Myriobiblion* of Photius (*PG* 103, col. 214 and note 28, 1860; McCulloch, 1960:142–143), and was depicted in various ways, *e.g.* with the head of a man or woman, and a tail like that of a scorpion. Ktesias, a Greek (Fifth century BCE), was a court physician for Darius II of Persia. Sirens were variously pictured as birdlike below the navel, or (after the second quarter of the Twelfth century) with fish-tails (Faral 1953; McCulloch 1960:167n151). Empedocles (Wright 1981, fragment 61) supposed that "human-faced bulls and again bull-headed humans", as well as other hybrid forms, were originally produced from the earth, but gradually died out. Paracelsus suggested that sea monsters may eventually "separate out" into their component beings (*Philosophy addressed to the Athenians* 1910, volume 2, text vi, page 251; and text xiv, page 255). Paracelsus was particularly explicit about the origin of hybrids, stating that "wandering night spirits" could carry sperm between humans and impure animals such as goats, dogs, worms, toads; from this illicit traffic, vile monsters would arise (Webster 1996:82). For lares, genii, etc. see Burton (1621:61–62).

50. *Iliad* 16:148–151; Virgil, *Georgics* 3; Pliny, *Naturalis historia* 8.166; L Ariosto, *Orlando Furioso* [1532] in Harington's translation (1591), Book 15, verse 29. See also Herodotus, *Histories* 3.16.3; Dalton (1872); Zirkle (1936); and Barber & Riches (1971).

51. But see McLennan (1869–1870). Carved seals in Harappa (Punjab) show a human-like god perched in a tree, sometimes accompanied by votives bearing plants (Allchin & Allchin 1968:313).

52. Sandars, *Gilgamesh* (1960:116); Veenker (1981).

53. Beseke (1797:88–90). So far as is known, Theophrastus alluded only once to the use of a mineral in human health (Scarborough 1978:384n174). See also Chapter 9.

54. Male and female variants of mandrake are shown in *The herbal of Rufinus* (edited by Thorndyke, 1946), and in the Straßburg *Hortus sanitatis* (*ca* 1497, cap. 275; woodcuts from the workshop of Johann Grüninger). Singer (1928:179) shows a tracing from a miniature in the Juliana Anicia Codex, a superb illuminated manuscript of Dioscorides's *On materia medica* that was presented in 512 CE as a wedding gift for Juliana Anicia, daughter of Anicius Olybrius, Emperor of the West, and of his wife Placidia, daughter of Valentinian III; it shows a very human-looking mandrake with eyes, fingers, and toes. Isidore likewise credits both male and female forms (*Etymologies* 17.9.30). The story of Jacob and Leah (*Genesis* 30:14–16) attests the fertilizing power of the mandrake, a property that has been "enlarged upon in rabbinical literature" (White 1954:26n1). For myths about the origin of mandrakes, see Pinto-Correia (1997:225 and 344n29). Quote from Evans (1896:110–111n1). Elephants and mandrakes: McCulloch (1960:115–117).

55. Paul, *1 Corinthians* 15:35–44; Origen, *De principiis* 3.5.4 (in Roberts & Donaldson 10:256–258, 1869); Jerome, *Epistle to Avitus*, cited at ibid. 10:256–257n4. We discuss the corresponding Neoplatonic tradition in Chapter 5.

NOTES   459

56. Bruno, *La cena de le ceneri* [1584], Dialogue 3: in *Opere Italiane* 1 (1907), page 77. See also Michel (1973:250–268), particularly page 262; and J Needham, *SCC* 2:540 (1956).

57. *Epinomis* 980c ff. (especially 981b,c and 984 b,d) as cited by Lloyd (1993:54n217).

58. Eliade (1978:53–64), especially pp. 54–55.

59. Ibn Waḥshiyya, *Nabatæan agriculture*; Maimonides, *Guide* 3, chapters 29 and 37; Tolkowsky (1938); Eliade (1949). Ibn Waḥshiyya claims to have translated *Kitāb al-Filahâ al-nabâtiyya* (*Nabatæan agriculture, ca* 904 CE) from earlier Babylonian texts. Sarton (*Introduction* 1:634, 1927) considered the *Nabatæan agriculture* to be a forgery; Fahd (1977:276–277) is more nuanced.

60. Braudel (2001:45). Langer (1972:17) dates Çatal Höyük to about 7000 BCE.

61. Inanna's journey to hell. In: Sandars (1971:117–167), pp. 156–157. This Eighteenth-century BCE poem is the earliest known that describes descent into the underworld.

62. Budge (1967) pp. xxxviii, 340, and the corresponding chapters.

63. Human into mineral, *Genesis* 19:26; vegetable (wood) into animal, *Exodus* 7:8–13. The sorcerers and magicians of Egypt successfully emulated Aaron, but his snake consumed theirs.

64. Empedocles, fragment 117D, quoted in de Santillana (1970:128). A slightly different translation is given by Coomaraswamy (1949:126), who presents derivative passages from William Blake, Māṇikka Vāçagar, Jalāl ad-Dīn Rūmī, and Aitareya Āraṇyaka. See also Birge (1937:124–125). The fragment from Empedocles attracted the scorn of Gregory of Nyssa (*De hominis opificio* 28:3, as translated by Moore & Wilson, in Schaff & Wace 5:419, 1893): "such doctrines as this of saying that one soul passed through so many changes are really fitting for the chatter of frogs or jackdaws, or the stupidity of fishes, or the insensibility of trees." Basil was no less uncharitable (*Hexaemeron*, Homily 8:2; ibid. 8:95–96, 1895): "avoid the nonsense of those arrogant philosophers who do not blush to liken their soul to that of a dog; who say that they have been formerly themselves women, shrubs, fish. Have they ever been fish? I do not know; but I do not fear to affirm that in their writings they show less sense than fish." Cf. Epimenides of Crete: "I have been an eagle, I have been a sea coracle . . . I have been a sword in the hand, I have been a shield in battle, I have been a string in a harp" (quoted by Burkert 1972:163). A coracle is a light boat.

65. Budge (1967), pp. lix and 176.

66. By my count, 112 transformations of body type, not including those of sex or colour. The quotation is from the Humphries translation (1957:3).

67. Julius Cæsar was killed the year before Ovid was born; Virgil, Horace, Propertius, and Tibullus were alive when he was a student. He was a friend of Propertius, is known to have heard Horace, and to have spoken with Virgil (Frazer, in Introduction to *Fasti*, 1931:ix–x). One message of *Metamorphoses* may have been that that not even "eternal Rome" could persist unchanged forever (Galinsky 1975:43–44). Colavito (1989, Appendix A) enumerates the metamorphoses slightly differently. At Book 15 the Humphries translation "the earth has something animal about it" (1960:375) seems too weak; Ovid's text reads *Nam sive est animal tellus, et vivit, habetque*. Ovid relates that men were produced from fungi which sprang up after the rains (Book 7). Seaweed to coral is at *Metamorphoses* 15, lines 416–417. Ovid may have patterned the *Metamorphoses* on the poem *Heteroeumena* by Antoninus Liberalis, of which only fragments survive; in it, people are turned into stone, insects, trees, statues, and other objects (Gow & Scholfield, *Nicander* 1953:205–208).

68. Lucius Apuleius built his *Metamorphoses* (*The golden ass*) on an earlier narrative, now lost, by Lucius of Patræ. Both in his lifetime and during the Renaissance, Apuleius was accused of magic. His stories posed problems for the Church, which considered his *Metamorphoses* part of the classical *corpus* but was troubled by claims of people being transformed into animals and vice-versa. Johannes Vincentii (*ca* 1475) circumvented this concern by supposing

460 NOTES

Apuleius to have been a victim of Satan, who forced dreams of metamorphoses upon him while he lay asleep (Kieckhefer 2000:32).

69. Isidorus Hispalensis, *Etymologiarum* 11, chapters 3 and 4. Isidore was canonized in 1598, and declared a Doctor of the Church in 1722. We return to the production of bees, locusts, and the like in Chapter 3.

70. Kieckhefer (2000:46). Various myths remained current describing women who, under more-or-less unnatural conditions, had given birth to animals.

71. Kieckhefer (2000:98–99) cites unnamed medieval fictional works that featured magicians who could turn animals into knights, or stones into cheese—or at least make onlookers believe that this had occurred.

72. Browne, *Religio medici* (1652:215–216); in English translation (1892:71). See, however, Browne at Chapter 8.

73. Petrus Bonus of Ferrara, *Pretiosa margarita novella* (*ca* 1330): Aldine edition by Janus Lacinius (1546: fol. 22r–22v); Waite translation 1894:93). Such descriptions, in innumerable variants, were common throughout antiquity; similar passages occur *e.g.* in Ovid's *Metamorphoses* (Book 15) and in *Outlines of Pyrrhonism* by Sextus Empiricus (Chapter 3).

74. Linnæus, Anmärkung (1740/1743); Latin epitome (1762). Swedes have long stereotyped their Norwegian cousins as unsophisticated "country bumpkins".

75. Theophrastus, crop plants: *Historia plantarum* 8.7.1 and 8.3 (Loeb Classical Library, *Enquiry into plants* at 2:182–183 and 192–195); ivy: *Historia* 3.18.7–10, *Enquiry* 1:273–278. Greene (1983) 1:205–209 and 470–471 (1909) for wheat, 1:228n85 and 478–479 for Nicon and Galen. Galen, *De alimentorum facultatibus*, in *Opera omnia* (1821–1833) volume 6, book 1, section 37, especially page 552. Scaliger, *Anamadversiones* (1584).

76. Greene (1:205) reports that even in the Twentieth century some farmers still believed that seeds of barley or wheat may germinate and grow into *Bromus secalinus*, "chess" or "cheat". When I was a child in rural Indiana in the mid-1950s my maternal grandfather, Russell Frank Hiatt (1905–1988), tried to convince me that if a horse's hair were left submerged in water, it would turn into a worm. I suspected that he might be pulling my leg, but he stuck by the story.

77. We discuss the alchemical tradition at length in Chapter 9. Patai (1994:307) alludes to instructions, set out in the alchemical work *Iggeret ha-Sodot* (probably of Sixteenth-century Spanish origin, falsely attributed to Maimonides), for producing a homunculus by melting metals at an astrologically determined hour. If I understand Patai's passage correctly, this is surely one of the few supposed transformations of metal into human.

78. Aristotle, *Metaphysica* 983b28–984a11. For Thales, see de Santillana (1961:22–26); Anaximenes (1961:41, 45); and Heraclitus (1961:43–51, especially fragments 24 and 25 at page 50). Also TS Hall (1950), pp. 345 and 350; and Braudel (2001:255 ff.), who makes the point that in early Miletus, the elements were considered living. The primary reference for these fragments is Diels (1934–1937 ff.).

79. Ibn Wahshiyya, *Nabatæan agriculture*; Maimonides, *Guide* 3, chapter 29; Fahd (1977). See also note [59] above.

80. Kieckhefer (2000:50–52).

81. For a taste of this literature, see Lang (1884).

82. Sark (1988:6–8).

83. Wang Ch'ung, a sceptical philosopher influenced by Confucianism and Taoism, wrote *Lun hêng* (*Discourses weighed in the balance*), probably in 82–83 CE: J Needham, *SCC* 2:374–376 and 420 (1956). Many more transformations (including grass into firefly) are described by the alchemist Ko Hung in *Pao phu tzu* (*ca* 320 CE): *SCC* 5(2):63 and note *c* at that page (1974). Ko Hung concluded that transformations are spontaneous in nature.

84. Needham, *SCC* 2:81.

NOTES 461

85. Chêng Ching-Wang, *Mêng chai pi than* (early Twelfth century), *SCC* 2:421 (1956). Chêng comments that "unfortunately, I fear that many people do not believe it". A Tang dynasty (618–907 CE) tomb sculpture shows the transformation of a man into a quadruped, that turns into a serpent, that turns into a woman: Rastelli (2009).
86. Mitford [AB Freeman-Mitford, Lord Redesdale] (1871), 1:255–258.
87. Westermarck (1933).
88. *Quran* 5:60. For a diversity of translations, see corpus.quran.com/translation. An interesting discussion is given by Pellat *et al.* (1996:306).
89. This stone was able to heal the eye: Ullmann (1972:25).
90. Kafka, *Die Verwandlung* (1915). Characters may change form voluntarily or involuntarily in Japanese *manga*, whereas in Western comics, change has typically been potentiated by exposure to an external force, *e.g.* radiation. Scientist Sven Larsen became "Animal-Vegetable-Mineral Man" (DC Comics, 1964) after falling into a vat of amino acids.
91. For linguistic evidence relating to extension to other animals, and the power of saints, see McLennan (1883). Demonic influence: Wier (1566); Bodin, *Démonomanie* (1580); Boguet (1603); Psellus (1838). Paracelsus not unreasonably credited Satan with similar powers (*Liber de nymphis* 1941). The oft-quoted statement of Aquinas, *Quod angeli boni et mali ex virtute naturae suae habent potestatem transmutandi corpora nostra sicut etiam alia corpora naturalia*, is from *Scripta super libris Sententiarum* 2.8.1.5. Sandars, *Gilgamesh* (1960:85–88); Ovid, *Metamorphoses* 1:163–239 (1817).
92. Relocation of the soul to a different body does not, of course, require that its nature be changed. Metempsychosis was sometimes seen as providing an opportunity for the soul to be purified or improved, but this was typically a change within type, not a transformation to a different type. More importantly for our argument, metempsychosis could involve the transgression of boundaries, often between animals and humans, and the establishment of a mixed being.
93. Herodotus, *Persian Wars* 2.123 (Rawlinson translation, 1942). Herodotus goes on to say that some Greeks (*i.e.* Pythagoras) adopted this belief as if it were their own. Pythagoras is said to have spent twenty-two years in Egypt, and to have been initiated into the mysteries (Diogenes Laërtius 8.1,3; Iamblichus, *The life of Pythagoras* (1905–1915 and 1918), page 42; and *On the Pythagorean life* (1989:8). Reincarnation was especially connected with the mysteries of Osiris. Pythagoras, who supposedly spent a further twelve years in Babylon, claimed to be the reincarnation of Euphorbus, the son of Panthus, who lived at the time of the Trojan War. Burkert (1972:126 and note 36) emphatically rejects this claim by Herodotus, stating that "the Egyptians never had such a doctrine". Vegetarianism was, however, a component of the ancient Orphic tradition in Greece (Burkert 1972:125 and note 31).
94. Burkert (1972:120–124). Iamblichus (*Life of Pythagoras* 1905–1915 and 1918, page 42; *On the Pythagorean life* 1989:38) suggests that Pythagoras may have believed that the human soul does not incarnate in those animals that is it "lawful to kill". Pythagoras also believed in anamnesis.
95. Iamblichus, *On the Pythagorean life* (1989:47–48) *et passim*; similarly in the *Life of Pythagoras* by Porphyry, and in that by Diogenes Laërtius. Flesh would not itself be ensouled, because the soul leaves the body at death; so their belief in metempsychosis might better provide justification for not *killing* animals. Some animals could, however, be killed as sacrifices to the gods. Porphyry suggests that the Pythagorean injunction against eating beans may have been related, in part, to supposed sympathies with parts of the human body (*Life of Pythagoras*, section 44).
96. For Empedocles and Hericleides, see Burkert (1972:133n74). Hericleides of Pontus, transmigration of soul into plants: Diogenes Laërtius, *Life of Pythagoras*, section 4 (Guthrie 1987:142).

462   NOTES

97. *Phaedrus* 249a–b. Although Plato refers to metempsychosis, the doctrine does not play a major role in his teaching, so (following Marsilio Ficino) this passage might be read as allegorical. Some even vaguer passages are cited by Ruderman (1988:213n32–33).

98. Numenius, Albinus, and Plotinus believed that a human soul could pass into the body of an animal if that soul had become too burdened with evil; Porphyry, Iamblichus, Sallustius, Proclus, and Nemesius of Emesa disagreed, or interpreted this possibility only figuratively: Iamblichus, *De Mysteriis* 8.6; Augustine, *City of God* 3, chapter 30; Ruderman (1988:130 and notes 35 and 37); Dillon (*The Middle Platonists* 1996:291–292 and 375–378). Numenius (Dillon 1996:375–378) believed that souls gather in the Milky Way, then migrate down through the planetary spheres, picking up influences on their way to inhabiting earthly bodies. For this, Proclus (*Commentary on the Republic* 2.128:26 ff) accused Numenius of astrology and mystery-cultism (Dillon 1996:375). There is no metempsychosis between man and animal in the *Chaldæan Oracles* (Dillon, *Iamblichi Chalcidensis* 2009:46 and note 1 at that page). If souls cannot transmigrate, one might ask how each person comes to have one. The Stoics held that the soul enters the body at the time of conception (von Arnim, *Stoicorum veterum fragmenta* 11:804 ff.); Albinus, that it entered the body at the moment the embryo was formed (*Didaskalikos*, chapter 25; Dillon 1996:293). Tertullian, Gregory of Nyssa, and others believed the soul to be traduced in the generative seed (Moore & Wilson, in Prolegomena to Schaff & Wace, *Nicene and post-Nicene Fathers* 5, at page 19). Alternatively, each soul might be created anew by God, although this raises other problems.

99. This was explicitly recognized by Nemesius of Emesa (*De natura hominis* section 51) and many other ancients; see *e.g.* Dillon, *Iamblichi Chalcidensis* 2009:45).

100. Tertullian, *De anima* chapters 28–35. Metempsychosis endangered the Christian belief in a single resurrection of body, at which point the soul re-enters its body and the resurrected person becomes subject to the Last Judgement. Tertullian devoted chapters 34–35 to various heresies involving the transmigration of souls. Metempsychosis was apparently accepted in pagan Rome: Persius, *Satires* 6:9; Horace, *Epistles* 2.1:52; Lucretius, *De rerum natura* 1.124; and Virgil, *Georgics* 4.219. Horace and Lucretius cite an apparently lost passage from the *Annals* of Ennius.

101. It is interesting to consider the relationship between metempsychosis and the idea of Purgatory, an intermediate state in which disembodied souls may be purified. The doctrine came to prominence in the second half of the Twelfth century and was well-entrenched by the Fifteenth, but came under attack from Martin Luther in the Sixteenth. The Church aggressively defended the concept following the Council of Trent. Popular belief in Purgatory remained widespread at the end of the Sixteenth century, thanks not only to the Church per se but also to literary works by Dante and others. Abraham Yagel equated Purgatory with the rabbinic *gehinom*, a twelve-month intermediate state for those sinners who were neither thoroughly righteous nor thoroughly wicked (Ruderman 1988:136–137).

102. Matteo Ricci read Pythagoras allegorically on this point, as did Marsilio Ficino Plato. Cornelius Agrippa and Giordano Bruno accepted that human souls could become trapped or imprisoned in the bodies of animals, but many other contemporary Christian Neoplatonists (*e.g.* Johann Reuchlin) rejected the idea. Spence (1986:251–252); Agrippa (1651:474 and 481–482); Reuchlin, *De arte cabalistica* (1983) 2:169; Pico della Mirandola, *De hominis dignitate* (1572:314–315).

103. Ruderman (1988:121–138); the dispute between da Spoleto and Yagel is detailed on pages 121–126. Quotation from *Beit Ya'ar ha-Levanon*, part 3, chapter 9, folios 131a-b as cited at 1988:126n21. Scholem (quoted at 1988:128) speculates that belief in metempsychosis spread following the expulsion of Jews from Spain in 1492, but appeared in the kabbalistic book *Temunah*, which may be as old as 1280 CE (Scholem 1987:461 and 467–468). See also note [101] above. A related idea—that the Messiah was a reincarnation of David, who in

NOTES   463

turn was a reincarnation of Adam—was current after the Thirteenth century (Scholem 1987:190–191).

104. Al-Bīrunī, *India* 1:43 and 1:65 as quoted by Wilczynski (1959:465).

105. It is found *e.g.* in the *Upanishads*. See also Needham, *SCC* 2:397–399 (1956).

106. Māṇikka Vāçagar, from Pope, *Tiruvāçagam* (1900) as quoted by Coomaraswamy (1949:125). Compare Empedocles above and note [64].

107. An interesting, if imperfect, recent analogue in Western popular culture is provided by the Trill character Jadzia in the 1990s science fiction television series *Star Trek: Deep Space Nine*. The Trill host carries a slug-like endosymbiont, Dax. The latter is of great antiquity, and its presence imparts to each successive host the experience accumulated in its previous incarnations.

108. Albinus, *Didaskalikos*, chapter 25, quoted in Dillon (1996:291).

109. Needham, *SCC* 5(2):11 and note *a* (1974).

110. TH Huxley applied the term incorrectly in his famous 1868 lecture on the physical basis of life. In illustrating the unity of protoplasm among all forms of life, he observed that the protoplasm in our food is changed into human protoplasm. According to Huxley, we "transubstantiate sheep into man" (*The Fortnightly Review*, 1869, at page 137). Although a defensible proposition for the *protoplasm* of sheep, the analogy does not extend to the entire animal.

111. According to Moore & Wilson (Prolegomena 1893, page 13), Gregory of Nyssa was the first to express the idea of transformation in the sacrament of the Eucharist (*Oratio catechetica*, chapter 37: "it at once is changed into the Body by the Word"). William of Ockham argued that when the bread is consecrated, it is changed into the substance of Christ's body, not his blood or his soul, nor to accidents inhering in his body; and indeed, into the complete body of Christ, not some portion thereof: Stump (1982:210–211).

112. Dampier (1948:87). Innocent III was then Pope.

Chapter 2

1. From the poem "Yr awdil vraith" ("Diversified song") attributed to Gwion son of Gwreang (*floruit* Sixth century CE), from the late Sixteenth-century Peniardd manuscripts as interpreted by Robert Graves (1961:151–155), verse 24 at page 154. For another interpretation (referring to Solomon and the "Tower of Babylon", but not specifically to "Asia"), see verse 25 of "The song of varieties" in Nash (1858:304).

2. Moorjani *et al.* (2013); Reich (2018). The migration at 2000 BCE is supported by bronze artefacts at Harappa and other sites: Allchin & Allchin (1968).

3. Allchin & Allchin (1968), pp. 297–298 and 313.

4. Some estimates date *Rigveda* to 1700 BCE. It was transcribed in writing only at 300–500 CE; the oldest extant (partial) manuscript dates to the Eleventh century CE. It comes to us organized in ten mandalas (books), of which the final mandala is considered later than the others: Horne (1917) 9:14. Mantra 90 of Mandala 10 describes how Purusa of the thousand heads, thousand eyes, and thousand feet was divided sacrificially to produce the sun, moon and stars, all worlds, cattle, horses, goats, and sheep. Mantras 129 and 130 of Mandala 10 discuss creation. For the primary text in translation: Griffith (1889–1892); second edition 1896.

5. *Brahadāraṇyaka* 5:5; *Chandogya* 7:10. Raju (1952); Subbarayappa (1966).

6. *Brahadāraṇyaka* 2:9,28. Bhalchandra Balte (2017).

7. *Aitareya Āraṇyaka* 1:5,1,9 and 3:61.5; *Manusaṃhitā* 1:49. Vishnu-Mittre (1970). Primary text: Dutt (1909).

464   NOTES

8. *Mahābhārata* may date to the Ninth century BCE, but the dialogue on plants is in the Santi Parva (chapter 184), regarded as a later addendum. See RFG Müller (1935), and Balchandra Balte (2017).
9. Surapala (1996).
10. Plants which bear fruits without flowers; those which bear fruits with flowers; those which spread with tendrils, *i.e.* creepers; and bushes.
11. *Vrikshayurveda*, verses 259–260.
12. Chapple (2007); Sridhar & Bilimoria (2007).
13. Sikdar (1992).
14. *Ākārānga Sūtra*, in Jacobi (1884) and at http://www.sacred-texts.com, accessed 30 April 2018. Interestingly, *animals* and *plants* are not the highest-level classes (unique beginners) in the *Ākārānga Sūtra*, where the six classes of beings are given as earth-bodied, fire-bodied, air-bodied, water-bodied, unmoving (*e.g.* plants and trees), and moving (insects, birds, animals, humans, hell-beings, and gods).
15. Darius I sent Skylax of Caryanda, a Greek, to reconnoitre in Sind in 519 BCE. See Filliozat (1963). As for the opposite direction, Aristoxenus of Tarentum relates a probably fictitious conversation between Socrates and an Indian philosopher who supposedly had been brought to Athens. An Indian delegation is supposed to have met with Augustus Cæsar around 13 CE.
16. Many but not all scholars now date the Buddha's birth and *parinirvana* somewhat later; I take these dates from Needham, *SCC* 2:398 (1956).
17. For a broader consideration of intellectual links between China and India, see Sen (2004).
18. For this I borrow heavily from Dampier (1948:8). See also Needham, SCC (1956) 2:402–403. The lexicon *Nāmalingānushasānam*, known as the *Amarakosha* (*Treasury of Amara*) compiled by Amarasimha, a Buddhist, lists animals and plants, but without a unified systematization. Aquatic and terrestrial plants are listed in separate chapters, while aquatic animals are described together with other water-related objects including rivers, boats, and fishermen (Sundara Rajan 2016). The lexicon has variously been dated to the mid-Fifth to Seventh centuries CE.
19. The Madhyamika school began in the First century BCE and was systematized around 120 CE. Its principal document is the *Maha-prajnaparamita sutra* (*The perfection of wisdom*): *SCC* 2:404–405 (1956).
20. Lao Tze. Until 2004 the authors of *SCC* used a modified Wade-Giles system to render Chinese characters; the series adopted *pinyin* from Volume 5, Part 11 (2008). Here I use *pinyin* for names of better-known persons, places, and texts, and to represent words, but (particularly in these endnotes) retain the form used in *SCC* for lesser-known persons and texts, in hopes of facilitating reference to this invaluable work. As a special case I retain *tao* (rather than *dao*) in deference to its wide usage.
21. de Groot (1918:5–55); Yosida (1973).
22. Chang-tzu, probably Third century BCE. See *On the unity of nature*, from Feng Yo-lan, *Translations of Chuang Tzu*, Shanghai 1933 as cited by Bodenheimer (1958:184).
23. Veith (1960:5) following *SCC* 2:265 and 273–278. Sarton (1927) 1:122 gives the title as *Huang ti nei Ching Su Wen*, as does Veith. Either translates as *Pure* [or *Simple*] *questions of the Yellow Emperor*; Veith adds *Canon of internal medicine*. Huang Ti was said to have lived 2698–2598 BCE, but the compilation itself is probably no older than the Fourth century BCE.
24. Tsou Yen. In the *Shih chi*, chapter 74; translated in *SCC* 2:232–234 (1956). See also *SCC* 2:232–244. We should not equate these with Greek or modern concepts of *elements*, but instead consider them as *powers* or *virtues*. *Metal* is sometimes replaced with *gold*. For more on how the first *chih* arose from space and time, were re-sorted into *chih* of earth and sky, met to form the *chih* of *yin* and *yang*, then formed fire and water, sun, moon, and stars, see Yosida (1973:77).

NOTES 465

25. *Mo Ching*: *Ching* and *Ching shuo* of *Ching hsia* 43, about 270 BCE; see *SCC* 2:259–260. For the Neo-Confucian statement of five elements and their cycle, see *SCC* 2:463. Johan Nieuhof, visiting China in 1655–1657 with a Dutch East-India Company legation, noted that although the Chinese had in earlier times recognized five elements, following the visit of Matteo Ricci they eagerly accepted the European doctrine of only four elements. He also claimed that they thought the world is an egg: Nieuhof (1665:77–78). Ricci was in China from 1583 until his death in 1610.

26. *SCC* 2:261–266. For example, water became associated with the kidneys, fire with the heart, metal with the lungs, wood with the liver, and earth with the stomach (Johnson 1928:102). See also Yosida (1973:76) and Veith (1960:7–8).

27. *Da Dai li ji*, or *Record of rites of the elder Dai*. By tradition, this was assembled by Dai De (Tai Tê) between 73 and 49 BCE, but is more properly dated to 80–105 CE (*SCC* 2:268). The quotation, as translated at *SCC* 2:271–272, is from chapter 81. For more on the interconversion of birds and mussels, see Chapter 10.

28. *SCC* 6(1):220 ff (1986). Ni Chu-Mo (1624) said that Shen Nong "had tested all the plants, so historically they had precedence over the mineral and animal kingdoms." He is said to have discovered tea in the process.

29. *SCC* 6(1):222, citing Lung Po-Chien (1957).

30. Hsieh (1947); *SCC* 6(1):224 and 601 (1986). Needham disputed that this arrangement was based in *fang shih* or alchemy, arguing that it was "primitively obvious". My discussion in this chapter, and in the previous one, undermines global claims of primitive obviousness.

31. *SCC* 6(1):224. Needham gives a slightly different arrangement at *SCC* 6(1):247.

32. *SCC* 6(1):266. Harada (1993:245–246) states (erroneously, I believe, based on a close reading of Needham) that the *Hsin hsiu* gives *pen tsao* and *Thang pen tsao* as identical.

33. *SCC* 6(1):248–264.

34. For example, at *SCC* 6(1):224 we read that Hu Shih-Kyo asked (in *Pen tsao ko kua*, 1295) "If (the physicians) do not read the pharmaceutical natural histories, how can they know the names, virtues, properties and active principles (of the minerals, plants and animals) which bring about health and longevity?" Footnote *a* credits a bibliography by Taki Mototane (1831, reprinted 1933 and 1936) (*SCC* 6(1):609). Do the words in parentheses appear in the original text, or were they added (rightly or wrongly) by Taki, or by Needham?

35. The primary reference is the *Shi ji* (*Records of the Grand Historian*) completed by Sima Qian in 94 BCE.

36. *SCC* 2:55–56, 410–419, and 493–495. Neo-Confucians such as Ch'en Shun (in his *Pei-ch'i tzi i*), as well as Taoists, recognized the challenge of Buddhism; see *SCC* 2:413–416 for extended quotations. Ch'en Shun (*floruit ca* 1200 CE) was a student of Chu Hsi (died *ca* 1200 CE). Buddhism (and other foreign ways of thinking) were severely set back by the persecution under Tang Emperor Wuzong (845 CE). A partial synthesis of Confucian, Taoist, and Buddhist philosophies was finally developed during the Eleventh and Twelfth centuries. See also Ching (1974).

37. *Shih ka mou ni*. *SCC* 2:396 footnote a, and page 400.

38. Around 700 CE. The Gautama Buddha is said to have attained enlightenment under a bodhi (pipal) tree in Bodhgaya.

39. The term *Silk Road* (in the plural *Seidenstrassen*) was coined by the German geographer [Baron] Ferdinand von Richthofen (1877).

40. *SCC* 2:3–15.

41. *Analects, ca* 500 BCE. Books 16–18 and 20 are later than the others and contain Taoist material (*SCC* 2:5 note *b*). Pagination refers to the Legge edition [1930]. Confucian philosophy became better known in Europe through the *Confucius sinarum philosophus* (1687) of Jesuit missionary Prospero Intorcetta (*SCC* 2:634).

466 NOTES

42. Presumably the *Shih ching*.

43. *Analects* 17.9.7. For "grasses and trees", see Legge's note 9.7 (1930:261). In analect 19.12.2 Confucius seems to refer to the assortment of plants into classes, but according to Legge's note 12.2 (1930:291), "classes" is a problematic translation.

44. Confucian school as recorded by Tsze-sze, *Chung yung* (*Doctrine of the mean*), chapter 9, states that the mountain holds the grass and trees, the birds and beasts. Aquatic animals (tortoises, iguanas, iguanodons, dragons, fishes, and turtles) are mentioned separately: Legge [1930], pp. 406–407. This passage is quoted somewhat differently at *SCC* 2:12. The *Chung yung* was compiled over many years, and contains some material by Confucius's grandson. The passage discussed here is probably mid-Third century BCE. Mencius (Fourth century BCE) referred to vegetation, birds, beasts, and men (Veith 1960:9–10).

45. Needham gives two slightly different dates for Xun Kuang (Hsün Ch'ing): 298 to 238 BCE (*SCC* 2:19) and *circa* 305 to 235 BCE (*SCC* 2:23). Knechtges & Shih (2004:1757) give *circa* 310–235 BCE and *circa* 314–217 BCE. The quotation appears at *SCC* 2:23. See the table at *SCC* 2:22 for later *scala naturae* teachings by Liu Chou (Sixth century CE) and Wang Khuei (Fourteenth century CE).

46. In *Liu Tzu*, Liu Chou (519 to 570 CE) wrote that plants *seng*, but animals both *seng* and *shi*; Needham (*SCC* 2:24) translates the latter as *consciousness*. Chĕng I added "good instinct" (*liang neng*) to the *zhi* (perception) of animals and man. See also the *Li hai chi* of Wang Khi (perhaps late Fourteenth century CE): *SCC* 2:23–24.

47. Zhu Xi (Chu Hsi) taught that inorganic things had only substance and qualities (*xing zhi chou wei*); plants further possessed *seng ji* (vital force), and animals and man further possessed the *ji* of blood, and perception-sensation (*xue ji zhi jue*). Zhu Xi must be considered a syncretist, as his Neo-Confucianism is strongly mixed with Taoism and other philosophical influences. The quotation is from *SCC* 2:569.

48. As quoted at *SCC* 2:33. For alchemy, see *SCC* 2:34; shamanism and magic, *SCC* 2:132–139; distrust of reason and logic, *SCC* 2:163.

49. *SCC* 2:35–36. Sarton (1927) 1:66–67 gives 604 BCE as the date of Laozi's birth, but most modern scholars consider Laozi to be a collective identity, or altogether fictitious.

50. *Zhuangzi* (*Chuang Tzu*), chapter 22 as quoted at *SCC* 2:38–39.

51. *Chuang Tzu*, chapter 18 as translated by Hu Shih; *SCC* 2:78–79. The character *chi* is unusual, but also occurs in the *I ching* (*Book of changes*), where it refers to the "minute embryonic beginnings of things" (*SCC* 2:80). A slightly different translation is given by Giles (1889:228). The concept was elaborated in the early Twelfth century by Chĕng Ching-Wang (*Mêng chai pi t'an*): *SCC* 2:421 and Chapter 1, note [85]. Ores and metals were likewise considered to undergo transformation: *SCC* 2:79–80 and footnote 1 at page 80.

52. *Kuan tzu*. Date uncertain but not earlier than the Fifth century BCE: *SCC* 2:42–43.

53. *Kuan yin tzu*, also known as *Wên shih chen ching* (*True classic of the word*), anonymous (perhaps Thien Thung-Hsiu, Eighth century) and edited by Chen Hsien-Wei (also known as Pao I Tzu), 1254 CE. See also the *Hua Shu* (*Book of transformations*) attributed to Than Chiao (Tenth century or perhaps slightly later): *SCC* 2:444–449.

54. Mozi (Mo Ti or Mo Tzu), who flourished in Song after Confucius but before Mencius, is considered the founder of Chinese logic. His *Canon* is composed of two chapters of canon per se (definitions, methods, propositions, and solutions) plus two of explanation and disputation.

55. *Mo ching*, in the *Ching shuo*, or Exposition, of *Ching shang*, the First or Upper Canon, section 85; see *SCC* 2:178.

56. *SCC* 2:196. The *Canon*, at least as it has come down to us, contains many unique and rare words, forms, and usages (Graham & Sivin 1973:108–110).

57. Reischauer (1990), pp. ix–x. See also Sugimoto & Swain (1978), and Reich (2018:198).

58. So far as can be determined, the arrangement of books in the *Honzō-wamyo* is identical to that in the *Hsin hsiu pen tsao*, texts and copies of which have been lost in China. Interestingly, entries in *Honzō-wamyo* include not only inorganic substances and living things, but also "natural phenomena, utensils, and imaginary creatures" (Harada, 1993:245–246).

59. And only twice in the Eighteenth century, in 1709 and 1775. In the late 1700s the philosopher Miura Baiyen set out a hierarchical division of the natural world into plant and animal branches, but it had little influence (Harada 1993).

60. Ueno (1964), at page 316[82].

61. Harada (1993:247–254), especially Table 6; Ueno (1964:329[95]).

62. The lone but critically important exception was a trading post built for Portuguese traders on a small artificial island (Dejima) at Nagasaki, and occupied by the Dutch East India company during 1641–1795. During these years, Japanese exposure to European natural history was through Dutch books, or books in Dutch editions; for examples, see Ueno (1964), pp. 320[86]–321[87]. Naturalists employed by the East India Company and posted to Dejima include Engelbert Kaempfer (1690–1692), Carl Peter Thunberg (later the successor to Linnæus at Uppsala: 1775–1776) and Philipp von Siebold (1823–1829). Natural-historical interactions between Japan and the outside world in the decades prior to the arrival of Commodore Perry in Edo Bay (1853) are described by Ueno (1964). The East India Company went bankrupt in 1795, and much of Dejima was destroyed in a fire in 1798.

63. Udagawa, *Botanikakyō* (1822) and *Shokugaku keigen* (1833). Iinuma, *Sōmokuzusetsu* (1856). The latter was still influenced by Dutch texts including Oskamp's *Afbeeldingen der artsenygewassen* (1796–1800) and Houttuyn's *Natuurlyke historie* (1761–1785); for details see Ueno (1964:331).

64. Greene (1983) 1:177 and 1:187.

65. We examine some of these in Chapter 10.

66. For this concept in *e.g.* Raymond of Seybonde, Saint Francis, George Herbert, and Henry Vaughan, see E Holmes (1932), pp. 45 ff. For the "good shepherd" metaphor, see *SCC* 2:576–578. See also EP Evans (1906); Hyde (1916); and P Singer (1975, 1979).

Chapter 3

1. Whitehead (1929:63); corrected edition (1978:39).

2. Taylor (1997:1–2). The terms *mythos* and *logos* were not distinguished before the Fifth century BCE: Osborne (1997:34).

3. According to Plutarch, Alexander strove to bring to reality what the Stoic philosopher Zeno could only dream of: a universal civil order in which Greeks were not favoured as "friends and domestics", nor barbarians scorned "as mere brutes and vegetables": Plutarch, *De Alexandri Magni fortuna aut virtute* 1.6 in the translation of Philips (1874:481), reproduced by de Santillana (1970:292–293). Alexander's encounter with Indian philosophers is a staple of the various versions of the Alexander romance, for which see *e.g.* Stoneman (1995).

4. From the majestic Lattimore translation (1951), but Henry Cary began his 1821 translation in the same way.

5. Sarton (1970:134–135); Osborne (1997:10–12).

6. Thales of Miletus in Ionia (*ca* 624 to *ca* 548–545 BCE) was said to have travelled to Egypt, where he studied geometry and presumably other sciences. O'Grady (*Internet Encyclopedia of Philosophy*) considers this possible, as by 620 BCE Miletus held a trading concession at Naucratis, at the mouth of the Nile, and had established colonies around the eastern Mediterranean and the Black Sea.

7. Diogenes Laërtius, *Lives* 1.24.

468 NOTES

8. Pindar, *Pythian Odes* 4.315. We return to Orpheus in Chapter 9.

9. Kahn (1960:3–4); de Santillana (1970); Sarton (1970:10–40).

10. Anaximander of Miletus, a key figure in Sixth-century Ionian thought, was a pupil of and/ or successor to Thales. Anaximander founded the scientific tradition of prose works with his *Peri physeos* (*On nature*), perhaps sharing this distinction with Pherecydes of Syros. The stories credited to Anaximander may be due to Theophrastus: Kirk, Raven & Schofield (1983:100–142). See also Censorinus, *De die natali* 4.7.

11. For original sources see Kahn (1960), pp. 68–71 and 109–110.

12. Empedocles is associated with Acragas (Agrigentum) in Sicily, then a wealthy and sophisticated centre of Hellenic culture. Some Pythagoreans had settled there after they were driven out of Croton (Sarton 1970:246–247). Empedocles set out the four elements in a poem *Physika*.

13. Anaximander also recognized a fifth, the Non-limited (*apeiron*), that served as "a kind of substratum of the others". The earlier Orphic tradition had recognized only three elements, a concept continued by Ion of Chios. Plato introduced the term *stoicheia* for the elements. Aristotle considered the *stoicheia* to be qualities—not earth, fire, air, and water, but rather the dry, the hot, the cold, and the moist—and assumed that they gave primal matter (*hyle*) its form (*eidos*): hence *hylomorphism*. Aristotle's elements could change into one another (*alloiosis*) and combine in different ways: by *synthesis*, *mixis*, or *krasis*. Philolaus of Croton added a fifth element, *holkas*, in analogy with the five geometric figures; Plato identified this fifth element with *æther*, and Aristotle said that *æther* occurred only in heavenly bodies. The author of the *Epinomis* (traditionally Philip of Opus, editor of Plato's *Laws*) also accepted *æther* as the fifth element: Kahn, *Anaximander* (1960:153); Dampier, *History of science* (1948:23); Needham, *Science and civilization in China* 2:245 (1956).

14. Hall (1950:342–344).

15. Hall cites Tyrtæus 15.5 as example. See also Chevalley (2014).

16. For Pythagoras, more-or-less untrustworthy *vitæ* come down to us from Diogenes Laërtius; Porphyry (*Life of Pythagoras*); Iamblichus (*Life of Pythagoras*); and an anonymous source preserved by Photius (Guthrie 1987:137–140). Origen (*Contra Celsum* 1:15) cites Hermippus (*On lawgivers*) to the effect that Pythagoras derived his philosophy from the Jewish people: Roberts & Donaldson, *Ante-Nicene Fathers* 10:412–413 (1869); and Hadas (1958:6–7). For a critical summary of sources, see Zhmud (2012) and Huffman (2014).

17. Iamblichus, *Life of Pythagoras*, sections 8 and 12; Diogenes Laërtius, *Life of Pythagoras*, section 6; and Guthrie (1987), pp. 66, 70, and 143.

18. There seems to have been some diversity of practise among his followers: see Huffman (2014).

19. Dampier (1948:17). A massacre of Pythagoreans is described in Iamblichus's *Life* (section 35). Pythagoras taught that "all nature is number", and according to Dampier (1948:18) "the Pythagorean idea of numbered order in geometry, arithmetic, music and astronomy made those four subjects the *quadrivium* of medieval instruction." See also Philo, *Quaestiones et solutiones in Genesim* 4.8; John of Salisbury, *De septem septenis* 7; and Nicholas of Cusa, *De docta ignorantia* 1.7.18.

20. Emanationist cosmologies posit that "all that exists (the universe and everything within it) has arisen through a process of flowing-out from, and willed by, a deity or First Principle. This flowing-out necessarily gives rise to a hierarchy or continuum of entities of which those closest to the First Principle are the most-perfect, while those farther away are increasingly material, embodied and imperfect. These systems are to be contrasted with those positing a (perfect) creator who stands outside his (less-perfect) creation" (Ragan 2009). Such teachings were common in the eastern Mediterranean lands at that time.

21. Alexander Polyhistor, as quoted by Diogenes Laërtius, *Lives* 8.24–33 (Dillon 1996:342). See also Aristotle, *Metaphysica* 985b23 ff; Sextus Empiricus, *Pros physikous* 2.261 ff; and Colavito

(1989), pp. v–vi and 39–44. Dillon (1996:373) notes that there was in fact an "age-old controversy among Pythagoreans" as to whether there are one or two First Principles: a Monad that produces from itself a Dyad; or alternatively, eternally opposed Monad and Dyad. Pythagoras is credited with discovering the mathematical basis of musical harmony, and Pythagoreans saw this as an example of the mathematical basis of harmony throughout nature. However, Zhmud (2012) argues that Pythagoras himself did not teach a numerical metaphysics, the association of which with Pythagoras appeared only in the Fourth century BCE in the context of debates specific to that time.

22. ML Kuntz (1987:253–265); Colavito (1989); Porphyry, *Life of Pythagoras*; Hussey (1997).

23. As did the author of the anonymous *Life of Pythagoras* preserved by Photius, at paragraph 6 (Guthrie 1987:138).

24. According to Aulus Gellius (*Noctes atticæ* 4.11.12–13), who claimed to quote Plutarch quoting Aristotle (*Attic nights*, translated by Beloe, 1795, 1:264); also Porphyry, *Life of Pythagoras*, section 45 (Guthrie 1987:132). Note that Aristotle considered sea-anemones to "dualize" (see later).

25. Pythagoreans had settled in Croton, a town in southern Italy. He may have been the teacher of Archytas of Tarentum, the last of the major Pythagorean philosophers. See CA Huffman, *Philolaus of Croton* (1993).

26. Burkert, *Lore and science in ancient Pythagoreanism* (1972); Huffman (1993).

27. From section 25.17 of the *Theologumena arithmeticae*, a work often but controversially attributed to Iamblichus, who (if indeed he was the author) may have drawn on earlier sources, notably Nicomachus of Gerasa (Second century CE). Burkert (1972) and Huffman (1993) accept this fragment as genuine, although other authorities differ (see Huffman 1993:17–35 and 307–314).

28. Democritus the atomist, of the same generation as Philolaus but probably younger by one or two decades. See Huffman (1993:311–314).

29. *De generatione animalium* 745b23.

30. Ocellus Lucanus (attrib.), *On the nature of the universe*, Chapter 3. In the translation of Taylor (1831), at page 19: "The origin, however, of the generation of man was not derived from the earth, nor that of other animals, nor of plants."

31. Hippodamus of Miletus was renowned as an architect and town planner. *On felicity*, in Guthrie (1987:215–217). Sextus the Pythagorean (*floruit ca* 200 CE) likewise uses *animal* in a high-level sense (*Sentences* 25: Guthrie 1987:268).

32. The anonymous *Life of Pythagoras* preserved by Photius (*ca* 820–891 CE) has Plato "the ninth in succession from Pythagoras" (Guthrie 1987:137).

33. *Phaedo*: Colavito (1989:5–6). Plato's discussion of microcosm and macrocosm (*Timaeus* 29d–47e) is sometimes considered Pythagorean, although the idea was relatively widespread in contemporary cosmologies. Perhaps more tellingly, the myths in *Timaeus* bear similarities to those in the Speech of Pythagoras (Book 15) of Ovid's *Metamorphoses*. Dillon (1996:3) considers that Plato took on a more Pythagorean position in his later dialogues. Apuleius of Madauros (*Florida* 15.26) goes farther: "our own Plato, deviating no whit, or hardly at all, from this sect, is in large part a Pythagorean" (Dillon 1996:306).

34. For the merger of Neopythagoreanism into Neoplatonism, see Guthrie (1987:38–43).

35. Parmenides distinguished between the worlds of reason and sense-perception, a distinction later extended by Plato. Melissus was admiral of the Samian fleet at its defeat in 440 BCE; unlike his predecessors, he argued that the cosmos could not be finite. Xenophanes of Colophon, regarded as a precursor to the Eleatics, interpreted fossils as animals and plants that had been transformed into stone (Diels 1951–52, at 21 A33).

36. Zeno as quoted by Cicero, *De natura deorum* 2.7.22, in the Rackham translation. See also Broadie (1981:10–11 and 21).

470   NOTES

37. Kretzmann (1982:7); de Santillana (1970:143). Little is known of Leucippus; by tradition he was born in Miletus, the city of Anaximander and Thales, and had once been an Eleatic, perhaps studying under Zeno. He flourished in the mid-Fifth century BCE, a generation before Democritus of Abdera, who was born during the Eightieth Olympiad (460–457 BCE) and died about 370 BCE. Democritus wrote a book, now lost, on Pythagoras.

38. *Stobaeus* 1.4.7c; Diels (1951–1952), fragment 67B2; Taylor (1999), fragment L1.

39. Diels (1967), fragment 68B9; Taylor (1999), with words in square brackets added for emphasis following de Santillana (1970:145).

40. Sarton (1970:254).

41. It was, however, accorded credence by the Mu'tazilite movement (Eighth to Tenth centuries CE) within Islam, on the grounds that denying that God could reduce matter to indivisible atoms would be denying His omnipotence: Wolfson (1976:467–470). Abū Isḥāq Ibrāhīm an-Naẓẓām rejected atomism nonetheless.

42. Epicurus of Samos campaigned against superstition, divination, and magic, but distrusted mathematics and logic, and (so far as we know) was uninterested in natural history: Bailey (1926).

43. It would be a mistake to imagine too strong a connection between the atomic theory of Leucippus and Democritus, and the theory we know by that name today, attributable to John Dalton: Greenaway (1966:8–31).

44. Empedocles was a philosopher, mystic, and magician (Kingsley 1995). He was the last Hellenic philosopher to write in verse. More than 450 lines of his poem *On nature* are preserved in more than 130 fragments. Quotation from MR Wright (1997:177).

45. MR Wright (1981) as fragment 9; and (1995) as fragment 13.

46. Wright (1995:22–30); *roots* at fragment 6 (1981) and fragment 7 (1995); quotation at fragment 21 (1981) and 14 (1995). A very similar construction is at fragment 23 (1981) and 15 (1995). Elsewhere Empedocles mentions shrubs, shellfish, buccinia, turtles, and "things and plants".

47. Wright (1995:228–229); fragments 62, 79 and 82 (1981); and 53, 65 and 71 (1995).

48. Wright: fragment 110 (1981) and 100 (1995). See also fragment 103 (1981) and 81 (1995).

49. Diogenes of Apollonia in Thrace (a Black Sea colony of Miletos, not the Apollonia in Crete), quoted in Barnes (1982:571) as fragment 515 (Diels-Kranz DK 64B2).

50. Cicero, *Tusculan disputations* 4.10. That is, Socrates abandoned the philosophy of nature, and instead founded the philosophy of morality.

51. Most of what we know about Socrates comes *via* his students Plato (*Apology, Crito, Phaedo, Symposium*) and Xenophon (*Memorabilia*), and *via* a caricature by Aristophanes (*Clouds*). The somewhat contrasting views of Socrates in these works have been synthesized by Vlastos (1991). Divine purpose: Drummond (1888) 1:52ff after Xenophon, *Memorabilia*. Benson (1997:323–324) reminds us that Pythagoras, as well as Socrates's contemporaries Protagoras, Gorgias, and Democritos, engaged in moral philosophy.

52. *Metaphysics* M4 1078b27.

53. Plato is thought to have been born in Athens.

54. *Phaedo* 65–78; *Republic* 485 ff.; *Phaedrus* 246; *Parmenides* 131–135.

55. *Republic* 495d–e and 522b; *Laws* 919c. See also Feyerabend (1996). Ironically, Plato's term *Demiurge* comes from a common word that originally meant *craftsman* or *artisan*.

56. de Santillana & Pitts (1951:117) called the *Timaeus* "practically a Pythagorean enclave in Plato's work".

57. Wetherbee (1973:10). He further argues that *Timaeus* was "of tremendous importance" in "seem[ing] to sanction the study of causality and analogy within the cosmic framework as manifestations of the 'aurea catena' or hierarchy of cosmic powers which descends from the Platonic Demiurge and expresses its will." Had Aristotelianism (or perhaps atomism) become

NOTES   471

predominant instead, such sanction might have been unnecessary. Sarton (1927) 1:352–353 dates Chalcidius with considerable uncertainty to the first half of the Fourth century CE; others place him in the Fifth or Sixth century. His translation of *Timaeus* covered only the first 53 chapters.

58. Critias and Hermocrates. The quotation about Timaeus is at 27a. For the most part I have used the translation of Bury (1929) in Loeb Classical Library 234.

59. *Timaeus* 28–33. Here and elsewhere, "living creature" is ζῷον (*zoon*), which could be translated *animal*. Where "beasts of the earth" is intended, Plato can use *theria* (*e.g. Timaeus* 76e). For Aristotle and most subsequent philosophers, a living creature endowed with intelligence would be man; but Plato so credited the universe itself. The stars too are living creatures (*Timaeus* 38e), each assigned a soul (*Timaeus* 41).

60. *Timaeus* 33–34. This is against the Eleatics, who taught that motion was illusory.

61. Ibid. 35–36, a very Pythagorean passage.

62. Ibid. 37–39.

63. Ibid. 39e–42. The Demiurge fashioned these four classes "as Reason perceives Forms existing in the Absolute Living Creature". This is often taken as a statement of the doctrine of *plenitude*, of which we shall have much to say in later chapters. The point is reinforced in the final sentence of the dialogue (ibid. 92). The number four is perhaps a Pythagorean touch.

64. Ibid. 41–42. Recall the spirits in the meadow, in Socrates's tale of Er the son of Armenius, concluding Book 10 of *Republic* (614–621).

65. *Timaeus* dwells at length on the mathematical foundations of "matter". At least in the perceptible world, fire, air, and water were built of elementary triangles, while earth was constructed of cubes (ibid. 55–57), all of particular dimensions. According to Aristotle, Plato's world-soul was constructed out of the four primal numbers (one, two, three, and four), and it is exactly at the level of the soul that these four numbers take on their respective aspects point, line, plane, and solid (*De anima* 404b16–26).

66. *Timaeus* 39–40, 91–92.

67. Ibid. 91–92. The three faculties of the human soul are the rational (or reasoning: *to logistikon*), the passionate (spirited, irascible: *to thymoeides*), and the appetitive (desirous, concupiscent: *to epithymetikon*). Each is localized differently in the human body: reason in the head, passion in the chest, and appetite between the diaphragm and navel (ibid. 69–70, 77, 90). The liver was thought to mediate between the rational and appetitive faculties, and livers of sacrificial animals were thus useful in divination (Thunberg 1965:189)—a practise of which Plato disapproved (*Timaeus* 71–72).

68. Socrates speaks at *Phaedrus* 245d: "Every soul is immortal. For that which is ever moving is immortal but that which moves something else or is moved by something else, which it ceases to move, ceases to live. Only that which moves itself, since it does not leave itself, never ceases to move." See also *Laws* 10:893. Aristotle criticized this in *De anima* 1:3. Origen (Chapter 6) later recognized a hierarchy of beings based on the origin and nature of their motion.

69. *Timaeus* 76e–77c. The soul was divided into immortal and mortal parts; the former was situated in the head. The mortal part was further divided, the superior part (responsible for courage and passion, and potentially under the rule of reason) was situated in the chest (thorax), whereas the lower part was situated at the navel "as though it were a creature which, though savage, they must necessarily keep joined to the rest and feed, if the mortal stock were to exist at all" (*Timaeus* 69–70). Diogenes Laërtius credits a very similar concept to the later Pythagoreans: see Colavito (1989:10–11).

70. *Timaeus* 77 for "trees and plants and seeds" (δένδρα καὶ φυτὰ καὶ σπέρματα), *Republic* 441 for animals including human children. Empedocles and Heracleides earlier had included plants among *zoa* (Burkert 1972:133n74).

71. *Timaeus* 92b–c.

472   NOTES

72. For example, in the description of man as "not an earthly but a heavenly plant up from earth towards our kindred in heaven" at *Timaeus* 90a–b.

73. Or more accurately, of the triangles of which the elements are generated (*Timaeus* 54–60). The four elements continually convert into one another (*Timaeus* 49–50). Gold, adamant, and copper are indeed not earth, but a fusile kind of water; vegetable juices and oils, the sap of trees, wine, honey, and vegetable acids are also kinds of water (*Timaeus* 58–59).

74. Plato does not use a technical term for "matter", referring to it instead as "that which is" or some similar construction. Aristotle was first to use the term *hyle* for a concept analogous to modern *matter*.

75. The chariot analogy at *Phaedrus* 246a–254e; and *Republic* Book 4.

76. *Phaedrus* 229c–230a.

77. Such beings would in any event be unreal, as only forms (ideas) are real. Among the potentially "intermediate organisms" would be the sponge, for which see note [82] below.

78. *Timaeus* 29d ff; *Philebus* 28d–30d. See also Holmes, Lyons & Linker (1935) 1:136–138.

79. Indeed, we today cannot count ourselves free of its embrace.

80. For example, in *Parmenides* 131–135.

81. *Sophist* 218e–221c. The argument is led by the Eleatic stranger, who has accompanied Theodorus and Theaetetus to visit Socrates.

82. Mair (1963), page xxxviii.

83. If the dichotomy is comprehensive, the Eleatic stranger requires hunted beings other than animals to be lifeless. Recall that Timaeus accepts that trees, plants, and seeds share the appetitive faculty; if Plato agreed with both the Eleatic stranger and Timaeus, then sponges are either attached beings that nonetheless are not alive, or are stones or minerals. Both *Sophist* and *Timaeus* are considered late works, with the latter likely later than *Sophist*.

84. Aristotle of Stagira, 384 to 322 BCE.

85. *Metaphysics* 987b1–10, 1078b30–32. The doctrine that everything is composed of immaterial form and material substrate is called *hylomorphism*: Manning (2013). Form also makes the material world eternal (*De caelo* 279b4–283b2).

86. Aristotle itemizes six meanings for *nature* (*Metaphysics* 1014b16–1015a19; see Weisheipl 1985:5–8), summarizing them (*Metaphysics* 1015a13–19) as: "nature in the primary and strict sense is the essence of things which have in themselves, as such, a source of movement; for the matter is called the nature because it is qualified to receive this, and processes of becoming and growing are called nature because they are movements proceeding from this. And nature in this sense is the source of the movement of natural objects, being present in them somehow, either potentially or in complete reality." Thus for Aristotle, as for other Greek thinkers through at least the late Fifth century, *nature* was a *principle* or *source*, not a collection of natural things. Gorgias (late Fifth century BCE) may have been among the first to use *physis* as more-or-less equivalent to *kosmos*, the aggregate of natural things: Weisheipl (1985), pp. 2–3 and footnote 3 at page 3.

87. *Physica* 231a19 ff. See also Kretzmann (1982:8). For Aristotle on Zeno's paradox, see FD Miller Jr (1982:96–98). Aristotle nonetheless taught (in *Physica*) that a body can move only by a series of jerks. Diodorus Cronus, who unlike Aristotle was an atomist, likewise accepted that movement in space involved a series of jerks: Sorabji (1982:59).

88. Thus through its combinations, base matter could take on different properties—a view not incompatible with later beliefs of alchemists. Aristotle also posited a fifth element, *æther*, of which heavenly bodies were composed (*De caelo*).

89. In organizing this passage I have followed Nordenskiöld (1932:36–37). More precisely, soul is the place of *potential* forms (*De anima* 429a26–29).

90. In the sense of referring to *psyche*, the soul.

91. *De anima* 434a22 ff.

NOTES 473

92. Ibid. 413a20–413b14. Translation by JA Smith, in the Oxford *Aristotle*.

93. Ibid. 413b25–26.

94. Ibid. 413b32–414a3.

95. Ibid. 414a29–414b10.

96. Although Aristotle considers heavenly bodies to move rationally and voluntarily, he never calls them *zoia* or living beings: Falcon (2005:91).

97. *De anima* 416b20–25.

98. Ibid. 415a23–30, 416a19.

99. It might be asked whether animate soul encompasses vegetative soul, or corresponds only to the powers necessarily superadded in animals—*i.e.* whether animals possess only animate soul, or a distinct and separate vegetative soul too. Aristotle considers this at *De anima* 413b13–24 and *De partibus animalium* 641a17 ff without reaching a generally applicable conclusion. The ambiguity may be a necessary consequence of requiring form to inhere in a body. Most authorities have interpreted Aristotle to refer to three types of soul—vegetative soul in all living beings, a distinct animate soul in animals and man, and rational soul in man alone. Keeping Aristotle's own inconclusiveness in mind, I follow the traditional schema here.

100. For a detailed comparison of human soul as understood by Plato, Aristotle, Galen, the Stoics, Posidonius, Proclus, Plotinus, Nemesius, the Church Fathers, and others, see Thunberg (1965).

101. *De anima* 414a29–415a13, especially 414b31 ff.

102. Mansion (1973) argues that Aristotle's concept of life evolved from his earlier to his more mature works, and that in the former he may have considered that plants, being alive, are necessarily sensate. See also Lloyd (1961), and Pellegin (1986) pp. 124 ff., and notes 15 and 16 at pp. 200–202.

103. For example, oysters and sponges: *Historia animalium* 487b8–10.

104. Aristotle appears inconsistent on self-motion, on the one hand denying that it exists in nature (*Physics*, Book 8) but repeatedly referring to it among animals. See Gill & Lennox (1994), and Berryman (2002).

105. *Historia animalium* 487b6–33 and 588b14–17.

106. *Metaphysica* 1038a8–24.

107. *De generatione animalium* 733a32–733b16.

108. *Historia animalium* 588b18–23.

109. *De caelo* 292b1–10: "We must, then, think of the action of the lower stars as similar to that of animals and plants. For on our earth it is man that has the greatest variety of actions—for there are many goods that man can secure; hence his actions are various and directed to ends beyond them—while the perfectly conditioned has no need of action, since it is itself the end, and action always requires two terms, end and means. The lower animals have less variety of action than man; and plants perhaps have little action and of one kind only. For either they have but one attainable good (as indeed man has), or, if several, each contributes directly to their ultimate good."

110. *Historia animalium* 487b6 ff. and 588b24 ff.

111. *De partibus animalium* 656a4–10.

112. Pellegrin (1986:63); Atran (1990).

113. *De partibus animalium* 642b10–644a10 is particularly clear. Aristotle recognized five classes of predicable things: *genos*, *eidos*, *diaphora*, *propria*, and accidents (*Topics* I 101b17–25). His theory of classes revolved around the first three of these, and "it is by understanding these terms as 'genus,' 'species,' and 'differentia' that systematicians of the modern period have made of Aristotle the inventor of taxonomy" (Pellegin, 1986:53). Some commentators believe that there coexist in Aristotle both technical and non-technical senses of these three terms. Pellegin (1986:50–112) considers these issues in detail.

474 NOTES

114. At *Historia animalium* 490b7–16 Aristotle identifies seven extensive genera of animals (birds, fishes, cetaceans, shellfish, soft-shelled animals, molluscs, and insects), but immediately makes it clear that further, less-speciose genera exist including (but not limited to) viviparous and oviparous wingless quadrupeds, serpents, and man. He also appreciated that unknown types of beings probably exist in faraway lands, but assumed that they would not greatly upset his analysis: Atran (1985:156–158 and 161).

115. Including Crombie (1953:127), Ley (1968:31–32), Mair (1963:xxviii), Peck (1937:22–23), Sarton (1970:536), and Singer (1921, 2:16 and 2:21; 1922:30 and 45–46; and 1957:18 and 26).

116. *De generatione animalium* 731a22–24.

117. *De iuventute et senectute, de vita et morte, de respiratione* 468a5, 8–12 (discussing the entry and discharge of food). Another example: *De longitudine et brevitate vitae* 467a32–467b3.

118. Keep in mind that Aristotle believed that plants reproduced only asexually: *De generatione animalium* 731a1–8.

119. Ibid. 731b3–14.

120. Ibid. 736a32–736b2.

121. Ibid. 736b12–13, 740b9–10. Aristotle goes on to consider whether, and how, the soul might enter the embryo. This problem was taken up later by Nicole Oresme and Thomas Browne (Chapter 8). Plant-animal analogies involving the embryo were known before Aristotle: Singer (1922:16) quotes an example from an unknown Greek writer of about 380 BCE, in a work *On generation*. The idea that the developing animal egg transits "from the life of a *Plant*, to the life of an *Animal*" persisted through the time of William Harvey and beyond (quotation from page 89 of the 1653 English translation of his 1651 *Exercitationes*, at page 49).

122. *De generatione animalium* 778b31–779a5.

123. Today *Pinna nobilis* is the Mediterranean pen-shell, a type of clam.

124. In the Oxford translation. According to Chevalley (2014:34), in this passage "the difficulty of translating *zôion* is . . . at its greatest".

125. *Historia animalium* 588b4–23. At *Historia animalium* 532b18–27 Aristotle noted that "there are some strange creatures to be found in the sea, which from their rarity we are unable to classify".

126. Presumably Aristotle is here simply emphasizing, or making a strong case for, continuity. I do not propose to enter the long-standing debate on whether some of Aristotle's works reflect an earlier "metaphysical" view inspired by Plato, while others reflect his more-mature thought (*e.g.* Jaeger 1923, 1955; in English translation 1934, 1948). Also note [102] above. A similar statement is at *De partibus animalium* 681a11–14.

127. Peck (1961:333) translates this "completely plants".

128. Epipetrum, literally *rockplant*. Translator Ogle suggests it may be a *Sedum*.

129. *De partibus animalium* 681a10–681b9 (within a broader treatment, 679b32–682a2). See also *Historia animalium* 531a7–531b19 and 546b15–549a14.

130. For identification of the "residuum" in the passages above as undigested food, not excrement, see *De partibus animalium* 681a4–6 and note 2 in Ogle's translation in the Oxford *Aristotle*.

131. In the 1504 Latin translation of Theodorus Gaza (681b:1 ff, at folio 73 *verso*): "Quas aût urticas appellant, non testa operiunt, sed exclusae oîno sunt iis, quae î genera diuisimus, ancipiti natura hoc genus est, ambigens & plantae, & aîali." The term *zoophyta* does not appear.

132. *Historia animalium* 548a31–549a14 (this quote 548b10–14). Perhaps the people of Torone (modern-day Toroni) were Platonists (see [83] above).

133. Building on Atran (1985:150–153).

NOTES 475

134. For usage of *ἐπαμφοτερίξειν* before Aristotle, see Lloyd (1993), pp. 44–53 and 44n172; and Pellegrin (1986:185n15).

135. *Historia animalium* 589a18–30 and 590a11–19; *De partibus animalium* 697a16–b5.

136. *Historia animalium* 588b16–17; *De generatione animalium* 731b8–13 and 761a15–31.

137. *Republic* 479b11.

138. *Phaedrus* 257b.

139. See Chapter 12 (Wotton, Aldrovandi). The Roman numinate god Janus was *ancipitus*. Unhelpfully, *anceps* could also mean *doubtful, undecided, equivocal, risky*, or *precarious*.

140. See Lloyd (1993:51–52). In *De partibus animalium* 697b1–5 Aristotle states that seals (similarly, bats) are half-way between two groups, "and so belong to both kinds or to neither". But "much the same may be said also of the Libyan ostrich" (*De partibus animalium* 697b14–26), which (to my reading) he retains among the birds, as the "peculiarities" that associate it with quadrupeds are explained by its bulk. Again, "intermediate in shape between man and quadrupeds is the ape, belonging therefore to neither or to both" (*De partibus animalium* 689b32–35); but surely Aristotle did not countenance that apes might partake of the rational soul.

141. For example, *zôion* could mean "living thing" (including plants), or "animal" (excluding plants). Both senses sometimes occur in the same sentence and must be understood from context, as in Simplicius, *On Aristotle's On the soul* 1.1–2.4, at 103, 19–22 (commenting on *De anima* 413b27–414a3). As translated by Urmson: "A difference between different kinds of living thing is made by either all forms of life [*zôion*] being present, as in man, or not all but most, as in other animals [*zôion*], or one only, the vegetative, as in plants."

142. I thank Dr Anitra Laycock (Classics, Dalhousie University) for confirming this by search of the Pandora CD-ROM of classical Greek texts. I also searched Lambros (1885) and related works, and (later) online resources.

143. Examples include the three works by Singer cited above at note [115], in turn reproduced by Sarton (1970:536); Cole (1949:28); and Pellegrin (1986:125). Singer's classification at 1922:46 is particularly egregious, with *zoophyta* written in Greek.

144. See *De longitudine et brevitate vitæ* 467b5.

145. English translation by ES Forster, in the Oxford *Aristotle* at 815a–830b.

146. Nicolaus's original Greek work has in turn been lost; a single manuscript in Syriac survives, as do later translations into Arabic, Hebrew, Latin, and Byzantine Greek. For critical discussion, see Lindberg (1992:350); Sarton (1970:546); Sarton, *Introduction* (1927) 1:226–227, 601, and 730; Greene (1983) 1:460–461 and note 24; Ullmann (1972:71–72); Baffioni, in Ikhwān al-Ṣafā (2013), pp. 328–329 note 50; and Kruk (1993:833). Kruk speculates that Nicolaus incorporated passages from Theophrastus. Albertus Magnus's *De vegetabilibus libri vii* incorporates a paraphrase of *De plantis*.

147. *De plantis* 815a, first sentence of text.

148. Ibid. 815b10–34.

149. Ibid. 817b24. The Latin text states that for this reason plants have "only part of a soul", a formulation that Forster finds to be "an Arabic turn of expression".

150. Ibid. 816a1–6.

151. Ibid. 816a8–17.

152. Ibid. 816a38–816b1, 816b6–9.

153. Ibid. 819a41–819b27. Some plants, like mallow and beet, "verge on" two different classes—*i.e.* dualize.

154. Ibid. 819a30–31.

155. Ibid. 822a28–822b25; the comparison with pottery is at 822a15–21.

156. Ibid. 827b18–19.

476   NOTES

157. Theophrastus, some fifteen years younger than Aristotle, was (like his older friend) origi-nally a student of Plato. For his life and works, see Greene (1983) 1:128–211. Eresos was the most important town on the island of Mytilene (Lesbos).

158. Including the remaining genus, that of minerals; for which see Caley & Richards (1956).

159. *Historia plantarum*, preserved in nine books with part of a tenth; *De causis plantarum*, orig-inally in eight books, of which six are extant. *De causis* was reportedly translated into Arabic by Ibn Bakkush in the Tenth century CE, but the translation appears to have been lost. A Latin translation by Theodorus Gaza, commissioned by Pope Nicholas V, was completed during winter 1453–1454 and first published in 1483. A number of misidentifications and mistranslations by Gaza, who was not a student of plants, were corrected in Johannes van Stapel's 1644 edition of the *Historia*, and further in the Loeb editions of 1916.

160. Greene (1983) 1:142.

161. Greene (1983) renders these as *ethical susceptibilities* and *the power of voluntary action*: 1:146 and note 52 at page 462.

162. *Historia plantarum* 1.1.1–4.

163. Ibid. 1.2.5–6.

164. Ibid. 1.3.1 ff. He notes that for some plants, "it might seem that our definitions overlap", and under cultivation might even "appear to become different and depart from their essential nature", *i.e.* when mallow grows tall and becomes tree-like. See also ibid. 1.3.4–1.4.4.

165. Ibid. 1.1.11.

166. The fullest description of seaweeds prior to (at least) the late Seventeenth century is at *Historia plantarum* 4.6.1–4.7.1. A lengthy description of aquatic and maritime flowering plants follows, ibid. 4.7.2–4.12.4. For the lichens *Usnea* and *Roccella*, see Hensson & Jahns (1974). Theophrastus reiterates at *Historia plantarum* 4.6.4 that seaweeds can perish, *i.e.* are alive.

167. Ibid. 4.7.1.

168. Ibid. 4.6.10. In a footnote, translator Arthur Hort identifies *aplysiae* as "some kind of sponge", but more probably Theophrastus is referring to other of Aristotle's "certain objects in the sea". Today, *Aplysia* is a marine opisthobranch mollusc ("sea-hare").

169. Ibid. 4.2.11; Greene (1983) 1:203 and note 189 at page 469. For later descriptions of sensitive plants, see Ritterbush (1964:144–156); also Sachs (1875:578–608), in English (1890:535–563). Webster (1966:7–9) notes descriptions of *Yerua biua* and *Yerua mimosa* by Christóbal Acosta (1585), and of *Yerva viva* by John Layfield (died 1619) after 1598. John Parkinson figured a *Mimosa* in *Theatrum botanicum* (1640:1617); I have been unable to find it in his *A garden of pleasant flowers* (1629). *Mimosa* was planted in England by 1638, and perhaps as early as 1634 (Webster 1966:9).

170. Cition was a settlement on Cyprus at the site of present-day Larnaca. Diogenes Laërtius (*Lives* 7.2) quotes Zeno's master Crates, the Cynic philosopher, referring to Zeno as a Phœnician. For Phœnician influences in Cition, see Sarton (1970:599–600 and note 44). Zeno taught at the *Stoa poikile* (frescoed portico) in the Agora at Athens; hence the name of the school. For more on Stoic physics, see Edelstein (1936:290–292); Dillon (1996:108); Clement of Alexandria, *Stromates* 1.11, in Roberts & Donaldson (1867) 4:385; Wolfson (1976:506); and Hager (1982).

171. Chrysippus of Soloi was the second successor to Zeno. See Hager (1982:102–103); and Long & Sedley (1987), 1:280–289.

172. Seemingly, including the elements as well: Hager (1982:102–103).

173. Chrysippus also held that there are eight parts of soul. In rational beings, soul resides in a central facility (the *hegemonikon*) which, in turn, has four powers, one of which (in man) is *logos*: Hahm (1977:164–165) citing von Arnim (1964) volume 2, fragments 458–460, 473, 634, 714–716, 804, and 1013.

NOTES   477

174. Sorabji (2000); Baltzly (2019).

175. According to Sextus Empiricus (*Against the Physicists* 1:79–81: RG Bury translation 1987), Posidonius departed from earlier Stoic teaching in arguing that only animals and plants (but not stones or wood) are unified bodies.

176. While *epithymia* is correctly translated as *desire*, here it refers to the appetitive powers of soul, more broadly encompassing *sensation*; *psyche* encompasses the passionate (but non-rational) powers of soul. See Edelstein (1936:296) and note 41 citing Diogenes Laërtius (*Lives* 7.86) and von Arnim (1964) 2:708–710; and Thunberg (1965:192–193) citing Pohlenz (1948:225). Posidonius has been viewed as the last Greek philosopher prior to the unification of Hellenic and Oriental thought in subsequent Neoplatonism: Edelstein (1936:287) citing Reinhardt (1926).

177. Telfer (1955:212–213).

178. For example, by Aëtius, Themistius, Galen, Alexander Aphrodisias, and Philo. See von Arnim (1964), fragments 708–737. Thus Philo (*Quod Deus sit immutabilis* 35–37): "Among the various kinds of bodies, the Creator has bound some by means of cohesion, others by growth, others by soul, and others yet by rational soul. Thus, in stones and timber that have been detached from its organic growth, he made cohesion a truly powerful bond. . . . Growth God allotted to plants, constituting it a blend of many faculties, nutritive, transformative, and augmentative."

179. The supposed journey of Pyrrho to India is described by Diogenes Laërtius, *Lives* 9.61.

180. Sextus Empiricus, *Outlines of Pyrrhonism*. I have used the Bury translation in Loeb Classical Library (1933/1993) unless otherwise indicated. The Greek text in Bury follows *Sextus Empiricus ex recensione Immanuelis Bekkeri* (1842).

181. Later Platonists argued that Socrates was a sceptic, as he put opposing views into play rather than forcing an opinion.

182. *Outlines of Pyrrhonism* 1.40–79.

183. Fittingly, this trope can be interpreted in different ways: see Vogt (2016).

184. *Outlines* 1.40–41 (1933/1993).

185. *Sextus Empirici opera* 1:iii–xvii (1912); Bury (1933/1993) 1, pp. xliii–xlv; and Floridi (1995 and 2002).

186. *Sexti philosophi Pyrrhoniarum hypotypwsewn libri iii* (Stephanus, 1562). Estienne Latinized his name as Henricus Stephanus. His translation reads "vt zwophyta quæ in caminis videmus" at page 18: *caminis* could equally well be translated *forges, fireplaces,* or *chimneys*; we would expect *fornaces* for *furnaces*.

187. *Sexti Empirici viri longe doctissimi adversvs mathematicos . . . eivsdem Sexti Pyrrhoniarvm hypotypwsewn libri tres* (Plantin, 1569); *Sextus Empiricus opera qvæ extant* (P Chouët & J Chouët, 1621); *Sexti Empirici opera Græce et Latine* (Fabricius, 1718); *Sexti Empirici opera Graece et Latine* (Fabricius, 1842); *Sextus Empiricus ex recensione Immanuelis Bekkeri* (1842).

188. Bibliothèque nationale, MS Fonds latin 14700, at folio 85v.

189. At Latin to Greek, folio 11v. See also Greek to Latin, folio 64r. We return to the dictionary tradition in Chapter 11.

190. Diogenes Laërtius, *Lives* 9.79. Forms of *pyribios* do not otherwise occur in the *Perseus* Greek corpus (1 June 2018). Diogenes Laërtius is not often accused of accuracy.

191. Sextus Empiricus, *Pyrrhoneïsche Grundzüge . . . Pappenheim* (1877); Patrick (1899); and Sextus Empiricus, *Outlines of Scepticism* (1994).

192. Pliny, *Naturalis historia* 11.119. The quoted passage is very similar to Aristotle at *Historia animalium* 552b10–13, where the creatures are considered *theria*.

193. Cicero, *De natura deorum* 1.103 (*bestia*); Apuleius, *De deo Socratis* 1.30 (*animalia*); Aelian, *De natura animalium* 2.2 (πυριγόνα or *pyrigones*, translated by AF Scholfield as *fire-flies* but with a cautionary note distinguishing them from the insects known today as fireflies).

478 NOTES

194. Philo of Alexandria [Philo Judæus]: *De gigantibus* 1.7; and in two works that may not be authentic: *De aeternitate mundi* 45, and *De plantatione* 12. Aelian, *De natura animalium* 2.2.

195. Porta (1597), pp. 52–53. I have not examined the first twenty-book edition (1589). So far as I can determine, Porta did not describe *pyrigones* in the earlier four-book editions. A delightful English translation, *Natural magick*, was published in 1658. Muffet (in Topsell 1658:1081) likewise accepts that "the creature called *Pyrigonus*" is generated in the flames.

196. Smelting made use of charcoal, *i.e.* of plant origin but mostly carbon.

197. Ross (1995).

198. Dampier (1948:36).

199. Andronicus of Rhodes was considered the tenth successor to Aristotle (Ammonius, *On interpretation* 5.24; Strabo, *Geography* 13.1.54; Plutarch, *Lives* at Sulla 26). The only subsequent successor of note was Alexander Aphrodisias. For the transmission of Aristotle's manuscripts see Grant (2004:34–36).

200. Wingate (1931). Boëthius intended to translate all of Aristotle's works into Latin, but most authorities doubt that he completed the zoological works. Aristotle's *Historia animalium*, *De partibus animalium*, and *De generatione animalium* were translated into Arabic by (or under the direction of) Yaḥyā ibn al-Baṭrīḳ in the early Ninth century (Dunlop 1959:144); Peck (1937:40) instead credits Yaḥyā ibn al-Baṭrīḳ's father, a physician and translator. Michael Scot then translated them from Arabic—probably from al-Baṭrīḳ's manuscript—as *De animalibus* no later than 1217 (Peck 1937:41). Scot's translation was used by Grosseteste and Albert; Aquinas instead relied on a translation, directly from Greek, by William of Moerbeke (before 1260). Later, George of Trebizond and Theodorus Gaza translated the zoological works (Peck 1937:42–43).

201. For Antiochus of Ascalon see Dillon (1996:52–105). Dillon also cites Varro to this effect, although without agreeing fully (1996:57–59).

202. Dillon (1996:53–61).

203. For Pamphilius and other authorities on folk medicine in these years, see Scarborough (2013).

Chapter 4

1. Varro, *Rerum rusticarum libri tres* 1.5.2, in Loeb Classical Library (1934).

2. Greene (1983) 1:461n26 (by FN Egerton) states baldly that "Greek science was largely non-utilitarian".

3. Ocellus Lucanus (Sixth century BCE) and Philolaus of Croton (Fourth century BCE).

4. Cato the Elder, *De agri cultura* (or in early printed editions, *De re rustica*) in Loeb Classical Library (1934).

5. Varro (1934). Some searches were conducted via the Roman Texts (now Perseus under PhiloLogic) webpage of Bill Thayer at perseus.uchicago.edu in April 2018. A notable exception, in which Varro explicitly refers to *animal* in its philosophical sense, is at *Rerum rusticarum* 2.1.3.

6. Columella, *De re rustica* in Loeb Classical Library (1941). For insects, see 1.6.15, 1.6.17, 2.9.10, and perhaps 1.6.13. He uses *animalia* in a general sense at 2.20.1, and refers to "omnia animalia, quae ruri sunt" (all farm animals) at 2.9.14. Elsewhere he uses *pecudum* and (at 2.14.7) *quadrupedum*.

7. Columella, *De arboribus* 10.1.

8. Cicero, *De natura deorum* 2.9–11 and 2.130 following Rackham (1933).

9. Ibid. 2.37. The examples Cicero provides of animals serving man are horses for riding, oxen for ploughing, and dogs for hunting and keeping guard. Interestingly, he does not mention animals as a direct source of meat.

NOTES   479

10. Ibid. 1.103. See also Chapter 3.
11. Norman, HistoryofInformation.com.
12. For the manuscript history of *De rerum natura*: Greenblatt (2011); Butterfield (2013); and Beretta (2016).
13. Gordon (1962) counted more than fifty manuscripts (some incomplete, or of wretched fidelity) in the first century after its rediscovery, and thirty print editions by 1600. For a possible thirty-first, see M Smith & Butterfield (2010).
14. *De rerum natura* 1:56–61. English terms from the Leonard translation (1916). In structuring this sentence I follow Kahn (1960:213).
15. *De rerum natura* 5:772 ff. Lucretius explicitly denied that animals owe their existence to a golden chain let down from above (ibid. 2:1149–1155). This did not prevent later authors from bolting spontaneous generation onto ideas of a Great Chain of Being.
16. Ibid. 5:797–798.
17. Ibid. 5:878 ff. John Ray (*The wisdom of God*, 1717) saw fit to quote Lucretius on this point, with a colourful translation.
18. Snyder (2011); Palmer (2012). For ownership, see also Greenblatt, cited above [12]. For classical versus Renaissance Epicurianism, see Yates (1964:223–225). Various Neoplatonic interpretations were likewise hung on Lucretius's *De rerum natura*, as on many other works from classical antiquity.
19. Seneca, *Naturales quaestiones*. Vegetable and animal life, Book 5 chapter 6. Originally in eight books, parts of two books have been lost, and the *Naturales quaestiones* is now presented in seven books.
20. von Hofsten (1958); Stearn (1959:14).
21. *Epistulae morales ad Lucilium* 58.8–24, words taken or paraphrased from the Gummere translation in the Loeb Classical Library (*Epistles*, vol. 1), 1917.
22. Ibid. 94.41 citing Phaedo (the person, not the Platonic dialogue); and possibly again at 123.16 (*Epistles*, vol. 3), 1925.
23. *Epistulae morales ad Lucilium* 124.14 (*Epistles*, vol. 3). See also Epistle 41.8 in vol. 1.
24. Syme (1969).
25. Beagon (1992). Seneca and Columella were members of the equestrian class as well.
26. Pliny, *Naturalis historia* 1.13–15, translation by MAR. Direct quotations in the text are from the Rackham, Jones & Eichholz translation (Loeb Classical Library, second edition 1983–1995) unless indicated otherwise. Citations to the text use the Giannini (section) rather than the Masseiana (chapter) system, as the former affords greater resolution and the latter does not apply to Book 1.
27. Gudger (1924) corrects this to 473 authors, 146 Roman and 327 Greek. Unusually for the time, Pliny identifies and lists the authors he consulted.
28. This apparatus was progressively improved over the centuries. Together with his reference at 1.14 to "the subjects the Greeks call *enkuklios paideias*", this has prompted later authors to refer to the *Naturalis historia* as an encyclopædia. According to Doody (2009), "it is not entirely clear what *enkuklios paideia* meant in antiquity". Other so-called encyclopædias by Varro (*Disciplinae*), Cato (*Ad filium*), and Celsus (*Artes*)—all now lost—differed from each other and from Pliny's *Naturalis historia*, and were intended to introduce "a well-defined curriculum of subjects that students at Rome would follow before more specialized study". Recent scholarship questions whether Cato's work existed in the sense discussed here, and Varro's may simply have consisted of lists of aphorisms. See also Doody (2010).
29. *Naturalis historia*, Books 7 (man), 8 (land animals), and 9 (aquatic animals). Pliny accepts that in foreign lands there are animals unknown to us, but at *Naturalis historia* 32.142–143 he gives his solemn word that all animals in the sea are known to us, precisely 144 species in all.

480 NOTES

Different manuscripts of the *Naturalis historia* give this number as 144, 164, or 176: see note 28 at Book 32, Chapter 53 (volume 6, page 59) of the Bostock & Riley translation (1855–1857).

30. Ibid. 9.143–145.

31. Ibid. 9.146 (translation by MAR): *Equidem et iis inesse sensum arbitror quae neque animalium neque fruticum sed tertiam quandam ex utroque naturam habent, urticis dico et spongeis.* Similar words appear in the *Summarium* referring to this section: *De his quae tertiam naturam habent animalium et fruticum.* Among the authors credited for Book 9 are Seneca, Cicero, Aristotle, Democritus, Melissus, and Theophrastus. Rackham translates *urticis* as *jellyfish*, but Warmington (*Index of fishes*, at end of volume 8) gives *sea anemone* or *sea nettle*. For *sea nettle*, see note 1 by Ogle at *De partibus animalium* 681a37 in the Oxford *Aristotle*.

32. *Naturalis historia* 9.148–149.

33. Ibid. 9.154.

34. Ibid. 13.135–138. He confirms at 32.154 that seaweeds are inanimate.

35. Ibid. 22.92–99.

36. Ibid. 25.130, 26.149, 27.42, and 27.56 (seaweeds), and 26.135 (fungi); but at 32.110–111 seaweeds are mentioned in a passage dealing otherwise with animals.

37. Ibid. 32.21–24. Some of the "marine shrubs" at 13.139–142 are probably corals, although he presumably did not realize this. He mentions the sea-lung (presumably jellyfish, but potentially also including ctenophores) at 32.111 in a section dealing with other marine productions, mostly animals (see previous note).

38. "Indeed, for my own part, I am strongly of opinion that there is a sense existing in those bodies which have the nature of neither animals nor vegetables, but a third which partakes of them both:—sea-nettles and sponges, I mean": ibid. 9.146 in the translation of Bostock & Riley (Volume 2, page 453).

39. Ibid. 31.123.

40. Ibid. 13.31.

41. At ibid. 32.86–87 Pliny describes four kinds of alcyoneum, a substance of uncertain origin found in the sea. It is probably a marine invertebrate: see notes [63] and [82] below.

42. McHam (2013). The reference to Alberti is at page 103.

43. Wethered (1937:284–287).

44. Doody (2010), particularly Chapter 2.

45. Kofoid (1940).

46. *Pline . . . semble avoir mesuré la Nature.* Buffon, *Premier discours* (1769:69).

47. Earlier authors included Menestor of Sybaris (*De causis plantarum* 6.3.5), Leophanes (ibid. 2.4.12) and Clidemus (*Historia plantarum* 3.1.4). The latter is said to have maintained that "plants are made of the same elements as animals, but that they fall short of being animals in proportion as their composition is less pure and as they are colder" (Theophrastus, *Enquiry* 3.1.4).

48. Theophrastus, *Historia plantarum* 1.1.11.

49. The work was unfinished on his death, and one book of *Historia plantarum* has been lost. At 4.7.2 he mentions in passing that mushrooms can be "turned to stone" by the sun; this may be a figure of speech, not a claim for actual transformation. See also 4.7.1. Theophrastus also describes two lichens, *Roccella tinctoria* (4.6.5) and a species of *Usnea* (3.8.6); see Watson (1758).

50. *Historia plantarum* 4.6.10.

51. Caley & Richards (1956), pp. 24–25 and 53; see also the editors' commentary at pp. 140–141.

52. Schmid, (1939); Nissen (1958); Stannard (1969); Einarson (1976); Scarborough (1978).

53. Arber (1912).

54. In any event, root-diggers (rhizotomists) had long been associated with unsavoury practises including sorcery and incantations. See *The book of Enoch the Prophet* (1883) 7:1–15 (particularly verse 10) and 8:3.

NOTES 481

55. One is attributed to Diocles of Carystus, a physician who practised in Athens in the Fourth century BCE. He also wrote on the structure of human bodies, and is regarded as the first to refer to this as *anatomy*.

56. Mithridates VI Eupator ruled from about 120 to 63 BCE. Dioscorides (*De materia medica*, Preface to Book 1) called Cratevas a *rizotómo*. Barbaro (1516) translated this as *Radiciseca*; Sprengel (in Kühn, *Medicorum Graecorum Opera* 25, 1829) as *herbalist*; and Osbaldeston (2000) as *rhizotomist*.

57. The illustrated edition is described by Pliny, *Naturalis historia* 25.8. See Deutsches Museum, accessed 2018–2020.

58. Pliny also cautioned that copyists would render the paintings inaccurately.

59. For Dioscorides I use the Greek and Latin texts in the Sprengel edition (1778–1830), and the English of Gunther (1959) following Goodyer (1655).

60. *Pneumonos thalassios* (Greek), *pulmo marina* (Latin), or sea-lungs (archaic English).

61. Mushrooms at 4.83; three types of *phykos thalassion* at 4.100, and the lichen 4.53.

62. At 2.175. Dioscorides does not specifically identify them as plants, but he calls them a "root" and describes them in the section on pot herbs.

63. Gunther (1959), 5.16 (alcyonium), 5.138 (sponges), and 5.139 (coral, called *lithodendron*, *i.e.* stone-tree). For alcyonium, see note [82] below.

64. Galen later criticized it for its descriptions of plants the author had never seen, and for "absurdities and superstitions": Singer (1928:178).

65. Everett (2012:5–9). Despite its name, the book was not written by the physician Galen.

66. Gow & Scholfield (1953:156–157).

67. Although raw or boiled shellfish are recommended for those poisoned by eating the thorn-apple: *Alexipharmaca* lines 376–396, in Gow & Scholfield (1953).

68. Here "Hippocrates" refers to Hippocrates of Cos, together with the school around him.

69. The idea of equality (*isonomia*) among humours is generally credited to Alcmaeon of Crotona (perhaps Fifth century BCE) but may be older. Empedocles taught much the same idea. For a broader sense of the Hippocratic approach, see Lyons & Petrucelli (1978:194–217).

70. Blood (air, spring), yellow bile (fire, summer), black bile (earth, autumn) and phlegm (water, winter). Health (harmony in man, the microcosm) was thus linked with harmony in the macrocosm (the cosmos). Another axis was blood (warm, moist), yellow bile (warm, dry), black bile (cold, dry), and phlegm (cold, moist); treatment was directed at bringing these back into balance. See Lyons & Petrucelli (1978:195). Galen identified the four humours as did Hippocrates, but other identifications eventually appeared, *e.g.* blood, choler, melancholy, and phlegm; and connections were drawn to the so-called Four Ages of Man, *e.g.* in the poems of Anne Bradstreet.

71. *Hippocrates with an English translation*, Loeb Classical Library (1923–2018); Lloyd (1978:9–60); Jouanna (1999).

72. Singer (1928:21–22).

73. *De medicina* 2.18–33.

74. Ibid. 2.24: *duri ex media materia pisces, ostrea, pectines, murices, purpuræ, cochleæ.* I use the Spencer (1935–1938) Latin text, but Daremberg (1891) and Marx (1915) give identical wording.

75. Galen, *De placitis Hippocratis et Platonicis* Books 2 and 3 (1984); Gill (2007).

76. *De placitis Hippocratis et Platonis* 457.2 ff. See also Edelstein (1936), especially pp. 296–297 and note 44.

77. A work entitled *De historia philosophia* in Galen's name attributes to Aristotle the ideas that plants are endowed with faculties of a lower degree than animals, and that a hierarchy of perception, sentience, and rationality is found among animals; but its attribution to Galen

482   NOTES

is considered false. *De historia philosophia liber*, in Charterius (1679) 2:28 (Chapter 6 at *De natura*) and 2:57 (Chapter 38 at *Vtràm stirpes sint animalia*). There is also Pseudo-Galen on plants and stones: see Thorndike (1963).

78. Emperor Marcus Aurelius was a Stoic.

79. Singer (1957:46–52). A brief but useful analysis of Galen's philosophical stance is at Dillon (1996:339–340). For Galen's complex views on Stoic philosophy, see Gill (2007).

80. *De simplicium medicamentorum temperamentis et facultatibus libri undecim* Books 6–8, 9, and 10–11.1; and marine and aquatic productions that are not animals, Book 11.2, all from Charterius 13:1–318 (1639). For snails, see also *De alimentorum facultatibus* 3:2.

81. *De simplicium* 7.12.25. See also 11.1.31 and *De alimentorum facultatibus* 2:66–67.

82. The word *alcimonium* (spellings vary) covers sponges and holothurians "which resist modern identification". For a definitive discussion, see Everett (2012:153). Galen's description at *De simplicium* 11.2.4 bears more than a passing resemblance to that of Dioscorides, *De materia medica* 5.136. Indeed, *De simplicium* has been called "largely an abridgement of Dioscorides" (Everett 2012:7–8 at note 6).

83. Kotrc & Walters (1979) count 2.6 million words (±10%) in extant works of Galen that are considered genuine.

84. Roberts & Skeat (1983).

85. Singer (1922:123–128); Temkin (1973); Nutton (2008).

86. *Articella*, based on Ḥunain ibn ʾIsḥāq's synthesis of Hellenic medicine (*Isagoge Ioanniti ad tegni Galieni*, or *Tegni* for short), which in turn was based on Galen's *Ars medicinalis*.

87. Randall (1961), especially pp. 19 ff.

88. Nutton (2008:355).

89. This example from Mair (1963:xxviii). For more examples see Chapter 3, note [115].

90. Pellegin has argued against ascribing to Aristotle "any taxonomic project, even an unformulated one" (1986, especially pp. 115–130 and 159).

91. Aristophanes of Byzantium became head of the library in Alexandria in about 197 BCE. There is evidence that his *Epitome* included material from sources other than Aristotle. The *Epitome* itself has been lost, but substantial portions were preserved in Byzantine works to which text from Aelian and others was added, further complicating the textual traditions. See Lambros (1885) 1:1–282; Hellmann (2006); and Falcon (2016).

92. Aristophanes, *Historiae animalium epitome* 2.1.3–5 (Lambros 1885:36) in the translation of Hellmann (2006:336).

93. Aristophanes, ibid. 2.3.13–15 (Lambros 1885:36). The order of presentation by Aristophanes is somewhat idiosyncratic, and information that might be thought necessary to identify animals is scant, perhaps because the audience for the *Epitome* might have been authors, poets, or others looking for succinct but not deeply philosophical information about specific animals (Hellmann 2006:354).

94. See Mair, Oppian (1963), pp. xxiii–xlviii.

95. Cuttlefish: *Halieutica* 1.638–639 and Mair's note *g* at page 267; bivalves and tiny fish: 1.762–768.

96. It is possible that Oppian of Apamea adopted, or was given, the name Oppian in tribute to the (slightly) earlier Oppian of Anazarbus. See Mair (1963), pp. xiii–xxiii. Sponges: *Cynegetica* 5.649–653.

97. Hatzimichali (2013), especially pp. 79–81; and Terian in Philo, *De animalibus* (1981:55).

98. Scholfield (1958), pp. xi–xii.

99. Scholfield points to "striking parallels" with the order of presentation in the *Halieutica* of Oppian of Anazarbus, and in the *De sollertia animalium* of pseudo-Plutarch.

100. *De natura animalium* 8.16. Compare *De sollartia animalium* 30.1 of [pseudo-]Plutarch, in Loeb Classical Library (1957:444–447).

NOTES   483

101. Curley (1980).

102. AS Cook (1919); Wellmann (1930); Gunther (1952) 3:147–148; TH White (1954); and McCulloch (1960).

103. Cook (1919:lix) lists similar allusions in the Bible.

104. Gunther (1952) 3:148 is correct that "the book *Physiologus* written by heretics and prefixed with the name of blessed Ambrose" was declared *apocryphus* in the so-called *Decretum Gelasanium*; however, internal evidence indicates that this decree could not have been formulated by Pope Gelasius I, but is instead a non-papal document of later date: see Burkitt (1913).

105. Aristotle, *Historia animalium* 552b15–18. Elsewhere (*Meteorologica* 382a6–8) Aristotle says that animals do not live in air or fire, and (*De generatione animalium* 737a1–3) that fire generates no animal.

106. Theophrastus, *De igne* 60 (see also van Raalte 2010); Nicander, *Alexipharmaca* line 539 and *Theriaca* line 818; Seneca, *Naturales quaestiones* 5.6; Pliny, *Naturalis historia* 10.188; Tertullian, *De anima* 32:3 (1844: col. 703); Aelian, *De natura animalium* 2.31; Isidore, *Etymologiarum* 12.4.36; Porta (1597:53); F Bacon (1730) 3:180 (Cent. 9, sec. 860); T Browne, *Pseudodoxia epidemica* 3.14 (1658); Topsell (1658) 2:747–750; *Talmud, Chagigah* 27a; *Suda*, Adler Σ 46; and Andrewe (1521). Window *en grisaille* from Château d'Écouen, probably early 1540s. See also McCulloch (1960:161–162); Mistele (2012); and Biederbick (2017:294–303). Nicander (Gow & Scholfield, 1953): *Theriaca*, pp. 28–93; *Alexipharmaca*, pp. 94–136.

107. Scanlan (1987:24–25).

108. *Daniel* 3:8–25.

109. In its somewhat forced allegory, the agate-stone that leads the diver to the pearl is John the Baptist, the pearl is Jesus, and the two shells of the oyster are the Old and New Testaments.

Chapter 5

1. Epistle 7 (The Seventh letter) attributed to Plato, *Epistles* 341C–D as translated by Bury (1929) 9:531. Bury considers Epistle 7 one of "the only two which we can with any confidence regard as genuine". The passage as translated reads "therewith": I take "Idea, the Real that lies behind all existence" from page 473.

2. The Academy may have been Pythagorean in an ethical sense too: both Xenocrates and Polemon, the second and third successors to Plato, refrained from eating meat. See Dillon (1996), pp. 35 and 40.

3. For a summary, see Dillon (1996:1–51).

4. And a very few women, *e.g.* the Neoplatonic philosopher Hypatia (d. 415 CE).

5. Grant (2004), pp. 97 ff.

6. Following JD Turner (1992:451) we could add passages in *Sophist*, the *Symposium*, *Theatetus*, and the *Republic* as well as Platonic teaching reflected in Aristotle's *Metaphysics*.

7. Turner (1992:451–452) enumerates these as "an ultimate ground of being beyond the transcendent realm of being itself, which latter properly begins with the realm of ideas and (ideal) numbers, followed by the World Soul as source of all movement, and finally by the sensible, corporeal world." See also Krämer (1967:193–369).

8. Cicero, *Academia priora* 11–12. The episode is retold by Dillon (1996: 53–55).

9. Dillon (1996:55).

10. Ptolemy, like Alexander a Macedonian nobleman, took on the title King in 305 BCE. He moved the government of Egypt from Memphis to Alexandria, and founded the Great Library. Scholars associated with Alexandria during the Ptolemaic dynasty included Herophilus (considered the first physician systematically to dissect human and animal bodies in public), Erasistratus (the father of physiology), Euclid, and Archimedes. See also Singer (1957).

484   NOTES

11. Oulton & Chadwick (1954:15); Elsee (1908:32 ff.); Singer (1957:35).

12. This is not to say that the different communities always got along splendidly. Riots in 38 and 40 CE were directed against the Jewish residents.

13. Also known as Philo Judæus.

14. Dillon (1996:140–142). See also Wolfson (1970), page v.

15. Dillon (1996:120).

16. Ibid. (1996:143). Broadie (1981:7) recognizes a second argument by Philo why Hellenic philosophy and the books of Moses should be consistent: philosophy had been "showered down by heaven" upon us (*De specialibus legibus* 3.33. 185).

17. Philo Judæus (*Legum allegoriae* 11.6, and *De specialis legibus* 1.333), supposedly following *Genesis* 9:4, *Leviticus* 7:11,14, and *Deuteronomy* 12:23: see Dillon (1996:174–175).

18. *Quaestiones et solutions in Genesim* II 59. The Aristotelian corpus had recently been assembled, but it is more likely that Philo's knowledge of Aristotle was indirect.

19. *Quod Deus sit immutabilis* 35–37 as translated by Winston (1981:110); for terms in square brackets, see note 124 at page 342. The passage is at pp. 26–29 in Loeb Classical Library 247 (1930).

20. Holmes (1935) 1:113 also mentions some apocryphal Hebrew writings. Here I follow the Colson & Whitaker translation, *On the account of the World's creation given by Moses* in Loeb 226 (1929).

21. *De opificio mundi* 13.43.

22. Ibid. 21.65–66. Philo summarizes similarly at ibid. 22.68.

23. Ibid. 20.62.

24. *De plantatione* 11–17 following *Timaeus* 90A and 91E. For the analogy between roots of plants and the head of animals, see also *Quod deterius potiori insidiari soleat* 85.

25. *De plantatione* 12. The fire-born beings are mentioned again at *De gigantibus* 7.

26. Philo, *Works* 4:185 (1855), in the translation of Yonge. In *Aristotelis operum volumen secundum . . . Philonis Iudaei de mundo liber unus* (1497) the passage appears in the (pseudo-) Philo *Peri khosmos* at folio 227v. In *De mvndo Aristotelis Lib. I. Philonis Lib. I. Gulielmo Budæo interprete* (1533) the passage appears (in "Philo") at page 39 as "ζωόφυτα, quasi plantaníma uocant, ab natura inter animal plantam'q[ue] ambigua." See also *De mundo* 3, in *Opera omnia cura Richter* (1828–1830), at 6:153.

27. Boëthius is cited at *De mundo* 14 (*Works* 4:198 in translation of Yonge), and in the spurious *De incorruptibilitate mundi* (*Works* 4:34). See also *Philonis Alexandrini opera qvae svpersvnt* edited by Cohn & Wendland (1897), at 2:vi–vii and 136–137.

28. Tiberius Julius Alexander was later Procurator of Judæa under Claudius (about 46–48 CE), then Prefect of Egypt under Nero (66–70 CE).

29. *De animalibus* 79, in the translation of Terian (1981:101–102).

30. *De animalibus* 100; Terian (1981:108).

31. Ammonius "of Athens" (not to be confused with Ammonius Saccas, nor with Ammonius Hermiae): Dillon (1996:184 and 191–192).

32. Plutarch, *On Isis and Osiris*. In: *Moralia* (Loeb 306, 1936). Dillon (1996:200–204).

33. *Timaeus* 28b, 30a, 34c, and 52d–53b.

34. Lucius Calvenus Taurus, *On the distinction between the theories of Plato and those of Aristotle*. This treatise is known only via the *Suda*. See Dillon (1996:237–247), at page 239.

35. Dillon (1996:242).

36. Dillon (1996:243) explains that *mesê* is a term from music. The μεση was the middle string on the Greek lyre, the νεατη the lowest string (with the highest pitch) and the υπατη the highest string (with the lowest pitch). Assuming he analogizes the two outer strings with animals and plants, then mixed beings can exist in equal mathematical proportion to them both—a very Pythagorean concept.

NOTES 485

37. Taurus as quoted by John Philoponus, *On the eternity of the world*; translation by Dillon (1996:243). Philoponus: *Ioannes Philoponus, De aeternitate mundi contra Proclum* (1899).

38. Porphyry, as quoted by Eusebius, *Ecclesiastikon istoria* 6.19.7; Dillon (1996:381). In English as *Ecclesiastical history* (1833), 6.19 (pp. 237–241).

39. Seeberg (1941); Benz (1951); Hindley (1964).

40. Cassius Longinus was the teacher of Porphyry. Later, as a counsellor to Queen Zenobia of Palmyra, he reportedly encouraged her to rebel against Rome. The revolution was defeated by Emperor Aurelian, and Longinus was executed.

41. Origen Adamantius, later a Christian theologian.

42. *Encyclopædia Britannica*, Ninth edition 1:743–744, article "Ammonius".

43. Elsee (1908:55–57); Wildberg, Neoplatonism (2016).

44. *Enneads* 5.2.1–2, as translated by MacKenna & Page. See also Plotinus, *Enneads* (1991:362). Where MacKenna has "the Intellectual-Principle" I substitute Plotinus's *psyche*.

45. Ibid. 3.2.7. Here Plotinus implicitly criticizes Aristotle, for whom soul had not ultimately emanated from The One.

46. Ibid. 1.4.3.

47. Ibid. 3.4.2 and 4.7.13.

48. Porphyry of Tyre. His Phoenician birth name was Malchos; Longinus gave him the Hellenic name by which he is known.

49. *Select works of Porphyry* (1823).

50. *Against the Christians*. Banned by successive imperial decrees ordering all copies to be burnt, the work is now lost.

51. Elsee (1908:62–65).

52. *The Epistle of Porphyry to the Egyptian Anebo*, in *Iamblichus on the mysteries* (1895).

53. See Todd's translation of Themistius, *On Aristotle's* On the soul (1996) 1:16.19 (pp. 31–32) and note 8 (page 160).

54. Porphyry wrote two commentaries on Aristotle's *Categories*, of which the shorter is extant. His *Isagoge* is an introduction to Aristotelian logic more generally. Boëthius wrote two commentaries on Porphyry's *Isagoge*, one using a translation by Marius Victorinus (written *ca* 504–505 CE), the second based on his own translation (*ca* 507–509 CE). His commentary on *In Categorias Aristotelis* dates to *circa* 509–511 CE. By this route, by the Sixth century Porphyry's *Isagoge* "had become a part of the *Organon*" and preserved a small but important part of Aristotelian teaching within "the curriculum of a thriving Neoplatonism" (Peters 1968:9). This irony and others are noted by Peters (1968:14–15).

55. Aristotle, *Topics* I 101b17–25.

56. Hence the Scholastic *universalia post rem, universalia ante rem*, and *universalia in re*. See Seth (1886:419). These questions exercised Medieval philosophers including Albert and Aquinas.

57. Porphyry is not known to have made a drawing of his Tree. Depictions of the Tree of Porphyry date to at least the year 1100 CE: Codex Vaticanus, Ottobonianus latinus 1406, fol. 11a as cited by Scholem (1987:447n194).

58. Spade (1994). I use an earlier draft: Spade, Porphyry, *Isagoge*, Oklahoma translations P-20-30s (1986), particularly lines 369–375. Porphyry also asserts that animals are sensitive, and plants insensitive, in *De abstentia* 3.19 (*Select works* 1823:97).

59. Porphyry (1992) 56.1–5 (page 30) and 116.21–27 (page 120). Porphyry's second, longer commentary (*Ad Gedalium*) on the *Categories* has been lost, apart from fragments.

60. Eunapius Sardianus (1568:37) refers to him as *secundum locum* to Porphyry. It is not known whether this is the Anatolius who later became (Christian) Bishop of Laodicea.

61. *The theology of arithmetic*, as translated by Waterfield (1988). The title is that of the other main identifiable source, the *Theology of arithmetic* by Nicomachus of Gerasa.

486   NOTES

62. *The theology of arithmetic* (1988:51). *Three* is the first odd number because to harmonize Pythagorean numerology with Neoplatonic metaphysics *one*, the *monad*, has to be simultaneously even, odd and even-odd (ibid. 1988:35).

63. Ibid. (1988:52–73).

64. Iamblichus of Chalcis was from a royal family in what is now northern Syria; one ancestor had perished in Marc Antony's fleet at Actium. Antioch and northern Syria were pillaged by Shāpūr I, King of Persia, in 256 CE. Unlike Porphyry, Iamblichus kept his Semitic name.

65. In any event, Iamblichus knew Porphyry's work well, and Porphyry dedicated his *On knowing oneself* to Iamblichus. Iamblichus returned to Syria, establishing himself in Apamea (where Numenius had taught in the Second century).

66. Shaw (2014), especially pp. 1–49 and 266–272.

67. Dillon, *Iamblichi Chalcidensis* (1973:49); Dillon (2010) 1:358–374.

68. Indeed, Neoplatonists after Plotinus (Amelius Gentilianus, Porphyry, Iamblichus, and Theodorus of Asine) sought to reconcile Plato not only with Pythagoras and Aristotle, but also with Moses, Homer, Hesiod, and the Orphic poems as well as the *Chaldæan Oracles*. See Dillon (1973:26–27), and Gersh (1978).

69. Iamblichus (1570), and (1895:17–365). The title *De mysteriis* is due to Ficino.

70. It is not known precisely what activities were involved, although Shaw (2014) mentions invocations, music, chanting, sacrifices, and worship of Helios (the Sun as god). Iamblichus is said to have laughed at the belief among his students that while meditating he could levitate and take on a golden colour.

71. Iamblichus, *Exhortation to philosophy* (1988), section 4 (page 31).

72. Ibid., section 9 (page 54).

73. Ibid., section 5 (page 42).

74. What we know of Iamblichus's commentary comes via Simplicius (*In Aristotelis Categorias Commentarium* 2.9 ff): see Dillon (1997). According to Simplicius, Iamblichus removed some of Porphyry's verbosity, and formalized the recognition of ten categories.

75. Dexippus, in Busse (1888); English translation by Dillon (1990).

76. Dexippus 48.27–29 (1990:89). Dillon points out that this part of Dexippus's argument closely follows Iamblichus.

77. Dexippus 49.15–19: ζωοφύτων at 49.16 (1888), *zoophytes* at 49.16 (1990:89–90).

78. Dexippus (1990:90n52). The third solution is that "the differentia is not only constitutive of the substance, but also a part of it".

79. Simplicius, *In Aristotelis Categorias Commentarium* (1907), 98.19–35 (*zoophyton* at 98.27). The Marcianus / Bessarion manuscript tradition reads *zoiphyton*. An English translation in four volumes is available from Cornell UP (2000–2003); Books 5 and 6 are in the translation of de Haas & Fleet (2001).

80. Simplicius acknowledges (at *Categorias* 2.25) that Dexippus wrote a commentary on *Categories*, but he devotes far more attention on Iamblichus and Porphyry. It is not out of the question that Iamblichus, in turn, might have taken *zoophyte* from Porphyry's lost commentary on the *Categories, Ad Gedalium*.

81. Dexippus (1549): ζωόφυτον at folio 34 *verso*.

82. A statesman and philosopher, Themistius of Paphlagonia was famous for his oratory. He led an independent school of philosophy in Constantinople from about 345–355 CE (Todd, *Themistius* 1996).

83. Blumenthal (1990).

84. Todd, *Themistius* (1996:62) 44.9, on *De anima* 413a20–31. The opposite directions refer to the roots growing downward, while the shoot grows upward.

85. Aristotle, *De anima* 410b26–411a2; Themistius, *In libros Aristotelis De anima* (Heinze, 1899) 35.6–25 (ζώφυτα at 35.20). William of Moerbeke brought this into Latin as *quae vocantur*

zoophyta et entoma (Verbeke 1957:84); Todd (1996:52). Todd (1996:167n25) bases his translation on ζωόφυτα in the Thirteenth-century Laurentianus folios, rather than ζώφυτα preferred by Heinze.

86. Aristotle, *De anima* 413a31–b9; Themistius (1899) 44.21–34 (ζώφυτα at 44.29, ζωφύτων at 44.32); Todd (1996:62). Moerbeke translates the first instance as *ostrea aut zoophyta aut etiam animalia*, and the second as *zoophytis*. Verbeke notes (1957:105) that other manuscripts give alternative spellings for these two instances: *zoofyca/zoofytis* (Toledo, late Thirteenth century), *zilophita/zilophitus* (Munich, late Thirteenth century; Oxford and Paris, Fourteenth century) and *zilofita/zilofotis* (Erfurt, Fourteenth century). Aquinas (*In Aristotelis De anima commentarium*) retained the testaceans (*sicut ostreae, quae non moventur motu progressivo*) but does not mention zoophytes. Moerbeke was travelling with Aquinas on papal business at the time of the translation (late 1267). See Verbeke (1957), pp. ix–xv; and De Corte (1932).

87. Aristotle, *De anima* 414a29–b6; Themistius (1899) 47.7–21 (ζωφύτοις at 47.16, and ζώφυτα at 47.22); Todd (1996:65–66). Heinze notes that the Laurentianus folios and a Fifteenth-century Parisian manuscript read instead ζωοφύτοις and ζωόφυτα respectively. Moerbeke again translates as *zoophytis / zoophyta*; as before, see Verbeke (1957:111) for alternative spellings of the Latin.

88. Aristotle, *De anima* 425a9–10; Themistius (1899) 81.4–11 (ζώφυτα at 81.6); Todd (1996:103). Heinze notes that a Fifteenth-century Parisian manuscript reads ζωόφυτα. Moerbeke translates as *zoophyta*, although four manuscripts read *zeophita* (Verbeke 1957:185).

89. Aristotle, *De anima* 434b18–24; Themistius (1899) 124.10–24 (ζώφυτα at 124.19); Todd (1996:151–152). Moerbeke again translates as *zoophyta*, noting *zeophyta* in four manuscripts (Verbeke 1957:277).

90. Lyons (1973), pp. vii–xvi.

91. Verbeke (1957) offers a critical translation (into French) of Moerbeke's translation (into Latin) of Themistius's paraphrase of Aristotle's *De anima*. Moerbeke also translated (entirely or in part) Aristotle's *Physica* and *Historia animalium*; the commentary on Book 3 of *De anima* by John Philoponus; commentaries by Alexander Aphrodisias and Ammonius Hermiae; the *Elementario theologia* of Proclus, and the latter's commentaries on *Parmenides* and *Timaeus*; and works by Hippocrates and Claudius Ptolemy. See also Dod (1982:45–79).

92. The 1534 Aldine, edited by Victor Trincavellus (Vettore Trincavelli), was based on a single manuscript. Moerbeke's translation seems to have been based on a Thirteenth-century manuscript later used by Heinze (1899). The 1866 Spengel edition was based on the 1534 Aldine, plus a Fifteenth-century text and Sixteenth-century notes. The 1899 Heinze edition (*Commentaria in Aristotelem Graeca* 5:3) is based on Eleventh- (Parisiensis Coislinianus 386) and Thirteenth-century manuscripts.

93. Elsee (1908:72).

94. Peters (1968:9–10).

95. Plutarch, *Quaestiones naturales* (1890) 2:1114.

96. Damascius was head of the Neoplatonic school at Athens when it was closed by Emperor Justinian I in 529 CE, perhaps in response to the appearance of Philoponus's *Against Proclus on the eternity of the world*.

97. Simplicius wrote his commentaries after being evicted from Athens. A treaty between Justinian and Khosrow I, signed in 533 CE, allowed the philosophers to return.

98. Simplicius, *In Aristotelis Physicorum libros quattuor commentaria* (Diels 1892–1895) 3.6 (*Prœmium*) and 1271.31 (*Physica* 261a14); *De anima commentaria* (Hayduck 1882) 180.21 (*De anima* 425a8); and *De caelo commentaria* (Heiberg 1894) 384.20 (*De caelo* 284b6), respectively. The attribution of *De anima commentaria* to Simplicius is disputed, with Priscian of Lydia, a colleague of Simplicius in Athens, the chief alternative.

488 NOTES

99. The respected Platonist Hypatia taught there until her death at the hands of a mob of Christians in 415 CE. Hierocles of Alexandria, a Neoplatonist, was active around 430 CE.

100. Hermias was a fellow-student with Proclus under Syrianus. It is tempting to date his move to Alexandria to about 437 CE, the year Syrianus died. Hermias's wife Aedesia was a relative of Syrianus, who had planned to wed her to Proclus until a divine warning precluded the match. Hermias and Aedesia had at least three sons, of whom the second was Ammonius Hermiæ (see note [103] below).

101. This is the traditional understanding, but in fact there is little evidence internal to their surviving works that Elias and David were Christian. Similar texts in Armenian are ascribed to one David the Invincible, a Christian, who may or may not be identical with the Alexandrian scholar: see Elias and David as edited by Gertz (2018:2–3).

102. Sorabji (1983:202).

103. Ammonius Hermiæ studied under Proclus in Athens, upon whose death in 485 CE Ammonius returned to Alexandria as head of the school of Neoplatonism there. His students included Asclepius, John Philoponus, Damascius, Simplicius, Olympiodorus and Origen. Zacharias, later Bishop of Mytilene, listened to his lectures.

104. The surviving commentaries are by Asclepius, John Philoponus, and one or more anonymous students.

105. *Ammonii in Porphyrii Isagogen* (1891) 70.16 (commenting on Porphyry 3.22); and 77.18 and 21 (on Porphyry 4.10–12).

106. Busse, in *Ammonii in Porphyrii Isagogen* (1891), pp. v–ix (1895); Blank (2017).

107. Translation from Greek by Dr Anitra Laycock, Department of Classics, Dalhousie University (1995).

108. John Philoponus "the Grammarian" (so called by Simplicius to make the point that John was employed to teach grammar, not philosophy). John is also credited with the discovery of the principle of inertia. He later turned his attention to Christian theology and apologetics.

109. *Philoponus on Aristotle's On the soul* 1.3–1.5 (van der Eijk 2006) pp. 108–109; on *De anima* 410b16–20 (Hayduck 1897), page 184.18–32. Philoponus returns to the same point in *In Aristoteles Analytica priora commentaria* (Wallies 1905), page 14.1–13.

110. *Philoponus on Aristotle's On the soul* 2.1–6 (Charlton 2005), page 14; on *De anima* 412b16–20 (Hayduck 1897), page 214.2–21.

111. Philoponus, *In Aristoteles analytica priora commentaria* (Wallies 1905), page 16.1–7. Translation from Greek by Dr Anitra Laycock.

112. *Philoponus on Aristotle's On the soul* 3.9–13 (Charlton 2013), pp. 54–55; on *De anima* 434b9–11 (Hayduck 1897), pp. 600.13–601.3. The Greek manuscript tradition for Book 3 is, however, problematic. Twelfth- and Fifteenth-century manuscripts say that Book 3 is "from the voice of" one Stephanus, and this was accepted by Hayduck (1897). This Stephanus may have been the last senior scholar in Alexandria (*ca* 581 to 610 CE), and may be identical with Stephanus of Athens, active after 610 CE in Constantinople after having been summoned there by Emperor Heraclius. Charlton (2013) agrees up to a point, allowing that the author may have been someone who attended lectures by Stephanus [in 3.9–13] (2013:6), *i.e.* the commentary would have been "from his voice".

113. *Philoponus on Aristotle's On the soul* 3.9–13 (Charlton 2013), page 27; on *De anima* 432b14–16 (Hayduck 1897), page 577.7–31.

114. *Philoponus on Aristotle's On the soul* 3.9–13 (Charlton 2013), pp. 42–43; on *De anima* 433b13–14 (Hayduck 1897), pp. 589.27–590.7.

115. *Philoponus on Aristotle's On the soul* 3.9–13 (Charlton 2013), page 49; on *De anima* 434a22–23 (Hayduck 1897), page 595.28–36.

116. Wildberg, in *Philoponus against Aristotle* (1987), page 4 and footnote 7.

117. Sorabji (1987:11).

NOTES 489

118. Philoponus, *Proœmium* to *On Aristotle's De anima* (Hayduck 1897), page 6.31–32, in the translation of Dudley (1974–1975), at page 70.
119. Philoponus, in Hayduck (1897), page 17.9–11; Dudley (1974–1975), page 82. We will see in Chapter 6 that Augustine makes a similar point, citing Varro.
120. Philoponus, in Hayduck (1897), page 18.26–28; Dudley (1974–1975), page 83.
121. Ammonius's immediate successor was probably the mathematician Eutocius, about whom little is known. For Olympiodorus and his school, see Elias and David as edited by Gertz (2018:1–9).
122. Westerink (1962), pp. x–xxv.
123. Texts that now bear the name Elias formerly circulated as anonymous manuscripts, so the attribution to Elias must have come later: Wildberg, Elias (2016).
124. Elias (Busse 1900), page 123.9–11.
125. Philoponus (van der Eijk 2006), page ix.
126. Elias (Busse 1900), pp. 63.24–28 and 64.18–19. A partial translation into French was made by Mueller-Jourdan (2007).
127. Elias (Busse 1900), pp. 64.24–65.7. Translated from Greek by Dr Anitra Laycock.
128. For manuscripts in the Armenian language by "David the Invincible", see Gertz (2018:203) and note [101] above.
129. David (Busse 1904), pp. 71.25–72.5. Translated from Greek by Dr Anitra Laycock. This passage has more recently been translated by Gertz (2018:155).
130. David (Busse 1904), page 148.8–35. Translated from Greek by Dr Anitra Laycock.
131. For its authorship and a partial translation see Mueller-Jourdan (2007), pp. xvi–xxvi; this example at 21.3 (2007:86).
132. Archytas of Tarentum. Most work that goes under his name is falsely ascribed. See Pseudo-Archytas (ed. Szlezák) 1972:19.
133. Szlezák (1972:172), in reference to the Greek text at 5.33 (page 64).
134. Joannes [Alexandrinus] (Dietz 1834), at 2:219–220.
135. For example, *Philebus* 29, and *Timaeus* 35–36. Macrocosm-microcosm is often claimed to be much older, but the evidentiary basis for this is spotty. Thunberg (1965:141) points to a fragment from Democritus, but it does not appear in CCW Taylor (1999) and may not be authentic. Needham (*SCC* 2:295) claims the universe-analogy was widespread in ancient India, pointing to *Rigveda* X:90; this hymn relates that the gods sacrificed Perusa, the Lord of Immortality, and the earth, sun, moon, and certain animals were formed from parts of his body. Needham (*SCC* 2:299–300) further states that "the universe-analogy was implicit in the whole world-outlook of the ancient Chinese" and notes that the *I ching* "likens heaven to the head and earth to the belly"; but the other evidence he cites is three to five centuries more-recent than Plato. Macrocosm-microcosm is a theme in the Kabbalah (notably in Sefirot), but its main elements date to the Second century BCE.
136. *Physica* 252b25–28. Thunberg (1965:141) points out that Aristotle seems to restrict this analogy to the human soul.
137. For example, Cicero (citing Chrysippus), *De natura deorum* 2.14.37–39. See also Thunberg (1965:140–144).
138. Galen, *De usu partium* 3.10.241.
139. Macrobius, *Commentarium in somnium Scipionis* 2.12.11; English translation by Stahl (Macrobius 1952:224).

Chapter 6

1. Tertullian, *The prescription against heretics*, in Roberts & Donaldson (1870) 15:9.

490　NOTES

2. Including the Edict of Milan in February 313 CE.

3. The most-important doctrinal issue facing the Council was the so-called Arian heresy, which concerned the relationship between God the Father and God the Son, including whether the latter was eternal or had been created. Ironically perhaps, Eusebius was an Arian.

4. His Jewish name was Saul of Tarsus.

5. *Acts* 22:3.

6. Kee & Young (1957:208).

7. Paul's phrase "through a glass darkly" (*1 Corinthians* 13:12) recalls *Phaedrus* 250b.

8. Wolfson (1970:9–11).

9. The first generation of Christian theologians (*ca* 90 to *ca* 160 CE) following the Twelve Apostles, including Clement (Bishop of Rome), Ignatius of Antioch, and Polycarp of Smyrna.

10. [Saint] Clement of Rome, *First epistle to the Corinthians*, in Lightfoot & Harmer (1912): compare 19.3 and 20.11 (1912:16–17). See also Wolfson (1970) 1:11n83.

11. This second generation included Aristides, Clement of Alexandria, Justin Martyr, Tatian, and Theophilus. At least the latter three were influenced by the works of Philo of Alexandria: Wolfson (1970) 1:11–12.

12. Wolfson (1970) 1:12–14 and 97–140.

13. Elsee (1908:43–44).

14. Justin accepted the perennial myth that Plato had spent time in Egypt, where he learned from various wise men, rabbis, and mystics: see *First apology* chapters 8, 20, 44, and 59–60, and *Hortatory address to the Greeks* 20, 26–27, 29, and 31–33. Justin also spoke well of the Stoics (*First apology* 20) and Homer (*Hortatory address* 28 and 30), while nonetheless accusing Plato of inconsistency (*Hortatory address* 5–7 and 23). In Roberts & Donaldson, volume 2 (1909).

15. Justin Martyr, *First apology* 5 and *Second apology* 10.

16. Justin Martyr, *Dialogue with Trypho* 4.

17. Justin Martyr, *Dialogue* 20.

18. See note [1] above, and Harnack (1888).

19. Tertullian, *Ad nationes* 2.4, in Roberts & Donaldson (1869) 11:473.

20. Tertullian, *De anima* 10, in Roberts & Donaldson (1870) 15:430–433. Clement of Alexandria briefly alludes to a similar argument at *Stromata* 7.6, in Roberts & Donaldson (1869) 12:428.

21. Tertullian, *De anima* 19, in Roberts & Donaldson (1870) 15:454–457.

22. *Stromata* (*Miscellanies*) 1.29 and 2.14, in Roberts & Donaldson 4:469–470 (1867) and 12:274–301 (1869) respectively.

23. *Exhortation to the heathen* 6–7, in Roberts & Donaldson (1867) 4:69–75.

24. *Proverbs* 5:20; at *Stromata* 1.5, in Roberts & Donaldson (1867) 4:367.

25. Ashwin-Siejkowski (2010:115–128).

26. Translated by Potter (*Klementos Alexandreos* 1715, at 2:866) as *per sanctam hebdomadem*. Wilson in Roberts & Donaldson (1869) 12:448 translates this as *through the holy septenniad [of heavenly abodes]*.

27. Clement of Alexandria, *Stromata* (*Miscellanies*) 7.10, in Roberts & Donaldson (1869) 12:446–449.

28. *Stromata* 7.2, in Roberts & Donaldson (1867) 4:412.

29. The *Recognitions* (preserved only in a Latin translation by Rufinus), together with a related Greek text (the *Homilies*), purport to have been written by Clement, supposedly a cousin of Emperor Domitian, later a companion of the Apostle Peter and eventually Pope Clement I. The archetype from which both the *Recognitions* and the *Homilies* likely descend is thought to date to the mid-Second to mid-Third century. See the introductions by Riddle (pp. 69–71) and T Smith (pp. 73–74) in Roberts & Donaldson 8 (1903); Gebhardt (2014); and Wikipedia, "Clementine literature" accessed 26 July 2018 and 19 October 2020.

30. *Recognitions* 8.22, in Roberts & Donaldson (1886) 8:171.

NOTES 491

31. *Recognitions* 8.25, in Roberts & Donaldson (1886) 8:172.
32. Leonides was martyred in 202 CE.
33. These and other details of his life from Crombie, in Roberts & Donaldson 23 (1872), pp. xxiii–xxxviii; Oulton & Chadwick (1954); and the introductions by de Lubac and Butterworth to Butterworth, *Origen* (1973).
34. Moore & Wilson, in Schaff & Wace (1893) 5:16.
35. Origen, *Contra Celsum* 4.39 and 6.19, in Roberts & Donaldson (1872) 23:203–205 and 357–358.
36. The *Logos alethes* (*On the true doctrine*) of Celsus was banned in 448 CE by Theodosius along with Porphyry's treatise against the Christians, and is no longer extant.
37. *Contra Celsum* 4.87, in Roberts & Donaldson (1872) 23:243, where Origen states that Celsus "regards the books of the Jews and Christians as exceedingly simple and commonplace".
38. McGuckin (2004) considers *Contra Celsum* as the "intellectual charter" of Christian Byzantium. Even so, his use of Platonic metaphysics, for example underlying his idea of "double creation" (first of souls; then as a consequence of sin, of bodies) led to charges of Arianism.
39. Origen, *De principiis* 1.1.6, 1.3.4, and 1.5.1–2, in Roberts & Donaldson (1869) 10:13, 36, and 44–46; *Contra Celsum* 5.11, in Roberts & Donaldson (1872) 23:278–279; *On prayer* 7, in Oulton & Chadwick (1954), pp. 254 and 339.
40. *De principiis* 2.1.3, in Roberts & Donaldson (1869) 10:74.
41. E Moore, in *Internet Encyclopedia of Philosophy*. See also Butterworth (1973:72n8 and 74n2), and Butterworth's Introduction, page xxxvii. Based on wording in Jerome (*Epistle to Avitus* 4), Koetschau reconstructed Origen's *De principiis* 1.8 with text stating that souls exist in a realm of their own, "carried round with the whirl of the universe". If a soul is inclined towards evil, it would pass into the body of man. If then virtuous, the soul could rise back to its heavenly realm, but if wicked it could continue to sink to "the life of an irrational animal", then further to "the insensate life of a plant". Corresponding text is absent from the Latin translation by Rufinus in 397 CE. Jerome famously charged Rufinus with having intentionally mistranslated or deleted passages in which Origen's teachings went against current orthodoxy.
42. *De principiis* 3.6.2, in Roberts & Donaldson (1869) 10:265.
43. *On prayer* 6.1, in Oulton & Chadwick (1954:250–251); *De principiis* 3.1.2, in Roberts & Donaldson (1869) 10:157–159.
44. *De principiis* 3.1.2, in Roberts & Donaldson (1869) 10:158.
45. Ibid. 3.1.3, in Roberts & Donaldson (1869) 10:160.
46. *Contra Celsum* 4.88–95, in Roberts & Donaldson (1872) 23:254–262.
47. *Phaedrus* 245d; see Chapter 3.
48. *De principiis* 3.4.1–5, in Roberts & Donaldson (1869) 10:244–252. At least in the Latin translation of Rufinus, Origen sets out the case for each solution, and asks the reader to decide. Philo (Chapter 5) earlier taught that there were two such souls.
49. *Leviticus* 17:14.
50. *De principiis* 2.8.1, in Roberts & Donaldson (1869) 10:118–120.
51. Ibid. 2.9.3, in Roberts & Donaldson (1869) 10:129–130.
52. *Commentary on the Song of Songs* 3; translation in Curley (1980:4–5). The words of Paul may be those in *Romans* 1:20.
53. *De principiis* 3.1. See Butterworth (1973:83n1). Even so, Origen argued against "determined cycles of existence": *Contra Celsum* 4.57–58, in Roberts & Donaldson (1872) 23:232–234.
54. *De principiis* 1.2–3, in Roberts & Donaldson (1869) 10:18–43, notably 1.2.2 (pp. 18–29); *Commentary on John* 1.22, in Menzies (1906) 9:307–308.
55. Origen died from injuries sustained under torture in the Decian persecution of 250–251 CE, occasioning Pope Dionysius of Alexandria to declare him a martyr.
56. [Saint] Athanasius, Patriarch of Alexandria.

492 NOTES

57. Telfer (1955), pp. 210–211 and note 24.

58. *Nemesii Episcopi Premnon physicon* in the Latin translation of Alfanus (1917). For many years the work was attributed to Gregory of Nyssa.

59. Nemesius, *On the nature of man* (1955); Telfer reviews the history of publication and translations.

60. Telfer (1955:213).

61. Nemesius, *On the nature of man* 1.3 (1955:232–234). Telfer (1955:233n5) points out that apart from the sentence about the ancients, this passage is "word for word identical" with one in Galen's *The agreement between Hippocrates and Plato* in which Galen, in turn, cites *On the passions* by Posidonius.

62. Giorgio Valla (in the Gryphius edition, 1538:11) translated this sentence *Quocirca hæc talia ueterum consuetudine sapientum animaliplata (ut noua in re nouum confingam mihi uerbum) quod Græci zoophyta dicunt, sunt uocitata* (On account of which, the wise men of ancient times called them animal-plants (so might the new word compose for me)). Nicasius Ellebodius (1565) accepted ζωόφυτα in Greek (page 12, bottom) and left it untranslated in his Latin text (page 4, line 9). Similarly, Matthaeus (1802) read ζωόφυτα (page 42, line 3) and left it untranslated in Latin (page 3, bottom line): *Quocirca haec omnia eruditi, juncto ex animali et stirpibus nomine ζωόφυτα nuncupant* (On account of which, all learned ones call these things *zoophyta*, the adjunction between animal and plant).

63. *On the nature of man* 1.12 (1955:264).

64. Ibid. 1.18 (1955:286–289 and note 6).

65. [Saint] Basil, Bishop of Cæsarea; [Saint] Gregory, Bishop of Nyssa; and [Saint] Gregory of Nazianzus, Archbishop of Constantinople.

66. For details, see Jackson in Schaff & Wace (1895) 8:xv–xvi.

67. Philo of Alexandria, *De opificio mundi* 3.13.

68. *Genesis* 1:3–19.

69. *Contra Celsum* 6.60–61, in: Roberts & Donaldson (1872) 23:402–404; *De principiis* 4.1.16, in Roberts & Donaldson (1869) 10:314–317. In Chapter 7 we shall see that in early Jewish rabbis read *Genesis* 1 and 2 as meaning that all things were created on the first day, and were thereafter revealed on the other days. The Scofield *Bible* (1967:1n6) offers a similar argument in a Christian tradition: clouds no longer obscure the sun.

70. Basil, *Hexaemeron* 2.7–8 and 6.2–3, in Schaff & Wace (1895) 8: 63–65 and 82–84 respectively. Another statement of his rejection of allegorical interpretation is at *Hexaemeron, Homily 3*, ibid. (1895) 8:71.

71. Ross, Gregory of Nyssa (at section "World"), in *Internet Encyclopedia of Philosophy*.

72. Ladner (1958:73n55). That is, Gregory presents the days as allegorical, *e.g.* as steps or phases rather than than 24-hour periods: *Liber in Hexaemeron* cols 71B (Latin) and 72B (Greek), in Migne, *Patrologiæ series Græca* 44 (1863). Augustine later took a similar position.

73. Basil, *Hexaemeron*, in Schaff & Wace (1895) 8:51–107. In this literature, *hexæmeron* is spelled in different ways.

74. Holmes (1935) 1:133–134.

75. Basil, *Hexaemeron* 6.2–4, in Schaff & Wace (1895) 8:82–84.

76. Ibid. 7.1, in Schaff & Wace (1895) 8:89–90.

77. Basil, *Letter 188 to Amphilochius* 15, in Schaff & Wace (1895) 8:228. The wife of Gregory of Nazianzus, Nonna, had a brother named Amphilochius, who in turn had a son of the same name (Amphilochius, Bishop of Iconium).

78. Basil, *Hexaemeron* 7.6, in Schaff & Wace (1895) 8:93–94.

79. Ibid. 7.7, in Schaff & Wace (1895) 8:99–100.

NOTES   493

80. Gregory of Nyssa, *De hominis opificio* 8.4–5 and 7, in Schaff & Wace (1893) 5:393–394; quotation at 8.7, page 394. See also 30.32 at page 427.

81. Ibid. 14.2, in Schaff & Wace (1893) 5:403.

82. Gregory of Nyssa, *On the soul and the resurrection*, in Schaff & Wace (1893) 5:433. The Teacher is Basil's sister (Saint) Macrina (*ca* 330 to 379 CE).

83. Gregory of Nazianzus, Oration 38 *On the Theophany, or the birthday of Christ* 9–10, in Schaff & Wace (1894) 7:347–348; very similar words are found in Oration 45 *Second oration on Easter* 5–6, in ibid. 7:424–425. See also page 212, footnote δ.

84. Gregory of Nazianzus, *Second theological oration* 23, in Schaff & Wace (1894) 7:297.

85. [Saint] Ambrose, Bishop of Milan.

86. Savage (1961), pp. ix–x.

87. Abbot Ælfric is thought to be the same as Ælfric Grammaticus, Archbishop of Canterbury from 996 to 1006 CE: see HW Norman (1849), pp. vii–xv.

88. Costache (2012); Gregory of Nyssa, *In Hexaemeron* (Drobner 2009).

89. Although Gregory of Nazianzus took a particularly hostile view towards Pyrrho and Sextus Empiricus: see his Oration 21 *On the great Athanasius* 12, in Schaff & Wace (1894) 7:272. In the Latin West, [Saint] Hilary of Poitiers attacked "the godless system of Epicurus": Sanday, in Schaff & Wace (1899) 9, page v.

90. Elsee (1908) pp. 71, 92, and 110–113.

91. Augustine, *Confessions* 7.9. Speculation on the authors of these books usually turns to Plotinus and Porphyry.

92. For example, in *City of God* 8.1–14.

93. Lindberg (1983:526–527).

94. Augustine, *De Genesi ad litteram* (written about 393 CE); English translation by Teske (1991:143–188). O'Toole (1944); see also *Confessions* 13. Even so, he was fascinated by numerology, notably the properties of the numeral six: *On the Trinity* 4.4(7)–4.5(9), in Schaff (1887) 3:73–74; and Bourke (1993:207–208).

95. *Confessions* 7.17.

96. For example, *Enchiridion* 3.9, in Outler (1955) 7:341–342, although he directs his criticism of its value to "matters of religion". Sharpe (1964:9 and note 16) makes a similar point about Augustine's condemnation of natural philosophy: astronomy could usefully be studied, but astrology (with which astronomy was then "almost inextricably intermixed") was to be condemned.

97. Augustine, *On Christian doctrine* 2.24, in Schaff (1887) 2:543. See also Lindberg (1983), especially page 521.

98. Keenan (1939).

99. Quotation from *On Christian doctrine* 1.8, in Schaff (1887) 2:524–525 *et passim*. The same division is in evidence in *City of God* 5.11 (mentioning stones), 11.16 *et passim*; and in *De Genesi ad litteram libri duodecim* 1.19(39), in Migne (1845) 34, column 261.

100. *City of God* 8.6.

101. *Morals of the Manichaeans* 17, paragraph 55 (Schaff 1887, Select Library 4:84).

102. According to Augustine (*City of God* 7.23, as translated by Healey), "Varro, in his book of the select gods, puts three degrees of the soul in all nature. One, living in all bodies unsensitive, only having life: this, he says, we have in our bones, nails, and hair; and so have trees living without sense. Secondly, the power of sense diffused through our eyes, ears, nose, mouth, and touch. Thirdly, the highest degree of the soul, called the mind, or intellect, confined unto man: wherein are that part of the world's soul he calls a god, and in us a genius. So divides he the world's soul into three degrees. First, stones and wood, and this insensible earth which we tread on, which are as it were its bones and nails, Secondly the sun, moon, and stars, of which we are sensible, and which are its senses. Thirdly the ether, which is its

494 NOTES

mind and which penetrates the stars, making them gods." Varro's "book of the select gods" was surely his *Antiquitates rerum divinarum* in sixteen books, now lost: see Jocelyn (1982).

103. *De vera religione* 109, in Migne (1845) 34, column 170 (1845). English translation by Burleigh in *Augustine. Early writings* (1953/2006), page 280. Eriugena (*Periphyseon* 736B, as translated by Sheldon-Williams (1987:373) later adopted this Stoic-like formulation.

104. *On Christian doctrine* 2, chapter 33 paragraph 51, in Schaff (1887) 2:551.

105. *Contra Faustus* 21.10, in Schaff (1887) 4:269.

106. *Confessions* 7.5.

107. *Confessions* 13.2. This formulation is due to Paul Kuntz (1987:51–52).

108. Bourke (1976); Sorabji (1983:163–172).

109. *City of God* 12 ff.

110. Sheed (1948), page v. Note however that Greek had not been part of Augustine's early education, and he never became fluent in the language (Wills 2016).

111. Chenu (1963:226–227).

112. [Saint] Dionysius the Areopagite was converted to Christianity by [Saint] Paul following Paul's defence of Christianity before jurors of the Athenian high court at the Areopagus, a large rock near the Acropolis (*Acts* 17:16–34).

113. Some find a connection with Edessa: see Elsee (1908:129–130). The writings of Dionysius were first mentioned in 533 CE.

114. Pseudo-Dionysius, *On the celestial hierarchy* 6–10, in Parker (1899) 2:23–42. According to Paul Kuntz (1987:41), Pseudo-Dionysius was the first to use the term *hierarchy*.

115. Elsee (1908:132) interprets this as referring to the Scriptures.

116. *On the celestial hierarchy* 1.3, in Parker (1899) 2:2–3.

117. Ibid. 15.7–8, in Parker (1899) 2:63–64.

118. *On the divine names* 2.7, in Rolt (1920:74n2).

119. Ibid. 5.3, in Rolt (1920:133–134 and 133n2); Gersh (1978:169–173). A ⊃ B means "A is a superset of B, but B cannot equal A".

120. See also Augustine, *Concerning the nature of good, against the Manichaeans* 1, in Schaff (1887) 4:351; and Aquinas, *Quaestiones disputatae de potentia Dei* 7.5.14.

121. Rolt (1920:134n1).

122. The Ostrogoths, like many northern Europeans at the time, were Arian Christians.

123. Courcelle (1935) argues the case that Boëthius studied under Ammonius Hermiæ.

124. Boëthius translated Aristotle's *Categories* and *On interpretation*, and apparently others; but there is no evidence that he translated any of Aristotle's treatises on animals. Boëthius also intended to translate Plato's dialogues. See McKeon (1929) 1:66; and Spade (2018).

125. Boëthius, *In Isagogen Porphyrii commentorum editionis secundae* 1.1, in Brandt (1906) 48:135–138; and in Migne (1847) 64: cols 71–72. Book 1 has been translated into English by McKeon (1929) 1:70–99. The quotation is from McKeon (1929:70–71).

126. Pliny, *Naturalis historia* 9.149.

127. Boëthius, *Philosophiae consolationis* 3.m9, in Weinberger (1934) 67:63–64.

128. Wetherbee (1973:10).

129. *Philosophiae consolationis* 4.m1, in Weinberger (1934) 67:79–80.

130. Dante Alighieri, *The divine comedy: Paradise* 10.121–129. Souls in the fourth (solar) sphere illuminate the world through their wisdom.

131. John of Damascus (John Damascene), born Yanah ibn Mansūr ibn Sarjum.

132. *De fide orthodoxa* 2.1–12, in Schaff & Wace (1899) 9, John of Damascus pp. 18–32.

133. Ibid. 2.7–9.

134. *De fide orthodoxa* 2.12, in the translation of Chase (1958) 37:237.

135. Chase (1958), page xxviii. The work exists in two variants, one longer than the other, both by John himself; these differ the most in chapters 6 and 9–14. The printed editions differ among themselves in other respects too, and this affects the text cited here at [137], which begins

chapter 47 in Lequien (hence Migne and Chase) but is found in chapter 30 in the 1548 and 1559 Petri editions cited at [139].

136. *Dialectica* 7–10, Chase (1958) 37:25–40. Angels are rational too: ibid. 8, Chase (1958) 37:27–28.

137. Ibid. 47, Chase (1958) 37:70 following Lequien in Migne (1864) 94, columns 619–620. Lequien (Paris 1711–1712) made a new translation from Latin, and a new edition of the Greek text. For further detail see Lequien in Migne (1864) 94.

138. *Dialectica. Version of Robert Grosseteste*, in Colligan (1953), at 4:6 (Chapter 3) and 4:26 (Chapter 30). See also Chapter 7 note [236], and Chapter 11, of the present work.

139. *Dialectica* 30, in Stapulensis (1548:590); and in *Opera* (1559:493).

140. John of Damascus, *Fragmenta*, in Migne (1860) 95: columns 229–230. Translation from the Greek by Dr Anitra Laycock. The Latin is very similar, although not identical. This fragment goes on to summarize: "The corporeal is divided into souled [ἔμψυχον] and non-souled [ἄψυχον]; the souled into the animal endowed with sensation and the insensate. The non-souled [ἄψυχον] into zoophyte and plant; the zoophyte into 'sea fig'; the plant into grass, shrub, tree. The animal, marking a new matter other than the natural, [is divided] into the rational and non-rational. The rational is mortal and immortal; the immortal [divided] into angel [and] dæmon, the mortal into man, cattle, dog and suchlike." For the underlined term *non-souled* we expect *insentient*. This is surely a copyist's error, as both the text and other Damascene fragments read *insentient* (Pandora, accessed 8 August 1995 by Anitra Laycock). Lequien translates ἄψυχον as *inanimum* without comment.

141. Aristotle, *De anima* 413b32–414a3, 415a23–30, and 416a19. See Chapter 3.

142. *Philoponus on Aristotle's On the soul* 3.9–13 (Charlton 2013), pp. 54–55; on *De anima* 434b9–11 (Hayduck 1897), pp. 600.13–601.3. See Chapter 5.

143. Peters (1968), page xviii.

144. Elsee (1908:76); O'Leary (1964:28–29); Sorabji (1983:199–200).

145. Peters (1968:17). Conflicting evidence suggests that Harran may have been a centre of the Sabian faith at about this time: TM Green (1992:106–112), and Churton (2002:26–27).

146. Basil II was co-emperor from 960, and sole emperor from 976 CE.

147. The *Quæstionum naturalium* of Michael Psellus hearkened back to Plutarch nine centuries prior. See Westerink (1948); Meeusen (2014); and Psellus in Migne (1889) 122: columns 783–810. Delatte (1939:5n1) following Gianelli (1939) claims that *Quæstionium* was actually written by (or modelled closely upon the *Conspectus rerum naturalium* of) Siméon Seth, a contemporary of Psellus in Constantinople.

148. Preus (1981:7–8).

149. Preus (1981). Michael also wrote commentaries, now lost, on the *Sophistical refutations*, the *Politics*, and the *Rhetoric*, and contributed to one on the *Nicomachean ethics*. See Sorabji (1987:5–6), reprinted (1995:188–189).

150. de Mowbray (2004).

151. Even Jerome, a fierce critic of pagan philosophy, accepted the distinction among animals, vegetables, and minerals: Jerome, *Against Jovinianus*, in Schaff & Wace (1893) 6:392, section 6. See also Lindberg (1983).

152. For example, John the Baptist as the reincarnation of Elijah (*Matthew* 17:1–13), or the transfer of the spirit of Elijah to Elisha (*2 Kings* 2:1–15).

Chapter 7

1. Jalāl ad-Dīn Muḥammad Rūmī, from "I died a mineral", in the translation of Arberry (1958:241).

2. Bairoch (1988).

496   NOTES

3. Existing cuneiform tablets date to the Twelfth century BCE, but the creation myth is much older: Mark (2018). For Asarre see Lambert (2007), Tablet 7, lines 1–2 at page 55.

4. I examined three English-language interpretations of these notoriously difficult Avedic verses: those of Mills (1900), Guthrie (1912), and Wadia (2017), as well as extensive secondary literature.

5. After wrestling with an angel, Jacob was given the name *Israel* (*Genesis* 32:28). We will discuss his eponymous ladder in Chapter 14.

6. M Smith (2002).

7. Broadie (1981:1–2).

8. According to Broadie (1981), the *Memar* shows influence of Hellenic philosophy in general, and Philo of Alexandria in particular.

9. Broadie (1981:19–20, 83, 149, and 157); and *Hymns* 1.131 and 2.213.

10. Also known as *Bereshith* ("In the beginning") Rabbah. English translation by Freedman & Simon (1961). Most *midrashim* date to 444–270 BCE and were originally oral, but began to be written down in the Third and Fourth centuries CE. *Genesis Rabbah* in its current form is usually ascribed to the Palestinian scholar (*amora*) Hoshaya in the Third century CE.

11. *Genesis Rabbah* 11.9.

12. Ibid. 7.5.

13. Ibid. 12.4 and 13.1. A similar interpretation was offered by Philo of Alexandria (Chapter 5).

14. Ibid. 10.7.

15. Ibid. 10.6.

16. Ibid. 8.11 and 14.3.

17. From about 293 CE (the establishment of the Tetrarchy to rule Rome) to the Muslim Arab conquests of the mid-Seventh century: see P Brown (1971).

18. Different authorities offer dates for completion of the Babylonian *Talmud* between 475 and about 700 CE.

19. Shāpūr I reigned *ca* 240 to 270/272 CE.

20. Peters (1968:45–49).

21. Khosrow I Anushirvān, reigned 531–579 CE.

22. Persecutions under Shāpūr II "The Great", from 339 to 379 CE, had been motivated by concerns that Christians were a fifth column supporting the empire's historic enemy, Rome. See Baum & Winkler (2003); Russell (2012); and Egger (2017), Chapter 1.

23. Its founder was Mani, 216–274 CE.

24. The Avesta is known directly only from Ninth-century commentaries on written texts (now lost) dating to the Fifth or Sixth century CE, and from partial copies dating to the Fourteenth century, but the tradition is much older.

25. Duchesne-Guillemin (1962). The relevant Pahlavi texts are *Mēnok ī xrat*, *Dātastān ī dēnīk*, and *Artāg vīrāf*.

26. A supposed connection to the Cathar (Albigensian) heresy that spread in southern France from the Eleventh to early Fourteenth centuries is less certain: see Weber (1907); Warner (1922); and van Schaik (2006).

27. Le Coq (1911), lines 79–84.

28. Wallis (1992:1).

29. Indeed, some regions were ruled by Christian or Jewish kings from about 400 to 620 CE. See Hoyland (2001), particularly pp. 139–150.

30. The word *Islam* connotes submission (to the will of God).

31. Later renamed Medina. The move was the *hijra* or *hegira*.

32. *Quran* 13:2–4.

33. Wolfson (1976), pp. 355–372 and 725–727.

34. *Quran* 16:4–11.

NOTES   497

35. Ibid. 50:38.
36. Ibid. 55:22.
37. Ibid. 5:95–96.
38. Pellat *et al.* (1986:306–307) (Pellat).
39. See Wolfson (1976:1–2) for this broader usage. More precisely, pre- and post-philosophical periods may be recognized for *kalām*: Wolfson (1976:19–20).
40. Regarding free will (as limiting God's power), and attributes (as calling into question the eternal pre-existence of the *Quran*); see Wolfson (1976:15–18).
41. Al-Māʾmūn reigned 813–833 CE in Baghdad.
42. Wolfson (1976:19–20 and 29–30) identifies syllogism (from the *Prior analytics*), and a new use of analogy.
43. See Chapter 6, John of Damascus.
44. Georr (1948) argues that the translation of Hellenic philosophic ideas into Arabic built on the scientific terminology developed by the earlier phase of translation into Syriac that began in the 530s CE.
45. Peters (1968:42).
46. Peters (1968:17); Walzer (1991).
47. Peters (1968:xx).
48. ʿAbbāsid Caliph Abū Jaʾfar ʿAbdallāh al-Manṣūr ibn Muḥammad reigned in Baghdad, 754–775 CE.
49. Lunde & Stone (1989:14) date this to the capture of Chinese papermakers at the Battle of Talas in 751 CE. Paper was made in Baghdad by 794 CE: Bloom (2001).
50. Wolfson, *Kalam* (1976:52); Treiger (2015:443–444).
51. Wolfson, *Kalam* (1976:53).
52. So far these are Dexippus, *In Aristotelis Categoriae*; Themistius, *In Aristotelis De anima*; Simplicius, *In Aristotelis Physica / De anima / De caelo*; "Ammonius", *In Porphyrii Isagogen*; John Philoponus, on Ammonius's lectures on *De anima*, and *In Aristotelis Analytica priora*; Elias, *In Porphyrii Isagogen*; David, *In Porphyrii Isagogen*; Nemesius of Emesa, *Premnon physicon / De natura hominis*; and John of Damascus, *Dialectica*; and the epitome misattributed to Philo, *Peri khosmos / De mundo*). We might add the lost commentaries by Porphyry (*Ad Gedalium*) and Iamblichus (*In Aristotelis Categoriae*), although if these works had been preserved in Arabic, they wouldn't be considered lost.
53. Ḥunain ibn ʾIsḥāq misattributed it to Gregory of Nyssa. See Graf (1944–1953) 1:319. The translation was known to al-Kindī and to Orthodox Christian ʿAbdallāh ibn al Faḍl in Antioch in the early Eleventh century (ibid. 2:58–59). For an overview of manuscripts, including in Syriac and Arabic, see Wicher (1960), especially pp. 37–38.
54. Lindberg (1992:168–170).
55. Lyons (1955); Lyons (1973).
56. Lyons (1973:344). I thank Dr Amin Mohamed (CSIRO) for translation from the Arabic.
57. Graf (1944–1953) 2:43; Treiger (2015:442–455).
58. Steinschneider (1869), page 157 and note 9.
59. Graf (1944–1953) 2:160–161; D'Ancona (2017). Lettinck (2014:3) states that Simplicius's commentary on Aristotle's *Physica* was "unknown in the Arabic world".
60. Ivry (2012, note 1) states that no Arabic translation exists of the commentary on *De anima* by John Philoponus.
61. Indeed, Ibn al-Ṭayyib refers at one point to "John Philoponus, Olympiodorus, Elias, and all the Alexandrian scholars" (Gyekye 1979:79, paragraph 148). See also Dunlop (1951); Stern (1957); and Peters (1968:82 and note 44).
62. If this refers to animal (but not vegetative animal), it presumably should read "from place to place".

498 NOTES

63. Gyekye (1979:71, paragraph 129).
64. Job of Edessa, as edited by Mingana (1935:23–24).
65. Job of Edessa, ibid. (1935:99–100); see also pp. 173 and 230.
66. Kruk (1993).
67. Pellat *et al.* (1986).
68. Levy (1957:460).
69. An important example is the *Kitāb al-Ḥayawān* of al-Jāḥiẓ: note [67]; Nasr (1992: 109–110); and J Miller (2013).
70. Nasr (1978:6n9).
71. Ullmann (1972:6–8).
72. *One thousand and one nights* was compiled over several centuries, building in part on Sasanian and perhaps Sanskrit prototypes. See DF Reynolds (2006:270–291).
73. Ullmann (1972:84–85).
74. Ad-Dīnawari, *Kitāb al-Nabāt.* See Kruk (1993); Silberberg (1910–1911); and Breslin (1986).
75. Ullmann (1972:54–56). Compare with Chapter 1.
76. ʿAbd al-Raḥīm ibn Muḥammad al-Ḥayyāt, *Kitāb al-Intiṣār* (full title in English: *The book of triumph and of the refutation of Ibn al-Rawandi the heretic*), quoted in Wolfson (1976:498–499). Wolfson (1976:500–504) distinguishes this from a more-specific theory of latency, referring only to properties released by human action, by Abū ʾl-Ḥasan al-Ashʿari.
77. *Kitāb al-Ḥayawān* (1949–1950).
78. Palacios (1930), especially pp. 26–27; Pellat *et al.* (1986); Miller (2013), pp. 143–144 and 151. Miller translates the relevant text (*Kitāb al-Ḥayawān* 1.26.1–1.37.8) at (2013:370–383).
79. Miller (2013), pp. 8–13, 17–22 and 183 ff.
80. Ibid. (2013:78–79).
81. Al-Muʿtaṣim, the eighth ʿAbbasid caliph (reigned 833–842), was the half-brother of his predecessor Caliph al-Māʾmūn.
82. Jolivet & Rashed (1986); and al-Kindī, *Œuvres* (1998).
83. Al-Kindī, *Prosternation du corps extrême*, in *Œuvres* (1998:173–199), at pp. 196–199.
84. Al-Fārābī, *On the perfect state*, in Walzer (1998:136–141).
85. Ibid. (1998:164–175); Bakar (1998:48–64).
86. Alternatively, "Herodotus of the Arabs"; in recognition of his extensive travels, and synthesis of the history and geography of foreign lands.
87. Khalidi (1974). The title of the *Tanbīh* is sometimes translated as *Book of warning and revision*, as in al-Masʿūdī, *Le livre de l'avertissement et de la revision* (1896).
88. Al-Masʿūdī, *Meadows of gold*, translated by Sprenger (1841:60–61); and *Les prairies d'or*, translated by Barbier de Meynard & Pavet de Courteille (1861–1877) 1:61.
89. For example, at *Prairies d'or* (1861–1877) 2:399, 3:31, 3:398, 3:437, 4:54, and 8:325.
90. *Prairies d'or* (1861–1877) 4:7–8.
91. Ibid. (1861–1877) 1:330.
92. Al-Masʿūdī (1896:14–15) as further translated by Sarton (1943) at page xvii. Sarton interprets al-Masʿūdī as inviting his listeners to "play with the ancient conceit of the ladder of nature", and elsewhere (*Introduction* 2:62 and 3:212) claims that the idea of a Great Chain of Being is expressed clearly in al-Masʿūdī. The texts I have seen do not support a strong claim. Describing the celestial sphere, the Ikhwān al-Ṣafā write that the planets and stars are "yoked to the sphere of the sun" (*Epistles* 22.26; Goodman & McGregor, 2009:237–238). Under the heading *Tanbīh*, Peters (1968:275–276) reconstructs an outline of al-Masʿūdī's *Various types of knowledge*, by which we may suppose (1968:273) he means the *Kitab Akhbār al-zamān* (for the lost works of al-Masʿūdī, see Khalidi 1974). This outline lists under *Corporalia* heavenly bodies, rational animals, non-rational animals, plants, minerals and metals, and the four elements.

NOTES   499

93. Also translated as *Brethren of Sincerity*; *floruit* in late Tenth-century Basra, with a branch in Baghdad. According to HR Turner (1997:190) they fell under suspicion by the religious authorities, and went underground in the Eleventh century.

94. Goodman & McGregor (2009), page xi.

95. *Epistle* 16.2 (Baffioni 2013:135–137) and 19.1 (2013:227). For macrocosm-macrocosm in the *Rasā'il*, see also Baffioni (2013:22–23 and 34), and Goodman & McGregor (2009:237n359, and 272–274 including note 446).

96. Sarton (1927) 1:660–661; Nasr (1978:25–104); Turner (1997:190–191). Their call on the Pythagorean tradition can be seen in *Epistle* 22, *Prologue* 16 (Goodman & McGregor 2009:91–94). Neoplatonic concepts may be found *e.g.* at *Epistle* 22, *Prologue* 6 (2009:71–72) and *Epistle* 22.18 (2009:189–190).

97. *Epistle* 22, *Prologue* 1. Goodman & McGregor (2009:63–64). Very similar wording is in *Epistle* 21.1 (Baffioni 2013:317) and 21.3 (2013:342–346).

98. For example, at *Epistle* 15.5–8 (Baffioni 2013:112–115).

99. *Epistle* 17.2 (Baffioni 2013:180).

100. Ibid. 22, *Prologue* 2–3 (Goodman & McGregor 2009:65–67).

101. Ibid. 22.35 (2009:274).

102. Ibid. 17.12–13 (Baffioni 2013:180–181), 19.3 (2013:229), and 19.9 (2013:247–248).

103. Ibid. 21.3 (2013:340).

104. Ibid. 21.1 (2013:326–329).

105. Ibid. 21.3 (2013:340–341).

106. Nasr (1978:93); *Epistle* 21.3 (Baffioni 2013:343).

107. *Epistle* 21.3 (Baffioni 2013:341–342). The date palm was widely appreciated to be sexually dimorphic.

108. Ibid. 19.3 (2013:230).

109. *Epistle* 22, *Prologue* 7. Goodman & McGregor (2009:74–76); Ullmann (1972:51–52).

110. *Epistle* 21.3 (Baffioni 2013:342–343 and note 108).

111. *Epistle* 22.11–18 (Goodman & McGregor 2009:150–199).

112. *Epistle* 21.3 (Baffioni 2013:344–345).

113. *Epistle* 22.36 (Goodman & McGregor 2009:277).

114. *Epistle* 19.11 (Baffioni 2013:257–258) and 22.38 (Goodman & McGregor 2009:300). At Epistle 19.3 (Baffioni 2013:230) they state that pearl is animal, but at *Epistle* 19.11 (2013:257–258) that it is mineral.

115. Nasr (1978:98). See also John Philoponus, notes on Ammonius on Aristotle's *De anima* (Chapter 5).

116. *Quran* 6.38 and 17.44.

117. *Epistle* 22.35 (Goodman & McGregor 2009:272).

118. Al-Bīrūnī opposed some of Aristotle's teaching, for example regarding the movement of the heavens (*De caelo*). For Platonic and Neoplatonic concepts in al-Bīrūnī, see Fakhry (1979:344–349). Regarding his lost philosophical works, see Nasr (1979:401).

119. Nasr (1978:147–148).

120. Al-Bīrūnī (1934), paragraphs 367–369, row 8, page 222.

121. Qadri (1979:588–589).

122. Ullmann (1972), pp. 7 and 69.

123. When Ibn Sīnā was born, Bukhara was capital of the Samanid empire (819–1004/1005 CE). See Pandita (1986) for a perspective on social, political, and intellectual currents in the region at that time.

124. Siraisi (1987).

125. Alternatively, *The cure*.

500 NOTES

126. For an outline and mapping to the books of Aristotle, Porphyry, Euclid, Ptolemy, Plato, and others, see Gutas (2016). In the late Twelfth century (with additions in the late Thirteenth century) the sections on metaphysics and natural history were translated into Latin as the *(Liber) sufficientia*.

127. Ibn Sīnā, *Compendium on the soul*, sections 2 and 4–8 (Landauer 1876, pp. 378, 384–411); and *Remarks and admonitions* (Inati 2014, pp. 102, 111–112, and 165).

128. For Ibn Sīnā, motion covers not only locomotion but also motion in situ, and indeed changes of quantity and quality: see Nasr (1978:217 and note 10).

129. Landauer (1876:387–388); Ibn Sīnā (van Dyck 1906:43–44). Ibn Sīnā claims to have carried out this experiment many times. Given that Ibn Sīnā defines motion broadly (see previous note), Landauer is incorrect in stating (1876:387n1) that Ibn Sīnā contradicts Aristotle with his claim that every being with the power of perception moves voluntarily, and vice-versa.

130. In the section of *al-Šifa'* dealing with animals, Ibn Sīnā fails to mention Aristotle's formulation on nature proceeding little by little from things lifeless to animal (*Historia animalium* 588b46). I examined Michael Scot's Latin translation in *Avicēne perhypatetici philosophi ac medicorum facile primi opera* (1508).

131. Tawara (2014).

132. See Chapter 3 and *De plantis* 816a1–17. Tawara attempts to connect Ibn Sīnā's new opinion to comments in his *al-Qānūn fī al-Ṭibb*.

133. In his *Manṭiq al-mashriqiyūn*; see Nasr (1978:185–196); these quotations from pp. 186–187.

134. Nasr (1978:191 and 263–274).

135. Aristotle, *De partibus animalium* 645a4–36, in Ogle's translation.

136. Shaker (2012:1–4) presents the end of *falsafah* as a liberation from "moribund" Aristotelianism, allowing a flowering of philosophy, including that of a mystical nature.

137. In his *al-Munqidh min al-ḍalāl*. See Rizvi (1986). Not only was Ibn Sīnā *falāsifa*, but his father and brother were *bāṭinyya*. For the influence of sufism on al-Ghazālī, see Politella (1964:180–183).

138. And to a lesser extent al-Fārābī, who like Ibn Sīnā accepted that the cosmos was eternal while disagreeing on some details of its emanation.

139. Or more precisely, not accepting that the matter can be settled only by revelation. See Griffel (2005).

140. Al-Ghazālī, *Incoherence* (2000:226–227).

141. Puig Montada (1992); Marmura (1995); Griffel (2009, 2011).

142. Marmura, Introduction (2000), pp. xv–xvi.

143. Rizvi (1986:278). See also Macdonald (1936:11).

144. Kruk (1993:833).

145. Baneth (1938:23–30); English summary at pp. iv–v.

146. Known in its Latin translation as *Logica et philosophica Algazelis Arabis* (1506). The sections on metaphysics and natural history were edited by Muckle (1933).

147. Muckle (1933:196). The idea can be found elsewhere in al-Ghāzalī as well.

148. The default position must be that in *Tahāfut*, al-Ghāzalī set out the philosophical positions he intended to refute (Macdonald 1936; Griffel 2005:273–296). However, some Jewish scholars have suspected that the *Tahāfut* was a cover for his espousal of philosophy (Griffel, 2019).

149. Wolfson (1976:593–600).

150. Dampier (1948:77).

151. Ibn Rushd, *Tahafut* (1987:113).

152. Indeed, this formed part of Ibn Rushd's argument for the existence of God: *Kitāb al-Kashf 'an manāhij al-adilla fī 'aqā'id al-milla*, translated by Najjar as *Faith and reason in Islam*.

NOTES   501

Averroës' exposition of religious arguments (2001). See also Hillier, Existence and attributes of God (*Internet Encyclopedia of Philosophy*).

153. Ibn Rushd, *Commentarium magnum* (Crawford, 1953), pp. 111 and 155; *Long commentary* (Taylor, 2009), pp. 94 and 125; *Middle commentary* (Ivry, 2002), pp. 39 and 49.

154. Indeed, plants and some sea animals may even have more than one soul, as a part remains alive (at least for some time) even after being cut off: Taylor (2009:102); Ivry (2002:49–50).

155. I searched fourteen books or commentaries by Ibn Rushd dealing with natural history or the soul, a number of his medical and other works, and extensive secondary literature.

156. von Hees (2002), pp. 103 and 107–108.

157. Al-Attas (2009).

158. Sarioğlu (2007).

159. Niẓamī Arūẓī, *Chahár maqála*, Persian text (1910); in Browne's translation (1921).

160. *Chahár maqála* (1921:5–6).

161. Ibid. (1921:6). Original terms in Persian have been deleted.

162. Ibid. (1921:7). Original terms in Persian have been deleted. Material in square brackets is from Browne's footnote 1.

163. Ibid. (1921:7–9). See also Sarton (1943), page xviii. This particular satyr was not treated by Tyson (1699).

164. Ibid. (1921:9–11). Muḥammad was the "most perfect of the Prophets" (1921:1).

165. As indeed Browne does at *Chahár maqála* (1921:10–11): "Human World", "Angelic World". Where Nizamī later refers to geopolitical kingdoms he uses *mulk* (*dominion, kingdom, country*). I thank Dr Atefeh Taherian Fard for expert assistance with the Persian text.

166. JV Carus (1872:166–169); al-Qazwīnī, *'Aja'ib al-makhlūqāt* 80, as cited by Ahmad (2008). The botanical parts of his *'Aja'ib al-makhlūqāt*, descriptive and arranged more-or-less alphabetically, were translated by Wiedemann (1916–1917), reprinted as *Beiträge* 54 (1970). Wiedemann also translated the zoological parts (reprinted as *Beiträge* 53).

167. Carus (1872:166–169). Al-Qazwīnī refers to "a worm that lives inside a mineralized tube" and can also be found on the seashore (*Afern*): page 168. For its identification as the *snail*, see also note [110] above.

168. Anonymous entry "Natural history" in Sharif, *A history of Muslim philosophy*; Ullmann (1972:32–33 and 52); von Hees (2002:102–103, 129–130).

169. Stephenson, Royal Asiatic Society (1928) and *Isis* (1928). The first nine text pages of the former are reproduced by Nasr (1992:119–125). See also the anonymous "Natural history" in Sharif (above).

170. Ibn Khaldūn, *Muqaddimah*, Sixth prefatory discussion (1958). Rosenthal notes that the term he translates as "the last stage" [of minerals] is literally "horizon".

171. Chittick (1998), pp. xxxi, 136–138, 280, 304, and 311; Murata (1992), pp. 28, 37–40, and (including *jinn*) 145. Chittick translates *muwallad* as *progeny*, Murata as *children*. I thank Professor Chittick for personal communication (2018) on this section.

172. Al-Qūnawī was a stepson of Ibn 'Arabī and a member of the spiritual circle around Jalāl ad-Dīn Muḥammad Rūmī.

173. *Miftāḥ al-ghayb al-jam' wa al-wujūd* 297–298, as translated by Chittick (2012:125–126). An earlier (almost identical) translation is at Chittick (1992:196–197). I have added [*al-muwalladāt al-thalātha*] for clarity.

174. Birge (1937), pp. 115–118 and 214–215, and note 1 at pp. 129–130.

175. Ibid. (1937:116) identifies the manuscript as *Devriyei Arṣiye* by Misrii El Niyazi, page 449.

176. Nasr (1992:112–113); Ullmann (1972:33–43).

177. Maimonides, *Guide for the perplexed* 1.71.

502 NOTES

178. See also Wolfson (1976:43–58). Earlier, Solomon Ibn Gebirol, usually considered a Neoplatonist, briefly gave the standard definition of vegetable and animal souls in his *Fons vitæ* 3.46: see the text edited Baeumker (1895:181–182).

179. Ibn Daud (1986), Introduction & treatises 2–4; Fontaine (2007).

180. The Arabic original (*Kitab al-'Aqīd al-rafi'a*), written in 1160, has been lost. It was translated into Hebrew as *Sefer ha-emunah ha-ramah* (*Book of exalted faith*) in 1391–1392 CE by Solomon ben-Lavi, and soon thereafter as *Sefer ha-emunah ha-nissa'ah* (*Book of sublime faith*) by Samuel ibn Motot. The former was in turn translated into German by Weil (1852), and into English by Samuelson (1986).

181. *Sefer ha-emunah ha-ramah*. In view of the comments of Ivry (1989), I have followed the Weil 1919 (*i.e.* 1852) edition at pp. 31–40, part of which was presented in English in Bodenheimer (1958:199–204).

182. Ibn Daud (1852:44).

183. The more one learns of God, the more one is perplexed. The highest level of knowledge is realization that one does not know Him. Lobel (2007:11) suggests that the concept in Maimonides, hence the title, may have a Sufi origin.

184. *Guide for the perplexed* 1:72 (Friedländer) 1904:114–119.

185. According to Maimonides, only man (not, for example, horses) can be said to be a microcosm of the universe; but even so, the parallel is imperfect. See *Guide for the perplexed* 1:72 (1904:117–119).

186. Friedländer (1904:liii), referring to *Guide* 2:30 (1904:217–218).

187. *Guide* 1:72 (Friedländer 1904:116).

188. *Ezekiel* 1:1–28.

189. Poncé (1973:13–15). Note that the sefirot of *Sefer yetzirah* are not identical with those of later *kabbalah*.

190. Kaplan (1997), pp. ix–xxvi.

191. Séd (1966). See also notes [8] and [9] above.

192. *Sefer yetzirah* 4.15 (Kaplan 1997:185).

193. Abulafia, *Otzar eden ha-ganuz* 75b, written in 1285/1286, as cited by Kaplan (1997:190). See also Kaplan (1997:325 and 373n84).

194. Weber (1907).

195. Scholem (1987:35–38).

196. Scholem (1987:49–198), summarized at pp. 196–198; quotation from page 198.

197. Poncé (1973:47); Scholem (1987:5–6); Tishby (1989:13–17, 91–96).

198. Tishby (1989:229–422); creation ex nihilo at pp. 279–281, 549–558, and 572–573.

199. Ibid. (1989:661–664).

200. Zonta (1996).

201. *Sefer magen 'avot* (Livorno 1785) 10a, as translated by Patai (1994:266). Comments in square brackets by Patai.

202. See Chapter 9. For more on Alemanno, see Idel (1988).

203. *Sefer sha'ar ha-hesheq* (Livorno 1790); see below [205].

204. Patai, *Jewish alchemists* (1994:293).

205. *Sefer sha'ar ha-hesheq* 55b–56a, as translated by Patai (1994:294). Comments in square brackets by Patai. The work under this title is the introduction (*Shir ha-ma'alot*) to Alemanno's *Heshek Shelomo*, which otherwise remains mostly unpublished. A partial translation of *Shir ha-ma'alot* is in Lesley (1976).

206. As translated by DR Blumenthal (1987:184). Blumenthal does not identify which work by Albotini is being translated, but it is almost certainly his *Sullam ha-aliyyah*. Comments in square brackets by Blumenthal.

207. Ruderman (1988:162).

NOTES 503

208. This is a close paraphrase of Ruderman (1988:5–6).

209. Randall (1961:19–20); Seth (1886:426–427).

210. Pasnau (2012:666).

211. Packard (1920:7–52).

212. Singer (1928:75–81); Wingate (1931); Minio-Paluello (1952); Haskins (1957:278–302); Grant (1974:35–41); Cranz (1984); Grant (2004:165–169).

213. Themistius, *Commentaire* (Verbeke, 1957); Thomas (1959:16–17); Dod (1982:63–64); Todd (2003: 61 and 78). See also Chapter 5, note [91].

214. Themistius (Verbeke, 1957), pp. 84, 105 (twice), 111 (twice), 185 and 277.

215. In Book 3, chapter 1. I examined the 1527, 1542, and 1559 editions, where *Zoophyta* appears at folio 73r, page 285, and page 186 respectively. For these editions see Todd (2003:72–73 and 78–79).

216. See Chapter 5, note [92].

217. Themistius (Schenkl, 1900).

218. In Books 1 and 8. In the 1527 and 1559 editions these occur at folio 14r and 52v, and pp. 31 and 134, respectively. In the 1542 edition the former appears at page 50; the copy I examined was defective, but the latter instance should appear at page 202.

219. Todd (2003:91–92).

220. [Themistius], *Hermolai Barbari . . . In paraphrasin Physices Themistii* (1559), Book 1, page 31.

221. This fragment is thought to date to after 1235 (Dod 1982).

222. This fragment is thought to date to after 1247 (Dod 1982).

223. Heiberg (1894:xii); Dod (1982); Chapter 5, note [98]. In the 1540 edition the term *zoophita* is found at folio 60r.

224. Dod (1982); Lautner (1995:5–6).

225. Thomas (1959:15); Dod (1982); Sorabji (1987:14); Charlton (2013:1–16); Wildberg, John Philoponus (2016).

226. Philoponus, *De anima* (Hayduck, 1897). The word *zoophyte* does not appear in the Moerbeke translation (Verbeke, 1966).

227. [Saint] Alfanus I was Archbishop of Salerno from 1058 to his death in 1085 CE.

228. Frederic Barbarossa, Holy Roman Emperor (reigned 1155–1190).

229. John of Salisbury, Bishop of Chartres; Peter Lombard, Bishop of Paris.

230. Telfer (1955:217–219); Chenu (1968:34n77); Chapter 6, notes [59] and [62]. Alfanus did not indicate an author for the *Premnon physicon*, perhaps because the Greek manuscript he worked from was likewise unattributed.

231. Apollonius of Tyana was known in the Islamic world as Balīnās, Balīnūs, or similar variants: Asl (2016:437).

232. Iskander (1981); Zimmermann (1981); Haq (1994:29–30). *Kitāb Sirr al-khalīqa* dates to the Eighth or first half of the Ninth century, and has been critically edited by Ursula Weisser (1979 in Arabic, 1980 in German).

233. Hudry (1997–1999, published 2000).

234. Chase (1958), pp. xxxvi–xxxviii. Chase does not indicate a year for this *Dialectica*, but Burgundio translated *De fide orthodoxa* in 1153–1154 (Buytaert 1955, pp. ix–xv and xlii). Buytaert, however, disputes that Burgundio translated the *Dialectica* (1955, page ix).

235. Chase (1958, pp. xxxvi–xxxviii); John of Damascus (Colligan, 1953). Buytaert (1955, page vii) gives 1135–1140 CE.

236. Colligan (1953:6), with "in sensibile animal" in the left-hand column and "et insensible et zoophyton et plantam" in the right-hand column. The same division is presented at page 26. For textual variants, see Colligan (1953:57).

237. Panetius, before 1497; De Billy, 1577; Lequien, 1712. The *Dialectica* was also taken into Russian in the mid-Sixteenth century.

504 NOTES

238. Dexippus, *In Aristotelis Categoriae*. Latin translation by Felicianus, published in 1546. According to Dillon (1990:16) this edition was based on an inferior manuscript. I have been unable to locate a 1546 Dexippus by this name, but have examined *Dexippi philosophi Platonici . . . In defensionem Praedicamentorum Aristotelis aduersus Plotinum . . . quaestionum libri 3. Nunc primùm in latinam linguam conuersi, atque in lucem editi, Ioanne Bernardo Feliciano autore*, supposedly 1549 (per Internet Archive, and Worldcat). However, the title page and colophon are absent.

239. The Greek text was printed by Vlastos in 1500. The Latin *Ammonius in qvinqve voces in Porphyrii per Pomponivm Gavricvm Neapolitanvm*, translated in 1502 and printed in Venice (Io Baptista Sessa, 1504), gives *zoophytō* (*i.e. zoophyton*) on the twenty-eighth (unnumbered) sheet; the second instance is translated "ostrea et spongae" (on the thirtieth sheet). Busse (*CAG* 4, 1891) points out that Gauricus's text includes his own interpretation; this was removed for a 1539 edition. According to Busse (1891:xli–xliii) the second and third Latin versions, both in the translation of Rasario, were published in 1547 and 1549; but a 1542 Rasario (Venice) is available at archive.org (accessed February 2018). ζωόφυτον is left in Greek in a 1545 Rasario second edition (Venice), while this section is absent altogether from a 1559 Rasario (front title page absent) at archive.org (accessed February 2018). The 1542 edition takes liberty with this passage, reading (in my translation): "body is animate or inanimate. The animate is animal, or plant, or truly a composite of both, which they call Zoophytum." Neither Barnes (2003) nor I could confirm the existence of a supposed 1494 translation by Gauricus; note that Gauricus was born in 1481 or 1482. See Barnes for further information on the Latin editions.

240. In the Henri Estienne edition (Paris, 1511), *zoophyton* (as a Latin word) appears at folio 15r, and *ostrea et spongiæ* at 16r.

241. Translated into Latin by Budæus for publication in 1533.

242. An anonymous Twelfth-century translation of *De partibus* is also known (Dod 1982). See also Sarton, *Introduction* 2:63; and William of Moerbeke, in Grant (1974:39–41).

Chapter 8

1. *Marius: On the elements*, as translated by Dales (1976:176–177).
2. Cassiodorus had earlier hoped to open a Christian school in Rome, as a counterpart of the pagan schools of Athens and Alexandria, but he had to change his plans with the death of Pope Agapetus in 536 CE: Sharpe (1964:11). See also Lejay & Otten (1908). Singer (1928:185) implies that the Benedictine monastery at Squillace was pre-existing and Cassiodorus merely "took great interest in its literary welfare".
3. The earliest is attributed to Saint Pachomius in 318 CE (Huddleston 1911).
4. The Monastery of Saint Athanasius in present-day Bulgaria dates to 344 CE, while monasteries were established near Cannes in 410 CE and near Marseilles around 415 CE. The movement reached Ireland in the late Fifth century, while the first monastery in England was founded by Augustine of Canterbury in 598 CE.
5. Millas-Vallicrosa (1963:129–130).
6. Turner (1908).
7. Walafrid Strabo, in Prologue to Einhard's *Life of Charlemagne*: AJ Grant (1905:1–3), and Thorpe (1969:50). Original as *Vita Karoli Magni*, date unknown but perhaps 817 and 833 CE: McKitterick (2004:30).
8. Gregory VII (later Saint) was Pope from 1073–1085 CE.

NOTES   505

9. The Carmelites, perhaps about 1155 CE (Zimmerman 1908); the Trinitarians, 1198 CE (Moeller 1912). The Franciscan, Dominican, and Augustinian orders date to the first half of the Thirteenth century.

10. The so-called *dispersal*, March 1229 to April 1231. See Seth (1886:427).

11. The Rabbinical and alchemical traditions were exceptions.

12. For which, see Seth (1886).

13. Seth (1886:425).

14. Cassiodorus, *De institutionibus divinarum litterarum*, and *De artibus ac disciplinis liberalium litterarum*, written 543–555 CE, respectively at columns 1105–1150 and 1149–1220 of Migne, *PL* 70 (1865). The two works are sometimes referred to collectively as *De institutiones divinarum et sæcularium litterarum*.

15. The *trivium* and *quadrivium* were first formalized by Varro (Grant 2004:91) and had been promoted by Jerome, Augustine (Lejay & Otten 1908), and Martianus Capella (Grant 2004:138–139).

16. Cassiodorus, *De anima* (1589:284–306); and in Migne, *PL* 70 (1865), columns 1279–1308. His *De anima* also quotes from the Bible, and concludes with a supplication to Jesus.

17. Cassiodorus, *Variarum* (1589:74); and in Migne, *PL* 69 (1865), column 602.

18. Singer (1928:185).

19. Sharpe (1964:6–7).

20. Barney *et al.* (2006:18).

21. Cassiodorus, *De artibus*, in *PL* 70 (1865), column 1152.

22. Barney *et al.* (2006:21–22) point out certain exceptions; others are at *Etymologies* 11.3 (2006:243–244).

23. *Etymologiæ* 12.4.4–9, 12.4.20–23, 12.6.3, 12.6.7–8, 12.6.21, and 12.6.58–62.

24. Ibid. 11.4.3.

25. Ibid. 16.8.1.

26. Isidore (Bekker 1967): pestilences at Chapter 39, pp. 67–68. In *Etymologiæ* 4.6.17, pestilence is due to "powers in the air".

27. Crombie (1953:10–11).

28. Alcuin, *Epistola* 116, in Migne, *PL* 100 (1863), column 418.

29. Alcuin, *Dialogus de rhetorica et virtutibus*, as represented in Migne, *PL* 101 (1863), columns 945–950.

30. The name Maurus was given him by his teacher Alcuin.

31. Earlier known as *De sermonum proprietate*, *Liber etymologiarum*, or *De naturis rerum*, but called *De universo libri XXII, sive etyologiarum opus* by Migne, *PL* 111 (1864), columns 9–614.

32. Sharpe (1964:19).

33. Hrabanus, *De universo* Book 8 (animals) closely follows Isidore, *Etymologiæ* Book 12; Hrabanus, Book 17 (stones and metals) follows Isidore, Book 16; and Hrabanus, Book 19 (agriculture, with a focus on plants) follows Isidore, Book 17 (rural matters).

34. Hrabanus, *De anima*, in Migne, *PL* 110 (1864), columns 1109–1120.

35. He signed his translation of Pseudo-Dionysius as *Eriugena*, *i.e.* born in Ireland; the combination John (Johannes) Scotus (Scottus) Eriugena is due to Archbishop Ussher in 1632 (O'Meara 1987:11).

36. It is not known for certain whether he was, or was not, a monk.

37. Capella's *De nuptiis* is remarkable not only for its Neoplatonism, heavy use of allegory, and embrace of the seven liberal arts, but also for positing that Mercury and Venus revolve around the sun.

38. O'Meara (1987:13). The latter is at *Annotationes in Marcianum* 57.15 (Lutz) 1939:64.

39. Eriugena is first attested there in 850/851 CE. Charles the Bald was grandson of Charlemagne.

506 NOTES

40. Eriugena, *Periphyseon*, in Floss, *PL* 122 (1853), column 441A; and in the Sheldon-Williams translation (1987:25). See also O'Meara (1981).

41. Eriugena, *Periphyseon*, in Floss, *PL* 122 (1853), cols 441B–442A; and Sheldon-Williams (1987:25–26).

42. Eriugena distinguishes two types of nothing—nihil *per privationem* and *nihil per excellentiam—and* argues that creation flows from God's superabundant nothingness. See O'Meara (1987:17–18), and Moran (2008).

43. *Periphyseon* (1853), cols 685B ff; and (1987:313 ff.).

44. Ibid. (1853), cols 727D–729C; and (1987:362–365). Quotation at column 728D and page 364 respectively.

45. Ibid. (1853), col. 710A; and (1987:341).

46. Ibid. (1853), col. 733A–B; and (1987:369). Largely repeated at cols 733D–734A, and pp. 369–370 respectively. See also cols 735C–736B, and pp. 372–373.

47. Ibid. (1853), cols 735C–736B; and (1987:372–373); also cols 751A ff, and pp. 391 ff. However, angels are not animals (col. 762A; page 389).

48. The *Clavis* summarizes the first four books of *Periphyseon*, and reproduces the fifth. The corresponding passages of the *Clavis* are at sections 219–223 and 225 (1974:172–177).

49. The so-called Medieval Climatic Optimum, from the Tenth through Fourteenth centuries in northern and central Europe: Mann *et al.* (2009).

50. Haskins (1927/1957); Chenu (1968); Bisson (2008).

51. Wetherbee (1973:2).

52. Seth (1886:418); McKeon (1929:142–143).

53. *Monologion* 4, in Anselm (2000:11).

54. *De grammatico* 3 ff., in Anselm (2000:133–134 ff.). He returns to this theme in *Monologion* 19 (page 21) and *Philosophical fragments* 27:26 (page 394).

55. *De veritate* 2, in Anselm (2000:168); see also *De grammatico* 7 (2000:139–140).

56. Dampier (1948:80); Seth (1886:425).

57. Quotations at McKeon (1929:207); see also Seth (1886:424–425).

58. Abelard, *Glosses on Porphyry*, in McKeon (1929), pp. 211, 224–226, 228, 230, 232, 234, 246 and 256, translated from Geyer (1919). For vegetative soul (in man), see Geyer (1919:105).

59. Peter Lombard, in Migne, *PL* 191 (1841), cols 11–454; *Sententiae* (1971–1981); and *The sentences* (2007–2010).

60. Peter Lombard, *Sentences* 2.12–15 (2008:49–68).

61. Ibid. 2.15.6 (2008:65–66).

62. Ibid. 2.15 (2008:63–68); quote at Distinction 15.4 (page 64) following Augustine, *De Genesi ad litteram* 3 (see Chapter 6).

63. Adelard was born about 1080. Bath was sacked during the Rebellion of 1088.

64. Adelard, *De eodem et diverso / On the same and the different* (1998:1–79).

65. Adelard, *Questiones naturales / Questions on natural science* (1998:81–235). See also Haskins (1924/1960), Chapter 2 (pp. 20–42); and *Dodi venechdi*, translated by Gollancz (1920:85–161). Gollancz (1920, page i) questions whether *Quaestiones naturales* was in fact the original title. Sarton (*Introduction* 2:169) calls the Gollancz translation "careless".

66. Burnett (1988, page xii) argues that in *Quaestiones naturales* Adelard draws on his experiences at Antioch.

67. *Questiones naturales* (1988) pp. 92–93, 118–121, 126–127, 150–151, and 176–177. On creation *versus* eternity, it is conceivable that Adelard read Philoponus or the Fathers in Arabic translation, and/or drew from *e.g.* Ibn Sīnā. For the question of whether Adelard was using "the Arabs" as insurance against a possible charge of heresy, see Burnett (1988), pp. xxii–xxx.

68. *Questiones naturales* (1988:178–179).

NOTES   507

69. Burnett (1988), pp. xxiii–xxiv. The instances are: *De eodem et diverso* (1988), pp. 46–47 and note 47, and pp. 70–71 and note 60; and *Questiones naturales* (1988), pp. 124–127 and note 31. For the *Premnon physicon*, see also Chapters 7 and 9.

70. *Questiones naturales* (1988:90–91).

71. Ibid. (1988:224–225).

72. Haskins (1924/1960 page 31, item 14) refers to a manuscript of miscellaneous notes, possibly attributable to Adelard, which contains the sentence *Animam composuit Deus ex substantia et ex eodem et diverso, id est ex indivuitate et vegetatione, ex mutabilitate et immutabilitate, anima ergo tercium genus nature [naturæ?—MR] est ex mutabilitate et immutabilitate mixtum.*

73. According to Sarton (*Introduction* 2:349), Rabbi Berachya [Berakya, Berechiah] ben Natronai Krespia ha-Naqdan was probably Benoît le Puncteur (Benedictus Punctuator), a French Jew who flourished in England toward the end of the Twelfth century. He is better known for his *Mishle shu'alim (Fox fables).*

74. Gollancz, *Dodi venechdi* (1920).

75. Ibid. (1920), pp. iv–vi.

76. Berakya in Gollancz (1920:20). According to *midrashim*, Lilith was Adam's first wife; after a dispute with Adam, she pronounced the Name of God and vanished into the air. See Witcombe (2000), chapter 7.

77. Berakya in Gollancz (1920:27–28).

78. For an overview of Hildegard's eventful life, see Sweet (2003).

79. Berger (1999), pp. ix–xi. *Liber subtilitatum* was written between 1151 and 1158 CE. Moulinier (1995) dates its division to between 1179 and 1223. The single manuscript of *Causæ et curæ* seems to have been "the result of fragmentation and compilation" (Berger 1999, page xi).

80. Hildegard, *Subtilitatum diversarum naturarum creaturarum libri novem*, in Migne, *PL* 197 (1882), cols 1117–1352. English translation by Throop (1998).

81. Sweet (2003:194–250).

82. Ibid. (2003:262–310); quotation at page 273.

83. Berger (1999), pp. 58, 60, 78, and 141; Sweet (2003:301–307).

84. *On the elements*, translated by Dales (1976). The unique, albeit partial, manuscript on which the translation is based probably dates to the 1190s. It was originally part of a codex that also contained two translations—one anonymous, and that of Alfanus—of the *De natura hominis* by Nemesius (ibid. 1976:7–9).

85. Ibid. (1976), pp. 106, 128, 170, *et passim.*

86. Ibid. (1976:156–163).

87. Ibid. (1976:172–173).

88. Ibid. (1976:174–175).

89. Here Dales (1976:174) footnotes Nicolaus of Damascus, *De plantis* 1.3. See also *On plants* in Aristotle (Barnes) 2:1251--1271 (1984); and Chapter 3.

90. *On the elements* (1976:174–177).

91. Ibid. (1976:182–183).

92. Fulbert was Bishop of Chartres from 1006 to 1028 CE. Gerbert of Aurillac (later Pope Sylvester II) is credited with introducing Arabic numerals into Europe.

93. McInerny (1963); Giacone (1974).

94. Chenu (1968:4–18); Grant (2010:17–18).

95. For example, by Copleston (1963), volume 3(2), page 11.

96. Wetherbee (1973:31).

97. Haring (1955:189); Wetherbee (1973:11).

98. Hugh of St Victor, *Didascalicon* (1961), 1.10 at page 57.

99. Sheridan (1980:35–45).

508 NOTES

100. John of Salisbury (*Metalogicon* 4.20, in Migne, *PL* 199 (1855), column 928C) refers his readers to *Phrenonphysicon*. Hall (*Metalogicon*, 1991:158) offers *prenonphysicon* without comment. McGarry (*Metalogicon*, 1955:234n244) and Chenu (1968:34n77) identify this as Alfanus's *Premnon physicon*. Lejeune (*Metalogicon*, 2009:305) asks whether the reference might instead be to *Physica animæ, i.e.* Book 2 of *De natura corporis et animæ* of Guillaume, Abbot of St-Theodore [St-Thierry], for which, see Migne, *PL* 180 (1902), cols 707–726. At column 710 Guillaume says that stones exist, herbs and trees partake of vegetative life, beasts have sensitive life, and man shares rational life with the angels; but he denies that herbs, trees, brute animals, and the like have soul.

101. Chenu (1968:14); Wetherbee (1973:11).

102. Jourdain (1862:76).

103. *Didascalicon* (1961), 1.3 at pp. 49–50.

104. William of Conches, *Dragmaticon* as retracting opinions expressed in his *Philosophia*. See McInerny (1963).

105. Wetherbee (1973), page 21.

106. Dronke (1978:7).

107. Stone (1996).

108. Silvestris, *The Cosmographia* (Wetherbee, 1973); and *Cosmographia* (Dronke, 1978). The work is also known as *De mundi universitate libri duo sive megacosmus et microcosmus* (Barach & Wrobel, 1876).

109. In the delightful words of Singer (1928:220).

110. *Cosmographia*, Megacosmus lines 422–425: Wetherbee (1973:85), Dronke (1978:115).

111. *Cosmographia* (1978:12–15).

112. Blund, *Treatise on the soul* (Dunne & Hunt, 2012); Hunt (1970), pp. x–xi.

113. The Condemnation of 1210 proscribed the works of Aristotle from being read in the Faculty of Arts at the University of Paris; as Prefect of Theology, Blund would have been exempt, at least in principle.

114. Blund, *Tractatus de anima*, paragraph 359 (Dunne & Hunt 2012:196–197).

115. Ibid., paragraphs 36–45 (2012:20–27).

116. Ibid., paragraph 55 (2012:32–35).

117. Ibid., paragraph 310 (2012:168–169).

118. *De anima* 410b26–411a2, not discussed; 413a31–b9, discussed at paragraphs 146 and 311; 414a29–b6, paragraph 146; 425a9–10, no direct discussion but neighbouring sentences discussed at paragraphs 96, 98, 196, and 239; 434b18–24, at paragraph 146 again, and at 210 and 221. In paragraph 314 he rejects "a continuity of diverse things" on the authority of Aristotle (*Physica*).

119. Robert Grosseteste translated works by John of Damascus and Pseudo-Dionysius, the *Nichomachean ethics* and *De caelo* of Aristotle, and the commentary of Simplicius on the latter.

120. Grosseteste, *Hexaëmeron*, translated by Martin (1996).

121. Ibid. 4.28 (1996:152).

122. Ibid. 4.30 (1996:155–158).

123. Ibid. 6.4 (1996:190).

124. Ibid. 4.14 (1996:141). This is in his section on the Fourth Day, when plants appeared.

125. Ibid. 6.7 (1996:192).

126. Ibid. 6.10 (1996:194–195).

127. Ibid. 7.13 (1996:214).

128. Ibid. 7.13 (1996:211–212).

129. Ibid. 9.2 (1996:273–275).

130. Ibid. 7.5 ff (1996:206 ff.); see also 7.14 (1996:217).

NOTES   509

131. Ibid. 7.14 (1996:218); Augustine, *City of God* 21.4 (Healey, 1968) 2:322.

132. Thomas Cantimpratensis, *Liber de natura rerum* (Boese, 1973). Dating according to Thorndike (*Isis*, 1963).

133. Gold, electrum, silver, copper, tin, lead, and iron.

134. van Maerlant, *Naturen bloeme* (1878).

135. See Chapter 12.

136. According to Salimbene de Adam, *Cronica* (1966) 1:134. The *Cronica* was written in 1283 CE (Steele, 1893:5–6).

137. Steele (1893:1). Steele (1893:5) dates *De proprietatibus rerum* to 1248–1260, probably closer to the former date.

138. Bartholomæus Anglicus (1505), *Liber de proprietatibus rerum* 3.7–3.9 (fol. unnumb.). In English as *Batman uppon Bartholome* (1582), folios 14r–14v.

139. Alphabetically by their Latin name. In the vernacular editions I have examined, the entries retain their original (Latin) order.

140. *Liber de proprietatibus* (1505) 13.29; *Batman* (1582), folios 198v–201r.

141. Ibid. (1505) 16.33; *Batman* (1582), folio 258v.

142. Ibid. (1505) 18.95; *Batman* (1582) 18.97, folios 380r–380v.

143. Ibid. (1505) 18.90; *Batman* (1582) 18.92, folio 379r.

144. One of the four books, *Speculum morale*, was compiled by fellow Dominicans half a century after his death: Archer (1888); and Franklin-Brown (2012:98).

145. Vincent of Beauvais, *Speculum naturale* 5.79 (1591), folio 60v; *Bibliotheca mvndi* (1624), [vol. 1], column 354. Vincent returns to sponges at 17.91 (1591), folio 220v; (1624), [vol. 1], cols 1295–1296.

146. *Speculum naturale* 8.56 (1591), fols 88v–89r; (1624), [vol. 1], cols 522–523.

147. Ibid. 12.59 (1591), folio 155v; (1624), [vol. 1], column 915.

148. Ibid. 8.24 (1591), folio 86r; (1624), [vol. 1], column 505; also 28.60 (1591), folio 347v; (1624), [vol. 1], column 2033.

149. In 1941 Pius XII proclaimed Albert patron saint of natural scientists.

150. Führer (2018).

151. Dominicans were forbidden to travel by horseback or cart.

152. Albert, *Physica Libri 1–4* (Hossfeld, 1987), at 1.1.1.

153. We find this at many places in Albert's works, *e.g.* in *Liber I Physicorum* 1.4 (*Opera omnia* 1890) 3:9; and *Parva naturalia, De intellectu et intelligibili* 1.3 and 1.5 (ibid.) 9:480–481 and 483–485). See also McKeon (1929), pp. 331 and 338.

154. Albert, *De animalibvs* 2.1.1 (1651) 6:96; *Opera omnia* (Borgnet, 1890) 11:162; and *On animals* (Kitchell & Resnick, 1999), page 287. I have replaced the ugly word *genuses* (Kitchell & Resnick) with *genera*.

155. Albert, *Quaestiones super De animalibus* 7.2; see Resnick & Kitchell (2011:227). According to Grant (1974:681n1) the questions were disputed in 1258 (or "sometime after 1257": Kitchell & Resnick 1999:35), and were edited after 1260 by one of Albert's students, Conrad of Austria.

156. Albert, *De vegetabilibus* 1.2.5 (1867:77). English translation in Grant (1974:695–696).

157. *De vegetabilibus* 1.2.5 (1867:75–77); Grant (1974:695).

158. *On animals* 7.1.1 (1999:588).

159. *Quaestiones super De animalibus* 7.2 (2011:227–228).

160. Albert, *De principiis motus processivi* 1 (1909:8).

161. *De vegetabilibus* 1.1.4 (1867:15).

162. Albert's *De animalibus* is in 26 books, the first 19 of which comment on Aristotle's *Historia animalium* (Books 1–10), *De partibus animalium* (11–14) and *De generatione animalium* (15–19) in the translation of Michael Scot, which in turn were based on the *Kitāb*

510   NOTES

al-Ḥayawān attributed to Abū Yaḥyā ibn al-Baṭrīḳ (Chapter 7). Albert's books 20 and 21 are his own, while the final five books, a "dictionary" of animals, come largely from the *De natura rerum* of Albert's student and Dominican brother Thomas of Cantimpré: Scanlan (1897:12–13); Kitchell & Resnick (1999:xviii).

163. *Quaestiones super De animalibus* 7.2; see Resnick & Kitchell (2011:226–229).

164. Seth (1886:427–428); quotation from Ueberweg (1872) 1:436.

165. Dampier (1948:86); Dewan (1991); Grant (2004:184–187); McInerny & O'Callaghan (2018). For Aquinas on creation, see the translation by Baldner & Carroll of Book 2, Distinction 1, Question 1 of *Scriptum super libros Sententiarum Petri Lombardi* (1997), particularly pp. 22–29 and Appendix D.

166. Aquinas, *De ente et essentia*. Translated by Maurer as *On being and essence* (1968), especially pp. 10–11. See also Maurer (1962:190), and Baldner & Carroll (1997:39n75).

167. Aquinas, *In Aristotelis librum De anima commentarium* 2.3.5.285 (Foster & Humphries 1951); *Quæstiones disputatæ de anima* 11 (Rowan 1949); and *Quodlibetal questions* 1.4.1 (Edwards 1983).

168. *In Aristotelis librum De anima commentarium* 2.2.3.255, 2.2.3.260, and 2.3.5.287 (1951).

169. Ibid. 2.2.3.255 (1951). See also *Summa contra gentiles* 2.68.6 (Anderson 1955–1957).

170. *In Aristotelis librum De anima commentarium* 2.2.3.259–260 (1951).

171. Ibid. 1.5.12.189 (1951).

172. Aquinas based his commentary on the Latin translation of *De anima* by his contemporary and fellow-Dominican William of Moerbeke. Aristotle did not use ζωόφυτον in *De anima*, nor did Moerbeke use *zoophyta* in his translation. As Moerbeke also translated the commentary of Themistius on *De anima*, he was aware of the term.

173. *Summa contra gentiles* 4.11.1–5 (1957). See also Aquinas, *Summa theologica* 1.91.4.3.

174. *Quæstiones disputatæ de anima* 1.7 (Rowan 1949), citing *De plantis* 815b35.

175. *In Aristotelis librum De anima commentarium* 2.3.5.288 (1955), citing Aristotle's *Metaphysics* 1043b33–1044b14.

176. Seth (1888:426–427); Peters (1968:221–237); Grant (2004:176–177).

177. Wippel (1977).

178. For an introductory study, see Cullen (2006).

179. Bonaventure, De reductione artium ad theologiam. *Opera omnia* (1889) 5:317–325. In English as: *On the reduction of the arts to theology* (Hayes 2006).

180. Houser (1999); Noone & Houser (2014).

181. For example at *Senteniarum* 2.1.2.3.1 (*Opera omnia* 1889, 2:47–49), 2.2.2.1.2 (2:73–75), and 2.9.1.1.3 (2:242).

182. Ibid. 2.15.2.2, conclusio 4 (2:385). Similar constructions are at *Itinerarium* 2.2 (*Opera omnia* 1889, 5:298–299); and *Breviloquii* 2.4 (ibid. 5:221).

183. Bonaventure, *Itinerarium mentis in Deum* 2.2 (1259), in the edition of Boehner (1956) at page 51; *Opera omnia* (1889) 5:300.

184. *Itinerarium* 1.2 (1259), Boehner (1956:39); *Opera omnia* (1889) 5:297.

185. Bonaventure, *Collationes in Hexaëmeron. Opera omnia* (1889) 5:327–454.

186. Étienne Tempier, Chancellor of the Sorbonne (1263–1268) and Bishop of Paris (1268–1279).

187. Grant (1979). Aquinas was canonized in 1323, and the articles specifically against him were declared null and void (by the then Bishop of Paris) in 1325: Grant (1974:47).

188. A range of opinion exists on the broader impact of the Condemnation of 1277, including (a) that by liberating enquiry from ancient Hellenic philosophy it marked the birth of modern science (Duhem, *Études* 2:412, 1909); (b) that by prohibiting any limitation on the power of God, it created intellectual space for "unphilosophical" ideas such as multiple worlds, or an infinite void (Grant, 1969; Dales, 1980); or (c) the condemnation was narrowly cast and largely ineffectual (Gooch, 2006).

NOTES 511

189. Glendinning (2015), citing the Dartmouth Dante Project. See also Seth (1888:430); CE Norton (1902), page viii; and Freeman (1921:76).

190. *Divine Comedy*, Canto 7 (1902) 3:55.

191. Wetherbee & Aleksander (2018).

192. *Divine Comedy*, Canto 10 (1902) 3:80–83; and Canto 12 (3:100–101). See also Gilson (1948:256 ff.).

193. Matthews & Briffa (2005); Mann et al. (2009).

194. Walshe (2009:4). The Dominicans were realists and considered intellect to be the highest power of soul, whereas the Franciscans, who included "extreme nominalists", regarded the will as the highest.

195. Blakney (1941), pp. xiv–xv.

196. Workman (1926); Robson (1961).

197. Quoting Maimonides, *Guide to the perplexed*. See the Colledge & McGinn edition of Eckhart (1981), pp. 28–29 and 93–95; quotation from page 95 (see note 83 at page 321).

198. Meister Eckhart (Kohlhammer, 1964 and 2015).

199. For this example, see his multiple commentaries on *Genesis* 1:11: *Prologi in Opus tripartitum*, *Expositio libri Genesis* (1964:95–97); *Liber parabolarum genesis* (1964:252–253; *Expositio libri Genesis, secundum recensionem* (2015:135–137 and 211); and *Liber parabolarum Genesis . . . recensio altera* (2015:410 and 431).

200. Eckhart (1964:215–215); and (2015:325–327). In another passage he expands his scope to "animals, grain-plants and metals" (2015:450).

201. *Commentary on John* 83 and 89 (Colledge & McGinn 1981, pp. 153, 155–156), and *Parables of Genesis* 151 (page 115). Eckhart considers the planetary spheres to be rational: *Parables of Genesis* 142 (page 111). At *Parables of Genesis* 141 (page 111) Eckhart cites the *Book of causes*: "the noble soul has three activities, for among its activities there is an animal one, an intellectual one and a divine one".

202. Eckhart, *Liber parabolarum Genesis . . . recensio altera* (2015:351); translation by MAR. Colledge & McGinn (1981:322n107) attribute this aphorism to the *Sentences from Aristotle* falsely attributed to Bede.

203. Eckhart, *Sermon 83* (1981:206–208); see also Fragment 42 in Blakney (1941:248).

204. *Parables of Genesis* 156 (1981:118), attributed by Colledge & McGinn (1981:326n220) to Aquinas, *Summa theologica* 1.91.3 response to Objection 1.

205. Seth (1888:418); Sarton, *Introduction* 3(1):552 (1946); Dampier (1948:94).

206. Murdoch (1982:184–189); and Stump (1982:207–230).

207. For example, at *Summæ logicæ* 3-3.18, in *Opera philosophica* 1 (1974), pp. 651–671 (1974); and *Summula philosophiae naturalis*, Praeambula, ibid. 6 (1984), page 154.

208. Buridan, *Summulae de dialectica* (2001).

209. *Quaestiones super libris quattuor de Caelo et Mundo* (Moody, 1942): stones, plants, and animals at pp. 46 and 271; man, animals, and plants at page 156.

210. Aristotle, *De anima* 412a27.

211. Oresme became Dean of the Cathedral of Rouen, and later Bishop of Lisieux.

212. Zupko (2018).

213. Taschow (2003).

214. Oresme, *De causis mirabilium* 3.519–524 (Hansen, 1985), pp. 232–233.

215. *De causis mirabilium* 3.559–575 and 592–603 (1985:234–239).

216. The Hippocratic corpus, Democritus, and Aristotle, Chapter 3; John Philoponus, Chapter 5. T Browne, *Religio medici* (1652:203); in English (1894:65–66).

217. Seth (1888:431) points out that cases could be made for Gabriel Biel (1420/1425–1495) or Francisco Suárez (1548–1617). Some historians recognize a second scholastic period in the Sixteenth and Seventeenth centuries which was particularly influenced by Jesuit thinkers such as Suárez.

512 NOTES

218. Nicholas of Cusa, *De coniecturis*. In: *Opera omnia*, volume 3 (1972). Available as *Nicholas of Cusa: metaphysical speculations*, volume 2 (Hopkins, 2000).

219. Dampier (1948:95).

220. Nicholas of Cusa, *De visione Dei* 24.106–107. In: *Dialectical mysticism* (1988:731–732). See also *Of learned ignorance* 2.5 (Heron, 1954:85; Hopkins, 1985:72).

221. As no instance can be at the maximum for its genus, it must have "contracted" back from that limit.

222. Nicholas of Cusa, *De docta ignorantia* 3.1 (Hopkins, 1985:112–114). Although Nicholas does not mention plants, he clearly considers these "oysters, sea mussels, and other things" (*in ostraeis et conchis marinis et aliis exempla reperiuntur*) to stand between plants and animals, as man stands in relation to animals and spirits (3.187, page 114).

223. The Council of Ferrara relocated to Florence in January 1439.

224. Basilios Bessarion was invested as Cardinal in 1439, and in 1463 became Latin Patriarch of Constantinople.

225. Georgius Gemistus took the name Plethon (Anglicized as Pletho) to show his respect for Plato.

226. Gemistus Pletho, *De Platonicæ et Aristotelicæ philosophiæ differentia* [1439], in Migne, *PG* 160 (1866), cols 889–932.

227. Gemistus Pletho, ibid. (1866), cols 893–894; *zoophyton* at 893D. See also Lagarde (1973:323), line 31.

228. Georgius Chariander: see Gemistus Pletho (1574), folio 8v. Nor is *zoophyta* used in the 1540 paraphrase (in the form of a dialogue) by Bernardino Donato (reprinted 1541); Lagarde (1973) finds it replete with errors and omissions, and "often unintelligible".

229. John Eugenikos of Trebizond. An opponent of reunification, he left the Council in 1438 to return home, only to be shipwrecked. He recounted the disaster, and gave thanks for his survival, in *Logos eucharisterios*.

230. Eugenikos of Trebizond (1832:372), at lines 27–28.

231. Cosimo de' Medici was de facto head of the Florentine government.

232. George of Trebizond became secretary to Pope Nicholas V.

233. Monfasani (1976); quotation from page 37.

234. Ibid. (1976), pp. 142–143, 162, and 167–169. The translation was probably commissioned in late 1458.

235. We meet Theodorus Gaza again in Chapter 11. His former pupil Pietro Barbo (1417–1471) became Pope Paul II in 1464 CE.

236. His library, donated in 1468 to the Republic of Venice, became the foundation of the Biblioteca Marciana.

237. Bessarion, *In calumniatorem Platonis* (1469).

238. Spade & Panaccio (2016).

239. Peters (1968), pp. 234–235 and note 38 at page 235.

Chapter 9

1. Marvell (1681:96).

2. Ficino studied arts and medicine, taught briefly at the University of Firenze, and enjoyed the patronage of Cosimo de'Medici. At age forty he was ordained as a priest. Pico was of a noble although fractious family. Agrippa held a succession of military and diplomatic positions, lectured at various universities, and for a time practised as a physician.

3. D Wright (1913); Burkert (1985:7); Mylonas (2015).

4. Parker (1995).

NOTES   513

5. Herodotus, *The Persian Wars* 2.123 (1942).
6. Iamblichus, *De mysteriis* (1570:5–178); *Iamblichus on the mysteries* (1895:17–365).
7. The *Oracles* have come down to us only in fragments. Here I use the reconstruction in Le Clerc [Clericus], *Philosophiæ orientalis*, Book 4. *In quo Chaldaïca Zoroastris, ejúsque discipulorum Oracula*. In: *Opera philosophica*, volume 2 (1710), pp. 321–377.
8. Julian the Theurgist (to distinguish him from his father, Julian the Chaldæan), *floruit* during the reign of Marcus Aurelius, *i.e.* 161–180 CE.
9. Dillon (1996), pp. 363 and 392–396; quotation from page 363.
10. Le Clerc (1710) 2:319.
11. Ibid. (1710) 2:340–343; animals at line 234, plants at line 282.
12. Hall (1928:94).
13. Scarborough (1988); Copenhaver (1992), pp. xiii–lix.
14. See also Chapter 6.
15. Clement of Alexandria, *Stromata* 6.4, in Roberts & Donaldson (1869) 12:323–324.
16. Hall (1928:97).
17. Nasr (1992:31), and (1978), pp. 13n28, 34–35, and 132.
18. Shumaker (1988:294).
19. Steele (1920), pp. xii–xiii. For al-Makīn, see Budge (1896:355).
20. Copenhaver (1992), page xiv.
21. Yates (1964:2–3).
22. Iamblichus, *On the mysteries* 1.1 (Gale, 1678), page 1 (as Mercurius); and (Taylor, 1895), page 17 (as Hermes).
23. Or if Seleucus and Manetho (as cited by Iamblichus) are to be believed, tens of thousands: *On the mysteries* 8.1 (1678:157) and (1895:300).
24. Yates (1964:2–3). The Eleventh-century authority is Michael Psellus, whom we met in Chapter 6.
25. Yates (1964:12–14). A fifteenth book from the Eleventh-century Byzantine *corpus*, missing from the manuscript available to Ficino, was translated by Ludovico Lazzarelli in the 1490s. For the remaining books see Yates (1964) and Copenhaver (1992).
26. They were first bound together in the 1505 Paris edition of *Pimander* and other Hermetic works in the translation of Ficino, edited by Jacques Lefèvre d'Étaples.
27. *Asclepius* 24, Copenhaver (1992:81). See also Copenhaver (1992), pp. xl and 238, and Yates (1964), pp. 9 and 170–172. For Augustine see *City of God* 8.23–26 (Dent 1968) 1:235–252.
28. The citation is in Lactantius, *Institutiones Divinæ* 4.6, Migne *PL* (1844) 6, cols 461–463; *The divine institutions*, in Roberts & Donaldson (1886) 7:105.
29. *Asclepius* 35: Copenhaver (1992:89).
30. Ibid. 6: Copenhaver (1992:70).
31. *Corpus Hermeticum* 1.15: Copenhaver (1992:3).
32. *Corpus Hermeticum* 9.1 and 10.23–24, and *Asclepius* 6, 32, 37, and 41: Copenhaver (1992) pp. 27, 35–36, 70, 87, 89, and 92 respectively.
33. *Corpus Hermeticum* 9.5 (quotation), 10.12, and 11.4: Copenhaver (1992) pp. 28, 33, and 38.
34. Ibid. 6.2: Copenhaver (1992:21–22).
35. [Hermes Trismegistus] *Tabvla Smaragdina* (1541), with commentary by Hortulanus (written *ca* 1350 CE), pp. 364–373 in some copies; Holmyard (1923); Colavito (1989), Appendix B, pp. 83–86; Agrippa (1997), Appendix I (pp. 709–711).
36. J Marshall (ed.), *Emerald tablet of Hermes* (undated).
37. Jābir in turn attributed it to Apollonius of Tyana (Chapter 7). For attributions, see Ullmann (1972), pp. 74 ff, 170–172, and 378–379. There is an ongoing debate concerning whether Jābir was a single person, or was alternatively the collective identity of a group, perhaps around the Ikhwān al-Ṣafā. In any event, some works attributed to Jābir (or in the Latinised form of his name, Geber) are medieval forgeries.

514 NOTES

38. Jennings (1884), quotation at pp. viii–ix. In a different version of the story, Apollonius of Tyana took the tablet from the hands of the dead Hermes.

39. Ficino, Argumentium Marsilij Ficini Florentini (1505), folio 2r; also Ficino (1576), 2:1836. See also Copenhaver (1992), page xlviii.

40. Scully (2003:322).

41. As Brahma, Vishnu, and Shiva, the three deities of the Trimūrti: Jennings (1884), page xii.

42. "J.F." (1650) [published 1649], at fols A3r–A4v. For this identification "J.F." cites "*Geber Paracel[sus] Henricus Nollius in theoria Philosophiæ Hermeticæ tractatu priimo*"; see below.

43. Ficino sometimes traces the supposed lineage further, to Zoroaster: Ficino, *Theologiæ Platonicæ*, in *Opera* (1576) 1:386 and 2:871. See also Kristeller (1943:25–26); Yates (1964:14); and Shumaker (1988:294).

44. Schmitt (1966).

45. Yates (1964), pp. 83 (quotation), 169, and 172–173. Evidence to the contrary, *e.g.* the idolatry that drew Augustine's scorn, could be blamed on later additions to the text supposedly made by the evil magician Apuleius.

46. *Corpus Hermeticum* 1.6 (Copenhaver 1992:2; Yates 1964:23) and, in the Greek original, *Asclepius* 8 (Yates 1964:7 and note 4). Elsewhere the cosmos is called the son of God (*Corpus Hermeticum* 9.8 and 10.14: Copenhaver 1992:29 and 33).

47. Ficino, *Opera* (1576) 2:1232.

48. *Harmony:* three is a harmonious number (Pythagoras). *Completeness:* God is complete in three Persons, and in knowledge, presence, and power; for some Gnostics, man is composed of *chous, psyche,* and *pneuma. Resurrection:* dry land, plants, and trees appeared on the Third Day of Creation; Jonah was in the fish for three days; Jesus rose from the dead on the third day.

49. Wolfson (1970:24–72).

50. Ibid. (1970:495–574); Wallis & Bregman (1992); Dillon (1996:384–389); Brakke (2010).

51. Wolfson (1970:497–503).

52. Nasr (1992:64–65).

53. Puig Montada (1992:117–118).

54. Wolfson (1970:501–502); *1 Corinthians* 2:10; *Phaedrus* 247C–D.

55. Brons (undated).

56. For example, Gnostic texts from Nag Hammadi, the *Gospel according to Mary Magdalene*, and texts attributed to Valentinus and members of his school, available online via The Gnostic Society Library (www.gnosis.org/library.html); *Pistis Sophia* (1921).

57. Conger (1922:2).

58. Aristotle, *Physica* 252b26.

59. Plato, *Timaeus* 29d ff; *Philebus* 28d–30d.

60. Nasr (1992:223–224).

61. Lyons & Petrucelli (1978:310). For the connection between gemstones and heavenly bodies, see Saif (2011), particularly at page 613; and (2015:79).

62. Plato, *Timaeus* 41e. Indeed, the stars themselves were living creatures (*Timaeus* 38e).

63. "We can imagine with what feelings gnostic men must have looked up to the starry sky. How evil its brilliance must have looked to them, how alarming its vastness and the rigid immutability of its courses, how cruel its muteness!"—Jonas (1963:261).

64. For Nieremberg, see Hendrickson (2015).

65. Presumably Theodorus Gaza.

66. Nieremberg (1635), Book 3, chapter 8, pp. 35–36, translated by Professor Denis Brosnan and MAR. Explanations in square brackets are mine; parentheses are in the original. Spacing into paragraphs is mine (single paragraph in the original). In the *Genesis* account, plants appeared on the third day; the sun, moon, and stars on the fourth; and animals on the fifth and sixth.

67. Eliade (1978:22).

NOTES 515

68. Ibid. (1978:19–26); Cobb & Goldwhite (1995:6–10); Braudel (2001:60–61).

69. Nasr (1992:243); Eliade (1978:53–78).

70. Pliny, *Naturalis historia* 34.49. See Eliade (1978), especially pp. 41–47 and 50–51.

71. Eliade (1978:51–52).

72. Here I focus mainly on the Islamic and European alchemical traditions. For Chinese alchemy, see Dubs (1947); Needham & Lu, *SCC* 5(2):1–304 (1974); and Eliade (1978:109–126). For Indian alchemy, see Stapleton (1962); P Ray (1967); and Eliade (1978:127–141). For the Arabic and Islamic traditions, see Nasr (1992:242–292) and Ullmann (1972:145–270). For the Jewish tradition, see Patai (1994).

73. Aristotle, *De generatione et corruptione* 329a1 ff.; quotations at 329a26 and 329b26–30. See also Lloyd (1964).

74. Aristotle, *Meteorologica* 378a12–378b4. See also Eichholz (1949) and Norris (2006).

75. Baeumker (1916:20); English translation by Longeway (1992) at al-Fārābī. See also Bakar (1998) page 30, and note 122 at page 41.

76. Holmyard (1931:56–58). For mercury and sulphur in later Medieval alchemy, see Newman (2014).

77. Norris (2006), particularly pp. 46–47.

78. Dating the beginning of chemistry to Robert Boyle's *The sceptical chymist* (1661).

79. Eliade (1978:149–152), and below.

80. Colavito (1989:64 and 67).

81. Alchemy Lab (undated).

82. C Burnett (2011).

83. Ullmann (1972:171–173); Haq (1994:29–30); Asl (2016); Colavito (2018). An alternative history has been suggested by Zimmermann (1981). See also Nau (1907).

84. de Sacy (1798–1799). The quotation (from 1798–1799:150) is in the English translation (from French) by Colavito (2018).

85. De Sacy (1799:152–153); Colavito (2018).

86. Ullmann (1972:74); de Sacy (1798–1799:150–155) translates this as *règne* (hence Colavito as *Kingdom*), while Weisser (1980) renders it as *Reich*. See Chapter 7, note [171]. Weisser uses *Übergangsstufen* for the transitional stages (1980:120).

87. Weisser (1979:371 in Arabic section; and 1980:120–121 and 211–212).

88. Hugo de Santalla, *Liber de secretis naturæ*. According to Colavito (2018), this translation "contained significant mis-readings and additions and redactions". Travaglia (2001) counterposes extended portions with an Italian translation of *Kitāb Sirr al-khaliqa*. See also Nau (1907) and Hudry (1997–1999).

89. Travaglia (2001:191 and 218).

90. Nau (1907), *Liber de secretis* 18v at page 103; alternatively, as *animalia, nascentia atque minerias* at 19v (Travaglia 2001:218), where we also find *mundi maioris*.

91. Haq (1994:228).

92. R Bacon, *Opera hactenus inedita*, volume 5 (1920), pp. 176–179.

93. SJ Williams (2000:79–94).

94. Steele (1894), pp. ix–x.

95. Steele (1920), pp. vii–lxiii, particularly pp. xiii–xvi.

96. Ibid. (1920), pp. xviii–xxiii; Williams (1991:173–174). This 1259 CE "official" recension was the root of perhaps nine separate Latin manuscript families distinguished by inclusion or exclusion of sections of text, and details of their subdivision into chapters. From (some of) these translations were made, often with significant further modification (*e.g.* into verse), into at least ten European languages, while a Hebrew translation was made directly from the Western Islamic form. French versions were also translated into English, and so on. As a consequence, versions of *Secretum* (or *Secreta*) *secretorum* may be scarcely recognizable as such. For some of these, see Steele (1894, 1898); and Manzalaoui (1977).

516 NOTES

97. Steele (1920), pp. 114, 157–158, and 161 (Latin), pp. 254, 256, and 261 (English).

98. Ibid. (1920), pp. 114–115 (Latin) and 261 (English).

99. Ashmole (1652:397–403).

100. For a translation of the discourses on substances and instruments, and a summary of and excerpts from that on operations, see Stapleton, Azo & Ḥusain (1927).

101. Stapleton *et al.* (1927:321–324); see also pp. 370, 394, and 396.

102. For a few among dispiritingly many such claims, see van Spronsen (1969:25); Singer (1996:146); Strathern (2000:46); Darian (2003:30); Doak (2010:112); and Moncrief (undated).

103. Frazer (1922:48–52). For other distinctions between magic and religion, see Hubert & Mauss (1902–1903), page 56.

104. Coles (1657:3).

105. Paracelsus, *Of the nature of things*, Book 9 (1650:100–145) and (1674:259–301). See also Waite (1910) 2:304–305, and Foucault (1970:25–28).

106. Böhme, *Signatura rerum* [1622], in *Sämmtliche Werke* (Schiebler) 4:269–462 (1842); translated as *The signature of all things* (1651). Also available in his *Works* (1781), 4:9–140. See especially Chapter 9 (1781) 4:59–69; quotations at page 59.

107. In many cases these were presumably occult names for bioactive plants, rather than actual body parts of bats or newts. Bat and newt are examples from *Macbeth*.

108. Institoris (Kramer), *Malleus maleficarum* (1487; I examined 1494 and 1507 editions).

109. Kieckhefer (2000:37).

110. For example, Godwin (1834).

111. Jābir ibn Ḥayyān (1935:481–482) as translated by Nasr (1992:264–265). See also Nasr (1992) pp. 163, 165, 166, 175, and 186. Again, these "kingdoms" are better expressed as "progeny" or "children".

112. Jābir, *Kitāb al-Aḥjār* as translated by Haq (1994:163–202).

113. At Haq (1994:252).

114. Ibid. (1994:208).

115. Stapleton *et al.* (1927), pp. 321 and 326.

116. Goodman (1995); Nasr (1992:268–269).

117. Nasr (1978:148–149).

118. Partington (1937); Nasr (1978:247–248); Needham & Lu, *SCC* 5(2):30–31 (1974); Ibn Sīnā, *Kitāb al-Šhifa*, in Grant (1974:569–573); and Ibn Sīnā, *De congelatione* (1927:33–41).

119. Ikhwān al-Ṣafā, *Epistle* 19 (Baffioni 2013:223–283), particularly pp. 228, 238–239, 248–250, 261, and 267.

120. Ikhwān al-Ṣafā, *Epistle* 22 (Goodman & McGregor 2009:65).

121. Nasr (1978:91–92).

122. Abu' l-Qāsim al-ʿIraqi, see Holmyard (1926).

123. Al-ʿIraqi as translated by Holmyard (1923:23–24).

124. Ullmann (1972:192–193); al-Hassan (2004). Robert of Chester's Latin translation of *Liber de compositione alchemiæ* was first brought into English in the Seventeenth century (see Holmyard 1925). More-recent translations have been made by Stavenhagen (1974) and McLean (2002). Khālid ibn Yazīd, grandson of Muʾawiyya, founder of the Umayyad dynasty, is supposed to have introduced alchemy to the Muslim world. According to legend, Maryānus (Morienus) was Khālid's teacher.

125. Ruska (1931); Plessner (1954). No Arabic (or Hebrew or Greek) manuscript is known. The Latin version exists in at least three main variants.

126. Waite (1896:18–22).

127. Albertus Magnus, *Liber mineralium* (1890), especially Book 4; in English as *Book of minerals* (1967:202–236); and Newman (2014). See also [Pseudo-Albertus Magnus], *Libellus de alchimia* (1958), excerpted in Grant (1974:586–603), for sulphur and mercury.

NOTES   517

128. See Patai (1994:141–143) and Suler (2007). Artephius wrote that "metals grow like plants, but whereas the plants are composed of water and dust, the metals are composed of sulphur and mercury" (Suler 2007:600).

129. Suler (2007).

130. Partington (1937:13–16). Pope John Paul XXII condemned and banned alchemy in 1317 (Sloane 1907).

131. Grant (1974:573–586), particularly pp. 581 and 583.

132. In his *De secretis naturae sive quinta essentia* incorrectly attributed to another Raymund (Llull): Patai (1994:175–195), at page 194.

133. Bornstein (1979) refers to 500 Latin manuscripts alone.

134. Kristeller (1944); Celenza (2017).

135. Ficino, in Kristeller(1944:288).

136. Ibid. (1944:316). Translation by MAR.

137. Porphyry, Iamblichus, Synesius, Proclus, pseudo-Dionysius, and Psellus. For Bishop Synesius, I consulted his *Œuvres* (1878).

138. "[Things called] sponges, marine snails [*Murex*, used for production of Tyrian purple], conches and oysters, which have no interior sense, and externally only two senses" (Ficino, *Opera* 1576, 3:1243). Elsewhere on page 1243 he refers to *ostreœ marinœ*. See also his *Praedicationes* (1576) 2:493.

139. Ficino, *Platonic theology*. English translation by Allen with Warden, Latin text edited by Hankins with Bowen (2001–2006) 3:114–119. See also 3:150–157.

140. Pico, *Asclepius* 1.6. In English translation of Wallis (1965:3).

141. Pico, *Heptaplus*, in *Opera omnia* (1557) 1:7. English translation of Carmichael (1965:78).

142. Pico's Second Proem was closely paraphrased, without attribution, by "Pierre de la Primaudaye" in *L'Academie Françoise* (1590); *Zoophyte(s)* occurs twice, at folio 26v and 27r. *L'Academie* was translated into English in 1577–1601 and 1618; in the 1618 edition *Zoophyta* occurs twice at page 671. It was also taken into Italian and German. For possible authorship by Francis Bacon, see Dawkins (2017).

143. Pico, *On the dignity of man*, and *Heptaplus* (1965), pp. 4–5 and 79 respectively. See also Kristeller (1953:119–123).

144. I do not know of it in the Kabbalistic literature either, although this is relatively little-studied and for the most part available, if at all, only in Hebrew. Here I treat *zoophyton* as Latin. Latin words of Greek origin ending in -on are declined as second-declension neuter, consistent with Pico's forms *Zoophytis* and *Zoophytu[m]*.

145. Kristeller (1965:46).

146. Kibre (1936), pp. 15, 29–30, and 36, and inventory items 575 (Themistius, *De anima*), 455 and 1020 (Simplicius, *De caelo*), 439, 446, 457, 464, 499, 514 and 745 (Simplicius, *Physica*), 447 (Simplicius, *De anima*), 449 and 1131 (Philoponus, *De anima*), 1589 (Philoponus, *Analytica priora*), 658 (Damascene, *Dialectica*), and 774 (Barbaro editions of Themistius). Pico corresponded with Barbaro (1936:13, 51, 53, 61, and 101).

147. Ficino, *De vita libri tres (De triplici vita)* (1489); in translation as *Three books of life* (1998).

148. Pico, *Disputationum, in astrologiam*, in *Opera* (1557), 1:411–732. See also Voss (2006), particularly pp. 25–48 and 215–216.

149. Pico, *Apologia*, in *Opera* (1557) 1:170; and *Conclusiones* [1486], at *Conclusiones magicæ* (1557: 104–106), especially Conclusion 3.

150. Pico, *Apologia*, in *Opera* (1557:172) as translated by Farmer (1998:127).

151. Pico, *Conclusiones* [1486]. Conclusiones magice numero XXVI secundum opinionem propriam (1557:104–106), Conclusion 9 (page 105) per Farmer (1998:497). The investigative commission established by Pope Innocent VIII deemed this thesis heretical (Farmer 1998:126).

518 NOTES

152. For Gianfrancesco's work as literary executor for his uncle, see Farmer (1998:152–176).

153. GF Pico, *De studio divinæ et humanæ philosophiæ*, in *Opera* (1573:37–38).

154. GF Pico, *Examen vanitatis doctrinæ gentium et veritatis discipline Christianæ*, in *Opera* (1573:1134). Translation by MAR.

155. The index (*per* the Olms reprint edition, Hildesheim 1969) erroneously indicates a third instance of *zoophyta* at page 853. At (1573:855) GF Pico mentions the *zophita* that appear in furnaces (see Chapter 3), as discussed by "many philosophers".

156. Agrippa, *De incertitvdine* (1531), cap. 7, 52 (n.p.); in translation as *The vanity of arts and sciences* (1676), pp. 40–43 and 133–142. The term *Aristotelian epistemology* is from Compagne (2017).

157. Agrippa, Liber de triplici ratione cognoscendi Deum. In: *Opervm pars posterior* (1600:480–501).

158. The 1533 edition of *De occulta philosophia* was a thorough revision of his 1510 first draft, which I have not examined. Here I refer to the 1533 edition in the Jung collection (archive. org) and its translation as *Three books of occult philosophy* (1997). In Agrippa's introduction (1997:lilii) he claims that his *De incertitudine* constituted a retraction "for the most part" of the 1510 version.

159. Marlowe (1604).

160. Goethe, *Faust. Eine Tragödie*. In: *Goethe's Werke*, Band 8 (1808:1–234) ("Part 1"); and *Faust. Der Tragödie zweyter Theil in fünf Acten*. In: *Vollständige Ausgabe letzter Hand* [VA], Band 41 (1832–1833).

161. Agrippa, *Three books* 1.24 (1997:81).

162. Ibid. 1.28 (1997:91).

163. Ibid. 1.35 (1997:106), presumably referring to the shell of oysters, the great density of ebony wood, and fossils (see also 1997:107).

164. Ibid. 1.37 (1997:110–111). I have reworded the first sentence of the second paragraph to follow more precisely the 1533 original (at pp. 43–44).

165. Agrippa 2.8 (1533:112); and (1997:263), as "plant-animal". The terms also appear in the corresponding text.

166. Quotation from *Das Buch Paragranum* [1529–1530], in *Sämtliche Werke* (1922–1933) Part 1, volume 8, pp. 182–184, as translated by Goodrick-Clarke (1999:75).

167. Paracelsus, ibid. 1.8:185. As can be the case with Paracelsus, he says the opposite in the spurious *Cœlum philosophorum*, in *Sämtliche Werke* 1.14:418–419; in English at *Hermetical and alchemical writings* (Waite/Laurence 1910) 1:16.

168. Goodrick-Clarke (1999:20).

169. Paracelsus, *Opus Paramirum* [1530–1531], in *Sämtliche Werke* 1.9:45–47; Goodrick-Clarke (1999:78). Depending on context, these terms may refer to the corresponding principles rather than to the elements or compounds.

170. Paracelsus, *De natura rerum* [1537], in *Sämtliche Werke* 1.11:318 ff.; Dorn (1584), especially pp. 153–168; *Hermetical and alchemical writings* (1910) 1:89–95; Goodrick-Clarke (1999:176–179). Gerhard Dorn was a follower of Paracelsus.

171. Paracelsus, *De natura rerum*, in *Sämtliche Werke* 1.11:360–373 ff.; *Of the nature of things* (1650:79–99), with quotation at page 126; *Hermetical and alchemical writings* (1910) 1:161–170. See also Dorn, *Avrora philosophorvm* (1584:32–48); *Aurora of the philosophers*, in *Hermetical and alchemical writings* (1910) 1:53–59.

172. Dorn (1584:81–82); and in Waite (1910) 1:64.

173. Dorn (1584:44); and in Waite (1910) 1:56.

174. Paracelsus, *Das Buch der Mineralibus*, in *Sämtliche Werke* 1.3:40; *A book about minerals*, in *Hermetical and alchemical writings* (1910) 1:243; *De natura rerum*, in *Sämtliche Werke* 1.11:331; *Of the nature of things* (1650:32), and (1910), 1:136; *A little book concerning the*

NOTES   519

Quintessence, in (1910) 2:138. See also *Cœlum philosophorum*, in *Sämtliche Werke* 1.14:419; and (1910) 1:17.

175. Paracelsus, *Philosophia ad Athenienses* 14 and 17, in *Sämtliche Werke* 1.13:395 and 397; *Philosophy addressed to the Athenians, Hermetical and alchemical writings* (1910) 2:255 and 257.

176. *Das Buch der Mineralibus*, in *Sämtliche Werke* 1.3:40 (*Pfifferling*, a chanterelle); *A book about minerals*, in *Hermetical and alchemical writings* (1910) 1:256. Also, *Philosophia de generationibus et fructibus quatuor elementorum*, in *Sämtliche Werke* 1.13:57 (*schwemmen*, *i.e.* wood-rot fungi); *The philosophy of the generation of the elements*, (1910) 1:227.

177. Paracelsus, *De gradibus et compositionibus receptorum et naturalium*, in *Sämtliche Werke* 1.4:26; *Concerning the alchemical degrees and compositions of recipes and of natural things*, (1910) 2:187.

178. See M Müller (1960:277).

179. von Meyer (1906:78); quote from Thorpe (1930:48).

180. More, *Enthusiasmus triumphatus* (1662:33).

181. For Flamel's reputed alchemical success and public charity, see Nasr (1992:285–286).

182. Norton (1477), reproduced (1928); also in *Musæum Hermeticum* (1678:432–532), as *Crede mihi sive ordinale*.

183. *Rosarium philosophorum* (1550); facsimile edition edited by Telle (1992).

184. Probably written by Johann Thölde: see Telle (2009). Translated into Latin by Michael Maier as part of *Tripvs avrevs* (1618) and included in *Musæum Hermeticum* (1625 and 1678). "Basil Valentine" is probably a pseudonym, perhaps used by more than one author.

185. Croll (1608).

186. *Novum lumen chymicum* (1614) and *Tractatvs de svlphvre* (1616) attributed to Michael Sendivogius (Michał Sędziwój), both also in *Musæum Hermeticum* (1678). Zink (www. livresanciens.com) and others have claimed that *Novum lumen* was actually written by Alexander Sethon (Seton), with its first edition in Prague (1604) and another at Paris (1608). For an entertaining tale of Sethon and Sędziwój see Waite (1888:171–181); for a more-rigorous examination see Prinke (2016).

187. Andreae, *Chymische Hochzeit* [1616], translated by Godwin as *The chemical wedding of Christian Rosenkreutz* (1991). Rosenkreutz was the probably mythical founder of the Rosicrucian movement.

188. Chrysogonus Polydorus (ed.), *De alchemia* (1541). "Chrysogonus Polydorus" is probably a pseudonym.

189. *De alchimia opuscula complura veterum philosophorum* (1550).

190. *Theatrum chemicum* (1602–1661) in six volumes.

191. *Musæum Hermeticum* (1625). A German translation was published the same year, as *Dyas chymica tripartita*.

192. *Musæum Hermeticum* (1678); translated by Waite as *The Hermetic museum restored and enlarged* (1893).

193. Ashmole (1652).

194. Manget [Mangetus], *Bibliotheca chemica curiosa* (1702).

195. For example, in Norton's *Ordinall of alchemy* (1477), in Maier, *Tripvs tractatus* (1618:160) and Ashmole (1652:82); and perhaps in the *Rosarium philosophorum* (1550), at page 117 (Latin) and page 101 (German) in the Telle (1992) edition.

196. Rosarium (1550), at page 43 (Latin) and page 43 (German) in Telle (1992).

197. Maier, *Atalanta fugens* (1617:138–139). I consulted the second edition (1618). Translation by MAR, making reference to an Old English translation at British Library MS Sloane 3645 transcribed by Hereward Tilton at alchemywebsite.com/atalanta.html (accessed 16 January 2019). Maier devotes Epigram 29, with the corresponding woodblock print and fugue, to the salamander, which like the Philosopher's stone lives in the fire.

520 NOTES

198. Mennens (1622), pp. 333 and 463 respectively.

199. Johnson [printed Iohnson], Lexicon chymicum. Section 1.3.1 of Manget (1702), pp. 266 and 287 respectively.

200. The authorial name attached, Divi Leschi Genus Amo, is an anagram of Michael Sendivogius, the Latin name of Michał Sędziwój. A second edition appeared in Frankfurt in 1611. I examined the (undated) version in *Theatrvm chymicvm*, volume 4 (1613), text pp. 471–502.

201. Ibid. (1613) 4:489. Also "trium Naturæ regnorum" at page 507.

202. Sendivogius, *Novum lumen chemicum* (1614), pp. 27–31 and 69; *Tractatvs de svlphvre* (1616:32–57). In English as *New light of alchymie* (1650), pp. 22–25, 57, 107–125 plus allegorical references.

203. *Musæum Hermeticum* (1678), pp. 17 (*Tractatus aureus*), 187 (*Via veritatis*) and 823, 846, 850, 860, and 862 (*Vitulus aureus*).

204. That is, in addition to *Duodecim tractatus* (above). One of the four, *Ænigma philosophicum*, is by Sendivogius (as Divi Leschi Genus Amo).

205. Including the *Vitulus aureus* (above), and the *Novum lumen* and *Tractatus de sulphura* attributed to Sendivogius. In some passages, the natural Kingdoms are explicitly contrasted with a Regnum Dei or Regnum cœlorum. The Hermetic *Tractatus aureus* (1:403) refers to Regnum Hominum, but this may be a reference to holy scripture. Kircher also argued (*Magneticum naturæ regnum*, 1667) that magnetism was the basis of all manner of sympathies and attractions, including the heliotropism of sunflowers; in it "regnum" is applied across all of nature.

206. "B Valentine", *Last vvill and testament* (1671), pp. 127, 136, 141, 265, and 483. For other references to kingdoms (and other divisions) of nature from the Seventeenth-century alchemical literature, see Isidore Geoffroy Saint-Hilaire, *Histoire* (1859) 2:18–22.

207. [R Boyle], *The sceptical chymist* (1661). The work as typically bound consists of two pamphlets, one brief (to page 34) and the other much longer, with title pages indicating different publishers but with consecutive pagination (apart from the second title page).

208. Ibid. (1661:429–430).

209. Ibid. (1661), pp. 34 (Animal Kingdom), 288 (Mineral Kingdom) and 332 (Vegetable Kingdom). Boyle mentions one or more of these kingdoms in nearly fifty passages in his written work, as judged by search of his collected *Works* in six volumes (1772).

210. Yates (1964:205–274); quotation at page 273.

211. Bruno, *De universo et immenso* 8.10, in *Opera latine conscripta* 1(2):313 (1879–1884).

212. Bruno, *Summa terminorum metaphysicorum*, in *Opera latine conscripta* 1(4):101; *Spaccio de la bestia trionfante* (1863:198); and *Expulsion of the triumphant beast* (1964:235–237).

213. Bruno, *Rationes articulorum physicorum adversus Peripateticos Parisiis propositorum, etc.*, in *Opera latine conscripta* 1(1):81.

214. Bruno, *Lampas triginta statuarum*, in *Opera latine conscripta* 3:177–181; zoophyta at page 180.

215. Some other passages are at *Opera latine conscripta* 1(2):396–399 and 464 (*De monade numero et figura*); 1(4):141 (*Lampas triginta statuarum*, in reference to Aristotle); 3:269, 376 and 390 (*Libri physicorum Aristotelis explanati*), 429 (*De magia*) and 552–553 (*De rerum principiis*).

216. *De monade numero et figura*, in *Opera latine conscripta* 1(2):464.

217. Fludd, *Pane et tritico in genere, simvl atqve de illorvm excellentia*, in: *Anatomiae amphitheatrum* (1623:25); *Philosophia Moysaica* (1638), particularly 3.3, 4.5, and 5.1 (n.p.); *Mosaicall philosophy* (1659), particularly pp. 47–48, 69–70, and 82–83. See also *Philosophicall Key* (1979:1–21).

218. Fludd, *Philosophia Moysaica* (1638), 3.3; *Mosaicall philosophy* (1659:48).

NOTES 521

219. Fludd, *In qvo venti orientalis in microcosmo afflatvs*, in: *Anatomiae amphitheatrum* (1623:264–265); *Pulsus seu nova et arcana pulsuum historia (ca 1631)*, page 11; Debus (1961), particularly pp. 378–382.

220. References to the animal, vegetable, and/or mineral genera are common throughout Fludd's *Utriusque cosmi* (1617–1621), *Anatomiae amphitheatrvm* (1623), and other works. For identification as Kingdoms, see (inter alia) *De primariis naturæ elementis* (pp. 94 and 98) and *De principiis microcosmi physicis seu secondarius* (pp. 159, 174–175) in *Utriusque cosmi* volume 2 (1619); *Pane et tritico*, in *Anatomiae amphitheatrum* (1623), pp. 9, 13–15 and 31; *Mosaicall philosophy* (1659) pp. 163, 167, 177, 215, 225, 228, 236, 239, 241, 243, 244, 259, 267, 275, 276, and 299; and *Philosophicall key* (1979), pp. 91–93, 96, and 149.

221. Fludd (1979), pp. 37, 142–144.

222. Fludd, *Monochordvm mvndi symphoniacvm, sev replicatio* (1623), in: *Anatomiae amphitheatrum* (1623), pp. 287–331, at pp. 322–323. For the Fludd-Kepler controversy see Yates (1964:440–444).

223. Casaubon is buried in Westminster Abbey.

224. Casaubon (1614).

225. Böhme, *Mysterium magnum* [1623], *passim*. I have examined only the 1640 edition, and those in *Works* volume 3 (1772), and *Sämmtliche Werke* volume 5 (1843). Böhme also contrasts the Kingdom of Glory from the Efflux of God's Unity with the Kingdom of the Properties of Nature: see his *Four tables of Divine revelation* [1654], Table 2 (Tetragrammaton), Fourth form: in *Works* (1772) 3:13–14, pag. sep.

226. Nolle (1617), pp. 16 and 60.

227. Thomas Vaughan, *Anima magica abscondica* (1650:30). *Three treatises* (under the pseudonym Eugenius Philalethes), in *Hermetic museum* (1678:235–241).

228. Quotation from "And do they so? Have they a sense", in *The works of Henry Vaughan* (1914) 2:432.

229. Kirchweger died in 1746.

230. Kirchweger, *Aurea catena Homeri* (1723:30–32). Translation by MAR.

231. The works of Paracelsus were variously censored during the 1580s and 1590s, and prohibited in their entirety from 1596 to 1598. Bruno was tried for heresy (1593 ff.) and burned at the stake (1600).

232. da Vinci, *Literary works* (1883), volume 2. Macrocosm-microcosm at section 929 (2:179), and *anima vegetatiua* at section 1000 (2:220–221).

233. Cardano, *Les Livres* (1556), folio 108. He stated that stones and metals (but not elements) are alive and are nourished (fols 106–108). Chapters 5–11 treat metals, stones, precious stones, plants, beasts arising from putrefaction, perfect beasts, and man. He discussed marine zoophytes (without using the term) at folio 227, in the chapter on perfect beasts.

234. Webster (1996:57–62).

235. Porta, *Magia naturalis* (editions from 1558). In English as *Natural magick* (1658): sympathies at 1.7–1.9, plant emotions at 1.7, lenses in Book 17, distillation in Book 10, tin into silver at 5.1, dogs from tigers at 2.6, beautiful children at 2.20, and party tricks at 18.2. I examined the 1560 Antwerp (Plantin) and numerous later editions.

236. Bacon, *Advancement and proficience of learning* 4.2 (1640:186), first published in 1605. Even so, Bacon accepted that iron and lead grow underground: *Natural history* 8.797, in *Works* (1730), 3.163.

Chapter 10

1. Sylvester (1641:58), translating *La semaine, ov creation dv monde* of Guillaume de Salluste, Seigneur du Bartas (1578). The immediately preceding paragraphs mention creatures bred "of

522 NOTES

live-less bodies", the salamander, and the "Fly *Pyrausta*" in furnaces. Sylvester's *Divine Weeks* was first published in full in 1602. Du Bartas also translated Ovid's *Metamorphoses* (1608).

2. Boethius, *Philosophiae consolationis*, Book 2.

3. *Genesis* 19.

4. *Exodus* 7.8–12. At *Numbers* 17:8 his rod bore buds, blossoms, and ripe almonds. In *Quran* 26.32–46 the staff of Moses (not that of Aaron) is changed into a serpent.

5. *Exodus* 14.21–30; *Quran* 26.60–67.

6. The term from Carl Sagan (1996).

7. Hohler (1989); Reed (1997); Langley (2000:61–62). Langley accepts a pagan origin for the dragon-carvings, while Reed interprets them as "national pride rather than a symptom of a dying culture".

8. Thurston (1909).

9. For delightful examples, see Benton (1992:59–64); quotation at page 62.

10. McCulloch (1960:21–44).

11. [Physiologus], in Wilhelm (1916) 8B:13–52.

12. Beer (2003:9). The pearl was a stone, so the oyster (which was of interest only in producing the pearl) makes no appearance.

13. Ibid. (2003:90–94).

14. Freeman (1976), describing the seven Unicorn Tapestries held by the Metropolitan Museum, New York, dating to 1495–1505. Also Taburet-Delahaye & de Chancel-Bardelot (2018), describing six tapestries held by *Musée national du Moyen Âge*, Paris, dating to about 1500.

15. McCrindle (1882). Ktesias's original text has been lost, but survives in an abridged version by Photios, patriarch of Constantinople (Ninth century CE).

16. This survives in excerpts by Diodorus Siculus, Strabo, Pliny, Arrian, Aelian, and others: see Schwanbeck (1846); Müller (1848) 2:39–439; McCrindle (1877); and Wittkower (1942:159–165).

17. Strabo, *Geography* 2.1.9 (1917) 1:262–263. Punctuation simplified somewhat.

18. Pliny, *Naturalis historiae* 7.2.21–22, in Loeb Classical Library 352, pp. 518–521. Pliny took many of these tales from Ktesias and Megasthenes.

19. St Jerome, The life of Paulus the first hermit. In Schaff & Wace (1893) 6:300–301. The *Life* was written in 374 or 375 CE. For other versions of the story, in which both the centaur and the satyr are clearly demons, see Bacchus (1911).

20. *Genesis* 1.11, 12, 21, and 24–26.

21. Isidore of Seville, *Etymologiae* 11.3.

22. Abelard (1929:250), translating Geyer (1919:27–28).

23. Aquinas, *Scriptum super libros Sententiarum* 1.19.5.1.co, at www.corpusthomisticum.org/snp1019.html.

24. Aquinas, *Commentaria in octo libros Physicorum* 2.14 (199a34–b33) (1963).

25. Hugolino of Orvieto, Article 43 (1932); translation by Idziak (1984).

26. For examples, see Benton (1996:147–165).

27. This paragraph draws heavily on Wittkower (1942), who provides citations for the sources mentioned, plus 45 well-chosen illustrations of "marvels" from illuminated manuscripts, codices, maps, tapestries, and cathedral walls. Also: Jourdain de Séverac (Frater Jordanus) at (1839) [original prior to 1305], and (1868) in translation; Orchard (1995). Mandeville (who also tells of the Dog-heads), in Pollard (1915), Haklyut (1965) 1:24–79, and Quinn & Skelton (1965), pp. xxvi–xxvii. The *Journal* [1331] of Friar Odoric, in Pollard (1915:344–346 *et passim*). We shall meet Gessner, Wotton, Aldrovandi, and Topsell in Chapter 12.

28. Herodotus, *Historia 3 Thalia, 106*. In the translation of Beloe (1791) 2:126–127, and (1806) 2:287–289.

NOTES 523

29. Larcher (1829) 1:597 at section 106, largely followed by H Lee (1887:46–51). The reports are by Ktesias (see above), Aristobulus of Cassandreia, and Nearchus (Fourth century BCE) via Strabo (*Geographia* 15.1.21); the elder Pliny (*Naturalis historiae* 19.2.14 *et passim*); Pomponius Mela (*De situ orbis* 3.7); Arrian (*Indica* 16, for which also see McCrindle 1877); and Julius Pollux (*Onomasticon* 7.75; in the 1542 Latin edition by Rudolf Gwalter, 7.17). For cotton clothing in ancient times: Gilroy (1845:315–332).

30. Theophrastus, *Historia plantarum* 4.7.7.

31. Liddell & Scott (1901:960); Woodhouse (1910:764).

32. Wolfson (1976:597–600), particularly page 598 and note 24.

33. *Leviticus* 19.31. This tradition is attested from (at least) the Twelfth through Seventeenth centuries: Duret (1605:322–341) (the paragraph translated by Lee is at pp. 323–324); and Lee (1887:5–8).

34. Laufer (1915:115–116); Lippmann (1933:46–47). Laufer (1915) does not accept Lee's thesis (1887) that the tale relates to the cotton plant.

35. The *Journal* [1331] of Friar Odoric, in Pollard (1915:353). See also Hakluyt (1810) 2:158–174; this passage at pp. 155 (*intùs invenitur vna bestiola similis vni agnello*) and 170 (English).

36. Mandeville (1915:174); for an older spelling see Lee (1887:5). See also Mandeville, in Hakluyt (1810) 2:123–124. Substantial portions of Mandeville's *Travels* have almost certainly drawn on Odoric's *Journal*.

37. Variants include *baromez, barometz, borames, boramnetz,* and *boranetz*. Kaempfer (1712:505–508) notes that the Persian word for sheep is *barrèh*. The claim of Lippmann (1933:47) that the word was first used by Clusius (de l'Éscluse) in 1558 is incorrect. The first appearance in print seems to be that of Sigismund, Freiherr von Herberstein, whose notes of his 1517–1518 experiences in Russia were later rewritten, and published as *Rerum Moscoviticarum commentarii* (1549 *et seq.*). In the extended translation provided by Lee (1887:11–13), Sigismund reports that Tartars call it *Boranetz*. In *Rerum Moscovitaricum* the term appears at folios 20v–21r (1549), pp. 99–100 (1556), folios 106v–107r (1557) and pp. 99–100 (1571).

38. Duret (1605:330); Parkinson (1629), frontispiece; Kircher (1641:730); Zahn (1696) volume 2, opposite page 235; La Croix (1728:1), see also page 15; and Lacroix (1791), frontispiece. Lee (1887:37, central image) reproduces the image from Lacroix (1791), which differs in detail from those of Kircher (1641) and Zahn (1696). See also Ritterbush (1964:145n13). A woodcut attributed to an elusive edition of Mandeville has been widely reproduced, including by Lee (1887:3) and Ley (1987, dust cover and page 60).

39. Worm (1655:189–191); Tradescant (1656:26). Bondeson (1999:208–210) relates other stories of pelts and barometz-skin coats.

40. Sigismund, Freiherr von Herberstein and diplomat of the Holy Roman Empire under Maximilian I, was posted to Muscovy in 1517–1526.

41. Lee (1887:13). Sigismund wrote *eamque esse ex animali instar plantae in terram defixo*; see note [37] above.

42. Du Bartas, *Edem*, in his unfinished *La seconde semaine* (1584), fols 1–14r, at folio 10v; as translated by Sylvester as *Eden* (1641:81–87), at page 86 [numb. sep.].

43. Cardano, *De rerum varietate* 6.22 (1557:216–217), as translated by Lee (1887:13–14).

44. Libavius, De agno vegetabili Scythiæ. In *Singularium* (1599) 2:289–314, at page 294: *Non est res mira nec ignota, quod zoophyta gramina pascantur. Item facit murex quidam.* Libavius makes several errors here: *murex* are obligate marine carnivores (*i.e.* do not eat grass). Very few authorities have considered snails zoophytes (one exception was Ibn al-Ṭayyib, although for the Brethren of Purity, Zakarīyā al-Qazwīnī, and Ibn Khaldūn they were the animals closest to plants: Chapter 7). Libavius, however, includes aquatic and terrestrial snails among zoophytes: De zoophytis, qvæstiones. In: *Singularium* 2:314–334, at pp. 324–326. For the relative perfection of terrestrial animals, see *Singularium* 2:300.

524   NOTES

45. Liceti, *De spontaneo viventium ortu* 4.67 (1618:311–313).

46. Kaempfer (1712:508).

47. Scaliger, *Exotericarum exercitationum* 181.29 (1557). Scaliger does not refer to the borametz as a zoophyte, but at 181.28 claims incorrectly that Aristotle called the sponge a zoophyte. For Scaliger's antagonism towards Cardano, see Lee (1887:13–15) and Bonderson (1999:204–206).

48. Liceti's book on monstrosities is *De monstrorum caussis, natura, & differentijs* (1616). He discusses uterine tumours at *De spontaneo* 3.34 (1618:225). For this usage of *mola*, see Pliny, *Naturalis historiae* 7.15.63, where Pliny refers to a uterine tumour, and 10.84.184 referring to an abortive fœtus or *mooncalf.*

49. Nieremberg, *Historia naturæ* 3.6 (1635:34–35). In the first paragraph Nieremberg notes that Liceti was not displeased by Scaliger's argument, then proceeds to append much of Liceti's text (from *De spontaneo* 4.6), however altering *ut olim probauimus convinire Zoophytis* to *quod convenit zoophytis*. This is a useful standardization, but one wonders if Liceti had intended a play on *probata* (from Greek *probaton, sheep*). Lee (1887:11) misstates the publication date of *De spontaneo* (1618, not 1518) and of Nieremberg's *Historia naturae* (1635, not 1605).

50. Duret (1605:322–341).

51. T Vaughan, *Anthroposophia theomagica* (1650:40).

52. Ritterbush (1964) identified Caspar Bose (1728) and Demetrius de Lacroix (1728) as the last to take the tale seriously. Francis Bacon remarked that "the *Figure* maketh the *Fable*" and offered rational explanations (*Sylva sylvarum* 7.609, at 1627:155), while John Ray called it "false and fabulous" (*Historia plantarum* 1686, 1:2).

53. E Darwin, *Botanic garden* (1791) 2.1.281–290; La Croix (1791) page xx, translated by Lee (1887:36 and 39). William Blake (in *Milton*, 1804) uses "Lamb" (the Saviour) and "Tartary" in the same sentence (1804:6–7): is this a coincidence?

54. See Chapter 1.

55. For example, Kepler (1606), chapter 24. See also Guthke (1990:80–82).

56. Æneas Silvius Bartolomæs Piccolomini travelled covertly to Scotland in late 1435 in hopes of persuading James I to attack England and thereby help France in the Hundred Years' War. The story of his quest for the barnacle-goose-tree is mentioned by F Max Müller (1868), at 2:590 and 2:599–600, following Münster, *Cosmographia* (1544), pp. xxxix–xl, where the tree and geese are nicely illustrated; and 1550:49 and 1554:49.

57. Only in 1803 was the barnacle goose (*Branta leucopsis*) formally recognized as distinct from the Brant goose, which Linnæus had named *Branta bernicula* in 1758. There are smaller populations in the Netherlands and elsewhere. It is now known that the Scottish and Irish birds breed in Greenland and Svalbard.

58. Charles Morton famously proposed that migrating birds fly to the moon: Harrison (1954).

59. Neckam, *De naturis rerum et De laudibus divinæ sapientiæ* (1863:99–100) in the translation of Heron-Allen (1928:10–11). Neckam returns to the barnacle goose in *De laudibus*, Dictinctio secunda, lines 481–484.

60. Giraldus Cambrensis (1905:36–37).

61. Heron-Allen (1928:10–108). Bondeson (1999:220–221) quotes a possible reference to the myth in the Eighth-century *Book of Exeter*.

62. Boece (1526/1527), folio 14v; and (1821) volume 1, pp. xlviii–l. See also FM Müller (1871) 2:583–604, with an extract from Boece at 2:591–593; and Heron-Allen (1928), pp. 22–25 and notes 74 and 78 at pp. 139–141.

63. Albertus Magnus, *On animals* (1999) 2:1563 (*Barliates*).

64. Peter Damian, *Opuscula* 36.11, in Migne, *PL* 144 (1867), column 614A. Damian wrote that *ex arborum ramis volucres prodeant, et ad pomerum similitudinem animate atque pennati*

NOTES 525

fructus erumpent (614B). Here as in the Garden, this may not be an apple—for which we expect *mālum*.

65. Heron-Allen (1928:20); Lippmann (1933:40–41); Needham, *SCC* (1956) 2:268–272.

66. Houssay (1895); Lankester (1915:100–141); Heron-Allen (1928:109–124).

67. The supposed origin of the barnacle geese likewise posed issues for Jewish dietary law: see FM Müller (1871) 2:593–594; Heron-Allen (1928:16–18); and *Zohar* 2:15b–16a, *Midrash ha-Neʾelam* (Lachower & Tishby 1989, 2:671–672).

68. Vincent of Beauvais remarked that geese, like trees, produce no excreta: *Speculum maius 1* (*Speculum naturæ*) 16.40 (1591), folio 200v.

69. Sayers (1994:17–21).

70. Thrupp (1867). Even so, goose became associated with the feast of Michaelmas: Thrupp (1867:162); and Clarkson (2014:913).

71. Clarkson (2014:476).

72. Thomas Cantimpratensis, *Liber de natura rerum* 5.23 (1973:186–187). Thomas studied under Albert from 1233 to 1240.

73. Vincent of Beauvais, *op. cit.* 16.40 (1591), folio 200v. Vincent went on to state that geese are too dry to eat; this recalls Albertus Magnus (1999) 2:1558, who reported that "a wild goose which was caught in our lands was not able to be softened even after three full days of continuous boiling. It was so tough that it could not be cut with a knife and no other beast even wanted to taste it. The flesh of the goose is unusually cold and dry, hard and melancholic, and indigestible."

74. Conrad von Megenberg ([14]75). In the 1481 Bämler edition, the *Bachad* is discussed at folio 65v; the pope is mistakenly identified as Innocent IV. Conrad mentions that he was at work on his *Buch* in 1349. In the Schulz translation (1897), the story is mentioned at 3B.11 (page 143).

75. Lippmann (1933:39–40) and note [56] above.

76. del Rio, *Disquisitionum magicarum* 2.14 (1599); I have examined only the 1603 and 1606 editions (quotation at 1:137–138, and 1:144–145 respectively). See also Heron-Allen (1928:154–155n152), and Lippmann (1933:44–45).

77. Schroeder (1937:236–296).

78. Innocent III, *Opera omnia* (1575); *Opera omnia*, in Migne, *PL* 214–217 (1889–1891). In addition to the sources cited in notes [78–84], I examined Baluze (1682) and nineteen further printed and electronic sources.

79. Hageneder (1972); Hageneder *et al.*, *Die Register Innozenz' III*, 1964 ff.

80. Cherubini (1655–1673) 1:79–89; Potthast (1874).

81. Pottast (1874) 1:1–467; see also Cheney (1952).

82. Deslile (1858); Kempf (1945); Cheney (1955).

83. Cheney (1952).

84. Valentini, *Historia simplicium reformata* 3.22 (1716), pp. 326 ff.

85. Paracelsus, *Of the nature of things*, Book 1 (1650:3) and (1674:163). A somewhat different version is attributed to Paracelsus by Porta, *Natural magick* 2.3 (1658:31).

86. Porta, *Natural magick* 2.3–2.4 (1658:31–33).

87. Purchas (1905) 2:110. The account is from Pigafetta's journal. Magellan himself had been killed about three months earlier, on 27 April 1521. The tree and leaves were illustrated by Duret (1605:319).

88. Bodin, *Vniversae naturæ theatrum* (1596:298–299); and *Le theatre de la natvre vniverselle* (1597:424–426).

89. W Turner, *Avivm praecipvarvm* (1544), sig. B3–B5; AH Evans (1903:24–29). Turner did not include the barnacle-goose tree in his *Libellus de re herbaria novus* (1538) or *A new herball* (1551–1568), but it appears as the final entry in Gerard's *Herball* (1597:1391–1392; and 1636:1587–1589).

526 NOTES

90. For the later history of the barnacle-goose-tree, see Bondeson (1999:226–231).

91. Mettinger (2001).

92. Achilles, Asclepius, Memnon, Romulus, and others.

93. *1 Kings* 17:17–24; *2 Kings* 4:8–37; *2 Kings* 13:21.

94. *John* 11:1–44.

95. *Matthew* 28:1–20; *Mark* 16:1–20; *Luke* 24:1–53; *John* 20:1–31. Some imagined that Jesus was John the Baptist raised from the dead (*Matthew* 14:2–12).

96. Delahaye (1909:45–76).

97. Paré, *Monstres et prodiges*. First edition 1573 (as the second book of chirurgie: 1573:365–581); "definitive edition" 1585; then in Paré, *Oevvres* (1628:1003–1081). His *Ouevres* were taken into Latin in by Guillemeau (1582) [not examined], and thence into a dodgy English edition by Thomas Johnson (1634). I have used the 1628 *Oevvres*, and a modern English translation by Pallister (1982).

98. Paré, "other stories" at (1628:1049–1050) and (1982:106–107); see also (1628:1030) and (1982:64–65).

99. Ibid. (1682:1079) and (1982:156–157).

100. Ibid. (1628:1057) and (1982:118–119); compare Rondelet, *Vniversæ aquatilium historiæ pars altera* (1555:130–131), where these are discussed in the book De insectis et zoophytis. Paré does Rondelet a disservice to infer that the aquatic organisms discussed in De insectis et zoophytis are somehow "insect fish".

101. Pallister (1982), page xxv.

102. Paré (1982:53–55). In his *Oevvres* (1628) this chapter is found at pp. 731–734, in a work titled De le petite verole, rovgeolle et vers des petits enfans, & de la Lepre.

103. Paré (1628:1050) and (1982:107).

104. Duret (1605:240).

105. Ibid. (1605:283).

106. Bosch, *Garden of earthly delights* (1490–1510); and in earlier studies.

107. Bronzino, *An allegory with Cupid and Venus* (*ca* 1545); also Moffitt (1996).

108. Blake, *Newton* (1795 to 1804/1805); also Fletcher (2015–2016). Blake also illustrated Milton's *Comus* (1801, and again *ca* 1815) showing animal-headed revellers.

109. Moreau, *Les chimères* (1884).

110. Linnæus, *Systema naturæ*, First (1735), Second (1740), Third (1740), Fourth (1744), and Fifth (1747) editions. The Second, Fourth, and Fifth editions further included the manticore, *lamia* (a human-animal chimæra), and siren. The group was called "Paradoxa" in First and Third editions, "Animalia paradoxa" in the Second, Fourth, and Fifth editions.

111. Watson (1763); see also Ritterbush (1964:138–139).

112. Ramsbottom (1941), citing "le Bossu" [JB Bossu] and "PA Engramelle, the entomologist" at pp. 323–324. Title pages of the multivolume *Papillons d'Europe* (1779–1792) and the corresponding plates (*Insectes d'Europe*, 1779) state that the descriptions are by "R.P. Engramelle Relig. Aug^tin. Q.S.G.". The *Catalogue of the Library of the British Museum (Natural History)* 2:533 (1904), and the *Dictionary of entomology* (Gordh & Headrick, 2011:508), credit Marie-Dominique-Joseph Engramelle (1727–1780/1). Schmitz & Ord-Hume (2001) state that Marie-Dominique-Joseph Engramelle (1727–1805), a builder of mechanical musical instruments, monk, and sometime prior of the convent of the Petits-Augustins in Paris, is occasionally confused with his younger brother, amateur entomologist Jacques Louis Florentin (1734–1814), who likewise was a monk. PA (in Ramsbottom) might mean *Père augustin* or *Petits-Augustins*, while RP (on the title pages) presumably means *Révérend Père*. Alas, neither distinguishes one brother from the other.

113. Gelli (1550); in English translation (1710).

114. Borges & Guerrero (1957, 1967); English translation (1969).

NOTES   527

Chapter 11

1. G Morrhy [Morrhius] (1530); English translation by MAR. Identical (or nearly identical) wording can be found in many subsequent reference works: see text.
2. Hellinga, *The introduction of printing in Italy* (undated).
3. Renouard (1803).
4. The Barbaro edition of Themistius, *In Aristotelis De anima*: see Chapter 7 note [215]; and of Themistius's paraphrase of Aristotle's *Physica* (via text borrowed from *Simplicius*): see Chapter 7 notes [218] and [220].
5. Philo [attrib.], In *Aristotelis operum volumen secundum* (1497). See also Chapter 5, note [26] and below, notes [7] and [45].
6. Ammonius: in Greek by Vlastos (1500), and in Latin by Gauricus (editions from 1503). Latin editions from 1539 followed the Greek original more closely than did Gauricus. A supposed 1494 Latin edition remains elusive: see Chapter 7 note [239], and Barnes (2003).
7. See Chapter 7. A mid-Fifteenth century Latin translation of *De fide orthodoxa* by John of Damascus by Giovanni Battista Panetti is known in a single incomplete manuscript (Backus, 1986) but was not published; for Panetti, see Bargellesi-Severi (1961). The translations of *De fide* by Jacques Lefèvre d'Étaples, published by the elder Henri Estienne in 1507 and 1512, did not contain the *Dialectica*.
8. Some knowledge of Greek persisted at the Université de Paris and similar institutions, but Greek texts were scarce: Dahan (1995).
9. Considine (2008:27).
10. Festus (1474): see also the editions of Festus edited by Thewrewk de Ponor (1889) and Lindsay (1913).
11. Nebrija, *Grammatica* (1492).
12. Nebrija, *Lexicon* (1492).
13. Hamann (2015:12–13).
14. Nebrija, *Dictionarium* (1495). *Dictionarium* is taken from the first line of the main text. A half-title uses the term *vocabulario*.
15. The Latin-Castilian section may not have been printed until 1517: see Hamann (2015), pp. 18–19 and note 29 at page 160. Nebrija actually used the Tironian *et*, but I replace it here with the ampersand for the sake of familiarity.
16. Apart from how the term itself was spelled: *zophyton* in 1516, 1536, and 1560 (*Lexicon latinocatalanum*); *zophitum* in 1520; *zoophiton* in 1560 (*Dictionarium latinohispanicum*), 1570, 1622, and 1638. The term is absent from 1513 and 1545 editions.
17. Morrhy (1530). I examined the copy in the British Library, shelfmark 623.m.5 (8 May 1997). See also Greswell (1833) 1:117–120.
18. Methodology: in February–March 2019 I searched archive.org for works bearing one or more of the following words in its title: *alphabetum*, *dictionarium*, *dictionnaire*, *lexicon*, *lexikon*, *nomenclator*, *onomasticon*, *onomastikon*, *ordbog*, *ordbok*, *thesaurus*, *vocabularia*, *vocabulista*, or *Wörterbuch*, ignoring works focused on the Bible, religion, morals, rhetoric, jurisprudence, geography, or mathematics. I then focused on those published before about 1650. I supplemented this list with relevant leads from the Indiana State University Library list of dictionaries before 1501, and with certain other reference works I knew of independently. This yielded 174 items with the following profile of publication dates: classical (1), 1400–1499 (16), 1500–1549 (41), 1550–1599 (47), 1600–1699 (59), 1700–1799 (8), and 1800–1899 (2). The language groups covered (other than Latin and Greek) were: Germanic (30), French (25), Hispanic (24), Italian (20), English (11), Flemish (9), Danish (4), Hebrew (2), Hungarian (2), Polish (2), Swedish (2), and other (8). Forty-four of the 174 mentioned *zoophyta*, and 39 of these had a separate entry for *zoophyta* (*zoophyta* was sometimes also mentioned at *ostrea*, *spongia*, and/or *urtica*).

528   NOTES

19. Considine (2008:27–28).

20. Crastone, *Dictionarium Græco-latinvm* (title varies), editions of 1497, 1510, 1519, 1524, and 1525.

21. I examined 16 dictionaries and lexicons from 1440 (manuscript) and 1472 (printed) up to 1500.

22. Considine (2008), pp. 29–31.

23. Calepino, *Dictionarium latinum* (title varies) editions of 1502, 1513, 1535, 1548 (Aldus), 1550, 1555, 1556, 1559 (Aldus), 1561, 1563, 1568, and 1573, and the multilingual editions of 1579 and 1625. The spelling *zoophita* was used in 1563 and 1573.

24. Mommsen, Krueger & Watson (1985).

25. Notably the Thirteenth-century *Gloss* of Franciscus Accursius of Bagnolo, against which Budé inveighs at every opportunity, including in the 1508 *zoophyta* passage discussed here.

26. Considine (2008:32).

27. Kelley (1970:53–86); quotation at page 57.

28. Budé, *Annotationes* (1541:394–397). The 1541 edition is printed in a more-legible script, but (apart from sigla) the text here is identical with that of 1508. The Aristotelian text cited in regard of ἐπαμφοτερίζειν is from *De generatione animalium* 772b1–b7. The quotation from *Mark* is at 16:15 (King James Version). See also Lloyd (1993:44–53).

29. See Chapter 5, note [26], and below at Brunfels.

30. See Chapter 5, notes [105] and [107].

31. Brunfels (1530); another edition 1532–1536. For the German "fathers of botany", see Greene (1983) 1:234–238 (Brunfels at 1:239–270).

32. Brunfels (1534), reprinted 1553.

33. Aristotle, *De mvndo* (1533), with ζωόφυτα at page 39, in the (pseudo-)Philonic *De mundo*, not in the (pseudo-)Aristotelian one. See also above and Chapter 5, note [26]. Budé left ζωόφυτα in Greek, and (true to form) offered *plantanima* in Latin. The introductory letter from Budé to Jacob Tusanus is dated April 1526.

34. For the context of this correspondence, see Plattard (1930:22–23).

35. H Brown (1976:22–25).

36. Dannenfeldt (1968:15–18).

37. The work was published in four volumes between about 1532 and 1552; a fifth appeared in 1562–1564 and is unlikely to have been written by Rabelais. The entire work (with the probably spurious fifth book) was published together only in 1567.

38. Brown (1976:20).

39. Rabelais (1546) in the transcription of Clouzot (1913) 2:22–23. The critical edition by Screech (1964:70) gives the original wording as "*Voyez comment nature, voulent les plantes, arbres, arbrisseaulx, herbes et zoophytes . . .*" This was brought rather expansively into English by Urquhart [1653] as "Behold how nature, having a fervent desire, after its production of plants, trees, shrubs, herbs, sponges, and plant-animals . . ." (*Works*, 1851, volume 1, page 510). The translation provided is by Screech (2006:439–440).

40. *Oxford English Dictionary* (1989) 20:825; *Trésor de la langue Française* (1994) 16:1436.

41. Brown (1976:29). The banquet's host, the scholar and printer Étienne Dolet, was strangled to death, and his body burnt, for heresy in 1546.

42. Dannenfeldt (1968:16). Pellicier had amassed these texts while ambassador to Venice (1540–1542).

43. In the Sixteenth century, a *Parlement* was a provincial appellate court, and that of Paris was the country's most-prestigious. Royal decrees had to be approved by a provincial *parlement* before coming into effect in that province. A *parlement* might (and often did) take on further roles, *e.g.* approving university faculty appointments.

44. HA Lloyd (2017).

NOTES 529

45. Bodin, *Démonomanie* (1580), Livre premier, preface de l'avthevr (pag. unnumb.); MLD Kuntz (1975), pp. xxxiv–xxxv.

46. The Latin version, *De republica libri sex* (1586), became widely known, and served as the main (but not only) basis for the English edition *Six bookes of a commonweale* (1606). The Latin version was a "complete redrafting and rewriting of the French text", with some material omitted and new text added, particularly in generalizing his argument beyond France and beyond monarchies: McRae (1962), page A29.

47. Bodin, *Les six livres* (1576:1–7).

48. He adds that mathematics might be carved off from the natural sciences as a fourth branch: *Methodvs* (1572:17–18). In English translation (1945:19).

49. *Colloquium* was completed in 1588 but, unpublishable in 1590s France, appeared first as *Ioannis Bodini Colloquium heptaplomeres de rerum sublimium arcanos adcitis* (1857). Bodin seems to have drawn on manuscripts left by Guillaume Postel: see Kuntz (1975), notes 121 and 122 at pp. lxi–lxii.

50. Bodin does not use the word *zoophyta* in *Methodus*, but in Chapter 3 (1572:42; and 1945:35) promises that his next book (*i.e. Theatrum*) will treat "the principles of nature . . . the elements and their nature; imperfect bodies; metals and stones; the types of plants; living things separated into three groups (*de animantibus triplici ordine distinctis*); the heavenly bodies; the size and shape of the world."

51. Bodin, *Six bookes* (1606:793–794). The wording "les Zoophytes, ou plante bestes" is at page 758 of the 1576 first French edition, and at page 797 of the more-readable 1577 edition from the same publisher. *Zoophyta* occurs on page 778 in the first (1586) Latin edition, and at page 1219 in a 1609 Latin edition printed at Frankfurt.

52. *Vniversæ natvræ theatrum* (1596), dedicatory epistle, second (unnumbered) page.

53. *Pectunculi* are small scallops or bivalves, but the 1597 French edition uses *les Nacres*, suggesting gastropods.

54. Bodin, *Theatrum* (1596), pp. 297–298 and 320; *Le theatre de la natvre vniverselle* (1597) at pages 317, 381 (index), 423, 425–426, 258 (spelled *zoophite*), 641, and in the fourth table. Oysters appear separately in the tenth table. Bodin also discusses whether barnacle-geese might be considered zoophytes (1596:298–299, and 1597:424–426). For *zoophyte* in an earlier explicit logical structure, see John of Damascus, *Dialectica* (Chapter 6, and below).

55. Lloyd (2017:225–226).

56. *Colloquium* (1857:2) and (1975:5).

57. Ibid. (1857:2) and (1975:4–5). The number 1296 is the fourth power of the Pythagorean "perfect number" six. See also Lloyd (2017:244).

58. *Démonomanie* (1580), folio 7v; and (1587), folio 8r.

59. Wiener (1936). It is said that Cesare Cremonini, the successor to Zabarella as Professor of Logic at Padua, was the friend of Galileo who famously refused to look into the telescope, lest it cause him to abandon the Aristotelian concept of the heavens: Copleson (1963), volume 3(2), page 32. Joachim Jung (Chapter 12) and William Harvey (Chapter 13) studied under Cremonini.

60. Randall (1961), particularly pp. 18–21; and HA Lloyd (2017:19–20).

61. Risse (1966), pp. vi–vii. Risse characterizes the 1483 Venice edition of Aristotle as Averroist, the 1495–1498 Aldine as old-style conservative (page vii).

62. Randall (1961:59–60).

63. Ibid. (1961:63). Mikkeli (2018) disputes Randall's assessment, allowing that Zabarella explained Aristotelian logic and natural philosophy particularly clearly, but denying that his method was a precursor to modern science.

64. In the second edition (1586) Ammonius is mentioned seven times in six passages, Philoponus twice in one passage, Simplicius five times in five passages, and Themistius 38 times in about 18 passages.

530   NOTES

65. Zabarella, *Opera logica* (1586), column 236; (1597), column 237; (1604:124); and in the translation (see text) of McCaskey (2013) 2:42–45.
66. Jean de Jandun (1473).
67. Tommaso de Vio ['Thomas de Vio Cardinalis Caietanus] (1514) and (1938–1939).
68. Zabarella, *Commentarij* (1605) Liber 2, textus 16, folio 41r–41v. Unlike in *De methodus*, Zabarella keeps ζωοφύτων in Greek in this passage. Translation by MAR.
69. Zabarella, *De anima* (1605) Liber 2, textus 27 (folio 63v), textus 31 (folio 68r) and textus 38 (folio 88v). Geometrical symmetry had been appreciated since the time of Pythagoras, but this may be one of the earliest descriptions of bilateral symmetry in animals.
70. Ramus (1560:80–96) and (1583:132–167); MacIlmaine (1581:54–61); Graves (1912:129–132).
71. Kneale & Kneale (1962).
72. According to Luccio (1963), Marko Marulić used the term *psichiologia* around 1520, but his work bearing that title has been lost; Freigius then used *Psychologia* in a prefix to his *Ciceronianus* (1575). I have been unable to locate any work of Freig under this title. Freig did edit a 1577 edition of Ramus's *Ciceronianvs*, but I do not find *psychologia* in the prefatory material (or main text). The term does appear, however, in Freig's *Quæstiones logicæ et ethicæ* (1576) in the bifurcating-tree diagram labelled *Typvs philosophiæ*, and Book 27 of his *Qvæstiones physicæ* (1579) bears the title *De psychologia*.
73. Freigius, *Quaestiones physicae* (1579:8). He elaborates slightly at page 888, noting the affinity (*cognatum*) between zoophytes and plants, and names sponges, sea-urchins, and holothurians as examples. See also page 916. He treats coral both as precious stones (page 753) and as marine plants that arise in the deep sea amongst rocks and stones, and have the appearance of a small bush (page 777). A 1585 printing reproduces this exactly. Note *Psychologia* in the figure.
74. John of Damascus, *Dialectica*: see Chapter 6, note [138].
75. R Burton [as Democritvs Iunior] (1621) and later editions.
76. Quotation from the 1621 first edition, at pp. 319–320. Burton continually added to the *Anatomy*. This passage remains intact in the 1628 edition at page 232 (although some material is transferred from marginal notes into the text). In the fifth edition (1638:243–244) and 1651:243–244 the passage reads "Many strange places ... where Cities have bin ruined or swallowed, battels fought, creatures, Sea-monsters, *Remora*, &c. minerals, vegetals. Zoophites were fit to bee considered ..."
77. For both, see Chapter 10.
78. Nieremberg, *Historia natvræ* (1635).
79. Pimentel (2009).
80. Pimentel (2009:96–97), reworded to present tense. *De la diferencia* was translated into Latin by 1651. I have examined only a 1762 edition of the Spanish version.
81. *Historia natvræ* (1635): condor at page 90; guardian angels at page 91; lost tribe of Israel at pp. 379–380.
82. Pimentel (2009:93–94). Ledezma (2010) situates the *Historia naturæ* "half way between the symbolic and allegoric natural history of Renaissance humanism and the morphological and taxonomical discipline that will impose itself during the Enlightenment".
83. *Historia natvræ* (1635:29), translation by Professor Denis Brosnan and MAR. The sponge apparently grasps the sensitive plants (aloë, scylla), which he treats in the same sentence as ancient reports of the Mycenæan sponge that contracts when touched (Book 3, Chapter 7, 1635:35).
84. Nieremberg's word *tramam* might be translated *woof* or *weft* (of a fabric), but *noduli* suggests a fishing-net. Pliny, whom Nieremberg cites two sentences later, used *nodulis* to describe lily stems tied together in "little knots" for smoke-drying (*Naturalis historia* 21.13, Loeb). *Naturalis historia* also contains the word *tramas* in connection with silk-moths preparing

themselves for winter (Jones translates it "woof-threads" at 11.27), but *in tramas* is a contested reading, and other authorities prefer *inter ramos* (among the branches). For *trama* my *Cassell's Latin dictionary* (undated, mid-1850s) also gives "applied to the spider's web" citing Pliny, but I have been unable to locate such a passage in *Naturalis historia*.

85. Ragan (2009).
86. *Historia natvræ* (1635:31–32).
87. Ibid. (1635:100–101). Book 13 (1635:286–293) is untitled in the text itself, but in the *Svmma totivs operis* facing page 1 its title is given as *De insectis & zoophytis similiter.*
88. Person, *Varieties* (1635), 1.36 (herring), 1.31–33 (adamant) and 1.37–39 (barnacle-geese).
89. Ibid. (1635) 1.39–40.
90. Ibid. (1635) 5.2–3 and 5.7.
91. Ibid. (1635) 1.2–3.
92. Henry (2016).
93. Quotation from Henry (2016).
94. See also Webster, *Isis* (1966), page 21 and footnote 63.
95. More, *Two choice and useful treatises* (1682).
96. Glanvill, *Lux orientalis* (1682), pp. 61–70 [pag. sep.].
97. More, *Annotations upon* Lux orientalis, in More (1682), page 53 [pag. sep.].
98. Another figurative use is in the phrase *zoophyte law*, a situation in which it is scarcely possible to determine "where the first law ends, or where [the] second begins" and as such ensures endless employment for lawyers: Francis (1869:25–26).

Chapter 12

1. Hegel, *Phenomenology of spirit* (1977), paragraph 247.
2. Aristotle, *De anima* 413a20–414b10 and 415a23–415b2.
3. Aristotle, *De partibus animalium* 643b810–644a10.
4. It was eventually challenged by Adam Zaluziansky à Zaluzian (1592) and Joachim Jung (1662, 1747): see Atran (1990:160), Ogilvie (2006:221 and note 51, page 323), and below.
5. Hort considers that what we know as Theophrastus's two works on plants were assembled by his students, based on lecture notes: Hort (1916) 1:xxi. Greene (1983) 2:636 notes that as the plants described by Theophrastus were mostly cultivated, it is right that his books bear the word *planta* in Latin translation. The Renaissance herbalists we discuss below focused much more on wild plants for which the more-inclusive word *stirps*, referring to the root and what is attached thereto, is arguably more appropriate.
6. In Chapters 4 and 8 respectively.
7. For Book 6 of Albert's *De vegetabilibus* as a herbal, see Reeds (1980) and Stannard (1980). For its date, see Poortman (2003), page vi and note 18.
8. Bartholomæus Anglicus (1505), Book 17.
9. Conrad von Megenberg, about 1350 (first printed 1475). Ogilvie (2006:97) notes that manuscripts of *Puch der Natur* were often bound with medical books, indicating an audience of "physicians, apothecaries, or literate lay healers".
10. In these years "alphabetical" often referred to the first letter only.
11. For a Nineteenth-century English translation of the illustrated Cotton Vitellus C iii (ff 11–85 in part) in the British Museum, see Cockayne (1864) 1:1–325. *Herbarium* was first printed, in Rome, about 1481.
12. Arber (1938:16–37); detailed bibliographic information in AC Klebs (1917–1918). *Gart der Gesundheit* became known as "the German *Herbarius*".

532 NOTES

13. Lonicer (1783). Its original title was *Vollständiges Kräuter-Buch* (1713 and earlier editions); *Buch der drey Reiche der Natur* is a late (Eighteenth-century) addition. *Lonicer* is sometimes spelled *Lonitzer*.

14. *Ortus sanitatis* (1491).

15. Andrewe (1521), facsimile edition 1954.

16. The work was printed in 1493: Barbaro, *Castigationes Plinianæ* (1493). Another edition appeared, in Cremona, in 1494.

17. For details of the protagonists and debate see Ogilvie (2006), pp. 30–33 and 37. Pandolfo Collenuccio, whose *Pliniana defensio* (1493) countered Leoniceno to some extent, is considered to have developed the first natural-science museum in Italy: Maffei (1853) 1:216.

18. Ogilvie (2006), pp. 31–32 and 137–138.

19. V Cordus, *In hoc volumine . . . annotationes* (1561); and *Stirpium . . . Italia* (1563). See also Greene (1983) 1:371–372 especially note 18.

20. Two other translations appeared at about the same time, by Ermolao Barbaro (posthumous) and Marcello Virgilio, but that of Ruel was considered superior and was reprinted much more often: Greene (1983) 2:602.

21. Ogilvie (2006:34–37).

22. The second generation of Renaissance naturalists ("the first phytographers, 1530–1560") in the classification of Ogilvie (2006:34–37). Many historians have recognized generations within Renaissance science including Sachs (1875) and Pinon (1995). For a broader perspective, see Bullough (1970).

23. Many of these woodcuts were re-used by later authors including Bock, D'Alechamps, Bauhin, Dodoens, Gerard, and Turner. They were idealized in the sense of ignoring plant-to-plant variation.

24. Examples are given by Greene (1983) 1:288–298.

25. Arber (1938:59) memorably calls its language the "plain, racy German of the people".

26. For this point I acknowledge Greene (1983) 1:307 ff.

27. Tragus, *De stirpivm* (1552), with a preface by Conrad Gessner. Regarding the illustrations see Ogilvie (2006:198).

28. Bock does not identify, much less justify, his criteria for arrangement, but they can be reconstructed ex post facto, as was done by Greene (1983) 1:327–347.

29. For a poetically inspired grouping, see Greene (1983) 1:345–346.

30. Tragus (1552:939–945). At pp. 687–688 we find "De Phyco Marino" with an introductory sentence stating that "many genera of algae are born in waters all about", but the genus is identified as *Potamogeton* (today commonly called pondweed); the illustration clearly depicts a flowering plant.

31. *Et facient lautas optata tonitrua cœnas.* Truffles have been associated with thunder and/or lightning-strikes since antiquity. The association is mentioned by Athanæus of Naucratis (Third century CE) in *Deipnosophistae* (1927) volume 1, excerpts from Book 2 (62A–B), where in a footnote the editor (Gulick) refers to Fragment 167 of Theophrastus as recognized by Friedrich Wimmer. Juvenal's *Satura* 5 reads in part: "Post hunc raduntur tubera, si ver tunc erit, & facient optata tonitrua cœnas majores" (lines 116–118). See also Plutarch's *Moralia. Quæstiones conviviales* 4.2 at 664B (Loeb 8:317), and Pliny's *Naturalis historia* 19.13.37. In keeping with the Renaissance nature of early scientific botany, Jean Ruel quotes this passage in *De natura stirpium* (1536:519, and 1537:392), with attribution to Aquinas the poet, as do Amato Lusitano [João Rodriguez de Castello Branco] translating the *De medica materia* of Dioscorides (1558) at 2.139, and Giambattista della Porta at the conclusion of his *Phytognomica* (1588). Other instances could be mentioned. However, I am at a loss in regard of the attribution (by Ruel, Lusitanius, and Bock) to "Aquinas Poëta". Saint Thomas Aquinas was an accomplished poet, but this subject matter would be atypical for him. Within the

NOTES 533

220 Thomistic works searchable at corpusthomsticum.org (3 April 2019) *tonitrua* occurs 47 times, *optata* 27, and *lautas* and *coenas* once each, none remotely related to fungi. The term *fungi* occurs three times (from the verb *fungor*); no instance of *tuber, tubera,* or *fungus* was retrieved. I am unaware of any classical poet named Aquinus or Aquinas; for the Thirteenth-century Sicilian poet Jacopo d'Aquino, see Langley (1913). Claudio Deodato (1628:285) attributes the passage only to "the Poet" (*Poeta*).

32. The 1530 edition did little more than reproduce the plant descriptions of classical and more-recent authors, together with high-quality illustrations. The liverwort *Marchantia* is the only non-flowering plant (1530:191) illustrated. A second edition (1532–1536) reproduced the 1530 edition as volume 1, while volume 2 added further herbs as well as lengthy excerpts from Dioscorides, Manardo, Leoniceno, Collenuccio, Fuchs, Bock, and others.

33. Matthiolius (1583:497–498). The work incorporates considerable material from Luca Ghini.

34. A second edition (1576) included a tract from Rondelet on internal medicine. Charles l'Écluse also studied with Rondelet, as did Felix Platter, Jacques d'Aléchamps, Jean Bauhin, and Jean Desmoulins: Arber (1938:85).

35. Arber (1938:91 and 174–178) makes a relatively strong claim for L'Obel having distinguished monocots and dicots. Greene (1983:894–921, particularly 910–911) fundamentally agrees, but adds nuance.

36. L'Obel and Pena discuss ferns (*Adversaria* pp. 358–364) immediately before thistles.

37. Ibid., pp. 353–355 and 361–363 respectively.

38. Ibid., page 449.

39. This *Litoxyla* seems to be petrified wood, although the name was later applied to a gorgonian: Pallas (1766:160, and 1787:199). The sponge, barnacle-goose-tree, and *Litoxyla* are present in the 1570–1571 *Adversaria* but do not appear in *Nova stirpium adversaria* (1576). The Appendix of *Nova stirpium* concludes (1576:471) with a seaweed, possibly a member of Phaeophyceae.

40. Plantin saw to it that woodblocks were shared among books by the three authors. Those from L'Obel's *Kruydtboeck* were published by themselves as a field manual, *Plantarvm sev stirpivm icones* (1581). Plantin's collection of woodcuts was later used to illustrate Johnson's editions of Gerard's *Herball* (1633 and 1636). For a history of woodblocks in botanical illustration, see Arber (1938) pp. 134, 176 and 185–243.

41. L'Obel, *Plantarvm seu stirpium historia* (1576), pp. 636 (*Agaricum*) and 643–654. The sponge, barnacle-goose-tree and *Litoxyla* (now spelled *Lithoxyla*) return (pp. 654–655); compare note [39] above. Further illustrations of these organisms were added in the *Kruydtboeck* (1581:278–294).

42. l'Écluse, *Rariorum aliquot stirpium per Hispanias* (1576) and *Rariorum aliquot stirpium, per Pannoniam* (1583).

43. The section itself is dated 1598 and reproduces an essay from the *Villæ* of Giambattista della Porta (1592:764–768). An accompanying set of colour illustrations (Codex BPL 303, Universitätsbibliothek Leiden) went missing from his publisher's premises (*Rariorvm*, 1601:292; Ogilvie, 2006:175–176) and could not be included in the 1601 publication. Many of the illustrations appeared later, without attribution, in the *Theatrum fungorum* of Frans van Sterbeeck (1675). The *Codex* was reunited with the *Fvngorvm* for an anniversary (re)issue (*Fungorum in Pannonis observatorum.* Budapest and Graz, 1983).

44. Ghini had studied under Leoniceno in Bologna.

45. See Findlen (1994:256–257) and (2017). Cordus died of malaria in 1544, the year Ghini moved from Bologna to become the founding director of the botanical garden at Pisa. Cosimo I de' Medici had offered the position to Fuchs, who declined. Ghini later became personal physician to Cosimo I.

46. Cesalpino moved with Ghini from Bologna to Pisa in 1544.

534 NOTES

47. Cesalpino, *Peripateticarum quaestionum* (1571), fols 1B, 38B, and elsewhere.

48. Linnæus, *Philosophia botanica* (1751), aphorism 54 at page 18.

49. Cesalpino, *De plantis* (1583), pp. 1–3 and 27. Sachs (1890:43–55) offers a useful paraphrase of *De plantis* Book 1, although his assessment of its value is not always generous.

50. Cesalpino (1583:27). See also Sloan (1972:12–13) and Atran (1990:151–157). Aristotle made theoretical provision for the effect of contingencies (locus, weather) on material form in individual species; by contrast, Cesalpino treated species as eternal and incorruptible (Atran 1990:138–140).

51. That is, the "genuine demonstrative knowledge of essences" (*epistēmē*), not just opinion (*doxa*): Sloan (1972:9).

52. Cesalpino's classification has been outlined outlined by Sachs (1890:56–57) and Bremekamp (1952–53); the latter is reproduced in Greene (1983) 2:812–814.

53. *De plantis* (1583:3–5).

54. Ibid. (1583:28). JF Wolff considers Lenticula palustris a synonym of the duckweed *Lemna minor* (1801:24).

55. And not only in regard of their shape, but of their substance as well, and the places where they grow: *De plantis* (1583:613).

56. Greene (1983:1018 note 3) also reminds us of *Ḥadīqat al-azhār fī, sarḥ māhīyat al-ʿushb wa al-ʿaq qār* [*Garden of flowers, or explanation of the characters of herbs and drugs*] of Qāsim ibn-Muḥammad al-Wazīr al-Ghassānī, which also appeared in 1586. No author is credited on the title page of the *Historia generalis plantarum*, but d'Aléchamps was senior author, while Jean Bauhin and Jean Desmoulins also contributed.

57. Many of the illustrations are based on the woodcuts used in the second edition of Mattioli's *Commentarii* on Dioscorides (1565). Insects were sometimes added, flying around flowers.

58. D'Aléchamps, *Historia generalis plantarvm* (1586–1587): horsetails, pp. 1069–1072; seaweeds, some marine vascular plants and corals, 1367–1376; barnacle-goose-tree, 1397–1398; petrified wood, 1398; truffles and fungi, 1585–1588; borametz, 1849.

59. *Historia generalis plantarvm* (1586–1587), page 1878.

60. Zaluziansky à Zaluzian, *Methodi herbariæ* (1592). Another edition appeared in 1604, and a facsimile edition in 1940.

61. *Methodi herbariæ* (1592), sig. B; and (1940:7). Arber (1938:144) gives an English translation of this passage.

62. Ogilvie (2006:226–227) makes the astute distinction between classifying plants, versus organizing pedagogy in botanical science. Zaluziansky's use of Ramist trees served the latter aim.

63. *Methodi herbariæ* (1592), sig. N2–N4; and (1940:62–66). He included ferns among the more-perfect plants: (1592), sig. T ff; and (1940:100–103). True mosses were apportioned between both, depending on the shape of their leaf.

64. The literature on plant nomenclature is considerable and not uniformly enlightening. As we saw in Chapter 1, since ancient Sumer multiple characters have been used to indicate that a particular type of plant is a member of a more-inclusive group. Under the influence of scholasticism, plant names sometimes became quite cumbersome. In many other instances, a single (genus) name was no longer sufficient once constituent species were recognized. The Bauhins were in the forefront of efforts to standardize plant names to two or a few words, but crisp binomials can occasionally be found in earlier herbals, encyclopædias, and other works.

65. To clarify the types of orchids (a notoriously difficult group), Bauhin added a detailed Ramist tree prepared by Cornelius Gemma: *Phytopinax* (1596:115–116). Gemma, professor of medicine at Leuven and amateur astronomer, was perhaps the first European "orchid hobbyist" (Jacquet, 1994).

NOTES   535

66. More than 5000 species were described in the posthumous *Historiæ plantarum universalis* (1650–1651) of Gaspard's elder brother Jean, who died in 1613.

67. Bauhin, *Pinax* (1623:368). The section does indeed end with a comma.

68. Adanson (1763), pp. xiii–xiv states that seventeen of Bauhin's seventy-two classes are natural, the greatest number and proportion to that date (surpassing Cesalpino: page x). According to Stafleu (1963:150), Adanson based this on the 1671 edition of *Pinax*.

69. *Pinax* (1623:15–16). Imahori (1954:30) states that Bauhin's "*Equisetum fœtidum sub aqua repens* 5. in Prod." (1623:16) is a charophyte, and the illustration in Bauhin's *Prodromus theatri botanici* (1620:25) supports its identification as a member of genus *Chara*.

70. *Pinax* (1623:513–514).

71. J Bauhin & Cherler, *Historiæ plantarum universalis* (1650–1651) 3: 809–810. One imagines that the delay was due, at least in part, to the Thirty Years' War (1618–1648).

72. Notably Camerarius, *De sexu plantarum* (1694); Grew, *The anatomy of vegetables begun* (1672) and *The anatomy of plants* (1682); Magnol, *Botanicum Monspeliense* (1676, 1686) and *Prodromus historiæ generalis plantarum* (1689); Rumphius, *D'Amboinsche Rariteitkamer* (1705); and Zwinger, *Theatrum botanicum* (1696) and (with his son Friedrich) again, 1744.

73. According to Kangro (2008) he is said to have left upon his death 75000 manuscript pages, of which two-thirds were lost in a fire in 1691. The anonymous entry "Jungius, Joachim" in Rees' *Cyclopædia* (1819) is also worthwhile.

74. Jung, *Opvscvla botanico-physica* (1747), sig. A–F3, pp. 1–46 with further annotation by Vaget (pp. 47–66). Chapter 1 is at sig. A3, pp. 5–6. Translation by MAR with reference to the German paraphrase in Sachs (1875:65–66), and (1890:60–61). At the sixth point, Vaget adds that Jung has more to say on this in the next section (signum B), but in any case the *Isagoge* teaches nothing about this *Abortus*; the wording might be modified to make it clear that Jung was referring to other speculation, not his own. A footnote provides Eighteenth-century references both opposed to (Micheli, Plaz, Wedel) and favouring (Lancisi) the idea, and notes that Linnæus (*Critica botanica* 1737) later replaced words in plant names that properly refer only to animals. Section 269 of *Critica* begins at page 219, and *abortiva* is mentioned at page 220.

75. Jung completed *Doxoscopiæ physicæ minores* in or by 1630. It appeared in two dissertations in 1642, then in an edition edited by Martin Fogel (1662). Fragments related to botany were then included in the *Opuscula* (1747), sig. I2–Z, pp. 67–178 along with further annotation by Fogel.

76. Jung, *De plantis doxoscopiæ physicæ minores*, in *Opvscvla botanico-physica* (1747), sig. T, page 145.

77. *De plantis doxoscopiæ* (1747:130–145): seeds in plants, including fungi following Scaliger and Porta, page 134; trees and herbs, pp. 70 and 72–74 crediting Rivinius at page 72. Cesalpino (see text) and Zaluziansky (*Methodus* 2.18, *De palmis*) had likewise spoken against the supposed distinctiveness of shrubs vis-à-vis trees.

78. *Physiologus*, Chapter 4; *De proprietatibus rerum* (Bartholomæus Anglicus) derived in part from *Physiologus*, Chapter 8. A few Medieval animal-books based on *Physiologus* were mentioned in Chapter 10.

79. McCulloch (1960), particularly pp. 34 ff.

80. Ibid. (1960:47–54); Wright (1841:20–131).

81. Wright, Preface (1841), page x.

82. Gervaise: McCulloch (1960:55–56); de Fournival: McCulloch (1960:46) and Chapter 10.

83. McCulloch (1960:47); Brunetto Latini (1993); Mayer (1890). Waldensian is a Francoprovençal dialect similar to Lyonnais.

84. Trevisa was important in the formation of English vocabulary (Fowler, 1995).

536   NOTES

85. The publication date is given as 1508, but it may have been printed as late as 1512 (JI Davis, 1958:vi). This *Libellus* may be the Latin source of *De las propriotas de las aminançes.* Surprisingly, the *Physiologus* itself was not printed until 1587.

86. The division is announced in the Incipit prefaciuncula at signum a ij, lines 1–9.

87. Hugo de Folieto, *Aviarum* (1992).

88. Frederick II, *De arte venanci cvm avibvs* (1788–1789); in English (1943). Haskins (1921:342) indicates 1248 as the most-likely date. The Arabic and Persian texts on falconry are mentioned (1943), page xlix.

89. For this and other early works on the veterinary care of horses, see Curth (2013:39–58).

90. P de' Crescenzi [Petrus de Crescentijs] (1471), Books 9 and 10. This was the first printed book on rural economy.

91. Hoffmann (1985:879).

92. Nazari (2014).

93. Turner, *Avivm praecipvarvm* (1544); AH Evans (1903).

94. Herr (1546). The unique copy is held by the Herzog-August-Bibliothek, Wolfenbüttel. A facsimile was printed by Dudelsack (1934), and an updated edition by Sollbach (1994). See Wickersheimer (1960); and Pinon (1995:74–75).

95. Agricola (1549:76–79). See also Hoover & Hoover (1912:217n26).

96. Lankester (1888:803–804); Pollard (1900); Bäumer (1990).

97. Wotton (1552), Praefatio (fol. unnumb.). [Referring to Aristotle:] *Hoc igitur ceu itíneris duce usus, (explicata tamen prius & declarata ratione qua animalium differentiæ inveniri possint) initium paulò altius ab ipso Animali repetere institui, & ita repetere, ut si quæ non illius propria, sed superiorum generum communia occurrerint, ea ipsa suo, hoc est primo loco, plana fecerim.* I have in part followed Bäumer (1990:16), who wrote *der Ausgangspunkt liege bei den Tieren selbst* to capture the sense of *initium paulò altius ab ipso Animali repetere institui.*

98. As we have seen (Chapter 3), Aristotle considered this at one point: *Metaphysica* 1038a8–24.

99. Wotton (1552), Praefatio: *Veterum scriptorum sententias in unum quasi cumulum coaceruaui, de meo nihil addidi.* Wotton lists the cited authors in a three-page Catalogus.

100. Ibid. (1552), sig. A, fol. 1E: *Erit itaque animal corpus animatum, quod sensu præditum est.*

101. Ibid. (1552), sig. L, fol. 198A: *Continet exanguium aquatilium differentias, scilicet mollium, crustaceoru[m], testatorum, & quę zoophyta appellantur, quę omnia à nonnullis etiam piscium nomine uocantur. Mollia* are Aristotle's μαλάκια (cephalopods: octopi, squid), crustacea his μαλάκόδερμα (crabs), and testata his οστρακόδερμα (hard-shelled animals, shell-fish). Wotton treats Aristotle's μαλάκια and μαλάκόδερμα together as of *natura mollium,* soft-bodied. While it is tempting to translate *mollia* as molluscs, the membership is not identical; and Cephalopoda as a taxon name dates only from the late Eighteenth century. I therefore leave the word in Latin. See also Peck (1937:23) and Mair (1963, page xxviii).

102. Wotton (1552), sig. Oo, fol. 217E.

103. Wotton also devotes chapter 91 of *De differentia* to "certain monstrous animals arising in India and Ethiopia".

104. This is particularly clear in his table of contents (*Summatim*) prior to Book 1.

105. Although he comes close, for example in Chapter 248, note 2, at folio 217E. See Chapter 3, note [115].

106. Just as Wotton (1552) followed his four chapters (248–251) on zoophyta with a chapter (252) on certain unworthy matters—he used the word *purgamenta*—so too Belon (1553) followed his chapter on Zoophyta (*De aquatilibus* Book 2.11, pp. 432–436) with one (2.12, pp. 436–445) titled On marine *deiectamenta.* In his *Summatim* Wotton included all five chapters under the heading ZOOPHYTA, but Belon does not correspondingly subdivide his *Index capitvm,* rendering it uncertain whether he includes these *deiectamenta* as zoophyta, or not.

Leuckart (1875:94–95 and 100) points out that in any case, Rondelet swept all these genera into his "insects and zoophytes".

107. Belon (1553:432). Pollicepeds are a type of barnacle. The *auris marina* depicted in the second part (see below) of Rondelet's *Libri de piscibus marinis* (1555) at page 5 appears to be an abalone.

108. Belon (1553:435–436).

109. The second part of Rondelet's *Libri de piscibus marinis* is under a different title, *Vniuersæ aquatilium historiæ pars altera, cum veris ipsorum inaginibus* (1555); the book on insects and zoophytes is at pp. 107–136. These are followed by books of fish of brackish waters, of lakes, and of rivers.

110. Rondelet, *Vniuersæ aquatilium* (1555:107–108).

111. His name in German (Swiss) vernacular was Conrad Gessner (or Cûnrat Geßner) with a double "s" (or an Eszett). Latinized, it became Conradus Gesner with a single "s" (Pyle 2000).

112. Or more memorably: "It was the object of these authors [Gessner and Aldrovandi] to amass every thing that had been said of animals by poets, shepherds, grammarians, philosophers, physicians, and old women. Their prolixity, of course, is insufferable. Their labours, however, may be regarded as rude quarries, from which some valuable materials may be dug up; but the expence of removing the rubbish will, perhaps, overbalance their intrinsic value" (Smellie, Preface to Buffon, *Natural history* (1780) 1:vi–vii). Gessner is known to have met both Belon and Rondelet (*e.g.* Romero, 2012).

113. Kusukawa (2010); Egmond (2013).

114. *Historiæ animalium* 1 (1551), Epistola nuncupatoria, second (unnumbered) page, at lines 14–20. Gessner lived to see four volumes in print (viviparous quadrupeds, 1551; amphibia and reptiles, 1554; birds, 1555; fish and aquatic animals, 1558), along with an appendix on quadrupeds (1554) and lightly annotated volumes of *Icones* (1553–1560); an incomplete volume on serpents and dragons appeared posthumously (1587) and contained a lengthy section on the scorpion that Gessner had intended for his sixth volume. An abbreviated German edition in four volumes (*Fischbůch, Thierbůch, Schlangenbüch*, and *Vogelbuch*) appeared a bit later (1557–1589).

115. He sometimes grouped very similar genera, thereby departing from strict alphabetical order.

116. Hippolito Salviani is credited as well, on the first (unnumbered) page of the Præfatio.

117. *Historiæ animalium* 4 (1558), pp. 1064–1079; zoophytes at page 1066; Ramist tree at page 1070. Five organisms are distinguished by this tree, two of which he decides are not true sponges.

118. Gessner, *Nomenclator aquatilum animantium* (1560); zoophytes at pp. 271–273. Translation by MAR from page 271. The phrase set in italics is in Greek in Gessner (*ta mesos kau epamphoteízouta*). Gessner also acknowledges a stepwise ascent in nature (*Historiæ animalium* 4:1066–1067).

119. The title page of Salviani's *Aquatilium animalium historia* bears the date 1554, but the first printing appeared only in 1557: Hendrikx (2015).

120. I intentionally avoid a running commentary on invertebrate classification as it evolved over these years, in part because it is not always clear what organism(s) should be matched to names used by earlier authors. For a dated but nevertheless useful attempt that engages the Sixteenth-century zoological works we discuss in this chapter, see Leuckart (1875). See also Voultsiadou *et al.* (2017).

121. See also Cole (1949:66).

122. Gessner, *Historiæ animalium* (1585) 3:766–772. I have been unable to examine the 1555 first edition of volume 3.

123. Ibid. (1585) 3:109–111.

538 NOTES

124. Belon (1553) illustrated the fabulous horse of Neptune (pp. 26–27) and the monk-fish (pp. 38–39); Rondelet (1554) the sea-lion, monk-fish, and bishop-fish, pp. 491–494; Gessner (*Historiæ animalium*) the unicorn (1:781–786), the fabulous horse of Neptune (4:433) and various *hominibvs marinis* (4:519–522, sometimes qualified with *rumor est*).

125. Belon (1553:89–93), coldness at page 89; Rondelet, *Liber de piscibus marinus* (1554:358–363), occult power at pp. 359 and 361 following Pliny and Galen respectively). See also Copenhaver (1991).

126. The famous comparison of human and bird skeleta is at (1555:40–42).

127. Publication dates: birds (1599–1603); insects (1602); other exsanginous animals (1605–1606); fish (1613); quadrupeds with non-cloven (1616) or cloven (1621) hoofs; serpents and dragons (1639–1640); monstrosities (1642); viviparous and oviparous quadrupeds (1645). In addition, works appeared posthumously on metals (1648) and trees (1667–1668). I examined some of these only in later editions (see below).

128. One need only examine the first chapter of Aldrovandi's *Serpentvm et draconvm historiæ* (1639–1640), pp. 2–108 to see this in full glory. See also Ashworth (1996).

129. Aldrovandi, *Serpentvm, et draconv[m] historiæ* (1639–1640): frog and cock with serpentine tails pp. 61–62, basilisk pp. 361–376, hydra pp. 386–401, winged dragons pp. 416–427; *De piscibus* (1613): sea-serpent page 368, sea-monkeys pp. 405–406, rays in the form of dragons pp. 443–444, cetacean with saw protruding from forehead page 695, monstrous orca page 700; *De reliquis* (1605–1606): the barnacle-goose-tree (as *Concha anatifera*), pp. 543–544.

130. Aldrovandi, *Monstrorvm historia* (1642): monk-fish illustrated at page 28; monstrous plants pp. 623–715; clouds, comets, solar halos, etc., pp. 716–748.

131. Aldrovandi, *De animalibus insectis* (1602:717–760). The Ramist tree dividing insects is on an unnumbered page immediately preceding the *Catalogvs avthorvm*.

132. Aldrovandi, *De reliquis animalibus* (1605–1606): zoophytes at pp. 563–593; the supposed marine fungus at page 587; sea-hand at page 593.

133. Ibid. (1605–1606), at page 563: "Of such kinds are [*zoophyta*], Urtica, Holothuria, Tethya, Mentula marina, Malum insanum, Cucumis, Pulmo and others of that sort, among which many enumerate the Sponges, which we shall place among the imperfect plants."

134. Topsell (1607), title page and pp. 262–263. For an account of Topsell's *History* in three volumes, see Ley (1967).

135. According to Ley (1967) these additional chapters were taken from [Thomas] Bonham. Raven (1947:223–226) argues that Topsell also borrowed from Wotton (1552) and others. The second edition of *Serpents* was used in the 1658 combined volume.

136. Mouffet died before completing *Insectorum sive minimorum animalium theatrum*; the work was edited by Théodore Turquet de Mayerne, and published (in Latin) in 1634. Its title page credits in part (*olim*) Edward Wotton, Conrad Gessner, and Thomas Penny. According to Ley (1967), Penny had worked extensively in Gessner's entomological collections.

137. Its title page gives the principal author's name as *Topsel* with a single "l", although his earlier 1607 and 1608 treatises spelled his name Topsell. Alternative spellings were not uncommon in those years in England, *viz.* Shakespeare, Shakspeare, Shakespear (likewise Mouffet, Moufet, Moffet, Muffet).

138. Hippopotamus at 1:256–257; manticore at 1:343–345.

139. Topsell (1658) 3:889.

140. Ibid. (1658) 3:1028.

141. *Historiæ naturalis De piscibus et cetis libri V* (title page undated, no colophon, Ad lectorem dated 1649), *De avibus libri VI* (1650), *Exangvibus aqvaticis libri IV* (1650), and *De insectis libri III, De serpentibus et draconibus libri II* (1653) were subsequently bound together.

NOTES 539

Although the title page of the latter promises three books of insects, four are present. *De quadrupedibus libri* followed (1657), and the 1650–1653 volumes were reissued in 1657, curiously with the figures mirror-imaged left-to-right. The volumes were translated into several vernacular languages, and were variously reissued until at least 1768.

142. Locy (1925:301).

143. It would be interesting to trace the intellectual lineage of semi-popular zoologies in about six parts from Jonston's *Historiæ naturalis* to (for example) Wood's *Animate Creation* (1898).

144. Jonston, *Historiæ naturalis de exanguibus aquaticus* (1650:77–78), and (1657:58). Another puzzling feature of this volume is the depiction of sea-turtles in Table 8, as sea-turtles are described in his *Quadrupedibus* (1657), pp. 147–148 and Table 80.

145. Imperato (1599). Cole (1949:346n1) cites Böhme that the first edition was 1593, and this 1599 is a second edition, but no evidence of the former has come to light. See also Accordi (1981). The catalogue was notable as well for the large engraving that depicts his cabinet as it was actually displayed. The work remained popular; a slightly expanded second edition, edited by GM Farro, was released in 1672, and a Latin translation of the latter in 1695.

146. Worm (1655).

147. Spener (1693).

148. Aldrovandi assembled an exceptionally large collection of plants and animals. In the Eighteenth century Rumph, Seba, Gersaint, Dávila, Mesny, and Favanne de Montcervelle among others assembled scientifically important collections, while the British Museum was built in significant part from the cabinet of Sir Hans Sloane. Buffon's *Histoire naturelle* (Chapter 13) was in part the description of the cabinet of the King of France.

149. Imperato (1599:713). He considered *Savaglia* (now considered to be a zoanthid) to be like a coral, and a plant (1599:724).

150. Ibid. (1599:713–724). Brook (1893:2) argues that *pora* in *madrepora, millepora*, etc. comes from the Greek πῶρος (*stone*), hence refers to the stony material within which the pores are found, not to the pores themselves.

151. Ibid. (1599:725–734). Note the marginal "*Spongia pia[n]ta animale affisa*" at page 727.

152. Ibid. (1599:735–750). Some of his mosses (pp. 744–750) may be vascular plants.

153. The garden is described in his beautifully produced *Hortvs Eystettensus* (1613), printed in both coloured and uncoloured versions, and reprinted in 1640 and 1713.

154. A second edition of *Gazophylacium* appeared in 1716, as did a somewhat more-complete overview of Besler's cabinet (*Rariorum Musei Besleriana*, 1716). Basilius Besler's illustrated *Fasciculus rariorum* [1616] divides the natural items into Animalia, Marina, Conchilla, Lapides, and Fructus. An octopus is pictured with Conchilla, while corals and various fossils are Lapides.

155. Charleton (1668), third (unnumb.) page of Præfatio: *nomina si nescis, perit cognitio rerum*. Charleton was physician to Charles I, and later to Charles II.

156. Charleton (1668:119).

157. Zoophyta is accorded its own header (ZOOPHYTA rather than PISCES), and although explicitly called a Class, the same is true for Serpentia (1558:30–34), which is unequivocally distinct from Quadrupedum or Insecta (see Roman numerals in the Index). I take this to mean that Charleton recognized only a single class within each of Serpentia and Zoophyta. He does not list Zoophyta or any zoophyte genus in the index.

158. Charleton (1668:193–195, mistakenly printed "165"). The same text appears in his *Exercitationes de differentiis & nominibus animalium* (1677), Pisces, pp. 68–69 with some new typefaces and typographical corrections.

159. JT Müller (1780) 1:238–256.

160. Willis (1672), and in English (1683).

540 NOTES

161. Lister, *Historiæ animalium Angliæ* (1678) and *Appendicis* (1685); *Historiæ conchyliorum* (1685-1691) and *Appendix* (1692); *Exercitatio anatomica* (1694, 1695); *Conchyliorum bivalvium* (1696); and in *Philosophical Transactions* (1697).

162. Cole (1949:256-270) and references therein.

163. Goedart (1662-1669).

164. Swammerdam (1737-1738), with preface by H Boerhaave.

165. Redi, *Esperienze intorno* (1668), translated into Latin (1671) and English (1909). See also his *Opuscvlorvm* (1685-1686).

166. Regarding animals Jung followed Zabarella, who in turn largely followed Aristotle. See Jung (ed. Vaget), *Doxoscopiæ physicæ minores* (1662) 2.1.2.24, especially point 5; and the critical edition edited by Meinel (1982:188).

167. Jung (ed. Vaget), *Historiæ vermium* (1691:28-30); "fabulosæ metamorphoses" at page 30. Jung identified "Primum Regnum Mineralium, hoc est Fossilium, sive Mistorum Inanimatorum, Secondum Regnum Vegetalium, sive Stirpium [et] Tertium Regnum Animalium" at *Doxoscopiæ* (1662) 1.2.2.1.24.1-2, and Mineralium regnum again at 2.2.7.2. Meinel (1982:12n17) calls the *Historia vermium* a "reine Materialsammlung". See also Meinel (1984), Einleitung, especially pp. xi, xiv, and xvi.

168. Mayerne (1658), fourth (unnumbered) page.

169. Alsted (1615:395). Very similar words appear in his *Scientiarum omnium encyclopædiæ* (1649) 2:144: *Corpora perfecti mixta in tria veluti regna sunt divisa: regnum videlicet minerale, vegetabile, & animale.* Alsted considered corals to be marine plants that turn to stone in the air (ibid. 2:409). See also his extensive Ramist trees outlining the study of natural history, at 2:210-225.

170. [Epistle] CLXXIXb. Descartes a Mersenne (1639), in *Oeuvres* (1913:97). Descartes was familiar with the work of JA Comenius, who had studied under Alsted.

171. Libavius, De zoophytis, qvæstiones. *Singularium* 2:12 (1599:314-334).

172. *Singularium* (1599): Scaliger at pp. 319, 325-329, and 332; Hector Boece, page 319; Francisco López [de Gomara] and Clusius, page 328; Filippo Mocenigo, Archbishop of Nicosia, page 331. Mocenicus wrote of two orders of perfection (1581); Libavius (1599:331) directly quotes Mocenigo (1581:343) in part.

173. *Singularium* (1599:327); sponges again at page 331.

174. Ibid. (1599:334).

175. In 1585 (*Physice* volume 1, *Primæ partis physicorum capita præcipua*) Michael Neander, a mathematician, astronomer, and professor of medicine in Jena, referred to a Regio Elementaris (fire, air, water, earth), then arranged composite bodies into five *ordines seu species ordinis*: Meteora, Metalla, Stirpes siue plantas, Animantia siue animalia, and Compositorum speciem hominem.

176. Lemery (1675), fourth (unnumbered) page of Preface, *et passim*.

177. In *Cours de chymie . . . nouvelle édition* (1756) the three groups are individually called *regnes* (always in lower case), and we find *trois regnes des corps naturels* at page 289, but only in the footnotes added by Théodore Baron for this edition. Baron also referred to the *regne des maladies épidémiques* (Lemery 1756:803na). The groups are called Kingdoms in the dedications of the 1725, 1737, and 1748 English translations of a book based in part on Lemery, Pierre Pomet's *Histoire générale des drogues* (1694), although in the original Pomet referred to them only as "les trois familles des Vegetaux, des Animaux, & des Mineraux". I have not examined the first English edition (1712).

178. Mentioned at the beginning of Chapter 1, as cited by Beseke (1797). König also edited *Thesaurus remediorum è triplici regno, vegetabili, animali, minerali* (1693). At Basel he lectured on ancient Greek and natural philosophy, before serving as professor of natural history (1706-1711) and of theoretical medicine (1711-1731).

NOTES   541

179. König, *Regnum animale* (1682:208–240); Zoophyta at pp. 227–228.
180. Hooke (1665).
181. Leeuwenhoek emphasized that magnifying power needed to be appropriate for the problem at hand. His most-powerful microscopes had a resolution approaching 1 μm.
182. Roger (1997:71).
183. Sloan, *Isis* (1976).
184. Rassemblons des faits pour nous donner des idées: Buffon, in *Histoire naturelle, générelle et particuliére,* tome 2 (1749), page 18.
185. Anderson (1982:9–10).
186. Ford (2009).
187. Antonovics & Hood (2018).

Chapter 13

1. Nietzsche, *Also sprach Zarathustra*, Vorrede, section 3 (1883) in the translation of Kaufmann (1954).
2. Leeuwenhoek recorded his observations in letters, written in Dutch, notably to Henry Oldenburg and subsequent Secretaries of the Royal Society, London, whence they were "very incompletely and imperfectly printed in the form of extracts or abstracts" in the *Philosophical Transactions.* Many letters relevant to this chapter have been translated in full by Dobell (1932); quotation at page 388. See also Corliss (1975). Leeuwenhoek's letters (in Dutch, with English translation) from 1673–1707 have been published in 15 volumes (1939–1999) and are available online at www.dbnl.org/tekst/leeu027alle00_01/. For the spelling of Leeuwenhoek's name, see Dobell (1932:300–305).
3. The Royal Society at first refused to publish Leeuwenhoek's communication on "little animals": Dobell (1932:171–173); Lane (2015).
4. *Timaeus* 28–33.
5. This formulation by Aquinas, *Summa theologica* 1.108 following Pseudo-Dionysius, *De cœlesti hierarchia* 6–10. For the latter, see *The heavenly hierarchy* in *The works of Dionysius the Areopagite* (1899) 2:23–42.
6. Epicurus to Herodotus 1.45, in *The extant remains* (1926:24–25). For other early Hellenic views on the plurality of worlds, see Kahn (1960:46–53); Duhem (1987:431–439); and Guthke (1990:36–37).
7. *Metaphysica* 1003a2 and 1071b13. Here I am of course indebted to Lovejoy (1936), pp. 50–58 and note 38 at page 338.
8. *De caelo* 276a18–279b4; see also Duhem (1987:431–435).
9. Duhem (1987:435–439).
10. Kahn (1960:46–53); Duhem (1987:434–435); and Guthke (1990:36). For the idea in Lucretius: *De rerum natura* 2.1043–1089. For Chinese Buddhist ideas of a plurality of worlds: J Needham, *SCC* (1956) 2:419–420.
11. Origen, *De principiis* 3.5.3 citing *Isaiah* 66:22 and *Ecclesiastes* 1:9, in Roberts & Donaldson (1869) 10:255.
12. Augustine, *The city of God* 12.11.
13. Albert, *De cœlo et mundo* 1.1.3.2 (Iammy, *Opera* vol. 2; Borgnet, *Opera omnia* vol. 4); Duhem (1987) pp. 446–447 and note 12 at page 544.
14. Aquinas, *Expositio super libros De caelo et mundo Aristotelis* 1.19; Grant (1979:219); Duhem (1987:448–449).
15. Aquinas, *Summa theologica* 1.15.2.4; Grant (1979:213 and note 2). Other early Scholastics broadly followed either Albert or Aquinas: see Duhem (1987:441–471) and Guthke (1990:37).

542 NOTES

16. *Quod prima causa non posset plures mundos facere*: Grant (1979:219).

17. Grant (1979); Duhem (1987).

18. McColley & Miller (1937); Grant (1994:168). Later Athanasius Kircher, a Jesuit cleric "well known as a voluminous and reckless writer on all manner of subjects" (Dobell 1932:365), wrote a highly popular story of an expedition into space (*Iterarivm exstaticvm* 1656; in editions from 1660 titled *Iter extaticum cœleste*) in which he reflects on whether a baptism in the waters of Venus would be valid (1656:91) (yes, it would).

19. Nicholas of Cusa, *De docta ignorantia* 2.12, 169–175 (Hopkins, 1981:96–98). See also Duhem (1987:509 ff.) and Guthke (1990:40). Cusanus completed the book in February 1440.

20. Bruno, De l'infinito universo et mondi. Dialogo terzo. In *Opere italiane* (1907) 1:332–361, particularly pp. 357–358.

21. *Entretiens* appeared in numerous editions—I have seen tallies of twenty-eight to thirty-three for the number during Fontenelle's (long) lifetime, while Gelbart (1990, page vii) counts "approximately one hundred" by 1990—and has been translated into at least ten other languages, despite being on the Church's *Index* from 1687 until well into the Nineteenth century, and again in the Twentieth. For Fontenelle in the context of the French Enlightenment see Marsak (1959).

22. Fontenelle, *Entretiens* (1687:146–147): [*liquids*] *full of little fish or little worms, which one would never have suspected to live there.* Editions from 1687 usually have this wording, whereas the 1686 editions at Paris and Amsterdam read *de petits Animaux*. As we shall see in Chapter 15, Linnæus in fact classified protozoa as Vermes. I have not examined a 1686 edition published at Lyon.

23. Huygens (1698), and in *Opera varia* (1724) 1:641–722. English editions appeared in 1698 and 1722. His description of extraterrestrial worlds occupies most of Book 1.

24. And not only the public: those embracing a multiplicity of worlds included Richard Bentley, Johann Elert Bode, William Derham, Nehemiah Grew, Edmund Halley, John Keill, Johann Heinrich Lambert, Pierre Louis de Maupertuis, Isaac Newton, Alexander Pope, Jacques-Henri Bernardin de Saint-Pierre, Emanuel Swedenborg, William Whiston, and Edward Young. See Guthke (1990), pp. 216–224 and 244–281; and Fara (2004).

25. Varro, *Rerum rusticarum libri tres* 12.2, in the translation of Storr-Best (1912:39).

26. Columella, *De re rustica* 1.5. *Nec paludem quidem vicinam esse oportet aedificiis, nec iunctam militarem viam; quod illa caloribus noxium virus eructat et infestis aculeis armata gignit animalia, quae in nos densissimis examinibus involant.*

27. Virgil, *Georgics* 3.558ff.

28. Lucretius, *De rerum natura* 6.1091–1095 (1712:369) as translated by Singer (1928:7).

29. Aristotle, *Historia animalium* 572a13; Varro, *Rerum rusticarum* 2.19 (1912:138–139); Columella, *De re rustica* 6.27; Virgil, *Georgics* 3.332–338; Pliny, *Naturalis historia* 8.67 (Loeb 353, 1983:116–117); Solinus, *De mirabilibus mundi* 23; Augustine, *City of God* 21.5. Ovid (*Metamorphosis* 15.364–394) is sometimes cited as referring to microscopic animals (OF Müller 1786, unnumb. page following title page; Entz 1888:224), but his *parva animalia* are worms, bees, hornets, frogs, and the like.

30. Philo, *The giants* 2, as translated by Winston (1981:62).

31. Fracastoro, *De contagione* (1546), folios 29r–33v. English translation by Wright (1930), reproduced in part by Brock (1961:69–75). See also Wolf (1935:443).

32. Harvey, *Exercitationes de generatione animalium* (1651:112), as translated by Willis (1847:321).

33. The idea that certain animals (partridge, tiger) could be impregnated (like the date palm) directly by the wind was reportedly widespread in the Arabic world: Ullmann (1972:55).

34. John Hill [as Abraham Johnson], *Lucina sine concubitu* (1750). A wickedly hilarious hoax, set out as a letter to the Royal Society, supposedly in response to having been rejected as a member. I examined only the second edition (also 1750).

NOTES 543

35. Leibniz, *Extraits de la théodicée* (1912:2) in the translation of Latta (*Monadology*, 1898:21).

36. Leibniz, *Réplique aux réflexions de Bayle* (1702) and *Epistola ad Bierlingium* (1712) respectively: Latta (1898:33n3).

37. Leibniz, in Latta (1898:213–271). The original manuscript of *Monadology*, in French, bears no title and was not published until 1839 (*Opera philosophica* 1839–1840, 2:705–712). See also Latta (1898:30–36); Clark (1947:265–268); and Look (2020), particularly sections 5.2 and 5.3.

38. Leibniz, *Monadology* [1714] 66–68 in the translation of Latta (1898:256). Needham (*SCC* 2:499, referring specifically to section 67) and others have remarked on the Buddhist-like sensitivity of this passage. From 1697 to 1702 Leibniz corresponded with Joachim Bouvet, a Jesuit missionary to China, and was familiar with Neo-Confucian thought: Needham, *SCC* 2:291–293, 340–345 and 496–505.

39. Quoted by Latta (1898:37–39). I have replaced Latta's word *inorganic* with Leibniz's *informes* (unorganized), and his *disputable territory* with Leibniz's *régions d'inflexion et de rebroussement* (*regions of inflection and of cusps*). We return in Chapter 14 to ideas of continuity among organisms. As for Budæus, Leibniz's biographer Guhrauer gives the name as Buddeus with a double *d*. Latta (1898:38n2) follows suit, supposing that Leibniz is referring to Johannes Franciscus Buddeus. This Buddeus discussed animals, plants, and minerals in *Institutionvm philosophiæ eclecticæ* [1706], but does not mention zoophytes or plant-animals even when touching on the plant-animal interface (Fifth edition, 1714–1715, 2:153–154). Lovejoy (1936:145) and Dawson (1987:130) spell the name with a single *d* indicating Guillaume Budé, who (as we saw in Chapter 11) *did* call zoophytes *plant-animals*. Dawson points out that in this passage Leibniz predicts that zoophytes would *someday* be discovered. He must have known that the term was widely applied to sponges, jellyfish, and the like; did he disagree with this usage?

40. *Monadology* 19 ff and 70; Latta (1898) pp. 30–36 and 108–112.

41. This passage follows Mackie (1845:134).

42. For preformationism versus epigenesis, see Pinto-Correia (1997).

43. *Monadology* 74.

44. Rádl (1905) 1:68–71 (quote at page 70); Latta (1898:114–116).

45. Leibniz mentions Leeuwenhoek in *Théodicée* (1710) 91, and in 1715–1716 exchanged five letters with Leeuwenhoek. For details of their content see Ehrenberg, *Rede* (1845), particularly pp. 10–16. See also Leibniz, *Principles of nature and grace* 6 (Latta 1898:412–414) and *New system* 6 (1898:304–306).

46. Needham, *SCC* 2:292.

47. Kant. Über die von der K. Akademie der Wissenschaften zu Berlin für das Jahr 1791 ausgesetzte Preisfrage (1804), in *Sämmtliche Werke* (1840) 1:521.

48. Fontenelle, *De l'Existence de Dieu*, in *Oeuvres* 3:231–242, at pp. 235–236, as translated by Marsak (1959:27). Fontenelle's essay is in the *Oeuvres* of 1742 (Brunet), not that of 1790–1792 (Bastien) as asserted by Marsak (1959:27n19, referring to 1959:5n1).

49. Sines & Sakellarakis (1987).

50. Aristophanes, *The clouds* (419 BCE).

51. Seneca, *Quæstiones naturales* 1.6.5–6.

52. According to Miall (1911:28–29), Roger Bacon mentioned crystal lenses for reading (1276). Dobell (1932:363–364) credits [the fictitious] Salvino d'Armato degli Armati with the invention of spectacles, and the monk Alessandro di Spino (*ca* 1300) with their subsequent popularization. Rosen (1956) argues that these claims are without substance.

53. D Barbaro (1569), although according to Price (1957, page vii) there may be an oblique reference in Cardano's *De subtilitate* (1550).

54. Porta, *Magia naturalis* (1597:571–619). Lens-making was not mentioned in editions earlier than 1589. See also Price (1957), page vii. Baruch Spinoza was a grinder of optical lenses.

544   NOTES

55. Istituto e Museo di Storia della Scienza. Galileo's microscope (undated); and Miall (1911:29).

56. Stelluti, *Melissographia*. See Istituto e Museo di Storia della Scienza. *Apiarium* [e] *Melissographia* (undated), and Bignami (2000). Stelluti's edition of the satires of Persius (*Persio*, 1630) includes two engraved plates, one showing bees (page 52), the second showing two weevils (page 127). The bee was the heraldic emblem of the Barberini family, to which Pope Urban VIII belonged.

57. In a letter from Giovanni Fabri [Faber] to Federigo Cesi, 13 April 1625: see Dobell (1932:364) citing G Govi (1888). Prince Cesi was a founder of the Accademia dei Lincei.

58. Borel (1656). Henry Power also published a book of micrographia (1663–1664) prior to Hooke's vastly better-illustrated *Micrographia* (1665). For the social context of *Micrographia*, see Bennett (1989). Needham (*SCC* 2:516) remarks on the book *Chhi chhi mu lüeh* (*Enumeration of strange machines*), written in 1683 by Tai Jung, which treats the inventions of Huang Lü-Chuang; among these were "microscopes and magnifying glasses".

59. Dobell (1932:109–111). Cole (1926:10–11) calls our attention to a few specialized exceptions: nummulitic limestones (containing fossilized foraminifera), known to the ancients and mentioned by Strabo (*Geographica* 17.1.34) and l'Écluse (letter to Johannes Crato von Krafftheim [April 1567], in *Epistolæ* 37 (1847:55–56)); the foraminiferan *Vaginulina*, referred to by Gessner (*De rervm fossilivm,* 1565) as a mollusc of genus *Strombus*; and a protozoan, possibly *Rotalia beccarii*, illustrated by Hooke (1665:80–81 and Scheme 5, Figure X following page 44).

60. Dobell (1932:109–299) and Cole (1926:12–14) and (1937) venture genus-level identification for several of these.

61. JR Baker (1953:421).

62. Cole (1937:9–11).

63. Dobell (1932:373–374); M Santer, *Perspectives in Biology and Medicine* 52:566–578 (2009).

64. Leeuwenhoek, "Eighteenth letter" dated 9 October 1676 (new style). It was received by Henry Oldenburg on 19 October; read in English translation to the Royal Society on 1, 15, and 22 February 1677 (old style); and published in part in *Philosophical Transactions* 12 (1677). The wretched creatures in question are ciliates of genus *Vorticella*. See Dobell (1932:112–166); quotation at page 118, from whence the title of this chapter is taken. The phrase "most wretched creatures" does not appear in Oldenburg's translation as published (1677:821). The entire letter, in Dutch and English, is at Leeuwenhoek, *Alle de brieven* 2:60–161 (1941), with the "wretched creatures" at page 67.

65. For reports by Leeuwenhoek, see Dobell (1932:388–397) and Cole (1937:185–235). Other letters appeared from [Sir] Edmund King (1694); [Rev.] John Harris (1698); and an anonymous Gentleman in the Country (1703), whose letter in issue 284 includes the first description and illustration of *Paramecium* (Figure E), and in issue 288 perhaps the first description and illustration of a diatom, *Tabellaria* (Figure VII): see Cole (1926:14–15) and Dobell (1932:371–372). Filippo Buonanni, a student of Kircher, depicted a ciliate (probably *Colpoda* or *Colpidium*) in *Observationes circa viventia* (1691), opposite page 175. For the probable identity of the Gentleman, see Dolan (2019).

66. Leeuwenhoek (Letter 18) and others describe the little animals "attempting to escape" from entrapment in filaments or particles.

67. King (1693:863–865). King specifically mentions their "sensibility" at page 865.

68. Leeuwenhoek, Brief 67 [Dobell/Cole 35] dated 3 March 1682, to Hooke. In: *Alle de brieven* (1948), 3:382–415, at page 397. Louis Joblot (below) agreed (1718) 1:33.

69. Leeuwenhoek, Brief 207 [Dobell/Cole 122] dated 2 January 1700 to Hans Sloane. In: *Alle de brieven* (1993) 13:2–23, at pp. 16–17; and (1701:515). As recently as 1916 *Volvox* was considered a "primitive metazoan" (Doflein 1916:455–456).

NOTES 545

70. Joblot, *Descriptions* (1718). The book was brought into English (with sections deleted, the remaining text translated or paraphrased, and new text added), together with Abraham Trembley's *Mémoires* (1744), by George Adams as *Micrographia illustrata* (1746). *Descriptions* was posthumously republished with additional material (1754–1755). See also Lechevalier (1976), and Ratcliff (2009:33–47).

71. For example, Carpenter, *The microscope and its revelations* (eight editions 1848–1901); Jabez Hogg, *The microscope* (15 editions 1854–1911). Gosse (*Evenings at the microscope*, at least 13 editions 1859–1915) and others popularized biological microscopy (and fanciful names for microscopic animals) but without the technical introduction.

72. *Descriptions* (1718), 2:79 (*Chausson*).

73. Ibid. (1718), 2:38–40; Adams (1746:122–123). Joblot conducted his experiments in June–October 1711; Spallanzani began his in May 1761 (Mancini *et al.*, 2007).

74. Chapter 12, note [165].

75. Adams (1746:110–111). I have been unable to locate a precisely corresponding passage in *Descriptions*, although important elements can be found at (1718) 2:3–4 and 2:45–46. Note that Joblot "uses the terms 'insects' and 'fishes' to designate almost anything" (Lechavalier 1976:248).

76. Leclerc had been made a member in 1734 based on several publications in mathematics. He was ennobled only in 1772. This section draws on publications by Lanessan (1884) and Roger (1997). We meet Buffon again in Chapter 14.

77. In 1745 Buffon appointed his relative Louis-Jean-Marie Daubenton to succeed Bernard de Jussieu as Guardian and Demonstrator of the Cabinet of Natural History (Roger 1997:60). Daubenton's description of the Cabinet forms part of Volume 3 (1749) of Buffon's *Histoire naturelle* (1749). Buffon also helped to shape the course of French science through his appointments to posts at the Garden including Antoine-Laurent de Jussieu, Lacépède, and Lamarck.

78. According to the anonymous prospectus in *Le Journal des Sçavans* (1748:639–640), the series would consist of 15 volumes in quarto: nine embracing the preface, methods, and the animal kingdom; three the vegetable kingdom; and three on minerals and precious stones.

79. Lanessan (1884), from whom I borrow *lumineuse* (*Notice biographique*, page 44*) and *un vaste spectacle* (*Introduction*, page 51*). The asterisks denote text by Lanessan, and are in the original.

80. Lanessan, *Notice* (1884:43*).

81. Lanessan argues that Buffon was "the true founder of the doctrine of transformism and evolution . . . the precursor of Darwin" (*Introduction*, 1884:411*), and strongly influenced Adanson, Bonnet, Goethe, Lamarck, von Baer, and others.

82. We return in Chapter 14 to Buffon's views on continuity in nature.

83. Buffon, *Histoire naturelle* (1749) 2:1–168. For the completion date, see Roger (1997:126).

84. Lanessan, *Introduction* (1884:317*–319*): *moule intérieur*.

85. Ibid. (1884:311*–317*).

86. Ibid. (1884:321*–322*).

87. *Histoire naturelle* (1749) 2:24–25, as translated by JS Barr in Lyon & Sloan (1981:173). In 1737 Buffon translated Newton's *Fluxions* into French, but after 1738 his writing came to reflect Leibniz's ideas: Sloan & Lyon (1981:20–24).

88. Lanessan, *Introduction*, pp. (1884:322*–327*).

89. JT Needham, *New microscopical discoveries* (1745:56–59); *Nouvelles observations microscopiques* (1750:65–70).

90. Roger (1997:140–142); Sloan (1992). The modified Wilson-Cuff instrument used by Needham is described by Baker (1742:14–15), and illustrated in Lyon & Sloan (1981:164) and Sloan (1992:423).

546   NOTES

91. Needham, *Philosophical Transactions* (1748), section 20, page 637.

92. Ibid. (1748), sections 19–22. Needham and Buffon conducted scores of such experiments. See also Roe (1983:161–162), and Pinto-Correia (1997:187–191). Needham's report was published separately (1749), and in greatly extended form in *Nouvelles observations* (1750:245–524).

93. Needham, *Philosophical Transactions* (1748), sections 16–17.

94. Ibid. (1748), section 27, pp. 646, 647, and 650.

95. Ibid. (1748), section 29, pp. 653–654. His observations of (male and female) spermatic animalcules are at sections 23–25, pp. 640–645; see particularly section 24, page 642.

96. Spallanzani, *Saggio* (1765); translated by Needham as *Nouvelle recherches* (1769). Needham appended extensive commentary on Spallanzani's text (pp. 139–235), plus remarks on the 1767 Linnæan thesis *Mundus invisibilis* (pp. 235–238: see Chapter 15); on "the passage from vegetal to vital" (pp. 239–261); and on the vegetative "propagation" of Eve from Adam (pp. 291–298). Spallanzani's 1765 tract has been reprinted (1914).

97. Spallanzani, *Opuscoli* (1776). See also Roe (1983:160–168); and Mazzolini & Roe (1986).

98. Farley (1982).

99. Needham (1748), sec. 26, page 645.

100. Ibid. (1748), section 15, page 631.

101. Ibid. (1748), section 31, page 657.

102. Ibid. (1748), section 15, page 630.

103. Note also that at the time, water moulds were not generally considered zoophytes.

104. Needham to Bonnet, 10 January 1761. Letter 7 in Mazzolini & Roe (1986:193–196), at page 195. See also Roe (1983:167) and Mazzolini & Roe (1986), pp. 41 and 73. Buffon had done so in volume 2 of *Histoire naturelle* (1749): see Chapter 14.

105. Haller (1753) had distinguished between irritability and sensitivity. His essay was soon translated into English, French, and Italian (1755). See also Mazzolini & Roe (1986:195n7).

106. JT Needham, Note de l'éditeur following the text of l'Abbé [Blaise] M[onestier], *La vraie philosophie* (1774:460–470); quotation from page 463. Needham wrote that "Végéto-Vital appartient aux Zoophites, qui, quoi-qu'ils aient pour la plupart la faculté de se mouvoir progressivement, sont purement vitaux sans être sensitifs; comme aussi aux parties organisées de tout corps animal quelconque, soit purement irritables, soit destinées à être les instrumens immédiats de la sensation" (page 461). See also his *Idée sommaire* (1781:4–9) (first edition 1776).

107. G Rousseau (2012).

108. Hill, *An history of animals. Containing descriptions of the birds, beasts, fishes, and insects, of the several parts of the world; and including accounts of the several classes of animalcules, visible only by the assistance of microscopes* (1752), pp. 1–12.

109. Cyclidium, Paramecium, Craspedarium, Brachurus, Macrocercus, Scelasius, Brachionus.

110. Hill, *A review* (1751:78) and *History of animals* (1752:87). With the former, he sought to cast ridicule on the Royal Society (supposedly for not having been invited into its Fellowship); he found worthy targets including John Bartram's claim (1744) that oysters and mussels, like plants, obtain nourishment through their "roots" (Hill 1751:84–88).

111. Hill (*The history of plants*, 1751) grouped sponges, corals, seaweeds, and algae into the first class of plants, Submarines (pp. 1–25), and recognized fungi as plants on the basis of their "absolute and perfect seeds" and roots "of the same nature with those of other vegetables" (pp. 26–72). Some species of fungi have flowers "consisting usually each of a single anthera" (page 26).

112. Bradbury (1968), particularly pp. 85–146.

113. Baker's *Microscope made easy* went through five editions in English, and was translated into Dutch (1744) and French (1754).

114. Bennett (1989); Ratcliff (2009); Hausmann & Machemer (2018).

NOTES   547

115. Baker (1753:402–403). *Noctiluca* was discovered by Joseph Sparshall (see Baker 1753:402) and first illustrated by Martinus Slabber (1778) at Plate 8, Figs 4–5.

116. *Employment* (1753:260–266) and Plate X, No. XI, 1–6. Baker observed the Proteus together with John Turberville Needham (page 265).

117. Virgil, *Georgics* 4:439 ff. Jabez Hogg (*The microscope*, 1854:134) pointed out that Proteus "could be either animal, vegetable, or elemental in his nature". Later authors have generally considered Baker's Proteus a ciliate.

118. Or worms, although by this he often refers to the larval stage of an insect, *e.g. Employment* (1753) pp. 249, 272–273, and 313.

119. Rösel came from a family of artists; the honorific "von Rosenhof" was bestowed on his uncle Wilhelm Rösel, whose paintings featured landscapes and animals. *Insecten-Belüstigüng* was also translated into Dutch (1764). Rösel also published a beautifully illustrated book of frogs (1758) with an introduction by Albrecht von Haller.

120. Rösel, *Insecten-Belüstigüng* 1 (1746), Vorrede, sig. C2r ff.

121. Ibid. 1 (1746), Vorrede, sig. Cv—C2r.

122. Ibid. 3 (1755), pp. 617–621 and Supplementary Table C1, Fig. 3.

123. Ibid. 3 (1755), pp. 621–624 and Supplementary Table C1, A-T.

124. Baker (1753:322–324) and Plate XII, Fig. 27.

125. In his report he has "mixed [them] in with" [*mit untermische*] insects: *Insecten-Belüstigüng* 3 (1755), page 617.

126. Ledermüller (1756, 1758).

127. Ledermüller (1760–1762); he added an *Anhang* in 1765.

128. Ledermüller (1760) 1:88 and Table 48 (1760) for "worms in hay-infusion"; and (1761) 2:174–176 and Table 88 for the trumpet-animal (*Stentor*) and "social polyps" (vorticellids). He also described nematodes of vinegar and paste (1:33–36 and Table 17); freshwater polyps (Tables 67, 71, 82 and 87); and "Schlammoos" (the green alga *Hydrodictyon reticulatum*), Table 72.

129. The section "Die Würmer im Heuwasser" begins *Diese Kreaturen gehören mit in die Classe der Infusions Thierlein* (1760, 1:88), but later in the same paragraph concluded "I therefore think that these little animals arise from eggs which are laid by certain insects upon fresh grass in the summer, and after the hay is gathered after lying [in the field], they then crawl out into the water. Perhaps with the rain or mildew their eggs may already be in the grass, because even in stagnant rainwater they can be found with other kinds of infusion-animalcules" (translation MAR).

130. Ledermüller, *Amusement microscopique* (1764) 1:118–119 and Table 48. Bütschli (in *Bronn's Klassen und Ordnungen* 1887:1129–1130), Entz (1888:224), Lankester (1888:806), Miall (1912:303) and Locy (1925:220) are among those incorrectly claiming that Ledermüller used the term *Infusoria* and/or established a Class for these animals.

131. At signum A (which, had it been numbered, would be page 1) Wrisberg gives the title more fully as *Observationvm de animalcvlorvm infvsoriorvm genesi et indole satvra*. That is, *Observations on the genesis and diverse nature of infusorial animalcules* (translation by MAR). In Roman times a *satura* was a dish filled with various sorts of fruit: thus, a mixture or medley. Wrisberg dedicated his book to Haller and to Haller's friend Paul Gottlieb Werlhof, a physician and poet.

132. Tubbs *et al.* (2014).

133. Wrisberg also reported observations on worms (1765:1–14), eggs (1765:14–16), and river-water (1765:52–54).

134. In addition, *animalicula spermatica* and *procreanda infusoria* occur once each, at pages 29 and 82 respectively. The words *animalcula* and *infusoria* are sometimes separated within a sentence.

548 NOTES

135. The supposition of Cole (1926:17) that Wrisberg was probably the first to employ the term *Infusoria* is incorrect.

136. Notable discoveries included foraminifera (Bianchi writing as Janus Plancus, 1739), cyano-bacteria including *Spirulina* (Corti, 1774), and the heliozoan *Actinosphærium* (Eichhorn, 1775, 1783). See also the historical overview by Ehrenberg (1838:519–520).

137. Brownian motion: Brown (1828, 1830) and in *Miscellaneous botanical works* (1866) 1:463–486; Pearle *et al.* (2010). For size comparison with Needham's atoms, see Lyon & Sloan (1981:166–167).

138. Ruestow (1996:275n62) points out that Wrisberg seems to have used a powerful single-lens instrument for some observations: see Wrisberg (1765:17).

139. Ruestow (1996), especially pp. 260–279 and 284–285.

Chapter 14

1. La Mettrie, *Man a machine*, in the translation of Bussey as revised by Calkins (1912:117), corresponding to *L'homme-machine* (1748:46).

2. Lovejoy (1936:56–66); quotations at pp. 59 and 58 respectively.

3. Carli-Rubbi (1750) also mentions "the Pyramid of nature".

4. PG Kuntz, Formal preface (1987), particularly note 9 at pp. 12–13.

5. Homer, *Iliad* (Lattimore) 8, lines 18–27. Later (*Iliad* 15, lines 18–22) Zeus relates having sus-pended Hera between earth and sky on a golden chain.

6. *Theaetetus* 153c–d. According to Nussbaum (1978:320–321), in this passage Plato (for Socrates) "claims that by the golden rope Homer actually means the sun, and that the point of the passage is to show that the continual regular motion of the heavens is necessary to preserve order in the universe."

7. *De motu animalium* 699b35–700a3.

8. For the period up to Macrobius, see Edelstein (1953:48–66). The unidirectionality of its threatened use was noted by Bodin (*Colloquium of the Seven*), who further has Salomon opine that "the Homeric chain is nothing other than the ladder represented in the nocturnal vision of Jacob the Patriarch" (MLD Kuntz, 1975:32). See also Tashiro (1965).

9. *Ex hæc est Homeri cathena aurea, quam pendere de cœlo in terras deum iussisse commemorat.* Macrobius, *Commentarii in somnium Scipionis* (1528), folio 37v; in the Stahl translation (1952:145) 1.14.15. The *editio princeps* of Macrobius's *Commentarium* was printed by Nicolaus Jenson (1472).

10. Porta (1658:7); Lovejoy (1936:63).

11. Kirchweger (1723), figure prior to title page. See also Chapter 9.

12. Kuntz & Kuntz (1987:324).

13. *Genesis* 28:12. Some versions (*e.g.* the New International Version) refer to a *staircase*. Philo (*On dreams* 1.133 ff) uses κλῖμαξ, which is usually translated *scale* or *staircase*.

14. Kuntz & Kuntz (1987:324) identify Origen, Gregory of Nyssa, John Chrysostom, John Climacus, Augustine, and Benedict of Nursia.

15. Ibid. (1987:323–327).

16. Budge (1895), pp. lxx–lxxi and cv. For dating, see Chapter 1 note [15].

17. Origen, *Contra Celsum* 6.22 (1953:334).

18. Kuntz & Kuntz (1987:319–334) argue that the tree is "the most universally widespread" symbol, offering examples from ancient India and Scandinavia through Ramon Llull to Max Ernst. However, tree-symbolism reaches unhelpfully far beyond connectivity between earth and heaven per se, encompassing *e.g.* mortality and death, and sin and redemption. See Frazer (1890); R Cook (1974); and Ladner (1979).

19. *The Epic of Gilgamesh* (Sandars) 1960:70–84.
20. Scholem (1987:71), referring to *Isaiah* 44:24 ("I am the Lord that maketh all things; that stretcheth forth the heavens alone; that spreadeth abroad the earth by myself."). For *Bahir* see also Chapter 7.
21. Scholem (1987:72–80), particularly page 76.
22. Pseudo-Chrysostom, Sermon 6, sec. 5, in Migne (1862) 59, cols 743 (Latin) and 745 (Greek); in English translation at de Lubac (1988:442–443) with "stairway of Jacob" at page 442.
23. Reuchlin, *De arte cabalistica* (1494). The copies I consulted were bound with G Pico della Mirandola, *Opera omnia* (1557, 1572), in which see pages 836 and 2:3104 respectively.
24. Porta, *Magiæ natvralis* 1.5 (1560). By the 1597 edition this chapter, now expanded, had moved to 1.6, where it remained for the 1658 English translation (cited here at page 8).
25. Milton (1668), Book 2, lines 1004–1005 and 1050–1051; and Book 3, lines 510–543.
26. Jean de Meun, *Roman de la rose* [*ca* 1275]. *Roman* was written in two stages: 4058 lines by Guillaume de Lorris *circa* 1230, then a further 17724 lines by Jean de Meun *circa* 1275. In the four-volume Méon edition (1814), the four degrees of nature (*pierres et metaulx, vegetaulx, tiers, homme*) are at volume 4, pp. 171–172.
27. Llull, *Liber de ascensu et decensu intellectus* [1304]. I have seen the work cited as 1305 Montpellier. The staircase diagram (which precedes the table of contents of *De ascensu et decensu* in the 1512 folio) is reproduced in Ragan (2009). Interestingly, fire (*flāma*) has its own step, above stones but below plants. Llull's argument is complex, involving not one but three interrelated ladders (Cofresi, 1987:154–155).
28. Raymond de Sebonde, *Theologia naturalis* [1434]. See also Tillyard (1943:25–26).
29. Fortescue, *De natura legis naturæ* 2.59 [1461–1463]; in English (1869) 1:322 (pag. var.).
30. Bouelles, *Liber de sapiente* (1510), folios 117v and 119v (figure reproduced in Kuntz & Kuntz, 1987:312); and *Physicorum elementorum* (1512), folios 23v and 73v (the latter reproduced in Ragan, 2009).
31. F Bacon, *Sylva sylvarum* (1627), layout and *e.g.* 601–609 (pp. 153–155).
32. Mersenne (1623), cols 325–326 (*graduum mundi*).
33. Bourguet (1729). See also Beseke (1797:110–111).
34. Nieremberg (1635:29).
35. Copernicus (1543).
36. Bruno (1950:313–314), English translation based on the "Venice" edition of 1584, which however was probably printed in London (McIntyre, 1903:358n4; Elton, 1907:322n20). See also (1950:239). Bruno went on to use the new cosmology to undercut Aristotle's ideas on animals (1950:315–318).
37. Guthke (1990:54–57) considers the Great Chain to have been compromised; Lovejoy (1936:99–143) disagrees.
38. Latta (1898:52–53); Hatfield (2018).
39. [Epistle] CLXXIXb. Descartes a Mersenne (1639), in *Oeuvres* (1913:97).
40. This logic led Réaumur to deny (at first) that the freshwater polyp (which lacked such internal structures) could be an animal, or that the option even existed for it to be intermediate between plants and animals: see Dawson (1987), particularly pp. 17–18 but also pp. 95 and 100–105.
41. By Ignace-Gaston Pardies, Noël Régnault, and others: see Rosenfield (1941), and Dawson (1987:32–34).
42. For this period, see Latta (1898); Cassirer (1932), in English (1951); Lovejoy (1936); Glass (1968); H Brown (1976); Shank (2008); and Feingold (2010).
43. Lovejoy (1936:183).
44. I exclude Kirchweger's alchemical-Hermetic *Aurea Catena Homeri* (1723) with its Spiritus Mundi, Universal Sperm and regenerated Chaos.

550 NOTES

45. Addison and Richard Steele co-founded *The Spectator* as a daily newspaper in 1711, catering to the emerging middle class. It ran until 1712, then was restarted by Addison in 1714.

46. Addison in *The Spectator* 121 (19 July 1711); in the [1907] reprint at volume 2, pp. 141–146; quotations from page 145.

47. Jackson (1886).

48. Bradley (1721:13–14). He also quotes (at page 10) from an article by William Stukeley FRS (1719) in which coral is given as a special case of the growth of stones.

49. These are the supposed ability of coral to strike root, its brittleness, and its absence of capillary vessels (Bradley 1721:15–17). For the latter, Bradley refers to "Enquiry with my best Microscopes" (page 16).

50. Bradley (1721:17–19).

51. Ibid. (1721:22).

52. Ibid. (1721:25–32). I find no evidence that this essay was published separately. See Dryander (1796) 2:459.

53. Bradley (1721:33–34).

54. Ibid. (1721:37–41).

55. Ibid. (1721:49–51).

56. Ibid. (1721:53–59). Bradley does not treat jellyfish, squid ,or octopi.

57. Ibid. (1721:71–75). At page 69 he seems to group these under Quadrupeds(!), but does not follow through.

58. Ibid. (1721:76–88).

59. Ibid. (1721:86–87).

60. Ibid. (1721:88–95).

61. Bradley assures us that some larger apes and monkeys can be taught to perform simple tasks on command; like men, they will drink strong liquor to excess, and love tobacco, such that they will smoke it abundantly if they can obtain it (1721:95).

62. Ibid. (1721:97).

63. Ibid. (1721:95–116), particularly pp. 101 ff.; quotation at page 103.

64. Ibid. (1721:104).

65. Ibid. (1721:105).

66. Ibid. (1721:116).

67. Ibid. (1721:147).

68. Ibid. (1721:155).

69. Ibid. (1721:155–157).

70. Ibid. (1721:159). One wonders if Bradley paused to consider if this claim could be reconciled with the possibility of life on other worlds.

71. Ibid. (1721:162).

72. Blumenbach (1806:107–108) accused Bradley of holding back for a second Scale (from man downward) those things which cannot be accommodated (*nicht füglich zu unterbringen lassen*) in the first. Thienemann (1910:241) called Bradley's Scale "highly peculiar" (*höchst sonderbar*).

73. Lovejoy (1936:186–207).

74. Theophrastus, respectively *Historia plantarum* 4.7.1 (see also Chapter 3) and *De lapidus* (Chapter 4, note [51]). For al-Bīrūnī, see Chapter 7.

75. Tournefort, *Élemens* (1694) 1:445–446; and *Institutiones* (1700) 1:572–575.

76. Réaumur writing anonymously (1712:70). Marsigli (1707) referred to a "bark" of corals and other so-called lithophytes, but without distinguishing it as the only vegetable portion of the organism. Leeuwenhoek (1708) noted that a mineral shell was left behind after coral was heated.

77. Marsigli (1725:126–127).

78. Finding coral largely composed of "fixed salts", Leeuwenhoek wrote that those who claim a medicinal value for coral merely "amuse common People with uncommon Medicines, and thereby get themselves a Name, whilst they are in the mean time only cheating the World" (1708:134).

79. Boyle, Experiment 42, in *Works* (1772) 1: 113–114; see also 1:435–441. Boyle considered coral to be a plant: (1772) 3:59–60.

80. This is substantially earlier than other marine research stations: Kofoid (1910); Murray & Hjort (1912:21); Egerton (2014).

81. Secondary sources for this section include JR Baker, *Abraham Trembley* (1952:118–129); and Gibson (2015:117–128). Unpublished letters and documents important to this narrative are discussed by McConnell (1990); and Vandersmissen (2012).

82. McConnell (1990:55–56).

83. Marsigli (1707); and (1725:115–116) and Tables 39 and 40. Jean-Paul Bignon was a member of the Académie Francaise and of the Académie des Sciences.

84. François Xavier Bon de Saint Hilaire was former president of the Société royale de Montpellier, and an early investigator on spider-silk.

85. McConnell (1990:58–60).

86. Ibid. (1990:64).

87. Réaumur writing anonymously, Sur le corail (1729); and reading work by Peyssonel (1729), especially pp. 274–275.

88. Peyssonel (1752:454) per Watson.

89. Peyssonel [Watson], *op. cit.* (1752:456–457); see also pp. 464–465 and 468–469. The view that the coral-animals (and indeed other marine organisms) secrete their mineral polypary much as wasps build up their nest was shared by Réaumur and others, although not by Trembley (Baker, *Trembley* 1952:127) or Parsons (1752).

90. The observations were made in 1741, published in 1744: Trembley (1744:208–220).

91. B de Jussieu: read in 1742, published in 1745. According to Johnston (1838:8), most of de Jussieu's observations of the 1740s were never published.

92. Réaumur, *Memoires* (1742) tome 6, pp. lxxiij–lxxx.

93. Donati (1750), page v. The Italian original (the text of which is dated 2 November 1745) was translated into German (1753) and French (1758). Excerpts read before the Royal Society on 7 February 1750 (published 1751) and 1 April 1756 (published 1757) are noted as having been translated from French. The German edition follows the outline of the Italian original, but appends new material including a Chain of Being according to Bonnet. In the French edition the textual material is divided differently, with Chapter 7 "The first degree by which Nature passes in the sea from plants to animals, or the Polyparies"; Chapter 8 "The second degree of this passage: or the true Zoophytes"' and Chapter 9 "The third degree of this passage, or, the Phyto-zoos, or Animal-plants".

94. In 1756, Donati wrote to Trembley that "I am now of opinion, that coral is nothing else than a real animal, which has a very great number of heads. I consider the polypes of coral only as the heads of the animal" (as translated by Birch): Trembley (1757:59).

95. Donati (1750), page xxi.

96. Ibid. (1750), Indice de paragrafi (no page number) and page xxii. The former passage, dated 1749, implies that *Della storia marina* will be divided into four books, recognizing marine plants plus three intermediate stages; the latter, written in 1745, mentions only two intermediates: *Poliparj*, and *Zoofiti o Animali-piante*. The monograph is set out in four books corresponding to the divisions in the Index.

97. Ibid. (1750), pp. xx and xxi respectively (translation as in Ragan 2009).

98. Ibid. (1750), pp. xx–xxii.

552 NOTES

99. The French edition uses *entrelacés* to describe the interconnection of rings between chains (1758:20). For Nieremberg's network, see Chapter 11.

100. Appreciation that the coral animal hosts a range of endosymbionts, notably dinoflagellates of family Symbiodiniaceae but also apicomplexa, fungi, bacteria, and archaea, would come much later.

101. Hill, *A general natural history* (1751) 2:12–14; Parsons (1752); Baker, *Employment* (1753:217–218, and 1764:217–218); Baster (1757); Pallas (1766:229–230); Ritterbush (1964:130–137); Ratcliff (2009:117–123). We return to this matter in Chapter 15.

102. Leeuwenhoek, Letter 149 (Dobell 1932:275°285) or 239 (*Alle brieven* 14:159–179); and (1703).

103. A Gentleman in the Country, Two letters (1703).

104. For details of Trembley's residence in The Netherlands, see Baker, *Trembley* (1952:12–19); and Lenhoff & Lenhoff (1986), Book 1, pp. 33–34. For Jussieu: Baker, *Trembley* (1952:37).

105. For the origin and evolution of the name, see Trembley (1744:11) and Baker, *Trembley* (1952:33–34). The genus name *Hydra* was assigned by Linnæus (*Fauna svecica* 1746:367) and included in the Tenth edition of *Systema naturæ* (1758:816–818). Trembley's three species were accorded Latinate binomials by Pallas (1766:16–17 and 25–32).

106. Trembley (1744:7–8), as translated by Lenhoff & Lenhoff (1986), Book 2, page 5.

107. Ibid. (1744: 9–11), and (1986) 2:6–7.

108. Ibid. (1744:11–12 and 66–69), and (1986) 2:7 and 37–39.

109. Ibid. (1744:12–13), and (1986) 2:7.

110. Ibid. (1744:12–17), and (1986) 2:7–10. Indeed, the basal portion regenerated an apical portion with the same number of arms as before.

111. Ibid. (1744:18–19), and (1986) 2:11. Leeuwenhoek had interpreted the same process as a mother (animal) giving birth to its young (1703:1307).

112. Ibid. (1744:20), and (1986) 2:11. These were of a different species.

113. Ibid. (1744:19–66), and (1986) 2:11–37.

114. Ibid. (1744:111), and (1986) 2:69–70.

115. Ibid. (1744:149–208), and (1986) 2:93–125.

116. M Trembley (1943); Baker, *Trembley* (1952), pp. xvii–xix, 33–40, and 43–48; and Ritterbush (1964:122–127).

117. Réaumur (1742), pp. xlix–lv and lxxvj–lxxvij.

118. Ibid. (1742), pp. lv–lxx.

119. Trembley (1744:302–304), and (1986) 2:182–183.

120. Ibid. (1744:306), and (1986) 2:184. Buffon made the same point in *Histoire naturelle* (1749) 1:260–262.

121. Ibid. (1744:306–307) and (1986) 2:184–185, citing Boerhaave (Leiden 1732:57 ff.): *Unde alimenta plantarum radicibus externis, animalium internis, hauriuntur; terra alens stirpi eterna semper, interna vero animali perpetuo habetur.* I have examined only the London (1732) and Paris (1733) editions, where the passage appears at 1:24 and 1:35 respectively. Boerhaave goes on to say that the same generalization applies to zoophytes: the animal is affixed inside, the shell fastened externally.

122. Trembley (1744:253 ff.), and (1986) 2:154 ff.

123. Ibid. (1744:307–308), and (1986) 2:185.

124. Buffon, *Histoire naturelle* (1749) 2:20–21.

125. Ibid. (1749) 2:261–263; *Oeuvres complètes* (Lanessan) 4:289–290 (1884–1886).

126. Rousseau (1946) page 131, and note 187 at pp. 215–216.

127. Vartanian (1950:260).

128. Voltaire, Chaîne des êtres créés, in *Questions sur l'Encyclopédie* (1770) 3:284–287, and in *Dictionnaire philosophique* (1822) 3:73–76. As translated by Lovejoy (1936:252): "This

NOTES 553

hierarchy pleases those good folk who fancy they see in it the Pope and his cardinals followed by archbishops and bishops; after whom come the curates, the vicars, the simple priests, the deacons, the subdeacons; then the monks appear, and the line is ended by the Capuchins."

129. Voltaire (1752). For another satire of La Mettrie's *L'homme plant*, see *The man-plant* by "Vincent Miller" (1752).

130. La Mettrie (1994).

131. Buffon, *Histoire naturelle* (1749) 2:8–9; Bonnet to Trembley (24 March 1741) cited by Dawson (1987:138).

132. La Mettrie, *L'homme machine* (1912:145), (1994:73), and *Œuvres* (1796) 3:192; and *L'homme plante* (1994:88), and *Œuvres* (1796) 2:67. Also Vartanian (1950:273). One "further property of animals" might be "fighting like polyps", perhaps a reference to competition among freshwater polyps for a worm (Trembley 1744:112), and (1986) 2:70.

133. Robinet, *De la nature* 4 (1766): no intermediates, pp. 4–5; matter is essentially organic, pp. 113–116; *Tout est animé*, page 289 as chapter summary. Robinet went on to contend that man is this single prototype: Robinet, *Considérations philosophiques* (1768). An almost identically titled work (*Vue philosophique . . .*) was published in Amsterdam (1768) as volume 5 of *De la nature*. See also Thienemann (1910:255–256); Lovejoy (1936:275–283); and Crocker (1968:134–136).

134. Diderot, *Rêve de D'Alembert* (1875) 2:128–132. *Rêve* was written in 1769, and distributed in correspondence from 1782, but published only in 1830.

135. La Mettrie touched on such issues in *Man a machine* (1912:136–138), (1994:66–67), and *Œuvres* (1796) 3:180–183. See also Vartanian (1950:271–272).

136. Mirabaud [actually d'Holbach] (1770).

137. For the distinction between Bonnet's *emboîtement* and the *enveloppement* of Leibniz discussed in Chapter 13, see Rieppel (1988:130–134).

138. Bonnet, *Essai analytique* (1760:488–493), polyps at page 493; and *Considérations* (1762) 1:28–39 (polyps at page 36).

139. For editions of the *Œuvres*, see Anderson (1982:150). Citations to the *Œuvres* are to the octavo edition.

140. And for all time, since (as we shall see in Chapters 15 and 16), the Great Chain did not outlive the Eighteenth century, or in some circles the first few years of the Nineteenth. Information on dates of material from Anderson (1976:45n1) and (1982:5–7).

141. According to Baker, *Trembley* (1952:17), their exact blood relationship is unknown but "Trembley regarded Bonnet as his cousin".

142. Their scientific views often clashed, but both were committed anti-materialists.

143. Anderson (1982:2–3) also credits *Spectacle de la nature* by the Abbé [Noël-Antoine] Pluche.

144. Dawson (1987:243–252).

145. Ibid. (1987), pp. 77–80, 88–90, and 93–95. Parthenogenesis is the development of offspring from unfertilized eggs.

146. Ibid. (1987), pp. 144–146 and 165.

147. Much of Bonnet's *Considérations* is given over to the Buffon-Needham theory, insect metamorphosis, his work on regeneration of earthworms, and Trembley's polyp. For Spallanzani, see *Considérations* (1789) 2:317–339 in footnotes.

148. *Contemplation* was also translated into English (1766) and German (1766); an abridgement was included in editions of John Wesley's *A survey of the wisdom of God in the creation* from 1809 (4:49–277).

149. Bonnet, *Insectologie*, in *Œuvres* (1779) 1, pp. xliv–xlviij; translation by MAR. For a slightly different interpretation, see Anderson (1982:6).

150. Bonnet, *Contemplation*, in *Œuvres* (1781) 7:54. This is generally consistent with the rest of *Contemplation*. Anderson (1982:146) asserts that "the chain of beings was not presented

554 NOTES

by Bonnet as a description of the actual order of nature; rather, it served as a model or as a metaphor, useful for taxonomic purposes as a means of envisaging a configuration of all the relations binding together everything in the world. It was these relations that were foundational for him. It was their nature and scope that had to be understood, both among the objects of the world and among the objects of the mind, in order that some semblance of the interrelated whole might be grasped." Bonnet—a man of the Enlightenment—certainly analyzed "the idea as a representative of the thing" (Voltaire) more explicitly than had (say) Nieremberg, but where Bonnet discusses the steps on his Scale (*e.g. Œuvres* 1:56–176) he speaks of actual kinds (often genera) of minerals, vegetables, and animals.

151. For fungi, liverworts, and mosses, see *Contemplation*, in *Œuvres* (1781) 7:86–91. The additional organisms include Byssus and Tremella (aquatic plants: *Œuvres* 7:86n1). More precisely, he identifies the French edition of Donati's *Della storia* at *Œuvres* 7:82n2. Bonnet's scale diagram *Idée d'un échelle des etres naturels* is bound at *Œuvres* (*Insectologie*) 1:1 (1779).

152. *Contemplation*, in *Œuvres* (1781) 7:83. At ibid. 7:46–47 Bonnet follows Aristotle (or Aquinas) in recognizing vegetative, sensitive, and reflective life in plants, brutes, and man respectively, but does little with this subsequently.

153. Ibid. (1781) 7:85 and 8:322–518; *Palingénésie philosophique*, in *Œuvres* (1783) 15:402–444; and elsewhere.

154. Ibid. (1781) 7:102 and note 1.

155. Ibid. (1781) 7:85 and note 1.

156. Ibid. (1781) 7:179.

157. Ibid. (1781) 8:137–138.

158. *Considérations*, in *Œuvres* (1779) 6:301–303. Translation by Georges Merinfeld.

159. *Contemplation*, in *Œuvres* (1781) 8:137–234.

160. The chapter (Book 8, Chapter 19) on infusion-animalcules (*Œuvres* 8:221–234) which draws upon "the beautiful observations of M. Spallanzani" was not present in earlier editions of *Contemplation*. It states that polyps *include* infusion-animals (*Œuvres* 7:221). Bonnet refers to work by OF Müller (presumably *Vermium terrestrium et fluviatilium*, 1773–1774) on the classification of these organisms. We shall meet Müller in this context again in Chapter 15.

161. *Contemplation*, in *Œuvres* (1781) 7:297–298 and note 2. Bonnet has little to say about sponges in his biological works.

162. Ibid. 7:203–204.

163. Ibid. 7:130–138.

164. Anderson (1976), particularly page 55; and Bonnet, *Contemplation*, in *Œuvres* (1781) 7:82: *la Nature semble faire ici un faut.*

165. *Contemplation*, in *Œuvres* (1781) 7:47–48.

166. *Palingénésie philosophique*, in *Œuvres* (1783) 15:218–220. Charles Perrault is credited with developing the fairy tale as a genre of literature. Sébastian Le Prestre de Vauban was a French military engineer under Louis XIV. Bonnet's point is that their contributions were of a lesser philosophical level than those of Newton or Leibniz.

167. *Considérations*, in *Œuvres* (1779) 6:313–314. Anderson (1976:54–55) argues that Bonnet has gotten ahead of himself here, as it is the pre-existing germs that enable this progression, but he believes these germs to be absent from stones.

168. Jenyns (1782:1–11); see also Lovejoy (1936:197–198).

169. JHB de Saint-Pierre (1784), *e.g.* 1:79–83.

170. Herder, *Ideen* (1785), for example at 5.1 (1:265).

171. Hunter, *Georgical essays* 1.3 (1777:37–46).

172. Peirson, *Georgical essays* 2.3 (1777:318–322), particularly at page 319. In the table of contents, the family name of the author is given variously as Peirson or Pierson, but as Peirson in the 1803 edition. A Rev. Robert Peirson, formerly of Jesus College Cambridge,

NOTES   555

was Archdeacon of Cleveland (in the prebendary of York), and Master of the Grammar School, Coxwold. Hunter's *Georgical essays* was printed in York.

173. Franklin, "An Arabian tale" [1779?], in *Writings* (1907).

174. R Watson (1787) 5:169–170.

175. C White (1799).

176. Anonymous, *Entwurf* [1780], transcribed Thienemann (1910:188–230).

177. Lamarck, Discours préliminaire, in *Flore françoise* (1778) 1, pp. lxxxvij–xvij, and table at pp. cxiv–cxvij. For Haüy's contribution, see Burkhardt (1977:52), and Stevens (1984) at pp. 79–80 and note 145.

178. Lamarck, Discours d'ouverture, in *Système des animaux sans vertèbres* (1801:16–18); in the translation of Elliot in Lamarck, *Zoological philosophy* (1984:416).

179. Bonnet continued to believe that a single scale encompasses all of living nature, even if we understand some parts of it imperfectly. In the passage quoted here, however, Lamarck does not quite argue that plants and animals are part of the same scale.

Chapter 15

1. *Psalm* 104:24, Revised Standard Version. Quoted (in Latin) by Linnæus in every edition of *Systema naturæ* from 1735 through 1789, except the (incomplete) Eighth (1753).

2. Greene (1983) 1:177.

3. Locke, *An essay concerning human understanding* (1689 but dated 1690) 3.3.13 and 3.6.30, at (1853) pp. 276 and 299–300 respectively; quotation at page 299.

4. Dewhurst (1963:8); Sloan (1972:21n50).

5. Locke (1689) 3.6.12, at (1853:294).

6. Bauhin, *Pinax* (1623); see Chapter 12.

7. Magnol, *Novus character* (1720:23). The calyx is the collective name for the outermost whorl of plant parts (sepals) that form and may support the flower.

8. Magnol, *Prodromus* (1689), Præfatio, sixth and seventh (unnumbered) pages. Translation by MAR.

9. Tournefort, *Élémens* (1694) and *Institutiones* (1700).

10. Tournefort (1694) 1:40. Translation by MAR.

11. Ibid. (1694) 1:29, and especially (1700) 1:61.

12. Ray, *Catalogus Cantabrigiam* (1660).

13. Ray, as reflected in Wilkins (1668:67–120). For Ray's contribution, see the Epistle to the Reader, third (unnumbered) page.

14. Ray (1757) 3:169–173; reprinted by Gunther (1928:77–83). JC Greene (1959:129–131 and note 2) dates this discourse to 1672.

15. Ray, *Historia plantarum* (1686) 1:40. Translation from Mayr (1982:256–257). Lazenby has translated Book 1 (1995).

16. Ray, *Methodus plantarum nova* (1682), Præfatio ad lectorem, fourth and fifth (unnumbered) pages.

17. For examples, see Sloan (1972:29).

18. Mayr (1982:192–193) notes that Aristotle, Bock, Gaspard Bauhin, Magnol, and Robert Morison had used non-fructification characters openly, as did Cesalpino, Tournefort, and (later) Linnæus "clandestinely".

19. Ray, *Methodus plantarum emendata et aucta* (1703), Præcognoscenda, sig. 4r–4v. The word *commodissimum* can carry the sense of *opportunely* or *conveniently*. Translation by MAR. For a somewhat less-literal translation, see Sloan (1972:46–47), who also provides important nuance on what Ray means by "accident" (1972:45–46 and note 115).

556 NOTES

20. Ray (1682), Præfatio ad lectorem, second (unnumbered) page.
21. Ray, *Wisdom of God* (1692), for example at 1:5–6.
22. Ray (1703), Præcognoscenda, sig. 4v.
23. Ray, *Synopsis methodica animalium* (1693:5); translation by MAR. Leeuwenhoek's animalcules also feature in *Wisdom* (1692) 1:159–160 and 2:90 as examples of God's wisdom, art and power.
24. Ray (1686) 1:59–118.
25. Ibid. (1686) 1:1.
26. Ray (1693:1–2).
27. Ray (1686) 1:2.
28. For example, at (1693:50–55).
29. Ray (1692) 2:76–81, in the context of criticizing the idea of spontaneous generation of worms and insects.
30. Xenophon, *Memorabilia* 1.4.4 ff., in Loeb 168 (1923:54 ff.).
31. Goodman (1971:72 ff.).
32. Aquinas, *Summa theologica* 3.2.
33. Raymond de Sebonde, *Theologia naturalis*.
34. Newton (1726), Liber 3, Scholium generale, particularly pp. 527–529. See also Newton (1953:65–67).
35. On the Continent, *Le spectacle de la nature* (1732–1751) by the Abbé Pluche was even at the time considered derivative from Derham's *Physico-theology* (1713), while specialized works such as Lesser's *Lithotheologie* (1732), *Insecto-theologia* (1738), and *Testaceo-Theologie* (1744) proliferated. We revisit this literature in Chapter 20.
36. Derham (1713: 406–408), citing Leeuwenhoek. I have corrected Derham's " . . . all those Parts have each of of them" [*sic*].
37. Mather (1721:151).
38. Edwards, Sinners in the hands of an angry God [1741], in *Basic writings* (1966:159). Edwards studied and wrote on spiders from an early age (1966:31–37). Was he aware, for example through Topsell (1658 at 2:749), of Albert's dictum that spiders are "cold" and can remain unharmed by a "hot burning iron"?
39. Wesley (1763). In the second edition (1770): microscopic animalcules at 1:188–189, corals and corallines at 1:204–205n1 and 1:267–268, freshwater polyps at 3:73–80, tapeworm at 3:62, barnacle-goose at 3:42, translation of plants into animals and vice versa at 1:268–270 and 3:34. Wesley mentions some but not all of these in the first edition (1763).
40. Wesley (1770) 3:73.
41. Lindroth (1983).
42. By far the best English-language summary of the Linnæan dissertations is provided by Pulteney (1805:350–490). The 171 dissertations are collected in the first eight volumes of *Amœnitates academicæ*; those in the ninth volume are not Linnæan, while the tenth presents (inter alia) Linnæus's own thesis, and several of his orations. Each dissertation was also published separately, usually with text identical to that in *Amœnitates*; for reasons of accessibility, I cite the latter.
43. Linnæus, *Philosophia botanica* (1751) and subsequent editions, Aphorism 77: *Natura non facit saltus*. Stafleu (1971:28) notes that these words were crossed out in Linnæus's own copy of the *Philosophia*. Even so, Linnæus believed that as more plants are discovered, the apparent gaps between orders and classes will tend to disappear.
44. Linnæus, *Nemesis Divina*: see Lepenies (1982). English translations by Petry (2001) and Miller (2002); quotation from Lindroth (1983:51).
45. Linnæus, *Clavis medicinæ* (1766); *Metamorphosis humana* (1769).
46. Chapter 16.

NOTES 557

47. Goethe, Letters to Zelter, 7 and 14 November 1816. At: www.zeno.org/Literatur/M/Goe the,+Johann+Wolfgang/Briefe/1816. English translation (1892:140–141). See also Benedikt (1945); and Wetzels (1985), although in my opinion Wetzels focuses too much on the nomenclatural component of Linnæus's botany and not enough on metamorphosis, cortex-medulla theory and prolepsis.
48. Lindroth (1983:50–51) referring particularly to his speculations on the medulla, and on leaves.
49. The British Museum *Catalogue* (1933) lists 3741 primary (plus many subsidiary) titles, not all by Linnæus himself.
50. Pulteney (1805:42–75).
51. In the former stream Linnæus set out the three kingdoms of nature (*Systema naturæ*, 1735); wrote the first monograph devoted to a single plant genus, the banana (*Musa Cliffortiana*, 1736); classified the plants of Lapland (*Flora Lapponica*, 1737) and of the garden and tropical greenhouses of Clifford's estate (*Hortus Cliffortianus*, 1737); and organized the vegetable kingdom at the ranks of genus (*Genera plantarum*, 1737) and class (*Classes plantarum*, 1738). The latter stream consists of *Fundamenta botanica* (1736), set out as 365 aphorisms; *Critica botanica* (1737), elaborating on the nomenclatural aphorisms (§210–§324) of *Fundamenta*; and *Bibliotheca botanica* (1736) on the history of botany, including a classification of botanists. In addition, he contributed to botanical works by Gronovius and others.
52. Also *Fundamenta botanica* (1736:1), Aphorism 2. He later erected Imperium Naturæ above the three kingdoms: *Systema naturæ* (1758) 1:5–8.
53. *Lapides* crescunt. *Vegetabilia* crescunt & vivunt. *Animalia* crescunt, vivunt & sentiunt. Hinc limites inter hæcce Regna constituta sunt. Also *Fundamenta botanica* (1736:1), Aphorism 3.
54. Lindroth (1983:16–17). Linnæus refers to the Chain at various places throughout his writings, *e.g. Animalia composita* (1760). Although in the First edition of *Systema naturæ* (1735) the text is arranged Vegetablia-Lapides-Animalia, thereafter the kingdoms were ordered Lapides-Vegetabilia-Animalia (Second through Fifth) or Animalia-Vegetabilia-Lapides (Sixth through the Gmelin Thirteenth editions).
55. Each was assigned a letter of the (Latin) alphabet: J and W were not used.
56. Literally "hidden marriage" or "clandestine nuptials". Clandestine marriages were then customary in Europe, and were outlawed in England only in 1753 (Schiebinger 1996:113).
57. Also a bryozoan (Eschara) and the green alga Acetabularia (as Acetabulum, following Tournefort). Linnæus also included *Ficus* (fig tree), but corrected the error in *Species plantarum* (1753) and the Ninth edition of *Systema naturæ* (1756).
58. Linnæus maintained the list of Paradoxa through the Fifth edition of *Systema naturæ* (1747) and added further examples, including the manticore and "antilope". Likewise, he retained some paradoxical Minerals including "Judaicos lapides": *Systema naturae* Twelfth edition (1770) 3:9, and Thirteenth edition (Gmelin 1793) 3:12.
59. Only the First (1735), Second (1740), Sixth (1748), Tenth (1758), and Twelfth (1767–1770) were written by Linnæus himself. The other editions include translations, extensions, and an epitome; among these the so-called Thirteenth edition (1788–1793) edited by Gmelin is valuable. For these I largely follow the Wikipedia page on *Systema Naturæ* (accessed 10 August 2019) but omit the Göttingen (1772) edition. The Eighth edition treats only Vegetabilia.
60. See also *Corallia Baltica* (1749); English translation by Brand (1781:457–480).
61. Linnæus included a few foraminifera with the molluscs, and classified vorticellids with the hydrozoa (genus *Vorticella*). He lumped all infusorial animals under *Chaos infusorium*. For *Furia infernalis*, see *Miracula insectorum* (1756:322); Pulteney (1805:516); and Jamieson (1827:39–43). Stevens & Cullen (1990:216) argue that *Furia* is and always was apocryphal "despite the fact that a vicar, no less, reported that one had fallen onto his plate". Linnæus attributed to *Furia infernalis* a mysterious swelling on his arm, while sunbathing near Lund

558 NOTES

in 1728, that became infected and nearly caused his death: Heller (1964:45). Nor are all species of his *Volvox* and *Chaos* protozoa. The bioluminescent dinoflagellate now known as *Noctiluca* appears in the Tenth edition of *Systema naturæ* (1:654) under genus *Nereis* (Vermes Mollusca): see *Noctiluca marina* (1756), and *Mundus invisibilis* (1769:389–390).

62. *Systema naturæ* Twelfth edition (1767) 1(2):1324–1327, in the translation of Dobell (1932:377).

63. My translation is informed by that of Turton: Linnæus, *A general system of nature* (1806) 4:4. The italics (other than quoting Pliny) indicate characteristics of plants. For Pliny (*Naturalis historia* 9.146), see Chapter 4.

64. *Systema naturæ* Twelfth edition (1767) 1(2):1071: *adeoque Vegetabilia naturali metamorphosi mutanda in Animalia.*

65. Lindroth (1979:13).

66. *Systema naturæ* (1735); *Fundamenta botanica* (1736), Aphorism 162; *Genera plantarum* (1737), Ratio operis §5–§6; *Philosophia botanica* (1751), Aphorism 157.

67. Linnæus's views on creation and hybridization evolved over time: see Hull (1985), pp. 38, 46–47 *et passim*.

68. *Gemma arborum* (1751); *Metamorphoses plantarum* (1759); *Fundamentum fructificationis* (1763). We have seen this distinction in Theophrastus (*Historia plantarum* 1.2.5–6) and Cesalpino (1583:11), although some subsequent details differ. Stevens & Cullen (1990:184) translate a key page of *Gemma arborum*.

69. I am indebted for the synopsis in this paragraph to Gunnar Broberg (1985:163–166). See also Farley (1982:24), and Stevens & Cullen (1990:186–188).

70. The details—numbers of stamens and pistils, their presence in the same or different flowers, and so forth—place each genus within a Class.

71. *Philosophia botanica* (1751), Aphorism 77.

72. *Metamorphoses plantarum* (1759); *Animalia composita* (1760); *Generatio ambigena* (1763); and Broberg (1985:164–165).

73. *Prælectiones* (1792:14): *Nam medulla nuda pro lubitu se extendere & contrahere potest, quod apparet in ultima Vermium species. (System. Natur.)* Volvox Chaos *dicta, quæ varias & quascunque figuras pro lubitu assumit.*

74. More precisely, those borne on the shoot.

75. Although this is unusual in animals other than Zoophyta.

76. *Gemma arborum* (1751); *Metamorphoses plantarum* (1759).

77. *Prolepsis plantarum* (Ullmark, 1760), and *Prolepsis plantarum* (Ferber, 1763), both in *Amœnitates* 6 (1763). For a clear if simplified explanation, see Carr (1837:105–107); for more detail, Stevens & Cullen (1990:190–194).

78. *Tænia* (1751); *Animalia composita* (1760); *Mundus invisibilis* (1769).

79. *Systema naturæ* (1758) 1:789.

80. *Philosophia botanica* (1751), Aphorism 76 citing *Corallia Baltica* (1749).

81. *Systema naturæ* (1758) 1:6, at Naturalia.

82. *Animalia composita, efflorescentia. Stirps vegetans* (*Systema naturæ* 1758, 1:799).

83. Stevens & Cullen (1990:199) translate the phrase "with an animated body". In the Twelfth edition (1766:1287) this became "Composite animals, efflorescing. *Stirps vegetans*, passing over by metamorphosis into Animal blossoms"; and in Gmelin's Thirteenth edition (1788:3753) "Composite animals, efflorescing in the manner of animals". The latter scarcely explains Turton's English translation (1806): "Composite animals, efflorescing like vegetables".

84. Linnæus to Ellis (16 September 1761), in JE Smith (1821) 1:151–152. Also quoted in Johnston (1838:19) and Ritterbush (1964:137).

85. By which Linnæus means the soft-bodied animal, not their calcareous substratum.

86. As do aphids, but (so far as Linnæus knew) not other animals.

NOTES 559

87. Stevens & Cullen (1990:203–204), citing *Mundum invisibilem* (Roos, 1767:20). In this section, *Mundum invisibilem* is nearly identical with the post-defence *Mundus invisibilis* as it appears in *Amoenitates academicæ* 7 (1769).

88. Porta, *Phytognomica* 6.2 (1588:240). Hooke (*Micrographia* 1665:121) observed possible seed-pods, but not seeds, in rose blight.

89. Micheli (1729) reported flowers and seeds in plants now known as liverworts, mosses, and lichens as well as diverse fungi.

90. Haller to Linnæus (21 December 1737), in JE Smith (1821) 2:306–307; also Ramsbottom (1941:291–292).

91. Among the sceptics was Sébastien Vaillant (1718:4), who wrote of the "cursed race" of "Flowers without flower".

92. Excerpts from the relevant correspondence between Linnæus and Münchhausen are reproduced in Ramsbottom (1941:364–365).

93. Münchhausen, *Der Hausvater* 1:329 ff, 2:751–752 and 3(3):899–901 (1765–1768); the latter section is missing from the copy at archive.org. According to *Allgemeine Deutsche Biographie* 23:7–8 (1886) he studied at Universität Göttingen, but was not (as claimed by Broberg) its chancellor: that was Gerlach Adolf von Münchhausen (1688–1770).

94. Münchhausen, *Hausvater* 1 §76 (1765:149–151).

95. *Hausvater* 2 §758 (1766:751–752), translation by MAR. This passage corresponds to *Mundus invisibilis* (1769:397–398), which in turn has been translated into English by "J.E." [John Ellis], *St James's Chronicle* no. 1012 (25–27 August 1767); and by Antonovics & Kritzinger (2016:372).

96. *Mundus invisibilis* (1769:399); Antonovics & Kritzinger (2016:373).

97. "J.E." [John Ellis], *St James's Chronicle* no. 1023 (19–22 September 1767); Ellis to Linnæus, in JE Smith (1821) 1:216; Ainsworth (1976:23–24); Antonovics & Hood (2018), particularly pp. 220 ff.

98. *Systema naturæ*, Twelfth edition (1766) 1:1327; Antonovics & Kritzinger (2016:358). See also note [111] below.

99. Pasteur (1857) two articles; Alba-Lois & Segal-Kischinevzky (2010).

100. For the "contagion of eruptive fevers" [Febrium Exanthematicarum *contagium*], see also Linnæus (1760); and DeLacy & Cain (1995).

101. Münchhausen to Linnæus (26 September 1754) as excerpted in Ramsbottom (1941:364–365). Münchhausen followed his suggestion with an expression of doubt.

102. *Hausvater* 2 (1766), Conspectus (seventh unnumbered page) and 2:253–259; quotation at 2:253; §679 at 2:719–720; and §679 at 2:745–747. Their supposed intermediacy among all three Kingdoms was reiterated by Beseke (1797:101).

103. Linnæus compared Münchhausen's discoveries with those of Harvey, Leeuwenhoek, and Trembley: *Mundus invisibilis* (1769:398).

104. Antonovics & Kritzinger (2016) point out the few differences between the thesis as submitted by Roos (*Mundum invisibilem*) and as later edited by Linnæus and published (*Mundus invisibilis*) in *Amœnitates Academiae* 7 (1769).

105. Ibid. (1769) 7:386.

106. Ibid. (1769) 7:402. Linnæus [Roos] does not offer a citation.

107. Ibid. (1769) 7:403–404; quote at page 404, in the translation of Antonovics & Kritzinger (2016:378).

108. Ibid. (1769) 7:404 (1769); quote at page 404, in the translation of Antonovics & Kritzinger (2016:378). The final sentence "In a word . . ." is not present in *Mundum invisibilem* (1767:20).

109. Linnæus, Manuscript X 505 in Royal Library Stockholm, as cited by Broberg (1985:170n23).

560   NOTES

110. Broberg (1985:171–172) summarizing the argument in Broberg (1975:104–149). For continuity I replace Broberg's word "marrow" with the Linnæan "medulla".

111. *Mundus invisibilis* includes one scholium (1769:408) absent from *Mundum invisibilem* (page 23), asking whether infusorial animalcules are disseminated with the seeds of mucors. This scholium was added by Linnæus after he set the thesis (1767) but before its inclusion in the *Amœnitates* (1769), and reflects his doubt in the full Münchhausen hypothesis, given Ellis's opinion that the infusorial animalcules are bystanders unrelated to the generation of fungi.

112. *Mundus invisibilis* (1769) 7:408, in the translation of Antonovics & Kritzinger (2016:382).

113. Stevens & Cullen (1990:203) suggest that Linnæus believed that fungi had little cortex, and for that reason were variable in form.

114. Ellis, for example: see Johnston (1838:23–24) and (1847:428–429).

115. *Systema naturæ* Tenth edition (1758) 1:5 [unnumbered]).

116. *Encyclopædia Britannica*, Botany (1771) 1:648; for authorship, see Smellie (1790:245–246 note). Regarding life, see Linnæus, *Sponsalia botanica* (1749:332), where life is indeed characterized in this way; and *Philosophia botanica* (1751), Aphorisms 3, 133, and 134. The definition of life as the propulsion of humours (*Vita est spontanea propulso humorum*) is credited to William Harvey. I can find no instance where Harvey defines life using these precise words, but it is a fair distillation of his *Exercitatio anatomica de motu cordis* (1628).

117. *Encyclopædia Britannica*, Botany (1771) 1:653.

118. Taiz & Taiz (2017:391).

119. Siegesbeck (1737); Buffon, *Histoire naturelle* (1749) 1:17–19; English translation in Lyon & Sloan (1981:105–106).

120. Sachs (1890:89–90). Sachs calls Linnæus a "dangerous guide for weak minds" (page 89).

121. Haeckel, *Generelle Morphologie* (1866) 1:193.

122. Bicheno (1827:481).

123. Beseke (1797:100 ff.) mentions systems by Münchhausen, Denso, Becher, Wallerius, Titius, Borowsky, and Stokkenstrand. A decade later, Thornton counted 52 systems of botanical classification (1807), Apology (unnumbered page at end of Volume 1), footnote. See also note [186] below.

124. A Botanical Society, at Lichfield. Linnæus, *A system of vegetables* (1783).

125. Hill, *Essays* (1752:28); DSA Büttner (unpublished, 1756, and 1760) as related by Weis (1770), pp. 1–2 and footnote at pp. 2–3; Ramsbottom (1941:302), who (incorrectly, so far as I can determine) gives the year of publication of Weis's monograph as 1769; and Villemet (1784).

126. Dillenius (1719); KA Rudolphi, "Afterorganismen" (1807:21–23); Lippmann (1933:80).

127. La Métherie (1780); Medicus (1789:247); Ramsbottom (1941:325–329).

128. Weis (1770) also argued that fungi do in fact move from place to place, as evidenced by their fruiting bodies appearing in different locations from year to year. Dryander replied that this supposed movement was due to dispersal of their seeds by the wind, and/or the thorough collection of fungi in one location by peasant women. See also Lütjeharms (1936:174–175).

129. See also Lidbeck [Dryander] (1776); and Kölreuter (1777).

130. Fungi are "terrestrial lithophytes" (Marchant [1711], published 1730); or transformations of "terrestrial zoophytes" (Lichtenstein, two articles, 1803). See also Hewitt *et al.* (2016:51–52).

131. Picco, *Melethemata* (1788). Picco is quoted as saying that fungi are animals (Ainsworth 1976:32), but I find no clear statement to that effect in *Melethemata*.

132. Durande (1785); Lütjeharms (1936); Ramsbottom (1941); Ainsworth (1976).

133. OF Müller (1768:48–49); page 48 mistakenly numbered 58.

134. Necker, *Physiologie des corps Organises* (1775); originally published in Latin as *Physiologia muscorum* (1774): *délire poétique* at (1775:242), *poetica* at (1774:234); Ovid and the barometz at (1775:244) and (1774:235–236).

NOTES 561

135. Necker, *Acta Academiæ* (1775), footnote x at pp. 286–287. He presents this as a riddle: see also Lütjeharms (1936:107 and note 1).

136. Necker, *Traité sur la mycitologie* (1783), note 78 at pp. 103–104.

137. Ibid. (1783), pp. 103–105 and note 79.

138. Villemet (1784); quotation at page 211 (translation by MAR).

139. Persoon (1793).

140. Chapters 19 and 21.

141. Alongside Plants, Animals, and Man: Nees von Esenbeck, *Handbuch* (1820) 1:12. His *Handbuch* is a work of *Naturphilosophie* (Chapter 18): fungi represent the northerly or earthly system of Earth, directly opposite the plants (1:18); they are "the living product of the objective polarity of – + or Earth" (1:14), and "individualised elements of the texture and structure of the plant-body" (1:26). In *System der Pilze und Schwämme* (1817) he recognized a *Luftreich*, *Lichtreich*, and *Wärmereich* (folding table at page 142), not to mention multiple *Reiche within* the *Schwammwelt*: three within Pilze (Regnum Primum, Regnum Secundum, and Regnum Tertium), and four within Schwämme (Regnum Primum through Regnum Quartum) (folding table at page 270; and Ueberblick des Systems, *passim*). The latter four (or seven) groups cannot be interpreted as kingdoms in any standard taxonomic sense.

142. John Hill's sprawling *The vegetable system* (1759–1775) introduced (from its second volume, 1761) Linnæan binomials to English-language botany, and used an idiosyncratic system based variously on floral characters, or the number of petals or leaves; but it had the element of *grand spectacle* about it, and did not influence subsequent botanical practise. See also Elliott (2018).

143. For his experiences in Sénégal, see *Histoire naturelle du Sénégal* (1757), translated into English (minus the section on shells) as *A voyage to Senegal* (1759); and Nicolas (1963:16–30).

144. Adanson, *Familles des plantes* (1763) 1, pp. clv–clvij. Stafleu (1963:178) has translated key paragraphs into English. The use of all available characters also underlies his treatment of mollusc-shells: *Histoire des coquillages* (1757), typically bound with *Histoire naturelle du Sénégal* (1757). Heywood (1985:9) comments that Adanson's own method was no more successful.

145. Adanson (1763) 1, page clxiv: *L'univers a pu n'être pas divisé, & il ne l'est peut-être pas relativement à la nature ou à l'Etre suprême; mais il est réelemant divisé en parties relativemant à nous, & cela sufit.*

146. Ibid. (1763) 1, pp. clviij–clxix.

147. Stafleu (1963:206–207) maps these where possible to the corresponding taxa of A-L de Jussieu (*Genera plantarum*, 1789) and others. According to Nicolas (1963:69–70), this *Tremella* was probably *Oscillatoria*, now recognized as a cyanobacterium.

148. For the reception and fate of *Familles*, see Stafleu (1963:240–246).

149. The second edition of *Flora Carniolica* (1772) was, by contrast, fully Linnæan.

150. Scopoli (1777:367–368).

151. Stafleu (1971:197).

152. Aristotle, *De partibus animalium* 643b810–644a10.

153. As did Gessner, Columna, Morison, and (corolla only) Tournefort: see A-L de Jussieu, *Genera plantarum* (1789), pp. xx–xxij and xxix–xxxj.

154. Linnæus, *Philosophia botanica* (1751), Aphorism 77.

155. A-L de Jussieu, *Genera plantarum* (1789), xxxv–xliv. An English translation by S Rosa is provided in PF Stevens, *The development of biological systematics. Antoine-Laurent de Jussieu, nature, and the natural system* (Columbia UP, 1994), pp. 355–364. Species reborn by continued generation, at 1789:xix and 1994:340.

156. A-L de Jussieu, *Genera plantarum* (1789), page ij: Regnum organicum and Regnum inorganicum.

562 NOTES

157. Ibid. (1789), pp. xxxv–xxxvj.

158. A-L de Jussieu, *Principes* (1824:39), from *Dictionnaire des science naturelles* (1824). See also Stevens (1994), page 166, and note 71 at page 457.

159. A-L de Jussieu, *Genera plantarum* (1789:2).

160. One approach was to burn a substance: animal (but not plant) matter was considered to release a smell of burnt bone, horn, or hair. This was tricky to scale-down to small samples, however, and its validity at the limit of the animal kingdom was unknown. Burnt animal matter might also leave a residue of "volatile salts" (ammonium carbonate) or lime (calcium oxides and hydroxides). See Linnæus, *Systema naturae* Twelfth edition (1766), 1:1304 footnote; his letter to John Ellis (20 July 1767), in JE Smith (1821) 1:208–209; Gibson (2015:59–63).

161. Pallas (1766), Præfatio, page xiv. Job Baster in particular (1761:111) had argued that corallines are plants; in this he was opposed by John Ellis.

162. Pallas (1766), Præfatio, page viij; organicorum IMPERIUM at page viij. Translation by MAR.

163. *Hydra, Eschara, Cellularia, Tubularia, Brachionus, Sertularia, Gorgonia, Antipathes, Isis, Millepora, Madrepora, Tubipora, Alcyonium, Pennatula,* and *Spongia.*

164. Pallas (1766:3–5); *Organicorum Corporum Imperium* at page 4.

165. Ibid. (1766:21–22). For *Lycoperdon,* see *Systema naturæ* (1766) 2:726.

166. Pallas (1766:23).

167. Ibid. (1766:375).

168. Ibid. (1766:23–24). Translated here by MAR, building on earlier translations by Genovese in Archibald (2009:563), and by Ragan (2009).

169. Pallas (1787:48). Translated by Ragan (2009).

170. Thienemann (1910:272): *Analogie—"superficieller und idealischer Anverwandtschaft"—und Homologie—"Struktur und Zeugungsart".*

171. Groner (1996:283).

172. OF Müller (1776), pp. xxvii–xxxii.

173. Corliss (1992:49). Müller also (unknowingly) included certain bacteria. The genera are *Brachionus, Bursaria, Cercaria, Cyclidium, Enchelis, Gonium, Himantopus, Kerona, Kolpoda, Leucophra, Monas, Paramecium, Proteus, Trichoda, Vibrio, Volvox,* and *Vorticella.*

174. Blainville, Infusoires (1822:416).

175. Bory de Saint-Vincent, Microscopiques (1824), 2:517.

176. Batsch (1789) 2:664–666.

177. *Die Pflanzenthiere* (1788–1830) has a complex publication history. Begun by Esper, it was continued by Frédéric-Louis Hammer but probably never completed. A *Fortsetzungen* (1794–1806) likewise remained incomplete. See Stafleu & Cowan (1976) 1:802–803; and Ott (1989).

178. *Systema naturæ,* Twelfth edition (1766).

179. Esper, *Die Pflanzenthiere,* Theil 1 (1791), Einleitung, page 1.

180. Ibid. (1791:21–22).

181. Ibid. (1791:22): *Wir können einmahl die Gränzen der Naturreiche nicht nach einzelnen Gattungen in ihren Uebergang, oder der genauesten Verbindung bestimmen.*

182. Cassirer (1951), pp. 13–16 and 23–24; Foucault (1970), pp. 46–77 and 268.

183. Lindroth (1979:11).

184. Wallerius, *Hydrologia* (1748); *Hydrologie* (1751).

185. Denso (1751), tenth (unnumbered) page.

186. Beseke (1797:101–102) cites Rozier (1776, 2:318–319) as stating that Becher termed atmospheric air *Chaos,* so the corresponding kingdom was *ein chaotisches Reich.* This is not an accurate representation of Rozier's text, but Rozier does not tell us what work by Becher he is referring to. In *Physica subterranea* (1703, 1738) Johann Joachim Becher, an early

NOTES 563

proponent of what would become phlogiston theory, discusses three kingdoms but does not (to my reading) accept a Chaotic Kingdom, nor a Kingdom of Air. Beseke (1797:103) briefly mentions a similar proposal by "Stokkenstrand"; Jourdan (1834, volume 1:125) credits "Stokenstrand" with a *règne atmosphérique*.

187. Tietz (1777:16–178).
188. Borowski (1779). See also Delafosse (1860), especially page 33.
189. Howard (1803). Goethe celebrated Howard's cloud classification in verse, in particular that the metamorphosis of one cloud type into another reveals an underlining "coherence and unity of forms" (Wetzels, 1985:140–141).
190. Linnæus, *Systema naturæ* First (1735), Second (1740), Sixth (1748), and Twelfth (1768) editions; Wallerius, *Mineralogia* (1747) and *Systema minerologicum* (1772–1775); Cronstedt (1758); Oken, *Lehrbuch der Naturgeschichte* 1 (1813); Whewell (1828); St. Clair (1965).
191. Haüy (1801), classification summarized at volume 5:1–10; Stevens (1984).
192. TO Bergman (1775); in English translation (1785), with his name spelled "Bergmann".
193. Linnæus, *Materia medica* (1749); and *Clavis medicinae* (1766).
194. Linnæus, *Genera morborum* (1763); Boissier de Sauvages de la Croix, *Nosologia* 1 (1763); Pulteney (1805:138–167).
195. Linnæus, *Bibliotheca botanica* (1736); and *Classes Plantarum* (1738).
196. Beseke (1797). For modern counterparts, see JM Carpenter (1987), and Ebach, Morrone & Williams (2008).
197. Linnæus, *Philosophia botanica* (1751), Aphorism 77; Whewell (1857) 3:267–272; Sloan (1972). I have not systematically distinguished *method* from *system* because the usage of these terms, never uniform, evolved over the century (and the Nineteenth): Stevens (1994), pp. 10–13 and 22.
198. Today a *polyhedron* is a flat-sided geometric object in three dimensions; a flat-sided geometric object in two dimensions is a *polygon*. Pallas did not draw this distinction.
199. Giseke, Tabula genealogico-geographica affinitatum plantarum, secundum ordines naturales Linnæi delineavit [1789], in Linnæus, *Prælectiones* (1792). According to Stafleu & Cowan (1976) 1:949–950, the Tabula was also published separately (1789).

Chapter 16

1. Augier-Favas (1809), as translated by Hellström, André & Philippe (2017:34).
2. Vicq d'Azyr (1774); reviewed in Anonymous, Table pour server à l'histoire anatomique & naturelle (1776). Excerpts in Thienemann (1910:254–255).
3. Daubenton, Histoire naturelle [Sur la nomenclature méthodique] (1800:434–435).
4. Blumenbach, Introductory letter to Sir Joseph Banks (1795), pp. v–xx; in the translation of Bendyshe (1865:150–151).
5. "Table of Pythagoras" usually refers to a multiplication table, but more generally can mean any rectangular table (matrix) in which each square (element) displays the result of a logical operation on the values of the corresponding column and row. Thus, in Duchesne (1795), the Reptilian features "brilliant colours in regions [of the body] and hot seasons" and "diverse shapes" are seen in Vegetables. Among the Worms, Lithophyta and Zoophyta display distinguishing characters of Vegetables.
6. Duchesne (1795), folding table "Rapports réels, *ou apparens*, des êtres de la nature, (rapprochés de leurs caractères propres et quelquefois exclusifs) dans les classes principales" opposite page 289.
7. It was rescued by Stevens (1994:194–195).
8. As translated by de Beer (1949:235).

564 NOTES

9. Linnæus, *Philosophia botanica* (1751), Aphorism 77.

10. Linnæus, *Prælectiones* (1792), facing page 623. Giseke's history of the map is at pp. xil–xl. An updated version appears in Hull, *Elements of botany* (1800) volume 2, following page 392.

11. Linnæus, *Prælectiones* (1792:21–622). Orders 16 (Calycifloræ) and 54 (Miscellaneæ) were omitted: Giseke (1792:627) and Hull (1800) 2:397. Broberg (1985:175) gives the number of Linnæan natural orders as 58 (Giseke) or 68 (JC Fabricius).

12. L'Héritier (1795:28–31).

13. Linnæus, *Philosophia botanica* (1751), Aphorisms 190, 206, 208, and 337.

14. Ibid. (1751), Aphorism 178.

15. A-L de Jussieu (1777:216); English translation by Stevens (1994:275). Quotation from Ragan (2009). See also ÉF Geoffroy (1741); Whewell (1847) 3:129; and Stevens (1994), pp. 38–39 and 183–198.

16. Macleay (1819–1821:362–366), (1823), and (1829); Virey (1825); Fries, *Systema mycologicum* (1821) 1:xvi, and *Systema orbis vegetabilis* (1825:55–56) *et passim*; Reichenbach (1837:108). All too often, however, *affinity* and *analogy* were left undefined (Stevens 1994:192–193).

17. Fries, *Systema mycologicum* (1821) 1, page xvii.

18. Conceivably, he might have meant affinities both upward and downward within a linear scale; or both within and orthogonal to a local section of the Chain: fragments, so to speak, of Donati's net. The map offers examples of the latter.

19. Linnæus (Giseke), *Prælectiones* (1792:623): *Dixi eam genealogicam, quia vox Affinitatis de familiis adhiberi solet pariter ac utrumque de Plantis, sed not ita, ut ab Avo ad nepotes deduci possit; potius eo sensu, quod patrueles & affines collocentur ita, ut vincula, quibus inter se nectantur, pateant.* I translate *avus* (grandfather) and *nepos* (grandson) generically as *ancestor* and *descendant*; in any case, these are contrasted with *patrueles & affines* (*cousins & kinfolk*).

20. Nor, for that matter, of the sort of genus-level progressivism of Bonnet or early Lamarck.

21. Richard, *Méthode* (1826:507); Bromhead (1836:247); Sharpe (1868–1871), frontispiece. Alphonse de Candolle (1841:148–168) gives step-by-step instructions for computing affinity among groups so they can be represented in such a map. A map encompassing six taxonomic levels within the vertebrates, with additional lines depicting direct affinities, is presented by Milne-Edwards (1844), facing page 98. See also O'Hara (1980) and (1991); and Stevens (1994), pp. 97–98 and 169–171.

22. Chapter 11.

23. Donati (1750), page xxi.

24. Ibid. (1750), pp. lxiiii–lxiv; "kingdom of the waters" (*regno dell'acque*) at page xxi.

25. Ibid. (1750), pp. xxi–xxii.

26. For example, by Treviranus (1802) 1:473–474.

27. Olivi (1792:68). Translation by MAR.

28. "Un officier du Roi" [JHB de Saint-Pierre] (1773) 1, Plate 3 *Idée d'un ordre sphérique* (folding table at page 146).

29. Oken, *Die Zeugung* (1805), pp. 188–216; figures at pp. 188 and 191.

30. Rüling (1774), folding plate opposite page 7.

31. Hermann, *Tabula affinitatum* (1783), folding plate opposite (unnumbered) page 1. The 1783 monograph was based on an earlier thesis (1777) defended by his student Georg Würtz: see Hermann (1783:27 note *k*). Further sub-diagrams are provided throughout the text, and in footnotes. According to Thienemann (1910:248), Hermann's system remained in textbooks for many years as the best expression of the so-called natural system.

32. Summarized at Hermann (1783:295). In the large folding diagram, *Raja* can be found about eight lines below AMPHIBIA, about one-third of the way down the left-hand side.

33. Batsch (1802), table at end.

NOTES 565

34. Treviranus (1802) 1:473–474. Translation by MAR.

35. Ragan (2009).

36. R Brown, *Prodromus* (1810), page v; and (1821), column 801.

37. Buffon, *Histoire naturelle, générale et particuliére* (1753) 5:223–229, and "Table de l'ordre des chiens" preceding page 301. Quotation at page 225. As translated by Smellie, the table bears the title "Genealogical table of the different races of dogs" (1791), table preceding page 49 in Volume 4. An English version is reproduced in Greene (1959:149); for Buffon on the fixity of species, see also Greene (1959:138–155).

38. National boundaries in Europe were redrawn by the Congress of Vienna in 1815 following the Napoleonic Wars. Rudolphi, Goldfuss, von Baer, and Eichwald worked in the Kingdom of Prussia and/or the Russian Empire, in lands now variously part of Germany, the Baltic countries, or Russia. Oken was based at Jena, in the Duchy of Saxe-Weimar-Eisenach.

39. Rudolphi, Über eine neue Eintheilung der Thiere, in *Beyträge* (1812:81–106) at page 95.

40. Ibid. (1812:103). He acknowledges Aristotle as his inspiration for the terms Myeloneura and Ganglioneura (1812:98).

41. Ibid. (1812:101).

42. In a related context (1812:95) Rudolphi distinguishes *Verwandtschaft* and *Hauptverwandtschaften*, suggesting that the lines in his figure represent or include affinity. Later he refers to Hermann's network as a *Verwandtschaftstafel* (1812:105). Mikulinskii (1961:347) argues persuasively that they are not lines of phylogeny or evolution.

43. For a biography, see TP Becker (1999).

44. Goldfuss and Nees von Esenbeck had been prime instigators of moving the library and natural-history collection of the Leopoldina, along with a number of gifted young researchers, from Erlangen to the new university at Bonn. Goldfuss eventually became Rector at the University of Bonn (Becker, 1999). Nees von Esenbech, who had studied under Batsch, later became president of the Leopoldina. His last official act as president was to admit Charles Darwin as a member (1857), the first non-British society to do so.

45. Recalling W Harvey (*Exercitationes de generatione animalium*, 1651): *Nos autem afferimus (ut ex dicendis constabit) omnia omnino animalia, etiam vivipara, atque hominem adeo ipsum ex ovo progigni* (1651:2).

46. Note that "north" (*Septemtrio*) is at the right of the figure, "east" (*Ortus*) at the bottom, "south" (*Meridies*) to the left, and "west" (*Occasus*) at the top.

47. Goldfuss (1817:13–14). Translation by MAR.

48. Ibid. (1817:17–18).

49. Ibid. (1817:18).

50. Ibid. (1817:21–22).

51. Ibid. (1817:56).

52. Ibid. (1817:58).

53. Ibid. (1817:22): *die einfachsten infusoriellen Gestalten.*

54. We return to *Naturphilosophie* in Chapter 18.

55. Goldfuss (1820) 1:xi–xii and 1:57–125. In this work Goldfuss updated his 1817 egg-diagram (volume 1, Table 1), and presented similar diagrams of Insecta (1, Table 2), Aves (2, Table 3) and Mammalia (2, Table 4).

56. Von Baer, unpublished [1819], as cited by Brauckmann (2012). See also Raikov (1951) 2:81–95 (diagram at pp. 91 and 95), and Mikulinskii (1961:341–343) (diagram at page 345), not least for von Baer's comments on the Goldfuss egg-diagram.

57. For a detailed explanation, see Raikov (1951) 2:90–92, who connects the taxa to find a tree. See also von Baer (1827:738 ff.); preface dated 20 August 1826. English translation in Henfrey & Huxley (1853:177 ff.).

566 NOTES

58. Von Baer, *Autobiography* (1986:375). The "geographical map" of affinities among mammals (Mikulinskii 1961:358) according to von Baer was, I believe, rendered by Mikulinskii, not by von Baer himself.

59. Successfully defended, this qualified him to serve as a Privatdocent in zoology, *i.e.* to teach (*venia legendi*) at the University. See Lindemann (1870:282). For Eichwald see also Raikov (1951) 2:321–389.

60. Eichwald (1821); *Historia animalium* 588b4–6 quoted (in Greek) opposite page 1.

61. Ibid. (1821:17).

62. Ibid. (1821:8–23).

63. Ibid. (1821:24–25).

64. Ibid. (1821), Table 2 following the text.

65. AL de Jussieu, *Genera plantarum* (1789:296–299).

66. Adrien de Jussieu (1825); quotation at page 393.

67. Ibid. (1825), Plate 29 (bottom). The French title quoted in the text is from the Explication des planches, page 542. The figure itself is labelled *Tentamen tabulæ genera Rutacearum secundum mutuas affinitates disposita exhibentis*. Two scans of this volume of the *Mémoires du Muséum* are available at archive.org. The dashed lines of affinity are not in evidence in the scan of the volume held at the British Museum, but are clear in the scan of the volume held at the University of Michigan (and others).

68. These groups correspond to the top-down divisions in his *Tabula analytica Rutacearum* (1825:522–523). He asserts that agreement with the plan of nature, not taxonomic rank, is important (1825:393–394).

69. Chapter 14 and Bonnet, *Contemplation*, in *Œuvres* (1781) 7:130–138.

70. Chapter 15 and Pallas (1766:23–24).

71. Bonnet was gracious in mentioning Donati's network: *Contemplation*, in *Œuvres* (1781) 7:82n2.

72. Rudolphi (1812:15).

73. Duchesne (1766), following page 228.

74. Pallas's tree metaphor was cited by Hermann (1783:25–26) and Rudolphi (1812:15–16) among others. I am unaware of any contemporary citations of Duchesne's genealogical tree of strawberries.

75. Augier (1801), pp. vi–vij as translated by Stevens (1983:206). Stevens translates *tranchants* as *striking*; I render it *decisive*.

76. Hellström *et al.* (2017:27).

77. Augier-Favas (1809) as translated by Hellström *et al.* (2017:29).

78. Ibid. (1809) per Hellström *et al.* (2017:35).

79. Hellström *et al.* (2017:18).

80. Lamarck, *Philosophie zoologique* (1809), especially chapter 5 (pp. 102–129), who mentions his own interest extending back to 1794. In Chapter 14 we mentioned Lamarck's linear arrangement of plants (*Flore françoise*, 1788).

81. Lamarck, *Histoire naturelle des animaux sans vertèbres* (1815) 1:83–84, as translated by Macleay (1819) 1:199.

82. Lamarck, *Philosophie* (1809) 2:463, and *Histoire naturelle* (1815) 1:457. We return to Lamarck in Chapter 17.

83. Eichwald, *Zoologia* (1829) 1:41–44 and figure opposite page 41. Translation by MAR.

84. Pallas (1766:23–24).

85. Earlier (Ragan, 2009) I stated that in 1961 (*Razvitie*, pp. 315–316) and 1972 (*Istoriya*, pp. 276 and 344) Mikulinskii presented Eichwald's diagram as a depiction of Pallas's tree. Mikulinskii stated that Eichwald's tree was the first after Pallas, but he did not specifically claim it as a representation thereof. For an interpretation of Pallas's tree, see Raikov (1952) 1:76.

NOTES 567

86. Eichwald (1829) 1:44.

87. Adrien de Jussieu (undated [1843]), at page 538. According to Stevens (1994:98 and 432n35), the work first appeared in 1843, and went through nine editions by 1864. The second part (pp. 277–728) is dated 1844.

88. Eichwald joined the faculty there in 1838, but the two may have met as early as 1832. For a biography of Horaninov, see Raikov (1951) 2:390–479. Horaninov translated Volume 5 (Zoology, 1835–1836) of Oken's *Allgemeine Naturgeschichte* into Russian (Raikov 1951, 2:467 ff.).

89. Horaninov (1834), folding figure at end.

90. Horaninov believed that there are 52 (13 × 4) chemical elements in all (1843:3). In *Primae lineae* he mentions selenium and thorium (1834:4), the fifty-first and fifty-second elements to be discovered (in 1826 and 1829 respectively). He does not mention the fifty-third, titanium, discovered in 1830.

91. Horaninov (1834:2–4) and (1843:3–7): quotations from (1843:19) and (1834:20) respectively.

92. Horaninov (1847), page iv.

93. G Fischer [von Waldheim], *Tableaux synoptiques de zoognosie* (1808), page 181.

94. John Venn (1834–1923) laid the groundwork for his eponymous diagrams in *Symbolic logic* (1881).

95. Fischer [von Waldheim] (1808), Tables 1 (unnumbered) and 12 at front of text, and page 183.

96. Delise, *Histoire* (1822:24–25) and Cercle lichénologique (also called Cercle methodique des genres dans la famille des lichens), following page 6 in the *Atlas* (1825); Choisy (1833:102); A[-P] de Candolle, *Collection de Mémoires* 9 (1838), page 9.

97. AP de Candolle, *Mémoires de Muséum d'histoire naturelle* (1828), Plate 1 following page 40; published separately as Revue (1829), Plate 1 following page 119.

98. AP de Candolle, *Collection* 1 (1828), pp. 4–5 and 9, and Plate 1.

99. Ibid. 2 (1828), Plate 2.

100. Cassini, Inulées (1822:577).

101. Cassini, Plate 83 (1816–1829).

102. Eichwald (1833), figure following page 374. The figure is reproduced by Mikulinskii (1961:319).

103. Mikulinskii (1961:318).

104. Holland (1988); see also chronology, pp. 149–155.

105. Cuvier's four-volume *Le règne animal* was published in 1817, and the first volume of Geoffroy's *Philosophie anatomique* in 1818. We return to these issues in Chapter 17.

106. Novick (2016), pp. 96 and 105–108.

107. Macleay (1823), notably pp. 48–49; Panchen (1992:23–24); Novick (2016:111–113).

108. Macleay (1819–1821), page 212.

109. Ibid. (1819–1821), pp. 210–212; quote at page 211.

110. Novick (2016:99–101) provides a particularly clear exposition; quotation from page 100.

111. Linnæus, *Systema naturæ* Twelfth edition (1766); Lamarck, *Système* (1801) pp. 8 and 35; *Philosophie* (1809); and *Extrait* (1812); Treviranus (1802) 1:175–398; Oken, *Übersicht des Grundrisses des Systems der Naturphilosophie, und der damit entstehenden Theorie der Sinne* [1802–1803], published as *Grundriss des Naturphilosophie* (1804); *Lehrbuch der Naturphilosophie* 1(3) (1811), pp. xix–xxiii and §2968–§3438; and *Isis* 1 (1817) cols 1153–1154; Cuvier, Sur un nouveau rapprochement (1812); and *Le Règne animal* (1817). For interrelationships among Oken's *Übersicht* [1802–1803], *Grundriss* (1804), two *Abriße* (1805) and *Die Zeugung* (1805), see Bach *et al.* (2007), pp. xxii–xxiii; Owen, Oken (1911); and Gambrotto (2017:333–334). The five classes are not in the version of *Übersicht* in *Laurentius Oken. Gesammelte Schriften* (1939:3–24).

568 NOTES

112. Macleay (1819–1821), pp. 201–203, 206–208 and 216–227. More precisely, Macleay appropriated the Polypi natantes, Polypi vaginati, and "Polypi rudes" of Lamarck. In *Animaux sans vertèbres* 2 (1816) Lamarck recognized five orders of *polypes*: Polypi ciliati, denudati, vaginati, natantes (2:17–18), and tubiferi (2:403–411). Macleay made the ciliati (rotifers) an osculant group, and ignored the tubiferi; his rudes are presumably Lamarck's denudati. According to Lamarck, Polypi denudati includes *Hydra*; vaginati includes corals, sertularia, bryozoa, millepores, gorgonia, sponges, corallines, and alcyonaria; and natantes the sea-pen *Pennatula*.

113. Macleay (1819–1821), pp. 213–215 and 273–274. His primary criterion was circulation of the blood.

114. Ibid. (1819–1821), page 223. Infusoria are presumably the "sketch" for Acrita itself.

115. Ibid. (1819–1821), page 320. This is puzzling, as Acrita lack a common type.

116. Ibid. (1819–1821), page 227.

117. The top-level figure shows all Animalia (ibid., page 318). One of the five animal groups, Annulosa, itself expands into a circle of five circles (page 390); and one of the five annulosan groups, Mandibulata, into a circle of five orders (page 439). Five ranks farther down (page 467), Macleay maps "affinities which appear to connect the described species of the genus Scarabeus"; five species are identified, but he does not draw the circles (page 521).

118. Ibid. (1819–1821), page 323.

119. Novick (2016:122–127) accepts his argument.

120. Macleay (1819–1821), page 395, note.

121. Virey (1835:215–217).

122. Linnæus, *Metamorphosis humana* (1769).

123. Fries, *Systema mycologicum* 1 (1821).

124. Kirby & Spence, *Introduction* (1826) 3:15 note *a*.

125. T Browne, *The garden of Cyrus* (1658).

126. Chapter 9.

127. Novick (2016:98) distinguishes between the quinarian system promoted by Macleay, and that by Swainson (1835:196–352). Swainson took Macleay to task for ignoring the osculant taxa: the fundamental number is more properly *ten*, not five (pp. 203–206). Swainson likewise admonished Fries for forgetting the centrum (thus his number is *five*, not four), and for not fully acknowledging that his groups are circles (pp. 213–214). Regarding taxonomic ranks, Swainson himself returned to the number *three* (pp. 268–271). We return to quinarian systems briefly in Chapter 20 in regard of Swainson.

128. Strickland (1841); Novick (2016:128–129).

129. Barber (1980:107–110); Knight (1986:93–97); Hull (1988:92–96); Novick (2016:129).

130. Chambers, *Vestiges* (1844:237–251). Chambers subsequently backed away to some extent from his support (*Explanations*, 1845:76–77). Hull (1988:96) notes that favourable treatment in the controversial *Vestiges* harmed the reputation of quinarian systematics.

131. H Miller, *Testimony* (1857:493).

132. Pritchard (1861:240). The figure does not appear in the first (1842) or third (1852) editions. The first edition was reissued (1845) with uncoloured figures.

133. Kaup, *Journal für Ornithologie* 2 (1854). Following issue 12 (November 1854) is a separately numbered section "Erinnerungsschrift zum Gedächnisse an die VIII. Jahresversammlung der deutschen Ornithologen-Gesellschaft, abgehalten im Gotha vom 17. bis 20. Juli 1854" (1855), within which Kaup's article appears at pp. xlvii–lvi; his quinarian diagram is Fig. 10 in Tafel II at the end of the added section, following page cxiv.

134. Following Swainson's *principle of evil* manifest in the animal kingdom (1835:245–249), it is tempting to identify the elements of Kaup's crow diagram as *pentagrams*.

135. Ruse (1996:258–259).

NOTES   569

136. Coggon (2002).

137. Well into the Nineteenth century, linearity was emphasized by Turpin, Mémoires (1819:429) [see Stevens 1994:157]; and Arnott (1831:59–60).

138. Pallas (1766:23–24): see Chapter 15.

139. Giseke, *Prælectiones* (1792) pp. 623–627, and note *a* beginning at page 2.

140. Strickland (1844:69); Jardine (1858), pp. ccii–ccv and folding plate following page cciv. The plate is reprinted in O'Hara (1991:260), Figure 3. At the same meeting, GR Waterhouse mapped the mammals as ten adjoined circles (1844:65).

141. Macleay (1819–1821), pp. 331–334. Macleay calls Lamarck's trees a *table* (page 332). The trees (which Lamarck called *séries*) are at *Histoire naturelle* (1815) 1:457. See also Macleay (1819–1821), p. 213 ("nature seems in the animal kingdom to have set out from inorganic matter by two different routes"). Macleay also found a branching relationship, like the letter V, to be compatible with his two circles of animals and vegetables (page 200).

142. Lamarck (1815) 1:458.

143. Barsanti (1988), pp. 81–83 and Figure 11; also published separately (1992), with this passage at pp. 46–47.

144. For examples, see R Cook (1974), plate 37 at (unnumbered) pp. 68–69; and Murdoch (1984), pp. 40, 44–45 and 49–50.

145. Providing that it does not contain a cycle.

146. For Ray and Linnæus on the *scala naturæ*, see Chapter 15 notes [20–22], and Chapter 15 note [54], respectively.

147. Von Baer, *Entwickelungsgeschichte* (1828), 1:219–232 and 1:242–262. The folding table itself follows 1:224; its text is repeated at 1:227. See also Raikov (1951) 2:112–113.

148. Brauckmann (2012), Figure 2.

149. Lindley, Exogens (1838:130). Two arms of the star contain nine orders each; in one case the "missing order" is marked by question marks, while in the other a single order spans both positions.

150. The innermost orders "exhibit in their own series a degree of organization equivalent to what occurs at the same point in other series", as do the orders at the points of the five arms (rays); Lindley does not state whether this constitutes an analogy. The orders along the sides of the arms, however, are connected to orders in the nearest arm by lines of analogy (Lindley 1838:130).

151. Bourdon, Animal (1822:369).

152. Ben-David (1970).

153. Waquet (1998) in the translation of Howe (2001:25).

154. Nyhart (1995:15).

155. Mendelsohn (1964); Richards (2002).

156. Barber (1980), especially pp. 27–44.

Chapter 17

1. Bichat (1799–1800:3). *On diroit que le végétal est l'ébauche, le canevas de l'animal, et que, pour former ce dernier, it n'a fallu que revêtir ce canevas d'un appareil d'organes extérieurs, propre à établir des relations.* The previous paragraph set up *exterior* as the antithesis of *interior*, using the term *l'environne*.

2. Chapter 15 and Burkhardt (1977:50–51).

3. Notably in Jussieu's *Genera plantarum* (1789): see Chapter 15.

4. For this and earlier precedents, see Shackleton (1970).

5. The *Encyclopédie* of d'Alembert and Diderot appeared in 28 volumes (17 volumes of text, 1751–1765; 11 volumes of plates, 1762–1772). A four-volume *Supplément* appeared in

570 NOTES

1776–1777, and a fifth supplementary volume and a two-volume index by 1780. By contrast, Panckoucke's *Encyclopédie méthodique* ran to about 216 volumes.

6. Gillespie (1959).

7. D'Alembert, Discours preliminaire des editeurs, in *Encyclopédie* (1751) 1, pp. i–xlv; Shackleton (1970:394).

8. For Daubenton and Buffon, see Farber (1975).

9. *Encyclopédie* (1751) 1, page l.

10. For example, at ibid. 2:195 and 2:340 (1752), 12:721 (1765), 16:869 (1765), and 17:744 (1765).

11. Ibid. 1:468–474. According to Shackleton (1970:394), Buffon had agreed to write the article on Nature, but failed to deliver it.

12. The unsigned entry *Plantes marines* (ibid. 12:721–722, 1765) is curious, equating the term solely with the organisms (corals and relatives) studied by Marsigli, Peyssonel, Bernard de Jussieu, Donati, and Ellis. They are "productions of the sea formed by insects, and are consequently part of *regne animal*" (12:721).

13. Ibid. 17:744 (1765). The author is not identified.

14. Diderot, *De l'interpretation de la nature* (1753). In *Œuvres complètes* Assézat (1875) 2:15–16. See also Gregory (2007:101–145), particularly pp. 101 and 120–122.

15. Diderot, *Entretien* (1769). English translation by Stewart & Kemp (1937:52).

16. Diderot, *Élements de physiologie* [1774–1780], in *Œuvres complètes* (Varloot) 17:295–303 (1987). For *nisus*, see 17:297n5. For Adanson on tremella, see *Familles des plantes* 1 (1763), II. Partie, Premiere famille, page 1; also Chapter 14, note [151], and Chapter 15, note [147]. See also Fontana (1776); and Ratcliff (2009), pp. 157, 207, 225, and 232–239. The quoted passage is at *Œuvres complètes* Assézat (1875) 9:253–260.

17. For Panckoucke's original plan for the *Encyclopédie méthodique*, and its final register of constituent volumes, see Watts (1958).

18. Daubenton (1782). Histoire naturelle de l'homme (pp. xix–lxxxxij) included extracts from Haller's *Elementa physiologiæ corporis humani* and Buffon's *Histoire naturelle*. An unsigned Avertissement (pp. v–viij) introduces the volume.

19. *Encyclopédie méthodique. Histoire naturelle des animaux* (1782) 1, pp. iv–v.

20. Ibid. (1782) 1, pp. xij–xv. Daubenton states that minerals are unorganized, or have only superficial organization: *il faut qu'il ait perdu toute organisation pour être minéral* (page xij). As we have seen, naturalists from Aristotle onward have considered stones and minerals to be organized. For a more-nuanced description of Pallas's views on Kingdoms, see Chapter 15.

21. Daubenton not only returned to earlier questions but also recycled his earlier text: compare, for instance, his text at *Séances des Écoles normales* 8:7–23 (1800) to that at *Encyclopédie méthodique. Animaux* (1782), pp. lxix–lxxvj. He is unlikely to have been alone in such practise. Multiple sources state that Daubenton, like Münchhausen, had regarded fungi as a kingdom intermediate between plants and animals: Delaméthrie (1799:51); AP de Candolle, *Théorie* (1815:9–10); Loiseleur-Deslongchamps & Marquis (1820:142–143); and Dunal (1838:3).

22. Daubenton, Histoire naturelle (*Séances . . . Débats* 1800, 1:97); undated, but from context presented between 31 January and 4 February 1795 inclusive). The first course of lectures at the new Écoles normales was held at the Muséum from 20 January 1795 to 19 May 1795, after which the Écoles were deemed unworkable, and no further courses were held until the school was reestablished in stages from 1808 to 1830. The first lectures were recorded by stenographers and published in two editions, of which I cite the much more-available *Nouvelle édition* (1800–1801). Although most material in these editions dates to 1795, some was added later.

23. Daubenton, Histoire naturelle (*Séances* 1800, 5:277). Again this lecture is undated, but the previous lecture, by Berthollet, is dated 12 Floréal [1 May 1795], and the subsequent one, by Mentelle, 13 Floréal [2 May 1795].

NOTES 571

24. *Séances* (1800), 6:3–11 [6 May 1795].

25. Barthélemy-Madaule (1982).

26. Chapter 14.

27. Corsi (1988:48). The Jardin du Roi was renamed the Jardin des plantes during the Revolution.

28. Burkhardt (1995:94–95).

29. Lamarck used *biologie* in *Hydrogéologie* (written 1800–1801, published January 1802) at pp. 8 and 188. At pp. 187–188 he writes, "Finally, the observations I made on living bodies, the main results of which I exposed in the opening discourse of my course in Year 9 [1800–1801] at the Muséum, will be the subject of my *Biologie*, the third and final part of Terrestrial Physics." This latter work was never published, but his *Recherches* (July 1802) includes the corresponding discourse from his Year 10 course (delivered May 1802). See also Corsi (1988:118–122) and (2006). The term *biology* was introduced into English by Beddoes (1799:4): "Physiology therefore—or more strictly *biology*—by which I mean *the doctrine of the living system in all its states*." *Biologie* (in German) was first used in this sense by Roose, introducing his *Grundzüge der Lehre von der Lebenskraft* (1797, page iii; second edition 1800); he was followed by Burdach (1800:62): "This knowledge can be understood under the name Biologie or life science [Lebenslehre] of Man." The note reads *Lehenslehre*, but I interpret it as *Lebenslehre*, as the corresponding text reads "Die Erscheinungen an dem lebenden Menschen". Treviranus titled his five-volume monograph *Biologie, oder die Philosophie der lebenden Natur* (1802–1822). For Hanov's use of *biologica* in 1766 (*Philosophiæ natvralis*, Tome 3), see McLaughlin (2002). For botanists as *biologi* in Linnæus (*Bibliotheca botanica*, 1736:148) and his translators, see Richards (2002:4n8). Huxley (On the study of biology [1876] (1893:268)) credits Bichat (1801, vol. 1, pp. xxxv–xl) with a concept of an integrated science of living things, but he did not use the term *biologie*.

30. Haeckel considered him the first (*Last words*, 1906:64). Lamarck first called attention to the fundamental distinction between vertebrates and invertebrates in his 1794 lecture series at the Muséum (*Philosophie* 1809, 1:118), and claimed priority for the idea (*Système* 1801:6). He first mentioned the distinction in print in his *Mémoires* (1797). In 1795 Latreille had used the absence of vertebrae as a distinguishing feature of insects (1801–1802:15–16). Even earlier, Lyonnet had drawn the same distinction, in annotating his 1742 translation of Lesser's *Insecto-theologia* (at volume 1:83n*).

31. Lamarck, *Système* (1801:6–7) *et passim*.

32. Lamarck, *Flore françoise* (1778) 1:4.

33. Ibid. (1778) 1, pp. iv–v. For Lamarck's subsequent views on Buffon's ideas more broadly, see Corsi (1988:45–46).

34. *Flore françoise* (1778) 1, pp. cxiv–cxvij. The final entry is "*Cyperus*, &c." The order of presentation in the text does not follow that shown in the scale.

35. Lamarck, *Encyclopédie méthodique. Botanique* 1–4 (1783–1796), continued by Poiret (1804–1808), with a historical *Discours préliminaire*; the illustrated *Tableau des trois règnes de la nature. Botanique* in three volumes (1791–1823), much of which reappeared in the 15-volume *Histoire naturelle des végétaux* (1803, with Mirbel); and the *Synopsis plantarum* (1806, with AP de Candolle). His *Flore françoise* went to three editions.

36. Burkhardt (1970); Barthélemy-Madaule (1982:7–8); Corsi (1988:47–48); Burkhardt (1995:29–33).

37. Ruse (1979:5–6); Outram (1984:164–165); Burkhardt (1995:117–120).

38. Lamarck, *Système des animaux sans vertèbres* (1801:16–17), in the translation of Elliot in Lamarck, *Zoological philosophy* (1984:416). This is the Discours d'ouverture (*Système* 1801:1–48), separately dated le 21 Floréal An 8 [11 May 1800], at *Zoological philosophy* (1984:409–430).

572 NOTES

39. Burkhardt (1995:51–52 and 57–58), notably an unpublished manuscript quoted by Burkhardt (as note 53) at pp. 57–58.

40. *Flore française* (1778) 1:17n1.

41. Ibid. (1778) 1:2–3.

42. *Système* (1801:5).

43. Ibid. (1801:10).

44. Ibid. (1801:34–35).

45. Ibid. (1801:358): *en quelque sorte les ébauches de l'animalisation.*

46. Ibid. (1801:41–42). Translation by MAR. In a footnote at 1801:16, Lamarck comments that *Mucor viridescens* may likewise represent the minimum of vegetality.

47. Ibid. (1801:357–397). His Polypes amorphes mostly follow OF Müller.

48. Ibid. (1801:391).

49. Ibid. (1801:391–392).

50. Ibid. (1801:392).

51. *Philosophie zoologique* (1809) 1:126, referring to his 1807 course at the Muséum.

52. Ibid. (1809) 2:87–90. As for infusoria, Lamarck mentions "certain vermin, which cause maladies of the skin, or pullulate there" (2:88–89), clearly following Linnæus (*Systema naturæ* Twelfth edition, 1767, 1:1327 under *obscura*).

53. *Philosophie zoologique* (1809); and Sloan, Evolutionary thought (2019).

54. *Philosophie zoologique* (1809) 1:126–128 *et passim.*

55. Ibid. (1809), Additions, 2:463–464; see also 2:88.

56. Ibid. (1809), Additions, 2:451–466. For the gap between molluscs and fishes, see 2:458; for the "Tableau Servant à montrer l'origine des différens animaux", see 2:463.

57. Ibid. (1809) 1:102–103, as translated by Elliot (*Zoological philosophy* 1984:56). Barbançois later presented a detailed tree of "filiation" in which most animals arise from polyps, not from worms (1816), table bound at end of the volume. For Barbançois as for Lamarck, polyps arise in turn from infusoria.

58. Lamarck, *Inédites* (1972:210), as cited by Burkhardt (1984), page xxxi.

59. *Philosophie* (1809) 1:126–128 in the translation of Elliot (1984:135).

60. Lamarck, *Histoire naturelle des animaux sans vertèbres* (1815) 1, pp. i–ij.

61. Ibid. (1815) 1:79–80 and 128–130; quotation at page 130.

62. Ibid. (1815) 1:8 and 79.

63. Ibid. (1815) 1:395–396 and 411.

64. Ibid. (1815) 1:8–11, 22, and 409–410.

65. Ibid. (1815) 1:380–381 (table at page 381).

66. Ibid. (1815) 1:382.

67. Ibid., Supplément (1815) 1:451–462.

68. Lamarck cites Desmarest as "Desmarets": *Histoire naturelle* (1815) 1:451. The publications to which Lamarck presumably refers are identified by Burkhardt (1995:251n76).

69. At Lamarck, *Histoire naturelle* (1815) 1:451, Lamarck provisionally places the new series (ascidia to acéphales to molluscs) after the Radiaria, but the diagram (page 457) shows it connected directly to the Polypes. At page 460 Lamarck apologizes that the tree is "a little disfigured", and states that the correct reading is in the text.

70. Ibid. (1815) 1:459.

71. Ibid. (1815) 1:456.

72. Ibid. (1815) 1:458–459.

73. Ibid. (1815) 1:454.

74. Ibid. (1815) 1:455.

75. Ibid. (1815) 1:346 and 452–453.

76. Lamarck, *Système analytique* (1820).

NOTES 573

77. Ibid. (1820:121–125). Briefly: composition from molecules resulting in an individual; concrete body parts containing fluids; internal vital motion; order and state conducive to vital motion; regeneration and ageing; acquisition and transformation of food; growth and development; reproduction from like; common faculties; and a discrete lifespan.

78. Ibid. (1820:126–128). Briefly: inability to move suddenly or iteratively; inability to locomote; inability of solid parts to move in response to external stimuli (although their fluids can so respond); absence of specialized internal organs; absence of internal digestion; inability to circulate fluids, apart from movement of sap; growth both upward (shoot) and downward (roots); tendency for upward growth perpendicular to the horizon; body usually formed of individuals, often with annual generations. It is curious that several of these characters pertain only to vascular plants, as Lamarck often counselled that such generalizations should be based on the simplest members of a taxon.

79. Ibid. (1820:130–132). For the possible spontaneous origin of lichens, see also *Philosophie* (1809) 2:88–89.

80. *Système analytique* (1820:136–138). Briefly: instant, iterative contractility; locomotion; repeated movement in response to repeated excitation; motion not entirely determined by its cause; both solid and fluid parts participate in vital motion; digestion of composed solids; great range of organization, including some with specialized organs; some only irritable, others also sensitive; no tendency to develop horizontally. None of these is characteristic of plants (1820:138, citing *Histoire naturelle* 1:111–113 (1815)).

81. *Système analytique* (1820:128–130).

82. Ibid. (1820:135 and 144).

83. Ibid. (1820:144–145). Lamarck now refers to *tuniciers*, not *ascidiens*. Today, ascidians are considered a (probably paraphyletic) group within Subphylum Tunicata.

84. Ibid. (1820:146–147); however, the origin of vertebrates remains obscure.

85. Ibid. (1820:148).

86. Ibid. (1820:22). For Augustine, see Chapter 6, specifically note [93]. See also Mayr (1972:70–71).

87. Georges was the name of his older brother, who died at age four in 1769, the year Jean-Léopold-Nicolas-Frédéric was born: Outram (1984:17).

88. Many accounts state that the Abbé Henri Tessier, whom Cuvier befriended in Normandy around 1793, was instrumental in introducing Cuvier to Parisian academic society: *e.g.* R Lee (1833:21–23); Appel (1987:30–32); but see Outram (1984:40–48). For his academic appointments in Paris, see Lee (1833:23–29) and Outram (1984:45–56).

89. Outram (1984:38–39 and note 50 at page 208); Appel (1987:32 and note 58 at page 250).

90. In *Genera plantarum*, A-L de Jussieu argued that supraspecific natural groups exist and, although defined by fewer characters than are species, can nonetheless be delineated using characters appropriate in each case. See also Chapter 15, note [145].

91. Geoffroy & [G] Cuvier, Mammalogie (1795), read to the Société d'histoire naturelle du Paris on 1 Floréal, An 3 [20 April 1795]; *le raisonnement et l'expérience* at pp. 168–169. Although Geoffroy is credited as co-author, the main ideas in this article have long been attributed to Cuvier (Appel 1987:33–34). See also G Coleman (1964:74–106); Foucault (1970:263–279); and Richards (1992:50–52).

92. Cuvier, Mémoire sur la structure (1795), read to the Société d'histoire naturelle on 21 Floréal, An 3 [10 May 1795]; Burkhardt (1995:20). The classes were molluscs, crustaceans, insects, worms, echinoderms, and zoophytes.

93. Cuvier, *Tableau élémentaire* (1798:20–21).

94. Cuvier, Sur un nouveau rapprochement (1812:76–77): *le système nerveux est au fond tout l'animal; les autres systèmes ne sont là que pour le servir ou pour l'entretenir; il n'est donc pas étonnant que ce soit d'après lui qu'ils se règlent*. In a footnote, Cuvier credits Virey with "strongly

574 NOTES

analogous ideas" (Animal, 1803). In 1795 Cuvier (with Geoffroy) had written that "the organs of sensation" exert the greatest influence on the entire organism (Mammalogie, 1795:171).

95. Cuvier, Sur un nouveau rapprochement (1812); *provinces ou embranchemens* at page 77.

96. Ibid. (1812:83–84).

97. Ibid. (1812:80–81).

98. Ibid. (1812:81–82). Translation by MAR, with the final clause informed by the translation by Winsor (1976:14).

99. Although dated 1817, *Le règne animal* appeared in early December 1816. Volume 3, on crustacea, spiders, and insects, was written by Latreille. The second edition (1829–1830) was in five volumes. The posthumous third edition (1836–1849), in 22 volumes, was prepared by twelve colleagues ("disciples") including Deshayes, d'Orbigny, Milne-Edwards, Valenciennes, and Quatrefages. Volumes of plates appeared separately. Translations were variously made into English, German, and other European languages.

100. *Le règne animal* (1817) 4:1–94.

101. Ibid. (1817) 4:89. Translation by MAR, with reference to the corresponding passage in Cuvier (1854:659–660).

102. Cuvier, *Recherches sur les ossemens* (1812), particularly its Discours préliminaire (volume 1:1–116); and *Discours sur les révolutions* (1826). See also Russell (1916:42–44).

103. The chair was divided in 1794, with Geoffroy carrying on as professor of mammals and birds, while Lacépède took on reptiles and fish (from 1795). Appel (1987:21) comments that in 1793 Geoffroy "knew next to nothing" about zoology.

104. Geoffroy, Mammiferes (1796).

105. Ibid. (1796:21) as translated by Appel (1987:28).

106. Buffon, *Histoire naturelle* (1766) 14:28–29. Robinet repeatedly made much the same point throughout *Vue philosophique* (1768); see also Russell (1916:23), and Lovejoy (1936:269–283). Appel (1987:29) contends that Geoffroy's comment could be taken as a "pleasant introductory remark" that "did not yet constitute a program of research".

107. Appel (1987:29–34). Their monograph on mammals was eventually written by Geoffroy and Georges-Frédéric Cuvier, younger brother of Georges (1824–1842).

108. Form interpreted narrowly, *e.g.* of a specific bone or organ.

109. Geoffroy, *Philosophie anatomique, Tome 1. Des organes respiratoirs sous le rapport de la détermination et de l'identité de leurs pièces osseuses* (1818). The second volume (1822) dealt with teratology.

110. Geoffroy, Mémoires sur l'organisation des insectes 3 (1820), reprinted in the Litterarischer Anzeiger to Oken's *Isis* (1820), cols 527–551, with an unsigned *Nota* not in the original (but seemingly by Geoffroy) at cols 551–552, followed by commentary signed by Oken (cols 552–559). The passage quoted (my translation) is from this *Nota*, column 552.

111. Appel (1987:143–155).

112. Geoffroy, *Principes* (1830), pp. 146, 178, 180.

113. Geoffroy, *Philosophie* (1818). For Geoffroy on analogues, see also Cahn (1962:60–72); Appel (1987), pp. 4–5, 84–90, 93, and 97–112; Rieppel (1994), especially pp. 70–72; and Hull (1988:90–92).

114. Dumortier (1832) drew precisely this analogy; at page 295 he depicted a V-shaped *echelle organique* arising from *Monade*. Each arm passes through three *degrés de structure*: asquelettés, exosquelettés, and *endosquelettés* on the animal arm, and correspondingly axylés, exoxylés, and *endoxylés* on the plant arm.

115. Appel (1987:3–5).

116. Also Buffon (1749–1766), Necker (1783), Saint-Pierre (1784), Villemet (1784), A-L de Jussieu (1789), Duchesne (1795), and Cassini (1816–1829). Necker proposed a *mésymale* kingdom, and Villemet one of *pseudo-zoo-litho-phytes* (both Chapter 15).

117. Vicq-d'Azyr (1786:5–6). This text can also be found in JL Moreau (1805) 4:17–18. Note that the text of *Oeuvres* does not faithfully follow that of *Traité d'anatomie* (1786), as Moreau combines, excerpts, and interpolates different works of Vicq-d'Azyr without warning (see 4:1–4). For contemporary definitions of life, see Outram (1986).

118. Vicq-d'Azyr (1792), page iv; Moreau, *Oeuvres* (1805) 4:232. *Règne vivant* also at *Oeuvres* 4:143 and 207, yet "three kingdoms" at 4:17–18 and 212.

119. Vicq-d'Azyr (1792), page xij; Moreau, *Oeuvres* (1805) 4:248. Moreau divides the sections differently.

120. Vicq-d'Azyr (1792), page xxvj; Moreau, *Oeuvres* (1805) 4:270. Moreau has omitted text, and altered the section heading.

121. This is explicit in not only in Aristotle but (among others) Ammonius Hermiae, Ibn Daud, many Scholastics, the *Sirr al-khalīqa*, and Ficino (Chapters 3, 5, and 7–9).

122. Bonnet: Chapter 14, note [164].

123. A-L de Jussieu: Chapter 15, note [156]; Pallas: Chapter 15, notes [162] and [164]. Rudolphi (1812:28, 82 and 107–172) agreed with Pallas.

124. Daubenton, *Séances* (1800) 1:427 (presented 10 February 1795). Instead of kingdoms, he said, we should simply refer to *minerals*, *vegetables*, or *animals*. Place Louis XV was renamed Place de la Révolution in 1789, then Place de la Concorde in 1795.

125. Bruguière (1792) 1, pp. vij–viij. OF Müller was the first to recognize Infusoria as a distinct taxon: *Vermium* 1 (1773); *Zoologiæ Danicæ prodromus* (1776), pp. xvii–xxviii and 202–212; and *Animalcula infusoria* (1786). See also Chapter 15. Gmelin (*Systema naturæ* Thirteenth edition, 1788) divided Vermes into Intestina, Mollusca (including *Medusa*, starfish and sea-urchins), Testacea, Zoophyta (including madrepores, corals, sponges, bryozoans, and *Hydra*), and Infusoria. Recognition of echinoderms as a taxon follows Klein (1734).

126. Sources disagree on authorship. The title page of Tome 1 (1792) mentions only Bruguière, but Sherborn & Woodward (1906) attribute the long entry on *Conus* (1:586–757) to Hwass and Deshayes, and state that Tome 1 was published in two parts (1789 and 1792). The collections of Hwass are cited extensively at 1:602–767. The title page of Tome 2 (1830) attributes authorship to Bruguière and Lamarck, continued by Deshayes, but Sherborn & Woodward deny Lamarck was involved. The (unsigned) Avertissement to Tome 2 praises Lamarck, but stops short of attributing authorship to him. Tome 3 (1832) was written by Deshayes. Bruguière joined an expedition to the Middle East (1792–1798), and died soon after returning to Europe (1798). Hwass died in 1803, and Lamarck in 1829. Three volumes of plates (*Tableau encyclopédique et méthodique*) were published in 1791 (title page: Bruguière), 1797 (*Vers testacées, a coquilles bivalves*; no author identified) and An VI [1797–1798] (*Mollusques testacés*, "by Citizen Lamarck"). A three-volume *Tableau encyclopédique et méthodique des trois règnes de la nature. Vers, coquilles, mollusques et polypiers* appeared in without attribution in 1827: see also Sherborn & Woodward, who state (1906:581n14) that Bory de Saint-Vincent succeeded Lamarck in this task (see *Tableau* 1:83–84).

127. Bruguière (1792) 1, page i.

128. Virey, Nature (1803) 15:380.

129. Ibid. (1803) 15:383. *La* nature *n'a produit d'abord qu'un animal, qu'un végétal très-simple, qu'elle a variés à l'infini, et compliqués par nuances jusqu'aux plus parfaites espèces.*

130. Virey (1835:262–263). Parallel lines cannot approach each other at only one end, of course, but Virey seems to have forgotten his Euclid.

131. Ibid. (1835:251–252).

132. Virey uses *evolution* in this sense at *Nouveau dictionnaire* (1803) 15:394, but in the sense of an ongoing renewal of body parts at 15:218, and of *development* at 15:406. According to Corsi, "Virey was probably the first, and for many years the only, European naturalist to

576   NOTES

suggest that the term 'evolution'—until then exclusively used to describe the phases of embryonic development—could be applied to the successive appearance of life forms on earth" (1988:172–173, citing Virey, Animal, 1816, 2:30).

133. Virey (1835:255).

134. Virey: the "great tree of life" (1835:410); species branching from a common trunk (1835:410–411); two kingdoms, Inorganic and Organized (1835:415); three *manières d'exister*—mineral, vegetable, and animal (1835:423); a single chain from minerals to man (1835:423).

135. Turpin illustrated works by Alexander von Humboldt and Aimé Bonpland (1808), Pierre Antoine Poiteau (1819–1820), and Augustin Saint-Hilaire (1825–1832), among others.

136. Turpin, *Essai* (1820).

137. Turpin, *Organographie végétale* (1827). He collected a large number of plants at the first stage or visible degree of the plant kingdom into genus *Globulina* (1827:13–15), and speculated on whether the green matter of Priestley and Ingen-Housz may have been a type of globuline or *Globulina* (1827:19–21).

138. In Poiret & Turpin, *Leçons de flore*, tome 3 (1820), following page 32 in the copy at Real Jardín Botánico de Madrid, but following page 185 in the copy at Université Lille; and at the front of tome 2 in the copy at Boston Public Library.

139. At (1820:28n1) Turpin states that the plant series could equally well be terminated by magnolia, or another plant of such complexity. As *gradation* includes both discrete and continuous cases, I do not offer a translation.

140. This may be the superb colour plate in *Organographie végétale* (1827, between pages 50 and 51) illustrating the "degrees of vegetable organization" [also at *Mémoires du Muséum* 14, following page 62]. See also Turpin, Aperçu, *Mémoires du Muséum* 16 (1828), plate 13 following page 344.

141. *Organographie végétale* (1827:3).

142. Turpin, Tables (1816–1830), *Planches 2ᵉ partie: Règne organisé. Botanique. 1. Végétaux acotylédones*, at page 1n1.

143. Turpin, ibid., first three (unnumbered) plates. Similar plates, but differing in detail, accompany his articles in *Mémoires du Muséum* volumes 14–16 and 18 (1827–1829).

144. Léman, Echinella (1819), Navicula (1825), and Bacillaria (1845). His corresponding entries in *DSN* 61 present multiple points of view.

145. Turpin, Surirelle (1827). He labelled the third plate in *DSN Planches Botanique* 1 (Surirelle) "*production microscopique, inerte*". A very similar plate in *Mémoires du Muséum d'histoire naturelle* 16 (1828), opposite page 368, is labelled "Production organisée, inerte, microscopique".

146. Appel (1987:66–67). Blainville edited the *Journal de physique, de chimie, d'histoire naturelle et des arts* from 1817 until its demise in 1823.

147. In this assessment Blainville included the needs of palæontologists and geologists: Blainville, *Manuel d'actinologie* (1834), page vij.

148. Blainville, Prodrome, *Journal de physique* (1816:245–246). The article also appeared in *Bulletin de la Société philomatique* (1816), although with a different layout and with Ramist trees instead of textual tables. Pagination in the latter volume is erroneous in part. See also Appel (1980).

149. Blainville, Prodrome, *Journal* (1816:247); *Bulletin* (1816:107). He had presented these ideas publicly since 1810: see *Journal* (1816:244) and *Bulletin* (1816:105).

150. Blainville, Prodrome, *Journal* (1816:247–248); *Bulletin* (1816:108).

151. Blainville, *De l'organisation des animaux; ou, Principes d'anatomie comparée* (Paris, 1822), pp. 217–218 and Sheets 1 and 10 of the folding Table.

152. Blainville (1822), pp. xl–xlij.

NOTES 577

153. Lamarck's chair was divided into two: Lamarck's understudy Latreille took on the section on articulates, while Blainville was assigned molluscs, worms, and zoophytes. See also Appel (1987), pp. 57–58 and 119–121.

154. Blainville, *Manuel d'actinologie*. The title page of the text-volume bears the date 1834, and refers to (a separate volume of) plates, otherwise undated. At page viij Blainville states that printing had been interrupted several times since 1830—the year not only of his accession to (part of) Lamarck's chair, but also of *le Révolution de Juillet*. Page 687 of the text-volume is dated December 1836. He considered the *Manuel* an update of his lengthy article on zoophytes in *Dictionnaire des sciences naturelles* (1830).

155. Blainville (1834:5). He discusses Amorphozoa in more detail at (1834:527–545).

156. Ibid. (1834:108–109). He followed Gaillon in considering oscillatoria, bacillaria, and certain other unicellular beings as plants. Blainville thought zoospermic animalcules not to be organized beings at all, but some sort of particle (1834:109–110).

157. Von Baer (1827) and (1828–1837); and Chapters 16 and 18.

158. Barry, On the unity of structure, and Further observations (1836–1837).

159. Milne-Edwards (1835) and (1844). For an excellent overview, see Ospovat (1995:124–129).

160. Milne-Edwards (1844:70–72).

161. Sarrut & Saint-Edme (1836); Role (1973); Ferrière (2009).

162. Bory de Saint-Vincent, Matière (1826) 10:250–251. These six states are not primitive, but arise from a further "multitude of other states which it is not given to us to perceive" (page 250).

163. Bory, Psychodiaire (1824) 2:657–663, particularly the table at page 659. This table is also reproduced in Bory, Histoire naturelle (1825) 8:247; and in his article on (his own) *Dictionnaire classique*, in *Revue encyclopédique* 28:53 (1825).

164. Bory, Chaodinées (1823) 3:12. The word is misspelled ("Cahodinées") and is consequently out of its proper alphabetical place.

165. Bory, Chaodinées (1823) 3:13; and Chaos (1823) 3:15.

166. Bory, Chaos (1823) 3:472.

167. Bory, Matière (1826) 10:254.

168. Bory, Matière (1826) 10:252–257, with further details through page 272.

169. Bory, Instinct (1825) 8:587.

170. Bory, Chaodinées (1823) 3:12–15.

171. Dobell (1932:381) believes that Bory borrowed the term from OF Müller (*Vermium terrestrium* 1773, 1:4): "microscopica *dicuntur, quod unice lenticulæ amplificantis ope videntur*"). According to Bory, *animalcule* is unsuitable, as this signifies the diminutive of *animal*, and these organisms are not diminished, but are as perfect as any other.

172. Bory, Microscopiques (1826), particularly the Tableau des ordres, des familles et des genres de microscopiques preceding page 533. See also Bory, Règne animal. Microscopiques (1831).

173. Bory, Matière (1826) 10:254–255; Brongniart (1824); Richard, Végétaux (1830).

174. Bory, Psychodiaire (1824) 2:660.

175. See plates 51–55 in *Dictionnaire classique* 17 (1831). In some but not all passages Bory refers specifically to the polyp of *e.g.* corals.

176. Bory, Psychodiaire (1824) 2:661–662.

177. Ibid. (1824): *insensibles nuances* at 2:663; *réseau* at 2:658 and 663.

178. Miall (1911:46) observes that where once theologians saw the Great Chain as a proof of the wisdom of God, after the scientific acceptance of Cuvier's four *embranchemens* they rejected the insensible gradations that Charles Darwin's theory of evolution seemed to require. Atran (1985:154) credits the four *embranchemens* as "cosmically, [raising] insects to the level of man and [reducing] man to an object on par with the lowliest bug".

578 NOTES

179. This sentence is a close paraphrase of Russell (1916:305).
180. Quatrefages (1854) 2:18–22.

Chapter 18

1. Schelling, *Über den Ursprung*, in *Werke* (1907) 1:596. Translation by MAR.
2. For the purposes of this chapter, *Germany* is approximately the territory of the member states and cities of the German Confederation (1815–1866) that had also been part of the Holy Roman Empire in the decades before 1806. I distinguish specific jurisdictions, *e.g.* Prussia, where required by context.
3. Late in life Schelling was recruited to the Friedrich-Wilhelms-University in Berlin (to the chair of philosophy formerly held by Hegel) to counter the dominance of Hegelianism, but eventually found it impossible to work there. His students there included Mikhail Bakunin, Friedrich Engels, and Søren Kierkegaard.
4. French was the official language of the Akademie through the mid-Eighteenth century.
5. Bahr, Ryan & Jaeger, German literature (2019).
6. Knight (1986:52–53); Ezekiel (undated, 2017).
7. Wolff also refused to submit his lecture for examination by the faculty of theology: Hettche & Dyck (2019). Under Frederick II, Wolff was reinstated at Halle, became Chancellor in 1743, and was ennobled in 1745.
8. Fichte equated God with the moral order of the universe, enraging Herder and others.
9. Waquet (2001), particularly pp. 72–90. Of the Latin monographs I cite in this book, the most recent (Gegenbaur's *De animalium plantarumque*, 1860) presents an address delivered at a German university. The most-recent journal article in Latin I cite is by Ferdinand Cohn, professor at the Königliche Universität zu Breslau [now Uniwersytet Wrocławski], 1872 (both cited in Chapter 21).
10. Scholars are at odds over Herder's contribution to the Enlightenment (or Anti-Enlightenment). See Haym (1880–1885); I Berlin (1976); Sternhell (2010); Forster (2019). Quote from Berlin (1998:56).
11. Herder, *Ideen* (1785) 1:285 [written 1784]. Translation by MAR. A fifth volume is represented by a plan, first published in 1820.
12. Ibid. (1785) 1:97, as beings having numerous roots and able to survive in harsh hot climates, and contrasted with large animals such as elephants and buffalo. The structure of the sentence at *Ideen* 1:285 suggests that *Pflanzenschöpfung* means *plants* here as well.
13. Ibid. (1785) 1:123.
14. Ibid. (1785): "zoophytes and insects" at 1:116–117, *Ubergang* (*sans* umlaut) at 1:136.
15. Ibid. (1785) 1:286. Elsewhere he refers to *die untern Reiche, das Erdreich, Menschenreich, Naturreiche, Gedankenreich, das ganze Reich der Ideen des Menschen, das Reich der Menschenorganisation, das Reich der Realität*, and so on.
16. Ibid. (1785) 1:286. For Herder's Chain of Beings in relation to other inhabited worlds, see Guthke (1990:218–221).
17. Magnus (1906:18–26); in English (1961:24–28).
18. *Werther* first appeared in 1774, and in a revised edition in 1787. For more on both editions, see Appelbaum (2004), pp. v–x. Satirist Christoph Friedrich Nicolai quickly published a parody (1775).
19. From 1815, Grand Duke of Saxe-Weimar-Eisenach.
20. Goethe first read Linnæus by 1785, and from no later than 1786 was guided in his study by August Batsch, whom we met briefly in Chapter 16. See Benedikt (1945); Arber (1946), particularly pp. 68–73; and Larson (1967). According to Goethe (*Geschichte meines botanischen*

NOTES   579

Studiems, MA 12:20–25, at pp. 21–22; *Der Verfasser teilt die Geschichte seiner botanischen Studien mit, HA* 13:148–169) he studied the *Termini botanici, Fundamenta botanica,* and *Philosophica botanica,* and JA Gesner's 1743 *Dissertationes de partium vegetationis* written under the direction of Linnæus. For Goethe's exposure to botany earlier than Weimar, see Magnus (1906:18–22) and (1961:24–26).

21. The Goethean *corpus* extends over a dozen major editions totalling more than 350 volumes. Goethe returned to some of his works, adding further sections, text, and commentaries over decades, with the result that they now exist in multiple versions. He also wrote retrospectives of the evolution of his own ideas, and an intellectual-artistic autobiography of his early life (*Aus meinem Leben*) that is selective and, perhaps, *ein wenig ausgestaltet.* His scientific ideas are sometimes prefigured in his poetry and dramatic works.

22. The intermaxillary bone in human is somewhat easier to resolve in the embryo than in the adult. One wonders to what extent this famous example informed Geoffroy Saint-Hilaire, who (Chapter 17) noted that characters were sometimes easier to observe in the vertebrate embryo. Goethe, Dem Menschen wie den Tieren ist ein Zwischenknochen der obern Kinnlade zuzuscheiben [early 1786], in: *LA* 1.9:154–161; and *MA* 12:156–190.

23. Steiner (1897), pp. 63–64 and 71–72; in English (1928), chapter 7. For the latter I have worked from an electronic edition that does not preserve pagination.

24. For the Linnæan theory of prolepsis, see Chapter 15 and note [77] there, and Arber (1946:76). Goethe introduced his *Versuch die Metamorphose der Pflanzen* (1790) with a quotation from the Linnæan thesis *Prolepsis plantarum* (Ullmark) 1763:341, and discussed *Antizipation* in sections 107–111.

25. Goethe, *Versuch* (1790), and *MA* 3.2:318–366. English translation (*Essay . . .*), 1863. See also *MA* 12:29–68. In notes not published at the time he called this a symbolic or transcendental leaf: Richards (2002:395–396).

26. Goethe, *Italienische Reise* (1816–1817); *HA* 11. The story of how Goethe came up with the idea of a plant archetype while contemplating a series of leaves of the fan palm (*Chamærops humilis*) in the botanical garden at Padua (and refined the idea later, while in Palermo) has been related by Steiner, *Einleitung* 1:14–39 (1884); Bielschowsky (1905) 1:373 and 398; Magnus (1906); Arber (1946) and (1950:40–45); and others. As pointed out by Richards (2002:376), Goethe's letter to Charlotte von Stein dated 9–10 July 1786 (in *Goethes Briefe, HA* 1:512–515, and note at page 760) shows that he had begun to formulate the idea no later than two months before his departure for Italy. He did not, however, use the term *Urpflanze* in this letter.

27. Charlotte von Stein to Karl Ludwig von Knebel, 1 May 1784. In: Knebel (1858:119–121).

28. Goethe, *Glückliches Ereignis* (1817–1822): *LA* 1.9:79–83 (*eine symbolische Pflanze* at page 81); and *MA* 12:86–90, at page 88. See also Arber (1950:59); and Richards (2018).

29. Goethe, Letter to Herder, 17 May 1787: *HA* 11:323–324; *MA* 15:392–394 at page 394.

30. Goethe, Bildung und Umbildung organischer Naturen. *VA* 58:1–18, at page 17: *den Begriff, die Idee des Thiers.*

31. Goethe, *Bryophyllum calycinum* I and II (1820): *LA* 1.10:211–213 and 228; *MA* 13.2, pp. 305–307 and 320–321. Quotation from Goethe, Tag- und Jahres-Hefte als Ergänzung meiner sonstigen Bekenntnisse (1820): *VA* 32:156–157, *MA* 14:280. *Bryophyllum* produces plantlets along the edges of its leaves. The species has since been named *B. pinnatum,* and is also known as *Kalanchoë pinnata.*

32. Goethe, Granit I and II (1784): *MA* 2.2, pp. 487–488 and 503–507, with other works on granite through page 515. See also *MA* 9:873–880; *MA* 13.2:267–268; Baldridge (1984); and Görner (1993, 1995). For the Neptunist classification of rock types: Werner (1787).

33. Ribe (1982) as cited by Amrine (1990) at page 202 and note 37. Goethe seems not, however, to have used the word *Urgestein.*

580 NOTES

34. Goethe, *Christus*, in *Sämmtliche Werke* (1840) 31:296–297.

35. Goethe, unpublished notes on botany, cited in Richards (2002), pp. 414–415 and 415n18, referring to *MA* 3.2:303. The awkward attempt of Turpin (*Atlas* 1837, Plate 3) to depict the *Urpflanze* attracted ridicule. Schleiden later presented a generalized flowering plant "as a teaching device devoid of poetic interpretation". For both, see Eyde (1975:432–433).

36. Goethe, *Morphologie* [title varies]. *MA* 4.2:188–204; and Richards (2002:453–457). Although Goethe did not set out a detailed programme for the science of morphology, Karl Friedrich Burdach did so in *Über die Aufgabe der Morphologie* (1817).

37. An idiosyncratic effort in this direction was made by Hayata (1921). Hayata's work is based on an early theory of genes, interpreted in a manner influenced by Buddhist scripture (1921:80–85).

38. Goethe followed the Cuvier-Geoffroy debate closely (Appel, 1987:1). Recall that Geoffroy extended the vertebrate plan to other *embranchemens* (Chapter 17).

39. Goethe, Einige Bermerkungen über die sogennante Tremella, and Infusions-Tiere, in *LA* 1.10:24–40; *MA* 2.2:562–580; and Becker (1999:36–50). See also Magnus (1906:158–160) and (1961:97); and Richards (2002:375).

40. Goethe, *Zur Morphologie* 1.1 (1817–1822): *JA* 39:254; *MA* 12:15–16. In the previous paragraph Goethe describes the Infusionstiere as the bottommost step of the animal kingdom. Haeckel excerpted the passage in *Protistenreich* (1878:67). My translation is informed by that of ER Lankester in Haeckel (*History of creation* 1876, 1:92). *Beweglich/Beweglichkeit* might be rendered *motile/motility*, but I have gone with *flexible/flexibility* to continue in the direction indicated by *starr* (rigid). Haeckel claimed that "this remarkable passage not only indicates most explicitly the genealogical relationship between the vegetable and animal kingdoms, but contains the germ of the monophyletic hypothesis of descent" (*History* 1:92). The passage is at pp. 82–83 in the German original (Haeckel, *Natürliche Schöpfungsgeschichte*, Fourth edition 1873).

41. Brittan (1978:3).

42. For an overview of Kant's argument, see Latta (1898:168–178); and Kant, *Prolegomena* 30–37 (1783:101–113). English translation by Friedrich (*The philosophy of Kant* 1949:84–91). Eco (2000:72) tabulates the categories. Many of Kant's works are online at https://korpora.zim.uni-duisburg-essen.de/kant/.

43. Kant introduced the problem in *Allgemeine Naturgeschichte* (1755). For self-cause and self-effect, see *Kritik der teleologischen Urtheilskraft* [1790], in *Gesammelte Schriften. Werke* 5 §§64–65 (1913:369–376). For an orientation into the problem, see Walsh (2006) and Ginsborg (2019).

44. Herder, *Ideen* (1785), 1:21–24 and 67–99. In reaction Kant, *Recensionen von JG Herders Ideen*, in *Gesammelte Schriften* 8:53–55 (1923). English translation by Wood, in Louden & Zöller, *Immanuel Kant* (2007:131–133). See also Lovejoy (1904:327–336) and (1959); Nisbet (1970:38–39); and Zammito (2018:180–185).

45. Kant, *Kritik der reinen Vernunft* [1781], in *Gesammelte Schriften* (1911) 3:441–442, in the translation by Guyer & Wood (1998:604); also Lovejoy (1936:241).

46. Kant, *Kritik der teleologischen Urtheilskraft* [1790], in *Gesammelte Schriften. Werke* 5 §80 (1913:418–419); and *Critique of judgement* (1914:337–338).

47. Kant, *Vorlesungen* (1821:97–98). Translation by MAR.

48. Kant, *Metaphysik K₂* [early 1790s] as translated by Ameriks & Naragon, in *Lectures on metaphysics* (1997:403); original in *Gesammelte Schriften* 28(1):762 (1970). In *Metaphysik vigilantius* (K₃) [1794–1795], *Lectures* (1997:499–500), Baumgarten reports a very similar passage; original in *Gesammelte Schriften* (1983) 29:1033.

49. Ignaz Born offered yet other names for this force (1789:54). See also Gigante (2009), pp. 20 and 251n31.

NOTES   581

50. Wolff, *Theoria generationis* (1759). For vitalism more broadly, see Bergson (1910, 1911); Driesch (1914); and Harris (1925).

51. Blumenbach, *De generis humani* (1795:84–85), in translation as *The anthropological treatises* (1865:194–196).

52. Blumenbach, *Über den Bildungstrieb* (1780, 1781). The term was also adopted by Schiller and Hölderlin, but to refer to the æsthetic impulse (Waibel, 2018).

53. *Kritik der teleologischen Urtheilskraft* [1790], in *Gesammelte Schriften* 5 §81 (1913:424); *Critique of judgement* (1914:345–346). For the somewhat different meanings of *Bildungstrieb* in Blumenbach and Kant, see Richards (2000).

54. Hegel's *Philosophie der Natur* was conceived during 1805–1806, first committed to text in 1817, then appeared as brief sections in his *Encyclopädie*, before its 1842 publication under the editorship of Michelet. English translations by Miller from the *Encyclopædia* (1970); Petry, distinguishing different versions (1970); and Taubeneck, from the 1817 text (1990). See also Redding (2020).

55. Frank (1997, 2004); I Berlin (2013), especially pp. 293–310.

56. *Kritik der teleologischen Urtheilskraft* [1790], in *Gesammelte Schriften* 5 §75 (1913:397–401); *Critique of judgement* (1914:309–313).

57. Novalis published very little before his untimely death. See K Gjesdal (2014), Ezekiel (undated, 2017).

58. The point is well made by Richards (2002:12–14).

59. Chamisso. *Peter Schlemihl* (1814). I examined the 1827 edition, in which *Zoophyten-Inseln* appears at page 114. Quotation in text from the translation of Bowring (1824:148).

60. Fichte, *Grundlage* (1794), and *Grundriss* (1795).

61. Recall that organisms are cause and effect (Kant).

62. Not only as physiology per se but also the as ability of the organism to interact with the surroundings, for example in locomoting to secure food.

63. Schelling, *Einleitung* (1799); English translation by Peterson as *First outline* (2004).

64. *Einleitung* (1799:51); *First outline* (2004:43). Translation by Peterson. Schelling worries that this conclusion is unsatisfactory, as the productive force must partition itself into male and female to enable generation; but there is only one force of production, and although its action is inhibited at different stages, it nonetheless strives to make a unitary product. Schelling decides that this is equivalent (1799:235–236; 2004:149).

65. *Einleitung* (1799:234–235); *First outline* (2004:149). On the question of sensibility in plants, see (1799:229–230) and (2004:146).

66. Ibid. (1799:226, 232–233); *First outline* (2004:144–145, 147–148). Schelling does not specify what he means by *zoophytes*.

67. Schelling, *Darlegung* (1806). We may, of course, question the wisdom of revitalizing Nature two decades after Spallanzani provided strong evidence against spontaneous generation (Chapter 13). According to Richards (2002:182), "Fichte's was an idealism of the ego, his [Schelling's] an idealism of nature".

68. Lovelock, *Gaia* (1995, 2000); Ruse (2013).

69. Cuvier, *Histoire* (1845) 5:313–376; Rupke (1993); Richards (2002:517–518) and (2018).

70. Richards (2002:179).

71. Kant, *Metaphysische Anfangsgründe* (1786). I have examined only the second edition (1787). English translation by Bennett (2017).

72. For this in Schelling, see his *Darstellung* (1801); English translation (*Presentation*) by Vater (2001), and in Vater & Wood (2012).

73. *Presentation* §141 (2012:199).

74. Goldfuss (1817), and (1820) Table 1; Chapter 16 and Figure 16.3. For these axes in Schelling: *Presentation* (2012:179 and 202).

582   NOTES

75. Horaninov (1834); Chapter 16 and Figure 16.9. For these axes in Schelling: *Presentation* (2012:177–182) *et passim*.

76. *Presentation* §153a (2012:202). Moreover, "Earth itself becomes animal and plant, and it is precisely the earth evolved into animal and plant that we now perceive in organised entities" (also §153a, page 202). In *Weltseele* [1798] he asserts that plants are not alive, but merely have the appearance of life (2000) 1.6, page 185.

77. Lovejoy (1936:320). The esteem was not reciprocal: Schelling could not accept Oken's identification of unfolding nature with God (Ford, 1987).

78. At Jena he was first Professor Extraordinary (1807–1812), then Professor Ordinary (1812–1819). After being released from Jena he taught at Munich, then in 1833 relocated to the new Universität Zürich, where he became the first rector. See Proß (1991). For a sympathetic review of Oken and *Naturphilosophie* see Gambarotto (2017).

79. Klein (2008); Butscher (2008).

80. Gould (1977:39).

81. Burdach (1817:19–25) anticipated two forms of *Wissenschaft*: *die reine Wissenschaft* that arises from self-knowledge, and *Erfahrungswissenschaft* that must confront the world of phenomenon. See also Richards (2008:463–464).

82. Williams (1972).

83. Galvanism is electricity produced by a chemical reaction, and/or present in living tissues such as muscle. For Luigi Galvani's popularization of his discovery, see Richards (2002:317–318). Mary Shelley identified Galvani's experiments as an influence on her novel *Frankenstein* (1818), although it was not until the 1931 Universal motion picture that Victor Frankenstein used electricity to bring the monster to life.

84. Oken, *Lehrbuch der Naturphilosophie* #885 (1843:149); in the translation of Tulk as *Elements of physiophilosophy* #884 (1847:182). At *Lehrbuch* #173 (1843:32) and *Elements* #173 (1847:40) Oken speaks of *alle Principien des Lebens, alle Zahlen* (all the principles of life, all numbers).

85. Oken, *Abriss des Systems der Biologie* (1805); also in *Gesammelte Werke* (2007) 1:17–86, at pp. 35–37. This is illustrated in Oken, *Die Zeugung* (1805), figure at page 188. The *Abriss* was also published separately as *Abriss der Naturphilosophie* (1805).

86. Oken, *Abriss des Systems* (1805), in *Gesammelte Werke* (2007) 1:36. When Oken wrote *Abriss* and *Die Zeugung* he was a Privatdozent at Universität Göttingen, lecturing in the same faculty as Heinrich August Wrisberg. In *Observationvm de animalcvlis infusoriis satvra* (1765) Wrisberg reported that small globules arising in animal- or plant-infusions congeal and fuse to form infusorial animalcula and, thereafter, polyps (Chapter 13).

87. Oken, *Die Zeugung* (1805:109): *Die Welt des Organischen ist geschieden in Infusorien, Pflanzen und Theire*.

88. *Die Zeugung* (1805:186). The diagrams at pages 188 and 191 accord Polyp (or Coral) and Plant equivalent stations in respect to Animal.

89. Among the features Oken identifies as distinguishing animal from plant, self-movement is the one first mentioned: *Allgemeine Naturgeschichte* 4:20–21 (1839).

90. Oken, *Lehrbuch der Naturphilosophie* #453 (1831:84); #472 (1843:78); and *Elements* #472 (1847:96). The first edition of *Lehrbuch* (1809–1811) seems not to contain this precise sentence, but its three volumes correspond generally to the three standard kingdoms. He identified animals, plants and minerals in *Werth der Naturgeschichte*: (1809:13 and 16) and (1939: 264 and 270).

91. *Lehrbuch* (1843), Vorwort, page v; *Elements* (1847), Preface, page xiii. Likewise, the major classes of plants "commence from below" and "ascend in a parallel series beside each other" (*Lehrbuch* #1755, 1843:264; *Elements* #1754, 1848:317).

92. *Lehrbuch* #1041 (1843:167); *Elements* #1040 (1847:204–205).

NOTES   583

93. *Lehrbuch* #1771–1773 (1843:268); *Elements* #1770–1772 (1847:322).

94. *Lehrbuch* #1779 and #1781 (1843:269); *Elements* #1778 and #1780 (1847:323–324).

95. *Lehrbuch* #1851 (1843:278); *Elements* #1850 (1847:335).

96. *Lehrbuch* #2134 (1843:308); *Elements* #2133 (1847:375).

97. *Die Zeugung* (1805:22). Translation by MAR. Or more colourfully, "the animal semen of the planet, the animal dissolved": *Lehrbuch* #3152 (1843:409), *Elements* #3151 (1847:512).

98. Oken, *Über das Universum* [1808], in *Gesammelte Werke* (2007) 1:379–403; quotation at page 401. Also *Gesammelte Schriften* (1939:97–144), quotation at page 141. His "theory of generation" is presumably *Die Zeugung*.

99. Oken, *Entstehung des erste Menschen* [1819], in *Gesammelte Schriften* (1939:145–159), quotation at page 156; and *Isis* (1819), at column 1122.

100. *Lehrbuch* #1058 (1843:170–171); *Elements* #1057 (1847:209).

101. *Lehrbuch* #3468 (1843:450); *Elements* #3467 (1847:570).

102. *Lehrbuch* (1843), page iii; *Elements* (1847), pp. xi–xii.

103. *Lehrbuch* #3089, (1843:398–399); #3100 (pp. 400–401); #3108 (pp. 402–403); #3143 (page 408); #3149 (page 409); and ##3463–3472 (pp. 449–450); respectively *Elements* #3088 (1847:498); #3099 (page 500); #3107 (page 503); #3142 and #3148 (page 511); and ##3462–3472 (pp. 570–571).

104. *Lehrbuch* ##3118–3119 (1843:404–405); *Elements* ##3117–3118 (1847:506). *Physalia* (the Portuguese man-o'-war) is "only [a] giant Infusori[um]": *Lehrbuch* #3484 (1843:452); *Elements* #3483 (1847:573).

105. *Lehrbuch* #3118, (1843:404); *Elements* #3117 (1847:506).

106. Liebig (1840:120–121). Mass murderers because physicians were trained in *Naturphilosophie*.

107. Treviranus, *Biologie* 1:447–448 (1802); *des Reichs der Zoophyten* at 1:399 and elsewhere. He recognizes a complex gradient within animals, and within plants (1:446–475). See also 5:82 (1818).

108. Ibid. 1:399–425 (1802).

109. Friedrich Tiedemann briefly suggested *ein drittes Reich unter den Namen Zoophyten* for "the most-simply organised animals the polyps, and the most-simply organised plants the fungi, seaweeds etc." (*Zoologie* 1:22, 1808), but did not pursue the matter. Achille Requin (Animal 1:557, 1836 [published 1834]) offered *plantanimal* as another name for Bory's *règne psychodiaire* (Chapter 17). Note also Oken's Kingdom of Corals (above).

110. Humboldt, *Floræ Fribergensis* (1793). The aphorisms were translated by Gotthelf Fischer (Humboldt 1794). Humboldt reported further experiments on the irritability of muscle fibres in *Versuche* (1797) 2:430–436.

111. For background on Humboldt's interest, see Richards (2002:316–321).

112. *Vis vitalis* in Latin, *Lebenskraft* in German.

113. *Aphorismi* 5, 6, and 9–11 (1793). Humboldt revealed his Romantic side in an allegory, published in Schiller's literary journal *Die Horen*, entitled *Die Lebenskraft, oder der rhodische Genius* (1795). He included the item in the second (1826) and third (1849) editions of his *Ansichten* (at 2:297–308 in the third edition); in *Aspects* (1849) it appears at 2:251–257.

114. *Floræ Fribergensis . . . accedunt Aphorismi* (1793) footnote at pp. 130–131; *Aphorismen* (1794), note 75 at pp. 46–48.

115. The Note follows the text of *Die Lebenskraft* in the third edition (*Ansichten* 2:309–314; *Aspects* 2:259–263).

116. *Kosmos* 1:367 (1845); *Cosmos* 1:340 (1877). Here I use the translation of Elise Otté.

117. *Kosmos* 1:368–369; *Cosmos* 1:341–342.

118. *Kosmos* 1:208 and 1:369–373; *Cosmos* 1:202 and 1:342–346.

584   NOTES

119. For example, his reference to the "green ovaries" of Coscinodiscæ at *Kosmos* 1:369 (*Cosmos* 1:342). *Coscinodiscus* was established by Ehrenberg in 1838 as a genus of microorganisms responsible for chalk deposits. Most species have subsequently been assigned to phylum Bacillariophyta (diatoms).

120. *Kosmos* 1:371–372; *Cosmos* 1:344.

121. Von Baer, *Nachrichten* (1866). An English translation of the second edition (1886), edited by Oppenheimer, was published as *Autobiography* (1986). For his early exposure to *Naturphilosophie*, and dissatisfaction with Schelling's philosophy as developed by Oken, see (1986) pp. 86, 121, and 203–205.

122. Von Baer, *Über Entwickelungsgeschichte* (1828) 1:224, as translated by Huxley (Fragments, 1853:214). Von Baer used the word *law* (*Gesetz*); a modern-day philosopher of biology would be more cautious.

123. For the longstanding debate between epigeneticists and preformationists, see Farley (1982) and Pinto-Correia (1997). The first two laws, read together, seem also to point to a common ground between the positions of Cuvier and Geoffroy, where the specialization of characters parallels the specialization of form, although von Baer goes out of his way to downplay the connection (*e.g.* at 1828, 1:225).

124. Raikov (1968:364–422); Gould (1977:52–63); Richards (2008:478–481).

125. As mentioned in Chapter 16, von Baer set out the hierarchy of embryological transformations that he considered to have given rise to the main groups of animals (of double-symmetric form) not only in a Ramist diagram (*Entwickelungsgeschichte* 1, following page 224) but also (in unpublished notes *circa* 1826) on a sketched (botanical) tree: Brauckmann (2012:655, figure 2). In *Entwickelungsgeschichte* 1:225–231 von Baer commented on the figure at some length, without the slightest hint that it might be interpreted to depict else but the progressive development of form by increased histological and morphological differentiation, from more-general towards specialized.

126. *Entwickelungsgeschichte* (1828) 1:220. See also Richards (2008:479–481). Huxley translated *Haupttypus*, and *Typus* where indicated by context, as *archetype*. Contrary to the claim of Brauckmann (2012:654), Huxley translated von Baer's *Grundform* as *fundamental form*. So far as I can determine, von Baer did not use the terms *Urthier*, *Urtypus*, or *Archetypus* in his publications.

127. Beiträge (1827:746–747); Fragments (1853:183). The articulates, radiates, and molluscs are primary provinces; the vertebrates "unite" the articulates and molluscs in their organs, and perhaps the radiates too (in their head). Cuvier called his fourth *embranchement Les zoophytes ou Animaux rayonnés*.

128. So-called affinities: Beiträge (1827:731–732); Fragments (1853:177).

129. Von Baer, Beiträge (1827:731–732 and 738–743); Fragments (1853:177–181). According to von Baer's *Autobiography* (1986:318), the seven constituent papers were written in 1824–1826.

130. Beiträge (1827:738–739); Fragments (1853:177–178).

131. Beiträge (1827:737–738). Lamarck had separated the genera *Brachionus* (*Histoire naturelle* 1816, 2:30–36) and *Vorticella* (ibid. 2:45–51) from the Vibratiles (ibid. 2:23–27).

132. Beiträge (1827:738); Fragments (1853:177). See, however, Chapter 19.

133. Beiträge (1827:738).

134. Ibid. (1827:736–739, 748 and 755); *Entwickelungsgeschichte* (1828) 1:211; Fragments (1853:177–178 and 199). Prototypes need not contain "special details" such as an intestinal or nervous system; only the "general character" needs to be recognized (Beiträge 1827:747; Fragments 1853:184).

135. Von Baer comments that Goldfuss's term *Protozoa* would be more appropriate than *Infusoria* or *Microscopica*, as it does not indicate a small size or infusory origin (Beiträge 1827:736–737).

NOTES 585

136. Ehrenberg, *Sylvæ mycologicæ* (1818). Oken's *mathesis* was a cosmogeny in which mathematical *Urformen* (numbers and geometrical figures) are paired with those of physics (electricity, heat, light, etc.) to form the inorganic world, of which the organic world is then an image or reflection (*Abbild*). Ehrenberg would have studied the version in *Übersicht des Grundrisses des Sistems der Naturfilosofie* [1804], for which see Oken, *Gesammelte Schriften* (1939:4–24), and *Gesammelte Werke* (2007) 1:1–14; or that in *Abriss des Naturphilosophie* (1805). For *exanthemata*, see Linnæus, *Systema naturæ* Twelfth edition (1767) 1(2):1327; *Exanthemata viva* (1760); *Genera morborum* (1763); *Lepra* (1769); *Morbi artificum* (1769); *Mundus invisibilis* (1769); and Chapter 15.

137. Ehrenberg, *Sylvæ mycologicæ* (1818).

138. Laue (1895); Jahn (1998).

139. One member stayed in Egypt; only Ehrenberg and his *Jäger* Heinrich Schulz, known as Falkenstein, returned to Europe. All others perished of disease. See Ehrenberg, *Naturgeschichtliche Reisen*, Band 1 (1828).

140. Siesser (1981); quotation at page 170.

141. Lazarus (1998). Ehrenberg recognized 22 families, 123 genera, and 553 species in his *Infusionsthierchen* (1838) alone.

142. Ehrenberg, *Infusionsthierchen* (1838), pp. v and xvii–xviii. Pritchard (1842:45) states that most of Ehrenberg's discoveries were made at magnification less than 380×; Siesser (1981:172–173) claims no greater than 300×. According to Pritchard, 250× suffices for "instructive amusement", whereas 800× is often necessary to observe details.

143. *Infusionsthierchen* (1838), plates 14 and 23–24 respectively.

144. Ehrenberg, *Organisation* (1830:57).

145. *Infustionsthierchen* (1838), unnumbered page following Foreword. Oken called infusoria *Magenthiere*: *Lehrbuch* (1843:449). Pritchard's *A history of infusoria* (1842, reprinted 1845) is not only "arranged according to" Ehrenberg's *Infusionsthierchen*, but presents extended passages in English translation.

146. Oken, *Lehrbuch* #3108 (1843:402); *Elements* #3107 (1847:503).

147. *Infustionsthierchen* (1838), unnumbered page following Foreword. Consult the original text for the complete diagnosis, which I have greatly condensed.

148. Chapter 13, notes [108] and [109].

149. Chapter 17, notes [120], [38] and [152] respectively, although Lamarck later changed his mind (Chapter 17, note [68]).

150. Audouin & Milne-Edwards (1828).

151. As inferred by Winsor (1976:28–32).

152. The eyespot or *stigma* is not an eye, but allows the organism (*e.g.* *Chlamydomonas* or *Euglena*) to detect the direction and intensity of a light source. Ehrenberg did not claim to have found a circulatory system, supposing its parts to be smaller than could be resolved with his microscope (1838:2).

153. *Infusionsthierchen* (1838), page xiv, point 28: *völlig deutliche Geistesfähigkeiten, wie andere Thiere*.

154. Derham (1713:406–408); Chapter 15, note [36].

155. *Infusionsthierchen* (1838), page xiv, point 29.

156. Ibid. (1838:99).

157. Dujardin (1835), quotation at page 354 (see also page 367); *vacuoles adventives* at page 371; groups of vacuoles mistaken for genital organs at pp. 371–372. He later refined the definition of *sarcode* (1838:258). Franz Meyen (1839) showed that food-vacuoles are not connected by an alimentary canal; it follows that infusoria are not complete animals.

158. Dujardin (1841:38). Nor is there any hint in Dujardin that sarcode might occur in plants.

159. Their objections are described by Stein (1859) 1:28–29.

586 NOTES

160. Wolff, *Theoria generationis* (1759); and Editio nova (1774). For a summary of Wolff's observations and interpretation, see Sachs (1890:249–253).

161. Treviranus (1805) 3:233. Treviranus cites Wolff §§2, 3, 26, and 93 (1774) but, so far as I can determine, does not cite or mention Oken in his six-volume *Biologie* (1802–1822).

162. Schleiden, *Beiträge* (1838); *Grundzüge* (1842:191–289); *The plant* (1848:43–55); and *Principles* (1849:31–110).

163. Eyde (1975:433).

164. Schwann, *Mikroskopische Untersuchungen* (1839), especially pp. 191–197; translated as *Microscopical researches* (1847), especially pp. 161–166.

165. Mohl (1846, 1852). The term *protoplasm* was first used in a cell-biological context (although in reference only to a primordial substance in animals, analogous with the cambium in plants) by Purkinje (1840); see also Baker (1949:90–91).

166. Schultze (1861:11). Schultze used the term *Kern*, which had typically been used in reference to cells of multicellular animals, and had been appropriated by Siebold for rhizopods and infusoria. See below and Siebold (1848:23): *Fast bei allen Infusorien, aber auch bei Rhizopoden kömmt im Inneren des Körpers ein scharf abgegrenzter Körper, eine Art Kern (nucleus) vor.*

167. Schultze (1865:399–400) considered diatoms neither animals nor plants, but rather *Urorganismen*. The context is organizational (he mentions their rigid membrane, nucleus, protoplasm, oil droplets, and the like), not systematic. It is easy to find in his words the implication that there exist *Urorganismen* in addition to diatoms, but he does not elaborate.

168. Geddes (1883–1884); Baker in *Quarterly Journal* (1948–1955); Hall (1969) 2:171–218; Geison (1969); Jacobs (1989); H Harris (1999); Mazzarello (1999); JA Mendelsohn (2003); and D Liu (2017). Hall (1950) attempts to trace the concept *protoplasm*, although not the word itself, to the Presocratics. The term *protoplast* is also to be found in the Slavonic mystical tradition, referring to the first humans as archetypes.

169. Cohn, *Nachträge* (1850). Following Mohl, Cohn believed that in plants, the sarcode and nucleus lie inside a cellulosic membrane, the *Primordialschläuche*, generally translated *primordial utricle* (1850:665–666). Note that his plant was the green alga *Protococcus pluvialis*, and his animals *Euglena* and Monades.

170. Virchow (1855:23). I cannot confirm the claim (by *e.g.* Tan & Brown, 2006; and Wright & Poulsom, 2012) that the aphorism *omnis cellula e cellula* was introduced by François-Vincent Raspail in 1825. See also Chvátal & Verkhratsky (2018:255).

171. Sudhoff (1922:5–7). For attribution to Virchow (1858,) see Temkin (1959), pp. 349–350 and notes 99 and 102.

172. Schwann, Author's preface (1847), page ix. For plants and animals see also pp. 161–163.

173. Baker (1948:106), propositions 6 and 7. For a particularly clear example, see Barry (1843).

174. See Baker, *Trembley* (1952:155–158); and (1953:422–423).

175. OF Müller (1786), pp. 56–57 and plate VII, figure 13; Baker (1953:423).

176. Barry (1839); Baker (1953:429–430).

177. Reichert (1840); Baker (1953:430). Reichert also mentions the *organische Reiche* (pp. 45, 78, and 147) and the *Their- und Pflanzenreichs* (page 96), so I believe we should read *Wirbelthier-Reich* in a non-taxonomic sense ("all the vertebrates" rather than "Kingdom Vertebrata").

178. C Bergmann (1841, and (1842:95); Baker (1953:430–431).

179. Kölliker (1847:12–13) as summarized by Baker (1953:431). Kölliker wrote that "*Furchungskugeln . . . wie Infusorien sich theilen.*" See also Richmond (2000:264–265).

180. Richmond (2000:257 and note 36) makes this point particularly well.

181. Siebold (1848:3). *Thiere, in welchen die verschiedenen Systeme der Organe nicht scharf ausgeschieden sind, und deren unregelmässige Form und einfache Organisation sich auf eine Zelle reduziren lassen.*

NOTES 587

182. Siebold (1848:25). In making this point, Siebold explicitly equated the terms *nucleus* and *Kern* (page 23). Churchill (1989:194n17) claims that "Siebold and many of his German contemporaries used 'Nucleus' when speaking of protozoa and 'Kern' when referring to the cells of higher organisms."

183. J Müller, Ueber eine eigenthümliche krankhafte parasitische Bildung (1841:493–494); he uses *Nucleus* and *Kern* interchangeably at page 494, as does Remak at page 511. See also Müller, Über einen krankhaften Hautausschlag (1841); and Beobachtung (1841).

184. At *Infusionsthierchen* (1838:24–25) Ehrenberg recognized a family of monads, the *Theilmonaden*, that multiplied in this way. In describing other taxa, he identified those in which he believes *Theilung* to occur. See also Harris (1999:61–62).

185. Owen, *Lectures* (1843), Figure 5 at page 17.

186. Ibid. (1843:25–26). In Lecture 2 (pp. 16–28) Owen follows Ehrenberg's interpretation of Polygastria at considerable length. Elsewhere he considers the analogy between infusoria and animal-cells "overstrained", *e.g.* at page 46.

187. Perty (1837–1846). Perty later wrote on the psychological character of infusoria, amœbæ and corals (1865:151–154), and on mysticism and the occult.

188. Perty (1852), pp. 51–54, 64–70, 76–77, *et passim*. Quotation at page 66: *eine gewisse Klasse von Blässchen und Körperchen*. For this work Perty used a Plössl microscope.

189. It is difficult to pinpoint the first such report, but Stein noted that "not a few" infusoria possess two or more nuclei, *Loxophyllum* up to 8–10, and *Loxodes* up to 12–20 (1859, 1:95). Ciliates have both macronuclei and micronuclei, each of which can be present in multiple copies.

190. Churchill (1989:203).

191. Huxley (1853).

192. Richmond (2000).

193. Huxley considered "unicellular" to be a contradiction in terms (1853:304). So far as I can determine, he did not use the term *acellular*.

194. Sedgwick (1894).

195. Whitman (1893).

196. Vines (1889).

197. Richmond (2000:282–284).

198. Dobell (1911:270).

199. Dobell (1911:272). Dobell concluded, inter alia, that "the various structures called cells certainly exist—but the cell theory is a myth" and "the cell theory must be abolished" (1911:285). For *protist*, see Chapter 21.

200. We return to Richard Owen in Chapter 20.

201. Not to be confused with Jules Victor Carus, zoologist and professor of comparative anatomy at Leipzig. The label "the last *Naturphilosoph*" comes from Baker, *Trembley* (1952:161), who may have followed Nordenskiöld (1932:290). For CG Carus as an idealist and Romantic, see Richards (2008:470–474).

202. Carus studied under Caspar David Friedrich. Today his works are on permanent display at national galleries in Europe, and at the Metropolitan Museum of Art in New York.

203. Carus, *Von den Ur-theilen* (1828). Blumenbach assembled one of the first research collections of human skulls (at Göttingen in the 1780s), and von Baer built a collection in St. Petersburg in the 1840s: Tammiksaar & Kalling (2018).

204. Carus, *Lehrbuch der vergleichenden Zootomie* §§42–48 (1818:25–28); in English as *An introduction to the comparative anatomy of animals* §§42–48 (1827:35–39). Both German editions (1818, 1834), although curiously not the English, display a brief verse by Goethe on the title page, and Goethe is recorded as having read Carus with joy and fulfilment: see Gibian (1955:372n4).

588   NOTES

205. *Lehrbuch* §§13–14 (1818:9–10); *Introduction* §§13–14 (1827:22).

206. *Lehrbuch* §29 (1818:16–17); *Introduction* §29 (1827:28).

207. Carus, *Lehrbuch*, figure at §36 (1834) 1:20. *Mensch* is presented as the eighth Class within *Eithiere*.

208. Carus, *Grundzüge* (1841); *Zwölf Briefe* (1841); *Psyche* (1846); *Physis* (1851); *Organon* (1856); *Lebensmagnetismus* (1857); *Symbolik* (1858); *Vergleichende Psychologie* (1866). In *Grundzüge* (1841), criminals lead a merely vegetative life.

209. Carus, *Natur und Idee* (1861:221–232).

210. Ibid. (1861:230).

211. Ibid. (1861:229).

212. For this earlier period, see Zammito (2018).

213. In the words of Klages (1926), page ii.

214. Kant, *Allgemeine Naturgeschichte* (1755, new edition 1798); and essays on the rotation of the earth [1754]; the age of the earth [1754]; the constitution and origin of the universe [1755]; earthquakes (three essays, [1756]); and the winds [1756], in *Gesammelte Schriften* 1 (1910).

215. Jenyns (1835), note at page 157.

216. Chapter 17. Règne Psychodiaire contained testate amœbæ, and sessile ciliates such as *Vorticella*.

Chapter 19

1. Shakespeare (1605), sig. D4v. "Heauen" given here as "heaven"; two commas added per the First Folio. The First Folio (1623) reads "our Philosophy". Since 1676, the passage is usually cited as 1.5.167–168. Also quoted in Unger (1843:2); and on the title page of Kützing, *Die Sophisten und Dialektiker* (1844).

2. Linnæus had died in 1778, but the animal-plant dichotomy was preserved in posthumous editions of *Systema naturæ* including the Leipzig (1788) and Lyon (1789–1796) Thirteenth edition edited by Gmelin, and its various translations.

3. *Bacillaria paradoxa* is now considered a colonial diatom, *Volvox* a colonial green alga.

4. OF Müller, *Animalcula infusoria* (1786), edited for publication by Otto Fabricius after Müller's death in 1784. Fourteen genera were recognized by both Müller and Gmelin. Gmelin recognized only one genus, *Bacillaria* (with one species), not in Müller. Müller recognized three genera (*Proteus, Kerona, Himantopus*) not in Gmelin.

5. According to Anker (1943:197–198), Müller had access to both simple and compound microscopes after the design of Watkins; but as the latter was uncorrected for chromatic aberration, he mostly used the former.

6. Linnæus, *Systema naturæ*. Thirteenth edition (1796) 2:1354–1397. The other four genera contain liverworts, hornworts, and lichens. Genus *Lichen* accommodated 314 species. Linnæus also recognized *Chara* (in Monandria Monogynia) and *Tremella* (in Cryptogamia Fungi).

7. This was perhaps most-clearly enunciated by Dillwyn, who wrote that "it will be necessary to reduce the present genera into one mass, and proceed in nearly the same manner as if nothing had been done before" (1809:4).

8. Siebold (1848).

9. Perty (1852).

10. Stein (1859–1883).

11. Tournefort (1694) 1:21–23, 1:428 and 1:438 ff. Tournefort included corals, madrepores, and sponges among such "plants". Ferns and mosses are without flowers, but "the feeling of those who believe that all plants have seeds is based on very probable conjectures" (1:22). See also note [21] below.

NOTES 589

12. Adanson (1763): overview at part 2, page 3; details at part 1, Systems 27–52 (pp. cclviij–ccxciij); membership in his families Bissus, Champiñons and Fucus at part 2, page 8.

13. AP de Candolle, *Théorie élémentaire* (1819), pp. 236–242 and 249–250.

14. Richard, *Nouveaux elémens* Third edition (1825), pp. 32–33, 444 and 466–470; and Fourth edition (1828), pp. 25, 386 and 399–403; *Elements of botany and vegetable physiology* (translated from the Fourth edition by Macgillivray, 1831), pp. 23, 355, and 368–371. I have not examined earlier editions.

15. Hooker (1821), part 2, page 3 note.

16. Lindley (1829:3).

17. Cellulares: de Candolle (1818), Lindley (1829), and (as Cellulaires) de Candolle (1819); Acotylédonés aphylles: de Candolle (1819), and (as Acotyledons) Hooker (1821); Agames: Richard (1825); Inembryonées: Richard (1828, 1833); Thallogens: Lindley (1846:5); Thallophyta: Endlicher (1836–1840), pp. i and liii. On the other hand, Lamarck (*Flore françoise*, 1778) and CF Brisseau-Mirbel (1815) followed Linnæus. See also Farley (1982:5–14).

18. Lamouroux (1813), summarized at page 291.

19. In part this is because colour is strongly correlated with photosynthetic pigments, which are highly conserved within each of these orders, and differ between them.

20. In Linnæus's cortico-medullary theory, a seed arose from the (female) medulla via fertilization, thereby involving the (male) cortex: see *Metamorphoses plantarum* (1759) and Chapter 15.

21. Tournefort (1694), pp. 22, 429–430, and 439.

22. J Lindsay (1794).

23. Stackhouse (1795–1801), pp. ix–xiv, particularly note § at page xi.

24. Lindley (1846), pp. xxvii and 3. Ramsbottom (1941:338–339) credits the term to Hedwig (1798); see note at 1798:64. Hedwig seems not to have intended any distinction between *spore* and *seed*, except that spores are formed in *sporangia*.

25. For example, from Joblot and Spallanzani: Chapter 13.

26. Treviranus (1803) 2:264–406; Creplin (1841), Sec. 1, 35:76–83; Pouchet (1859); Taschenberg (1882); Lippmann (1933).

27. Chapters 16 (Eichwald), 17 (Lamarck) and 18 (the others mentioned).

28. Farley (1972).

29. The most-comprehensive source by far is Hans Kniep's highly technical *Die Sexualität der niederen Pflanzen* (1928), which covers fungi as well. Link (1929) deals mostly with fungi. For individual taxa: Engler & Prantl, *Pflanzenfamilien. Algen* (1896); Oltmanns (1904–1905), second edition (1922–1923); and Fritsch (1935 and 1945).

30. For example, meiosis has not been directly observed in symbiotic dinoflagellates, despite highly suggestive evidence from genomic and transcriptomic sequencing: Morse (2019); and Shah *et al.* (2020).

31. Priestley (1772).

32. Priestley, *Experiments and observations* (1779) 1:341–342. The publication history of his *Experiments and observations* is complicated; I cite the three-volume Pearson & Rollason (Birmingham) edition (1779–1786) for simplicity. See also Schofield (2004).

33. Priestley, *Experiments and observations* (1781) 2:16–17. He noted that "my own eyes having always been weak, I have, as much as possible, avoided the use of a microscope" (2:17). Priestley carried out several curious experiments to confirm its vegetable nature (2:17 ff.), but was more interested in the origin of the dephlogisticated air.

34. Ibid. (1781) 2:32–33. It had long been known that green material sometimes formed in vessels, and could release bubbles of air; likewise, it was known that green material sometimes appeared on the walls of thermal baths. For the early observations of Leeuwenhoek, La

590 NOTES

Hire (1690), Homberg (1710), Adanson (1757), Fontana (1783), Scherer (1787), and others, see Senebier (1799) 48:155–162. In 1691–1692 Woodward (1700) observed green material on glass vessels containing water and plants, and considered it "vegetable material" (as opposed to "terrestrial matter"), but not a plant in its own right. Woodward separately remarked on "a fine thin *Conferva*" in some of his vessels (1700:203).

35. Priestley (1809). In this article he also denies that animals might change into plants, or vice-versa.

36. Ingen-Housz (1789) 2:24–26. For an overview of his research, see Wiesner (1905).

37. Chapter 14. The *polypiers coralligènes flexibles* of Lamouroux (1816) include a few genera of seaweeds.

38. Ingen-Housz (1789) 2:31 cites a personal demonstration by the Abbé Fontana, and subsequent personal communication from him (2:52). See also Fontana (1783:660–664). According to Tilloch writing anonymously in his *Philosophical Magazine* (1800:312), Fontana believed the green material to be a kind of polypier, produced by the motile insects.

39. Ingen-Housz compared this with fertilized frog-eggs breaking free from their jelly-like egg mass (1789) 2:131.

40. Ibid. (1789) 2:113–114, translation by MAR.

41. Ibid. (1789) 2:120.

42. Ibid. (1789) 2:51–52 and 93–102 ("The elaboration of dephlogisticated air is no proof whatsoever that it is a plant").

43. Girod-Chantrans (1802:7–8). Translation by MAR. See also Ritterbush (1964:141).

44. Senebier (1799). Stony case at 48:426; *Ulva* or *Nostoc* at 49:218; locomotion throughout, with particularly clear instances at 48:362–363 and 49:366–367.

45. B Thompson (1787).

46. Gruithuisen (1808): *So entsteht die priestley'sche grünen Infusorien, die an sich selbst schon eine Pflanzennatur insich tragen, und die ein wahres Mittelding zwischen einem Infusinonsthiere und einer Pflanze ist* (cols 607–608). *Cf.* Goethe (Chapter 18). Granite and cryptogams, (1808) column 607. Further (regarding the supposed loss of arbitrariness by animals upon procreation): *der Mensch wird selbst zur Pflanze, wenn er den Beyschlaf ausübt. Dieses waren für jeden Naturforscher von jeher sehr auffallende Momente* (column 607).

47. Ingen-Housz (1789), see text; Girod-Chantrans (1802), Plate 12, Figures 25′′′ and 26′′. See also Wiesner (1905:141–142).

48. Ingen-Housz (1789:41–42) noted that the insects produced by green matter differed from those produced by decomposing animal matter.

49. Ibid. (1789:56–57).

50. Ibid. (1789), pp. 29, 42, 128, *et passim*.

51. Treviranus later repeated some of Priestley's work (1803, 2:297–354). He described the green matter as arising from infusion animals (page 350), and regarded the green matter, like Needham's observations, as evidence for the transformation of animal forms into vegetable, and vice-versa (pp. 378–379).

52. Kniep (1928:2); Farley (1982); Drews (2000); and Chapters 18 and 21.

53. Farley (1982:25–31). Spallanzani called spermatozoa "parasites".

54. Vaucher (1841). Vaucher held that the stamen provided a fluid, which he called *nectaire*, to the ovary. See also Alphonse de Candolle (1842).

55. Vaucher contrasted the wondrous fecundity of nature "at the final limits of the vegetable kingdom" with "entire Linnæan classes, in which the sexual organs show almost no difference", and remarked that further modes of reproduction (hence families) doubtlessly exist among the marine confervas (1801), pp. 353 and 357. For the six families (containing eleven orders, not all of which are formally described), see Vaucher (1803).

56. His Conjuguées included the well-known *Spirogyra*.

NOTES 591

57. Vaucher (1803), pp. 135 (quote) and 140. In considering conjugation a sexual process he followed Hedwig (1798). Conjugation in *Spirogyra* (as *Conferva jugulis*) resulting in a zygote was pictured by OF Müller (*Flora Danica* 5(15), Table 883, 1782). Hedwig (1798) depicts the "flower", "seed", and mature "fruit" of *Chara vulgaris* at Table 35. See also Kniep (1928:1).

58. Vaucher (1803), Plate 7, Figure 1e,f.

59. His criteria for animality were self-movement; division of the body into distinct parts; or an interior organization "that could be related to that of some animal" (1803:143–146). He insisted that confervas are plants.

60. Vaucher's identification of sexual isomorphism was confirmed by (among others) CG Carus, Beitrag (1823:517–522). Carus remarks that this and three other genera of convervas are *zu höherer Vollkommenheit gediehen, zugleich pflanzlich und thierisch bewegt, als die höchste Evolution des infusoriellen oder primitiven Lebens in der Richtung zur Vegetation hervor* (pp. 516–517); Nees von Esenbeck, who commented throughout Carus's paper, jokingly (one imagines) calls this *Ausschweifung*, and admits that "the arrangement attempted here" (*i.e.* by Carus) is "more physiological than descriptive, and cannot exist in any system of algae" (page 522).

61. Kniep (1928:2) points particularly to his identification of antheridia and oogonia.

62. Trentepohl (1807): release at pp. 191–194, transition to animal at pp. 195–196. He observed the cycle through three generations.

63. Nees von Esenbeck (Bamberg, 1814); release at page 40. Nees von Esenbeck remarked that, as with Priestley's green matter, the infusorum relapses to the level of a plant; and finds it odd that this relapse usually takes place in the morning (*i.e.* given the associations of sunrise and East: see our discussion of Goldfuss's Easter-egg in Chapter 16).

64. For progress in lenscraft, microscope design, and technique during these years, see the early pages of volumes by Jabez Hogg (*The microscope*, first edition 1854) or Carpenter (*The microscope*, first edition 1848). Carpenter specifically discusses the role of improved microscopes in understanding the reproduction of cryptogams (1856:43–47).

65. In 1836 Unger was appointed professor of botany at the University of Graz, and from 1850 professor of botany at the University of Vienna.

66. *Conferva dilatata β clavata* Roth, or equivalently *Ectosperma clavata* Vaucher; now *Vaucheria bursata* (OF Müller) C. Agardh, a member of class Xanthophyceae.

67. Unger (1827); pages 791–792 are a Foreword by Nees von Esenbeck; *thierische Geschöpfe* at page 794.

68. Strongly implying that the conferva is a plant, and physiologically active.

69. In this work Unger used a compound microscope (he does not identify it further), at an unspecified magnification.

70. Unger called them *Kolben* or *Kölbchen*. An ear of maize can be called a *Kolbe*. They are now known to be sporangia (spore-bearing bodies).

71. Vaucher (1803:144).

72. Unger (1827:800). He looked back on this work in *Botanische Briefe* (1852:148–156), translated by Paul as *Botanical letters to a friend* (1853:110–116).

73. Oken, [Review of:] Die Metamorphose (1829).

74. Ehrenberg, Dritter Beitrag (1835), note at pp. 157–158, without specifically citing Unger. For other commentary, see Unger (1843:9–12).

75. Flotow (1844), with commentary by Nees von Esenbeck at pp. 566–574 (who in turn cites Morren & Morren, 1841). For other commentary, see Unger (1843:4–12).

76. JG Agardh (1836).

77. Dumortier (1832); Mohl (1835, 1845).

78. Unger (1838). Also Schleiden (1845–1846); English translation (in part) by Lankester as *Principles of scientific botany* (London, 1849), in which see especially pp. 145–150.

592 NOTES

79. Schleiden, *Principles* (1849:529–532); Farley (1982:47–54).

80. Schleiden, Einige Blicke (1837:312–315); Some observations (1838:243–245). See also Farley (1982:48–49).

81. Endlicher, *Grundzüge* (1838:15–20); Farley (1982:51).

82. Farley (1982:52–54).

83. Simon Plössl introduced an achromatic objective. Entz (1888:30 and 248) gives his name as Plössel.

84. Unger (1843), pp. 36, 39–42, and 93 (item 9). These are now known as *flagella*.

85. Ibid. (1843), item 16 at page 95, echoing Nees von Esenbeck (1814:40).

86. Ibid. (1843:96–97).

87. Endlicher an Unger [16 July 1842], in *Briefwechsel* (1899:133–134). Unger did not take up the offer.

88. Schleiden, *Principles* (1849:101–102).

89. Bornet (1875); Farlow (1876).

90. Thuret, Recherches sur les zoospores, in *Annales* (1850–1851).

91. Thuret (1853); and *Annales* (1854–1855). Thuret did not observe the smaller gametes ("antherozoids") fuse with the larger one; this remained for Pringsheim (1855).

92. Thuret, *Recherches* (1851:5–35). Most of this work was derived from an 1847 prize-winning *Mémoire*, with the delay in publication due to the time required for the plates to be engraved (see 1851:5).

93. Ibid. (1851:42–43). Translation by MAR.

94. For fungi, see Prévost (1807) for the white rust *Cystopus*; Gruithuisen (1821), whose *Conferva ferax* is now the water-mould *Saprolegnia ferax*; Kniep (1928); and Waterhouse (1962).

95. CA Agardh held the professorship at Lund from 1812–1835, then served as Bishop of Lund until his death in 1859. His son Jacob Georg Agardh, also a phycologist, was professor of botany at Lund from 1854–1879.

96. CA Agardh, *Synopsis algarum* (1817), page vi.

97. CA Agardh, *Aphorismi* (1817–1826), Aphorism 126 at page 72; *Species algarum* (1820–1828).

98. *Systema* (1824), one scale facing pp. [1], and two between pp. 168 and 169. The work is organized in a curious way, and pagination is sometimes problematic.

99. *Systema* (1824), pp. xvi–xvii. Page xvi is mistakenly numbered xv.

100. Koster (1965); Stafleu (1966). Agardh's six orders within class Algæ were Diatomeae (including desmids), Nostochinae, Confervoideae (including *Oscillatoria*, *Chara*, several red algae, and a few fungi), Ulvaceae (including *Vaucheria* and *Caulerpa*), Florideae, and Fucoideae. At *Systema* (1824), page vii, he credited Roth with the name *Algae* (presumably as distinct from *Cryptogamia Algæ*). Roth referred to *cryptogamischen Wassergewächsen (Algae)* in *Neue Beyträge* (1802), pp. viii–ix, in turn citing Mertens, *Algæ aquaticæ brevibus descriptionibus et iconibus illustratæ* [*Fasc.* 1, Bremæ, 1801]. I have been unable to locate the latter. Mertens illustrated the algae in the third volume of Roth's *Catalecta botanica* (1806), which nonetheless followed the Linnæan system (*Catalecta* 1806, 3:84–350).

101. CA Agardh, *Dissertatio* (1820); also in Oken's *Isis* (1820).

102. CA Agardh, *Dissertatio* (1820:5–6); quote at page 6. He called the new alga *Ulva Bullosa*; what Roth called *Ulva Bullosa* was moved to *Monostroma bullosum* by Thuret. In general, I do not attempt to trace synonymy; for those braver than myself, I heartily recommend Michael Guiry's *AlgaeBase* (www.algaebase.org).

103. CA Agardh, *Dissertatio* (1820:4–14), in order of appearance.

104. Ibid. (1820:14–18).

NOTES 593

105. Ibid. (1820:18). Translation by Professor Denis Brosnan and MAR. References to Leibniz in other works (*e.g. Classes Plantarum* 1825:3) suggest that by "monads", Agardh means the supposed elemental entities, not individuals of genus *Monas*. Agardh returned to plant metamorphosis in *Lärobok* (1829–1830) 1:211–213, and throughout the second volume, which I have examined only in the German translation of Creplin (1832). Agardh further identified simple algae (*e.g. Protococcus*) with *organs* of plants, although his precise meaning is sometimes obscure. *Elementar-Organe* (*Elementarorgane*) meant *fundamental structural units*, that is, *cells*; but bear in mind that Agardh predated much of cell-theory.

106. *Ulva* can be induced to grow as "long thin *Enteromorpha*-like tubes" by marine bacteria: Provasoli & Pintner (1980). Pleomorphism in fungi and lichens was notably studied by LR Tulasne (1851); Tulasne & Tulasne (1861–1865); and others.

107. Lichtenstein (two articles, 1803); see also Hewitt *et al.* (2016:51–52) and Chapter 15, note [130].

108. For example, Richard, *Nouveaux élémens* (1825:8–9).

109. Link (1820:5). The letter is dated 1 May 1819 (see page 8).

110. Hornschuch(?) (1819). The article is presented under the heading "III. Botanische Notizen. *Hornschuchiana*", but it is not clear whether Hornschuch is the author. See Ruppel (2019:140n180).

111. Meyen (1827).

112. Ironically, Jacob Georg Agardh, son of Carl Adolph and himself a professor of botany at Lund (1853–1879), argued against algal metamorphosis: see note [76] above, particularly note 3 at (1836:202–203).

113. Chamisso, *De Salpa* (1819). See also *Reise um die Welt* (1836), butterfly analogy at 1:50–51. Bodenheimer (1958:360–361) translates a *mélange* of paragraphs; Farley (1982:72–82) discusses whether the alternation is of generations, or of individuals.

114. C Darwin, On the ova of Flustra [March 1827], in *Collected papers* (1977) 2:285–291. Darwin mentions having read the report to the Wernerian and Plinian Societies. See also Ashworth (1935).

115. Sars (1846); Sars, Koren & Danielssen (1856); and Koren & Danielssen (1877).

116. Steenstrup, *Om Forplantning* (1842). The manuscript was translated into German by CH Lorenzen (1842), then from German into English by Busk (1845). A French translation was also made.

117. Gaillon, Aperçu microscopique (1821:371–372).

118. Gaillon, Observations microscopiques et physiologiques (1822); metamorphosis at pp. 35 and 36. The report was delivered by August Le Prévost.

119. Gaillon, Observations microscopiques sur le Conferva (1824:311–315).

120. Ibid. (1824:314): that is, members of Bory's *Bacillariées*. Gaillon did not find the species in OF Müller, but their form and movement were analogous to his *Vibrio bipunctatus* and *V. tripunctatus*. Note that Gaillon identified his oyster-animalcule as *Vibrio ostrearius* (see text).

121. Gaillon, ibid. (1824:314–316).

122. Bory, Némazoaires (1827:502). See also Léman, Navicula (1825); and Blainville, Némazoaires (1825), and Némazoones (1825).

123. Gaillon's observations appeared as short notes, including in *Bulletin des sciences par la Société philomatique* (presented by Blainville, 1820); *Journal de physique* (1820); and *Précis analytique* (1821). See also Briée (2010).

124. Bory, Navicule (1824). In 1974 the diatom was renamed *Haslea ostrearia*.

125. Gaillon, Aperçu d'histoire naturelle (1833, 1834). The latter is reproduced in part in Tableaux synoptiques (1834).

594 NOTES

126. Gaillon, Aperçu d'histoire naturelle (1834:47–56).

127. Gaillon does not specify the taxonomic rank of this group.

128. Ibid. (1834:48).

129. Kützing was self-taught in subjects such as Latin required for matriculation to university. Around 1833 he served briefly as assistant to Franz Wilhelm Schweigger-Seidel, head of a pharmaceutical institute attached to the University of Halle, while attending classes and lecturing in medical botany (amidst a cholera epidemic). From 1835–1883 Kützing taught at the *Realschule* at Nordhausen. His (unfinished) autobiography, *Aufzeichnungen und Erinnerungen*, was edited by Müller & Zaunick (1960).

130. Kützing, Beitrag (1833): *nun wie mit einem Zauberschlage* at page 353.

131. Ibid. (1833:359–361).

132. The full question is presented in Dutch in *Natuurkundige Verhandelingen van de Hollandsche Maatschappij der Wetenschappen te Haarlem* 12(1), 1841, at page vi. A German translation, which adds a sense of *returning to* the original organization in the absence of favourable conditions, is given by Kützing at page xix of his prize-winning essay Die Umwandlung (1841), in this same issue; and in *Aufzeichnungen* (1960:235).

133. On 11 May 1839 he received notification that he had won the prize: *Aufzeichnungen* (1960:236).

134. Kützing (1841:4). By 1834 he had access to an early Schiek microscope (see Kützing, *Verwandlung*, 1844:7); after 1837 at least, Schiek microscopes had a reputation equal to those by made Plössl (see note [83] above). Schleiden, Schwann, and Johannes Müller used Schiek microscopes: Mappes (undated), and Goren (undated).

135. Kützing (1841:9–11); I translate *Spielereien* as *frivolities* (1841:10). Kützing spells out this law of nature in detail at (1841:13–14).

136. Ibid. (1841:11–14).

137. Ibid. (1841:14–76).

138. Ibid. (1841:109–115). By contrast, mosses are more constant.

139. Kützing, *Phycologia generalis* (1843), page xiii.

140. Ibid. (1843:4).

141. Schleiden, as quoted by Kützing (1843:4n2), perhaps in reference to *Grundzüge* (1842, 1:263–266), where Schleiden describes the interpretations of Ingen-Housz, Agardh, *et al.* in these terms, and worse. See also note [88] above.

142. Kützing, *Sophisten* (1844:13–14).

143. In *Species algarum* (1849), Kützing broke apart the Linnæan genus *Conferva*; but Kützing's critique of species clearly went deeper than this.

144. Kützing, *Sophisten* (1844:16–18).

145. Variously called *Sphaerella nivalis, Protococcus viridis, P. nivalis, Enchelys sanguinea, Astasia nivalis, Giges sanguineus*, etc.: Kützing, *Verwandlung* (1844:8–9).

146. Kützing, *Verwandlung* (1844), pp. 9–13 and 23–24.

147. Ibid. (1844:13–20). I have modernized the spelling of his *Chlamidomonas* and *Stygeoclonium*.

148. Ibid. (1844:20).

149. Describing the fermentation of sugar in making wine, Adamo Fabbroni wrote that sugar is decomposed by a "*sostanza vegeto-animale*" in grapes or maize (1787). Kützing associated this process with yeast, and in 1834 sent a manuscript describing his observations to Poggendorff, editor of *Annalen der Physik*, who however neither published the work, nor would return the manuscript. Kützing eventually re-wrote it (Microscopische Untersuchungen, 1837), but by then had been scooped by Cagniard de la Tour, and by Schwann. The discovery was later exploited by Pasteur. One type of Turpin's *globulines*

NOTES 595

(Chapter 17) was "Globuline du vin, de la bière" (see the anonymous listing in *Dictionnaire des sciences naturelles*, 1845). See also Schützenberger (1876:34–44).

150. Kützing, Synopsis diatomearum (1833).

151. According to Prescott (1951:4–5), Kützing described more genera than any other phycologist.

152. Kuhlbrodt (undated).

153. Kützing, *Verwandlung* (1844:21–24).

154. Something *complete*, or *developing*. The distinction, often formulated as *being* versus *becoming*, reflects a theme that can be traced back through Aristotle (*Metaphysics*) and Plato (Bolton, 1975) to the Presocratics.

155. Kützing, *Verwandlung* (1844:22–23). As an example of this physiology, see his proposal for how the outer cell-envelope and cell-membrane direct an infusorian to develop as a filamentous plant (pp. 16–17).

156. Ibid. (1844:23).

157. Ibid. (1844:24).

158. Leeuwenhoek, Brief 239 [Dobell/Cole 149] dated 25 December 1702 (Dobell 1932:276, 278); *Philosophical Transactions* (1703), Figure 8, between L and K; *Alle de brieven* 14:158 (1948).

159. A Gentleman in the Country (1703): see Chapter 13, note [65].

160. CA Agardh made *Desmidium* one of nine genera within order Diatomeæ (*Systema Algarum*, 1824, page xv). Ehrenberg set Desmidiacea as one of four Sections within family Bacillaria (1838:140–165). See also Ehrenberg, *Zur Erkenntniss* (1832:66 ff.). According to Jahn (1995:114), Ehrenberg did not use the term *diatom* because he considered it to imply that these organisms are plants. Kützing eventually divided family Diatomeen into two *Hauptgruppe* of equivalent rank, Diatomaceae and Desmidiaceae: note [150] above.

161. Corti (1774), pp. 111–115 and Tav. 2, Fig. 17. He called it *corpicetti a Baccello*, a small body resembling a bean-pod. Vibratory movement, and movement from place to place, at (1774:113). His *Osservazioni* also contains the first description of cytoplasmic streaming in *Chara*.

162. Eichhorn (1775), page 48, Tab. 5, Fig. c. The more-widely available 1781 edition appears identical, apart from spelling of the title and (in the copies I examined) location of the figures. There is also a *Zugabe* (1783). Eichhorn called *Closterium* the "half moon".

163. Ehrenberg (1838:139 and 175). He applied the same analogy to the amœba *Cyphidium* (1838:135). See also Kützing, *Die kieselschaligen Bacillarien* (1844), pp. 4 and 26.

164. Kützing, *Die kieselschaligen Bacillarien* (1844:26). Likewise, physician and histologist John Dalrymple (1840) found no protrusion or foot.

165. Ralfs (1848:20–22).

166. Ehrenberg (1838), pp. 136 (*Darmkanal*), 139 (diatom stomachs) and 142 (desmid stomachs), the latter interpretation marked as uncertain. Eckhard, on the other hand, claimed to find internal alimentary and intestinal canals (1846:218–226); in English (1847:440–445).

167. Chapter 18, note [157].

168. Kützing, *Kieselschaligen Bacillarien* (1844:25–29); Meneghini, *Sulla animalità* (1846), in English (1853); Eckhard (1846), in English (1847); Ralfs (1848:17–34). For *Staurastrum*, see Ralfs (1848:18).

169. Meyen (1835:208); Dalrymple (1840); Kützing, *Kieselschaligen Bacillarien* (1844:25–29); Ralfs (1848:17–37); Thuret, *Recherches* (1851:42 and note 2).

170. Kützing, *Kieselschaligen Bacillarien* (1844:28). The ability of diatoms to produce significant quantities of oxygen in the sunlight, "like all decided plants", was the one line of evidence he did not immediately qualify, except to point out that oxygen production by green monads

596 NOTES

and *Euglena* "does not prove anything for the animal nature of the diatoms, but makes it very doubtful that those infusoria are animals".

171. Ralfs (1848:35).

172. For example, the presence of starch could make desmids plants, without implication for diatoms: Ralfs (1848), pp. 31–34 and 36.

173. Link (1820:4), *Bacillaria*. See also note [109] above.

174. His construction *bald . . . bald* might suggest alternating rather than simultaneous plant- and animal-elements (*Kieselschaligen Bacillarien*, 1844:28).

175. Ibid. (1844:28–29).

176. Meneghini criticized Kützing for claiming that diatoms represent a combination of inorganic and organic kingdoms: *Sulla animalità* (1845:5–7); in English (1853:347–349).

177. Pritchard (1852:701–704). Pritchard's genera include *Rotifer* and *Tardigrada*. Pritchard often follows Ehrenberg, but includes genera from CA Agardh, Bailey, de Brèbisson, Dujardin, Kützing, and Ralfs. Siebold (1848, 1:10–11) lists only representative families and genera.

178. Kützing, *Species algarum* (1849). Both Pritchard (1852) and Kützing (1849) include diatoms and desmids. Nägeli (*Die Neuern Algensysteme*, 1847) mentions only common or representative genera. Perty (1852:221–228) lists 229 genera by my count, but focuses on those reported from Switzerland.

179. Nägeli, *Neuern Algensysteme* (1847:116). Pagination is identical in the freestanding and *Neue Denkschriften* editions.

180. Ibid. (1847:118).

181. Kützing had required algae to be aquatic: *Phycologia generalis* (1843:145).

182. See note [91] above. Pringsheim subsequently observed sexual reproduction in other algae, notably the conjugation of isogamous zoospores in the green alga *Pandorina* (1870).

183. Cohn, Nachträge (1850: 634–636); in English (1853:523–524).

184. An important step in this direction was taken by WH Harvey, who established brown, red, and green algae first as Divisions (1836), then as Series (1841). The latter made it natural to include unicellular forms, *e.g. Protococcus* in Chlorospermeæ (1841:181–183). He was nonetheless careful to avoid referring to solitary forms as *unicellular*. Harvey established Diatomaceæ as a fourth Division (1836:249–254) or Series (1841:194–216) with diatoms and desmids as members, but later (1849) labelled the three series as sub-classes, and brought diatoms into Sub-Class Chlorospermeæ or Confervales (1849:185).

185. Sachs (1875:212–218) and (1890:197–203). Also Hofmeister (1851), in English (1862); and Neue Beiträge (1859 and 1861).

186. For example, Schleiden, *Grundzüge* §80 (1845) 1:28–29; *Principles* §80 (1849:145–146); and WH Harvey (1841), pp. vii and 181–183.

187. Chlorophyll was extracted from leaves and named by Pelletier & Caventou (1817). By the 1840s JG Agardh, Kützing, Nägeli, Schleiden, and others were using *Chlorophyll* to mean a chemical substance or pigment in algae. Even so, different structures were described as *Chlorophyllbläschen* or *Chlorophyllkörnchen*. Sorby (1873) later studied the distribution of so-called blue chlorophyll (now known as chlorophyll *a*), yellow chlorophyll (chlorophyll *b*) and chlorofucine (chlorophyll *c*) in brown, red and green algae, and concluded that "as far as their constituent colouring-matters are concerned, the green *Algæ* are therefore perfectly typical plants" (1873:474). The trace of chlorophyll *c* he observed in red algae was undoubtedly due to epiphytic diatoms. The same connection had been drawn by Stokes (1864) but with little explanation. Haeckel (*Generelle Morphologie* 1866, 1:210) allowed that a few animals (*Borellia*, *Hydra*, *Stentor*, Turbellaria) do produce chlorophyll.

NOTES   597

188. As detected by the iodine test introduced by Colin & Gaultier de Claubry (1814) and by Stromeyer (1815), and first applied to plant material by Schleiden (Einige Bemerkungen, 1838).

189. Bergson (1910:122–123); in English (1911:112–113).

Chapter 20

1. E Darwin, *The temple of nature* (1803:25–26).

2. By Britain, I refer to the Kingdom of England (1536–1707), the Kingdom of Great Britain (1707–1801), the United Kingdom of Great Britain and Ireland (1801–1922), and/or the United Kingdom of Great Britain and Northern Ireland (1922 ff.) as appropriate. By Scotland, I refer to the Kingdom of Scotland (until 1707), or the corresponding part of Britain (after Union).

3. London's population, about 710,000 persons (1775), surpassed that of Paris and Istanbul: Chandler (1987).

4. University College London was founded in 1826, but received a Royal charter (allowing it to award degrees) only in 1836/1837. King's College London was founded in 1829.

5. Royal College of Physicians (chartered 1518); The Royal Society (1660); Royal Society for the Encouragement of Arts, Manufactures and Commerce (1754, chartered 1847); Linnean Society of London (1788, chartered 1802); Royal Institution (1799); Royal Society of Medicine (1805); Geological Society of London (1807); Zoological Society of London (1826); Royal Geographical Society (1830).

6. For a quantitative analysis, see Sorrenson (1996).

7. The Corporation Act (1661), the Test Act (1673), and the related Act of 1678 were in place, if not always rigorously enforced, until their repeal in 1828–1829.

8. The volumes on the natural history of birds appeared later, in 1792–1793. Further editions of the *Natural history* appeared until 1817. See also Loveland (2004).

9. Smellie (1790:245–263). See Chapter 15, notes [116] and [117] for similar views in the first edition of *Encyclopædia Britannica*. Smellie continued his attack on Linnæus as a "nomenclator, without philosophy, though he may be useful by mechanically marking distinctions, is incapable of enriching our minds with general ideas" (1790:523).

10. Smellie (1790:520–526); quote at page 526.

11. Ibid. (1790:1–14). At page 12 he allows a "vast chasm in the chain of being" between minerals and plants.

12. Eddy (2008); Walker (1812).

13. Chapter 13, including notes [112] and [113].

14. See also Armstrong (2000).

15. Including Carl Agardh, Bonaventura Corti, Johann Conrad Eichhorn, René-Just Haüy, Johann Gottfried Herder, Pierre Latreille, John Turberville Needham, Michael Sars, Franz von Paula Schrank, Jean Senebier, Lazzaro Spallanzani, Johann Trentepohl, and Jean-Pierre Vaucher.

16. Including Miles Joseph Berkeley, Stephen Hales, John Stevens Henslow, Leonard Jenyns, William Kirby, David Landsborough, John Lightfoot, and Joseph Priestley; and William Buckland, Adam Sedgwick, and William Whewell in natural philosophy.

17. Our Gilbert White was born at the home of his grandfather (also named Gilbert White), who was vicar of Selborne. Gilbert White (the grandson) took over the home in 1758 on the death of his father, but was ineligible to serve as vicar in Selborne because he had studied at Oriel College, while the office of vicar was in the gift of Magdalen College. He instead served as curate of Selborne until his death.

598   NOTES

18. Contrast White's *Natural history of Selborne* with other works of natural history that appeared in 1789, *e.g.* Jussieu's *Genera plantarum*, Batsch's *Versuch einer Anleitung, zur Kenntniss und Geschichte der Thiere und Mineralien*, Lavoisier's *Traité élémentaire de chimie*, or the second volume of Ingen-Housz's *Nouvelles expériences*; or, for that matter, with Erasmus Darwin's *Loves of the plants*, published anonymously at first (below, and J Browne, 1989).

19. G White, *Selborne* (1789), pp. 3, 14, and 288–289. At page 48 he finds an analogous "degree of dubiousness and obscurity attending the propagation" of reptiles and eels on one hand, and "the cryptogamia in the sexual system of plants" on the other; but he does not pursue the analogy.

20. That is, rather than alphabetically. Ray called his high-level taxa *genera*, but their rank is similar to that of Linnæan classes or orders. In 1724 Ray's *Synopsis methodica stirpium Britannicarum* was updated (particularly its cryptogams) by Johann Dillen, professor of botany at Oxford. For the history of this edition, see Stearn (1973:23–28).

21. Chapter 15, notes [20] (*Methodus plantarum nova*) and [23] (*Synopsis methodica animalium quadrupedem*) respectively.

22. Hill, *An history of animals* (1752:87). A second edition appeared in 1772–1773. Hill passed up few opportunities to criticize Linnæus: see Stafleu (1971:207–209).

23. H Baker, *Employment* (1753), particularly Part 2 at pp. 233 ff. See also Chapter 13. *Employment* was reprinted in 1764.

24. Ibid. (1753:217–221).

25. Ellis (1755), pp. v–xvii.

26. Stafleu (1971:211). "Perhaps [the British] were more interested in growing their plants than in talking about them; a commendable attitude."

27. A second edition appeared in 1765, a third in 1776, and a fourth in 1788. An 1810 edition appearing under the name of James Lee, son of the original author (but actually by Robert Thornton), bills itself the "Fourth edition, corrected and enlarged". According to the *Catalogue* of the British Museum (Natural History) 3:1078 (1910), an 1811 edition (revised by Stewart) may be the ninth.

28. Hope was also Regius Keeper of the Royal Botanical Garden at Edinburgh, and King's Botanist. He published very little, but was influential through his position and teaching.

29. Withering, *A botanical arrangement* (1776) 1, page v. Second (1787) and third (1796) editions appeared under slightly different titles.

30. Robert Waring Darwin of Elston, brother of Erasmus Darwin.

31. Martyn also translated Jean-Jacques Rousseau's *Lettres élémentaires sur la botanique*, appending further "letters" setting out the Linnæan system (1785).

32. Thornton lost a considerable sum of money on his *New illustration*, and had to scale back the large plates to fewer than half of the originally intended number. This suggests that the high-end market for botanies had become saturated by 1807.

33. A second edition appeared in 1824.

34. Geldart (1914:648) says that Rose wrote *Elements of botany* with the assistance of the Rev. Henry Bryant. Bryant is mentioned briefly in Rose's *Elements* (1775:450).

35. This Botanic Society was co-founded by [Sir] Brooke Boothby and Erasmus Darwin: see King-Hele (1963:24), and Uglow (2002:379–383). The translation is of the Thirteenth edition. See also Chapter 15, note [124].

36. Novelist and poet Oliver Goldsmith drew heavily on Buffon for his eight-volume popular *History of the earth, and animated nature* (1774): see Gibson (2012:467–472), and the exchange between Lynskey (1945) and Lovejoy (1946).

37. Kerr also translated Lavoisier's *Traité élémentaire de chimie*, and Cuvier's *Recherches sur les ossemens fossiles de quadrupedes* (the latter published posthumously as *Essay on the theory of the earth*, 1822).

NOTES 599

38. Anonymous, *A new system* (1791) 1:1–16.
39. For the travels of Linnæus's students, see Stafleu (1971), especially pp. 143–155.
40. Banks was accompanied by Daniel Solander, a pupil of Linnæus: see also Stafleu (1971), pp. 151 and 153.
41. Sara Elisabeth Moraea, known as Sara Lisa (1716–1806). Linnæus's library and collections first passed to Carl Linnæus the younger, while Sara Lisa kept the herbarium and eventually locked it in an unheated building, where it was badly damaged by mould and vermin. In 1784 Smith purchased the library, collections, and herbarium for 1000 guineas sterling, and had them shipped to London: see Stearn & Bridson (1978:14–16).
42. Pleasance [Reeve] Smith (1773–1877). The purchase sent the Linnean Society into debt, from which it emerged only in 1861.
43. Also *Introduction to physiological and systematical botany* (1807), and the explanatory text to the 36-volume *English botany* illustrated by Sowerby (1790–1814). In volumes 1–4 of *The English flora* (first edition 1824–1828, second edition 1828–1830) Smith himself treated Linnæan Classes 1–23 plus ferns (part of Class 24), while WJ Hooker contributed volume 5(1) on mosses, liverworts, lichens, Characeæ, and algæ (1833), and MJ Berkeley volume 5(2) on fungi (1836). The volumes by Hooker and Berkeley also formed volume 2 (*Cryptogamia*) of the fourth (1838) and fifth (1842) editions of Hooker's own *British flora*.
44. Smith, *Introduction* (1807:1–9); quotation at pp. 4–5. He speaks favourably (pp. 5–6) of the observation by Brisseau-Mirbel (1801–1802, 1:19–20) that only plants can derive nourishment from inorganic matter, whereas animals feed on "what is or has been organized matter, either of a vegetable or animal nature".
45. *Encyclopædia Britannica*, Botany (1771, 1:648–653); see above and Chapter 15, notes [116] and [117]. The First edition was reprinted in 1773.
46. *Encyclopædia Britannica*, Second edition, Botany (1778, 2:1304). The article "Kingdoms, in natural history" states that philosophers affirm that "we may descend from the most perfect animal to the rudest mineral by insensible degrees, and without finding any interval from which a division might be made"; but the opinions of naturalists are divided on the subject, and "if we avoid investigating extremes . . . the distinctive marks must be acknowledged sufficiently obvious to justify the triple division" (1780, 6:4040). The article "Natural history" (1781) is arranged according to the three kingdoms of nature; the brief entry "Zoology" (1783) is not further informative.
47. *Encyclopædia Britannica*, Third edition, Botany (1797): Smith's translation at pp. 454–457. Linnæus glorified: Fourth edition (1810), 4(1):63. See also the corresponding entries in editions through the Sixth (1823).
48. Smith, Botany (1824) 2:376–422.
49. Stevens (1994), pp. 100 and 102–103. Brown also described and named the cell nucleus (1833:710–713) and (1866) 1:511–514); and described the eponymous motion (1828; 1830; and 1866 1:463–486). For SF Gray's non-Linnæan *A natural arrangement of British plants* (1821), see Stearn (1989).
50. In the Seventh edition (1842, 5:73) Walker-Arnott called the Linnæan system "unquestionably the best and most simple of all the artificial systems attempted either before or since", but personally preferred Jussieu's system. In the Eighth (1854) and Ninth editions (1876) Balfour was likewise fair in assessing their strengths and weaknesses.
51. Bailey (1726). A very similar meaning is assigned to the Latin *fossilis* (Pliny) by Gouldman (1678).
52. Hill, *General natural history* (1748–1752), final (unnumbered) page of the Preface. In the first volume (1748) the volume title is given (second title page) as *A history of fossils*; in the second volume (1751, unnumbered page opposite the Preface), *The history of fossils*; and in the third volume (1752, unnumbered page opposite the Preface), *A general history of minerals and other fossils* (1752).

600 NOTES

53. *Encyclopædia Britannica*, Second edition, Natural history (1781) 7:5300; and Third edition (1797) 12:653. At least the former may have been written by James Tytler.

54. In *Organic remains* (1808–1811), Parkinson refers three times to the Fossil kingdom (2:44, 84, 263), each time citing or paraphrasing others; elsewhere he calls it the Mineral kingdom. The Second edition (1833) reprinted the First, with identical text and pagination.

55. Krause (1879); King-Hele (1963); and Uglow (2002). Specifically: poetry (King-Hele 1963:133–152); age of the earth (1963:72); embryonic forms (1879:141, 1963:73); survival of the strongest (1879:113–114); nerves and electricity (1963:50); self-improvement of animality (1963:71); and Buffon (2002:269). The quotation on taking all knowledge as one's province (King-Hele 1963:1) echoes Francis Bacon, in a 1592 letter to his uncle William Cecil, the first Baron Burghley (*Letters and life* 1:108–109, 1861).

56. For the main editions and translations of Darwin's books, see King-Hele (ed.), *Essential writings* (1968:205–206). Bound volumes of *The botanic garden* often mix-and-match editions of *Economy of vegetation*, and *Loves of the plants*, with a volume title page appropriate for one but not the other; here I cite by internal title page only. *Zoonomia* means "The laws of animal life" (Garnett 1804); Eichwald (1829) 1:89.

57. In Additional notes following *Economy of vegetation*, Second edition (1791), Note 5, page 10 [also Third edition (1796), Note 5, page 11]; and in Additional notes to *Loves of the plants*, Fourth edition (1794:191) [not in the Third edition, 1791]. I find no kingdoms in *Temple of nature* (1803), but Darwin mentions vegetables and animals as collective groups, *e.g.* in the notes at page 28.

58. *Loves of the plants* 1.427 note (Third edition 1791:43; Fourth edition 1794:43.

59. Additional notes following *Economy of vegetation*, Second edition (1791), Note 36, pp. 98 [also Third edition (1796), Note 36, page 99]; *Phytologia* (1800), pp. 5, 239, 485–490, *et passim*.

60. Additional notes following *Economy of vegetation*, Second edition (1791), Notes 35–39, pp. 96–112 [also Third edition (1796), Notes 35–39, pp. 97–113]; *Phytologia* (1800), pp. 5–9, 132–139, *et passim*; *Zoonomia* (1794) 1:101–107. Pleasurable sensation: *Phytologia* (1800), pp. 557 and 559.

61. Ritterbush argues that a scale may be inferred upon close reading (1964:162n15).

62. *Zoonomia* (1794) 1:102; *Phytologia* (1800:1); *Temple of nature* 4.66 note (1803:134–135).

63. *Temple of nature* (1803) 2.71 note (1803:48–49); *Phytologia* (1800:301).

64. *Phytologia* (1800:487).

65. *Zoonomia* (1794) 1:105.

66. *Zoonomia* (1803) 1:435–436. This passage is part of an Appendix (1803, 1:416–438) that does not appear in the first London edition (1794). See also the first Additional note following *Temple of nature* (1803). Darwin accepts that Priestley's green matter is vegetable.

67. *Temple of nature* 1.247–250 (1803:22).

68. *Zoonomia* (1794) 1:502–507, quotation at page 507; and (1803) 1:395–399, quotation at page 399. The filament is fundamental for Darwin: all filaments are contractile (1794 1:465), the embryo of animals and man begins as a living filament (1794, 1:480 and 498), and organic bodies are built of living filaments (1:492).

69. "Poetic enthusiasm" following Ritterbush (1964:164), who continues: "We cannot claim Darwin as an exponent of the more rigorous scientific faculty which was to characterise early nineteenth-century natural science."

70. According to the *Oxford English Dictionary* (accessed online 23 May 2020), *Economy of vegetation* (1791) was the first work in English to mention *hydrogene*, and the first non-technical work in English to mention *oxygene*. *Economy* also mentions *carbone* (Second edition, 1791:106 note); the *OED* reports only one earlier use in English, again in a technical publication.

71. *Economy of vegetation*, Second edition 2.361–394 (1791:91–93); Third edition (1795:91–93).

72. This period is well-captured by Uglow (2002:435–463).

NOTES 601

73. Coleridge famously broke with the French Revolution in his poem *The recantation: an ode* (1798), later renamed *France: an ode*.

74. Canning tied his political career to that of Pitt, eventually becoming Foreign Secretary (twice) and Leader of the House of Commons. Frere became a Member of Parliament, diplomat and member of Privy Council. Their *Loves of the triangles. A mathematical and philosophical poem* first appeared in the Tory political newspaper (co-founded by Canning) *Anti-Jacobin*, issues of 16 and 23 April and 7 May 1798. For the poem, see *Poetry of the Anti-Jacobin* (1807:108–129) compiled by Gifford (first edition 1799).

75. T Brown (1798). "To reason from analogy is, in most cases, to mislead" (page 247).

76. Good (1826) maintains the three standard kingdoms of nature, but criticizes Erasmus Darwin implicitly at 1:178–183 (calling the idea that plants possess irritability, a heart, brain, and stomach "mere fancy"), while extending the "loves and intermarriages" to atoms, magnets, gases, chemicals, ferments, and contagions; and by name at 2:86, for the "absurdity" of deriving man from "the race of oysters".

77. Scholars delineate natural theology differently. Here I focus on its intent and arguments in the natural-historical and related philosophical traditions. Chignell & Pereboom (2020) emphasize its logical and cognitive contexts. Barber (1980) unhelpfully conflates natural theology with allegorical animal-tales, Victorian moralization, and broader ideas of the "economy of nature" in *e.g.* Linnæus, Buffon, and Treviranus.

78. For example, Brockes, *Irdisches Vergnügen in Gott* (1721–1748); JA Fabricius, *Pyrotheologie* (1732) and *Hydrotheologie* (1734); Lesser, *Lithotheologie* (1732, 1735), *Insecto-theologia* (1738), and *Testaceo-Theologia* (1744); Richter, *Ichthyotheologie* (1754); and at least a dozen similar works focused on other areas of animate and inanimate nature, surveyed by Michel (2008). Lesser's publications in particular were variously paraphrased in, and translated into, French and/or English.

79. *Luke* 12:27, King James Version. Similarly sparrows (ibid. 12:6–7), ravens (12:24), and field grass (12:28).

80. See also Barnouw (1981), and Grant, *History* (2007:293–307). The boundary between faith and reason was patrolled with particular attentiveness by the Royal Society: Newton, its president from 1703–1727, prohibited any mention of religion (Grant 2007:296–297).

81. Grant (2007:299–300).

82. Boyle, *Excellency* [1665], in *Works* (1772) 4:1–72, at page 38. According to the Publisher's advertisement (1772, 4:1) Boyle wrote *Excellency* in 1665 while in the countryside (to avoid the plague), but thereafter suppressed the work for some years, for fear it might be used against the study of natural philosophy.

83. Chapter 9, note [209].

84. Knight (1986:32). This was to be done in a generic manner, distinguishing Christianity from other religions.

85. A list of the known Boyle Lectures is available at Wikipedia (accessed 17 May 2020). The Reverend [later, Bishop] Alfred Barry presided at the funeral of Charles Darwin (1882).

86. The analogy can be traced through Bernard de Fontenelle (*Entretiens* 1686, Premier soir), Joseph Glanvill (*Scepsis scientifica* 1665:32) and Nicole Oresme (*Le livre du ciel et du monde* 2:288 [1377]) to Cicero (*De natura deorum* 2.38 (1933:216–217).

87. Including the astronomer [Rev.] John Brinkley, Buffon via Goldsmith, Derham, Erasmus Darwin, the Abbé Fontana, the surgeon Everard Home, Linnæus, the Dutch natural theologian Bernard Nieuwentyt, Ray, Jacques-Henri Bernardin de Saint-Pierre, Smellie, Spallanzani, and Withering.

88. Paley, *Natural theology* (1802:318–321). At page 306 Paley quotes Priestley (*Letters* 1787, 1:153) in the argument that a complex, well-ordered contrivance is evidence that the Designer is a person, not an impersonal principle *e.g.* "nature".

602 NOTES

89. Paley (1802), pp. 311 and 314.
90. Contributors included David Brewster, John Stevens Henslow, John Herschel, James Macintosh, Walter Scott, and Mary Shelley. Reform of the voting system in 1832 helped define the middle class, at least in the Boroughs.
91. Swainson (1834:107–116); "every branch" at page 110. He was little interested in plants, and distinguished between botany and natural history (1834:108).
92. Swainson (1835:343). The title page depicts Linnæus in an engraving by William Finden, after a painting by Henry Corbould. In the passage quoted, a footnote at "mineral kingdom" refers the reader to Swainson & Richardson (1831), page liv, where Swainson considers this point at greater length, albeit without resolution. A footnote at page liii indicates that Macleay had, at one point, considered that all "matter" might be organized into five groups: two "normal groups" (Animals and Vegetables), and three as-yet undetermined "aberrant groups" comprising Inorganic matter. See also Chapter 16.
93. As taken from the Notice published at the front of each Treatise.
94. Davies Gilbert PRS, William Howley, and Charles James Blomfield respectively.
95. Kirby (1835) 1:145–148; quote at pp. 146–147. Kirby concludes that "these doubtful forms [have] no just claim to be considered as animals". Other editions (*e.g.* Philadelphia, 1836) insert commas in the title, and have different pagination. Kirby (with co-author Spence) referred to the three kingdoms of nature in the earlier *Introduction to entomology* (1815–1817) 1:1, 1:4–5, and 2:138.
96. Kidd (1837) refers explicitly to the atmospherical kingdom at pp. 8 and 340, and to four kingdoms of nature at pp. 7 and 8. However, at page 81 he refers to the four "kingdoms or divisions of nature", and at page 187 to "the atmosphere, and the vegetable, and animal kingdoms, being three out of the four general departments of the external world". At page 169 he identifies three kingdoms (animal, vegetable, and mineral), although the limitation is probably required by context.
97. Kidd (1837:52).
98. Notably his *Lectures on the comparative anatomy and physiology of the vertebrate animals* (1846), which however covered only the fishes, and the three-volume *Anatomy of vertebrates* (1866–1868). Owen also introduced the term *Dinosauria* (in 1841: see *Report* 1842:103).
99. Officially the British Museum (Natural History) from its establishment (1881) until 1992.
100. Rupke (1994:112–123). Buckland had earlier described a "great fossil lizard", which he called Megalosaurus—the first fossil dinosaur. For Cuvier, see Chapter 17.
101. Owen (1835).
102. Rudolphi (1808–1810).
103. Rudolphi (1819:572).
104. *Radiata* in Latin.
105. Rudolphi (1819:572). Owen (1835:388) adds dryly that Rudolphi's assessment was just.
106. Chapter 16.
107. Owen (1835:390).
108. Owen cited Cuvier's *Le règne animal, Nouvelle édition*, Tome 3 (1830). This edition maintained the five classes of Zoophytes from 1817, with different extents of revision.
109. Owen (1835) and Acrita (1836).
110. Owen's term for "the classes which present the diffused condition of the nervous globules", *i.e.* polygastric infusoria, sponges, polyps, acalephs, and parenchymous entozoa (1835:390).
111. Owen, Acrita (1836) 1:48–49.
112. Chapter 18, notes [133] through [135].
113. Chapter 17, notes [112] and [113].
114. Appel (1987:5); Rupke (1994:115).

NOTES **603**

115. Ospovat (1995:129–130). Ospovat made the further point that this distinction was important to morphologists, but unintelligible to teleologists (*e.g.* natural theologians). The definitions are from Owen, *Lectures* (1843), pp. 374 and 379. See also Owen, *Report on the archetype* (1847:173–176; definitions at page 175); excepted as *On the archetype* (1848), with definitions on page 7.

116. Owen, *Lectures* (1843:370).

117. Ibid. (1843:16–28); true animals at page 20.

118. Ibid. (1843:45).

119. Ibid. (1843:94).

120. *Lectures* (1855:8).

121. Owen, Palæontology, in *Encyclopædia Britannica* (1858) 17:91–92.

122. Owen, *Palæontology* (1860:4).

123. Ibid. (1860), pp. v and 5. He retained Kingdom Protozoa in the second edition (1861), page vii, although at page 6 the heading reads "Acrita or Protozoa", perhaps in response to Hogg's criticism of *-zoa* (below). As usual, Owen does not consider the plants.

124. Chapter 18. Perhaps we should not read too much into the *-zoa* of his Protozoa, as Owen presumably felt bound by prior usage, and was not much interested in the vegetable kingdom.

125. *Encyclopædia Britannica* (1858) 17:91. This statement does not appear in *Palæontology* (1860), which however adds a brief consideration of the hypotheses of Buffon, Lamarck, the *Vestiges*, Wallace, and Charles Darwin (1860:403–406).

126. Owen, *Anatomy* (1868) 3:786–825. See also Rupke (1994:220–258).

127. Owen, *Anatomy* (1868) 3:818. The previous paragraph mentions "genetic descent from a germ or cell", but it is unclear whether Owen viewed these roots in a genetic sense.

128. Ibid. (1868) 3:817.

129. Notably Buckland (1836). Buckland was later appointed Dean of Westminster.

130. Chambers revised *Vestiges* through the Eleventh edition (1860). A sequel, *Explanations* (1845), appeared in two editions, after which it was usually bound with *Vestiges*. By 1884 more than two dozen editions had appeared in English; translations were made into German (twice), Dutch, Hungarian, Italian and Russian. To mark the 150th anniversary of *Vestiges*, University of Chicago Press released a facsimile of the First (1844) and Tenth (1853) editions.

131. Insects: *Vestiges* (1844:185–187); Tenth edition (1853:135–136); Twelfth edition (1884:174–175); and *Explanations* (1845:119–120, 189–195). Fungi ("electro-vegetation"): *Explanations* (1845:195–198).

132. [Chambers], Preface to *Vestiges*, Tenth edition (1853), pp. v–viii.

133. Alexander Ireland affixed Chambers's name (and image) to the Twelfth edition (1884).

134. Hodge (1972).

135. *Vestiges* (1844:204), and with slightly different wording at (1853:156).

136. *Vestiges* (1844:191). Chambers later re-worded the "obvious gradation" section, removing the quinary circles but retaining much the same sense (1853:185–186).

137. *Vestiges* (1844:237).

138. Animalcules (*Vestiges* 1844:191,199) and (1853:187); sponges (1844:171) and (1853:128,187); fungi (1844:83,179) and (1853:54,168,174); lichens (1844:83,191) and (1853:54,168).

139. Chapters 15 and 16.

140. Forbes (1843:4–5). The examples of analogies he offers are highly curious.

141. For an introduction to the immense Darwin literature, see Sloan, Darwin (2019); and van Wyhe, darwin-online.org.uk.

142. C Darwin, *Origin of species* (1859); tree diagram between pages 116 and 117. In *Origin*, Darwin refers once to the "inextricable web of affinities" among members of a class

604   NOTES

(1859:434). He considered natural selection "the main but not exclusive means of modification" (1859:6).

143. C Darwin to Jenyns, 7 January 1860. In F Darwin (1887) 2:57–58.

144. C Darwin to Lyell, 11 October 1859. In ibid. (1887) 2:6. For Darwin on microbial life, see O'Malley (2009).

145. Darwin, *Origin,* Third edition (1861:518–519). He continued: "When the views advanced by me in this volume, and by Mr. Wallace in the *Linnean Journal,* or when analogous views on the origin of species are generally admitted, we can dimly foresee that there will be a considerable revolution in natural history." The identical passage appeared in the Fourth edition (1866:570–572), while Darwin slightly revised the wording for the Fifth (1869:572–573; New York, 1871:432), and again for the Sixth (1872:424–425). Only the first sentences appear in the First (1859:484) and Second (1860:484–485) editions. In the London Fifth edition (Murray), the first sentence reads "farther", whereas in the New York Fifth edition (Appleton) it reads "further".

146. Darwin read German only with difficulty. Asa Gray held that "*Natural selection [is] not inconsistent with natural theology*", to quote (in part) the title of his 1861 book.

147. Darwin sometimes used *kingdom* to mean *taxon of high rank,* as in the *Articulate* (or *articulated) kingdom* (*A monograph on Cirripedia,* 1851–1854, 1:212, 2:9, 2:12, and 2:22) and the *Molluscous kingdom* (ibid. 2:9).

148. John Hogg, *On the natural history of the vicinity* (1829), Appendix 2 (separately numbered pp. 1–94); for attribution to Hogg see (unnumbered) page iv, opposite the Contents. In 1840 (On the tentacular classification) Hogg states that he wrote this sketch in 1825, and that it was published in 1827. Hogg arranges this Appendix by animal, vegetable, and mineral, and lists polyps, corals, and sponges among animals (pp. 31–39). The first edition of the *Parochial history* was published in 1796.

149. John Hogg, On the nature (1824), as *Nerita glaucina* L. See also Fleming (1828:319). *Natica glaucina* Lamarck 1822 is now *Neverita didyma* Röding 1798.

150. Hogg, *op. cit.* (1840:376).

151. Hogg, *op cit.* (1839:404–405).

152. Hogg, *op cit.* (1839:405–406).

153. John Hogg, Address (1857:166). Later, quoting himself, Hogg adds "and nervous" immediately after "muscular" (*Edinburgh New Philosophical Journal* 1860:217).

154. Ibid. (1860). Lankester reading the paper: [Anonymous], *Athenæum* No. 1706 (1860:26). As failed approaches Hogg mentioned "*animal-vegetable*" theory (*i.e.* Unger and Kützing), and criteria based on locomotion, chemical composition, or oxygen and carbonic acid. According to the anonymous report in *Athenæum,* Lankester "could not agree with the author as to the necessity of a fourth kingdom in nature".

155. Hogg, *op. cit.* (1860:220). He went on to clarify some of these terms and concepts (1860:221).

156. Hogg's paper was read in the Botany section by Edwin Lankester, senior secretary of "Section D, Zoology and Botany, including Physiology". JS Henslow was President of Section D, and Owen one of three vice-presidents: see *Report of the Thirtieth Meeting of the British Association* (1861), page xxix. For an extended abstract of Hogg's paper, see this *Report* under the separately paginated Notices and Abstracts, pp. 111–112. Owen's presence: Rupke (1994:272). Daubeny's paper, mentioned by Rupke, is abstracted at *Report of the Thirtieth Meeting of the British Association* (1861:109–110). Daubeny was likewise a vice-president of Section D.

157. Hogg (1860:223) citing Owen, *Lectures* (1855:2).

158. Hogg (1860:223). I have omitted (after the mention of Blainville) a paragraph in which Hogg states that "Amorphoctista" applies only to "the living beings of Sponges in their fresh state", not to "the Spongiaries (*Spongiaria*), or skeletons, or remains of the sponge after the

NOTES 605

death and decomposition of the live jelly, or living being,—*Spongioctiston*", as many of the latter are not amorphous. From this we see that Hogg continued to view sponges as a sort of polypary.

159. Above, and *Palæontology* (1860:5–8). Blainville's *amorphozoaires* included sponges, false alcyonia, and some of OF Müller's infusoria (*De l'organisation* 1822, pp. 26, 217–218, *et passim*; and *Manuel* 1834, vol. 1, pp. 6, 527–545, and figure opposite page 110).

160. Hogg rightly noted that -zoa means "living thing" rather than "animal", but sensibly conceded to established usage. Hogg had an interest in classical antiquity, and in earlier publications had combined this with his natural history.

161. Hogg (1860:225) reiterated that Protoctista includes both Protophyta and Protozoa, without mentioning Amorphoctista.

162. The standard reading is that Hogg meant to include Amorphoctista within Protoctista; and that *Protoctista*, despite its capital *P*, was only a name for the members of Primigenum, not a taxon in its own right (Rothchild, 1989). Even so, Protoctista has been treated as a taxon: Margulis (1989), notably "Kingdom Protoctista" at page xvi.

163. By "have been well compared" Hogg seems not to take credit for the idea. He does not identify a precedent, but *cf.* Bourdon, Animal (1822) 1:369. The entry was also published separately (1822).

164. Hogg (1860:224) and Plate 3.

165. Hitchcock, *Elementary geology* (1840). See Archibald (2009), particularly note 2 at page 562. The coloured chart is present in my copy of the Third edition (1844).

166. Owen, On the characters (1858), pp. 19–20 and 37; *On the classification* (1859), pp. 25–26 and 52.

167. Owen, ibid. (1859:50–51); Rupke (1994:260–270).

168. *Palæontology* (1860:401–403).

169. Wilson & Cassin (1864); their manuscript had been presented for publication on 19 May 1863, and ordered to be published on 26 May 1863. This is the only published work (co-)authored by Wilson, whose natural-historical interests lay primarily in minerals, insects, and birds. He was elected president of the Academy in December 1863. Apart from this article, Cassin published only on birds, but also made observations on insects. For Wilson, see Stone, in *Cassinia* (1909); and Day (1984). For Cassin, see Fuller (undated).

170. Leidy (1853). He went on to publish important work on diatoms (1875), rhizopods (1879), and protozoan parasites of termites (1881). Cassin had nominated Leidy for membership in Academy (1845). See also Warren (1998).

171. Wilson supported the Confederate side, and in 1864 resigned from the presidency of the Academy owing to conflicts with most members on the issue. Cassin volunteered for the Union side, and was captured and held prisoner by the Confederates.

172. Wilson & Cassin (1864:113–114). The volume is recognized as volume 15, although the title page indicates only that the Proceedings are for 1863, and were printed in 1864. The volumes were formally numbered (other than by year) from volume 53 (1901).

173. Wilson & Cassin emphasize this difference by referring to "positive or relative" characters (ibid., pp. 114 and 115).

174. Ibid. (1864:115–116).

175. Ibid. (1864:116). They do not call their group Protozoa, arguing (page 118) that Goldfuss has priority on the term.

176. Ibid. (1864:117). This formulation does not help their argument, as it is perfectly consistent with (inter alia) a linear chain of being.

177. Ibid. (1864:118).

178. Ibid. (1864:118–119).

606 NOTES

179. Of the organisms identified by Owen (1858, 1860) as Protozoa, sponges obviously map to Wilson & Cassin's Spongiæ, and (for lack of alternatives) rhizopods and polygastria to Conjugata. It is not known how Wilson & Cassin considered diatoms and desmids to map. See also Rothschild (1984:286).

180. In a Darwinian context, this became a central tenet of so-called cladistic systematics, for which a leading reference is Hennig (1950). The broader area, including theory and methods, is highly contested.

181. They consider "first specialisation" or "first development" in reference to Primalia being older than, and/or of "inferior organisation" to, Vegetabilia and/or Animalia (1864, especially page 116).

182. Richard Owen introduced Joseph Leidy to Charles Darwin in 1848, and Darwin became a member in 1860. See also Academy of Natural Sciences of Drexel University, Darwin's chair (undated).

183. Victoria was Queen from 1837–1901, but here we focus on the period to about 1880.

184. This literature has itself given rise to a large number of academic studies, including Allen (1969); Barber (1980); Merrill (1989); and the series of papers in *Victorian Literature and Culture* edited by Gates (2007).

185. Carpenter (1848); Eighth edition (with modifications by Dallinger), 1901. Carpenter illustrated later editions of the *Vestiges*.

186. Jabez Hogg (1854); Eleventh edition, 1886.

187. For example, Jabez Hogg's comment that diatoms are one of "the border tribes that occupy a sort of neutral, and yet not undisputed, ground between the confines of the animal and vegetable kingdoms" (1856:43); but Hogg considered them animals (1856:148–150). His "border tribes" statement remained to the Eleventh edition (1886:68).

188. Anonymous. *Minerals and metals* (1835), note at page 1. The author does not mention quinarian systems, Macleay, or Swainson. Founded in 1698, the SPCK was active in the popularization of natural history.

189. Miller, *Cruise* (1858:88). His *Footprints of the Creator* (1849) was a reaction to the *Vestiges*.

190. Needless to say, these do not constitute a random sample of Victorian popular natural histories. For titles of the forty-four works and further comments on methodology, see the Appendix.

191. Forty by direct mention, and two by clear implication. The remaining two do not mention kingdoms of nature. Mantell (1846:105–106; identically in 1850:105–106) identified the kingdoms as animal, plant, and inorganic.

192. Mivart (1876): Human, Animal, and Vegetable kingdoms at page 181 (quoting Charles Darwin's *Descent of man*, where these kingdoms appear at page 162 of the 1874 New edition); man a separate kingdom (1876:181 and 183); and mineral, vegetable, animal, and rational kingdoms (1876:358). Mivart, a convert to Roman Catholicism, sought to reconcile Darwin's theory with religion; he later emigrated to Belgium, and as a professor at Université Catholique de Louvain fell further afoul of the Church, and was excommunicated in 1900.

193. Kingsley (1855), at pp. 40 and 160. Rev. Charles Kingsley was a friend of Charles Darwin.

194. Harper (1860), pp. 23, 65, 70, 167 and 278.

195. Buckley (1880:36). The book was reprinted in multiple editions. Arabella Buckley was secretary to Charles Lyell from 1864 until Lyell's death in 1875.

196. Landsborough (1849:331–333); Second edition (1851:348–349); Third edition (1857:348–349). Also Gosse (1857:30); Fifth edition (1878:30).

197. Gilbert & Sullivan (1879).

198. Evidence for the Flood: Buckland (1823); on glaciation: Buckland (1840–1841). See also Davies (1968), and GW White (1970). In 1845, Buckland succeeded Samuel Wilberforce as Dean of Westminster, when the latter was appointed Bishop of Oxford.

NOTES   607

199. If we allow that Primalia (Wilson and Cassin) was a direct reaction to Owen's Protozoa, and to Hogg's Primigenum.

200. Regnum neutrum (Münchhausen), Regnum chaoticum (Linnæus), Regnum mesymale (Necker), Phytozoa (Horaninov), and Némazoaires (Gaillon) had few if any exponents by 1864.

Chapter 21

1. Scheffauer (1910:217). Scheffauer had been deeply affected by Haeckel's *Die Welträthsel* (1899).

2. *Häckel ist ein Narr. Das wird sich schon noch herausstellen.* Virchow as quoted by Schleich (1922:190) as translated by Miall (1935:159). The quotation itself dates to between 1884 and 1889.

3. For his early life, see Bölsche [1906], pp. 29–50; Di Gregorio (2005:26–35); and Richards (2008:19–24). Humboldt and Bonpland's *Voyage* (1815–1825) was brought into German as *Reise* (1815–1832); Darwin's *Journal of researches* (1839) appeared in German as *Naturwissenschaftliche Reisen* (1844).

4. Bölsche [1906], pp. 51–81; Di Gregorio (2005:35–50); Richards (2008:26–30 and 39–44). General (as opposed to mechanical) anatomy took in the study of cells and tissues using techniques of microscopy and histology: Nyhart (1995:84).

5. For this period, see Bölsche [1906], pp. 82–93; Di Gregorio (2005:56–58); and Richards (2008:55–65). The second war of Italian independence was fought (in the north) between April and July 1858.

6. J Müller (1834–1840), translated by Baly as *Elements of physiology* (1837–1843). In the German original: organisms and life, 1:18–39; animals and plants, 1:39–62; Bory, 1:41; the other matters mentioned, 1:40–44; Ehrenberg on infusoria, 1:467–468; alcyonia as animals, 1:468. In English translation: organisms and life, 1:18–40; animals and plants, 1:40–63; Bory, 1:42n2; other matters, 1:41–44; Ehrenberg, 1:487–488; alcyonia, 1:488.

7. Kölliker (1852), pp. 7 and 11; in English (1854), pp. 39 and 44.

8. Virchow (1858:3–6); the second German edition (1859) in English translation (1860:4–7).

9. The Latin aphorism appears only in the second German edition (1859:25) and its English (1860:27) and French (1861:23–24) translations, but the concept is clearly stated in the first German edition (1858:25). Spontaneous generation: Virchow (1858:25), (1859:25), (1860:27) and (1861:23). See also Chapter 18.

10. Leydig (1857:4–7).

11. A Braun, *Betrachtungen* (circulated 1850, published 1851: see page i). English translation by Henfrey as *Reflections* (1853). The work argues that rejuvenation distinguishes the organic from the inorganic, and offers a framework to understand progression in nature and society.

12. Braun, ibid. (1851:131–145) and (1853:123–136). Braun returned to these issues in *Algarum unicellularium* (1855, especially pp. 4–11), a work he dedicated to Nägeli. At (1855:69–70) he acknowledged the dispute over diatoms and desmids.

13. Braun, ibid. (1851:227) and (1853:212); see also (1851:219n3) and (1853:205n3).

14. Nordenskiöld (1932:551–552).

15. Kölliker (1845), particularly the note beginning at page 97; and (1848). See also Siebold (1849).

16. Braun, ibid. (1851:198n3) and (1853:185n10). He did not mention the interpretations of Kützing and Unger.

17. Braun, Über Chytridium (1856:23). Today *Chytridium* is recognized as a fungus (Chytridiomycota), whereas *Achlya* and *Saprolegnia* are classified as chromists (Oomycota).

608 NOTES

18. Schenk (1858); quote at page 20. In 1852–1853 Haeckel attended Schenk's course on crypto-gamic botany (Richards 2008:30).

19. Nyhart (1995:146–160) describes the parallels between Gegenbaur and Haeckel. Gegenbaur's studies in Messina resulted in *Lehre vom Generationswechsel* (1854). For more on Gegenbaur, see the minisymposium edited by Hoßfield, Olsson & Breidbach (2003).

20. Gegenbaur (1859:7–8).

21. Gegenbaur (1859), quotation at page 8. He admitted an animal *Grundtype* in the sense of Saint-Hilaire or Owen, as opposed to that of Schelling or Goethe. For the Second edition of *Grundzüge* (1870), Gegenbaur based his argument instead on Darwin's theory of descent.

22. Recall that infusoria (ciliates) are often multinucleate, and were thought by some to repre-sent a fusion of cells (Chapter 18). Some gregarines exhibit a superficial transverse septum that demarcates a "head" (*protomerite*). Gregarine life cycles often involve the enclosure of two individuals within a common cyst, segmentation within that cyst, and emergence of a string of oocysts—forms that invite over-interpretation.

23. Gegenbaur [1860], particularly pp. 14–16. In *Grundzüge* (1859:45) Gegenbaur also in-cluded sponges among the protozoa. So far as I can determine, Gegenbaur did not use the tree-trunk metaphor for the plant or animal series, or refer to branches. Earlier, JV Carus (1853:44) held (in regard of *Euglena*) that if a simple being displays non-animal characters, the simplest conclusion is that it is not an animal.

24. Claus (1863); quotation at page 10 (translation by MAR). Claus completed his doctoral studies under Leuckart, and after an appointment at Würzburg took up the professorship of zoology at Marburg. Later, as professor at Vienna and head of the experimental station at Trieste, Claus supervised the early zoological studies of Sigmund Freud (1877); see also Bernfeld (1949).

25. Claus (1863); quotation at page 24. Even so, Claus ended up agreeing with Gegenbaur that flagellates, peridinians, and astasiæ should be excluded from the animals, and protozoa re-stricted to infusoria, rhizopods, sponges, and gregarines (1863:24).

26. Haeckel, Ueber die Gewebe (1857), *e.g.* pp. 529–532. His inaugural dissertation was *De telis quibusdam Astaci fluviatilis* (1857).

27. Haeckel, *Die Radiolarien* (1862).

28. Haeckel considered the digestive mechanism of radiolaria an "interesting connecting-step (*Verbindungsstufe*) between animals and plants": *Radiolarien* (1862), 1:161.

29. Here Haeckel cites Gegenbaur [1860].

30. Haeckel, *Radiolarien* (1862:162).

31. Ibid. (1862:162–165): *trefflich*[es] at (1862:163); neglected developmental states at (1862:164); complexes of fused cells and sarcode at (1862:165).

32. Ibid. (1862:163n1). See note [22] above.

33. The first German edition of Darwin's *Origin of species* appeared in 1860 (*Über die Entstehung*, translated by HG Bronn from the second English edition).

34. Haeckel, *Radiolarien* (1862:231–234 and note 1). Haeckel named a *Coccodiscus* in honour of Darwin, citing his "classic studies 'on the origin of species'" (1862:486, 558, and plate 28:11,12).

35. Haeckel, *Über die Entwickelungstheorie Darwins* (1863). Comparison with Newton, 1:27–28; quotation at 1:27.

36. Anna Haeckel (neé Sethe) died on 16 February 1864, Ernst Haeckel's thirtieth birthday; Gegenbaur's wife Emma (neé Streng) died after childbirth on 1 August 1864, at which point Haeckel had begun work on *Generelle Morphologie*.

37. Haeckel, *Generelle Morphologie* (1866). The first volume is dedicated to Carl Gegenbaur; the second to Charles Darwin, Goethe, and Lamarck. Translations from *Generelle Morphologie* by MAR.

NOTES 609

38. Haeckel, Die Gastraea-Theorie (dated 1873, published 1874), page 9. English translation (Wright): The gastraea-theory (1874:149).

39. *Generelle Morphologie* (1866) 1:3.

40. Ibid. (1866) 1:60. By contrast, special morphology involves the extension of these explanations to individual groups and subgroups of organisms.

41. Ibid. (1866) 2:8. Haeckel draws a parallel with the introduction of generation-theory by Wolff (1759), which lay embryonic until its application to the alimentary canal by Oken & Kieser (1806–1807), and its popularization by Meckel (1812) via his translation of Wolff's *De formatione intestinorum* (1768–1769). Haeckel returns to the history of descent-theory at 2:150–166, and to Darwin's theory of selection at 2:166–170. For an extended discussion, see Haeckel, *Die Naturanschauung* (1882); in English translation (minus the Foreword, an introductory poem, and lengthy notes) in *Nature* (1882).

42. *Generelle Morphologie* (1866) 2, pp. xvii–xviii. The allusion is to Goethe's *Metamorphose der Pflanzen*.

43. Ibid. (1866) 1:24–30, summarized at 1:30. Further on tectology and promorphology at 1:46–49; tectology, 1:239–374; promorphology, 1:375–558; ontogeny, 2:1–300; and phylogeny, 2:301–422. An organic *Stamm* or *phylon* is "the sum of all the indicated organisms that derive their common origin from one-and-the-same *Stammform*" (1:57; see also 2:xix). Each *Stamm* is a third-level genealogical individual (1:57).

44. In this context, *Entwickelungsgeschichte* is often translated *biogenesis*. It can also be translated *evolution*, but Haeckel uses the term *Evolution* separately.

45. Ibid. (1866) 1:179–190. Haeckel carefully distinguished *Autogonie* from creation, and from various versions of *Urzeugung* (spontaneous generation).

46. Ibid. (1866) 1:179.

47. Ibid. (1866) 1:180–185.

48. Ibid. (1866) 1:187 and 2:407–408.

49. Ibid. (1866) 1:188, 1:202 and 2:xxi.

50. Ibid. (1866) 1:135–136.

51. Ibid. (1866) 1:185.

52. Ibid. (1866) 1:199–202. In any case, it would be nearly impossible to tell whether the original stem-forms were identical or not (1:197–199; 2:411).

53. Ibid. (1866) 1:37, 1:195–196, 2:xvii, and 2:374–391. Haeckel also introduced the term *Genealogema* as a synonym of *Stammbaum* (2:xvi–xx), but used it only three times in *Generelle Morphologie*.

54. Ibid. (1866) 1:50–60 and 2:xviii–xix.

55. His three levels of genealogical individuality are the generation-cycle, the species, and the stem or phylum (2:304–305). For physiological individuals (*bionts*), see 1:332–363 and 2:4–5.

56. Ibid. (1866) 1:210–211 and 1:269–331, and slightly differently at 2:367–369.

57. Ibid. (1866) 1:112.

58. Haeckel, ibid. (1866) 1:269–289; and *Krystallseelen* (1917), page vii.

59. *Generelle Morphologie* (1866) 2, page xix.

60. In Chapter 1, I argued that it was not so simple as this.

61. Ibid. (1866) 1:192. Haeckel identifies two types of error: the attempt to find definitive characters, and claims that the animal and plant kingdoms merge into a single great kingdom of organisms (1:192–193). As for the claim (by Gegenbaur and himself) that cells retain greater independence in plants than in animals, Haeckel argued that this was never put forward as an absolute difference between the two kingdoms. In any case, he is now approaching the question from the standpoint of descent-theory (1:193–194).

62. Ibid. (1866) 1:198–201.

610 NOTES

63. Ibid. (1866) 1:202–203. A footnote at page 203 derives *Protisten* from πρώτιστον, the very first [*Allereste*] or primordial [*Ursprüngliche*].

64. Haeckel's protoplasts (protoplasta) are amœbæ and gregarines. At ibid. (1866) 2:406–407 he offers the possibility of unfolding the phycophyte *Stamm* into three separate *Stämme*: archephytes, Florideen (red algae), and Fucoideen (brown algae), referring to his systematic introduction (2:xxxii). Haeckel designed his *Stammbaum* of all organisms to be read against alternative baselines, resulting in different numbers of animal and plant phyla (2:417n1 and Table 1); note that the number of protistan phyla remains eight regardless of whether the baseline is s-t, x-y, or m-n.

65. Indeed, possibly from several monera: Ibid. (1866) 1:198–206; phyla at 1:206. The system is elaborated in some detail at 2, pp. xx–clx.

66. And to Diatomea, if diatoms are considered algae.

67. Ibid. (1866) 1:215–226, 2:xx–xxii, and 2:404. Goethe: *Analyse und Synthese* (1829), in *Werke*, HA 13:49–52. For animals as oxidation-organisms and plants as reduction-organisms, see ibid. (1866) 1:210 and 1:221 respectively.

68. Haeckel considered it "a complete certainty" that the protistan kingdom was polyphyletic (ibid. 2:404). He further held that kingdoms could not be diagnosed in a sharp and definitive manner except by ignoring the lowest and simplest forms (1:208–209). This would apply all the more to Protista, where all forms are "low and simple".

69. Followed by Haeckel himself, in *Radiolaria* (above). At *Generelle Morphologie* 2:408n1, Haeckel characterized Protozoa as a *Rumpelkammer* (junk-room), like the earlier Vermes.

70. Ibid. (1866) 2:lxxvii–lxxviii. In the first paragraph I translate *Gliederung* as *arrangement*, but it can mean *structure, organization, disposition, articulation*, or *segmentation*. See also ibid. 2:408–409.

71. Suctoria feed on prey by use of a specialized organelle ("tentacle"). Like his contemporaries, Haeckel thought that prey was drawn in by suction: hence *Saugröhren* (suction-tubes). See Rudzinska (1973). Suctoria are now classified within the ciliates.

72. *Generelle Morphologie* 2, page lxxviii. My translation assumes that for *actinetenartigen Larven* Haeckel intended *acinetenartigen Larven*.

73. Ibid. (1866) 2, page lxxix: *Von allen Würmern stehen diese ohne Zweifel den Infusorien am nächsten, und sind selbst durch einige so zweifelhafte Uebergangs-Formen mit den Ciliaten verbunden, dass ihre Abstammung von diesen nicht geleugnet werden kann.* One expects *bestimmte* in place of *zweifelhafte*.

74. Ibid. (1866) 2, page lxxxv.

75. Schmidt (1849). Schmidt's report had been publicized by Friedrich Stein (1859 1:33n9). Schmidt, who published on Goethe (1853), sponges (1862–1864), and Darwinism (1873), had taught at Jena before Haeckel arrived.

76. J Müller, Wurmlarve (1850); similarity to infusoria at page 487. One of these larval types is now known as "Müller's larva". In the 1840s and 1850s Müller himself, Louis Agassiz, Charles Girard, and others described ciliated larvæ of other worms, and of polyps, medusæ, echinoderms, bryozoa, and rotifers. See also Lacalli (1982); and Nielsen (2008, 2018).

77. *Generelle Morphologie* 2:412. Alternatively, acinetes might be directly linked with rhizopods or protoplasts, *i.e.* not through the common infusorian ancestor (2, page lxxix).

78. For example, he transferred sponges from Protista to Cœlenterata in 1868, but the supporting evidence appeared only in 1870: see notes [95] and [96] below. Volume 2 of his monograph on calcispongæ appeared more than a year after Volume 1, but both are dated 1872 (see Haeckel, *Kalkschwämme* 1872 1:339).

79. Häckel, Monographie der Moneren (1868:71–107); in English as Monograph of Monera (1869), pp. 34–42, 113–134, and 219–222. He had mentioned *Protamœba primitiva* in *Generelle Morphologie* (1866) 1:133.

NOTES 611

80. Haeckel, Ueber den Sarcodekörper (1865:360). Max Schultze had described it as *Amoeba porrecta* (1854:8).

81. Cienkowski (1865). The article was communicated from Dresden (1865:232). Haeckel renamed the former *Protomonas amyli* (*Generelle Morphologie* 2, page xxiii) because "the term *Monas* has many significations" (Monera, 1868:69 and 1869:32).

82. In Appendix A to Dayman (1858:64).

83. Wallich (1861:52–53). See also Rehbock (1975:510–513). For Wallich on *Bathybius*, see Wallich (1875).

84. Huxley (1868), quotation at page 210.

85. Haeckel, Beiträge (1870:499–510). For his later views on *Bathybius* and the surrounding controversy, see his *Protistenreich* (1878:68–85); and Rehbock (1975), pp. 522–523 and 531.

86. The story is engagingly told by Rehbock (1975).

87. Cienkowski (1867).

88. Haeckel, Monera, 1868:121–122 and 127; and 1869:232, 330–331, and 335.

89. Haeckel discovered *Magosphæra* in 1869, mentioned it in the Second edition of *Natürliche Schöpfungsgeschichte* (1870:373 and 383–384), and published a full report in 1871 (Die Catallacten). He derived *Catallacta* from χαταλλάχτοζ ("mediator"), in reference to the organism's change of form through its life cycle.

90. Reynolds & Hülsmann (2008).

91. Haeckel, *Natürliche Schöpfungsgeschichte*, Eighth edition (1889). From his description, these are the organisms now known as coccolithophorids (class Prymnesiophyceæ). Coccolithophorids were first described by Ehrenberg (Weitere Nachrichten, 1836).

92. Grant (1826).

93. Carter (1847, 1848). Carter considered the sponge body to be an aggregation of amœboid cells that can, at times, disaggregate into individual motile cells (1848:309).

94. The connection presumably being with foraminifera, although diatoms too have a silicaceous shell.

95. Haeckel, Monera (1868:118–120) and (1869:327–328).

96. Haeckel, Ueber den Organismus der Schwämme (1870); *entoderm* and *ectoderm* at page 212. In English as On the organization of sponges (1870); *entoderm* and *ectoderm* at page 6. For the origin and acceptance of these terms, see MacCord (2013). *Entoderm* is now more usually *endoderm*; *ectoderm* is occasionally *exoderm*.

97. Haeckel, Schwämme (1870:213–214). The taxon Coelenterata was introduced by Frey & Leuckart (1847:137), and Leuckart (1848:13–31). Haeckel responded to critics at *Kalkschwämme* (1872) 1:453–473.

98. And not only of Darwin's theory, but of monism too. Haeckel, Schwämme (1870:233); Sponges (1870:118); *Kalkschwämme* (1872) 1:22 and 1:474–484. Haeckel returned to this point over coming decades (Reynolds, 2019). In *Kalkschwämme*, Haeckel named several species and varieties in honour of Darwin, and others for Gegenbaur, Lamarck, and Goethe.

99. Haeckel, Schwämme (1870:228–233); Sponges (1870:113–118); *Kalkschwämme* (1872) 1:338–340.

100. Now *Leucosolenia*. *Olynthus* additionally has spicules, but removing them by treatment with acid reveals the proto-sponge form (*Kalkschwämme* 1:338, 1872).

101. Latinized from Greek: γάστήρ or γάστρα can mean a *hollow* (of a shield), *paunch, belly*, or *womb* (Liddell & Scott, 1940); the *-ula* is diminutive. However, at *Kalkschwämme* (1872) 1:333–334 Haeckel wrote: "By the name *gastrula* or gastric larva [*Magenlarve*] I attest to the juvenile developmental stage that initially emerges from the planula in the calcisponges and which, in my opinion, due to its wide distribution in different animals has *an extraordinarily great importance for the general phylogeny of the animal kingdom*. By *gastrula* I mean a spherical or spheroidal, egg-shaped or elongated round body, which contains an inner cavity with

612 NOTES

an outer opening (*primordial stomach-cavity [Magenhöhle] with mouth-opening*); the wall of this cavity consists of *two different cell-layers or lamellæ*: an outer light layer with flagella, and an inner cloudy layer without flagella; the former corresponds to the *exoderm* or the outer (animal, sensory or *dermal*) germ-layer, the latter to the *entoderm* or inner (vegetative, trophic or *gastric [gastralen]*) *germ-layer* of the higher animals." *Magen* means *stomach*, and was used by Oken in that sense. In Gastraea-Theorie (1874:10) Haeckel refers to metazoa (*i.e.* animals descending from a gastrulated ancestor) as *Darmthiere*.

102. Chapter 18, particularly note [146].

103. Schwämme (1870:233–234); Sponges (1870:118–119 (1870); *Kalkschwämme* (1872) 1:22. Haeckel renamed this sponge *Ascetta blanca*, and it has since been referred to genus *Clathrina*. The specimens he described almost certainly were of sponges of different species growing atop each other. See also Miklucho-Maclay (1868).

104. Using the terminology he introduced in *Generelle Morphologie*, he generalized the development of sponge (and coral) morphology as an enrichment of antimers, metamers, and persons: *Kalkschwämme* (1872) 1:96–130.

105. *Kalkschwämme* (1872) 1:467.

106. Haeckel, Gastraea-Theorie (1874:10–11); in English as The gastraea-theory (1874:150). I have corrected Wright's [Urdarm] to [*Darmanlage*], as in Haeckel's text. Note that Haeckel now speaks of four (not six) orders of individuality, and divides the animal kingdom into seven (in the quotation) or eight phyla (Gastraea-Theorie 1874:10–11; Gastraea-theory 1874:246).

107. Siebold had not mentioned gregarines. Siebold recognized two classes within Protozoa, Infusoria and Rhizopoda, while excluding vibrios, most unicellular green algae, diatoms, desmids, and sponges. His Infusoria included *Euglena, Chlorogonium, Peridinium, Opalina*, and ciliates (1848:7–25). See also Chapter 18.

108. Haeckel, Gastraea-Theorie (1874:27–28); Gastraea-theory (1874:223–224). Confusingly, here and above [106] Haeckel includes (some?) monera amongst the animals, and places monothalamia (single-chambered foraminifera) at the tectological level of monera. Yet in Monera he stated that "Monera are in fact protista. They are neither animals nor plants" (1868:65 and 1869:29).

109. Cohn had demonstrated sexuality in the unicellular alga *Sphæroplea* (Über die Fortpflanzung, 1855) and in *Volvox* (Observations, 1856). See also his review on sexuality in algae (Ueber das Geschlecht, 1855).

110. Phycochromaceæ (Rabenhorst), previously Myxophyceæ (Wallroth), later became known as cyanophyceæ or blue-green algae; they are now cyanobacteria.

111. Haeckel, Monera (1868:119–121) and (1869:328–330). Haeckel termed reproduction solely by asexual processes *monogony*, and organisms which reproduce in this way *monogenetic*. He anticipated many such revisions over time, as our knowledge of simple organisms improved. He was in error, however, in claiming that fungi reproduce only asexually (see below).

112. Haeckel, ibid. (1868:122) and (1869:330–331). His "Kingdom of Protists or the monogenetic organisms" was now Monera, Flagellata (minus *Volvox*), Labyrinthulea, Diatomea, Phycochromacea, Fungi, Myxomycetes, Protoplasta (amœbæ and gregarines), Noctilucæ, and Rhizopoda.

113. Fungi had long been considered cryptogams (plants with hidden sexuality), although even in the Eighteenth century, Hedwig and others claimed that fungi were sexual: see Lütjeharms (1936:163–173). The situation was slowly resolved by Tulasne and Pringsheim in the 1850s, and by de Bary in the 1860s.

114. Haeckel, *Natürliche Schöpfungsgeschichte*, Second edition (1870), pp. 404 and 415–416.

NOTES 613

115. Haeckel, *Das Protistenreich* (1878:52–53). In the Eighth edition of *Natürliche Schöpfungsgeschichte* (1889) he included phycochromaceæ (and bacteria), but not fungi.

116. Haeckel introduced the idea of a moneran in his 1863 Stettin lecture: note [35] above.

117. Haeckel, Zur Morphologie (1873:558–559); *Natürliche Schöpfungsgeschichte*, Second edition (1870), pp. 407–408 (protophytes) and 459–460 (protozoa); Fourth edition (1873), pp. 407–408 (protophytes) and 468 (protozoa), and in English as *The history of creation* (1876) 2:85 and 2:154 respectively.

118. Haeckel, Monera (1868:65–66) and (1869:29–30).

119. Haeckel appreciated that this point cannot be decided, because monera are without structure, and if they differed from one another (*e.g.*, chemically) we could not detect it: *Generelle Morphologie* (1866) 2:411–412.

120. Ibid. (1866), 1:203 and 2:403–406.

121. *Natürliche Schöpfungsgeschichte* (1868), pp. 347, 382, and 392; Second edition (1870), pp. 398, 399, and 441; Fourth edition (1873), pp. 398, 399, and opposite 440.

122. Ibid. (1868:328–329); and Fourth edition (1873:378), as translated (1876) 2:50–52.

123. Haeckel based *Natürliche Schöpfungsgeschichte* on a series of popular lectures he delivered at Jena during 1867–1868; and *Anthropogenie* on lectures he presented during summer 1873. For the latter as complementary to the former, see Haeckel's Vorwort to *Anthropogenie* (1874), page xv.

124. Near the end of *Natürliche Schöpfungsgeschichte* was a section titled Thierische Vorfahrenkette oder Ahnenreihe des Menschen (The chain of animal ancestors, or ancestral line of man), or a variant thereof. The series remained at 22 stages from the first (1868) through Seventh editions (1879), with revisions and increasing detail; Haeckel added three stages in the Eighth edition (1889). He also presented truncated versions focusing on the series from monera to gastraea, *e.g.* in Die Gastrula (1875:406), reprinted in *Biologische Studien* 2 (1877:66).

125. *Natürliche Schöpfungsgeschichte* (1868), figure at page 347 (see also page 556). At 1868:330 the running head refers to "neutral Amœbae of the present day", and at 1868:335 he supposes that *Noctiluca* may be a "neutral protist".

126. *Natürliche Schöpfungsgeschichte*, Second edition (1870), figures at pp. 398 and 399. See also pp. 405 and 441. He retained the two figures in the fourth (1873), sixth (1875), and seventh (1879) German editions, but replaced it for the eighth (1889).

127. In *Natürliche Schöpfungsgeschichte* (1870:579) he called amœbæ "*einzellige Urthiere* (Protozoa unicellularia)", and "animal amœbæ" appear under Protozoa in the animal *Stammbaum* (1870:449). A more-comprehensive statement was at Zur Morphologie (1873:556–559), and "amœbæ" appear under Protozoa in animal *Stammbäume* in the Fourth (1873:449) and Sixth editions (1875:449). Recall his earlier view (above) that certain protists are closer to plants, or to animals: note [64] above.

128. *Natürliche Schöpfungsgeschichte* (1870), compare pp. 383 and 460–462.

129. Haeckel presented two versions: one that does not identify (most) Protista as Protozoa (Gastraea-Theorie, 1874:53), and one that does (Zur Morphologie 1873:560). The former article, dated 29 September 1873, was reprinted in volume 2 of his *Biologische Studien* (1877), with the figure appearing at page 55. The tree at Zur Morphologie (1873:560) bears the title (in translation) "Phylogenetic table of the stem-relationships of the phyla of the animal kingdom", and distinguishes 'Monera (neutralia)' from 'Monera (animalia)'.

130. Haeckel, Nachträge (1877:71–73); reprinted in *Biologische Studien* 2 (1877:243–244). Haeckel had earlier sought to downplay the contradiction: Zur Morphologie (1873:557–559).

131. Haeckel, *Protistenreich* (1878), at pp. 2 (in Latin) and 3 (German). *Urthiere* (or a variant thereof) appears twelve times, while *Protist* (or a variant) appears ninety-eight times in

614   NOTES

the main text (including the title page, but excluding the running head and the taxonomic *Anhang*).

132. Ibid. (1878:16–18).

133. Ibid. (1878:64).

134. Ibid. (1878:61). A similarly anachronistic concept, still to be found in textbooks, identifies so-called *animal* and *vegetable* poles of the (animal) embryo.

135. Ibid. (1878:63). Similarly if less sharply, the thallus (or prothallus) separates plants from protists (not from protophytes).

136. *Generelle Morphologie* (1866) 1:232–233. In his translation of Haeckel's *Kristallseele* (1917), Mackay translates *Seele* as *psyche* (1999:15).

137. *Generelle Morphologie* (1866) 1:234; *Natürliche Schöpfungsgeschichte* (1868:341) and (1870:392–393).

138. Summarized in *Kristallseelen. Studien über das anorganische Leben* (1917).

139. *Protistenreich* (1878:29). The protistan soul was probed further by Max Verworn in *Psycho-physiologische Protisten-Studien* (1889), dedicated to Haeckel.

140. Haeckel, *Plankton-Studien* (1890), page 18 and note 1. In English translation (Field) as Planktonic studies (1893:578).

141. *Natürliche Schöpfungsgeschichte*, Eighth edition (1889:401–422). Continuing his analogy with the state, he characterizes histones as a "state community" (*eine staatliche Gemeinschaft*): *Anthropogenie*, Fourth edition (1891:128). Haeckel's concept is unrelated to the current meaning of *histone* (a positively charged protein that packages and orders DNA in the nucleus).

142. *Natürliche Schöpfungsgeschichte* (1889), pp. 420 and 421.

143. Ibid. (1889:454); and (with some changes in membership among plant and animal monera), Ninth edition (1898:456).

144. Ibid. (1889:419).

145. Haeckel does not mention labyrinthulæ, or any residual "neutral monera". At ibid. (1889:434) he states that flagellates, amœbæ, gregarines, diatoms, and myxozoa belong to the neutral monera; this is surely an error, as they are not monera at all. At the corresponding point in the Ninth edition (1898:435) he refers instead to "neutral groups of atypical or asemic protists" [for which see below]. For *Protista vegetalia* and *Protista animalia* see the tables at (1889:452 and 453).

146. Although *Volvox* and relatives, which he removed from Protista to the plant kingdom in 1874 (above), now appear on the animal side of the ledger (ibid. 1889:499 and 697).

147. In describing the morphologically based classification, he· refers to Protista and Histones as the two *Haupt-Reiche* (main kingdoms); each contains two *Reiche*, one of beings with a plant-like plasm and metabolism, the other with an animal-like plasm and metabolism (*Natürliche Schöpfungsgeschichte* 1889:419). Protophyta and Protozoa are accordingly shown as kingdoms at (1889:452–453). He also refers, seemingly informally, to an *Urpflanzen-Reiche* (1889:466) and a *Metazoen-Reiche* (1889:507 and 520). A generous explanation would be that *Natürliche Schöpfungsgeschichte* is a work of popularization.

148. *Natürliche Schöpfungsgeschichte* (1889:427).

149. Respectively *plasmogens* and *plasmophages* (*Natürliche Schöpfungsgeschichte* 1889:433); or *plasmodomes* and *plasmophages* (*Systematische Phylogenie* 1894, 1:36 and 1:43–46).

150. Haeckel, *Anthropogenie*, Fourth edition (1891) 2:491–492 and 2:512 (see also 2:513).

151. Ibid. (1891), 2:524 and 2:616 (with *Volvox*), and Table 15 following 2:472 (as Blastaeaden).

152. Ibid. (1891) 2:493–495.

153. Richards (2009:321).

154. Haeckel, *Systematische Phylogenie* (1894) 1:40–41.

NOTES 615

155. Ibid. (1894) 1:46.

156. Ibid. (1894) 1:46–47.

157. Ibid. (1894) 1:47. *Asemic* means *carrying no meaning*, although Haeckel offers *atypische* as a synonym.

158. Ibid. (1894) 1:49–51.

159. Ibid. (1894) 1:47–48.

160. Ibid. (1894) 1:47–49 and 52–53. Haeckel cautions that we must take care to exclude swarm-spores of algae, sponges, and cnidaria.

161. Ibid. (1894) 1:49–50, 1:52 and 1:93–251. On the plant side of the border are archephytes (phytomonera, chromaceæ) and mastigotes (phytomonads, volvocinæ, dinoflagellates); on the animal side, archezoa (zoomonera, bacteria), flagellates (zoomonads, catallacts, noctilucæ), gregarines, phycomycetes, and zygomycetes.

162. Ibid. (1894) 1:53 (protists), 1:91 (the organic world), 1:97 (protophytes), and 1:139 (protozoa). Regarding interpretation, Chromacea (asemic protists and protophytes per 1:52) are, as expected, on the protophyte side of the (dashed) boundary-line in the *Stammbaum* of the protistan kingdom (1:53), but on (or below?) the (solid) boundary-line in the *Stammbaum* of the organic world (1:91). The former (1:53) illustrates Haeckel's point that (plant-like) plasma-building preceded (animal-like) plasma-consuming metabolism.

163. Ibid. (1894) 1:52.

164. Haeckel acknowledges the polyphyletic basis at ibid. (1894) 1:95 (for protophytes), 1:135 (protozoa), 1:295 (thallophytes), and elsewhere.

165. Siebold (1848:3). See Chapter 18, note [181].

166. J Müller, Einige Beobachtungen (1856).

167. Balbiani (1861).

168. A spermatic body according to Müller; an ovum according to Balbiani. For this episode, see Churchill (1989) and Jacobs (1989).

169. Stein (1859) 1:96–98.

170. Bütschli had trained in mineralogy, chemistry, and palæontology, but also passed the examination in zoology at Heidelberg; after a brief stint with Leuckart, and service during the Franco-Prussian war (1870–1871), he worked mostly in his private laboratory in Frankfurt. His *Habilitationsschrift* at the Polytechnische Hochschule in Karlsruhe was published as *Studien* (1876). On the recommendation of Gegenbaur, he was appointed Professor of Zoology at Heidelberg (1878). See Fokin (2013).

171. *Studies on the early developmental processes of the egg-cell, (on) cell division and (on) conjugation in the infusoria.* Also published in *Abhandlungen der Senckenburgischen Naturforschenden Gesellschaft* (1876).

172. Bütschli, *Studien* (1876:219); *Abhandlungen* (1876:431).

173. *Studien* (1876:147); *Abhandlungen* (1876:359).

174. Weismann (1883); in English translation (1889). For a broader consideration of the role of the nucleus in inheritance (including the contributions of Haeckel, Nägeli, and others), see W Coleman (1965).

175. Bütschli, Protozoa (1880–1889). Pagination is continuous across all three *Abtheilungen*. He wrote the introduction in 1888 (see page i).

176. Ibid. (1880–1889), page ix.

177. Ibid. (1880–1889), page xii.

178. Bütschli derived the sponges from a common ancestor with choanoflagellates, undercutting Haeckel's gastraea-theory (ibid., pp. xvi and 877–879). He also considered it possible that the conjugate algae (*e.g. Spirogyra*) may have originated separately from holophytic mastigotes rather than via Protococcoidea (pp. xv–xvi).

179. Ibid. (1880–1889), pp. xiii–xiv

616 NOTES

180. *Ein consequentes Bestreben nach möglichst natürlicher, der Genealogie entsprechender Gruppirung der Organismen führt uns so zur Anerkennung des Mittelreiches, der Häckel'schen Protisten in modificirtem Sinne* (ibid., page xvi). He does not give the (intended) kingdom any further name.

181. On grounds of their physiological similarities: ibid., page xvii.

182. Reduced to "nucleic granules": ibid. (1880–1889), pp. xi–xiii; Schmitz (1880:196–198). See also Chapter 22.

183. Bütschli (1880–1889), pp. xvii–xviii.

184. Saville Kent (1880–1882), 1:50–56 and 74. Volume 1 is dated 1881–1882.

185. Ibid. (1880–1882) 1:44–45.

186. Ibid. (1880–1882) 1:143–194.

187. Ibid. (1880–1882) 1:31–44; quotation at 1:34.

188. Ibid. (1880–1882) 1:37.

189. Lanessan (1882).

190. Hartog, Sollas, Hickson & MacBride (1906).

191. Hickson, Lister, Gamble, Willey, Woodcock, Weldon & Lankester (1909).

192. Hickson *et al.* (1909:155–158).

193. Lang (1901:33).

194. Calkins (1909), pp. 28–33, 86–88. Calkins contrasted the "distributed" nucleus of bacteria with the compact "morphological" nucleus of "higher cells" (*e.g.* 1909:221 *et passim*).

195. Minchin (1912:5). It is unclear why the same charge could not be held against animals, or plants. Doflein (1909:311–313) acknowledged that while it can be useful to consider some potential protozoa as monera, plants, or animals, it is also illuminative to consider many of these organisms as protozoa.

196. Minchin (1912:5). From time to time, Haeckel pointed to the purely practical benefits of recognizing Kingdom Protista (*e.g.*, *Generelle Morphologie* 1:229–230, 1866), and (as we have seen) later integrated most protists into plants and animals, albeit not on a fully phylogenetic basis.

197. Theophrastus, *Historia plantarum* 4.6.1–4.7.1.

198. Linnæus, *Systema naturæ*, Tenth edition (1759) 2:1337–1348. As discussed in Chapter 15, Linnæus adjusted the composition of Cryptogamia Algæ several times; in the Tenth edition it also included liverworts, lichens, and sponges.

199. Lamouroux (1813), summarized at page 291.

200. CA Agardh, *Systema algarum* (1824), particularly pp. xi–xii.

201. See Chapter 19, note [184]. Agardh's diatoms included desmids.

202. For example, WH Harvey (1841), pp. xvii–xxi.

203. Brown algae (*Phæophyceæ*) do not include unicellular forms; red unicells were known to CA Agardh and Nägeli but were not classified as red algae until much later, and even then were suspected to be degenerate filaments. Earlier authors recognized series among diatoms, from single cells to filaments. For mature statements of parallel form-series see Fritsch (1935) 1:27; and Chapman & Chapman (1973:330–336).

204. Nägeli, *Neuern Algensysteme* (1847:116).

205. Cohn, Nachträge (1850:634–636); in English translation (1853:523–525).

206. Pringsheim (1858).

207. Cohn, Conspectus (1872a).

208. Sachs, *Lehrbuch* (1874:235–249); in English translation (1882:231–244). Farlow (1881:9–21) arranged algae in much the same way, with putatively asexual genera at the bottom, followed by zygosporic, oosporic, and carposporic genera.

209. de Bary (1881).

210. Bennett (1888).

211. Wille (1897), particularly the figure at page 26. Green algae are still used to study the transition from unicellularity to multicellularity: see (for example) Herron & Michod (2008); and Shelton & Michod (2010).

212. Pascher (1914), summarized at page 158.

213. Examples include Chodat (1897); Blackman (1900); and Oltmanns (1905) 2:3–23. For earlier, if less well-developed, concepts, see Cienkowsky (1870) and Woronin (1880).

214. Nägeli (1884:378–405): law of union at page 378; form-series illustrated, page 401.

215. Ibid. (1884). Nägeli helpfully summarized his argument at pp. 524–552, which in turn has been translated into English by Clarke & Waugh (1898). I examined the second edition (1914).

216. Nägeli (1884:90). He remarks that he might have called them *Protobien*, but *Protisten* had already been applied to a group of later-appearing beings of higher organization.

217. Ibid. (1884), pp. 163–164, 196, 206, 338–339, 341, 346, 347, 349, 351, 367, 421, 526, and 547.

218. He referred to "the two kingdoms" (ibid. 1884:170), and to "the two organic kingdoms" (1884:18).

219. Ibid. (1884:547). Given that he considered life to have arisen on multiple occasions at different times, one might imagine probias, plants, and animals had coexisted, or might still coexist: but Nägeli does not remark upon this.

220. Oltmanns (1904–1905), *e.g.* 1:3, 2:5 (and more in the second edition, 1922–1923).

221. Klebs (1896), pp. ix–x. The second (general?) part of the work did not appear.

222. Klebs (1893).

223. Klebs uses the term *Verwandtschaft* throughout, and finds it conceivable that the *netzförmig verlaufenden Verwandtschaftslinien* could be explained on the basis of *Transmutationslehre* (1893:436).

224. Ibid. 1893:430).

225. Ibid. (1893:429). He also allows that bacteria might be reduced forms (page 430).

226. Klebs is quite clear that the network depicts multiple, if uncertain, origins of higher plants: ibid. (1893:430–437). These green forms populate the lower right-hand quadrant of the figure (page 428).

227. Leeuwenhoek, Brief 26 [Dobell/Cole 18], 9 October 1676: see note [64] in Chapter 13; Brief 76 [Dobell/Cole 39], 17 September 1683, in *Alle de brieven* (1952) 4:118–155; and others.

228. Admixed with other organisms, as worms in OF Müller, *Vermium* (1773) 1:25–27 (*Monas*) and 1:39–49 (*Vibrio*); and as infusoria in *Animalcula* (1786), pp. 1–9 (*Monas*) and 43–77 (*Vibrio*).

229. Ehrenberg, *Animalia evertebrata* (1828).

230. Ehrenberg, *Infusionsthierchen* (1838:75–86). *Spirodiscus* has not been recorded since. Ehrenberg also recognized a genus (and family) *Monas*, but for organisms now recognized as chrysophytes, green algae, and bodonids (1838:10–38). Migula (1897:11) claimed that *Ophidomonas*, placed by Ehrenberg among the Cryptomonadina (1838:44), is a bacterium.

231. Dujardin (1841:212–226). However, he merged *Spirochæte* into *Spirillum*. He also recognized a non-bacterial genus *Monas* (pp. 279–286).

232. Nägeli, *Die neuern Algensysteme* (1847:152–155), *Nostoc*; and *Gattungen* (1849:44–60), Chroococcaceæ.

233. Caspary (1857), column 760.

234. Cohn, Untersuchungen (1872b:201–204). In 1871 Cohn had proposed the taxon name Schizosporeæ (Conspectus, 1872a), but now considered it inappropriate because not all bacteria produce spores (Untersuchungen 1872b:201).

235. Sachs, *Lehrbuch* (1874:248–249); and in English (1882:244). His class Protophyta contained Cyanophyceæ, Palmellaceæ (in part), Schizomycetes, and Saccharomycetes.

236. Haeckel, *Generelle Morphologie* (1866) 2, page xxiii *et passim*.

618   NOTES

237. Ibid. (1866) 2, pp. xxxiii–xxxiv *et passim.*

238. Haeckel, *Protistenreich* (1878), pp. 59–60 and 87.

239. Haeckel, *Systematische Phylogenie* (1894) 1:39, 1:52, 1:90–91, 1:101–103, 1:138, and 1:140–144.

240. Ibid. (1894) 1:39. He introduces, but scarcely uses, the term *Bactromonera* (1:137 and 1:140).

241. Ibid. (1894) 1:53 and 1:139. The *Stammbaum* at 1:91 is difficult to interpret on this point (as on others). Polyphyly at 1:143.

242. Ibid. (1894) 1:101–103 and 1:140–144. In *Systematische Phylogenie* he called these pigment-bodies *Chromatellen*, but in *Die Lebenswunder* (1904) refers to them as *chromoplasts (e.g.* pp. 157, 162, 397, and 410), *chromatophores* (page 37), or *chloroplasts* (pp. 161–162 and 195). Throughout *Lebenswunder* he refers to chromacea and bacteria together, sometimes as monera.

243. Chapter 15. During 1714–1721 Richard Bradley (Chapter 14) developed a theory of infectious disease based on living agents (Santer, 2009).

244. *The Economist* (1849), as cited in *The Economist* (2020). Wainwright (2001) briefly reviews studies on the microbiology of disease prior to Pasteur.

245. Rayer (1850). Rayer described the bacillus, while Davaine had observed the disease in sheep, and later (two articles, 1863) demonstrated its transmissibility. See also Théodoridès (1966).

246. Koch (1876).

247. As tabulated by Brock (1999:290); reprinted in Blevins & Bronze (2010:e750).

248. I surveyed 19 bacteriology books published between 1878 and 1899. Of these, 18 stated outright, or strongly implied (*e.g.* by describing growth as *vegetation*), that bacteria are plants. Three of these 18 commented that the line separating the plant and animal kingdoms is uncertain or somewhat arbitrary, and one mentioned "certain unicellular organisms whose relations to the plant and animal kingdoms are doubtful" (Hueppe, 1899:10). The remaining book did not comment on the place of bacteria in nature.

249. For Pasteur's experiments disproving spontaneous generation, see Porter (1961).

250. Griffiths (1893), page xi.

251. Entz (1888). For the years of composition, see page vii. Entz (with Török Aurei) translated Darwin's *Descent of man, and selection in relation to sex* into Hungarian (1884). I thank Professor Miklós Müller for the gracious gift of a copy of Entz's *Studien*. Entz refers to "*Protista animalia seu Protozoa*" (1888:272), a concept developed by Haeckel in *Natürliche Schöpfungsgeschichte* (1889): see note [126] above. Note also that Entz credits *Protorganismen* to JV Carus (1888:272n*) when in fact it was introduced by CG Carus (Chapter 18).

252. Entz, Bevezető. In: G Entz & S Mágocsy-Dietz (eds), *Az élők világa növény- és állatország.* Athenaeum, Budapest (1907), pp. 1–60, at page 2. In this paragraph I translate *véglényeknek* as *protists: vég*, end, terminus, or (figuratively) border; *lény*, being, organism (Miklós Müller, personal communication 27 August 2020; Török Júlia, personal communication 21 September 2020).

253. Entz (1907:50).

254. Leclercq (1890); list of protistan classes at page 80, quotation at page 85, diagram at page 89. For a biographical sketch, see the anonymous entry at www.bestor.be.

255. Strasburger (1913:52); Hertwig (1913); Poll (1913:39); Roux (1915:173); Hartmann (1915). At page 300 Hartmann states that there is no comprehensive work on "microbiology (biology of plantlike and animal protists)".

256. Hertwig (1893:145–146).

257. Singer (1931:342).

258. The prospectus has been translated by Mollenhauer (2000:286). I have not seen the German original. *Protist* continues the *Archiv für Protistenkunde.*

NOTES 619

259. Apart from the medically oriented *Centralblatt für Bakteriologie und Parasitenkunde* (1887), which in 1901 was renamed *Centralblatt für Bakteriologie, Parasitenkunde und Infektionskrankheiten*; and the *Journal of Pathology and Bacteriology* (1892). The medical focus of these bacteriological journals led to the launch of *Journal of Bacteriology* (1916) as a forum for applied, but not necessarily medical, papers: see WT Sedgwick (1916). For other early journals in cognate fields, see Corliss (1998).

260. Hertwig (1902).

261. Bütschli (1902); Schaudinn (1902).

262. Doflein (1902). Doflein wrote not only his respected *Lehrbuch der Protozoenkunde* (first edition 1909), but also *Probleme der Protistenkunde* (1909).

263. Exclusive of the volume title, running heads, and reference lists.

264. Prowazek, who had survived malaria as a youth, died of typhus during World War I.

265. Dobell later wrote the authoritative *Antony van Leeuwenhoek and his "little animals"* (1932).

266. Dobell (1911); primitivity at page 272. Dobell wrote that "there is no cell 'theory' . . . this may perhaps be called a theory, but it is more accurately called a misconception" (1911:277). "The cell must be defined in terms of the organism, not vice-versa. . . . A correct interpretation of Protista can never be reached until the cell theory has disappeared" (1911:284–285).

267. Huxley used *Protista*, *Protistæ*, and *Protistic* in his review of Haeckel's *Natürliche Schöpfungsgeschichte* (1869:41); reprinted in *Critiques & Addresses* (1873): see pp. 312–314. Wyville Thomson used *Protista* in *Philosophical Transactions* (1869:716). According to the *OED*, *Protist* appeared as a noun in 1873, and as an adjective in 1879 (*Oxford English Dictionary* online, accessed 22 August 2020).

268. Haeckel, *Die Welträthsel* (1899), page vii.

269. Incongruously labelled *Radix communis Organismorum* in *Generelle Morphologie* (1866) 2, Table 1.

Chapter 22

1. Dallinger (1872:300). The Reverend William Henry Dallinger conducted research in microbiology, and was an early supporter of Charles Darwin's work on evolution, but challenged Haeckel's kingdom Protista (Dallinger 1872; Haas 2000).

2. Russett (1976); Glick (1988); Engels & Glick (2008).

3. For the latter, Wainwright (1997) and below.

4. For a stark characterization of mid-Twentieth-century microbiology, see Woese (1987:222–224), and Woese (1994b).

5. To these evils we may add Lysenkoism, which admixed ideas from Lamarck and Darwin while writing out Mendel; and state imperialism particularly in tropical lands, which benefited from the study of protistan parasites.

6. Chapter 3, at note [27].

7. Porphyry, Ammonius, and Philophonus, Chapter 5; John of Damascus, Chapter 6.

8. Chapter 9.

9. Linnæus, *Systema naturæ*, Tenth edition (1758) 1:5–8. Imperium Naturæ included Minerals.

10. Pallas, *Elenchus zoophytorum* (1766), pages viij and 3–5.

11. AL de Jussieu, *Genera plantarum* (1789), page ij. Note also the *Règne organique* of Turpin (1820).

12. Chapter 17, note [118].

13. Such a *supertaxon* would rank above (animals + plants) together, *i.e.* alongside the four examples provided in the text, but exclude plants and animals. Haeckel's distinction of

620 NOTES

Protista from *Histones* does not provide a counterexample, as each contained beings that he called plants and animals: *Natürliche Schöpfungsgeschichte*, Eighth edition (1889:420), and Chapter 21.

14. Chapter 21, notes [232] and [233]. Nägeli considered fungi and algae to be plants.

15. Animals: Chapter 21, notes [229] through [231]; Plants: Cohn (below), and Migula (1900); Fungi: Nägeli (above). Haeckel: *Natürliche Schöpfungsgeschichte*, Eighth edition (1889): blue-green algae as Protophytes (Protista vegetalia), page 452; colourless bacteria as Protozoa (Protista animalia), page 453; and *Systematische Phylogenie* (1894), 1:96 and 1:138 respectively.

16. Cohn did not elaborate on his reasoning, apart from pointing to "morphological and evolutionary relationships" (Cohn, *Beiträge* 1872b), at page 191. He did not refer to their common lack of a nucleus, or assign any particular rank to Schizophyta.

17. Schleiden and Schwann, Chapter 18; Remak (1855:164–169).

18. Haeckel, *Systematische Phylogenie* (1894) 1:101. He is very clear that Chromaceen have no Zellkern and are therefore Cytodes, not true cells. Bacteria are Protozoen: Archezoa (1894, 1:137–138, 140–144). He nonetheless kept the taxa separate.

19. Clark (1928:39).

20. Ward (1928:7); Sharp (1934:192–194).

21. JR Baker (1955:474).

22. For example, EG Pringsheim (1949:87–91).

23. The concern runs deep in bacteriology: Cohn, *Beiträge* (1875:201): *Veilleicht möchte sich die Bezeichnung Schizophytae für diese erste und einfachste Abtheilung lebender Wesen empfehlen, die mir, den höheren Pflanzengruppen gegenüber, natürlich abgegrenzt erscheint, wenn auch die Merkmale, durch welche sie charakterisirt ist, mehr negativer als positiver Art sind.*

24. At the time Enderlein, an entomologist, was director of the Zoological Institute at Universität Berlin. His *Bakterien-Cyclogenie* (1925) is said to contain 200 purpose-built neologisms.

25. The *Mychit* remains singular in micrococci (Enderlein 1925:236). As noted by Henrici (1928:5–7) and others, Enderlein described (and presented drawings of) subcellular structures of dimensions far beneath the limit of resolution of the light microscope.

26. Enderlein (1925:223–235) *et passim*. At 1925:353 he defines Mychota (bacteria) as *Regnum der Protomychota*. By contrast, Protota (*e.g.* bacteriophage) lack cytoplasm and are noncellular (1925:327–329).

27. Enderlein went on to produce and sell naturopathic remedies against the resulting diseases including tuberculosis, arthritis, and cancer.

28. *Stamm* can usually be translated as *phylum*, but at (1925:236) he identifies Mychota as a kingdom (Regnum). Earlier he wrote that "the bacteria stand in fully the same relation [*Verhältnis*] to the animals as to the plants" (1917:309). At (1925:305–309), however, he describes a *Stammbaum* with chemoautotrophic bacteria as the common base, photoautotrophic bacteria as the deepest branch along the plant-series, and the remaining colourless bacteria on the line leading to fungi, protozoa, and metazoa.

29. HF Copeland (1947:351).

30. Chatton (1907). Chatton lists this work as 1906 (*Titres et travaux* 1938:6), but it was read in the session of 12 January 1907, and the volume is dated 1907.

31. Chatton (1925). Sapp (2005:293) states that Chatton "wrote the paper in 1923 at the University of Strasbourg", but Chatton describes observations up to December 1924, and cites articles published in 1924. Chatton had published at length on the nucleus and mitosis of amœbæ (Chatton 1910).

32. He later called Sporamœbidæ "a family or a tribe" (Chatton 1938:160).

33. *Procaryote* was certainly understood as implying that these organisms are older and more-primitive than eukaryotes, and gave rise to them: see Woese in Schleifer & Stackebrandt (1985), especially pp. 2–4; and Pace (2009), Figure 1 at page 2009.

34. Chatton (1925:76). The table bears the title *Essai de classification des Protistes*, and a footnote cautions that "it was not a question of drawing up here a complete and detailed classification of Protists, but of summarizing and illustrating the views set out in this memoir and showing how they differ from classical notions."

35. Chatton (1925:77), *Essai sur la phylogenie des Protistes*.

36. The text (Chatton 1925) offers no evidence that Chatton considered algae or sponges to be protists. Fungi are mentioned in passing (1925:45). Chatton (1925:38) cited without comment a passage in which Léger & Duboscq (1910:220) referred to the *règne des Protistes*. The latter authors, in turn, followed Hartmann (1907), Doflein (1909), and others in a standard delineation of protists. For archegoniates (liverworts, mosses, ferns, and perhaps gymnosperms) see Underwood (1895) and Davis (1909). Although Chatton (writing in French) spelled *procaryote* and *eucaryote* with a *c*, today these terms are usually spelled with a *k* in recognition of the Greek root κάρυον (kernel).

37. For example, Sapp (2005:292–294).

38. Lwoff (1932:3). Lwoff then distinguished Leuophytes from pigmented Chlorophytes. Chatton's article was published in 1925, not 1926. Lwoff later considered Chatton "the greatest protozoologist of all time" (Lwoff 1971:3).

39. Knight (1936:156), who (as we see) attributed the division into procaryotic and eucaryotic protists to Lwoff, rather than to Chatton. The further subdivision of Eucaryotes is correctly credited to Lwoff. At page 175 Knight thanks Lwoff "for having read the MS. and offer[ing] valuable criticisms and emendations". Knight co-founded the Microbiology Society (1943), and was a founding co-editor of *Journal of General Microbiology* (1946). An avid Francophile, he was highly regarded for his translations of Stendahl.

40. Lwoff (1938:194). Very similar wording is at Lwoff (1944:71).

41. Chatton (1918).

42. Chatton (1938). According to Katscher (2004:257), some copies bear a main title page dated 1937 that lists his academic positions at the time, but also (following page 60) have a second title page dated 1938 that states his new positions. My copy, previously owned by phycologist Jean Feldmann (1905–1978), bears only a single title page (replicated as the front cover) dated 1938.

43. Chatton (1938:50); translated by MAR. Note that in his earlier phylogenetic network (1925:77), the transition from flagellates (cryptomonads) to brown (and thence red) algae does not take place "at the base of the vegetable kingdom"; nor is the vegetable kingdom fully in evidence. His 1938 formulation included angiosperms (at page 50).

44. Recall (above) that Chatton (1925:38) cited Léger & Duboscq (1910) referring to *règne des Protistes*. Chatton succeeded Octave Duboscq as director of Laboratoire Arago at Banyuls.

45. I know of no evidence that Chatton ever returned to this question (nor does Marie-Odile Soyer-Gobillard, *pers. commun.* 2021). Chatton was not entirely disinterested in taxonomy; he (in part with Lwoff) had established several higher taxa. We might instead point to his new directorial responsibilities from 1938, war (1939–1945), and occupation (1940–1944). Chatton died in 1947 following a lengthy illness.

46. Novák (1930), pp. 71–76; categorical ranks at page 8. Aphanobionta ("unseen bionts"), page 79; Akaryonta, pp. 80–89; overview of Karyonta, page 90. Bacteria and cyanophytes might be elevated to phyla (*kmeny*) or even kingdoms, if it were proven that their morphological similarity is convergent (page 80). Flagellates are on the "edge" of animals; akaryonts have no phylogenetic relationship with karyonts (page 90).

47. Novák (1930), page 80.

622 NOTES

48. Harms (1946) [first edition]. The latter group included not only multicellular animals, but also *Zellverbandstiere* (page 37). He left Jena after the [DDR] "Cultural Advisory Committee for the Publishing Industry" withheld permission for a second edition of *Zoobiologie* until he discussed "modern authorities" (*e.g.* Michurin, Lysenko) rather than Mendel. From 1949 he was associated with Universität Marburg. See also Pflugfelder (1966).

49. Rothmaler (1948), summarized at pp. 248–249. His Anucleobionta included bacteria and blue-green algae.

50. Rothmaler (1951). He now recognizes Novák's Aphanobiointa, and subsumes blue-green algae (Cyanoschizeae) into Schizophyta (page 259).

51. Haeckel, *Generelle Morphologie* (1866) 1:112.

52. Haeckel, Monographie der Monera (1868:65–66); Monograph of Monera (1869:29–30).

53. Haeckel, *Systematische Phylogenie* (1894) 1:39, 1:52, 1:90–91, 1:101–103, 1:138, and 1:140–144.

54. Haeckel, *Die Lebenswunder* (1904:217–238) *et passim*; in English as *The wonders of life* (1905:190–209) *et passim*.

55. HF Copeland (1938:386).

56. Stanier & van Niel (1941:456–458). They recognized two divisions, Myxophyta (blue-green algae) and Schizomycetae (other bacteria including spirochætes). Their 1941 article did not commit to the number or identities of other kingdoms.

57. For example, by Barkley (1939) and (1949:88).

58. Marton (1940); Zworykin (1940); Marton (1941); Mudd & Lackman (1941); Knaysi & Baker (1947); Mudd (1948); Hillier *et al.* (1949).

59. An important early article is Chapman & Hillier (1953); for a review, see Brock (1988).

60. Stanier & van Niel (1962), pp. 21, 23–25, 28–29, and 33. The criterion involved the presence or absence of organelles and vacuoles, not only of discrete nuclei. We might speculate that they settled on *procaryotic* and *eucaryotic* for reasons of familiarity: Chatton, Lwoff, van Niel, and Stanier were microbiologists with connections to marine research stations, whereas Novák, both Copelands, Barkley, and Rothmaler were botanists. Ellsworth Dougherty (1957) had contrasted *eukaryon* (the nucleus of "higher organisms") with *prokaryon* (the corresponding feature of monera), but Stanier & van Niel (1962:20) explicitly credited the terminology to Chatton.

61. Stanier *et al.* (1963:85). More-detailed statements appeared in the Third (1970:33–34) and Fifth (1986:47) editions. By contrast, the first edition allows only that "the bacteria and blue-green algae are an isolated subgroup of the microbial world, characterized by a relatively simple cell structure" (1957:100), with a central region that corresponds chemically (DNA) but not structurally to "typical nuclei as we find them in the cells of the higher protists" (1957:101).

62. Stanier *et al.* (1963), three kingdoms at page 56, and lower protists at pp. 65 and 139–191; Third edition (1970), three kingdoms at page 33, prokaryotic protists at pp. 48 and 133; Fifth edition (1986), three kingdoms (now explicitly attributed to Haeckel) at page 47. In the Fifth edition they recognized Eucaryotes, Eubacteria, and Archaebacteria as taxonomic "groups" (1986:43 and 47–48), and downplayed the concept of protists (pp. 48–49).

63. Stanier began using the term *cyanobacteria* in his publications from about 1973. For a history of the integration of cyanobacteria into bacterial taxonomy, see Rippka & Cohen-Bazire (1983). *Blue-green algae* remained in use in the phycological community into the 1980s and beyond.

64. Copeland (1938:412–414), using as criterion the exclusive presence in Chlorophyceae and Embryophyta of *all* the following: chloroplasts, chlorophylls *a* and *b*, carotin, xanthophyll, true starch, and true cellulose. He accepted that some of these occur among protists, but held that the full set is found only in green algae and plants. Subsequent biochemical studies

proved him largely correct, although "carotin" and "xanthophyll" proved to be mixtures, and other characters must sometimes be argued narrowly. Molecular trees (Chapter 23) are more readily supportive.

65. Whittaker, *Science* (1969).

66. In regard of *convenience*, Copeland mentioned familiarity (*i.e.* tradition), avoidance of too many or too few taxa, and feasibility of description and delineation (1938:383–384). He reflected on the practicalities of teaching elementary botany (1938:384–385), and thanked his father Edwin for having taught him the principles of classification (1938:417). The article by EB Copeland (1927) he cites is heavily focused on the use of texts in teaching elementary botany.

67. HF Copeland (1938:385) cited his father EB Copeland (1927:389), who in turn cited an unnamed introductory botany text in which Schizophytes was "a distinct kingdom of living things". The senior Copeland quotes at length from the earlier text, but I have not been able to identify it.

68. Copeland (1938:393–412); quotation at page 416. Copeland held that although Protista is "undeniably heterogeneous", it is nonetheless "a natural group, having genetic continuity by the fact that [it] include[s] the original form of nucleate life and all of its descendants except those two specialized secondary developments, the familiar kingdoms of plants and animals. [Protista] could legitimately be treated as several kingdoms, but it is not possible to present knowledge definitively to delimit these, and several of them would probably turn out to be ludicrously inconsiderable in this rank. The entire assemblage, then, is to be treated as a single kingdom. It is expected that familiarity will make it acceptable" (Copeland 1947:344).

69. *Protamœba*, the presumed type of Haeckel's Monera, did not (in the opinion of Schaeffer 1926:6) exist as an independent organism, rendering the taxon invalid (Copeland 1947:350–351); Protoctista, he claimed, is "the same group as Protista Haeckel Gen. Morph. 1:203. 1866" (1947:351) but was published earlier. As we saw in Chapter 20, Hogg's "fourth kingdom" was Primigenum, its members Protoctista; it is far from obvious that Hogg held Protoctista to be a taxon at all. Nor did Hogg mention red algae, brown algae, or fungi: he explicitly included in Primigenum only Owen's Protozoa, *i.e.* sponges, rhizopods, diatoms, desmids, gregarines, and infusorial animalcules (Hogg 1860:221–223).

70. Copeland (1956).

71. Barkley (1939) and (1949).

72. Rothmaler (1948).

73. Whittaker (1957:537). Hagen (2012) provides important background.

74. Whittaker (1957:537).

75. Where he felt confident about the derivation he indicated "lines of evolution" (Whittaker 1959:219), later called "phyletic lines" (Whittaker 1969:157). He believed Rhodophyta, Myxomyetes, and Mesozoa to be derived separately, albeit from unknown protists (1959:217 *et passim*). His later five-kingdom diagram (Whittaker 1969) indicates a greater degree of polyphyly for Fungi. Chlorophyta (green algae) might equally well be placed within Protista (1969:157). At (1969:158): "Each [polyphyletic higher kingdom] includes a dominant evolutionary line to higher organisms as its major subkingdom, and minor subkingdoms which are independent experiments in multicellular or multinucleate organization in one of the three nutritive directions. In each case these minor subkingdoms are less widely successful than the principal subkingdom, represent somewhat lower and different organization from it, and may to some degree depart from the typical nutritive mode of the kingdom."

76. Margulis (1971). Some authorities worried, however, that the inclusion of sponges in Animalia might make the kingdom diphyletic.

624   NOTES

77. Margulis (1974), retaining *protist* rather than *protoctist* (1974:56); Whittaker & Margulis (1978), recognizing *protist* and *protoctist* as "valid alternatives" but indicating a sense of the former as "primarily unicellular forms" (1978:5). See also Margulis (1998:79–81). The 1978 article was based on Whittaker's keynote address to the Second international meeting of the Society for Evolutionary Protistology (22–24 June 1977).

78. For example, the two- and nineteen-kingdom schemes described by Leedale (1974), at pp. 267–268 and 268–269 respectively. Ideas of multiple (non-animal, non-plant) kingdoms had long been mooted informally: Copeland (1927); Martin (1932); GM Smith (1950:11); and others.

79. RC Moore (1954) recognized Schizomycetes and Myxophyceae as phyla within Monera, and twelve phyla (including Eumycophyta, and a reduced Protozoa) within Protoctista. Monera lack a "definitely organized nucleus", while Protoctista are "cells with distinct nucleus" (1954:594). Quote in text at (1954:592).

80. Dodson (1971) recognized kingdoms Mychota (including viruses), Plantae and Animalia; fungi and eukaryotic algae were plants, while protozoa and sponges were animals.

81. Jeffrey (1971). Jeffrey (like Novák) recognized superkingdoms of Acytota (viruses); Procytota (with two kingdoms, Bacteriobionta and Cyanobionta); and Eucytota (with five kingdoms: Rhodobionta (red algae), Chromobionta (golden and brown algae, diatoms, oömycetes), Zoobiota (protozoa and metazoa), Mycobiota (fungi and unicellular relatives), and Chlorobionta (metaphytes, green algae, euglenoids, perhaps trypanosomes). These were demarcated by thin combinations of pigmentation, biochemical, ultrastructural, and nutritional characters.

82. Leedale's "pteropod" scheme (1974:268), which however increases the degree of polyphyly of the animals, plants, and fungi.

83. Edwards (1976), based on cytological and biochemical criteria. Six were eukaryotic (Chlorobionta, Ochrobionta, Myxobionta, Erythrobionta, and Fungi 1 and 2); one was prokaryotic (Cyanochlorobionta). He provisionally treated bacteria and animals as single kingdoms.

84. Dougherty & Allen (1958; 1960:137–138). These levels represent "major evolutionary steps". Christensen (1962) recognized Procaryota and Eucaryota, and divided the latter into Aconta (without flagella or basal bodies: red algae) and Contophora (other eukaryotes). He quickly abandoned this system, however (Christensen 1964). See also Round, two articles (1963).

85. Dodge (1966:113), at kingdom level.

86. LS Dillon (1962), page 114 *et passim*. Many (perhaps most) of his interpretations of cytological states were incorrect (Heath, 1983).

87. LS Dillon (1962:115). Kingdom Plantae, Subkingdom Chrysophytaria, Province Metaphaeophyta, Subprovince Metazoa (Dillon 1963:76–82).

88. Margulis & Taylor (1975); Corliss (1986:15–18). According to the ISEP website (accessed 1/2021), a society was proposed by Lynn Margulis, Max Taylor, and Howard Whisler at ICSEB-1 (1973). The proposal was formalized at a symposium on the evolution of mitosis in eukaryotic microorganisms (Boston, 23–24 June 1975); this meeting was subsequently considered SEP-1. "International" was added after SEP-4, attended by the current author. ISEP was incorporated in 1982.

89. Six kingdoms: Cavalier-Smith (1981, 1983, 1998, 2004a); seven: Ruggiero *et al.* (2015); eight: Cavalier-Smith (1981, 1993); nine and ten: Cavalier-Smith (1981). He also formalized Prokaryota and Eukaryota as superkingdoms (1981).

90. Chapter 21, note [35]. Haeckel seems to have envisioned a gradual autogenous process, as he derived cells (with nuclei) from cytodes (without nuclei), and did not require an exogenous origin of chromatophores.

NOTES   625

91. Alternatively described as granules, molecules, molecular granules, etc. Tyson (1878:21) dates globule-theory to 1779–1842. Schleiden, *Beiträge* (1838); Schwann, *Microscopische Untersuchungen* (1839). Globule-theory superseded an earlier fibre-theory that can be traced to Haller (1757), who however had little concept of cells.

92. For the globule theory, see Virchow, *Cellularpathologie* (1858:23–24) and *Cellular pathology* (1860:25–27); Tyson (1870:20–25) and (1878:21–28); Altmann (1890:1–55); Baker (1948:114–121); Pickstone (1973); Wolpert (1995:228–229); and Schickore (2009). For Baumgärtner and Arnold, see Virchow (1858:23–24) and (1860:26), and Tyson (1878:26–28). Baumgärtner (1842), *Bildungskugeln, Molecularkügelchen* and many variants; Arnold (1836), *Kügelchen* or *Bläschen*. Baumgärtner (1842), see especially pp. 82–83 and the lengthy note at pp. 83–84, comparing his theory with that of Schwann.

93. Chapter 18.

94. Wilson (1925:57); *einem lebendingen Baue*, Brücke (1861:386).

95. Brücke (1861). Sigmund Freud studied anatomy under Brücke from 1877 to 1883.

96. Brücke (1861), note 1 at page 381; quoted by Altmann (1890), page 4.

97. Béchamp (1883); quotation at page iv. Béchamp was professor of medical chemistry and pharmacy at Montpellier (1856–1876), and afterward professor of organic and biological chemistry, and dean of the Faculty of Medicine at Université Lille. It did not help Béchamp's cause that he pursued a bitter personal campaign against Pasteur, accusing him of plagiarism and fraud. For a view from the Béchard camp, see Hume (1923). Enderlein (above) modelled his *protits* on Béchamp's microzymas.

98. Flemming (1882), page 77 & note 1. Flemming distinguished *fila* (alternatively *mitome*) and *interfilar mass* (*paramitome*) from metabolic products and inclusion-bodies, calling the latter "intruders in the cell".

99. Altmann (1890:124–125) and (1894:141–143). Intracellular fibres were also bioblasts. *Im Bioblast scheint jene morphologische Einheit der lebenden Materie gefunden zu sein.* "Protoplasm can be defined as a colony of bioblasts" (1894:143). For more on bioblasts, see Wilson (1925:74–78) and O'Rourke (2010).

100. Altmann (1890:8–9). Pages 1–16 were first published as a lecture "Zur Geschichte der Zelltheorien" (Leipzig 1889). A second edition of *Elementarorganismen* appeared in 1894; this quotation appears therein unchanged, also at pp. 8–9.

101. Benda (1898:397). See also Cowdry (1918) and (1953), and Wilson (1925:45–48).

102. Chapter 21 and note [247] there.

103. Sapp (1994:89–130).

104. Portier (1918:79–95). Portier claimed that he and others had cultured plant and insect mitochondria; "true microorganisms" at (1918:79). Mitochondria from vertebrates, however, resisted culture (1918:81–93). From 1911–1918 Portier was chair of marine physiology at the recently formed (1906) Institut océanographique. See also Sapp (1994:89–90).

105. Portier (1918), pp. 101–109, 131–132; Sapp (1994:90–92).

106. Reinke 1880:62). Details of the observation are in Schimper (1883), column 112, note 2. Reinke later became an active opponent of Haeckel's Monist League, which he considered atheistic.

107. Schmitz (1883).

108. Schimper (1883), note 2 at column 112. Translation by MAR. If plastids are not formed anew each generation, they must exist continuously across generations, as would a lineage of any organism. Schimper did not explicitly identify a free-living counterpart. In this 1883 article he coins the word *Chloroplastiden* (which he uses only in the plural); the clearest definition is at column 123.

109. Schimper (1885:202). Translation by MAR. He says much the same elsewhere in the article, *e.g.* at page 4.

626 NOTES

110. Sapp (1994), especially pp. 3–34.

111. Schwendener described the duality in an 1867 address to the Schweizerische Naturforschende Gesellschaft (Anonymous, Protokoll 1867), and later elaborated on their nutritional relationship (Schwendener 1869), referring inter alia to the alga as servant (*Diener*), and the fungus as ruler (*Beherrscher*, 1869:3) and parasite (1869:10). See also Honegger (2000). The duality mentioned by Schwendener was not entirely revolutionary at the time: see WL Lindsay (1856:115).

112. He introduced the term on 14 September 1878, in an address to the 51st congress of the Deutscher Naturforscher und Aerzte, meeting in Cassel (de Bary 1878:121–126). The address was printed the following year: de Bary (1879:5), *der Erscheinungen des Zusammenlebens ungleichnamiger Organismen*. Oulhen *et al.* (2016) offer an English translation. Albert Frank (1876:195–197) had earlier introduced the term *Symbiotismus* in a similarly neutral sense, and anticipated more-specific terms for relationships of increasing dependency.

113. Famintsin & Boranetzky (1867). According to Khakhina (1992:23n§), Famintsyn later claimed the work as exclusively his own. For details of his scientific career, see Khakhina (1992:17–33). Both Famintzyn and Baranetzky had studied under de Bary.

114. Famintzin, *Biologische Centralblatt* (1907:355–356).

115. Famintsin, *Zapiski* (1907:11), as translated in Khakhina (1992:33). The paper was submitted on 25 October (8 November) 1906, and published in 1907.

116. According to Famintsin, the sexual processes of plants and animals are likewise examples of symbiosis (1918, col. 282).

117. For alternative spellings of his family name, see Sapp *et al.* (2002):413n1. For his life and work more generally, see Khakhina (1992:34–50); Sapp (1994:51–59); and Sapp *et al.* (2002). His interest in the evolution of the cell dates to 1903, and seems to have grown out of his work on the classification of diatoms: Khakhina (1992:34–37).

118. He introduced the term in Russian in 1909, and in German in 1910.

119. Mereschkowsky (1905); Mereschkowsky (1910); Mérejkovsky (1920). For a high-level summary of his argument, see (1920:24); Sapp *et al.* (2002:420–423) reconstruct its argument somewhat differently.

120. Khakhina (1992:38–50).

121. Sapp (1994:51–59); Sapp *et al.* (2002:418–427).

122. *Chromatophores* include (green) *chloroplasts* in green algae and higher plants; their red, golden-brown (etc.) counterparts in the corresponding groups of algae; and colourless *leucoplasts*. Depending on context, some or all may simply be called *chloroplasts*. Today we use the term *organelles* ("little organs") for chloroplasts (*i.e.* chromatophores) and mitochondria (chondriosomes), and (in a broader sense) for other discrete (but not self-perpetuating) structures in the cell including the endoplasmic reticulum, Golgi apparatus, centrioles, and flagella. Merezhkowsky regarded the cilia of infusoria to be organs, *i.e.* not symbionts (1905:595).

123. Mereschkowsky (1905:596–598); Mérejkovsky (1920:22–31).

124. Merezhkowsky mentions their size, form, even dispersion of colour, lack of true nuclei, ability to take up carbon dioxide in the light, and reproduction by division (1905:599–601; 1920:31–52). Unaware that chloroplasts contain nucleic acids (1905:600)—this was, after all, confirmed only in 1961—he imagined that the *Nukleinkörner* of blue-green algae had been transformed into pyrenoids, and eventually lost.

125. Ibid. (1905:598–599); (1920:52–56). Merezhkowsky ignored that zoochlorellae are nucleate; for this he was rightly taken to task by Famintzin and others (Khakhina 1992:40).

126. According to Khakhina (1992:42), Merezhkowsky introduced the idea in *Teoriya dvukh plasm* (1909); I have been unable to locate a copy. Sapp *et al.* (2002:425) describe his 1910

article as the German translation of the 1909 book; the introduction to Mereschkowsky (1910) is dated 11 January 1909. For the two plasma types, see Mereschkowsky (1910); Khakhina (1992:42–46); Sapp (1994:53–54); and Sapp *et al.* (2002:424–427). Mycoplasm as the bearer of heredity at (1910:358).

127. Haeckel, *Systematische Phylogenie* (1894) 1:36 and 1:43–46; see Chapter 21 and notes [149] and [162] there. Merezhkowsky did, however, acknowledge Haeckel's comment that the plant cell evolved as "a symbiosis between plasmodomous green and plasmophagous not-green companions" (*Lebenswunder*, 1904:225; in English as *The wonders of life*, 1905:195–196). *Wonders* was copyrighted in 1904, but appeared in January 1905.

128. Although not in phycomycetes: Mereschkowsky (1910:364–366).

129. Mereschkowsky (1910:363).

130. Ibid. (1910:364). Had there been no symbioses, the organic world would have been dominated by a great Kingdom of Fungi, and "uniform, petty [*geringfügig*] Monera" (1910:363). Altmann (1890:138) had imagined a four-step systematics based on type of bioblast colony, but did not develop the idea.

131. Ibid. (1910:366): *Noch eine Schlussfolgerung meiner Theorie ist die Aufhebung des Reiches der Protisten—dieser Zoophyten des 19. Jahrhundert, welche ein Reich von Übergangsorganismen vorstellen sollen, die sich noch night in echte Tiere oder echte Pflanzen differenziert hätten.*

132. Ibid. (1910:366). Later he likened the groups of infusoria to "the grass of a lawn on which the great tree of animals grows" (Sporozoa alone gets credit as a "small shrub"): (1920:62n1). Metazoa, however, are strictly monophyletic (1920:64–65).

133. Ibid. (1905:602–603); six to nine origins (1910:363–366); fifteen or more origins (1920:58–62). In contrast to the "lawn" of minor animal groups, the plant kingdom resembles a grove (*un bosquet*): (1920:62n1).

134. Ibid. (1920:57–59).

135. Ibid. (1905:601).

136. Ibid. (1920:44). That is, they still possess a cell wall.

137. Ibid. (1920:82). To these we could add the evidence of the absent lady (1920:76–77).

138. Ibid. (1920:94).

139. Ibid. (1905:406). He recalled the imagery at (1920:95–97).

140. Taylor (1974:230); and Taylor (1976). Dillon's scenario (above) was autogenous, although nucleocentric.

141. Boveri (1904:90–91).

142. Kozo-Polyansky (1924:56–57); English translation (2010). See also Fet (2021). Khakhina (1992:80–94) presents other views in the USSR in the period 1920-1950. *Natura facit saltum*: Kozo-Polyansky (1921), reproduced in Fet (2021) at Figure 3, point 1. Johannsen (1911:158) wrote that *natura facit saltus* in the context of genetic mutations. *Saltum* is singular, *saltus* plural.

143. Lumière 1919). For this episode, see Sapp (1994:103–106).

144. Guilliermond (1919), two articles. Ivan Wallin agreed that plastids arose from mitochondria (1927:108–111). Altmann earlier held that bioblasts could change into Chlorophyllkörner: (1890:137–138) and (1894:150–151).

145. Wallin (1927). For Wallin's life and references to primary literature, see Mehos (1992).

146. Wilson (1925:738–739) specifically referred to Merezhkowsky's symbiosis-theory as an "entertaining fantasy" and as "flights of the imagination", but allowed that "it is within the range of possibility that they may some day call for more serious consideration". The third edition was reprinted in 1928, 1934, 1937, 1940, 1947, 1953, and 1959.

147. Béchamp and Enderlein claimed that bacteria are involved in the ætiology of cancer and other diseases, and can be used in therapy. Béchamp pursued such a bitter vendetta against Pasteur and his "germ theory of disease" that efforts were made to have his (Béchamp's)

628   NOTES

writings listed on the *Index librorum prohibitorum*. Portier found roles for symbionts in immunity (1918:269–272) and cancer (1918:272–281), and thought that symbionts could travel through space on comets (1918:147–149). Pyotr Kropotkin's analysis of cooperative behaviour (1902) was seen by some as linking symbiosis to anarcho-communism (his text does not mention cells or symbiosis). For Merezhkowsky's appalling life story and suicide, see Sapp *et al.* (2002:427–435).

148. Altenburg (1946), two articles; Kingdom Archetista at (1946:562). Altenburg held that these viroids cause "all" cancers. Interestingly, he supposed that plastids are *not* modified viroids, but rather a separate line of symbionts.

149. *The Journal of Biophysical and Biochemical Cytology* was rechristened *The Journal of Cell Biology* in 1962.

150. That is, two central microtubules, surrounded by nine pairs of microtubules. These include centrioles, flagellar and ciliary basal bodies, and certain specialized structures. For a review, see MJ Chapman *et al.* (2000).

151. Brody & Vatter (1959); Novikoff (1961:335–338).

152. Wilson (1925), pp. 253, 916–979 *et passim*; quotation at page 916.

153. Baur (1909). For the contribution of Carl Correns, see Hagemann (2000).

154. These, together with a further cytoplasmic *Plasmon*, comprise the idioplasm, a "specific living substance or hereditary substance" (Renner 1934:265).

155. Granick (1955:558–560).

156. Ris & Plaut (1962); see also Ris (1961), particularly page 115.

157. That is, free from the proteins characteristic of chromosomes in the nucleus.

158. Ris & Plaut (1962:388–390).

159. Nass & Nass (1963:627).

160. We have her own brief words in Khakhina (1992), page xvi; see also Sapp (1994:175–178), and Martin *et al.* (2017:4). An abstract (Sagan & Scher, 1961) reports labelling experiments showing the uptake of thymidine into the cytoplasm (*i.e.* chloroplast) of *Euglena gracilis*, consistent with DNA synthesis.

161. Sagan (1967).

162. Sagan (1967:228) and Table 1 facing page 248.

163. Sagan (1967:228–229).

164. Sagan (1967), page 231 *et passim*.

165. Sagan (1967), Table 1 facing 228. Different numbers are indicated elsewhere, *e.g.* ten in Margulis (1970:62), fifteen in ibid. (1970:63).

166. In fairness, this was not definitively settled until the 1990s. See Johnson & Rosenbaum (1991), and Hall & Luck (1995).

167. Li & Wu (2003); see also Li & Wu (2005).

168. Sagan (1967:247 and 252–254).

169. Desmids (step 2); Schizogoniaceae (step 2); Charales and higher plants (step 4); Oedogoniales (step 4); Siphonales and Ulvacea (step 5); and Volvocales (step 6): Sagan (1967), Figure 1. In Margulis (1970:62–63), a lineage of red and green algae branches immediately after fungi; charophytes branch with oömycetes, *Vaucheria* and brown algae; chytrids and diatoms are sister-groups; choanoflagellates and sponges become close relatives of green plants.

170. Margulis (1974); Whittaker & Margulis (1978); Margulis & Schwartz (1982).

171. Margulis *et al.* (2000:6955–6957); Margulis *et al.* (2005); Margulis *et al.* (2006:13081–13082). She credits the term to Janicki (1915), who introduced it in 1911 (at pp. 327–328).

172. Margulis *et al.* (2000:6957–6958); Dolan *et al.* (2002); Margulis *et al.* (2005); Margulis *et al.* (2006:13083–13084). The five occasions correspond to archamœbæ, calonymphids, chlorophyte green algae, ciliates, and foraminifera (Margulis *et al.* 2000). Given the mismatch

in taxon names, it is difficult to map these onto Figure 1 of (Sagan 1967), but they represent several of her six steps in the evolution of mitosing cells.

173. Margulis *et al.* (2000:6956); Margulis *et al.* (2006:13081).

174. Margulis (1996); Margulis *et al.* (2000); Margulis *et al.* (2006:13082–13083).

175. These include certain amœbæ (*e.g.*, *Pelomyxa*), diplomonads (*Giardia*), and trichomonads. For a time, Cavalier-Smith classified these as *Archezoa*, descendants of the first (pre-mitochondrial) eukaryotes (*Nature*, 1987b). Some were eventually found to contain much-reduced mitochondria, or putatively homologous bodies; all have nuclear genes that probably entered the eukaryotic lineage with the proto-mitochondrial symbiont (PMS). It is now widely accepted that the most-recent common ancestor of present-day eukaryotes was mitochondriate. It remains a formal possibility that the PMS entered an even earlier ancestor that was already eukaryotic; however, no evidence favours this sequence of events. See also Cavalier-Smith (1983, 2004a, 2010); and Eme *et al.* (2017).

176. Indeed, in some unspecified but substantial number of events; and that the flagellum arose yet separately. Merezhkowsky (1920:84–90) held that chromatophores and mitochondria are unrelated, and did not claim the latter as symbionts. Guilliermond (above) and Wallin (1927:107–111) held that plastids develop from mitochondria.

177. Thermal muds: Margulis *et al.* (2000:6956) and (2006:13084). Termite guts: Kirby (1941, 1994); Cleveland & Grimstone (1964). Bog creatures: Duval & Margulis (1995).

178. Margulis emphasized this herself at (2005:179–180) and (2006:13084).

179. Margulis & Schwartz call their separation of Siphonales, Charales, prasinophytes, and chlorophytes (on one hand) from conjugating green algae *somewhat arbitrary* (1982:102). The same can be said vis-à-vis the higher plants.

180. Even in 2010 (page 56) he called symbiogenetic theories "a 40-year distraction from the core problems of how a bacterium was transformed into a eukaryote".

181. Cavalier-Smith (1975:463). Rudi Raff and Henry Mahler (1972) likewise argued against the symbiotic origin of mitochondria.

182. Robertson (1959, 1960).

183. By contrast, vacuoles and endosomes (*e.g.* lysosomes and phagosomes) have a single membrane. Mitochondria are enclosed by two membranes, and their establishment from an endosymbiotic α-proteobacterium can be argued similarly (Roger *et al.* 2017).

184. Whatley *et al.* (1979).

185. That is, they are not located in the cytosol (as are plastids in red and green algae), but are instead within the lumen of the (host's) endoplasmic reticulum. The two outer membranes, collectively called the chloroplast endoplasmic reticulum (CER) by Bouck (1965), were given separate names (periplastidal rough endoplasmic reticulum, PRER; and periplastid membrane, PPM) by Cavalier-Smith (1989:386–389). In membrane nomenclature, "rough" refers to the presence of ribosomes.

186. The tubular construction makes the mastigonemes rigid, and rigid mastigonemes make a flagellum efficient in propelling the cell in a forward direction: Cavalier-Smith (1986:317–318). Therefore, tubular mastigonemes are likely to be selectively advantageous to organisms (such as these) that feed and reproduce in aqueous environments.

187. Cavalier-Smith (1986). Some chromistan lineages (oömycetes) have secondarily lost their plastids, and one (Haptophyta) has lost tubular mastigonemes. See also Cavalier-Smith (1989, 2018).

188. Greenwood *et al.* (1977).

189. Archibald & Keeling (2004); Sommer *et al.* (2006).

190. Sarai *et al.* (2019). Dinoflagellates may also engage in tertiary symbioses, which are beyond the scope of this discussion.

191. Cavalier-Smith (2004b).

630   NOTES

192. Archibald & Keeling (2004:67).

193. He sometimes grouped Eukaryota with Archaea (as "Neomura"), to the exclusion of Bacteria. See (for example) Cavalier-Smith (1987a).

194. In this context, in a primary endosymbiosis, the invading organism was a blue-green alga; in a secondary endosymbiosis, the invader was a eukaryote (a green or red alga). In a tertiary endosymbiosis (not discussed here), the invader already contains a secondarily endosymbiotic plastid.

195. This paragraph is based on Ruggiero *et al.* (2017), and looks back to earlier classifications as necessary. Quotation on animals from Cavalier-Smith (1998:208). In earlier articles he sometimes referred to Prokaryota and Eukaryota as *empires* (*e.g.* Cavalier-Smith 1998, 2004a).

196. Cavalier-Smith (2002b, 2003) versus (2010). In the latter he derives the uniciliate condition from the biciliate.

197. See note [175] above, and Chapter 23.

198. Into Fungi, Chromista, Protozoa, and Plantae respectively: Cavalier-Smith (1981).

199. Cavalier-Smith (1998:210–215). Paraphyletic taxa do not include all their descendants. A familiar example is reptiles, which are not normally construed to include birds, although birds arose from within the reptiles.

200. Cavalier-Smith (1981:477).

201. As we shall discuss in Chapter 23, this sort of multiple ancestry does not render eukaryotes polyphyletic.

202. Kirk *et al.* (2008). The photobiont might be given a Linnæan binomial, but lichen holobionts are not treated in algal or cyanobacterial systematics.

203. As (arguably) are others by its absence: protozoa are not chromists only because they show no evidence of an ancestral secondary symbiosis.

Chapter 23

1. Bather (1927), page c. Bather appreciated that "the cytoplasm exerts considerable influence on the bodily form (the phænotype)" and suggested that "the chromosomes convey all the minor points of difference ... [but] the cytoplasm conveys the established form" (1927, page c). He worried that "the whole of our System, from the great Phyla to the very unit cells, is riddled through and through with polyphyly and convergence", and held that the species stands as a last bulwark against the complete unravelling of the genetic, hence classificatory, enterprise (1927, page ci).

2. Correns (1900); de Vries (1900). For a forensic analysis of priority, see R Moore (2001).

3. W Bateson, letter to Adam Sedgwick (1854–1913), University of Cambridge MS Add. 8634:B42 G5, page 19 dated 18 April 1905; Bateson (1906). Bateson and Reginald Punnett founded the *Journal of Genetics* in 1910.

4. In German: *das Gen*. Johannsen (1909), derivation at page 125. His 1909 German text was an extension and expansion of a 1905 Danish publication that built his lectures at the University of Copenhagen in 1903 (see 1909, page iii). Peirson (2012) states that Johannsen contrasted *genotype* and *phenotype* in 1905. I have not located the Danish original.

5. Johannsen (1911). The *genotype* is "the sum total of all the 'genes' in a gamete or in a zygote", and the *phenotype* "all 'types' of organisms, distinguishable by direct inspection or only by finer methods of measuring or description" (1911:133–134). See also Peirson (2012).

6. Miescher (1871). He also characterized *protamine* as a constituent of the nucleus. For an overview of Miescher's research, see Dahm (2008).

7. Kossel (1884).

8. For details, see Morgan *et al.* (1915). For sake of argument, they supposed that a chromosome is "a chain of chemically complex substances (*e.g.*, proteins)" and that a mutation might be a "slight addition, loss or re-arrangement of the atoms in the molecules of a bead in such a chain" (1915:169). As we saw above, some heredity might be determined by factors external to the nucleus, notably in plastids.

9. Levene & Bass (1931).

10. Gulick (1941:449–450); quote at page 449.

11. Gulick (1941); DNA uniformity at page 450.

12. Fischer 1902; Hofmeister 1902 two articles; see also Fruton (1985).

13. For example, Gulick (1938:156–160).

14. Griffith (1928). Bacteria were, after all (in a sense), pleomorphic.

15. Avery *et al.* (1944). Remaining doubts that DNA is the genetic material were removed by the labelling experiments of Hershey & Chase (1952).

16. Dobzhansky (1941:47).

17. Beadle & Tatum (1941). The "one gene, one protein" concept was soon appreciated to be a simplification.

18. Watson & Crick (1953).

19. Watson (1970).

20. Fred Sanger and colleagues sequenced the A and B chains of insulin in 1951–1952. See Sanger (1988); Stretton (2002).

21. Šorm *et al.* (1957); Šorm & Keil (1963). For their (imperfect) ideas on evolution: (1963:202–205).

22. Fox & Homeyer (1955); quote at page 165.

23. Globins were later characterized and sequenced from plants (leghæmoglobins) and prokaryotes, but are paralogues of animal hæmoglobins and myoglobins.

24. Zuckerkandl & Pauling (1965). The article was written in 1963 and published in Russian translation in 1964; this is the original text. See also Zuckerkandl & Pauling (1965), notably pp. 97–102.

25. Margoliash (1963).

26. Margoliash *et al.* (1968). Three-kingdom tree (with twenty-three of these sequences) at page 274; where the identical sequence is present in multiple species (*e.g.* in man and chimpanzee), only one is used.

27. Ibid. (1968:273).

28. Not only data-dependent, but model-dependent as well. Fitch and Margoliash were pioneers in the inference of phylogenetic trees from molecular sequences. Here, the unit is minimal mutation distance. Values at inferred ancestral nodes are averaged sums of mutations in the subtended branches: Fitch & Margoliash (1967). Statistical methods for inference of phylogenetic trees have developed greatly since their work. The best treatment is that of Felsenstein (2004).

29. Fitch (1973).

30. McLaughlin & Dayhoff (1973), Figure 2 at page 105.

31. Ibid. (1973:104).

32. Sherman & Stewart (1971:257–258). Most of the early localization data were obtained in studies on *Saccharomyces cerevisiae*.

33. Dickerson, in Sigman & Brazier (1980).

34. In modern terminology, become "saturated" with change: Meyer *et al.* (1986).

35. Ambler *et al.* (1979), two articles. Their interpretation was disputed by Dickerson (*Nature*, 1980) and Woese, Gibson & Fox (1980).

36. Pechman & Woese (1972:230).

37. Claude (1943); Rheinberger (1995); Bąkowska-Żywicka & Tyczewska (2009).

632   NOTES

38. The rRNA in the large subunit (lsu) sediments at 23S (bacteria, plastids, and mitochondria) or 28S (eukaryote non-organelle); that in the small subunit (ssu) sediments at 16S (bacteria and organelles) or 18S (eukaryote non-organelle). S (Svedberg) is a unit of sedimentation velocity in a centrifugal field. Sedimentation velocity, in turn, is a function of volume, mass (hence density) and shape.

39. This section paraphrases Ragan *et al.* (2014). Articles in this issue of *RNA Biology* are dedicated to the memory of Carl Woese.

40. Alanine transfer RNA: Holley *et al.* (1965); 5S RNA: Brownlee *et al.* (1967).

41. Woese, Stackebrandt *et al.* (1985). The acronym *ssu* refers to the small ribosomal subunit (see note [38] above).

42. Woese (1987); CR Woese. Lecture at Nobel Symposium 70 (Karlskoge, Sweden) from notes of MAR, 29 August 1988. The corresponding article is Woese (1989).

43. The precise number is not recorded, but lies between thirty and perhaps fifty or sixty: see Woese & Fox (1977), and Ragan *et al.* (2014:177) and references therein.

44. Woese & Fox (1977).

45. Ibid. (1977:5089). The name had been suggested by Linda Magrum and David Nanney (1977:5090).

46. Woese, Kandler & Wheelis (1990).

47. DeLong (1998).

48. Brosius *et al.* (1978).

49. Woese (1987:231). His classic review also presents trees for purple (1987:238) and Gram-positive bacteria (1987:254). Throughout, he refers to the three major groups as *kingdoms*.

50. For example, Pace *et al.* (2012).

51. Yap *et al.* (1999) offer a rare counterexample.

52. Douglas & Turner (1991); Bhattacharya & Medlin (1995).

53. Yang *et al.* (1985); Gray (1988).

54. Pace (2009).

55. Lyons, 3 November 1977. See also *New York Times*, 6 November 1977, page E1. The *Times* continues to cover archaea and the origin of life.

56. The rank of *domain* was introduced by Woese, Kandler & Wheelis (1990) to avoid entanglement in either the kingdoms of Copeland or Whittaker, or the prokaryote-eukaryote dichotomy after Chatton (1990:4576).

57. Woese (1987), pp. 232–235 *et passim*. Quotation at page 232; *kingdom* (not *domain*) in the original. Other signatures likewise define phyla (1987:236).

58. Midpoint rooting made archaea and eukaryotes "quite specific relatives of one another" and eukaryotic rRNA a fast-evolving lineage—an outcome he found "intuitively unappealing" (Woese 1987:261). He rooted the kingdom-level trees by use of an outgroup, *e.g.* an archaeal sequence for the bacterial ssu-rRNAs.

59. He referred to ssu-rRNA as a *compound, non-linear chronometer*: CR Woese. Lecture at Nobel Symposium 70 from notes of MAR, 29 August 1988. Covariance sets, mappable onto folded structure, were key to his chronometric model. See also note [42] above.

60. Iwabe *et al.* (1989). Gogarten *et al.* (1989) reached a similar conclusion.

61. Woese, Kandler & Wheelis (1990:4578).

62. Golding & Gupta (1995).

63. JR Brown & Doolittle (1997).

64. Ibid. (1997:495).

65. It is tempting to cite the quantity-into-quality argument in Hegel, *Wissenschaft der Logik* (1812–1816). One of several English translations is that of di Giovanni (2010).

66. The US Department of Energy invited suggestions from Carl Woese (and others). See also Kyrpides *et al.* (2014).

NOTES   633

67. Gilbert (1991).
68. Fleischmann *et al.* (1995). The genome is a 1.83 million-basepair circle.
69. Beiko *et al.* (2005); Dagan *et al.* (2008); Ragan & Beiko (2009).
70. Andersson (2005); Keeling & Palmer (2009); Ku & Martin (2016); Van Etten & Bhattacharya (2020).
71. Chan *et al.* (2009), two articles.
72. Ragan & Beiko (2009:2243).
73. It was appreciated that DNA content per cell can vary greatly even if organism size or morphological complexity does not, but such comparisons were typically directed to eukaryotes: CA Thomas (1971); Cavalier-Smith (1985); Eddy (2012).
74. Horesh *et al.* (2021); includes *Shigella*.
75. The literature is huge, and the values quoted depend somewhat on features of the dataset (*e.g.*, proportions of biomedical, soil, or aquatic isolates). These numbers were drawn from Touchon *et al.* (2020) and Horesh *et al.* (2021).
76. Karberg *et al.* (2011).
77. Rouli *et al.* (2015); Park *et al.* (2019); Costa *et al.* (2020). So far as I am aware, no large-scale studies have been reported for archaea.
78. Virus types shared across domains nonetheless have restricted taxonomic distribution *within* Archaea: Prangishvili *et al.* (2017).
79. More precisely: for 453 *Oryza sativa* isolates with high-quality genome data, the pangenome consists of 23,876 gene families, while their actual genomes contain 12,770 to 14,826 gene families. Families observed in most or all of the 453 genomes are, on average, larger (have more members) than those found more occasionally: Wang *et al.* (2018).
80. Stewart (2012).
81. Pace *et al.* (1986); Pace (1997); DeLong & Pace (2001).
82. It is particularly difficult to assign rRNA genes to genomic "bins".
83. Hug *et al.* (2016), Supplementary Figure 4.
84. Mao *et al.* (2012); Anantharaman *et al.* (2016).
85. Lake *et al.* (1984). They retained Archaebacteria for *Halobacterium*, *Halococcus*, and any other former archaea that lack the ribosomal features of eocytes.
86. Cox *et al.* (2008).
87. Brochier-Armanet *et al.* (2008).
88. Schleper *et al.* (2005).
89. Spang *et al.* (2015); quote from title.
90. Eme *et al.* (2017); Zaremba-Niedzwiedzka (2017).
91. Hartman & Fedorov (2002).
92. Forterre (2015); Da Cunha *et al.* (2017).
93. Garg *et al.* (2021).
94. Even if present, they may not produce any recognizable eukaryotic phenotype: Dey *et al.* (2016).
95. BJ Baker *et al.* (2010); Imachi *et al.* (2020). It is possible to sequence the genome of an isolated single cell, although this approach brings other difficulties and potential artefacts: Rinke *et al.* (2013); and Saw *et al.* (2015).
96. The literature is immense, but Panchen (1992) is a good place to start.
97. Mayr (1981).
98. Taxa might be recognized directly from the topology (nodes) of an inferred tree, or via the distribution of character states. Regarding the latter, cladists use only shared derived (*synapomorphic*) character states in recognizing taxa.
99. For entry into the literature, see Wiley (1981:84) and Vanderlaan *et al.* (2013).

634 NOTES

100. Vanderlaan *et al.* (2013) distinguish phylogenetic (genealogical) from systematic (kinship-group) contexts; here I follow their definition for the former. In systematic contexts, a group is monophyletic if all its taxa are more closely related to each other than to any taxa not in the group. Cladists further assert that, to be monophyletic, a group must include all descendants of the common ancestor, *i.e.* not be paraphyletic. The term *holophyletic* is sometimes used to emphasize this inclusivity (Ashlock 1971).

101. Carpenter (1987); Hull (1988); Ebach *et al.* (2008).

102. Chapter 22. Barkley (1970) included a Kingdom Vira in addition to Monera, Protista, Plantae, and Animalia. Alexander & Bridges (1928:52–56) established Ultrabiontia for viruses, but as a grade or rank, not a taxon. Their "isobiontic" ranks are (in ascending order): Abiontia, Ultrabiontia, Symbiontia, Bacteriobiontia, Cytofungi, Cytophyta, Cytozoa, Metafungi, Metaphyta, Metazoa, Sociophyta, and Sociozoa (1928:56).

103. Smarda (1987).

104. Raoult & Forterre (2008); Schulz *et al.* (2017). These large viruses have accumulated genes from multiple sources: Moreira & López-García (2009). By contrast, the bacterial endosymbiont *Carsonella ruddii* encodes only 182 protein-coding genes: Nakabachi *et al.* (2006).

105. Forterre & Prangishvili (2009); Moreira & López-García (2009).

106. Trifonov & Kejnovsky (2016).

107. Croal *et al.* (2004).

108. CT Brown *et al.* (2015); see also Rinke *et al.* (2013).

109. Euryarchaeota and Crenarchaeota: Woese, Kandler & Wheelis (1990); Winker & Woese (1991).

110. Krieg & Garrity (2001).

111. Shay & Wright (2019).

112. Will & Lührmann (2011). For other processes and molecular complexes variously thought to be characteristic of eukaryotes, see Eme *et al.* (2017), especially page 715.

113. Since about 2010, concatenated protein sequences have been used, alongside or in place of ssu-rRNA sequences, for cross-domain phylogenetic inference. The proteins used typically include selected ribosomal proteins, elongation factors, and certain others. For a defence of the approach in application to the bacterial domain, see Ramulu *et al.* (2014). For a critique, see Forterre (2015). According to Ku *et al.* (2015), 1933 individual protein trees present eukaryotes as monophyletic, representing nearly 74.8% of proteins in their analysis for which the question could be posed.

114. Skejo & Franjević (2020) distinguish between their multiple roots prior to the last eukaryotic common ancestor, and their subsequent descent from that last common ancestor.

115. Chapter 22, note [175]. The archezoa hypothesis (that eukaryotes on the most-basal branches are ancestrally amitochondriate) was abandoned because genes of probable mitochondrial origin were found in their genomes, because homologous or reduced structures are present, and/or because methodological biases had artefactually placed their lineages deep in molecular-sequence trees. For a review, see Shiflett & Johnson (2010).

116. Chapter 22, note [151]. The hypothesis may not be entirely dead: Simpson *et al.* (2017).

117. Adl *et al.* (2012).

118. Burki (2014); Simpson *et al.* (2017).

119. Some of these supergroups, in turn, are sometimes joined into even-broader assemblages said to be of heuristic value, but with very thin support in primary data. For details, see the references cited in the previous note.

120. Nor is there evidence of specialized genetic machinery that might facilitate it: Ku *et al.* (2015), particularly pp. 429–430; and Ku & Martin (2016). The extent of LGT among eukaryotes remains controversial: see the exchange between Martin (2017) and Leger *et al.* (2018).

NOTES   635

121. Ku *et al.*, *Nature* (2015). From 7.37 million protein sequences from 55 eukaryotes, 1847 bacteria and 134 archaea, Ku *et al.* generated 2585 disjoint clusters, each of which contains representatives from at least five prokaryotes *and* at least two eukaryotes. In inferred trees, the eukaryote sequences were recovered as monophyletic for 1933 clusters (74.8%), while monophyly could not be rejected for 329 (12.7%) others. These are not necessarily orthogroups, but may be serviceable for large-scale protein-type comparison across and within domains.

122. That is, members of the same cluster.

123. And of course, genes in the cyanobacterial proto-plastid were shared with bacteria in other phyla through LGT and/or overlapping pan-genomes.

124. For the α-proteobacterial origin, see above and Thiergart *et al.* (2012). Bacterial lineages now extinct were probably also sources of genetic transfer: Fourner *et al.* (2009).

125. By contrast, the fungal and algal components of lichens preserve their respective cellular integrity, but (by convention) the association is considered a new species within the genus of the fungal component. See also Chapter 22.

126. Chapter 22.

127. For an alternative scenario also based on phagotrophy, see Cavalier-Smith (2002).

128. Zillig *et al.* (1989); Martin & Müller (1998).

129. Ku *et al.* (2015:430).

130. Sogin (1991); Kurland *et al.* (2006). Hartman (1984) offered a variant on this idea.

131. Reviewed by JR Brown & Doolittle (1997:462–465); and Keeling (2014).

132. Assuming eocytes remain Archaea, and the paralogue root is retained.

133. Brown & Doolittle (1997), summary at page 463; Rivera *et al.* (1998). Terminology (in italics) from Rivera *et al.* (1998).

134. Spang *et al.* (2015); Eme *et al.* (2017), especially page 715; Zaremba-Niedzwiedzka *et al.* (2017). Most show a spotty distribution, although this might indicate insufficiently deep sequencing, multiple gene losses, and/or misassembly.

135. The presence of individual genes would not mean that Asgard archaea (for example) can make cellular structures like those in eukaryotes. For an instructive case among simple animals, see Erwin (2009). Fournier & Poole (2018) also allow the possibility that the nucleocytoplasm and the Asgard metagenomes are sister groups (together *Eukaryomorpha*), but their common ancestor diverges outside the (other) archaea. It could also be argued that, were that the case, the Asgard metagenomes would draw the nucleocytoplasm into Archaea (or Arkarya).

136. Forterre (2015). Cavalier-Smith's *Neomura* is inappropriate because it implies a bacterial origin: Cavalier-Smith (1987a, 2002a). Forterre (2015:717) credits David Prangishvili for the name *Arkarya*.

137. Mayr (1998). "To sweep all this [diversity among eukaryotes] under the rug and claim that the difference between the two kinds of bacteria is of the same weight as the difference between the prokaryotes and the extraordinary world of the eukaryotes strikes me as incomprehensible" (1998:9723).

138. He may also have overestimated the species-richness of eukaryotes. According to Mayr (1998:9722), some ten thousand eubacteria have been named; there may be "thousands" of groups of archaea, and thirty-plus million species of eukaryotes. Mora *et al.* (2011) report that 1.43 million species of eukaryotes have been catalogued, and estimate that 10.95 million species exist; and that 11400 prokaryotes have been catalogued. Using ssu-rRNA sequences, Louca *et al.* (2019) estimate that between 0.8 and 1.6 million prokaryotic OTUs (operational taxonomic units) may exist. See references cited by Louca *et al.* for less-conservative estimates.

139. Mayr (1998); Woese (1998). The episode has been retold by Sapp (2009:267–281).

636   NOTES

140. Hennig (1966), pp. 139, 155, *et passim*.

141. Mayr (1998:9721) also accused Woese of ignoring autapomorphic characters—by which Mayr meant phenotype, not the (autapomorphic) oligonucleotide signatures that Woese actually used. As described above, Woese recognized Archaebacteria as a urkingdom in 1977 based on distinctiveness of oligonucleotide catalogues. He accepted a root for the rRNA tree at some point between 1987 (Woese 1987:231) and 1990 (Woese, Kandler & Wheelis 1990); archaea and eukaryotes became sister groups only upon this rooting. Based on extensive conversations I held with him in late August 1988, I greatly doubt that Woese had heard of Hennig or phylogenetic systematics in 1977; in any event, he had little time for formal systematics of any stripe. See also Sapp (2009:273–274).

142. Woese (1987), especially pp. 222–226; Woese (1994a,b).

143. Woese (1998:11046). I have corrected "Mayr's" to "Mayr", and omitted two reference citations.

144. Cavalier-Smith (1998:216) continued to maintain that "Archaebacteria are not really very different at all from *E. coli.*" He classified archaea not only within kingdom Bacteria, but alongside Gram-positive bacteria in subkingdom Unibacteria (2002a, 2004a). Ruggiero *et al.* (2015) later recognized Superkingdom Prokaryota with kingdoms Archaea and Bacteria.

145. Gupta (1998) recognized the same grouping, calling it Subdomain Monodermata.

146. At minimum, because it violates Hennig's dictum about the equivalent ranking of sister groups: see note [140] above.

147. Doolittle (2020), page R178 (where the two halves of this quote appear in reverse order to that given here). Doolittle holds that trees necessarily bifurcate, so there can be only two taxa at the highest (*i.e.* domain) level (pers. commun. to MAR, 4/2021). As we saw in Chapter 5, some Neoplatonists allowed trifurcations in trees of logic.

148. Doolittle (2020), page R179.

149. Raymann *et al.* (2015); Embley & Williams (2016); Marriott & Allers (2016); Zhou *et al.* (2018)

150. van der Gulik *et al.* (2017), particularly at third page: *a truly new type of cell [has] emerged.*

151. Hennig held that a species becomes extinct when it undergoes speciation; we might, for instance, consider that a lineage goes extinct when it irreversibly surrenders its genetic identity into a symbiosis. What happens beyond the point of extinction can be of no concern to the earlier taxon. Fusion and admixture at cellular and genomic levels have received little specific consideration in phylogenetic systematics; for an exception (although in my opinion not a tenable one), see Mindell (1992).

152. Nasir *et al.* (2016). The authors also present concerns about sample size, and about highly technical issues of phylogenetic inference.

153. O'Malley *et al.* (2010), introducing a further fourteen articles.

154. Doolittle & Bapteste (2007:2048). For multifaceted pattern pluralism as the common view, they cite Gould (1994).

155. See also Franklin-Hall (2010).

References

Abelard P. The glosses of Peter Abailard on Porphyry. In: R McKeon (ed.), *Selections from Medieval philosophers. 1. Augustine to Albert the Great.* New York: Scribner's (1929), pp. 208–258.

Academy of Natural Sciences of Drexel University. Darwin's chair. https://ansp.org/exhibits/online-exhibits/stories/darwins-chair/, undated. Accessed 5 June 2020 and 9 October 2020.

Accordi B. Ferrante Imperato (Napoli, 1550–1625) e il suo contributo alla storia della geologia. *Geologica Romana* 20:43–56 (1981).

Acosta C. *Tractado de las drogas y medicinas de las Indias Orientales.* Bvrgos: Martin de Victoria (1578).

Acosta. C. *Trattato di Christoforo Acosta . . . Della historia, natvra, et virtv delle droghe medicinali . . . .* Venetia: Francesco Ziletti (1585).

Adams G. *Micrographia illustrata, or, the knowledge of the microscope explain'd: together with an account of a new invented universal, single or double, microscope . . . to which is added a translation of Mr. Joblott's observations on the animalcula, that are found in many different sorts of infusions; and a very particular account of that surprising phænomenon, the fresh water polype, translated from the French treatise of Mr. Trembley.* London: printed for the author (1746).

Adanson M. *A voyage to Senegal, the Isle of Goree, and the River Gambia.* London: Nourse & Jonhston [*sic*] (1759).

Adanson M. *Familles des plantes.* In two volumes. Paris: Vincent (1763). Facsimile edition by Cramer, Lehre (1966).

Adanson M. *Histoire naturelle du Sénégal. [Histoire des] Coquillages.* Paris: Pauche (1757).

Adanson M. Mémoire sur un mouvement particulier découvert dans une plante appelée *Tremella. Mémoires de Mathématique et de Physique . . . de l'Académie Royale des Sciences* 1767:564–572 (1770).

Addison J. *The Spectator,* number 121 (19 July 1711). Reprinted as *The Spectator by Joseph Addison. Richard Steele & others.* (GG Smith, ed.). Eight volumes in four. Everyman's Library 164–168. London: Dent, London (undated [1907]).

Adelard of Bath. *Conversations with his nephew. On the same and the different, Questions on natural science, and On birds.* C Burnett (ed. and trans.). Cambridge UP (1998).

Adelard of Bath. *Quæstiones naturales.* In: H Gollancz (trans.), *Dodi ve-nechdi (Uncle & nephew). The work of Berachya Hanakdan . . . to which is added the first English translation from the Latin of Adelard of Bath's Quaestiones naturales.* London: Humphrey Milford; and Oxford UP (1920), pp. 85–161.

Adl SM, Simpson AGB, Lane CE, et al. The revised classification of eukaryotes. *Journal of Eukaryotic Microbiology* 59(5):429–493 (2012).

Aelian. *De natura animalium. On the characteristics of animals.* AF Scholfield (trans.). Loeb Classical Library 446, 448, and 449 (1959). Cambridge MA: Harvard UP (1958–1959).

Agardh CA. *Aphorismi botanici.* Lundæ: Berlingianis (1817–1826).

Agardh CA. *Classes plantarum . . . praeside CA Agardh . . . p.p. LP Holmberg.* Lundæ: Berlingiana (1825).

Agardh CA. De metamorphosi algarum. *Isis von Oken* 1820, Band 1(10), cols 644–654 (1820).

Agardh CA. *Dissertatio de metamorphosi algarum . . . praeside CA Agardh . . . pro gradu philosophico p.p. J Åkerman.* Lundæ: Berlingiana (1820).

Agardh CA. *Lärobok i Botanik.* In two volumes. Malmö: Thomson (1829–1830).

Agardh CA. *Lehrbuch der Botanik. Abtheilung 1: Organographie der Pflanzen.* L Meyer (trans.). Kopenhagen: Gyldendal (1831). *Abt. 2: Allgemeine Biologie der Pflanzen.* FCH Creplin (trans.). Griefswald: Koch (1832).

638 REFERENCES

Agardh CA. *Species algarum rite cognitae, cum synonymis, differentiis specificis et descriptionibus succinctis.* In two volumes (Volume 1 issued in two parts). Gryphiswaldiæ: Ernest Mauritius (1820–1828).

Agardh CA. *Synopsis algarum Scandinaviæ, adjecta dispositione universali algarum.* Lundæ: Berlingiana (1817).

Agardh CA. *Systema algarum.* Lundæ: Berlingianis (1824).

Agardh JG. Observations sur la propagation des algues. *Annales des Sciences Naturelles, 2 Série. Botanique* 6:193–212 (1836). [Pages 209–212 mistakenly numbered 109–112.]

["A Gentleman in the Country"]. An extract of some letters sent to Sir C.H. relating to some microspocal [*sic*] observations. *Philosophical Transactions* 23(284):1357–1372 (1703).

"A Gentleman in the Country". Two letters from a gentleman in the country, relating to Mr Leuwenhoeck's letter in Transaction, No. 283. *Philosophical Transactions* 23(288):1494–1501 (1703).

Agricola G. *De animantibvs subterraneis liber.* Basileæ: Froben (1549).

Agricola G. *De re metallica libri XII.* Basileæ: Froben (1556).

Agricola G. *De re metallica.* HC Hoover & LH Hoover (trans.). London: The Mining Magazine (1912). Facsimile edition by Dover, New York (1950).

Agrippa HC. *De incertitudi[n]e & vanitate scientiarum declamario inuectiue, qua vniuersia illla sophorum gigantomachia plus quam Herculea impugnatur audacia.* Coloniæ: Melchior Novesianus (1531).

Agrippa HC. *De occulta philosophia libri tres.* [Coloniæ]: [n.p.] (1533).

Agrippa HC. Liber de triplici ratione cognoscendi Deum. In: *Opervm pars posterior.* Lugduni: Beringos Fratres (1600), pp. 480–501.

Agrippa HC. *The vanity of arts and sciences.* London: JC for S Speed (1676).

Agrippa HC. *Three books of occult philosophy.* J Freake (trans.). London: RW (1651).

Agrippa HC. *Three books of occult philosophy.* J Freake (trans.). D Tyson (ed.). St Paul MN: Llewellyn (1997).

Ahmad SM. *Al-Qazwīnī,* Zakariyā ibn Muḥammad ibn Maḥmud, Abū Yaḥyā. *Complete Dictionary of Scientific Biography* 11:230–233 (2008).

Ainsworth GC. *Introduction to the history of mycology.* Cambridge UP (1976).

Ākārānga Sūtra. In: *Jaina Sutras,* Part 1 (H Jacobi, trans.). Sacred Books of the East 22. Oxford: Clarendon (1884), pp. 1–213.

Akers BP, Ruiz JF, Piper A & Ruck CAP. A prehistoric mural in Spain depicting neurotropic *Psilocybe* mushrooms? *Economic Botany* 65(2):121–128 (2011).

Alba-Lois L & Segal-Kischinevzky C. Yeast fermentation and the making of beer and wine. *Nature Education* 3(9):17 (2010).

Alberch P. The logic of monsters: evidence for internal constraint in development and evolution. *Geobios, Mémoire spécial* 12:21–57 (1989).

Albertus Magnus. *Beati Alberti Magni . . . Opera quæ hactenus habere potuerunt . . .* P Iammy (ed.). In 20 volumes. Tomus 6. *De animalibvs lib. XXVI.* Lvgdvni: Prost et al. (1651).

Albertus Magnus. *Book of minerals.* D Wyckoff (trans.). Oxford: Clarendon (1967).

Albertus Magnus. *De animalibus libri XXVI nach der Cölner Urschrift.* H Stadler (ed.). In two volumes. Beiträge zur Geschichte der Philosophie des Mittelalters. Texte und Untersuchungen 15–16. Münster: Aschendorff (1916 and 1920).

Albertus Magnus. *De vegetabilibus libri VII, historiae naturalis pars XVIII.* Editionem criticam ab Ernesto Meyero cœptam absolvit Carolus Jessen. Berolini: Reimer (1867). Reprinted by Minerva, Frankfurt am Main (1982).

Albertus Magnus. *Liber de principiis motus processivi ad fidem Coloniensis archetypi.* H Stadler (ed.). München: Straub (1909).

Albertus Magnus. *Liber mineralium.* In: A Borgnet (ed.), *Opera omnia, ex editione Lugdunensi religiose castigate,* volume 5. Paris: Vivès (1890), pp. 1–116.

Albertus Magnus. *Man and the beasts. De animalibus (books 22–26).* JJ Scanlan (trans.). Binghamton NY: Medieval & Renaissance Texts & Studies (1987).

REFERENCES   639

Albertus Magnus. *On animals. A medieval Summa zoologica*. KF Kitchell Jr & IM Resnick (trans.). In two volumes. Baltimore: Johns Hopkins UP (1999).

Albertus Magnus. *Opera omnia, ex editione Lugdunensi religiose castigate*. A Borgnet (ed.). In 38 volumes. Paris: Vivès (1890–1899).

Albertus Magnus. *Physica Libri 1–4*. In: P Hossfeld (ed.), *Sancti doctoris ecclesiæ Alberti Magni . . . Opera omnia*. Institutum Alberti Magni Coloniense, volume 4. Münster: Aschendorff (1987).

Albertus Magnus. *Questions concerning Aristotle's* On animals. IM Resnick & KF Kitchell Jr (trans.). In series: The Fathers of the Church. Mediaeval continuation, volume 9. Washington: Catholic University of America Press (2011).

Albertus Magnus. *The short natural treatises on the intellect and the intelligible*. In: R McKeon (ed.), *Selections from Medieval philosophers. 1. Augustine to Albert the Great*. New York: Scribner's (1929), pp. 326–375.

Alchemy Lab. What is alchemy? (undated). https://www.alchemylab.com/what_is_alchemy.htm, accessed 26 December 2018 and 16 September 2020.

Alcuin [Albinus BF]. Opera omnia. In: JP Migne (ed.), *Patrologiæ cursus completus, Series Latina prior* 100–101. Parisiis: Migne (1863).

Aldrovandi U. *De animalibvs insectis libri septem*. Bonon[iae]: Ioan. Bapt. Bellagamba (1602).

Aldrovandi U. *De piscibvs libri V. et de cetis lib. vnvs*. Bononiæ: Io. Baptista Bellagamba (1613).

Aldrovandi U. *De qvadrvpedib[us] digitatis viviparis libri tres et De qvadrvpedib[us] digitatis oviparis libro dvo*. Bononiæ: Nicolai Tebaldini for M. Antonij Berniæ (1645).

Aldrovandi U. *De qvadrvpedibvs solidipedibvs volvmen integrvm*. Bononiæ: Victori Bonatius (1616).

Aldrovandi U. *De reliquis animalibus exangibus libri quatuor post mortem eius editi: nempè De mollibvs, crvstaceis, testaceis, et zoophytis*. Bononiæ: Io. Baptista Bellagamba (1606). Colophon dated 1605.

Aldrovandi U. *Dendrologiæ natvralis scilicet arborvm historiæ libri dvo*. Bononiæ: Io. Baptistæ Ferronii (1668).

Aldrovandi U. *Monstrorvm historia. Cvm paralipomenis historiæ omnivm animalivm*. Bononiæ: Nicolai Tebaldini (1642).

Aldrovandi U. *Mvsaevm metallicvm in libros IIII distribvtvm*. Bononiæ: Io. Baptistæ Ferronij (1648).

Aldrovandi U. *Ornithologiae hoc est de avibvs historiae libri XII*. In three volumes. Bononiae: Franciscum de Francicsis et Io. Baptistam Bellagamba (1599–1603).

Aldrovandi U. *Qvadrvpedvm omniv[m] bisvlcorv[m] historia*. Bononiæ: Sebastiani Bonhommius (1621).

Aldrovandi U. *Serpentvm, et draconv[m] historiæ libro dvo*. Bononiæ: Clement Ferrini for M. Antonij Bernia (1640). Colophon dated 1639.

Alemanno J. *Sefer Sha'ar haHesheq*. [*The book of the gate of desire*]. Ḥayyim J Baruch ben Moses (ed.). Livorno (1790). Republished, Halberstadt: [*s.n.*] (*circa* 1860).

Alexander J & Bridges CB. Some physico-chemical aspects of life, mutation, and evolution. In: J Alexander (ed.), *Colloid chemistry. Theoretical and applied*, volume 2. *Biology and medicine*. New York: Chemical Catalog Company (1928), pp. 9–58.

Allchin B & Allchin R. *The birth of Indian civilization. India and Pakistan before 500 BC*. Harmondsworth UK: Penguin (1968).

Allen DE. *The Victorian fern craze*. London: Hutchinson (1969).

Alsted JH. *Scientiarvm omnivm encyclopaediæ*. In four volumes. Lvgdvni: Hvgvetan Ravavd (1649).

Alsted JH. *Theologia natvralis*. [n.c.]: Antonivm Hvmmivm (1615).

Altenburg E. The symbiont theory in explanation of the apparent cytoplasmic inheritance in *Paramecium*. *American Naturalist* 80(795):661–662 (1946).

Altenburg E. The "viroid" theory in relation to plasmagenes, viruses, cancer and plastids. *American Naturalist* 80(794):559–567 (1946).

Altmann R. *Die Elementarorganismen und ihre Beziehungen zu den Zellen*. Leipzig: Veit (1890).

640 REFERENCES

Altmann R. *Die Elementarorganismen und ihre Beziehungen zu den Zellen.* Second edition. Leipzig: Veit (1894).

Ambler RP, Daniel M, Hermoso J, *et al.* Cytochrome c_2 sequence variation among the recognised species of purple nonsulphur photosynthetic bacteria. *Nature* 278(5705):659–660 (1979).

Ambler RP, Meyer TE & Kamen MD. Anomalies in amino acid sequences of small cytochromes c and cytochromes c' from two species of purple photosynthetic bacteria. *Nature* 278(5705):661–662 (1979).

Ambrose. *Hexameron, Paradise, and Cain and Abel.* JJ Savage (trans.). Fathers of the Church. A new translation, volume 42. New York: Fathers of the Church (1961).

Ammonius [Hermeae]. *Ammonius in Aristotelis De interpretatione commentarius.* In: A Busse (ed.), *Commentaria in Aristotelem Graeca.* Academiae Litterarum Regiae Borussicae 4(5). Berolini: Reimer (1897).

Ammonius [Hermeae]. *Ammonius in Porphyrii Isagogen sive V voces.* In: A Busse (ed.), *Commentaria in Aristotelem Graeca.* Academiae Litterarum Regiae Borussicae 4(3). Berolini: Reimer (1891).

Ammonius [Hermeae]. *Ammonivs in qvinqve voces Porphyrii per Pomponivm Gavricvm Neapolitanvm.* Venetiis: Sessa (1504).

Ammonius [Hermeae]. *Hammonii Hermeae in qvinqve Porphyrii voces commentarivm. Ioannis Baptista Rasarivs á Vallevzia e Græco in Latinvm vertebat.* Venetiis: Scotum (1542).

Ammonius [Hermiae]. *Hypomnēma eis tas pente phōnas apophōnēs Ammōniou mikrou tou Hermeiou.* [Venezia]: Nikolaos Vlastos (1500).

Ammonius [Hermiae] and others. *Logices adminicvla his contenta, Ammonius in p[re]dictabilia. Boetij I eadem predicabilia . . . .* Parisiis: Henrici Stephani (1511).

Anantharaman K, Brown CT, Hug LA, *et al.* Thousands of microbial genomes shed light on interconnected biogeochemical processes in an aquifer system. *Nature Communications* 7:13219 (2016).

Anderson L. *Charles Bonnet and the order of the known.* Dordrecht: Reidel (1982).

Anderson L. Charles Bonnet's taxonomy and chain of being. *Journal of the History of Ideas* 37(1):45–58 (1976).

Andersson JO. Lateral gene transfer in eukaryotes. *Cellular and Molecular Life Sciences* 62(11):1182–1197 (2005).

Andreae JV. *Chymische Hochzeit: Christiani Rosencreutz. Anno 1459. Arcana publicata vilescunt; & gratiam prophanata amittunt. Ergo: ne margaritas objice porcis, seu asino substerne rosas.* Straßburg: Lazari Zetzners S. Erben (1616). *Published anonymously, but later claimed by Andreae.*

Andrewe L. *The noble lyfe & natures of man, of bestes, serpentys, fowles & fishes yt be moste knoweu* (Antwerp: Johñ of Doesborowe, *ca* 1521). Translated from *Der dieren palleys* (Antwerp: van Doesborgh, 1520), which in turn was copied from the 1517 *Hortus sanitatis*, the Tractatus de animalibus of the Mainz (1491) *Hortus sanitatis*. Facsimile: N Hudson (ed.), *An early English version of Hortus sanitatis. A recent bibliographical discovery.* London: Quaritch (1954).

Anker J. *Otto Friedrich Müller. Et bidrag til den biologiske forsknings historie i det attende aarhundrede.* In series: Acta Historica Scientiarum Naturalium et Medicinalium, volume 2. København: Munksgaard (1943).

Anonymous. *A new system of the natural history of quadrupeds, birds, fishes,—and insects.* In three volumes. Edinburgh: Hill; and London: Cadell (1791–1792).

Anonymous. *Entwurf einer nach der mutmaßlichen Stufen-Folge eingerichteten allgemeinen Naturgeschichte 1780.* In: A Thienemann. Die Stufenfolge der Dinge, der Versuch einer natürlichen Systems der Naturkörper aus dem achtzehnten Jahrhundert. *Zoologische Annalen* 3:185–230 (1910).

Anonymous. Globuline du vin, de la bière. *Dictionnaire des sciences naturelles* 61:21–23 (1845).

Anonymous. Jungius, Joachim. In: A Rees (ed.), *The cyclopædia; or, universal dictionary of arts, sciences, and literature*, volume 19 [pag. unnumb.] (1819).

Anonymous. Leclercq, Emma (1851–1933). https://www.bestor.be/wiki/index.php/Leclercq,_Emma_(1851-1933), accessed 20 August 2020 and 11 October 2020.

REFERENCES 641

Anonymous. *Libellus de natura animalium*. Monteregali: Vincentium Berruerius (1508). Facsimile edition (JI Davis, ed.) by Dawson, London (1958).

Anonymous. *Minerals and metals; their natural history and uses in the arts; with incidental accounts of mines and mining*. Published under the direction of the Committee of General Literature and Education, appointed by the Society for Promoting Christian Knowledge. London: Parker (1835).

Anonymous. Natural history. In: MM Sharif (ed.), *A history of Muslim philosophy*, Volume 2, Book 5, Chapter 66. http://www.al-islam.org/ar/node/39596, accessed 28 August 2018 and 12 October 2020.

Anonymous. Nouvelles litteraires. *Le Journal des Sçavans* 1748(10):630–640 (1748).

Anonymous. Protokoll der botanischen Sektion [of the meeting in Rheinfeld, 8–10 September 1867. Session of 10 September 1867]. *Actes de la Société Helvétique des Sciences Naturelles. Verhandlungen der Allgemeinen Schweizerischen Naturforschenden Gesellschaft* 51:88–91 (1867); report on the längeren Vortrag of Schwendener, pp. 88–90.

Anonymous. Science. British Association. *The Athenæum Journal of English and Foreign Literature, Science, and the Fine Arts* No. 1706. London (7 July 1860), pp. 18–32.

Anonymous. [Review of:] Table pour server à l'histoire anatomique & naturelle des corps vivans ou organiques, publiée le 12 Novembre 1774, à la séance publique de l'Académie Royale des Sciences; par Mr. Felix Vicq d'Azir. *Commentarii de Rebvs in Scientia Natvrali et Medicina Gestis* 22(2):263–271 (1776).

Anselm of Canterbury. *Complete philosophical and theological treatises*. J Hopkins & H Richardson (trans.). Minneapolis: Banning (2000).

Antonovics J & Hood ME. Linnaeus, smut disease and living contagion. *Archives of Natural History* 45(2):213–232 (2018).

Antonovics J & Kritzinger J. A translation of the Linnaean dissertation *The invisible world*. *British Journal for the History of Science* 49(3):353–382 (2016).

Appel TA. Henri de Blainville and the animal series: a Nineteenth-century Chain of Being. *Journal of the History of Biology* 13(2):291–319 (1980).

Appel TA. *The Cuvier–Geoffroy debate. French biology in the decades before Darwin*. Oxford UP (1987).

Appelbaum S. Introduction. In: S Appelbaum (ed. and trans.), *The sorrows of young Werther / Die Leiden des jungen Werther*. Mineola NY: Dover (2004), pp. v–x.

Apuleius. *Apulei Platonici Madaurensis De deo Socratis liber*. C Lütjohann (ed.). Griefswald: Kunike (1878).

Apuleius. *The works of Apuleius. A new translation comprising the Metamorphoses, or Golden Ass. The God of Socrates, the Florida and his Defence, or a Discourse on magic*. London: Bell (1914).

Aquinas T. *Aquinas on Creation. Writings on the "Sentences" of Peter Lombard 2.1.1*. SE Baldner & WE Carroll (trans.). Mediaeval Sources in Translation, volume 35. Toronto: Pontifical Institute of Mediaeval Studies (1997).

Aquinas T. Commentary on Aristotle's De anima. In: K Foster & S Humphries (trans.), *Aristotle's De anima in the version of William of Moerbeke and the Commentary of St. Thomas Aquinas*. New Haven: Yale University Press (1951).

Aquinas T. *Commentary on Aristotle's Physics by St. Thomas Aquinas*. RJ Blackwell, RJ Spath & WE Thirkel (trans.). New Haven: Yale UP (1963).

Aquinas T. *Corpus Thomisticum S. Thomae de Aquino Opera omnia*. https://www.corpusthomasti cum.org/opera.html, accessed 2018–2020.

Aquinas T. *Doctoris angelici divi Thomæ Aquinatis ... Opera omnia*. SE Fretté & P Maré (eds). In 34 volumes. Parisiis: Ludovicum Vivès (1872–1895).

Aquinas T. *On being and essence*. A Maurer (trans.). Second edition. Toronto: Pontifical Institute of Mediaeval Studies (1968).

Aquinas T. *Opera omnia of St. Thomas Aquinas*. In 60 volumes [anticipated]. Aquinas Institute. https://aquinasinstitute.org/operaomnia/, accessed 10 September 2020.

Aquinas T. *Quaestiones disputatae de anima. The soul*. JP Rowan (trans.). St Louis: Herder (1949).

642   REFERENCES

Aquinas T. *Quodlibetal questions 1 and 2*. S Edwards (trans.). Toronto: Pontifical Institute of Mediaeval Studies (1983).

Aquinas T. *Sancti Thomae de Aquino opera omnia: issue impensaque, Leonis XIII PM edita*. In 50 volumes. Romae: SC de Propaganda Fide [and others] (1882–).

Aquinas T. *Scripta super libris Sententiarum*. http://www.corpusthomisticum.org, accessed 5 September 2020.

Aquinas T. *Summa contra gentiles*. In four volumes. Book 1: *God*. AC Pegis (trans.); Book 2: *Creation*. JE Anderson (trans.); Book 3: *Providence*. VJ Bourke (trans.); Book 4: *Salvation*. CJ O'Neill (trans.). Garden City NY: Hanover (1955–1957). Reprinted as *On the truth of the Catholic faith*. Notre Dame IN: University of Notre Dame Press (1975–1977).

Aquinas T. *Summa theologiæ*. In 60 volumes. Cambridge: Blackfriars; and New York: McGraw-Hill (1964–1973).

Arber A. Goethe's botany. The *Metamorphosis of plants* (1790) and Tobler's *Ode to Nature* (1782). With an introduction and translations. *Chronica Botanica* 10(2):63–126 (1946).

Arber A. *Herbals. Their origin and evolution. A chapter in the history of botany 1470–1670*. Cambridge UP (1912). Revised edition (1938); reprinted by Hafner, Darien CT (1970).

Arber A. *The natural philosophy of plant form*. Cambridge UP (1950).

Archer TA. Vincent of Beauvais. *Encyclopædia Britannica*, Ninth edition 24:235–236 (1888).

Archibald JD. Edward Hitchcock's pre-Darwinian (1840) "tree of life". *Journal of the History of Biology* 42(3):561–592 (2009).

Archibald JM & Keeling PJ. The evolutionary history of plastids: a molecular phylogenetic perspective. In: RP Hirt & DS Horner (eds), *Organelles, genomes and eukaryote phylogeny. An evolutionary synthesis in the age of genomics*. Systematics Association Special Volume Series 68. Boca Raton FL: CRC Press (2004), pp. 55–74.

Ariosto L. *Orlando furioso*. First published in complete form, 1532. In English heroic verse by J Haringto[n]. London: Richard Field, for John Norton and Simon VVaterson (1591).

Aristophanes. The clouds (419 BCE). In: *The comedies of Aristophanes*, volume 1. WJ Hickie (trans.). London: Bohn (1853). http://www.perseus.tufts.edu, accessed 23 September 2020.

Aristophanes. *The comedies of Aristophanes*. WJ Hickie (trans.). Two volumes in one. London: Bohn (1853).

Aristotle. *Aristotle's De anima in the version of William of Moerbeke and the commentary of St. Thomas Aquinas*. K Foster & S Humphries (trans.). New Haven: Yale UP (1959).

Aristotle. *De mvndo. Aristotelis lib. I. Philonis lib. I* (G Budæus, trans.). *Cleomedis lib. II.* (G Valla, trans.). Basileæ: Ioan. Valdervm (1533).

Aristotle. *Habentvr hoc volvmine haec Theodoro Gaza interprete. Aristotelis de natura animalium. lib. IX. Eiusdem de partibus animalium. lib. IIII. Eiusdem de generatione animalium. lib. V. Theophrasti de historia plantarum*. Venetiis: Aldus (1503–1504).

Aristotle. *Hæc Aristotelis uolumina in hoc libro impressa continentur* [title varies]. With works of Galen, "Philo", Theophrastus, and Alexander Aphrodisias and others. Five volumes bound in six. Venetiis: Aldus (1495–1498).

Aristotle. *The complete works of Aristotle*. J Barnes (ed.). The revised Oxford translation. In two volumes. Princeton UP (1984).

Aristotle. *The works of Aristotle*. WD Ross (ed.). In twelve volumes. Volumes 4, 5, and 8: JA Smith & WD Ross (eds). Oxford UP (1908–1952).

Armstrong P. *The English parson-naturalist: a companionship between science and religion*. Leominster UK: Gracewing (2000).

Arnold F. *Lehrbuch der Physiologie des Menschen*. Teil 1 (1836) of F Arnold & JW Arnold, *Die Erscheinungen und Gesetze des lebenden menschlichen Körpers im gesunden und kranken Zustande*. In two volumes. Zürich: Orell & Füssli (1836–1842).

Arrian. *The Indica of Arrian*. W McCrindle (trans.). Reprinted from the Indian Antiquary, Vol. 5. Bombay: Education Society (1876).

Ashlock PD. Monophyly and associated terms. *Systematic Zoology* 20(1):63–69 (1971).

Ashmole E (ed.). *John Lydgate Monke of St. Edmundsbury, in his translation of the second epistle that King Alexander sent to his Master Aristotle*. In: *Theatrum chemicam britannicum*. London: Grismond (1652), pp. 397–403.

REFERENCES 643

Ashmole E (ed.). *Theatrum chemicum Britannicum. Containing severall poeticall pieces of our famous English Philosophers, who have written the Hermetique Mysteries in their owne ancient language.* London: Grismond (1652).

Ashwin-Siejkowski P. Clement of Alexandria on trial. The evidence of "heresy" from Photius' [*sic*] *Biblioteca.* In: J den Boeft *et al.* (eds), *Texts and studies of early Christian life and language* 101. Leiden: Brill (2010).

Ashworth JH. Charles Darwin as a student in Edinburgh, 1825–1827. *Proceedings of the Royal Society of Edinburgh* 55(2):97–113 (1935).

Ashworth WB Jr. Emblematic natural history of the Renaissance. In: N Jardine, JA Secord & EC Spary (eds), *Cultures of natural history.* Cambridge UP (1996), pp. 17–37.

Asl MKZ. Sirr al-khalīqa and its influence in the Arabic and Persianate world: 'Awn b. al-Mindhir's commentary and its unknown Persian translation. *Al-Qanṭara* 37(2):435–473 (2016).

Athanæus of Naucratis. *Deipnosophistiæ.* CB Gulick (ed.). Loeb Classical Library 204, 208, 224, 235, 274, 327, and 345. London: Heinemann; and Cambridge MA: Harvard UP (1927–1941).

Atkinson EG, Audesse AJ, Palacios JA, *et al.* No evidence for recent selection at *FOXP2* among diverse human populations. *Cell* 174(6):1424–1435 (2018).

Atran S. *Cognitive foundations of natural history. Towards an anthropology of science.* Cambridge UP (1990).

Atran S. Pre-theoretical aspects of Aristotelian definition and classification of animals: the case for common sense. *Studies in History and Philosophy of Science A* 16(2):113–163 (1985).

al-Attas SAT. *A guide to philosophy. The Hidāyat al-ḥikmah of Athīr al-Dīn al-Mufaḍḍal ibn 'Umar al-Abharī al-Samarqandī.* Subang Jaya (Malaysia): Pendaluk (2009).

Aubert M, Lebe R, Oktaviana AA, *et al.* Earliest hunting scene in prehistoric art. *Nature* 576(7787):442–445 (2019).

Audouin [V] & Milne-Edwards [H]. Résumé des recherches sur les animaux sans vertèbres, failtes aux îles Chausey. *Annales des Sciences Naturelles* 15:5–19 (1828).

Augier A. *Essai d'une nouvelle classification des végétaux, conforme à l'ordre que la nature paroît avoir suivi dans le règne vegetal; d'où résulte une méthode qui conduit à la connoissance des plantes & de leurs rapports naturels.* Lyon: Bruyset Ainéet (An 9, 1801).

Augier-Favas [A]. Observations sur l'ordre naturel des végétaux. *Journal Administratif, Judiciaire, Littéraire, etc. du Départment de la Drôme* 3(60):2–3 (28 Juillet 1809).

Augustine. [Works]. In: P Schaff (ed.), *A select library of the Nicene and post-Nicene Fathers of the Christian Church,* volumes 1–8. Buffalo: Christian Literature (1886–1890).

Augustine. *Confessions and Enchiridion.* AC Outler (ed. and trans.). Library of Christian Classics, volume 7. Philadelphia: Westminster (1955).

Augustine. *De Genesi ad litteram libri duodecim.* In: JP Migne (ed.), *Patrologiæ cursus completus* 34. Parisiis: Migne (1845), cols 215–486.

Augustine. *De vera religione.* In: JHS Burleigh (trans.), *Augustine. Early writings.* Library of Christian Classics. Ichthus edition. Philadelphia: Westminster (1953), pp. 218–283. Reprinted by Westminster John Knox, Louisville KY (2006).

Augustine. *De vera religione.* In: JP Migne (ed.), *Patrologiæ cursus completus* 34. Parisiis: Migne (1845), cols 121–172.

Augustine. *Saint Augustine on Genesis. Two books on Genesis against the Manichees, and On the literal interpretation of Genesis: an unfinished book.* RJ Teske (trans.). In: The Fathers of the Church. A new translation (TP Halton, ed. dir.), volume 84. Washington: Catholic University of America Press (1991).

Augustine. *The city of God.* John Healey (trans.), RVG Tasker (ed.). In two volumes. London: Dent; and New York: Dutton (1968).

Augustine. *The Confessions of St. Augustine.* FJ Sheed (trans.). New York: Sheed & Ward (1943).

Aulus Gellius. *Attic nights.* W Beloe (trans.). London: Johnson (1795).

Averroës—see Ibn Rushd

Avery OT, MacLeod CM & McCarthy M. Studies on the chemical nature of the substance inducing transformation of pneumococcal types: induction of transformation by a deoxyribonucleic acid fraction isolated from Pneumococcus Type III. *Journal of Experimental Medicine* 79(2):137–159 (1944).

644   REFERENCES

Avicenna—see Ibn Sīnā

Bacchus FJ. Paul the Hermit, Saint. *The Catholic Encyclopedia* 11:590–591 (1911).

Bach T, Breidbach O & von Engelhardt D (eds.). *Lorenz Oken. Gesammelte Werke*, Band 1. *Frühe Schriften und Naturphilosophe*. Weimar: Böhlaus (2007), pp. xi–xxvi.

Backus I. John of Damascus. *De fide orthodoxa*: translations by Burgundio (1153/54), Grosseteste (1235/40) and Lefève d'Etaples (1507). *Journal of the Warburg and Courtauld Institutes* 49:211–217 (1986).

Bacon F. *Natural history*. In: *The works of Francis Bacon*, volume 3. London: Knapton, Knaflock *et al.* (1730), pp. 1–209.

Bacon F. *Of the advancement and proficience of learning; or, The partitions of sciences, IX bookes*. G Wats (trans.). Oxford: Young & Forrest (1640).

Bacon F. *The letters and the life of Francis Bacon*. J Spedding (ed.). In seven volumes. Longman, Green *et al.* (1861–1874).

Bacon F. *Sylva sylvarvm: or, a naturall historie. In ten centuries*. London: J.H. for William Lee (1627).

Bacon R. *Opera hactenus inedita Rogeri Baconi*. Fasc. V. *Secretum secretorum cum glossis et notulis*. R Steele (ed.). Oxford: Clarendon (1920).

Baer KE von. *Account of the life and works of Dr Karl Ernst von Baer*. Second edition. Braunschweig: Viewig (1886).

Baer KE von. *Autobiography of Dr. Karl Ernst von Baer*. JM Oppenheimer (trans.). Canton MA: Science History Publications (1986).

Baer KE von. Beiträge zur Kenntniss der niedern Thiere. *Verhandlungen der Kaiserlichen Leopoldinisch-Carolinischen Akademie der Naturforscher / Nova Acta Physico-Medica Academiæ Cæsareæ Leopoldino-Carolinæ Naturæ Curiosorum* 13(2):523–762 (1827).

Baer KE von. Fragments relating to philosophical zoology. Selected from the works of KE von Baer. TH Huxley (trans.). In: A Henfrey & TH Huxley (eds), *Scientific memoirs, selected from the transactions of foreign academies of science, and from foreign journals. Natural history*. London: Taylor & Francis (1853), pp. 176–238.

Baer KE von. *Nachrichten über Leben und Schriften des Herrn Geheimraths Dr Karl Ernst von Baer, mitgetheilt von ihm selbst. Veröffentlicht bei Gelegenheit seines fünfzigjährigen Doctor-Jubiläums am 29. August 1864 von der Ritterschaft Ehstlands*. St Petersburg: Schmitzdorff (1866). Second edition, 1886.

Baer KE von. *Über Entwickelungsgeschichte der Thiere. Beobachtung und Reflexion. In two volumes*. Königsberg: Borntrager (1828 and 1837).

Baer KE von. *Über künstliche und natürliche Classification der Lebewesen und der Entwicklung. Teil I. Der Unterschied von künstlicher und natürlicher Classification bei den Naturkörpern. Teil II. Wie stellt man ein natürliches zoologisches System auf?* Unpublished manuscripts [1819], Russian Academy of Sciences (St Petersburg) as cited by S Brauckmann, *International Journal of Developmental Biology* 56(9):653–660 (2012).

Baeumker C. *Alfarabi. Über den Ursprung der Wissenschaften (De ortu scientiarum). Eine Mittelalterliche Einleitungsschrift in die philosophischen Wissenschaften*. In series: Beiträge zur Geschichte der Theologie und Philosophie des Mittelalters 19(3). Münster: Aschendorff (1916).

Bahr E, Ryan J & Jaeger CS. German literature. *Encyclopædia Britannica* (2019), www.britannica.com/art/German-literature, accessed 8 January 2020 and 3 October 2020.

Bailey C. *Epicurus. The extant remains*. Oxford: Clarendon (1926).

Bailey N. *An universal etymological English dictionary*. Third edition. London: Darby, Bettesworth *et al.* (1726).

Bairoch P. *Cities and economic development: from the dawn of history to the present*. University of Chicago (1988).

Bakar O. *Classification of knowledge in Islam. A study in Islamic philosophies of science*. Cambridge UK: Islamic Texts Society (1998).

Baker BJ, Comolli LR, Dick GJ, *et al.* Enigmatic, ultrasmall, uncultivated Archaea. *PNAS* 109(19):8806–8811 (2010).

REFERENCES 645

Baker H. *Employment for the microscope. In two parts. I. An examination of salts and saline substances . . . II. An account of various animalcules never before described.* London: Dodsley (1753).

Baker H. *Employment for the microscope.* Second edition. London: Dodsley (1764).

Baker H. *Het microscoop gemakkelyk gemaakt.* Te Amsterdam: Isaak Tirion (1744).

Baker H. *Le microscope a la portée de tout le monde.* Paris: Jombert (1754).

Baker H. *The microscope made easy: or, I. The nature, uses, and magnifying powers of the best kinds of microscopes described, calculated, and explained . . . II. An account of what surprizing discoveries have been already made by the microscope: with useful reflections on them.* London: Dodsley (1742).

Baker JR. *Abraham Trembley of Geneva. Scientist and philosopher 1710–1784.* London: Edward Arnold (1952).

Baker JR. The cell-theory: a restatement, history, and critique. Part I. *Quarterly Journal of Microscopical Science* 89(1):103–125 (1948).

Baker JR. The cell-theory: a restatement, history, and critique. Part II. *Quarterly Journal of Microscopical Science* 90(1):87–108 (1949).

Baker JR. The cell-theory: a restatement, history, and critique. Part III. The cell as a morphological unit. *Quarterly Journal of Microscopical Science* 93(2):157–190 (1952).

Baker JR. The cell-theory: a restatement, history, and critique. Part IV. The multiplication of cells. *Quarterly Journal of Microscopical Science* 94(4):407–440 (1953).

Baker JR. The cell-theory: a restatement, history, and critique. Part V. The multiplication of nuclei. *Quarterly Journal of Microscopical Science* 96(4):449–481 (1955).

Bąkowska-Żywicka K & Tyczewska A. The structure of the ribosome—short history. *Biotechnologia* 1(84):14–23 (2009).

Balbiani É-G. Recherches sur les phénomènes sexuels des infusoires. *Journal de Physiologie de l'Homme et des Animaux* 4(13):102–130, 4(14):194–220, 4(15):431–448, and 4(16):465–520 (1861).

Baldridge WS. The geological writings of Goethe. *American Scientist* 72(2):163–167 (1984).

Balfour JH. Botany. *Encyclopædia Britannica*, Eighth edition 5:65–239 (1854).

Balfour JH. Botany. *Encyclopædia Britannica*, Ninth edition 4:79–163 (1876).

Baltzly D. Stoicism. *The Stanford Encyclopedia of Philosophy* (Spring 2019 edition), EN Zalta (ed.), https://plato.stanford.edu/archives/spr2019/entries/stoicism/, accessed 4 March 2020.

Baluze E [Balvzivs S]. *Epistolarvm Innocentii III. Romani pontificis libri vndecim.* In two volumes. Paris: Muguet (1682).

Baneth DH. The common teleological source of Bahye Ibn Paqoda and Ghazzali. In: [*Sefer Magnes*]. *Magnes anniversary book. Contributions by members of the academic staff of the Hebrew University.* FI Baer *et al.* (eds). Jerusalem: Hebrew University Press (1938), pp. 23–30. English summary at pp. iv–v.

Barbançois [CH] de. Observations sur la filiation des animaux, depuis le polype jusqu'au singe. *Journal de Physique, de Chimie, d'Histoire Naturelle et des Arts* 82:444–448 (1816), with folding table bound at end of volume.

Barbaro D. *La practica della perspettiva.* Venetia: Borgominieri (1569).

Barbaro E [Barbarus H]. *Castigationes Hermolai in Plinium castigatissimæ.* Cremonæ: Carolus Darlerius (1494).

Barbaro E [Barbarus H]. *Castigationes Plinianæ.* Romæ: Eucharius Argenteus (1493).

Barber L. *The heyday of natural history. 1820–1870.* London: Cape (1980).

Barber R & Riches A. *A dictionary of fabulous beasts.* Woodbridge UK: Boydell (1971).

Bargellesi-Severi A. Due Carmelitani a Ferrara nel rinascimento Battista Panetti e Giovanni M. Verrati. Documenti inediti delle famiglie Panetti, Verrati, Severi. *Carmelus* 8(1):63–131 (1961).

Barkley FA. *Keys to the phyla of organisms, including keys to the orders of the plant kingdom.* Privately published. Missoula MT (1939). As cited by HF Copeland (1956:238).

Barkley FA. *Outline classification of organisms.* Third edition. Providence RI: Hopkins (1970).

Barkley FA. Un esbozo de clasificación de los organismos. *Revista Facultad Nacional de Agronomía Medellín* 10(34):83–103 (1949).

646 REFERENCES

Barnes J. [Review of:] Ammonius, *Commentaria in quinque voces Porphyrii*, translated by Pomponius Gauricus, and *In Aristotelis Categorias*, translated by Ioannes Baptista Rasarius, Commentaria in Aristotelem Graeca: versions latinae temporis resuscitatarum litterarum 9 (Stuttgart-Bad Canstatt: Frommann-Holzboog, 2002). *International Journal of the Classical Tradition* 10(2):307–309 (2003).

Barnes J. *The Presocratic philosophers*. Revised edition. London: Routledge & Kegan Paul (1982).

Barney SA, Lewis WJ, Beach JA & Berghof O. Introduction. In: SA Barney *et al.* (eds), *The Etymologies of Isidore of Seville*. Cambridge UP (2006), pp. 1–28.

Barnouw J. The separation of reason and faith in Bacon and Hobbes, and Leibniz's theodicy. *Journal of the History of Ideas* 42(4):607–628 (1981).

Barry M. Further observations on the unity of structure in the animal kingdom, and on congenital anomalies, including "hermaphrodites"; with some remarks on embryology, as facilitating animal nomenclature, classification, and the study of comparative anatomy. *Edinburgh New Philosophical Journal* 22:345–364 (1836–1837).

Barry M. On fissiparous generation. *Edinburgh New Philosophical Journal* 35:205–220 (1843).

Barry M. On the unity of structure in the animal kingdom. *Edinburgh New Philosophical Journal* 22:116–141 (1836–1837).

Barry M. Researches in embryology.—Second series. *Philosophical Transactions* 129:307–380 (1839).

Barsanti G. Le immagini della natura: scale, mappe, alberi 1700–1800. *Nuncius* 3(1):55–125 (1988).

Barsanti G. *Le immagini della natura: scale, mappe, alberi 1700–1800*. Firenze: Sansoni (1992).

Barthélemy-Madaule M. *Lamarck: ou le mythe de précurseur*. Paris: du Seuil (1979).

Barthélemy-Madaule M. *Lamarck the mythical precursor. A study of the relations between science and ideology*. Cambridge MA: MIT Press (1982).

Bartholomæus Anglicus. *Liber de proprietatibus rerum*. Argentoratum: Georg Husner (1505).

Bartram J. Extract of a letter from Dr. John Bartram, to Mr. Peter Collinson, FRS containing some observations concerning the salt-marsh muscle, the oyster-banks, and the fresh-water muscle, of Pensylvania [*sic*]. *Philosophical Transactions* 43(474):157–159 (1744).

Basil. *Hexaemeron*. B Jackson (trans.). In: P Schaff & H Wace (eds), *A select library of Nicene and post-Nicene Fathers of the Christian Church*. Second series, volume 8. New York: Christian Literature Company (1895), pp. 51–107.

Basil. *Letters and select works*. In: P Schaff & H Wace (eds), *A select library of Nicene and Post-Nicene Fathers of the Christian church*. Second series, volume 8. New York: Christian Literature (1895).

Basmachi F. The votive vase from Warka. *Sumer* 3:118–127 (1947). Also pp. 193–201 in Arabic.

Baster J. Dissertationem hanc de zoophytis. *Philosophical Transactions* 52:108–118 (1761).

Baster J. Observationes de Corallinis, iisque insidentibus Polypis, aliique animalculis marinis. *Philosophical Transactions* 50:258–280 (1757).

Bateson W. An address on Mendelian heredity and its application to man. *Brain* 29(2):157–179 (1906).

Bateson W. *Mendel's principles of heredity. A defence . . . With a translation of Mendel's original papers on hybridisation*. Cambridge UP (1902).

Bather FA. Biological classification: past and future. *Quarterly Journal of the Geological Society of London* 83(2): lxii–civ (1927).

Batman [S]. *Batman vppon Bartholome, his booke De proprietatubus rerum, newly corrected, enlarged and amended*. London: Thomas East (1582).

Batsch AJGC. *Tabula affinitatum regni vegetabilis, quam delineavit, et nunc ulterius adumbratam*. Vinariae: Landes-Industrie-Comptior (1802).

Batsch AJGC. *Versuch einer Anleitung, zur Kenntniss und Geschichte der Thiere und Mineralien*. In two volumes. Jena: Akademischen Buchhandlung (1788–1789).

Bauhin C. *Pinax theatri botanici*. Basileæ: Ludovici Regis (1623).

Bauhin C. Φυτοπιναξ *seu envmeratio plantarvm ab herbarijs nostro seculo descriptarum, cum earum differentijs*. Basileæ: Sebastianvm Henricpetri (1596).

Bauhin C. Πρόδρομος *theatri botanici*. Francofurti: Ioannis Treudelii (1620).

Bauhin C. *Theatri botanici*. Basileæ: Joannis Regis (1671).

Bauhin J & Cherler JH. *Historia plantarvm vniversalis, nova, et absolvtissima*. In three volumes. Ebŕodvni: [n.p.] (1650–1651).

Baum W & Winkler DW. *The Church of the East: a concise history*. London: Routledge (2003).

Bäumer Ä. Das erste zoologische Kompendium in der Zeit der Renaissance: Edward Wottons Schrift "Über die Differenzen der Tiere". *Berichte zur Wissenschaftsgeschichte* 13(1):13–29 (1990).

Baumgärtner KH. *Beiträge zur Physiologie und Anatomie*. Stuttgart: Scheible, Rieger & Sattler (1842).

Baur E (1909). Das Wesen und die Erblichkeitsverhältnisse der "Varietates albomarginatae hort." von *Pelargonium zonale*. *Zeitschrift für induktiv Abstammung- und Vererbungslehre* 1(4):330–351 (1909).

Beadle GW & Tatum EL. Genetic control of biochemical reactions in Neurospora. *PNAS* 27(11):499–506 (1941).

Beagon M. *Roman nature. The thought of Pliny the Elder*. Oxford: Clarendon (1992).

Béchamp A. *Les microzymas dans leurs rapport avec l'héterogénie, l'histogénie, la physiologie et la pathologie. Examen de la panspermie atmosphérique continue ou discontinue, morbifère ou non morbifère*. Paris: Balliere (1883).

Becher JJ. *Physica subterranea*. Editio novissima. Lipsiæ: Gleditsch (1703).

Becher JJ. *Physica subterranea*. Editio novissima. Lipsiæ: Wiedmann (1738).

Becker HJ. *Goethes Biologie. Die wissenschaftlichen und die autobiographischen Texte*. Würzburg: Könighausen & Neumann (1999).

Becker TP. Georg August Goldfuß und die Begründung der Naturwissenschaften in Bonn. In: *Chronik des Akademischen Jahres 1997/98*. Bonn: Rheinische Friedrich-Wilhelms-Universität (1999), pp. 182–189.

Beddoes T. *Contributions to physical and medical knowledge, principally from the West of England*. Briston: Biggs & Cottle (1799).

Bede. *De natura rerum*. In: JA Giles (ed.), *The complete works of Venerable Bede, in the original Latin*, volume 6. London: Whitaker (1843), pp. 99–122.

Beer J. *Beasts of love: Richard de Fournival's Bestiaire d'amour and a woman's response*. University of Toronto Press (2003).

Beiko RG, Harlow TJ & Ragan MA. Highways of gene sharing in prokaryotes. *PNAS* 102(40):14332–14337 (2005).

Bellenden J. Biographical introduction. In: H Boece. *The history and chronicles of Scotland*. J Bellenden (trans.), volume 1 (of 2). Edinburgh: Tait (1821), pp. xiii–liv.

Belon P. *De aquatilibus, libri duo: cum eiconibus ad viuam ipsorum effigiem, quoad eius fieri potuit, expressis*. Parisiis: Carolus Stephanus (1553).

Belon P. *L'histoire de la natvre des oyseavx, avec levrs descriptions, & naïfs portraicts retirez dv natvrel: escrite en sept livres*. Paris: Av Roy (1555).

Ben-David J. The rise and decline of France as a scientific centre. *Minerva* 8(2):160–179 (1970).

Benda C. Ueber die Spermatogenese der Vertebrten und höherer Evertebraten. II. Theil. Die Histiogenese der Spermien. *Archiv für Anatomie und Physiologie, Physiologische Abtheilung* 1898:393–398 (1898).

Benedikt E. Goethe und Linné. *Svenska Linné-Sällskapets Årsskrift* 28:49–54 (1945).

Bennett AW. On the affinities and classification of algae. *Journal of the Linnean Society. Botany* 24:49–61 (1888).

Bennett JA. The social history of the microscope. *Journal of Microscopy* 155(3):267–280 (1989).

Benson HH. Socrates and the beginnings of moral philosophy. In: CCW Taylor (ed.), *From the beginning to Plato*. Routledge history of philosophy (GHR Parkinson & SG Shanker, gen. eds), volume 1. London: Routledge (1997), pp. 323–355.

Benton JR. Gargoyles: animal imagery and artistic individuality in Medieval art. In: NC Flores (ed.), *Animals in the middle ages*. New York: Routledge (1996), pp. 147–165.

648 REFERENCES

Benton JR. *The medieval menagerie: animals in the art of the Middle Ages*. New York: Abbeville (1992).

Benz E. Indische Einflüsse auf die frühchristliche Theologie. *Abhandlungen der Geistes- und Sozialwissehschaftlichen Klasse* 1951(3):1–34 (1951).

Beretta M (ed.). *Lucrezio. De rerum natura: editio princeps (1472–73)*. Bologna: Bononia UP (2016).

Berger M. Preface. In: M Berger (ed. and trans.). *Hildegard of Bingen. On natural philosophy and medicine. Selections from* Cause et cure. Cambridge UK: Brewer (1999), at pp. ix–xvii.

Bergman T. Disquisitio de attractionibus electivis. *Nova Acta Regiæ Societatis Scientiarum Vpsaliensis* 2:161–250 (1775).

Bergmann [C]. Die Zerklüftung und Zellenbildung im Froschdotter. *Archiv für Anatomie, Physiologie und wissenschaftliche Medicin* 1841:89–102 (1841).

Bergmann [C]. Zur Verständigung über die Dotterzellenbildung. *Archiv für Anatomie, Physiologie und wissenschaftliche Medicin* 1842:92–101 (1842).

Bergmann T. *A dissertation on elective attractions*. London: Murray (1785).

Bergson H. *Creative evolution*. A Mitchell (trans.). New York: Holt (1911).

Bergson H. *Lévolution créatrice*. Paris: Félix Alcan (1910).

Berkeley MJ. *The English flora of Sir JE Smith, volume 5, part 2 (or Volume 2 of Hooker's British Flora). Class 24. Cryptogamia . . . Comprising the fungi*. London: Longman, Rees *et al.* (1836).

Berkenhout J. *Outlines of the natural history of Great Britain and Ireland, containing a systematic arrangement and concise description of all the animals, vegetables and fossils which have hitherto been discovered in these kingdoms*. In three volumes. London: Elmsly (1769–1772).

Bernfeld S. Freud's scientific beginnings. *American Imago* 6(3):163–196 (1949).

Berlin B, Breedlove DE & Raven PH. General principles of classification and nomenclature in folk biology. *American Anthropologist* 75(1):214–242 (1973).

Berlin B, Breedlove DE & Raven PH. *Principles of Tzeltal plant classification*. New York: Academic (1974).

Berlin I. My intellectual path. *New York Review of Books* 45(8):53–60 (14 May 1998).

Berlin I. *The roots of Romanticism*. Second edition. H Hardy (ed.). AW Mellon Lectures in the Fine Arts. National Gallery of Art, Washington DC: Bollingen Series XXXV:45. Princeton UP (2013).

Berlin I. *Vico and Herder: two studies in the history of ideas*. Princeton UP (1976).

Berryman S. Aristotle on *pneuma* and animal self-motion. *Oxford Studies in Ancient Philosophy* 23:85–97 (2002).

Beseke JMG. *Versuch einer Geschichte der Hypothesen über die Erzeugung der Thiere wie auch einer Geschichte des Ursprungs der Eintheilung der Naturkörper in drey Reiche*. Mitau: Steffenhagen (1797).

Besler B. *Fascicvlvs rariorvm et aspectv dignorvm varii generis*. [n.c.]: [n.p.] [1616 or 1622].

Besler B. *Hortvs Eystettensis*. [Nürnberg]: [n.p.] (1613).

Besler MR. *Gazophylacium rerum naturalium e regno vegetabili, animali et minerali depromptarum*. [Lipsiæ]: [Wittigau] (1642).

Besler MR. *Rariorum Musei Besleriana quae olim Basilivs et Michael Rvpertvs Besleri collegerunt*. [n.c.]: [n.p.] (1716).

Bessarion. *Bessarionis Cardinalis . . . capitula libri p[ri]m[um] aduersus calu[m]niatore[m] Plato[n]is incipiu[n]t feliciter*. N Perottus (trans.). [Romæ: Konrad Sweynheym & Arnold Pannartz] (1469).

Bhalchandra Balte B. Plant physiology in Sanskrit literature. *International Journal of Sanskrit Research* 3(5):109–111 (2017).

Bhattacharya D & Medlin L. The phylogeny of plastids: a review based on comparisons of small-subunit ribosomal RNA coding regions. *Journal of Phycology* 31(4):489–489 (1995).

Bianchi G. [Janus Plancus]. *De conchis minvs notis liber*. Venetiis: Pasqvali (1739).

Bichat X. *Anatomie générale, appliquée a la physiologie et a la médecine*. In four volumes. Paris: Brosson, Gabon (An 10, 1801).

Bichat X. *Recherches physiologiques sur la vie et la mort*. Paris: Brosson, Gabon (An 8 [1799–1800]).

Bicheno JE. On systems and methods in natural history. *Transactions of the Linnean Society of London* 15(2):479–496 (1827).

Biederbick MC. Tradition and empirical observation—nature in Giovio's and Symeoni's *Dialogo Dell'Imprese* from 1574. In: K Enenkel & PJ Smith (eds), *Emblems and the natural world*. Leiden: Brill (2017), pp. 271–318.

Bielschowsky A. *The life of Goethe*. WA Cooper (trans.). In three volumes. New York: Putnam (1905–1911).

Bignami GF. The microscope's coat of arms . . . or, the sting of the bee and the moons of Jupiter. *Nature* 405(6790):999 (2000).

Birge JK. *The Bektashi order of dervishes*. London: Luzac (1937).

al-Bīrūnī. *The book of instruction in the elements of the art of astrology. Kitab al-tafhim li awa'il sina'at al-tanjim*. RR Wright (trans.). London: Luzac (1934).

Bisson TM. *The crisis of the Twelfth century: power, lordship, and the origins of European government*. Princeton UP (2008).

Blackman FF. The primitive algae and the flagellata. An account of modern work bearing on the evolution of the algae. *Annals of Botany* 14(56):647–688 (1900).

Blainville HMD de. *De l'organisation des animaux, ou, principes d'anatomie comparée*. Paris: Levrault (1822).

Blainville HMD de. Des huîtres vertes, et des causes de cette coloration; par M. Benjamin Gaillon. *Bulletin des Sciences par la Société Philomatique de Paris* 1820:129–130 (1820).

Blainville HMD de. Infusoires. *Dictionnaire des sciences naturelles* [Second edition] 23:416–420 (1822).

Blainville HMD de. *Manuel d'actinologie, ou de zoophytologie*. Two volumes (text and plates). Paris: Levrault (1834–1836).

Blainville HMD de. Némazoaires. *Dictionnaire des sciences naturelles* 34:346–365 (1825).

Blainville HMD de. Némazoones. *Dictionnaire des sciences naturelles* 34:365–379 (1825).

Blainville H[MD] de. Prodrome d'une nouvelle distribution systématique du règne animal. *Journal de Physique, de Chimie, d'Histoire Naturelle et des Arts* 83:244–267 (1816).

Blainville H[MD] de. Prodrome d'une nouvelle distribution systématique du règne animal. *Bulletin de la Société Philomatique de Paris* 1816:105–124 (1816). *Pagination is problematic.*

Blainville HMD de. Zoophytes, *Zoophyta. Dictionnaire des sciences naturelles* 60:1–546 (1830).

Blake W. Eight illustrations for J Milton, *Comus*. Pen and water colour on wove paper (1801). Huntington Library and Collections, San Marino CA.

Blake W. Eight illustrations for J Milton, *Comus*. Water colour on paper (*ca* 1815). Museum of Fine Arts, Boston.

Blake W. *Milton: a poem in 2 books*. British Museum 1859,0625.1 (1804–1811). Text in: ERD Maclagan & AGB Russell (eds), *The prophetic books of William Blake*. London: Bullen (1907).

Blake W. *Newton*. Three prints (1795 to 1804/1805). In: J Fletcher, *Blake / An illustrated quarterly* 49(3), pag. unnumb. (2015–2016).

Blakney RB. Introduction. In: RB Blakney (trans.), *Meister Eckhart. The celebrated 14th Century mystic and scholastic*. New York: Harper & Row (1941), pp. xiii–xxviii.

Blank D. Ammonius. *The Stanford Encyclopedia of Philosophy* (Winter 2017 edition), EN Zalta (ed.), https://plato.stanford.edu/archives/win2017/entries/ammonius/, accessed 11 July 2018.

Blevins SM & Bronze MS. Robert Koch and the "golden age" of bacteriology. *International Journal of Infectious Diseases* 14:e744–e751 (2010).

Bloom J. *Paper before print: the history and impact of paper in the Islamic world*. New Haven: Yale UP (2001).

Blumenbach JF. *Beyträge zur Naturgeschichte*. In two volumes. Göttingen: Dieterich (1790–1811).

Blumenbach JF. *Beyträge zur Naturgeschichte*. Second edition. Theil 1. Göttingen: Dieterich (1806).

Blumenbach JF. *De generis humani varietate nativa*. Third edition. Gottingae: Vandenhoek & Rvprecht (1795).

650 REFERENCES

Blumenbach JF. Introductory letter to Sir Joseph Banks. In: T Bendyshe (ed. and trans.), *The anthropological treatises of Johann Friedrich Blumenbach*. London: Longman, Green *et al.* (1865), pp. 149–154.

Blumenbach JF. On the natural variety of mankind, third ed. 1795. In: T Bendyshe (ed. and trans.). *The anthropological treatises of Johann Friedrich Blumenbach*. London: Longman, Green *et al.* (1865), pp. 145–176.

Blumenbach JF. *The anthropological treatises of Johann Friedrich Blumenbach*. T Bendyshe (ed. and trans.). London: Longman, Green *et al.* (1865).

Blumenbach [JF]. Über den Bildungstrieb (Nisus formativus) und seinen Einfluß auf die Generation und Reproduction. *Göttingisches Magazin der Wissenschaften und Litteratur* 1(5):247–266 (1780).

Blumenbach JF. *Über den Bildungstrieb und das Zeugungsgeschäfte*. Göttingen: Dieterich (1781).

Blumenthal DB. Lovejoy's Great Chain of Being and the Medieval Jewish tradition. In: ML Kuntz & PG Kuntz (eds), *Jacob's Ladder and the Tree of Life. Concepts of hierarchy and the Great Chain of Being*. Revised edition. New York: Peter Lang (1987), pp. 179–190.

Blumenthal HB. Themistius: the last Peripatetic commentator on Aristotle? In: R Sorabji (ed.), *Aristotle transformed. The ancient commentators and their influence*. Ithaca NY: Cornell UP (1990), pp. 113–123.

Blund J. *Treatise on the soul*. M Dunne & RW Hunt (eds), MW Dunne (trans.). In series: Auctores Britannici Medii Aevi 2. New edition. British Academy, and Oxford UP (2012).

Bodenheimer FS. *The history of biology: an introduction*. London: Dawson (1958).

Bodin J. *Colloquium heptaplomeres de rerum sublimum arcanis abditis*. L Noack (ed.). Megaloburg: Bærensprung (1857).

Bodin J. *Colloquium of the seven about secrets of the sublime. Colloquium heptaplomeres de rerum sublimium arcanis abditis*. MLD Kuntz (trans.). Princeton UP (1975).

Bodin J. *De la demonomaine des sorciers*. Paris: du Puys (1580).

Bodin J. *De la demonomaine des sorciers*. Paris: du Puys (1587).

Bodin J. *De la démonomanie des sorciers*. V Krause, C Martin & E McPhail (eds). Travaux d'humanisme et Renaissance 559. Geneva: Droz (2016).

Bodin J. *De repvblica libri sex, Latine ab avtore redditi mvlto qvam antea locupletiores*. Parisiis: Iacobvm Dv-pvys (1586).

Bodin J. *De repvblica libri sex, Latine ab avtore redditi mvlto qvam antea locupletiores*. Francofvrti: Fischer (1609).

Bodin J. *Les six livres de la repvbliqve*. Paris: Iacques du Puys (1576).

Bodin J. *Les six livres de la repvbliqve*. Paris: Iacques du Puys (1577).

Bodin J. *Le theatre de la natvre vniverselle*. Lyon: Iean Pillenotte (1597).

Bodin J. *Method for the easy comprehension of history*. B Reynolds (trans.). New York: Columbia UP (1945). Reprinted by Norton, New York (1969).

Bodin J. *Methodvs, ad facilem historiarvm cognitionem*. Parisiis: Martinum Iuuenem (1572).

Bodin J. *The six bookes of a commonweale . . .* done into English, by Richard Knolles. London: Bishop (1606).

Bodin J. *Vniversæ natvræ theatrvm*. Lvgdvni: Iacobvm Rovssin (1596).

Boece H. *Scotorum historiæ a prima gentis origine*. [Parisiis]: Iodocus Badius Ascensius (1526/1527).

Boece H. *The history and chronicles of Scotland*. J Bellenden (trans.). In two volumes. Edinburgh: Tait (1821).

Boerhaave H. *Elementa chemiæ*. In two volumes. Parisiis: Guillelmum Cavelier (1733).

Boerhaave H. *Elementa chemiae*. Two volumes in one. Londini: SK et JK (1732).

Boëthius. *Anicii Manlii Severini Boethii Opervm Pars I: in Isagogen Porphyrii commenta*. In: S Brandt (ed.), *Corpvs scriptorvm ecclesiasticorvm Latinorvm. Editvm consilio et impensis Academiae Litterarvm Caesareae Vindobonensis*, volume 48. Vindobonae: Tempsky; et Lipsiae: Freytag (1906).

Boëthius. *Commentaria in Porphyrium a se trànslatum*. In: PL Migne (ed.), *Patrologiæ cursus completus. Series prima*, volume 64. Parisiis: Migne (1847), cols 71–158.

REFERENCES 651

Boëthius. *Philosophiae consolationis libros qvinqve.* In: G Weinberger (ed.), *Corpvs scriptorvm ecclesiasticorvm Latinorvm.* Editvm consilio et impensis Academiae Litterarvm Vindobonensis, volume 67. Vindobonae: Hoelder-Pichler-Tempsky; et Lipsiae: Akademische Verlagsgesellschaft (1934).

Boëthius. The second edition of the commentaries on the *Isagoge* of Porphyry. In: R McKeon (ed. and trans.), *Selections from medieval philosophers* 1. New York: Scribner's (1929), pp. 70–99.

Boguet H. *Discovrs execrable des sorciers.* Roven: de Beavvais (1603).

Böhme J. *De signatvra rervm: Das ist, Bezeichnung aller dingen, wie das Innere vom Eusseren bezeichnet wird . . . 1622. Durch Jacob Böhmen, sonst Teutonicum Philosophvm.* [n.c.]:[n.p.] (1635).

Böhme J. *Jakob Böhme's sämmtliche Werke.* KW Schiebler (ed.). In seven volumes. Leipzig: Barth (1831–1847).

Böhme J. *Mysterium magnum or an exposition of the first book of Moses called Genesis.* J Sparrow (trans). CJ Barker (ed.). In two volumes. London: Watkins (1924 and 1947). Reprinted 1965.

Böhme J. *Mysterivm magnvm, oder Erklärung uber das Erste Buch Mosis von der Offenbarung Göttlichen Worts. Verfasset in zwey Theil. Berichten Anno 1623.* Gedruckt den Liebhabern (1640).

Böhme J. *Signatura rerum: or the signatvre of all things . . . 1622. By Jacob Behmen, aliàs Teutonicus Phylosophus.* London: John Macock (1651).

Böhme J. *The works of Jacob Behman.* In four volumes. London: M Richardson (1764–1781).

Boissier de Sauvages de la Croix F de. *Nosologia methodica sistens morborum classes, genera et species.* In three volumes. Amstelodami: De Tournes (1763).

Bölsche W. *Haeckel. His life and work.* J McCabe (trans.). Philadelphia: Jacobs (undated [1906]).

Bolton R. Plato's distinction between being and becoming. *Review of Metaphysics* 29(1):66–95 (1975).

Bonaventure. *Doctoris seraphici S. Bonaventurae . . . Opera omnia.* In 10 volumes. Quaracchi (Clara Aqua): Collegii S. Bonaventurae (1882–1902).

Bonaventure. *Itinerarium mentis in Deum.* P Boehner (trans.). St Bonaventure NY: Franciscan Institute (1956). Reprinted 1990.

Bonaventure. *St. Bonaventure's On the reduction of the arts to theology.* Z Hayes (trans.). In: *Works of Saint Bonaventure* (FE Coughlin, series ed.), volume 1. St Bonaventure NY: Franciscan Institute (2006).

Bondeson J. *The Feejee mermaid and other essays in natural and unnatural history.* Ithaca NY: Cornell UP (1999).

Bonnet C. *Betrachtung über die Natur* von Herrn Karl Bonnet. Leipzig: Junius (1766).

Bonnet C. *Considérations sur les corps organisés, ou l'on traite de leur origine, de leur development, de leur réproduction, &c.* In two volumes. Amsterdam: Rey (1762).

Bonnet C. *Contemplazione della natura* del Signor Carlo Bonnet. In two volumes. Modena: Giovanni Montanari (1769–1770).

Bonnet C. *Essai analytique sur les facultés de l'ame.* Copenhague: Philibert (1760).

Bonnet C. *Œuvres d'histoire naturelle et de philosophie.* Eighteen volumes bound in nine: 1, *Traité d'Insectologie;* 2, *Observations diverses sur les insectes;* 3, *Mémoires d'histoire naturelle;* 4, *Recherches sur l'usage des feuilles;* 5–6, *Corps organisés;* 7–9, *Contemplation de la nature;* 10, *Ecrits d'histoire naturelle;* 11, *Ecrits et lettres d'histoire naturelle;* 12, *Lettres sur divers sujets d'histoire naturelle;* 13–14, *Essai analytique sur les facultés de l'ame;* 15–16, *La palingénésie philosophique;* 17, *Essai de psychologie;* 18, *Écrits divers.* Neuchatel: Fauche (1779–1783).

Bonnet C. *The contemplation of nature.* In two volumes. London: Longman, Becket and de Hondt (1766).

Borel P[ierre]. *Historiarvm, et observationvm medicophysicarum, Centuriæ IV.* Parisiis: Ioannem Billaine, et Mathvrini Dvpvis (1656).

Borges JL & Guerrero M. *El libro de los seres imaginarios.* Buenos Aires: Kier (1967).

Borges JL & Guerrero M. *Manual de zoologia fantastica.* México: Fondo de Cultura Económica (1957).

Borges JL & Guerrero M. *The book of imaginary beings.* NT di Giovanni (trans.). New York: Dutton (1969).

652 REFERENCES

Born I. Zwote Abhandlung über die Nutritionskraft. In: [JF] Blumenbach, [I] Born & [CF] Wolff. *Zwo Abhandlung über die Nutritionskraft*. St Petersburg: Kayserl[ichen] Akademie der Wissenschaften (1789), pp. 17–63.

Bornet Ed. M. Gustave-Adolphe Thuret. Esquisse biographique. *Annales des Sciences Naturelles, 6 Série. Botanique* 2:308–360 (1875).

Bornstein D. [Review of:] MA Manazlaoui (ed.), "Secretum secretorum": nine English versions, I: text (Early English Text Society 276), 1977. *Speculum* 54(2):402–406 (1979).

Borowski GH. *Abriß einer Naturgeschichte des Elementarreichs, zum Gebrauche seiner Vorlesungen.* Mannheim: der neuen Buchhandlung; und Berlin: Decker (1779).

Bory [de] Saint-Vincent [JB]. Dictionnaire classique d'histoire naturelle. *Revue encyclopédique* 28:46–61 and 369–378 (1825).

Bory de Saint-Vincent JB. Chaodinées [misspelled *Cahodinées*]. *Dictionnaire classique d'histoire naturelle* 3:12–15 (1823).

Bory de Saint-Vincent JB. Chaos [misspelled *Cahos*]. *Dictionnaire classique d'histoire naturelle* 3:15–16 (1823).

Bory de Saint-Vincent JB. Chaos. *Dictionnaire classique d'histoire naturelle* 3:472 (1823).

Bory de Saint-Vincent JB. Histoire naturelle. *Dictionnaire classique d'histoire naturelle* 8:244–252 (1825).

Bory de Saint-Vincent JB. Instinct. *Dictionnaire classique d'histoire naturelle* 8:585–588 (1825).

Bory de Saint-Vincent JB. Matière. *Dictionnaire classique d'histoire naturelle* 10:248–281 (1826).

Bory de Saint-Vincent JB. Microscopiques. *Dictionnaire classique d'histoire naturelle* 10:533–546 (1826). With *Tableau des ordres, des famillies et des genres de microscopiques* preceding page 533.

Bory de Saint-Vincent JB. Microscopiques. *Encyclopédie méthodique. Histoire naturelle des zoophytes, ou animaux rayonnés* 2:515–543 (1824).

Bory de Saint-Vincent JB. Navicule. *Encyclopédie méthodique. Histoire naturelle des zoophytes, ou animaux rayonnés* 2:562–565 (1824).

Bory de Saint-Vincent JB. Némazoaires. *Dictionnaire classique d'histoire naturelle* 11:501–503 (1827).

Bory de Saint-Vincent JB. Psychodiaire (Règne). *Encyclopédie méthodique. Histoire naturelle des zoophytes, ou animaux rayonnés* 2:657–663 (1824).

Bory de Saint-Vincent JB. Règne animal. Microscopiques. *Dictionnaire classique d'histoire naturelle* 17:50–110 and Plates 56–60 (1831).

Bosch H. *Garden of earthly delights*. Oil on wood (1490–1510). Museo del Prado, Madrid.

Bose C & Bose GM. *Dissertatio botanico-philosophica de motu plantarum sensus æmulo.* Lipsiæ: Breitkopf (1728).

Bouck GB. Fine structure and organelle associations in brown algae. *Journal of Cell Biology* 26(2):523–537 (1965).

Bouelles C de [Bovillus]. *Liber de intellectu. Liber de sensu. Liber de nichilo. Ars oppositorum. Liber de generatione. Liber de sapiente. Liber de duodecim numeris. Epistolę complures.* Parisiis: Henrici Stephani (1510).

Bouelles C de [Bovillus]. *Physicoru[m] eleme[n]torum Caroli Bouilli Samarobrini veromādui libri decem.* Parrhisiis: Ascensianis (1512).

Bourdon I. Animal. *Dictionnaire classique d'histoire naturelle* 1:369–381 (1822).

Bourdon I. *Considérations sur les animaux en générale*. Paris: Baudouin (1822).

Bourguet L. *Lettres philosophiques sur la formation des sels et des crystaux et sur la génération & le mécanisme organique des plantes et des animaux.* Amsterdam: Francois l'Honoré (1729).

Bourke VJ. Augustine of Hippo: the approach of the soul to God. In: ER Elder (ed.), *The spirituality of Western Christendom*. Kalamazoo MI: Cistercian (1976), pp. 1–12 and 189–191.

Bourke VJ. *Augustine's quest of wisdom. His life, thought and works*. Albany NY: Magi (1993).

Boveri T. *Ergebnisse über die Konstitution der chromatischen Substanz des Zellkerns*. Jena: Fischer (1904).

Bowen TJ. *Grammar and dictionary of the Yoruba language*. Washington: Smithsonian (1858).

REFERENCES   653

Boyle R. *The excellency of theology compared with natural philosophy* [1665]. In: T Birch (ed.), *The works of Robert Boyle. New edition*, volume 1. London: Rivington, Davis *et al.* [varies] (1772), pp. 1–72.

Boyle R. *The sceptical chymist: or chymico-physical doubts & paradoxes, touching the spagyrist's principles commonly call'd hypostatical, as they are wont to be propos'd and defended by the generality of alchymists.* London: Cadwell (1661).

Boyle R. *The works of the honourable Robert Boyle. A new edition.* In six volumes. London: (1772).

Bozic S & Marshall A. *Aboriginal myths.* Melbourne: Gold Star (1972).

Bradbury S. *The microscope past and present.* Oxford: Pergamon (1968).

Bradley R. *A philosophical account of the works of nature. Endeavouring to set forth the several gradations remarkable in the mineral, vegetable, and animal parts of the Creation. Tending to the composition of a scale of life.* London: Mears (1721).

Bradshaw Foundation. www.bradshawfoundation.com, accessed 4 September 2020.

Bradstreet A. *The poems of Mrs. Anne Bradstreet (1612–1672) together with her prose remains.* [n.c.]: Duodecimos (1897).

Brakke D. *The Gnostics: myth, ritual, and diversity in early Christianity.* Cambridge MA: Harvard UP (2010).

Brauckmann S. Karl Ernst von Baer (1792–1876) and evolution. *International Journal of Developmental Biology* 56(9):653–660 (2012).

Braudel F. *Les mémoires de la Méditerranée.* Paris: Fallois (1998).

Braudel F. *Memory and the Mediterranean.* New York: Vintage (2001).

Braun A. *Algarum unicellularium genera nova et minus cognita, praemissis observationibus de algis unicellularibus in genere.* Lipsiae: Engelmann (1855).

Braun A. *Betrachtungen über die Erscheinung der Verjüngung in der Natur, insbesondere in der Lebens- und Bildungsgeschichte der Pflanze.* Leipzig: Engelmann (1851).

Braun A. Reflections on the phenomenon of rejuvenescence in nature, especially in the life and developments of plants. A Henfrey (trans.). In: *Botanical and physiological memoirs.* London: Ray Society (1853), pp. 1–341.

Braun A. Über *Chytridium*, eine Gattung einzelliger Schmarotzergewächse auf Algen und Infusorien. *Physikalische Abhandlungen der Königlichen Akademie der Wissenschaften zu Berlin* 1855:21–83 (1856). Also published separately.

Bremekamp CEB. A re-examination of Cesalpino's classification. *Acta Botanica Neerlandica* 1(4):580–592 (1952–1953).

Breslin CAY. *Abu Hanifah al-Dinawari's Book of plants: an annotated English translation of the extant alphabetical portion.* MA thesis, University of Arizona (1986).

Briée C. *Le verdissement des huîtres: deux siècles de transformation d'un problème biologique.* PhD thesis, Université de Nantes (2010).

Brisseau-Mirbel CF. *Élémens de physiologie végétale et de botanique.* In two volumes. Paris: Magimel (1815).

Brisseau-Mirbel CF. *Traite d'anatomie et de physiologie végétales.* In two volumes. Paris: Dufart (An 10 [1801–1802]).

British Museum (Natural History). *Catalogue of the books, manuscripts, maps and drawings.* In five volumes. London: Longmans; and others (1903–1915).

British Museum. *A catalogue of the works of Linnæus. Second edition.* In two volumes. London: British Museum (1933–1936).

Brittan GG Jr. *Kant's theory of science.* Princeton UP (1978).

Broadie A. *A Samaritan philosophy. A study of the Hellenistic cultural ethos of the Memar Marqah.* In series: Studia Post-Biblica Instituta a PAH de Boer, volume 31. Leiden: Brill (1981).

Broberg G. Homo sapiens L. *Studier i Carl von Linnés naturuppfattning och människolära.* [Uppsala]: Borgströms, and Almquist & Wiksell (1975).

Broberg G. Linné's systematics and the new natural history discoveries. In: J Weinstock (ed.), *Contemporary perspectives on Linnaeus.* Lanham MD: University Press of America (1985), pp. 153–181.

654   REFERENCES

Brochier-Armanet C, Boussau B, Gribaldo S & Forterre P. Mesophilic crenarchaeota: proposal for a third archaeal phylum, the Thaumarchaeota. *Nature Reviews Microbiology* 6(3):245–252 (2008).

Brock TD (ed.). *Milestones in microbiology*. London: Prentice-Hall (1961).

Brock TD. *Robert Koch: a life in medicine and bacteriology*. Washington: American Society of Microbiology Press (1999).

Brock TD. The bacterial nucleus: a history. *Microbiological Reviews* 52(4):397–411 (1988).

Brockes BH. *Irdisches Vergnügen in Gott, bestehend in physicalisch- und moralischen Gedichten*. Nine volumes, bound in ten. Hamburg: Schiller & Kißner [varies] (1721–1748).

Brody M & Vatter AE. Observations on cellular structures of *Porphyridium cruentum*. *Journal of Biophysical and Biochemistry Cytology* 5(2):289–294 (1959).

Bromhead EF. Remarks on the arrangement of the natural botanical families. *Edinburgh New Philosophical Journal* 20(2):245–254 (1836).

Brongniart A. Cryptogamie. *Dictionnaire classique d'histoire naturelle* 5:155–159 (1824).

Brons D. A brief summary of Valentinian theology (undated). In: The Gnostic Society Library. http://gnosis.org/library/valentinius/Brief_Summary_Theology.htm, accessed 17 December 2018 and 16 September 2020.

Bronzino A. *An allegory with Venus and Cupid [Exposure of luxury]*. Oil on wood (*circa* 1545). National Gallery, London.

Brook G. *Catalogue of the Madreporarian corals in the British Museum (Natural History)*. In six volumes. *Volume 1. The genus Madrepora*. London: British Museum (Natural History) (1893).

Brosius J, Palmer ML, Kennedy PJ & Noller HF. Complete nucleotide sequence of a 16S ribosomal RNA gene from *Escherichia coli*. *PNAS* 75(10):4801–4805 (1978).

Brown CT, Hug LA, Thomas BC, *et al*. Unusual biology across a group comprising more than 15% of domain Bacteria. *Nature* 523(7559):208–211 (2015).

Brown H. *Science and the human comedy. Natural philosophy in French literature from Rabelais to Maupertuis*. University of Toronto Press (1976).

Brown JR & Doolittle WF. *Archaea* and the prokaryote-to-eukaryote transition. *Microbiology and Molecular Biology Reviews* 61(4):456–502 (1997).

Brown P. *The world of late antiquity*. AD 150–750. New York: Harcourt Brace Jovanovich (1971).

Brown R. A brief account of microscopical observations made in the months of June, July, and August 1827, on the particles contained in the pollen of plants; and on the general existence of active molecules in organic and inorganic bodies. *Edinburgh New Philosophical Journal* 5:358–371 (1828).

Brown R. A brief account of microscopical observations made in the months of June, July, and August, 1827, on the particles contained in the pollen of plants; and on the general existence of active molecules in organic and inorganic bodies. In: JJ Bennett (ed.), *The miscellaneous botanical works of Robert Brown*, volume 1 [of 3]. Ray Society. London: Robert Hardwicke (1866), pp. 463–486.

Brown R. Additional remarks on active molecules. *Edinburgh New Philosophical Journal* 8:41–46 (1830).

Brown R. On the organs and mode of fecundation in Orchideæ and Asclepiadeæ. *Transactions of the Linnean Society* 16:685–738 (read 1831, published 1833).

Brown R. *Prodromus floræ Novæ Hollandiæ et Insulæ Van-Diemen, exhibens characteres plantarum*, volume 1 [only volume published]. Londini: Taylor (1810).

Brown R. *Prodromus florae Novae-Hollandiae et Insulae Van-Diemen. Exhibens characteres plantarum*. [Second edition]. [London?]: Isidis (1821?).

Brown R. *The miscellaneous botanical works of Robert Brown*. In three volumes. London: Hardwicke, for the Ray Society (1866–1868).

Brown T. *Observations on the* Zoonomia *of Erasmus Darwin, MD*. Glasgow: Mundell; and London: Johnson & Wright (1798).

Browne J. Botany for gentlemen: Erasmus Darwin and "The loves of the plants". *Isis* 80(4):593–621 (1989).

Browne T. *Hydriotaphia. Urne-buriall, or, a discourse of the sepulchrall urnes lately found in Norfolk. Together with the Garden of Cyrus, or the quincunciall lozenge, or net-work plantations of the ancients, artificially, naturally, mystically considered.* London: Hen. Brome (1658).

Browne T. *Pseudodoxia epidemica: or, enquiries into very many received tenets, and commonly presumed truths.* Fourth edition. London: Edward Dod (1658).

Browne T. *Religio medici cum annotationibus.* Argentorati: Spoor (1652).

Browne T. *Religio medici with the "Observations" of Sir Kenelm Digby.* London: Cassell (1892).

Brownlee GG, Sanger F & Barrell BG. Nucleotide sequence of 5S-ribosomal RNA from *Escherichia coli. Nature* 215 (5102):735–736 (1967).

Brücke EW [von], Die Elementarorganismen. *Sitzungsberichte der Kaiserlichen Akademie der Wissenschaften in Wien, Mathematisch-naturwissenschaftliche Klasse* 44(2):381–406 (1861).

Bruguière [JG]. *Encyclopédie méthodique. Histoire naturelle des vers.* Tome 1. Paris: Panckoucke (1792).

Bruguière [JG] & Lamarck [JB de], continued by Deshayes GP. *Encyclopédie méthodique. Histoire naturelle des vers.* Tomes 2 and 3. Paris: Agasse (1830–1832).

Bruguière [JG] & Lamarck [JB de]. *Tableau encyclopédique et méthodique des trois règnes de la nature.* In three volumes: 1, Bruguière. *L'helminthologie, ou les vers infusoires, les vers intestins, les vers mollusques, &c.* Paris: Panckoucke (1791); 2, no author identified. *Vers testacées, a coquilles bivalves.* Paris: Agasse (An 5, 1797); 3, Lamarck. *Mollusques testacés.* Paris: Agasse (An 6 [1797–1798]).

Bruguière [JG], Lamarck [JB] de & Deshayes GP. *Encyclopédie méthodique. Histoire naturelle des vers,* volumes 2–3. Paris: Agasse (1830–1832).

Brumm A, Oktaviana AA, Burhan B, et al. Oldest cave art found in Sulawesi. *Science Advances* 7(3):eabd4648 (2021).

Brunetto Latini. *The book of the treasure (Li livres dou tresor).* P Barette & S Baldwin (trans.). Garland Library of Medieval Literature Series B, volume 90. New York: Garland (1993).

Brunfels O. [*Onomastikon*] *medicinæ.* Argentorati: Ioannem Schottum (1534).

Brunfels O. [*Onomastikon*] *seu lexicon medicinæ simplicis.* Argentorati: Ioannes Schottvs (1553).

Brunfels O. *Herbarvm vivæ eicones ad naturæ imitationem.* Argentorati: Ioannem Schottum (1532–1536).

Brunfels O. *Herbarvm vivæ eicones ad naturę imitationem.* Argentorati: Ioannem Schottum (1530).

Bruno G. *De l'infinito vniuerso et mondi.* Venetia: [n.p.] (1584).

Bruno G. *Expulsion of the triumphant beast.* A Imerti (trans.). Lincoln: University of Nebraska Press (1964).

Bruno G. *Jordani Bruni Nolani Opera Latine conscripta publicis sumptibus edita.* Three volumes bound in eight. Volume 1(1–2): F Fiorentino (ed.), Neapoli: Morano (1879–1884); volume 1(3–4): F Tocco & H Vitelli (eds), Florientiae: Le Monnier (1889); volume 2(1): V Imbriani & CM Tallarigo (eds), Neapoli: Morano (1886); volumes 2(2–3) and 3: F Tocco & H Vitelli (eds), Florentiae: Le Monnier (1889–1891).

Bruno G. *La cena de le ceneri.* In: *Opere Italiane* (1907–1908). *1. Dialoghi metafisici.* Bari: Laterza (1907), numb. sep.

Bruno G. *On the infinite universe and worlds.* In: DW Singer (trans.), *Giordano Bruno: his life and thought.* New York: Henry Schuman (1950), pp. 225–378.

Bruno G. *Opere Italiane.* In two volumes. *1. Dialoghi metafisici. 2. Dialoghi morali.* Nuovamente ristampati. Bari: Laterza & Figli (1907–1908).

Bruno G. *Spaccio de la bestia trionfante.* Nuova editione. Milano: Daelli (1863).

Buckland W. *Geology and mineralogy considered with reference to natural theology.* In two volumes. London: William Pickering (1836).

Buckland W. Memoir on the evidences of glaciers in Scotland and the north of England, first part. *Proceedings of the Geological Society of London* 3(2):332–337 (read 1840, published 1841).

Buckland W. *Reliquiæ diluvianæ; or, observations on the organic remains contained in caves, fissures, and diluvial gravel, and of other geological phenomena, attesting the action of an universal deluge.* London: John Murray (1823).

Buckley A. *Life and her children: glimpses of animal life from the amoeba to the insects.* London: Edward Stanford (1880).

656 REFERENCES

Buddæus JF. *Institutionvm philosophiæ eclecticæ*. In two volumes. 1. *Elementa philosophiæ instrvmentalis*. 2. *Elementa philosophiæ theoreticæ*. Halæ Saxonvm: Orphanotrophius (1706). I examined the fifth edition (1714–1715).

Budé G [Budæus]. *Annotationes . . . in qvatvor et viginti Pandectarvm libros*. [Parisiis]: Iododo Badio Ascensio (1508).

Budé G [Budæus]. *Annotationes . . . in XXIIII. Pandectarvm libros*. Lvgdvni: Gryphivm (1541).

Budge EAW. Introduction. In: EAW Budge (trans.), *The Egyptian Book of the dead. The Papyrus of Ani in the British Museum*. London: British Museum (1895), reprinted New York: Dover (1967), pp. ix–clv.

Budge EAW (trans.). *The Egyptian Book of the dead. The Papyrus of Ani in the British Museum*. London: British Museum (1895), reprinted New York: Dover (1967).

Budge EAW. *The life and exploits of Alexander the Great*. London: Clay (1896).

Buffon [GLL]. *Histoire naturelle, générelle et particuliére, avec la description du Cabinet du Roi*. In 18 volumes. Paris: l'Imprimerie Royale (1749–1775).

Buffon [GLL]. Initial discourse. J Lyon (trans.). In: J Lyon & PR Sloan (eds), *From natural history to the history of nature: readings from Buffon and his critics*. Notre Dame IN: University of Notre Dame Press (1981), pp. 97–128.

Buffon [GLL]. *Natural history, general and particular*. W Smellie (trans.). Third edition. In nine volumes. London: Strahan & Cadell (1791).

Buffon [GLL]. *Œuvres complètes de Buffon. Nouvelle édition annotée et précédée d'une introduction sur Buffon et sur les progrès des sciences naturelles depuis son époque par J-L de Lanessan . . . suivie de la correspondance générale de Buffon*. In 14 volumes. Paris: Abel Pilon (1884–1886).

Buffon [GLL]. Premier discours. In: *Histoire naturelle, générale et particulière*. Nouvelle édition, Tome premier. Paris: Imprimerie Royale (1769), pp. 1–90.

Buffon [GLL]. The generation of animals [selections]. JS Barr (trans.). In: J Lyon & PR Sloan (eds), *From natural history to the history of nature: readings from Buffon and his critics*. Notre Dame IN: University of Notre Dame Press (1981), pp. 170–209.

Buffon GLL. *Natural history, general and particular*. W Smellie (trans.). In nine volumes. Edinburgh: William Creech (1780).

Buffon GLL. *The natural history of birds*. W Smellie (trans.). In nine volumes. London: Strachan, Cadell & Murray (1793).

Bullough VL (ed.). *The scientific revolution*. New York: Holt, Rinehart & Winston (1970).

Buonanni F. *Obseruationes circa viuentia, quæ in rebus non viuentibus reperiuntur: cum micrographia curiosa siue rerum minutissimarum obseruationibus, quæ ope microscopij recognitæ ad viuum exprimuntur*. Romæ: Antonij Herculis (1691).

Burdach KF. *Propädeutik zum Studiem der gesammten Heilkunst. Ein Leitfaden akademischer Verlesungen*. Leipzig: Breitkopf und Härtel (1800).

Burdach KF. *Ueber die Ausgabe der Morphologie*. Leipzig: Dok'sche (1817).

Buridan J. *Iohannis Buridani. Quaestiones super Libris quattuor De caelo et mundo*. EA Moody (ed.). Mediaeval Academy of America Publication 40. Cambridge MA: Mediaeval Academy of America (1942).

Buridan J. *Summulae de dialectica. An annotated translation, with a philosophical introduction by G Kluma*. New Haven: Yale UP (2001).

Burkert W. *Greek religion*. Harvard UP (1985).

Burkert W. *Griechische Religion der archaischen und klassischen Epoche*. Stuttgart: Kohhammer (1977).

Burkert W. *Lore and science in ancient Pythagoreanism*. EL Minar Jr (trans.). Cambridge MA: Harvard UP (1972).

Burkhardt RW Jr. Lamarck, evolution, and the politics of science. *Journal of the History of Biology* 3(2):275–298 (1970).

Burkhardt RW Jr. *The spirit of system. Lamarck and evolutionary biology*. Cambridge MA: Harvard UP (1977). Reprinted 1995.

Burkhardt RW Jr. The Zoological Philosophy of J.B. Lamarck. In: JB [de] Lamarck, *Zoological philosophy*. H Elliot (trans.). University of Chicago Press (1984), pp. xv–xxxix.

REFERENCES 657

Burki F. The eukaryotic tree of life from a global phylogenomic perspective. *Cold Spring Harbor Perspectives in Biology* 6(5):a016147 (2014).

Burkitt FC. The Decretum Gelasianum. Review of: Das Decretum elasianum de libris recipiendis et non recipiendis in kritischem Text herausgegeben und untersucht von Ernst von Dobschütz. *Journal of Theological Studies* 14(55):469–471 (1913).

Burnett C. Apollonius of Tyana. In: H Lagerlund (ed.), *Encyclopedia of Medieval Philosophy. Philosophy between 500 and 1500*, volumes 1–2. Dordrecht: Springer (2011), pp. 82–83.

Burton R [as Democritvs Iunior]. *The anatomy of melancholy, what it is. Vvith all the kindes, cavses, symptomes, prognostickes, and severall cvres of it*. Oxford: Lichfield and Short (1621).

Burton R [as Democritvs Iunior]. *The anatomy of melancholy. What it is, with all the kinds causes, symptomes, prognostickes, and several cures of it*. Oxford: Henry Cripps (1628).

Burton R [as Democritvs Iunior]. *The anatomy of melancholy*. Oxford: Henry Cripps (1638). Anatomy spelled *Anatomie* on the half-title. Title page missing absent from the two copies I examined.

Burton R [as Democritvs Iunior]. *The anatomy of melancholy. What it is, with all the kinds causes, symptomes, prognostickes, and several cures of it*. Oxford: Henry Cripps (1651). *Anatomy* spelled *Anatomie* on the half-title.

Bury RG. Introduction. In: Sextus Empiricus. *Outlines of Pyrrhonism*. RG Bury (trans.). Sextus Empiricus 1. Loeb Classical Library 273. Cambridge MA: Harvard UP (1933, reprinted 1993), pp. vii–xlv.

Butscher HB. Oken (or Okenfuss), Lorenz. *Complete Dictionary of Scientific Biography* 23:331–335 (2008).

Bütschli O. Bemerkungen über Cyanophyceen und Bacteriaceen. *Archiv für Protistenkunde* 1(1):41–58 (1902).

Bütschli O. Protozoa. In: *Dr HG Bronn's Klassen und Ordnungen der Thier-Reichs*, Band 1. Abtheilung 1, *Sarkodina und Sporozoa* (1880–1882); Abtheilung 2, *Mastigophora* (1883–1887); and Abtheilung 3, *Infusoria und System der Radiolaria* (1887–1889). Leipzig & Heidelberg: Winter.

Bütschli O. *Studien über die ersten Entwicklungsvorgänge der Eizelle, die Zelltheilung und die Conjugation der Infusorien*. Frankfurt: Winter (1876).

Bütschli O. Studien über die ersten Entwicklungsvorgänge der Eizelle, die Zelltheilung und die Conjugation der Infusorien. *Abhandlungen der Senckenburgischen Naturforschenden Gesellschaft* 10(3–4):213–457 (1876).

Butterfield D. *The early textual history of Lucretius'* [sic] De rerum natura. Cambridge UP (2013).

Butterworth GW. Introduction. In: Origen. *On first principles. Being Koetschau's text of the De principiis translated into English*. GW Butterworth (ed.). Gloucester MA: Peter Smith (1973), pp. xxiii–lviii.

Buytaert EM. Introduction. In: EM Buytaert (ed.), *Saint John Damascene. De fide orthodoxa. Versions of Burgundio and Cerbanus*. Bonaventure NY: Franciscan Institute (1955), pp. vi–liv.

Cahn T. *La vie et l'œuvre d'Étienne Geoffroy Saint-Hilaire*. Paris: Presses Universitaires de France (1962).

Cajetan T—see Tommaso de Vio

Calepino A [Calepinus]. *Dictionarivm* [*latinvm*]. Rhegius: Dionysius Berthochus (1502).

Calepino A [Calepinus]. *Dictionarivm lingvarvm septem*. Basileæ: Henric. Petrina (1579).

Calepino A [Calepinus]. *Dictionarivm septem lingvarvm*. Venetiis: Georgium Valentium (1625).

Calepino A [Calepinus]. *Dictionarivm, in quo restituendo atque exornando hæc præstitimus*. Venetiis: Aldus (1548).

Calepino A [Calepinus]. *Dictionarivm, in qvo restivendo atqve exornando hæc præstitimus. Additamenta Pauli Manutii*. Venetiis: Manutianis (1573).

Calepino A [Calepinus]. *Dictionarivm. Post omnes alias œditiones à mvltis vtrivsqve linguae*. Venetiis: Gryphivs (1550).

Calepino A [Calepinus]. *Dictionu*[m] *latinarum e greco*. Argentiniæ: Joa[n]nis Grüninger (1513).

Caley ER & Richards JFC. *Theophrastus on stones. Introduction, Greek text, English translation, and commentary*. Columbus: Ohio State University (1956).

658 REFERENCES

Calkins GN. *Protozoölogy.* New York & Philadelphia: Lea & Febiger (1909).

Camerarius RJ. *Epistola ad D. Mich. Bern. Valentini de sexu plantarum.* Tuebingæ: [n.p.] (1694).

Candolle, Alphonse de. On the life and writings of JPE Vaucher. *Annals and Magazine of Natural History* 10(64):161–168, and 10(65):241–248 (1842).

Candolle, Alphonse de. Troisième mémoire sur la famille des Myrsinéacées. *Annales des Sciences Naturelles, 2 Série. Botanique* 16:129–176 (1841).

Candolle AP de. *Collection de mémoires pour servir a l'histoire du régne végétal.* Ten mémoires bound in nine volumes. Paris: Treuttel et Würtz (1828–1838).

Candolle AP de. *Regni vegetabilis systema naturale, sive ordines, genera et species plantarum secundum methodi naturalis normas digestarum et descriptarum.* In two volumes. Paris: Treuttel et Würtz (1818–1821).

Candolle AP de. Revue de la famille des cactées. *Mémoires de Muséum d'histoire naturelle* 17:1–119 (1828). Also published separately as: *Revue de la famille des cactées avec des observations sur leur végétation et leur culture.* Paris: Belin (1829).

Candolle AP de. *Théorie élémentaire de la botanique, ou exposition des principes de la classification naturelle et de l'art de décrire et d'etudier les végétaux.* Paris: Déterville (1815).

Candolle AP de. *Théorie élémentaire de la botanique, ou exposition des principes de la classification naturelle et de l'art de décrire et d'etudier les végétaux.* Second edition. Paris: Déterville (1819).

[Canning G & Frere JH]. Loves of the triangles. A mathematical and philosophical poem. Anti-Jacobin (16 and 23 April, and 7 May 1798). Available in: W Gifford (ed.), *Poetry of the Anti-Jacobin.* Fifth edition. London: Miller & Hatchard (1807), pp. 108–129.

Cardano G. *Les livres de Hierome Cardanvs* .... R le Blanc (trans.). Paris: Charles l'Angelier (1556).

Cardano H. *De rervm varietate libri XVII.* Basileæ: Henrichvm Petri (1557).

Cardano H. *De svbtilitate libri XXI.* Norìmbergæ: Ioh. Petreius (1550).

Carli-Rubbi G. [Dedication to Maupertuis]. In: V Donati, *Della storia naturale marina dell'Adriatico.* Venezia: Francesco Storti (1750), pag. unnumb.

Carpenter JM. Cladistics of cladists. *Cladistics* 3(4):363–375 (1987).

Carpenter WB. *The microscope and its revelations.* London: Churchill (1848).

Carpenter WB. *The microscope: and its revelations.* [First American, based on the Third London edition (1857)]. Philadelphia: Blanchard & Lea (1856).

Carpenter WB. *The microscope and its revelations.* Sixth edition. In two volumes. New York: William Wood (1883).

Carpenter WB & Dallinger WH. *The microscope and its revelations.* Eighth edition. London: Churchill; and Philadelphia: Blakiston (1901).

Carr DC. *The life of Linnæus, the celebrated Swedish naturalist. To which is added, a short account of the botanical systems of Linnæus and Jussieu, with a slight glance at the discoveries of Goëthe.* Holt: James Shalders (1837).

Carter HJ. Notes on the species, structure, and animality of the freshwater sponges in the tanks of Bombay. (Genus *Spongilla*). *Annals and Magazine of Natural History, 2 Series* 1(4):303–311 (1848).

Carter HJ. Notes on the species, structure, and animality of the freshwater sponges in the tanks of Bombay. (Genus; *Spongilla*). *Transactions of the Medical and Physical Society of Bombay* 8:101–107 (1847).

Carus CG. *An introduction to the comparative anatomy of animals; compiled with constant reference to physiology.* RT Gore (trans.). In two volumes of text, plus one volume of plates [n.p.]. London: Longman, Rees *et al.* (1827).

Carus CG. Beitrag zur Geschichte der unter Wasser an verwesenden Thierkörpern sich erzeugenden Schimmel- oder Algengattungen. *Verhandlungen der Kaiserlichen Leopoldinisch-Carolinischen Akademie der Naturforscher / Nova Acta Physico-Medica Academiæ Cæsareæ Leopoldino-Carolinæ Naturæ Curiosorum* 11(1):491–522 (1823).

Carus CG. *Grundzüge einer neuen und wissenschaftlich begründeten Cranioscopie (Schädellehre).* Stuttgart: Balz (1841).

Carus CG. *Lehrbuch der vergleichenden Zootomie. Mit stäter Hinsicht auf Physiologie ausgearbeitet.* Leipzig: Fleischer (1818).

REFERENCES 659

Carus CG. *Lehrbuch der vergleichenden Zootomie. Mit stäter Hinsicht auf Physiologie ausgearbeitet.* Second edition. In two volumes. Leipzig: Fleischer; and Wein: Gerold (1834).

Carus CG. *Natur und Idee oder das Werdende und sein Gesetz. Eine philosophische Grundlage für die specielle Naturwissenschaft.* Wien: Braumüller (1861).

Carus CG. *Organon der Erkenntniss der Natur und des Geistes.* Leipzig: Brockhaus (1856).

Carus CG. *Physis. Zur Geschichte des lieblichen Lebens.* Stuttgart: Scheitlin (1851).

Carus CG. *Psyche. Zur Entwicklungsgeschichte der Seele.* Pforzheim: Flammer & Hoffmann (1846).

Carus CG. *Symbolik der menschlichen Gestalt. Ein Handbuch zur Menschenkenntniss.* Second edition. Leipzig: Brockhaus (1858).

Carus CG. *Ueber Lebensmagnetismus und über die magischen Wirkungen überhaupt.* Leiptzig: Brockhaus (1857).

Carus CG. *Vergleichende Psychologie oder Geschichte der Seele in der Reihenfolge der Thierwelt.* Wien: Braumüller (1866).

Carus CG. *Von den Ur-theilen des Knochen- und Schalengerüstes.* Leipzig: Fleischer (1828).

Carus CG. *Zwölf Briefe über das Erdleben.* Stuttgart: Balz (1841).

Carus JV. *Geschichte der Zoologie bis auf Joh. Müller und Charl. Darwin.* München: Oldenbourg (1872).

Carus JV. *System der thierischen Morphologie.* Leipzig: Wilhelm Engelmann (1853).

Casaubon I. *De rebvs sacris & ecclesiasticis, exercitationes XVI.* Londini: Billium (1614).

Caspary R. Bericht über die Verhandlungen der botanischen Sektion der 33 Versammlung deutscher Naturforscher und Aerzte, gehalten in Bonn vom 18 bis 24 September 1857. *Botanische Zeitung* 15: cols 749–776 and 784–792 (1857).

Cassini A[H] de. Inulées. *Dictionnaire des sciences naturelles* 23:559–582 (1822).

Cassini A[H] de. Tableau exprimant les affinités des tribus naturelles de la famille des Synanthérées. *Dictionnaire des sciences naturelles. Planches, 2.e partie: Règne organisè. Botanique, végétaux dicotylédons* [volume 3 of plates], plate 83. Paris: Levrault (1816–1829).

Cassiodorus MA. De anima. In: *Magni Avrelii Cassiodori Senatoris VC. Variarum libri XII. De anima liber I* . . . . Parisiis: Sebastianum Niuellium (1589), pp. 284–306.

Cassiodorus MA. *Magni Avrelii Cassiodori Senatoris VC. Variarum libri XII. De anima liber I. De institutione diuinarum scripturarum libri II. . . . Iordani Episcopi Ravennatis.* . . . Parisiis: Sebastianum Niuellium (1589).

Cassiodorus MA. Opera omnia. In: JP Migne (ed.), *Patrologiæ cursus completus, Series Latina* 69 (cols 421–1334) and 70. Parisiis: Migne (1865).

Cassirer E. *Die Philosophie der Aufklärung.* Tübingen: Mohr (1932).

Cassirer E. *The philosophy of the Enlightenment.* FCA Koelln & JP Pettegrove (trans.). Princeton UP (1951).

Cato the Elder. *De agri cultura* (or in early printed editions, *De re rustica*). In: WD Hooper & HB Ash (trans.) *Cato and Varro on agriculture.* Loeb Classical Library 283. London: Heinemann (1934).

Cavalier-Smith T. The origin of nuclei and of eukaryotic cells. *Nature* 256(5517):463–468 (1975).

Cavalier-Smith T. Eukaryote kingdoms: seven or nine? *BioSystems* 14(3–4):461–481 (1981).

Cavalier-Smith T. A 6-kingdom classification and a unified phylogeny. In: HEA Schenk & WS Schwemmler (eds), *Endocytobiology 2: intracellular space as an oligogenetic ecosystem.* Berlin: de Gruyter (1983), pp. 1027–1034.

Cavalier-Smith T (ed.). *The evolution of genome size.* Chichester UK: Wiley (1985).

Cavalier-Smith T. The kingdom Chromista: origin and systematics. *Progress in Phycological Research* 4:309–347 (1986).

Cavalier-Smith T. The origin of eukaryotic and archaebacterial cells. *Annals of the New York Academy of Sciences* 503(1):17–54 (1987a).

Cavalier-Smith T. Eukaryotes with no mitochondria. *Nature* 326(6111):332–333 (1987b).

Cavalier-Smith T. The kingdom Chromista. In: JC Green, BSC Leadbeater & WL Diver (eds), *The chromophyte algae: problems and perspectives.* Systematics Association Special Volume 38. Oxford: Clarendon (1989), pp. 381–407.

660 REFERENCES

Cavalier-Smith T. Kingdom protozoa and its 18 phyla. *Microbiological Reviews* 57(4):953–994 (1993).

Cavalier-Smith T. A revised six-kingdom system of life. *Biological Reviews* 73(3):203–266 (1998).

Cavalier-Smith T. The neomuran origin of archaeabacteria, the negibacterial root of the universal tree and bacterial megaclassification. *International Journal of Systematic and Evolutionary Microbiology* 52(1):7–76 (2002a).

Cavalier-Smith T. The phagotrophic origin of eukaryotes and phylogenetic classification of Protozoa. *International Journal of Systematic and Evolutionary Microbiology* 52(2):297–354 (2002b).

Cavalier-Smith T. Protist phylogeny and the high-level classification of Protozoa. *European Journal of Protistology* 39(4):338–349 (2003).

Cavalier-Smith T. Only six kingdoms of life. *Proceedings of the Royal Society B: Biological Sciences* 271(1545):1251–1262 (2004a).

Cavalier-Smith T. The membranome and membrane heredity in development and evolution. In: RP Hirt & DS Horner (eds), *Organelles, genomes and eukaryote phylogeny. An evolutionary synthesis in the age of genomics.* Systematics Association Special Volume Series 68. Boca Raton FL: CRC Press (2004b), pp. 335–351.

Cavalier-Smith T. Origin of the cell nucleus, mitosis and sex: roles of intracellular coevolution. *Biology Direct* 5:7 (2010).

Cavalier-Smith T. Kingdom Chromista and its eight phyla: a new synthesis emphasising periplastid protein targeting, cytoskeletal and periplastid evolution, and ancient divergences. *Protoplasma* 255(1):297–357 (2018).

Celenza CX. Marsilio Ficino. *The Stanford Encyclopedia of Philosophy* (Fall 2017 edition), EN Zalta (ed.), plato.stanford.edu/archives/fall2017/entries/ficino/, accessed 5 January 2019.

Celsus. *A. Cornelii Celsi De medicina libri octo.* C Daremberg (ed.). Lipsiae: Teubneri (1891).

Celsus. *A. Cornelii Celsi quæ supersunt.* F. Marx (ed.). Lipsiae: Teubneri (1915).

Celsus. *De medicina.* WG Spencer (trans.). Loeb Classical Library 292, 304, and 336. London: Heinemann (1935–1938).

Cesalpino A [Cæsalpinus]. *De plantis libri XVI.* Florentiæ: Georgius Marescottus (1583).

Cesalpino A [Cæsalpinus]. *Peripateticarum quæstionum libri quinque.* Venetiis: Iuntas (1571).

[Chambers R]. *Explanations: a sequel to "Vestiges of the natural history of Creation".* London: John Curchill (1845).

[Chambers R]. *Vestiges of the natural history of Creation.* [First edition]. London: John Churchill (1844).

[Chambers R]. *Vestiges of the natural history of Creation.* Tenth edition. London: John Churchill (1853).

[Chambers R]. *Vestiges of the natural history of Creation.* Eleventh edition. London: John Churchill (1860).

Chambers R. *Vestiges of the natural history of Creation.* Twelfth edition. *With an introduction . . . by A Ireland.* London & Edinburgh: Chambers (1884).

Chambers R. *Vestiges of the natural history of Creation and other evolutionary writings.* JA Secord (ed.). University of Chicago Press (1994).

Chamisso A von. *A voyage around the world with the Romanzov exploring expedition in the years 1815–1818 in the brig Rurik, Captain Otto von Kotzebue.* H Kratz (ed. and trans.). Honolulu: University of Hawaii Press (1986).

Chamisso A de. *De animalibus quibusdam e classe Vermium Linnaeana in circumnavigatione Terrae,* volume 1. *De Salpa.* Berolini: Dümmler (1819).

Chamisso A von [as Lamotte Fouqué]. *Peter Schlemihl: from the German of Lamotte Fouqué.* [J Bowring, trans.]. London: Whittaker (1824).

Chamisso A von. *Peter Schlemihl's wundersame Geschichte mitgetheilt von Adelbert von Chamisso und herausgegeben von Friedrich Baron de la Motte Fouqué.* Nürnberg: Schrag (1814).

Chamisso A von. *Peter Schlemihl's wundersame Geschichte mitgetheilt von Adelbert von Chamisso. Zweite mit den Liedern und Balladen des Verfassers vermehrte Ausgabe.* Nürnberg: Schrag (1827).

Chamisso A von. *Reise um die Welt mit der Romanzoffischen Entdeckungs-Expedition in den Jahren 1815–18 auf der Brigg Rurik. Kapitain Otto v. Kotzebue.* In: *Werke* 1–2. Volume 1: *Tagebuch*; volume 2: *Anhang. Bemerkungen und Ansichten.* Leipzig: Wiedmann (1836).

Chan CX, Beiko RG, Darling AE & Ragan MA. Lateral transfer of genes and gene fragments in prokaryotes. *Genome Biology and Evolution* 1:429–438 (2009).

Chan CX, Darling AE, Beiko RG & Ragan MA. Are protein domains modules of lateral genetic transfer? *PLoS ONE* 4(2):e4524 (2009).

Chandler T. *Four thousand years of urban growth.* Second edition. Lewiston NY: St David's UP (1987).

Chapman GB & Hillier J. Electron microscopy of ultra-thin sections of bacteria. I. Cellular division in *Bacillus cereus. Journal of Bacteriology* 66(3):362–373 (1953).

Chapman MJ, Dolan MF & Margulis L. Centrioles and kinetosomes: form, function, and evolution. *Quarterly Review of Biology* 75(4):409–429 (2000).

Chapman VJ & Chapman DJ. *The algae.* Second edition. London: Macmillan (1973).

Chapple CK. Purgation and virtue in Jainism: toward an ecological ethic. In: P Bilimoria, J Prabhu & R Sharma (eds), *Indian ethics. Classical traditions and contemporary challenges* 1. Aldershot UK: Ashgate (2007), pp. 217–227.

Charleton W. *Exercitationes de differentiis & nominibus animalium.* Oxoniæ: Sheldonianum (1677). *Sections numbered individually.*

Charleton W. *Onomasticon zoicon, plerorumque animalium differentias & nomina propria pluribus linguis exponens.* London: Jacobum Allestry (1668).

Charlton W. Introduction. In: W Charlton (trans.), *Philoponus: On Aristotle* On the soul *3.9–13.* [With:] Stephanus: *On Aristotle* On interpretation. London: Bloomsbury (2013), pp. 1–16.

Charterius R (ed.). *Hippocratis Coi et Claudii Galeni Pergameni opera.* In 13 volumes. Lutetiæ Parisorum: Andræum Pralard (1639–1649). Revised edition (1679–1689). Lutetiæ Parisorum: Joannem Guignard / Andræum Pralard.

Chase FH. Introduction. In: FH Chase Jr (trans.), *Saint John of Damascus. Writings.* Fathers of the Church. A new translation 37. New York: Fathers of the Church (1958), pp. v–xxxviii.

Chatton E. Essai sur la structure du noyau et la mitose chez les Amœbiens. Faits et théories. *Archives de Zoologie Expérimentale et Générale, 5 Série* 5(5):267–337 (1910).

Chatton É. *Pansporella perplexa.* Amoebien a spores protégées parasite des daphnies. Réflexions sur la biologie et la phylogénie des protozoaires. *Annales des Sciences Naturelles, 10 Série. Zoologique* 8:5–84 (1925).

Chatton É. *Titres et travaux scientifiques (1906–1918).* Tunis: Rapide (1918).

Chatton É. *Titres et travaux scientifiques (1906–1937).* Sète: Sottano (1938).

Chatton É. Un protiste nouveau *Pansporella perplexa* nov. gen., nov. sp., parasite des Daphnies (Note préliminaire). *Comptes Rendus Hebdomadaires des Séances et Mémoires de la Société de Biologie* 62:42–43 (1907).

Cheney CR. Decretals of Innocent III in Paris, B.N. MS Lat. 3922A. *Traditio* 11:149–162 (1955).

Cheney CR. The letters of Pope Innocent III. *Bulletin of the John Rylands Library* 35(1):23–43 (1952).

Chenu MD. Nature and man at the School of Chartres in the Twelfth century. In: GS Métraux & F Crouzet (eds), *The evolution of science. Readings from the history of mankind.* Toronto: Mentor (1963), pp. 220–235.

Chenu MD. Nature and man. The renaissance of the Twelfth century. In: J Taylor & LK Little (eds and trans.). *Nature, man and society in the Twelfth century: essays on new theological perspectives in the Latin West.* University of Chicago Press (1968), pp. 1–48.

Chenu MD. *Nature, man, and society in the Twelfth Century. Essays on new theological perspectives in the Latin West.* J Taylor & LK Little (eds and trans.). University of Chicago Press (1968).

Cherubini L. *Magnvm bvllarivm Romanvm... Edition novissima.* In five volumes. Lvgdnvi: Philippi Borde, Laur. Arnaud, & Cl. Rigaud (1655–1673).

Chevalley C. Animal. In: *Dictionary of untranslatables: a philosophical lexicon.* Princeton UP (2014), pp. 34–37.

Chignell A & Pereboom D. Natural theology and natural religion. *The Stanford Encyclopedia of Philosophy* (Fall 2020 edition), EN Zalta (ed.), https://plato.stanford.edu/archives/fall2020/entries/natural-theology/), accessed 9 October 2020.

662 REFERENCES

Childe VG. The prehistory of science; archaeological documents. In: GS Métraux & F Crouzet (eds), *The evolution of science. Readings from the history of mankind.* Toronto: Mentor (1963), pp. 34–76.

Ching J. Truth and ideology: the Confucian way (Tao) and its transmission (Tao-T'ung). *Journal of the History of Ideas* 35(3):371–388 (1974).

Chittick WC. *In search of the lost heart. Explorations in Islamic thought.* M Rustom, A Khalil & K Murata (eds). Albany: SUNY Press (2012).

Chittick WC. The circle of spiritual ascent according to al-Qūnawī. In: P Morewedge (ed.). *Neoplatonism and Islamic thought.* Albany: SUNY Press (1992), pp. 179–209.

Chittick WC. *The self-disclosure of God: principles of Ibn Arabī's cosmology.* Albany: SUNY Press (1998).

Chodat R. On the polymorphism of the green algae and the principles of their evolution. *Annals of Botany* 11(41):97–121 (1897).

Choisy [JD]. Description des Hydroléacées. *Memoires de la Société de Physique et d'Histoire Naturelle de Genève* 6(1):95–122 (1833).

Christensen T. *Systematisk botanik Nr. 2: Alger.* In: TW Böcher, M Lange & T Sørensen (eds), *Botanik,* Bind 2. København: Munksgaard (1962).

Christensen T. The gross classification of algae. In: DF Jackson (ed.), *Algae and man. Based on lectures presented at the NATO Advanced Study Institute July 22–August 11, 1962 Louisville, Kentucky.* New York: Plenum (1964), pp. 59–64.

Churchill FB. The guts of the matter: infusoria from Ehrenberg to Bütschli: 1838–1876. *Journal of the History of Biology* 22(2):189–213 (1989).

Churton T. *The golden builders: alchemists, Rosicrucians, and the first Freemasons.* Lichfield: Signal (2002).

Chvátval A & Verkhratsky A. An early history of neuroglial research: personalities. *Neuroglia* 1(1):245–281 (2018).

Chymische Hochzeit: Christiani Rosencreutz. Anno 1459. Arcana publicata vilescunt; & gratiam prophanata amittunt. Ergo: ne margaritas objice porcis, seu asino substerne rosas. Straßburg: Lazari Zetzners (1616). *The chemical wedding of Christian Rosenkreutz,* trans. J Godwin. Grand Rapids MI: Phanes Press (1991).

Cicero. *De natura deorum. Academica.* H Rackham (trans.). Cicero 19. Loeb Classical Library 268. London: Heinemann (1933).

Cicero. *Tusculan disputations.* JE King (trans.). Loeb Classical Library 141. Cambridge MA: Harvard UP (1927).

Cienkowski L. Beiträge zur Kenntniss der Monaden. *Archiv für mikroskopische Anatomie* 1:203–232 (1865).

Cienkowski L. Ueber den Bau und die Entwickelung der Labyrinthuleen. *Archiv für mikroskopische Anatomie* 3:274–310 (1867).

Cienkowski [L]. Ueber Palmellaceen und einige Flagellaten. *Archiv für mikroskopische Anatomie* 6:421–438 (1870).

Clark GN. *The Seventeenth Century.* Second edition. Oxford: Clarendon (1947).

Clark PF. Morphological changes during the growth of bacteria. In: EO Jordan & IS Falk (eds), *The newer knowledge of bacteriology and immunology.* University of Chicago Press (1928), pp. 38–45.

Clarkson J. *Food history almanac. Over 1,300 years of world culinary history, culture, and social influence.* Lanham MD: Rowman & Littlefield (2014).

Claude A. The constitution of protoplasm. *Science* 97(2525):451–456 (1943).

Claus C. *Ueber die Grenze des thierischen und pflanzlichen Lebens.* Leipzig: Wilhelm Engelmann (1863).

Clement of Alexandria [attrib.]. *Recognitions.* In: A Roberts & J Donaldson (eds), *The Ante-Nicene Fathers,* volume 8. New York: Scribner (1903), pp. 75–211.

Clement of Alexandria. *Klementos Alexandreos ta Euriskomena. Clementis Alexandrini opera, quæ extant . . . per Joannem Potterum.* In two volumes. Oxonii: Sheldoniano (1715).

REFERENCES 663

Clement of Alexandria, *The miscellanies; or, stromata*. In: A Roberts & J Donaldson (eds), *Ante-Nicene Christian Library* volume 4, pp. 349–470 (1867), and volume 12, pp. 1–514 (1869). Edinburgh: Clark.

Clement of Alexandria. *Writings*. In: A Roberts & J Donaldson (eds), *Ante-Nicene Christian Library* 4 (1867) and 12 (1869). Edinburgh: Clark.

Cleveland LR & Grimstone AV. The fine structure of the flagellate *Mixotricha paradoxa* and its associated micro-organisms. *Proceedings of the Royal Society B. Biological Sciences* 159(977):668–686 (1964).

Cobb C & Goldwhite H. *Creations of fire. Chemistry's lively history from alchemy to the atomic age.* New York: Plenum (1995).

Cockayne TO. *Leechdoms, wortcunning, and starcraft of early England. Being a collection of documents, for the most part never before printed, illustrating the history of science in this country before the Norman conquest.* In three volumes. Rerum Britannicarum Medii Ævi Scriptores, volume 35. London: Longman, Green *et al.* (1864–1866).

Cofresi LL. Hierarchical thought in the Spanish Middle Ages: Ramon Lull and Don Juan Manuel. In: ML Kuntz & PG Kuntz (eds), *Jacob's Ladder and the Tree of Life. Concepts of hierarchy and the Great Chain of Being.* Revised edition. New York: Peter Lang (1987), pp. 153–159.

Coggon J. Quinarianism after Darwin's *Origin*: the circular system of William Hincks. *Journal of the History of Biology* 35(1):5–42 (2002).

Cohn F. Conspectus familiarum cryptogamarum secundum methodum naturalem dispositarum. *Hedwigia* 11(2):17–20 (1872a).

Cohn F. Nachträge zur Naturgeschichte des *Protococcus pluvialis* Kützing (*Haematococcus pluvialis* Flotow, *Chlamidococcus versatilis* A. Braun, *Chlamidococcus pluvialis* Flotow u. A. Braun.). *Verhandlungen der Kaiserlichen Leopoldinisch-Carolinischen Akademie der Naturforscher / Novorum Actorum Academiæ Cæsareæ Leopoldino-Carolinæ Naturæ Curiosorum* 22(2):605–764 (1850).

Cohn F. Observations sur les Volvocinées, et spécialement sur l'organisation et la propagation du *Volvox globator. Annales des Sciences Naturelles, 4 Série. Botanique* 5:323–332 (1856).

Cohn F. On the natural history of Protococcus pluvialis. G Busk (abstr. and trans.). In: A Henfrey (ed.), *Botanical and physiological memoirs.* London: Ray Society (1853), pp. 515–564.

Cohn F. Über die Fortpflanzung von *Sphaeroplea annulina. Bericht über die zur Bekanntmachung geeigneten Verhandlungen der Königlichen Preussischen Akademie der Wissenschaften zu Berlin* 1855:335–351 (1855).

Cohn F. Ueber das Geschlecht der Algen. *Jahres-Bericht der Schlesischen Gesellschaft für vaterländische Kultur* 33:95–105 (1855).

Cohn F. Untersuchungen über Bacterien. *Beiträge zur Biologie der Pflanzen* 1(2):127–224 (1872b).

Cohn F. Untersuchungen über Bacterien. II. *Beiträge zur Biologie der Pflanzen* 1(3):141–224 (1875).

Colavito J. *Book of the secret of creation and the Art of nature* (Kitāb sirr al-ḥalīqa), 2018. http://www.jasoncolavito.com/the-secret-of-creation, accessed 28 December 2018 and 16 September 2020.

Colavito MM. *The Pythagorean intertext in Ovid's* Metamorphoses. *A new interpretation.* Studies in Comparative Literature 5. Lewiston NY: Mellen (1989).

Cole FJ. *A history of comparative anatomy from Aristotle to the Eighteenth Century.* London: Macmillan (1949).

Cole FJ. Leeuwenhoek's zoological researches.—Part I. *Annals of Science* 2(1):1–46 (1937).

Cole FJ. Leeuwenhoek's zoological researches—Part II. Bibliography and analytical index. *Annals of Science* 2(2):185–235 (1937).

Cole FJ. *The history of protozoology.* University of London Press (1926).

Coleman G. *Georges Cuvier, zoologist. A study in the history of evolution theory.* Cambridge MA: Harvard UP (1964).

Coleman W. Cell, nucleus, and inheritance: an historical study. *Proceedings of the American Philosophical Society* 109(3):124–158 (1965).

664 REFERENCES

Coleridge WT. France: an ode [1798]. In: EH Coleridge (ed.), *The poems of Samuel Taylor Coleridge*. Oxford: Humphrey Milford, and Oxford UP (1921), pp. 243–247.

Coles W. *Adam in Eden: or, natures paradise. The history of plants, fruits, herbs and flowers*. London: Streater (1657).

Colin [JJ] & Gaultier de Claubry H[F]. Sur les combinaisons de l'iode avec les substances végétales et animales. *Annales de Chimie* 90:87–100 (1814).

Collenuccio P. *Pliniana defensio adversus Nicolai Leoniceni accusationem*. Ferrariæ: Andreas Bellfortis (1493).

Columella. *De re rustica*. Trans. by HB Ash (vol. 1) and ES Foster (vols 2–3). Loeb Classical Library 361 (1941), 407 (1954) and 408 (1955). London: Heinemann; and Cambridge MA: Harvard UP.

Compagne VP. Heinrich Cornelius Agrippa von Nettesheim. *The Stanford Encyclopedia of Philosophy* (Spring 2017 edition), EN Zalta (ed.), plato.stanford.edu/archives/spr2017/entries/agrippa-nettesheim, accessed 7 January 2019.

[Confucius]. *Analects*. In: J Legge (trans.). *The four books. Confucian Analects, The great learning, The doctrine of the mean, and The works of Mencius*. Shanghai: Chinese Book Company [1930], pp. 1–306.

Conger GP. *Theories of macrocosms and microcosms in the history of philosophy*. New York: Columbia UP (1922).

Conrad NJ. A female figurine from the basal Aurignacian of Hohle Fels Cave in southwestern Germany. *Nature* 459(7244):248–252 (2009).

Conrad von Megenberg. *Bůch der Natur*. [Third printed edition]. Augsburg: Johann Bämler (1481). US Library of Congress Incun. 1481 .K6, digital images at https://www.loc.gov/item/48035378/, accessed 31 March 2019 and 15 October 2020.

Conrad von Megenberg. *Buch der Natur*. Fifteenth century. Universitätsbibliothek Heidelberg, Codex Pal. germ. 300. Digital images at https://digi.ub.uni-heidelberg.de/digit/cpg300, accessed 31 March 2019 and 15 October 2020.

Conrad von Megenberg. *Das Buch der Natur von Conrad von Megenberg. Die erste Naturgeschichte in deutscher Sprache. In Neu-Hochdeutscher Sprache bearbeitet*. H Schulz (trans.). Griefswald: Julius Abel (1897).

Conrad von Megenberg. *Hye nach volget das půch der natur, das Innhaltet. Zu dem ersten von eygenschafft und natur des menschen, darnach von der natur un[d] eygenschafft des himels, der tier des gefügels, der kreuter, der steÿn und vo[n] vil ander natürliche[n] dingen . . . Welches půch meyster Cůnrat von Megenberg von latein in teütsch transsferiert un[d] geschribe[n] hat*. Augspurg ([14]75). Münchener StaatsBibliothek 2 Inc.c.a. 347. Digital images at https://www.digitale-sammlungen.de, accessed 31 March 2019 and 15 October 2020.

Considine J. *Dictionaries in early modern Europe. Lexicography and the making of heritage*. Cambridge UP (2008).

Cook AS (ed.). *The old English Elene, Phoenix, and Physiologus*. New Haven: Yale UP (1919).

Cook J. *Ice Age art: arrival of the modern mind*. London: British Museum Press (2013).

Cook R. *The tree of life. Image for the cosmos*. New York: Avon (1974).

Coomaraswamy AK. *The bugbear of literacy*. London: Dennis Dobson (1949).

Copeland EB. What is a plant? *Science* 65(1686):388–390 (1927).

Copeland HF. Progress report on basic classification. *American Naturalist* 81(800):340–361 (1947).

Copeland HF. *The classification of lower organisms*. Palo Alto CA: Pacific Books (1956).

Copeland HF. The kingdoms of organisms. *Quarterly Review of Biology* 13(4):383–420 (1938).

Copenhaver BP. A tale of two fishes: magical objects in natural history from antiquity through the Scientific Revolution. *Journal of the History of Ideas* 52(3):373–398 (1991).

Copenhaver BP. *Hermetica. The Greek Corpus Hermeticum and the Latin Asclepius in a new English translation with notes and introduction*. Cambridge UP (1992).

Copenhaver BP. Introduction. In: *Hermetica. The Greek Corpus Hermeticum and the Latin Asclepius in a new English translation with notes and introduction*. BP Copenhaver (ed.). Cambridge UP (1992), pp. xiii–lxi.

Copernicus N. *De revolvtionibvs orbium coelestium, libri VI*. Norimbergæ: Petreium (1543).

Copleston F. *A history of philosophy*. In nine volumes. Garden City NY: Image (1962–1975).

Cordus E. *Botanologicon*. Coloniae: Ioannem Gymnicum (1534).

Cordus V. *In hoc volumine continentur Valerii Cordi Simesusij annotationes in Pedacij Discoridis Anazarbei De medica materia libros V*. Argentorati: Rihelius (1561).

Cordus V. *Stirpium descriptionis liber quintus, qua in Italia sibi visas describit*. Argentorati: Rihelius (1563).

Corliss JO. Historically important events, discoveries, and works in protozoology from the mid-17th to the mid-20th Century. *Revista de la Sociedad Mexicana de Historia Natural* 42: 45–81 (1991, published 1992).

Corliss JO. Progress in protistology during the first decade following reemergence of the field as a respectable interdisciplinary area in modern biological research. *Progress in Protistology* 1:11–63 (1986).

Corliss JO. The protists deserve attention: what are the outlets providing it? *Protist* 149(1):3–6 (1998).

Corliss JO. Three centuries of protozoology: a brief tribute to its founding father, A. van Leeuwenhoek of Delft. *Journal of Protozoology* 22(1):3–7 (1975).

Correns C. G. Mendel's Regel über das Verhalten der Nachkommenschaft der Rassenbastarde. *Berichte der Deutschen Botanischen Gesellschaft* 18(4):156–168 (1900).

Corsi P. Biologie. In: P Corsi *et al*. (eds), *Lamarck, philosophe de la nature*. Paris: Presses Universitaires de France (2006), pp. 37–64.

Corsi P. *Oetre il mito. Lamarck e le scienze naturali del suo tempo*. Bologna: Il Mulino (1983).

Corsi P. *The age of Lamarck. Evolutionary theories in France 1790–1830*. Berkeley: University of California Press (1988).

Corti B. *Osservazioni microscopiche sulla Tremella e sulla circulazione del fluido in una pianta acquajuola*. Lucca: Guiseppe Rocchi (1774).

Costa SS, Guimarães LC, Silva A, *et al*. First steps in the analysis of prokaryotic pan-genomes. *Bioinformatics and Biology Insights* 14:1–9 (2020).

Costache D. Approaching *An apology for the Hexaemeron*: its aims, method and discourse. *Phronema* 27(2):53–81 (2012).

Courcelle P. Boèce et l'école d'Alexandrie. *Mélanges de l'école française de Rome* 52:185–223 (1935).

Cowdry EV. Historical background of research on mitochondria. *Journal of Histochemistry and Cytochemistry* 1(4):183–187 (1953).

Cowdry EV. The mitochondrial constituents of protoplasm. *Contributions to Embryology* 8(25):39–160 (1918). Carnegie Institution of Washington publication 271.

Cox CJ, Foster PG, Hirt RP, *et al*. The archaebacterial origin of eukaryotes. *PNAS* 105(51):20356–20361 (2008).

Cranz FE. *A bibliography of Aristotle editions 1501–1600*. Bibliotheca Bibliographica Aureliana 38. Second edition. Baden-Baden: Koerner (1984).

[Crastone G] [Crastonus]. *Dictionarivm Græcum copiosissimum secu[n]dum ordinem alphabeti cum interpretatione Latina*. Venetiis: Aldus Manutius (1497).

[Crastone G] [Crastonus]. *Dictionarivm Graecvm cum interpretatione latina*. Venetiis: Aldus, et Andreae Asvlani Soceri (1524).

[Crastone G] [Crastonus]. *Dictionarivm Græcvm, vltra Ferrariensem æditionem locupletatum locis infinitis*. Basileæ: Andreas Cartander (1519).

[Crastone G] [Crastonus]. *Dictionvm Graecarvm thesavrvs copiosus quantum nunq[am] antea . . . Dictionum latinarum thesaurus unuq[am] alias' impressus cum græca interpretatione*. [n.c.]: Giovanni Mazzocchi (1510).

Creplin FCH. Enthelminthologie, Endozoologie, Entozoologie. In: *Allgemeine Encyclopädie der Wissenschaften und Künste in alphabetischer Folge*. Section 1, volume 35. JS Ersch & JG Gruber (eds). Leipzig: Gleditsch (1841), pp. 76–83.

Croal LR, Gralnick JA, Malasarn D & Newmann DK. The genetics of geochemistry. *Annual Review of Genetics* 38:175–202 (2004).

666 REFERENCES

Crocker LG. Diderot and Eighteenth century French transformism. In: B Glass, O Temkin & WL Straus Jr (eds), *Forerunners of Darwin: 1745-1859*. Baltimore: Johns Hopkins UP (1968), pp. 114-143.

Croll O. *Osualdi Crollii Veterani Hassi Basilica chymica continens. Philosophicam propriâ laborum experientiâ confirmatum descriptionem et usum remediorum chymicorum selectissimorum é lumine gratiæ et naturæ desumptorum*. Francofurti: Godefridi Tampâchii (1608).

Crombie AC. *Augustine to Galileo. The history of science AD 400-1650*. Cambridge MA: Harvard UP (1953).

Crombie F. Life of Origen. In: A Roberts & J Donaldson (eds), *Ante-Nicene Christian Library* 23. Edinburgh: Clark (1872), pp. xxiii-xxxviii.

[Cronstedt AF]. *Försök till en mineralogie, eller Mineral Rikets upställning*. Stockholm: Wildiska (1758).

Cullen CM. *Bonaventure*. Oxford UP (2006).

Curley MJ. "Physiologus", Φυσιολογία and the rise of Christian nature symbolism. *Viator, Medieval and Renaissance Studies* 11(1):1-10 (1980).

Curth LH. *"A plaine and easie waie to remedie a horse": equine medicine in Early Modern England*. Leiden: Brill (2013), pp. 39-58.

Cuvier G. Biographical memoir of M. de Lamarck. In: *Zoological philosophy; an exposition with regard to the natural history of animals*. H Elliot (trans.). University of Chicago Press (1984), pp. 434-453.

Cuvier [G]. Discours préliminaire. In: [G] Cuvier, *Recherches sur les ossemens fossiles de quadrupèdes*. Tome 1. Paris: Deterville (1812), pp. 1-16.

Cuvier G. *Discours sur les révolutions de la surface du globe, et sur les changements qu'elles ont produits dans le règne animal*. Paris: Dufour & d'Ocagne (1826).

Cuvier [G]. *Essay on the theory of the earth*. R Kerr (trans.). Edinburgh: Blackwood (1813).

Cuvier G. *Histoire des sciences naturelles, depuis leur origine jusqu'à nos jours, chez tous les peuples connus*. In five volumes. Paris: Fortin, Masson (1841-1845). Facsimile by Culture et Civilisation, Bruxelles (1969).

Cuvier [G]. *Le règne animal distribué d'après son organisation, pour servir de base a l'historie naturelle des animaux et d'introduction a l'anatomie comparée*. In four volumes. Tome 3 (*Les crustacés, les arachnides et les insectes*) by Latreille [PA]. Paris: Deterville (1817).

Cuvier [G]. *Le règne animal distribué d'après son organisation. Nouvelle [second] édition, revue et augmentée*. In five volumes. Tomes 4-5 by Latreille [PA]. Paris: Deterville & Crochard (1829-1830).

Cuvier G. *Le règne animal distribué d'après son organisation . . . par une réunion de disciples de Cuvier*. [Third edition]. In 11 volumes, with accompanying plates. Paris: Fortin, Masson (1836-1849).

Cuvier G. Mémoire sur la structure interne et externe, et sur les affinités des animaux auxquels on a donné le nom de vers, lu à la société d'Histoire-Naturelle, le 21 floréal de l'an 3. *La Décade philosophique, littéraire et politique* 40:385-396 (An 3 [1795]).

Cuvier [G]. *Recherches sur les ossemens fossiles de quadrupèdes, ou l'on rétablit les caractères de plusieurs espècies d'animaux que les révolutions du globe paroissent avoir détruites*. In four volumes. Paris: Deterville (1812).

Cuvier G. Sur un nouveau rapprochement à établir entre les classes qui composent le Régne animal. *Annales du Muséum d'Histoire Naturelle* 19:73-84 (1812).

Cuvier G. *Tableau élémentaire de l'histoire naturelle des animaux*. Paris: Baudouin (An 6, 1798). Facsimile by Culture et Civilisation, Bruxelles (1969).

Cuvier [G]. *The animal kingdom, arranged after its organization . . . A new edition, with additions by WB Carpenter . . . and JO Westwood*. London: Orr (1854).

Da Cunha C, Gaia M, Gadelle D, *et al*. Lokiarchaea are close relatives of Euryarchaeota, not bridging the gap between prokaryotes and eukaryotes. *PLoS Genetics* 13(6):e1006810 (2017).

Dagan T, Artz-Randrup Y & Martin W. Modular networks and cumulative impact of lateral transfer in prokaryote genome evolution. *PNAS* 105(29):10039-10044 (2008).

Dahan G. La connaissance et l'étude des langues bibliques. In: M Fumaroli (ed.), *Les origines du Collège de France (1500–1560): actes du colloque international, Paris* (1995). Paris: Collège de France; et Klincksieck (1995), pp. 327–355.

Dahm R. Discovering DNA: Friedrich Miescher and the early years of nucleic acid research. *Human Genetics* 122(6):565–581 (2008).

[d'Aléchamps J, with Bauhin J & Desmoulins J]. *Historia generalis plantarvm, in libros XVIII. per certas classes artificiose digesta.* In two volumes. Lugduni: Gvlielmvm Rovillivm (1586–1587).

d'Alembert [J LeR]. Discours préliminaire des editeurs. *Encyclopédie, ou, dictionnaire raisonné des sciences, des arts et des métiers* 1:i–xlv (1751).

d'Alembert [J LeR]. *Preliminary discourse to the Encyclopedia of Diderot.* RN Schwab (trans.). University of Chicago Press (1995).

Dales RC. The de-animation of the heavens in the Middle Ages. *Journal of the History of Ideas* 41(4):531–550 (1980).

Dallinger WH. Should the naturalist recognise a fourth kingdom in nature? *Proceedings of the Liverpool Literary and Philosophical Society* 26:279–300 (1872).

Dalton ET. *Descriptive ethnology of Bengal.* Calcutta: Office of the Superintendent of Government Printing (1872).

[Dalrymple J]. Upon the family of Closterinæ. *Annals of Natural History; or, Magazine of Zoology, Botany, and Geology* 5(33):415–417 (1840).

Dampier WC. *A history of science and its relation with philosophy & religion.* Fourth edition. Cambridge UP (1948).

D'Ancona C. Greek sources in Arabic and Islamic philosophy. *The Stanford Encyclopedia of Philosophy* (Winter 2017 edition), EN Zalta (ed.), https://plato.stanford.edu/archives/win2 017/entries/arabic-islamic-greek/, accessed 21 August 2018.

Dannenfeldt KH. *Leonhard Rauwolf. Sixteenth-century physician, botanist, and traveler.* Cambridge MA: Harvard UP (1968).

Dante Alighieri. *La commedia secondo l'antica vulgata.* G Petrocchi (ed.). In four volumes. Milano: Mondadori (1966–1967).

Dante Alighieri. *The Divine Comedy of Dante Alighieri: Inferno, Purgatory, Paradise.* HF Cary (trans.), G Doré (illus.). New York: Union Library Association (1935).

Dante [Alighieri]. *The Divine Comedy of Dante Alighieri.* CE Norton (trans.). Revised edition. In three volumes. Boston: Houghton Mifflin; and Cambridge MA: Riverside (1902).

Darian S. *Understanding the language of science.* Austin: University of Texas Press (2003).

Dartmouth Dante Project. https://dante.dartmouth.edu, accessed 6 November 2018 and 14 September 2020.

Darwin C. *A monograph on the sub-class Cirripedia.* In two volumes. London: Ray Society (1851–1854).

Darwin C. *Az ember származása és ivari kiválás.* In two volumes. Török A & Entz G (trans.). Budapest: Kiadja (1884).

Darwin C. *Journal of researches into the geology and natural history of the various countries visited by MHS Beagle.* London: Henry Colburn (1839).

Darwin C. *Naturwissenschaftliche Reisen.* E Dieffenbach (trans.). In two volumes. Braunschweig: Vieweg (1844).

Darwin C. *On the origin of species by means of natural selection, or the preservation of favored races in the struggle for life.* [First edition]. London: Murray (1859).

Darwin C. *On the origin of species by means of natural selection.* [Second edition]. London: Murray (1860).

Darwin C. *On the origin of species by means of natural selection.* Third edition. London: Murray (1861)

Darwin C. *On the origin of species by means of natural selection.* Fourth edition. London: Murray (1866).

Darwin C. *On the origin of species by means of natural selection.* Fifth edition. London: Murray (1869).

668   REFERENCES

Darwin C. *On the origin of species by means of natural selection*. Fifth edition. New York: Appleton (1871).

Darwin C. *On the origin of species by means of natural selection*. Sixth edition. London: Murray (1872).

Darwin C. On the ova of Flustra, or, early notebook, containing observations made by CD. when he was at Edinburgh, March 1827. In: PH Barrett (ed.), *The collected papers of Charles Darwin*, volume 2. University of Chicago Press (1977), pp. 285–291.

Darwin C. *The descent of man, and selection in relation to sex*. [First edition]. In two volumes. London: John Murray (1871).

Darwin C. *The descent of man, and selection in relation to sex*. New edition, revised and augmented. New York: Hurst (undated, 1874).

Darwin C. *The formation of vegetable mould, through the action of worms, with observations on their habits*. London: Murray (1881).

Darwin C. *Über die Entstehung der Arten im Thier- und Pflanzen-Reich durch natürliche Züchtung, oder Erhaltung der vervollkommneten Rassen im Kampfe um's Daseyn*. HG Bronn (trans.). Stuttgart: Schweizerbart (1860).

Darwin E. *Phytologia; or the philosophy of agriculture and gardening*. London: Johnson (1800).

[Darwin E]. *The botanic garden. Part 1. Containing the economy of vegetation. A poem. With philosophical notes*. [First edition]. London: Johnson (1791).

[Darwin E]. *The botanic garden. Part 1. Containing the economy of vegetation. A poem. With philosophical notes*. Second edition. London: Johnson (1791).

[Darwin E]. *The botanic garden. Part 1. Containing the economy of vegetation. A poem. With philosophical notes*. Third edition. London: Johnson (1795).

[Darwin E]. *The botanic garden. Part 2. Containing The loves of the plants. A poem. With philosophical notes*. [First edition]. Lichfield: Jackson (1789).

[Darwin E]. *The botanic garden. Part 2. Containing The loves of the plants. A poem. With philosophical notes*. Third edition. London: Johnson (1791).

[Darwin E]. *The botanic garden. Part 2. Containing The loves of the plants. A poem. With philosophical notes*. Fourth edition. London: Johnson (1794).

Darwin E. *The essential writings of Erasmus Darwin*. D King-Hele (ed.). London: MacGibbon & Kee (1968).

Darwin E. *The temple of nature; or, the origin of society: a poem, with philosophical notes*. London: Johnson (1803).

Darwin E. *Zoonomia; or, the laws of organic life*. In two volumes. London: Johnson (1794–96).

Darwin E. *Zoonomia; or, the laws of organic life*. Second American, from the Third London edition. In three volumes. Boston: Thomas & Andrews (1803).

Darwin F (ed.). *The life and letters of Charles Darwin*. In two volumes. New York: Appleton (1887).

[Darwin RW]. *Principia botanica: or, a concise and easy introduction to the sexual botany of Linnæus*. Newark: Allin (1787).

Daubenton [LJM]. Histoire naturelle. [Leçon sur l'homme, etc.]. *Séances des Écoles Normales, Nouvelle édition* 8:3–31 (An 9, 1800).

Daubenton [LJM]. Histoire naturelle. [Sur la nomenclature méthodique de l'histoire naturelle]. *Séances des Écoles normales, Nouvelle édition* 1:425–444 (An 9, 1800).

Daubenton [LJM]. Histoire naturelle. [Sur la physiologie des végétaux, comparée à celle des animaux]. *Séances des Écoles Normales, Nouvelle édition* 5:269–278 (An 9, 1800).

Daubenton [LJM]. Histoire naturelle. *Séances des Écoles Normales, Nouvelle édition. Débats* 1:91–97 (An 9, 1800).

Daubenton [LJM]. Histoire naturelle. *Séances des Écoles Normales, Nouvelle édition* 6:3–11 (An 9, 1800).

Daubenton [LJM]. Introduction a l'histoire naturelle (pp. i–x); Les trois regnes de la nature (xj–xv); Regne animal (xvj–xviij, with a folding table); and Histoire naturelle de l'homme (pp. xix–lxxxxij). In: *Encyclopédie méthodique. Histoire naturelle des animaux*. Tome 1. Paris: Panckoucke; Liège: Plomteux (1782).

Daubeny CJB. Remarks on the final causes of the sexuality of plants, with particular reference to Mr. Darwin's work "On the origin of species by natural selection". *Report of the Thirtieth Meeting of the British Association for the Advancement of Science, Oxford, June-July 1860.* Notices and abstracts. London: John Murray (1861), pp. 109–110.

Davaine C. Nouvelles recherches sur les infusoires du sang dans la maladie connue sous le nom de *sang de rate. Comptes Rendus Hebdomadaires des Séances de l'Academie des Sciences* 57:351–353 and 386–387 (1863).

Davaine C. Recherches sur les infusoires du sang dans la maladie connue sous le nom de *sang de rate. Comptes Rendus Hebdomadaires des Séances de l'Academie des Sciences* 57:220–223 (1863).

Davaine C. Sur la nature des maladies charbonneuses. *Archives générales de Médecine, 6 Série* 10:144–148 (1868).

David [attrib.]. *Davidis prolegomena et in Porphyrii* Isagogen. In: A Busse (ed.), *Commentaria in Aristotelem Graeca.* Academiae Litterarum Regiae Borussicae 18(2). Berolini: Reimer (1904).

Davidson I. Images of animals in rock art: not just "good to think". In: *Oxford handbook of the archaeology and anthropology of rock art* (B David & IJ McNiven, eds), 2017 (online).

Davies GL. The tour of the British Isles made by Louis Agassiz in 1840. *Annals of Science* 24(2):131–146 (1968).

da Vinci L. *Scritti letterari di Leonardo da Vinci cavati dagli autographi e pubblicati. The literary works of Leonardo da Vinci compiled and edited from the original manuscripts by JP Richter.* In two volumes. London: Sampson Low, Marston, Searle & Rivington (1883).

Davis BM. The origin of the Archegoniates. *American Naturalist* 43(506):107–111 (1909).

Davis JI. Preface. In: Libellus de natura animalium. *A fifteenth century bestiary.* London: Dawson (1958), pp. v–vii.

Dawkins P. *The French Academy. Francis Bacon's connection with and possible authorship of "Academie Francoise" by Pierre de la Primaudaye* (undated, 2017). http://www.fbrt.org.uk/wp-content/uploads/2020/06/The_French_Academy.pdf, accessed 16 September 2020.

Dawson J. *Australian Aborigines. The languages and customs of several tribes of Aborigines in the Western District of Victoria, Australia.* Robertson: Melbourne (1881).

Dawson VP. *Nature's enigma. The problem of the polyp in the letters of Bonnet, Trembley and Réaumur.* Memoirs of the American Philosophical Society, volume 174. Philadelphia: American Philosophical Society (1987).

Day WH. TB Wilson, MD, a founder and benefactor of the American Entomological Society, and his family: our first Newark, Delaware-Philadelphia connection. *Entomological News* 95(4):137–147 (1984).

Dayman J. *Deep sea soundings in the North Atlantic Ocean between Ireland and Newfoundland, made in HMS Cyclops, Lieut.-Commander Joseph Dayman, in June and July 1857.* London: Admiralty, HMSO (1858). Reprinted by British Library (2011).

De alchimia opuscula complura veterum philosophorum. Francoforti: Cyriacus Iacobus (1550).

de Bary A. *Die Erscheinung der Symbiose. Vortrag, gehalten auf der Versammlung Deutscher Naturforscher und Aertze zu Cassel.* Strassburg: Trübner (1879).

de Bary A. Über Symbiose. *Tageblatt der 51. Versammlung Deutscher Naturforscher und Ärzte in Cassel.* Cassel: Baier & Lewalter (1878), pp. 121–126.

de Bary A. Zur Systematik der Thallophyten. *Botanische Zeitung* 39: cols 1–17 and 33–36 (dated 1880, published 1881).

de Beer GR. The correspondence between Linnaeus and Johann Gesner. *Proceedings of the Linnean Society of London* 161(2):225–241 (1949).

Debus AG. Robert Fludd and the circulation of the blood. *Journal of the History of Medicine and Allied Sciences* 16(4):374–393 (1961).

De Corte M. Themistius et Saint Thomas d'Aquin. Contribution à l'étude des sources et de la chronologie du *Commentaire* de Saint Thomas sur le *De anima. Archives d'Histoire doctrinale et littéraire du Moyen Age* 7:47–83 (1932).

de Groot JJM. *Universismus: die Grundlage der Religion und Ethik, des Staatswesens und der Wissenschaften Chinas.* Berlin: Reimer (1918).

670 REFERENCES

DeLacy ME & Cain AJ. A Linnean thesis concerning *contagium vivum*: the "Exanthemata viva" of John Nyander and its place in contemporary thought. *Medical History* 39:159–185 (1995).

Delafosse [G]. *Nouveau cours de minéralogie contenant la description de toutes les espèces minérales avec leurs applications directes aux arts.* Tome second. Paris: Roret (1860).

Delahaye H. *Les légends grecques des saints militaires.* Paris: Alphonse Picard (1909).

Delaméthrie JC. Discours préliminaire. *Journal de Physique, de Chimie, d'Histoire Naturelle et des Arts* 48:3–99 (An 7, 1799).

Delatte A (éd.). *Anecdota Atheniensia. Tome 2. Textes Grecs relatifs à histoire des sciences.* Bibliothèque de la Faculté de Philosophie et Lettres de l'Université de Liége, Fasc. 88. Liége: Faculté de Philosophie et Lettres; et Paris: Droz (1939).

Delise D. *Histoire des lichens. Genre sticta.* In two volumes (text plus atlas). Caen: Chalopin (1822–1825).

DeLong EF. Everything in moderation: archaea as "non-extremophiles". *Current Opinion in Genetics & Development* 8(6):649–654 (1998).

DeLong EF & Pace NR. Environmental diversity of bacteria and archaea. *Systematic Biology* 50(4):470–478 (2001).

del Paso y Troncoso F. Primer estudio. La botánico entre los nahuas. *Anales del Museo Nacional de México* 1(3):140–235 (1886).

del Rio M. *Disquisitionum magicarum libri sex.* In three volumes. Mogvntiæ: Ioannem Albinvm (1603).

del Rio M. *Disquisitionvm magicarvm libri sex.* In three volumes. Venetiis: de Franciscis (1606).

de Lubac H. *Catholicsm: Christ and the common destiny of man.* LC Sheppard & E Englund (trans.). San Francisco: Ignatius (1988).

de Lubac H. Introduction to the Torchbook edition. In: Origen. *On first principles. Being Koetschau's text of the* De principiis *translated into English.* GW Butterworth (ed.). Gloucester MA: Peter Smith (1973), pp. vii–xxii.

de Mowbray M. Philosophy as handmaid of theology: biblical exegesis in the service of scholarship. *Traditio* 59:1–37 (2004).

Denso JD. Vorrede des Uebersetzers. In: JG Wallerius, *Hydrologie oder Wasserreich.* Berlin: Nicolai (1751), pag. unnumb.

Deodato C. *Pantheum hygiasticvm Hippocratico-Hermeticvm, De hominis vita.* Brvntrvti: Wilhelmus Darbellay (1628).

Derham W. *Physico-theology or, a demonstration of the being and attributes of God, from his works of creation.* London: Innys (1713).

de Sacy S. Kitab sirr alkhalikat libelinous alhakim. Le livre du secret de la creature, par le sage Bélinous. *Notices et extraits des manuscrits de la Bibliothèque nationale, lus au Comité établi dans la ci-devant Académie des Inscriptions & Belles-Lettres* 4:107–158. Paris: Imprimerie de la République (An 7 [1798–1799]).

de Santillana G & Pitts W. Philolaos in Limbo, or: what happened to the Pythagoreans? *Isis* 42(2):112–120 (1951).

de Santillana G. *The origins of scientific thought. From Anaximander to Proclus. 600 BC to 300 AD.* University of Chicago Press (1970).

Descartes R. CLXXIXb. Descartes a Mersenne (1639). In: *Oeuvres de Descartes publiées par C Adam & P Tannery sous les auspices du Ministére de l'Instruction publique. Supplément* [to 1894 edition]. Paris: Vrin (undated, probably 1913), pp. 97–102.

Deslile L. Mémoire sur les actes d'Innocent III. *Bibliothèque de l'École des Chartes* 19 [*4 Série*, 4]:1–73 (1858).

Deutsches Museum. Die Botanik in der Antike. https://www.deutsches-museum.de/bibliot hek/unsere-schaetze/medizin/dioscurides/botanik-der-antike, accessed 21 May 2018 and 9 September 2020.

de Vries H. Sur la loi de disjonction des hybrides. *Comptes Rendus Hebdomadaires des Séances de l'Academie des Sciences* 130(13):845–847 (1900).

Dewan L. St. Thomas, Aristotle, and creation. *Dionysius* 15: 81–90 (1991).

Dewhurst K. *John Locke (1632–1704): physician and philosopher.* London: Wellcome (1963).

REFERENCES 671

Dexippus. *Dexippi philosophi Platonici Iamblichi discipuli, in defensionem Praedicamĕtorum Aristotelis aduersus Plotinŭ Porphyrij praeceptorĕ Platonicae disciplinae grauissimum & defensorem & propugnatore, Quaestionum libri III*. IB Feliciano (ed.). Parisiis: Vascosanum (1549).

Dexippus. *Dexippus in* Categorias. In: A Busse (ed.), *Commentaria in Aristotelem Graeca*. Edita consilio et auctoritate Academiae Litterarum Regiae Borussicae 4(2). Berolini: Reimer (1888).

Dexippus. *On Aristotle's* Categories. J Dillon (trans.). Ithaca NY: Cornell UP (1990).

Dey G, Thattai M & Baum B. On the archaeal origins of eukaryotes and the challenges of inferring phenotype from genotype. *Trends in Cell Biology* 26(7):476–485 (2016).

d'Huy J. La distribution des animaux à Lascaux reflèterait leur distribution naturelle. *Bulletin de la Société Historique et Archéologique du Périgord* 138:493–502 (2011).

Dickerson RE. Evolution and gene transfer in purple photosynthetic bacteria. *Nature* 283(5743):210–212 (1980).

Dickerson RE. The cytochromes *c*: an exercise in scientific serendipity. In: DS Sigman & MAB Brazier (eds), *The evolution of protein structure and function*. UCLA Forum in Medicine and Science 21. New York: Academic (1980), pp. 173–202.

Diderot D. Conversation between d'Alembert and Diderot. In: J Stewart & J Kemp (trans.), *Diderot, interpreter of nature. Selected writings*. London: Lawrence & Wishart (1937), pp. 49–63.

Diderot D. *Éléments de physiologie* [1774–1780]. In: J Assézat (ed.), *Œuvres complètes de Diderot*. Tome 9. Paris: Garnier (1875), pp. 235–440.

Diderot D. *Éléments de physiologie* [1774–1780]. In: J Varloot (ed.), *Œuvres complètes*. Tome 17. Paris: Hermann (1987), pp. 262–544.

Diderot D. Entretien entre d'Alembert et Diderot [1769]. In: J Assézat (ed.), *Œuvres complètes de Diderot*. Tome 2. Paris: Garnier (1875), pp. 101–121.

Diderot D. De l'interprétation de la nature [1753]. In: J Assézat (ed.), *Œuvres complètes de Diderot*. Tome 2. Paris: Garnier (1875), pp. 9–63.

Diderot D. *Rêve de D'Alembert*. In: *Œuvres* complètes de Diderot. J Assézat (ed.), volume 2 (1875). Paris: Garnier (1875–1879), pp. 101–181.

Diderot [D] & d'Alembert [J LeR] (eds). *Encyclopédie, ou, dictionnaire raisonné des sciences, des arts et des métiers*. Text: volumes 1–7, Paris: Briasson, David, Le Breton & Durand (1751–1757); volumes 8–17, Neufchastel: Faulche (1765). *Recueil de planches* (11 volumes), Paris: Briasson *et al.* [varies] (1762–1772). With: *Supplément à L'Encyclopédie* (five volumes); and *Table analytique* (two volumes, by P Mouchon). Amsterdam: Rey (1776–1780). Text volumes 1–17 and plates, http://enccre.academie-sciences.fr/encyclopedie/, accessed 2 October 2020.

Diels H. *Die Fragmente der Vorsokratiker. Griechisch und Deutsch*. W Kranz (ed.). Fifth edition. Berlin: Wiedmann (1934–1937).

Diels H. *Die Fragmente der Vorsokratiker*. Sixth edition. W Kranz (ed.). Berlin: Wiedmann (1951–1952).

Diels H. *Die Fragmente der Vorsokratiker*. W Kranz (ed.). Twelfth edition. Dublin: Wiedemann (1966–1967).

Di Gregorio MA. *From here to eternity. Ernst Haeckel and scientific faith*. Göttingen: Vandenhoek & Ruprecht (2005).

Dillenius JJ. *Catalogus plantarum sponte circa Gissam nascentium*. Francofurti: Maximilianum à Sande (1719).

Dillon J. Introduction. In: *Dexippus. On Aristotle's Categories*. Ithaca NY: Cornell UP (1990), pp. 7–17.

Dillon J. *The Middle Platonists. 80 BC to AD 220*. Revised edition. Ithaca NY: Cornell UP (1996).

Dillon J. Iamblichus of Chalcis and his school. In: LP Gerson (ed.), *Cambridge history of philosophy in Late Antiquity*, volume 1 (2010), pp. 358–374.

Dillon JM. *Iamblichi Chalcidensis in Platonis dialogos commentariorum fragmenta*. Philosophia Antiqua 23. Leiden: Brill (1973). New edition: Westbury UK: Prometheus (2009).

Dillon JM. Iamblichus' Νοερὰ Θεωρία of Aristotle's Categories. *Syllecta Classica* 8:65–77 (1997). Separate bibliography (8:219–235).

672 REFERENCES

Dillon LS. A reclassification of the major groups of organisms based upon comparative cytology. *Systematic Zoology* 12(2):71–82 (1963).

Dillon LS. Comparative cytology and the evolution of life. *Evolution* 16(1):102–117 (1962).

Dillwyn LW. *British confervae; or colored figures and descriptions of the British plants referred by botanists to the genus Conferva.* London: Phillips ([1802]–1809).

Diogenes Laërtius. *Lives of eminent philosophers.* In two volumes. RD Hicks (ed.). Loeb Classical Library 184 and 185. London: Heinemann (1925).

Dioscorides [Pedanius of Anazarbos]. *De medica materia libros qvinqve, amati Lvsitani doctoris medici ac philosophi celeberrimi enarrationes eruditissimæ.* Lvgdvni: Gulielmum Rouillium (1558).

Dioscorides. *De materia medica. Being an herbal with many other medicinal materials written in Greek in the first Century of the Common Era. A new indexed version in modern English by TA Osbaldeston and RPA Wood.* Johannesburg: Ibidis (2000).

Dioscorides. *In hoc volvmine hæc continentvr Ioannis Baptistæ Egnatii Veneti in Dioscoridem ad Hermolao Barbaro tralatvm annotamenta, qvibvs morborvm et remediorvm vocabvla obscvriora . . . explicantvr. Pedacii Dioscoridis Anazarbei de medicinali materia ab eodem Barbaro latinitate primum Donati libri quinque. Eiusdem de noxiis venenis ut caueri vitarique possint. Liber I. Eiusdem de venenatis animalibus & rabioso cane. Liber I. Eiusdem de [notis] eorum [sic] quos animalia venenata momorderint. Liber I. Hermolai Barbari . . . corollarium libris quinque àbsolutum.* Venetiis: de Gregoriis (1516).

Dioscorides. *Pedanii Dioscoridis Anazarbei De materia medica libri quinque. Ad fidem codicum manuscriptorum.* KPF Sprengel (ed.). In two volumes. Leipzig: Cnobloch (1778–1834 and 1829–1830).

Dioscorides. *The Greek Herbal of Dioscorides. Illustrated by a Byzantine AD 512. Englished by John Goodyer AD 1655. Edited and first printed AD 1933 by Robert T. Gunther.* New York: Hafner (1959).

Doak R. *Empire of the Islamic world.* Revised edition. New York: Chelsea House (2010).

Dobell C. *Antony van Leeuwenhoek and his "little animals". Being some account of the father of protozoology and bacteriology and his multifarious discoveries in these disciplines.* London: Staples (1932). Reprinted by Dover, New York (1960).

Dobell CC. The principles of protistology. *Archiv für Protistenkunde* 23(3):269–310 (1911).

Dobzhansky T. *Genetics and the origin of the species.* Second edition. New York: Columbia UP (1941).

Dod BG. Aristoteles Latinus. In: N Kretzmann, A Kenny & J Pinborg (eds), *Cambridge history of later medieval philosophy.* Cambridge UP (1982), pp. 45–79.

Dodge JD. The Dinophyceae. In: MBE Godward (ed.), *The chromosomes of the algae.* New York: St Martin's Press (1966), pp. 96–115.

Dodson EO. The kingdoms of organisms. *Systematic Zoology* 20(3):265–281 (1971).

Doflein F. Das System der Protozoen. *Archiv für Protistenkunde* 1(1):169–192 (1902).

Doflein F. *Lehrbuch der Protozoenkunde. Eine Darstellung der Naturgeschichte der Protozoen mit besonderer Berücksichtigung der parasitischen und pathogenen Formen.* Second edition of *Protozoen als Parasiten und Krankheitserreger.* Jena: Gustav Fischer (1909).

Doflein F. *Lehrbuch der Protozoenkunde.* Fourth edition. Jena: Gustav Fischer (1916).

Doflein F. *Probleme der Protistenkunde. I. Die Trypanosomen. Ihre Bedeutung für Zoologie, Medizin und Kolonialwirtschaft.* Jena: Gustav Fischer (1909).

Dolan JR. Unmasking "The eldest son of the Father of Protozoology": Charles King. *Protist* 170(4):374–384 (2019).

Dolan MF, Melnitsky H, Margulis L & Kolnicki R. Motility proteins and the origin of the nucleus. *Anatomical Record* 268(3):290–301 (2002).

Donati V. An account of a work published in Italian by Vitaliano Donati, MD containing, An essay towards a natural history of the Adriatic Sea. *Philosophical Transactions* 49:585–592 (1757).

Donati V. *Auszug seiner Natur-Geschichte des Adriatischen Meers den Boden des Meers zu untersuchen, nebst Instrumenten in solcher Tiefe zu fischen; von Classen der Meerpflanzen, der*

Polyparen, der Thierpflanzen und Pflanzenthiere, oder Uebergang der Natur vom Pflanzenreiche zum Thierreiche. Halle: Franckens (1753).

Donati V. *Della storia naturale marina dell'Adriatico.* Venezia: Francesco Storti (1750).

Donati V. *Essai sur l'histoire naturelle de la mer Adriatique . . . sur une nouvelle espece de plante terrestre.* La Haye: Pierre de Hondt (1758).

Donati V. New discoveries relating to the history of coral. *Philosophical Transactions* 47:95–108 (1751).

Donato B. *De Platonicæ atqve Aristotelicæ philosophiæ differentia, libellus nuper in lucem editus.* Venetiis: Scotus (1540). Reprinted 1541.

Doody A. *Pliny's encyclopedia: the reception of the* Natural History. Cambridge UP (2010).

Doody A. Pliny's *Natural History: Enkuklios paideia* and the ancient encyclopedia. *Journal of the History of Ideas* 70(1):1–21 (2009).

Doolittle WF. Evolution: two domains of life or three? *Current Biology* 30(4):R177–R179 (2020).

Doolittle WF & Bapteste E. Pattern pluralism and the Tree of Life hypothesis. *PNAS* 104(7):2043–2049 (2007).

Dorn G. *In Theophrasti Paracelsi Auroram Philosophorum, Thesaurum, & Mineralem oeconomiam, commentaria.* Francoforti: [C Rab] (1584).

Dougherty EC. Neologisms needed for structures of primitive organisms. 1. Types of nuclei. *Journal of Protozoology* 4(supp. 3):14, Abstract 55 (1957).

Dougherty EC & Allen MB. Is pigmentation a clue to protistan phylogeny? In: MB Allen (ed.), *Comparative biochemistry of photoreactive systems.* New York: Academic (1960), pp. 129–144.

Dougherty EC & Allen MB. The words "protist" and "protista". *Experientia* 14(2):78 (1958).

Douglas SE & Turner S. Molecular evidence for the origin of plastids from a cyanobacterium-like ancestor. *Journal of Molecular Evolution* 33(3):267–273 (1991).

Drews G. The roots of microbiology and the influence of Ferdinand Cohn on microbiology of the 19th century. *FEMS Microbiology Reviews* 24(3):225–249 (2000).

Driesch H. *The history & theory of vitalism.* CK Ogden (trans.). London: Macmillan (1914).

Dronke P. Introduction. In: P Dronke (ed.), *Bernardus Silvestris. Cosmographia.* Leiden: Brill (1978), pp. 1–69.

Drummond J. *Philo Judaeus; or, the Jewish-Alexandrian philosophy in its development and completion.* In two volumes. London: Williams and Northgate (1888).

Dryander J. *Catalogus bibliothecæ historico-naturalis Josephi Banks.* In five volumes. London: Bulmer (1796–1800).

Du Bartas [de Salluste G]. *La metamorphose d'Ovide, contenant l'Olympe des histoires poëtiques, traduits de Latin en François.* Roven: Reinsart (1608).

Du Bartas [de Salluste G]. *La seconde semaine ov enfance dv monde.* Paris: l'Huillier (1584).

Du Bartas [de Salluste G]. *La sepmaine, ov creation dv monde.* Paris: Feurier (1573).

Du Bartas [de Salluste G]. *The works of Guillaume de Salluste, Sieur Du Bartas.* UT Holmes, JC Lyons & RW Linker (eds). In three volumes. Chapel Hill: University of North Carolina Press (1935–1940).

Dubs HH. The beginnings of alchemy. *Isis* 38(1/2):62–86 (1947).

Duchesne [AN] fils. *Histoire naturelle des fraisiers, contenant les vues d'économie réunies à la botanique; & suivie de remarques particulieres sur plusieurs points qui ont rapport à l'histoire naturelle générale.* Paris: Didot & Panckoucke (1766).

Duchesne AN. Sur les rapports entre les Êtres naturels. *Magasin Encyclopédique* 1795(6):289–294 (1795) with folding table opposite page 289.

Duchesne-Guillemin J. Fire in Iran and Greece. *East and West* 13(2):198–206 (1962).

Dudley J. Johannes Grammaticus Philoponus Alexandrinus, "In Aristotelis De anima, Proemion" translated from the Greek. *Bulletin de la Société International pour l'Étude de la Philosophie Médiévale (Louvain)* 16–17:62–85 (1974–1975).

Duhem P. *Études sur Leonard de Vinci.* In three volumes. Paris: Hermann (1906–1913).

Duhem P. *Medieval cosmology. Theories of infinity, place, time, void, and the plurality of worlds.* R Ariew (ed. and trans.). University of Chicago Press (1987).

674 REFERENCES

Dujardin F. *Histoire naturelle des Zoophytes. Infusoires, comprenant la physiologie et la classification de ces animaux, et la manière de les étudier a l'aide du microscope.* In two volumes (text and atlas). Paris: Roret (1841).

Dujardin F. Mémoire sur l'organisation des infusoires. *Annales des Sciences Naturelles, 2 Série, Zoologie* 10:230–315 (1838).

Dujardin F. Recherches sur les organismes inférieurs. *Annales des Sciences Naturelles, 2 Série, Zoologie* 4:343–377 (1835).

Dumortier BC. Recherches sur la structure comparée et le développement des animaux et des végétaux. *Verhandlungen der Kaiserlichen Leopoldinisch-Carolinischen Akademie der Naturforscher / Nova Acta Academiæ Cæsareæ Leopoldino-Carolinæ Naturæ Curiosorum* 16(1):217–312b (1832).

Dunal [MF]. Végétal, végétation. *Encyclopédie du dix-neuvième siècle* 25:1–21 (1838).

Dunlop DM. The translations of al-Biṭrīq and Yaḥyā (Yuḥannā) b. al-Biṭrīq. *Journal of the Royal Asiatic Society of Great Britain and Ireland* 91(3–4):140–150 (1959).

Dunlop EM. The existence and definition of philosophy. From an Arabic text ascribed to al-Fārābī. *Iraq* 13(2):76–94 (1951).

Duran S ben Ṣemaḥ. *Sefer magen 'avot.* Livorno: Bi-befus ha-shutafim AY Kasṭilo ve-'Eli'ezer Sa'adon (1785).

Durande JF. Mémorie sur le champignon ridé, & sur les autres plantes de la même famille. *Noveaux Mémoires de l'Academie de Dijon, pour la Partie des Sciences et Arts* 1785(2):302–324 (1785).

Duret C. *Histoire admirable des plantes et herbes esmerueillables & miraculeuses en nature: mesmes d'aucunes qui sont vrays zoophytes, ou plant'-animales, plantes & animaux tout ensemble, pour auoir vie vegetatiue, sensitiue & animale.* Paris: Nicolas Bvon (1605).

Durkheim E & Mauss M. De quelques formes primitives de classification: contribution à l'étude des représentation collectives. *Année Sociologique* 6:1–72 (1901–1902).

Duval B & Margulis L. The microbial community of *Ophyrydium versatile* colonies: endosymbionts, residents, and tenants. *Symbiosis* 18(3):181–210 (1995).

Dutt MN. *Manu Samhita (English translation).* Calcutta: Elysium (1909).

Ebach MC, Morrone JJ & Williams DM. A new cladistics of cladists. *Biology & Philosophy* 23(1):153–156 (2008).

Eckhard C. Die Organisationsverhältnisse der polygastrischen Infusorien mit besonder Rücksicht auf die kürzlich durch Herrn v Siebold ausgesprochenen Ansichten über diesen Gegenstand. *Archiv für Naturgeschichte* 12(1):209–235 (1846).

Eckhard C. On the organization of the polygastric infusoria. *Annals and Magazine of Natural History* 18 (suppl.):433–452 (1847).

Eco U. *Kant and the platypus. Essays on language and cognition.* New York: Harcourt (2000).

Eddy MD. *The language of mineralogy. John Walker, chemistry and the Edinburgh Medical School, 1750–1800.* Farnham UK: Ashgate (2008).

Eddy SR. The C-value paradox, junk DNA and ENCODE. *Current Biology* 22(21):R898–R899 (2012).

Edelstein L. The Golden Chain of Homer. In: G Boas *et al.* (eds), *Studies in intellectual history.* Baltimore: Johns Hopkins UP (1953), pp. 48–66.

Edelstein L. The philosophical system of Posidonius. *American Journal of Philology* 57(3): 286–325 (1936).

Edwards J. Sinners in the hands of an angry God [1741]. In: OA Winslow (ed.), *Jonathan Edwards. Basic writings.* New York: Signet (1966), pp. 150–167.

Edwards P. A classification of plants into higher taxa based on cytological and biochemical criteria. *Taxon* 25(5–6):529–542 (1976).

Egerton FN. History of ecological sciences, part 51: formalizing marine ecology, 1870s to 1920s. *Bulletin of the Ecological Society of America* 95(4):349–430 (2014).

Egger VO. *A history of the Muslim world to 1750. The making of a civilization.* Second edition. New York: Routledge (2017).

REFERENCES   675

Egmond F. A collection within a collection. Rediscovered animal drawings from the collections of Conrad Gessner and Felix Platter. *Journal of the History of Collections* 25(2):149–170 (2013).

Ehrenberg CG. *Animalia evertebrata exclusis insectis*. In: [FG] Hemprich & [CG] Ehrenberg, *Symbolæ Physicæ, seu, icones et descriptiones corporum naturalium novorum aut minus cognitorum. Zoologia* 2. Berolini: Officina Academica (1828), pag. unnumb.

Ehrenberg CG. *Die Infusionsthierchen als vollkommene Organismen. Ein Blick in das tiefere organische Leben der Natur*. In two volumes (text and atlas). Leipzig: Voss (1838).

Ehrenberg CG. Dritter Beitrag zur Erkenntniss grosser Organisation in der Richtung des kleinsten Raumes. *Abhandlung der Königlichen Akademie der Wissenschaften zu Berlin* 1833:145–336 (1835).

Ehrenberg CG. *Naturgeschichtliche Reisen durch Nord-Afrika und West-Asien in den Jahren 1820 bis 1825 von Dr WF Hemprich und Dr CG Ehrenberg. Band 1. Historische Theil*. Berlin: Mittler (1828).

Ehrenberg CG. *Organisation, Systematik und geographisches Verhältniss der Infusionsthierchen*. In two volumes (text, plates). Berlin: Königlichen Akademie der Wissenschaften (1830).

Ehrenberg CG. *Rede zur Feier des Leibnitzischen Jahrestages über Leibnitzens Methode Verhältniss zur Natur-Forschung und Briefwechsel mit Leeuwenhoek in der öffentlichen Sitzung der Königlich-Preuss. Akademie der Wissenschaften am 3. Juli 1845*. Berlin: Königlichen Akademie der Wissenschaften (1845).

Ehrenberg CG. *Sylvæ mycologicæ Berolinensis*. Berolini: Bruschcke (1818).

Ehrenberg CG. Weitere Nachrichten über das Vorkommen fossiler Infusorien. *Bericht über die zur Bekanntmachung geeigneten Verhandlungen der Königlich Preussischen Akademie der Wissenschaften zu Berlin* 1:83–85 (1836).

Ehrenberg CG. *Zur Erkenntniss der Organisation in der Richtung des kleinsten Raums. Zweiter Beitrag. Entwickelung, Lebensdauer und Structur der Magenthiere und Räderthiere, oder sogenannten Infusorien, nebst einer physiologischen Characteristik beider Klassen und 412 Arten derselben*. Berlin: Akademie der Wissenschaften (1832).

Eichholz DE. Aristotle's theory of the formation of metals and minerals. *Classical Quarterly* 43(3–4):141–146 (1949).

Eichhorn JC. *Beyträge zur Natur-Geschichte der kleinsten Wasser-Thiere die mit keinem blossen Auge können gesehen werden und die sich in den Gewässern in und umb Danzig befinden*. Danzig: Müller (1775).

Eichhorn JC. *Beyträge zur Naturgeschichte der kleinsten Wasserthiere die mit blossen Auge nicht können gesehen werden und die sich in den Gewässern in und um Danzig befinden*. Berlin & Stettin: Nicolai (1781).

Eichhorn JC. *Zugabe zu meinen Beyträge zur Natur-Geschichte der kleinsten Wasser-Thiere die mit keinem blossen Auge können gesehen werden, mit zwey neuentdeckten Wasser-Thieren*. Danzig: Müller (1783).

Eichwald E. *De regni animalis limitibus atque evolutionis gradibus*. Dorpat: Schünmann (1821).

Eichwald E. Ueber eine neue Eintheilung der Thiere. *Zeitschrift für die organische Physik* 3(3):261–326 (1833).

Eichwald E. *Zoologia specialis quam expositis animalibus tum vivis, tum fossilibus potissimum Rossiae in universum, et Poloniae in species*. In three volumes. Vilnae: Zawadzki (1829–1831).

Einarson E. The manuscripts of Theophrastus' *Historia plantarum*. *Classical Philology* 71(1):67–76 (1976).

Eliade M. *Forgerons et alchimistes*. Paris: Flammarion (1956).

Eliade M. *The forge and the crucible*. Second edition. University of Chicago Press (1978).

Eliade M. *Traité d'histoire des religions*. Paris: Payot (1949).

Elias. *Eliae in Porphyrii* Isagogen *et Aristotelis* Categorias *Commentaria*. In: A Busse (ed.), *Commentaria in Aristotelem Graeca*. Academiae Litterarum Regiae Borussicae 18(1). Berolini: Reimer (1900).

Elias and David. *Introductions to philosophy. With Olympiodorus, Introduction to logic*. S Gertz (trans.). In series: Ancient commentators on Aristotle (R Sorabji & M Griffin, General eds). London: Bloomsbury (2018).

676   REFERENCES

Elliott B. Sir John Hill as a botanist: *The vegetable system.* In: C Brant & G Rousseau (eds), *Fame and fortune. Sir John Hill and London life in the 1750s.* London: Palgrave Macmillan (2018), pp. 267–290.

Ellis J. *An essay towards a natural history of the Corallines, and other marine productions of the like kind, commonly found on the coasts of Great Britain and Ireland.* London: printed for the author (1755).

Ellis J [as "J.E."]. [Untitled.] In: *The St James's Chronicle or, The British Evening-Post* no. 1012 (25–27 August 1767), second (unnumbered) page.

Ellis J [as "J.E."]. [Untitled.] In: *The St James's Chronicle or, The British Evening-Post* no. 1023 (19–22 September 1767), fourth (unnumbered) page.

Ellis J. Ellis to Linnæus (30 October 1767). In: JE Smith (ed.), *A selection from the correspondence of Linnæus,* volume 1 [of 2]. London: Longman, Hurst *et al.* (1821), pp. 216–219.

Elsee C. *Neoplatonism in relation to Christianity.* Cambridge UP (1908).

Elton O. *Modern studies.* London: Edward Arnold (1907).

Embley TM & Williams TA. Only two domains, not three: changing view on the tree of life. *Microbiology Today* 43(2):70–73 (2016).

Eme L, Spang A, Lombard J, *et al.* Archaea and the origin of eukaryotes. *Nature Reviews Microbiology* 15(12):711–723 (2017). Erratum at *ibid.* 16(2):120 (2018).

Empedocles. *The extant fragments* (MR Wright, ed.). Second edition. London: Hart (1995).

Encyclopædia Britannica—unattributed articles:
   Ammonius. *Encyclopædia Britannica,* Ninth edition 1:743–744 (1885).
   Botany. *Encyclopædia Britannica,* [First edition] 1:627–653 (1771).
   Botany. *Encyclopædia Britannica,* Second edition 2:1283–1318 (1778).
   Botany. *Encyclopædia Britannica,* Third edition 3:417–473 (1797).
   Botany. *Encyclopædia Britannica,* Fourth edition 4:62–332 (1810).
   Botany. *Encyclopædia Britannica,* Fifth edition 4:62–332 (1815).
   Kingdoms, in natural history. *Encyclopædia Britannica,* Second edition 6:4040–4042 (1780). [Page 4040 is mistakenly numbered 3040.]
   Natural history. *Encyclopædia Britannica,* Second edition 7:5299–5312 (1781).
   Natural history. *Encyclopædia Britannica,* Third edition 12:651–670 (1797).
   Natural history. *Encyclopædia Britannica,* Fourth edition 14:628–640 (1810).
   Zoology. *Encyclopædia Britannica,* Second edition 10:8991–8993 (1783).

Encyclopedia of Art Education. A–Z of prehistoric art. http://www.visual-arts-cork.com/site/prehistoric.htm, accessed 4 September 2020.

Enderlein G. *Bakterien-Cyclogenie. Prolegomena zu Untersuchungen über Bau, geschlehtliche und ungeschlechtliche Fortpflanzung und Entwicklung der Bakterien.* Berlin: de Gruyter (1925).

Enderlein G. Ein neues Bakteriensystem auf vergleichend morphologischer Grundlage. *Sitzungsberichte der Gesellschaft Naturforschender Freunde zu Berlin* 1917(4):309–319 (1917).

Endlicher S. [Brief 100]. Endlicher an Unger, 16 Juli 1842. In: G Haberlandt (ed.), *Briefwechsel zwischen Franz Unger und Stephan Endlicher.* Berlin: Borntraeger (1899), pp. 133–134.

Endlicher S. *Genera plantarum secundum ordines naturales disposita.* Vindobonae: Beck (1836–1840).

Endlicher S. *Grundzüge einer neuen Theorie der Pflanzenzeugung.* Wien: Beck (1838).

Engels EM & Glick TF (eds). *The reception of Charles Darwin in Europe.* In two volumes. London: Continuum (2008).

Engler A & Prantl K (eds). *Die natürlichen Pflanzenfamilien. 1. Teil, Abteilungen 1a-2.* Leipzig: Engelmann (1896–1900).

Engramelle "RP". *Papillons d'Europe, peints d'aprés nature par M. Ernst . . . D'ecrits par le R.P. Engrammele.* Eight parts in four volumes. Paris: Delaguette [et] Basan & Poignant (1779–1792). With three volumes of plates: *Insectes d'Europe, peints d'apres nature par M. Ernst.* [Paris] (1779).

[Enoch]. *The Book of Enoch the Prophet. Translated from an Ethiopic ms. in the Bodleian Library by the late Richard Laurence.* London: Kegan Paul, Trench (1883).

REFERENCES   677

Entz G. Bevezető. In: G Entz & S Mágocsy-Dietz (eds), *Az élők világa növény- és állatország*. Budapest: Athenaeum (1907), pp. 1–60.

Entz G. *Tanulmányok a véglények köréből. A kir. Magyar természettudományi társulat megbizásából. Studien über Protisten. Im Auftrage der Kön. Ung. Naturwissenschaftlichen Gesellschaft. I. Theil. Entwickelung der Kenntniss der Protisten—Ein historisch-kritisch Überblick.* Budapest: Magyar Természettudományi Társukat (1888).

Epicurus. *The extant remains.* C Bailey (ed. and trans.). Oxford: Clarendon (1926).

Eriugena [Johannes Scotus Erigena]. *Annotationes in Marcianum.* CE Lutz (ed.). Mediaeval Academy of America 34. Cambridge MA: Mediaeval Academy of America (1939).

Eriugena [Johannes Scotus Erigena]. *Periphyseon (The division of nature).* IP Sheldon-Williams (trans.), JJ O'Meara (rev.). Montréal: Bellarmin; and Washington: Dumbarton Oaks (1987).

Eriugena [Johannes Scotus Erigena]. *Periphyseon.* In: HJ Floss (ed.), *Patrologiæ cursus completus, Series secunda* 122. Parisiis: Migne (1853), cols 439–1022.

Erwin DH. Early origin of the bilaterian developmental toolkit. *Philosophical Transactions of the Royal Society B* 364(1527):2253–2261 (2009).

Esper EJC. *Die Pflanzenthiere in Abbildungen nach der Natur mit Farben erleuchtet nebst Beschreibungen.* Seventeen parts in three volumes (parts 16–17 by FL Hammer). Nürnberg: Raspe (1788–1830).

Esper EJC. *Fortsetzung der Pflanzenthiere in Abbildungen nach der Natur mit Farben erleuchtet nebst Beschreibungen.* Ten parts in two volumes. Nürnberg: Raspe (1794–1806).

Eugenikos of Trebizond. Του αύτου νομοφύλαχο; του Εύγενικου τη Τραπεζουντίων πόλει έγκωμιαστικὴ ἔκφρασις. In: *Eustathii Metropolitae Thessalonicensis Opuscula. Accedunt Trapezuntinae historiae scriptores Panaretus et Eugenicus . . .* TLF Tafel (ed.). Francofurti: Schmerber (1832), pp. 370–373.

Eunapius Sardianus. *De vitis philosophorum et sophistarum.* HJ Hornanum (trans.). Antverpiæ: Platinus (1568).

Eusebius. *The ecclesiastical history of Eusebius Pamphilius . . . in ten books.* CF Crusé (trans.). Philadelphia: R Davis (1833).

Evans AH (ed.). *Turner on birds: a short and succinct history of the principal birds noticed by Pliny and Aristotle, first published by Doctor William Turner, 1544.* Cambridge UP (1903).

Evans EP. *Animal symbolism in ecclesiastical architecture.* New York: Henry Holt (1896).

Evans EP. *The criminal prosecution and capital punishment of animals.* London: Heinemann (1906).

Everett N. *The Alphabet of Galen. Pharmacy from antiquity to the Middle Ages. A critical edition of the Latin text with English translation and commentary.* University of Toronto Press (2012).

Eyde RH. The foliar theory of the flower. *American Scientist* 63(4):430–437 (1975).

Ezekiel A. Novalis (Georg Philipp Friedrich von Hardenberg) (1772–1801). *Internet Encyclopedia of Philosophy*, undated [2017]. https://www.iep.utm.edu/novalis, accessed 7 January 2020 and 3 October 2020.

Fabbroni A. *Dell'arte di fare il vino.* Firenze: Giuseppe Tofani (1787).

Fabbroni A. *Dell'arte di fare il vino.* Second edition. Firenze: Jacopo Grazioli (1790).

Fabricius J[A]. *Hydrotheologie oder Versuch, durch aufmerksame Betrachtung der Eigenschafte, reichen Austheilung und Bewegung der Wasser, die Menschen zur Liebe und Bewunderung ihres gütigsten, weisesten, mächtigsten Schöpfers zu ermuntern.* Hamburg: König & Richter (1734).

Fabricius J[A]. *Pyrotheologie, oder Versuch durch nähere Betrachtung des Feuers, die Menschen zur Liebe und Bewunderung ihres gütigsten, weisesten, mächtigsten Schöpfers anzuflammen.* Hamburg: Wittwe (1732).

Fahd T. Matériaux pour l'histoire de l'agriculture en Irak: *al-Filâḥa n-nabaṭiyya.* In: B Spuler (ed.), *Handbuch der Orientalistik,* Abteilung 1 *(der Nahe und der mittere Osten),* Band 6 *(Geschichte der islamischen Länder),* Abschnitt 6 *(Wirtschaftsgeschichte des vörderen Orients in islamischer Zeit)* 1:276–377. Leiden: Brill (1977).

Fakhry M. Al-Bīrūnī and Greek philosophy: an essay in philosophical erudition. In: HM Said (ed.), *Al-Bīrūnī commemorative volume.* Karachi: Hamdard (1979), pp. 344–349.

Falcon A. *Aristotle and the science of nature: unity without uniformity.* Cambridge UP (2005).

678   REFERENCES

Falcon A (ed.). *Brill's companion to the reception of Aristotle in antiquity*. Leiden: Brill (2016).

Famintzin AS. Chto takoe lishainiki? [What are lichens?]. *Priroda* 1918 (April–May): cols 265–282 (1918).

Famintzin AS. Die Symbiose als Mittel der Synthese von Organismen. *Biologisches Centralblatt* 27(12):353–364 (1907).

Famintzin AS. O roli simbioza v' evolutsii organismov' [On the role of symbiosis in the evolution of organisms]. *Zapiski Imperatorskoi Akademii Nauk' po Fiziko-matematicheskomu Otdeleniu 8 Série / Mémoires de l'Académie impériale des sciences de St-Pétersbourg, Classe physico-mathématique. 8e Série* 20(3):1–14 (1907).

Famintzin A & Boranetzky J. Zur Entwickelungsgeschichte der Gonidien und Zoosporenbildung der Flechten. *Mémoires de l' Académie impériale des sciences de Saint-Pétersbourg, 7 Série*, 11(9):1–7 (1867).

Fara P. Heavenly bodies: Newtonianism, natural theology and the plurality of worlds debate in the Eighteenth century. *Journal of the History of Astronomy* 35(2):143–160 (2004).

al-Fārābī AN. *On the perfect state (Mabādi' ārā' ahl al-madīnat al-fāḍilah)*. R Walzer (trans.). Oxford: Clarendon (1998).

al-Fārābī. *The book of Al Farabi on the origin of the sciences. That is, letter assigning the cause from which the philosophical sciences have arisen, and their order in teaching*. Translated from the edition of Cl. Baeumker (1916). JL Longeway (trans.). University of Oklahoma Translation Clearing House A-30-451 (1992).

al-Fārābī. *Über den Ursprung der Wissenschaften (De ortu scientiarum). Eine Mittelalterliche Einleitungsschrift in die philosophischen Wissenschaften*. In series: Beiträge zur Geschichte der Theologie und Philosophie des Mittelalters 19(3). Münster: Aschendorff (1916).

Faral E. La queue de poisson des sirènes. *Romania* 74(296):433–506 (1953).

Farber PL. Buffon and Daubenton: divergent traditions within the *Histoire naturelle*. *Isis* 66(1): 63–74 (1975).

Farley J. *Gametes & spores. Ideas about sexual reproduction 1750–1914*. Baltimore: Johns Hopkins UP (1982).

Farley J. The spontaneous generation controversy (1700–1860): the origin of parasitic worms. *Journal of the History of Biology* 5(1):95–125 (1972).

Farlow WG. Gustave Thuret. *Journal of Botany, British and Foreign* 14:4–9 (1876).

Farlow WG. *Marine algæ of New England and adjacent coast*. Washington: Government Printing Office (1881).

Farmer SA. *Syncretism in the West: Pico's 900 theses (1486). The evolution of traditional religious and philosophical systems*. Tempe AZ: Medieval & Renaissance Texts & Studies (1998).

Feingold M. The war on Newton. *Isis* 101(1):175–186 (2010).

Felsenstein J. *Inferring phylogenies*. Sunderland MA: Sinauer (2004).

Ferrière H. *Bory de Saint-Vincent. L'évolution d'un voyageur naturaliste*. Paris: Syllepse (2009).

Festus, Sextus Pompeius. *De verborum significatione*. [Venezia]: Johannis de Colonia & Johannis Manthen (1474).

Festus, Sextus Pompeius. *De verborum significatu quae supersunt cum Pauli epitome. Thewrewkianis copiis usus edidit WM Lindsay*. Lipsiae: Teubner (1913).

Festus, Sextus Pompeius. *De verborum significatu quæ supersunt com Pauli epitome*. Æ Thewrewk de Ponor (ed.). Budapestini: Academiæ Litterarum Hungaricæ (1889).

Fet V. Lynn Margulis and Boris Kozo-Polyansky: how the *Symbiogenesis* was translated from Russian. *BioSystems* 199:104316 (2021).

Feyerabend P. Theoreticians, artists and artisans. *Leonardo* 29(1):23–28 (1996).

Fichte JG. *Grundlage der gesammten Wissenschaftslehre als Handschrift für seine Zuhörer*. Leipzig: Gabler (1794).

Fichte JG. *Grundriss des Eigenthümlichen der Wissenschaftslehre in Rüksicht auf das theoretische Vermögen als Handschrift für seine Zuhörer*. Jena & Leipzig: Gabler (1795).

Ficino M. Argumentum Marsilij Ficini Florentini: in libru[m] Mercurij Trismegisti: ad Cosmum Medicem. In: [Hermes Trismegistus]. *Pimander. Asclepius*. Parisiis: Henrici Stephani (1505), folios 2r–3v.

REFERENCES 679

Ficino M. Argumentvm Marsilii Ficini Florentini, in librum Mercurij Trismegistium, ad Cosmum Medicem, patriæ patrem. In: *Opera, & quæ hactenus extitêre*, volume 2. Basileæ: Henric Petrina (1576), page 1836.

Ficino M. *De vita libri tres (De triplici vita). Apologia. Quod necessaria sit ad vitam securitas.* Firenze: Antonius Mischominus (1489).

Ficino M. *Opera, & quae hactenus extitêre.* In three volumes. Basileæ: Henric Petrina (1576).

Ficino M. *Platonic theology.* MJB Allen & J Warden (trans.), J Hankins & W Bowen (eds). In six volumes. Cambridge MA: Harvard UP (2001–2006).

Ficino M. *Three books on life.* A critical edition. CV Kaske & JR Clark (eds and trans.). Binghampton: SUNY Press (1998).

Filliozat J. India and scientific exchanges in antiquity. In: GS Métraux & F Crouzet (eds), *The evolution of science.* Toronto: Mentor (1963), pp. 88–105.

Findlen P. *Possessing nature. Museums, collecting, and scientific culture in early modern Italy.* Berkeley: University of California Press (1994).

Findlen P. The death of a naturalist: knowledge and community in late Renaissance Italy. In: G Manning & C Klestinec (eds), *Professors, physicians and practices in the history of medicine.* New York: Springer (2017), pp. 155–195.

Fischer E. Über die Hydrolyse der Proteinstoffe. *Chemiker-Zeitung* 26:939–940 (1902).

Fischer [von Waldheim] G. *Tableaux synoptiques de zoognosie.* Moscou: l'Université Impériale (1808).

Fitch WM. Is the fixation of observable mutations distributed randomly among the three nucleotide positions of the codon? *Journal of Molecular Evolution* 2(2–3):123–136 (1973).

Fitch WM & Margoliash E. Construction of phylogenetic trees. *Science* 155(3760):279–284 (1967).

Fleischmann RD, Adams MD, White O, *et al.* Whole-genome random sequencing and assembly of *Haemophilus influenzae* Rd. *Science* 269(5223):496–512 (1995).

Fleming J. *A history of British animals.* Edinburgh: Bell & Bradfute; and London: Duncan (1828).

Flemming W. *Zellsubstanz, Kern und Zelltheilung.* Leipzig: Vogel (1882).

Floridi L. Sextus Empiricus. The transmission and recovery of Pyrrhonism. American Classical Studies 46 (American Philological Association). Oxford UP (2002).

Floridi L. The diffusion of Sextus Empiricus's works in the Renaissance. *Journal of the History of Ideas* 56(1):63–85 (1995).

Flotow J von. Über *Haematococcus pluvialis. Verhandlungen der Kaiserlichen Leopoldinisch-Carolinischen Akademie der Naturforscher / Novorum Actorum Academiæ Cæsareæ Leopoldino-Carolinæ Naturæ Curiosorum* 12(2):411–606 (1844).

Fludd R. *Anatomiæ amphitheatrvm effigie triplici, more et conditione varia, designatvm.* Francofúrti: de Brÿ (1623).

Fludd R. *Mosaicall philosophy: grounded upon the essentiall truth 'or eternal sapience.* London: Humphrey Moseley (1659).

Fludd R. *Philosophia Moysaica.* Govdæ: Petrus Rammazenius (1638).

Fludd R. *Pvlsvs seu nova et arcana pvlsvvm historia.* [Francofurti: n.p.] (*ca* 1631).

Fludd R. *Robert Fludd and his Philosophicall Key being a transcription of the manuscript at Trinity College, Cambridge, with an introduction by AG Debus.* New York: Science History Publications (1979).

Fludd R. *Utriusque cosmi maioris scilicet et minoris metaphysica, physica atqve technica historia.* In two volumes. Oppenhemii: de Bry (1617–1621).

Fokin SI. Otto Bütschli (1848–1920): where we will genuflect? *Protistology* 8(1):22–35 (2013).

Fontaine R. Ibn Daud, Abraham ben David Halevi. *Encyclopaedia Judaica*, Second edition 9:662–665 (2007).

Fontana F. Lettera del Sig. Felice Fontana . . . al Sig. Adolfo Murray . . . scritta il dì 20 di Ottobre 1781. *Memorie di Matematica e Fisica della Società Italiana* 1:648–706 (1783).

Fontana F. Sur l'ergot & le tremella. *Observations sur la Physique, sur l'Histoire Naturelle et sur les Arts* 7:42–52 (1776).

Fontenelle B le B de. *Conversations on the plurality of worlds.* HA Hargreaves (trans.). Berkeley: University of California Press (1990).

680   REFERENCES

Fontenelle B le B de. *Entretiens sur la pluralité des mondes*. Amsterdam: Mortier (1686).

Fontenelle B le B de. *Entretiens sur la pluralité des mondes*. Paris: Blageart (1686).

Fontenelle B le B de. *Entretiens sur la pluralité des mondes. Nouvelle edition, augmentée d'un nouvel Entretien*. Paris: Michel Guerout (1687).

Fontenelle B le B de. *Œuvres de Fontenelle. Nouvelle édition*. In eight volumes. Paris: Bastien (1790–1792).

Fontenelle B le B de. *Œuvres de Monsieur de Fontenelle. Nouvelle edition augmentée*. In eight volumes. Paris: Michel Brunet (1742).

Forbes E. *An inaugural lecture on botany, considered as a science, and as a branch of medical education*. London: Van Voorst (1843).

Ford BJ. The microscope of Linnaeus and his blind spot. *The Microscope (Surrey)* 57(2):65–72 (2009).

Ford LS. The last link in *The Great Chain of Being*. In: ML Kuntz & PG Kuntz (eds), *Jacob's Ladder and the Tree of Life. Concepts of hierarchy and the Great Chain of Being*. New York: Peter Lang (1987), pp. 299–311.

Forster M. Johann Gottfried von Herder. *The Stanford Encyclopedia of Philosophy* (Summer 2019 edition), EN Zalta (ed.), https://plato.stanford.edu/archives/sum2019/entries/herder/, accessed 17 January 2020 and 3 October 2020.

Forterre P. The universal tree of life: an update. *Frontiers in Microbiology* 6:717 (2015).

Forterre P & Prangishvili D. The origin of viruses. *Research in Microbiology* 160(7):466–472 (2009).

Fortescue J. *The works of Sir John Fortescue*. In two volumes. London: printed for private distribution (1869).

Foucault M. *The order of things. An archaeology of the human sciences. A translation of* Les mots et les choses. In series: World of man (RD Laing, ed.). New York: Pantheon (1970).

Fournier GP & Poole AM. A briefly argued case that Asgard archaea are part of the eukaryote tree. *Frontiers in Microbiology* 9:1896 (2018).

Fournier GP, Huang J & Gogarten JP. Horizontal gene transfer from extinct and extant lineages: biological innovation and the coral of life. *Philosophical Transactions of the Royal Society B* 364(1527):2229–2239 (2009).

Fowler DC. *The life and times of John Trevisa, medieval scholar*. Seattle: University of Washington Press (1995).

Fox SW & Homeyer PG. A statistical evaluation of the kinship of protein molecules. *American Naturalist* 89(846):163–168 (1955).

Fracastoro G. *Contagion, contagious diseases and their treatment*. WC Wright (trans.). New York: Putnam (1930).

Fracastoro G. *De sympathia et antipathia rervm liber vnvs. De contagione et contagiosis morbis et cvratione Libri III*. Venetiis: Florentini (April 1546).

Francis H. *The present and future government of the Colony of New South Wales*. Sydney: Cunningham (1869).

Frank AB. Ueber die biologischen Verhältnisse des Thallus einiger Krustenflechten. *Beiträge zur Biologie der Pflanzen* 2(2):123–200 (1876).

Frank M. *"Unendliche Annährung": die Anfänge der philosophischen Frühromantik*. Frankfurt: Suhrkamp (1997).

Frank M. *The philosophical foundations of early German Romanticism*. E Millan-Zaibert (trans.). Albany: SUNY Press (2004).

Frankenstein. J Whale (dir.). Universal City CA: Universal Pictures (1931), motion picture.

Franklin B. An Arabian tale [Paris, 1779?]. In: *The writings of Benjamin Franklin* (AH Smith, ed.), volume 10. New York: Macmillan (1907), pp. 123–124.

Franklin-Brown M. *Reading the world: encyclopedic writing in the scholastic age*. University of Chicago Press (2012).

Franklin-Hall LR. Trashing life's tree. *Biology and Philosophy* 25(4):689–709 (2010).

REFERENCES 681

Frazer JG. Creation and evolution in primitive cosmogonies. In: *Creation and evolution in primitive cosmogonies and other pieces.* Freeport NY: Books for Libraries (1935, reprinted 1967), pp. 3–34.

Frazer JG. *The golden bough. A study in comparative religion.* London: Macmillan (1890).

Frazer JG. *The golden bough. A study in magic and religion.* Abridged edition. New York: Macmillan (1922).

Frazer JG. Totemism. *Encyclopædia Britannica*, Ninth edition 23:467–476 (1904).

Frederick II. *Reliqva librorvm Friderici II. Imperatoris: De arte venandi cvm avibvs. Cvm Manfredi Regis additionibvs.* IG Schneider (ed.). In two volumes. Lipsiæ: Mvlleri (1788–1789).

Frederick II. *The art of falconry, being the* De arte venandi cum avibus *of Frederick II of Hohenstaufen.* CA Wood & FM Fyfe (trans. & ed.). Boston: Branford (1943).

Freeman MB. *The Unicorn Tapestries.* New York: Metropolitan Museum of Art (1976).

Freeman WFXR. Sources of Dante's inspiration. *The Fordham Monthly* 40:74–77 (December 1921).

Freig JT [Freigius]. *Quæstiones . . . seu logicæ & ethicæ.* Basileæ: Henricpetri (1576).

Freig JT [Freigius]. *Qvæstiones physicæ . . . Libris XXXVI.* Basileæ: Henricpetri (1579).

Freig JT [Freigius]. *Qvæstiones physicæ . . . Libris XXXVI.* Basileæ: Henricpetri (1585).

French J [as "J.F."]. To the Reader. In: *The Divine Pymander of Hermes Mercurius Trismegistus, in XVII. books.* London: Robert White (1650), folios A2r–A7v (numb. sep.).

Freud S. Arbeiten aus dem zoologisch-vergleichend-anatomischen Institute der Universität Wien. VII. Beobachtung über Gestaltung und feineren Bau der als Hoden beschriebenen Lappenorgane des Aals. *Sitzungsberichte der Mathematisch-Naturwissenschaftliche Classe der Kaiserlichen Akademie der Wissenschaften (Wien)* 75(1):419–431 (1877).

Frey H & Leuckart R. *Beiträge zur Kenntniss wirbelloser Thiere mit besonderer Berückschtigung der Fauna des Norddeutschen Meeres.* Branuschweig: Vieweg (1847).

Friedländer M. Analysis of the Guide for the perplexed. In: M Maimonides, *The guide for the perplexed.* M Friedländer (trans.). Fourth edition. New York: Dutton (1904), pp. xxxix–lix.

Fries E[M]. *Systema mycologicum, sistens fungorum ordines, genera et species, huc usque cognitas, quas ad normam methodi naturalis determinavit, disposuit atque descripsit.* Volumes 1–2, Lundæ: Berlingiana (1821–1823); volume 3, Gryphiswaldae: Mauritius (1829–1832).

Fries E[M]. *Systema orbis vegetabilis. Primas lineas novæ constructionis.* Lundæ: Typographia Academica (1825).

Fritsch FE. *The structure and reproduction of the algae.* In two volumes. Cambridge UP (1935 and 1945).

Fruton JS. Contrasts in scientific style. Emil Fischer and Franz Hofmeister: their research groups and their theory of protein structure. *Proceedings of the American Philosophical Society* 129(4):313–370 (1985).

Führer M. Albert the Great. *The Stanford Encyclopedia of Philosophy* (Summer 2018 edition), ER Zalta (ed.), plato.stanford.edu/archives/sum2018/entries/albert-great, accessed 31 October 2018.

Fuller H. Cassin of Philadelphia. http://www.towhee.net/history/cassin.html, undated. Accessed 3 June 2020 and 9 October 2020.

Gaillon B. *Aperçu d'histoire naturelle et observations sur les limites qui séparent le Règne Végétale du Règne Animal. (Lu à la Société d'Agriculture, du Commerce et des Arts, de Boulogne-sur-mer, dans sa séance publique du 19 septembre 1832.)* Boulogne: Le Roy-Mabille (1833).

Gaillon B. Aperçu d'histoire naturelle, ou observations sur les limites qui séparent le Règne Végétale du Règne Animal. *Annales des Sciences Naturelles, 2 Série. Botanique* 1:44–56 (1834).

Gaillon B. Aperçu microscopique et physiologique de la fructification des thalassiophytes symphysistées. *Annales Générales des Sciences Physiques* 8:362–372 (1821).

Gaillon B. Des huîtres vertes, et des causes de cette coloration. *Journal de Physique, de Chimie, d'Histoire Naturelle et des Arts* 91:222–225 (1820).

Gaillon [B]. Observations microscopiques et physiologiques sur l'Ulva intestinalis. *Precis Analytique des Travaux de l'Académie Royale des Sciences, Belles-Lettres et Arts de Rouen* 1821:33–37 (1822).

682 REFERENCES

Gaillon B. Observations microscopiques sur le Conferva comoïdes, Dillw. *Annales des Sciences Naturelles* 1:309–321 (1824).

Gaillon B. Sur les causes de la couleur verte que prennent les huîtres des pures à certaines époques de l'année. *Précis Analytique des Travaux de l'Académie Royale des Sciences, Belles-Lettres et Arts de Rouen* 1820:90–96 (1821).

Gaillon B. Tableaux synoptiques et méthodiques des genres des Némazoaires. *Mémoires de la Société Royale d'Émululation d'Abbeville* 1833:469–482 (1834).

Galen C. *De usu partium*. In: Clavdii Galeni opera omnia. CG Kühn (ed.), volumes 3 and 4. Lipsiae: Cnoblochii (1822).

Galen C. *On the usefulness of the parts of the body (Peri chreias morion.) (De usu partium.)* MT May (trans.). Ithaca NY: Cornell UP (1968).

Galen. *De alimentorum facultatibus*. In: *Clavdii Galeni opera omnia*. CG Kühn (ed.). Lipsiae: Cnobloch (1821–1833), volume 6 (1825), pp. 453–748. English translation as: Galen. *On the properties of foodstuffs (De alimentorum facultatibus)*. O Powell (trans.). Cambridge UP (2003).

Galen. *On the doctrines of Hippocrates and Plato*. P de Lacy (ed. and trans.). Third edition. Corpvs Medicorvm Graecorvm edidervnt Academiae Berolinensis Havniensis Lipsiensis Volume 4, 1,2 Galeni. De placitis Hippocratis et Platonis. In three volumes. Berlin: Akademie-Verlag (1984).

Galinsky GK. *Ovid's* Metamorphoses. *An introduction to the basic aspects*. Berkeley: University of California Press (1975).

Gambarotto A. Lorenz Oken (1779–1851): *Naturphilosophie* and the reform of natural history. *British Journal for the History of Science* 50(2):329–340 (2017).

Garg SG, Kapust N, Lin W, *et al.* Anomalous phylogenetic behavior of ribosomal proteins in metagenome-assembled Asgard archaea. *Genome Biology and Evolution* 13(1):1–12 (2021).

Garnett T. *Popular lectures on zoonomia, or the laws of animal life, in health and disease*. London: Royal Institution (1804).

Gart der Gesundheit. [Moguntiæ]: [Peter Schöffer] (1485).

Gates BT. Introduction: why Victorian natural history? *Victorian Literature and Culture* 35(2):539–549 (2007). Minisymposium, pp. 539–694.

Gebhardt JG. *The Syriac Clementine Recognitions and Homilies: the first complete translation of the text*. Nashville: Grave Distractions (2014).

Geddes P. A re-statement of the cell theory, with applications to the morphology, classification, and physiology of protists, plants, and animals. Together with an hypothesis of cell-structure, and an hypothesis of contractility. *Proceedings of the Royal Society of Edinburgh* 12(115):266–292 (1883–1884).

Gegenbaur C. *De animalium plantarumque regni terminis et differentiis. Programma quo ad orationem pro loco in medicorum ordine ienensi*. Lipsiae: Breitkopf & Haertel [1860].

Gegenbaur C. *Grundzüge der vergleichenden Anatomie*. Leipzig: Wilhelm Engelmann (1859).

Gegenbaur C. *Grundzüge der vergleichenden Anatomie*. Second edition. Leipzig: Wilhelm Engelmann (1870).

Gegenbaur C. *Zur Lehre vom Generationswechsel und der Fortpflanzung bei Medusen und Polypen*. Würzburg: Stahel (1854).

Geison GL. The protoplasmic theory of life and the vitalist-mechanist debate. *Isis* 60(3):272–292 (1969).

Gelbart NR. Introduction. In: B Fontenelle, *Conversations on the plurality of worlds*. Berkeley: University of California Press (1990), pp. vii–xxxii.

Geldart AM. Sir James Edward Smith and some of his friends. *Transactions of the Norfolk and Norwich Naturalists' Society* 9:645–692 (1914).

Gelli G. *La Circe*. Vinegia [Venezia]: Agostino Bindoni (1550).

Gelli GB. *The Circe of Signior Giovanni Battista Gelli . . . Consisting of ten dialogues between men transform'd into beasts*. T Brown (trans.). Second edition. London: J.N. (1710).

Gemistus Pletho G. *De Platonicæ et Aristotelicæ philosophiæ differentia* (1439). In: JP Migne (ed.), *Patrologia cursus completus. Series Græca posterior* 160. Parisiis: Migne (1866), cols 889–934. With preface by G Chariander (1574), cols 865–888.

REFERENCES 683

Gemistus Pletho G. *Georgii Gemisti Plethonis De Platonicae atqve Aristotelicæ philosophiæ differentia, libellus, ex Græca lingua in Latinam conuersus.* G Chariander (trans.). Basileæ: Pernam (1574).

Genesis Rabbah. In: H Freedman & M Simon (eds), *Midrash Rabbah.* In ten volumes. Volume 1. H Freedman (trans.). Third edition. London: Soncino (1961).

Geoffroy l'Aîné [ÉF]. De differents rapports observés en chimie entre differentes substances. *Mémoires de Mathématique et de Physique de l'Académie Royale des Sciences (Paris)* Année 1718:202–212 (1741).

Geoffroy [Saint-Hilaire É]. Mammiferes. Mémoire sur les rapports naturels des Makis Lemur, L. et description d'une espèce nouvelle de mammifère. *Magasin Encyclopédique* 1796(1):20–50 (1796).

Geoffroy Saint-Hilaire [É]. Mémoires sur l'organisation des insectes. Troisième mémoire, sur une colonne vertébrale et ses côtes dans les insectes apiropodes; lu à l'Académie des sciences, le 12 février 1820. *Journal Complémentaire du Dictionaire des Sciences Médicales* 6:138–158 (1820).

Geoffroy Saint-Hilaire [É]. Mémoires sur l'organisation des insectes. Troisième mémoire, sur une colonne vertébrale et ses côtes dans les insectes apiropodes; lu à l'Académie des Sciences, le 12 février 1820. *Isis von Oken* 1820, Litterarischer Anzeiger 1: cols 527–552 (1820). With commentary by Oken, cols 552–559.

Geoffroy Saint-Hilaire [É]. *Principes de philosophie zoologique, discutés en Mars 1830, au sein de l'Académie Royale des Sciences.* Paris: Pichon & Didier; et Rousseau (1830).

Geoffroy-Saint-Hilaire [É]. *Philosophie anatomique.* In two volumes. Paris: Ballière (volume 1, 1818); and l'Auteur (volume 2, 1822).

Geoffroy-Saint-Hilaire [É] & Cuvier F. *Histoire naturelle des mammifères.* In seven volumes. Paris: Belin (volumes 1–6, 1824); and Blaise (volume 7, 1842).

Geoffroy [Saint Hilaire É de] & Cuvier [G]. Mammalogie. Mémoire sur une nouvelle division des mammifères et sur les principes que doivent servir de base dans cette sorte de travail, lu à la société d'Histoire naturelle, le 1 floréal de l'an 3. *Magasin Encyclopédique* 1795(2):164–190 (An 3, 1795).

Geoffroy Saint-Hilaire I. *Histoire naturelle générale des règnes organiques.* In three volumes. Paris: Masson (1854–1862).

Georr Kh. *Les Catégories d'Aristote dans leurs versions syro-arabes.* Beirut: Institut français de Damas (1948).

Gerard J. *The herball or generall historie of plantes.* London: John Norton, London (1597).

Gerard J. *The herball or generall historie of plantes. Gathered by John Gerarde of London . . . Very much enlarged and amended by Thomas Johnson.* London: Islip, Norton & Whitakers (1633).

Gerard J. *The herball or generall historie of plantes. Gathered by John Gerarde of London . . . Very much enlarged and amended by Thomas Johnson.* London: Islip, Norton & Whitakers (1636).

Gersh S. *From Iamblichus to Eriugena. An investigation of the prehistory and evolution of the Pseudo-Dionysian tradition.* Leiden: Brill (1978).

Gesner J. Dissertationes physicæ de vegetabilibus [1743]. In: C von Linné (ed.), *Oratio de necessitate perergrinationum intra patriam.* [Bound with:] J Browallius, *Examen epicriseos Siegesbeckianæ in systema plantarum sexuale.* [And:] J Gesner . . . *Dissertationes de partium vegetationis et fructificationis structura, differentia et usu, in quibus Elementa Botanica dilucide explicantur.* Lugduni Batavorum: Haak (1743), pp. 55–108 [pag. sep. in part.].

Gessner C. *Appendix historiæ quadrupedum uiuiparorum & ouiparorum.* Tigvri: Froschover (1554).

Gessner C. *De rervm fossilivm, lapidvm et gemmarvm maximè, figvris & similitudinibus liber.* Tiguri: [Iacobus Gesnerus] (1565).

Gessner C. *De scorpione.* Tigvri: Froschoviana (1587).

Gessner [Gäßner] C. *Fischbůch das ist ein kurtze, doch vollkom[m]ne beschrybung aller Fischen so in dem Meer unnd süssen wasseren.* Zürich: Christoffel Froschower (1575).

Gessner C. *Historiæ animalium lib. V, qui est de serpentium natura.* Tigvri: Froschoviana (1587).

Gessner C. *Historiæ animalium liber II. Qui est ide quadrupedibus ouiparis.* Francofvrdi: Ioannis Wechel (1586).

684   REFERENCES

Gessner C. *Historiæ animalium. Lib. I. De quadrupedibus uiuiparis* (1551). *Lib. II. De quadrupedibus ouiparis* (1554). *Lib. III. Qui est auium natura* (1555). *Lib. IIII. Qui est de piscium & aquatilium animantium natura* (1558). Tigvri: Froschover.

Gessner C. *Icones animalivm qvadrvpedvm viviparorvm et oviparorvm.* Tigvri: Froschover (1553).

Gessner C. *Icones animalivm qvadrvpedvm viviparorvm et oviparorvm.* Editio secunda . . . [With:] *Icones avivm omnivm.* [Also:] *Nomenclator aqvatilivm animantivm. Icones animalivm aquatilium in mari & dulcibus aquis degentium.* Tigvri: Froschover (1560).

Gessner [Geßner] C. *Schlangenbüch das ist ein grundtliche und vollkom[m]ne Beschreybung aller Schlangen.* Zürych: Froschow (1589).

Gessner [Gäßner] C. *Thierbůch das ist ein kurtze beschrybung aller vier füssigen Thieren.* Zürich: Christoffel Froschouwer (1583).

Gessner [Geßner] C. *Vogelbuch darinn die art, natur unnd eigenschafft aller vöglen, sampt jrer waren Contrafactur, angezeigt wirdt.* Zürych: Christoffel Froschouer (1581).

Gessner [Geßner] C. *Vogelbüch oder Ausführliche beschreibung und lebendige ja auch eygentlĭche Controfactur und Abmahlung aller und jeder Vögel.* Franckfurt: Johann Saurn (1600).

Geyer B (ed.). Peter Abelards philosophische Schriften. I. Die Logica Ingredientibus. 1. Die Glossen zur Porphyrius. *Beiträge zur Geschichte der Philosophie des Mittelalters* 21(1). Münster: Aschendorff (1919), pp. 1–32.

al-Ghassānī, Qāsim ibn-Muḥammad al-Wazīr. *Ḥadīqat al-azhār fī, sarḥ māhīyat al-ʿushb wa al-ʿaqqār.* Fez(?): [n.p.] (1586).

al-Ghazālī. *Maqāṣid al-falāsifa (Doctrines of the philosophers).* In Latin translation [by Dominicus Gundissalinus and Magister Johannes] as *Logica et philosophica Algazelis Arabis.* Veneto: Petri Liechtensteyn (1506). Reprinted by Minerva, Frankfurt (1969).

al-Ghazālī. *The incoherence of the philosophers. A parallel English-Arabic text.* ME Marmura (trans.). Second edition. Provo UT: Brigham Young UP (2000).

Giacone R. Masters, books and library at Chartres according to the cartularies of Notre-Dame and Saint-Père. *Vivarium* 12:30–51 (1974).

Gianelli C. Di alcune versioni e rielaborazioni serbe delle "Solutiones breves quaestionum naturalium" attribute a Michele Pselle. *Atti del V Congresso Internazionale di Studi Bizantini: Roma, 20–26 Settembre 1936. Studi Bizantini e Neoellenici* 5:445–468 (1939).

Gibian G. CG Carus' [sic] *Psyche* and Dostoevsky. *The American Slavic and East European Review* 14(3):371–382 (1955).

Gibson S. *Animal, vegetable, mineral? How eighteenth-century science disrupted the natural order.* Oxford UP (2015).

Gibson S. On being an animal, or, the Eighteenth-century zoophyte controversy in Britain. *History of Science* 50(4):453–476 (2012).

Gigante D. *Life: organic form and Romanticism.* Yale UP (2009).

Gilbert W. Towards a paradigm shift in biology. *Nature* 249(6305):99 (1991).

Gilbert WS & Sullivan A. *The Major General's Song,* from: *The Pirates of Penzance; or, The Slave of Duty* (1879). Versions of Gilbert's libretto are available at https://gsarchive.net/pirates/html/bond_intro.html, accessed 9 October 2020.

Giles HA. *Chuang Tzŭ: mystic, moralist, and social reformer.* London: Bernard Quaritch (1889).

Gilgamesh (The epic of). NK Sandars (trans.). Revised edition. Harmondsworth: Penguin (1960).

Gill C. Galen and the Stoics: mortal enemies or blood brothers? *Phronesis* 52(1):88–120 (2007).

Gill ML & Lennox JG (eds). *Self-motion: from Aristotle to Newton.* Princeton UP (1994).

Gillespie CC. Science in the French Revolution. *PNAS* 45(5):677–684 (1959).

Gilroy CG. *The history of silk, cotton, linen, wool, and other fibrous substances.* New York: Harper (1845).

Gilson E. *Dante the philosopher.* London: Sheed & Ward (1948).

Ginsborg H. Kant's aesthetics and teleology. *The Stanford Encyclopedia of Philosophy* (Winter 2019 edition), EN Zalta (ed.), http://plato.stanford.edu/archives/win2019/entries/kant-aesthetics/, accessed 9 January 2020 and 3 October 2020.

Giraldus Cambrensis. *The historical works of Giraldus Cambrensis containing The topography of Ireland, and the history of the conquest of Ireland.* T Forester (trans.). *The itinerary through*

REFERENCES 685

Wales, and the description of Wales. RC Hoare (trans.). T Wright (ed.). London: George Bell (1905).

Girod-Chantrans J. *Recherches chimiques et microscopiques sur les conferves, bisses, tremelles, etc.* Paris: Bernard (An 10, 1802).

Giseke PD. In mappam affinitatum genealogico-geographicam commentarius. In: C Linné von. *Prælectiones in ordines naturales plantarum.* PD Giseke (ed.). Hamburgi: Hoffmanni (1792), pp. 623–627. With: Tabula genealogico-geográphica affinitatum plantarum. Secundum ordines naturales Linnæi delineavit (1789).

Gjesdal K. Georg Friedrich Philipp von Hardenberg [Novalis]. *The Stanford Encyclopedia of Philosophy* (Fall 2014 edition), EN Zalta (ed.), https://plato.stanford.edu/archives/fall2014/entries/novalis/, accessed 12 January 2020 and 3 October 2020.

Glanvill J. *Lux orientalis, or an enquiry into the opinion of the Eastern sages concerning the præexistence of souls.* In: H More (ed.), *Two choice and useful treatises.* London: Collins & Lowndes, London (1682), pag. sep.

Glanvill J. *Scepsis scientifica: or, contest ignorance, the way to science.* London: Cotes (1665).

Glass B. Maupertuis, pioneer of genetics and evolution. In: B Glass, O Temkin & WL Straus Jr (eds), *Forerunners of Darwin: 1745–1859.* Baltimore: Johns Hopkins UP (1968), pp. 51–83.

Glendinning P. *The* Summa *in verse* (2015). https://summainverse.wordpress.com, accessed 6 November 2018 and 14 September 2020.

Glick TF. *The comparative reception of Darwinism.* Second edition. University of Chicago Press (1988).

Gnostic Society Library (The). http://gnosis.org, accessed 18 December 2018 and 16 September 2020.

Godwin W. *Lives of the necromancers.* London: Frederick J Mason (1834).

Goedart J. *Metamorphosis et historia naturalis insectorum.* In three volumes. Medioburgi: Jacobum Fierensium (1662–1669).

Goethe [JW von]. *Aus meinem Leben. Dichtung und Wahrheit.* In three volumes. Stuttgart & Tübingen: Cotta (1811–1814).

Goethe JW von. Briefe an Zelter, 7 and 14 November 1816. At: www.zeno.org/Literatur/M/Goethe,+Johann+Wolfgang/Briefe/1816, accessed 26 August 2019 and 28 September 2020.

Goethe JW von. *Christus nebst zwölf alt- und neutestamentlichen Figuren den Bildhauren vorgeschlagen. IX. Matthäus der Evangelist.* In: *Goethe's sämmtliche Werke in vierzig Bänden. Vollständige, neugeordnete Ausgabe* 31. Stuttgart & Tübingen (1840), pp. 292–300 (1840).

[Goethe JW von]. *Die Leiden des jungen Werthers.* Two parts in one volume. Leipzig: Weygand (1774).

Goethe JW von. Essay on the metamorphosis of plants. EM Cox (trans.). *Journal of Botany, British and Foreign* 1:327–345 and 360–374 (1863).

Goethe W [von]. *Faust. Eine Tragödie.* In: *Goethe's Werke,* Band 8. Tübingen: Cotta (1808), pp. 1–234.

Goethe W von. *Faust. Der Tragödie zweyter Theil in fünf Acten.* In: *Vollständige Ausgabe letzter Hand* [VA] Band 41. Tübingen: Cotta (1832), pp. 1–344.

Goethe JW von. *Goethe. Die Schriften zur Naturwissenschaft.* D Kuhn (ed.). In 21 volumes. Weimar: Böhlaus (1977). [LA]

Goethe JW von. *Goethe's letters to Zelter.* AD Coleridge (ed.). London: George Bell (1892).

Goethe JW von. *Goethe's Werke. Vollständiger Ausgabe letzter Hand.* In 61 volumes. Stuttgart & Tübingen: Cotta (1827–1835). Volumes 55–60 (Nachgelassene Werke 16–20, and index), 1842. [VA]

Goethe JW von. *Goethes Briefe. Hamburger Ausgabe.* KR Mandelkow (ed.). In four volumes. München: Wegner (1962–1967). Reprinted by Beck, München (1972 ff.).

Goethe JW von. *Goethes Italienische Reise.* Leipzig: Insel (1914).

Goethe JW von. *Goethes Sämtliche Werke: Jubiläums-Aufgabe in 40 Bänden.* E von der Hellen (gen. ed.). In 41 volumes including index. Stuttgart: Cotta (undated, 1902–1907). [JA]

Goethe JW von. *Goethes Werke. Hamburger Ausgabe in 14 Bänden.* E Trunz (ed.). Hamburg: Wegner (1948–1960). Reprinted by Beck, München (1981). [HA]

686 REFERENCES

Goethe [J]. *Leiden des jungen Werther*. Leipzig: Göschen (1787).

Goethe JW von. *Sämtliche Werke nach Epochen seines Schaffens, Münchner Ausgabe*. K Richter *et al*. (eds). In 21 volumes. München: Hanser (1985–1998). [MA]

Goethe JW von. *Versuch die Metamorphose der Pflanzen zu erklären*. Gotha: Ettinger (1790).

Gogarten JP, Kibak H, Dittrich P, *et al*. Evolution of the vacuolar H+-ATPase: implications for the origin of eukaryotes. *PNAS* 86(17):6661–6665 (1989).

Goldfuss GA. *Handbuch der Zoologie*. In two volumes. Nürnberg: Schrag (1820).

Goldfuss GA. *Ueber die Entwicklungsstufen des Thieres. Omne vivum ex ovo, ein Sendschreiben an Herrn Dr. Nees v. Esenbeck*. Nürnberg: Schrag (1817).

Golding GB & Gupta RS. Protein-based phylogenies support a chimeric origin for the eukaryotic genome. *Molecular Biology and Evolution* 12(1):1–6 (1995).

Goldsmith O. *An history of the earth, and animated nature*. In eight volumes. London: Nourse (1774).

Gollancz H (trans.). *Dodi Ve-nechdi (Uncle & nephew). The work of Berachya Hanakdan, now edited from MSS. at Munich and Oxford, with an English translation, introduction etc. to which is added the first English translation from the Latin of Adelard of Bath's Quaestiones naturales*. London: Humphrey Milford; and Oxford UP (1920).

Gooch J. The effects of the Condemnation of 1277. *The Hilltop Review* 2(1):34–44 (2006).

Good JM. *Book of nature*. In three volumes. London: Longman, Rees *et al*. (1826).

Goodman LE & R McGregor (trans.). *The case of the animals versus man before the King of the Jinn. A translation from the Epistles of the Brethren of Purity*. Oxford UP (2009).

Goodman LE. al-Rāzī. *The encyclopaedia of Islam*, New [Second] edition 8:474–477. Leiden: Brill (1995).

Goodman LE. Ghazâlî's argument from creation. *International Journal of Middle East Studies* 2(1):67–85 and 2(2):168–188 (1971).

Gordh G & Headrick D. *A dictionary of entomology*. Second edition. Wallingford UK: CABI (2011).

Gordon CA. *A bibliography of Lucretius*. London: Rupert Hart-Davis (1962).

Goren Y. Goren collection of the history of the microscope. http://microscopehistory.com/schi eck-in-berlin (undated), accessed 12 April 2020 and 7 October 2020.

Görner R. Granit: zur Poesie eines Gesteins. *Aurora (Jahrbuch der Eichendorff-Gesellschaft für die klassische-romantische Zeit)* 53:126–138 (1993). Also published in: R Görner, *Goethe. Wissen und Entsagen—als Kunst*. Munich: Iudicium (1995), pp. 50–62.

Gosse PH. *Evenings at the microscope; or, researches among the minuter organs and forms of animal life*. London: Society for Promoting Christian Knowledge (undated, 1859).

Gosse PH. *Life in its lower, intermediate, and higher forms*. [First edition]. London: James Nisbet (1857).

Gosse PH. *Life in its lower, intermediate, and higher forms*. Fifth edition. London: James Nisbet (1878).

Gould SJ. *Ontogeny and phylogeny*. Cambridge MA: Belknap, Harvard UP (1977).

Gould SJ. Tempo and mode in the macroevolutionary reconstructionism of Darwinism. *PNAS* 91(15):6764–6771 (1994).

Gouldman F. *A copious dictionary in three parts*. Fourth edition. Cambridge: John Hayes (1678).

Govi G. Il microscopio composto inventato da Galileo. *Atti della Reale Accademia delle Scienze Fisiche e Matematiche, 2 Ser*. 2(1):117 (1888). English translation as: The compound microscope invented by Galileo. *Journal of the Royal Microscopical Society, London* 1889(4):574–598 (1889).

Graf G. *Geschichte der christlichen arabischen Literatur*. In five volumes. Città del Vaticano: Bibliotheca Apostolica Vaticana (1944–1953).

Graham AC & Sivin N. A systematic approach to the Mohist optics (ca. 300 BC). In: S Nakayama & N Sivin (eds), *Chinese science. Explorations of an ancient tradition*. Cambridge MA: MIT Press (1973), pp. 105–152.

[Grandjean de Fouchy JP]. Sur un mouvement spontané observé dans la plante appelée *Tremella*. *Histoire de l'Académie Royale des Sciences* 1767:75–78 (1770).

Granick S. Plastid structure, development and inheritance. In: W Ruhland (ed.), *Handbuch der Pflanzenphysiologie. Encyclopædia of plant physiology*, volume 1. Berlin: Springer (1955), pp. 507–563.

Grant AJ (trans.). *Early lives of Charlemagne by Eginhard and the Monk of St Gall*. London: Chatto & Windus (1905).

Grant E. *A history of natural philosophy. From the ancient world to the Nineteenth Century*. Cambridge UP (2007).

Grant E. *A source book in Medieval science*. Cambridge MA: Harvard UP (1974).

Grant E. Medieval and Seventeenth-century conceptions of an infinite void space beyond the cosmos. *Isis* 60(1):39–60 (1969).

Grant E. *Planets, stars, and orbs: the medieval cosmos, 1200–1687*. Cambridge UP (1994).

Grant E. *Science & religion, 400 BC to AD 1550. From Aristotle to Copernicus*. Baltimore: Johns Hopkins UP (2004).

Grant E. The Condemnation of 1277, God's absolute power, and physical thought in the late Middle Ages. *Viator* 10: 211–244 (1979).

Grant E. *The nature of natural philosophy in the late Middle Ages*. Washington: Catholic University of America Press (2010).

Grant RE. Observations on the spontaneous motions of the ova of the Campanularia dichotoma, Gorgonia verrucosa, Caryophyllea calycularis, Spongia panicea, Sp. papillaris, cristata, tomentosa, and Plumularia falcata. *Edinburgh New Philosophical Journal* 1:150–156 (1826).

Graves FP. *Peter Ramus and the educational reformation of the Sixteenth century*. PhD thesis, Columbia University (1912).

Graves R. *The white goddess. A historical grammar of poetic myth*. Amended and enlarged edition. London: Faber & Faber (1961).

Gray A. *Natural selection not inconsistent with natural theology. A free examination of Darwin's treatise on the origin of species, and of its American reviewers*. London: Trübner; and Boston: Ticknor & Fields (1861).

Gray MW. Organelle origins and ribosomal RNA. *Biochemistry & Cell Biology* 66(5):325–348 (1988).

Gray SF. *A natural arrangement of British plants, according to their relations to each other, as pointed out by Jussieu, De Candolle, Brown, &c*. Two volumes. London: Baldwin, Cradock & Joy (1821).

Green T. *The universal herbal; or, botanical, medical and agricultural dictionary. Containing an account of all the known plants in the world, arranged according to the Linnean system*. In two volumes. Liverpool: Caxton [1816].

Green T. *The universal herbal; or, botanical, medical, and agricultural dictionary*. Second edition. In two volumes. London: Caxton (1824).

Green TM. *The city of the Moon God. Religious traditions of Harran*. In series: Religions in the Graeco-Roman world (R van den Broek *et al.*, eds) 114. Leiden: Brill (1992).

Greenaway F. *John Dalton and the atom*. Ithaca NY: Cornell UP (1966).

Greenblatt S. *The swerve: how the world became modern*. New York: Norton (2011).

Greene EL. *Landmarks of botanical history*. FN Egerton (ed.). Volumes 1 (1909, reprinted 1983) and 2 (1983). Stanford UP.

Greene JC. *The death of Adam. Evolution and its impact on Western thought*. Ames: Iowa State University Press (1959).

Greenwood AD, Griffiths HB & Santore UJ. Chloroplasts and cell compartments in Cryptophyceae [abstract]. *British Phycological Journal* 12(2):119 (1977).

Gregory ME. *Diderot and the metamorphosis of species*. New York: Routledge (2007).

Gregory of Nazianzus. Select orations and letters. In: P Schaff & H Wace (eds), *A select library of Nicene and Post-Nicene Fathers of the Christian Church*. Second series, volume 7. New York: Christian Literature (1894), pp. 203–482.

Gregory of Nyssa. *De hominis opificio*. In: JP Migne (ed.), *Patrologiæ cursus completa. Series Græca prior* 44. Parisiis: Migne (1863), cols 123–256.

688 REFERENCES

Gregory of Nyssa. *De hominis opificio*. W Moore & HA Wilson (trans.), in: P Schaff & H Wace (eds), *A select library of Nicene and post-Nicene Fathers of the Christian Church*. Second series, volume 5. New York: Christian Literature Company (1893), pp. 386–427.

Gregory of Nyssa. *Explicato apologetica ad Petrum Fratrem, in Hexaemeron*. In: JP Migne (ed.). *Patrologiæ cursus completa. Series Græca prior* 44. Parisiis: Migne (1863), cols 61–124.

Gregory of Nyssa. *Gregorii Nysseni In Hexaemeron. Opera exegetica in Genesim*, pars I. HR Drobner (ed.). Leiden: Brill (2009).

Gregory of Nyssa. *On the making of man*. In: P Schaff & H Wace (eds), *A select library of Nicene and Post-Nicene Fathers of the Christian Church*. Second series, volume 8. New York: Christian Literature (1895), pp. 386–427.

Greswell WP. *A view of the early Parisian Greek press: including the lives of the Stephani*. In two volumes. Oxford: Collingwood (1833).

Greville RK. *Flora Edinensis: or a description of plants growing near Edinburgh, arranged according to the Linnean System. With a concise introduction to the natural orders of the Class Cryptogamia, and illustrative plates*. Edinburgh: Blackwood; and London: Cadell (1824).

Grew N. *The anatomy of plants. With an idea of a philosophical history of plants*. [London]: Rawlins (1682).

Grew N. *The anatomy of vegetables begun*. London: Spencer Hickman (1672).

Griffel F. Al-Ghazali. *The Stanford Encyclopedia of Philosophy* (Winter 2019 edition), EN Zalta (ed.), https://plato.stanford.edu/archives/win2019/entries/al-ghazali/, accessed 11 March 2020.

Griffel F. *Al-Ghazali's philosophical theology*. Oxford UP (2009).

Griffel F. *Taqlīd* of the philosophers: al-Ghazālī's initial accusation in his *Tahāfut*. In: S Günther (ed.), *Ideas, images, and methods of portrayal. Insights into classical Arabic literature and Islam*. Leiden: Brill (2005), pp. 273–296.

Griffel F. The Western reception of al-Ghazālī's cosmology from the Middle Ages to the 21st century. *Dîvân: Disiplinerarasi Çalişmalar Dergisi* 16(30):33–62 (2011).

Griffith F. The significance of pneumococcal types. *Journal of Hygiene* 27(2):113–159 (1928).

Griffith RTH (trans.). *The Hymns of the Rig Veda*. In four volumes. Benares: Lazarus (1889–1892.

Griffith RTH (trans.). *The Hymns of the Rig Veda*. Second edition. In two volumes. Benares: Lazarus (1896).

Griffiths AB. *A manual of bacteriology*. London: Heinemann (1893).

Groner J. Pallas, Peter Simon, MD. In: J Groner & PFS Cornelius, *John Ellis*. Pacific Grove CA: Boxwood (1996), pp. 282–283.

Grosseteste R. *On the Six Days of Creation. A translation of the Hexaëmeron*. CFJ Martin (trans.). In series: Auctores Britannici Medii Aevi 6(2). British Academy, and Oxford UP (1996).

Gruithuisen [F von P]. [Untitled]. *Oberdeutschen Literatur-Zeitung* Nr 110 (4 Okt 1808), cols 588–592; Nr 111 (6 Okt 1808), cols 606–608; and Nr 112 (8 Okt 1808), cols 619–624.

Gruithuisen F v[on] P. Die Branchienschnecke und eine aus ihren Ueberresten hervorwachsende lebendig-gebaehrende conferve. *Verhandlungen der Kaiserlichen Leopoldinisch-Carolinischen Akadademie der Naturforscher / Nova Acta Physico-Medica Academiæ Cæsareæ Leopoldino-Carolinæ Naturæ Curiosorum* 10(2):437–452 (1821).

Gudger EW. Pliny's *Historia naturalis*. The most popular natural history ever published. *Isis* 6(3):269–281 (1924).

Guillaume, Abbot of St-Theodore. *De natura corporis et animæ libri duo*. In: JP Migne (ed.), *Patrologiæ cursus completus. Series secunda* 180. Parisiis: Migne (1855), cols 695–726.

Guillaume de Lorris & Jean de Jandun. *Roman de la rose*. Nouvelle édition. M Méon (ed.). In four volumes. Paris: Didot (1814).

Guilliermond A. Mitochondrie et symbiotes. *Comptes Rendus Hebdomadaires des Séances et Mémoires de la Société de Biologie* 82:309–312 (1919).

Guilliermond A. Sur l'origine mitochondriale des plastides a propos d'un travail de M. Mottier. *Annales des Sciences Naturelles. Botanique, 10 série* 1(1):225–246 (1919).

Guiry M. AlgaeBase. https://www.algaebase.org.

Gulick A. The chemistry of the chromosomes. *Botanical Review* 7(9):433–457 (1941).

REFERENCES 689

Gulick A. What are the genes? II. The physico-chemical picture; conclusions. *Quarterly Review of Biology* 13(2):140–168 (1938).

Gunn BG. *The wisdom of the East: the instruction of Ptah-Hotep and the instruction of Ke'Gemni: the oldest books in the world*. London: Murray (1906).

Gunther RT. *Early science in Oxford*, volume 3. London: Dawsons (1952).

Gunther RWT (ed.). *Further correspondence of John Ray*. London: Ray Society (1928).

Gupta RS. Life's third domain (*Archaea*): an established fact or an endangered paradigm? *Theoretical Population Biology* 54(2):91–104 (1998).

Gutas D. Ibn Sina [Avicenna]. *The Stanford Encyclopedia of Philosophy* (Fall 2016 edition), EN Zalta (ed.), https://plato.stanford.edu/archives/fall2016/entries/Ibn-sina/, accessed 31 August 2018.

Guthke KS. *Der Mythos der Neuzeit*. Bern: Francke (1983).

Guthke KS. *The last frontier. Imagining other worlds, from the Copernican revolution to modern science fiction*. H Atkins (trans.). Ithaca NY: Cornell UP (1990).

Guthrie KS. *The Pythagorean sourcebook and library. An anthology of ancient writings which relate to Pythagoras and Pythagorean philosophy*. Grand Rapids MI: Phanes (1987).

Gyekye K. *Arabic logic: Ibn al-Tayyib's commentary on Porphyry's Eisagoge*. Albany: SUNY Press (1979).

Haas JW Jr. The Reverend Dr William Henry Dallinger, FRS (1839–1909). *Notes and Records of the Royal Society of London* 54(1):53–65 (2000).

Hadas M. Plato in Hellenistic fusion. *Journal of the History of Ideas* 19(1):3–13 (1958).

Haeckel E. *Anthropogenie oder Entwickelungsgeschichte des Menschen*. Second edition. Leipzig: Engelmann (1874).

Haeckel E. *Anthropogenie oder Entwickelungsgeschichte des Menschen*. Fourth edition. In two volumes. Leipzig: Engelmann (1891).

Haeckel E. Beiträge zur Plastidentheorie. *Jenaische Zeitschrift für Medicin und Naturwissenschaft* 5:492–550 (1870).

Haeckel E. *Biologische Studien. Heft 2. Studien zur Gastraea-Theorie*. Jena: Hermann Dufft (1877).

Haeckel E. Crystal souls—studies of inorganic life. AL Mackay (trans.). *Forma* 14(1–2):1–146 (1999).

Haeckel E. *Das Protistenreich. Eine populäre Uebersicht über das Formengebiet der niedersten Lebewesen. Mit einem wissenschaftlichen Anhange: System der Protisten*. Leipzig: Ernst Günther (1878).

Haeckel E. *De telis quibusdam Astaci fluviatilis. Dissertatio inauguralis histologica*. Berolini: Schade (1857).

Haeckel E. Die Catallacten, eine neue Protisten-Gruppe. *Jenaische Zeitschrift für Medicin und Naturwissenschaft* 6:1–22 (1871).

Haeckel E. Die Gastraea-Theorie, die phylogenetische Classification des Thierreichs und die Homologie der Keimblätter. *Jenaische Zeitschrift für Naturwissenschaft* 8:1–55 (1874).

Haeckel E. Die Gastrula und die Eifurchung der Thiere. *Jenaische Zeitschrift für Naturwissenschaft* 9:402–508 (1875).

Haeckel E. *Die Kalkschwämme. Eine Monographie*. Two volumes of text, plus atlas. Berlin: Reimer (1872).

Haeckel E. *Die Lebenswunder. Gemeinverständliche Studien über Biologische Philosophie. Ergänzungsband zu dem Buche über die Welträthsel*. Stuttgart: Alfred Kröner (1904).

Haeckel E. *Die Naturanschauung von Darwin, Goethe und Lamarck*. Jena: Fischer (1882).

Haeckel E. *Die Radiolarien. (Rhizopoda radiaria.) Eine monographie*. Text plus atlas. Berlin: Reimer (1862). Haeckel added three further volumes in 1887–1888.

Haeckel E. *Die Welträthsel. Gemeinverständliche Studien über Monistische Philosophie*. Bonn: Strauß (1899).

Haeckel E. *Generelle Morphologie der Organismen. Allgemeine Grundzüge der organischen Formen-wissenschaft, mechanisch begründet durch die von Charles Darwin reformirte Descendenz-Theorie. Band 1: Allgemeine Anatomie der Organismen. Band 2: Allgemeine Entwickelungsgeschichte der Organismen*. Berlin: Reimer (1866).

690 REFERENCES

Haeckel E. *Kristallseelen. Studien über das anorganische Leben.* Leipzig: Alfred Kröner (1917).

Haeckel E. *Last words on evolution. A popular retrospect and summary.* Translated from the second [German] edition by J McCabe. London: Owen (1906).

Haeckel [Häckel] E. Monograph of Monera. WF Kirby & EP Wright (trans.). *Quarterly Journal of Microscopical Science, New Series* 9:27–42, 113–134, 219–232, and 327–342 (1869).

Haeckel [Häckel] E. Monographie der Moneren. *Jenaische Zeitschrift für Medicin und Naturwissenschaft* 4:64–137 (1868).

Haeckel E. Nachträge zur Gastraea-Theorie. *Jenaische Zeitschrift für Naturwissenschaft* 11:55–98 (1877).

Haeckel E. *Natürliche Schöpfungsgeschichte. Gemeinverständliche wissenschaftliche Vorträge über die Entwickelungslehre im Allgemeinen und diejenige von Darwin, Goethe und Lamarck im Besonderen.* [First edition]. Berlin: Reimer (1868).

Haeckel E. *Natürliche Schöpfungsgeschichte.* Second edition. Berlin: Reimer (1870).

Haeckel E. *Natürliche Schöpfungsgeschichte.* Fourth edition. Berlin: Reimer (1873).

Haeckel E. *Natürliche Schöpfungsgeschichte.* Sixth edition. Berlin: Reimer (1875).

Haeckel E. *Natürliche Schöpfungsgeschichte.* Seventh edition. Berlin: Reimer (1879).

Haeckel E. *Natürliche Schöpfungsgeschichte.* Eighth edition. Berlin: Reimer (1889).

Haeckel E. *Natürliche Schöpfungsgeschichte.* Ninth edition. Berlin: Reimer (1898).

Haeckel [Häckel] E. On the organization of sponges, and their relationship to the corals. *Annals and Magazine of Natural History, 4 Series* 5(25):1–13 and 5(26):107–120 (1870).

Haeckel E. *Plankton-Studien. Vergleichende Untersuchungen über die Bedeutung und Zusammensetzung der pelagischen Fauna und Flora.* Jena: Gustav Fischer (1890).

Haeckel [as Hæckl] E. *Planktonic studies: a comparative investigation of the importance and constitution of the pelagic fauna and flora.* GW Field (trans.). In: *Report of the U.S. Commissioner of Fish and Fisheries for 1889 to 1891.* Washington: Government Printing Office (1893), pp. 565–641. Also published separately.

Haeckel E. Professor Haeckel on Darwin, Goethe, and Lamarck. *Nature* 26(674):533–541 (1882).

Haeckel E. *Systematische Phylogenie. Entwurf eines Natürlichen System der Organismen auf Grund ihrer Stammesgeschichte.* Volume 1: *Systematische Phylogenie der Protisten und Pflanzen* (1894); volume 2: *Systematische Phylogenie der wirbellosen Thiere (Invertebrata)* (1896); volume 3: *Systematische Phylogenie der Wirbelthiere (Vertebrata)* (1895). Berlin: Reimer.

Haeckel E. The gastraea-theory, the phylogenetic classification of the animal kingdom and the homology of the germ-lamellæ. EP Wright (trans.). *Quarterly Journal of Microscopical Science, New Series* 14:142–165 and 223–247 (1874).

Haeckel E. *The history of creation.* [First English, from the fourth German edition]. ER Lankester (trans. rev.). In two volumes. London: King (1876).

Haeckel E. *The wonders of life. A popular study of biological philosophy* (J McCabe, trans.). New York: Harper (1905).

Haeckel E. *Über die Entwickelungstheorie Darwins* [1863]. In: *Gesammelte populäre Vorträge aus dem Gebiete der Entwickelungslehre*, volume 1 [of 2]. Bonn: Emil Strauss (1878), pp. 1–28.

Haeckel E. Ueber den Organismus der Schwämme und ihre Verwandtschaft mit den Corallen. *Jenaische Zeitschrift für Medicin und Naturwissenschaft* 5:207–235 (1870).

Haeckel E. Ueber den Sarcodekörper der Rhizopoden. *Zeitschrift für wissenschaftliche Zoologie* 15:342–370 (1865).

Haeckel E. Ueber die Gewebe des Flusskrebses. *Archiv für Anatomie, Physiologie und wissenschaftliche Medicin* 1857:469–568 (1857).

Haeckel E. Zur Morphologie der Infusorien. *Jenaische Zeitschrift für Medizin und Naturwissenschaft* 7:516–560 (1873).

Hagemann R. Edwin Baur or Carl Correns: who really created the theory of plastid inheritance? *Journal of Heredity* 91(6):435–440 (2000).

Hagen JB. Five kingdoms, more or less: Robert Whittaker and the broad classification of organisms. *BioScience* 62(1):67–74 (2012).

Hageneder O. Die Edition der Register Papst Innocenz' III. https://geschichtsforschung.univie.ac.at/forschung/laufende-forschungsprojekte-des-ioeg/die-edition-der-register-pabst-innocenz-iii/ (undated), accessed 18 September 2020.

Hageneder O. Die päpstlichen Register des 13. und 14. Jahrhunderts. *Annali della Scuola Speciale per Archivisti* 12:45–76 (1972).

Hager P. Chrysippus' [sic] theory of pneuma. *Prudentia* 14(2):97–108 (1982).

Hahm DE. *The origins of Stoic cosmology.* Columbus: Ohio State University Press (1977).

Hakluyt R. *Hakluyt's collection of the early voyages, travels, and discoveries, of the English nation. New edition.* In five volumes. London: Evans, Mackinlay, and Priestley [varies] (1809–1812).

Hall JL & Luck DJL. Basal body-associated DNA: *in situ* studies in *Chlamydomonas reinhardtii. PNAS* 92(11):5129–5133 (1995).

Hall MP. *The secret teachings of all ages. An encyclopedic outline of Masonic, Hermetic, Qabbalistic and Rosicrucian symbolical philosophy.* San Francisco: Crocker (1928).

Hall TS. *Ideas of life and matter. Studies in the history of general physiology, 600 BC–1900 AD.* In two volumes. University of Chicago Press (1969).

Hall TS. The scientific origins of the protoplasm problem (based upon a re-examination of the relations between the concept of life and the doctrine of matter in pre-Socratic Greek science). *Journal of the History of Ideas* 11(3):339–356 (1950).

Haller A de. Partibvs corporis hvmani sensilibvs et irritabilibvs. *Commentarii Societatis Regiae Scientiarum Gottingensis* 1752, 2:114–158 (1753).

Haller A [von]. *A dissertation on the sensible and irritable parts of animals.* London: Nourse (1755).

Haller A [von]. *Dissertation sur les parties irritables et sensibles des animaux.* Lausanne: Bousquet (1755).

Haller A [von]. *Dissertazione intorno le parti irritabili, e sensibili, degli animali.* Napoli: Benedetto Gessari (1755).

Haller A [von]. *Elementa physiologiæ corporis humani.* Eight volumes in four. Volume 1, Lausannæ: Bousquet (1757); volume 2, Lausannæ: d'Arnay (1760); volumes 3–5, Lausannæ: Grasset (1762–1766); and volumes 6–8, Bernæ: Societatis Typographicæ (1766).

Haller A von. Haller to Linnæus (21 December 1737). In: JE Smith (ed.), *A selection from the correspondence of Linnæus,* volume 2 [of 2]. London: Longman, Hurst *et al.* (1821), pp. 305–308.

Hamann BE. *The translations of Nebrija. Language, culture, and circulation in the early modern world.* Amherst: University of Massachusetts Press (2015).

Hanov MC. *Philosophiæ natvralis sive physica dogmaticæ.* In four volumes. Halæ Magdebvrgicæ: Rengeriana (1762–1768).

Hansen B. *Nicole Oresme and the marvels of nature. A study of his* De causis mirabilium. Studies and Texts 68. Toronto: Pontifical Institute of Medieval Studies (1985).

Haq SN. *Names, natures and things. The alchemist Jābir ibn Ḥayyān and his* Kitāb al-Aḥjār *(Book of stones).* Boston Studies in the Philosophy of Science 158. Dordrecht: Kluwer (1994).

Harada E. Carcinology in classical Japanese works. In: F Truesdale (ed.), *History of carcinology.* Rotterdam: Balkema (1993), pp. 243–258.

Haring N [Häring NM]. The creation and Creator of the world according to Thierry of Chartres and Clarenbaldus of Arras. *Archives d'histoire doctrinale et littéraire du Moyen Age* 22:137–216 (1955).

Harmon DW. *A journal of voyages and travels in the interiour [sic] of North America.* Andover: Flagg and Gould (1820).

Harms JW. *Zoobiologie für Mediziner und Landwirte.* Jena: Fischer (1946).

Harnack A. Tertullian. *Encyclopædia Britannica*, Ninth edition 23:196–198 (1888).

Harper J. *Glimpses of ocean life; or, rock-pools and the lessons they teach.* London: Nelson (1860).

Harris F. Physiology and "vital force". *Nature* 115(2895):608–610 (1925).

Harris H. *The birth of the cell.* New Haven: Yale UP (1999).

692 REFERENCES

Harris J. Some microscopical observations of vast numbers of animalcula seen in water. *Philosophical Transactions* 19(220):254–259 (read 1696, published 1698).

Harrison AJ. *Savant of the Australian seas. William Saville-Kent (1845–1908) and Australian fisheries*. Hobart: Tasmanian Historical Research Association (1997).

Harrison AJ. Saville-Kent, William (1845–1908). *Australian Dictionary of Biography. Supplement 1580–1980*. C Cunneen (ed.). Melbourne UP (2005). http://adb.au.edu.au/biography/saville-kent-william-13185, accessed 10 November 2020.

Harrison TP. Birds in the moon. *Isis* 45(4):323–330 (1954).

Hartman H. The origin of the eukaryotic cell. *Speculations in Science and Technology* 7(2):77–81 (1984).

Hartman H & Fedorov A. The origin of the eukaryotic cell: a genomic investigation. *PNAS* 99(3):1420–1425 (2002).

Hartmann M. Das System der Protozoen. Zugleich vorläufige Mitteilung über *Proteosoma* [Labbé]. *Archiv für Protistenkunde* 10(1):139–158 (1907).

Hartmann M. Mikrobiologie. Allgemeine Biologie der Protisten. In: *Die Kultur der Gegenwart* (ed. P Hinneberg, ed.). Teil 3, Abteilung 4. *Organische Naturwissenschaften*. Band 1. *Allgemeine Biologie* (C Chun, W Johannsen & A Günthart, eds). Leipzig & Berlin: Teubner (1915), pp. 283–301.

Hartog M, Sollas IBJ, Hickson SJ & MacBride EW. *Protozoa, porifera (sponges), coelenterata & ctenophora, echinodermata*. In: SF Harmer & AE Shipley (eds), *The Cambridge natural history* 1. London: Macmillan (1906).

Harvey W. *Anatomical exercises on the generation of animals*. R Willis (trans.). In: *The works of William Harvey*. London: Syndenham Society (1847), pp. 143–518.

Harvey W. *Anatomical exercitations, concerning the generation of living creatures*. London: James Young (1653).

Harvey W. *Exercitatio anatomica de motv cordis et sangvinis in animalibvs*. Francofvrti: Gvilielmius Fitzerus (1628).

Harvey W. *Exercitationes de generatione animalium. Quibus accedunt quædam De partu: de membranis ac humoribus uteri: & de conceptione*. Londini: Pulleyn (1651).

Harvey W. *The anatomical exercises of Dr. William Harvey*. London: Francis Leach (1653).

Harvey WH. *A manual of the British algæ: containing generic and specific descriptions of all the known British species of sea-weeds, and of confervæ, both marine and fresh-water*. [First edition]. London: van Voorst (1841).

Harvey WH. *A manual of the British marine algæ: containing generic and specific descriptions of all the known British species of sea-weeds*. [Second edition.] London: van Voorst (1849).

Harvey WH. Algæ. In: JT MacKay, *Flora Hibernica, Part second*. Dublin: Curry (1836), pp. 157–254.

Haskins CH. *Studies in the history of mediaeval science*. Cambridge MA: Harvard UP (1924). Reprinted by Frederick Ungar, New York (1960).

Haskins CH. The "De arte venandi cum avibus" of the Emperor Frederick II. *English Historical Review* 36(143):334–355 (1921).

Haskins CH. *The renaissance of the Twelfth Century*. Cambridge MA: Harvard UP (1927).

Haskins CH. *The renaissance of the Twelfth Century*. New York: Meridian (1957).

al-Hassan AY. The Arabic original of *Liber de compositione alchemiae*. The Epistle of Maryānus, the Hermit and Philosopher, to Prince Khālid ibn Yazīd. *Arabic Sciences and Philosophy* 14(2):213–231 (2004).

Hatfield G. René Descartes. *The Stanford Encyclopedia of Philosophy* (Summer 2018 edition), EN Zalta (ed.), https://plato.stanford.edu/archives/sum2018/entries/descartes/, accessed 24 July 2019 and 26 September 2020.

Hatzimichali M. Encyclopaedism in the Alexandrian library. In: J König & G Woolf (eds), *Encyclopaedism from antiquity to the Renaissance*. Cambridge UP (2013), pp. 64–83.

Haughton S. *The three kingdoms of nature briefly described*. London: Cassell, Petter & Galpin [1869].

REFERENCES 693

Hausmann K & Machemer H. The microcosm under the microscope: a passion of amateurs and experts. *Denisia* 41:1–46 (2018).

Haüy [RJ, as Citoyen]. *Traité de minéralogie.* In five volumes. Paris: Louis (An 10, 1801).

Hayata B. An interpretation of Goethe's *Blatt* in his "Metamorphose der Pflanzen", as an explanation of the principle of natural classification. In: Hayata B, *Icones plantarum Formosanarum nec non et contributiones ad floram Formosanam* 10. Taihoku: Bureau of Productive Industries, Government of Formosa (1921), pp. 75–95.

Hayata B. The natural classification of plants according to the dynamic system. In: Hayata B, *Icones plantarum Formosanarum nec non et contributiones ad floram Formosanam* 10. Taihoku: Bureau of Productive Industries, Government of Formosa (1921), pp. 97–234.

Hayes Z. Introduction. In: Z Hayes (trans.), *St. Bonaventure's On the reduction of the arts to theology.* St Bonaventure NY: Franciscan Institute (2006), pp. 1–10.

Haym R. *Herder nach seinem Leben und seinem Werken.* In two volumes. Berlin: Rudolph Gaertner (1880–1885).

Heath IB. Review of "Ultrastructure, macromolecules and evolution" by L.S. Dillon. *ISEP Newsletter* 2(2), insert (1983).

Hedwig J. *Theoria generationis et frvctificationis plantarvm cryptogamicarvm Linnaei. Retractata et avcta.* Lipsiae: Breitkopf-Haertel (1798).

Hegel GWF. *Encyclopädie der philosophischen Wissenschaften im Grundrisse.* Theil 1: *Die Logik.* L von Hennig (ed.). In: P Marheineke, *et al.* (eds), *Werke. Vollständige Ausgabe.* Band 6. Berlin: Dunder & Humblot (1840).

Hegel GWF. *Encyclopädie der philosophischen Wissenschaften im Grundrisse.* Theil 3: *Philosophie des Geistes.* L Boumann (ed.). In: P Marheineke, *et al.* (eds), *Werke. Vollständige Ausgabe.* Band 7(2). Berlin: Dunder & Humblot (1845).

Hegel GWF. *Hegel's* Philosophy of nature. AV Miller (trans.). Oxford: Clarendon (1970).

Hegel GWF. *Hegel's* Philosophy of nature. MJ Petry (trans.) In three volumes. London: Allen & Unwin (1970).

Hegel GWF. Hegel's *Philosophy of Nature.* SA Taubeneck (trans.) from the Heidelberg text of 1817. In: E Behler (ed.), *Encyclopedia of the philosophical sciences in outline and critical writings* (ed. E Behler). London: Continuum (1990). https://www.marxists.org/reference/archive/hegel/natin dex.htm, accessed 16 January 2020 and 5 October 2020.

Hegel GWF. *Phenomenology of spirit.* Revised edition. AV Miller (trans.). Oxford UP (1977).

Hegel GFW. *The science of logic* (G di Giovanni, trans. and ed.). Cambridge UP (2010).

Hegel GWF. *Vorlesungen über die Naturphilosophie als der Encyclopädie der philosophischen Wissenschaften im Grundrisse.* Theil 2. CL Michelet (ed.). In: P Marheineke, *et al.* (eds), *Werke. Vollständige Ausgabe.* Band 7(1). Berlin: Dunder & Humblot (1842).

Hegel GFW. *Wissenschaft der Logik.* In two volumes. Nürnberg: Schrag (1812–1816).

Heller JL. The early history of binomial nomenclature. *Huntia* 1:33–70 (1964).

Hellinga L. *The introduction of printing in Italy: Rome, Naples and Venice.* The John Rylands University Library, Manchester (undated). http://www.library.manchester.uk/firstimpressi ons/, accessed 8 March 2019.

Hellmann O. Peripatetic biology and the *Epitome* of Aristophanes of Byzantium. In: WW Fortenbaugh & SA White (eds), Aristo of Ceos: text, translation, and discussion. Rutgers University Studies in Classical Humanities 13. New Brunswick NJ: Transaction (2006), pp. 329–359. Reprinted by Routledge, London (2017).

Hellström NP, G André & M Philippe. Augustin Augier's botanical tree: transcripts and translations of two unknown sources. *Huntia* 16(1):17–38 (2017).

Hendrickson DS. *Jesuit polymath of Madrid. The literary enterprise of Juan Euseubio Nieremberg (1595–1658).* In series: Jesuit Studies 4. Leiden: Brill (2015).

Hendrikx S. *Ippolito Salviani (1514–1572)* (2015). http://www.rarefishbooks.com/k2-categories/ Books/311-salviani-ippolito-151401572.html, accessed 21 April 2019 and 22 September 2020.

Henfrey A & Huxley TH (eds). *Scientific memoirs, selected from the transactions of foreign academies of science, and from foreign journals. Natural history.* London: Taylor & Francis (1853).

694 REFERENCES

Hennig W. *Grundzüge einer Theorie der phylogenetischen Systematik*. Berlin: Deutscher Zentralverlag (1950). Reprinted by Koeltz, Koenigstein (1980).

Hennig W. *Phylogenetic systematics*. Urbana: University of Illinois Press (1966).

Henrici AT. *Morphologic variation and the rate of growth of bacteria*. In series: Microbiology Monographs (RE Buchanan, EB Fred & SA Waksman, eds) 1. Springfield IL: Thomas (1928).

Henry J. Henry More. *The Stanford Encyclopedia of Philosophy* (Winter 2016 edition), EN Zalta (ed.), https://plato.stanford.edu/archives/win2016/entries/henry-more/, accessed 22 March 2019 and 20 September 2020.

Henshilwood CS, d'Errico F, Yates R, *et al*. Emergence of modern human behavior: Middle Stone Age engravings from South Africa. *Science* 295(5558):1278–1280 (2002).

Hensson A & Jahns HM. *Lichenes. Eine Einführung in die Flechtenkunde*. Stuttgart: Georg Thieme (1974).

Herbarius Maguntie impressus, also known as *Aggregator practicus de simplicibus*. [Moguntiæ]: [Peter Schöffer] (1484).

Herder JG. *Ideen zur Philosophie der Geschichte der Menschheit*. In four volumes. Riga & Leipzig: Hartknoch (1785–1792).

Hermann J. *Affinitatum animalium tabulam brevi commentario illustratam . . .* proponit GC Würtz. Argentorati: Heitz (1777).

Hermann J. *Tabula affinitatum animalium olim academico specimine edita*. Argentorati: Treuttel (1783).

[Hermes Trismegistus]. *Pimander. Mercurij Trismegisti liber de sapienta et potestate dei. Asclepius. Eiusdem Mercurij de voluntate divina. Item Crater Hermetis a Lazarelo Septempedano*. Parisiis: Henrici Stephani (1505).

[Hermes Trismegistus]. *Tabvla Smaragdina Hermetis Trismegisti*. In: Chrysogonus Polydorus (ed.), *De alchemia*. Norimbergiæ: Johannes Petreium (1541), pp. 363–373.

Herodotus. *The History of Herodotus, translated from the Greek. With notes*. W Beloe (trans.). In four volumes. London: Leigh and Sotheby (1791).

Herodotus. *Herodotus, translated from the Greek, with notes*. Second edition, corrected and enlarged. W Below (trans.). In four volumes. London: Leigh and Sotheby, *et al*. (1806).

Herodotus. *The Persian wars*. G Rawlinson (trans.). New York: Modern Library (1942).

Heron-Allen E. *Barnacles in nature and myth*. Oxford UP; and London: Humphrey Milford (1928).

Herr M. *Gründtlicher Underricht, wahrhaffte und eygentliche Beschreibung wunderbarlicher Seltzamer Art, Natur, Krafft und Eygenschafft aller vierfüssigen Thier*. Straßburg: Beck (1546). Facsimile as: *Das Thierbuch des Michael Herr*. Kötzschenbroda: Dudelsack (1934).

Herron MD & Michod RE. Evolution of complexity in the volvocine algae: transitions in individuality through Darwin's eye. *Evolution* 62(2):436–451 (2008).

Hershey A & Chase M. Independent functions of viral protein and nucleic acid in growth of bacteriophage. *Journal of General Physiology* 36(1):39–56 (1952).

Hertwig R. Die Protozoen und die Zelltheorie. *Archiv für Protistenkunde* 1(1):1–40 (1902).

Hertwig R. *Lehrbuch der Zoologie*. Second edition. Jena: Gustav Fischer (1893).

Hertwig R von. Die einzelligen Organismen. In: *Die Kultur der Gegenwart* (ed. P Hinneberg, ed.). Teil 3, Abteilung 4. *Organische Naturwissenschaften*. Band 2(2). *Zoologischer Teil* (O Hertwig, ed.). Leipzig & Berlin: Teubner (1913), pp. 1–38.

Hettche M & Dyck C. Christian Wolff. *The Stanford Encyclopedia of Philosophy* (Winter 2019 edition), EN Zalta (ed.), https://plato.stanford.edu/archives/win2019/entries/wolff-christian/, accessed 8 January 2020 and 3 October 2020.

Hewitt DA, Amram P, Schmull M & Karakehian JM. An early mycota: Johannes Baptista von Albertini and Lewis David von Schweinitz's *Conspectus fungorum in Lusatiae superioris agro Niskiensi crescentium*, with a translation of the Latin introduction into English. *Bartonia* 69:47–61 (2016).

Heywood VH. Linnaeus—the conflict between science and scholasticism. In: J Weinstock (ed.), *Contemporary perspectives on Linnæus*. Lanham MD: University Press of America (1985), pp. 1–15.

REFERENCES 695

Hickson SJ, Lister JJ, Gamble FW, *et al*. Introduction and protozoa. First fascicle. In: [E]R Lankester (ed.), *A treatise on zoology*. In nine volumes. London: Black (1909).

Hildegard of Bingen. *Hildegard von Bingen's Physica. The complete English translation of her classic work on health and healing*. P Throop (trans.). Rochester VT: Healing Arts (1998).

Hildegard of Bingen. *On natural philosophy and medicine. Selections from* Cause et cure. M Berger (ed. and trans.). In series: Library of Medieval Woman (J Chance, series ed.). Cambridge UK: Brewer (1999).

Hildegard of Bingen. Opera omnia, ad optimorum librorum fidem edita. In: JP Migne (ed.), *Patrologiæ cursus completus*. Series Latina prior 197. Parisiis: Garnier (1882).

Hill J. *A general natural history: or, new and accurate descriptions of the animals, vegetables, and minerals, of the different parts of the world*. [Volume 1] *A history of fossils* (1748); [Volume 2] *A history of plants* (1751); [Volume 3] *An history of animals* (1752). Volume 3 lacks the series title page. London: Thomas Osborne.

Hill J. *A general natural history: or, new and accurate descriptions of the animals, vegetables, and minerals, of the different parts of the world*. [Volume 1] *A history of fossils* (1772); [Volume 2] *A history of plants* (1773); [Volume 3] *An history of animals* (1773). [Second edition]. London: printed for the author.

Hill J. *A review of the works of the Royal Society of London; containing animadversions on such of the papers as deserve particular observations. In eight parts: under the several heads of arts, antiquities, medicine, miracles, zoophytes, animals, vegetables, minerals*. London: Griffiths (1751).

Hill J. *An history of animals. Containing descriptions of the birds, beasts, fishes, and insects, of the several parts of the world; and including accounts of the several classes of animalcules, visible only by the assistance of microscopes*. London: Thomas Osborne (1752).

Hill J. *Essays in natural history and philosophy. Containing a series of discoveries, by the assistance of microscopes*. London: Whiston, White, Vaillant & Davis (1752).

Hill J. *The vegetable system: or, a series of experiments, and observations tending to explain the internal structure, and the life of plants; their growth, and propagation; the number, proportion, and disposition of their constituent parts; with the true course of their juices; the formation of the embryo, the construction of the seed, and the encrease from that state to perfection*. Five volumes bound in six. London: for the author (1759–1763).

Hill J. [as Johnson A]. *Lucina sine concubitu*. Second edition. London: Cooper (1750).

Hillier HC. Ibn Rushd (Averroes) (1126–1198). *Internet Encyclopedia of Philosophy*, undated. https://www.iep.utm.edu/ibnrushd, accessed 6 September 2018.

Hillier J, Mudd S & Smith AS. Internal structure and nuclei in cells of *Escherichia coli* as shown by improved electron microscopic techniques. *Journal of Bacteriology* 57(3):319–338 (1949).

Hindley JC. Ammonios Sakkas. His name and origin. *Zeitschrift für Kirchengeschichte* 75:332–336 (1964).

Hippocrates. *Hippocrates with an English translation*. WHS Jones, ET Withington, P Potter & WD Smith (trans). Loeb Classical Library 147–150, 472–473, 477, 482, 509, 520, and 538. London: Heinemann; and Cambridge MA: Harvard UP (1923–2018).

Hitchcock E. *Elementary geology*. [First edition]. Amherst MA: Adams (1840).

Hitchcock E. *Elementary geology*. Third edition. New York: Newman (1844).

Hodge MJS. The universal gestation of nature: Chambers' *Vestiges* and *Explanations. Journal of the History of Biology* 5(1):127–151 (1972).

Hoffmann RC. Fishing for sport in Medieval Europe: new evidence. *Speculum* 60(4):877–902 (1985).

Hofmeister F. Über Bau und Gruppierung der Eiweisskörper. *Ergebnisse der Physiologie* 1(1):759–802 (1902).

Hofmeister [F]. Ueber den bau des Eiweißmoleküls. *Naturwissenschaftliche Rundschau* 17(42):529–533 and (43):545–549 (1902).

Hofmeister W. Neue Beiträge zur Kenntniss der Embryobildung der Phanerogamen. *Abhandlungen der Königlichen Sächsischen Gesellschaft der Wissenschaften* 6:533–672 (1859) and 7:629–760 (1861).

696 REFERENCES

Hofmeister W. *On the germination, development, and fructification of the higher Cryptogamia, and on the fructification of the Coniferæ.* F Currey (trans.). London: Hardwicke, for the Ray Society (1862).

Hofmeister W. *Vergleichende Untersuchungen der Keimung, Entfaltung und Fruchtbildung höherer Kryptogamen.* Leipzig: Hofmeister (1851).

Hogg J[abez]. *The microscope: its history, construction, and application.* [First edition]. London: Orr (1854).

Hogg J[abez]. *The microscope: its history, construction, and application.* Second edition. London: Ingram (1856).

Hogg J[abez]. *The microscope[.] Its history, construction, and application.* New [Seventh] edition. London: Routledge (undated, 1867/1868).

Hogg J[abez]. *The microscope: its history, construction, and application.* Eleventh edition. London & New York: Routledge (1886).

Hogg J[ohn]. Address to the members. *Transactions of the Tyneside Naturalists' Field Club* 3:163–188 (read 1857, published 1858).

Hogg J[ohn]. Further observations on the *Spongilla fluviatilis*; with some remarks on the nature of the *Spongiæ marinæ. Transactions of the Linnean Society of London* 18(3):368–407 (read 1838 and 1839, published 1840).

Hogg J[ohn]. Observations on the *Spongilla fluviatilis. Transactions of the Linnean Society of London* 18(3):363–367 (read 1838, published 1840).

Hogg J[ohn]. On the distinctions of a plant and an animal, and on a fourth kingdom of nature. *Edinburgh New Philosophical Journal, New Series* 12:216–225 (1860).

Hogg J[ohn]. On the distinctions of a plant and an animal, and on a fourth kingdom of nature. *Report of the Thirtieth Meeting of the British Association for the Advancement of Science, Oxford, June–July 1860.* Notices and abstracts. London: John Murray (1861), pp. 111–112.

Hogg J[ohn]. On the natural history of the vicinity [of Stockton-on-Tees]. In: J Brewster, *The parochial history and antiquities of history of Stockton-upon-Tees.* Second edition. Stockton-upon-Tees: Jennett (1829), Appendix 2 (numb. sep.).

Hogg J[ohn]. On the nature of the marine production commonly called Flustra arenosa. *Transactions of the Linnean Society of London* 14(2):318–321 (read 1823, published 1824).

Hogg J[ohn]. On the tentacular classification of zoophytes. *Annals of Natural History; or, Magazine of Zoology, Botany, and Zoology* 4(26):364–367 (1840).

Hohler EB. Norwegian stave church carving, an introduction. *Arte Medievale* 3:77–116 (1989).

Holland J. Alexander Macleay (1767–1848) and William Sharp Macleay (1792–1865). In: P Stanbury & J Holland (eds), *Mr Macleay's celebrated cabinet.* University of Sydney (1988), pp. 9–37.

Holley RW, Apgar J, Everett GA, *et al.* Structure of a ribonucleic acid. *Science* 147(3664):1462–1465 (1965).

Holmes E. *Henry Vaughan and the Hermetic philosophy.* New York: Russell & Russell (1932).

Holmes UT Jr. *Guillaume De Salluste Sieur Du Bartas. A biographical and critical study.* In: UT Holmes Jr, JC Lyons & RW Linker (eds), *The works of Guillaume de Salluste Sieur Du Bartas. A critical edition with introduction, commentary, and variants,* volume 1. Chapel Hill: University of North Carolina Press (1935), pp. 3–228.

Holmes UT, Lyons JC & Linker RW. *The works of Guillaume De Salluste Sieur Du Bartas. A critical edition with introduction, commentary, and variants.* In three volumes. Chapel Hill: University of North Carolina Press (1935–1940).

Holmyard EJ. A romance of chemistry. [Transcription of British Museum Sloane MS 3697, *De compositione alchemiæ* translated from Arabic supposedly by Robert of Chester in 1144]. *Journal of the Society of Chemical Industry* 1925:75–77, 105–108, 272–276, 300–301, and 327–328 (1925).

Holmyard EJ. Abu' l-Qāsim al-'Irāqī. *Isis* 8(3):403–426 (1926).

Holmyard EJ. *Makers of chemistry.* Oxford: Clarendon (1931).

Holmyard EM. The emerald table. *Nature* 112(2814):525–526 (1923).

REFERENCES 697

Homer [attrib.]. *The first six books of the* Iliad *of Homer, literally translated into English prose*. H Cary (trans.). Cambridge: Hall *et al*. (1821).

Homer [attrib.]. *The Iliad*. R Lattimore (trans.). University of Chicago Press (1967).

Honegger R. Great discoveries in bryology and lichenology. Simon Schwendener (1829–1919) and the dual hypothesis of lichens. *The Bryologist* 103(2):317–312 (2000).

Honorius Augustodunensis. *Clavis physicae*. P Lucentini (ed.). In series: Temi e Testi (E Massa, ed.), volume 21. Roma: Storia e Letteratura (1974).

Hooke R. *Micrographia: or some physiological descriptions of minute bodies made by magnifying glasses with observations and inquiries thereupon*. London: Martyn & Allestry (1665).

Hooker WJ. *The British flora*. Fourth edition. In two volumes. London: Longman, Rees *et al*. (1838).

Hooker WJ. *The British flora*. Fifth edition. In two volumes. London: Longman, Rees *et al*. (1842).

Hooker WJ. *Flora Scotia; or a description of Scottish plants, arranged both according to the artificial and natural methods*. London: Taylor (1821).

Hooker WJ. *The English flora of Sir JE Smith. Volume 5, Part 1 (or Volume 2 of Hooker's British Flora). Class 24. Cryptogamia . . . Comprising the mosses, hepaticæ, lichens, characæ and algæ*. London: Longman, Rees *et al*. (1833).

Horaninov P. *Characteres essentiales familiarum ac tribuum regni vegetabilis et amphorganici ad leges tetractydis naturae conscripti*. Petropoli: Wienhöberianis (1847).

Horaninov P. *Primae lineae systematis naturæ, nexui naturali omnium evolutionique progressivae per nixus reascendentes superstructi*. Petropoli: Karoli Krajanis (1834).

Horaninov P. *Tetractys naturae seu systema quadrimembre omnium naturalium, quod primis lineis systematis naturae, a se editis*. Petropoli: Wienhöberianis (1843).

Horesh G, Blackwell GA, Tonkin-Hill G, *et al*. A comprehensive and high-quality collection of *Escherichia coli* genomes and their genes. *Microbial Genomics* 7:000499 (2021).

Horne CF (ed.). *The sacred books and early literature of the East. With historical surveys of the chief writings of each nation*. In 14 volumes. New York: Parke, Austin (1917).

[Hornschuch F]. Einige Beobachtungen über das Entstehen der Algen, Flechten und Daubmoose. *Flora oder Botanische Zeitung* 2, Bd 1(9):140–144 (1819).

Hort AF. Introduction. In: Theophrastus, *Enquiry into plants*. Loeb Classical Library 70. London: Heinemann; and Cambridge MA: Harvard UP (1916), pp. ix–xxi.

Hortus sanitatis. Straßburg (*ca* 1497). Bayerische StaatsBibliothek 2 Inc.s.a. 676. Digital image at https://www.digitale-sammlung.de/, accessed 5 September 2020.

Hoßfield U, Olsson L & Breidbach O. Carl Gegenbaur (1826–1903) and his influence on the development of evolutionary morphology. *Theory in Biosciences* 122(2–3):105–108 (2003). Minisymposium, pp. 105–301.

Houser RE. Bonaventure's threefold way to God. In: RE Houser (ed.), *Medieval masters: essays in memory of Msgr. EA Synan*. Thomistic Papers 7. Houston: Center for Thomistic Studies (1999), pp. 91–145.

Houssay F. Les théories de la genèse a Mycènes et le sens zoologique de certains symboles du culte d'Aphrodite. *Review Archéologique, 3 Série* 26:1–27 (1895).

Houttuyn M. *Natuurlyke historie: of, uitvoerige beschryving der dieren, planten en mineraalen*. Three parts in 37 volumes. Amsterdam: Houttuyn [publ. var.] (1761–1785).

Howard L. On the modification of clouds, and on the principles of their production, suspension, and destruction. [*Tilloch's*] *Philosophical Magazine* 16(62):97–107, 16(64):344–357, and 17(65):5–11 (1803).

Hoyland RG. *Arabia and the Arabs: from the Bronze Age to the coming of Islam*. Routledge (2001).

Hrabanus Maurus. Bibliotheca Augustana. Hrabani Mauri De rerum naturis. www.hs-augsburg. de/~harsch/ hra_rn00.html, accessed 16 October 2018 and 14 September 2020.

Hrabanus Maurus. Opera omnia. In: JP Migne (ed.), *Patrologiæ cursus completus, Series Latina* [*varies*] 107 (1864), 108 (1851), 109–111 (1864) & 112 (1878). Parisiis: Migne (1851–1864) & Garnier (1878).

Hsieh Sung-Mo. An investigation of the authenticities of ancient and medieval Chinese medical books: 1, the pharmaceutical natural histories. *Chinese Journal of Medical History* 1(1):57–66 (1947).

698   REFERENCES

Hubert H & Mauss M. Esquisse d'une théorie générale de la magie. *Année Sociologique* 7:1–146 (1902–1903).

Huddleston G. Monasticism. *The Catholic Encyclopedia* 10:459–464 (1911).

Hudry F. Le *De secretis nature* du Ps. Apollonius de Tyane, traduction latine par Hugues de Santalla du *Kitāb sirr al-ḥalīqa*. *Chrysopoeia* 6:1–154 (1997–1999), published 2000.

Hudson W. *Flora Anglica, exhibens plantas per Regnum Angliæ sponte crescentes, distributas secundum systema sexuale*. Londini: printed for the author (1762).

Hueppe F. *The principles of bacteriology*. EO Jordan (trans.). Chicago: Open Court (1899).

Huffman C. Pythagoras. *The Stanford Encyclopedia of Philosophy* (Summer 2014 edition), EN Zalta (ed.), https://plato.stanford.edu/archives/sum2014/entries/pythagoras/, accessed 3 May 2018.

Huffman CA. *Philolaus of Croton. Pythagorean and Presocratic. A commentary on the fragments and testimonia with interpretative essays*. Cambridge UP (1993).

Hug LA, Baker BJ, Anantharaman K, *et al*. A new view of the tree of life. *Nature Microbiology* 1(1):1–6 (2016).

Hugh of St Victor. *The Didascalicon of Hugh of St. Victor. A medieval guide to the arts*. J Taylor (trans.). New York: Columbia UP (1961).

Hugo de Folieto. *The medieval book of birds. Hugh of Fouilloy's Aviarium*. WB Clark (ed. & trans.). SUNY Binghamton Medieval & Renaissance Texts & Studies 80. Binghamton NY: Medieval & Renaissance Texts & Studies (1992).

Hugolino of Orvieto. Articles prohibited for the theological faculty of the University of Bologna, 1364. In: *I più antichi statuti della Facoltà Teologica dell'Università de Bologna. Contributo alla storia della scholastica medievale*. F Ehrle (ed.). Bologna: L'Istituto per la Storia dell'Università (1932). As translated by JM Idziak. Oklahoma State Translation Clearing House ref. HO-25–20i (1984).

Hull DL. Linné as an Aristotelian. In: J Weinstock (ed.), *Contemporary perspectives on Linnæus*. Lanham MD: University Press of America (1985), pp. 37–54.

Hull DL. *Science as a process. An evolutionary account of the social and conceptual development of science*. University of Chicago Press (1988).

Hull J. *Elements of botany*. Volume 1. *Introduction to the sexual system of Linnæus. Botanical terms and definitions*. Volume 2. *Characters of the genera of British plants. Lectures on the natural orders*. Manchester: Dean (1800).

Hull J. *The British flora, or a Linnean arrangement of British plants*. Manchester: Dean (1799).

Humboldt A von. *Ansichten der Natur, mit wissenschaftlichen Erläuterungen*. Second edition. In two volumes. Stuttgart & Tübingen: Cotta (1826).

Humboldt A von. *Ansichten der Natur, mit wissenschaftlichen Erläuterungen*. Third edition. In two volumes. Stuttgart & Tübingen: Cotta (1849).

Humboldt FA von. *Aphorismen aus der chemischen Physiologie der Pflanzen*. G Fischer (trans.). Leipzig: Voss (1794).

Humboldt A von. *Aspects of nature, in different lands and different climates*. [EJL] Sabine (trans.). In two volumes. London: Longman, Brown *et al*. (1849).

Humboldt A von. *Cosmos: a sketch of a physical description of the universe*. EC Otté (trans.). In five volumes. London: Bohn (1849–1858).

Humboldt A von. Die Lebenskraft, oder der Rhodische Genius. Eine Erzählung. *Die Horen* 1795(5):90–96 (June 1795).

Humboldt FA von. *Floræ Fribergensis specimen plantas cryptogamicas præsertim subterraneas exhibens. Accedunt Aphorismi ex doctrina physiologiæ chemicæ plantarum*. Berolini: Rottmann (1793).

Humboldt A von. *Kosmos. Entwurf einer physischen Weltbeschreibung*. In five volumes. Stuttgart: Cotta (1845–1862).

Humboldt FA von. *Versuche über die gereizte Muskel and Nervenfaser, nebst Vermuthungen über den chemischen Process des Lebens in der Their und Pflanzenwelt*. In two volumes. Posen: Decker; Berlin: Rottmann (1797).

Humboldt A von & Bonpland A. *Voyage aux régions équinoxiales du nouveau continent, fait en 1799, 1800, 1801, 1802, 1803 et 1804.* In nine volumes. Paris: Maze (1815–1825).

Humboldt A von & Bonplandt [*sic*] A. *Reise in die Aequinoctial-Gegenden des neuen Continents in den Jahren 1799, 1800, 1801, 1802, 1803 und 1804.* Six volumes in seven. Stuttgart & Tübingen: Cotta (1815–1832).

Hume E. *Béchamp or Pasteur? A lost chapter in the history of biology.* Chicago: Covici-McGee (1923).

Hunn ES. *Tzeltal folk zoology. The classification of discontinuities in nature.* New York: Academic (1977).

Hunt RW. Introduction. In: J Blund, *Tractatus de anima.* DA Callus & RA Hunt (eds). In series: Auctores Britannici Medii Aevi 2. British Academy, and Oxford UP (1970), pp. vii–xx.

Hunter A (ed). *Georgical essays.* York: Ward for Dodsley, Cadell *et al.* (1777).

Hunter A. On vegetation, and the analogy between plants and animals. In: A Hunter (ed.), *Georgical essays.* York: Ward (1777), pp. 37–46.

Hussey E. Pythagoreans and Eleatics. In: CCW Taylor (ed.), *From the beginning to Plato. Routledge history of philosophy* (GHR Parkinson & SG Shanker, gen. eds), volume 1. London: Routledge (1997), pp. 128–174.

Hutton J. Theory of the earth; or an investigation of the laws observable in the composition, dissolution, and restoration of land upon the globe. *Transactions of the Royal Society of Edinburgh* 1(2):209–304 (1788).

Huxley [TH]. On some organisms living at great depths in the North Atlantic Ocean. *Quarterly Journal of Microscopical Science, New Series* 8:203–212 (1868).

Huxley TH. On the physical basis of life. *The Fortnightly Review, New Series* 5:129–145 (1869).

Huxley TH. On the study of biology [1876]. In: *Collected essays 3. Science and education.* London: Macmillan (1893), pp. 262–293.

Huxley TH. [Review of:] The natural history of creation—by Dr E Haeckel. *The Academy* 1:13–14 (9 October 1869) and 40–43 (13 November 1869).

Huxley TH. The cell-theory. *British and Foreign Medico-Chirurgical Review* 12(2):285–314 (1853).

Huxley TH. The genealogy of animals [1869]. In: *Critiques & Addresses.* London: Macmillan (1873), pp. 303–319.

Huygens C. *Christiani Hugenii Zulichemii . . . Opera varia*, volume 1. Lugduni Batavorum: Vander Aa (1724).

Huygens C. *The celestial worlds discover'd: or, conjectures concerning the inhabitants, plants and productions of the worlds in the plants.* London: Timothy Childs (1698).

Huygens C. *The celestial worlds discover'd: or, conjectures concerning the inhabitants, plants and productions of the worlds in the plants.* Second edition. London: James Knapton (1722).

Huygens C. *Κοσμοθεωρος, sive De terris cœlestris.* Hagæ-Comitum: Adrianum Mortjens (1698).

Hyde WW. The prosecution and punishment of animals and lifeless things in the Middle Ages and modern times. *University of Pennsylvania Law Review & American Law Review* 64(7):696–730 (1916).

Iamblichus [attrib.]. *The exhortation to philosophy. Including the letters of Iamblichus and Proclus' Commentary on the Chaldean Oracles.* TM Johnson (trans.). Grand Rapids MI: Phanes (1988).

Iamblichus [attrib.]. *The theology of arithmetic.* R Waterfield (trans.). Grand Rapids MI: Phanes 1988).

Iamblichus. *De mysteriis Ægyptiorvm, Chaldæorum, Assyriorum.* With works by Proclus, Porphyry, Psellus, and Mercurius Trismegistus. Lvgdvni: Tornaesivm (1570), pp. 5–178.

Iamblichus. [*Iamblichou Chalkideōs tes Koilēs Syrias Peri mysteriōn logos.*] *Iamblichi Chalcidensis ex Coele-Syria, De Mysteriis liber. Præmittitur Epistola Porphyrii ad Anebonem Ægyptium, eodem argumento.* T Gale (trans.). Oxonii: Sheldoniano (1678).

Iamblichus. *Iamblichi Chalcidensis In Platonis dialogos commentariorum fragmenta.* Edited with translation and commentary by JM Dillon. Westbury UK: Prometheus (2009).

Iamblichus. *On the mysteries of the Egyptians, Chaldeans, and Assyrians.* T Taylor (trans.). Second edition. London: Bertram Dobell & Reeves and Turner (1895).

Iamblichus. *On the Pythagorean life.* G Clark (trans.). Liverpool UP (1989).

700   REFERENCES

Iamblichus. *The life of Pythagoras.* T Taylor (trans.). Krotona CA: Theosophical Publishing House (1905–1915 and 1918).

Ibn Daud A [Ben David Halevi A]. *Das Buch Emunah Ramah oder: Der erhabene Glaube.* S Weil (trans.). Frankfurt: Typographischen Anhalt (1852).

Ibn Daud A. *The exalted faith.* NM Samuelson (trans.), G Weiss (ed.). London: Fairleigh Dickinson UP (1986).

Ibn Gebirol [Abencebrolis]. Fons vitae. Ex Arabico in Latinvm translatvs ab Iohanne Hispano et Dominco Gvndissalino. In: C Baeumker (ed.), *Beiträge zur Geschichte der Philosophie des Mittelalters. Texte und Untersuchungen* 1(2–3). Münster: Aschendorff (1895).

Ibn Khaldūn. *The Muqaddimah. An introduction to history.* F Rosenthal (trans.). In three volumes. Bollingen 43. Princeton UP (1958).

Ibn Rushd [Averroës]. *Averroes (Ibn Rushd) of Cordoba. Long commentary on the De anima of Aristotle.* RC Taylor (trans.). New Haven: Yale UP (2009).

Ibn Rushd [Averroës]. *Averroës Middle commentary on Aristotle's De anima.* AL Ivry (ed.). Provo UT: Brigham Young UP (2002).

Ibn Rushd [Averroës]. *Corpvs commentariorvm Averrois in Aristotelem.* HA Wolfson, D Baneth & FH Fobes (eds). *Versionum Latinarum 6(1). Commentarivm magnvm* (FS Crawford, ed.). Mediaeval Academy of America Publication 59. Cambridge MA: Mediaeval Academy of America (1953).

Ibn Rushd [Averroës]. *Faith and reason in Islam. Averroes' exposition of religious arguments.* I Najjar (trans.). Oxford: Oneworld (2001).

Ibn Rushd [Averroës]. *Tahafut al-tahafut (Incoherence of the incoherence).* S van den Bergh (trans.). In two volumes. EJW Gibb Memorial Trust. London: Luzac (1954). Reprinted in one volume (1987).

Ibn Sina [Avicenna]. *A compendium on the soul.* EA van Dyck (trans.). Verona: Nicola Paderno (1906). Facsimile by Antioch Gate, Birmingham UK (2007).

Ibn Sīnā [Avicenna]. *Avice[n]ne perhypatetici philosophi: ac medicorum facile primi opera in luce[m] redacta: ac nuper quantum ars niti potuit per canonicos emendata.* Venetiis: Bonetus Locatellus Borgamensis (1508). Facsimile by Minerva, Frankfurt (1961).

Ibn Sīnā. *Kitab al-Shifa.* [Section:] On the formation of minerals and metals and the impossibility of alchemy. EJ Holmyard & DC Mandeville (trans.). In: E Grant (ed.), *Source book in medieval science.* Cambridge MA: Harvard UP (1974), pp. 569–573.

Ibn Sīnā. *Kitab al-Shifa. The book of the remedy* [in part]. In: *Avicennae De congelatione et conglutinatione lapidum. Being sections of the Kitâb al-Shifâ'.* EJ Holmyard & DC Mandeville (trans.). Paris: Paul Geuthner (1927).

Ibn Wahshiyya [Wahshih, Ahmad bin Abubekr bin]. *Kitab al-filahâ al-nabâtiyya* [*Nabatæan agriculture*], *ca* 904 CE. English translation in: J Hammer. *Ancient alphabets and hieroglyphic characters explained; with an account of the Egyptian priests, their classes, initiation, and sacrifices.* London: Bulmer (1806).

Idel M. The anthropology of Yohann Alemanno: sources and influences. *Topoi* 7(3):201–210 (1988).

Iinuma Yokusai. *Sōmokuzusetsu.* [*An iconography of herbs and woody plants*]. In 20 volumes. Ōgaki: Heirinshō (1856).

Ikhwān al-Ṣafā. *Epistles of the Brethren of Purity. On the natural sciences. An Arabic critical edition and English translation of Epistles 15–21.* C Baffioni (ed. and trans.). Oxford UP (2013).

Ikhwān al-Ṣafā. *The case of the animals versus man before the King of the Jinn. A translation from the Epistles of the Brethren of Purity.* Goodman LE & R McGregor (trans.). Oxford UP (2009).

Imachi H, Nubu MK, Nahakara N, *et al.* Isolation of an archaeon at the prokaryote-eukaryote interface. *Nature* 577(7791):519–525 (2020).

Imahori K. *Ecology[,] phytogeography and taxonomy of the Japanese Charophyta.* Kanazawa University (1954).

Imperato F. *Dell'historia natvrale di Ferrante Imperato napolitano libri XXVIII.* Napoli: Constantio Vitale (1599).

REFERENCES 701

Imperato F. *Historia naturale . . . In questa Seconda impressione aggiontouí da Gio: Maria Ferro . . . alcune annotationi alle piante nel libro vigesimo ottauo.* Venetia: Combi, & La Noù (1672).

Imperato F. *Historiæ naturalis libri XXIIX. Accesserunt nonnullæ Johannis Mariæ Ferro adnotationes ad librum vigesmum octavum.* Coloniæ: Saurmann (1695).

Inati S. *Ibn Sina's* Remarks and admonitions: physics and metaphysics. *An analysis and annotated translation.* Columbia UP (2014).

Ingen-Housz J. *Nouvelles expériences et observations sur divers objets de physique.* In two volumes. Paris: Barrois (1785–1789).

Innocent III [Pontifex Romanus]. *D. Innocentii Pontificis Maximi eivs nominis III. Viri ervdit issimi simul atq[ve] grauissimi opera.* Two volumes in one. Coloniæ: Maternvm Cholinvm (1575).

Innocent III [Pontifex Romanus]. *Opera omnia.* In: JP Migne (ed.) *Patrologiæ cursus completus. Series Latina* 214–217. In four volumes. Parisiis: Garnier (1889–1891).

Institoris [Kramer H]. *Malleus maleficarum.* Nurbergen: Antonium Koberger (1494).

Institoris [Kramer H]. *Malleus maleficarum.* Parisiis: Jehan Petit (1507).

al-'Iraqi. *Kitab al-'ilm al-muktasab fi zira'at adh-dhahab, The book of knowledge acquired concerning the cultivation of gold.* EJ Holmyard (trans.). Paris: Paul Geuthner (1923).

Isidore Hispalensis. *De natura rerum liber.* G Becker (ed.). Berolini: Weidmann (1857). Reprinted by Hakkert, Amsterdam (1967).

Isidore of Seville. *The* Etymologies *of Isidore of Seville.* SA Barney *et al.* (eds). Cambridge UP (2006).

Isidorus Hispalensis. *Etymologiarum libri XX.* In: JP Migne (ed.), *Patrologiæ cursus completus . . . Series secunda* 87. *Sancti Isidori Hispalensis* 3–4. Paris: Migne (1850), cols 74–1059.

Iskander AZ. [Review of:] Ursula Weisser (editor), "Buch über Geheimnis der Schöpfung und die Darstelling der Natur" von Pseudo-Apollonius von Tyana. *Medical History* 25(4):438–439 (1981).

Istituto e Museo di Storia della Scienza, Accademia Nazionale dei Lincei. Apiarium [e] Melissographia. Edizione critica digitale (undated). https://brunelleschi.imss.fi.it/apiarium/, accessed 23 September 2020.

Istituto e Museo di Storia della Scienza, Accademia Nazionale dei Lincei. Galileo's microscope. Firenze (undated). Online at brunelleschi.imss.fi.it/esplora/microscopio/dswmedia/risorse/complete_texts.pdf, accessed 19 May 2019 and 23 September 2020.

Ivry A. Arabic and Islamic psychology and philosophy of mind. *The Stanford Encyclopedia of Philosophy* (Summer 2012 edition), EN Zalta (ed.), https://plato.stanford.edu/archives/sum2 012/entries/Arabic-islamic-mind, accessed 23 August 2018 and 11 September 2020.

Ivry AL. [Review of:] *The exalted faith by Abraham ibn Daud,* [by] NM Samuelson and G Weiss. *Speculum* 64(3):721–722 (1989).

Iwabe N, Kuma K, Hasegawa M, *et al.* Evolutionary relationship of archaebacteria, eubacteria, and eukaryotes inferred from phylogenetic trees of duplicated genes. *PNAS* 86(23):9355–9359 (1989).

Jābir ibn Ḥayyān. *Muktār rasā'il. Essai sur l'histoire des idées scientifiques dans l'Islam.* P Kraus (ed.). Cairo: al-Khānjī (1354 [1935]).

Jackson B. Sketch of the life and works of Saint Basil. In: P Schaff & H Wace (eds), *A select library of Nicene and Post-Nicene Fathers of the Christian church.* Second series, volume 8. New York: Christian Literature (1895), pp. xiii–lxxvii.

Jackson BD. Richard Bradley. In: L Stephen (ed.), *Dictionary of National Biography* 6:172. London: Smith, Elder (1886).

Jacobs NX. From unit to unity: protozoology, cell theory, and the new concept of life. *Journal of the History of Biology* 22(2):215–242 (1989).

Jacquet P. History of orchids in Europe, from antiquity to the 17th century. In J Arditti (ed.), *Orchid biology: reviews and perspectives* 6. Dordrecht: Kluwer (1994), pp. 33–102.

Jaeger W. *Aristoteles. Grundlegung einer Geschichte seiner Entwicklung.* Berlin: Wiedemann (1923), second edition 1955. Translated by R Robinson as *Aristotle: fundamentals of the history of his development.* Oxford: Clarendon (1934), second edition 1948.

702   REFERENCES

Jahn R. CG Ehrenberg's concept of the diatoms. *Archiv für Protistenkunde* 146(2):109–116 (1995).

Jahn R. Christian Gottfried Ehrenberg: the man and his contribution to botanical science. In: DM Williams & R Huxley (eds), *Christian Gottfried Ehrenberg (1795–1876). The man and his legacy. The Linnean*, Special issue 1 (1998), pp. 15–28.

[Jamieson R]. On the rein-deer. 1. Its naturalization in Scotland. 2. Its food. 3. Rein-deer milk, and preparations made from it. 4. Speed of the rein-deer. 5. Rein-deer eats the lemming. 6. On the Furia infernalis. *Edinburgh New Philosophical Journal* 3:30–43 (1827).

Janicki C. Untersuchungen an parasitischen Flagellaten. II. Teil. Die Gattungen Devescovina, Parajoenia, Stephanonympha, Calonympha.—Über den Parabasalapparat. Über Kernkonstitution und Kernteilung. *Zeitschrift für wissenschaftliche Zoologie* 112(4):573–691 (1915).

Janicki C. Zur Kenntnis des Parabasalapparats bei parasitischen Flagellaten. *Biologisches Centralblatt* 31(11):321–330 (1911).

Jardine W. *Memoirs of Hugh Edwin Strickland, M.A.* London: van Voorst (1858).

"J.E." [John Ellis]. [Untitled.] In: *The St James's Chronicle or, The British Evening-Post* no. 1012 (25–27 August 1767), second (unnumbered) page.

"J.E." [John Ellis]. [Untitled.] In: *The St James's Chronicle or, The British Evening-Post* no. 1023 (19–22 September 1767), fourth (unnumbered) page.

Jean de Jandun. *Quaestiones super tres libros Aristotelis De anima.* Venetiis: F de Hailbrun & N de Franckfordia (1473).

Jean de Meun—see Guillaume de Lorris

Jeffrey C. Thallophytes and kingdoms—a critique. *Kew Bulletin* 25(2):291–299 (1971).

Jennings H. Introduction. In: *The Divine Pymander of Hermes Mercurius Trismegistus.* Translated from the Arabic by Dr. Everard. Originally published at London: Robert White (1650). Reprint, London: George Redway (1884), pp. i–xiv.

Jenyns L. Report on the recent progress and present state of zoology. In: *Report of the Fourth Meeting of the British Association for the Advancement of Science; held at Edinburgh in 1834.* London: John Murray (1835), pp. 143–251.

Jenyns S. *Disquisitions on several subjects.* London: Dodsley (1782).

Jerome. *Against Jovinianus.* In: P Schaff & H Wace (eds), *A select library of Nicene and Post-Nicene Fathers of the Christian Church.* Second series, volume 6. New York: Christian Literature (1893), pp. 346–416.

Jerome. *Letter 124. To Avitus.* In: P Schaff & H Wace (eds), *A select library of Nicene and Post-Nicene Fathers of the Christian church.* Second series, volume 6. New York: Christian Literature (1893), pp. 238–244.

Jerome. The life of Paulus the first hermit. In: P Schaff & H Wace (eds), *A select library of Nicene and post-Nicene fathers of the Christian church.* Second series, volume 6. New York: Christian Literature (1893), pp. 299–303.

"J.F." [John French]. To the Reader. In: *The Divine Pymander of Hermes Mercurius Trismegistus, in XVII. books.* London: Robert White (1650), folios A2r–A7v (numb. sep.).

Joannes [Alexandrinus]. Σχολια Ιπποκρατους εις το Περι παιδιου φυσεως απο Φωνης Ιωαννου. [*Commentarii in Hippocratis librum de natura pueri.*] In: FR Dietz (ed.), *Scholia in Hippocratem et Galenum* 2:205–235 (1834). Reprinted by Hakkert, Amsterdam (1966).

Joblot L. *Descriptions et usages de plusieurs nouveaux microscopes, tant simples que composez.* Paris: Jacques Collombat (1718).

Joblot L. *Observations d'histoire naturelle faites avec le microscope. Sur un grand nombre d'insectes, & sur les animalcules qui se trouvent dans les liqueurs préparées.* Paris: Briasson (1754–1755).

Job of Edessa. *Encyclopædia of philosophical and natural sciences as taught in Baghdad about A.D. 817, or, Book of treasures. Syriac text edited and translated with a critical apparatus by A. Mingana.* Cambridge: Heffer (1935).

Jocelyn HD. Varro's *Antiquitates rerum diuinarum* and religious affairs in the late Roman Republic. *Bulletin of the John Rylands Library* 65(1):148–205 (1982).

Johannsen W. *Elemente der exakten Erblichkeitslehre.* Jena: Fischer (1909).

REFERENCES 703

Johannsen W. The genotype conception of heredity. *American Naturalist* 45(531):129–159 (1911).

John of Damascus. *De fide orthodoxa. In hoc opere contenta.* JF Stapulensis (trans.). Parisiis: Henricvm Stephanvm (1512).

John of Damascus. *De fide orthodoxa.* In: P Schaff & H Wace (eds), *A select library of Nicene and Post-Nicene Fathers of the Christian church,* Second series, volume 9. New York: Scribner's (1899), pp. 1–101.

John of Damascus. *Saint John Damascene: De fide orthodoxa. Versions of Burgundio and Cerbanus.* St Bonaventure NY: Franciscan Institute (1955).

John of Damascus. *Saint John of Damascus. Writings.* FH Chase Jr (trans.). In series: Fathers of the Church. A new translation, volume 37. Fathers of the Church, New York (1958).

John of Damascus. *Sancti patris nostri Joannis Damasceni . . . Opera omnia quæ exstant.* PM Lequien (trans.). In three volumes. In: JP Migne (ed.), *Patrologiæ cursus completes. Series Græcæ prior,* volumes 94 (1864), 95 and 96 (1860). Parisiis: Migne.

John of Damascus. *St. John Damascene. Dialectica. Version of Robert Grosseteste.* OA Colligan (ed.). In: Franciscan Institute Publications Text Series, volume 4 (EM Buytaert, ed.). St Bonaventure NY: Franciscan Institute (1953).

John of Damascus. *Ta tou makariou Ioannou tou Damaskēnou herga. Beati Ioannis Damasceni opera.* Basileæ: Henrichvm Petri (1559).

John of Damascus. *Theologia Damasceni. De orthodoxa fide.* JF Stapulensis (trans.). Parisiis: Henricvm Stephanvm (1507).

John of Damascus. *Tou en hagiois patros hēmōn Iōannou tou Damaskēnou . . . Sancti patris nostri Joannis Damasceni . . . Opera omnia quæ exstant.* In two volumes. Parisiis: Delespine (1711–1712).

John of Damascus. *Tou makariou Ioannou tou Damaskinou ekdosis akribis tis orthodoxou piseōs . . . Beati Ioannis Damasceni Orthodoxæ fidei accvrata explicatio.* IF Stapulensis (ed. and trans.). Basileæ: Henrichvm Petri (1548). . . . *Series secunda* 199. *Joannis cognomine Saresberiensis . . . Opera omnia.* Paris: Migne (1855), cols 945–964.

John of Salisbury. *Ioannis Saresberensis. Metalogicon.* JB Hall (ed.). In series: Corpvs Christianorvm. Continuatio Mediaeualis 98. Tvrnholti: Brepols Editores Pontificii (1991).

John of Salisbury. *Metalogicon. Jean de Salisbury.* J Lejeune (trans.). In series: Collection Zêtêsis: Textes et essais. Québec: Université Laval; et Paris: Vrin (2009).

John of Salisbury. *Metalogicus.* In: PL Migne (ed.), *Patrologiæ cursus completus . . . Series secunda* 199. *Joannis cognomine Saresberiensis . . . Opera omnia.* Paris: Migne (1855), cols 823–946.

John of Salisbury. *The Metalogicon of John of Salisbury. A Twelfth-century defense of the verbal and logical arts of the trivium.* DD McGarry (trans.). Berkeley: University of California Press (1955).

Johnson G. Lexicon chymicum. In: JJ Manget (ed.), *Bibliotheca chemica curiosa.* Genevæ: Chouet, De Tournes *et al.* (1702), pp. 217–291.

Johnson KA & Rosenbaum JL. Basal bodies and DNA. *Trends in Cell Biology* 1(6):145–149 (1991).

Johnson OS. *A study of Chinese alchemy.* Shanghai: Commercial Press (1928).

Johnston G. *A history of the British zoophytes.* Edinburgh: Lizars (1838).

Johnston G. *A history of the British zoophytes.* Second edition. In two volumes. London: Van Voorst (1847).

Jolivet J & Rashed R. al-Kindī. *The encyclopaedia of Islam,* New [Second] edition 5:121–123. Leiden: Brill (1986).

Jonas H. *The Gnostic religion.* Second edition. Boston: Beacon (1963).

Jonston J. *Historiæ naturalis de auibvs libri VI.* (1650). [With:] *Historiæ naturalis de piscibus et cetis libri V.* (undated; *Ad lectorem* 1649). [Also:] *Historiæ naturalis de exangvibus aqvaticis libri IV.* (1650). [And:] *Historiæ natvralis de insectis. Libri III. De serpentibus et draconib[us], libri II.* (1653). Francofvrti: Merianoru[m].

Jonston J. *Historiæ naturalis de exanguibus aquaticis libri IV.* Amstelodami: Johannis Jacobi Schipper (1657).

Jonston J. *Historiæ naturalis de quadrupedibus libri.* [With:] *De insectis libri III* [actually IV]. [And:] *De serpentibus et draconibus libri II.* Amstelodami: Schipper (1657).

704 REFERENCES

Jonston J. *Historiæ naturalis de quadrupedibus libri*. Amstelodami: Schipper (1657).

Jouanna J. *Hippocrates*. Baltimore: Johns Hopkins UP (1999).

Jourdan AJL. *Atmosphérique*. *Dictionnaire raisonné, étymologique, synonymique et polyglotte, des termes usités dans les sciences naturelles* 1:125 (1834).

Jourdain C. *Des commentaires inédits de Guillaume de Conches et de Nicolas Triveth sur la Consolation de la philosophie de Boèce*. *Notices et extraits des manuscrits de la Bibliothèque Impériale et autres bibliothèques* 20(2). Paris: Impériale (1862), pp. 40–82.

Jourdain de Séverac [Frater Jordanus]. *Mirabilia descripta*. *Recueil de voyages et de mémoires, Société de Géographie (Paris)* 4:37–68 (1839).

Jourdain de Séverac [Frater Jordanus]. *Mirabilia descripta*. *The wonders of the East, by Friar Jordanus*. H Yule (trans.). London: Hakluyt Society (1868).

Jung J [Jungius]. *Doxoscopiæ physicæ minores, sive Isagoge physica doxoscopica . . . Ex recensione et distinctione M.F.H., cujus annotationes quædam accedunt*. Hamburgi: Johannis Naumanni. Typis Pfeifferianis (1662).

Jung J [Jungius]. *Historia vermium*. J Vagetius (ed.). Hamburgi: Brendekianis (1691).

Jung J [Jungius]. *Opvscvla botanico-physica ex recensione et distinctione Martini Fogelii . . . et Ioh. Vagetii . . . cvm eorvndem annotationibvs accedit Iosephi de Aromatariis . . . ad Bartholomevm Nanti epistola De generatione plantarvm ex seminibvs omnia collecta*. IoS Albrecht (ed). Cobvrgi: Georgii Ottonis (1747).

Jung J [Jungius]. *Praelectiones Physicae. Historisch-kritische edition*. C Meinel (ed.). Göttingen: Vandenhoeck & Ruprecht (1982).

Jussieu A[drien] de. *Cours élémentaire d'historire naturelle . . . conformément au Programme de l'Université du 14 september 1840 . . . Botanique*. Paris: Langlois & Leclerq, and Fortin, Masson (undated, [1843] and 1844).

Jussieu A[drien] de. *Mémoire sur le groupe de Rutacées*. *Mémoires du Muséum d'Historie Naturelle* 12:384–542 plus 29 plates (1825).

Jussieu AL de. An introduction to the history of plants [*Genera plantarum* (1789), pp. i–lxxii]. S Rosa (trans.). In: PF Stevens, *The development of biological systematics. Antoine-Laurent de Jussieu, nature, and the natural system*. New York: Columbia UP (1994), pp. 319–384.

Jussieu AL de. *Examen de la famille des Renoncules*. *Mémoires de Mathématique et de Physique de l'Académie Royale des Sciences* 1773:214–240 (1777).

Jussieu AL de. *Genera plantarum secundum ordines naturales disposita*. Parisiis: Herissant & Barrois (1789).

Jussieu AL de. *Méthode naturelle des végétaux*. *Dictionnaire des science naturelles* [Second edition] 30:426–468 (1824). Also published separately as *Principes des méthode naturelle des végétaux*. Paris et Strasbourg: FG Levrault (1824).

Jussieu B de. De quelques productions marines qui ont été mises au nombre des plantes, & qui sont l'ouvrage d'une sorte d'insectes de mer. *Mémoires de Mathématique & de Physique . . . de l'Académie Royale des Sciences* 1742:290–302 (1745).

Justin Martyr. *Writings*. In: A Roberts & J Donaldson (eds), *Ante-Nicene Christian Library*, volume 2. Edinburgh: Clark (1909).

Juvenal [Juvenalis, Decimus Junius]. *D. Junii Juvenalis et A. Persii Flacci. Satiræ*. L Prateus (trans.). Londini: Rivington *et al*. (1823).

Kaempfer E. *Amœnitatum exoticarum politico-physico-medicarum fasciculi V, quibus continentur variæ relationes, observationes & descriptiones rerum Persicarum & ulterioris Asiæ*. Lemgoviæ: Meyer (1712).

Kafka F. *Die Verwandlung*. Leipzig: Kurt Wolfe (1916).

Kahn CH. *Anaximander and the origins of Greek cosmology*. New York: Columbia UP (1960).

Kangro H. Jungius, Joachim. *Complete Dictionary of Scientific Biography* 7:193–196 (2008).

[Kant I]. *Allgemeine Naturgeschichte und Theorie des Himmels, oder Versuch von der Verfassung und dem mechanischen Ursprunge des ganzen Weltgebäudes nach Newtonischen Grundsätzen abgehandelt*. Königsberg & Leipzig: Petersen (1755).

Kant I. *Allgemeine Naturgeschichte und Theorie des Himmels, oder Versuch von der Verfassung und dem mechanischen Ursprunge des ganzen Weltgebäudes nach Newtonischen Grundsätzen abgehandelt*. Neue Auflage. Zeitz: Wilhelm Webe (1798).

REFERENCES 705

Kant I. *Critique of judgement.* JH Bernard (trans.). London: Macmillan (1914).

Kant I. *Critique of pure reason.* P Guyer & AW Wood (eds and trans.). Cambridge UP (1998).

Kant I. *Das Bonner Kant-Korpus.* https://korpora.zim.uni-duisburg-essen.de/kant/, accessed January 2020 and 1 November 2020.

Kant I. *Immanuel Kant. Anthropology, history, and education.* RB Louden & G Zöller (eds). Cambridge UP (2007).

Kant I. *Immanuel Kant's Sämmtliche Werke.* K Rosenkranz & FW Schubert (eds). In 14 volumes, bound in 12. Leipzig: Leopold Voss (1838–1842).

Kant I. *Kant's gesammelte Schriften.* Abteilung 1. *Werke* (Bände 1–9). Abt. 2. *Briefwechsel* (Bde 10–13). Abt. 3 *Handschriftlicher Nachlass* (Bde 14–23). Abt. 4. *Vorlesungen* (Bde 24–29). Berlin: Reimer (*Werke* 1–7, 1910–1917); and de Gruyter (remaining volumes, 1923–1983).

Kant I. *Lectures on metaphysics.* K Ameriks & S Naragon (trans.). Cambridge UP (1997).

Kant I. *Metaphysical foundations of natural science.* J Bennett (trans.). https://earlymoderntexts. com/assets/pdfs/kant1786.pdf (2017), accessed 15 January 2020 and 5 October 2020.

Kant I. *Metaphysische Anfangsgründe der Naturwissenschaft.* Riga: Hartknoch (1786).

Kant I. *Metaphysische Anfangsgründe der Naturwissenschaft.* Second edition. Riga: Hartknoch (1787).

Kant I. *Prolegomena zu einer jeden künftigen Metaphysik, die als Wissenschaft wird auftreten können.* Riga: Hartknoch (1783).

Kant I. Review of JG *Herder's Ideas for the philosophy of the history of humanity. Parts 1 and 2* (1785). AW Wood (trans.). In: *Immanuel Kant. Anthropology, history, and education.* RB Louden & G Zöller (eds). Cambridge UP (2007), pp. 121–142.

Kant I. *The philosophy of Kant. Immanuel Kant's moral and political writings.* CJ Friedrich (ed. and trans.). New York: Modern Library (1949).

Kant I. Über die von der K. Akademie der Wissenschaften zu Berlin für das Jahr 1791 ausgesetzte Preisfrage: welches sind Die wirklichen Fortschritte, die Die Metaphysik seit Leibnitz's und Wolf's Zeiten in Deutschland gemacht hat (1804). In: K Rosenkranz & FW Schubert (eds), *Immanuel Kant's Sämmtliche Werke,* volume 1 (1840). Leipzig: Leopold Voss (1838–1840), pp. 483–578.

Kant I. *Universal natural history and theory of the heavens.* SL Jaki (trans.). Edinburgh: Scottish Academic Press (1981).

Kant I. *Vorlesungen über die Metaphysik.* Erfurt: Keyser (1821).

Kaplan A. *Sefer Yetzirah. The Book of Creation.* Revised edition. York Beach ME: Samuel Weiser (1997).

Karberg KA, Olsen GJ & Davis JJ. Similarity of genes horizontally acquired by *Escherichia coli* and *Salmonella enterica* is evidence of a supraspecies pangenome. *PNAS* 108(50):20154–20159 (2011).

Katscher F. The history of the terms prokaryotes and eukaryotes. *Protist* 155(2):257–263 (2004).

Kaup J. Einige Worte über die systematische Stelling der Familie der Raben, Corvidae. In: J Cabanis (ed.), *Erinnerungsschrift zum Gedächnisse an die VIII. Jahresversammlung der deutschen Ornithologen-Gesellschaft, abgehalten im Gotha vom 17 bis 20 Juli 1854.* Cassel: Theodor Fischer (1855), xlvii–lvi. *Erinnerugsschrift* also appeared as a separately numbered part of *Journal für Ornithologie* 2 (1854).

Kee H & Young FW. *Understanding the New Testament.* Englewood Cliffs NJ: Prentice-Hall (1957).

Keeling PJ. The impact of history on our perception of evolutionary events: endosymbiosis and the origin of eukaryotic complexity. *Cold Spring Harbor Perspectives in Biology* 6:a016196 (2014).

Keeling PJ & Palmer JD. Horizontal gene transfer in eukaryotic evolution. *Nature Reviews Genetics* 9(8):605–618 (2009).

Keenan ME. St. Augustine and biological science. *Osiris* 7:588–608 (1939).

Kelley DR. *Foundations of modern historical scholarship. Language, law, and history in the French Renaissance.* New York: Columbia UP (1970).

Kempf F. *Die Register Papst Innocenz' III. Eine paläographisch-diplomatische Untersuchung.* Miscellanea Historiæ Pontificiæ, volume 9. Romæ: Pontificia Universita Gregoriana (1945).

706 REFERENCES

Kepler J. *De stella nova in pede serpentarii . . . Libellus astronomicis, physicis, metaphysicis, meteorologicis & astrologis disputationibus.* Pragæ: Pauli Sessii (1606).

Khakhina LN. *Concepts of symbiogenesis. A historical and critical study of the research of Russian botanists.* Translated from Russian original (1979) by S Merkel & R Coalson; edited by L Margulis & M McMenamin. New Haven: Yale UP (1992).

Khalidi T. Mas'ūdī's lost works: a reconstruction of their content. *Journal of the American Oriental Society* 94(1):35–41 (1974).

Kibre P. *The library of Pico della Mirandola.* Morningside Heights NY: Columbia UP (1936).

Kidd J. *On the adaptation of external nature to the physical condition of man.* [Fifth edition]. London: William Pickering (1837).

Kieckhefer R. *Magic in the Middle Ages.* Cambridge UP (2000).

al-Kindī. *Œuvres philosophiques et scientifiques d'al-Kindī,* volume 2. *Métaphysique et cosmologie.* R Rashed & J Jolivet (eds and trans.). Leiden: Brill (1998).

King BJ. Animal images in prehistoric rock art: looking beyond Europe. npr.org/sections/13.7/2017/08/03/540805776/animal-images-in-prehistoric-rock-art-looking-beyond-europe, accessed 4 September 2020.

King E. Several observations and experiments on the animalcula, in pepper-water, &c. *Philosophical Transactions* 17(203):861–865 (read 1693, published 1694).

King-Hele D. *Erasmus Darwin. Grandfather of Charles Darwin.* New York: Scribners (1963).

King-Hele D. *The essential writings of Erasmus Darwin. Chosen and edited with a linking commentary.* London: MacGibbon & Kee (1968).

Kingsley C. *Glaucus; or, the wonders of the shore.* Boston: Ticknor & Fields (1855).

Kingsley P. *Ancient philosophy, mystery, and magic: Empedocles and Pythagorean tradition.* Oxford: Clarendon (1995).

Kirby H (annotated by Margulis L). Harold Kirby's symbionts of termites: karyomastigont reproduction and calomnymphid taxonomy. *Symbiosis* 16(1):7–63 (1994).

Kirby H Jr. Organisms living on and in protozoa. In: GH Calkins & FM Summers (eds), *Protozoa in biological research.* New York: Columbia UP (1941), pp. 1009–1113.

Kirby W. *On the power wisdom and goodness of God as manifested in the creation of animals and in their history habits and instincts.* In two volumes. London: William Pickering (1835).

Kirby W. *On the power, wisdom and goodness of God, as manifested in the creation of animals, and in their history, habits and instincts.* Philadelphia: Carey, Lea & Blanchard (1836).

Kirby W & Spence W. *An introduction to entomology: or elements of the natural history of insects.* In two volumes. London: Longman, Hurst *et al.* (1815–1817).

Kirby W & Spence W. *An introduction to entomology.* Fourth edition. In four volumes. London: Longman *et al.* [varies] (1822–1826).

Kircher A. *Itinerarivm exstaticum cœleste, quo mundi opificium.* Herbipoli: Endter (1660).

Kircher A. *Itinerarivm exstaticvm qvo mvndi opificivm.* Romæ: Vitalis Mascardi (1656).

Kircher A. *Magnes sive de arte magnetica opvs tripartitvm.* Romae: Ludouici Grignani (1641).

Kircher A. *Magneticum naturæ regnum sive disceptatio physiologica de triplici in Natura rerum magnete, juxta triplicem ejusdem Naturæ gradum digesto inanimato: animato: sensitivo.* Amstelodami: Johannis Janssonii à Waesberge & Elizei Weyerstraet (1667).

Kirchweger AJ. *Aurea catena Homeri oder eine Beschreibung von dem Ursprung der Natur und natürlichen Dingen.* Franckfurt und Leipzig: Böhme (1723).

Kirk GS, Raven JE & Schofield M. *The Presocratic philosophers. A critical history with a selection of texts.* Second edition. Cambridge UP (1983).

Kirk PM, Cannon PF, Minter DW & Stalpers JA. *Dictionary of the fungi.* Tenth edition. Wallingford: CABI (2008).

Kitāb al-ḥayawān. Nala hadha al-kitab al-ja'iza al-ula li-an-nashr wa at-tahqiq al-'imi fi an-musabaqat al-adabiyya allati nazamaha al-mujamma' al-lughawi. In eight volumes. Cairo: Matba'at wa maktabat Mustafa al-Babi al-Halabi wa auladih bi-Misr (1949–1950).

Klages L. Einleitung. In: CG Carus, *Psyche. Zur Entwickelungsgeschichte der Seele.* In series: Gott-Nature, Schriftenreihe zur Neubegründung der Naturphilosophie 1. Jena: Eugen Diederichs (1926), pp. i–xx.

REFERENCES 707

Klebs AC. Incunabula lists I. Herbals. *Papers of the Bibliographic Society of America* 11(3):75–92 (1917) and 12(1–2):41–53 (1918).

Klebs G. Flagellatenstudien. *Zeitschrift für wissenschaftliche Zoologie* 55(3):266–351 and 55(4):353–445 (1893).

Klebs G. *Ueber die Fortpflanzungs-Physiologie der niederen Organismen, der Protobionten. Specieller Theil. Die Bedingungen der Fortpflanzung bei einigen Algen und Pilzen.* Jena: Gustav Fischer (1896).

Klein JT. *Naturalis dispositio echinodermatum. Accesseit lucubratiuncula de aculeis echinorum marinorum.* Gedani: Schreiber (1734).

Klein M. Oken (or Okenfuss), Lorenz. *Complete Dictionary of Scientific Biography* 10:194–196 (2008).

Kleiner FS & Mamiya CJ. *Gardner's art through the ages, volume 1: The Western perspective.* [n.c.]: Wadsworth/Thompson (2006).

Knaysi G & Baker RF. Demonstration, with the electron microscope, of a nucleus in *Bacillus mycoides* grown in a nitrogen-free medium. *Journal of Bacteriology* 53(5):539–553 (1947).

Kneale W & Kneale M. *The development of logic.* Oxford: Clarendon (1962).

Knebel KL von. *Zur deutschen Literatur und Geschichte.* Nürnberg: Bauer & Raspe (1858).

Knechtges DR & Shih H. Xunzi. In: DR Knechtges & T Chang (eds), *Ancient and early medieval Chinese literature: a reference guide*, volume 3/4. Leiden: Brill (2004), pp. 1757–1765.

Kniep H. *Die Sexualität der niederen Pflanzen. Differenzierung, Verteiling, Bestimmung und Vererbung des Geschlects bei den Thallophyten.* Jena: Fischer (1928).

Knight BCJG. *Bacterial nutrition. Material for a comparative physiology of bacteria.* London: HMSO (1936).

Knight D. *The age of science. The scientific world-view in the Nineteenth century.* Oxford: Blackwell (1986).

Koch R. Untersuchungen über Bacterien. V. Die Aetiologie der Milzbrand-Krankheit, begründet auf die Entwicklungsgeschichte des Bacillus Anthracis. *Beiträge zur Biologie der Pflanzen* 2(2):277–310 (1876).

Kofoid CA. Review of *Natural History* by Pliny (H Rackham). *Isis* 31(2):433–434 (1940).

Kofoid CA. The biological stations of Europe. *Bulletin of the United States Bureau of Education* 440 (1910 no. 4). Washington: Government Printing Office (1910).

Kölliker A [von]. Beiträge zur Kenntniss niederer Thiere. *Zeitschrift für wissenschaftliche Zoologie* 1(1):1–37 (1848).

Kölliker A [von]. Die Lehre von der thierischen Zelle und den einfacheren thierischen Formelementen, nach den neuesten Fortschritten dargestellt. *Zeitschrift für wissenschaftliche Botanik* 2:46–102 (1845).

Kölliker A [von]. *Handbuch der Gewebelehre des Menschen für Aerzte und Studirende.* Leipzig: Wilhelm Engelmann (1852).

Kölliker A [von]. *Manual of human microscopical anatomy.* G Busk & T[H] Huxley (trans.). J da Costa (ed.). Philadelphia: Lippincott, Grambo (1854).

Kölliker A. Zur Lehre von den Furchungen. *Archiv für Naturgeschichte* 13(1):9–22 (1847).

Kölreuter JG. *Das entdeckte Geheimniß der Cryptogamie.* Carlsruhe: Michael Maklot (1777).

König E (ed.). Κεραζ αμαλθειαζ, *seu thesaurus remediorum è triplici regno, vegetabili, animali, minerali.* Basileæ: König (1693).

König E. *Regnum vegetabile speciale plantarum descriptionem.* Basileæ: Regum (1696).

König E. *Regnvm animale, sectionibus III.* Coloniae Munatianae: König (1682).

König E. *Regnvm minerale.* Basileae Rauracorum: König (1686).

Königlichen Akademie der Wissenschaften. *Allgemeine Deutsche Biographie.* In 56 volumes. Leipzig: Dunder & Humblot (1875–1912).

Koren J & Danielssen DC (eds). *Fauna littoralis Norvegiæ.* Heft 3. Bergen: Beyer (1877).

Kossel A. Ueber einen peptonartigen Bestandteil des Zellkerns. *Zeitschrift für physiologische Chemie* 8(6):511–515 (1884).

Koster JTh. Preface. In: CA Agardh, *Systema algarum* (1824). Reprint by Asher, Amsterdam (1965), two pages between pp. iv and v.

708 REFERENCES

Kotrc RF & Walters KR. A bibliography of the Galenic corpus. A newly researched list and arrangement of the titles of the treatises extant in Greek, Latin, and Arabic. *Transactions & Studies of the College of Physicians in Philadelphia,* 5 Series 1(4):256–304 (1979).

Kozo-Polyansky BM. *Noviy printsip biologii. Ocherk teorii simbiogeneza. [New principle of biology. An outline of the theory of symbiogenesis.]* Moscow: Puchina (1924).

Kozo-Polyansky BM. *Symbiogenesis. A new principle of evolution.* V Fet (trans.); V Fet & L Margulis (eds). Cambridge MA: Harvard UP (2010).

Kozo-Polyansky BM. Theory of symbiogenesis and, "Pangenesis, a provisional hypothesis". Abstract, All-Russian Congress of Russian Botanists, Petrograd (1921).

Krämer HJ. *Der Ursprung der Geistmetaphysik: Untersuchungen zur Geshichte des Platonismus zwischen Platon und Plotin.* Second edition. Amsterdam: Grüner (1967).

Krause E. *Erasmus Darwin.* WS Dallas (trans.). *With a preliminary notice by Charles Darwin.* London: John Murray (1879).

Krause J, Lalueza-Fox C, Orlando L, *et al.* The derived *FOXP2* variant of modern humans was shared with Neandertals. *Current Biology* 17(21):1908–1912 (2007).

Kretzmann N. *Infinity and continuity in ancient and medieval thought.* Ithaca NY: Cornell UP (1982).

Krieg NR & Garrity GM. On using the *Manual.* In: DR Boone & RW Castenholz (eds). *Bergey's manual of systematic bacteriology.* Second edition (GM Garrity, editor-in-chief). Volume 1. The *Archaea* and deeply branching and phototrophic bacteria. New York: Springer (2001), pp. 15–19.

Kristeller PO. Giovanni Pico della Mirandola and his sources. In: *L'opera e il pensiero di Pico della Mirandola nella storia dell'Umanesimo. Convegno Internazionale per il V Centenario della nascita di Giovanni Pico della Mirandola . . . Mirandola: 15–18 Settembre 1963, volume 1. Relazioni.* Firenze: Istituto nazionale di studi sul Rinascimento (1965), pp. 35–142.

Kristeller PO. *Il pensiero filosofico di Marsilio Ficino.* Firenze: Samsoni (1953).

Kristeller PO. *The philosophy of Marsilio Ficino.* New York: Columbia UP (1943). Reprinted 1964.

Kristeller PO. The scholastic background of Marsilio Ficino: with an edition of unpublished texts. *Traditio* 2:257–318 (1944).

Kropotkin P. *Mutual aid: a factor in evolution.* New York: McClure Phillips (1902).

Kruk R. Nabāt, plants. *The encyclopaedia of Islam,* New [Second] edition 7:831–834. Leiden: Brill (1993).

Ku C & Martin WF. A natural barrier to lateral gene transfer from prokaryotes to eukaryotes revealed from genomes: the 70% rule. *BMC Biology* 14:89 (2016).

Ku C, Nelson-Sathi S, Roettger M, *et al.* Endosymbiotic origin and differential loss of eukaryotic genes. *Nature* 524(7566):427–432 (2015).

Kuhlbrodt P. Friedrich Traugott Kützing—ein bedeutender Naturforscher des 19. Jahrhunderts. Friedrich Traugott Kützing als Nordhäuser Bürger (undated). http://www.geschichtsportal-nordhausen.de/biografien/biografie-friedrich-traugott-kuetzing/, accessed 9 April 2020 and 7 October 2020.

Kuhlwilm M, Gronau I, Hubisz JM, *et al.* Ancient gene flow from early modern humans into Eastern Neanderthals. *Nature* 530(7591):429–433 (2016).

Kühn CG (ed.). *Medicorum Græcorum opera quæ exstant.* In 26 volumes. Lipsiæ: Cnoblochii (1821–1833).

Kuntz ML. Pythagorean cosmology and its identification in Bodin's Colloquium heptaplomeres. In: ML Kuntz & PG Kuntz (eds), *Jacob's Ladder and the Tree of Life. Concepts of hierarchy and the Great Chain of Being.* Revised edition. New York: Peter Lang (1987), pp. 253–265.

Kuntz MLD. Introduction. In: *Colloquium of the seven about secrets of the sublime. Colloquium heptaplomeres de rerum sublimium arcanis abditis.* MLD Kuntz (trans.). Princeton UP (1975), pp. xv–lxxxi.

Kuntz PG & Kuntz ML. The symbol of the tree interpreted in the context of other symbols of hierarchical order, the Great Chain of Being and Jacob's Ladder. In: ML Kuntz & PG Kuntz (eds), *Jacob's Ladder and the Tree of Life. Concepts of hierarchy and the Great Chain of Being.* Revised edition. New York: Peter Lang (1987), pp. 319–334. [Author order per Table of Contents].

Kuntz PG. "From the angel to the worm": Augustine's hierarchical vision. In: ML Kuntz & PG Kuntz (eds), *Jacob's Ladder and the Tree of Life. Concepts of hierarchy and the Great Chain of Being*. Revised edition. New York: Peter Lang (1987), pp. 41–53.

Kuntz PG. A formal preface and an informal conclusion to The Great Chain of Being: the necessity and universality of hierarchical thought. In: ML Kuntz & PG Kuntz (eds), *Jacob's Ladder and the Tree of Life. Concepts of hierarchy and the Great Chain of Being*. New York: Peter Lang (1987), pp. 3–14.

Kurland CG, Collins LJ & Penny D. Genomics and the irreducible nature of eukaryotic cells. *Science* 312(5776):1011–1014 (2006).

Kusukawa S. The sources of Gessner's pictures for the *Historia animalium. Annals of Science* 67(3):303–328 (2010).

Kützing FT. *Aufzeichnungen und Erinnerungen*. RHW Müller & R Zaunick (eds). Leipzig: Barth (1960).

Kützing FT. Beitrag zur Kenntniss über die Entstehung und Metamorphose der niedern vegetabilischen Organismen, nebst einer systematische Zusammenstellung der hieher gehörigen niedern Algenformen. *Linnaea* 8:335–384 (1833).

Kützing FT. *Die kieselschaligen Bacillarien oder Diatomeen*. Nordhausen: Köhne (1844). Reprinted by Koeltz, Koenigstein (1983).

Kützing FT. *Die Sophisten und Dialektiker, die gefährlichsten Feinde der wissenschaftlichen Botanik. (Zugleich als Erwiderung auf eine Note in "Schellings's und Hegel's Verhältniss zur Naturwissenschaft" von Dr. M.J. Schleiden, Prof. zu Jena)*. Nordhausen: Förstemann (1844).

Kützing FT. Die Umwandlung niederer Algenformen in höhere, so wie auch in Gattungen ganz verschiedener Familien und Klassen höhere Cryptogamen mit zelligem Bau. *Natuurkundige Verhandelingen van de Hollandsche Maatschappij der Wetenschappen te Haarlem. 2 Verzamling* 1:xix–xxiv and 1–120 (1841).

Kützing F[T]. Microscopische Untersuchungen über die Hefe und Essigmutter, nebst mehreren andern dazu gehörigen vegetabilischen Gebilden. *Journal für praktische Chemie* 11(7):385–409 (1837).

Kützing FT. *Phycologia generalis oder Anatomie, Physiologie und Systemkunde der Tange*. Leipzig: Brockhaus (1843).

Kützing FT. *Species algarum*. Lipsiae: Brockhaus (1849). Reprinted by Asher, Amsterdam (1969).

Kützing FT. Synopsis diatomearum oder Versuch einer systematischen Zusammenstellung der Diatomeen. *Linnaea* 8:529–620 (1833).

Kützing FT. *Über die Verwandlung der Infusorien in niedere Algenformen*. Nordhausen: Köhne (1844).

Kyrpides NC, Hugenholz P, Eisen JA, *et al*. Genomic encyclopedia of bacteria and archaea: sequencing a myriad of type strains. *PLoS Biology* 12(8):e1001920 (2014).

Lacalli TC. *The nervous system and ciliary band of Müller's larva. Proceedings of the Royal Society of London. Series B, Biological Sciences* 217(1206):37–58 (1982).

Lachower P & Tishby I (eds). *The wisdom of the Zohar*. In three volumes. London: Littman Library of Jewish Civilization (1989).

La Croix D de. *Connubia florum Latino carmine demonstrata*. Parisiis: Theobustea (1728).

Lacroix D. *Connubia florum Latino carmine demonstrata*. R Clayton (ed.). Bathoniæ: Hazard (1791).

Lactantius, Lucius Caecilius Firmianus. *Institutiones Divinæ*. In: JP Migne (ed.), *Patrologiæ cursus completus. Series prima* 6. Parisiis: Sirou (1844), cols 111–1094.

Lactantius, Lucius Caecilius Firmianus. *The divine institutes*. In: A Roberts & J Donaldson (eds), The Ante-Nicene fathers 7. Buffalo: Christian Literature (1886), pp. 9–223.

Ladner GB. Medieval and modern understanding of symbolism: a comparison. *Speculum* 54(2):223–256 (1979).

Ladner GB. The philosophical anthropology of Saint Gregory of Nyssa. *Dumbarton Oaks Papers* 12:61–94 (1958).

Lagarde B. Le "De differentiis" de Pléthon d'après l'autographe de la Marcienne. *Byzantion* 43:312–343 (1973).

710 REFERENCES

Lahontan LA. *New voyages to North-America*. London: Bonwicke (1703).

Lake JA, Henderson E, Oakes M & Clark MW. Eocytes: a new ribosome structure indicates a kingdom with a closer relationship to eukaryotes. *PNAS* 81(12):3786–3790 (1984).

Lamarck [JB] de. *Encyclopédie méthodique. Botanique*. Volumes 1–4 (1783–1796); with JLM Poiret, volumes 5–8 (1804–1808), and *Supplément* in five volumes (1810–1817). Paris: Panckoucke; and Liège: Plomteux (vols 1–3); Paris: Agasse (vols 4–8, and *Supplément*).

Lamarck [JB] de. *Extrait du cours de zoologie du Muséum d'histoire naturelle. Sur les animaux sans vertèbres*. Paris: D'Hautel & Gabon (1812).

Lamarck [JB] de. *Flore françoise*. In three volumes. Paris: L'Imprimerie Royale (1778).

Lamarck [JB de] [as: le C, presumably *le Citoyen*]. *Flore françoise*. Second edition. In three volumes. Paris: Agasse (An 3, 1795).

Lamarck [JB] de. *Histoire naturelle des animaux sans vertèbres, présentant les caractères généraux et particuliers de ces animaux, leur distribution, leurs classes, leurs familles, leurs genres, et la citation des principales espèces qui s'y rapportent*. In seven volumes. Paris: Verdière [varies] (1815–1822).

Lamarck JB [de]. *Hydrogéologie ou recherches sur l'influence qu'ont les eaux sur la surface du globe terrestre*. Paris: L'Auteur, Agasse & Maillard (An 10, 1802).

Lamarck JB [de]. *Inédites de Lamarck, d'après les manuscrits conservés à la bibliothèque centrale du Muséum National d'Histoire Naturelle de Paris*. M Vachon, G Rousseau & Y Laissus (eds). Paris: Masson (1972).

Lamarck JB [de]. *Mémoires de physique et d'histoire naturelle, établis sur des bases de raisonnement indépendendantes de toute théorie*. Paris: L'Auteur (An 5, 1797).

Lamarck JBPA [de]. *Philosophie zoologique, ou exposition des considérations relatives à l'histoire naturelle des animaux*. In two volumes. Paris: Dentu; and the author (1809).

Lamarck JB [de]. *Recherches sur l'organisation des corps vivans*. Paris: L'auteur; et Maillard (undated, 1802).

Lamarck [JB]. *Système analytique des connaissances positives de l'homme, restreintes a celles qui proviennent directement ou indirectement de l'observation*. Paris: L'auteur, [et] Belin (1820).

Lamarck JB [de]. *Systême des animaux sans vertèbres*. Paris: L'auteur; et Deterville (An 9, 1801).

Lamarck [JB] de. *Tableau encyclopédique et méthodique des trois règnes de la nature. Botanique*. In three volumes of text, and 1000 plates bound in four volumes. Paris: Panckoucke (volumes 1–2, 1791–1793); and Agasse (volume 3 and plates, 1823).

Lamarck JB [de]. *Zoological philosophy. An exposition with regard to the natural history of animals*. [With:] Introductory lecture for 1800 (from *Système des animaux sans vertèbres*). [And:] Biographical memoir of M. de Lamarck (*Eloge de M. Lamarck*, by G Cuvier . . . as published in *The Edinburgh New Philosophical Journal*, 1836). H Elliot (trans.). University of Chicago Press (1984).

Lamarck [JB] de & Candolle [AP] de. *Flore françoise*. Third edition. Four volumes (bound in five) originally issued in 1805, reissued in 1815 with new title pages; plus a new Tome 5 under the sole authorship of de Candolle. Paris: Desray (1815).

Lamarck JB de & Candolle AP de. *Synopsis plantarum in flora Gallica descriptarum*. Paris: Agasse (1806).

Lamarck JB [de] & Mirbel B. *Histoire naturelle des végétaux, classés par familles, avec la citation de la classe et de l'ordre de Linné . . . avec des renvois aux familles naturelles de AL de Jussieu*. In 15 volumes. Paris: Deterville (An 11, 1803).

Lambert WG. Mesopotamian creation stories. In: MJ Geller & M Schipper (eds), *Imagining creation*. IJS Studies in Judaica 5. Leiden: Brill (2007), pp. 15–59.

Lambros SP (ed.). *Excerptorum Constantini De natura animalium libri duo. Aristophanis. Historiae animalium epitome*. In: Academiae Litterarum Regiae Borussicae (ed.), *Supplementum Aristotelicum* 1. Berlin: Reimer (1885).

La Métherie [JC] de. *Vues physiologiques sur l'organisation animale et végétale*. Amsterdam: Didot (1780).

[La Mettrie JO de]. *L'homme machine*. Leide: Luzac (1748).

La Mettrie JO de. *Man a machine* and *Man a plant*. RA Watson & M Rybalka (trans.). Indianapolis: Hackett (1994).

La Mettrie JO de. *Man a machine*. GC Bussey (trans.), MW Calkins (rev.). La Salle IL: Open Court (1912).

La Mettrie JO de. *Œuvres philosophiques de La Mettrie. Nouvelle édition*. In three volumes. Berlin: Tutot (1796).

Lamouroux JVF. Essai sur les genres de la famille des Thalassiophytes non articulés. *Annales du Muséum d'Histoire Naturelle* 20:21–47, 115–139 and 267–293 (1813).

Lamouroux JVF. *Histoire des polypiers coralligènes flexibles, vulgairement nommés zoophytes*. Caen: Poisson (1816).

Lamouroux [JVF], Bory de Saint-Vincent [JB] & Eudes-Deslongchamps [JA]. *Encyclopédie méthodique. Histoire naturelle des zoophytes, ou animaux rayonnés*. Tome 2. Paris: Agasse (1824).

Landauer S. Die Psychologie des Ibn Sînâ. *Zeitschrift der Deutschen Morgenländischen Gesellschaft* 29(3/4):335–418 (1876).

Landsberger B. Die Fauna des alten Mesopotamien. *Abhandlung der philologisch-historischen Klasse der Sächsische Akademie der Wissenschaften* 42(6):1–144 (1934).

Landsborough D. *A popular history of British sea-weeds*. [First edition]. London: Reeve, Behnam & Reeve (1849).

Landsborough D. *A popular history of British sea-weeds*. Second edition. London: Reeve & Benham (1851).

Landsborough D. *A popular history of British sea-weeds*. Third edition. London: Lovell Reeve (1857).

Lane N. The unseen world: reflections on Leeuwenhoek (1677) "Concerning little animals". *Philosophical Transactions of the Royal Society* B 370:20140344 (2015).

Lanessan JL de. Introduction. In: *Œuvres de Buffon. Nouvelle édition* 1:51*–452*. Paris: Pilon [et] Le Vasseur (1884).

Lanessan JL de. Notice biographique. In: *Œuvres de Buffon. Nouvelle édition* 1:1*–50*. Paris: Pilon [et] Le Vasseur (1884).

Lanessan JL de. *Traité de zoologie. Protozoaires*. Paris: Octave Doin (1882).

Lang A. *Lehrbuch der vergleichenden Anatomie der wirbellosen Thiere. Protozoa*. Second edition. Jena: Gustav Fischer (1901).

Lang A. Mythology. *Encyclopædia Britannica*, Ninth edition 17:143–146 (1884).

Langer WL (ed.). *An encyclopedia of world history. Ancient, medieval, and modern: chronologically arranged*. Fifth edition. Boston: Houghton Mifflin (1972).

Langley EF. The extant repertory of the early Sicilian poets. *PMLA* 28(3):454–520 (1913).

Langley MM. *Sacred wood. A study of Norwegian stave churches*. MA thesis, University of Louisville (2000).

Lankester [E]R. *Diversions of a naturalist*. London: Methuen (1915).

Lankester ER. Zoology. *Encyclopædia Britannica*, Ninth edition 24:799–820 (1888).

La Primaudaye P de. *L'Académie Françoise*. Troisième tome. Paris: G Chaudiere (1590).

La Primaudaye P de. *The French Academie*. London: Thomas Adams (1618).

Larcher PH. *Larcher's notes on Herodotus. Historical and critical remarks on the Nine books of the History of Herodotus; with a chronological table. Translated from the French*. In two volumes. London: JR Priestley (1829). Second edition: Whittaker *et al.*, London (1844).

Larson JL. Goethe and Linnæus. *Journal of the History of Ideas* 28(4):590–596 (1967).

Latreille PA. Premier discours. De la nature des insectes, et de leur ordre dans la série des animaux. In: [GLL] Buffon, *Histoire naturelle, générelle et pariculière* (CS Sonnini, ed.). *Des crustacées et des insectes*, tome 1 (An 10 [1801–1802]), pp. 15–51.

Latta R. Introduction. In: [GW] Leibniz, *The Monadology and other philosophical writings*. R Latta (trans.). Oxford: Clarendon (1898), pp. 1–211. Reprinted 1948.

Laue M. *Christian Gottfried Ehrenberg. Ein Vertreter deutscher Naturforschung im neunzehnten Jahrhundert. 1795–1876*. Berlin: Julius Springer (1895).

712   REFERENCES

Laufer B. The story of the pinna and the Syrian lamb. *The Journal of American Folk-Lore* 28(108):103–128 (1915).

Lautner P. Introduction III. In: *Simplicius*. On Aristotle's On the soul 1.1–2.4. JO Urmson (trans.). Ithaca NY: Cornell UP (1995), pp. 4–10.

Lavoisier [AL]. *Elements of chemistry*. R Kerr (trans.). Edinburgh: William Creech (1790).

Lavoisier AL. *Traité élémentaire de chimie*. In three volumes (Volume 3 as *Nomenclature chimique*). Paris: Cuchet (1789).

Lawson AJ. *Painted caves: Paleolithic rock art in Western Europe*. Oxford UP (2012).

Lazarus D. The Ehrenberg collection and its curation. In: DM Williams & R Huxley (eds), *Christian Gottfried Ehrenberg (1795–1876). The man and his legacy. The Linnean*, Special issue 1 (1998), pp. 31–48.

Lazenby EM. *The* Historia plantarum generalis *of John Ray*. In three volumes. PhD thesis, University of Newcastle upon Tyne (1995).

Lechevalier H. Louis Joblot and his microscopes. *Bacteriological Reviews* 40(1):241–258 (1976).

Le Clerc J [Clericus]. *Philosophiæ orientalis*. In: *Opera philosophica*, volume 2 (of 4). Amstelodami: Wetstenios (1710), pp. 167–375.

Leclercq E. Les microorganismes intermèdiaires aux deux règnes. *Bulletin de la Société Belge de Microscopie* 16:70–131 (1890).

L'Écluse C [Clusius]. *Caroli Clusii Atrebatis ad Thomam Redigerum et Joannem Cratonem epistolae; accedunt Remberti Dodonaei, Abrahami Ortelii, Gerardi Mercatoris et Ariae Montani ad Eumdem Cratonem epistolae*. PFX de Ram (ed.). Bruxelles: Hayez (1847).

L'Écluse C de [Clusius]. *Fungorum in Pannoniis observatorum: brevis historia et Codex Clusii*. Budapest: Akadémiai Kiadó; and Graz: Akademische Druck- u. Verlagsanstalt (1983).

L'Écluse C de [Clusius]. *Rariorum aliquot stirpium per Hispanias obseruatarum historia, libris dvobvs espressa*. Antverpiæ: Christophiri Plantini (1576).

L'Écluse C de [Clusius]. *Rariorum aliquot stirpium, per Pannoniam, Austriam, & vicinas quasdam Prouincias obseruaterum historia, qvatvor libris expressa*. Antverpiae: Christophorus Plantinus (1583).

L'Écluse C de [Clusius]. *Rariorvm plantarvm historia*. Antverpiæ: Plantiniana (1601).

Le Coq A v[on]. Dr. Stein's Turkish Khuastuanift from Tun-Huang, being a confession-prayer of the Manichæn auditores. *Journal of the Royal Asiatic Society of Great Britain and Ireland* 1911:277–314 (1911).

Ledermüller MF. *Amusement microscopique tant pour l'esprit, que pour les yeux*. In three volumes. [Nürnberg]: Winterschmidt (1764–1768).

Ledermüller MF. *Mikroskoopische vermaaklykheden*. In four parts. Te Amsterdam: Houttuyn (1776).

Ledermüller MF. *Mikroskopische Gemüths- und Augen-Ergötzung*. In three volumes (Volume 3 as *Nachlese*). [Nürnberg]: Christian de Launoy (1760–1762).

Ledermüller MF. *Physicalische Beobachtungen derer Saamenthiergens, durch die allerbesten Vergrösserungs-Gläser und bequemlichsten Microscope betrachtet*. Nürnberg: Monath (1756).

Ledermüller MF. *Versuch zu einer gründlichen Vertheidigung derer Saamenthiergen; nebst einer kurzen Beschreibung derer Leeuwenhoeckischen Mikroscopien und einem Entwurf zu einer vollständigern Geschichte des Sonnenmikroskops*. Nürnberg: Monath (1758).

Ledermüller MF. *Vertheidigung; als in Anhang seiner Mikroskopische Gemüths- und Augen-Ergötzung*. Nürnberg: Christian de Launoy (1765).

Ledezma D. Interpreting New World nature: Niermberg's [sic] Historia Naturæ as a palimpsest of fantastic literature. In: J Funke, S Riekeles & A Broeckmann (eds), *Proceedings of the 16th International Symposium on Electronic Art, 20–29 August 2010, Ruhr* (2010), pp. 297–299.

Lee H. *The vegetable lamb of Tartary; a curious fable of the cotton plant*. London: Sampson Low, Marston, Searle, & Rivington (1887).

Lee J. *An introduction to botany. Containing an explanation of the theory of that science; extracted from the works of Dr. Linnæus*. [First edition]. London: Tonson (1760).

Lee J. *An introduction to botany. Containing an explanation of the theory of that science; extracted from the works of Dr. Linnæus*. Fourth edition. London: Rivington et al. (1788).

REFERENCES   713

Lee J. *An introduction to the science of botany, chiefly extracted from the works of Linnæus; to which are added, several new tables and notes, and a life of the author. By the late James Lee . . . Fourth edition, corrected and enlarged, by James Lee, son and successor to the author.* London: Rivington et al. (1810).

Lee R [Mrs]. Memoirs of Baron Cuvier. London: Longman, Rees *et al.* (1833).

Leedale GF. How many are the kingdoms of organisms? *Taxon* 23(2–3):261–270 (1974).

Leeuwenhoek A van. *Alle de brieven.* In 15 parts. Editors n/a (Parts 1–4), LC Palm (5 and 10–15), JJ Swart (6–8), and J Heniger (9). Amsterdam: Swets & Zeitlinger (Parts 1–9 and 12); and Lisse (10–11 and 13–15) (1939–1993). Online at https://www.dbnl.org/tekst/leeu027alle00_01/, accessed 22 May 2019 and 23 September 2020.

Leeuwenhoek A van. Microscopical observations on red coral. *Philosophical Transactions* 26(316):126–134 (1708).

Leewenhoeck A van. Observations, communicated to the publisher by Mr. Antony van Leewenhoeck, in a Dutch letter of the 9th of Octob. 1676, here English'd: concerning little animals by him observed in rain- well- sea- and snow-water; as also in water wherein pepper had lain infused. *Philosophical Transactions* 12(133):821–831 (1677).

Leeuwenhoek [A van]. Part of a letter from Mr Anthony Van Leeuwenhoek, concerning the worms in sheeps livers, gnats, and animalcula in the excrements of frogs. *Philosophical Transactions* 22(261):509–518 (read 1700, published 1701).

Leeuwenhoek A van. Part of a letter from Mr Antony van Leeuwenhoek, FRS concerning green weeds growing in water, and some animalcula found about them. *Philosophical Transactions* 23(283):1304–1311 (1703).

Le Fèvre de La Boderie N. *L'introduction.* In: F Georgio, *L'harmonie dv monde, divisee en trois cantiqves.* G Le Fèvre (trans.). [With:] G Pico della Mirandola, *L'heptaple.* N Le Fèvre (trans.). Paris: Macé (1579), pag. unnumb. (20 pages).

Léger L & Duboscq O. *Selenococcidium intermedium* Lég. et Dub. et la systématique des Sporozoaires. *Archives de Zoologie Expérimentale et Générale, 5 Série* 5(5):187–238 (1910).

Leger MM, Eme L, Stairs CW & AJ Roger. Demystifying eukaryote lateral gene transfer. *BioEssays* 2018:1700242 (2018).

Leibniz GW. *Essais de théodicée sur la bonté de Dieu, la liberté de l'homme, et l'origine du mal. Nouvelle edition.* In two volumes. Amsterdam: François Changuion (1747).

Leibniz [GW]. *Extraits de la théodicée. Essais sur la bonté de Dieu, la liberté de l'homme et l'origine du mal.* P Janet (ed.). Fourth edition. Paris: Hachette (1912).

Leibniz GW. *Opera philosophica quæ extant Latina Gallica Germanica omnia.* In two volumes. JE Erdmann (ed.). Berlolini: Eicherl (1839–1840).

Leibniz GW. *The Monadology and other philosophical writings.* R Latta (trans.). London: Oxford UP (1898). Reprinted, 1948.

Leidy J. *A flora and fauna within living animals.* Smithsonian Contributions to Knowledge 5(2). Washington: Smithsonian (1853).

Leidy J. *Fresh-water rhizopods of North America. Report of the United States Geological Survey of the Territories 12.* Washington: Government Printing Office (1879).

Leidy J. On the motive power of diatoms. *Annals and Magazine of Natural History, 4 Series* 15(87):234–235 (1875).

Leidy J. The parasites of the termites. *Journal of the Academy of Natural Sciences of Philadelphia. Second series* 8(4):425–447 (1881).

Lejay P & Otten J. Cassiodorus. *The Catholic Encyclopedia* 3:405–407 (1908).

Léman DS. Bacillaria. *Dictionnaire des sciences naturelles* 3 (Supplément): 158–159 (1816). See also *ibid.* 61:5–6 (1845).

Léman DS. Echinella. *Dictionnaire des sciences naturelles* 14:185 (1819).

Léman DS. Navicula, *Navicule. Dictionnaire des sciences naturelles* 34:318–319 (1825). See also *ibid.* 61:30–33 (1845).

Lemery N. *Cours de chymie contenant la maniere de faire les operations qui sont en usage dans la medecine, par une methode facile.* Paris: l'Autheur (1675).

Lemery [N]. *Cours de chymie . . nouvelle édition.* Paris: Herisant (1756).

714 REFERENCES

Lenhoff SG & Lenhoff HM. *Hydra and the birth of experimental biology—1744. Abraham Trembley's Mémoires concerning the polyps. Book I. Some reflections on Abraham Trembley and his Mémoires. Book II. A translation from the French of Mémoires, pour servir à l'histoire d'un genre de polypes d'eau douce, à bras en forme de cornes.* Pacific Grove CA: Boxwood (1986).

Lepenies W. Linnaeus's *Nemesis divina* and the concept of divine retaliation. *Isis* 73(1):11–27 (1982).

Lesley AM Jr. The *Song of Solomon's virtues* by Yohanan Alemanno: love and human perfection according to a Jewish associate of Giovanni Pico della Mirandola. PhD thesis, University of California at Berkeley (1976).

Lesser [FC]. *Insecto-theology: or, a demonstration of the being and perfections of God, from a consideration of the structure and economy of insects.* Edinburgh: Creech; and London: Cadell (1799).

Lesser [FC]. *Theologie des insectes, ou demonstration des perfections de Dieu dans tout ce qui concerne les insectes.* P Lyonnet (trans.). In two volumes. La Haye: Jean Swart (1742).

Lesser FC. *Insecto-Theologia, oder: Vernunfft- und Schrifftmäßige Versuch, wie ein Mensch durch aufmercksame Betrachtung derer sonst wenig geachteten Insecten zu lebendiger Erkänntniß und Bewunderung der Allmacht, Weißheit, der Güte und Gerechtigkeit des grossen Gottes gelangen könne.* Franckfurt und Leipzig: Blochberger (1738).

Lesser FC. *Kurzer Entwurff einer Lithotheologie, oder eines Versuches durch natürliche und geistliche Betrachtung derer Steine, die Allmacht, Güte, Weißheit und Gerechtigkeit des Schöpffers zuerkennen, und die Menschen zur Bewunderung, Lobe und Dienste desselben aufzumuntern.* Nordhausen: by the author, and Cöler (1732).

Lesser FC. *Lithotheologie, das ist: Natürliche Historie und geistliche Betrachtung derer Steine, also abgefaßt, das daraus die Allmacht, Weißheit, Güte und Gerechtigkeit des grossen Schöpffers gezeuget wird.* Hamburg: Brandt (1735).

Lesser FC. *Testaceo-Theologie, oder: Gründlicher Beweis des Daseyns und der vollkommensten Eigenschaften eines göttlichen Wesens, aus natürlicher und geistlicher Betrachtung der Schnecken und Muscheln, zur gebührenden Verherrlichung des grossen Gottes und Beförderung des ihm schuldigen Dienstes ausgefertiget.* Leipzig: Blochberger (1744).

Lettinck P. Introduction [to Philoponus]. In: P Lettinck & HO Urmson (eds), *Philoponus On Aristotle* Physics 5–8 *with Simplicius On Aristotle on the void.* London: Bloomsbury (2014), pp. 3–18.

Leuckart R. Die Zoophyten. Ein Beitrag zur Geschichte der Zoologie. *Archiv für Naturgeschichte* 41(1):70–110 (1875).

Leuckart R. *Ueber die Morphologie und die Verwandtschaftsverhältnisse der wirbellosen Thiere.* Braunschweig: Vieweg (1848).

Levene PA & Bass LW. *Nucleic acids.* American Chemical Society Monograph Series. New York: Chemical Catalog Company (1931).

Levy R. *The social structure of Islam.* Cambridge UP (1957).

Lewin R. An ancient cultural revolution. *New Scientist* 83(1166):352–355 (1979).

Lewis IM. The cytology of bacteria. *Journal of Bacteriology* 5(3):181–230 (1941).

Ley W. *Dawn of zoology.* Englewood Cliffs NJ: Prentice-Hall (1968).

Ley W. *Exotic zoology.* New York: Viking (1959). Reprinted by Bonanza, New York (1987).

Ley W. Introduction. In: E Topsell, *The history of four-footed beasts and serpents and insects,* volume 1 (of 3). Facsimile edition. New York: Da Capo (1967), pag. unnumb.

Leydig F. *Lehrbuch der Histologie des Menschen und der Thiere.* Frankfurt: Meidinger (1857).

L'Héritier [de Brutelle] CL. Mémoire sur un nouveau genre de plante appelé *Cadia. Magasin Encyclopédique* 1795(5):20–31 (An 4, 1795).

Li JY & Wu CF. New symbiotic hypothesis on the origin of eukaryotic flagella. *Naturwissenschaften* 92(7):305–309 (2005).

Li JY & Wu CF. Perspectives on the origin of microfilaments, microtubules, the relevant chaperonin system and cytoskeletal motors—a commentary on the spirochaete origin of flagella. *Cell Research* 13(4):219–227 (2003).

Libavius A. *Singvlarivm Andreæ Libavii.* In four parts. Francofvrti: Ioannis Savri (1599–1601).

REFERENCES 715

Liceti F. *De monstrorum caussis, natura, & differentijs libri dvo.* Patavii: Crivellarium (1616).

Liceti F. *De spontaneo viventivm ortv libb: quatuor, in quibus de generatione animantium, quæ vulgo ex putri exoriri dicuntur, accurante aliorum opiniones omnes primum exanimantur.* Vicentiæ: Dominici Amadei (1618).

Lichetenstein AAH. Ueber die außerhalb des Wassers lebenden Pflanzenthiere, wie auch deren merkwürdige Verwandlung in mancherlei Erdschwämme und andere ähnliche, bisher gewöhnlich zum Gewächsreiche gerechnete organische Natur-Erzeugnisse. *(Voigt's) Magazin für den neuesten Zustand der Naturkunde* 6(1):493–510 (1803).

Lichtenstein AAH. Ueber die Luftzoophyten. *(Voigt's) Magazin für den neuesten Zustand der Naturkunde* 6(1):45–56 (1803).

Lidbeck EG. *Dissertatio gradualis fungos regno vegetabili vindicans . . . publico examini subjicit J Dryander* (1776). Londini Gothorum: Berlinglanis (1776).

Liddell HG & Scott R. *A Greek-English lexicon.* Eighth edition. New York: American (1901).

Liddell HG & Scott R. *A Greek-English lexicon.* Oxford: Clarendon (1940). Accessed at http://www.perseus.tufts.edu.

Liebig J von [as "J.L."]. Der Zustand der Chemie in Preussen. *Annalen der Chemie und Pharmacie* 34:97–136 (1840).

Lightfoot J. *Flora Scotia: or, a systematic arrangement, in the Linnæan method, of the native plants of Scotland and the Hebrides.* In two volumes. London: White (1777).

Lightfoot JB & Harmer JR (eds). *The apostolic Fathers.* London: Macmillan (1912).

Lindberg DC. Science and the early Christian church. *Isis* 74(4):509–530 (1983).

Lindberg DC. *The beginnings of western science. The European scientific tradition in philosophical, religious, and institutional context, 600 BC to AD 1450.* University of Chicago Press (1992).

Lindemann E von. Das fünfzigjährige Doktorjubiläum Eduard von Eichwald's. *Verhandlungen der Russisch-Kaiserlichen Mineralogischen Gesellschaft zu St. Petersburg, 2 Serie* 5:278–358 (1870).

Lindley J. *A synopsis of the British flora; arranged according to the natural orders: containing vasculares, or flowering plants.* London: Longman, Rees *et al.* (1829).

Lindley J. Exogens. *Penny Cyclopædia of the Society for the Diffusion of Useful Knowledge* 10:120–131 (1838).

Lindley J. *The vegetable kingdom: or, the structure, classification, and uses of plants, illustrated upon the natural system.* London: Bradbury & Evans (1846).

Lindroth S. Linnaeus in his European context. In: G Broberg (ed.), *Svenska Linnésällskapets Årsskrift Årgång 1978. Yearbook of the Swedish Linnaeus Society. Commemorative Volume.* Uppsala: Almqvist & Wiksell (1979), pp. 9–17.

Lindroth S. The two faces of Linnaeus. In: T Frängsmyr (ed.), *Linnaeus. The man and his work.* Berkeley: University of California Press (1983), pp. 1–62.

Lindsay J. Account of the germination and raising of ferns from the seed. *Transactions of the Linnean Society of London* 2:93–100 (1794).

Lindsay WL. *A popular history of British lichens, comprising an account of their structure, reproduction, uses, distribution, and classification.* London: Lovell Reeve (1856).

Link GKK. Reproduction in thallophytes, with special reference to fungi. *Botanical Gazette* 88(1):1–37 (1929).

Link HF. Epistola ad virum celeberrimum Nees ab Esenbeck . . . De algis aquaticis in genera disponendis. In: CG Nees von Esenbeck (ed.), *Horae physicae Berolinenses.* Bonnae: Adophus Marcus (1820), pp. 2–8.

Linné C von [Linnæus]. *A dissertation on the sexes of plants.* JE Smith (trans.). Dublin: White (1786).

Linné C von [Linnæus]. *A general system of nature, through the three grand kingdoms of animals, vegetables, and minerals.* W Turton (trans.). In seven volumes. London: Lackington & Allen (1806).

Linné C von [Linnæus]. *A generic and specific description of British plants, translated from the Genera et Species Plantarum of the celebrated Linnæus.* J Jenkinson (trans.). Kendal: Ashburner (1775).

716   REFERENCES

Linné C von [Linnæus]. *A selection of the correspondence of Linnæus, and other naturalists, from the original manuscripts.* By JE Smith. In two volumes. London: Longman, Hurst *et al.* (1821).

Linné C von [Linnæus]. *A system of vegetables, according to their classes orders genera species with their characters and differences . . . Translated from the Thirteenth edition (as published by Dr. Murray) of the Systema vegetabilium of the late Professor Linneus; and from the Supplementum Plantarum of the present Professor Linneus.* A Botanical Society, at Lichfield (trans.). In two volumes. Lichfield: Leigh and Sotheby (1783).

Linné C von [Linnæus]. *Animalia composita . . . proposuit A Bäck* (1759). In: C Linnæus (ed.), *Amœnitates academicæ* 5:343–352. Holmiæ: Laurentius Salvius (1760).

Linné C von [Linnæus]. Anmårkning öfwer de diuren, som sågas komma neder utur skyarnI i Norrige. *Konglige Svenska Vetenskaps Academiens Handlingar* 1:320–325 (1740, published 1743). A Latin epitome (Annotationes de animalibus, quæ in Norwegia ex nubibus decidere dicuntur) appeared in *Analecta Transalpina* 1:68–73 (1762).

Linné C von [Linnæus]. *Bibliotheca botanica recensens libros plus mille de plantis huc usque editos, secundum systema auctorum naturale in classes, ordines, genera & species dispositos . . . cum explicatone Fundamentorum Botanicorum pars 1ma.* Amstelodami: Salomonem Schouten (1736).

Linné C von [Linnæus]. *Classes plantarum. Seu systemata plantarum omnia a fructificatione desumta . . . secundum classes, ordines et nomina generica cum clave cujusvis methodi et synonymis genericis.* Lugduni Batavorum: Conradum Wishoff (1738).

Linné C von [Linnæus]. *Clavis medicinæ duplex, exterior & interior.* Holmiæ: Laurentius Salvius (1766).

Linné C von [Linnæus]. *Corallia Baltica . . . proposita H Fougt* (1745). In: C Linnæus (ed.), *Amœnitates academicæ* 1:74–106. Holmiæ et Lipsiæ: Kiesewetter (1749). English translation: FJ Brand (trans.), *Select dissertations from the Amœnitates academicæ* (1781), pp. 457–480.

Linné C von [Linnæus]. *Critica botanica in qua nomina plantarum generica, specifica, & variantia examini subjiciuntur.* Lugduni Batavorum: Conradum Wishoff (1737).

Linné C von [Linnæus]. *Disquisitio de sexu plantarum* (1759). Manuscript GB-110/LM/LP/BOT/3/7/2, at http://linnean-online/144137, accessed 8 October 2020.

Linné C von [Linnæus]. *Exanthemata viva . . . proposuit JC Nyander (1757).* C Linnæus (ed.), *Amœnitates academiae* 5:92–105. Holmiæ: Laurentius Salvius (1760).

Linné C von [Linnæus]. *Fauna Svecica sistens animalia Sveciæ regni.* Stockholmiæ: Laurentius Salvius (1746).

Linné C von [Linnæus]. *Flora Lapponica exhibens plantas per Lapponiam crescentes, secundum systema sexuale.* Amstelædami: Salomonem Schouten (1737).

Linné C von [Linnæus]. *Fundamenta botanica quæ majorum operum prodromi instar theoriam scientiæ botanices per breves aphorismos tradunt.* Amstelodami: Salomonen Schouten (1736).

Linné C von [Linnæus]. *Fundamenta botanica in qvibvs theoria botanices aphoristice traditvr.* [With:] J Gesner, *Dissertationes physicæ in qvibvs celeb. Linnæi Elementa Botanica dilvcide explicantvr.* Halæ: Bierwirth (1747).

Linné C von [Linnæus]. *Fundamentum fructificationis . . . proposuit JM Gråberg (1762).* In: C Linnæus (ed.), *Amœnitates academicæ* 6:279–304. Holmiæ: Laurentius Salvius (1763).

Linné C von [Linnæus]. *Gemma arborum . . . propositæ P Löfling (1749).* In: C Linnæus (ed.), *Amœnitates academicæ* 2:182–224. Holmiæ: Laurentius Salvius (1751).

Linné C von [Linnæus]. *Genera morborum . . . proposuit J Schröder (1759).* In: Linnæus C (ed.), *Amœnitates academicæ* 6:452–486. Holmiæ: Laurentius Salvius (1763).

Linné C von [Linnæus]. *Genera plantarum eorumque characters naturales secundum numerum, figuram, situm, & propotionem omnium fructificationis partium. Fundamentorum botanicorum pars II.* Lugduni Batavorum: Conradum Wishoff (1737).

Linné C von [Linnæus]. *Generatio ambigena . . . proposuit C Ramström (1759).* In: C von Linné (ed.), *Amœnitates academicæ* 6:1–16. Holmiæ: Laurentius Salvius (1763).

Linné C von [Linnæus]. *Hortus Cliffortianus plantas exhibens.* Amstelædami [n.p.] (1737).

REFERENCES 717

Linné C von [Linnæus]. *Indelning i ort-riket, efter Systema naturæ, på Svenska öfversatt af Johan J. Haartman.* [Eighth edition]. Stockholm: Lars Salvius (1753).

Linné C von [Linnæus]. *Institutions of entomology: being a translation of Linnaeus's Ordines et genera insectorum.* TP Yeats (trans.). London: Horsfield (1782).

Linné C von [Linnæus]. *Lachesis Lapponica, or a tour in Lapland,* now first published from the original manuscript journal of the celebrated Linnæus; by JE Smith. In two volumes. London: White & Cochrane (1811).

Linné C von [Linnæus]. *Lepra . . . proposuit I Uddman (1763).* In: C von Linné (ed.), *Amœnitates academicæ* 7:94–108. Holmiæ: Laurentius Salvius (1769).

Linné C von [Linnæus]. Linnæus to Ellis (16 September 1761). In: JE Smith (ed.), *A selection from the correspondence of Linnæus,* volume 1 [of 2]. London: Longman, Hurst *et al.* (1821), pp. 148–152.

Linné C von [Linnæus]. Linnæus to Ellis (20 July 1767). In: JE Smith (ed.), *A selection from the correspondence of Linnæus,* volume 1 [of 2]. London: Longman, Hurst *et al.* (1821), pp. 208–209.

Linné C von [Linnæus]. Manuscript X 505, Kungliga Biblioteket, Stockholm. As cited by G Broberg, in: J Weinstock (ed.), *Contemporary perspectives on Linnæus* (1985), page 170.

Linné C von [Linnæus]. *Materia medica, liber I. de plantis.* Holmiæ: Laurentius Salvius (1749).

Linné C von [Linnæus]. *Metamorphoses plantarum . . . proposita NE Dahlberg (1755).* In: C Linnæus (ed.), *Amœnitates academicæ* 4:368–386. Holmiæ: Laurentius Salvius (1759).

Linné C von [Linnæus]. *Metamorphosis humana . . . proposuit JA Wadström (1767).* In: C von Linné (ed.), *Amœnitates academicæ* 7:326–344. Holmiæ: Laurentius Salvius (1769).

Linné C von [Linnæus]. *Miracula insectorum . . . proposuit GE Avelin (1752).* In: C Linnæus (ed.), *Amœnitates academicæ* 3:313–334. Holmiæ: Laurentius Salvius (1756).

Linné C von [Linnæus]. *Morbi artificum . . . proposuit N Skragge (1764).* In: C Linné von (ed.), *Amœnitates academicæ* 7:84–93. Holmiæ: Laurentius Salvius (1769).

Linné C von [Linnæus]. *Mundum invisibilem . . . submittit censuræ JC Roos (1767).* BL 146, in: *Linnæ Dissertationes* 121–150. Upsaliæ: [n.p.] [1767]. In Linnæus's personal library, Linnean Society of London.

Linné C von [Linnæus]. *Mundus invisibilis . . . proposuit JC Roos (1767).* In: C von Linné (ed.), *Amœnitates academicæ* 7:385–408. Holmiæ: Laurentius Salvius (1769).

Linné C von [Linnæus]. *Musa Cliffortiana florens Hartecampi.* Lugduni Batavorum: [n.p.] (1736).

Linné C von [Linnæus]. *Nemesis divina.* E Miller (trans.). Lanham MD: University Press of America (2002).

Linné C von [Linnæus]. *Nemesis divina.* MJ Petry (ed. and trans.). In series: Archives Internationales d'Histoire des Idées 177. Dordrecht: Springer (2001).

Linné C von [Linnæus]. *Noctiluca marina . . . publice ventilandam sistit CF Adler (1752).* In: Linnæus C (ed.), *Amœnitates academicæ* 3:202–210. Holmiæ: Laurentius Salvius (1756).

Linné C von [Linnæus]. *Philosophia botanica, in qua explicantur botanices fundamenta, studio Curtii Sprengel; editio aucta et emendata.* Tornaci Nervioru: Casterman-Dieu (1824).

Linné C von [Linnæus]. *Philosophia botanica in qva explicantur fundamenta botanica cum definitionibus partium, exemplis terminorum, observationibus rariorum, adjectus figuris æneis.* Stockholmiae: Kiesewetter (1751).

Linné C von [Linnæus]. *Prælectiones in ordines naturales plantarum . . . Accessit uberior palmarum et scitaminum expositio praeter plurium novorum generum reductiones cum mappa geographico-genealogica affinitatum ordinum.* PD Giseke (ed.). Hamburgi: Hoffmanni (1792).

Linné C von [Linnæus]. *Prolepsis plantarum . . . auctore JA Ferber (1763).* In: C von Linné (ed.), *Amœnitates academicæ* 6:365–383. Holmiæ: Laurentius Salvius (1763).

Linné C von [Linnæus]. *Prolepsis plantarum . . . proposuit H Ullmark (1760).* In: C von Linné (ed.), *Amœnitates academicæ* 6:324–341. Holmiæ: Laurentius Salvius (1763).

Linné C von [Linnæus]. *Select dissertations from the Amœnitates academicae, a supplement to Mr. Stillingfleet's Tracts relating to natural history.* FJ Brand (trans.), volume 1 [only volume published]. London: Robson (1781).

718 REFERENCES

Linné C von [Linnæus]. *Species plantarum, exhibentes plantas rite cognitas, ad genera relatas, cum differentiis specificis, nominibus trivialibus, synonymis selectis, locis natalibus, secundum systema sexuale digestas.* In two volumes. Holmiæ: Laurentius Salvius (1753).

Linné C von [Linnæus]. *Sponsalia botanica . . . proposita JG Wahlbom (1746).* In: Linnæus C (ed.), *Amœnitates academicæ* 1:327–380. Holmiæ et Lipsiæ: Kiesewetter (1749).

Linné C von [Linnæus]. *Systema naturæ, sive regna tria naturæ systematice proposita per classes, ordines, genera, & species.* [First edition]. Lugduni Batavorum: Theodorum Haak (1735).

Linné C von [Linnæus]. *Systema natvræ in quo natvræ regna tria, secvndvm classes, ordines, genera, species, systematice proponuntur. Editio secunda, auctior.* Stockholmiæ: Gorrfr. Kiesewetter (1740).

Linné C von [Linnæus]. *Systema natvræ, sive regna tria natvrae systematice proposita per classes, ordines, genera et species. Natur-Systema, oder die in ordentlichem Zusammenhange vorgetragene drey Reiche der Natur, nach ihren Classes, Ordningen, Geschlechtern und Arten, in die Deutsche Sprache übersetzt.* Johann Joachim Langen (trans.). [Third edition]. Halle: Gebauer (1740).

Linné C von [Linnæus]. *Systema naturæ in quo proponuntur naturæ regna tria secundum classes, ordines, genera & species. Editio quarta ab auctore emendata & aucta.* Parisiis: Michaelis-Antonii David (1744).

Linné C von [Linnæus]. *Systema natvræ in qvo natvræ regna tria, secvndvm classes, ordines, genera, species, systematice proponvntvr . . .* Mich. Gottl. Agnethlero (ed.). Editio altera avctior et emendatior. [Fifth edition]. Halæ Magdebvrgicæ: [n.p.] (1747).

Linné C von [Linnæus]. *Systema natvræ sistens regna tria naturae, in classes et ordines genera et species redacta tabulisque æneis illustrata. Editio sexta, emendate et aucta.* Stockholmiae: Kiesewetter (1748).

Linné C von [Linnæus]. *Systema naturæ sistens regna tria naturæ, in classes et ordines genera et species redacta tabulisque æneis illustrata. Secundum sextam Stockholmiensem emendatam & auctam editionem.* [Seventh edition]. Lipsiae: Kiesewetter (1748).

Linné C von [Linnæus]. *Systema naturæ sistens regna tria naturæ in classes et ordines genera et species redacta, tabulisque æneis illustrata. Accedunt vocabula Gallica. Editio multi auctior & emendatior.* [Ninth edition]. Lugduni Batavorum: Theodorum Haak (1756).

Linné C von [Linnæus]. *Systema naturæ per regna tria naturæ, secundum classes, ordines, genera, species, cum characteribus, differentiis, synonymis, locis. Editio decima, reformata.* In two volumes, pages numbered consecutively. Holmiæ: Laurentius Salvius (1758–1759).

Linné C von [Linnæus]. *Systema natvrae per regni tria natvrae, secvndvm classes, ordines, genera, species, cvm characteribvs, differentiis, synonymis, locis. Ad editionem decimam reformatam Holmiensem.* [Eleventh edition]. Halae Magdebvrgicae: Cvrt. Volumes 1–2 (1760); volume 3 (1770).

Linné C von [Linnæus]. *Systema naturae per regna tria natura, secundum classes, ordines, genera, species, cum characteribus, differentiis, synonymis, locis. Editio duodecima, reformata.* In three volumes; Tomus I in two parts. Holmiae: Laurentius Salvius (1766–1768).

Linné C von [Linnæus]. *Systema naturae, per regna tria naturae, secundum classes, ordines, genera, species cum characteribus, differentiis, synonymis, locis. Editio decima tertia ad editionem duodecimam reformatam Holmiensem.* In three volumes; Tomus I in two parts. ["Edition 12a"]. Vindobonae: Thomae (1767–1770).

Linné C von [Linnæus]. *Systema naturae per regna tria naturae, secundum classes, ordines, genera, species, cum characteribus, differentiis, synonymis, locis. Editio decima tertia, aucta, reformata. Cura Jo. Frid. Gmelin.* Three volumes in ten parts. [The standard Thirteenth edition]. Lipsiae: Beer (1788–1793).

Linné C von [Linnæus]. *Systema naturæ per regna tria naturæ, secundum classes, ordines, genera, species; cum characteribus, differentiis, synonymis, locis. Editio decima tertia, aucta, reformata.* JF Gmelin (ed.). Three volumes bound in ten. [Exact reprint of the Leipzig Thirteenth edition]. Lugduni: Delamolliere (1789–1796).

Linné C von [Linnæus]. *Tænia . . . publico examini submisit G Dubois (1748).* In: C Linnæus (ed.), *Amœnitates academicæ* 2:59–99. Holmiæ: Laurentius Salvius (1751).

Linné C von [Linnæus]. *Termini botanici . . . proposuit J Elmgren (1762)*. In: C von Linné (ed.), *Amœnitates academicæ* 6:217–246. Holmiæ: Laurentius Salvius (1763).

Linné C von [Linnæus]. *The animal kingdom, or zoological system . . . Class I. Mammalia . . . being a translation of that part of the Systema Naturae, as lately published, with great improvements, by Professor Gmelin of Goettingen.* R Kerr (trans.). London: Strahan & Cadell; and Edinburgh: Creech (1792).

Linné C von [Linnæus]. *The elements of botany . . . being a translation of the Philosophia botanica, and other treatises.* H Rose (trans.). London: Cadell & Hingeston (1775).

Lippmann EO von. *Urzeugung und Lebenskraft. Zur Geschichte dieser Probleme von den ältesten Zeiten an bis zu den Anfängen des 20. Jahrhunderts.* Berlin: Julius Springer (1933).

Lister M. *Conchyliorum bivalvium utriusque aquæ exercitatio anatomica tertia.* Londini: printed for the author (1696).

Lister M. *Exercitatio anatomica altera, in qua maximè agitur De buccinis fluviatilibus & marinis.* Londini: Sam. Smith & Benj. Walford (1695).

Lister M. *Exercitatio anatomica. In qua de cochleis, maximè terrestribus et limacibus, agitur.* Londini: Sam. Smith & Benj. Walford (1694).

Lister M. *Historiae animalium Angliae tres tractatus. Unus de araneis. Alter de cochleis tum terrestribus tum fluviatilibus. Tertius de cochleis marinis. Quibus adjectus est quartus de lapidibus ejusdem Insulae ad cochlearum quandam imaginem figuratis.* Londini: Joh. Martyn (1678).

Lister M. *Historiæ conchyliorum* (1685–1691). [With:] *Appendix ad historiæ conchyliorum librum IV* (1692). Londini: printed for the author.

Lister M. *Johannes Goedartius De insectis . . . cum notularum additione operâ M. Lister . . . item appendicis ad Historiam animalium Angliæ, ejusdem M. Lister.* Londini: S. Smith (1685).

Lister M. The anatomy of the scallop. *Philosophical Transactions* 19(229):567–570 (1697).

Liu D. The cell and protoplasm as container, object, and substance, 1835–1861. *Journal of the History of Biology* 50(4):889–925 (2017).

Lloyd GER. The development of Aristotle's theory of the classification of animals. *Phronesis* 6(1):59–81 (1961).

Lloyd GER. Introduction. In: *Hippocratic writings.* GER Lloyd (ed.). Harmondsworth UK: Penguin (1978), pp. 9–60.

Lloyd GER. *Science, folklore and ideology. Studies in the life sciences in ancient Greece.* Cambridge UP (1993).

Lloyd GER. The hot and the cold, the dry and the wet in Greek philosophy. *Journal of Hellenic Studies* 84:92–106 (1964).

Lloyd HA. *Jean Bodin, "This pre-eminent man of France": an intellectual biography.* Oxford UP (2017).

Llull R. *Liber de ascensu et decensu intellectus* (1304). In: *Raymundi Lullij Doctoris illuminati de noua logica. de correllatiuis. necnon [et] de ascencu [et] descensu intellectus: quibus liquide[m] tribus libellis pbreui ad facili artificio.* A de Proaza (ed.). [Valencia]: [Jorge Costilla] (1512), fol. sep.

Lobel D. *A Sufi-Jewish dialogue. Philosophy and mysticism in Baḥya ibn Paqūda's Duties of the heart.* Philadelphia: University of Pennsylvania Press (2007).

L'Obel M de. *Kruydtboeck oft beschrÿuinghe van allerleye ghewassen, kruyderen, hesteren, ende gheboomten.* t'Antwerpen: Christoffel Plantyn (1581).

L'Obel M de. *Plantarvm sev stirpivm historia.* Antverpiae: Christophori Plantini (1576).

[L'Obel M de]. *Plantarvm sev stirpivm icones.* Antverpiæ: Christophori Plantini (1581).

Locke J. *Locke's essays. An essay concerning human understanding. And a treatise on the conduct of the understanding. Complete, in one volume: with the author's last additions and corrections.* Philadelphia: Troutman & Hayes (1853).

Locy WA. *The growth of biology. Zoölogy from Aristotle to Cuvier, botany from Theophrastus to Hofmeister, physiology from Harvey to Claude Bernard.* New York: Henry Holt (1925).

Loiseleur-Deslongchamps JLA & Marquis AL. Plante. *Dictionaire des Sciences Médicales* 43:141–162 (1820).

720 REFERENCES

Long AA & Sedley DN. *The Hellenistic philosophers*. In two volumes. Cambridge UP (1987).

Lonicer A. *Adams Lonicers . . . vollständiges Kräuter-Buch, oder das Buch über der drey Reiche der Natur*. Augsburg: Josef Wolff (1783).

Look BC. Gottfried Wilhelm Leibniz. *The Stanford Encyclopedia of Philosophy* (Spring 2020 edition), EN Zalta (ed.), https://plato.stanford.edu/archives/spr2020/entries/leibniz/, accessed 23 September 2020.

Louca S, Mazel F, Doebeli M & Parfrey LW. A census-based estimate of Earth's bacterial and archaeal diversity. *PLoS Biology* 17(2):e3000106 (2019).

Lovejoy AO. Goldsmith and the Chain of Being. *Journal of the History of Ideas* 7(1):91–98 (1946).

Lovejoy AO. Kant and evolution. In: B Glass, O Temkin & WL Straus Jr (eds), *Forerunners of Darwin 1745–1859*. Baltimore: Johns Hopkins UP (1959), pp. 173–221.

Lovejoy AO. Some Eighteenth century evolutionists. *Popular Science Monthly* 65:238–251 and 323–340 (1904).

Lovejoy AO. *The Great Chain of Being. A study of the history of an idea*. Cambridge MA: Harvard UP (1936). Reprinted (identical pagination) by Harper, New York (1960).

Loveland J. Georges-Louis Leclerc de Buffon's *Histoire naturelle* in English, 1775–1815. *Archives of Natural History* 31(2):214–235 (2004).

Lovelock J. *Gaia. A new look at life on Earth*. Oxford UP (2000).

Lovelock J. *The ages of Gaia. A biography of our living earth*. New York: Norton (1995).

Luccio R. Psychologia: the birth of a new scientific context. *Review of Psychology* 20(1–2):5–14 (1963).

Lucretius Carus, Titus. *De rerum natura libri sex*. London: Tonson (1712).

Lucretius Carus, Titus. *Of the nature of things*. WE Leonard (trans.). London: Dent; and New York: Dutton (2000).

Lum P. *Fabulous beasts*. London: Thames & Hudson (1944).

Lumière A. *Le mythe des symbiotes*. Paris: Masson (1919).

Lunde P & Stone C. Introduction. In: *The meadows of gold* by Mas'ūdī. P Lunde & C Stone (ed. and trans.). London: Kegan Paul (1989), pp. 11–20.

Lütjeharms WJ. *Zur Geschichte der Mykologie das XVIII. Jahrundert*. Mededeelingen van de Nederlandsche Mycologische Vereeniging 23. Gouda: Koch & Knuttel (1936).

Lwoff A. From protozoa to bacteria and viruses. Fifty years with microbes. *Annual Review of Microbiology* 25:1–26 (1971).

Lwoff A. *L'évolution physiologique. Étude des pertes de fonctions chez es microorganismes*. Paris: Hermann (1944).

Lwoff A. *Recherches biochimiques sur la nutrition des protozoaires*. Collections des Monographies de l'Institut Pasteur. Paris: Masson (1932).

Lwoff A. Remarques sur la physiologie comparée des Protistes eucaryotes. Les Leucophytes et l'oxytrophie. *Archiv für Protistenkunde* 90(2):194–209 (1938).

Lynskey W. Goldsmith and the Chain of Being. *Journal of the History of Ideas* 6(3):363–374 (1945).

Lyon J & Sloan PR (eds). *From natural history to the history of nature: readings from Buffon and his critics*. Notre Dame IN: University of Notre Dame Press (1981).

Lyons AS & Petrucelli RJ II. *Medicine. An illustrated history*. New York: Abrams (1978).

Lyons MC. An Arabic translation of the commentary of Themistius. *Bulletin of the School of Oriental and African Studies, University of London* 17(3):426–435 (1955).

Lyons MC (ed.). *An Arabic translation of Themistius Commentary on Aristoteles De anima*. Thetford: Cassirer (1973).

Lyons RD. Scientists discover a form of life that predates higher organisms. *The New York Times*, 3 November 1977, pp. A1 and A20.

MacCord K. Germ layers. *Embryo Project Encyclopedia* (2013), http://embryo.asu.edu/ handle/ 10776/6273, accessed 19 July 2020 and 11 October 2020.

Macdonald DB. The meanings of the philosophers by al-Ghazzāli. *Isis* 25(1):9–15 (1936).

MacIlmaine R [Makylmenæum Scotum]. *The logike of the most excellent philosopher P. Ramvs Martyr: newly translated, and in diuers places corrected, after the minde of the author*. London: Thomas Vautrollier (1581).

Mackie JM. *Life of Godfrey William von Leibnitz. On the basis of the German work of Dr. G.E. Guhrauer.* Boston: Gould, Kendall & Lincoln (1845).

Macleay WS. A reply to some observations of M. Virey in the "Bulletin des Sciences Naturelles, 1825". *Zoological Journal* 4:47–71 (1829).

Macleay WS. *Horæ entomologicæ: or essays on the annulose animals.* Two parts in one volume [all published]. London: Bagster (1819–1821).

Macleay WS. Remarks on the identity of certain general laws which have been lately observed to regulate the natural distribution of insects and fungi. *Transactions of the Linnean Society of London* 14(1):46–68 (1823).

Macrobius [Ambrosius Theodosius]. *Commentary on the dream of Scipio.* WH Stahl (trans.). Records of Civilization (AP Evans, ed.) 48. New York: Columbia UP (1952). Reprinted 1990.

Macrobius [Ambrosius Theodosius]. *In somnium Scipionis exposito.—Saturnalia.* Venezia: Nicolaus Jenson (1472).

Macrobius. *Macrobii in somnivm Scipionis ex Ciceronis VI libro.* Venetiis: Aldus et Soceri (1528).

Maffei G. *Storia della letteratura italiana.* In two volumes. Third edition, corrected. Firenze: Felice Le Monnier (1853).

Magnol P. *Botanicvm Monspeliense sive plantarvm circa Monspelium nascentium.* Monspeliensis: Francisci Bovrly (1676).

Magnol P. *Botanicvm Monspeliense sive plantarvm circa Monspelium nascentium index.* Monspelii: Danielis Pech, impensis Pauli Marret (1686).

Magnol P. *Novus character plantarum, in duos tractatus divisus.* Monspelii: Pech (1720).

Magnol P. *Prodromus historiæ generalis plantarum in quo familiæ plantarum per tabulas disponuntur.* Monspelij: Gabrielis & Pech (1689).

Magnus R. *Goethe als Naturforscher.* Leipzig: Barth (1906).

Magnus R. *Goethe as a scientist.* New York: Schuman (1949). In paperback, Collier (1961).

Maier M. *Atalanta fugiens.* https://alchemywebsite.com/atalanta.html, accessed 16 January 2019 and 17 September 2020.

Maier M. *Atalanta fvgiens, hoc est, emblemata nova de secretis naturæ chymica.* Oppenheimii: de Bry (1617).

Maier M. *Atalanta fvgiens, hoc est, emblemata nova de secretis naturæ chymica.* Second edition. Oppenheimii: de Bry (1618).

Maier M. *Tripvs avrevs, hoc est, tres tractatvs chymici selectissimi.* Francofvrti: Lvcæ Iennis (1618).

Maimonides M. *The guide for the perplexed.* M Friedländer (trans.). Fourth edition. New York: Dutton (1904). Reprinted by Shalom, Brooklyn NY (1969).

Mair AW. Introduction [to Oppian]. In: *Oppian. Colluthus. Tryphiodorus.* AW Mair (trans.). Loeb Clasical Library 219. London: Heinemann (1963), pp. xiii–lxxx.

Mancini R, Nigro M & Ippolito G. Lazzaro Spallanzani e la confutazione della teoria della generazione spontanea. Lazzaro Spallanzani and his refutation of the theory of spontaneous generation. *Le Infezioni in Medicina* 15(3):199–206 (2007).

Mandeville J [attrib.]. *Libri Ioannis Mandevil.* In: R Hakluyt, *The principall navigations, voiages and discoveries of the English nation.* London: George Bishop and Ralph Newberie (1589), pp. 24–79. Reprinted (facsimile) in two volumes by Cambridge UP (1965).

Mandeville J [attrib.]. *The travels of Sir John Mandeville. The version of the Cotton Manuscript in modern spelling.* Library of English classics. London: Macmillan (1915).

Manget JJ [Mangetus] (ed.). *Bibliotheca chemica curiosa.* Genevæ: Chouet et al. (1702).

Mann ME, Zhang Z, Rutherford S, et al. Global signatures and dynamical origins of the Little Ice Age and Medieval Climate Anomaly. *Science* 326(5957):156–1260 (2009).

Manning G. The history of "hylomorphism". *Journal of the History of Ideas* 74(2):173–187 (2013).

Mansion S. Deux définitions différentes de la vie chez Aristote? *Revue Philosophique de Louvain. Quatrième Série* 71(11):425–450 (1973).

Mantell GA. *The invisible world revealed by the microscope; or, thoughts on animalcules.* A new edition. London: John Murray (1850).

Mantell GA. *Thoughts on animalcules; or, a glimpse of the invisible world revealed by the microscope.* London: John Murray (1846).

722 REFERENCES

Manzalaoui MA (ed.). *Secretum secretorum. Nine English versions*, volume 1. Early English Text Society, volume 276. Oxford UP (1977).

Mao DP, Zhou Q, Chan CY & Quan ZY. Coverage evaluation of universal bacterial primers using the metagenomic datasets. *BMC Microbiology* 12:66 (2012).

Mappes T. FW Schiek: Kleines Stangen-Mikroscop. http://www.museum-optischer-instrumente.de/Schiek_229 (undated), accessed 12 April 2020 and 7 October 2020.

Marchant [J]. Observations touchant la nature des plants, & de quelques-unes de leurs parties cachées ou inconnuës. *Mémoires de l'Academie Royal des Sciences* 1711:99–108. See also: Sur une vegetation singulieré. *Histoire de l'Academie Royal des Sciences 1711*:42–43 (published 1730).

Margoliash E. Primary structure and evolution of cytochrome *c*. *PNAS* 50(4):672–679 (1963).

Margoliash E, Fitch WM & Dickerson RE. Molecular expression of evolutionary phenomena in the primary and tertiary structures of cytochrome *c*. *Brookhaven Symposia in Biology* 21(2):259–302 (1968), with discussion pp. 302–305.

Margulis L. Archaeal-eubacterial mergers in the origin of Eukarya: phylogenetic classification of life. *PNAS* 93(3):1071–1076 (1996).

Margulis L. Five-kingdom classification and the origin and evolution of cells. In: T Dobzhansky, MK Hecht & WC Steele (eds), *Evolutionary biology* 7. New York: Plenum (1974), pp. 45–78.

Margulis L. Introduction. In: L Margulis, JO Corliss, M Melkonian & DJ Chapman (eds), *Handbook of Protoctista*. Boston: Jones & Bartlett (1989), pp. xi–xxiii.

Margulis L. *Origin of eukaryotic cells. Evidence and research implications for a theory of the origin and evolution of microbial, plant, and animal cells on the Precambrian Earth*. New Haven: Yale UP (1970).

Margulis L. *The symbiotic planet. A new look at evolution*. London: Phoenix (1998).

Margulis L. Whittaker's five kingdoms of organisms: minor revisions suggested by consideration of the origin of mitosis. *Evolution* 25(1):242–245 (1971).

Margulis L & Schwartz KV. *Five kingdoms. An illustrated guide to the phyla of life on earth*. San Francisco: Freeman (1982).

Margulis L & Taylor FJR. Symposium on the evolution of mitosis in eukaryotic microorganisms. *BioSystems* 7(3–4):295–297 (1975).

Margulis L, Chapman M, Guerrero R & Hall J. The last eukaryotic common ancestor (LECA): acquisition of cytoskeletal motility from aerotolerant spirochetes in the Proterozoic Eon. *PNAS* 103(35):13080–13085 (2006).

Margulis L, Dolan MF & Guerrero R. The chimeric eukaryote: origin of the nucleus from the karyomastigont in amitochondriate protists. *PNAS* 97(13):6954–6959 (2000).

Margulis L, Dolan MF & Whiteside J. "Imperfections and oddities" in the origin of the nucleus. *Paleobiology* 31(2) Supplement:175–191 (2005).

Marius. *On the elements*. RC Dales (ed. and trans.). Centre for Medieval and Renaissance Studies, UCLA. Berkeley: University of California Press (1976).

Mark JJ. Enuma Elish, the Babylonian epic of creation. Full text (2018). *Ancient History Encyclopedia*, https://www.ancient.eu/article/225/, accessed 16 August 2018.

Marlowe C. *The tragicall history of D. Faustus. As it hath bene acted by The Right Honorable The Earle of Nottingham his seruants*. London: Thomas Bushell (1604).

Marmura ME. Ghazālian causes and intermediaries. *Journal of the American Oriental Society* 115(1):89–100 (1995).

Marmura ME. Translator's introduction. In: Al-Ghazālī, *The incoherence of the philosophers. A parallel English-Arabic text*. Second edition. Provo UT: Brigham Young UP (2000), pp. xv–xxvii.

Marriott H & Allers T. Archaea and the meaning of life. *Microbiology Today* 43(2):74–77 (2016).

Marsak LM. Bernard de Fontenelle: the idea of science in the French Enlightenment. *Transactions of the American Philosophical Society, New Series* 49(7):1–64 (1959).

Marshall J (ed.). *Emerald tablet of Hermes* (translations undated). http://www.levity.com/alchemy/emerald.html, accessed 16 December 2018 and 16 September 2020. Also at http://archive.org/.

Marsigli LF. Extrait d'une lettre ecrite de Cassis, prés de Marseille, le 18. de Decembre 1706. à Monsieur l'Abbé Bignon, par Monsieur le Comte Marsilli, touchant quelques branches de corail qui ont fleuri. *Journal des Sçavans* 1707 *(Supplement)*:59–66 (1707).

Marsigli [Marsilli] LF de. *Histoire physique de la mer.* Amsterdam: De'pens de la Compagnie (1725).

Martianus Capella MF. *De nuptiis philologiæ et Mercurii.* Mittellateinische Studien und Texte (HJ Westra *et al.*, eds) 1. Leiden: Brill (1994).

Martianus Capella MF. *De nvptiis philologiæ et Mercvrii.* Basileæ: Henricus Petrus (1532).

Martin B. *Bibliotheca technologica: or, a philological library of literary arts and sciences.* London: Idle (1737).

Martin GW. Systematic position of the slime molds and its bearing on the classification of the Fungi. *Botanical Gazette* 93(4):421–435 (1932).

Martin WF. Too much eukaryote LGT. *BioEssays* 2017:1700115 (2017).

Martin W[F] & Müller M. The hydrogen hypothesis for the first eukaryote. *Nature* 392(6671):37–41 (1998).

Martin WF, Tielens AGM, Mentel M, Garg SG & Gould SB. The physiology of phagocytosis in the context of mitochondrial origin. *Microbiology and Molecular Biology Reviews* 81(3):e00008–17 (2017).

Marton L. A new electron microscope. *Physical Reviews* 58(1):57–60 (1940).

Marton L. The electron microscope. A new tool for bacteriological research. *Journal of Bacteriology* 41(3):397–413 (1941).

Martyn T. *Thirty-eight plates, with explanations; intended to illustrate Linnæus's system of vegetables, and particularly adapted to the letters on the elements of botany.* London: White (1799).

Marvell A. Upon Appleton House. In: *Miscellaneous poems* (M Marvell, ed.). London: Robert Boulter (1681), pp. 76–103.

al-Mas'ūdī. *El-Mas'ūdī's historical encyclopædia entitled "Meadows of gold and mines of gems".* A Sprenger (trans.). Volume 1 [only volume published]. London: Oriental Translation Fund of Great Britain and Ireland (1841).

al-Mas'ūdī. *Maçoudi. Le livre de l'avertissement et de la revision.* B Carra de Vaux (trans.). Paris: L'Imprimerie nationale (1896).

al-Mas'ūdī. *Maçoudi. Les prairies d'or.* C Barbier de Meynard & P de Courteille (trans.). In nine volumes. Paris: L'Imprimerie Impériale (1861–1877).

Mateu TE. Representations of women in Spanish Levantine rock art. *Journal of Social Archaeology* 2(1):81–108 (2002).

Mather C. *The Christian philosopher: a collection of the best discoveries in nature, with religious improvements.* London: Matthews (1721).

Matthews JA & Briffa KR. The "Little Ice Age": re-evaluation of an evolving concept. *Geografiska Annaler* 87A:17–36 (2005).

Matthiolus PA. *Commentarij in VI. libros Pedacij Dioscoridis Anazarbei de medica materia . . . Adjectis magnis, ac nouis plantarum, ac animalium iconibus.* Venetijs: Felicem Valgrisium (1583).

Mattiolus PA. *Commentarii in sex libros Pedacii Dioscoridis.* Venetiis: Valgrisana (1565).

Maurer AA. *Medieval philosophy.* New York: Random House (1962).

Mayer A. Der waldensische Physiologus. *Romanische Forschungen* 5:392–418 (1890).

Mayerne T. Epistle Dedicatory (1634). In: T Mouffet, *The theater of insects: or, lesser living creatures.* London: "E.C." [E. Couts] (1658), pag. unnumb.

Mayr E. Biological classification: toward a synthesis of opposing methodologies. *Science* 214(4520):510–516 (1981).

Mayr E. Lamarck revisited. *Journal of the History of Biology* 5(1):55–94 (1972).

Mayr E. *The growth of biological thought. Diversity, evolution, and inheritance.* Cambridge MA: Harvard UP Belknap (1982).

Mayr E. Two empires or three? *PNAS* 95(17):9720–9723 (1998).

724   REFERENCES

Mayr E, Linsley EG & Usinger RL. *Methods and principles of systematic zoology.* New York: McGraw-Hill (1953).

Mazzarello P. A unifying concept: the history of cell theory. *Nature Cell Biology* 1(1):E13–E15 (1999).

Mazzolini RG & Roe SA. *Science against the unbelievers: the correspondence of Bonnet and Needham, 1760–1780.* Oxford: Voltaire Foundation (1986).

McColley G & HW Miller. Saint Bonaventure, Francis Mayron, William Vorilong, and the doctrine of a plurality of worlds. *Speculum* 12(3):386–389 (1937).

McConnell A. The flowers of coral—some unpublished conflicts from Montpellier and Paris during the early 18th Century. *History and Philosophy of the Life Sciences* 12(1):51–66 (1990).

McCrindle JW. *Ancient India as described by Ktêsias the Knidian; being a translation of the abridgement of his "Indika" by Phôtios, and of the fragments of that work preserved in other writers. Reprinted (with additions) from the "Indian Antiquary," 1881.* Calcutta: Thacker, Spink; Bombay: Thacker; and London: Trübner (1882).

McCrindle JW. *Ancient India as described by Megasthenês and Arraian; being a translation of the fragments of the Indika of Megasthenês collected by Dr. Schwanbeck, and of the first part of the Indika of Arrian. Reprinted (with additions) from the "Indian Antiquary," 1876–77.* Calcutta: Thacker, Spink; Bombay: Thacker; and London: Trübner (1877).

McCulloch F. *Mediaeval Latin and French bestiaries.* University of North Carolina Studies in the Romance Languages and Literatures 33. Chapel Hill: UNC Press (1960).

McGuckin JA. The scholarly works of Origen. In: JA McGuckin (ed.), *The Westminster handbook to Origen.* Louisville KY: Westminster John Knox Press (2004), pp. 25–44.

McHam SB. *Pliny and the artistic culture of the Italian Renaissance.* New Haven: Yale UP (2013).

McInerny R. The School of Chartres. In: AR Caponigri & R McInerny (eds), *A history of Western philosophy,* volume 2. Notre Dame IN: University of Notre Dame Press (1963), pp. 157–188.

McInerny R & O'Callaghan J. Saint Thomas Aquinas. *The Stanford Encyclopedia of Philosophy* (Summer 2018 edition), EN Zalta (ed.), plato.stanford.edu/archives/sum2018/entries/aquinas/, accessed 1 November 2018.

McIntyre JL. *Giordano Bruno.* London: Macmillan (1903).

McKeon R (ed.), *Selections from Medieval philosophers. 1. Augustine to Albert the Great.* New York: Scribner's (1929).

McKitterick R. *History and memory in the Carolingian world.* Cambridge UP (2004).

McLaughlin P. Naming biology. *Journal of the History of Biology* 35(1):1–4 (2002).

McLaughlin PJ & Dayhoff MO. Eukaryote evolution: a view based on cytochrome c sequence data. *Journal of Molecular Evolution* 2(2–3):99–116 (1973).

McLean A. *The book of the composition of alchemy.* Glasgow: Author (2002).

McLennan JF. Lycanthrophy. *Encyclopædia Britannica,* Ninth edition 15:89–92 (1883).

McLennan JF. The worship of animals and plants. *Fortnightly Review, New Series* 6:407–427, 562–582 (1869) and 7:194–216 (1870).

McRae KD. Introduction. In: J Bodin. *The six bookes of a commonweale. A facsimile reprint of the English translation of 1606.* KD McRae (ed.). Cambridge MA: Harvard UP (1962), pp. A1–A67.

Meckel JF (ed. and trans.). *Caspar Friedrich Wolff über die Bildung des Darmkanals im bebrüteten Hühnchen. Uebersetzt und mit einer enleitenden Abhandlung und Anmerkung versehen von JF Meckel.* Halle: Renger (1812).

Meckel JF. *Handbuch der pathologischen Anatomie.* Three volumes bound in two. Leipzig: Reclam (1812–1818).

Medicus FC. Sur l'origine et la formation des champignons. JAL Reynier (ed. and trans.). *Observations sur la Physique, sur l'Histoire Naturelle et sur les Arts* 34:241–247 (1789)

Meeusen M. Salt in the holy water: Plutarch's *Quaestiones naturales* in Michael Psellus' *De omnifaria doctrina.* In: LR Lanzillotta et al. (eds), *Plutarch in the religious and philosophical discourse of late Antiquity.* Leiden: Brill (2014), pp. 101–121.

Mehos DC. Appendix: Ivan E. Wallin and his theory of symbionticism. In: LN Khakhina, *Concepts of symbiogenesis.* L Margulis & M McMenamin (eds). New Haven: Yale UP (1992), pp 149–163.

REFERENCES 725

Meinel C. *Der handschriftliche Nachlaß von Joachim Jungius in der Staats- und Universitätsbibliothek Hamburg.* In: Katalog der Handschriften der Staats- und Universitätsbibliothek Hamburg, Band 9. Stuttgart: Hauswedell (1984).

Meister Eckhart. *Die deutschen und lateinischen Werke. Herausgegeben im Auftrage der Deutschen Forschungsgemeinschaft. Die lateinischen Werke. Band 1, Teile [1] and 2.* Stuttgart: Kohlhammer (1964 and 2015).

Meister Eckhart. *The complete mystical works of Meister Eckhart.* MO'C Walshe (trans.). Revised edition. New York: Herder & Herder / Crossroad (2009).

Meister Eckhart. *The essential sermons, commentaries, treatises and defense.* E Colledge & B McGinn (trans.). New York: Paulist; and Toronto: Ramsey (1981).

Mendelsohn E. The biological sciences in the Nineteenth century: some problems and sources. *History of Science* 3(1):39–59 (1964).

Mendelsohn JA. Lives of the cell. *Journal of the History of Biology* 36(1):1–37 (2003).

Meneghini G. On the animal nature of [the] Diatomeæ, with an organographical revision of the genera, established by Kützing. C Johnson (trans.). In: A Henfrey (ed.), *Botanical and physiological memoirs.* London: Ray Society (1853), pp. 341–513.

Meneghini G. *Sulla animalità delle diatomee e revisione organografica dei generi di diatomee stabiliti dal Kützing.* Venezia: Naratovich (read 1845, published 1846).

Mennens G. *Avrei Velleris sive Sacræ Philosophiæ vatvm selectæ ac vnicæ mysteriorvmqve Dei, natvræ, & artis admirabilium, libri tres.* In: *Theatrum chemicum* 5 (1622), pp. 267–470.

Mérejkovsky C de. La plante considéreé comme un complexe symbiotique. *Bulletin de la Société des Sciences naturelles de l'Ouest de la France, 3 Série* 6:17–98 (1920). Preface dated 25 October 1918.

Mereschkowsky C. Theorie der zwei Plamaarten als Grundlage der Symbiogenesis, einer neuen Lehre von der Entstehung der Organismen. *Biologisches Centralblatt* 30(9):278–303, (10):321–347, and (11):353–367 (1910).

Mereschkowsky C. Über Natur und Ursprung der Chromatophoren im Pflanzenreich. *Biologisches Centralblatt* 25(18):593–604; Nachtrag, 25(21):689–691 (1905).

Merezhkowsky KS. *Teoriya dvukh plasm kak osnova simbiogenezisa, novogo ucheniya o proiskhozhdenii organismov [Theory of two plasms as the basis of symbiogenesis, a new study on the origins of organisms].* Kazan: Imperial University (1909). Not examined.

Merrill LL. *The romance of Victorian natural history.* Oxford UP (1989).

Mersenne M. *Qvæstiones celeberrimæ in Genesim.* Lvtetiæ Parisiorvm: Sebastiani Cramoisy (1623).

Mettinger TND. *The riddle of resurrection. "Dying and rising gods" in the ancient Near East.* Coniectanea Biblica Old Testament Series 50. Stockholm: Almqvist & Wiksell (2001).

Meyen [F]J. Einige Bemerkungen über den Verdauungs-Apparat der Infusorien. *Archiv für Anatomie, Physiologie und wissenschaftliche Medicin* 1839:74–79 (1839).

Meyen [F]J. Ueber die Priestleysche grüne Materie, wie über die Metamorphose des Protococcus viridis in Priestleya botryoides und in Ulva terrestris. *Linnaea* 2:388–409 (1827).

Meyen J. Jahresbericht über die Resultate der Arbeiten im Felde der physiologischen Botanik von dem Jahre 1834. *Archiv für Naturgeschichte* 1(1):133–251 (1835).

Meyer TE, Cusanovich MA & Kamen MD (1986). Evidence against use of bacterial amino acid sequence data for construction of all-inclusive phylogenetic trees. *PNAS* 83(2):217–220 (1986).

Miall LC. *History of biology.* London: Watts (1911).

Miall LC. *The early naturalists. Their lives and work (1530–1789).* London: Macmillan (1912).

Michel P. *Physikotheologie. Ursprünge, Leistung und Niedergang einer Denkform.* Zürich: Gelehrten Gesellschft (2008).

Michel PH. *The cosmology of Giordano Bruno.* REW Maddison (trans.). Paris: Hermann (1973).

Micheli PA. *Nova plantarvm genera ivxta Tovrnefortii methodvm disposita.* Florentiæ: Bernardi Paperinii (1729).

Miescher F. Ueber die chemische Zusammensetzung der Eiterzellen. *Medicinisch-chemische Untersuchungen. Aus dem Laboratorium für angewandte Chemie zu Tübingen* 4:441–460 (1871).

726 REFERENCES

Migula W. Schizomycetes (Bacteria, Bacterien). In: A Engler (ed.), *Die natürlichen Pflanzenfamilien nebst ihren Gattungen und wichtigeren Arten ... begrünget von A Engler und K Prantl.* Teil 1, Abteilung 2. Leipzig: Engelmann (1897), pp. 2–13.

Migula W. Schizophyta (Spaltpflanzen). In: A Engler (ed.), *Die natürlichen Pflanzenfamilien nebst ihren Gattungen und wichtigeren Arten ... begrünget von A Engler und K Prantl.* Teil 1, Abteilung 1a. Leipzig: Engelmann (1900), pp. 1–13.

Mikkeli H. Giacomo Zabarella. *The Stanford Encyclopedia of Philosophy* (Spring 2018 edition), EN Zalta (ed.), https://plato.stanford.edu/archives/spr2018/entries/zabarella, accessed 14–15 March 2019 and 19 September 2020.

Miklucho-Maclay N. Beiträge zur Kenntniss der Spongien I. *Jenaische Zeitschrift für Medicin und Naturwissenschaft* 4:221–240 (1868).

Mikulinskii SR (ed.). *Istoriya biologii s dreveishiky vremen do nachala XX veka.* Moskva: Nauka (1972).

Mikulinskii SR. *Razvitie obshchikh problem biologii v Rossii. Pervaya polovina XIX veka.* Moskva: Akademii Nauk SSSR (1961).

Millar J. *A guide to botany, or, a familiar illustration of the principles of the Linnæan classification.* Edinburgh: Hill (1818).

Millas-Vallicrosa J. Translations of oriental scientific works (to the end of the Thirteenth century). In: GS Métraux & F Crouzet (eds), *The evolution of science. Readings from the history of mankind.* Toronto: Mentor (1963), pp. 128–169.

Miller FD Jr. Aristotle against the atomists. In: PD Kretzman (ed.), *Infinity and continuity in ancient and medieval thought.* Ithaca NY: Cornell University Press (1982), pp. 87–111.

Miller H. *Footprints of the Creator: or, the Asterolepis of Stromness.* London: Johnstone & Hunter (1849).

Miller H. *The cruise of the Betsey; or, a summer ramble among the fossiliferous deposits of the Hebrides. With Rambles of a geologist; or, ten thousand miles over the fossiliferous deposits of Scotland.* Edinburgh: Constable; and London: Hamilton & Adams (1858).

Miller H. *The testimony of the rocks; or, geology in its bearings on the two theologies, natural and revealed.* Boston: Gould & Lincoln (1857).

Miller J. *More than the sum of its parts: animal categories and accretive logic in Volume One of al-Jāḥiẓ's Kitāb al-ḥayawān.* PhD thesis, New York University (2013).

"Miller V" . *The man-plant: or, scheme for increasing and improving the British breed.* London: Cooper (1752).

Milne-Edwards H. Considérations sur quelques principes relatifs a la classification naturelle des animaux, et plus particulièrement sur la distribution méthodique des mammifères. *Annales des Sciences Naturelles, 3 Série, Zoologie* 1:65–99 (1844).

Milne-Edwards H. Observations sur les changemens de forme que divers crustacés éprouvent dans le jenue áge. *Annales des Sciences Naturelles, 2 Série, Zoologie* 3:321–334 (1835).

Milton J. *Paradise lost. A poem in ten books.* London: Simmons (1668).

Minchin EA. *An introduction to the study of the protozoa with special reference to the parasitic forms.* London: Edward Arnold (1912).

Mindell DP. Phylogenetic consequences of symbioses: Eukarya and Eubacteria are not monophyletic taxa. *BioSystems* 27(1):53–62 (1992).

Minio-Paluello L. Iacobus Veneticus Grecus: canonist and translator of Aristotle. *Traditio* 8:265–304 (1952).

Mirabaud [JB] de [Thiry PHT, Baron d'Holbach]. *Système de la nature. Ou des loix du monde physique & du monde moral.* In two volumes. Londres: [n.p.] (1770).

Mistele WR. *Mermaids, sylphs, gnomes, and salamanders. Dialogues with the kings and queens of nature.* Berkeley CA: North Atlantic (2012).

Mitford AB. *Tales of old Japan.* In two volumes. London: Macmillan (1871).

Mivart St. G. *Lessons from nature, as manifested in mind and matter.* London: John Murray (1876).

Mocenigo P [Mocenius]. *Vniversales institutiones ad hominum perfectionem.* Venetiis: Aldus (1581).

Moeller C. Trinitarians, Order of. *The Catholic Encyclopedia* 15:45–47 (1912).

Moffitt JF. An exemplary humanist hybrid: Vasari's "Fraude" with reference to Bronzino's "Sphinx". *Renaissance Quarterly* 49(2):303–333 (1996).

Mohl H von. *Principles of the anatomy and physiology of the vegetable cell.* A Henfrey (trans.). London: Van Voorst (1852).

Mohl H von. Ueber die Saftbewegung im Innern der Zellen. *Botanische Zeitung* 4(5–6): columns 73–78 and 89–94 (1846). English translation as: On the circulation of the sap in the interior of cells. *Annals and Magazine of Natural History* 18(116):1–10 (1846).

Mohl H von. *Ueber die Vermehrung der Pflanzen-Zellen durch Theilung.* Inaugural-Dissertation... der öffentlichen Prüfung vorlegt AW Winter. Tübingen: Fues (1835).

Mohl H von. Ueber die Vermehrung der Pflanzenzellen durch Theilung. In: H von Mohl, *Vermischte Schriften botanischen Inhalts.* Tübingen: Fues (1845), pp. 362–371. Page 362 mistakenly numbered 623.

Mollenhauer D. Founder of "Archiv für Protistenkunde": Fritz Schaudinn—his unfinished life. *Protist* 151(3):283–287 (2000).

Mommsen T, Krueger P & Watson A (eds). *The Digest of Justinian.* In four volumes. Philadelphia: University of Pennsylvania Press (1985).

Moncrief JW. Abu Bakr ar-Razi. Science and its times: understanding the social significance of scientific discovery (undated). At: encyclopedia.com, accessed 30 December 2018 and 16 September 2020.

Monfasani J. *George of Trebizond: a biography and a study of his rhetoric and logic.* In series: Columbia Studies in the Classical Tradition 1. Leiden: Brill (1976).

Moore E. Origen of Alexandria. *Internet Encyclopedia of Philosophy,* undated. https://iep.utm.edu/origen-of-alexandria/, accessed 30 July 2018.

Moore R. The "rediscovery" of Mendel's work. *Bioscene* 27(2):13–24 (2001).

Moore RC. Kingdom of organisms named Protista. *Journal of Paleontology* 28(5):588–598 (1954).

Moore W & Wilson HA. Prolegomena. In: P Schaff & H Wace (eds), *A select library of Nicene and post-Nicene Fathers of the Christian Church.* Second series, volume 5. New York: Christian Literature Company (1893), pp. 1–32.

Moorjani P, Thangaraj K, Patterson N, *et al.* Genetic evidence for recent population mixture in India. *American Journal of Human Genetics* 93(1):1–17 (2013).

Mora C, Tittensor DP, Adl S, *et al.* How many species are there on earth and in the ocean? *PLoS Biology* 9(8):e1001127 (2011).

Moran D. John Scottus Eriugena. *The Stanford Encyclopedia of Philosophy* (Fall 2008 edition), EN Zalta (ed.), plato.stanford.edu/archives/fall2008/entries/scottus-eriugena/, accessed 19 October 2018.

More H (ed.). *Two choice and useful treatises: the one Lux orientalis, or an enquiry into the opinion of the Eastern sages concerning the præexistence of souls. The other, a Discourse of truth, by the late Reverend Dr. Rust. With annotations on them both.* London: Collins & Lowndes, London (1682). *Tracts are paginated individually.*

More H. Enthusiasmus triumphatus; or, a brief discourse of the nature, causes, kinds, and cure of enthusiasm (1662). In: *A collection of several philosophical writings of Henry More.* Second edition. London: James Flesher (1662). Each tract numbered individually.

Moreau G. *Les chimères.* Oil on canvas (1884). Musée national Gustave Moreau, Paris.

Moreira D & López-García P. Ten reasons to exclude viruses from the tree of life. *Nature Reviews Microbiology* 7(4):306–311 (2009).

Moreau JL. *Œuvres de Vicq-d'Azyr, recueillies et publiées avec des notes et un discours sur sa vie et ses ouvrages.* In six volumes. Paris: Duprat-Duverger (1805).

Morgan TH, Sturtevant AH, Muller HJ & Bridges CJ. *The mechanism of Mendelian heredity.* New York: Henry Holt (1915).

Morren A & Morren Ch. Recherches physiologiques sur les hydrophytes de Belgique. Troisième mémoire. *Nouveaux Mémoires de l'Académie Royale des Sciences et Belles-Lettres de Bruxelles* 14 (1841), numb. sep.

Morrhy G [Morrhius]. *Lexicon Graecolatinvm cui praeter omneis omnium additiones hactenvs.* Parisiis: Gerardus Morrhius (1530).

728 REFERENCES

Morse D. A transcriptome-based perspective of meiosis in dinoflagellates. *Protist* 170(4):397–403 (2019).

Mouffet T. *Insectorvm sive minimorum animalium theatrvm: olim ab Edoardo Wottono. Conrado Gesnero. Thomaqve Pennio inchoatum: tandem Tho. Movfeti Londinâtis operâ sumptibusq[ve]*. London: Thom. Cotes (1634).

Moulinier L. *Le manuscript perdu à Strasbourg. Enquête sur l'oeuvre scientifique de Hildegarde*. Paris: Sorbonne (1995).

Mozzi A, Forni D, Clerici M, *et al*. The evolutionary history of genes involved in spoken and written language: beyond *FOXP2*. *Scientific Reports* 6:22157 (2016).

Muckle JT. *Algazel's* Metaphysics. *A medieval translation*. Toronto: Pontifical Institute of Mediaeval Studies (1933).

Mudd S. Submicroscopic structure of the bacterial cell, as shown by the electron microscope. *Nature* 161(4087):302–303 (1948).

Mudd S & Lackman DB. Bacterial morphology as shown by the electron microscope. I. Structural differentiation within the streptococcal cell. *Journal of Bacteriology* 41(3):415–420 (1941).

Mueller-Jourdan P. *Une initiation à la philosophie de l'antiquité tardive. Les leçons du Pseudo-Elias*. Fribourg: Academic (2007).

Muhlenberg H. *Catalogus plantarum Americæ Septentrionalis, huc usque cognitarum indigenarum et cicurum: or, a catalogue of the hitherto known native and naturalized plants of North America, arranged according to the sexual system of Linnæus*. Lancaster PA: Hamilton (1813).

Müller FM. *Lectures on the science of language*. Sixth edition. In two volumes. London: Longmans, Green (1871).

Müller [J]. Beobachtung über die Psorospermien (Fortsetzung). *Bericht über die zur Bekanntmachung geeigneten Verhandlungen der Königlichen Preussischen Akademie der Wissenschaften zu Berlin* 1841:246–250 (1841).

Müller [J]. Einige Beobachtungen an Infusorien. *Monatsberichte der Königlichen Preussische Akademie der Wissenschaften zu Berlin* 1856:389–393 (1856).

Müller J. *Elements of physiology*. W Baly (trans.). In three volumes. London: Taylor & Walton (1838–1842).

Müller J. *Handbuch der Physiologie des Menschen für Vorlesungen*. In two volumes. Coblenz: Hölscher (1834–1840).

Müller J. Ueber eine eigenthümliche krankhafte parasitische Bildung mit specifisch organisirten Samenkörperchen. *Archiv für Anatomie, Physiologie und wissenschaftliche Medicin* 1841:477–496 (1841).

Müller J. Ueber eine eigenthümliche Wurmlarve, aus der Classe der Turbellarien und aus der Familie der Planarien. *Archiv für Anatomie, Physiologie und wissenschaftliche Medicin* 1850:485–500 (1850).

Müller [J]. Über einen krankhaften Hautausschlag mit specifisch organisirten Samenkörperchen. *Bericht über die zur Bekanntmachung geeigneten Verhandlungen der Königlichen Preussischen Akademie der Wissenschaften zu Berlin* 1841:212–222 (1841).

Müller JT. *Einleitung in die oekonomische und physikalische Bücherkunde und in die damit verbundenen Wissenschaften bis auf die neuesten Zeiten*. In three Bände. Leipzig: Schwickert (1780–1784).

Müller K (ed.). Megasthenes. In: *Fragmenta historicorum Graecorum* 2. Parisiis: Firmin Didot (1848), pp. 397–439.

Müller K, Langlois V & Müller T. *Fragmenta historicorum Graecorum*. Five volumes in six. Parisiis: Firmin Didot (1841–1873). Digital at: http://www.dfhg-project.org, accessed 13 October 2020.

Müller M. *Theophrast von Hohenheim genannt Paracelsus Sämtliche Werke Erste Abtheilung: medizinische, naturwissenschaftliche und philosophische Schriften hsg. K Sudhoff, Registerband. Nova Acta Paracelsica, Supplementum*. Einsiedeln: Eberle (1960).

Müller OF. [V]on der Entdeckung eines neuen Geschlechts von Thierpflanzen. *Berlinische Sammlung zur Beförderung der Arzneywissenschaft, der Naturgeschichte, der Haushaltungskunst, Kameralwissenschaft und der dahin einschlagenden Litteratur* 1(1):41–52 (1768).

REFERENCES 729

Müller OF. *Abbildungen der Pflanzen, welche in der Königreichen Dännemark und Norwegen . . . wild wachsen, zu Erläuterung des unter dem Titel Flora Danica. Band 5 enthaltend das 13–15 Heft oder* Tab. 721–900. Kopenhagen: Hallager (1778–1782).

Müller OF. *Animalcula infusoria fluviatilia et marina, quæ detexit, systematice descripsit et ad vivum delineari curavit . . . opus hoc posthumum.* O Fabricius (ed.). Hauniæ: Nicolai Möller (1786).

Müller OF. *Vermivm terrestrium et fluviatilium, seu animalium infusoriorum, helminthicorum et testaceorum, non marinorum, succincta historia.* In two volumes. Havniæ et Lipsiæ; Heineck et Faber (1773–1774).

Müller OF. *Zoologicæ Danicæ prodromus, seu animalium Daniae et Norvegiae indigenarum characteres, nomina, et synonyma imprimis popularium.* Havniæ: Hallageriis (1776).

Müller RFG. Natur- und Medizingeschichtliches aus dem Mahābhārata. *Isis* 23(1):25–53 (1935).

Münchhausen O von. *Der Hausvater.* Volumes 1 (1765), 2 (1766), 3 (1768), 4 (1769), 5 (1770), and 6 (1773). Internal title page gives 1764 for Volume 1, Erstes Stück. Hanover: Försters.

Münster S. *Cosmographiae uniuersalis Lib. VI.* Basileæ: Henrichvm Petri (1550). Another printing 1554.

Münster S. *Cosmographia.* Basel: Heinrichum Petri (1544).

Murata S. *The tao of Islam. A sourcebook on gender relationships in Islamic thought.* Albany: SUNY Press (1992).

Murdoch JE. *Album of science. Antiquity and the Middle Ages.* In series: Album of science (IB Cohen, gen. ed.), volume 1. New York: Scribner (1984).

Murdoch JE. William of Ockham and the logic of infinity and continuity. In: N Kretzmann (ed.), *Infinity and continuity in ancient and medieval thought.* Ithaca NY: Cornell UP (1982), pp. 165–206.

Murray J & Hjort J. *The depths of the ocean. A general account of the modern science of oceanography based largely on the scientific researches of the Norwegian steamer Michel Sars in the North Atlantic.* London: Macmillan (1912).

Musæum Hermeticum, omnes Sopho-Spagyricæ artis discipulos fidelissime erudiens. Francofurti: Lucæ Jennisis (1625). In German translation as *Dyas chymica tripartita, das ist: Sechs herzliche teutsche philosophische Tractätlein.* Franckfurt: Lvca Jennis (1625).

Musæum Hermeticum reformatum et amplificatum. Francofurti: Hermannum à Sande (1678). Later translated into English by AE Waite as *The Hermetic museum restored and enlarged.* In two volumes. London: James Elliott (1893).

Myer I. *Oldest books in the world.* New York: Dayton (1900).

Mylonas GE. *Eleusis and the Eleusinian mysteries.* Princeton UP (1961). Reprinted 2015.

Nägeli C von. *A mechanico-physiological theory of organic evolution.* VA Clarke & FA Waugh (trans.). Chicago: Open Court (1898).

Nägeli C von. *A mechanico-physiological theory of organic evolution. Summary.* VA Clarke & FA Waugh (trans.). Second edition. Chicago: Open Court (1914).

Nägeli [Nægeli] C. Die neuern Algensysteme und Versuch zur Begrüngung eines eigenen Systems der Algen und Florideen. *Neue Denkschriften der Allgemeinen Schweizerischen Gesellschaft für die Gesammten Naturwissenschaften / Nouveaux Mémoires de la Société Helvétique des Sciences Naturelles* 9:1–275 (1847), numb. sep.

Nägeli C. *Die neuern Algensysteme und Versuch zur Begründung eines eigenen Systems der Algen und Florideen.* Zürich: Friedrich Schulthess (1847).

Nägeli C. *Gattungen einzeliger Algen physiologisch und systematisch bearbeitet.* Zürich: Schulthess (1849).

Nägeli C von. *Mechanisch-physiologische Theorie der Abstammungslehre. Mit einem Anhang: 1. Die Schranken der naturwissenschaftlichen Erkenntniss, 2. Kräfte und Gestaltungen im molecularen Gebiet.* München & Leipzig: Oldenbourg (1884).

Nakabachi A, Yamashita A, Toh H, *et al.* The 160-kilobase genome of the bacterial endosymbiont *Carsonella. Science* 214(5797):267 (2006).

Nash DW (ed.). *Taliēsyn; or, the bards and druids of Britain. A translation of the remains of the earliest Welsh bards, and an examination of the bardic mysteries.* London: Smith (1858).

730   REFERENCES

Nasr SH. Al-Bīrūnī as philosopher. In: HM Said (ed.), *Al-Biruni commemorative volume*. Karachi: Hamdard (1979), pp. 400–406.

Nasir A, Kim KM, Da Cunha V & Caetano-Anollés G. Arguments reinforcing the three-domain view of diversified cellular life. *Archaea* 2016:1851865 (2016).

Nasr SH. *An introduction to Islamic cosmological doctines. Conceptions of nature and methods used for its study by the Ikhwān al-Ṣafa, al-Bīrūnī, and Ibn Sīnā*. Revised edition. London: Thames and Hudson (1978).

Nasr SH. *Science and civilization in Islam*. New York: Barnes & Noble (1992).

Nass S & Nass MMK. Intramitochondrial fibers with DNA characteristics. II. Enzymatic and other hydrolytic treatments. *Journal of Cell Biology* 19(3):613–618 (1963).

Nau F. Une ancienne traduction latine du Bélinous Arabe (Apollonius de Tyane) faite par Hugo Sanctelliensis et conserve dans un Ms. du XIIe siècle. *Revue de l'Orient Chrétien, 2 Série* 2(1):99–106 (1907).

Nazari V. Chasing butterflies in Medieval Europe. *Journal of the Lepidopterists' Society* 68(4):223–231 (2014).

Neander M. *Physice, siue potius syllogæ physicæ rerum eruditarum* [title varies]. In two volumes. Lipsiæ: Desner (1585).

Nebrija EA de. *Dictionariu[m] Aelij Antonij Nebrissen[sis] nu[n]c [de]mu[m] auctu[m] & recognitu[m]*. Salamanca: Joannis Varele (1516).

Nebrija EA de. *Dictionarium ex hispaniensi in latinum sermone[m]*. Salamanca (1495).

Nebrija EA de. *Dictionarivm latinohispanicvm, et vice versa hispanicolatinvm, Ælio Antonio Nebrissensi interprete*. Antverpiæ: Ioannis Steelsij (1560).

Nebrija EA de. *Gramãtica de la le[n]gua castellana*. Salamanca: [de Porras] (1492).

Nebrija EA de. *Lexicon hoc est dictionarium ex sermone latino in hispaniense[m]*. Salamanca: [n.p.] (1492).

Nebrija EA de. *Lexicon latinocatalanvm, sev dictionarium Aelij Antonij Nebrissensis ipsius auctoris opera primum concinnatum*. Barcinone: Clavdius Bornatius (1560).

Nebrija EA de. *Vocabularium Nebrissense: ex latino sermone in siciliensem & hispaniensez denuo traductum*. Venetiis: Bernardinum Benalium (1520).

Neckam A. *De naturis rerum libri duo. With the poem of the same author, De laudibus divinæ sapientiæ*. T Wright (ed.). Rerum Britannicarum Medii Ævi scriptores. London: Longman, Green et al. (1863).

Necker [NJ] de. Eclaircissemens sur la propagation des filicées en général. *Acta Academiæ Theodoro-Palatinæ (Physicvm)* 3:275–318 (1775).

Necker NJ de. *Physiologia muscorum per examen analyticum de corporibus variis naturalibus inter se collatis continuitatem proximamve animalis cum vegetabili concatenationem indicantibus*. Manheimii: Schwan (1774).

[Necker NJ de]. *Physiologie des corps organisés, ou examen analytique des animaux & des végétaux comparés ensemble, à dessein de démontrer la chaîne de continuité qui unit des différens regnes de la nature*. Bouillon: Société Typographiques (1775).

Necker NJ de. *Traité sur la mycitologie ou discourse historique sur les champignons en général, dans lequel on démontre leur véritable origine & leur génération*. Mannheim: Mathias Fontaine (1783).

Needham J. *Science and civilization in China*. In multiple volumes and parts [authorship varies]. Cambridge UP (1954 ff.). [SCC]

Needham JT. *Idée sommaire, ou vüe générale du systeme physique, et metaphysique de Monsieur Needham sur la génération des corps organisés*. Second edition. Bruxelles: Lemaire (1781).

Needham JT. *New microscopical discoveries*. London: Needham (1745).

Needham JT. Note de l'éditeur. In: L'Abbé M*** [Monestier B], *La vraie philosophie*. Bruxelles: Valade (1774), pp. 460–470.

Needham JT. *Nouvelles observations microscopiques, avec des découvertes intéressates sur la composition & la décomposition des corps organisés*. Paris: Louis-Étienne Ganeau (1750).

REFERENCES   731

Needham JT. Observations upon the generation, composition, and decomposition of animal and vegetable substances. *Philosophical Transactions* 45(490):615–666 (1748). Also published separately (1749).

Needham R. Introduction. In: E Durkheim & M Mauss, *Primitive classification* (R Needham, ed. and trans.). Originally published as: De quelques formes primitives de classification. *Année Sociologique* 6:1–72, 1901–1902. University of Chicago Press (1963), pp.vii–xlviii.

Nees von Esenbeck CG. *Das System der Pilze und Schwämme. Ein Versuch.* In two volumes. With: *Uebersicht des Systems der Pilze und Schwämme zur Erklärung der Kupfertafeln*, numb. sep. Würzburg: Stahel (1817).

Nees von Esenbeck CG. *Die Algen des süssen Wassers nach ihren Entwicklungsstufen.* Bamberg: CF Kunt (copy at NHM London); or Würzburg: [n.p.] (copy at BSB München) (1814).

Nees von Esenbeck CG. *Handbuch der Botanik.* In two volumes, together Theil 4 of GH Schubert (ed.), *Handbuch der Naturgeschichte, zum Gebrauch bei Vorlesungen.* Nürnberg: Schrag (1820).

Nemesius. *Nemesii Episcopi Premnon Physicon sive Peri physeōs anthrōpou liber a N. Alfano Archiepiscopo Salerni.* C Burkhard (trans.). Lipsiae: Teubneri (1917).

Nemesius. *Nemesii philosophi clarissimi De natvra hominis liber utilissimvs.* Georgio Valla Placentino (trans.). Lvdgvni: Gryphivm (1538).

Nemesius. *Nemesiou episkopou kai philosophou Peri phuseōs anthrōpou, biblion en. Nemesii episcopi et philosophi De natvra hominis, lib. vnvs.* N Ellebodio Casletano (ed. and trans.). Antverpiæ: Plantini (1565).

Nemesius. *Nemesius Emesenus De natura hominis Graece et Latine.* CF Matthaeus (ed.). Halae Magdeburgicae: Gebauer (1802).

Nemesius. *On the nature of man.* In: W Telfer (ed.), *Cyril of Jerusalem and Nemesius of Emesa.* Library of Christian Classics 4. London: SCM Press (1955), pp. 224–453.

Newman RW. Mercury and sulphur among the high Medieval alchemists: from Rāzī and Avicenna to Albertus Magnus and Pseudo-Roger Bacon. *Ambix* 61(4):327–344 (2014).

Newton I. On universal design. In: HS Thayer (ed.), *Newton's philosophy of nature: selections from his writings.* New York: Hafner (1953), pp. 65–67.

Newton I. *Philosophiæ naturalis principia mathematica. Editio tertia aucta & emendata.* Londini: Innys (1726).

Nicander. *The poems and poetical fragments.* ASF Gow & AF Scholfield (eds and trans.). Cambridge UP (1953).

Nicholas of Cusa. *De docta ignorantia. Die belehrte Unwissenheit*, volumes 1–2: P Wilpert & HG Senger (ed. and trans.); volume 3: R Klibansky (ed.), HG Senger (trans.). Philosophische Bibliothek, vols 264a–c. Hamburg: Felix Meiner (1994–1999).

Nicholas of Cusa. *Dialectical mysticism.* J Hopkins (trans.). Minneapolis: Banning (1988).

Nicholas of Cusa. *Metaphysical speculations.* J Hopkins (trans.). In two volumes. Minneapolis: Banning (1997–2000).

Nicholas of Cusa. *Nicolai de Cusa Opera omnia.* E Hoffmann, R Klibansky *et al.* (eds.). In 20 volumes. Academiae Litterarum Heidelbergensis. Hamburg: Meiner (1932–2006).

Nicholas of Cusa. *Of learned ignorance.* F Heron (trans.). London: Routledge & Kegan Paul (1954).

Nicholas of Cusa. *Of learned ignorance (De docta ignorantia).* J Hopkins (trans.). Minneapolis: Banning (1981).

Nicholas of Cusa. *Of learned ignorance (De docta ignorantia).* J Hopkins (trans.). Second edition. Minneapolis: Banning (1985).

Nicholas of Damascus. *Nicolai Damasceni de Plantis libri duo Aristoteli vulgo adscripti. Ex Isaac ben Honain. Versione Arabica Latine vertit Alfredus.* EHF Meyer (ed.). Leipzig: Leopoldi Voss (1841).

Nicolai [C]F. *Freuden des jungen Werthers.* Berlin: Friedrich Nicolai (1775).

Nicolas JP. Adanson, the man. In: GHM Lawrence (ed.), *Adanson. The bicentennial of Michel Adanson's "Familles des plants"*, volume 1 [of 2]. Pittsburgh: Hunt Botanical Library (1963), pp. 1–121.

Nielsen C. Origin of the trochophora larva. *Royal Society Open Science* 5(7):180042 (2018).

732 REFERENCES

Nielsen C. Six major steps in animal evolution: are we derived sponge larvae? *Evolution & Development* 10(2):241–257 (2008).

Nieremberg JE. *De la diferencia entre lo temporal y lo eterno: crisol de desengaños*. Madrid: Manuel Martin (1762).

Nieremberg JE. *Historia natvræ, maxime peregrinæ, libris XVI. Distincta. In quibus rarissima Naturæ arcana, etiam astronomica, & ignota Indiarum animalia, quadrupedes, aues, pisces, reptilia, insecta, zoophyta, plantæ, metalla, lapides, & alia mineralia . . . describuntur.* Antverpiæ: Balthasaris Moreti (1635).

Nietzsche FW. *Also sprach Zarathustra. Ein Buch für Alle und Keinen.* Chemnitz: Schmeitzer (1883–1885).

Nietzsche FW. *The portable Nietzsche.* W Kaufmann (sel. and trans.). New York: Viking (1954).

Nieuhof J. *Het gezantschap der Neêrlandtsche Oost-Indische Compagnie, aan den grooten Tartarischen Cham.* Amsterdam: van Meure (1665).

Nisbet HB. *Herder and scientific thought.* Cambridge UK: Modern Humanities Research Association (1970).

Nissen C. *Herbals of five centuries.* Zurich: L'Art Ancien (1958).

Niẓami Arūẓi. *Chahár maqála ("The four discourses") of Aḥmad ibn 'Alí an-Niẓámí al-' Arúḍí as-Samarqandí, edited . . . by Mírzá Muḥammad.* EJ Gibb Memorial Series 11(1). Leyden: Brill; and London: Luzac (1910).

Niẓami Arūẓi. *Revised translation of the Chahár maqála ("Four discourses") of Niẓámí-i-'Arúḍí of Samarquand, followed by an abridged translation of Mírzá Muḥammad's notes to the Persian text.* EG Browne (trans.). EJ Gibb Memorial Series 11(2). Cambridge UP; and London: Luzac (1921).

Nolle H [Nollius]. *Theoria philosophiæ Hermeticæ, septem tractatibvs.* Hanoviæ: Petrvm Antonivm (1617).

Noone T & Houser RE. Saint Bonaventure. *The Stanford Encyclopedia of Philosophy* (Winter 2014 edition), EN Zalta (ed.), plato.stanford.edu/archives/win2014/entries/bonaventure/, accessed 6 November 2018.

Nordenskiöld E. *The history of biology. A survey.* LB Eyre (trans.). New York: Knopf (1932).

Norman HW. Preface. In: HW Norman (trans.). *The Anglo-Saxon version of the Hexameron of St. Basil, or, Be Godes six daga weorcum.* Second edition. London: John Russell Smith; and Oxford: Francis Macpherson (1849), pp. vii–xx.

Norman J. Jeremy Norman's HistoryofInformation.com. The Codex Vossanius Oblongus: earliest & best surviving test of Lucretius's *De rerum natura*. Circa 825. Accessed 17 May 2018 and 8 September 2020.

Norris JA. The mineral exhalation theory of metallogenesis in pre-modern mineral science. *Ambix* 53(1):43–65 (2006).

Norton CE. Aids to the study of the "Divine comedy". In: CE Norton (trans.), *The Divine Comedy of Dante Alighieri.* Revised edition, volume 1 [of 3]. Boston: Houghton Mifflin (1902), pp. v–viii.

Norton T. *The ordinall of alchimy . . . being a facsimile reproduction from* Theatrum Chemicum Britannicum *with annotations by Elias Ashmole.* London: Edward Arnold (1928).

Norton T. *The ordinall of alchymy.* [Bristol]. British Library Add. MS 10302 (1477).

Norton T. *Tractatus chymicus dictus Crede mihi sive ordinale. Tractatus secundus.* In: *Musæum Hermeticum reformatum et amplificatum.* Francofurti: Hermannum à Sande (1678), pp. 432–532.

Novák FA. *Systematická botanika. Díl první.* In series: Rostlinopis (S Prǎt, ed.), volume 8. Praha: Aventinum (1930).

Novick A. On the origins of the quinarian system of classification. *Journal of the History of Biology* 49(1):95–133 (2016).

Novikoff AB. Mitochondria (chondriosomes). In: J Brachet & AE Mirsky (eds), *The cell. Biochemistry, physiology, morphology.* Volume 2: *Cells and their component parts: biochemistry, physiology, morphology.* New York: Academic (1961), pp. 299–421.

Nussbaum MC. *Aristotle's* De motu animalum. *Text with translation, commentary, and interpretative essays.* Princeton UP (1978).

Nutton V. The fortunes of Galen. In: RJ Hankinson (ed.), *The Cambridge companion to Galen.* Cambridge UP (2008), pp. 355–390.

Nyhart LK. *Biology takes form. Animal morphology and the German universities, 1800–1900.* University of Chicago Press (1995).

Ocellus Lucanus (attrib.). *On the nature of the universe.* T Taylor (trans.). London: Taylor (1831).

Ockham. *Guillelmi de Ockham Opera philosophica et theologica ad fidem codicum manuscriptorum edita. Opera philosophica.* P Boehner, G Gál, S Brown *et al.* (eds). In seven volumes. St Bonaventure NY: Instituti Franciscani (1974–1988).

Odoric. Itinerarium fratris Odorici. The journall of Frier Odoricus. In: *Hakluyt's collection of the early voyages, travels, and discoveries, of the English nation.* New edition, volume 2. London: RH Evans, J Mackinlay, and R Priestley (1810), pp. 142–158 (Latin) and 158–174 (English).

Odoric. *The journal of Friar Odoric.* In: AW Pollard (ed.), *The travels of Sir John Mandeville.* London: Macmillan (1915), pp. 326–362.

Ogilvie BW. *The science of describing. Natural history in Renaissance Europe.* University of Chicago Press (2006).

O'Grady P. Thales of Miletus. *Internet Encyclopedia of Philosophy,* undated. https://iep.utm.edu/thales/, accessed 2 May 2018.

O'Hara RJ. Diagrammatic classifications of birds, 1819–1901: views of the natural system i[n] 19th-century British ornithology. *In: Acta XIX Congressus Internationalis Ornithologici 2* (H Ouellet, ed.). Ottawa, 22–29 VI 1986. University of Ottawa Press (1988), pp. 2746–2759.

O'Hara RJ. Representations of the natural system in the Nineteenth century. *Biology and Philosophy* 6(2):255–274 (1991).

Oken [L]. *Abriss der Naturphilosophie. Bestimmt zur Grundlage seiner Vorlesungen über Biologie.* Göttingen: Vandenhoek & Ruprecht (1805).

Oken [L]. *Abriss des Systems der Biologie. Zum Behufe seiner Vorlesungen.* Göttingen: Vandenhoek & Ruprecht (1805).

Oken [L]. *Allgemeine Naturgeschichte für alle Stände.* In seven volumes plus Register. Stuttgart: Hoffmann (1833–1842).

Oken L. Cuviers und Okens Zoologien neben einander gestellt. *Isis von Oken 1817,* Band 1(8), sections 144–148 (1817). *Column numbering is absent in part, and unreliable.*

Oken L. *Die Zeugung.* Bamberg und Wirzburg: Goebhardt (1805).

Oken L. *Elements of physiophilosophy.* A Tulk (trans.). London: Ray Society (1847).

Oken L. Entstehung des ersten Menschen. *Isis von Oken 1819,* Band 2(7): cols 1117–1123 (1819).

Oken L. *Grundriss des Naturphilosophie der Theorie der Sinne und der darauf gegründeten Classification der Thiere.* Frankfurt (1804).

Oken L. *Laurentius Oken. Gesammelten Schriften.* J Schuster (ed.). Berlin: Kieper (1939).

Oken L. *Lehrbuch der Naturgeschichte. Erster Theil. Mineralogie.* Leipzig: Reclam (1813).

Oken [L]. *Lehrbuch der Naturphilosophie.* [First edition]. In three volumes. Jena: Frommann (1809–1811).

Oken [L]. *Lehrbuch der Naturphilosophie.* Second edition. Jena: Frommann (1831).

Oken [L]. *Lehrbuch der Naturphilosophie.* Third edition. Zürich: Friedrich Schultheß (1843).

Oken L. *Lorenz Oken—Gesammelte Werke.* T Bach, O Breidbach & D von Engelhardt (eds). In four volumes. Weimar: Böhlaus (2007–2013).

[Oken L]. [Review of:] Die Metamorphose der *Ectosperma clavata* Vaucher, von F Unger. *Isis von Oken* 22(2): col. 136 (1829).

Oken [L]. *Ueber den Werth der Naturgeschichte, besonders für die Bildung der Deutschen.* Jena: Frommann (1809). In: Gesammelten Schriften. J Schuster (ed.). Berlin: Kieper (1939), pp. 255–274.

Oken [L]. *Vseobschaya estestvennaya istoriya dlya vsekh sostoyanij,* volume 5. *Zoologii chast vtoraya.* St Petersburg: Christian Gintse (1836).

734 REFERENCES

Oken [L] & Kieser [DG]. *Beiträge zur vergleichende Zoologie, Anatomie und Physiologie*. In two volumes. Bamberg: Göbhardt (1806–1807).

Oken [L], in part with Walchner FA. *Allgemeine Naturgeschichte für alle Stände*. In seven volumes. Stuttgart: Hoffmannn (1833–1841).

[*Old English herbal, illustrated*]. British Library: Cotton Vitellius C III (ff 11–85 in part). Copied in early Eleventh century. http://www.bl.uk/manuscripts, accessed 1 April 2019 and 21 September 2020.

O'Leary DeL. *How Greek science passed to the Arabs*. London: Routledge & Kegan Paul (1949).

Olivi G. *Zoologia adriatica*. Bassano: Remondini (1792).

Oltmanns F. *Morphologie und Biologie der Algen*. In two volumes. Jena: Gustav Fischer (1904–1905).

Oltmanns F. *Morphologie und Biologie der Algen*. Second edition. In three volumes. Jena: Gustav Fischer (1922–1923).

O'Malley MA. What *did* Darwin say about microbes, and how did microbiology respond? *Trends in Microbiology* 17(8):341–347 (2009).

O'Malley MA, Martin W & Dupré J. The tree of life: introduction to an evolutionary debate. *Biology & Philosophy* 25(4):441–453 (2010).

O'Meara DJ. The concept of natura in John Scottus Eriugena (De divisione naturae Book I). *Vivarium* 19(2):126–145 (1981).

O'Meara JJ. Introduction. In: Eriugena, *Periphyseon*. IP Sheldon-Williams (trans.), JJ O'Meara (rev.). Montréal: Bellarmin; and Washington: Dumbarton Oaks (1987), pp. 11–21.

Oppian. In: AW Mair (trans.), *Oppian. Colluthus. Tryphiodorus*. Loeb Clasical Library 219. London: Heinemann (1963).

Orchard A (ed.). *Pride and prodigies: studies in the monsters of the Beowulf-manuscript*. University of Toronto Press (1995).

Oresme N. *Le livre du ciel et du monde* [1377]. AD Menut & AJ Denomy (eds), AD Menut (trans.). Madison: University of Wisconsin Press (1968).

Oresme N. *Nicole Oresme and the marvels of nature. A study of his* De causis mirabilium. B Hansen (ed. and trans.). In series: Studies and Texts, volume 68. Toronto: Pontifical Institute of Medieval Studies (1985).

Origen. *Against Celsus*. In: A Roberts & J Donaldson (eds), *The Ante-Nicene Fathers*, Volume 10 (1869), pp. 391–478, and Volume 23 (1872), pp. 1–559. Edinburgh: Clark.

Origen. *Commentary on the Gospel of John*. In: A Menzies (ed.), *The Ante-Nicene Fathers*. Fifth edition, volume 9. New York: Scribner (1906), pp. 297–408.

Origen. *Contra Celsum*. H Chadwick (trans.). Cambridge UP (1953).

Origen. *De principiis*. In: A Roberts & J Donaldson (eds), *Ante-Nicene Christian Library* 10:1–365 (1869).

Origen. *On first principles. Being Koetschau's text of the* De principiis *translated into English*. GW Butterworth (ed.). Gloucester MA: Peter Smith (1975).

Origen. *On prayer*. JEL Oulton (trans.) In: JEL Oulton & H Chadwick (eds), Alexandrian Christianity. Selected translations of Clement and Origen. Library of Christian Classics. Ichthus edition. Philadelphia: Westminster (1954), pp. 180–387.

Origen. *Writings*. In: A Roberts & J Donaldson (eds), *Ante-Nicene Christian Library*, volumes 10 and 23. Edinburgh: Clark (1869 and 1872).

O'Rourke B. From bioblasts to mitochondria: ever expanding roles of mitochondria in cell physiology. *Frontiers in Physiology* 1:7 (2010).

Ortus sanitatis. Moguntiæ: Jacob Meydenbach (1491). Bayerische StaatsBibliothek 2 Inc.c.a. 2576. Digital image at https://www.digitale-sammlungen.de/, accessed 14 October 2020.

Osborne R. The polis and its culture. In: CCW Taylor (ed.), *From the beginning to Plato*. Routledge history of philosophy (GHR Parkinson & SG Shanker, gen. eds), volume 1. London: Routledge (1997), pp. 9–46.

Oskamp DL. *Afbeeldingen der artseny-gewassen met derzelver Nederduitsche en Latynsche beschryvingen*. In six volumes. Amsterdam: Sepp en Zoon (1796–1800).

REFERENCES   735

Ospovat D. *The development of Darwin's theory. Natural history, natural theology, and natural selection, 1838–1859.* Cambridge UP (1995).

O'Toole CJ. *The philosophy of creation in the writings of St. Augustine.* Washington: Catholic University of America Press (1944).

Ott FD. A bibliographical contribution to our appreciation of EJC Esper's *Die Pflanzenthiere in Abbildungen nach der Natur* (1788–1830) and its *Fortsetzungen* (1794–1806). *Taxon* 38(2):204–215 (1989).

Oulhen N, Schultz BJ & Carrier TJ. English translation of Heinrich Anton de Bary's 1878 speech, "Die Erscheinung der Symbiose" ("*De la symbiose*"). *Symbiosis* 69(3):131–139 (2016).

Oulton JEL & Chadwick H (eds). *Alexandrian Christianity. Selected translations of Clement and Origen.* Library of Christian Classics, volume 2. Ichthus edition. Philadelphia: Westminster (1954).

Outram D. *Georges Cuvier. Vocation, science and authority in post-revolutionary France.* Manchester UP (1984).

Outram D. Uncertain legislator: Georges Cuvier's laws of nature in their intellectual context. *Journal of the History of Biology* 19(3):323–368 (1986).

Ovidus Naso, Publius [Ovid]. *Fasti.* JG Frazer (trans.). London: Heinemann (1931).

Ovidius Naso, Publius [Ovid]. *Metamorphoseon libri XV interpretatione et notis.* Novi-Eboraci: Eastburn (1817).

Ovidus Naso, Publius [Ovid]. *Metamorphoses.* R Humphries (trans.). Bloomington: Indiana UP (1957).

Owen R. Acrita. [*Todd's*] *Cyclopædia of Anatomy and Physiology* 1:47–49 (1836).

Owen R. *Lectures on the comparative anatomy and physiology of the invertebrate animals, delivered at the Royal College of Surgeons of England, in 1843.* London: Longman, Brown *et al.* (1843).

Owen R. *Lectures on the comparative anatomy and physiology of the invertebrate animals, delivered at the Royal College of Surgeons.* Second edition. London: Longman, Brown *et al.* (1855).

Owen R. *Lectures on the comparative anatomy and physiology of the vertebrate animals, delivered at the Royal College of Surgeons of England, in 1844 and 1846.* London: Longman, Brown *et al.* (1846).

Owen R. Oken, Lorenz. *Encyclopædia Britannica*, Eleventh edition 20:55–57 (1911).

Owen R. *On the anatomy of vertebrates.* In three volumes. London: Longmans, Green (1866–1868).

Owen R. *On the archetype and homologies of the vertebrate skeleton.* London: Van Voorst (1848).

Owen [R]. On the characters, principles of division, and primary groups of the class Mammalia. *Journal of the Proceedings of the Linnean Society. Zoology* 2:1–37 (read 1857, published 1858).

Owen R. *On the classification and geographical distribution of the Mammalia.* London: Parker (1859).

Owen R. *Palæontology or a systematic summary of extinct animals and their geological relations.* Edinburgh: Black (1860). Reprinted by Arno, New York (1980).

Owen R. *Palæontology or a systematic summary of extinct animals and their geological relations.* Second edition. Edinburgh: Black (1861).

Owen R. Palæontology. *Encyclopædia Britannica*, Eighth edition 17:91–176 (1858).

Owen R. Remarks on the Entozoa, and on the structural differences existing among them: including suggestions for their distribution into other classes. *Transactions of the Zoological Society of London* 1(4):387–394 (1835).

Owen R. Report on British fossil reptiles, Part 2. In: *Report of the Eleventh Meeting of the British Association for the Advancement of Science, Plymouth, July 1841.* London: John Murray (1842), pp. 60–204.

Owen [R]. Report on the archetype and homologies of the vertebrate skeleton. In: *Report of the Sixteenth Meeting of the British Association for the Advancement of Science, Southampton, September 1846.* London: John Murray (1847), pp. 169–340.

Oxford English Dictionary. https://www.oed.com.

Oxford English Dictionary. Second edition. In 20 volumes. Oxford: Clarendon (1989).

736 REFERENCES

Pace NR. A molecular view of microbial diversity and the biosphere. *Science* 276(5313):734–740 (1997).

Pace NR. Problems with "procaryote". *Journal of Bacteriology* 191(7):2008–2010 (2009).

Pace NR, Sapp J & Goldenfeld N. Phylogeny and beyond: scientific, historical, and conceptual significance of the first tree of life. *PNAS* 109(4):1011–1018 (2012).

Pace NR, Stahl DA, Lane DJ & Olsen GJ. The analysis of natural microbial populations by ribosomal RNA sequences. *Advances in Microbial Ecology* 9:1–55 (1986).

Packard FR. History of the School of Salernum. In: J Harington, *The School of Salernum. Regimen Sanitatis Salernitanum*. New York: OB Hoeber (1920), pp. 7–52. Reprinted by AM Kelley, New York (1970).

Palacios MA. El "Libro de Los Anaimales" de Jâḥiẓ. *Isis* 14(1):20–54 (1930).

Paleolithic & Neolithic rock art. Cave paintings & rock engravings. http://paleolithic-neolithic. com (Lascaux, updated 02/2015), accessed 4 September 2020 and 14 October 2020.

Paley W. *Natural theology: or, evidences of the existence and attributes of the Deity. Collected from the appearances of nature*. Philadelphia: John Morgan (1802).

Pallas PS. *Charakteristik der Thierpflanzen, worin von den Gattungen derselben allgemeine Entwürfe, und von denen dazugehörigen Arten kurze Beschreibungen gegeben werden . . . Aus dem Lateinischen übersetzt und mit Anmerkungen versehen von CF Wilkens . . . und von JFW Herbst*. Nürnberg: Kaspisch (1787).

Pallas PS. *Elenchus Zoophytorum sistens generum adumbrationes generaliores et specierum cognitarum succinctas descriptiones cum selectis auctorum synonymis*. Hagæ Comitum: Petrum van Cleef (1766).

Pallister JL. Introduction. In: A Paré, *On monsters and marvels*. JL Pallister (trans.). University of Chicago Press (1982), pp. xv–xxxii.

Palmer A. Reading Lucretius in the Renaissance. *Journal of the History of Ideas* 73(3):395–416 (2012).

Panchen AL. *Classification, evolution, and the nature of biology*. Cambridge UP (1992).

Pandita KN. Central Asian society in Ibn Sīnā's time. *Indian Journal of History of Science* 21(3):251–256 (1986).

Paracelsus. *Essential readings*. N Goodrick-Clarke (trans.). Western Esoteric Masters series. Berkeley: Atlantic (1999).

Paracelsus. *Liber de nymphis*. English translation by HE Sigerist as: A book on nymphs, sylphs, pygmies, and salamanders, and on the other spirits. In: HE Sigerist (ed.), *Four treatises of Theophrastus von Hohenheim called Paracelsus*. Baltimore: Johns Hopkins UP (1941), pp. 213–253.

Paracelsus. *Of the nature of things. Nine books*. [Bound with:] M Sandivogius. *A new light of alchymie*. J.F. [J French] (trans.). London: Richard Cotes (1650), pp. 1–145 (numb. sep.).

Paracelsus. *Of the nature of things. Nine books*. [Bound with:] M Sandivogius. *A new light of alchymy*. J.F. [J French] (trans.). London: Andrew Clark[e] (1674), pp. 153–301.

Paracelsus. *Philosophy addressed to the Athenians*. In: LW de Laurence (ed.), *The hermetic and alchemical writings of Aureolus Philippus Theophrastus Bombast of Hohenheim, called Paracelsus, the Great*. From the London edition (1894) edited by AE Waite. Chicago: De Laurence, Scott (1910). Reprinted by Kessinger, Kila MT, undated (1990s), volume 2, pp. 249–281.

Paracelsus. *Sämtliche Werke*. K Sudhoff (ed.). In 14 volumes. München: Oldenbourg (1929–1933). With: Müller M. Registerband zu Sudhoffs Paracelsus-Gesamtausgabe. *Nova Acta Paracelsica*, Supplementum. Einsiedeln: Eberle (1960).

Paré A. *Des monstres & prodiges*. In: *Les Œvvres d'Ambroise Paré. Quatriesme edition*. Paris: Gabriel Buon (1585), pp. 1020–1097.

Paré A. *Des monstres et prodiges*. In: *Deux livres de chirurgie*. Paris: André Wechel (1573), pp. 365–580.

Paré A. *Des monstres et prodiges*. In: *Les Œvvres d'Ambroise Paré*. Paris: Nicolas Bvon (1628), pp. 1004–1081.

Paré A. *Des monstres et prodiges*. In: *Œuvres complètes*. JF Malgaigne (ed.). In three volumes. Volume 3 (1841). Paris: Baillière (1840–1841), pp. 1–68.

Paré A. *Of monsters and prodigies*. In: *The workes of that famous chirurgion Ambrose Parey*. T Johnson (trans.). London: Cotes & Young (1634), pp. 961–1026.

Paré A. *On monsters and marvels*. JL Pallister (trans.). University of Chicago Press (1982).

Park SC, Lee K, Kim YO, *et al*. Large-scale genomics reveals the genetic characteristics of seven species and importance of phylogenetic distance for estimating pan-genome size. *Frontiers in Microbiology* 10:834 (2019).

Parker R. Early Orphism. In: *The Greek world* (A Powell, ed.). London: Routledge (1995), pp. 483–510.

Parkinson J. *Organic remains of a former world. An examination of the mineralized remains of the vegetables and animals of the antediluvian world; generally termed extraneous fossils.* Volume 1: *The vegetable kingdom*. Volume 2: *The fossil zoophytes*. Volume 3: *The fossil starfish, echini, shells, insects, amphibia, mammalia, &c.* London: Sherwood, Neely & Jones, *et al*. (1808–1811).

Parkinson J. *Organic remains of a former world. An examination of the mineralized remains of the vegetables and animals of the antediluvian world; generally termed extraneous fossils.* Volume 1: *The vegetable kingdom*. Volume 2: *The fossil zoophytes*. Volume 3: *The fossil starfish, echini, shells, insects, amphibia, mammalia, &c.* London: Nattali (1833).

Parkinson J. *Paradisi in sole paradisus terrestris. Or a garden of all sorts of pleasant flowers which our English ayre will permitt to be noursed up.* London: Lownes and Young (1629). Facsimile edition by Dover, New York (1976).

Parkinson J. *Theatrum botanicvm: the theater of plants. Or, an herball of a large extent.* London: Tho. Cores (1640).

Parsons J. A letter from James Parsons . . . concerning the formation of corals, corallines, &c. *Philosophical Transactions* 47: 505–513 (1752).

Partington JR. Albertus Magnus on alchemy. *Ambix* 1(1):3–20 (1937).

Pascher A. Über Flagellaten und Algen. *Berichte der deutschen Botanischen Gesellschaft* 32:136–160 (1914).

Pasnau R. The Latin Aristotle. In: C Shields (ed.), *The Oxford handbook of Aristotle*. Oxford UP (2012), pp. 665–689.

Pasteur L. Mémoire sur la fermentation alcoolique. *Comptes Rendus Séances de l'Academie des Sciences* 45:1032–1036 (1857).

Pasteur L. Mémoire sur la fermentation appelée lactique. *Comptes Rendus Séances de l'Academie des Sciences* 45:913–916 (1857).

Patai R. *The Jewish alchemists. A history and source book*. Princeton UP (1994).

Patrick MM. *Sextus Empiricus and Greek Skepticism*. Thesis, University of Bern. Accompanied by a translation from the Greek of the first book of the "Pyrrhonic sketches" by Sextus Empiricus. London: Deighton Bell (1899).

Pearle P, Collett B, Bart K, *et al*. What Brown saw and you can too. *American Journal of Physics* 78(12):1278–1289 (2010).

Pechman KJ & Woese CR. Characterization of the primary structural homology between the 16S ribosomal RNAs of *Escherichia coli* and *Bacillus magaterium* by oligomer cataloging. *Journal of Molecular Evolution* 1(3):230–240 (1972).

Peck AL. Introduction [to *Parts of animals*]. In: Aristotle. *Parts of animals. Movement of animals. Progression of animals.* Aristotle 12. Loeb Classical Library 323. Cambridge MA: Harvard UP (1937), pp. 8–50.

Peirson BRE. Wilhelm Johannsen's genotype-phenotype distinction. *Embryo Project Encyclopedia* (2012-12-07), http://embryo.asu.edu/handle/10776/4206, accessed 27 February 2021.

Peirson R. On the analogy between plants and animals. In: A Hunter (ed.), *Georgical essays*. York: Ward (1777), pp. 318–322.

Pellat C, Sourdel-Thomine J & Boratav PN. Ḥayawān "the animal kingdom". *The Encyclopaedia of Islam*, New [Second] edition 3:304–315. Leiden: Brill (1986).

Pellegrin P. *Aristotle's classification of animals. Biology and the conceptual unity of the Aristotelian corpus.* A Preus (trans.). Berkeley: University of California Press (1986).

738   REFERENCES

Pelletier [J] & Caventou [J]. Sur la matière verte des feuilles. *Journal de Pharmacie et des Sciences Accessoires* 3:486–491 (1817).

Pena P & L'Obel M de. *Nova stirpivm adversaria, perfacilis vestigatio, luculentaqve accessio ad priscorum, præsertim Dioscoridis, et recentiorum materiam medicam. Additis Gvillielmi Rondelletii aliquot Remediorum formulis.* Antverpiæ: Christophorum Plantinum (1576).

Pena P & L'Obel M de. *Stirpivm adversaria nova, perfacilis vestigatio, luculentaque accessio ad priscorum, præsertim Dioscoridis, et recentiorum materiam medicam.* Londini: Purfoot (1570–1571).

Persius [Persius Flaccus, Aulus]. *Persio. Tradotto in verso sciolto e dichiarato da Francesco Stellvti.* Roma: Giacomo Mascardi (1630).

Person D. *Varieties: or, a svrveigh of rare and excellent matters, necessary and delectable for all sorts of persons.* London: Richard Badger for Thomas Alchorn (1635).

Persoon CH. Was sind eigentlich die Schwäme? *(Voigt's) Magazin für das Neueste aus der Physik und Naturgeschichte* 8(4):76–85 (1793).

Perty M. *Allgemeine Naturgeschichte, als philosophische und Humanitätswissenschaft für Naturforscher, Philosophen und das hoher gebildete Publikum.* In four volumes. Bern: Fischer (1837–1846).

Perty M. *Ueber das Seelenleben der Thiere. Thatsachen und Betrachtungen.* Leipzig & Heidelberg: Winter (1865). A second edition appeared in 1876.

Perty M. *Zur Kenntniss kleinster Lebensformen nach Bau, Funktionen, Systematik, mit Specialverzeichniss der in der Schweiz beobachteten.* Bern: Jent & Reinert (1852).

Peter Damian. *Opusculum 36. De divina omnipotentia in reparatione corruptæ, et factis infectis reddendis.* In: JP Migne (ed.), *Patrologiæ cursus completus, Series Latina* 145. Parisiis: Migne (1867), cols 595–622.

Peter Lombard. *Opera omnia.* In: JP Migne (ed.), *Patrologiæ cursus completus . . . Series secunda* 101–102. Paris: Migne (1854–1855).

Peter Lombard. *Sententiae in IV libris distinctae,* Editio tertia. Spicilegium Bonaventurianum, volumes 4 and 5. Roma: Collegii S. Bonaventurae (1971 and 1981).

Peter Lombard. *The sentences.* G Silano (trans.). In four volumes. Toronto: Pontifical Institute of Mediaeval Studies, Toronto (2007–2010).

Peters FE. *Aristotle and the Arabs. The Aristotelian tradition in Islam.* New York: New York UP (1968).

Petrus Bonus of Ferrara. *Pretiosa margarita novella (ca* 1330). In the edition of Janus Lacinius. Venetiis: Aldus (1546).

Petrus Bonus of Ferrara. *The new pearl of great price.* AE Waite (trans.). London: Elliott (1894).

Petrus de Crescentijs. *Epistola in libru[m] commodoru[m] ruralium.* Augstensem: Johann Schüszler (1471).

Peyssonel [JA]. An account of a manuscript treatise, presented to the Royal Society, intituled, Traité de corail . . . by the Sieur de Peyssonnel, MD. . . . Extracted and translated from the French by Mr. William Watson, FRS. *Philosophical Transactions* 47:445–469 (1752).

Pflugfelder O. Harms, Jürgen. *Neue Deutsche Biographie* 7:685. Berlin: Duncker & Humblot (1966).

Philo. *Philo with an English translation.* FH Colson & GH Whitaker (trans.). Loeb Classical Library 226–227, 247, 261, 275, 289, 320, 341, 363 and 379. Supplements: 380 and 401 (1953). R Marcus (trans.). Cambridge MA: Harvard UP (1929–1962).

Philo. *Philonis Alexandrini De animalibus: the Armenian text with an introduction, translation, and commentary by A Terian.* Studies in Hellenistic Judaism. Supplements to *Studia Philonica.* E Hilgert & BL Mack (eds), volume 1. Chico CA: Scholars Press (1981).

Philo. *Philonis Alexandrini Opera qvae svpersvnt.* L Cohn & P Wendland (eds). Berolini: Reimer (1896–1906).

Philo. *Philonis Judaei opera omnia.* CE Richter (ed.). In eight volumes. Lipsiae: EB Schwickerti (1828–1830).

Philo [attrib.]. *Philonis Iudaei de mundo liber unus.* In: *Aristotelis operum volumen secundum.* Venetiis: Aldus (1497).

REFERENCES 739

Philo. *Quaestiones et solutiones in Genesim. English translation as Questions and answers on Genesis. Translated from the ancient Armenian version of the original Greek by R Marcus.* Loeb Classical Library 380, Philo Supplement 1. Cambridge MA: Harvard UP (1953).

Philo. *Quod Deus immutabilis sit. English translation as On the unchangeableness of God.* FH Colson & GH Whittaker (trans.). Loeb Classical Library 247. Cambridge MA: Harvard UP (1968).

Philo. *The contemplative life, The giants, and selections.* D Winston (trans.). New York: Paulist (1981).

Philo. *The works of Philo Judæus, the contemporary of Josephus.* CD Yonge (trans.). In four volumes. London: Henry G. Bohn (1854–1855).

Philoponus J. *Ioannes Philoponus, De aeternitate mundi contra Proclum* (H Rabe, ed.). Lipsiae: Teubneri (1899).

Philoponus J. *Ioannis Philoponi In Aristotelis Analytica priora commentaria.* In: M Wallies (ed.), *Commentaria in Aristotelem Graeca.* Academiae Litterarum Regiae Borussicae 13(2). Berolini: Reimer (1905).

Philoponus J. *Ioannis Philoponi In Aristotelis De anima libros commentaria.* In: M Hayduck (ed.), *Commentaria in Aristotelem Graeca.* Academiae Litterarum Regiae Borussicae 15. Berolini: Reimer (1897).

Philoponus J. *Jean Philopon. Commentaire sur le De anima d'Aristote. Traduction de Guillaume de Moerbeke. Édition critique.* G Verbeke (ed.). Louvain: Publications Universitaires (1966).

Philoponus J. *Philoponus Against Aristotle, On the eternity of the world.* C Wildberg (trans.). Ithaca NY: Cornell UP (1987).

Philoponus J. *Philoponus: On Aristotle On the soul 1.3–5.* PJ van der Eijk (trans.). London: Bloomsbury (2006).

Philoponus J. *Philoponus: On Aristotle On the soul 3.9–13.* With Stephanus: *On Aristotle On interpretation.* W Charlton (trans.). London: Bloomsbury (2013).

Photius. *Myriobiblion sive Bibliotheca* (codd. 1–249). In: Photii, Constantinopolitani Patriarchæ, *Opera omnia.* Tomus 3. In: *Patrologiæ cursus completus . . . Series Græca* (ed. JP Migne) 103. Lutetiæ Parisiorum: Migne (1860).

[Physiologus]. Dicta Chrysostomi. In: F Wilhelm (ed.), *Denkmäler deutscher Prosa des 11. und 12. Jahrhunderts.* Münchener Texte 8A (Text), pp. 4–28; and 8B (Kommentar), pp. 13–52. München: Callwey (1914–1916). Reprinted by Hüber, München (1960).

Pickstone JV. Globules and coagula: concepts of tissue formation in the early Nineteenth century. *Journal of the History of Medicine and Allied Sciences* 28(4):336–356 (1973).

Picco V [Picus]. *Melethemata inauguralia.* Augustæ Taurinorum: Briolus (1788).

Pico della Mirandola Io. *De hominis dignitate.* In: *Opera omnia, Ioannis Pici, Mirandvlæ . . . Tomo II.* Basileæ: Henricpetrina (1572), pp. 313–331.

Pico della Mirandola G. *On the dignity of man.* CG Wallis (trans). *On being and the one.* PJW Miller (trans.). *Heptaplus.* D Carmichael (trans.). Library of Liberal Arts. Indianapolis: Bobbs-Merrill (1965).

Pico della Mirandola G. *Opera omnia Ioannis Pici, Mirandvlæ.* Basileæ: Heinricvm Petri (1557).

Pico GF. *Opera omnia, Ioannis Francisci Pici, Mirandvlæ . . . Item, tomo II.* Basileæ: Henricpetrina (1573).

Pimentel J. *Baroque nature. Juan E. Nieremberg, American wonders, and preterimperial natural history.* In: D Bleichmar *et al.* (eds), *Science in the Spanish and Portuguese empires 1500–1800.* Stanford CA: Stanford UP (2009), pp. 93–111.

Pindar. *Pythian odes.* In: *The odes of Pindar including the principal fragments.* J Sandys (trans.). Loeb Classical Library 485. London: Heinemann (1915), pp. 151–311.

Pinon L. *Livres de zoologie de la Renaissance*[:] *une anthologie (1450–1700).* Paris: Klincksieck (1995).

Pinto-Correia C. *The ovary of Eve. Egg and sperm and preformation.* University of Chicago Press (1997).

Pistis Sophia. GRS Mead (trans.). London: JM Watkins (1921).

Plato. Epistle VII. In: *Plato in twelve volumes,* volume 9. RG Bury (trans.). Loeb Classical Library 234. Cambridge MA: Harvard UP (1929), pp. 463–565. Reprinted 1981.

740 REFERENCES

Plato. *Timaeus*. In: *Plato in twelve volumes*, volume 9. RG Bury (trans.). Loeb Classical Library 234. Cambridge MA: Harvard UP (1929), pp. 17–105. Reprinted 1981.

Plattard J. *The life of François Rabelais*. London: Routledge (1930). Reprinted by Cass, London (1968).

Plessner M. The place of the *Turba philosophorum* in the development of alchemy. *Isis* 45(4):331–338 (1954).

Pliny the Elder [Gaius Plinius Secundus]. *Natural history*. In ten volumes. Trans. H Rackham (volumes 1–5 and 9), WHS Jones (volumes 6–7), and DE Eichholz (volume 9). Loeb Classical Library 330, 352–353, 370–371, 392–394, and 418–419. Cambridge MA: Harvard UP (1938–1963).

Pliny the Elder. *Plinivs secundus nouocomensis . . . Libros Natvralis Historiæ*. Venetis: Spira (1469).

Pliny the Elder. *The Natural History of Pliny*. J Bostock & HT Riley (trans.). In six volumes. London: HG Bohn (1855–1857).

Plotinus. *The Enneads*. S Mackenna (trans.). Abridged by J Dillon. London: Penguin (1991).

Plotinus. The six Enneads. S Mackenna & BS Page (trans.). http://classics.mit.edu/Plotinus/enne ads.html, accessed 2 July 2018 and 9 September 2020.

Pluche A. *Le spectacle de la nature, ou entretiens sur les particularités de l'histoire naturelle qui ont paru les plus propres à rendre les jeunes-gens curieux, & à leur former l'esprit*. Eight volumes in nine. Paris: Estienne (1732–1751).

Plutarch. *De Alexandri Magni fortuna aut virtute*. J Philips (trans.), WW Goodwin (rev.), *Plutarch's Morals*, volume 1. Boston: Little, Brown (1874), pp. 475–516.

Plutarch. *De sollertia animalium. The cleverness of animals*. W Helmbold (trans.). Loeb Classical Library 406. Cambridge MA: Harvard UP (1957), pp. 309–486.

Plutarch. *Moralia*. FC Babbitt *et al.* (eds). Loeb Classical Library 197, 222, 245, 305–306, 321, 337, 405–406, 424–429, 470, and 499. Cambridge MA: Harvard UP (1927–2004).

Plutarch. *Plutarch Chæronensis Scripta moralia. Græce et Latine*. F Dübner (ed.). In two volumes. Parisiis: Firmin Didot (1866 and 1890).

Pohlenz M. *Die Stoa. Geschichte einer geistigen Bewegung*. In two volumes. Göttingen: Vandenhoeck & Ruprecht (1948).

Poiret JLM & Turpin PJF *Leçons de flore. Cours complet de botanique*. In series: Flore médicale (FP Chaumeton, ed.). Paris: Panckoucke (1819–1820).

Politella J. Al-Ghazālī and Meister Eckhart. Two giants of the spirit. *The Muslim World* 64(3):180–194 and 64(4):233–244 (1964).

Poll H. Zellen und Gewebe des Tierkörpers. In: *Die Kultur der Gegenwart* (ed. P Hinneberg, ed.). Teil 3, Abteilung 4. *Organische Naturwissenschaften*. Band 2(2). *Zoologischer Teil* (O Hertwig, ed.). Leipzig & Berlin: Teubner (1913), pp. 39–93.

Pollard AF. Wotton, Edward (1492–1555). In: S Lee (ed.), *Dictionary of National Biography* 63:48–49. London: Smith, Elder (1900).

Pollard AW (ed.). *The travels of Sir John Mandeville*. London: Macmillan (1915).

Pollux J. *Iulii Pollucis Onomasticon, hoc est instructissimum rerum et synonymorum dictionarium, nunc primum latinitate donatum*. R Gualtero (trans.). Basileæ: Robertvm Vvinter (1542).

Polydorus C (ed.). *De alchemia*. Norimbergiæ: Johannes Petreium (1541).

Pomet [P]. *A compleat history of druggs, written in French by Monsieur Pomet . . . to which is added, what is further observable on the same subject, from Mess. Lemery and Tournefort, divided into three classes, vegetable, animal and mineral*. Second edition. London: Bonwicke (1725).

Pomet [P]. *A compleat history of druggs*. Third edition. London: Bonwicke (1737).

Pomet [P]. *A complete history of drugs*. Fourth edition. London: Bonwicke (1748).

Pomet P. *Histoire generale des drogues, traitant des plantes, des animaux, & des mineraux*. Paris: Lotson & Pillon (1694).

Pomponius Mela. *Pomponii Melæ De sitv orbis libri tres, accvratissime emendati*. Lvtetiæ: Vascosanum (1551).

Poncé C. *Kabbalah. An introduction and illumination for the world today*. San Francisco: Straight Arrow (1973).

REFERENCES 741

Poortman ELJ. *Petrus de Alvernia, Sententia super librum "De vegetabilibus et plantis".* Leiden: Brill (2003).

Porphyry. *Life of Pythagoras.* In: DR Fideler (ed.), *The Pythagorean sourcebook and library. An anthology of ancient writings which relate to Pythagoras and Pythagorean philosophy.* KS Guthrie (comp. and trans.). Grand Rapids MI: Phanes (1987), pp. 123–135.

Porphyry. *On Aristotle's Categories.* SK Strange (trans.). Ithaca NY: Cornell UP (1992).

Porphyry. *Select works of Porphyry; containing his four books On abstinence from animal food; his treatise On the Homeric cave of the nymphs; and his Auxiliaries to the perception of intelligible natures.* T Taylor (trans.). London: Thomas Rodd (1823).

Porphyry. *The Epistle of Porphyry to the Egyptian Anebo.* In: *Iamblichus on the mysteries of the Egyptians, Chaldeans, and Assyrians.* T Taylor (trans.) Second edition. London: Dobell, and Reeves & Turner (1895), pp. 1–16. Online at http://www.tertullian.org/fathers/porphyry_an ebo_02_text.htm, accessed 3 July 2018 and 9 September 2020.

Porta JB. *Magiæ natvralis libri viginti, in qvibvs scientiarum naturalium diuitiæ, & deliciæ demonstrantur.* Francofvrti: Wecheli, Marnium & Aubrium (1597).

Porta JB. *Magiæ natvralis, sive de miracvlis rervm natvralivm libri IIII.* Antverpiæ: Christophorus Plantinus (1560).

Porta JB. *Natural magick by John Baptista Porta, a Neapolitane: in twenty books . . . wherein are set forth all the riches and delights of the natural sciences.* London: Young & Speed (1658).

Porta JB. *Phytognomonica . . . Octo libris contenta.* Neapoli: Horatium Saluianum (1588).

Porta JB. *Villæ . . . Libri XII.* Francofvrti: Wecheli, Marnium, & Aubrium (1592).

Porter JR. Louis Pasteur. Achievements and disappointments, 1861. *Bacteriological Reviews* 25(4):389–403 (1961).

Portier P. *Les symbiotes.* Paris: Masson (1918).

Potthast A. *Regesta Pontificum Romanorum inde ab a. post Christum natum MCXCVIII ad a. MCCCIV.* In two volumes. Berolini: de Decker (1874). Reprinted by Akademische Druck-u. Verlagsanstalt, Graz (1957).

Pouchet FA. *Hétérogenie, ou traité de la génération spontanée, basé sur de nouvelles expériences.* Paris: Baillière (1859).

Power H. *Experimental philosophy, in three books: containing new experiments microscopical, mercurial, magnetical.* London: Martin and Allestry (1664).

Prangishvili D, Bamford DH, Forrterre P, *et al.* The enigmatic archaeal virosphere. *Nature Reviews Microbiology* 15(12):724–739 (2017).

Prescott GW. History of phycology. In: GM Smith (ed.), *Manual of phycology.* New York: Ronald (1951), pp. 1–11.

Preus A. (trans.). Aristotle. *On the movement and progression of animals.* Aristotle, *De motu animalium* and *De incessu animalium.* With Michael of Ephesus, *Commentaria in De motu et De incessu animalium.* Studien und Materialien zur Geschichte der Philosophie 22. Hildesheim and New York: Georg Olms (1981).

Prévost B. *Mémoire sur la cause immédiate de la carie ou charbon des blés, et de plusieurs maladies des plantes, et sur les préservatifs de la carie.* Paris: Bernard (1807).

Price DJ. Giambattista della Porta and his Natural magick. In: [Facsimile edition of:] JB Porta, *Natural magick.* London: Thomas Young and Samuel Speed (1658). New York: Basic (1957), pp. v–ix.

Priestley J. *Experiments and observations relating to various branches of natural philosophy; with a continuation of the observations on air.* [Volume 1], London: Johnson (1779); Volume 2, Birmingham: Pearson & Rollason, for Johnson (1781); and Volume 3, Birmingham: Pearson & Rollason, sold by Johnson (1786).

Priestley J. *Letters to a philosophical unbeliever.* Second edition. In two parts. Birmingham: Pearson & Rollason (1787).

Priestley J. Observations and experiments relating to equivocal, or spontaneous, generation. *Transactions of the American Philosophical Society* 6:119–129 (1809).

Priestley J. Observations on different kinds of air. *Philosophical Transactions* 62:147–252; with an Appendix by W Hey, pp. 253–264 (1772).

742 REFERENCES

Pringsheim EG. The relationship between bacteria and myxophyceae. *Bacteriological Reviews* 13(2):47–98 (1949).

Pringsheim N. Beiträge zur Morphologie und Systematik der Algen. *Jahrbücher für wissenschaftliche Botanik* 1:284–306 (1858).

Pringsheim N. Über die Befruchtung der Algen. *Bericht über die zur Bekanntmachung geeigneten Verhandlungen der Königl[ichen] Preuss[ischen] Akademie der Wissenschaften zu Berlin* 1855:133–164 (1855).

Pringsheim [N]. Über Paarung von Schwärmsporen, die morphologische Grundform der Zeugung im Pflanzenreiche. *Monatsberichte der Königlich Preussischen Akademie der Wissenschaften zu Berlin* 1869:721–738 (1870).

Prinke RT. New light on the alchemical writings of Michael Sendivogius (1566–1636). *Ambix* 63(3):217–243 (2016).

Pritchard A. *A history of infusoria, living and fossil.* [First edition]. London: Whittaker (1842).

Pritchard A. *A history of infusorial animalcules, living and fossil.* New [Third] edition. London: Whittaker (1852).

Pritchard A. *A history of infusoria, including the Desmidiaceæ and Diatomaceæ, British and foreign.* Fourth edition. London: Whittaker (1861).

Pritchard JB. *The ancient Near East in pictures. Relating to the Old Testament. Second edition with supplement.* Princeton UP (1969). Second printing 1974.

Proß W. Lorenz Oken—Naturforschung zwischen Naturphilosophie und Naturwissenschaft. In: *Die deutsche literarische Romantik und die Wissenschaften* (N Saul, ed.). University of London, Institute of Germanic Studies, volume 47. München: Iudicium (1991), pp. 44–71.

Provasoli L & IJ Pintner. Bacteria induced polymorphism in an axenic laboratory strain of *Ulva lactuca* (Chlorophyceae). *Journal of Phycology* 16(2):196–201 (1980).

Psellus [Psellos] M. *De operatione daemonum cum notis Gaulmini.* JF Boissonade (ed.). Norimbergae: Campe (1838). Originally written (in Greek) mid-Eleventh century.

Psellus M. *Επιαυσεις συντομοι φυσικων ζητηματων / Sqlutiones* [sic] *breves quæstionum naturalium.* In: JP Migne (ed.), *Patrologiæ cursus completus . . . Series Græca* 122. Parisiis: Garnier (1889), columns 783–810.

[Pseudo-Albertus Magnus]. *Libellus de alchimia ascribed to Albertus Magnus. Translated from the Bornet Latin edition.* V Heines (trans.). Berkeley: University of California Press (1958). Excerpts in: E Grant (ed.), *Source book in medieval science* (1974), pp. 586–603.

Pseudo-Archytas. *Über Die Kategorien. Texte zur griechischen Aristotles-Exegese.* TA Szlezák (ed. and trans.). Berlin: de Gruyter (1972).

Pseudo-Chrysostom. [Sermon 6 for Holy Week]. In: JP Migne (ed.), *Patrologiæ cursus completus* 59. Parisiis: Migne (1862), cols 735–746.

Pseudo-Dionysius. *Dionysius the Areopagite On the divine names and the Mystical theology.* CE Rolt (trans.). London: Society for Promoting Christian Knowledge; and New York: Macmillan (1920).

Pseudo-Dionysius. *The works of Dionysius the Areopagite.* J Parker (trans.). In two volumes. London: James Parker (1897–1899).

Puig Montada J. Ibn Rushd versus al-Ghazāli: reconsideration of a polemic. *The Muslim World* 82(1–2):113–131 (1992).

Pulteney R. *A general view of the writings of Linnæus.* London: Payne & White (1781).

Pulteney R. *A general view of the writings of Linnæus.* Second edition. London: Mawman (1805).

Purchas S. *Pvrchas his pilgrimage. Or relations of the world and the religions obserued in all ages and places discouered, from the Creation vnto this present.* Fourth edition. London: William Stansby for Henrie Fetherstone (1626).

Purchas S. *Hakluytus posthumus or Purchas his pilgrimes. Contayning a history of the world in sea voyages and lande travells by Englishmen and others.* In twenty volumes. Glasgow: MacLehose; and New York: Macmillan [varies] (1905–1907).

Purkinje [Purkyně] J. [Ueber die Analogieen in den Struktur-Elementen des thierischen und pflanzlichen Organismus]. *Uebersicht der Arbeiten und Veränderungen der schlesischen Gesellschaft für vaterländische Kultur* 16:81–82 (1840 for 1839).

REFERENCES 743

Pyle CM. Conrad Gessner on the spelling of his name. *Archives of Natural History* 27(2):175–186 (2000).

Qadri MAH. *Kitāb al-Jamāhir fi ma'rifah al-jawāhir': al-Bīrūnī's contribution to biological studies and concepts*. In: HM Said (ed.), *Al-Bīrūnī commemorative volume*. Karachi: Hamdard (1979), pp. 587–593.

Quatrefages A de. *Souvenirs d'un naturaliste*. In two volumes. Paris: Charpentier (1854).

Quinn DB & Skelton RA. Introduction. In: R Hakluyt, *The principall navigations, voiages and discoveries of the English nation*. Cambridge UP (1965), pp. ix–lx.

Rabelais F. *Gargantua and Pantagruel*. MA Screech (trans.). London: Penguin (2006).

Rabelais F. *Gargantua et Pantagruel. Texte transcrit et annoté par Henri Clouzot*. In three volumes. Paris: Larousse (1913).

Rabelais F. *Le tiers livre. Édition critique commentée par MA Screech*. Droz: Genève; et Paris: Minard (1964).

Rabelais F. *The works of Francis Rabelais. A new edition*. T Urquhart (trans. Books 1–3); PA Motteux (trans. Books 4–5). In two volumes. London: Bohn (1851).

Rabelais F. *Tiers liure des faictz et dictz heroïques du noble Pantagruel*. Paris: Wechel (1546).

Rádl E. *The history of biological theories*. EJ Hatfield (trans.). Oxford UP; and London: Humphrey Milford (1930).

Rádl EM. *Geschichte der biologischen Theorien seit dem Ende des siebzehnten Jahrhunderts*. In two volumes. Leipzig: Wilhelm Engelmann (1905–1909).

Raff RA & Mahler HR. The non symbiotic origin of mitochondria. *Science* 177(4049):575–582 (1972).

Ragan MA. Trees and networks before and after Darwin. *Biology Direct* 4:43 (2009).

Ragan MA & Beiko RG. Lateral genetic transfer: open issues. *Philosophical Transactions of the Royal Society B* 364(1527):2241–2251 (2009).

Ragan MA, Bernard G & Chan CX. Molecular phylogenetics before sequences. Oligonucleotide catalogs as *k*-mer spectra. *RNA Biology* 11(3):176–185 (2014).

Raikov BE. *Karl Ernst von Baer 1792–1876. Sein Leben und sein Werk*. In series: Acta Historica Leopoldina, volume 5. Lss: Barth (1968).

Raikov BE. *Russkie biologi-evolyutsionisty do Darvina. Materialy k istorii evolyutsionnoi idei v Rossii*. In three volumes. Moskva: Akademii Nauk SSSR (1951–1955).

Raju PT. The development of Indian thought. *Journal of the History of Ideas* 13(4):528–550 (1952).

Ralfs J. *The British Desmidieæ*. London: Reeve, Benham & Reeve (1848). Reprinted by Cramer, Weinheim (1962).

Ramsbottom J. Presidential address. *Proceedings of the Linnean Society of London* 151(4):280–367 (1941).

Ramulu HG, Groussin M, Talla E, *et al*. Ribosomal proteins: toward a next generation standard for prokaryotic systematics? *Molecular Phylogenetics and Evolution* 75:103–117 (2014).

Ramus P. *Ciceronianvs, et Brvtinæ qvæstiones*. Basileæ: Pernam (1577).

Ramus P. *Dialecticæ libri duo*. Parisiis: Andream Wechelum (1560).

Ramus P. *Dialecticæ libri duo*. Francofurti: Wecheli (1583).

Randall JH Jr. *The School of Padua and the emergence of modern science*. Padova: Antenore (1961).

Raoult D & Forterre P. Redefining viruses: lessons from mimivirus. *Nature Reviews Microbiology* 6(4):315–319 (2008).

Rastelli S. *China at the court of the emperors: unknown masterpieces from Han tradition to Tang elegance, 25–907*. Milan: Skira (2009).

Ratcliff J. *The quest for the invisible: microscopy in the Enlightenment*. Farnham UK: Ashgate (2009).

Raven CE. *English naturalists from Neckham to Ray: a study in the making of the modern world*. Cambridge UP (1947).

Ray J. A discourse on the specific differences of plants (1672). In: T Birch, *History of the Royal Society of London*, volume 3. London: Millar (1757), pp. 169–173.

[Ray J]. *Catalogus plantarum circa Cantabrigiam nascentium*. Cantabrigiæ: Field (1660).

744 REFERENCES

Ray J. *Historia plantarum: species hactenus editas aliasque insuper multas noviter inventas & descriptas complectens*, volumes 1 (1686), 2 (1688) and 3 (1704). London: Faithorne & Kersey [varies].

Ray J. *Methodus plantarum emendata et aucta*. Londini: Smith & Walford (1703).

Ray J. *Methodus plantarum nova, brevitatis & perspicuitatis causa synoptice in tabulis exhibita*. Londini: Faithorne & Kersey (1682).

Ray J. *Synopsis methodica animalium quadrupedum et serpentini generis* Londini: Smith & Walford (1693).

Ray J. *Synopsis methodica stirpium Britannicarum Editio tertia 1724*. [Bound with:] C Linnæus. *Flora Anglica 1754 & 1759*. London: Ray Society (1973).

Ray J. *Synopsis methodica stirpium Britannicarum*. Third edition. Londini: Innys (1724).

Ray J. *The wisdom of God manifested in the works of the Creation*. Second edition. London: Samuel Smith (1692).

Ray J. *The wisdom of God manifested in the works of the Creation*. Seventh edition. London: Harbin (1717).

Ray P. Origin and tradition of alchemy. *Indian Journal of History of Science* 2(1):1–21 (1967).

Rayer [PFO]. Inoculation du sang de rate. *Comptes Rendus des Séances et Mémoires de la Societé de Biologie* 2:141–144 (1850).

Raymann K, Brochier-Armanet C & Gribaldo S. The two-domain tree of life is linked to a new root for the Archaea. *PNAS* 112(21):6670–6675 (2015).

Raymond de Sebonde. *Theologia naturalis siue liber creatur[um]*. [Deventer]: [Ricardus Pafraet] (1480–1485).

Raymond de Sebonde. *Theologia naturalis, sive liber creaturarum* [1434]. [Lyon]: [Guillaume Balsarin] (*circa* 1488).

Réaumur RAF de. *Memoires pour server a l'histoire des insectes*. In six volumes. Paris: L'Imprimerie Royale (1734–1742).

Réaumur [RAF] de [actually Peyssonel JA de]. Observations sur la formation du corail, & des autres productions appellées plantes pierreuses. *Mémoires de Mathématique & de Physique . . . de l'Académie Royale des Sciences* 1727:269–281 (1729).

[Réaumur RAF de]. Sur le corail. *Histoire de l'Académie Royale des Sciences* 1727:37–39 (1729).

[Réaumur RAF de]. Sur les plantes de la mer. *Histoire de l'Académie Royale des Sciences* 1710:69–78 (1712).

Redding P. Georg Wilhelm Friedrich Hegel. *The Stanford Encyclopedia of Philosophy* (Spring 2020 edition), EN Zalta (ed.), https://plato.stanford.edu/archives/spr2020/entries/hegel/, accessed 23 January 2020 and 5 October 2020.

Redi F. *Esperienze intorno alla generazione degl' insetti*. Firenze: [All' insegna della stella] (1668).

Redi F. *Experimenta circa generationem insectorvm*. Amstelodami: Andreæ Frisii (1671).

Redi F. *Experiments on the generation of insects*. M Bigelow (trans.). Chicago: Open Court (1909).

Redi F. *Opvscvlorvm pars prior, sive experimenta circa generationem insectorum . . .* [et] *Opusculorum Tomus alter. Experimenta circa varias res naturales*. Amstelædami: Henricum Wetstenium (1685–1686).

Reed MF. Norwegian stave churches and their pagan antecedents. *RACAR: revue d'art canadienne / Canadian Art Review* 24(2):3–13 (1997).

Reeds K. Albert on the natural philosophy of plant life. In: JA Weisheipl (ed.), *Albertus Magnus and the sciences. Commemorative essays 1980*. Toronto: Pontifical Institute of Mediaeval Studies (1980), pp. 341–354.

Rehbock PF. Huxley, Haeckel, and the oceanographers: the case of *Bathybius haeckelii*. *Isis* 66(4):504–533 (1975).

Reich D. *Who we are and how we got here. Ancient DNA and the new science of the human past*. New York: Pantheon (2018).

Reichenbach HGL. *Handbuch des natürlichen Pflanzensystems nach allen seinen Classen, Ordnungen und Familien*. Dresden und Leipzig: Arnold (1837).

Reichert KB. *Das Entwickelungsleben im Wirbelthier-Reich*. Berlin: August Hirschwald (1840).

REFERENCES 745

Reinhardt K. *Kosmos und Sympathie. Neue Untersuchungen über Poseidonios*. München: Beck (1926).

Reinke J. *Lehrbuch der allgemeinen Botanik mit Einschluß der Pflanzenphysiologie*. Berlin: Wiegandt, Hempel & Parey (1880).

Reischauer EO. Foreword. In: M Watanabe, *The Japanese and Western science*. OT Benfey (trans.). Philadelphia: University of Pennsylvania Press (1990), pp. ix–x.

Remak [R]. Anatomische Beobachtungen über das Gehirn, das Rückenmark und die Nervenwurzeln. *Archiv für Anatomie, Physiologie und wissenschaftliche Medicin* 1841:506–522 (1841).

Remak R. *Untersuchungen über die Entwickelung der Wirbelthiere*. Berlin: Reimer (1855).

Renner O. Die pflanzlichen Plastiden als selbständige Elemente der genetischen Konstitution. *Berichte über die Verhandlungen der königlich sächsischen Akademie der Wissenschaften zu Leipzig. Mathematisch-physische Klasse* 86:241–266 (1934).

Renouard AA. *Annales de l'imprimerie des Aldes, ou l'histoire des trous Manuce et de leurs éditions*. In two volumes. Paris: Renouard (An 12, 1803).

Report of the Thirtieth Meeting of the British Association for the Advancement of Science; held at Oxford in June and July 1860. London: John Murray (1861).

Requin AP. Animal. *Encyclopédie nouvelle: Dictionnaire philosophique, scientifique, littéraire et industriel, offrant le tableau des connaissances humaines au XIXe siècle* 1:554–564 (1836).

Reuchlin J. *De arte cabalistica*. [Bound with:] Pico della Mirandola J. *Opera omnia Ioannis Pici, Mirandvlæ*. Basileæ: Henricvm Petri (1557), pp. 733–899.

Reuchlin J. *De arte cabalistica . . . Libri III* (1494). [Bound with:] Pico della Mirandola Io. *Opera omnia, Ioannis Pici, Mirandvlæ . . . Item, Tomo II*. Basileæ: Henricpetrina (1572), pp. 3001–3164 (pag. sep.).

Reuchlin J. *On the art of the Kabbalah: De arte cabalistica*. M Goodman & S Goodman (trans.). New York: Abaris (1983).

Reynolds A & Hülsmann N. Ernst Haeckel's discovery of *Magosphaera planula*: a vestige of metazoan origins? *History & Philosophy of the Life Sciences* 30(3–4):339–386 (2008).

Reynolds AS. Ernest Haeckel and the philosophy of sponges. *Theory in Biosciences* 138(1):133–146 (2019).

Reynolds DF. A thousand and one nights: a history of the text and its reception. In: R Allen & DS Richards (eds), *Arabic literature in the post-classical period. Cambridge history of Arabic literature* 6. Cambridge UP (2006), pp. 270–291.

Rheinberger H-J. From microsomes to ribosomes: "strategies" of "representation". *Journal of the History of Biology* 28(1):49–89 (1995).

Ribe NM. Science and symbol in Goethe's geology. Paper presented at annual meeting of the History of Science Society, Ann Arbor MI (1982) as cited by F Amrine, *Goethe Yearbook* 5:187–202 (1990).

Richard A. *Elements of botany and vegetable physiology*. W Macgillivray (trans.). Edinburgh: Blackwood; and London: Cadell (1831).

Richard A. Méthode. *Dictionnaire classique d'histoire naturelle* 10:493–511 (1826).

Richard A. *Nouveaux élémens de botanique et de physiologie végétale*. Third edition. Paris: Béchet (1825).

Richard A. *Nouveaux élémens de botanique et de physiologie végétale, quatrième édition, revue, corrigée et augmentée du caractère des familles naturelles du Règne végétal*. Fourth edition. Paris: Bechet (1828).

Richard A. *Nouveaux élémens de botanique et de physiologie végétale; cinquième édition, revue, corrigée et augmentée du caractère des familles naturelles du Règne végétal*. Fifth edition. Paris: Béchet (1833).

Richard A. Végétaux. *Dictionnaire classique d'histoire naturelle* 16:518–531 (1830).

Richards RJ. Kant and Blumenbach on the *Bildungstrieb*: a historical misunderstanding. *Studies in the History and Philosophy of Science C* 31(1):11–32 (2000).

Richards RJ. The foundations of archetype theory in evolutionary biology: Kant, Goethe, and Carus. *Republic of Letters* 6(1), 4 March 2018.

746 REFERENCES

Richards RJ. *The meaning of evolution. The morphological construction and ideological reconstruction of Darwin's theory*. University of Chicago Press (1992).

Richards RJ. *The romantic conception of life. Science and philosophy in the age of Goethe*. University of Chicago Press (2002).

Richards RJ. *The tragic sense of life. Ernst Haeckel and the struggle over evolutionary thought*. University of Chicago Press (2008).

Richmond ML. TH Huxley's criticism of German cell theory: an epigenetic and physiological interpretation of cell structure. *Journal of the History of Biology* 33(2):247–289 (2000).

Richter JGO. *Ichthyotheologie, oder: Vernunft- und Schriftmäßiger Versuch die Menschen aus Betrachtung der Fische zur Bewunderung, Ehrfurcht und Liebe ihres großen, liebreichen und allein weisen Schöpfers zu führen*. Leipzig: Erben (1754).

Riddle MB. Introductory notes to the Pseudo-Clementine Literature. In: A Roberts & J Donaldson (eds), *The Ante-Nicene Fathers*, volume 8. New York: Scribner (1903), pp. 69–71.

Rieppel O. Homology, topology, and typology: the history of modern debates. In: BK Hall (ed.), *Homology: the hierarchical basis of comparative biology*. San Diego: Academic (1994), pp. 63–100.

Rieppel O. The reception of Leibniz's philosophy in the writings of Charles Bonnet (1720–1793). *Journal of the History of Biology* 21(1):119–145 (1988).

Rink H. *The Eskimo tribes. Their distribution and characteristics, especially as regards language, with a comparative vocabulary and sketch-map*. London: Williams and Norgate (1887).

Rinke C, Schwientek P, Sczyrba A, et al. Insights into the phylogeny and coding potential of microbial dark matter. *Nature* 499(7459):431–437 (2013).

Rippka R & Cohen-Bazire G. The Cyanobacteriales: a legitimate order based on the type strain *Cyanobacterium stanieri? Annales de Microbiologie (Institut Pasteur)* 134B(1):21–36 (1983).

Ris H. Ultrastructure and molecular organization of genetic systems. *Canadian Journal of Genetics and Cytology* 3(2):95–120 (1961).

Ris H & Plaut W. Ultrastructure of DNA-containing areas in the chloroplast of *Chlamydomonas. Journal of Cell Biology* 13(3):383–391 (1962).

Risse W. Einführung. In: *Jacobi Zabarellae: Opera logica*. Hildesheim: Georg Olms (1966), pp. v–xii.

Ritterbush PC. *Overtures to biology. The speculations of Eighteenth-century naturalists*. New Haven: Yale UP (1964).

Rizvi SAA. Ibn Sīnā's impact on the rational and scientific movements in India. *Indian Journal of History of Science* 21(3):276–284 (1986).

Rivera MC, Jan R, Moore JE & Lake JA. Genomic evidence for two functionally distinct gene classes. *PNAS* 95(11):6239–6244 (1998).

Roberts CH & Skeat TC. *The birth of the codex*. London: British Academy and Oxford UP (1983).

Robertson JD. The molecular structure and contact relationships of cell membranes. *Progress in Biophysics and Biophysical Chemistry* 10:343–418 (1960).

Robertson JD. The ultrastructure of cell membranes and their derivatives. *Biochemical Society Symposium* 16:3–43 (1959).

Robinet JB. *Considerations philosophiques sur la gradation naturelle des formes de l'etre, ou les essais de la nature qui apprend a faire l'homme*. Paris: Charles Saillant (1768).

[Robinet JB]. *De la nature*. In four volumes. Amsterdam: van Harrevelt (1761–1766).

Robinet JB. *Vue philosophique de la gradation naturelle des formes de l'etre, ou Les essais de la nature qui apprend a faire l'homme*. Amsterdam: van Harrevelt (1768).

Robson JA. *Wyclif and the Oxford schools. The relation of the "Summa de ente" to scholastic debates at Oxford in the later Fourteenth century*. Cambridge UP (1961).

Roe SA. John Turberville Needham and the generation of living organisms. *Isis* 74(2):158–183 (1983).

Roger AJ, Muñoz-Gómez SA & Kamikawa R. The origin and diversification of mitochondria. *Current Biology* 27(21):R1177–R1192 (2017).

Roger J. *Buffon. A life in natural history*. SL Bonnefoi (trans.). Ithaca NY: Cornell UP (1997).

REFERENCES 747

Roger J. *Buffon, un philosophe au Jardin du Roi*. Paris: Fayard (1989).

Role A. *Un destin hors série: la vie aventureuse d'un savant. Bory de St Vincent 1778–1846*. Paris: La Pensée Universelle (1973).

Romero A. When whales became mammals: the scientific journey of cetaceans from fish to mammals in the history of science. In: A Romero & EO Keith (eds), *New approaches to the study of marine mammals*. Rijeka: InTech (2012), pp. 223–232.

Rondelet G. *Libri de piscibus marinis, in quibus veræ piscium effigies expressæ sunt*. [With:] *Vniuersæ aquatilium historiæ pars altera, cum veris ipsorum imaginibus*. Lugduni: Matthiam Bonhomme (1555).

Roose TGA. *Grundzüge der Lehre von der Lebenskraft*. Braunschweig: Thomas (1797).

Roose TGA. *Grundzüge der Lehre von der Lebenskraft*. Second edition. Göttingen und Braunschweig: Thomas (1800).

Rosarium philosophorum. Secunda pars alchimiæ de lapide philosophico vero modo præparando. In: *De alchimia opuscula complura veterum philosophorum*. Part 2. Francoforti: Cyriacus Iacobus (1550).

Rose H. *The elements of botany . . . being a translation of the* Philosophia botanica, *and other treatises of the celebrated Linnæus*. London: Cadell & Hingeston (1775).

Rösel van Rosenhof A. *De natuurlyke histoirie der insecten*. In eight volumes (Deel 1–4, each in two parts). Haarlem & Amsterdam: Bohn & de Wit [Deel 1], and Bohn & Gartman [Deel 2–4] (1764).

Rösel von Rosenhof A. *Der monatlich-herausgegebenen Insecten-Belustigung*, volumes 1 (1746), 2 (1749), 3 (1755), and 4 (1761). Nürnberg: Fleischmann.

Rösel von Rosenhof A. *Historia natvralis Ranarvm nostrativm . . . Die natürliche Historie der Frösche heisigen Landes*. Nürnberg: Fleichmann (1758).

Rosen E. The invention of eyeglasses. *Journal of the History of Medicine and Allied Sciences* 11(1):13–46 and 11(2):183–218 (1956).

Rosenfield LC. *From beast-machine to man-machine: animal soul in French letters from Descartes to La Mettrie*. Oxford UP (1941).

Ross D. *Aristotle*. With an introduction by JL Ackrill. Sixth edition. London and New York: Routledge (1995).

Ross DL. Gregory of Nyssa. *Internet Encyclopedia of Philosophy*, undated. https://www.iep.utm.edu/gregoryn/, accessed 8 March 2020.

Rösslin E. *Kreutterbůch von allem Erdtgewächs. Anfenglich von Doctor Johan Cuba zusamen bracht . . . Mit warer Abconterfeitung aller Kreuter*. Franckfurt: Christian Egenolph (1533).

Rösslin E. *Kraüterbůch von aller Kraüter, Baüm, Gestaüd, und Frücht*. Franckfúrt (1546).

Roth AW. *Catalecta botanica*. In three volumes. Lipsiae: Müller (volume 1, 1797) and Gleditsch (volumes 2–3, 1800–1806).

Roth AW. *Neue Beyträge zur Botanik. Erster Theil* [only part published?]. Frankfurth, Wilmans (1802).

Rothmaler W. Die Abteilungen und Klassen der Pflanzen. *Feddes Repertorium* 54(2–3):256–266 (1951).

Rothmaler W. Über das natürliche System der Organismen. *Biologisches Zentralblatt* 67:242–250 (1948).

Rothschild LJ. Protozoa, Protista, Protoctista: what's in a name? *Journal of the History of Biology* 22(2):277–305 (1989).

Rouli L, Merhaj V, Fournier PE & Raoult D. The bacterial pangenome as a new tool for analysing pathogenic bacteria. *New Microbes and New Infections* 7:72–85 (2015).

Round FE. [Review of] T Christensen, *Botanik 2. British Phycological Bulletin* 2(2):269–270 (1963).

Round FE. Taxonomy of the Chlorophyta. *British Phycological Bulletin* 2(4):224–235 (1963).

Rousseau G. *The notorious John Hill: the man destroyed by ambition in the era of celebrity*. Bethlehem PA: Lehigh UP (2012).

Rousseau JJ. *Discours sur les sciences et les arts. Edition critique*. GR Havens (ed.). New York: Modern Language Association of America; and London: Oxford UP (1946).

748 REFERENCES

Rousseau JJ. *Letters on the elements of botany . . . Translated into English, with notes, and twenty-four additional letters, fully explaining the system of Linnæus.* T Martyn (trans.). London: White (1785).

Rousseau JJ. *Lettres élémentaires sur la botanique.* In: *Œuvres complètes de JJ Rousseau.* Tome 6. Paris: Hachette (1909), pp. 30–65.

Roux W. Das Wesen des Lebens. In: *Die Kultur der Gegenwart* (P Hinneberg, ed.). Teil 3, Abteilung 4. *Organische Naturwissenschaften.* Band 1. *Allgemeine Biologie* (C Chun, W Johannsen & A Günthart, eds). Leipzig & Berlin: Teubner (1915), pp. 173–187.

Rozier F. *Sammlung brauchbarer Abhandlungen aus des Herrn Abt Rozier Beobachtungen über die Natur und Kunst,* volume 2. Leipzig: Weidemanns Erben & Reich (1776).

Ruderman DB. *Kabbalah, magic, and science.* Cambridge MA: Harvard UP (1988).

Rudolph RC. The jumar in China. *Isis* 40:35–37 (1949).

Rudolphi CA. *Entozoorum synopsis cui accedunt mantissa duplex et indices locupletissimi.* Berolini: Augustus Rücker (1819).

Rudolphi CA. *Entozoorum, sive vermium intestinalium historia naturalis.* In two volumes, bound in three. Amstelaedami: Tabernae Librariae et Artium (1808–1810).

Rudolphi KA. *Anatomie der Pflanzen.* Berlin: Mylius (1807).

Rudolphi KA. *Beyträge zur Anthropologie und allgemeinen Naturgeschichte.* Berlin: Haude & Speuner (1812).

Rudzinska MA. Do suctoria really feed by suction? *BioScience* 23(2):87–94 (1973).

Ruel J. *De natura stirpium libri tres.* Basileæ: Frobeniana (1537).

Ruel J. *De natura stirpium libri tres.* Parisiis: Simonis Colinæis (1536).

Ruestow EG. *The microscope in the Dutch Republic. The shaping of discovery.* Cambridge UP (1996).

Rufinus. *The Herbal of Rufinus edited from the unique manuscript* (L Thorndyke, ed.). Corpus of Mediaeval scientific texts 1. University of Chicago Press (1946).

Ruggiero MA, Gordon DP, Orrell TM, *et al.* A higher level classification of all living organisms. *PLoS ONE* 10(4):e0119248 (2015).

Rüling JP. *Ordines natvrales plantarvm commentatio botanica.* Goettingae: Vandenhoeck (1774).

Rūmī, Jalāl ad-Dīn Muḥammad. Az jamadi murdam (I died a mineral). In: AJ Arberry (trans.), *Classical Persian literature.* London: Allen & Unwin (1958), page 241.

Rumph GE [Rumphius]. *D'Amboinsche rariteitkamer.* T'Amsterdam: François Halma (1705).

Rupke NA. *Richard Owen. Victorian naturalist.* New Haven: Yale UP (1994).

Rupke NA. Richard Owen's vertebrate archetype. *Isis* 84(2):231–251 (1993).

Ruppel S. *Botanophilie. Mensch und Pflanze in der auflkärerisch-bürgerlichen Gesellschaft um 1800.* Wien: Böhlau (2019).

Ruse M. *Monad to man. The concept of progress in evolutionary biology.* Cambridge MA: Harvard UP (1996).

Ruse M. *The Darwinian revolution. Science red in tooth and claw.* University of Chicago Press (1979).

Ruse M. *The Gaia hypothesis. Science on a pagan planet.* University of Chicago Press (2013).

Ruska J. *Turba philosophorum: ein Beitrag zur Geschichte der Alchemie.* Berlin: Julius Springer (1931).

Russell ES. *Form and function. A contribution to the history of animal morphology.* London: John Murray (1916).

Russell JR. Christianity I. In pre-Islamic Persia: literary sources. *Encyclopædia Iranica* 5(5): 327–328 (2012). https://iranicaonline.org/articles/christianity-i, accessed 20 August 2018.

Russett CE. *Darwin in America. The intellectual response 1865–1912.* San Francisco: Freeman (1976).

Sachs J von. *Geschichte der Botanik vom 16. Jahrhundert bis 1860.* München: Oldenbourg (1875).

Sachs J von. *History of botany (1530–1860).* HEF Garnsey (trans.), IB Balfour (rev.). New York: Russell & Russell (1890), reprinted 1967.

Sachs J. *Lehrbuch der Botanik nach dem gegenwärtigen Stand der Wissenschaft.* Fourth edition. Leipzig: William Engelmann (1874).

Sachs J (ed.). *Text-book of botany. Morphological and physiological.* Second edition. Oxford: Clarendon (1882).

Sagan C. *The demon-haunted world. Science as a candle in the dark.* New York: Ballantine (1996).

Sagan L. On the origin of mitosing cells. *Journal of Theoretical Biology* 14(3):225–274 (1967).

Sagan L & Scher S. Evidence for cytoplasmic DNA in *Euglena gracilis. Journal of Protozoology* 8 (supplement 4):8, Abstract 20 (1961).

Said HM (ed.). *Al-Bīrūnī commemorative volume. Proceedings of the International Congress held in Pakistan on the occasion of millenary of Abū Rāihan Muhammad ibn Ahmad al-Bīrūnī (973–ca 1051 AD) November 26, 1973 thru' December 12, 1973.* Karachi: Hamdard (1979).

Saif L. *The Arabic influences on early modern occult philosophy.* London: Palgrave Macmillan (2015).

Saif L. The Arabic theory of astral influences in early modern medicine. *Renaissance Studies* 25(5):609–626 (2011).

Saint-Pierre JHB de. *Études de la nature.* In three volumes. Paris: Didot (1784).

Saint-Pierre JHB de [as "Un officier du Roi"]. *Voyage à l'Isle de France, à l'Isle de Bourbon, au Cap de Bonne-Espérance, &c., avec des observations nouvelles sur la nature & sur les hommes.* Volume 1, Neuchatel: Société Typographique; volume 2, Amsterdam: Merlin (1773).

Salimbene. *Salimbene de Adam Cronica.* Nuova edizione. G Scalia (ed.). In two volumes. Bari: Laterza & Figli (1966).

Salviani I. *Aquatilivm animalivm historiæ, liber primvs cvm eorvmdem formis, ære excvsis.* Romæ: Hippolyto Salviano (1554).

Samorini G. The oldest representations of hallucinogenic mushrooms in the world (Sahara Desert, 9000–7000 BP). *Integration* 2(3):69–78 (1992).

Sandars NK (trans.). *Poems of heaven and hell from ancient Mesopotamia.* London: Penguin (1971).

Sandars NK (trans.). *The epic of Gilgamesh.* Revised edition. Harmondsworth: Penguin (1960).

Sanday W. Introduction [to St Hilary of Poitiers]. In: P Schaff & H Wace (eds), *A select library of Nicene and Post-Nicene Fathers of the Christian church.* Second series, volume 9. New York: Christian Literature (1899), pp. i–xcvi.

Sanger F. Sequences, sequences, and sequences. *Annual Review of Biochemistry* 57:1–28 (1988).

Santer M. Richard Bradley: a unified, living-agent theory of the cause of infectious diseases of plants, animals, and humans in the first decades of the 18th Century. *Perspectives in Biology and Medicine* 52(4):566–578 (2009).

Sapp J. *Evolution by association. A history of symbiosis.* Oxford UP (1994).

Sapp J. The prokaryote-eukaryote dichotomy: meanings and mythology. *Microbiology and Molecular Biology Reviews* 69(2):292–305 (2005).

Sapp S. *The new foundations of evolution: on the tree of life.* Oxford UP (2009).

Sapp J, Carrapiço F & Zolotonosov M. Symbiogenesis: the hidden face of Constantin Merezhkowsky. *History and Philosophy of the Life Sciences* 24(3–4):413–440 (2002).

Sarai C, Tanifuji G, Nakayama T, *et al.* Dinoflagellates with relic endosymbiont nuclei as models for elucidating organellogenesis. *PNAS* 117(10):5364–5375 (2019).

Sarioğlu H. Abharī. In: T Hockey *et al.* (eds), *The biographical encyclopedia of astronomers.* New York: Springer (2007), pp. 7–8.

Sark JJ. The birth of Minegoo. In: Lennox Island Band Council (ed.), *Micmac legends of Prince Edward Island.* Charlottetown & Lennox Island PEI: Lennox Island Band Council & Ragweed Press (1988).

Sarrut G & Saint-Edme B. Bory de Saint-Vincent. In: *Biographie des hommes du jour.* Tome 2(1). Paris: Krabbe (1836), pp. 279–285.

Sars M. *Fauna littoralis Norvegiae oder Beschreibung und Abbildungen neuer oder wenig bekannten Seetheire.* Heft 1. Christiana: Johann Dahl (1846).

Sars M, Koren J & Danielssen DC. *Fauna littoralis Norvegiæ.* Heft 2. Bergen: Beyer (1856).

Sarton G. *Ancient science through the Golden Age of Greece.* New York: Norton (1970), republished by Dover (1993). First published as *A history of science, volume 1: Ancient science through the Golden Age of Greece.* Cambridge MA: Harvard UP (1952).

750 REFERENCES

Sarton G. Foreword. In: MF Ashley Montagu (author). *Edward Tyson MD, FRS 1650–1708 and the rise of human and comparative* anatomy *in England*. A study in the history of science. *Memoirs of the American Philosophical Society* 20 (1943), pp. xv–xx.

Sarton G. *Introduction to the history of science*. Three volumes bound in four. Carnegie Institution of Washington Publication 376. Baltimore: Williams & Wilkins (1927–1948).

Savage JJ. Introduction. In: *Saint Ambrose. Hexameron, Paradise, and Cain and Abel*. JJ Savage (trans.). Fathers of the Church. A new translation. New York: Fathers of the Church (1961), pp. v–xi.

Saville Kent W. *A manual of the infusoria: including a description of all known flagellate, ciliate, and tentaculiferous protozoa, British and foreign, and an account of the organization and affinities of the sponges*. In three volumes. London: Bogue (1880–1882).

Saw JH, Spang A, Zaremba-Niedzwiedzka K, *et al*. Exploring microbial dark matter to resolve the deep archaeal ancestry of eukaryotes. *Philosophical Transactions of Royal Society B* 370(1678):20140328 (2015).

Sayers J. *Innocent III: leader of Europe 1198–1216*. London: Longman (1994).

Scaliger JC. *Anamadversiones in historias Theophrasti*. Lvgdvni: Iuntæ (1584).

Scaliger JC. *Exotericarvm exercitationvm liber qvintvs decimvs*. Lvtetiæ: Michaelis Vascosani (1557).

Scanlan JJ. Introduction. In: *Albert the Great. Man and the beasts. De animalibus (Books 22–26)*. JJ Scanlan (trans). Binghamton NY: Medieval & Renaissance Texts & Studies (1987), pp. 1–56.

Scarborough J. Adaptation of folk medicines in the formal "materia medica" of classical antiquity. *Pharmacy in History* 55(2/3):55–63 (2013).

Scarborough J. Hermetic and related texts in classical antiquity. In: *Hermeticism and the Renaissance. Intellectual history and the occult in early modern Europe*. I Merkel & AG Debus (eds). Washington: Folger (1988), pp. 19–44.

Scarborough J. Theophrastus on herbals and herbal remedies. *Journal of the History of Biology* 11(2):353–385 (1978).

Schaeffer AA. Taxonomy of the amebas with descriptions of thirty-nine new marine and freshwater species. Papers from the Department of Marine Biology of the Carnegie Institution of Washington 24. *Carnegie Institution of Washington Publication* 345:1–116 (1926).

Schaudinn F. Beiträge zur Kenntnis der Bakterien und verwandter Organismen. I. Bacillus bütschlii n. sp. *Archiv für Protistenkunde* 1(2):306–342 (1902).

Scheffauer H. Haeckel, a Colossus of science. *The North American Review* 192(657):217–226 (1910).

Schelling FWJ. *Darlegung des wahren Verhältnisses der Naturphilosophie zu der verbesserten Fichte'schen Lehre*. Tübingen: Cotta (1806).

Schelling FWJ. Darstellung meines Systems der Philosophie. *Zeitschrift für Spekulative Physik* 2(2): i–xiv and 1–127 (1801).

Schelling FWJ. *Einleitung zu seinem Entwurf eines Systems der Naturphilosophie*. Jena & Leipzig: Gabler (1799).

Schelling FWJ. *Erster Entwurf eines Systems der Naturphilosophie. Zum Behuf seiner Vorlesungen*. Jena & Leipzig: Gabler (1799).

Schelling FWJ. Fernere Darstellungen aus dem Systems der Philosophie. *Neue Zeitschrift für Spekulative Physik* 1(1):1–77 and 1(2):1–180 (1802).

Schelling FWJ. *First outline of a system of the philosophy of nature*. KR Peterson (trans.). SUNY series in Contemporary Continental Philosophy (DJ Schmidt, ed.). Albany: State University of New York Press (2004).

Schelling FWJ. Further presentations from the system of philosophy (1802). MG Vater (trans.). *The Philosophical Forum* 32(4):373–397 (2001).

Schelling FWJ. Presentation of my system of philosophy (1801). MG Vater (trans.). *The Philosophical Forum* 32(4):339–371 (2001).

REFERENCES 751

Schelling FWJ. *Presentation of my system of philosophy*. In: MG Vater & DW Wood (ed. and trans.), *The philosophical rupture between Fichte and Schilling: selected texts and correspondence (1800-1802)*. Albany: SUNY Press (2012), pp. 141-225.

Schelling FWJ. *Schellings Werke*. In three volumes. Leipzig: Fritz Eckhardt (1907).

Schelling FWJ. *Über den Ursprung des allgemeinen Organismus*. In: *Schellings Werke, Band 1. Schriften zur Naturphilosophie*. Leipzig: Fritz Eckardt (1907), pp. 587-679.

Schelling FWJ. Von der Weltseele—eine Hypothese der höhern Physik zur Erklärung des allgemeinen Organismus (1798). J Jantzen (ed.). In series: Schelling. Historisch-kritische Ausgabe im Auftrag der Schelling-Kommission der Bayerischen Akademie der Wissenschaften. Reihe 1: *Werke*. Band 6. Stuttgart: Fromann-Holzboog (2000).

Schelling FWJ. *Von der Weltseele, eine Hypothese der höhern Physik zur Erklärung des allgemeinen Organismus*. Hamburg: Friedrich Perthes (1798).

Schenk A. *Über das Vorkommen contractiler Zellen im Pflanzenreiche*. Würzburg: Thien (1858).

Schickore J. Error as historiographical challenge: the infamous globule hypothesis. In: G Hon, J Schickore & F Steinle (eds), *Going amiss in experimental research*. Boston Studies in the Philosophy of Science 267 (2009), pp. 27-45.

Schiebinger L. The loves of the plants. *Scientific American* 274(2): 110-115 (1996).

Schimper AFW. Ueber die Entwickelung der Chlorophyllkörner und Farbkörper. *Botanische Zeitung* 41(7-10), cols 105-112, 121-131, 137-146, and 153-162 (1883).

Schimper AFW. Untersuchungen über die Chlorophyllkörner und die ihnen homologen Gebilde. *Jahrbucher für wissenschaftliche Botanik* 16:1-247 (1885).

Schleich CL. *Besonnte Vergangenheit. Lebenserinnerungen (1859-1919)*. Berlin: Ernst Rowohlt (1922).

Schleich CL. *Those were good days! Reminiscences*. B Miall (trans.). London: Allen & Unwin (1935).

Schleiden MJ. Beiträge zur Phytogenesis. *Archiv für Anatomie, Physiolologie und wissenschaftliche Medicin* 1838:137-176 (1838).

Schleiden MJ. *Die Botanik als induktive Wissenschaft. Grundzüge der Wissenschaftlichen Botanik nebst einer Methodologischen Einleitung als Anleitung zum Studium der Pflanze*. Second edition. In two volumes. Leipzig: Wilhelm Engelmann (1845-1846).

Schleiden MJ. Einige Bemerkungen über den vegetabilischen Faserstoff und sein Verhältniss zum Stärkemehl. *Annalen der Physik und Chemie* 43(2):391-397 (1838).

Schleiden MJ. Einige Blicke auf die Entwicklungsgeschichte des vegetabilischen Organismus bei den Phanerogamen. *Archiv für Naturgeschichte* 3(1):289-320 and 414 (1837).

Schleiden MJ. *Grundzüge der wissenschaftlichen Botanik . . . Erster Theil: Methodologische Einleitung. Vegetabilische Stofflehre. Die Lehre von der Pflanzenzelle*. Leipzig: Engelmann (1842).

Schleiden MJ. *Grundzüge der wissenschaftlichen Botanik . . . Erster Theil: Methodologische Grundlage. Vegetabilische Stofflehre. Die Lehre von der Pflanzenzelle*. Second edition. Leipzig: Engelmann (1845-1846).

Schleiden MJ [printed "JM"]. *Principles of scientific botany; or, botany as an inductive science*. E Lankester (trans.). London: Longman, Brown *et al.* (1849).

Schleiden MJ. Some observations on the development of the organization in phænogamous plants. *London & Edinburgh Philosophical Magazine and Journal of Science* 12(73):172-189 and 12:(74):241-249 (1838).

Schleiden MJ. *The plant; a biography*. A Henfrey (trans.). London: Balliere (1848).

Schleper C, Jurgens G & Jonuscheit M. Genomic studies of uncultivated archaea. *Nature Reviews Microbiology* 3(6):479-488 (2005).

Schmid A. *Über alte Kräuterbücher*. Bern: Haupt (1939).

Schmidt O. *Deszendenzlehre und Darwinismus*. Leipzig: Brockhaus (1873).

Schmidt O. *Die Spongien des adriatischen Meers*. Text (1862) plus three supplements. Leipzig: Engelmann (1862-1868).

752   REFERENCES

Schmidt O. Einige neue Beobachtungen über die Infusorien. In: MJ Schleiden & R Froriep (eds), *Notizen aus dem Gebiete der Natur- und Heilkunde, Dritter Reihe* 9(1), cols 5–7. Weimar: Landes-Industrie-Comptoirs (1849).

Schmidt O. *Goethes Verhältnis zu den organischen Naturwissenschaften.* Berlin: Hertz (1853).

Schmitt CB. Perrenial [*sic*] philosophy: from Agostino Steuco to Leibniz. *Journal of the History of Ideas* 27(4):505–532 (1966).

Schmitz F. Die Chromatophoren der Algen. *Verhandlungen des naturhistorischen Vereins der preußischen Rheinlande und Westphalens (Decheniana)* 40:1–180 (1883).

Schmitz [F]. Untersuchungen über die Struktur des Protoplasmas und der Zellkerne der Pflanzenzellen. *Sitzungsberichte der Niederrheinischen Gesellschaft für Natur- und Heilkunde zu Bonn* 1880:159–198 (1880).

Schmitz H-P & Ord-Hume, AWJG. Engramelle, Marie Dominique Joseph (2001). In: *Oxford music online.* https://www.oxfordmusiconline.com, accessed 11 February 2019 and 18 September 2020.

Schofield RE. *The enlightened Joseph Priestley. A study of his life and work from 1773 to 1804.* University Park: Pennsylvania State UP (2004).

Scholfield AF. Introduction. In: *Aelian. On the characteristics of animals.* AF Scholfield (trans.). Loeb Classical Library 446. Cambridge MA: Harvard UP (1958), pp. xi–xxix.

Scholem G. *Origins of the Kabbalah.* RJ Zei Werblowsky (ed.), A Arkush (trans.). Princeton UP (1987).

Scholem G. *Ursprung und Anfänge der Kabbala.* Berlin and New York: de Gruyter (1962).

Schroeder HJ. *Disciplinary decrees of the General Councils. Text, translation, and commentary.* St Louis MO: Herder (1937).

Schultze M[S]. Die Bewegung der Diatomeen. *Archiv für mikroskopische Anatomie* 1:376–402 (1865).

Schultze MS. *Über den Organismus der Polythalamien (Foraminiferen) nebst Bermerkungen über die Rhizopoden im Allgemein.* Leipzig: Wilhelm Engelmann (1854).

Schultze M[S]. Ueber Muskelkörperchen und das, was man eine Zelle zu nennen habe. *Archiv für Anatomie, Physiologie und wissenschaftliche Medicin* 1861:1–27 (1861).

Schulz F, Yutin N, Ivanova NN, *et al.* Giant viruses with an expanded complement of translation system components. *Science* 356(6333):82–85 (2017).

Schützenberger P. *On fermentation.* International Scientific Series 20. New York: Appleton (1876).

Schwanbeck EA. *Megasthenis Indica.* Bonnae: Pleimseii (1846).

Schwann T. Author's preface. In: *Microscopical researches into the accordance in the structure and growth of animals and plants.* H Smith (trans.). London: Sydenham Society (1847), pp. ix–xviii.

Schwann T. *Microscopical researches into the accordance in the structure and growth of animals and plants.* H Smith (trans.). London: Sydenham Society (1847).

Schwann T. *Mikroskopische Untersuchungen über die Uebereinstimmung in der Struktur und dem Wachsthum der Thiere und Pflanzen.* Berlin: Sander (Reimer) (1839).

Schwendener S. *Die Altentypen der Flechtengonidien. Programm für die Rectoratsfeier der Universitæt.* Basel: Schultze (1869).

Scofield CI (ed.). *Holy Bible. Authorized King James version.* New York: Oxford UP (1967).

Scopoli JA. *Flora Carniolica exhibens plantas Carniolæ indigenas et distributas in classes naturales.* Viennæ: Trattner (1760).

Scopoli JA. *Flora Carniolica exhibens plantas Carnioliae indigenas et distribvtas in classes, genera, species, varietates, ordine Linnaeano.* Second edition. In two volumes. Vindobonæ: Kravss (1772).

Scopoli JA. *Introdvctio ad historiam natvralem sistens genera lapidvm, plantarvm, et animalivm.* Pragæ: Wolfgangvm Gerle (1777).

Scully N. *Alchemical healing: a guide to spiritual, physical, and transformational medicine.* Rochester NY: Bear (2003).

Séd N. Le Mêmar samaritain, le Sêfer Yeṣīrā et les trente-deux sentiers de la Sagesse. *Revue de l'Histoire des Religions* 170(2):159–184 (1966).

REFERENCES 753

Sedgwick A. On the inadequacy of the cellular theory of development, and on the early development of nerves, particularly of the third nerve and of the sympathetic in elasmobranchii. *Quarterly Journal of Microscopical Science, New Series* 37(145):87–101 (1894).

Sedgwick WT. Foreword. The genesis of a new science,—bacteriology. *Journal of Bacteriology* 1(1):1–4 (1916).

Seeberg E. Ammonius Sakas. *Zeitschrift für Kirchengeschichte* 61:136–170 (1942).

Sen A. Passage to China. *New York Review of Books* 51(19):61–65 (2 December 2004).

Sendivogius M. *A new light of alchymie*. In: J.F. [J French] (trans.), *A new light of alchymie* (1650). London: Richard Cotes (1650), pp 1–74.

Sendivogius M. *A treatise of svlphvr*. In: J.F. [J French] (trans.), *A new light of alchymie* (1650). London: Richard Cotes (1650), pp. 75–147.

Sendivogius M. *Novvm lvmen chymicvm, e natvræ fonte et manvali experientia depromptum, & in duodecim tractatus diuisum*. Coloniæ: Antonium Boëtzerum (1614).

Sendivogius M. *Tractatvs de svlphvre: altero natvræ principio, ab avthore eo, qvi et primum conscripsit principium*. Coloniæ: Ioannem Crithium, Coloniæ (1616).

Senebier J. Sur la matière verte qu'on trouve dans les vases remplis d'eau, lorsqu'ils sont exposés à la lumière, de même que sur les conferves et tremelles considérées relativement à leur nature, et à leur propriété de donner du gaz oxigène au soleil. *Journal de Physique, de Chimie, d'Histoire Naturelle et des Arts* 48:155–162, 202–210, 294–301, 361–368, and 417–431; and 49:3–8, 135–141, 213–219, and 357–368 (1799).

Seneca. *Epistles*. RM Gummere (trans.). Loeb Classical Library 75–77. London: Heinemann (1917–1925). Volume 1, 1917; volume 2, 1920; volume 3, 1925.

Seneca. *Naturales quaestiones*. TH Cororan (trans.). Loeb Classical Library 450 and 457. Cambridge MA: Harvard UP (1971–1972).

Seth A. Scholasticism. *Encyclopædia Britannica*, Ninth edition 21:417–431 (1886).

[Sextus Empiricus]. Bibliothèque nationale, MS Fonds latin 14700, at folio 85v. Accessed online (BnF), 30 May 2018.

Sextus Empiricus. *Des Sextus Empiricus Pyrrhoneïsche Grundzüge. Aus dem Griechischen übersetzt und mit einer Einleitung und Erläuterungen versehen von Eugen Pappenheim*. Philosophische Bibliothek. Band 89. Leipzig: Dürr (1877).

Sextus Empiricus. *Outlines of Pyrrhonism*. RG Bury (trans.). Sextus Empiricus 1. Loeb Classical Library 273. Cambridge MA: Harvard UP (1933, reprinted 1993).

Sextus Empiricus. *Outlines of Scepticism*. J Annas & J Barnes (trans.). Cambridge UP (1994).

Sextus Empiricus. *Pros physikous. Against the physicists*. RG Bury (trans.). Sextus Empiricus 3, in Loeb Classical Library 311. Cambridge MA: Harvard UP (1987).

Sextus Empiricus. *Sexti Empirici opera Græce et Latine. Pyrrhoniarum institutionum libri III. cum Henr. Stephani versione et notis. Contra mathematicos, sive disciplinarum professores, libri VI. Contra philosophos libri V. Cum versione Gentiani Herveti. Græca ex mss. codicibus castigavit, versiones emendavit supplevitqve, et toti operi notas addit Jo. Albertus Fabricius*. Lipsiæ: Gleditschii (1718).

Sextus Empiricus. *Sexti Empirici opera Graece et Latine. Pyrrhoniarum institutionum libri III. cum Henrici Stephani versione et notis. Contra mathematicos, sive disciplinarum professores, libri VI. Contra philosophos libri V. Cum versione Gentiani Herveti. Græca ex mss. codicibus castigavit, versiones emendavit supplevitque et toti operi notas addit Io. Albertus Fabricius . . . Editio emendatior. Tomus I. Pyrrhonias institutions continens*. Lipsiae: Kuehnianae (1842).

Sextus Empiricus. [*Sextou Empeirikou Ta Sōzomena.*] *Sexti Empirici opera qvæ extant. Magno ingenii acvmine scripti, Pyrrhoniarvm hypotypωsewn libri III. Qvibus in tres Philosophiæ partes acerrimè inquiritur, Henrico Stephano interprete: Aduersus mathematicos, hos est, eos qui disciplinæ profitentur, Libri X. Gentiano Herveto Avrelio interprete, Græcè nunc primùm editi. Adiungere visum est Pyrrhonis Eliensis Philosophi vitam: nec non Clavdii Galeni Pergameni de Optimo docendi genere librum, quo aduersus Academicos Pyrrhoniosqve disputat*. Avrelianæ: Chouët (1621).

Sextus Empiricus. *Sexti Empirici opera recensuit Hermannus Mutschmann*, volume 1. Lipsiae: Teubneri (1912).

754 REFERENCES

Sextus Empiricus. *Sexti Empirici viri longe doctissimi Adversvs mathematicos . . . Gentiano Herveto Avrelio interprete. Eivsdem Sexti Pyrrhoniarvm Hypotypωsewn libri tres.* Antverpiæ: Plantini (1569).

Sextus Empiricus. *Sextus Empiricus ex recensione Immanuelis Bekkeri.* Berolini: Reimer (1842).

Sextus Empiricus. *Sexti Philosophi Pyrrhoniarum hypotypωsewn libri III . . . Interprete Henrico Stephano.* [Genève]: Henricus Stephanus (1562).

Shackleton R. The "Encyclopédie" as an international phenomenon. *Proceedings of the American Philosophical Society* 114(5):389–394 (1970).

Shah S, Chen Y, Bhattacharya D & Chan CX. Sex in Symbiodiniaceae dinoflagellates: genomic evidence for independent loss of the canonical synaptonemal complex. *Scientific Reports* 10:9792 (2020).

Shaker AF. *Thinking in the language of reality. Ṣadr al-Dīn Qūnawī (1207–74 CE) and the mystical philosophy of reason.* XLibris (2012).

Shakespeare W. The tragedie of Hamlet, Prince of Denmarke. In: *Mr. William Shakespeares comedies, histories, & tragedies. Published according to the true originall copies.* London: I Iaggard & E Blount, for W Iaggard, E Blount, I Smithweeke & W Aspley (1623). First Folio, Bodleian Arch. G c.7. Digital facsimile at https://firstfolio.bodleian.ox.ac.uk, accessed 14 March 2020 and 7 October 2020.

Shakespeare W. *The tragicall historie of Hamlet, Prince of Denmarke.* London: I.R for N[icholas] L[ing] (1605). Second variant quarto, Sig. D4v. Garrick copy, British Library. Digital facsimile at https://www.bl.uk/collection-items/second-quarto-of-hamlet-1605, accessed 14 March 2020 and 7 October 2020.

Shank JB. *The Newton wars and the beginning of the French Enlightenment.* University of Chicago Press (2008).

Sharp LW. *Introduction to cytology.* Third edition. New York: McGraw-Hill (1934).

Sharpe RB. *A monograph of the Alcedinidæ: or, family of kingfishers.* London: published by the author (1868–1871).

Sharpe WD. Isidore of Seville: the medical writings. An English translation. *Transactions of the American Philosophical Society. New Series* 54(2):1–75 (1964).

Shaw G. *Theurgy and the soul. The Neoplatonism of Iamblichus.* Second edition. Kettering OH: Angelico (2014).

Shay JW & Wright WE. Telomeres and telomerase: three decades of progress. *Nature Reviews Genetics* 20(5):299–309 (2019).

Sheed FJ. Foreword. In: FJ Sheed (trans.). *The Confessions of St. Augustine.* New York: Sheed & Ward (1943), pp. v–vi.

Shelley M. *Frankenstein: the 1818 text.* New York: Penguin Classics (2018).

Shelley M. *Frankenstein; or, the modern Prometheus.* In three volumes. London: Lackington, Hughes *et al.* (1818).

Shelton DE & Michod RE. Philosophical foundations for the hierarchy of life. *Biology & Philosophy* 25(3):391–403 (2010).

Sherborn CD & Woodward BB. On the dates of publication of the natural history portions of the "Encyclopédie Méthodique". *Annals and Magazine of Natural History, 7 Series* 17(102):577–582 (1906).

Sheridan JJ. Introduction. In: Alan of Lille, *The plaint of Nature.* JJ Sheridan (trans.). Toronto: Pontifical Institute of Medieval Studies (1980), pp. 1–64.

Sherman F & Stewart JW. Genetics and biosynthesis of cytochrome *c. Annual Review of Genetics* 5:257–296 (1971).

Shiflett AM & Johnson PJ. Mitochondrion-related organelles in eukaryotic protists. *Annual Review of Microbiology* 64:409–429 (2010).

Shumaker W. Literary Hermeticism: some test cases. In: *Hermeticism and the Renaissance. Intellectual history and the occult in early modern Europe.* I Merkel & AG Debus (eds). Washington: Folger (1988), pp. 293–301.

Siebold CTh v[on]. *Lehrbuch der vergleichenden Anatomie der wirbellosen Thiere.* In: [CTh] v[on] Siebold & [H] Stannius, *Lehrbuch der vergleichenden Anatomie. Erster Theil. Wirbellosen Thiere.* Berlin: Veit (1848).

REFERENCES   755

Siebold CTh v[on]. Ueber einzellige Pflanzen und Thiere. *Zeitschrift für wissenschaftliche Zoologie* 1(2–3):270–294 (1849).

Siegesbeck JG. *Botanosophiæ verioris brevis sciagraphia in vsvm discentivm adornata: accedit ob argvmenti analogiam, epicrisis in clar. Linnæi nvperrime evulgatvm systema plantarum sexvale, et hvic svperstrvctam methodvm botanicam.* Petropoli: Academiæ [Scientiarum] (1737).

Siesser WG. Christian Gottfried Ehrenberg: founder of micropaleontology. *Centaurus* 25(2):166–188 (1981).

Sigismund, Freiherr von Herberstain [*sic*]. *Rervm Moscoviticarvm comentarii.* Viennæ: [n.p.] (1549).

Sigismund, Freiherr von Herberstain [*sic*]. *Rervm Moscoviticarum commentarij Sigismundi Liberi Baronis in Herberstain, Neyperg, & Guettenhag: Rvssiæ, & quæ nunc eius metropolis est, Moscouiæ, breuissima descriptio.* Basileæ: Ioannem Oporinum (1556).

Sigismund, Freiherr von Herberstein. *Rervm Moscoviticarvm commentarij, Sigismundo Libero authore. Rvssiæ breuissima descriptio, & de religione eorum varia inserta sunt.* Antverpiæ: Ioannis Steelsij (1557).

Sigismund, Freiherr von Herberstein. *Rervm Moscoviticarvm commentarij Sigismundi Liberi Baronis in Herberstain, Neyperg, & Guettenhag.* Basileæ: Oporiniana (1571).

Sikdar JC. Fabric of life: paryāpti prānāpana in Jaina agama. *Indian Journal of History of Science* 27(3):223–229 (1992).

Silberberg B. Das Pflanzenbuch des Abû Hanîfa Ahmed ibn Dâ'ûd ad-Dînawarî. Ein Beitrag zur Geschichte der Botanik bei den Arabern. *Zeitschrift für Assyriologie und Vorderasiatische Archäologie* 24:225–265 (1910) and 25:39–88 (1911).

Silvestris B. *Cosmographia.* P Dronke (ed.). Leiden: Brill (1978).

Silvestris B. *De mundi universitate libri duo sive megacosmus et microcosmus.* CS Barach & J Wrobel (trans.). Innsbruck: Wagner (1876).

Silvestris B. *The Cosmographia of Bernardus Silvestris.* W Wetherbee (trans.). New York: Columbia UP (1973).

Simplicius. *In Aristotelis De caelo commentaria.* JL Heiberg (ed.). *Commentaria in Aristotelem Graeca.* Academiae Litterarum Regiae Borussicae 7. Berolini: Reimer (1894).

Simplicius. *In Aristotelis Physicorum libros quattuor commentaria.* H Diels (ed.). *Commentaria in Aristotelem Graeca.* Academiae Litterarum Regiae Borussicae 9 (1882) and 10 (1895). Berolini: Reimer.

Simplicius. *In libros Aristotelis De anima commentaria.* In: M Hayduck (ed.), *Commentaria in Aristotelem Graeca.* Academiae Litterarum Regiae Borussicae 11. Berolini: Reimer (1882).

Simplicius. *In Aristotelis Categorias commentarium.* C Kalbfleisch (ed.). In: *Commentaria in Aristotelem Graeca.* Academiae Litterarum Regiae Borussicae 8. Berolini: Reimer (1907).

Simplicius. *On Aristotle, Categories 1–4.* M Chase (trans.). Ithaca NY: Cornell UP (2003).

Simplicius. *On Aristotle, Categories 5–6.* FAJ de Haas & B Fleet (trans.). Ithaca NY: Cornell UP (2001).

Simplicius. *On Aristotle, Categories 7–8.* B Fleet (trans.). Ithaca NY: Cornell UP (2002).

Simplicius. *On Aristotle, Categories 9–15.* R Gaskin (trans.). Ithaca NY: Cornell UP (2000).

Simplicius. *On Aristotle's On the soul 1.2–2.4.* JO Urmson (trans.). Ithaca NY: Cornell UP (1995).

Simpson AGB, Slamovits CH & Archibald JM. Protist diversity and eukaryote phylogeny. In: AGB Simpson, CH Slamovits & JM Archibald (eds), *Handbook of the protists.* Second edition. Springer (2017), pp. 1–21.

Sines G & YA Sakellarakis. Lenses in antiquity. *American Journal of Archaeology* 91(2):191–196 (1987).

Singer C. *A history of biology to about the year 1900.* Oxford: Clarendon (1931). Reprinted by Iowa State UP, Ames (1989).

Singer C. *A history of scientific ideas.* New York: Barnes and Noble (1996).

Singer C. *A short history of anatomy from the Greeks to Harvey.* Second edition. New York: Dover (1957).

Singer C. *From magic to science. Essays on the scientific twilight.* New York: Boni and Liveright (1928).

756   REFERENCES

Singer C. *Greek biology & Greek medicine*. Oxford: Clarendon (1922).

Singer C. Greek biology and its relation to the rise of modern biology. In: C Singer (ed.), *Studies in the history and method of science*, volume 2. Oxford: Clarendon (1921), pp. 1–101.

Singer P. *Animal liberation. A new ethics for our treatment of animals*. New York: Random House (1975).

Singer P. *Practical ethics*. Cambridge UP (1979).

Siraisi NG. *Avicenna in Renaissance Italy: the Canon and medical teaching in Italian universities after 1500*. Princeton UP (1987).

Skejo J & Franjević D. Eukaryotes are a holophyletic group of polyphyletic origin. *Frontiers in Microbiology* 11:1380 (2020).

Slabber M. *Natuurkundige verlustigingen, behelzende microscopise waarneemingen van in- en uitlandse water- en land-dieren*. Te Haarlem: Bosch (1778).

Sloan P. Darwin: from *Origin of species* to *Descent of man*. *The Stanford Encyclopedia of Philosophy* (Summer 2019 edition), EN Zalta (ed.), https://plato.stanford.edu/archives/sum2019/entries/origin-descent/, accessed 30 May 2020 and 9 October 2020.

Sloan P. Evolutionary thought before Darwin. *The Stanford Encyclopedia of Philosophy* (Winter 2019 edition), EN Zalta (ed.), https://plato.stanford.edu/archives/win2019/entries/evolution-before-darwin, accessed 26 September 2019 and 1 October 2020.

Sloan PR. John Locke, John Ray, and the problem of the natural system. *Journal of the History of Biology* 5(1): 1–53 (1972).

Sloan PR. Organic molecules revisited. In: J-C Beaune *et al.* (eds), *Buffon 88: actes du Colloque international pour le bicentenaire de la mort de Buffon. Paris, Montbard, Dijon, 14–22 juin 1988*. Paris: J Vrin (1992), pp. 415–438.

Sloan PR. The Buffon-Linnaeus controversy. *Isis* 67(3): 356–375 (1976).

Sloan PR & Lyon J. Introduction. In: J Lyon & PR Sloan (eds), *From natural history to the history of nature: readings from Buffon and his critics*. Notre Dame IN: University of Notre Dame Press (1981), pp. 1–32.

Sloane TO'C. Alchemy. *The Catholic Encyclopedia* 1:272–273 (1907).

Smarda J. Viroids: molecular infectious agents. *Acta Virologica* 31(6):506–524 (1987).

Smellie W. Preface by the translator. In: [GLL] Buffon, *Natural history, general and particular*. Volume 1. Edinburgh: William Creech (1780), pp. v–xxi.

Smellie W. *The philosophy of natural history*. Edinburgh: Elliot *et al.* (1790).

Smith GM. *The fresh-water algae of the United States*. Second edition. New York: McGraw-Hill (1950).

Smith JE (ed.). *A selection from the correspondence of Linnæus, and other naturalists, from the original manuscripts*. In two volumes. London: Longman, Hurst *et al.* (1821).

Smith JE. *An introduction to physiological and systematical botany*. London: Longman, Hurst *et al.* (1807).

Smith JE. Botany. *Encyclopædia Britannica*, Supplement to the Fourth, Fifth and Sixth editions 2:376–422 (1824).

Smith JE. *English botany; or, coloured figures of British plants... figures by J Sowerby*. In 36 volumes. London: Wiles & Taylor (1790–1814).

Smith JE. *Flora Britannica*. In three volumes. Londini: Davis [varies] (1800–1804).

Smith JE. *The English flora*. In four volumes. London: Longman, Rees *et al.* (1824–1828). Second edition, 1828–1830. For the fifth volume (to both editions) see Hooker WJ (1833) and Berkeley MJ (1836).

Smith M. *The early history of God. Yahweh and other deities of ancient Israel*. Second edition. Grand Rapids MI: Eerdmans (2002).

Smith MF & Butterfield D. Not a ghost: the 1496 Brescia edition of Lucretius. *Aevum* 84(3):683–696 (2010).

Smith T. Introductory notice to the Recognitions of Clement. In: A Roberts & J Donaldson (eds), *The Ante-Nicene Fathers*, volume 8. New York: Scribner (1903), pp. 73–74.

Smith WR. *Kinship and marriage in early Arabia*. Cambridge UP (1885).

Snyder JG. Marsilio Ficino's critique of the Lucretian alternative. *Journal of the History of Ideas* 72(2):165–181 (2011).

Sogin ML. Early evolution and the origin of eukaryotes. *Current Opinion in Genetics and Development* 1(4):457–463 (1991).

Solinus, Caius Julius. *Ad aduentu[m] polihistor siue de situ orbis ac mu[n]di mirabilibus liber.* G Tardivus (ed.). [Parisiis]: [Louis Symonel] (1474–1475).

Sollbach GE. *Das neue Tier- und Arzneibuch des Doktor Michael Herr AD 1546.* Würzburg: Königshausen & Neumann (1994).

Sommer MS, Gould SB, Kawach O, *et al.* Photosynthetic organelles and endosymbionts. In: LA Katz & D Bhattacharya (eds), *Genomics and evolution of microbial eukaryotes.* Oxford UP (2006), pp. 94–108.

Sorabji R. Atoms and time atoms. In: PD Kretzman (ed.), *Infinity and continuity in ancient and medieval thought.* Ithaca NY: Cornell UP (1982), pp. 37–86.

Sorabji R. *Emotion and peace of mind: from Stoic agitation to Christian temptation.* Oxford UP (2000).

Sorabji R. General introduction. In: C Wildberg (trans.). *Philoponus against Aristotle,* On the eternity of the world. Ithaca NY: Cornell UP (1987), pp. 1–17. Reprinted as: *Appendix: the commentators.* In: JO Urmson (trans.), *Simplicius on Aristotle's On the soul 1.2–2.4* (1995), at pp. 185–194.

Sorabji R. Appendix: the commentators. In: Simplicius. On Aristotle's *On the soul 1.1–2.4.* JO Urmson (trans.). Ithaca NY: Cornell UP (1995), pp. 185–194.

Sorabji R. *Time, creation and the continuum. Theories in antiquity and the early Middle Ages.* Ithaca NY: Cornell UP (1983).

Sorby HC. On comparative vegetable chromatology. *Proceedings of the Royal Society* 21(146):442–483 (1873).

Šorm F & Keil B. Regularities in the primary structure of proteins. *Advances in Protein Chemistry* 17:167–207 (1963).

Šorm F, Keil B, Holeyškový V, *et al.* On proteins. XXXIX. Structural resemblance in certain proteins. *Collection of Czechoslovak Chemical Communications* 22(4):1310–1329 (1957).

Sorrenson R. Towards a history of the Royal Society in the Eighteenth century. *Notes and Records of the Royal Society of London* 50(1):29–46 (1996).

Spade PV. *Five texts on the medieval problem of universals: Porphyry, Boethius, Abelard, Duns Scotus, Ockham.* Indianapolis: Hackett (1994).

Spade PV. Medieval philosophy. *The Stanford Encyclopedia of Philosophy* (Summer 2018 edition), EN Zalta (ed.), https://plato.stanford.edu/archives/sum2018/entries/medieval-philosophy/, accessed 7 August 2018.

Spade PV & Panaccio C. William of Ockham. *The Stanford Encyclopedia of Philosophy* (Winter 2016 edition), EN Zalta (ed.), plato.stanford.edu/archives/win2016/entries/ockham, accessed 18 November 2018.

Spallanzani L. *Nouvelles recherches sur les découvertes microscopiques, et la génération des corps organisés.* JT Needham (trans.). Londres et Paris: Lacombe (1769).

Spallanzani L. *Opuscula di fisica animale e vegetabile.* In two volumes. Modena: Societa Tipografica (1776).

Spallanzani L. *Saggio di osservazioni microscopiche concernenti il sistema della generazione de' signori di Needham e Buffon.* In: B Vergine (ed.), *Dissertazioni due dell' Abate Spallanzani.* Modena: Soliani (1765), pp. 1–87.

Spallanzani L. *Saggio di osservazioni microscopiche concernenti il sistema della generazione de' signori di Needham e Buffon.* In: A Mieli & E Troilo (eds), *Classici delle Scienze e della Filosofia.* Serie Scientifica 2. Bari: Barese (1914).

Spang A, Saw JH, Jørgensen SL, *et al.* Complex archaea that bridge the gap between prokaryotes and eukaryotes. *Nature* 521(7551):173–179 (2015).

Spence JD. *The memory palace of Matteo Ricci.* Markham ON: Penguin (1986).

Spener JJ. *Museum Spenerianum, sive catalogus rerum tam artificiosarum, quam naturalium, tam antiquarum, quam recentium, tam exoticarum, quam domesticarum, quas clarissimus ... collegit.*

758   REFERENCES

JM Michaelis (ed.). *Das Spenerische Kabinet, oder kurtze Beschreibung aller sowol künstlich-als natürlicher, alter, als neuer, fremder, als einheimischer curiösen Sachen*. Leipzig: Fleischer (1693).

Sridhar MK & Bilimoria P. Animal ethics and ecology in classical India—reflections on a moral tradition. In: P Bilimoria, J Prabhu & R Sharma (eds), *Indian ethics. Classical traditions and contemporary challenges* 1. Ashgate, Aldershot UK (2007), pp. 297–328.

Stackhouse J. *Nereis Britannica*. Bathoniæ: Hazard, for the author (1795–1801).

Stafleu FA. Adanson and the "Familles des plantes". In: GHM Lawrence (ed.), *Adanson. The bicentennial of Michel Adanson's "Familles des plantes"*, volume 1 [of 2]. Pittsburgh: Hunt Botanical Library (1963), pp. 123–264.

Stafleu FA. Agardh's *Systema algarum*. *Taxon* 15(7):276–277 (1966).

Stafleu FA. *Linnaeus and the Linnaeans. The spreading of their ideas in systematic botany, 1735–1789*. Utrecht: Oosthoek (1971).

Stafleu FA & Cowan RS. *Taxonomic literature*. Second edition. In seven volumes, with supplements. Utrecht: Bohn, Scheltema & Holkema [varies] (1976).

Stanbury P & J Holland (eds). *Mr Macleay's celebrated cabinet. The history of the Macleays and their museum*. University of Sydney (1988).

Stanier RY & van Niel CB. The concept of a bacterium. *Archiv für Microbiologie* 42(1):17–35 (1962).

Stanier RY & van Niel CB. The main outlines of bacterial classification. *Journal of Bacteriology* 42(4):437–466 (1941).

Stanier RY, Doudoroff M & Adelberg EA. *The microbial world*. [First edition]. Englewood Cliffs NJ: Prentice-Hall (1957).

Stanier RY, Doudoroff M & Adelberg EA. *The microbial world*. Second edition. Englewood Cliffs NJ: Prentice-Hall (1963).

Stanier RY, Doudoroff M & Adelberg EA. *The microbial world*. Third edition. Englewood Cliffs NJ: Prentice-Hall (1970).

Stanier RY, Ingraham JL, Wheelis ML & Painter PR. *The microbial world*. Fifth edition. Englewood Cliffs NJ: Prentice-Hall (1986).

Stannard J. Albertus Magnus and medieval herbalism. In: JA Weisheipl (ed.), *Albertus Magnus and the sciences. Commemorative essays 1980*. Toronto: Pontifical Institute of Mediaeval Studies (1980), pp. 345–377.

Stannard J. The herbal as a medical document. *Bulletin of the History of Medicine* 43(3):212–220 (1969).

Stapleton HE, Azo RF & Hidāyat Ḥusain M. Chemistry in 'Irāq and Persia in the Tenth century AD. *Memoirs of the Asiatic Society of Bengal* 8(6):317–417 (1927).

Stapleton HE, Azo RF, Hidāyut Ḥusain M & Lewis GL. Two alchemical treatises attributed to Avicenna. *Ambix* 10(2):41–82 (1962).

Stavenhagen L (ed. and trans.). *A testament of alchemy being the revelations of Morienus, ancient adept and hermit of Jerusalem to Khalid ibn Yazid ibn Muʿawiyya, King of the Arabs of the divine secrets of the magisterium and accomplishment of the alchemical art*. Hanover NH: University Press of New England (1974).

St. Clair CS. *The classification of minerals: some representative mineral systems from Agricola to Werner*. PhD thesis, University of Oklahoma (1965).

Stearn WT. Introduction. In: J Ray. *Synopsis methodica stirpium Britannicarum. Editio tertia. 1724.* [With:] Linnaeus C. *Flora Anglica 1754 & 1759*. London: Ray Society (1973), pp. 1–90.

Stearn WT. S.F. Gray's "Natural arrangement of British plants" (1821). *Plant Systematics and Evolution* 167(1):23–34 (1989).

Stearn WT. The background of Linnaeus's contributions to the nomenclature and methods of systematic biology. *Systematic Zoology* 8(1): 4–22 (1959).

Stearn WT & Bridson G. *Carl Linnaeus (1707–1778). A bicentenary guide to the career and achievements of Linnaeus and the collections of the Linnean Society*. Linnean Society of London (1978).

REFERENCES 759

Steele R. Forewords. In: R Steele (ed.), *Lydgate and Burgh's Secrees of old philosoffres. A version of the "Secreta secretorum".* Early English Text Society Extra Series, volume 66. London (1894), pp. vii–xxi.

Steele R. Introduction. In: R Steele (ed.), *Opera hactenus inedita Rogeri Baconi.* Fasc. 5. *Secretum secretorum cum glossis et notulis.* Oxford: Clarendon (1920), pp. vii–lxiii.

Steele R (ed.). *Lydgate and Burgh's Secrees of old philosoffres. A version of the "Secreta secretorum".* Early English Text Society Extra Series, volume 66. London (1894).

Steele R (ed.). *Medieval lore: an epitome of the science, geography, animal and plant folk-lore and myth of the Middle Age: being classified gleanings from the encyclopedia of Bartholomew Anglicus On the properties of things.* London: Elliot Stock (1893).

Steele R (ed.). *Three prose versions of the Secreta Secretorum,* volume 1. Early English Text Society Extra Series, volume 74. London: Kegan Paul, Trench, Trübner (1898).

Steenstrup JJS. *Om forplantning og udvikling gjennem vexlende generationsrækker, en særegen form for opfostringen i de lavere dyrklasser.* Kjøbenhavn: [Reitzel] (1842).

Steenstrup JJS. *On the alternation of generations, or, the propagation and development of animals through alternate generations.* G Busk (trans.). London: Ray Society (1845).

Steenstrup JJS. *Ueber den Generationswechsel oder die Fortpflanzung und Entwickelung durch abwechselnde Generationen.* CH Lorenzen (trans.). Copenhagen: Reitzel (1842).

Stein F. *Der Organismus der Infusionsthiere nach eigenen Forschungen in systematischer Reihenfolge bearbeitet.* Three volumes bound in four. Leipzig: Engelmann (1859–1883).

Steiner R. *Einleitung zu Goethes Naturwissenschaftlichen Schriften (1884). Zugleich eine Grundlegung der Geisteswissenschaft (Anthroposophie).* Fourth edition. Dornach: Steiner (1987). Originally published as introductions to each of the four volumes of *Goethes Naturwissenschaftliche Schriften.* In: J Kürschner (ed.), *Deutsche National-Literatur* 114–117. Stuttgart: Spemann [varies] (1883–1897).

Steiner R. *Goethe's conception of the world.* E Collinson (ed.). London: Anthroposophical Publishing (1928).

Steiner R. *Goethes Weltanschauung.* Weimar: Emil Felber (1897).

Steinschneider M. Al-Farabi (Alpharabius), des arabischen Philosophen Leben und Schriften, mit besonderer Rücksicht auf die Geschichte der griechischen Wissenschaft under den Arabern. Nebst Anhängen: Joh. Philoponus bei den Arabern, Darstellung der Philosophie Plato's, Leben und Testament des Aristoteles von Ptolemaeus. *Mémoires de l'Académie Impériale des Sciences de St.-Pétersbourg, 7 Série* 13(4):1–268 (1869).

Stephenson J. *The zoological section of the Nuzhatu-l-Qulūb of Ḥamdullāh al-Mustaufī al-Qazwīnī.* London: Royal Asiatic Society (1928).

Stephenson J. *The zoological section of the Nuzhatu-l-Qulūb. Isis* 11(2):285–315 (1928).

Sterbeeck F van. *Theatrum fungorum, oft het toonsel der campernoelien.* T'Antwerpen: Ioseph Iacobs (1675).

Stern SM. Ibn al-Ṭayyib's commentary on the *Isagoge. Bulletin of the School of Oriental and African Studies, University of London* 19(3):419–425 (1957).

Sternhell Z. *The anti-Enlightenment tradition.* D Maisel (trans.). New Haven: Yale UP (2010).

Stevens PF. Augustin Augier's "Arbre botanique" (1801), a remarkable early botanical representation of the natural system. *Taxon* 32(2): 203–211 (1983).

Stevens PF. Haüy and A.-P. Candolle: crystallography, botanical systematics, and comparative morphology, 1780–1840. *Journal of the History of Biology* 17(1):49–82 (1984).

Stevens PF. *The development of biological systematics. Antoine-Laurent de Jussieu, nature, and the natural system.* New York: Columbia UP (1994).

Stevens PF & SP Cullen. Linnaeus, the cortex-medulla theory, and the key to his understanding of plant form and natural relationships. *Journal of the Arnold Arboretum* 71(2):179–220 (1990).

Stewart EJ. Growing unculturable bacteria. *Journal of Bacteriology* 194(16):4151–4160 (2012).

Stokes GG. On the supposed identity of biliverdin with chlorophyll, with remarks on the constitution of chlorophyll. *Proceedings of the Royal Society of London* 13:144–145 (1864).

Stone DM. Bernardus Silvestris, "Mathematicus": edition and translation. *Archives d'histoire doctrinale et littéraire du Moyen Age* 63:209–283 (1996).

Stone W. Thomas B Wilson, MD. *Cassinia. Proceedings of the Delaware Valley Ornithological Club* 13:1–6 (1909).

Stoneman R. Naked philosophers: the Brahmins in the Alexander historians and the Alexander Romance. *Journal of Hellenic Studies* 115:99–114 (1995).

Strabo. *The Geography of Strabo*. HL Jones (trans.). In eight volumes. Loeb Classical Library 49–50, 182, 196, 211, 223, 241, and 267. London: Heinemann; and Cambridge MA: Harvard UP (1917–1930).

Strasburger E. Pflanzliche Zellen- und Gewebelehre. In: *Die Kultur der Gegenwart* (P Hinneberg, ed.). Teil 3, Abteilung 4. *Organische Naturwissenschaften*. Band 2(1). *Botanischer Teil* (E Strasburger & W Benecke, eds). Leipzig & Berlin: Teubner (1913), pp. 1–174.

Strathern P. *Mendeleyev's dream: the quest for the elements*. New York: St Martin's (2000).

Stretton AOW. The first sequence. Fred Sanger and insulin. *Genetics* 162(2):527–532 (2002).

Strickland HE. Description of a chart of the natural affinities of the insessorial order of birds. *Report of the Thirteenth Meeting of the British Association for the Advancement of Science held at Cork in August 1843, Notices and Abstracts of Communications*. London: John Murray (1844), page 69.

Strickland HE. On the true method of discovering the natural system in zoology and botany. *Annals and Magazine of Natural History* 6(36):184–194 (1841).

Stromeyer [F]. Ein sehr empfindliches Reagens für Jodine, aufgefunden in der Stärke (Amidon). *Annalen der Physik* 49(1–2):146–153 (1815).

Stukely W. An account of the impression of the almost entire sceleton [*sic*] of a large animal in a very hard stone, lately presented the Royal Society, from Nottinghamshire. *Philosophical Transactions* 30(360):963–968 (1719).

Stump E. Theology and physics in *De sacramento altaris*: Ockham's theory of indivisibles. In: N Kretzmann (ed.), *Infinity and continuity in ancient and medieval thought*. Ithaca NY: Cornell UP (1982), pp. 207–230.

Subbarayappa BV. The Indian doctrine of five elements. *Indian Journal of the History of Science* 1(1):60–67 (1966).

Suda. Suda On Line, http://www.stoa.org/, accessed April 2018 and 8 September 2020.

Sudhoff K. *Rudolf Virchow und die Deutschen Naturforscherversammlungen*. Leipzig: Akademische Verlagsgesellschaft (1922).

Sugimoto M & Swain DL. *Science and culture in traditional Japan, AD 600–1854*. Cambridge MA: MIT Press (1978).

Suler B. Alchemy. *Encyclopaedia Judaica* (M Berenbaum & F Skolnik, eds), Second edition 1:599–603. Detroit: Macmillan Reference (2007).

Sundara Rajan S. Amarakośa—a biological assessment. *Indian Journal of History of Science* 51(3):548–555 (2016).

Surapala. *Vrikshayurveda* (the science of plant life). N Sadhale (trans.). *Asian Agri-History Foundation Bulletin* 1. Secundurabad: Asian Agri-History Foundation (1996).

Swainson W. *A preliminary discourse on the study of natural history. Lardner's Cabinet Cyclopædia* 59. London: Longman, Rees *et al*. (1834).

Swainson W. *A treatise on the geography and classification of animals. Lardner's Cabinet Cyclopædia* 66. London: Longman, Rees *et al*. (1835).

Swainson W & Richardson J. *Fauna boreali-Americana; or the zoology of the northern parts of British America . . . Part second, the birds*. London: John Murray (1831).

Swammerdam J. *Bybel der natuure . . . Of Historie der insecten*. Te Leyden: Severinus, van der Aa & van der Aa (1737–1738).

Sweet V. *Body as plant, doctor as gardener. Premodern medicine in Hildegard of Bingen's Causes and cures*. PhD thesis, University of California San Francisco (2003).

Sylvester J (trans.). *Du Bartas his Diuine weekes and workes*. London: Young (1641).

Syme R. Pliny the Procurator. *Harvard Studies in Classical Philology* 73:201–236 (1969).

Synesius. *Œuvres de Synésius. Évêque de Ptolémais, dans la Cyrénaïque au commencement du Ve siècle*. H Duron (trans.). Paris: Hachette (1878).

Tableau encyclopédique et méthodique des trois règnes de la nature. Vers, coquilles, mollusques et polypiers. In three volumes. Paris: Agasse (1827).

Taburet-Delahaye É & de Chancel-Bardelot B. *La Dame à la licorne.* Paris: RMN (2018). In English (A Keens & S Heft, trans.) as: *The lady and the unicorn.* Paris: RMN (2018).

Tabvla Smaragdina Hermetis Trismegisti. In: Chrysogonus Polydorus (ed.), *De alchemia.* Norimbrgiæ: Johannes Petreium (1541), pp. 363–373.

Taçon PSC, May SK, Lamilami R, *et al.* Maliwawa figures—a previously undescribed Arnhem Land rock art style. *Australian Archaeology* 86(3):208–225 (2020).

Taiz L & Taiz L. *Flora unveiled. The discovery and denial of sex in plants.* Oxford UP (2017).

Tammiksaar E & Kalling K. "I was stealing some skulls from the bone chamber when a bigamist chemist stopped me." Karl Ernst von Baer and the development of physical anthropology in Europe. *Centaurus* 60(4):276–293 (2018).

Tan SY & Brown J. Medicine in stamps. Rudolph Virchow (1821–1902): "pope of pathology". *Singapore Medical Journal* 47(7):567–568 (2006).

Taschenberg O. *Die Lehre von der Urzeugung sonst und jetzt. Ein Beitrag zur historischen Entwicklung derselben.* Halle: Max Neimeyer (1882).

Taschow U. *Nicole Oresme und der Frühling der Moderne.* Four books in two volumes. Halle: Avox (2003).

Tashiro TT. Three passages in Homer, and the Homeric legacy. *The Antioch Review* 25(1):63–89 (1965).

Tauxe D. L'organisation symbolique du dispositif pariétal de la grotte de Lascaux. *Préhistoire du Sud-Ouest* 15:177–266 (2007).

Tawara A. Avicenna's denial of life in plants. *Arabic Sciences and Philosophy* 24(1):127–138 (2014).

Taylor CCW. Introduction. In: CCW Taylor (ed.), *From the beginning to Plato.* Routledge history of philosophy (GHR Parkinson & SG Shanker, gen. eds), volume 1 . London: Routledge (1997), pp. 1–8.

Taylor CCW. *The atomists: Leucippus and Democritus. Fragments, a text and translation with commentary.* University of Toronto Press (1999).

Taylor FJR. Autogenous theories for the origin of eukaryotes. *Taxon* 25(4):377–390 (1976).

Taylor FJR. Implications and extensions of the serial endosymbiosis theory of the origin of the eukaryotes. *Taxon* 23(2–3):229–258 (1974).

Telfer W (ed.). *Cyril of Jerusalem and Nemesius of Emesa.* Library of Christian Classics, volume 4. London: SCM Press (1955).

Telfer W. General introduction [to Nemesius of Emesa]. In: W Telfer (ed.), *Cyril of Jerusalem and Nemesius of Emesa.* Library of Christian Classics 4. London: SCM (1955), pp. 203–223.

Telle J. (ed.). *Rosarium philosophorum. Ein alchemisches Florilegium des Spätmittelalters.* In two volumes. Weinheim: VCH (1992).

Telle J. "Die prima materia lapidis philosophici". Zu einer deutschen Lehrdictung im Basilius-Valentinus-Alchemicacorpus. *Zeitschrift für Germanistik, Neue Folge* 19(1):12–40 (2009).

Temkin O. *Galenism: rise and fall of a medical philosophy.* Ithaca NY: Cornell UP (1973).

Temkin O. The idea of descent in post-Romantic German biology. In: B Glass, O Temkin & WL Straus Jr (eds), *Forerunners of Darwin 1745–1859.* Baltimore: Johns Hopkins UP (1959), pp. 323–355.

Tertullian [Quintus Septimus Florens Tertullianus]. *De anima.* English translation in: A Roberts & J Donaldson (eds), *Ante-Nicene Christian library*, volume 15. Edinburgh: Clark (1870), pp. 410–541.

Tertullian. *De anima.* In: JP Migne (ed.), *Patrologiæ cursus completus . . . Series prima 2. Quinti Septimii Florentis Tertulliani . . . Opera omnia 2.* Parisiis: Migne (1844), cols 641–752.

Tertullian. *The writings of Tertullian.* In: A Roberts & J Donaldson (eds), *Ante-Nicene Christian Library*, volumes 11, 15 and 18. Edinburgh: Clark (1869–1870).

Thayer B. Roman Texts. At: Perseus under PhiloLogic, 2018 edition. http://perseus.uchicago.edu, accessed April 2018.

762 REFERENCES

Theatrum chemicum, præcipuos selectorum auctorum tractatus de chemiæ et lapidis philosophici. In six volumes: vol. 1, Argentorati: Zetzner (1602); vols 2 and 3: Ursellis: Zetzner (1602); vol. 4, Argentorati: Zetzner (1613); vol. 5, Argentorati: Zetzner (1622); vol. 6, Argentorati: "Zetzner" [Heilman?] (1661). Volumes 1–3 were republished in 1613 and 1659; volume 4 in 1659; and volume 5 in 1660.

The Economist (1849), as cited in: Cleanliness is next to growth. *The Economist* 436(9205), 1–7 August 2020, pp. 65–67. Online under the title "How hand-washing explains economic expansion" (https://www.economist.com/books-and-art/2020/08/01/how-hand-washing-explains-economic-expansion), accessed 18 August 2020.

The Epic of Gilgamesh (trans. NK Sandars). Harmondsworth: Penguin Classics (1960).

Themistius. *An Arabic translation of* Themistius [*sic*] Commentary on Aristoteles De anima. MC Lyons (ed.). Thetford UK: Cassirer (1973).

Themistius. *Libri paraphraseos Themistii . . . Interprete Hermolao Barbaro.* Venetiis: Scotus (1527).

Themistius. *Omnia Themistii opera, hoc est paraphrases, et orationes. Alexandri Aphrodisiensis libri dvo De anima, et De fato vnvs.* Venetiis: Aldus (1534).

Themistius. *On Aristotle's* On the soul. RB Todd (trans.). Ithaca NY: Cornell UP (1996).

Themistius. *Themistii in Aristotelis Physica paraphrasis.* H Schenkl (ed.). *Commentaria in Aristotelem Graeca.* Academiae Litterarum Regiae Borussicae 5(2). Berolini: Reimer (1900).

Themistius. *Themistii In libros Aristotelis De anima paraphrasis.* Heinze R (ed.). In: *Commentaria in Aristotelem Graeca.* Academiae Litterarum Regiae Borussicae 5(3). Berolini: Reimer (1899).

Themistius. *Themistii paraphrases Aristotelis librorum quae supersunt.* L von Spengel (ed.). In two volumes. Berolini: Teubneri (1866).

Themistius. *Themistii peripatetici lvcidissimi paraphrasis in Aristotelis . . . Hermolao Barbaro . . . interprete.* Venetiis: Scotus (1542).

Themistius. *Themistii peripatetici lvcidissimi, paraphrasis in Aristotelis . . . Hermolao Barbaro . . . interprete.* Venetiis: Scotus (1559).

Themistius. *Thémistius. Commentaire sur le* Traité de l'ame d'Aristote. *Traduction de Guillaume de Moerbeke. Édition critique.* G Verbeke (ed.). In series: Centre de Wulf-Mansion, Corpus Latinum Commentariorum in Aristoteles Graecorum 1. Louvain: Publications Universitaires (1957).

Théodoridès J. Casimir Davaine (1812–1882): a precursor of Pasteur. *History of Medicine* 10(2):155–165 (1966).

Theophrastus. *De causis plantarum.* B Einarson & GKK Link (trans.). In three volumes. Loeb Classical Library 471 (1976) and 474–475 (1990). London: Heinemann; and Cambridge MA: Harvard UP.

Theophrastus. *De igne. A post-Aristotelian view of the nature of fire.* V Courant (ed. and trans.). Assen: Van Gorcum (1971).

Theophrastus. *Enquiry into plants, and minor works on odours and weather signs.* A Hort (trans.). In two volumes. Loeb Classical Library 70 and 79. London: Heinemann; and Cambridge MA: Harvard UP (1916 and 1926).

Theophrastus. *Historia plantarvm libri decem. Græcè et Latinè.* IB à Stapel, IC Scaliger & R Constantins (eds). Amstelodami: Henricum Laurentium (1644).

Theophrastus. *On stones.* ER Caley & JFC Richards (trans.). Columbus: Ohio State University (1956).

Theophrastus. *Theophrasti De historia plantarvm liber primvs* [*–decimvs*] *per Theodorvm Gazam in Latinvm ex Craeco* [*sic, Graeco*] *sermone versvs.* Taruisium: Bartholomaevm Confalonerivm de Salodio (1483).

Theophrastus. *Theophrasti Eresii opera quae supersunt omnia. Ex recognitione Friderici Wimmer.* In three volumes. Lipsiæ: Teubner (1854–1862).

Thienemann A. Die Stufenfolge der Dinge, der Versuch eines natürlichen Systems der Naturkörper aus dem achtzehnten Jahrhundert. Eine historische Skizze. *Zoologische Annalen (Würzburg)* 3:185–274 (1910).

Thiergart T, Landan G, Schenk M, *et al.* An evolutionary network of genes present in the eukaryote common ancestor polls genomes on eukaryotic and mitochondrial origin. *Genome Biology and Evolution* 4(4):466–485 (2012).

Thomas CA Jr. The genetic organization of chromosomes. *Annual Review of Genetics* 5:237–256 (1971).

Thomas I. Introduction. In: K Foster & S Humphries (trans.), *Aristotle's De anima in the version of William of Moerbeke and the Commentary of St. Thomas Aquinas.* New Haven: Yale UP (1959), pp. 13–37.

Thomas Cantimpratensis [Thomas of Cantimpré]. *Liber de natura rerum. Editio princeps secundum codices manuscriptos.* T Boese (ed.). Berlin: de Gruyter (1973).

Thompson B. Experiments on the production of dephlogisticated air from water with various substances. *Philosophical Transactions* 77:84–124 (1787).

Thompson RC. *A dictionary of Assyrian botany.* London: British Academy (1949).

Thompson RC. *A dictionary of Assyrian chemistry and geology.* Oxford: Clarendon (1936).

Thompson RC. *The Assyrian herbal.* London: Luzac (1924).

Thomson W. On *Holtenia*, a genus of vitreous sponges. *Philosophical Transactions* 159:701–720 (1869).

Thorndike L. More manuscripts of Thomas of Cantimpré, *De naturis rerum. Isis* 54(2):269–277 (1963).

Thorndike L. The Pseudo-Galen, *De plantis* (with Latin text of chapters on stones and those of chemical interest). *Ambix* 11(1):87–94 (1963).

Thornton RJ. *New illustration of the sexual system of Carolus von Linnaeus.* Three parts in two volumes. London: Bensley (1807).

Thorpe E. *History of chemistry.* Two volumes bound in one. London: Watts (1930).

Thorpe LGM (trans.). *Two lives of Charlemagne.* Penguin Classics L213. Harmondsworth: Penguin (1969).

Thrupp J. British superstitions as to hares, geese, and poultry. *Transactions of the Ethnological Society of London* 5:162–167 (1867).

Thunberg L. *Microcosm and mediator. The theological anthropology of Maximus the Confessor.* Lund: Gleerup; and Copenhagen: Munksgaard (1965).

Thuret G. Recherches sur la fécondation des Fucacées, suivies d'observations sur les anthéridies des algues. *Annales des Sciences Naturelles, 4 Série. Botanique* 2:197–214 (1854), and 3:5–28 (1855).

Thuret G. Recherches sur les zoospores des algues et les anthéridies des cryptogames. *Annales des Sciences Naturelles, 3 Série. Botanique* 14:214–260 (1850), and 16:5–39 (1851).

Thuret G. *Recherches sur les zoospores des algues et les anthéridies des cryptogames.* Paris: Masson (1851).

Thuret G. Sur la fécondation des Fucacées. *Comptes Rendus Hebdomadaires des Séances de l'Académie des Sciences* 36:745–748 (1853).

Thurston H. George, Saint. *The Catholic Encyclopedia* 6:453–455 (1909).

Tiedemann F. *Zoologie. Zu seinen Vorlesungen entworfen.* In three volumes. Landshut: Weber (1808–1814).

Tietz JD [Titius]. *Lehrbegriff des Naturgeschichte zum ersten Unterrichte.* Leipzig: Herteln (1777).

[Tilloch A]. A cursory view of some of the late discoveries in science. *Philosophical Magazine* 6:126–132, 243–251 and 304–312 (1800).

Tillyard EMW. *The Elizabethan world picture.* London: Chatto & Windus (1943).

Tishby I. General introduction. In: P Lachower & I Tishby (eds), *The wisdom of the Zohar*, volume 1. London: Littman Library of Jewish Civilization (1989), pp. 1–126.

Todd RB. Themistius. *Catalogus Translationum et Commentariorum* 8:57–102 (2003).

Tolkowsky S. *Herperides. A history of the culture and use of citrus fruits.* London: Bale & Curnow (1938).

Tommaso de Vio [Thomas de Vio Cardinalis Caietanus]. *Commentaria in De anima Aristotelis.* PI Coquelle (ed.). In two volumes. *Scripta Philosophica.* Romæ: Angelicum (1938–1939).

Tommaso de Vio [Thomas de Vio Cardinalis Caietanus]. *Commentaria . . . in libros Aristotelis De anima.* Venetiis: Georgius Arriuabenus (1514).

Topsell E. *The historie of fovre-footed beastes.* London: William Iaggard (1607).

764   REFERENCES

Topsell E. *The history of four-footed beasts and serpents . . . Whereunto is now added, the Theater of insects; or, lesser living creatures: as bees, flies, caterpillars, spiders, worms, &c. A most elaborate work: by T. Muffet*. In three volumes. London: Cotes (1658).

Touchon M, Perrin A, Moura de Sousa JA, *et al*. Phylogenetic background and habitat drive the genetic diversification of *Escherichia coli*. *PLoS Genetics* 16(6):e1008866 (2020).

Tournefort JP de. *Élemens de botanique, ou méthod pour connoître les plantes*. In three volumes. Paris: L'Imprimerie royale (1694).

Tournefort JP de. *Institutiones Rei Herbariæ. Editio altera, Gallica longe auctior*. In three volumes. Parisiis: Typographia Regia (1700–1703).

Tradescant J. *Musæum Tradescantianum, or, a collection of rarities preserved at South-Lambeth neer London*. London: J Grismond (1656).

Tragus H [Bock J]. *De stirpivm, maxime earvm, qvæ in Germania nostra nascvntvr*. Argentorati: Vuendelinus Rihelius (1552).

Travaglia P. *Una cosmologia ermetica. Il Kitab sirr al-ḫalīqa / De secretis naturae*. Napoli: Liguori (2001).

Treiger A. The Fathers in Arabic. In: K Parry (ed.), *The Wiley Blackwell companion to patristics*. Chichester: Wiley (2015), pp. 442–455.

Trembley A. Extract of a letter of Mr. Abraham Trembley, FRS to Tho. Birch. *Philosophical Transactions* 50:58–62 (1757).

Trembley A. *Mémoires pour server à histoire d'un genre de polypes d'eau douce, à bras en forme de cornes*. Leide: Verbeek (1744).

Trembley A. Memoirs concerning the natural history of a type of freshwater polyp with arms shaped like horns. SG Lenhoff & HM Lenhoff (trans.). In: SG Lenhoff & HM Lenhoff. *Hydra and the birth of experimental biology—1744. Book 2*. Pacific Grove CA: Boxwood (1986), pag. sep.

Trembley M. *Correspondance inédite entre Réaumur et Abraham Trembley comprenant 113 lettres recueillies et annotées*. Genève: Georg (1943).

Trentepohl JF. Beobachtungen über die Fortpflanzung der Ectospermen des Herrn Vaucher, insonderheit der Conferva bullosa Linn. nebst einigen Bemerkungen über die Oscillatorien. In: AW Roth (ed.), *Botanische Bemerkungen und Berichtigungen*. Leipzig: Joachims literarischem Magazin (1807), pp. 180–216.

Trésor de la Langue Française. In 16 volumes. Paris: Centre National de la Recherche Scientifique (1994).

Treviranus GR. *Biologie, oder Philosophie der lebenden Natur für Naturforscher und Aerzte*. In six volumes. Göttingen: Röwer (1802–1822).

Trifonov EN & Kejinovsky E. Acytota—associated kingdom of neglected life. *Journal of Biomolecular Structure and Dynamics* 34(8):1641–1648 (2016).

Tubbs RS, Padmalayam D, Shoja MM & Loukas M. Heinrich August Wrisberg (1736–1808). *Clinical Anatomy* 27(1):10–13 (2014).

Tulasne LR & Tulasne C. *Selecta fungorum carpologia*. In three volumes. Parisiis: Jussu (1861–1865).

Tulasne LR. Note sur l'appareil reproducteur dans les lichens et les champignons. *Comptes Rendus Hebdomadaires des Séances de l'Académie des Sciences* 32:427–430 and 470–475 (1851).

Turner HR. *Science in medieval Islam. An illustrated introduction*. Austin: University of Texas Press (1997).

Turner JD. Neoplatonism and Gnosticism. In: RT Wallis (ed.), *Neoplatonism and Gnosticism. Studies in Neoplatonism: ancient and modern* (RB Harris, general ed.) 6. Albany: SUNY Press (1992), pp. 425–459.

Turner W. *A new herball, wherin are conteyned the names of herbes in Greke, Latin, Englysh, Duch Frenche, and in the poteraries and herbaries Latin* [title varies]. Volume 1, London: Steven Mierdman (1551); volumes 2 and 3, Collen [Köln]: Arnold Birckman (1562 and 1568).

Turner W. *Avivm praecipvarvm, qvarvm apvd Plinivm et Aristotelem mentio est, breuis & succincta historia*. Coloniæ: Ioan. Gymnicus (1544).

Turner W. Carlovingian schools. *The Catholic Encyclopedia* 3:349–351 (1908).

Turner W. *Libellus de re herbaria novvs, in quo herbarium aliquot nomina greca, latina & Anglica habes.* London: Byndell (1538).

Turpin PJF. Aperçu organographique sur le nombre deux. *Mémoires du Muséum d'Histoire Naturelle* 16:295–344 (1828).

Turpin PJF. *Atlas contenant deux planches d'anatomie comparée, trois de botanique et deux de géologie.* [With:] Goethe [JW von]. *Œuvres d'histoire naturelle de Goethe comprenant divers mémoires d'anatomie comparée de botanique et de géologie.* CF Martins (trans.). Paris: Cherbuliez [1837].

Turpin PJF. *Essai d'une iconographie élémentaire et philosophique des végétaux.* In: JLM Poiret & PJF Turpin, *Leçons de flore. Cours complet de botanique* (1819–1820). Tome 3 [of 3]. Paris: Panckoucke (1820).

Turpin PJF. Mémoires sur l'inflorescence des Graminées et des Cypérées, comparée avec celles des autres végétaux sexiféres; suivi de quelques observations sur les disques. *Mémoires du Muséum d'Histoire Naturelle* 5:426–492 (1819).

Turpin PJF. Observations sur le nouveau genre Surirella. *Mémoires du Muséum d'Histoire Naturelle* 16:361–368 (1828).

Turpin PJF. Observations sur quelques productions marines, qui avoient été considérées, les unes, comme des animalcules isolés; les autres, comme des agrégations filamenteuses d'animalcules analogues aux premiers. *Mémoires du Muséum d'Histoire Naturelle* 15:299–328 (1827).

Turpin PJF. Organographie microscopique, élémentaire et comparée des végétaux. Observations sur l'origine ou la formation primitive du tissue cellulaire. *Mémoires du Muséum d'Histoire Naturelle* 18:161–211 (1829).

Turpin PJF. Organographie végétale. Observations microscopiques sur l'organisation tissulaire, l'accroissement et le mode de reproduction de la truffe comestible, comparée aux tissus, à la production de la globuline, et de tous les corps reproducteurs des autres végétaux. *Mémoires du Muséum d'Histoire Naturelle* 15:343–376 (1827).

Turpin PJF. Organographie végétale. Observations sur quelques végétaux microscopiques, et sur le rôle important que leurs analogues jouent dans la formation et l'accroissement du tissue cellulaire. (Lues à l'Académie des Sciences de l'Institut, en sa séance du 12 juin 1826.) *Mémoires du Muséum d'Histoire Naturelle* 14:15–67 (1827). Also published separately by Belin, Paris (1827).

Turpin PJF. Surirelle, *Surirella. Dictionnaire des sciences naturelles* 51:408–410 (1827).

Turpin PJF. Tables des planches. In: *Dictionnaire des sciences naturelles. Planches 2e partie: Règne organisé. Botanique... Végétaux acotylédons.* Paris: Levrault (1816–1829), pp. 1–8.

Tyson E. *Orang-outang, sive homo sylvestris: or, The anatomy of a pygmie compared with that of a monkey, an ape, and a man. To which is added, A philological essay concerning the pygmies, the cynocephali, the satyrs, and sphinges of the ancients.* London: Bennett and Brown (1699).

Tyson J. *The cell doctrine: its history and present state.* [First edition]. Philadelphia: Lindsay & Blakiston (1870).

Tyson J. *The cell doctrine: its history and present state.* Second edition. Philadelphia: Lindsay & Blakiston (1878).

Udagawa Yōan. *Botanikakyō.* [*Botany sutra*]. Edo: Udagawajuku (1822).

Udagawa Yōan. *Shokugaku keigen.* [*The principles of botany*]. In three volumes. Edo: Bosatsurô (1833).

Ueberweg F. *A history of philosophy, from Thales to the present time. From the fourth German edition.* GS Morris (trans.), N Porter (add.). In two volumes. London: Hodder & Stoughton (1872–1874).

Ueno M. The Western influence on natural history in Japan. *Monumenta Nipponica* 19(3/ 4):315[81]–339[105] (1964).

Uglow J. *The lunar men. The friends who made the future. 1730–1810.* London: Faber & Faber (2002).

Ullmann M. *Die Natur- und Geheimwissenschaften im Islam.* Handbuch der Orientalistik (B Spuler, ed.), Abteilung 1, Ergänzungsband 6, Abschnitt 2. Leiden: Brill (1972).

Underwood LM. The classification of the Archegoniates. *Bulletin of the Torrey Botanical Club* 22(3):124–129 (1895).

766 REFERENCES

Unger F. *Botanical letters to a friend*. DB Paul (trans.). London: Samuel Highley (1853).

Unger F. *Botanische Briefe*. Wien: Gerold (1852).

Unger F. Die Metamorphose der Ectosperma clavata Vaucher. *Verhandlungen der Kaiserlichen Leopoldinisch-Carolinischen Akademie der Naturforscher / Nova Acta Physico-Medica Academiæ Cæsareæ Leopoldino-Carolinæ Naturæ Curiosorum* 13(2):789–808 (1827).

Unger F. *Die Pflanze im Momente der Thierwerdung*. Wien: Beck (1843).

Unger F. Mikroskopische Beobachtungen. *Verhandlungen der Kaiserlichen Leopoldinisch-Carolinischen Akademie der Naturforscher / Nova Acta Academiæ Cæsareæ Leopoldino-Carolinæ Naturæ Curiosorum* 18(2):685–710 (1838).

"Un officier du Roi" [Saint-Pierre JHB de]. *Voyage à l'Isle de France, à l'Isle de Bourbon, au Cap de Bonne-Espérance, &c., avec des observations nouvelles sur la nature & sur les hommes*. Volume 1, Neuchatel: Société Typographique; volume 2, Amsterdam: Merlin (1773).

Vaillant S. *Discours sur la structure des fleurs, leurs differences et l'usage de leurs parties*. Leide: Vander Aa (1718).

Valentine B [attrib.]. *Ein kurtz summarischer Tractat von dem grossen Stein der Vralten*, hrsg. Johann Thölde. Eisleben: Johann Hörnig (1599).

Valentine B [attrib.]. *Ein kurtz summarischer Tractat von dem grossen Stein der Vralten*. With *Anhang*. Leipzig: Apel (1602).

Valentine B [attrib.]. *Practica cvm dvodecim clavibvs et appendice, De magno lapide antiqvorvm sapientvm, scripta & relicta*. M Maier (trans.). In: M Maier (ed.), *Tripus aureus, hoc est, tres tractatvs chymici selectissimi*. Francofvrti: Paulus Iacobus (1618), pp. 7–75.

Valentine B [attrib.]. *The last vvill and testament of Basil Valentine, monke of the Order of St. Bennet*. London: S.G. and B.G. for Edward Brewster (1671).

Valentini MB. *Historia simplicium reformata, sub musei museorum . . . Auctoris D. Joh. Conrado Beckero*. Francofurti: Zunneriana (1716).

van der Gulik PTS, Hoff WD & Speijer D. In defence of the three-domains of life paradigm. *BMC Evolutionary Biology* 17(1):218 (2017).

Vanderlaan TA, Ebach MC, Williams DM & Wilkins JS. Defining and redefining monophyly: Haeckel, Hennig, Ashlock, Nelson and the proliferation of definitions. *Australian Systematic Biology* 26(5):347–355 (2013).

Vandersmissen J. Le débat sur la véritable nature du corail au XVIIIe siècle. In: *Neuvième Congrès de l'Association des Cercles Francophones d'Histoire et d'Archéologie de Belgique. LVIe Congrès de la Fédération des Cercles d'Archéologie et d'Histoire de Belgique, Liège, 23–26 août 2012, 6e Section*. Liège (2012), pp. 1–9.

Vandiver PB, Soffer O, Klima B & Svoboda J. The origins of ceramic technology at Dolni Věstonice, Czechoslovakia. *Science* 246(4933):1002–1008 (1989).

Van Etten J & Bhattacharya D. Horizontal gene transfer in eukaryotes: not if, but how much? *Trends in Genetics* 36(12):915–925 (2020).

van Maerlant J. *Jacob van Maerlant's Naturen bloeme, uitgegeven door Dr. Eelco Verwijs. Eerste Deel*. In series: Bibliotheek van Middelnederlandsche Letterkunde (HE Moltzer & J te Winkel, eds). Groningen: Wolters (1878).

van Raalte M. The nature of fire and its complications: Theophrastus' *De igne* 1–10. *Bulletin of the Institute of Classical Studies* (University of London) 53(1):47–97 (2010).

van Schaik JML. Cathars, Albigensians, and Bogomils. *Encyclopædia Iranica* (2006), https://iranicaonline.org/articles/cathars-albigensians-and-bogomils, accessed 19 August 2018.

van Spronsen JB. *The periodic system of chemical elements*. Amsterdam: Elsevier (1969).

van Wyhe J (ed.). The complete work of Charles Darwin online. http://darwin-online.org.uk.

Varro, M Terentius. *Rerum rusticarum libri tres*. In: WD Hooper & HB Ash (trans.), *Cato and Varro on agriculture*. Loeb Classical Library 283. London: Heinemann (1934).

Varro, M Terentius. *Varro on farming. M. Terenti Varronis Rerum rusticarum libri tres*. L Storr-Best (trans.). London: Bell (1912).

Vartanian A. Trembley's polyp, La Mettrie, and Eighteenth-century French materialism. *Journal of the History of Ideas* 11(3):259–286 (1950).

Vaucher [J]P. Mémoire sur les graines des conferves. *Journal de Physique, de Chimie, d'Histoire Naturelle et des Arts* 52:344–359 (An 9, 1801).

Vaucher JP. *Histoire des conferves d'eau douce, contenant leurs différens modes de reproduction, et la description de leurs principales espèces, suivie de l'histoire des trémelles et des ulves d'eau douce.* Genève: Paschoud (An 11, 1803).

Vaucher JP. *Histoire physiologique des plantes d'Europe ou exposition des phénomènes qu'elles presentent dans les diverses periodes de leur développement.* In four volumes. Paris: Aurel (1841).

Vaughan H. *The works of Henry Vaughan.* LC Martin (ed.). In two volumes. Oxford: Clarendon (1914).

Vaughan T [as Eugenius Philalethes]. *Anima magica abscondita; or a discourse of the universall spirit of nature, with his strange, abstruse, miraculous ascent, and descent.* London: T.W. for H.B. (1650).

Vaughan T [as Eugenius Philalethes]. *Anthroposophia theomagica: or a discourse of the nature of man and his state after death.* London: T.W[.] for H. Blunden (1650).

Vaughan T [as Philalethes]. *Philalethæ tractatus tres.* In: *Musæum Hermeticum reformatum et amplificatum.* Francofurti: Hermannum à Sande (1678), pp. 741–814.

Vaughan T [as Philalethes]. *The three treatises of Philalethes.* In: AE Waite (ed.), *The Hermetic Museum restored and enlarged,* volume 2. London: James Elliott (1893), pp. 225–269.

Veenker RA. Gilgamesh and the magic plant. *The Biblical Archaeologist* 44(4):199–205 (1981).

Veith I. Creation and evolution in the Far East. In: S Tax & C Callender (eds), *Evolution after Darwin,* volume 3. University of Chicago Press (1960), pp. 1–17.

Venn J. *Symbolic logic.* London: Macmillan (1881).

Verbeke G. Thémistius et le commentaire de S. Thomas au *De anima* d'Aristote. In: G Verbeke (ed.), *Thémistius. Commentaire sur le* Traité de l'ame *d'Aristote. Traduction de Guillaume de Moerbeke. Édition critique.* Centre de Wulf-Mansion, Corpus Latinum Commentariorum in Aristoteles Graecorum 1. Louvain: Publications Universitaires de Louvain (1957), pp. ix–xxxviii.

Verworn M. *Psycho-physiologische Protisten-Studien. Experimentelle Untersuchungen.* Jena: Gustav Fischer (1889).

Vicq-d'Azyr F. *Encyclopédie méthodique. Systême anatomique. Quadrupèdes.* Tome 2. Paris: Panckoucke; and Liège: Plomteux (1792).

Vicq d'Azyr F. Table pour server à l'histoire naturelle & anatomique des corps organiques ou vivans, présentee dans la séance publique de l'Académie Royale des Sciences, le 12 Novembre 1774. *Observations sur la Physique, sur l'Histoire Naturelle et sur les Arts* 4:479, with large folding table opposite (1774).

Vicq d'Azyr [F]. *Traité d'anatomie et de physiologie.* Tome 1 [only volume published]. Paris: Didot l'Aîné (1786).

Villemet [PR]. Essai sur l'histoire naturelle du champignon vulgaire. *Academie des Sciences, Arts et Belles-Lettres, Dijon, (Nouveau) Mémoires* 1783(2):195–211 (1784).

Vincent of Beauvais. *Bibliotheca mvndi . . . Specvlvm qvadrvplex, natvrale, doctrinale, morale, historiale.* In four volumes. Dvaci: Belleri (1624).

Vincent of Beauvais. *Specvli maioris Vincentii Bvrgvndi præsvlis Belvacensis.* In four volumes. Venetiis: Dominicum Nicolinum (1591).

Vines SH. An examination of some points in Prof. Weismann's theory of heredity. *Nature* 40:621–626 (1889).

Virchow R. *Cellular pathology as based upon physiological and pathological histology.* F Chance (trans.). London: John Churchill (1860).

Virchow R. Cellular-Pathologie. *Archiv für pathologische Anatomie und Physiologie und für klinische Medicin* 8(1):3–39 (1855).

Virchow R. *Die Cellularpathologie in ihrer Begründung auf physiologische und pathologische Geweblehre.* Berlin: August Hirschwald (1858).

Virchow R. *Die Cellularpathologie in ihrer Begründung auf physiologische und pathologische Geweblehre.* Second edition. Berlin: August Hirschwald (1859).

768   REFERENCES

Virchow R. *La pathologie cellulaire basée sur l'étude physiologique et pathologique des tissus*. P Picard (trans.). Paris: Baillière (1861).

Virey JJ. Animal. *Nouveau dictionnaire d'histoire naturelle* 1:419–466 (An 11, 1803).

Virey JJ. Animal. *Nouveau dictionnaire d'histoire naturelle. Nouvelle édition* 2:1–81 (1816).

Virey JJ. Mue. *Nouveau dictionnaire d'histoire naturelle* 15:217–220 (An 11, 1803).

Virey JJ. Nature. *Nouveau dictionnaire d'histoire naturelle* 15:358–414 (An 11, 1803).

Virey JJ. *Philosophie de l'histoire naturelle ou phènomènes de l'organisation des animaux et des végètaux*. Paris: Baillière (1835).

Virey JJ. Remarques sur l'identité de certaines lois générales observées dans une distribution naturelle des insectes et des champignons; par WS Macleay, esq. *Bulletin des Sciences Naturelles et de Géologie* 4:275–278 (1825).

Virgil [Vergilius Maro, Publius]. *The Georgics*. RD Blackmore (trans.). London: Sampson Low & Marston (1871).

Virgil [Vergilius Maro, Publius]. *Virgil 1 and 2*. Loeb Classical Library 63–64. London: Heinemann; and Cambridge MA: Harvard UP (1916–1918).

Vishnu-Mittre. Biological concepts and agriculture in ancient India. *Indian Journal of the History of Science* 5(1):144–161 (1970).

Vlastos G. *Socrates. Ironist and moral philosopher*. Cambridge UP (1991).

Vogt K. Ancient Skepticism. *The Stanford Encyclopedia of Philosophy* (Winter 2016 edition), EN Zalta (ed.), https://plato.stanford.edu/archives/win2016/entries/skepticism-ancient/, accessed 28 May 2018.

Voltaire [Arouet FM]. Chaîne des êtres créés. In: *Dictionnaire philosophique* 3:73–76 (1822).

[Voltaire] [Arouet FM]. Chaîne des êtres créés. In: *Questions sur l'Encyclopédie, par des amateurs*. 3:284–287 (1770).

Voltaire [Arouet F-M]. *Dictionnaire philosophique*. In eight volumes. Paris: Touquet (1822).

Voltaire [Arouet FM]. *Le micromégas*. Londres: [n.p.] (1752).

[Voltaire] [Arouet FM]. *Questions sur l'Encyclopédie, par des amateurs*. In nine volumes. [Vols 1–8: Genève: Cramer] (1770–1772).

von Arnim J. *Stoicorvm vetervm fragmenta collegit Ioannes ab Arnim*. In four volumes. Stvtgardiae: Tevbneri (1964). Reproduction of first edition (1905).

von Hees S. *Enzyklopädie als Spiegel des Wiltbildes. Qazwīnīs Wunder des Schöpfung—eine Naturkunde des 13. Jahrhunderts*. Wiesbaden: Harrassowitz (2002).

von Hofsten N. Linnaeus's conception of Nature. *Kungliga Svenska Vetenskaps-Societetens Årsbok for 1957* (1958), pp. 65–105.

von Lippmann EO. *Urzeugung und Lebenskraft. Zur Geschichte dieser Probleme von den ältesten Zeiten an bis zu den Anfängen des 20. Jahrhunderts*. Berlin: Springer (1933).

von Meyer E. *A history of chemistry from earliest times to the present day*. London: Macmillan (1906).

von Richthofen [F]. Ueber die centralasiatischen Siedenstrassen bis zum 2. Jahrhundert n. Chr. *Verhandlungen der Gesellschaft für Erdkunde zu Berlin* 4:96–122 (1877).

Voss A. (ed.) *Marsilio Ficino*. Western Esoteric Masters series. Berkeley: North Atlantic (2006).

Voultsiadou E, Gerovasileiou V, Vandepitte L, et al. Aristotle's scientific contributions to the classification, nomenclature and distribution of marine organisms. *Mediterranean Marine Science* 18(3):468–478 (2017).

Waibel VL. Hölderlin's idea of "Bildungstrieb": a model from yesteryear? *Educational Philosophy and Theory* 50(6–7):640–651 (2018).

Wainwright M. Extreme pleomorphism and the bacterial life cycle: a forgotten controversy. *Perspectives in Biology and Medicine* 40(3):407–414 (1997).

Wainwright M. Microbiology before Pasteur. *Microbiology Today* 28(2):19–21 (2001).

Waite AE. *Lives of alchemystical phyilosophers . . . to which is added a bibliography of alchemy and Hermetic philosophy*. London: George Redway (1888).

Waite AE (ed.). *The hermetic and alchemical writings of . . . Paracelsus*. In two volumes. Chicago: Laurence, Scott (1910). Reprinted by Kessinger, Kila MT (undated, *circa* 1992).

REFERENCES   769

Waite AE (trans.). *The Hermetic Museum restored and enlarged*. In two volumes. London: James Elliott (1893).

Waite AE. *The Turba Philosophorum or Assembly of the sages. Called also the Book of truth in the art and the Third Pythagorical Synod*. London: George Redway (1896).

Walker J. *Essays on natural history and rural economy*. London: Longman, Hurst *et al.* (1812).

Walker-Arnott GA. Botany. *Encyclopædia Britannica*, Seventh edition 5:30–141 (1842). Pages 74–89 written by JE Smith.

Wallerius JG. *Hydrologia, eller Wattu-riket indelt och beskrifvit*. Stockholm: Lars Salvius (1748).

Wallerius JG. *Hydrologie, oder Wasserreich*. Berlin: Nicolai (1751).

Wallerius JG. *Mineralogia, eller Mineral-riket*. Stockholm: Lars Salvius (1747).

Wallerius JG. *Systema mineralogicum, qvo corpora mineralia in classes, ordines, genera et species, suis cum varietatibus divisa, describuntur*. In two volumes. Holmiæ: Laurentius Salvius (1772–1775).

Wallich GC. On the true nature of the so-called "Bathybius," and its alleged function in the nutrition of the protozoa. *Annals and Magazine of Natural History, 4 Series* 16(95):322–339 (1875).

Wallich GC. Remarks on some novel phases of organic life, and on the boring powers of minute annelids, at great depths in the sea. *Annals and Magazine of Natural History, 3 Series* 8(43):52–58 (1861).

Wallin IE. *Symbionticism and the origin of species*. Baltimore: Williams & Wilkins (1927).

Wallis RT. Introduction, part 1. In: RT Wallis & J Bregman (eds), *Neoplatonism and Gnosticism*. In series: Studies in Neoplatonism: ancient and modern (RB Harris, gen. ed.), volume 6. Albany: SUNY Press (1992), pp. 1–3.

Wallis RT & Bregman J (eds). *Neoplatonism and Gnosticism*. Studies in Neoplatonism: ancient and modern (RB Harris, general ed.), volume 6. Albany: SUNY Press (1992).

Walsh DM. Organisms as natural purposes: the contemporary evolutionary perspective. *Studies in History and Philosophy of Biological and Biomedical Sciences* 37(4):771–791 (2006).

Walshe MO'C. Introduction to Part One. In: MO'C Walshe (trans. and ed.), *The complete mystical works of Meister Eckhart*. Revised edition. New York: Herder & Herder/Crossroad (2009), pp. 3–28.

Walzer R. al-Fārābī. *The encyclopaedia of Islam*, New [Second] edition 2:778–781. Leiden: Brill (1991).

Wang W, Mauleon R, Hu Z, *et al*. Genomic variation in 3,010 diverse accessions of Asian cultivated rice. *Nature* 557(7703):43–49 (2018).

Waquet F. *Latin or the empire of a sign. From the sixteenth to the twentieth centuries*. J Howe (trans.). London: Verso (2001).

Waquet F. *Le latin ou l'empire d'un signe*. Paris: Albin Michel (1998).

Ward HC. The newer knowledge of the morphology of bacteria. In: EO Jordan & IS Falk (eds), *The newer knowledge of bacteriology and immunology*. University of Chicago Press (1928), pp. 1–13.

Warmington EH. Index of fishes. In: Pliny. *Natural history 8*. Loeb Classical Library 418. Cambridge MA: Harvard UP (1963), pp. 585–596.

Warner HJ. *The Albigensian heresy*. London: Macmillan (1922).

Warren L. *Joseph Leidy. The last man who knew everything*. Yale UP (1998).

Waterhouse GM. Presidential address. The zoospore. *Transactions of the British Mycological Society* 45(1):1–20 (1962).

Waterhouse GR. On the classification of the mammalia. *Report of the Thirteenth Meeting of the British Association for the Advancement of Science held at Cork in August 1843, Notices and Abstracts of Communications*. London: John Murray (1844), pp. 65–67.

Watson JD. *Molecular biology of the gene*. Second edition. New York: Benjamin (1970).

Watson JD & Crick FHC. Molecular structure of nucleic acids: a structure for deoxyribose nucleic acid. *Nature* 171(4356):737–738 (1953).

Watson R. *Chemical essays*. In five volumes. London: Evans (1781–1787).

Watson W. An account of the insect called the vegetable fly. *Philosophical Transactions of the Royal Society* 53:271–274 (1763).

770 REFERENCES

Watson W. An historical memoir concerning a genus of plants called lichen, by Micheli, Haller, and Linnaeus: and comprehended by Dillenius under the terms Usnea, Coralloides, and Lichenoides: tending principally to illustrate their several uses. *Philosophical Transactions* 50:652–688 (1758).

Watts GB. The *Encyclopedie methodique*. *PMLA* 73(4):348–366 (1958).

Weber NA. Albigenses. *The Catholic Encyclopedia* 1:267–269 (1907).

Webster C. *From Paracelsus to Newton. Magic and the making of modern science*. New York: Barnes & Noble (1996).

Webster C. The recognition of plant sensitivity by English botanists in the Seventeenth century. *Isis* 57(1):5–23 (1966).

Weis FG. *Plantæ cryptogamicæ*. Gottinge: Vandenhoek (1770).

Weisheipl JA. *Nature and motion in the Middle Ages*. WE Carroll (ed.). Washington: Catholic University of America Press (1985).

Weismann A. *Über die Vererbung*. Jena: Gustav Fischer (1883). Translated as: On heredity [1883]. In: EB Poulton, S Schönland & AE Shipley (eds), *Essays upon heredity and kindred biological problems*. In series: Translations of foreign biological memoirs 4. Oxford: Clarendon (1889), pp. 67–105.

Weisser U (ed.). *Buch über das Geheimnis der Schöpfung und die Darstellung der Natur (Buch der Ursachen) von Pseudo-Apollonius von Tyana*. Aleppo: Institute for the History of Arabic Science (1979).

Weisser U. *Das "Buch über das Geheimnis der Schöpfung und die Darstellung der Natur" von Pseudo-Apollonius von Tyana*. In series: Ars medica. Texte und Untersuchungen zur Quellenkunde der Alten Medizin 3(2). A Dietrich & O Spies (eds). Berlin: de Gruyter (1980).

Wellmann M. *Der Physiologos: Eine religionsgeschichtlich-naturwissenschaftliche Untersuchung*. *Philologus* Supplementband 22.1. Leipzig: Dieterich (1930).

Werner AG. *Kurze Klassifikation und Beschreibung der verschiedenen Gebirgsarten*. Dresden: Walter (1787).

Wesley J. *A survey of the wisdom of God in the Creation: or a compendium of natural philosophy*. In two volumes. Bristol: William Pine (1763).

Wesley J. *A survey of the wisdom of God in the Creation: or, a compendium of natural philosophy*. In three volumes. The second edition. Bristol: William Pine (1770).

Wesley J. *A survey of the wisdom of God in the Creation: or, a compendium of natural philosophy*. A new edition. In five volumes. London: Maxwell & Wilson; and Williams & Smith (1809).

Westerink LG. *Anonymous prolegomena to Platonic philosophy*. Amsterdam: North-Holland (1962).

Westerink LG. *Michael Psellus De omnifaria doctrina. Critical text and introduction*. Nijmagen: Centrale (1948).

Westermarck E. *Pagan survivals in Mohammedan civilization*. London: Macmillan (1933).

Wetherbee W & Aleksander J. Dante Alighieri. *The Stanford Encyclopedia of Philosophy* (Fall 2018 edition), EN Zalta (ed.), plato.stanford.edu/archives/fall2018/entries/dante, accessed 7 November 2018.

Wetherbee W. Introduction. In: *The Cosmographia of Bernardus Silvestris*. New York: Columbia UP (1973), pp. 1–62.

Wethered HN. *The mind of the ancient world. A consideration of Pliny's Natural history*. London: Longmans, Green (1937).

Wetzels W. Some observations on Goethe and Linné. In: J Weinstock (ed.), *Contemporary perspectives on Linnæus*. Lanham MD: University Press of America (1985), pp. 135–151.

Whatley JM, John P & Whatley FR. From extracellular to intracellular: the establishment of mitochondria and chloroplasts. *Proceedings of the Royal Society B: Biological Sciences* 204(1155):165–187 (1979).

Whewell W. *An essay on mineralogical classification and nomenclature; with tables of the orders and species of minerals*. Cambridge: J. Smith (1828).

REFERENCES 771

Whewell W. *History of the inductive sciences, from the earliest to the present time.* New [Second] edition. In three volumes. London: Parker (1847).

Whewell W. *History of the inductive sciences, from the earliest to the present time.* Third edition. In three volumes. London: Parker (1857).

White C. *An account of the regular gradation in Man, and in different animals and vegetables; and from the former to the latter . . . Read to the Literary and Philosophical Society of Manchester, at different meetings, in the year 1795.* London: Dilly (1799).

White G. *The natural history and antiquities of Selborne, in the county of Southampton.* London: Bensley (1789).

White GW. Early discoverers XXVII. Announcement of glaciation in Scotland. William Buckland (1784–1856). *Journal of Glaciology* 9(55):143–145 (1970).

White TH. *The book of beasts. Being a translation from a Latin bestiary of the Twelfth century.* London: Cape (1954).

Whitehead AN. *Process and reality.* New York: Macmillan (1929).

Whitehead AN. *Process and reality.* Corrected edition. DR Griffin & DW Sherburne (eds). New York: Free Press (1978).

Whitehurst J. *An inquiry into the original state and formation of the earth; deduced from facts and the laws of nature.* London: Cooper (1778).

Whitman CO. The inadequacy of the cell-theory of development. *Journal of Morphology* 8(3):639–658 (1893).

Whittaker RH. New concepts of kingdoms of organisms. *Science* 163(3863):150–160 (1969).

Whittaker RH. On the broad classification of organisms. *Quarterly Review of Biology* 34(3):210–226 (1959).

Whittaker RH. The kingdoms of the living world. *Ecology* 38(3):536–538 (1957).

Whittaker RH & Margulis L. Protist classification and the kingdoms of organisms. *BioSystems* 10(1–2):3–18 (1978).

Whorf BL. *Language, thought, and reality* (JB Carroll, ed.). Cambridge MA: MIT Press (1956).

Wicher HB. Nemesius Emesensus. *Catalogus Translationum et Commentariorum* 6:31–72 (1960).

Wickersheimer E. Le livre des quadrupèdes de Michel Herr, médicin Strasbourgeois (1546). In: A Koyré (ed.), *La science au XVIe siècle. Colloque international de Royaumont, 1–4 juillet 1957.* Paris: Hermann (1960), pp. 267–283.

Wiedemann E. Über die Kriechtiere nach al Qazwînî nebst einigen Bemerkungen über die zoologischen Kenntnisse der Araber. *Sitzungsberichte der physikalisch-medizinischen Societät zu Erlangen* 48–49: 228–285 (1916–1917). Reprinted as *Beiträge* 53 of E Wiedemann, *Aufsätze zur arabischen Wissenschaftsgeschichte* 2. Hildesheim: Olms (1970), pp. 314–371.

Wiedemann E. Übersetzung und Besprechung des Abschnittes über die Pflanzen von Qazwînî. *Sitzungsberichte der physikalisch-medizinischen Societät zu Erlangen* 48–49: 286–321 (1916–1917). Reprinted as *Beiträge* 54 of E Wiedemann, *Aufsätze zur arabischen Wissenschaftsgeschichte* 2. Hildesheim: Olms (1970), pp. 372–407.

Wiener PP. The tradition behind Galileo's methodology. *Osiris* 1:733–746 (1936).

Wier J. *De præstigiis dæmonvm, et incantationibus ac ueneficijs, libri v.* Tertia editione aucti. Basileæ: Ioannem Oporinum (1566).

Wiesner J. *Jan Ingen-Housz: sein Leben und sein Wirken als Naturforscher und Arzt.* Wien: Carl Konegen (1905).

Wilczynski JZ. On the presumed Darwinism of Alberuni eight hundred years before Darwin. *Isis* 50(4):459–466 (1959).

Wildberg C. Elias. *The Stanford Encyclopedia of Philosophy* (Fall 2016 edition), EN Zalta (ed.), https://plato.stanford.edu/archives/fall2016/entries/elias/, accessed 12 July 2018.

Wildberg C. John Philoponus. *The Stanford Encyclopedia of Philosophy* (Spring 2016 edition), EN Zalta (ed.), https://plato.stanford.edu/archives/spr2016/entries/philoponus, accessed 17 September 2018.

Wildberg C. Neoplatonism. *The Stanford Encyclopedia of Philosophy* (Spring 2016), EN Zalta (ed.), https://plato.stanford.edu/archives/spr2016/entries/neoplatonism/, accessed 2 July 2018.

Wiley EO. *Phylogenetics. The theory and practice of phylogenetic systematics.* New York: Wiley (1981).

772 REFERENCES

Wilkins J. *An essay towards a real character, and a philosophical language.* London: Gellibrand & Martyn (1668).

Will CL & Lührmann R. Spliceosome structure and function. *Cold Spring Harbor Perspectives in Biology* 3(7):a003707 (2011).

Wille N. Chlorophyceae. In: A Engler (ed.), *Die natürlichen Pflanzenfamilien nebst ihren Gattungen und wichtigeren Arten . . . begrünget von A Engler und K Prantl.* Teil 1, Abteilung 2. Leipzig: Engelmann (1897), pp. 24–161.

William of Moerbeke. A list of translations made from Greek and Latin in the Thirteenth century. Arranged, introduced and annotated by Edward Grant. In: E Grant (ed.), *Source book in medieval science.* Cambridge MA: Harvard UP (1974), pp. 39–41.

Williams LP. [Review of:] Hegel's *Philosophy of nature. Isis* 63(2):290–291 (1972).

Williams SJ. Philip of Tripoli's translation of the pseudo-Aristotelian Secretum secretorum viewed within the context of intellectual activity in the Crusader Levant. In: I Draelants, A Tihon & B van der Abeele (eds), *Occident et Proche-Orient: contacts scientifiques au temps des Croisades: Actes du colloque de Louvain-la-Neuve, 23 et 25 mars 1997* (2000), pp. 79–94.

Williams SJ. *The scholarly career of the pseudo-Aristotelian Secretum secretorum in the Thirteenth and early Fourteenth century.* PhD thesis, Northwestern University (1991).

Willis T. *De anima brutorum quæ hominis vitalis ac sensitiva est, exercitationes duæ.* Osonii: Sheldoniano (1672).

Willis T. *Two discourses concerning the soul of brutes, which is that of the vital and sensitive of man.* London: Thomas Dring, Ch. Harper, and John Leigh (1683).

Wills G. Reading Augustine's mind. *New York Review of Books* 63(1):71–72 (14 January 2016).

Wilson EB. *The cell in development and heredity.* Third edition. New York: Macmillan (1925).

Wilson TB & Cassin J. On a third kingdom of organized beings. *Proceedings of the Academy of Natural Sciences of Philadelphia* 15:113–121 (presented 1863, published 1864).

Windels F. *The Lasceaux cave paintings.* New York: Viking (1950).

Wingate SD. *The Mediaeval Latin versions of the Aristotelian scientific corpus, with special reference to the biological works.* London: Courier (1931).

Winker S & Woese CR. A definition of the domains Archaea, Bacteria and Eucarya in terms of small subunit ribosomal RNA characteristics. *Systematic and Applied Microbiology* 14(4):305–310 (1991).

Winsor MP. *Starfish, jellyfish, and the order of life. Issues in Nineteenth-century science.* New Haven: Yale UP (1976).

Wippel JF. The condemnations of 1270 and 1277 at Paris. *Journal of Medieval and Renaissance Studies* 7(2):169–210 (1977).

Witcombe CLCE. Eve and the identity of women (2000). http://witcombe.sbc.edu/eve-women, accessed 24 October 2018.

Withering W. *A botanical arrangement of all the vegetables naturally growing in Great Britain. With descriptions of the genera and species, according to the system of the celebrated Linnæus.* In two volumes. Birmingham: Swinney (1776).

Withering W. *A botanical arrangement of British plants; including the uses of each species, in medicine, diet, rural œconomy and the arts.* Second edition. In three volumes. Birmingham: Swinney (1787).

Withering W. *An arrangement of British plants; according to the latest improvements of the Linnæan system.* Third edition. In four volumes. London: Robinson (1796).

Wittkower R. Marvels of the East. A study in the history of monsters. *Journal of the Warburg and Courtauld Institutes* 5:159–197 (1942).

Woese CR. Archaebacteria and the nature of their evolution. In: B Fernholm, K Bremer & H Jörnvall (eds), *The hierarchy of life. Molecules and morphology in phylogenetic analysis. Proceedings from Nobel Symposium 70 . . . August 29–September 2, 1988.* Amsterdam: Excerpta Medica (1989), pp. 119–130.

Woese CR. Bacterial evolution. *Microbiological Reviews* 51(2):221–271 (1987).

Woese CR. Default taxonomy: Ernst Mayr's view of the microbial world. *PNAS* 95(19):11043–11046 (1998).

Woese CR. Microbiology in transition. *PNAS* 91(5):1601–1603 (1994a).

Woese CR. There must be a prokaryote somewhere: microbiology's search for itself. *Microbiological Reviews* 58(1):1–9 (1994b).

Woese CR. Why study evolutionary relationships among bacteria? In: KH Schleifer & E Stackebrandt (eds), *Evolution of prokaryotes*. London: Academic (1985), pp. 1–30.

Woese CR & Fox GE. Phylogenetic structure of the prokaryotic domain: the primary kingdoms. *PNAS* 74(11):5088–5090 (1977).

Woese CR, Gibson J & Fox GE. Do genealogical patterns in purple photosynthetic bacteria reflect interspecific gene transfer? *Nature* 283(5743):212–214 (1980).

Woese CR, Kandler O & Wheelis ML. Towards a natural system of organisms: proposal for the domains Archaea, Bacteria, and Eucarya. *PNAS* 87(12):4576–4579 (1990).

Woese CR, Stackebrandt E, Macke TJ & Fox GE. A phylogenetic definition of the major eubacterial taxa. *Systematic and Applied Microbiology* 6(2):143–151 (1985).

Wolf A. *A history of science, technology, and philosophy in the 16th & 17th centuries*. London: Allen & Unwin (1935).

Wolff CF. De formatione intestinorvm observationes, in ovis incvbatis institvtae [title varies]. *Novi Commentarii Academiæ Scientiarum Imperialis Petropolitanæ* 12:43–47, 403–448, 449–507 (1768), and 13:478–530 (1769).

Wolff CF. *Theoria generationis. Editio nova aucta et emendata*. Halae: Hendel (1774).

Wolff CF. *Theoria generationis*. Halæ: Hendel (1759).

Wolff CF. *Über die Bildung des Darmkanals im bebrüteten Hühnchen*. JF Meckel (trans.). Halle: Renger (1812).

Wolff JF. *Commentatio de Lemna*. Altorfii et Norimbergæ: Lechner (1801).

Wolfson HA. *The philosophy of the Church Fathers*, volume 1. Faith, trinity, incarnation. Third edition. Cambridge MA: Harvard UP (1970).

Wolfson HA. *The philosophy of the kalam*. Cambridge MA: Harvard UP (1976).

Wolpert L. Evolution of the cell theory. *Philosophical Transactions of the Royal Society of London B. Biological Sciences* 349(1329):227–233 (1995).

Wood JG. *Animate creation. Revised and adapted to American zoology by JB Holder*. In six volumes. New York: Selmar Hess (1898).

Woodhouse SC. *English-Greek Dictionary. A vocabulary of the Attic language*. London: Routledge & Kegan Paul (1910).

Woodward J. Some thoughts and experiments concerning vegetation. *Philosophical Transactions* 21(253):193–227 (1700).

[Wordsworth W & Coleridge ST]. *Lyrical ballads: with a few other poems*. London: Arch (1798).

Workman HB. *John Wyclif: a study of the English medieval church*. In two volumes. Oxford: Clarendon (1926).

Worm O. *Museum Wormianum. Seu historia rerum rariorum*. Lugduni Batavorum: Elseviriorum (1655).

Woronin M. Chromophyton Rosanoffi. *Botanische Zeitung* 38, columns 625–631 and 641–648 (1880).

Wotton E. *De differentiis animalivm libri decem*. Lvtetiae Parisiorvm: Vascosanvm (1552).

Wright D. *The Eleusinian mysteries & rites*. London: Theosophical (1913).

Wright MR (ed.). *Empedocles: the extant fragments*. New Haven: Yale UP (1981).

Wright MR (ed.). *Empedocles: the extant fragments*. Second edition. London: Duckworth (1995).

Wright MR. Empedocles. In: CCW Taylor (ed.), *From the beginning to Plato*. Routledge history of philosophy (GHR Parkinson & SG Shanker, gen. eds), volume 1. London: Routledge (1997), pp. 175–207.

Wright NA & Poulsom R. *Omnis cellula e cellula* revisited: cell biology as the foundation of pathology. *Journal of Pathology* 226(2):145–147 (2012).

Wright T (ed.). *Popular treatises on science written during the Middle Ages, in Anglo-Saxon, Anglo-Norman, and English*. London: Historical Society of Science (1841).

774   REFERENCES

Wright T. Preface. In: T Wright (ed.), *Popular treatises on science written during the Middle Ages, in Anglo-Saxon, Anglo-Norman, and English*. London: Historical Society of Science (1841), pp. vii–xvi.

Wrisberg HA. *Observationvm de animalcvlis infvsoriis satvra*. Goettingae: Vandenhoeck (1765).

Xenophon. *Memorabilia. Oeconomicus. Symposium. Apology*. EC Marchant (trans.). Loeb Classical Library, volume 168. Cambridge MA: Harvard UP; and London: Heinemann (1923).

Yang D, Oyaizu Y, Oyaizu H, *et al*. Mitochondrial origins. *PNAS* 82(13):4443–4447 (1985).

Yap WH, Zhang Z & Wang Y. Distinct types of rRNA operons exist in the genome of the actinomycete *Thermomonospora chromogena* and evidence for horizontal transfer of an entire rRNA operon. *Journal of Bacteriology* 181(17):5201–5209 (1999).

Yates FA. *Giordano Bruno and the Hermetic tradition*. University of Chicago Press (1964).

Yosida M. The Chinese concept of nature. In: S Nakayama & N Sivin (eds), *Chinese science. Explorations of an ancient tradition*. Cambridge MA: MIT Press (1973), pp. 71–89.

Zabarella J. *On methods*. JP McCaskey (ed. and trans.), volumes 1 (Books 1–2) and 2 (Books 3–4 and *On regressus*). The I Tatti Renaissance Library. Cambridge MA: Harvard UP (2013).

Zabarella J. *In tres Aristotelis Libros de Anima commentarij*. Venetiis: Franciscvm Bolzettam (1605).

Zabarella J. *Opera logica . . .* editio tertia. Coloniæ: Lazarus Zetzner (1597).

Zabarella J. *Opera logica . . .* secunda editione. Venetiis: Paulum Meietum (1586).

Zabarella J. *Opera logica . . .* editio sextadecima. Tarvisii: Robert Meietl (1604).

Zahn J. *Speculæ physico-mathematico-historicæ*. In three volumes. Norimbergæ: Lochner (1696).

Zaluziansky à Zaluzian A. *Methodi herbariæ, libri tres*. Pragæ: Georgij Dacziceeni (1592).

Zammito JH. *The gestation of German biology: philosophy and physiology from Stahl to Schelling*. University of Chicago Press (2018).

Zaremba-Niedzwiedzka E, Caceres ER, Saw JH, *et al*. Asgard archaea illuminate the origin of eukaryotic cellular complexity. *Nature* 541(7637):353–358 (2017).

Zhou Z, Liu Y, Li M & Gu JD. Two or three domains: a new view of tree of life in the genomics era. *Applied Microbiology and Biotechnology* 102(7):3049–3058 (2018).

Zhmud L. *Pythagoras and the early Pythagoreans*. K Windle & R Ireland (trans.). Oxford UP (2012).

Zillig W, Klenk HP, Palm P, *et al*. Did eukaryotes originate by a fusion event? *Endocytobiosis and Cell Research* 6(1):1–25 (1989).

Zimmerman B. The Carmelite Order. *The Catholic Encyclopedia* 3:354–370 (1908).

Zimmermann FW. [Review of:] Ursula Weisser (editor), "Buch über Geheimnis der Schöpfung und die Darstelling der Natur" von Pseudo-Apollonius von Tyana. *Medical History* 25(4):439–44 (1981).

Zirkle C. Animals impregnated by the wind. *Isis* 25(1):95–130 (1936).

Zirkle C. The jumar or cross between the horse and the cow. *Isis* 33(4):486–506 (1941).

Zonta M. Mineralogy, botany and zoology in medieval Hebrew encyclopaedias: "descriptive" and "theoretical" approaches to Arabic sciences. *Arabic Sciences and Philosophy* 6(2):263–315 (1996).

Zoroaster [in part]. *The holy Gathas (sacred hymns) of Zarathustra . . . Presented as a recitable prayer in English*. J Wadia (ed.). Online document at zoroastriansnet.files.wordpress.com/ 2017], accessed 15 August 2018.

Zoroaster. *The Gâthas of Zarathushtra (Zoroaster) in metre and rhythm*. LH Mills (ed.). Leipzig: Brockhaus (1900).

Zoroaster. *The hymns of Zoroaster usually called the Gathas*. KSL Guthrie (ed.). No. Yonkers NY: Platonist (1912).

Zuckerkandl E & Pauling L. Evolutionary divergence and convergence in proteins. In: V Bryson & HJ Vogel (eds), *Evolving genes and proteins*. New York: Academic (1965), pp. 97–181.

Zuckerkandl E & Pauling L. Molecules as documents of evolutionary history. *Journal of Theoretical Biology* 8(2):357–366 (1965).

Zupko J. John Buridan. *The Stanford Encyclopedia of Philosophy* (Fall 2018 edition), EN Zalta (ed.), plato.stanford.edu/archives/fall2018/entries/buridan/, accessed 9 November 2018.

Zwinger T. *Theatrum botanicvm, das ist: Neu vollkommenes Kräuter-Buch*. Basel: Jacob Bertsche (1696).

Zwinger T. *Theatrum botanicvm, das ist: Neu vollkommenes Kräuter-Buch . . . durch Friedrich Zwinger*. Basel: Hans Jacob Bischoffs (1744).

Zworykin VK. An electron microscope for the research laboratory. *Science* 92(2377):51–53 (1940).

Index of Persons

For the benefit of digital users, indexed terms that span two pages (e.g., 52–53) may, on occasion, appear on only one of those pages.

Endnotes are indexed only if they refer to persons not mentioned in the corresponding text, or offer substantive additional information. Greek names are Latinized (Chrysippus) or Anglicized (Plato). Mythological, legendary and fictional figures are, in general, not indexed. Years of birth and death (not *e.g.* of papacy, reign etc.) are given or estimated.

Tables and figures are indicated by *t* and *f* following the page number.

al-Abharī, Athīr al-Dīn al-Mufaḍḍal ibn ʿUmar al-Samarqandī, also known as al-Munajjim (*ca* 1190/1200 to 1262/1265 CE), 103

Abulafia, Rabbi Abraham ben Samuel (b. 1240, d. after 1291), 109

Accursius, Franciscus [Accorso di Bagnolo, Francesco] (*ca* 1182 to 1260/1263), 528n.25

Acosta, Christóbal [da Costa, Christóvão] (*ca* 1525 to *ca* 1594), 476n.159

Adams, George [the elder] (*ca* 1709 to 1773), 216, 221–22

Adanson, Michel (1727–1806), 259, 260, 291, 301–2, 336, 535n.68

Addison, Joseph (1672–1719), 227–28

Adelard [Athelard] of Bath (*ca* 1080 to *ca* 1152 CE), 119–20, 122

Adeliza de Louvaine (*ca* 1103 to 1151), 197–98

Ælfric, Archbishop (d. 1005), 82

Aelian [Aelianus, Claudius] (*ca* 170/175 to 235 CE), 46, 56–57, 81, 134, 198, 482n.91, 522n.16

Aelius Nicon [father of Galen] (*fl.* Second century CE), 9–10

Aenesidemus of Knossos (*fl.* First century BCE), 44–45, 46

Aesop (*ca* 620 to *ca* 564 BCE), 57

Agardh, Carl Adolph, Bishop of Karlstad (1785–1859), 344–46, 347, 353, 363, 403, 594n.141, 595n.160, 597n.15

Agardh, Jacob Georg (1813–1901), 592n.95, 593n.112, 596n.187

Agassiz, Jean Louis Rodolphe FRS(For) (1807–1873), 610n.76

"A Gentleman in the Country" (poss. King, Charles, b. 1654, d. after 1703), 233–34, 350, 544n.65

Agricola, Georgius. *See* Pawer, Georg

Agrippa, Heinrich Cornelius, of Nettesheim (1486–1535), 136, 152–53, 158, 160, 173, 226, 284–85, 462n.102

Åkerman, Joachim (1798–1876), 345

Alain de Lille [Alan of Lille] (before 1128 to 1202/1203 CE), 121–22

Alberti, Leon Battista (1404–1472), 51–52

Albertus Magnus [Albert the Great], Saint (before 1200 to 1280 CE), 57, 80, 102–3, 112, 115, 124–27, 133–34, 138, 147, 149, 154, 168, 169, 173, 189, 190, 194, 198, 211, 226, 249, 287, 475n.146, 478n.200, 556n.38

Albinus [philosopher] (*fl.* 149–157 CE), 12, 462n.98

Albotini [al-Botini], Rabbi Judah ben Moses (d. 1519), 110–11

Alcmaeon of Croton (perhaps Fifth century BCE), 25–26, 481n.69

Alcuin of York [Flaccus Albinus Alcuinus] (*ca* 735 to 804 CE), 111, 114–15, 116

Aldrovandi, Ulisse [Ulysse] (1522–1605), 165, 193, 200, 203–5, 206, 226, 243, 537n.112, 539n.148

Alemanno, Yohanan (*ca* 1435 to 1504 or later), 110

Alexander [of] Aphrodisias (*fl. ca* 200 CE), 173–74, 477n.178, 478n.199, 487n.91

Alexander III ["the Great"] of Macedon (356–323 BCE), 16, 24, 25, 44–45, 59, 60, 91, 138, 144–46

Alexander of Hales (*ca* 1185 to 1245 CE), 115

778 INDEX OF PERSONS

Alexander, Tiberius Julius (*ca* 10/14 to 70 CE), 62

Alexius [Alexios] I Comnenus, Emperor (1048/1057 to 1118), 90

Alfanus I, Archbishop of Salerno, Saint (*ca* 1015/1020 to 1085), 80, 112, 119, 120–21, 492n.58, 508n.100

Alighieri, Durante di Alighiero degli, known as Dante (*ca* 1265 to 1321), 87, 129–30, 462n.101

Allen, Mary Belle (1922–1973), 419–21

Alsted, Johann Heinrich (1588–1638), 206–7, 208

Altenburg, Edgar (1888–1967), 426

Altmann, Richard (1852–1900), 422, 627n.130, 627n.144

Amarasiṃha (*fl. ca* 375 CE), 464n.18

Ambrogini, Agnolo [Angelo], known as Poliziano (1454–1494), 175

Ambrose of Milan, Saint (*ca* 340 to 397), 7, 82, 83, 123, 125, 197–98, 483n.104

Ammonius "of Athens" (*fl.* First century CE), 62

Ammonius Hermiæ (*ca* 435/440 to *ca* 520 CE), 69–70, 71, 73, 89, 102–3, 112, 113, 173–74, 176, 180, 412, 487n.91, 488n.103, 494n.123, 497n.52, 575n.121

Ammonius Saccas (d. 241/242 CE), 63, 77

Anatolius [? Bishop of Laodicea, d. 283], 65–66

Anaxagoras of Clazomenæ (*ca* 500 to *ca* 428 BCE), 42, 68–69

Anaximander of Miletus (*ca* 610 to *ca* 546 BCE), 25–26, 211, 470n.37

Anaximenes of Miletus (*ca* 585/586 to *ca* 525/526 BCE), 10, 25–26, 140

Andreae, Johann Valentin (1586–1654), 154

Andronicus of Rhodes (*fl. ca* 70–30 BCE), 46–47

Anselm, Archbishop of Canterbury, Saint (1033/1034 to 1109), 117, 118–19

al-Antaki, ʿAbdallāh ibn al Faḍl (*fl.* mid-Eleventh century), 497n.53

Antiochus of Ascalon (*ca* 130/125 BCE to *ca* 68 BCE), 47, 60

Antoninus Liberalis (*fl.* Second or Third century CE), 459n.67

Antonius [Antonios] of Antioch, Abbott (*fl.* first half of Eleventh century), 95, 113

Apollonius of Tyana [Balīnās or Balīnūs] (*ca* 15 to *ca* 97 CE), 112–13, 143, 513n.37, 514n.38

Apuleius Madaurensis, Lucius (*ca* 123/124 to *ca* 170 CE), 9, 46, 469n.33, 514n.45

Apuleius Platonicus [Apuleius Barbarus; Pseudo-Apuleius] (Fourth century CE), 53, 189–90

Archimedes of Syracuse (*ca* 287 to *ca* 212 BCE), 111, 483n.10

Archytas of Tarentum (*fl.* early Fourth century BCE), 24, 73, 469n.25

Aristides of Athens (*fl.* Second century CE), 490n.11

Aristobulus of Cassandreia (*ca* 375 to 301 BCE), 523n.29

Aristophanes of Athens (*ca* 446 to *ca* 386 BCE), 215, 470n.51

Aristophanes of Byzantium (*ca* 257 to *ca* 185/180 BCE), 55

Aristotle of Stagira (384–322 BCE), 2, 19, 24–25, 27, 28–29, 30–31, 34–41, 42–43, 46–47, 49–50, 51, 55, 57–58, 59, 60, 61, 63, 64–65, 66, 67–68, 69, 70, 71, 73–74, 77, 79–80, 81, 87, 94–95, 97, 101–3, 107, 111–12, 113, 114, 115, 117, 118, 122–23, 125–27, 128–29, 131–32, 134, 137, 140, 142–43, 144–46, 147, 152, 166–67, 171–72, 173–75, 176–77, 178–79, 180, 181–82, 184–85, 187, 188–89, 193, 194–95, 198–201, 202, 207, 208, 210–11, 212, 214, 223, 224, 226, 241, 243, 247, 248, 260, 273, 287, 302–3, 344, 350, 385, 390, 412, 468n.13, 469n.24, 471n.59, 471n.65, 471n.68, 472n.74, 477n.192, 481–82n.77, 484n.18, 499n.118, 500n.129, 500n.130, 509–10n.162, 511n.202, 524n.47, 534n.50, 549n.36, 554n.152, 555n.18, 565n.40, 570n.20

Aristoxenus of Tarentum (b. *ca* 375, d. after 322 BCE), 464n.15

Arnold, Friedrich (1803–1890), 421

Arouet, François-Marie. *See* Voltaire

Arrian of Nicomedia (*ca* 86/89 to perhaps 146/160 CE), 522n.16, 523n.29

Artaxerxes II Mnemon [Arses] of Persia (*ca* 453/445 to 359/358 BCE), 164

Artephius (*fl.* mid-Twelfth century CE), 149

al-Ashʿari, Abū ʾl-Ḥasan ʿAlī ibn Ismāʿil (*ca* 874 to 935/936 CE), 498n.76

Ashmole, Eilas FRS (1617–1692), 145–46, 154

Ashoka "the Great", Emperor (*ca* 304 to 232 BCE), 4

Athanæus of Naucratis (*fl.* late Second to early Third century CE), 532–33n.31

Athanasius I, Bishop of Alexandria, Saint (*ca* 296/298 to 373 CE), 79, 504n.4

Audouin, Jean Victor (1797–1841), 585n.150

INDEX OF PERSONS 779

Augier de Favas, François Augustin Marie (1758–1825), 265, 276–78, 279, 301–2
Augustine, Bishop of Hippo [Aurelis Augustinus Hipponensis], Saint (354–430 CE), 83–85, 87, 93, 111, 114, 117, 118, 119, 123, 124, 129, 130, 137–38, 211, 297–98, 462n.98, 489n.119, 492n.72, 505n.15, 514n.45, 542n.29, 574n.102
Augustine of Canterbury (b. early Sixth century, d. 604), 504n.4
Augustus Cæsar [Gaius Octavius; Octavian] (63 BCE to 14 CE), 464n.15
Aurelian [Lucius Domitius Aurelianus], Emperor (214–275 CE), 485n.40
Averroës. See Ibn Rushd
Avery, Oswald Theodore Jr ForMemRS (1877–1955), 432–33
Avicenna. See Ibn Sīnā

Bachmann [Backmann], August [Rivinius] (1652–1723), 535n.77
Bacon, Francis, Viscount St Alban [Lord Verulam] (1561–1626), 57, 161, 226, 318, 355, 358–59, 361–62, 517n.142, 524n.52, 600n.55
Bacon, Roger (1214/1220 to ca 1292), 144–45, 147, 543n.52
Baer, Karl Ernst, Ritter von [Ber, Karl Maksimovich] ForMemRS (1792–1876), 273, 286, 287, 307, 324–25, 335, 366, 386–87, 545n.81, 565n.38, 587n.203
Baker, Henry FRS (1698–1774), 209, 220–22, 356, 357, 545n.90
Bakunin, Mikhail Alexandrovich (1814–1876), 578n.3
Balbiani, Édouard-Gérard (1823–1899), 400
Banks, Sir Joseph PRS (1743/1744 to 1820), 358, 563n.4
Baranetzky [Boranetzky], Josif (1843–1905), 423
Barbançois-Villegongis, Charles-Hélion de (1760–1822), 572n.57
Barbaro, Ermolao [Barbarus, Hermolaus] (1454–1493), 111–12, 175, 190, 481n.56, 517n.146, 527n.4
Barkley, Fred Alexander (1908–1989), 417, 444, 622n.57, 622n.60
Baron d'Hénouville, Théodore (1715–1768), 540n.177
Barry, Alfred, Bishop of Sydney (1826–1910), 362
Barry, Martin FRS (1802–1855), 307, 329
Bartholomæus Anglicus (before 1203 to 1272), 125, 189, 197–98

Bartram, John (1699–1777), 546n.110
Basil II Porphyrogenitus, Byzantine Emperor (ca 958 to 1025), 90
Basil of Caesarea, Saint (329/330 to 379 CE), 80–81, 82, 117, 165, 459n.64
al-Baṣrī, Abū ʿUthman ʿAmr ibn Baḥr al-Kinānī, known as al-Jāḥiẓ (776 to 868/869 CE), 96–97, 102, 105, 498n.69
Baster, Job [Hiob] FRS (1711–1775), 562n.161
Bateson, William FRS (1861–1926), 432
Bather, Francis Arthur FRS (1863–1934), 432
Batsch, August Johann Georg Carl (1761–1802), 263, 268–69, 272, 565n.44, 578–79n.20, 598n.18
Bauhin, Gaspard [Caspar] (1560–1624), 195–96, 244, 555n.18
Bauhin, Jean [Johann] [botanist] (1541–1613), 178, 196, 533n.34, 534n.56, 534n.64
Baumgärtner, Karl Heinrich (1798–1886), 421
Baur, Erwin (1875–1933), 427
Beadle, George Wells ForMemRS (1903–1989), 432–33
Béchamp, Pierre Jacques Antoine (1816–1908), 422, 627–28n.147
Becher, Johann[es] Joachim (1635–1682), 263–64, 560n.123
Beddoes, Thomas (1760–1808), 571n.29
Bede, the Venerable, Saint (ca 672/673 to 735 CE), 114–15, 116, 133–34
Belon, Pierre (1517–1564), 199–200, 201, 202, 203, 207, 219, 226, 537n.112
Benda, Carl (1857–1932), 422
Benedict of Nursia, Saint (ca 480 to ca 547 CE), 114, 548n.14
Ben-Lavi, Solomon (fl. 1391–1392), 502n.180
Bennett, Alfred William (1833–1902), 403–4
Benoit le Puncteur [Benedictus Punctuator]. See Berachya ha-Naqdan
Bentinck, Imperial Count Willem [William] van Rhoon (1707–1774), 234
Bentley, Rev. Richard FRS (1662–1742), 362, 542n.24
Berachya [Berechiah] ben Natronai ha-Naqdan (fl. late Twelfth to Thirteenth centuries), 119–20
Bergman, Torbern Olaf [Olof] (1735–1784), 563n.192
Bergmann, Carl Georg Lucas Christian (1814–1865), 329
Bergson, Henri-Louis (1859–1941), 354, 581n.50
Berkeley, George, Bishop (1685–1753), 180, 315
Berkeley, Rev. Miles Joseph FRS (1803–1889), 597n.16, 599n.43

780 INDEX OF PERSONS

Berkenhout, John (1726–1791), 359
Bernard of Chartres (d. after 1124), 121–22
Bernard of Clairvaux, Saint (1090–1153), 118–19
Bernard Silvester [Bernardus Silvestris] (fl. Twelfth century), 122, 226
Bernardin de Saint-Pierre, Jacques-Henri (1737–1814), 268–69, 542n.24, 601n.87
Berthollet, Claude Louis (1748–1822), 570n.23
Beseke, Johann Melchior Gottlieb (1746–1802), 2, 263–64, 559n.102, 560n.123
Besler, Basilius (1561–1629), 205–6, 208
Besler, Michael Rupert (1607–1661), 205–7, 208
Bessarion, Cardinal Basilios, Latin Patriarch of Constantinople (1403–1472), 134, 486n.79
Bianchi, Giovanni Paolo Simone [Plancus, Janus] (1693–1775), 548n.136
Bichat, Marie François Xavier (1771–1802), 289, 380, 571n.29
Biel, Gabriel (1420/1425 to 1495), 511n.217
Bignon, Abbé Jean-Paul FRS (1662–1743), 232, 551n.83
al-Bīrūnī, Abū 'l-Rayḥān Muḥammad ibn Aḥmad (973 to after 1050 CE), 12, 100–1, 148, 231
Blainville, Henri Marie Ducrotay de ForMemRS (1777–1850), 262–63, 306–7, 308, 309, 326, 335, 366, 372, 605n.159
Blake, William (1757–1827), v, 172, 459n.64, 524n.53
Bloch, Marcus Elieser (1723–1799), 337
Blomfield, Charles James, Bishop of London (1786–1857), 363
Blumenbach, Johann Friedrich ForMemRS (1752–1840), 263, 265–66, 313, 316, 322–23, 550n.72, 581n.53, 587n.203
Blund, John (ca 1175 to 1248), 122–23
Boccacio, Giovanni (1313–1375), 122
Bock, Hieronymus [Jerome] [Tragus, Hieronymus] (1497/1498 to 1554), 191, 194, 532n.23, 533n.32, 555n.18
Bode, Johann Elert (1747–1826), 542n.24
Bodin, Jean [of Angers] (1529/1530 to 1596), 169–70, 178–79, 226, 461n.90, 548n.8
Boece, Hector [Bœcius, Boethius] (1465–1536), 168, 169, 183
Boerhaave, Herman FRS (1668–1738), 235, 249, 540n.164
Boëthius, Anicius Manlius Severinus (ca 477 to 524 CE), 62, 74, 87, 114, 119, 125–26, 162, 226, 478n.200, 485n.54

Böhme, Jakob (1575–1624), 146–47, 159, 311–12
Boissier de Sauvages de la Croix [de Lacroix], François FRS (1706–1767), 563n.194
Bombastus von Hohenheim, Philippus Aureolus Theophrastus, known as Paracelsus (1493/1494 to 1541), 146–47, 153–54, 157, 160, 161, 169–70, 194, 226, 311–12, 316, 458n.49, 461n.91
Bon de Saint Hilaire, François Xavier (1678–1761), 231
Bonaventure [Fidanza, Giovanni di], Saint (1221–1274), 115, 129–30, 226
Bonham, Thomas (ca 1564 to 1628/1629), 548n.135
Bonnet, Charles FRS (1720–1793), 209, 236–41, 244, 251–52, 261–62, 264, 266, 269, 276, 286, 290–91, 301–3, 354, 545n.81, 551n.93, 564n.20
Bonpland, Aimé (1773–1858), 323, 576n.135
Boothby, Sir Brooke (1744–1824), 598n.35
Borel, Pierre [Borellus, Petrus] (ca 1620 to 1671), 544n.58
Borges, Jorge Luis (1899–1986), 172
Born, Edler Ignaz [Ignatius] [von] FRS (1742–1791), 316
Borowski [Borowsky], Georg Heinrich (1746–1801), 263–64, 560n.123
Bory de Saint-Vincent, Jean-Baptiste Geneviève Marcellin (1778–1846), 262–63, 290, 307–9, 333–34, 335, 336, 346–48, 354, 378, 380, 386, 575n.126, 583n.109
Bosch, Hieronymus [van Aken, Jheronimus] (ca 1450 to 1516 CE), 172
Bouelles [Bovelles], Charles de (ca 1475 to after 1566), 226
Bourdon, Jean Baptiste Isidor[e] (1795–1861), 287, 605n.163
Bourguet, Louis (1678–1742), 226
Bouvet, Joachim (1656–1730), 543n.38
Boveri, Theodor Heinrich (1862–1915), 426
Boyle, Robert FRS (1627–1691), 157, 231, 361–62, 515n.78
Bradley, Richard FRS (1688–1732), 227–31, 236–37, 238, 244, 262, 264, 269, 286, 618n.243
Bradstreet, Anne, née Dudley (1612–1672), 481n.70
Braun, Alexander Carl Heinrich (1805–1877), 379, 380–81
Bremser, Johann Gottfried (1767–1827), 337
Brethren of Purity. See Ikhwān al-Ṣafā
Brewster, Sir David FRS PRSE (1781–1868), 602n.90

INDEX OF PERSONS 781

Brinkley, Rev. John Mortimer (1763/1766 to 1835), 601n.87

Brisseau-Mirbel CF. *See* Mirbel, CF Brisseau de

Brockes, Barthold Heinrich (1680–1747), 601n.78

Bronn, Heinrich Georg (1800–1862), 401, 608n.33

Bronzino [Cosimo, Agnolo di] (1503–1572), 172

Brown, Robert FRS (1773–1858), 222, 269, 359, 363

Brown, Thomas FRSE (1778–1820) [poet, philosopher], 361

Browne, Sir Thomas (1605–1682), 9, 57, 132, 284–85, 474n.121

Brücke, Ritter Ernst Wilhelm von (1819–1892), 421–22

Bruguiére, Jean Guillaume (1749–1798), 303

Brunetto Latini (*ca* 1220 to 1294), 197–98

Brunfels, Otto (1488? to 1534), 176–77, 190–92, 194

Bruno, Giordano [the Nolan] (1548–1600), 8, 158, 160, 173, 211, 226–27, 284–85, 462n.101

Bryant, Rev. Henry (1721–1799), 598n.34

Buckland, Rev. William FRS (1784–1856), 364, 377, 597n.16, 603n.129

Buckley, Arabella (1840–1929), 606n.195

Buddeus, Johann Frans [Buddeus, Johannes Franciscus] (1667–1729), 543n.39

Budé, Guillaume [Budæus] (1467–1540), 62, 160, 173, 175–77, 178, 213–14, 504n.241, 528n.39, 549n.26

Buffon, Comte de. *See* Leclerc, Georges-Louis

Bulwer, John (1606–1656), 165

Buonanni [Bonanni], Filippo SJ (1638–1723), 544n.65

Burchard of Worms (*ca* 950/965 to 1025), 9

Burdach, Karl Friedrich (1776–1847), 318–19, 324, 571n.29, 580n.36

Burgundio of Pisa (d. 1193), 80, 112, 113

Buridan, Jean [John] (before 1301 to 1358/1362), 130, 131

Burman, Johannes (1707–1780), 249

Burton, Robert (1577–1640), 183, 458n.49

Bütschli, Johann Adam Otto (1848–1920), 400–2, 402f, 408, 409, 421, 547n.130

Cajetan, Thomas. *See* de Vio, Tommaso

Calepino, Ambrogio [Calepinus] (*ca* 1440 to 1510), 175, 178, 185

Calvenus Taurus, Lucius (*ca* 105 to *ca* 165 CE), 63

Camerarius [Camerer], Rudolf Jakob (1665–1721), 535n.72

Candolle, Alphonse Louis Pierre Pyrame de (1806–1893), 423, 564n.21, 590n.54

Candolle, Augustin Pyramus [Pyrame] de ForMemRS (1778–1841), 274–75, 281–82, 336

Canning, George FRS (1770–1827), 361

Capella, Martianus Minneus Felix (*ca* 360 to *ca* 428 CE), 116, 505n.15

Cardano, Gerolamo [Girolamo] [Cardan, Jérôme] (1501–1576), 161, 166–67, 171–72, 226, 543n.53

Carli-Rubbi, Gianrinaldo [Carli, Gian Rinaldo] (1720–1795), 548n.3

Carpenter, William Benjamin FRS (1813–1885), 376, 545n.71, 591n.64

Carus, Carl Gustav (1789–1869), 317, 322, 331–34, 335, 337, 354, 367, 378, 591n.60, 618n.251

Carus, Jules Victor (1823–1903), 105, 587n.201, 608n.23, 618n.251

Casaubon, Isaac (1559–1614), 159

Cassin, John (1813–1869), 373–75, 377–78

Cassini, Comte Alexandre Henri Gabriel de (1781–1832), 281–82

Cassiodorus Senator, Flavius Magnus Aurelius (*ca* 485/490 to *ca* 583/585 CE), 114, 115, 116

Cato, Marcus Porcius ["the Elder"] (234–149 BCE), 47, 48, 479n.28

Cavalier-Smith, Thomas FRS (1942–2021), 421, 428–29, 430, 447–48, 629n.175, 635n.127, 635n.136

Cellini, Benvenuto (1500–1571), 57

Celsus (*fl.* Second century CE), 7, 77–78, 225

Celsus, Aulus Cornelius (*ca* 25 BCE to *ca* 50 CE), 47, 54–55, 479n.28

Censorinus (*fl.* Third century CE), 25–26

Cesalpino, Andrea [Cæsalpinius, Andreas] (1524/1525 to 1603), 190, 193–95, 226, 243, 245, 246, 248, 260, 336–37, 535n.77, 558n.68

Cesi, Federico Angelo (1585–1630), 544n.57

Chambers, Ephraim (*ca* 1680 to 1740), 289–90

Chambers, Robert FRSE [publisher] (1802–1871), 368–69, 568n.130

Chamisso, Adelbert von (1781–1838), 317, 325, 346

Chandragupta Maurya, Emperor (*ca* 350 to 297/293 BCE), 16, 164

Charlemagne, Holy Roman Emperor (747–814), 111, 114–15, 116, 505n.39

782 INDEX OF PERSONS

Charles IX Maximilien, King of France (1550–1574), 170–71

Charles the Bald, Emperor (823–877 CE), 117

Charleton, Walter [Charletonus, Gualteri] FRS (1619–1707), 206

Chatton, Édouard Pierre Léon (1883–1947), 414–15, 422, 436, 622n.60, 632n.56

Chaucer, Geoffrey (*ca* 1343 to 1400), 122, 130

Ch'en Shun (*fl. ca* 1200 CE), 465n.36

Chêng Ching-Wang (early Twelfth century), 461n.85

Cherler, Jean-Henri (*ca* 1570 to *ca* 1610), 196

Choisy, Jacques Denys [Denis] (1799–1859), 567n.96

Chrysippus of Soloi (*ca* 280/279 to *ca* 208/206 BCE), 44, 48, 54

Chu Hsi. *See* Zhu Xi

Chuang Tzu. *See* Zhuang Zhou

Cicero, Marcus Tullius (106–43 BCE), 46

Cienkowski, Leon. *See* Tsenkovsky, Lev

Claude, Albert (1899–1983), 435–36

Claus, Carl Friedrich Wilhelm (1835–1899), 381–82

Clement VI, Pope [Roger, Pierre] (1291–1352), 169

Clement XIV, Pope [Ganganelli, Giovanni Vincenzo Antonio] (1705–1774), 257

Clement of Alexandria (*ca* 150 to *ca* 215 CE), 76–78, 137, 490n.11

Clement, Bishop of Rome [Pope Clement I], Saint (*ca* 35 to 99 CE), 76, 490n.29

Clericus, Johannes. *See* Le Clerc, Jean

Clifford, George III (1685–1760), 249

Clusius, Carolus. *See* l'Écluse, Charles de

Cohn, Ferdinand Julius ForMemRS (1828–1898), 328, 352–53, 403–4, 405–7, 412, 578n.9, 612n.109, 620n.23

Coleridge, Samuel Taylor (1772–1834), 361

Collenuccio, Pandolfo (1444–1504), 532n.17, 533n.32

Colonna, Fabio [Columna] (1567–1640), 561n.153

Columbus, Christopher (1451–1506), 157–58

Columella, Lucius Junius Moderatus (4 CE to *ca* 70 CE), 47, 48, 198–99, 212, 479n.25

Columna. *See* Colonna

Comnena [Komnene], Anna (1083 to at least 1148), 90

Condillac, Étienne Bonnot de (1714–1780), 311–12

Confucius. *See* Kǒng Fūzǐ

Conrad von Megenberg [Cunrat or Konrad von Alemann] (1309–1374), 124–25, 165, 169, 189

Constantine I, Emperor [Flavius Valerius Constantinius] (*ca* 272 to 337), 65–66, 75

Conti di Segni, Lotario dei. *See* Innocent III, Pope

Cook, James FRS (1728–1779), 358

Cope, Edward Drinker (1840–1897), 285–86

Copeland, Edwin Bingham (1873–1964), 622n.60, 623n.66, 623n.67

Copeland, Herbert Faulkner (1902–1968), 413–14, 416, 417, 419–21, 428, 622n.60, 632n.56

Copernicus, Nicolaus (1473–1543), 157–58, 212, 226–28, 230–31

Corbould, Henry (1787–1844), 602n.92

Cordus, Euricius [Ritze, Heinrich] (1486–1535), 190–91

Cordus, Valerius (1515–1544), 190–91, 193, 199–200

Corliss, John Orzo (1922–2014), 541n.2

Correns, Carl Erich (1864–1933), 628n.153, 630n.2

Corti, Abate Bonaventura SJ (1729–1813), 350–51, 548n.136, 597n.15

Crastone, Giovanni [Crastonus, Johannes] (d. after 1497), 45–46, 175

Cratevas [Crateuas] (111–64 BCE), 52–53

Crato von Krafftheim, Johannes [Krafft, Johannes] (1519–1585), 544n.59

Cremonini, Cesare (1550–1631), 529n.59

Crick, Francis Harry Compton FRS (1916–2004), 432–33

Croll, Oswald (*ca* 1563 to 1609), 154

Cromwell, Oliver (1599–1658), 147

Cuvier, Georges-Frédéric (1773–1838), 574n.107

Cuvier, Jean Léopold Nicolas Frédéric [Baron Georges] ForMemRS (1769–1832), 263, 266, 278, 282, 283–84, 290, 295, 298–302, 306, 307–8, 309, 318, 323, 324–25, 335, 337, 358, 364–65, 366, 373, 386–87, 580n.38

Cyril, Patriarch of Alexandria, Saint (*ca* 376 to 444 CE), 92

d'Aléchamps [Daléchamps], Jacques (1513–1588), 178, 194–95, 532n.23, 533n.34

d'Alembert, Jean-Baptiste Le Rond FRS (1717–1783), 289–91, 311–12

Dallinger, Rev. William Henry FRS (1839–1909), 411, 606n.185

Dalrymple, John FRS (1803–1852), 595n.164

Damascius (*ca* 458 to at least 538 CE), 69, 89–90, 488n.103

Dandini, Girolamo [Dandinus, Hieronymus] SJ (1554–1634), 185

INDEX OF PERSONS 783

Darius I ["the Great"], King and Pharaoh (*ca*
550 to 486 BCE), 16
Darius II [Ochus], King (d. 404 BCE), 458n.49
Darwin, Charles Robert FRS (1809–1882),
245, 325, 346, 369–70, 375, 376, 377, 379,
383–85, 390, 399–400, 409, 411, 423,
443, 449, 545n.81, 565n.44, 577n.178,
601n.85, 606n.192, 606n.193, 608n.21,
608n.37, 618n.251, 619n.1
Darwin, Erasmus FRS (1731–1802), 257, 355,
359–61, 368, 377, 598n.35, 601n.87
Darwin, Robert Waring [of Elston] (1724–
1816), 357–58
da Spoleto, Rabbi Barukh Abraham (*fl.* late
Sixteenth century), 11–12
Daubeny, Charles Giles Bridle FRS (1795–
1867), 604n.156
Daubenton, Louis-Jean-Marie FRS (1716–
1800), 265, 290, 291–93, 298, 302–3,
545n.77
Davaine, Casimir-Joseph (1812–1882), 406–7
David (*fl.* late Sixth or early Seventh century
CE), 69, 71, 72–73, 95, 97, 113, 497n.52
Dávila, Pedro Franco (1711–1786), 539n.148
da Vinci, Leonardo (1452–1519), 57, 157–58
Dayhoff, Margaret Belle, née Oakley (1925–
1983), 631n.30
de Bary, Heinrich Anton ForMemRS (1831–
1888), 403–4, 422–23, 612n.113
del Rio, Martin Anton SJ (1551–1608), 169
de Vio, Tommaso [Cajetan, Thomas;
Gaetanus], Cardinal (1469–1534), 181
Delise, Domenec François (1780–1841),
567n.96
Democritus of Abdera (*ca* 460 to *ca* 370 BCE),
27, 28–29, 35, 42, 68–69, 132, 158, 211,
470n.43, 480n.31
Denso, Johann Daniel (1708–1795), 263–64,
560n.123
Derham, William FRS, Canon of Windsor
(1657–1735), 247–48, 326–27, 356, 362,
542n.24, 601n.87
Descartes, René [Cartesius, Renatus] (1596–
1650), 171–72, 186, 206–7, 214, 222,
227–28, 230–31
Deshayes, Gérard Paul (1795–1875), 303,
574n.99
Desmarest, Anselme Gaëtan (1784–1838), 297
Desmoulins, Jean (1580–1620), 533n.34, 534n.56
Dexippus [philosopher] (*fl.* mid-Fourth
century CE), 66–67, 69, 80, 113, 150–51,
497n.52
Diderot, Denis (1713–1784), 51–52, 236,
289–91

Dillen [Dillenius], Johann Jacob FRS (1687–
1747), 598n.20
Dillon, Lawrence Samuel (1910–1999), 419–
21, 627n.140
Dillwyn, Lewis Weston FRS (1778–1855),
588n.7
ad-Dīnāwari, Abū Ḥanīfa Āḥmad ibn Dawūd
(*ca* 815/820 to 895/896 CE), 96
Diocles of Carystus (*ca* 375 to 295 BCE),
481n.55
Diocletian [Gaius Aurelius Valerius
Diocletianus], Emperor (242/245 to 311/
312), 65–66, 75
Diodorus Cronus (d. *ca* 284 BCE), 472n.87
Diodorus Siculus (*fl.* First century BCE), 522n.16
Diogenes Laërtius (*fl.* Third century CE), 25, 46
Diogenes of Apollonia (*fl.* latter half of Fifth
century BCE), 10, 30
Dionysius the Areopagite, Bishop of Athens,
Saint (*fl.* First century CE), 85, 87
Dionysius "the Great", Saint, Pope of
Alexandria (d. 264), 85, 491n.55
Dioscorides. *See* Pedanius Dioscorides
Dobell, Cecil Clifford FRS (1886–1949), 330,
409, 541n.2, 543n.52, 544n.59, 544n.64,
577n.171, 587n.199
Dobzhansky, Theodosius Grygorovych
ForMemRS (1900–1975), 432–33
Dodge, John David (b. 1935), 419–21
Dodoens, Rembert [Joenckema, Rembert van]
(1517–1585), 192, 532n.23
Dodson, Edward Ottway (1916–2002),
419–21
Doflein, Franz Theodor (1873–1924), 409,
616n.195
Dolet, Étienne (1509–1546), 528n.41
Donati, Vitaliano FRS (1717–1762), 232–33,
238, 262, 264, 266, 268, 269, 276, 290
Doolittle, W. Ford (b. 1942), 439, 447–48, 449
d'Orbigny, Alcide Charles Victor Marie
Dessalines (1802–1857), 574n.99
Dorn, Gerhard (*ca* 1530 to 1584), 518n.171
Dougherty, Ellsworth Charles (1921–1965),
419–21, 622n.60
Dryander, Jonas Carlsson (1748–1810),
560n.128
Duboscq, Octave Joseph (1868–1943),
621n.36, 621n.44
Duchesne, Antoine Nicolas (1747–1827), 266,
276, 277*f*
Dujardin, Félix (1801–1860), 301–2, 327–28,
351, 366, 405
Dumortier, Comte Barthélémy Charles Joseph
(1797–1878), 574n.114

784 INDEX OF PERSONS

Duns, John [Duns Scotus] (*ca* 1265/1266 to 1308), 115, 134–35
Duran, Rabbi Simeon ben Ẓemaḥ (1361–1444), 109–10
Duret, Claude (*ca* 1570 to 1611), 167, 171, 173, 196, 203, 525n.87

Eckhart von Hochheim. *See* Meister Eckhart
Edwards, Jonathan [theologian] (1703–1758), 247–48, 362
Egerton, Rev. Francis Henry FRS, [the Eighth] Earl of Bridgewater (1756–1829), 363
Ehrenberg, Christian Gottfried ForMemRS (1795–1876), 323–24, 325–28, 329, 330, 332–33, 335, 336, 342, 348–49, 351, 353, 364–65, 366, 367, 369, 380, 382, 400, 405, 406, 595n.160
Eichhorn, Pastor Johann Conrad (1718–1790), 350–51, 597n.15
Eichwald, Karl Eduard [von] [Eykhvald, Eduard Ivanovich] (1795–1876), 273, 274f, 278–79, 279f, 280, 282, 286, 287, 337, 354, 565n.38
Elias (*fl.* mid-Sixth century CE), 69, 71–72, 95, 97, 113, 497n.52
Ellis, John FRS (*ca* 1710 to 1776), 252, 254, 255, 257, 291–92, 339, 357, 390, 562n.161
Empedocles of Acragas [Agrigento] (*ca* 490/494 to *ca* 434 BCE), 8–9, 11, 29–30, 42, 158, 165, 457n.38, 458n.49, 468n.12, 471n.70, 481n.69
Enderlein, Günther (1872–1968), 413–14, 417, 625n.97, 627–28n.147
Endlicher, Stephan Ladislaus [István László] (1804–1849), 342, 343, 589n.17
Engels, Friedrich (1820–1895), 578n.3
Engramelle, Jacques Louis Florentin (1734–1814) (*), 526n.112
Engramelle, Marie-Dominique-Joseph (1727 to 1780/1 or 1805) (*), 526n.112
Entz, Géza (1842–1919), 407–8, 592n.83
Epicurus of Samos (341–270 BCE), 24, 29, 77, 210–11, 493n.89
Epimenides of Crete [Epimenides of Cnossos], Seventh or Sixth century BCE, 459n.64
Erasistratus of Ceos (*ca* 304 to *ca* 250 BCE), 483n.10
Erasmus, Desiderius [Geeris, Geert] (*ca* 1466/1469 to 1536), 160, 175
Eriugena, John [Johannes] Scotus (*ca* 800 to *ca* 877 CE), 116–17, 118–19, 549n.26
Ernst, Max (1891–1976), 548n.18
Esper, Eugen[ius] Johann Christoph (1742–1810), 263

Estienne, Henri [the Elder] (1460/1470 to 1520), 504n.240, 527n.7
Estienne, Henri [Stephanus, Henricus] (1528/1531 to 1598), 45–46
Estienne, Robert [Stephanus, Robertus] (1503–1559), 175
Euclid (*fl. ca* 300 BCE), 94–95, 119, 483n.10, 500n.126, 575n.130
Eugenius Philalethes. *See* Vaughan, Thomas
Euler, Leonhard FRS (1707–1783), 311–12
Eunapius Sardianus (347 to at least 402 CE), 485n.60
Eusebius, Bishop of Nicomedia [later Archbishop of Constantinople] (d. 341), 75, 485n.38

Fabbroni, Adamo (1748–1816), 594–95n.149
Faber, Giovanni [Fabri, Johann] (1574–1629), 544n.57
Fabricius, Johan Christian (1745–1808), 564n.11
Fabricius, Johann Albert[us] (1668–1736), 45–46, 601n.78
Fabricius, Otto (1744–1822), 588n.4
Famintsyn [Famintzin], Andrei Sergeyevich (1835–1918), 423, 427, 430–31
al-Fārābī, Abū Naṣr Muḥammad ibn Muḥammad ibn Ṭarkhān (*ca* 870/872 to 950/951 CE), 94, 97–98, 101, 102, 108, 109–10, 111, 143, 500n.138
Favanne de Montcervelle, Jacques-Guillaume de (1716–1770), 539n.148
Feldmann, Jean (1905–1978), 621n.42
Festus [Sextus Pompeius Festus] (*fl.* late Second century CE), 174
Fichte, Johann Gottlieb (1762–1814), 311–12, 316–17, 318
Ficino, Marsilio (1433–1499), 49, 134, 136, 137–39, 149–50, 151–52, 157–58, 226, 284–85, 461n.96, 462n.102, 486n.69
Finden, William (1787–1852), 602n.92
Fischer von Waldheim, Gotthelf [Grigorij Ivanovich] (1771–1853), 281, 282f
Fischer, Gustav Paul Dankert (1845–1910), 408–9
Fischer, Hermann Emil Louis ForMemRS (1852–1919), 432
Flamel, Nicholas (*ca* 1330/1340 to 1418), 154
Flemming, Walther (1843–1905), 422
Flotow, Major Julius Christian Gottlieb von (1788–1856), 342, 386
Fludd, Robert (1574–1637), 158
Focke, Gustav Woldemar (1810–1877), 327
Fogel, Martin (1634–1675), 535n.75

Fontana, Abbé Felice [Félix] (1730–1805), 291, 338–39, 340, 601n.87

Fontenelle, Bernard Le Bovier de FRS (1657–1757), 211–12, 214–15, 221–22, 289, 601n.86

Forbes, Edward FRS (1815–1854), 327, 369

Fortescue, Sir John (ca 1394 to 1479), 226

Foulcher [Fulcher] de Chartres (ca 1059 to after 1128), 165

Fox, George Edward (b. 1945), 436

Fox, Sidney Walter (1912–1998), 433

Fracastoro, Girolamo [Fracastorius, Hieronymus] (ca 1476/1478 to 1553), 212

Francis of Assisi, Saint [di Bernardone, Giovanni di Pietro] (1181/1182 to 1226), 467n.66

Frank, Albert Bernhard (1839–1900), 581n.55

Franklin, Benjamin FRS FRSE (1706–1790), 240–41

Frederic[k] Barbarossa, Holy Roman Emperor (1122–1190), 112

Frederick II "the Great", King of Prussia (1712–1786), 311–12

Frederick II of Hohenstaufen, Holy Roman Emperor (1194–1250), 198

Frederick Augustus [Friedrich August] II, King (1797–1854), 331

Freig [Frey] [Freigius], Johann Thomas (1543–1583), 182–83, 226

Frere, John Hookham (1769–1846), 361

Freud, Sigmund [Sigismund Schlomo] (1856–1939), 608n.24, 625n.95

Friedrich, Caspar David (1774–1840), 587n.202

Fries, Elias Magnus FRS FRSE (1794–1878), 267–68, 284–85, 568n.127

Fuchs, Leonhard [Leonhart] (1501–1566), 190–91, 192, 193, 194, 533n.32, 533n.45

Fulbert, Bishop of Chartres (ca 960 to 1028), 121–22

Gaillon, [François] Benjamin (1782–1839), 301–2, 346–47, 351, 353, 354, 577n.156, 607n.200

Galen, Aelius [Galenus, Claudius] (ca 129 to ca 210/217 CE), 9–10, 44, 47, 53, 54–55, 56, 57, 73–74, 79, 88, 94–96, 111, 115, 119, 120, 140, 153, 154, 158, 171–72, 173–74, 177, 190, 191, 473n.100, 481n.69, 492n.61, 538n.125

Galilei, Galileo di Vincenzo Bonaiuti de' (1564–1642), 157–58, 171–72, 180, 211–12, 215

Gallego, Pedro [Peter] González (ca 1197 to 1267), 113

Galvani, Luigi (1737–1798), 582n.83

Gauricus, Pomponius (1481/1482 to 1530), 504n.239, 527n.6

Gautama Siddhārtha, known as the Buddha (563–483 BCE), 16–17, 18–19

Gaza, Theodorus (ca 1398 to ca 1475), 134, 140–41, 174–75, 474n.131, 476n.159, 478n.200

Geber. See Jābir ibn Hayyān, Abu Mūsā

Gegenbaur, Karl [Carl] ForMemRS (1826–1903), 379, 381–82, 386, 421, 578n.9, 608n.36, 608n.37, 609n.61, 611n.98, 615n.170

Gelli, Giambattista (1498–1563), 172

Gellius, Aulus (ca 125 to ca 180 CE), 469n.24

Gemistus Pletho, Georgius (1355/1360 to 1452/1454), 134, 150–51, 173

Gemma, Cornelius [Cornelio] (1535–1578), 534n.65

Geoffroy, Étienne François [l'Aîné, "the Elder"] FRS (1672–1731), 564n.15

Geoffroy Saint-Hilaire, Étienne (1772–1844), 282, 290, 300–2, 307, 318

Geoffroy Saint-Hilaire, Isidore (1805–1861), 2, 520n.206

George III [George William Frederick], King (1738–1820), 358

George of Trebizond (1395/1396 to 1486), 134, 478n.200

Gerald de Barri [Gerald of Wales; Giraldus Cambrensis] (ca 1146 to ca 1223), 122, 168–69

Gerard(e), John (ca 1545 to 1612), 525n.89, 532n.23, 533n.40

Gerbert of Aurillac, Pope Sylvester II (ca 946 to 1003), 121–22

Gersaint, Edme-François (1694–1750), 539n.148

Gershon ben Schlomoh [Solomon] of Arles (fl. late Thirteenth century), 109–10

Gervaise [de la Puette] of Normandy (fl. first half of Thirteenth century), 197–98

Gesner, Johan [Gessner, Johannes] (1709–1790), 266, 578–79n.20

Gessner, Conrad [Gesnerus, Conradus] (1516–1565), 57, 165, 190–91, 201–4, 207, 287, 532n.27, 544n.59, 549n.26, 561n.153

al-Ghassānī, Qāsim ibn-Muḥammad al-Wazīr (ca 1548 to ca 1611), 534n.56

al-Ghazālī, Abū Ḥāmid Muḥammad ibn Muḥammad [Algazel] (ca 1056/1058 to 1111 CE), 102–3, 106, 129, 139, 247

786 INDEX OF PERSONS

Ghini, Luca (1490–1556), 193, 533n.33
Gilbert [Giddy], Davies PRS (1767–1839), 363
Gilbert of Poitiers [Gilbert de la Porrée] (after 1085 to 1154), 121–22
Gilbert, Sir William Schwenck (1836–1911), 606n.197
Giovio, Paulo (1483–1552), 198–99
Giraldus Cambrensis. *See* Gerald de Barri
Girard, Charles Frédéric (1822–1895), 610n.76
Girod-Chantrans, Justin (1750–1841), 339–40, 341
Giseke, Paul Dietrich (1741–1796), 264, 266–68, 286
Glanvill, Joseph FRS (1636–1680), 186, 601n.86
Gmelin, Johann Friedrich (1748–1804), 250*t*, 256, 257, 262–63, 358, 558n.83, 575n.125, 588n.2, 588n.4
Goedart, Johannes [Jan] (1617–1668), 206
Goethe, Johann Wolfgang von (1749–1832), 152, 248, 312–15, 317, 318–19, 322–23, 327–28, 331, 333, 335, 337, 345–46, 350, 354, 355, 361, 364, 379, 384, 386, 545n.87, 563n.189, 608n.21, 608n.37, 611n.98
Goeze, Johann August Ephraim (1731–1793), 337
Goldfuss, Georg August (1782–1848), 270–72, 271*f*, 286, 318, 565n.38, 584n.135, 591n.63, 605n.175
Goldsmith, Oliver (1728–1774), 598n.36, 601n.87
Good, John Mason (1764–1827), 361
Gordian III [Marcus Antonius Gordianus], Emperor (225–244 CE), 63
Grant, Robert Edmond FRS FRSE (1793–1874), 390
Gray, Asa ForMemRS (1810–1888), 370, 373–74, 375
Green, Thomas (active 1816–1820), 357–58
Gregory VII, Pope [Hildebrand of Sovana] [Saint] (*ca* 1015 to 1085), 115
Gregory XIII, Pope [Boncompagni, Ugo] (1502–1585), 117
Gregory of Nazianzus, Saint (*ca* 329 to 390 CE), 80–81, 82, 493n.89
Gregory of Nyssa, Saint (*ca* 335 to *ca* 395 CE), 80–82, 116, 459n.64, 462n.98, 463n.111, 492n.58, 497n.53, 548n.14
Greville, Robert Kaye FRS FRSE (1794–1866), 357–58
Grew, Nehemiah FRS (1641–1712), 327, 535n.72, 542n.24
Griffith, Frederick (1877–1941), 432–33
Gronovius, Jan [Johann] Frederik (1690–1762), 249

Grosseteste, Bishop Robert (*ca* 1168 to 1253), 89, 112, 113, 123–24, 173, 247, 478n.200
Gruithuisen, Baron Franz von Paula (1774–1852), 340
Guettard, Jean-Étienne (1715–1786), 232
Guillaume de Lorris (*ca* 1200 to *ca* 1240), 549n.26
Guillaume de Salluste, Sieur du Bartas (1544–1590), 162, 166
Guillaume of St-Thierry [St-Theodore], Abbé (*ca* 1085 to 1148), 508n.100
Guilliermond, Alexandre (1876–1945), 426, 629n.176
Gundisalvo, Domingo (*ca* 1115 to at least 1190), 123
Gutenberg, Johannes (*ca* 1400 to 1468), 173–74
Gwion Bach ap Gwreang (*fl.* Sixth century CE), 14

Haeckel [Häckel], Ernst Heinrich Philipp August (1834–1919), 317, 324–25, 379–80, 381, 382–403, 394*f*, 398*f*, 404–5, 406, 407, 408, 409–10, 411, 412–13, 416, 417, 421, 424, 426, 443, 571n.30, 578n.4, 596n.187, 608n.18, 619–20n.13, 622n.62, 625n.106
Hales, Rev. Stephen (1677–1761), 597n.15
Haller, Albrecht von FRS (1708–1777), 253–54, 333, 546n.105, 547n.119, 547n.131, 570n.18, 625n.91
Halley, Edmund FRS (1656–1742), 542n.24
Hammer, Frédéric-Louis (1762–1837), 562n.177
Han Mingdi [Liu Zhuang], Emperor (28–75 CE), 16
Hanov [Hanow], Michael Christoph (1695–1773), 571n.29
Hardenberg, Georg Philipp Friedrich, Freiherr von, known as Novalis (1772–1801), 316–17
Harms, Jürgen Wilhelm (1885–1956), 416
Harris, Rev. John FRS (1666/1667 to 1719), 544n.65
Hartmann, Max (1876–1962), 404–5, 409, 421, 618n.255
Harvey, William [physician] (1578–1657), 212, 316, 355, 474n.121, 559n.103, 560n.116, 565n.45
Harvey, William Henry FRS [botanist] (1811–1866), 403, 596n.184
Haüy, Abbé René-Just ForMemRS (1743–1822), 241, 597n.15
Hayata Bunzō (1874–1934), 580n.37
al-Ḥayyāt, ʿAbd al-Raḥīm ibn Muḥammad (d. 902 CE), 498n.76

INDEX OF PERSONS   787

Hedwig, Johann[es] FRS (1730–1799), 589n.24, 591n.57, 612n.113

Hegel, Georg Wilhelm Friedrich (1770–1831), 188, 315, 316–17, 322, 578n.3, 632n.65

Heidenberg, Johann. *See* Trithemius, Johannes

Hennig, Emil Hans Willi (1913–1976), 447, 606n.180, 636n.146, 636n.151

Henri III [Alexandre Édouard], King of France (1551–1589), 170–71

Henry I, King of England (*ca* 1068 to 1135), 197–98

Henslow, Rev. John Stevens (1796–1861), 597n.16, 602n.90, 604n.156

Heracleides [Heraclides] of Pontus (*ca* 390 to *ca* 310), 11, 24, 471n.70

Heraclitus of Ephesus (*fl. ca* 500 BCE), 10, 25–26, 44, 101–2

Heraclitus of Tyre (*fl.* First century BCE), 47, 60

Heraclius I, Emperor (*ca* 575 to 641), 488n.112

Herberstein, Sigismund, Freiherr von (1486–1556), 166, 183, 230

Herbert, Rev. George (1593–1633), 467n.66

Herder, Johann Gottfried [von] (1744–1803), 240–41, 312–13, 315, 578n.8, 597n.15

Hermann [Herrmann], Johann [Jean] (1738–1800), 268–69, 272, 565n.42, 566n.74

Hermias [philosopher] (*ca* 410 to *ca* 450 CE), 69

Hermippus of Smyrna (*fl.* mid-Third century BCE), 468n.16

Hernández de Toledo, Francisco (1514–1587), 183–84

Herodotus (*ca* 485/484 to *ca* 425 BCE), 7, 10–11, 56–57, 136–37, 165–66, 498n.86

Herophilus [Herophilos] (335–280 BCE), 483n.10

Herr, Michel [Michael] (*ca* 1490/1495 to *ca* 1550), 198–99

Herschel, Sir John FRS (1792–1871), 602n.90

Hertwig, Richard Wilhelm Karl Theodor, Ritter von (1850–1937), 408, 409, 421

Hesiod (*fl.* 750 BCE), 486n.68

Hiatt, Russell Frank (1905–1988), 460n.76

Hierocles of Alexandria (*fl. ca* 430 CE), 63, 488n.99

Hilary of Poitiers, Saint (310/315 to 367/368), 493n.89

Hildebert, Archbishop (*ca* 1055 to 1133), 122

Hildebrand of Sovana. *See* Gregory VII, Pope

Hildegard of Bingen, Saint (1098–1179), 120–21, 146–47, 189

Hill, "Sir" John (1714–1775), 219–20, 221–22, 226, 326, 357, 542n.34, 561n.142

Hill, Peter (1754–1837), 358

Hincks, Rev. William (1794 to 1871), 285–86

Hippasus of Metapontum (*ca* 530? to *ca* 450? BCE), 10

Hippocrates of Cos (*ca* 460 to *ca* 370 BCE), 47, 53–54, 55, 73, 94–95, 111, 115, 154, 177, 492n.61

Hippodamus of Miletus ["The Thurian"] (498–408 BCE), 27–28

Hitchcock, Edward (1793–1864), 373

Hofmeister, Franz (1850–1922), 432

Hofmeister, Wilhelm Friedrich Benedikt (1824–1877), 596n.185

Hogg, Jabez (1817–1899), 376, 545n.71, 547n.117, 591n.64

Hogg, John FRS (1800–1869), 370–74, 373*f*, 377–78, 417, 603n.123

Holbach, Baron d'. *See* Thiry, Paul-Henri

Hölderlin, Johann Christian Friedrich (1770–1843), 581n.52

Homberg, Wilhelm (1652–1715), 589–90n.34

Home, Everard FRS (1756–1832), 601n.87

Homer [legendary author of *Iliad*] (*fl.* perhaps late Eighth century BCE), 26, 60–61, 76, 159, 164, 224, 225–26, 486n.68, 490n.14, 548n.5, 548n.6, 548n.8

Honorius III, Pope [Savelli, Cencio] (*ca* 1150 to 1227), 117

Honorius Augustodunensis of Regensburg [Honorius of Autun] (*ca* 1080 to 1154?), 117

Hooke, Robert FRS (1635–1703), 209, 210, 216, 327, 544n.58, 544n.59

Hooker, Sir William Jackson FRS (1785–1865), 336, 599n.43

Hope, John FRS (1725–1786), 357

Horace [Quintus Horatius Flaccus] (65–8 BCE), 459n.67, 462n.100

Horaninov, Paul [Gorianinov, Pavel Feodorovich] (1796–1865), 280–81, 280*f*, 284–85, 286, 287, 318, 607n.200

Hornschuch, Christian Friedrich (1793–1850), 345–46, 347

Howard, Luke FRS (1772–1864), 263–64

Howley, William FRS, Archbishop of Canterbury (1766–1848), 363

Hrabanus [Rabanus] Maurus Magnentius, Archbishop of Mainz (*ca* 776 to 856 CE), 114, 116

Hudson, William FRS (1730–1793), 357

Hugh of Saint Victor (*ca* 1096 to 1141), 122

Hugo de Folieto [Hugh of Fouilloy] (1096/1111 to *ca* 1172), 198

Hugo de Santalla (active 1141–1145), 112–13, 144

788 INDEX OF PERSONS

Hugolino of Orvieto (after 1300 to 1373), 522n.25

Hui Neng [the "Sixth Patriarch"] (638–713 CE), 18–19

Hull, John [physician] (1761–1843), 357–58, 564n.10

Humboldt, Friedrich Wilhelm Heinrich Alexander von ForMemRS (1769–1859), 307–8, 317, 322–24, 325, 333, 379, 576n.133

Hume, David (1711–1776), 159, 315

Ḥunain ibn 'Isḥāq al-'Ibādī (808/809 to 873/877 CE), 95–96, 112–13, 482n.86

Hunter, Alexander FRS (1729–1809), 240–41

Hus, Jan (ca 1370 to 1415), 157–58

Hutton, James FRSE (1726–1797), 359

Huxley, Thomas Henry FRS (1825–1895), 330, 371, 377, 379, 389, 401–2, 463n.110, 571n.29, 619n.267

Huygens, Christiaan FRS (1629–1695), 211–12

Hwass, Christian Hee (1731–1803), 575n.126

Hypatia (perhaps 350/370 to 415), 483n.4, 488n.99

Iamblichus of Chalcis (ca 240/245 to ca 325 CE), 11, 26, 65–67, 68–69, 73, 76, 80, 85, 90, 91, 136–38, 148, 462n.98, 497n.52

Ibn al-'Arabī, Abū 'Abd Allāh Muḥammad al-Hātimī aṭ-Ṭā'ī, known as al-Qushayrī (1165–1240 CE), 106

Ibn al-Baṭrīḳ [al-Biṭrīq], Abū Zakariyā' Yaḥyā [Yuḥannā] (fl. 796–806 CE), 94–95, 144–45, 478n.200, 509–10n.162

Ibn Daud, Abraham ben David Halevi, known as Rabad I (ca 1110 to ca 1180 CE), 108, 575n.121

Ibn Falaquera, Shem Tov ben Joseph (ca 1225 to after 1290), 109–10

Ibn Gebirol, Solomon (1021/1022 to perhaps 1070), 502n.178

Ibn Khaldūn al-Ḥaḍramī, Abū Zayd 'Abd ar-Raḥmān ibn Muḥammad (1332–1406 CE), 501n.169, 501n.170, 523n.44

Ibn Matkah. See Yehudah ben Schlomoh

Ibn Motot, Samuel Ben Saadiah (fl. ca 1390/1395), 502n.180

Ibn Rushd, Abū l-Walīd Muḥammad ibn 'Aḥmad [Averroës] (1126–1198 CE), 68, 102–3, 109–10, 111–12, 113, 129, 180, 210–11

Ibn Sīda. See al-Mursi

Ibn Sīnā, Abū 'Alī al-Ḥusayn ibn 'Abd Allāh ibn al-Ḥasan ibn 'Alī [Avicenna] (ca 980 to 1037 CE), 68, 101–2, 108, 109–10, 111–12, 113, 122–23, 125, 140, 148, 154, 165, 287

Ibn al-Ṭayyib, Abū al-Faraj 'Abd Allāh, known as al-'Irāqī (d. 1043 CE), 95, 96, 113

Ibn Waḥshiyya (d. 930/931 CE), 459n.59

Ibn Yazīd, [Abū Hāshim] Khālid (ca 668 to 704/709 CE), 148–49

Ignatius of Antioch, Saint (b. second half of First century, d. first half of Second century), 490n.9

Iinuma Yokusai (1783–1865), 22

Ikhwān al-Ṣafā [Brethren of Purity] (fl. late Tenth century), 98–100, 137, 148, 498n.92, 513n.37

Imperato, Ferrante (ca 1525? to ca 1615), 185, 196, 205, 207, 549n.26

Ingen-Housz [Ingenhousz], Jan FRS (1730–1799), 338–39, 340, 341, 347, 353, 576n.137, 594n.141, 598n.18

Innocent III, Pope [dei Conti di Segni, Lotario] (1160/1161 to 1216), 168–70, 463n.112

Innocent VIII, Pope [Cybo or Cibo, Giovanni Battista] (1432–1492), 517n.151

Intorcetta, Prospero [Yin Duoze] (1626–1696), 465n.41

Ion of Chios (ca 490/480 to ca 420 BCE), 468n.13

al-'Iraqi al-Simāwi, Abu' al-Qāsim Ahmad ibn Muḥammad (fl. Thirteenth century CE), 516n.122, 516n.123

Ireland, Alexander (1810–1894), 603n.133

Ishāq ibn Ḥunain, Abū Ya'qūb (ca 830 to ca 911/911 CE), 68, 95, 111–12

Isidore of Seville [Isidore Hispalensis], Saint (ca 560 to 636 CE), 9, 57, 114, 115–16, 125, 165, 197–98, 458n.54

Jābir ibn Hayyān, Abu Mūsā (ca 721 to ca 815 CE), 138, 143, 147, 148–49

Jacopo d'Aquino (fl. Thirteenth century), 532–33n.31

Jacques de Vitry [Jacobus de Vitriaco], Cardinal (ca 1160/1170 to 1240), 165

Jacquin, Nikolaus Joseph, Freiherr von (1727–1817), 307–8

al-Jāḥīẓ. See al-Baṣrī, Abū 'Uthman

James of Venice (fl. 1125–1150), 113

Jean de Jandun (ca 1285 to 1328), 181

Jean de la Rochelle (ca 1200 to 1245), 115

Jean de Meun [Meung] (ca 1240 to ca 1305), 226

Jeanne d'Arc [Joan of Arc], Saint (ca 1412 to 1431), 147

INDEX OF PERSONS 789

Jefferson, Thomas (1743–1826), 49

Jeffrey, Charles (1934–2022), 419–21, 444

Jenkinson, James, First Earl of Liverpool (*ca* 1739 to 1808), 357–58

Jenyns [Blomefield], Rev. Leonard (1800–1893), 597n.16, 604n.143

Jenyns, Soame (1704–1787), 240–41

Jerome of Stridon [Hieronymus, Eusebius Sophronius], Saint (340/342, or *ca* 347, to 420 CE), 8, 114, 164–65, 491n.41, 495n.151

Jesus of Nazareth (*ca* 4 BCE to 30/33 CE), 75, 76, 130, 139–40, 170, 458n.47, 483n.109, 514n.48

Job of Edessa [ar-Ruhāwi, Ayyūb] (*ca* 760 to at least 832 CE), 95–96

Joblot, Louis (1645–1723), 216, 545n.75

Johann Konrad von Gemmingen, Prince-Bishop (1561–1612), 205–6

Johannsen, Wilhelm (1857–1927), 432, 620n.14

John VIII Palaiologos, Byzantine Emperor (1392–1448), 134

John XXI, Pope [Julião, Pedro] (*ca* 1215 to 1277), 128–29

John, Lord of Ireland [later King of England] (1166–1216), 168

John Chrysostom, Saint (347–407 CE), 163–64, 225–26, 548n.14

John Climacus ["John of the Ladder"], Saint (*ca* 579 to 649), 548n.14

John Eugenikos of Trebizond (b. after 1394, d. after 1454/1455), 134

John Hyrkanus [Hyrcanus] [Yōḥānān Cohen Gadol] (164–104 BCE), 92

John of Damascus [John Damascene] (*ca* 675/676 to 749 CE), 87–89, 90, 91, 95, 113, 150–51, 182–83, 412, 497n.52, 508n.119, 527n.7, 529n.54

John of Italy [John Italus] (*ca* 1025 to after 1082), 90, 125–26

John of Salisbury, Bishop of Chartres (*ca* 1115/1120 to 1180), 80, 112, 121–22

John of Stobi. *See* Stobæus, Joannes

John Paul XXII, Pope [Duèze or d'Euse, Jacques] (*ca* 1244 to 1334), 517n.130

John the Baptist (*ca* First century BCE to *ca* 30 CE), 84–85, 483n.109, 495n.152, 526n.95

John the Gael. *See* Eriugena

Johnson, Samuel (1709–1784), 257

Johnson, Thomas (1595/1600 to 1644), 526n.97, 533n.40

Jones, Thomas Rymer FRS (1810–1880), 327

Jonson, Benjamin [Ben] (1572–1637), 49

Jonston, John [Jan] (1603–1675), 204–5, 206, 226, 243

Jourdain [Jordanus] de Séverac (*fl.* 1280 to *ca* 1330), 165

Jovian [Iovianus], Emperor (331–364), 65–66

Judah ben Solomon. *See* Yehudah ben Schlomoh

Julian "the Apostate" [Flavius Claudius Julianus], Emperor (331–363), 65–66, 79, 80–81

Julian the Theurgist (*fl.* 161–180 CE), 137

Julius Cæsar [Gaius Julius Cæsar] (100–44 BCE), 459n.67

Jung[e] [Jungius], Joachim (1587–1657), 196–97, 206, 247, 336–37, 531n.4

Jussieu, Adrien-Henri [-Laurent] de (1797–1853), 274–76, 275*f*, 279

Jussieu, Antoine de FRS (1686–1758), 228–29

Jussieu, Antoine-Laurent de ForMemRS (1748–1836), 260, 267–68, 278, 290, 298, 302–3, 358, 359, 412, 545n.77, 598n.18

Jussieu, Bernard de FRS (1699–1777), 232, 234, 241, 252, 255, 259, 293, 294, 545n.77, 570n.12

Justin Martyr of Neapolis (*ca* 100 to 163/167 CE), 76, 77–78

Justinian I ["the Great"], Byzantine Emperor (482–565 CE), 68–69, 71, 89–90, 92, 175–76, 487n.96

Juvenal [Decimus Junius Juvenalis] (*fl.* late First century to after 127 CE), 532–33n.31

Kaempfer, Engelbert (1651–1716), 167, 467n.62, 523n.37

Kafka, Franz (1883–1924), 10

Kaibara Ekiken [Ekken] (1630–1714), 22

Kant, Immanuel (1724–1804), 159, 214, 312, 315–17, 318, 333, 337, 384

Karl August, [Grand] Duke (1757–1828), 312–13

Kaup, Johann Jakob von (1803–1873), 285–86

Keill, John FRS (1671–1721), 542n.24

Kepler, Johannes (1571–1630), 158

Kerr, Robert FRSE (1757–1813), 358

Khosrow I Anushirvān, Sasanian King (512/514 to 579 CE), 69, 89–90, 92

Kidd, John FRS (1775–1851), 363–64

Kielmeyer, Carl Friedrich (1765–1844), 333

Kierkegaard, Søren Aabye (1813–1855), 578n.3

al-Kindī, Abu Yūsuf Yaʻqūb ibn ʻIshaq aṣ-Ṣabbāḥ (*ca* 801 to *ca* 873 CE), 97, 497n.53

King, Sir Edmund FRS (1629/1630 to 1709), 544n.65

790 INDEX OF PERSONS

Kirby, Rev. William FRS (1759–1850), 284–85, 363, 597n.16

Kircher, Athanasius SJ (1602–1680), 156–57, 542n.18, 544n.65

Kirchweger, Anton Josef (d. 1746), 159, 226

Klebs, Georg Albrecht (1857–1918), 404–5, 405f

Knebel, Karl Ludwig von (1744–1834), 313

Knight, Bert Cyril James Gabriel (1904–1981), 414, 621n.39

Koch, Heinrich Hermann Robert (1843–1910), 406–7

Ko Hung [Ge Hong] (283–343), 460n.83

Kölliker, Rudolf Albert [von] ForMemRS (1817–1905), 329, 379, 380–81, 386

Kǒng Fūzǐ [Confucius] (551–479 BCE), 19, 20, 466n.54

König, Emanuel (1658–1731), 2, 208

Kossel, Ludwig Karl Martin Leonhard Albrecht (1853–1927), 432

Kozo-Polyansky, Boris Mikhailovich (1890–1957), 426

Kropotkin, Pyotr Alexeyevich (1842–1921), 627–28n.147

Ktesias [Ctesias] of Knidos [Cnidus] (fl. Fifth century BCE), 164

Kūkai [Saeki no Mao] (774–835), 22

Kützing, Friedrich Traugott (1807–1893), 343, 347–50, 351–52, 353, 354, 370, 376–77, 380, 403, 588n.1, 595n.160, 596n.187

Lacépède, Bernard Germain Étienne de (1756–1825), 307–8, 545n.77, 574n.103

Lactantius, Lucius Caecilius Firmianus (ca 250 to ca 325 CE), 137–39

Lagrange, Joseph-Louis [Lagrangia, Giuseppe Luigi] FRS (1736–1813), 311–12

La Hire, Philippe de (1640–1718), 589–90n.34

Lake, James Albert (b. 1941), 442

Lamarck, Jean-Baptiste Pierre Antoine de Monet, Chevalier de (1744–1829), 241, 278, 283–84, 286, 290, 293–98, 300, 303, 307–8, 309, 325, 326, 335, 336, 337, 384, 443, 545n.77, 545n.81, 571n.29, 608n.37, 611n.98, 619n.5

Lambert, Johann Heinrich (1728–1777), 542n.24

La Mettrie, Julien Offray de (1709–1751), 223, 236, 311–12

Lamouroux, Jean Vincent Félix (1779–1825), 301–2, 336, 352–53, 403, 590n.37

Landsborough, Rev. David (1779–1854), 597n.16

Lankester, Edwin FRS (1814–1874) [father of Edwin Ray Lankester FRS], 371, 604n.156

Lankester, Edwin Ray FRS (1847–1929), 547n.130, 580n.40

Laozi [Lao Tze] (supposedly b. 604 BCE) (**), 17, 20

Laplace, Pierre-Simon, Marquis de (1749–1827), 384

Lardner, Dionysius FRS (1793–1859), 362

La Tour [Latour], Charles Cagniard de (1777–1859), 594–95n.149

Latreille, Pierre André (1762–1833), 307–8, 571n.30, 574n.99, 577n.153, 597n.15

Lavoisier, Antoine-Laurent de FRS (1743–1794), 290, 361

Leclerc, Georges-Louis FRS, Comte de Buffon (1707–1788), 159, 209, 216–18, 221–22, 226, 234, 235, 236, 237, 257, 259, 269, 270f, 276, 289, 290, 293, 298, 300–1, 307–8, 309–10, 316, 320, 355–56, 358, 359–60, 362, 364, 539n.148, 545n.81, 571n.33, 574n.116, 601n.77, 603n.125

Le Clerc, Jean [Clericus, Johannes] (1657–1736), 513n.7

Leclercq, Emma (1851–1933), 408

L'Écluse, Charles de [Clusius, Carolus] (1526–1609), 178, 192–93, 194–95, 533n.34

Ledermüller, Martin Froben[ius] (1719–1769), 221–22

Lee, James [nurseryman] (1715–1795), 357

Leedale, Gordon Frank (b. 1932), 419–21

Leeuwenhoek, Antony [Antonj] Philips van FRS (1632–1723), 206, 209, 210, 211–12, 214, 215–16, 221, 233–34, 237, 250, 251–52, 253, 350, 405, 550n.76, 551n.78, 552n.111, 559n.103, 589–90n.34

Le Fèvre de La Boderie, Nicolas (1550–1613), 150, 151f

Lefèvre d'Étaples, Jacques (ca 1455 to ca 1536), 160, 513n.26, 527n.7

Leibniz, Gottfried Wilhelm [von] FRS (1646–1716), 213–15, 222, 226, 227, 240, 315, 345–46, 545n.87, 553n.137, 593n.105

Leidy, Joseph Mellick (1823–1891), 373–74, 375

Lemery [Lémery], Nicolas (1645–1715), 208

Leoniceno, Niccolò (1428–1524), 190–91, 198, 533n.44

Leonides of Alexandria, Saint (d. 202), 76, 491n.32

Lesser, Friedrich Christian (1692–1754), 556n.35, 572n.39, 601n.78

Lesueur, Charles Alexandre (1778–1846), 297

Leucippus (fl. perhaps mid-Fifth century BCE), 28–29, 35, 211, 222, 470n.43

Leuckart, Karl Georg Friedrich Rudolf (1822–1898), 536–37n.106, 537n.120, 611n.97, 615n.170

Lewis, Isaac McKinney (1878–1943), 412–13

Leydig, Franz [von] (1821–1908), 379, 380

L'Héritier de Brutelle, Charles Louis (1746–1800), 266–67, 268
Libavius, Andreas (*ca* 1550 to 1616), 167, 207–8, 226
Liceti, Fortunio (1577–1657), 167
Lichtenstein, Anton August Heinrich (1753–1816), 345–46, 560n.130
Liebig, Justus, Freiherr von (1803–1873), 322
Lightfoot, Rev. John FRS (1735–1788), 357–58, 597n.16
Lindley, John FRS (1799–1865), 287, 336
Lindsay, John (d. 1803), 336–37
Link, Johann Heinrich Friedrich FRS (1767–1851), 325, 345–46, 351–52
Linnæus, Carl (the younger) (1741–1783), 358
Linnæus, Carolus [Linné, Carl von] FRS (1707–1778), 2, 9, 22, 49, 159, 172, 193, 204–5, 209, 233, 245, 248–57, 250*t*, 259, 260, 261, 263, 264, 266, 267–68, 267*f*, 274–75, 283–85, 286, 287, 291–92, 293–94, 303, 304–5, 312–13, 316, 335, 344–46, 350, 352, 353, 355–56, 357–60, 371, 373, 375, 377, 385, 390, 403, 406, 412, 552n.105, 555n.1, 555n.18, 571n.29, 585n.136, 601n.87, 602n.92, 607n.200
Linné, Carl von. *See* Linnæus, Carolus
Lister, Martin FRS (1639–1712), 206
Livy [Titus Livius] (59 BCE to 17 CE), 176
Llull, Ramon (*ca* 1232 to *ca* 1315/1316), 154, 226, 548n.18
L'Obel, Mathias de [Lobelius, Matthæus] (1538–1616), 178, 192, 193
Locke, John FRS (1632–1704), 180, 243–44, 311–12, 315, 358–59
Loder, Justus Christian (1753–1832), 313, 322–23
Longinus, Cassius (*ca* 213 to 273 CE), 63, 64
Lonicer [Lonitzer], Adam (1528–1586), 189–90
López de Gomara, Francisco (1511–1566), 540n.172
Loudon, John Claudius (1783–1843), 359
Louis XVI [Louis-Auguste], King of France (1754–1793), 361
Lovejoy, Arthur Oncken (1873–1962), 223, 227–28, 541n.7, 552–53n.128
Lucretius Carus, Titus (*ca* 99 BCE to *ca* 55 BCE), 29, 48–49, 212, 541n.10
Lucullus, Lucius Licinius (118 to 57/56 BCE), 60
Lumière, Auguste Marie Louis Nicolas (1862–1954), 426
Luther, Rev. Martin (1483–1546), 134–35, 147, 157–58, 379, 462n.101

Lwoff, André Michel ForMemRS (1902–1994), 414–15, 622n.60
Lyell, Sir Charles FRS (1797–1875), 369, 606n.195
Lyonnet [Lyonet], Pierre FRS (1706–1789), 237, 571n.30
Lysenko, Trofim Denisovich (1898–1976), 619n.5, 622n.48

Machiavelli, Niccolò (1469–1527), 49
Macintosh, Sir James FRS FRSE (1765–1832), 602n.90
Macleay, William Sharp (1795–1865), 282–86, 283*f*, 285*f*, 287, 364–65, 378, 602n.92
Macrobius [Macrobius Ambrosius Theodosius] (*fl. ca* 400 CE), 73–74, 224
Maerlant, Jacob van (1230/1240 to 1288/1300), 124–25
Magellan, Ferdinand [Magallanes, Fernando de] (1480–1512), 169–70
Magnol, Pierre (1638–1715), 243, 244–46, 535n.72, 555n.18
Maier, Michael (1568–1622), 154–56, 173, 519n.184
Maimon, Moses ben [Maymūn, Mōšeh běn; Maymūn, Mūsā bin] [Maimonides] (1135–1204), 6, 94–95, 107, 108–10
Maimonides. *See* Maimon, Moses ben
al-Makīn, Girgis ibn Abī al-Yāsir [Elmacin, George] (*ca* 1203/1205 to 1273/1274), 513n.19
Malpighi, Marcello FRS (1628–1694), 237, 327
al-Māʾmūn. *See* al-Rashīd, Abū al-Abbas ibn Hārūn
Manardo [Manardi], Giovanni (1462–1536), 533n.32
Mandeville, Sir John (pseudonym; *fl.* early-to mid-Fourteenth century), 165, 166, 523n.38
Manget, Jean-Jacques [Mangetus, Johann Jacob] (1652–1742), 154, 156–57
Mani (216 to 274 CE), 496n.23
al-Manṣūr, Abū Jaʿfar ʿAbd Allāh ibn Muḥammad, Caliph (*ca* 709/714 to 775), 94–95
Manutius, Aldus Pius (1449/1452 to 1515), 55, 173–74
Marcgrave [Marggraf], Georg (1610–1644), 204
Marcus Aurelius Antoninus, Emperor (121–180 CE), 44
Margoliash, Emanuel (1920–2008), 433–34, 631n.28
Margulis [Sagan], Lynn Petra, née Alexander (1938–2011), 418–21, 420*f*, 427–29, 446, 605n.162, 624n.88

792 INDEX OF PERSONS

Maria Theresa Walburga Amalia Christina, Empress (1717–1780), 338

Marie Antoinette Josèphe Jeanne, Queen (1755–1793), 302

Marius (*fl.* Twelfth century), 114, 120–21, 226

Marlowe, Christopher (1564–1593), 152

Marquah (*fl.* Fourth century CE), 92

Marsigli [Marsili], Comte Luigi Ferdinando FRS (1658–1730), 231–32, 290, 550n.76

Martin, Benjamin (*ca* 1704 to 1782), 359

Martyn, Thomas FRS (1735–1825), 357–58

Marulić Splićanin, Marko (1450–1524), 530n.72

Marvell, Andrew (1621–1678), 136

al-Mas'ūdī, Abu al-Ḥasan 'Alī ibn al-Ḥusayn ibn 'Alī (*ca* 893/896 to 956 CE), 98

Mather, Cotton FRS (1663–1728), 247–48, 362

Mattioli, Pierandrea [Pietro Andrea] Gregorio (1501 to *ca* 1577), 190–92, 534n.57

Maupertuis, Pierre Louis Moreau de FRS (1698–1759), 227, 236, 311–12, 542n.24

Maximus the Confessor, Saint (*ca* 580 to 662), 116

Mayerne, Sir Théodore Turquet de (1573–1655), 204, 206–7

Mayr, Ernst Walter ForMemRS (1904–2005), 447–48, 555n.18

Meckel, Johann Friedrich ForMemRS (1781–1833), 324–25, 609n.41

Medici, Cosimo di Giovanni de' ["the Elder"] (1389–1464), 137–38, 193

Medicus [Medikus], Friedrich Casimir [Kasimir] (1738–1808), 316

Megasthenes (*ca* 350 to *ca* 290 BCE), 164

Mehmed II "The Conquerer", Sultan (1432–1481), 134

Meister Eckhart [Eckhart von Hochheim] (*ca* 1260 to 1328/1329), 130–31, 134–35

Melanchthon, Philip [Schwartzerdt, Philipp] (1497–1560), 160

Melissus of Samos (*ca* 470 to *ca* 430 BCE), 28, 480n.31

Mencius. *See* Mengzi

Mendel, Abbot Gregor Johann (1822–1884), 245, 426–27, 432, 619n.5, 622n.48

Mengzi [Mencius] (372–289 BCE), 466n.44, 466n.54

Mentelle, Edme (1730–1816), 570n.23

Merezhkowsky [Mereschkowsky], Konstantin Sergeevich (1855–1921), 423–26, 425*f*, 427, 429, 430–31, 627n.146, 627–28n.147, 629n.176

Merian, Matthäus [the Younger] (1621–1687), 204

Mersenne, Rev. Marin (1588–1648), 157–58, 206–7, 226

Mertens, Franz Carl (1764–1831), 592n.100

Mertrud, Jean-Claude (1728–1802), 298

Mesny, Bartolomeo (1714–1787), 539n.148

Meyen, Franz Julius Ferdinand (1804–1840), 345–46, 347–48, 349, 351, 369, 585n.157

Michael of Ephesus (*ca* 1050 to 1129), 90

Micheli, Pier Antonio (1679–1737), 253–54, 535n.74

Michurin, Ivan Vladimirovich (1855–1935), 622n.48

Miescher, Johannes Friedrich (1844–1895), 432

Millar, James [physician] FRSE (1762–1827), 357–58

Miller, Hugh (1802–1856), 285–86, 376

Milne-Edwards, Henri ForMemRS (1800–1885), 307, 309, 326, 335, 366, 564n.21

Milton, John (1608–1674), 226

Minchin, Edward Alfred FRS (1866–1915), 402–3

Mirbel, Charles-François Brisseau de (1776–1854), 571n.35, 589n.17, 599n.44

Mithridates VI Eupator (135–63 BCE), 47, 52–53, 60

Miura Baiyen [Miura Susumu] (1723–1789), 467n.59

Mivart, St. George Jackson FRS (1827–1900), 606n.192

Mocenigo [Mocenicus], Filippo, Archbishop of Nicosea (1524–1586), 540n.172

Mohl, Hugo von ForMemRS (1805–1872), 327–28

Montaigne, Michel de [Michel Eyquem, Sieur de Montaigne] (1533–1592), 49

Moore, Raymond Cecil (1892–1974), 419–21

More, Henry FRS (1614–1687), 154, 186–87

Moreau, Gustave (1828–1898), 172

Morgan, Thomas Hunt ForMemRS (1866–1945), 432

Morison, Robert (1620–1683), 286, 555n.18, 561n.153

Morrhius Campensis, Gerard[us] [Morrhy, Gérard] (active 1529–1532), 173, 175

Morton, Rev. Charles (1627–1698), 524n.58

Moses (perhaps Thirteenth century BCE), 60–62, 76, 78, 110–11, 124, 137, 140–41, 147, 158, 187, 224–25, 313–14, 486n.68, 522n.4

Moses de León [Moshe ben Shen-Tov], Rabbi (*ca* 1240 to 1305), 109

Mouffet [Moufet, Moffet, Muffet], Thomas (1553–1604), 204, 476n.159

Mozi [Mo Ti; Mo Tzu] (*ca* 470 to *ca* 391 BCE), 21–22

Muḥammad (*ca* 570 to 632 CE), 93–94, 106, 139

Muhlenberg, Rev. Henry Melchior [Mühlenberg, Heinrich Melchior] (1711–1787), 357–58

Müller, Johannes Peter ForMemRS (1801–1858), 329, 379–80, 381, 382, 388, 400, 421, 594n.134

Müller, Miklós (b. 1931), 618n.251

Müller, Otto Friderich (1730–1784), 258, 262–63, 325, 329, 335–36, 349, 352, 353, 405, 542n.29, 554n.160, 575n.125, 577n.171

al-Munajjim. *See* al-Abharī, Athīr al-Dīn al-Mufaḍḍal

Münchhausen, Otto II, Freiherr von (1716–1774), 209, 250, 253–55, 258, 345–46, 354, 560n.111, 560n.123, 570n.21, 607n.200

Münster, Sebastian (1488–1552), 165, 524n.56

al-Mursi, Abū al-Ḥasan ʿAlī ibn Ismāʿīl ibn Sīda (1007–1066 CE), 100–1

al-Muʿtaṣim. *See* al-Rashīd, Abū Isḥaq Muḥammad

Nägeli, Carl Wilhelm von ForMemRS (1817–1891), 327–28, 352–53, 403, 404, 405, 412, 423, 607n.12

Napoleon I Bonaparte [Napoleone Buonaparte] (1769–1821), 307–8, 318–19, 565n.36

an-Naẓẓām, Abū Isḥaq Ibrāhīm ibn Sayyār ibn Hāniʿ (*ca* 775 to 845/846 CE), 96–97, 470n.41

Neander [Neumann], Michael (1529–1581), 540n.175

Nearchus [naval officer under Alexander] (*ca* 360 to *ca* 300 BCE), 523n.29

Nebrija, Antonio de [Martínez de Cala, Antonio] (1444–1522), 174–75, 185

Neckam, Abbot Alexander (1157–1217), 122, 168

Necker, Natalis [Noël] Martin Joseph [de] (1729–1793), 258, 574n.116, 607n.200

Needham, Rev. John Turbervill[e] FRS (1713–1781), 209, 217–20, 221–22, 226, 227, 234–35, 237, 239, 247–48, 254, 290–91, 314, 320, 327, 345–46, 354, 356, 590n.51, 597n.15

Needham, Noel Joseph Terence Montgomery FRS (1900–1995), 18, 214, 465n.31, 489n.135

Nees von Esenbeck, Christian Gottfried Daniel (1776–1858), 258–59, 270, 272, 324, 341–42, 345, 591n.60

Nemesius, Bishop of Emesa [now Homs] (*fl.* very late Fourth century CE), 79–80, 82–83, 90, 95, 112–13, 143, 462n.98, 497n.52

Nero Claudius Cæsar Augustus Germanicus [Lucius Domitius Ahenobarbus], Emperor (37–68 CE), 49, 484n.28

Nestorius, Patriarch of Constantinople (*ca* 386 to *ca* 450 CE), 92

Newton, Sir Isaac PRS (1642/1643 to 1726/1727), 171–72, 227, 240, 247, 311–12, 355, 358–59, 383, 542n.24, 545n.87, 601n.80

Nicander of Colophon (*fl.* Second century BCE), 53, 57

Nicholas V, Pope [Parentucelli, Tommaso] (1397–1455), 476n.159, 512n.232

Nicholas of Autrecourt (*ca* 1299 to 1369), 102–3

Nicholas of Cusa [Nikolaus von Kues] [Cusanus], Cardinal (1401–1464), 132–34, 211, 226–27

Nicolai, Christoph Friedrich (1733–1811), 578n.18

Nicolaus of Damascus (64 BCE to at least 4 BCE), 41–42

Nicomachus of Gerasa (*ca* 60 to *ca* 120 CE), 28, 469n.27, 485n.61

Nicon. *See* Aelius Nicon

Nieremberg y Otin, Juan Eusebio (1595–1658), 140–42, 167, 173, 183–85, 226, 233, 238, 264, 266, 268

Nietzche, Friedrich Wilhelm (1844–1900), 210

Nieuhof, Johan (1618–1672), 465n.25

Nieuwentyt [Nieuwentijt], Bernard (1654–1718), 601n.87

Niẓamī-i Arūẓī-i Samarqandī [Aḥmad ibn ʿUmar ibn ʿAlī], known as ʿArūḍī (*fl. ca* 1110–1161), 103–5, 106

Nolle, Heinrich (*ca* 1583 to 1626), 159

Norton, Thomas (before 1436 to *ca* 1513), 154

Novák, František Antonín (1892–1964), 416, 444, 622n.60, 624n.81

Novalis. *See* Hardenberg, Georg Philipp Friedrich

Numenius of Apamea (*fl.* second half of Second century CE), 28, 462n.98

Ocellus Lucanus (*fl.* Sixth century BCE), 27–28

Odoric of Pordenone [Odorico Mattiussi], Friar (1286–1331), 165, 166

Oken [Okenfuß], Lorenz (1779–1851), 268–69, 283–84, 318–22, 323, 325, 326, 327, 331, 333, 335, 337, 342, 343, 380, 389, 390, 565n.38, 585n.145

794   INDEX OF PERSONS

Oldenburg, Henry [Heinrich] FRS (*ca* 1618–1677), 541n.2

Olivi, Abate Giuseppe (1769–1795), 268, 269

Olympiodorus the Younger (*ca* 495 to 570 CE), 71, 72, 488n.103

Ono Ranzan [Ono Motohiro] (1729–1810), 22

Oppian of Anazarbus (*fl. ca* 180 CE; d. after 211 CE), 47, 56, 198, 202

Oppian of Apamea (*fl. ca* 200/215 CE), 56

Oresme, Nicole (1320/1325 to 1382), 131, 474n.121

Origen Adamantius [Origen of Alexandria] (184/185 to 253/254 CE), 8, 63, 77–79, 81, 163, 211, 224–25

Ovidius Nāsō, Publius [Ovid] (43 BCE to 17/18 CE), 7, 9, 93, 248, 258, 346, 457n.34, 521–22n.1, 542n.29

Owen, Sir Richard FRS (1804–1892), 318, 324–25, 329–30, 364, 365–67, 371–72, 373–75, 377–78, 623n.69

Pachomius "the Great", Saint (292–348), 504n.3

Paley, Rev. William (1743–1807), 248, 362, 377

Pallas, Peter Simon FRS (1741–1811), 261–62, 264, 266, 276, 279, 286, 291–92, 302–3, 369, 390, 412

Pamphilius [Greek physician in Rome] (*fl.* end of First century CE), 47, 53, 56

Panckoucke, Charles-Joseph (1736–1798), 291–92

Panetti, Giovanni Battista [Panetius] [the Carmelite] (1439/1440 to 1497), 503n.237

Pantænus ["the Philosopher"], Saint (d. *ca* 200), 76

Paracelsus. *See* at Bombastus von Hohenheim, Philippus

Paré, Ambroise (1510 to 1590), 170–72, 203, 226

Parkinson, James (1755–1824), 359

Parkinson, John (1567–1650), 476n.169

Parmenides of Elea (*ca* 515 to after 450 BCE), 25–26, 28–29, 31, 409

Pascher, Adolf Alois (1881–1945), 403–4

Pasteur, Louis ForMemRS (1822–1895), 254, 406–7, 594–95n.149, 625n.97, 627–28n.147

Paul [Saul of Tarsus], Saint (*ca* 5 to 64/67 CE), 8, 74, 76, 77–79, 80–81, 91, 118–19, 139–40, 494n.112

Paul II, Pope [Barbo, Pietro] (1417–1471), 512n.235

Pauling, Linus Carl ForMemRS (1901–1994), 631n.24

Paulus Diaconis [Paul the Deacon; Paulus Warnefredus] (*ca* 730 to 796/799 CE), 174

Pawer [Bauer], Georg [Agricola, Georgius] (1494–1555), 198–99

Pedanius Dioscorides of Anazarbus (*ca* 40 to 90 CE), 47, 53, 57, 115, 125, 189–92, 458n.54, 482n.82, 532–33n.31

Peirson, Rev. Robert (1738–1803), 240–41

Pellicier, Guillaume (*ca* 1490 to 1568), 178

Pena, Pierre (1520/1535 to 1600/1605), 192

Penny, Thomas (1532–1589), 538n.136

Perrault, Charles (1628–1703), 240

Persius [Aulus Persius Flaccus] (34–62 CE), 544n.56

Person, David (active 1635), 185–86

Persoon, Christiaan Hendrik (1761–1836), 258–59

Perty, Josef Anton Maximilian (1804–1884), 330, 336

Peter [Simon Peter], Saint (*ca* 1 to 64/68 CE), 169, 490n.29

Peter Abelard (*ca* 1079 to 1142), 118–19, 165

Peter Damian [Pietro Damiani], Saint (*ca* 988 to 1072/1073), 168

Peter Lombard, Bishop of Paris (*ca* 1096 to 1160), 80, 112, 119, 129

Peter of Cornwall (1139/1140 to 1221), 197–98

Petrus Bonus [of Ferrara] (*fl.* 1330s), 9, 149

Peyssonel [Peyssonal, Peyssonnel], Charles ["the Elder"] (1640–1720), 232

Peyssonel [Peyssonal, Peyssonnel], Jean-André FRS (1694–1759), 232, 252, 255, 290, 338

Pherecydes of Syros (*ca* 580 to *ca* 520 BCE), 468n.10

Philip II, King of Spain (1527–1598), 183–84

Philip of Tripoli [Philip of Foligno] (*fl.* 1218–1269), 144–45

Philippe de Thaun [de Thaon] (*fl.* early Twelfth century), 197–98

Philo of Alexandria [Philo Judæus] (*ca* 20 BCE to *ca* 50 CE), 6, 46, 60–62, 73, 81, 91, 113, 163, 173–74, 176–77, 212, 490n.11, 496n.8, 497n.52

Philo of Larisa [Larissa] (159/158 to 84/83 BCE), 60

Philolaus of Croton [alternatively of Tarentum] (*ca* 470 to *ca* 385 BCE), 27–28, 138–39, 412, 468n.13

Philoponus, John [the Grammarian] (*ca* 490 to *ca* 570 CE), 70–71, 72, 73, 89, 94–95, 97, 107, 112, 132, 150–51, 180, 412, 497n.52

Philostratus [Lucius Flavius Philostratus "the Athenian"] (*ca* 170 to *ca* 250 CE), 7, 56

Photius [Photios I of Constantinople], Saint (*ca* 810 to *ca* 893), 458n.49, 468n.16, 469n.23, 469n.33

INDEX OF PERSONS 795

Piccolomini, Æneas Silvius Bartolomæs [Enea Silvio Bartolomeo] [Pope Pius II] (1405–1464), 49, 167

Pico della Mirandola, Giovanni (1463–1494), 11, 110, 134, 136, 150–52, 151*f*, 157–58, 160, 173–74, 176, 225–26

Pico della Mirandola, Giovanni Francesco [Gianfrancesco] (1470–1533), 152, 226

Pietro de' Crescenzi [Petrus de Crescentiis] (*ca* 1230/1235 to *ca* 1320), 198

Pigafetta, Antonio (*ca* 1491 to *ca* 1531), 169–70

Pitt, William [the Younger] (1759–1806), 361

Pius II, Pope. *See* Piccolomini, Æneas Silvius

Plancus, Janus. *See* Bianchi, Giovanni

Plantin, Christophe [Christoffel van] (*ca* 1520 to 1589), 39, 192

Plato of Athens (428/427 or 424/423 to 348/347 BCE), 11, 24–25, 26, 28, 29, 31–35, 37, 40–41, 42, 44–45, 47, 49–50, 54, 59–61, 63, 64–66, 67, 68–69, 73–74, 76, 77–78, 81, 83, 87, 90, 94–95, 114, 117, 118–19, 134, 136–37, 138–40, 147, 150, 152, 157–58, 159, 179, 180, 204–5, 210–11, 223, 224, 226, 468n.13

Plaut, Walter Sigmund (b. 1923), 427

Pletho. *See* Gemistus Pletho, Georgius

Pliny the Elder [Caius or Gaius Plinius Secundus] (23/24 to 79 CE), 2, 46, 47, 50–53, 56–57, 76, 83, 87, 114, 116, 125, 164, 171–72, 173–74, 178–79, 185, 189–90, 191, 198–99, 200, 202, 207, 208, 212, 250–51, 524n.48, 530–31n.84

Plössl [Plössel], Simon (1794–1868), 342–43, 587n.188, 594n.134

Plotinus of Lycopolis (203/205 to 270 CE), 28, 63–64, 66, 71, 73, 83, 462n.98, 473n.100

Pluche, Abbé Noël-Antoine (1688–1761), 289, 309, 556n.35

Plutarch [of Athens], son of Nestorias (*ca* 350 to 430/432 CE), 68–69

Plutarch of Chæroneia (45/50 to 120/125 CE), 62

Poggendorff, Johann Christian (1796–1877), 594–95n.149

Poggio Bracciolini, Gian Francesco (1380–1459), 48–49

Poiret, Jean Louis Marie (1755–1834), 571n.35

Polemon of Athens (*ca* 314 to *ca* 270 BCE), 483n.2

Poliziano. *See* Ambrogini, Agnolo

Poll, Heinrich Wilhelm (1877–1939), 408

Pollux, Julius (d. 238? CE), 523n.29

Polo, Marco (*ca* 1254 to 1324), 19

Polycarp of Smyrna, Bishop, Saint (69–155), 490n.9

Pomet, Pierre (1658–1699), 540n.177

Pomponius Mela (d. *ca* 45 CE), 523n.29

Pope, Alexander (1688–1744), 542n.24

Porphyry of Tyre (232/234 to 305/309 CE), 63, 64–66, 69, 71, 72, 77, 87, 90, 91, 95, 111, 113, 115, 125–26, 147, 149, 173–74, 176, 180, 181–82, 412, 461n.95, 462n.98, 491n.36, 497n.52

Porta, Giambattista [Giovanni Battista] della (*ca* 1535 to 1615 CE), 46, 57, 161, 169–70, 205, 226, 253–54, 532–33n.31, 533n.43, 535n.77

Portier, Paul (1866–1962), 422, 426, 446, 627–28n.147

Posidonius [Poseidonius] of Apameia (*ca* 135 to *ca* 51 BCE), 44, 54

Postel, Guillaume (1510–1581), 529n.49

Power, Henry FRS (1623–1668), 544n.58

Priestley, Joseph FRS (1733–1804), 338–39, 340, 341, 345, 347, 353, 359–60, 361, 403, 576n.137, 591n.63, 597n.16, 601n.88

Pringsheim, Nathanael (1823–1894), 352–53, 380–81, 612n.113

Pritchard, Andrew FRSE (1804–1882), 285–86, 585n.142, 596n.177

Proclus Lycaeus ["the Successor"] (412–485 CE), 68–69, 85, 112–13, 462n.98, 473n.100, 488n.100

Prowazek, Stanislaus Josef Mathias von [Provázek z Lanova, Stanislav] (1875–1915), 409

Psellus [Psellos], Michael (1017/1018 to 1078/1079), 90

Pseudo-Dionysius (*fl.* late Fifth to early Sixth century), 76, 85–86, 116, 508n.119

Ptolemy I Soter, Pharaoh (*ca* 367 to 282 BCE), 483n.10

Ptolemy, Claudius (*ca* 100 to 168/170 CE), 94–95, 111

Pulteney, Richard FRS (1730–1801), 359, 556n.42

Punnett, Reginald Crundall FRS (1875–1967), 630n.3

Pyrrho of Elis (*ca* 360 BCE to *ca* 270 BCE), 24, 44–45

Pythagoras of Samos (*ca* 570 to *ca* 495/490 BCE), 11, 26–27, 28–29, 30, 47, 60–61, 66, 76, 91, 136–37, 138–39, 147, 148–49, 159, 266

al-Qazwīnī, Abū Yahya Zakarīyā ibn Muḥammad (1203–1283 CE), 105, 523n.44

al-Qazwīnī, Ḥamdallāh al-Mustaufī (1281/1282 to 1340/1349), 105

796 INDEX OF PERSONS

Quatrefages de Bréau, Jean Louis Armand de ForMemRS (1810–1892), 574n.99
al-Qūnawī, Ṣadr al-Dīn Muḥammad ibn Isḥāq ibn Muḥammad ibn Yūnus (1207–1274), 106

Rabanus Maurus. *See* Hrabanus Maurus
Rabelais, François (1483/1494 to 1553), 177–78
Raleigh, Sir Walter (*ca* 1552 to 1618), 165
Ralfs, John (1807–1890), 351, 596n.172
Ramus, Petrus [La Ramée, Pierre de] (1515–1572), 181–83
al-Rashīd, Abū al-Abbas ibn Hārūn, Caliph, known as al-Māʾmūn (786–833 CE), 94–96, 97
al-Rashīd, Abū Isḥaq Muḥammad ibn Hārūn, Caliph, known as al-Muʿtaṣim (796–842 CE), 97
Raspail, François-Vincent (1794–1878), 586n.170
Ratramnus (d. *ca* 868 CE), 165
Rauwolf, Leonhard [Leonhart] (1535–1596), 178
Ray [earlier Wray], Rev. John FRS (1627–1705), 209, 243, 245–47, 248, 259, 287, 289, 355, 356–57, 362, 377, 524n.52, 601n.87
Rayer, Pierre François Olive (1793–1867), 406–7
Raymond de Sebonde [Sebunde] (*ca* 1385 to 1436), 226, 247, 467n.66
Raymund de Tarrega (*fl.* Fourteenth century), 149
al-Rāzī, Abū Bakr Muḥammad ibn Zakariyyā [Rhazes] (854 to *ca* 925/932 CE), 140, 146, 148
ar-Rāzī, Šahmardān ibn Abi al-Ḥayr (*fl. ca* 1095–1119), 103
Réaumur, René-Antoine Ferchault de FRS (1683–1757), 232, 234–35, 237, 255, 259, 338
Redi, Francesco (1626–1697), 206, 216, 237
Reichenbach, Heinrich Gottlieb Ludwig (1793–1879), 564n.16
Reichert, Karl [Carl] Bogislaus (1811–1883), 329
Reinke, Johannes (1849–1931), 421, 422–23
Remak, Robert (1815–1865), 412–13, 587n.183
Renner, Otto ForMemRS (1883–1960), 427
Requin, Achille Pierre (1803–1854), 583n.109
Reuchlin, Johann[es] (1455–1522), 225–26, 462n.101
Ricci, Matteo [Mattheus] SJ (1552–1610), 462n.102, 465n.25
Richard, Achille (1794–1852), 336
Richard de Fournival (1201 to 1259/1260), 163–64, 197–98

Richter, Johann Gottfried Ohnefalsch (1703–1765), 601n.78
Richthofen, Ferdinand Freiherr von (1833–1905), 465n.39
Ris, Hans (1914–2004), 427
Ritter, Johann Wilhelm (1776–1810), 333
Rivinius [Augustus Quirinus Rivinius]. *See* Bachmann, August
Robert of Chester [Robertus Castrensis] (*fl.* mid-Twelfth century), 148–49
Robinet, Jean-Baptiste [-René] (1735–1820), 236
Roger, Pierre. *See* Clement VI, Pope
Rondelet, Guillaume (1507–1566), 170–71, 178, 185, 192, 193, 194–95, 200–1, 202–3, 204–5, 207, 219, 226
Roos, Johannes Carolus (1745–1828), 559n.87, 559n.104
Roose, Theodor Georg August (1771–1803), 571n.29
Rose, Hugh [apothecary] (*ca* 1717 to 1792), 357–58
Rösel von Rosenhof, August Johann (1705–1759), 220–21, 226, 253
Ross, Captain James Clark FRS (1800–1862), 323–24
Rösslin, Eucharius (*ca* 1470 to 1526), 189–90
Roth, Albrecht Wilhelm (1757–1834), 592n.100
Rothmaler, Werner Walter Hugo Paul (1908–1962), 416, 417, 444, 622n.60
Rousseau, Jean-Jacques (1712–1778), 236, 290
Roux, Wilhelm (1850–1924), 408
Rudolphi, Karl [Carl] Asmund (1771–1832), 270, 276, 279, 337, 364–65, 565n.38
Ruel, Jean (1474–1537), 190–92, 199, 532–33n.31
Rufinus [Italian monk and botanist] (*fl.* 1287–1300), 458n.54
Rufinus, Tyrannius [of Aquileia] (344/345 to 411 CE), 490n.29, 491n.41, 491n.48
Rufo [Ruffo], Giordano (*fl.* 1200–1256), 198
Rüling, Johann Philipp (b. 1741), 268–69
Rūmī, Jalāl ad-Dīn Muḥammad (1207–1273), 91, 459n.64, 501n.172
Rumph [Rumphius], Georg Eberhard (1627–1702), 535n.72, 539n.148
Rust, Bishop George (*ca* 1628 to 1670), 186

Sachs, Julius von (1832–1897), 403–4, 405–6, 532n.22
Sagan, Lynn. *See* Margulis, Lynn
Saint Hilaire. *See* at Bon and at Geoffroy
Saint-Pierre, Jacques-Henri Bernardin de (1737–1814), 240–41, 268–69, 542n.24, 601n.87

INDEX OF PERSONS 797

Salimbene di Adam (1221 to *ca* 1290),
509n.136
Sallust [Gaius Sallustius Crispus] (86 to *ca* 35
BCE), 462n.98
Salviani, Hippoloto [Ippolito] (1514–1572),
202, 207, 537n.116
Sanger, Frederick FRS (1918–2013), 435–36,
631n.20
Sars, Michael (1805–1869), 346, 597n.15
Sauvages de la Croix, François Boissier de. *See*
Boissier de Sauvages de la Croix, François
Savigny, Marie Jules César Lelorgne de (1777–
1851), 297
Saville Kent (***), William (1845–1908), 401–2
Savonarola, Rev. Girolamo (1452–1498), 157–58
Scaliger, Julius Cæsar [Scala, Giulio Cesare
della] (1484–1558), 9–10, 167, 196, 207,
226, 535n.77
Schaudinn, Fritz Richard (1871–1906), 408–9
Scheffauer, Herman George [Hermann Georg]
(1876–1927), 379
Schelling, Friedrich Wilhelm Joseph [von]
(1775–1854), 311, 314, 316–19, 333, 380,
608n.21
Schenk, Joseph August [von] (1815–1891),
379, 380–81
Scherer, Johann Baptist Andreas, Ritter von
(1755–1844), 589–90n.34
Schiller, Johann Christoph Friedrich von
(1759–1805), 325, 581n.52, 583n.113
Schimper, Andreas Franz Wilhelm (1856–
1901), 422–23
Schleiden, Matthias Jakob (1804–1881),
327–28, 342, 343, 344, 348–49, 350,
352–53, 380–81, 421, 580n.35, 596n.187,
597n.188
Schmidt, Eduard Oscar (1823–1886), 388
Schmitz, Carl Johann Friedrich (1850–1895),
422–23
Schrank, Franz von Paula (1747–1835), 597n.15
Schultze, Max Johann Sigismund (1825–
1874), 327–28, 354
Schwann, Theodor Ambrose Hubert
ForMemRS (1810–1882), 327–28, 329,
330, 412–13, 421, 594n.134, 594–95n.149
Schweigger-Seidel, Franz Wilhelm (1795–
1838), 594n.129
Schwendener, Simon ForMemRS (1829–
1919), 423
Scopoli, Giovanni Antonio (1723–1788),
259–60
Scot [Scotus], Michael (1175 to *ca* 1232),
113, 123, 198, 478n.200, 500n.130,
509–10n.162
Scott, Sir Walter FRSE (1771–1832), 602n.90

Seba, Albertus (1665–1736), 539n.148
Sedgwick, Rev. Adam FRS (1785–1873), 330,
597n.16
Seleucus I Nicator (*ca* 358 to 281 BCE), 164
Sendivogius, Michael [Sędziwój, Michał]
(1566–1636), 154, 156–57, 160, 208
Senebier, Jean (1742–1809), 338–39, 340,
597n.15
Seneca, Lucius Annæus ["the Younger"] (*ca*
4 BCE to 65 CE), 49–50, 56, 57, 215,
480n.31
Serres, Antoine Étienne Renaud Augustin
(1786–1868), 324–25
Seth, Siméon (*ca* 1035 to *ca* 1110), 495n.147
Sethon [Seton], Alexander "the
Cosmopolitan" (d. 1603/1604), 519n.186
Sextus Empiricus (*ca* 160 to *ca* 210 CE), 44–
46, 47, 460n.73, 477n.175, 493n.89
Sextus the Pythagorean (*fl. ca* 200 CE),
469n.31
Shakespeare, William (1564–1616), 51–52,
312–13, 335, 538n.137
Shāpūr I (*ca* 215 to 270/272 CE), 92
Shāpūr II ["The Great"] (309–397 CE), 496n.22
Shelley, Mary Wollstonecraft, née Godwin
(1797–1851), 582n.83, 602n.90
Shen Xiu [Yuqian Shenxiu] (perhaps 606/607
to 706), 18–19
Siebold, Karl Theodor Ernst von ForMemRS
(1804–1885), 329, 330, 333–34, 335, 336,
380–81, 386, 391, 400, 586n.166
Siebold, Philipp Franz von (1796–1866),
467n.62
Siger of Brabant (*ca* 1240 to 1284), 129
Sima Qian (*ca* 145/135 to *ca* 86 BCE), 465n.35
Simplicius of Cilicia (*ca* 490 to *ca* 560 CE), 67,
69, 89–90, 95, 112, 150–51, 173–74, 180,
210–11, 475n.141, 488n.103, 488n.108,
497n.52, 508n.119
Skylax of Caryanda (*fl.* late Sixth to early Fifth
centuries BCE), 464n.14
Slabber, Martinus (1740–1835), 547n.115
Sloane, Sir Hans PRS (1660–1753), 539n.148,
544n.69
Smellie, William FRSE (1740–1795), 355–56,
358, 565n.37, 601n.87
Smith, James Edward FRS (1759–1828), 358–
59, 377
Socrates (*ca* 470 to 399 CE), 27, 28, 30–31,
33, 60, 65, 76, 78, 147, 247, 361, 464n.15,
471n.68, 472n.81, 477n.181, 548n.6
Solander, Daniel Carlsson FRS (1733–1782),
599n.40
Solinus, Gaius Julius (*fl.* mid-Third century
CE), 197–98, 212

798 INDEX OF PERSONS

Šorm, František (1913–1980), 433, 435–36
Sowerby, James (1757–1822), 599n.43
Spallanzani, Rev. Lazzaro FRS (1729–1799), 216, 219, 226, 237, 248, 340, 341, 350–51, 597n.15, 601n.87
Spence, William FRS (ca 1783 to 1860), 284–85
Spener, Johann Jakob [Johannes Jacobus] (1669–1692), 205
Speusippus (ca 408 to 339/338 BCE), 28
Spina [di Spino], Alessandro (d. 1313), 543n.52
Spinzoa, Baruch [de] (1632–1677), 186, 315, 543n.54
Stackhouse, John (1742–1819), 336–37
Stahl, Georg Ernst (1659–1734), 316, 333
Stanier, Roger Yate FRS (1916–1982), 416–17, 622n.60
Steele, Sir Richard (1672–1729), 550n.45
Steenstrup, Johannes Japetus Smith ForMemRS (1813–1897), 346
Stein, Charlotte Albertine Ernestine von, née von Schardt (1742–1827), 313
Stein, Samuel Friedrich Nathaniel, Ritter von (1818–1885), 336, 400
Stelluti, Francesco (1577–1652), 544n.56
Stephanus of Alexandria (fl. ca 580 to ca 640 CE), 71, 112, 488n.112
Sterbeeck, Johannes Franciscus [Frans] van (1630–1693), 533n.43
Stobæus, Joannes [John of Stobi] (fl. Fifth century CE), 27–28
Strabo (64/63 BCE to 23/24 CE), 164, 544n.59
Strabo, Walafrid (ca 808 to 849 CE), 114–15
Strasburger, Eduard Adolf ForMemRS (1844–1912), 408
Strato of Lampsacus (ca 335 to ca 269 BCE), 46–47
Strickland, Hugh Edwin FRS (1811–1853), 285–86, 287
Stukeley, William FRS (1687–1765), 550n.48
Sturtevant, Alfred Henry (1891–1970), 432
Suárez, Francisco SJ (1548–1617), 511n.217
Sulla [Lucius Cornelius Sulla] (138–78 BCE), 47, 59, 60
Sullivan, Sir Arthur Seymour (1842–1900), 606n.197
Surapala (perhaps Tenth century CE), 15
Swainson, William John FRS (1789–1855), 285–86, 362–63
Swammerdam, Jan (1637–1680), 206, 237, 327
Swedenborg, Emanuel (1688–1772), 542n.24
Sylvester II, Pope. See Gerbert of Aurillac
Sylvester, Josuah (1563–1618), 162, 166

Synesius, Bishop of Ptolemais (ca 373 to ca 414 CE), 517n.137
Syrianus "the Great" (d. ca 437 CE), 68–69, 488n.100

Tacitus [Publius Cornelius Tacitus] (ca 56 to ca 120 CE), 7
Tanabe Fubito (fl. early Eighth century), 8
Tatian of Adiabene (ca 120 to ca 185 CE), 490n.11
Tatum, Edward Lawrie (1909–1975), 432–33
Taylor, Frank John Rupert ["Max"] (b. 1939), 624n.88
Tempier, Bishop [of Paris] Étienne (1210–1279), 129, 211
Tertullianus, Quintus Septumius Florens [Tertullian] (ca 155/160 to after 220 CE), 7, 11, 57, 75, 76, 462n.98
Tessier, Abbé Henri-Alexandre (1741–1837), 573n.88
Thales of Miletus (ca 624/623 to ca 548/545 BCE), 10, 25–26
Thao Hung-Ching [Tao Hongjing] (456–536), 18
Themistius of Paphlagonia (317 to ca 388/390 CE), 67–68, 70, 80, 95, 102–3, 109–10, 111–12, 113, 122–23, 127, 150–51, 173–74, 180, 497n.52
Theoderic [Theodoric] "the Great", King of the Ostrogoths (454–526), 114
Theophrastus of Eresos (ca 371/370 to ca 287/286 BCE), 9–10, 24, 42–43, 46–47, 48, 52, 53, 57, 94–95, 96, 134, 165–66, 168, 173–75, 178–79, 189, 190, 194–95, 231, 243, 403, 468n.10, 475n.146, 480n.31, 532–33n.31, 558n.68
Thevet, Rev. André (1516–1590), 165
Thierry of Chartres (d. before 1150/1155), 121–22
Thiry, Paul-Henri, Baron d'Holbach (1723–1789), 290
Thölde, Johann (ca 1565 to 1614), 519n.184
Thomas Aquinas [Tommaso d'Aquino], Saint (1225–1274), 68, 82, 112, 115, 126–28, 129–30, 133–34, 147, 150, 165, 173, 180, 181, 194, 211, 214, 226, 247, 461n.91, 487n.86, 532–33n.31, 554n.151
Thomas of Brabant. See Thomas of Cantimpré
Thomas of Cantimpré [Thomas of Brabant] (1201–1272), 88, 165, 169, 189
Thompson, Sir Benjamin [Count Rumford] FRS (1753–1814), 340
Thornton, Robert John (1768–1837), 357–58, 560n.123

INDEX OF PERSONS   799

Thunberg, Carl Peter (1743–1828), 467n.62
Thuret, Gustave Adolphe (1817–1875),
    343–44
Tiedemann, Friedrich FRS(For) (1781–1861),
    583n.109
Tietz[e] [Titius], Johann Daniel (1729–1796),
    263–64, 560n.123
Titus Cæsar Vespasianus [Titus], Emperor
    (39–81 CE), 50
Titus Flavius Vespasianus [Vespasian],
    Emperor (9–79 CE), 50
Tomitano, Bernardino [Tomitanus,
    Bernardinus]·(1517–1576), 180
Topsell, Edward (ca 1572 to 1625), 57, 165, 204
Tournefort, Joseph Pitton de (1656–1708),
    231, 243, 245–46, 289, 290, 336–37,
    555n.18, 561n.153
Tradescant, John [the Younger] (1608–1662),
    523n.39
Tragus, Hieronymus. See Bock, Hieronymus
Trembley, Abraham FRS (1710–1784), 232,
    233, 234–36, 237, 239, 255, 322–23, 329,
    339
Trentepohl, Johann Friedrich (1748–1806),
    341–42, 345, 597n.15
Treviranus, Gottfried Reinhold (1776–1837),
    269, 283–84, 322, 327, 337, 571n.29,
    590n.51, 601n.77
Trevisa, John [of] (fl. 1341/1342 to 1402),
    197–98
Trincavelli, Vettore or Vittore [Trincavellus,
    Victor] (1496–1568), 487n.92
Trithemius, Johannes [Heidenberg, Johann]
    (1462–1516), 153
Trogus, Gnaeus Pompeius (fl. First century
    BCE), 176
Tsenkovsky, Lev Semyonovich [Cienkowski,
    Leon] (1822–1887), 389, 390
Tulasne, Louis René Étienne [Edmond]
    (1815–1885), 593n.106
Turner, William [naturalist] (1509/1510 to
    1568), 169–70, 193, 198–99, 356
Turpin, Pierre Jean François (1775–1840),
    Frontispiece, 305–6, 309, 354, 580n.35,
    594–95n.149
Turton, William (1762–1835), 358, 558n.63,
    558n.83

Udagawa Yōan (1798–1846), 22
"Un officier du Roi". See Saint-Pierre, Jacques-
    Henri Bernardin de
Unger, Franz Joseph Andreas Nicolaus (1800–
    1870), 342–44, 353, 354, 363, 370, 376–
    77, 588n.1, 604n.154

Urban VIII, Pope [Barberini, Maffeo
    Vincenzo] (1568–1604), 544n.56

Vaget, Johann (1633–1691), 196, 206
Vaillant, Sébastien (1669–1722), 559n.91
Valenciennes, Achille (1794–1865), 574n.99
Valla, Georgio (1447–1500), 175
Valla, Lorenzo [Laurentius] (ca 1407 to 1457),
    492n.62
Vallisnieri, Antonio (1661–1730), 237
van Helmont, Jan Baptist [Johann Baptista]
    (1579/1580 to 1644), 316
van Niel, Cornelis Bernardus (1897–1985),
    416–17
Varro, Marcus Terentius (116–27 BCE), 47, 48,
    84, 198–99, 212, 479n.28, 505n.15
Vauban, Sébastian Le Prestre, Marquis de
    (1633–1703), 240
Vaucher, Rev. Jean-Pierre Étienne (1763–
    1841), 341–42, 345, 353, 597n.15
Vaughan, Henry (1621–1695), 159
Vaughan, Thomas ["Eugenius Philalethes"]
    (1621–1666), 159, 167, 173
Venn, John FRS (1834–1923), 281, 407–8,
    567n.94
Ventenat, Étienne Pierre (1757–1808), 278
Vergilius Maro, Publius [Vergil, Virgil] (70–19
    BCE), 67, 114, 147, 212, 248, 459n.67
Verrius Flaccus, Marcus (ca 55 BCE to 20 CE),
    174
Verworn, Max Richard Constantin (1863–
    1921), 614n.139
Vesalius, Andreas (1514–1564), 157–58
Vicq d'Azyr, Félix (1748–1794), 265, 302–3,
    326, 412
Villaneuva, Arnaldo de [Villa Nova, Arnaldus
    de] (ca 1240 to ca 1311 CE), 149
Villemet [Willemet], Pierre-Rémi[s]-François
    de Paule (1762–1790), 258
Vincent of Beauvais [Vincentius Burgundus]
    (perhaps 1184/1194 to ca 1264), 125, 149,
    168–69
Vines, Sydney Howard FRS (1849–1934), 330
Virchow, Rudolf Ludwig Carl ForMemRS
    (1821–1902), 328, 379, 380, 625n.92
Virey, Julien-Joseph (1775–1846), 284–85,
    303–5, 309, 573–74n.94
Virgil. See Vergilius Maro, Publius
Virgilio Adriani, Marcello (1464–1521), 532n.20
Vlastos, Nikolas [Vlasto, Nicolaus] (1430–
    1500), 504n.239, 527n.6
Voltaire [Arouet, François-Marie] FRS
    (1694–1778), 227, 236, 290, 311–12,
    553–54n.150

800 INDEX OF PERSONS

Vorilong, William [Willem of Verolon] (*ca* 1390 to 1463), 211

Walker, Rev. John FRSE (1731–1803), 356
Wallace, Alfred Russel FRS (1823–1913), 369
Wallerius, Johan Gottschalk (1709–1785), 263–64, 560n.123
Wallich, George Charles (1815–1899), 389
Wallin, Ivan Emanuel (1883–1969), 426, 446, 629n.176
Wang Ch'ung (27 to *ca* 97 CE), 10
Waterhouse, George Robert (1810–1888), 569n.140
Watson, James Dewey ForMemRS (b. 1928), 432–33
Watson, Richard FRS, Bishop of Llandaff (1737–1816), 240–41
Watson, William FRS (1715–1787), 232
Weiditz, Hans [the Younger] (1495 to *ca* 1537), 190–91
Weiss, Friedrich Wilhelm [Weis, Fridericus Guilielmus] (1744–1826), 560n.125, 560n.128
Werlhof, Paul Gottlieb (1699–1767), 547n.131
Werner, Abraham Gottlob (1749–1817), 322–23
Wesley, Rev. John (1703–1791), 247–48, 362, 553n.148
Whewell, Rev. William FRS (1794–1866), 364, 377, 597n.16
Whisler, Howard Clinton (1931–2007), 624n.88
Whiston, William (1667–1752), 542n.24
White, Charles FRS (1728–1813), 240–41
White, Rev. Gilbert (1720–1793), 356, 377
Whitehead, Alfred North FRS (1861–1947), 24, 33
Whitehurst, John FRS (1713–1788), 359
Whitman, Charles Otis (1842–1910), 330
Whittaker, Robert Harding (1920–1980), 417–21, 418*f*, 419*f*, 428–29, 430, 632n.56
Wilberforce, Samuel FRS, Bishop of Winchester (1805–1873), 371, 377
Willdenow, Carl Ludwig (1765–1812), 307–8, 333
Wille, Johan Nordal Fischer (1858–1924), 403–4
Willemet, Pierre-Rémi-François. *See* Villemet, Pierre-Rémi-François
William of Conches (*ca* 1090/1091 to after 1154), 121–22
William of Moerbeke, Archbishop of Corinth (*ca* 1215 to *ca* 1286), 68, 111–12, 478n.200, 486–87n.85, 510n.172

William of Ockham [Occam] (*ca* 1287 to 1347), 131, 133–35, 361–62, 463n.111
Willis, Thomas FRS (1621–1675), 206
Willughby, Francis FRS (1635–1672), 245–46
Wilson, Edmund Beecher FRS (For) FRSE (1856–1939), 426
Wilson, Thomas Bellerby (1807–1865), 373–75, 377–78
Withering, William FRS (1741–1799), 357, 359–60, 601n.87
Woese, Carl Richard ForMemRS (1928–2012), 435–37, 438*f*, 441, 447–48, 632n.56
Wolff, Caspar Friedrich (1733–1794), 316, 327, 609n.41
Wolff, Christian, Freiherr von FRS (1679–1754), 227, 311–12
Woodward, John FRS (1665–1728), 589–90n.34
Woodward, Rev. Josiah (1657–1712), 362
Wordsworth, William (1770–1850), 361
Worm, Ole (1588–1654), 205, 523n.39
Wotton, Edward [zoologist] (1492–1555), 165, 199–200, 202, 226, 243, 287, 538n.135, 538n.136
Wrisberg, Heinrich August (1736–1808), 221–22, 548n.138, 582n.86
Würtz [Wurtz], Georg Christoph (1756–1823), 564n.31
Wuzong [Li Chan], Emperor of Tang (814–846), 465n.36
Wyclif [Wycliffe], John (before 1330 to 1384), 130

Xenocrates of Chalcedon (396/395 to 314/313 BCE), 28, 483n.2
Xenophanes of Colophon (*ca* 570 to *ca* 478 BCE), 25–26, 469n.35
Xenophon of Athens (*ca* 430 to 355/354 BCE), 24, 470n.51, 556n.30
Xun Kuang [Xunzi; Hsün Ch'ing] (*ca* 310 to *ca* 235 BCE) (****), 19

Yagel, Abraham ben Hananiah (1553 to *ca* 1623), 11–12, 111, 462n.101
Yeats, Thomas Pattinson FRS (d. 1782), 358
Yehudah ben Schlomoh [Judah ben Solomon] ha-Cohen [ibn Matkah] (*ca* 1215 to *ca* 1274), 109–10
Young, Edward (1683–1765), 542n.24

Zabarella, Giacomo [Jacopo] (1533–1589), 180–82, 226, 540n.166
Zacharias of Mytilene, Bishop (b. *ca* 465, d. after 536), 488n.103

INDEX OF PERSONS 801

Zaluziansky à Zaluzian, Adam (1555–1613), 195, 535n.77

Zarathustra [Zoroaster] (*fl.* uncertain but perhaps late Second millennium BCE), 91, 137, 150, 199, 210, 514n.43

Zeno of Cition [Citium] (perhaps *ca* 334 to *ca* 262 BCE), 44, 467n.3

Zeno of Elea (*ca* 495/490 to *ca* 430 BCE), 28, 409, 472n.87

Zenobia [Semptima Zenobia], Queen of Palmyra (b. *ca* 240, d. after 274 CE), 485n.40

Zhang Qian (d. 114 BCE), 18–19

Zhu Xi [Chu Hsi; Hui'an] (1130–1200), 19–20, 465n.36

Zhuang Zhou [Zhuangzi; Chuang Tzu] (*ca* 369 to *ca* 286 BCE), 17, 20–21

Zou Yan [Tsou Yen] (*ca* 350 to *ca* 270 BCE; alternatively 305–240 BCE), 17

Zuckerkandl, Émile (1922–2013), 631n.24

Zwinger, Theodor III (1658–1724), 535n.72

(*) The brothers Engramelle: see 526n.112

(**) Laozi: see 466n.49

(***) His British scientific contemporaries often cited his name as Kent, but sometimes as Saville Kent. In *Savant of the Australian seas* (1997; Second edition 2005), biographer AJ Harrison states that he was named William Savill Kent by his parents, and first used Saville Kent in a personal letter dated November 1870. The hyphen is first attested on a report to the Tasmanian government in 1884. Harrison provides further examples of both forms; Saville-Kent used the hyphenated form in his will (1892). In *The Australian Dictionary of Biography* (2005), Harrison adds that "On 5 January 1876 at the parish church, Prestwich, Lancaster, he married Many Ann Livesay, giving his surname as Saville Kent. Later he added a hyphen."

(****) Xun Kuang: for dates, see also 466n.45

Index of Subjects

For the benefit of digital users, indexed terms that span two pages (e.g., 52–53) may, on occasion, appear on only one of those pages.

Endnotes are indexed only if they refer to persons not mentioned in the corresponding text, or offer substantive additional information.

Tables and figures are indicated by *t* and *f* following the page number.

agriculture, rural affairs, 8, 47, 48, 52, 56, 91, 115–16, 118, 198, 240–41, 289–90, 356
   Nabatæan agriculture, 10, 459n.59
air / wind in fecundation
   of animals, 96, 212
   of date palm, 96
   of humans, 212
alchemy, 9–10, 20, 33–34, 57, 138, 142–43, 144, 147–48, 149, 150–51, 153–57, 224, 231, 311–12
   alchemical treatises, 112–13, 138, 145–46, 148–49, 154–57, 549n.44
   banned by the Church, 149
   kingdoms (biological) in, 156–57, 205–7, 208
   lead to gold, 12–13, 143, 153, 154
   in medicine, 18, 153–54, 157–58
   philosopher's stone, 142, 145–46, 148–49, 154–56
   salamander in, 57, 519n.197
   superseded by chemistry, 157
algae
   as alternatively animal and plant, 341–43, 344, 345, 349, 354
   blue-green, cyanophyceae (*see* bacteria)
   class Algæ (Agardh), 344–45
   classified by colour (Lamouroux), 336, 430–31
   coralline algae, 233, 244, 247–48, 261, 263, 295, 299–300, 308–9, 357, 358, 568n.112
   Cryptogamia Algae, 249–50, 592n.100, 616n.198
   definition of (Nägeli), 403
   filaments as basis of plant body, 345–46
   form-series, 345–46, 347–48, 349–50, 403–4, 405–6, 411, 412
   life cycles / histories, 338–40, 341–46, 347–50, 352–53

   metamorphosis of, 344–46, 348–50, 353, 363
   pleomorphy, 345–46
   as polyparies, 338–40
   Priestley's green matter (*see* at microscopic beings)
   red algae as mesoprotists, 419–21
   reproductive structures, sexuality, 336–37, 340–44, 352–53, 403–4
   seaweeds (*see* marine plants / seaweeds)
   as sister-group to animals, 419–21
   Thierwerdung (Unger), 343, 344
   unicells classified into form-series, 352–53, 403–4, 412
   zoospores of, 341–44, 351, 596n.182
allegory, allegorical literature, 9, 11, 33, 57–58, 60–61, 62, 75, 79, 81, 90, 98–99, 115–16, 122, 124, 139, 148, 156–57, 161, 163–64, 171, 203, 224–25, 461n.96, 505n.37, 526n.107, 583n.113
alternation of generations, 308, 354
   in algae, 345, 346
   in marine invertebrates, 346, 608n.19
amber, 156
anamnesis, 12, 461n.94
angels
   as airborne souls, 212
   as consisting of fire, 148–49
   direct creation of, 120, 129–30
   as intellectual / rational beings, 78, 83, 85–86, 99, 104–5, 117, 121, 122–23, 128, 130–31, 153, 176, 495n.136, 508n.100
   as transporting animals to Noah's ark, 183–84
   as watching over earthly beings, 100
   See also hierarchy, celestial; soul, angelic
angling (example of logical division), 34

804 INDEX OF SUBJECTS

animalcules, 45–46, 175, 216, 217–20, 221–22,
230, 242, 243, 247–48, 249–50, 251, 254,
255, 256, 257, 295, 304, 314–15, 325,
326, 329, 338–39, 340, 341–42, 344–45,
346–47, 362, 366, 367, 368–69, 554n.160,
556n.23, 577n.156, 577n.171, 593n.120
born in furnaces, 45–46, 48, 62, 518n.155,
521–22n.1
See also microscopic beings; zoospores
Animal Kingdom, 103–4, 106, 138, 156–57,
161, 206–7, 208, 237–38, 244, 249, 253,
254–55, 258, 260, 261–62, 263, 272, 278,
282, 284, 285f, 287, 290–92, 296, 299–
300, 305, 306–7, 308, 316, 321, 322, 324–
25, 331, 339, 344–45, 347, 348, 355–56,
358, 362–63, 365, 366, 367, 368, 370, 372,
373, 373f, 375, 380–81, 382, 383, 385–86,
387–88, 391–92, 393–95, 394f, 397, 398f,
401–2, 404, 407–8, 408f, 417, 418–21,
418f, 419f, 420f, 424, 428, 430, 465n.28,
520n.209, 540n.167, 545n.78, 557n.54,
569n.141, 570n.12, 580n.40, 602n.96,
606n.187, 606n.192, 609n.61, 621n.43,
630n.198, 634n.102
first called a kingdom, 2, 156–57
Le régne animal (Cuvier), 299–300
True and Doubtful subkingdoms
(Blainville), 306
animals
all nature is animal (Robinet), 236
analogies with vegetables, 15, 18–19, 25–26,
30, 37–38, 43, 62, 120, 123, 228–29, 236,
238, 255, 260, 272, 282–83, 296, 301,
302–3, 320, 340, 342, 360, 382, 586n.165
archetype, 314, 317, 318, 329, 333, 384,
584n.126, 586n.168
chemical properties, 231, 307, 336, 351, 353,
370, 371, 381–82, 562n.160
composite (Linnæus), 250–51, 252, 365
defined, 36, 99, 181, 188, 294, 371, 430
as deities, 2, 6–7
as "fleeting vegetables", 159
fœtus, 203, 207
infusoria (*see* microscopic beings)
instinct, 79–80, 308–9, 466n.46, 577n.169
intrinsic worth, 100
as inverted vegetables, 37, 62
as irritable, 261, 289, 296, 297, 318
microscopic (*see* microscopic beings)
morphology / fine structure distinguishes
from plants, 382
motion, movement, motility (*see* motion)
nutrition, 35–36, 37–38, 67–68, 72–73, 98,
101, 181, 239, 343, 360, 367, 397, 417–18

as (polar) opposites to vegetables, 66–67,
188, 271–72, 280–81, 314–15, 318, 319
protozoa (*see* at microorganisms)
ranked within classes, 100
reason, intellect, 33, 50, 62, 64, 66, 78, 85–
86, 87, 99, 101, 117, 239–40, 294
senses, sensate / insensate, 8, 15–16, 27,
35–40, 41, 42, 45, 51, 62, 65, 66, 67–68,
69–70, 71, 72–73, 79–80, 83, 85–86,
87, 88–89, 99–101, 105, 108, 113, 117,
120–21, 123–24, 126, 127–28, 129, 132,
141–42, 148, 149–50, 153, 186, 188, 199,
205, 207, 219, 227, 246–47, 249, 250–51,
285–86, 294–95, 296, 308, 318, 371, 374–
75, 380, 382
soul of (*see* at soul)
spirits, dæmons, jinn as animals, 10, 105
subterranean, 198–99
transformed into vegetables (Kützing), 349,
590n.51
unicellular, 329–31, 353, 355, 356, 382,
388, 390, 391, 393–95, 400–1, 408,
613n.127
Urthier (Goethe, Oken), 313–14, 320–22,
330, 391, 396, 399–400, 584n.126,
613n.127, 613–14n.131
as variants on a prototype, 290–91, 303–4,
325, 366, 369–70
as vegetables, 62, 320, 419–21
vegetative animals, 95, 96
veterinary medicine, 198
animals: classification
in Aldrovandi, 203–4
blood / bloodlessness in, 55, 56, 80, 179,
199, 202, 204, 205, 206, 208–9, 298
bones / hair / nails as sentient or insentient,
44, 70–71, 84
"difficult" groups, 206, 534n.65
four *embranchemens*, 266, 283–84, 299, 301,
324–25
in Gessner, 201–3
in Lamarck, 293–98, 303, 325
nervous system in, 238, 247, 250–51, 252–
53, 255, 270, 297, 298–99, 301, 302–3,
309, 320, 326–27, 331, 359–60, 364–65,
366, 371, 396, 584n.134
Oozoa (egg-animals), 321
in Owen, 364–67, 372, 373
protozoa / metazoa (Haeckel), 391, 393,
396, 399–400
purgamenta, 201, 202, 536–37n.106
Ramist tree in, 182–83, 204, 208, 584n.125
skeletal anatomy in, 203, 299, 313–14,
538n.126

INDEX OF SUBJECTS 805

spinal column / vertebræ in, 270, 283–84,
293, 294, 295, 296, 299, 300–1, 305, 314,
326, 331, 386
in Topsell, 204
use of multiple characters in, 36–37, 260
animals (selected genera / taxa)
amphibians (*see* at dualizers)
bats (*see* at dualizers)
chamæleon, 203, 229, 362, 456n.19
choanoflagellates, 401–2, 430, 615n.178,
628n.169
corals (*see* corals)
Hydra (*see Hydra* [freshwater polyp])
infusoria (*see* microscopic beings)
insects (*see* insects)
invertebrates (*see* marine invertebrates)
protozoa (*see* at microorganisms)
rotifers, 215, 219–20, 260, 284, 295, 299–
300, 308, 325, 330, 336, 352, 365, 388,
568n.112, 610n.76
salamander, 6, 54, 57, 115, 119–20, 124–25,
147, 197–98, 519n.197, 521–22n.1
sea horse, 53, 54, 201
slime moulds, myxomycetes, 382, 386, 393–
95, 623n.75
snails (*see* snails)
whale, 115–16, 120, 125, 197–98, 200, 203,
276–77
worms (*see* worms)
zoophytes (*see* Zoophyta)
archaea
Asgard, 442–43, 446, 447–48
distinguished from bacteria, 436, 438*f*, 440
eukaryotes as, 446–48
genomes of, 441
major types / superphyla of, 442, 444
methanogens, 436
recognition of, 436
studied by metagenomics, 442–43, 446,
447–48
archaea: classification
in Cavalier-Smith, 630n.193
Kingdom Archaea, 636n.144
as prokaryotes (or not), 437, 442, 447–48
in Stanier *et al.*, 622n.62
as an urkingdom, domain or empire, 436,
437, 438*f*, 442, 444
archaebacteria. *See* archaea
archetype, 96, 312
See also at animals; vegetables
Aristotle: logic and philosophy
alternative readings in, 181, 473n.113
being versus becoming, 333, 350
classification of animals, 35–38, 55

logic, logical division, 37, 49–50, 65, 94
peripatetic school, 24–25, 27–28, 42–43, 44,
46–47, 59, 114
theory of intelligible forms (ideas), 34–35
Aristotle: works / corpus
Analytica priora, 70, 95, 111, 112, 150–51,
497n.52
Andronicus as editor, 46–47
banned by the Church (1210-1254),
128–29
Categoriæ, 65, 66–67, 69, 111, 113, 115,
485n.54, 494n.124, 497n.52
De anima, 35–36, 67–68, 69, 70, 95, 101,
102–3, 111–12, 113, 122–23, 127, 150–
51, 181, 188, 497n.52
De caelo, 69, 101, 112, 150–51, 473n.109,
497n.52, 508n.119
De generatione animalium, 94–95, 101, 113,
478n.200, 509–10n.162
De generatione et corruptione, 101
De partibus animalium, 94–95, 101, 113,
478n.200, 509–10n.162
De plantis (attrib.), 41–42, 101, 128, 194
De progressu animalium, *De motu
animalium*, 113
Historia animalium, 37, 38–40, 64, 94–95,
101, 113, 174–75, 181, 199, 478n.200,
487n.91, 509–10n.162
Metaphysica, 127
Meteorologica, 101
Organon, 101, 128–29, 131, 149, 180,
485n.54
Physica, 35, 111, 112, 128–29, 131, 150–51,
180, 487n.91, 497n.52
Politica, 127
rediscovery in Twelfth century, 111–12,
113, 118, 122–23, 125, 128–29, 131, 134
translation into Arabic, 94–95, 497n.52,
503n.226
ascent (spiritual). *See* mysticism
astrology, 14, 33–34, 50, 54, 98–99, 100–1, 103,
137–38, 145–46, 151–52, 325, 406
See also sun, stars, and planets
atomic theory
in Islam, 470n.41
in Leucippus and Democritus, 28–29, 35,
470n.43
in Lucretius, 49
augury, 63–64, 78, 145–46

bacteria
as causative agents of disease, 406–7, 413–
14, 627–28n.147
first observed, 215, 405, 548n.136

806 INDEX OF SUBJECTS

bacteria (*cont.*)
  as free-living organelles (*see*
    eukaryogenesis)
  genus *Bacterium*, 405
  life history of (Koch), 406–7
  not cells (Haeckel), 406
  as nucleate / anucleate, 401, 406, 412–14,
    416–17, 419–21, 616n.194
  pleomorphy, 411, 413–14, 631n.14
  as sexual / asexual, 406, 413–14
  as symbionts / endosymbionts (*see*
    eukaryogenesis; symbiosis)
bacteria: classification
  Akaryonta, Anucleobionta, 416
  as animals, 406, 412, 416
  Chroococcaceæ, 405
  cyanobacteria, Cyanophyta, 308, 416, 417,
    612n.110
  domain Bacteria, 438*f*, 442, 444
  empire Bacteria, 447
  as infusoria, 308, 405, 562n.173
  Kingdom Bacteria, 636n.144
  as "lower protists" (Stanier & van Neil), 417
  as monera / protists, 395, 396–97, 406, 412,
    414–15, 416, 418*f*, 419–21, 419*f*, 420*f*
  Mychota, 413–14
  as plants, 405–7, 412, 416, 419–21
  polyphyletic origin, 406, 413
  as prokaryotes, 414–15, 417, 437, 447, 448
  as protozoa, 409, 608n.18
  as Psychodiaire (Bory), 308–9
  Schizomycetes, 405–6, 412, 416, 418*f*
  Schizophyta, 405–6, 412
barnacle-goose. *See* chimæric beings
Bathybius, 389–90, 416
bestiaries, 55, 56–57, 96–97, 163–64
  See also *Physiologus*
Black Death, plague, 114, 130, 134, 153, 160,
  212, 322, 601n.82
Boëthius: works
  Philosophiæ consolationis, 87
borametz. *See* chimæric beings
botanic gardens. *See* institutions
Buffon: works
  Histoire naturelle, générale et particulière,
    216–17, 269, 270*f*, 289, 290, 355–56,
    539n.148

cabinet (of curiosities), 166, 205–6, 216–17,
  293–94, 312–13
  royal Cabinet (Paris), 216–17
cathedral schools
  Chartres, 121–22
  Notre Dame, 119

  Paris, 115
cave art, 3
cells, cell theory
  cell as elementary organism (Brücke),
    421–22
  cell as unit of physiology, 327–28, 421
  cell as unit of structure, 327–28, 384, 385
  cell-republics, cell-monarchies (Haeckel),
    395
  cells analogous in animals and plants, 327–
    28, 351, 380, 424–25
  cell theory, 327–31, 409, 413
  cellulosic membrane / wall (plant cells),
    327–28, 382
  cytodes (Haeckel), 385, 393, 396, 406,
    412–13
  early descriptions of cells, 327, 421
  in Enderlein, 413–14
  endosymbiotic origins of (*see*
    eukaryogenesis)
  mitochondria transformed into
    chloroplasts (Guillermond), 426
  mycoplasm / amœboplasm
    (Merezhkowsky), 424
  omnis cellula e cellula (Virchow), 328, 380
  opposed by Dobell, 330, 409
  opposed by Huxley, 330, 401–2
  origin of cells (*see* eukaryogenesis)
  origin of mitosis (Margulis), 427–28,
    624n.88
  plastid theory (Haeckel), 385
  protoplasma, protoplasm, 327–28, 413–14,
    415, 421, 625n.99
  sarcode (Dujardin), 327–28
  tissues, 321–22, 327, 381–82, 421
  unicellularity, 328–31, 353, 380–82, 386,
    388, 391, 400–1, 403, 408, 409, 596n.184
cellular ultrastructure
  contractile vacuoles, 382
  DNA in organelles, 427
  early ideas about, 305, 422
  flagella, cilia, 9 + 2 structures (*see* flagella)
  granules, vesicles, 352, 422
  macro- and micronuclei, 330, 401, 608n.22
  membranes, 421, 429–30, 445, 446, 448
  mitochondria, 414–15, 422, 426 (*see also*
    eukaryogenesis)
  nuclei in bacteria (*see* at bacteria)
  nucleomorph, 429–30
  nucleus as a gonad / sexual organ, 400
  nucleus, nuclear membrane, 327–28, 412–13,
    415, 416–17, 426, 445, 587n.182, 599n.49
  organelles cultured outside cell, 422, 423,
    426

INDEX OF SUBJECTS   807

plasmids, 439, 440, 444
plastids, chloroplasts, 406, 414, 416–17, 422–23 (*see also* eukaryogenesis)
ribosomes, 427, 429, 432–33, 435–37, 442, 445
Chaldæan Oracles, 66, 93, 137, 462n.98
chimæric beings, 33, 164–65, 203
  in Abelard, 165
  Animal-Vegetable-Mineral Man, 461n.90
  in Aquinas, 165
  in arts and architecture, 165, 172
  barnacle-goose(-tree), 167–70, 185, 192, 194–95, 196, 203, 247–48, 249, 529n.54
  borametz (Scythian lamb, vegetable lamb), 165–67, 168, 171, 172, 185, 194–95, 196
  Dog-heads, 165
  fauns, 7, 164–65
  Hippo-centaur, 164–65
  jeduah, 165–66
  in Lucretius, 49
  in Paré, 170–71
  satyr, 7, 104, 164–65, 172, 249
  See also cave art
chromosomes. *See* genetics
Church (Christian)
  Byzantine, 134
  Condemnations of 1210-1277, 122–23, 128–29, 211
  Crusades, 115
  Edict of Milan, 65–66, 490n.2
  Edict of Thessalonica, 75
  of England, 158, 355
  Fathers (Cappadocian, Greek, Latin), 29, 48–49, 76–79, 80–83, 87–88, 90, 94–95, 118–19, 224–25, 473n.100
  after French Revolution, 290, 309
  Index librorum prohibitorum, 117, 128–29, 521n.231, 627–28n.147
  Inquisition, 118
  Protestant / Lutheran, 462n.101
  Reformation, 134–35, 157–58, 160
  Roman Catholic, 114–19, 120–34, 606n.192
  Vatican, 126 (*see also* Popes, in Index of persons)
  wary of pagan learning, 129
  See also councils; Popes; religions
classification: methodology
  affinity (*see* comparative biology)
  analogy (*see* comparative biology)
  anatomy (comparative) in, 298, 299–300, 306, 329, 381, 384, 387, 399
  Archezoa hypothesis, 430, 445, 629n.175
  in Aristotle, 35–37, 188–89

embryology (comparative) in, 307, 324–25, 368, 384–85, 393
endosymbiosis as basis for (Merezhkowsky), 424, 425*f*
genealogical / genetic basis (*see* comparative biology)
hierarchical, 4–5, 21, 55, 194, 204–5, 206, 208, 216–17, 243–44, 248, 264, 284–85, 287, 314, 317, 324–25, 411, 414, 443, 449, 467n.59
in Linnæus, 248–50, 256–57
morphological basis, 364–65, 381–82, 383–85, 390, 396–400, 405–6, 424–25
natural system, 248, 260, 266, 359, 383, 384–85, 564n.32
nomenclature (binomial, descriptive), 191, 195, 197, 206, 243, 248, 256, 259, 260, 289, 357, 417, 557n.51, 563n.3
nutritional mode / ecological role in, 417–18, 428–29
observation versus experiment in, 248
observers versus classifiers, 209
physiological basis, 299–300, 329, 381–82, 396–97, 399–400, 424–25
in Plato, 32, 34
principle of character subordination, 193, 298, 301
principle of connections (Geoffroy), 300–1
sexual system, 249, 256, 260, 357–58, 595n.155, 598n.19
superkingdoms, supertaxa, 430, 432–39, 624n.81, 624n.89, 627n.144
in Theophrastus, 43, 189
unikont / bikont criterion, 430, 445
unique beginner, 5, 52, 464n.14
universality and scalability, 244, 358
See also species
classification (modern schools of)
  cladistic, phylogenetic, 443–44, 447–48, 606n.180
  evolutionary, 443–44
  numerical, 443
classificatory systems (selected modern)
  Cavalier-Smith (six to ten kingdoms), 430
  Copeland (four kingdoms), 417
  Dillon (one kingdom / superkingdom), 419–21
  Edwards (seven kingdoms), 419–21
  Leedale (two- and nineteen-kingdom), 419–21
  Margulis (five kingdoms), 418–19, 420*f*
  Moore (three kingdoms), 419–21
  Whittaker 1959 (four kingdoms), 417–18, 418*f*

808 INDEX OF SUBJECTS

classificatory systems (selected modern)
(*cont.*)
Whittaker 1969 (five kingdoms), 417–18,
419*f*
Whittaker & Margulis (five kingdoms),
418–19
comparative biology
affinity, 95–96, 238–39, 244–45, 261, 266–
68, 269, 272, 275–76, 281–82, 286, 287,
298, 530n.73, 564n.21, 565n.42
analogy, analogue, 262, 267–68, 282–83,
287, 300–1, 313, 359–60, 366, 367, 368–
69, 370, 380–81, 390, 574n.113
genealogical / genetic basis, 245–46, 259,
266–67, 268, 269, 270*f*, 276, 277*f*, 384–
85, 387, 397–99, 401, 411, 432, 443–44,
580n.40, 623n.68
homology, homologue, 262, 366, 384, 390
Stammverwandtschaft (Haeckel), 384
See also phylogenetics, molecular
continuum / gradient / scale of nature
analogized with a Commonwealth of man,
179
in Anselm of Canterbury, 118
in Aquinas, 127–28
in Aristotle, 38–39
in Arūzī, 103–5
in Augustine, 84
in Confucianism (Xunzi), 19
in Descartes, 206–7, 227
gaps in continuum, 184, 199, 259, 261–62,
265–66, 295, 297, 556n.43
in the Ikhwān al-Ṣafā, 99
Jacob's ladder, 224–26
in Jainism, 15–16
in Kant, 315–16
in Kirchweger, 159, 160*f*
in Linnæus, 248, 249
in Lucretius, 49
in Macrobius, 210–11
in al-Masʿūdī, 98
metaphors (chain, cord, ladder, stairway,
tree), 223–26
Natura non facit saltus, 207, 261–62,
556n.43
in Nemesius, 79–80
in Nicholas of Cusa, 132–33
observed only in portions, 223–24, 233,
238, 239–40, 244, 248, 260, 262, 564n.18
in Pallas, 261–62, 266, 269, 276
in Plato, 32
principle of continuity (*synecheia*), 38–39
in Pythagoras, 27
in al-Qazwīnī, 105

in Ray, 246
in Stoic philosophy, 44, 66
corals
in alchemy, chemistry and medicine, 231
in Ambrose and Basil, 82
as animals, 51, 100–1, 108, 231, 252, 262–
63, 289, 290, 291–92, 295, 306–7, 326,
357, 390, 551n.94, 568n.112, 604n.148
as animals, vegetables and/or minerals, 252,
253, 357
bark and core classified differently, 231
coral industry and uses, 231
coralline algae classified with, 233, 261, 352,
590n.37
coral reefs / islands, 317, 369
in curiosity-cabinets, 150–51, 152, 179
in Dioscorides, 53
flowers of, 231–32
in herbals, 189–90, 191
as intermediate between animate and
inanimate, 144, 358
as intermediate between minerals and
plants, 103–4, 106–7, 121, 144, 179, 207,
230–31, 240–41, 256
kingdom of, 319, 583n.109
as lithophytes (Linnæus), 249, 252, 253,
254–55
as living or lifeless, 197
as minerals or stones, 125, 153, 154, 231,
233, 550n.48
in Münchhausen (Regnum neutrum), 254–
55, 256
in Oken, 319, 321, 582n.88
in *Physiologus*, 57
as plants or trees, 51, 53, 68–69, 99, 125,
154, 156, 192, 194–95, 196, 197, 205,
219–20, 228, 231, 233, 244, 246–47, 403,
530n.73, 551n.79
as plants that can turn to stone, 9, 43, 81,
125, 208–9, 231, 540n.169
in Pliny, 51
as polyparies / nests for insects, 232, 233,
338, 339, 340, 360
as precious stones, 100–1, 115–16, 530n.73
as Psychdiare (Bory), 308–9, 333–34
in the *Quran*, 94
recognition of polyps, 231–33
seaweed to coral (Ovid), 9
supposed parasitic insects of, 231–32
in Theophrastus, 43, 52, 231
as zoophytes or similar, 218, 233, 239, 263,
299–300, 302, 317, 322
See also Golden Chain; Great Chain of
Being

INDEX OF SUBJECTS 809

correspondence (occult)
  in Agrippa, 153
  in Hildegard, 120
  in Person, 185
  in Plato, 33–34
  with words or letters, 109, 147
  See also signatures (occult); sympathetic
    medicine
cosmos / universe
  as created versus eternal, 63, 68–69, 92, 94,
    103, 109, 356–57, 490n.3
  creation myths, 6, 17, 31–33, 59–60, 91 (*see
    also Genesis*; hexæmeral literature)
  emanation from the One, 26, 63–64, 73–74,
    79, 106–7, 108–9, 110, 117, 127–28, 133–
    34, 139, 186, 223–24, 500n.138
  as a living being, 26–27, 28, 31, 33, 44, 108,
    138, 210–11, 227, 316
  temporal order of appearance, 49, 63, 92,
    96, 99, 109, 211
Councils
  Council of Ferrara, 134
  Council of Pisa, 160
  Council of Trent, 462n.101
  First Council of Ephesus, 92
  First Council of Nicæa, 75
  Fourth Council of the Lateran, 168–69
  Fourth Synodal Council of Toledo, 115–16
  Second Council of Constantinople, 77
creation, creation myths. *See* cosmos /
  universe
Cuvier-Geoffroy debate, 301–2, 323, 580n.38,
  584n.123

Darwin, Charles: books
  Descent of man, 369, 411, 618n.251
  Origin of species, 369, 373, 377, 383, 384,
    390, 409, 411, 608n.33
Demiurge, 31–32, 54, 73, 76, 79, 138–39, 140,
  142, 470n.55
dictionaries, lexicons, 45–46, 53, 56, 115–
  16, 156, 173, 174–75, 178, 456n.19,
  464n.18
divination, 33–34, 64, 78, 471n.67
domain (classificatory rank)
  introduced (Woese), 437, 442
  three domains, 437, 438*f*, 442, 446–48
  two domains, 447–48
  for specific domains (*see* high-level taxa)
dream-interpretation, 33–34, 53–54
dualizers: concept
  in Aristotle, 39–40, 475n.134
  in Calvenus Taurus (*mesê*), 63
  Greek word, 40–41, 176

in al-Jāḥiẓ, 97
Latin word, 40–41, 51
in Plato, 40–41
dualizers: organisms
  ape, 40–41
  bat, 5, 40–41, 81, 96–97, 147, 203, 229,
    276–77
  dolphin, 40–41, 199–200, 203
  hermit crab, 40–41
  man as dualizer, 176
  molluscs, "sea shells", 40–41, 42
  ostrich, 40–41, 96–97, 100, 124
  plants (beet, mallow), 475n.153
  sea anemone (sea nettle), 39, 40–41, 51, 65,
    79–80, 81, 126, 167, 171, 176, 184, 195–
    96, 198–99, 200, 203–4, 469n.24
  seal, 40–41, 97
  sea turtle, 40–41
  zoophytes, 124, 184, 200, 207, 247, 261,
    356–57

earth
  generative force of, 167, 279, 316
  as mother to mankind (Gaia), 6, 73, 148,
    315, 360, 581n.68
elements
  chemical elements (modern), 267–68, 280–
    81, 341
  in classical Hellas, 25–27, 30, 33, 35, 142–
    43, 476n.172
  in classical Rome, 48
  creation limited to (Thierry of Chartres),
    122
  in early China, 17
  four (earth, air, fire, water), 25–27, 30, 33,
    35, 48, 88, 92, 95–96, 98, 99, 101, 102,
    104, 106–7, 110–11, 117, 121, 124–25,
    128, 140, 142–43, 147, 149, 154, 157, 158,
    293, 320–21, 465n.25, 498n.92
  in Hildegard, 120–21
  as living, 460n.78
  mercury-sulphur theories, 99, 143, 144,
    146–47, 148–49, 154–55, 156, 157, 158
  On the elements (Marius), 114, 120–21
  oxygen in air, 338, 339, 340, 342, 351, 353,
    367, 600n.70, 634n.108
  phlogiston theory, 338, 340, 562–63n.186,
    590n.42
  silicon / silica, 332, 349–52, 390
  See also metals
embryo, embryology
  analogized with plant / plant germ, 27, 38,
    222, 238, 342
  animal and vegetable poles of, 614n.134

810 INDEX OF SUBJECTS

embryo, embryology (*cont.*)
    animal and vegetative layers (Haeckel),
        395–96
    characters useful in systematics, 307, 309,
        336, 579n.22
    embryo not an animal (Philoponus), 70
    embryonic forms analogize earlier forms of
        life (E Darwin), 359–60
    as ensouled / living, 38, 70, 462n.98,
        474n.121
    epigenetics versus preformationism, 214,
        236, 324–25
    as having vegetative life, 38, 100
    infusoria interpreted as animal embryos,
        342–43
    laws of embryology (von Baer), 324–25
    as successively fungal, zoophyte, then
        animal stages, 132
    as successively vegetable, animal, then
        human, 70, 100
    as successively vegetable, zoophyte, then
        animal, 73
    successive stages of, 133–34
    See also ontogeny
Emerald tablet. See Hermetic corpus
Empire (classificatory rank)
    Empire of Organic Bodies (*see*
        Organicorum [corporum] Imperium)
    Empires of Organized and Unorganized
        beings, 306
    Imperium Naturæ, 256, 412, 557n.52
    organic empire, 261–62, 306
    Organicorum [corporum] Imperium, 261–
        62, 302–3, 412
    Prokaryota / Eukaryota as empires, 447,
        630n.195
    See also at high-level taxa
Empires (political)
    Byzantine, 19, 71, 87–88, 92, 93–94, 134
    Ghaznavid, 103
    Holy Roman, 112, 523n.40, 578n.2
    Maurya, 16
    Ottoman, 106–7, 134
    Roman, 46–47, 48, 54–55, 59, 60–61, 65–66,
        74, 75
    Russian, 323, 325, 565n.38
    Samanid, 499n.123
    Sasanian, 92, 93–94
    See also Emperors (Index of Persons)
encyclopædias (selected)
    Book of treasures (*The*), 95–96
    Cabinet cyclopædia (Lardner), 362
    Cyclopædia (Chambers), 289–90
    De universo (Hrabanus Maurus), 116

    Dictionnaire classique d'histoire naturelle,
        289–90
    Dictionnaire des sciences naturelles, 289–90
    in early Greece and Rome, 54, 479n.28
    in early Japan, 22
    Encyclopædia Britannica, 256, 355–56,
        358–59
    Encyclopédie méthodique, 289–90, 291–92,
        303
    Encyclopédie, ou dictionnaire raisonné,
        289–90
    Etymologiæ (Isidore), 115–16
    Hebrew-language, 109–10
    Nouveau dictionnaire d'histoire naturelle,
        289–90
    Speculum maius (Vincent of Beauvais), 125,
        149, 169
    See also Encyclopædists; Ikhwān al-Ṣafā (in
        Index of Persons)
Encyclopædists, 203–5, 206–7
endosymbiosis. *See* eukaryogenesis
Enlightenment (The), 289–90, 311–12, 355–
    56, 530n.82, 542n.21, 553–54n.150,
    578n.10
eukaryogenesis
    analogies between bacteria and organelles,
        422, 423, 424, 426, 427, 428
    archezoa hypothesis, 430, 445, 629n.175
    autogenic (non-symbiotic) theories, 426,
        624n.90
    as basis for classification, 424, 428, 429, 430
    DNA in organelles as evidence for, 427, 428,
        437, 445
    early proposals, 421–26
    early steps / stages of (Margulis), 427–28
    host lineage for, 446
    karymastigont scenario, 428
    membranes as markers of, 429–30
    in Merezhkowsky (symbiogenesis),
        423–26
    molecular evidence of, 437, 445
    origin of flagella, 426, 427, 428
    origin of mitochondria, 422, 426, 427, 428–
        29, 430–31, 435, 437, 445, 446
    origin of nuclei, 424, 426, 428, 429, 430–31,
        446–47
    origins of plastids / chloroplasts, 422–23,
        426, 427, 428–30, 435, 437, 444, 445
    proteins as markers of, 435, 445, 446
    secondary / tertiary endosymbiosis, 429–
        31, 437, 439, 445–46
    serial endosymbiosis theory, 423–24,
        427–29
    symbiosis versus fusion, 446, 448

INDEX OF SUBJECTS   811

eukaryotes: classification
  contrasted with prokaryotes, 414, 417–19,
    430–31, 434–35
  Empire Eukaryota, 630n.195
  formalized as kingdom, 430, 624n.89
  introduced as high-level taxon, 414–15
  major groups / supergroups within, 428,
    430, 445, 449, 624n.83, 624n.84
  phylogenetic "big bang", 445
  superkingdom Eukaryota, 636n.144
evolution, evolutionary theory
  Buffon as precursor to Darwin, 545n.81
  Darwinian, 324–25, 369–70, 383–85, 386,
    399–400, 409, 423, 449, 608n.21, 609n.47
  in Lamarck, 293, 295
  mechanical-physiological (Nägeli), 404
  opposed by Koch and Pasteur, 406–7
  in Owen, 367
  process pluralism, 449
  in Virey, 303–5

falconry, 198
fermentation, 250, 254, 257, 349–50
fishing, 34, 47, 56, 198
flagella, cilia
  9 + 2 structure, 426, 427–28, 430, 445
  bacterial, 428
  early observations of, 216, 342–43, 344, 390
  mastigonemes, 429
folk taxonomy, 4–9, 23, 40, 96, 188–89, 191,
  287
fortune-telling, 33–34, 63–64
fowling, 34, 47, 56
French Revolution, 290, 293–94, 300, 302–3,
  309, 312, 359–60, 361
fungi, mushrooms, truffles
  as animals, 257, 262–63
  aquatic, 203–4, 205, 347, 380–81, 403–4,
    592n.94
  in Bonnet, 238
  component of lichens, 345–46, 423, 430–31,
    635n.125
  as connecting animals and vegetables, 140–
    41, 232, 233, 261–62, 268, 291, 302, 360
  Cytofungi, Metafungi, 634n.102
  in Dioscorides, 53
  in early rock art, 3
  as an evil / putrescent ferment, 53, 257
  evolved from bacteria (Merezhkowsky),
    424, 425f
  as excrescences or vapours, 126, 191, 192–
    93, 196, 197, 253–54
  filamentous, 297, 355, 403–4
  first monograph of fungi, 192–93

  flowers of, 336, 344–45, 546n.111
  in Galen, 54
  genera as form-series, 345–46
  generated by use of electricity, 368
  as giving rise to men, 459n.67
  growth promoted by rain or thunder, 191,
    532–33n.31
  as intermediate between minerals and
    plants, 126, 150, 194, 207, 228, 230–31
  as irritable, 258–59
  kingdom of, 245–46, 418f, 419f, 420f,
    561n.141, 627n.130
  in Linnæus, 249, 255–56, 257, 335, 403
  man a fungus, 255–56
  as members of a third, fourth or
    intermediate kingdom, 254–55, 258–59,
    280, 281, 322, 375, 391–92, 395, 417–
    21, 418f, 419f, 420f, 424, 430, 570n.21,
    607n.200, 620n.14
  men produced from (Ovid), 459n.67
  metamorphosis, 162, 254, 345–46
  as minerals, 257
  in Needham, 218, 219, 239, 355
  in Nees von Esenbeck, 258–59
  as non-living, 197
  as plants, 42–43, 51, 52, 53, 54, 96, 100–1,
    120, 154, 179, 191–93, 194–95, 197, 199,
    205, 208–9, 219–20, 238, 244, 246–47,
    249, 257, 258–59, 260, 268–69, 282–83,
    297, 302–3, 306–7, 308, 319, 325, 327,
    335, 344–45, 368–69, 371, 403–4, 405–6,
    612n.113
  in Pliny, 51
  as protists, 391–92, 395, 417
  seeds, spores, sexuality, 197, 253–54, 257,
    325, 336–37, 344, 360, 589n.29, 612n.111,
    612n.113
  smut, rust (mucor, ustilago), 209, 253–54,
    255, 336, 560n.111
  spontaneous generation of, 167, 337, 349–
    50, 352–53, 360, 368, 380
  as a stage in animal ontogeny, 132
  stamens analogized with, 228–29
  as thallophytes, 403–4, 405–6
  in Theophrastus, 42–43, 52
  as "vegetable crystals", 257
  "vegetable fly", 172
  as zoophytes, 322

Genesis (book of)
  animals in, 4, 90
  creation, 88, 92, 108–9, 122, 123, 125, 211,
    225
  Deluge / flood, 4, 183–84, 377

812 INDEX OF SUBJECTS

Genesis (book of) (*cont.*)
  Genesis Rabbah, 92
  Jacob's ladder (*see* at continuum)
  Six Days of Creation (*see* hexæmeral
    literature)
genetics, genes
  applied (plant- and animal-breeding), 269,
    270*f*, 276, 277*f*
  as basis of classification, 630n.1 (*see also*
    comparative biology)
  central dogma of molecular biology, 432–33
  chromatin, chromosomes, 413, 419–21,
    424, 426, 432, 445, 630n.1
  DNA, discovery / characterisation of, 432
  DNA in organelles, 427, 437, 445
  eugenics, 411
  genes, chemical nature of, 432–33
  genes, early ideas about, 630n.4
  genomics, genome (*see* genomics)
  genotype / phenotype, 432, 435
  lateral genetic transfer (LGT), physical basis
    of, 439, 440, 441, 444
  laws of inheritance (Mendel), 426, 432–33
  molecular sequences (*see* phylogenetics;
    proteins)
  organelles as non-Mendelian, 427, 631n.8
  plasmids, 439, 440, 444 (*see also* cellular
    ultrastructure)
  RNA in phylogenetics (*see* phylogenetics,
    molecular)
  terminology, first use of, 432
  transformation (genetic), 432–33
genomics, genomes
  of archaea, bacteria, eukaryotes, 440–41,
    442–43, 444–45, 446, 448
  environmental, 441–42
  first genomes sequenced, 440
  Genom, Plastom, 427
  metagenomics, 441–43
  "model organisms", 440
  pan-genomes, 440–41
  phylogenomics (*see* phylogenetics,
    molecular)
  signature proteins, 442, 446
  variation in genome size, 440–41
geographical locations (selected)
  Alexandria, 46–47, 56–57, 60, 63, 68–69, 71,
    76, 77, 89–90, 93–94, 504n.2
  Athens, 23, 44, 47, 59, 62–63, 64, 68–69, 75,
    80–81, 85, 87–88, 89–90, 111, 212, 504n.2
  Australia, 3, 5, 6, 274, 285–86, 317, 359
  Babylon, 4, 6–7, 28–29, 91, 101, 107, 139,
    140
  China, 16, 17–22, 94–95, 168

Constantinople, 67, 89–90, 94, 111, 134,
    180, 597n.3
Damascus, 87–88, 89–90, 93–94, 137
Egypt, 10–11, 14, 25, 26, 28–29, 56–57, 59,
    60, 62, 66, 77–78, 101, 114, 136–38, 139,
    158, 199–200, 457n.34, 467n.6, 490n.14,
    594n.139
Florence (Firenze), 134, 149, 157–58,
    173–74
Harran, 89–90, 94, 137
India, 6–7, 10, 14–16, 25, 28–29, 44–45, 56–
    57, 60, 63, 100–1, 164, 165–66, 231
Japan, 10, 22
Köln (Cologne), 126, 173–74
Lapland, 10, 269, 557n.51
Montpellier, 177, 178, 190–91, 192, 231
Padua, 55, 111, 173–74, 180, 190–91, 199,
    212, 579n.26
Paris, 45–46, 111, 115, 118, 119, 121–23,
    125, 126, 128–30, 131, 170–71, 173–74,
    175, 178, 211, 215, 216–17, 245, 259, 274,
    278, 282, 289, 291–92, 293, 298, 309–10,
    323, 597n.3
Persia, 7, 14–15, 28–29, 63, 69, 139, 146,
    164, 198, 224–25
Rome, 46–47, 48, 53–54, 59, 60, 63, 64, 67,
    75, 76, 83, 84–85, 130, 173–74, 459n.67,
    504n.2
Strasbourg, 173–74, 312–13, 414, 415
Tylos, Island of (Bahrain), 165–66, 168
See also cathedral schools; Councils;
    Empires (political); institutions; learned
    societies; universities; and Index of
    Persons
Gilgamesh (*Epic of*), 4–5, 6, 7, 10, 211
gods (Egyptian, Greek, Roman), 6–7, 9, 25–26,
    65–66, 91, 92, 93–94, 122, 137–39, 185–
    86, 224, 225, 226, 486n.70
Golden Chain [of Homer], 470–71n.57,
    479n.15
  in the *Iliad*, 224
  in Kirchweger, 159, 160*f*
  in Macrobius, 224
  in Nieremberg, 184, 226
  in Porta, 226
  in Reuchlin, 225–26
  See also continuum; Great Chain of Being
Great Chain of Being, 24, 63–64, 209, 213–14,
    220, 223, 224, 225–31, 232–33, 234–35,
    236–41, 243–44, 249, 256, 259, 262, 263,
    264, 265–66, 268, 269, 276, 278, 279, 284,
    289, 290–92, 293, 294, 296, 297, 298, 303,
    304, 305, 309, 312, 355–56, 364–65, 373–
    74, 383, 409, 479n.15, 498n.92, 576n.134

INDEX OF SUBJECTS   813

in Bonnet, 209, 236–40, 266, 551n.93
in Bradley, 227–31
branched, 239–40, 241, 244, 262, 276
in Lamarck, 241, 278, 293, 294, 296, 297
in Leibniz, 213–14, 227
See also continuum; Golden Chain

Haeckel: works (selected)
Anthropogenie, 393–95, 397
Das Protistenreich, 395–96, 406, 407–8,
   580n.40
Die Radiolarien, 382–83
Generelle Morphologie, 383–86, 387–89,
   392, 396, 397, 399, 406, 407–8, 409
Natürliche Schöpfungsgeschichte, 393–95,
   396, 398f, 399, 407–8
Systematische Phylogenie, 397–400, 406,
   407–8, 409, 416
Hanseatic League, 118
herbals, 47, 52–53, 169–70, 189–92, 194–95,
   196, 198–99, 243–44, 357–58, 458n.54,
   533n.40
first printed illustrated herbal, 189
Hermetic corpus / texts, 57, 76, 93, 137–39,
   143, 150
Asclepius, 137–38
Corpus Hermeticum, 137–39, 150, 159
Emerald tablet (*Tabula smaragdina*), 112–
   13, 138, 142, 143, 144–45
Musæum Hermeticum, 154, 156–57
hexæmeral literature, 61–62, 75, 81–83, 117,
   123–24, 150, 197–98
hierarchy, celestial, 76, 84, 85–86, 90, 93, 210–
   11, 230–31, 239, 468n.20, 470–71n.57
high-level classification
   animals, vegetables, minerals encompass all
      earthly beings, 4–5, 24, 100–1, 114, 121
   as artificial or arbitrary, 5, 243–45, 260,
      261–62, 265, 268, 291–92, 293, 350, 411,
      599n.50
   inferred from symbolic language, 4
high-level taxa (selected)
   Aconta (Christensen), 624n.84
   Acrita (Macleay), 283–84, 364–65, 378,
      603n.123
   Acytota (Jeffrey), 444, 624n.81
   Akaryobionta (Rothmaler), 416–17
   Akaryonta (Novák), 413, 416
   Amorphoctista (Hogg), 372
   Animalia, Animals (*see* Animal Kingdom)
   Anucleobionta, Anucleobionten, 416
   Aphanobionta (Novák), 416, 444, 622n.50
   Archaea, Archaebacteria (*see* archaea:
      classification)

Archetista (Altenburg), 426
Archezoa (Cavalier-Smith), 430, 445,
   629n.175
Archezoa (Haeckel), 615n.161, 620n.18
Archezoa (Perty), 330
Arkarya (Forterre), 446–47, 448
articulate kingdom, 604n.147
atmospherical kingdom / department, 363–
   64, 562–63n.186
Bacteria (*see* bacteria: classification)
chaotische Reich, 562–63n.186
Chromista (Cavalier-Smith), 429, 430–31
Contophora (Christensen), 624n.84
Elementarreich, 263–64
Eocyta (Lake), 442, 446
Eucaryota, Eukaryota (*see* eukaryotes:
   classification)
Eucytota (Jeffrey), 624n.81
Eukaryomorpha (Fournier & Poole),
   635n.135
Feuerreich, 263–64
fire kingdoms, 376–77
Fungi / Pilze, 258–59, 418f, 419f, 420f,
   627n.130
Histones, kingdom (Haeckel), 396–97, 399–
   400, 402–3
Human / rational kingdom, 376–77
Imperium Naturæ (Linnæus), 256, 412,
   557n.52
Infusoria, 221, 260, 262–63, 272, 299–
   300, 303, 367, 386, 387–88, 584n.135,
   612n.107
inorganic kingdom, 321, 596n.176,
   606n.191
Karyonta (Novák), 416
Karyophyta (Novák), 416
Karyozoa (Novák), 416
Kingdom of Corals, 319, 583n.109
kingdom of heaven, 376–77
kingdom of imponderable agents, 376
kingdom of nature, 376–77
Kingdom of the Properties of Nature,
   521n.225
Lichtreich, 561n.141
Lithophyta, 249–50, 252, 253, 254–55, 258,
   261, 263, 555n.5
Lithozoa, 272, 321
Luftreich, 263–64, 561n.141
Man, Kingdom of (*see* at Man)
Materialienreich, 263–64
Mesocaryota (Dodge), 419–21
Mesoprotists (Doughery & Allen), 419–21
Metaphyta, metaphytes (Haeckel), 396,
   399–400

814 INDEX OF SUBJECTS

high-level taxa (selected) (*cont.*)
  metaprotists (Dougherty & Allen), 419–21
  Metazoa (Haeckel), 396, 614n.147
  Minerals (*see* Mineral Kingdom)
  Mittelreich, 254–55
  molluscous kingdom, 604n.147
  Monera, 386, 416–18, 419–21, 419*f*, 420*f*,
    634n.102
  Mychota, 413–14, 416–17
  Mykoidenreich (Merezhkowsky), 424
  Némazoaires, 347, 607n.200
  Neomura (Cavalier-Smith), 630n.193,
    635n.136
  Oozoa (Oken), 321
  organic kingdom, 586n.177, 596n.176
  Organicorum [corporum] Imperium
    (Pallas), 261–62, 412
  Phytozoa, 255, 263, 272, 280, 282, 318, 321,
    607n.200
  Plantæ, plants (*see* Vegetable Kingdom)
  Primalia, kingdom (Wilson & Cassin), 375,
    378, 607n.199
  Primigenum, Regnum (Hogg), 372–73,
    373*f*, 375, 378, 607n.199, 623n.69
  Probienreich (Nägeli), 404
  Procaryota, prokaryotes (*see* prokaryotes)
  Procytota (Jeffrey), 624n.81
  Protista (Haeckel) (*see* Protista)
  Protobionta (Rothmaler), 416
  Protobionten (Klebs), 404–5
  Protoctista (Hogg, Margulis), 372, 417,
    419–21, 624n.77
  Protophyta, 372, 392–95, 396–97, 399–400,
    406, 407, 620n.15
  Protorganismen (Carus), 331–34, 337,
    618n.251
  Protozoa (Goldfuss), 272, 584n.135,
    605n.175
  Protozoa (Siebold), 325, 329, 333–34,
    612n.107
  Protozoa, kingdom (Owen), 367, 372, 374–
    75, 376–77, 378, 607n.199
  Protozoa (other), 372, 391, 392–95, 396–97,
    399–400, 406, 407, 620n.15
  Pseudozoolithophyta (Class), 258, 264
  Psychodiare, Règne (Bory), 308–9, 333–34,
    346–47, 378, 583n.109
  Regio Elementaris, 540n.175
  Règne atmospherique, 562–63n.186
  Règne Ethéré, 308
  Règne inorganique ou minéral (Virey),
    576n.134
  Règne organique, Frontispiece, 305,
    619n.11

  Règne organisé (végétal ou animal) (Virey),
    576n.134
  Règne vivant (Vicq-d'Azyr), 302–3, 412
  Regnum ampho-anorganicum /
    amphorganicum, 280–81
  Regnum chaoticum, 248–49, 256, 264, 304–
    5, 364–65, 607n.200
  Regnum inorganicum, 260, 276, 279
  Regnum mesymale, 258, 264, 607n.200
  Regnum neutrum, 248–49, 254–55, 264,
    607n.200
  Regnum organicum, 260, 280, 302–3, 412
  Sponge-kingdom, 376–77
  star kingdoms, 376–77
  Thallophyta, 336, 403–4, 405–6
  Urorganismen, 354, 595n.167
  Urpflanzen-Reiche, 614n.147
  Vegetabilia, Vegetables (*see* Vegetable
    Kingdom)
  végéto-animaux, Frontispiece, 305
  Vertebrate Kingdom (*Wirbelthier-Reich*),
    586n.177
  Vira, kingdom, 634n.102
  Wärmereich, 561n.141
  Wasserreich, 263–64
  Wirbelthier-Reich, 586n.177
  Zoophyta (*see* at Zoophyta)
homology. *See* comparative biology
humours (Galenic), 53–54, 88, 119, 120, 140,
  146–47, 153
hunter-gatherers, 1–2, 23, 91
hunting, 1–3, 8, 34, 47, 56, 198
Hydra (freshwater polyp)
  animal, vegetable or zoophyte, 234–35,
    249–50, 254–55, 260, 261, 262–63, 289,
    295, 322–23, 575n.125, 596n.187
  described by Leeuwenhoek, 215, 233–34
  as evidence of / challenge to continuity in
    nature, 230, 233, 234–36
  genus named by Linnæus, 552n.105
  in Ledermüller, 221
  philosophical problems arising from, 236
  in Rösel, 220–21
  as sensate / insensate, 234–35, 238
  studied by Trembley, 233–36

Industrial Revolution, 406
infusoria. *See* at microscopic beings
insects
  in Bonnet, 236–37, 239
  as a branch off the Great Chain, 239, 262,
    276
  butterfly collecting, 198
  as carriers of disease, 212

in folk taxonomies, 5
generated by use of electricity, 368
insect-fish, 170–71
metamorphosis, 206, 229, 553n.147
parthenogenesis, 237, 251–52
in Priestley's green matter, 338–39
in Rösel, 220–21
spiders as "loathsome insects", 247–48
spontaneous generation of, 167, 169–70,
185, 206, 212
institutions
Aldine press, 55, 173–74, 175
Bodleian Library (Oxford), 15
British Museum (London), 364, 539n.148
Château d'Écouen, 483n.106
Dutch East India Company, 249, 465n.25,
467n.62
Écoles Normales (Paris), 292, 309–10, 570n.22
Institut océanographique de Monaco, 422
Iraq Museum, 4–5
Jardin des Plantes / Jardin du Roi (Paris),
216–17, 245, 293
Laboratoire Arago à Banyuls-sur-Mer, 415,
621n.44
Metropolitan Museum (New York),
522n.14, 587n.202
Muséum d'histoire naturelle (Paris), 287–
88, 290, 292, 293–94, 295, 298–99, 300,
301–2, 307–8
Natural History Museum (London), 364,
367
Orto botanico di Padova (Padua), 190–91,
579n.26
Orto e Museo botanico (Pisa), 190–91, 193
Petersburg Medical-Chirurgical Academy,
280
Royal Botanical Garden (Edinburgh), 598n.28
Royal Botanic Gardens, Kew, 358
Royal menagerie (London), 206
Scuola medica Salernitana (Salerno), 111
Station Biologique de Villefranche-sur-Mer,
415
See also cathedral schools; learned societies;
monasteries; universities
Islamic philosophy (schools of)
falsafah, 94, 97, 101–3
kalām, 94, 97, 102, 107
Muʿtazilite, 94, 96, 470n.41
ṣufiyya, 102, 106–7, 502n.183
Israel, lost tribe of, 183–84

Jacob's ladder. See at continuum

kingdom (classificatory rank)

in alchemical literature (1604 ff), 156–57
in Alsted (1615), 206–7
in Besler (1642), 205–7
in chemical literature, 157
in Descartes (1639), 206–7
in König (1682-1693), 2, 208
in Mayerne (1634), 206–7
superkingdoms, 412, 413, 414–16, 417,
419–21, 430, 624n.89, 636n.144
urkingdoms (Woese), 436, 636n.141
word as inappropriate, 302–3
word not in al-ʾArabī, 106, 144
word not in Arūẓī, 103–5, 144

Lamarck: works
Flore française, 241, 293, 294
Histoire naturelle des animaux sans
vertèbres, 296–97
Philosophie zoologique, 295–96, 297
Système analytique des connaissances
positives de l'homme, 297–98
Système des animaux sans vertèbres,
294–95
languages (selected mentions)
Algonquin, 5
ancient Middle Eastern, 4
Anglo-Norman, 197–98
Arabic, 96, 140, 147, 150–51
Armenian, 57, 488n.101, 489n.128
Australian aboriginal, 5
Aztec, 5
Castilian, Spanish, 172, 174–75
Cree, 5
Dutch, 215, 467n.62, 541n.2
Egyptian, 137
English, 172, 183, 204, 571n.29, 600n.70
Ethiopian, 57
French, 171, 172, 177–78, 183, 197–98, 289,
578n.4
German, Germanic, 198–99, 532n.25,
571n.29, 604n.146
Greek, 41, 54–55, 59, 77, 87–88, 114, 134,
137, 150–51, 173–74, 175, 178, 180, 190,
191, 194–95, 201, 228
Hebrew, 150–51
Latin, 46–47, 48, 54–55, 83, 118, 191–92,
228, 257, 287–88, 311–12, 509n.139,
594n.129
Mayan (Tzeltal), 5
Ojibwa, 6
Russian, 503n.237, 631n.24
Syriac, 57
vernacular, 118, 149, 174–75, 178, 183, 187,
190–91, 202, 289

816 INDEX OF SUBJECTS

languages (selected mentions) (*cont.*)
    Waldensian, 197–98
    Yoruba, 5
learned / professional societies
    Académie [Royale] des Sciences (Paris),
        232, 234, 237, 260, 290, 297, 301, 302
    Académie Royal de Peinture et Sculpture
        (Paris), 216
    Académie Française, 171–72, 302
    Académie Parisienne, 157–58
    Academy of Natural Sciences of
        Philadelphia, 373–74, 375
    Accademia dei Lincei, 544n.57
    British Association for the Advancement of
        Science, 371
    Geological Society of London, 369, 432,
        597n.5
    Gesellschaft Deutscher Naturforscher und
        Ärtze, 318–19, 383
    Imperial Leopoldinian-Carolinian Society
        of Naturalists, 270
    [International] Society for Evolutionary
        Protistology, 421, 624n.77
    Königliche Akademie der Wissenschaften
        zu Berlin, 311
    Koninklijke Hollandsche Maatschappij der
        Wetenschappen, 347–48
    Linnean Society (London), 282, 358, 597n.5
    Microbiology Society, 621n.39
    Microscopical Society of London, 364
    Royal College of Physicians, 597n.5
    Royal College of Surgeons (London), 364
    Royal Geographical Society (London),
        597n.5
    Royal Holland Society of Sciences (*see*
        Koninklijke Hollandsche Maatschappij
        der Wetenschappen)
    Royal Institution (London), 597n.5
    Royal Society (London), 209, 216, 222, 227–
        28, 232, 233–34, 237, 247, 287–88, 355,
        356, 358, 363, 376–77
    Royal Society for the Encouragement of
        Arts (London), 597n.5
    Royal Society of Medicine (London),
        597n.5
    Schweizerische Naturforschende
        Gesellschaft, 626n.111
    Société Royale de la Médecine, 260, 302
    Société Royale de Montpellier, 231
    Zoological Society of London, 597n.5
legendary / mythical creatures (selected)
    animated leaves, 169–70, 171, 196
    bishop-fish, monk-fish, 203
    centaur, 49–50, 163–65, 207

    dragon, 54, 115–16, 120, 124–25, 163–64, 172,
        197–98, 203, 204–5, 466n.44, 537n.114
    gargoyle, 165
    gorgon, 7, 204
    griffin, 7, 183–84
    Hellmouth, 163
    Hydra, 115–16, 163–64, 172, 197–98, 233,
        547n.129
    mermaid, 125, 204–5, 458n.49
    Neptune's horse, 203
    Phœnix (*bennu*), 4, 7, 8–9, 21, 57, 163–64,
        172
    sawfish (*serra*), 163–64, 197–98
    on stave churches, 163
    unicorn, 120, 163–64, 172, 197–98, 204–5,
        249, 458n.45, 538n.124
lichens, 191, 192, 194, 195, 238, 240, 246–47,
    254, 258, 259, 268–69, 297, 306–7, 308,
    315–16, 321, 322–23, 342, 352, 368–69
    algal and fungal components of, 345–46,
        423, 430–31, 635n.125
    classified with fungi, 259, 430–31
    in Dioscorides, 53
    as examples of symbiosis, 423, 424, 430–31
    as intermediate between plants and
        animals, 292–93
    as members of an intermediate kingdom,
        254–55, 322, 375
    metamorphoses involving, 343–44, 345–46,
        347–48
    pleomorphy of, 593n.106
    as polyphyletic, 424
    sexuality, flowers, seeds, 336, 344–45,
        559n.89
    spontaneous origin of, 573n.79
    in Theophrastus, 43, 480n.49
    as zoophytes, 322
life / the animate
    in Aristotle, 35–37
    as breath, 11, 25–26, 67, 76, 130, 212
    features distinguishing living from
        inorganic bodies, 297
    grades, modes or scale of life (Aquinas),
        127–28, 133–34, 149–50
    insentient, 84, 89, 180–81, 201, 371
    in John of Damascus, 88–89
    Lebenskraft, 316, 337, 380, 571n.29,
        583n.112, 583n.113
    matter a product of life, 311
    Omne vivum ex ovo, 270–71
    origin of, 278–79, 297, 315, 316, 320, 321,
        359–61, 368, 369, 384–85, 392–93, 395–
        96, 413–14 (*see also* cosmos, creation
        myths)

INDEX OF SUBJECTS 817

in Philoponus, 70
in the Presocratics, 25–26, 27, 28
as self-movement, 32, 78, 141–42
terminology (*zoë, zôion, zoon*), 26, 471n.59,
474n.124, 475n.141, 605n.160
Linnæus: works / corpus
Amœnitates academicæ, 248, 251
Mundus invisibilis, 255–56, 557–58n.61
Philosophia botanica, 251, 259, 267–68,
280–81, 357–58, 556n.43
purchased by Smith, 599n.41
reception of, 256–57, 263, 312–13, 355–56,
357–59
Species plantarum, 248, 357–58
Systema naturae, 22, 172, 243, 248, 249–51,
250*t*, 252, 254, 256, 259, 261, 262–63,
335, 357, 358, 585n.136
logical division
arbore (Medieval), 286
in Aristotle, 37, 49–50, 65, 94
differentiæ as mediating entity, 66
distributio (Ramus), 181–83
in Plato, 34, 204–5
Ramism, Ramist trees, 181–83, 192,
195, 201, 204, 208, 534n.65, 538n.131,
540n.169, 584n.125, 585n.148
regressus (Tomitanus, Zabarella), 180
Tree of Porphyry, 65, 181–82
tripartite, 69–70, 72–73, 88–89, 95–97
Lucretius: works
De rerum natura, 48–49, 212, 541n.10
lycanthropy, 10

macrocosm-microcosm, 54, 63–64, 73–74,
78–79, 82–83, 88, 90, 120, 121, 122,
129, 138, 139, 140, 143, 145–47, 149,
153, 154–55, 158, 159, 161, 320, 333,
481n.70
in Aristotle, 73–74, 140
in Islamic literature, 96, 98–99, 106–7, 109,
140, 502n.185
in Kabbalah, 489n.135
as literary device, 171
in Neoplatonism, 73–74
in Plato, 33–34, 73–74, 140, 469n.33
in Pythagoreanism, 27
in Stoicism, 73–74
magic, 9, 33–34, 50, 53–54, 56, 111, 137–38,
142, 146, 147, 150–51, 152, 153, 157–58,
159, 170
natural magic, 151–52, 161
persons considered magicians, 147, 459–
60n.68, 470n.44, 514n.45
as sister to religion, 146

sorcery, black magic, 147, 149, 161, 169,
459n.63, 514n.45
magnet, magnetism, 79–80, 120, 186, 293, 318,
362, 376, 520n.205, 601n.76
in *Naturphilosophie*, 318, 588n.208
man, humankind
Ages of Man, 197–98, 481n.70
analogized with vegetables, 15, 18–19, 30,
50, 62, 236, 472n.72
apes / monkeys approach form / nature
of man, 105–7, 108, 132, 157, 179, 229,
475n.140
children intermediate between animals and
humans, 38, 126
congenital anomalies / defects, 170–71,
203
contrasted with animals, vegetables (and
minerals), 27–28, 35, 36, 50, 66, 79–80,
83, 88, 90, 91, 92, 93, 98, 99, 101, 102, 103,
105–7, 108, 109, 110, 112–13, 118–20,
130–31, 133–34, 144, 147, 148–50, 153,
156, 176, 226, 227, 239–40, 312, 313, 331,
368, 373, 412, 561n.141, 606n.192
created in the image of God, 78–79, 260,
457n.35
defined by rational / intellective soul, 27, 36,
44, 61, 65, 66, 70–71, 72, 79–80, 81, 88–
89, 98, 101, 108, 117, 118, 121, 122–23,
125, 126, 127, 129, 149–50, 152, 182–83,
193, 206, 294, 376–77, 471n.67
emotions, 183
as the flower of the animal kingdom, 272
formed from clay or earth, 6, 73, 223
as fungus (Linnæus), 255–56
hierarchy within mankind, 102, 104–5,
552–53n.128
as a high-level category (modern), 280, 312,
331, 373, 448–49, 540n.175, 561n.151
(*see also* kingdom of, below)
as a high-level category (pre-modern), 27–
28, 36, 79–80, 83, 88–89, 90, 91, 92, 93,
98–99, 101, 103, 108, 109, 110, 112–13,
118–20, 130–31, 133–34, 144, 147, 148–
50, 153, 156, 159, 176, 185–86, 188, 219,
226, 227, 412, 508n.100
intermaxillary bone, 313
intermediate between beasts and angels,
176, 179
kingdom of, 520n.205 (*see also* at high-level
taxa: Human / rational kingdom)
Menschen, 578n.15
Menschenreich, 373, 606n.192
Regnum Hominum, 561n.141
as microcosm (*see* macrocosm-microcosm)

818 INDEX OF SUBJECTS

man, humankind (*cont.*)
    not an animal, 5
    not distinct osteologically from animals
        (Goethe), 292–93
    not the most-perfect animal, 230–31
    Perfect Man, 106–7
    spirit of (*see* Romanticism)
    women intermediate between beasts and
        humans, 32, 179
    wondrous forms of, 164
marine invertebrates
    alcimonium, 54
    alcyonia, soft coral, 195, 205, 244, 246–47,
        249, 263, 306, 322, 380, 403, 605n.159
    ascidia, tunicates, tethya, 39–40, 55, 140–
        41, 199, 200–2, 203–4, 205, 249–50, 268,
        284, 297–98, 308–9, 325, 346
    bryozoa, 203–4, 205, 220–21, 232, 263, 308–
        9, 326, 346, 369, 370, 568n.112, 575n.125,
        610n.76
    crab, lobster, 20–21, 40–41, 51, 54, 56, 65,
        115–16, 125, 166, 195, 200, 202, 206, 229,
        239, 536n.101
    holothuria, sea cucumbers, 39, 51, 55, 140–
        41, 183, 199, 201–2, 203–4, 205, 262–63,
        299, 322, 380, 530n.73
    jellyfish, medusæ, 53, 200, 201, 203–4, 208,
        230, 249–50, 252, 262–63, 270, 272, 278,
        299–300, 322, 325, 346, 381
    larvae of, 346, 388, 389, 390
    octopus, squid, cuttlefish, 51, 53, 54, 56,
        124–25, 198–99, 200, 202, 205, 206, 208,
        249–50, 301, 536n.101, 539n.154
    pinna, 38–39, 79–80
    sea anemone, acalephae, actinia, sea-nettle,
        27, 39, 40–41, 51, 55, 65, 79–80, 81, 101,
        126, 167, 171, 176, 184, 195, 198–99, 200,
        203–4, 280, 281, 295, 299–300, 321, 364–
        65, 480n.38
    sea-hare, sea-slug, 53, 184, 191–92, 199,
        200, 201, 249–50, 252, 476n.168
    sea-lungs, comb jellies, 39, 51, 171, 184, 199,
        201, 480n.37, 481n.60
    sea-pen, 201, 203–4, 263, 308–9, 322,
        568n.112
    sea-snail (*see* at snails)
    sea-urchin, 51, 65, 122, 184, 195, 200, 201,
        205, 206, 207, 208, 249–50, 263, 299, 322,
        325, 530n.73
    shellfish, oyster, mussel, testacea, 17–18,
        20–21, 32, 33, 37–41, 51, 54, 55, 57, 65,
        69–70, 71, 78, 79–80, 82, 84, 94, 95–96,
        98, 100, 101, 102–3, 105, 120–21, 122,
        123, 125, 126, 127, 133–34, 144, 149–50,

        153, 169–70, 172, 173, 176, 179, 187,
        198–201, 202, 203–4, 205, 206, 208,
        229, 238, 240, 246–47, 262–63, 276, 303,
        346–47, 351, 356–57, 470n.46, 473n.103,
        474n.114, 546n.110
    starfish, 51, 124–25, 184, 185, 199, 200–1,
        202, 203–4, 205, 219, 229, 234–35, 249–
        50, 263, 299, 322, 346, 456n.19
    See also corals; dualizers; snails; sponges;
        zoophytes
marine plants / seaweeds, 53, 168, 205, 232,
    252–53, 259, 308, 322, 346, 348, 390,
    480n.34, 480n.36, 546n.111, 551n.94
    coralline algae, 233–34, 244, 247–48, 261,
        263, 295, 299–300, 302, 306, 308–9, 357,
        358, 369, 568n.112
    flowers, sexuality, 336, 343–44
    included in herbals, 191–92, 194–95, 196,
        208–9
    in Pliny, 52
    as polyparies, 338, 339–40, 344
    in Theophrastus, 52, 403
    as zoophytes, 322, 593n.109
    See also algae; coral
materia medica, 7, 47, 53, 54, 176–77, 189,
    190–92, 563n.193
    Dioscorides, 47, 53, 115, 189–91, 460n.75,
        532–33n.31
    pen tsao texts, 18, 22
medical texts, 17, 18, 22, 47, 53–55, 115, 116,
    173–74, 189–90, 331, 501n.155, 531n.9
    See also *materia medica*; and Celsus,
        Dioscorides, Galen, Hippocrates in Index
        of Persons
metals and alloys
    bronze, 7, 142, 225, 463n.2
    copper, 46, 142, 143, 472n.73, 509n.133
    gold, 12–13, 50, 99, 110, 128, 142, 143, 144,
        148, 153, 154, 156–57, 162, 164, 177–78,
        225, 464n.24, 472n.73, 509n.133 (*see also*
        Golden Chain)
    iron, 79, 142, 143, 148, 156, 225, 318,
        509n.133, 521n.236
    lead, 12–13, 142, 143, 225, 509n.133,
        521n.236
    mercury, 362 (*see also* at elements)
    silver, 50, 110, 143, 148, 156, 157, 161, 225,
        509n.133
    tin, 142, 148, 161, 225, 509n.133
metamorphosis
    of algae (*see* at algae)
    in the ancient / classical world, 8–9, 10
    in ancient China / Japan, 10, 21–22
    in Arabic literature, 10

INDEX OF SUBJECTS 819

in cave art, 8–9
of clouds, 263–64
connecting plants and animals, 9, 10, 254,
    256, 257
in folk-tales / legends, 9–10
of frogs, 8, 10, 21–22, 229, 230
of fungi (*see* at fungi)
in Goethe, 313–14, 384, 579n.24
of insects (*see* at insects)
in Linnæus, 251–52, 253, 254, 256, 257–58,
    556n.45
metamorphoses (Ovid), 8–9, 258, 346,
    469n.33, 521–22n.1, 542n.29
in modern literature / popular culture, 10
in *Naturphilosophie*, 319, 333–34, 342
of Priestley's green matter, 339, 340
of zoophytes (*see* at zoophytes)
See also alchemy; lycanthropy; ontogeny;
    transubstantiation
metempsychosis (transmigration of souls),
    10–12, 14–15, 76, 80, 82–83, 90, 93, 108,
    136–37
    Augustine rejects, 83–84, 93
    in Plato, 32
    in Plotinus, 64
    in Pythagoras, 11, 27
microorganisms (selected)
    algae (*see* algae; marine plants)
    Amoeba, amœbæ, 220–21, 249–50, 254,
        263, 307, 308–9, 327–28, 330–31, 332–34,
        336, 381–82, 391, 392, 393–95, 396–97,
        409, 414, 416, 419–21, 424–25
    archaea, archaebacteria (*see* archaea)
    Bacillaria (*see* diatoms)
    bacteria (*see* bacteria)
    Chlorarachnion, 429–30
    Chlorella, 445–46
    Chytridium, 380–81
    ciliates, 215, 216, 219–20, 239, 308–9,
        325–26, 329, 330, 333–34, 387–88, 391,
        393, 396–97, 400, 401, 409, 414, 445–46,
        544n.65, 547n.117
    coccolithophorids, 390, 409
    desmids, 309, 329, 349–51, 366–67, 371,
        372, 376–77, 380, 382, 396–97
    diatoms, 305, 308–9, 325–26, 327–28, 329,
        332–34, 346–47, 349–52, 353, 366–67,
        371, 372, 376–77, 380, 382, 384–85, 386,
        392, 395, 396–97, 401–2, 403–4, 409, 416,
        417, 424–25, 429, 544n.65
    dinoflagellates, 220, 308, 416, 419–21, 429–
        30, 437, 557–58n.61, 589n.30, 615n.161
    Euglena, euglenoids, 215, 308, 343–44, 393–
        95, 429–30, 586n.169, 608n.23

flagellates, 215, 330, 386, 387–88, 391–92,
    393–95, 396–97, 400–1, 403, 404–5, 408,
    409, 414, 415, 416, 419–21, 424, 609n.46
foraminifera, 215, 285–86, 327, 367, 389,
    409, 548n.136, 557–58n.61, 611n.94
gregarines, 367, 380–82, 391, 393, 395, 396–
    97, 399, 409
Labyrinthula, 390, 393, 395, 408
Magosphæra, catallacta, 390, 393–95, 396–
    97, 399, 400–1, 408
Monas, monads, 283–84, 347, 351, 352, 355,
    366, 369, 382, 403, 404–5, 406, 586n.169,
    587n.184
monera (Haeckel), 383, 384–85, 386, 387–
    88, 389, 391, 392–97, 401, 402–3, 406,
    407, 409, 413, 416–18, 419–21, 424, 429
Noctiluca, 220, 386, 393, 395, 399, 557–
    58n.61, 612n.112
opalinids, 330
Pansporella, 414
Plasmodium (malarial parasite), 430
Protamœba, 389, 401–2, 408, 416, 623n.69
Proteus (Baker), 220–21, 355
Proteus, the lesser (Rösel), 220–21, 253,
    255–56, 295, 588n.4
protoplasta (Haeckel), 325, 386, 387–88,
    393
protozoa, 239, 262–63, 272, 325, 329–30,
    333–34, 336, 367, 371–72, 374–75, 380,
    386, 387, 390, 391, 392, 393–97, 399–400,
    401–4, 406, 407, 409, 410, 414–15, 416,
    423, 428, 430, 445, 542n.22, 608n.23,
    608n.25, 630n.203
radiolaria, 379–80, 381, 382–83, 414
Symbiodinium, 445–46
Volvox, volvocines, 216, 220–21, 249–50,
    251–52, 253, 261–62, 263, 295, 296, 332–
    33, 335, 336, 343–44, 391–92, 614n.146
Vorticella, 220–21, 239, 263, 295, 299–300,
    308–9, 335, 355, 400–1, 544n.64, 557–
    58n.61, 562n.173, 588n.216
microscopes
    bead-lens, 215
    chromatic aberration, 220, 222, 588n.5
    compound, 209, 588n.5, 591n.69
    electron, 416–17, 421, 426, 427, 429, 430,
        439
    optical resolution, 222, 325–26, 335, 341–
        42, 350, 585n.152, 620n.25
    Plössl (Plössel), 342–43, 587n.188
    Ross, 389
    Schieck, 594n.134
    simple (single-lens), 217–18, 588n.5
    spherical aberration, 220, 222

820   INDEX OF SUBJECTS

microscopes (*cont.*)
  technological improvement, 220, 221, 380–81, 591n.64
  See also microscopy
microscopic beings
  as able to develop into animal or into plant, 304–5, 308–9, 314–15, 332–33
  as alternatively animal and plant, 219, 254, 255, 308, 347, 349, 354
  as animal united with plant, 273, 278, 308, 347, 349, 351–52, 354
  Blastien-Theorie, 330, 332
  as complete / perfect animals, 303, 326–27, 329–30, 366
  exemplifying the wisdom and power of God, 247–48
  eyespot, 326–27, 349
  first classified in separate taxon, 219
  first comprehensive classification, 262–63
  as free-living cells, 328–31
  infusionary, infusoria, 210, 215, 216, 217–19, 220, 221–22, 230, 243, 253, 254, 255, 263, 272, 281, 292–93, 295, 304, 314, 319, 320–22, 326–27, 333–34, 340, 344–45, 366, 367, 572n.61
  as irritable / not irritable, 249
  in Lamarck, 295, 296, 297
  in Linnæus (*Mundus invisibilis*), 255–56
  in Linnæus (*Systema naturæ*), 249–50
  microscopiques (Bory), 308, 333–34
  as "moving atoms", 217–18, 222
  multiple nuclei in ciliates and opalinids, 330
  not unanticipated, 210, 211–12
  in Owen, 365–67
  polygastric theory, 325–27, 332–33, 351, 365, 366, 367, 382, 587n.186
  Priestley's green matter, 338–40, 345–46, 353, 360, 362, 576n.137
  produced from granite, 340
  as prototypes of higher animals, 325, 366
  responsible for disease, 212, 215, 250, 253–54, 256, 325, 406–7
  as seeds of macroscopic beings, 212
  as sensate / insensate, 216, 219, 296, 396
  spermatozoa, 221–22
  spontaneous generation of, 273, 337
  terminology describing, 215–16, 239, 337
  See also air / wind in fecundation; spontaneous generation
microscopy
  books / illustrations of objects at microscopic scale, 216, 546n.113
  introduction of word *microscopio*, 215
  as leisure activity, 220

  observation of living corpuscles / globules, 217–19, 221–22, 237, 253–54, 337, 421, 582n.86
  optical lenses, 161, 215
  See also microscopes
Mineral Kingdom, 22, 104–5, 106, 138, 144, 208, 227, 260, 291–92, 308, 316, 319, 320, 355–56, 359, 360, 362, 363–64, 368, 373, 373f, 376, 377, 455n.3, 465n.28, 540n.167
  first called a kingdom, 2, 104–5, 156–57, 161, 206–7, 208
  Fossil / Fossile Kingdom, 359, 540n.167
  in the Ikhwān al-Ṣafā, 99
  inorganic kingdom, 606n.191
  in Linnæus, 249, 256, 557n.54
  in Needham, 219
  Regnum Lapideum (Scopoli), 260
  Stone Kingdom, 313
minerals, stones
  analogized with plants (trees), 148
  as animals, 236
  as "fixed vegetables", 159
  granite, 313–14, 340
  as growing (or not) within the earth, 100–1, 142, 148, 154, 179, 228
  as lacking soul, 36, 101, 149
  as living or lifeless, 8, 21, 33, 35, 88, 100–1, 110, 126, 144, 150, 154, 217, 236, 371, 521n.233
  as mixtures of elements, 99, 143, 150
  as organized / unorganized beings, 36, 238, 239–40, 244, 306, 308, 362, 477n.175, 570n.20
  origins of, 313–14, 323
  ranked by nobility or perfection, 99, 105, 128, 142, 144
  as sensate / insensate, 15–16, 148
  use / non-use in medicine, 4, 7, 18, 140, 153, 154, 208
  vegetal mineral, mineral plant, 99, 100
  See also metals
miracles, 75, 85, 156, 203, 361–62, 383
  resurrection / return from the dead, 85, 102, 139, 143, 170, 462n.100
monad (simple substance or particle)
  in Agardh, 345
  in Leibniz, 213–15, 227, 593n.105
  in Oken, 321
  in Pythagoreanism, 26–27
  for the microorganism (form-type or genus) (*see* at microorganisms)
monasticism, monasteries, 80–81, 83, 114, 115, 116, 120, 130, 153, 176–77
  at Mar Saba, 87–88
  Montanist, 76

INDEX OF SUBJECTS 821

at Monte Cassino, 114, 126
Petit-Augustins, 526n.112
monera (Haeckel). *See* high-level taxa
monsters
  in Aldrovandi, 203
  congenital anomalies, fœtuses, tumours, 170–71, 207, 363–64
  in Gessner, 201
  in literature / mythology, 7, 8–9, 77, 124–25, 172
  Monstres et prodiges (Paré), 170–71
  in Salviani, 202
  as a step toward animals, 49
motion, movement (as criterion for classification)
  amœboid, 424
  arbitrary, 272
  Brownian, 222, 363
  contractile, 40, 71, 89, 296
  goal-directed, purposeful, 37, 306, 329
  locomotion (place to place, transitive), 22, 36–37, 42, 62, 68, 69–70, 72–73, 79–80, 89, 99, 101, 108, 120–21, 126, 141–42, 149–50, 152, 166–67, 188, 219–20, 238, 239, 250–51, 261, 305, 307, 308, 326–27, 338, 340, 347, 350–51, 353, 360, 366, 371, 573n.78, 573n.80, 604n.154
  motility (of diatoms, infusoria, zoospores), 216, 254, 257, 258–59, 288, 336, 339, 340, 341–44, 345, 347, 350, 351, 353, 380–81
  in place (local, in situ), 35, 56, 95, 101, 123, 158, 167
  as a property of soul, 32, 78, 79, 110
  as a property of stars, 141–42
  reflex, 396
  self-motion, 32–33, 41, 219, 336, 340, 344, 345, 351, 353, 473n.104, 582n.89, 591n.59
  spontaneous, 88, 141–42, 278, 294, 302
  tremulous, vibratory, 293, 595n.161
  voluntary, 22, 99–100, 101, 104, 108, 122–23, 252–53, 296, 327, 346, 350, 380, 382
multiple worlds
  in Bonnet, 239–40
  in Bruno, 158, 211
  cosmological, 79, 133–34, 153, 210–12, 226–27, 510n.188
  in Fontenelle, 211–12
  human inhabitants of, 211–12, 239–40
  microscopic, 211–12
mystery-books
  De mysteriis (Iamblichus), 66, 136–37, 148
  Kitab al-Asrār (*Liber secretorum*), 146
  Sirr al-asrār (*Secretum secretorum*), 144–46, 148–49

Sirr al-khalīqa (*Liber de secretis naturae*), 112–13, 138, 143–44, 147, 148–49
mysticism
  Babylonian mysteries, 101
  Egyptian mysteries, 101, 136–37, 461n.93
  Eleusinian mysteries, 14, 136–37
  Jewish mysteries, 77–78
  Kabbalah, 108–9, 110, 149, 157–58, 159, 225–26
  nature-mysticism, 157–59
  in Nieremberg, 183–84
  number-mysticism, numerology, 33–34, 59–60, 62, 109, 145–46, 157–58, 159, 185–86, 248, 284–85, 486n.62, 493n.94
  Orphic mysteries, 14, 101, 136–37
  Persian mysteries, 225
  spiritual ascent to God, 17, 73, 79, 84–85, 86f, 87, 106–7, 129, 152, 179, 223, 224–26
  Taoism (*see* at religions)

natural selection (Charles Darwin), 369, 370, 404
natural theology, 77, 82, 96–97, 98, 102, 103–4, 210, 240–41, 243, 246–48, 265–66, 309–10, 361–64, 377, 577n.178, 604n.146
  Boyle Lectures, 362
  Bridgewater treatises, 248, 363–64, 368, 377
  Natural theology (Paley), 248, 362
  The wisdom of God manifested in the works of the Creation (Ray), 246–47
nature
  all nature is animal (Robinet), 236
  early treatises on (*Peri physeos*), 25–26, 27–28, 29
  as expression of Spirit (Schelling), 317
  five kingdoms of, 375, 417–19, 419f, 420f, 428–29, 430, 434–35
  formative / vitalistic force in, 316
  four genera / kingdoms of (selected), 98–99, 149–50, 281, 397–400, 409, 417–18, 418f, 602n.96
  God manifest in (*see* natural theology)
  made cohesive by intermediate organisms, 179
  as model or teacher, 78–79, 90
  natura naturans, 333
  personification of, 122
  as spectacle (Buffon), 216–17, 257, 289, 309, 553n.143, 561n.142
  Spirit of Nature, 186
  three genera / kingdoms of (selected), 2, 22, 98–99, 112, 138, 156–57, 179, 206–7, 227, 249, 256, 259, 261–62, 265, 266–67, 276–77, 293, 319, 363, 417–18, 419–21, 575n.118, 622n.62

822 INDEX OF SUBJECTS

nature: plan or design of
  based on number or symmetry, 269, 273,
    282, 284–85, 286 (*see also* at mysticism;
    Pythagorean teachings)
  as bundles of sticks, 260
  as circle(s) or rings, 226, 281–82, 282*f*, 287
  as Easter-egg, 270–73, 271*f*
  genealogy as basis for, 266–67, 268, 269,
    276, 384, 385
  linear (*see* continuum; Great Chain of
    Being; Golden Chain)
  as map, 260, 266–68, 267*f*
  metaphors grade into each other, 225–26,
    286–87
  as multiple series, 232–33, 265, 267–69,
    276–77
  as network, web or fabric, 184, 232–33, 238,
    262, 264, 266, 268–69, 270*f*
  as one or more pyramids, 287, 373, 373*f*,
    548n.3
  pattern pluralism, 449
  as polygon, 269–70, 273–76, 273*f*, 274*f*,
    275*f*, 287
  as polyhedric figures, 262, 264
  quinarian arrangements, 282–86, 283*f*, 285*f*,
    287, 363, 368–69
  as spiral, 280–81, 280*f*
  as tetractys, 281
  as thallus / prothallus, 408
  as tree, 223–24, 225–26, 239, 241, 244, 262,
    265, 266, 268, 276–79, 277*f*, 279*f*, 286–87,
    307, 309, 369, 409, 565n.57, 584n.134,
    608n.23
  as two diverging branches or scales,
    Frontispiece, 276–77, 278, 296, 303–6,
    344, 368–69, 373
  See also Great Chain of Being
Naturphilosophie, 248, 272, 287–88, 309–10,
  317–22, 324, 325, 326, 330, 331–33, 353,
  561n.141
  Carus as "the last *Naturphilosoph*", 331
  compared with the Black Death (Liebig),
    322
  rejected by von Baer, 324
New York Times, The, 437

ontogeny
  defined (Haeckel), 384
  gastraea theory (Haeckel), 390–91, 393–95,
    402–3
  law of ascension (Virey), 304–5
  laws of embryology (von Baer), 324–25
  ontogeny parallels geological series, 368,
    575–76n.132

ontogeny recapitulates phylogeny, 324–25,
    404, 411
  ontogeny reveals vertebrate plan, 301,
    579n.22
  tree-metaphor implicit in (von Baer), 287
  vitalistic force underlies (Kant), 319
  See also embryology
organisms (bodies of)
  as assembles of corpuscles / living particles,
    213–15, 320–22, 347
  cortico-medullary theory (Linnæus),
    251–52
  as pneumatic machines, 217–18
  theory of moule interieur (Buffon), 217, 316
  tripartite (Philoponus), 70–71

palingenesis (Bonnet), 236
palm tree
  date palm as animal plant, 99
  date palm as engendered by wind, 96,
    542n.33
  date palm as highest / most-perfect plant,
    99, 103–4, 105, 106–7, 373, 374*f*
  date palm: reproduction, sexuality, 99
  fan palm (Goethe), 579n.26
  lion and palm tree (Merezhkowsky), 426
  in Theophrastus, 43
papermaking, 94–95, 118
parson-naturalist, 247, 356, 363, 597n.15,
    597n.16
pearl, pearls, 100, 110, 120, 146, 149, 156, 207,
    483n.109, 522n.12
pharmacopœia, 7, 57–58
  in early China, 18
  in early Japan, 22
  Theophrastus, 52
philosophy
  in accord / unity with religion, 77–78, 79,
    80–81, 116, 118–19
  as aid to the study of theology, 88, 89
  as autonomous / distinct from faith, 129,
    134–35, 361–62
  combined with Hermeticism and Kabbalah,
    158
  as handmaiden to theology, 76, 77–78,
    82–83, 90
  modern, beginnings of, 315
  personified as goddess, 87, 119, 122, 162
  philosophia perennis, 157–58
  as separate from theology, 131
  as subordinate to faith, 118, 129
philosophy, schools of / stances
  atomism, 28–29, 30–31, 35, 38, 44, 46–47,
    211

INDEX OF SUBJECTS   823

dualism (Cartesian), 186, 227
Eleatic, Eleatic paradoxes, 28–31, 44,
    471n.60
empiricism, 132, 315
Epicurean, 29, 44, 48–49, 88, 210–11,
    493n.89
humanism, 55, 80–81, 121–22, 134, 149–53,
    163, 171, 173–76, 177, 178, 190, 530n.82
idealism, transcendental, 312, 314, 315–17,
    322, 327–28, 331, 333
Islamic (*see* Islamic philosophy [schools
    of])
middle Platonism (including Philo), 59–63,
    224
Mohism, 17, 21–22
Monism (Haeckel), 400, 611n.98,
    612n.106
nature-philosophy (see *Naturphilosophie*)
Neoplatonism, 8, 11, 12–13, 28, 33–34,
    41, 49, 57–58, 60, 63–74, 82–83, 85, 87,
    89–90, 92, 94, 98–99, 101–2, 106–7, 116,
    134, 137, 143, 172, 178, 180, 210–11, 224,
    225–26
nominalism, 64–65, 115, 131, 133–34,
    511n.194
Pythagoreanism (*see* Pythagorean teachings
    / belief)
rationalism, 19, 122, 148, 186, 213, 315
Scepticism, 44–45, 47, 59, 60
scholasticism (*see* scholarly traditions)
Stoicism, 43–44, 49–50, 54, 56, 59, 60–61,
    63, 66, 73–74, 76, 88, 92, 94, 141, 247,
    462n.98
for individual philosophers (*see* at Index of
    Persons)
phylogenetics, molecular
    before sequences, 433, 435–36, 437
    disagreement among phylogenetic trees,
        438–39
    eocyte tree, 442, 446
    holophyly, 634n.100
    homology, homologue, 428, 446
    inference of phylogenetic trees, 437, 438–
        39, 442, 631n.28
    lateral genetic transfer (LGT) in
        phylogenetic inference, 435, 439, 440,
        444
    monophyly, 443–44, 448
    oligonucleotide catalogues, 435–36, 441
    orthology, orthologue, 435, 436–37
    paralogy, paralogue, 438–40, 631n.23
    paraphyly, 443–44, 448, 630n.199
    phylogenomics, 411–12, 440
    polyphyly, 443–44, 448

protein sequences in, 433–35, 434*f* (*see also*
    proteins)
    RNA in phylogenetics, 435–37, 438*f*, 442
    root (of phylogenetic tree), 435, 437, 447–
        48, 636n.141
    signature (Woese), 437, 636n.141
    Tree of Life (rRNA), 435, 436–37, 447–48
    See also classification; comparative biology
phylogeny. *See* comparative biology
Physiologus, 7, 56–57, 81, 125, 163–64, 171,
    197–98, 208
plants. *See* vegetables
plant-animal. *See* Zoophyta, zoophyte
Plato: philosophy
    Academy, 24–25, 44–45, 47, 59, 60, 75, 89–
        90, 92, 114
    Accademia Platonica (Florence), 134, 138–
        39, 149
    cosmology, 31–33
    logical division, 34, 204–5
    as Pythagorean, 28, 59, 60–61
    theory of intelligible forms (ideas), 31, 34,
        37, 59–60
Plato: works / corpus
    Epistle 7, 59
    Meno, 31
    Parmenides, 59–60, 134
    Phaedo, 28, 31
    Phaedrus, 11, 33, 78
    reintroduction to Europe, 118, 134
    Sophist, 472n.81
    Timaeus, 28, 31–33, 59–60, 62–63, 73, 83,
        87, 140
    translation into Arabic, 94–95
    See also macrocosm-microcosm
plenitude in nature, 223, 226, 227, 243, 309,
    481n.63
Pliny the Elder: works
    Naturalis historia, 50–52, 53, 530–31n.84
    Naturalis historia (supposed errors in), 190
Porphyry
    See also Tree of Porphyry
Porphyry: works (selected)
    Ad Gedalium, 485n.54, 485n.59, 486n.80,
        497n.52
    Epistle to Anebo, 64, 66
    Isagoge, 64–65, 69, 71, 72–73, 87, 95, 101,
        111, 113, 115, 149, 173–74, 176, 180,
        497n.52
    Life of Pythagoras, 461n.95, 468n.16
printing-press, early printed works, 54–55,
    125, 173–74, 189–90, 198, 202, 208,
    504n.239, 532n.16, 536n.85, 536n.90
prisca theologia, 138, 150, 157–58, 159, 178

824 INDEX OF SUBJECTS

prokaryotes, 414–15, 418–19, 427–29, 430–31, 434–35, 437, 442, 447, 622n.62, 632n.56, 635n.136
    Empire Prokaryota, 447–48, 630n.195
    formalized as a kingdom, 430, 624n.89
    Superkingdom Prokaryota, 636n.144
    term inappropriate, 437
    See also eukaryotes
proteins
    abiogenesis (Nägeli), 404
    cytochromes, 433–35
    early concepts of structure, 432
    folded structure / domains, 433, 448
    informational versus operational, 446
    as markers of eukaryogenesis, 442–43, 445, 446
    in molecular phylogenetics, 433–35, 434*f*
    nuclear, 419–21, 432, 614n.141
    presumed role in heredity, 432
    sequences of, 433–35
Protista, protists
    asemic, 399, 406, 614n.145
    controversies over, 407–9
    and Histones, 396–97, 399–400, 402–3
    as informal term, 409
    initial definition of, 385–86, 392
    Kingdom Protista, 386, 388–89, 390, 392, 393–99, 398*f*, 407, 408, 409, 414–15, 417–21, 418*f*, 419*f*, 420*f*, 424, 621n.44, 630n.1, 634n.102
    lack of sexuality, 391–92, 395–96
    membership, 386, 387–90, 391–400
    and monera, 386, 388, 391, 392–97, 406
    and protophyta / protozoa, 392, 393–95, 396–97, 399–400
protists (selected). *See* microorganisms
Protophyta, protophytes. *See* vegetables: classification
Protozoa, protozoa. *See* high-level taxa; microorganisms
pyrallis, pyrigones. See animalcules in furnaces
Pythagorean teachings / beliefs
    cosmology, 26–27, 28, 211
    dietary prohibitions, 11, 26, 27
    in later philosophers, 59, 60–61, 62, 65, 66, 76, 77, 88, 98–99, 101, 138–39, 143, 150–51, 157–58, 529n.57
    Loves of the triangles, 361
    metempsychosis, 11, 27
    number as basis of nature, 27, 59–60, 62, 65, 150–51, 266, 468–69n.21, 484n.36, 514n.48, 529n.57
    reincarnation, 12

Table of Pythagoras, 266

religions, denominations, and similar traditions
    Akkadian, 91
    Buddhism, 12, 16–17, 18–19, 22, 93, 541n.10, 543n.38, 580n.37
    Canaanite, 91
    Catholicism, 12–13, 178, 217–18, 311–12, 355, 606n.192 (*see also* Church)
    Christianity, 25, 27, 28, 29, 54, 65–66, 71, 75, 92, 114, 118, 125–26, 138, 149, 150, 157–58, 163, 170, 178, 210–11, 225, 601n.84
    Confucianism, 17, 18–20, 21–22, 460n.83, 543n.39
    Gnosticism, 66, 76, 93, 109, 137, 139–40, 143, 514n.48
    Hinduism, 12, 15
    Hittite, 91
    Islam, 19, 90, 93–94, 96, 106, 111, 178
    Jainism, 15–16
    Judaism, 60–61, 88, 90, 92, 99, 107, 108–10, 111, 178
    Local / pagan religions, 44, 59, 65–66, 71, 75, 76, 79, 80–81, 89–90, 91, 92, 122, 129, 147, 163, 165
    Lutheran, 311–12
    Manichæism, 19, 83, 93
    Methodist, 355
    Nestorian Christianity, 19, 92, 94, 95–96
    Nicene Christianity, 75, 87
    Quaker, 355
    Roman religion, 59, 137–38, 475n.139
    Rosicrucian, 154–55, 157–58, 311–12, 519n.187
    Sabian, 495n.145
    Samaritan, 92
    Shintō, 22
    Sumerian, 91
    Taoism, 10, 17, 18–19, 20–22, 465n.41
    Unitarianism, 355
    Valentinian, 139–40
    Zoroastrianism, 19, 91, 92, 150
religious orders
    Augustinians, 505n.9
    Benedictines, 114, 116, 120, 126, 153, 177
    Carmelites, 505n.9
    Dominicans, 115, 124–26, 130, 134–35, 158, 509–10n.162, 510n.172
    Franciscans, 115, 125, 129, 130, 131, 166
    Jesuits, 140, 169, 183, 185, 465n.41, 511n.217, 542n.18, 543n.38

INDEX OF SUBJECTS   825

Trinitarians, 505n.9
religious texts
  Avesta, 93
  Bible, 125, 146–47, 173–74, 361, 483n.103,
    492n.69, 505n.16, 555n.1
  Genesis Rabbah, 92
  "hidden scrolls", 109
  midrashim, 92, 107, 225, 507n.76, 525n.67
  Pahlavi texts, 93
  Pentateuch, 60–61, 92, 459n.63, 522n.4
  Quran, 10, 94, 96, 98–99, 497n.40, 522n.4
  Rigveda, 15, 16, 489n.135
  Talmud, 57, 92, 107–8, 165–66
  Vedic literature, 15
  Vrikshayurveda, 15
  See also *Chaldæan Oracles*; *Genesis*; Kabbalah
Renaissance
  Italian, 149, 157–58, 162, 172, 173–74, 180,
    190, 193, 198–99, 212, 231, 243, 244, 252,
    361, 406, 459–60n.68, 532n.22
  Northern, 181–82
rhizotomists, 52, 481n.56
Romanticism (German), 272, 287–88, 316–
    17, 322, 323, 325, 331, 333, 380, 385,
    583n.113

scale of nature. *See* continuum; Great Chain
    of Being
scholarly traditions
  etymology, 115–16, 201, 204
  lexicography, 96–97, 154, 174, 175–76
  quadrivium (arithmetic, music, geometry,
    astronomy), 115, 121–22, 468n.19
  rhetoric, 72, 73, 77, 115, 121–22, 181–82
  scholasticism, 85–86, 118–19, 120–34, 136,
    149, 172, 173, 175–76, 180, 181–82, 251,
    256, 257, 485n.56, 534n.64, 541n.15,
    575n.121
  trivium (grammar, rhetoric, dialectic), 115,
    121–22, 181–82
scientific disciplines (selected)
  anatomy, 196, 203, 215, 266, 288, 291, 298,
    299–300, 306, 309, 312–13, 322–23,
    327–28, 329, 330, 364, 381, 384, 387, 399,
    408–9, 481n.55, 587n.201, 607n.4
  anthropology, 5, 100–1, 248, 303, 368, 411
  archæology, 14–15, 325
  bacteriology, 400, 405–7, 412, 422, 620n.23
  biology, 16–17, 269, 287–88, 289, 293, 324–
    25, 327, 331, 414, 571n.29
  botany, 22, 42–43, 52, 161, 172, 174, 176–77,
    187, 190–97, 198, 206, 228, 240, 248, 257,
    280, 293–94, 322–23, 324, 325, 327–28,

    344–45, 348–49, 358–59, 368, 369, 385–
    86, 389, 397, 408–9, 557n.51
  cell biology, 309–10, 320, 321–22, 327–31,
    353, 426, 430, 435, 447
  chemistry, 143, 148, 157, 267–68, 293, 309,
    312–13, 355–56
  developmental biology, 307, 348, 381–82,
    384 (*see also* embryology; ontogeny)
  embryology, 307, 309, 324–25 (*see also*
    ontogeny)
  evolution, evolutionary biology, 293, 309,
    575–76n.132
  genetics (*see* genetics)
  geography (physical), 323
  geology, 293, 322–23, 355–56, 368, 376
  linguistics, 14–15, 411, 461n.91
  microbiology, 309–10, 323–24, 411, 416,
    437, 441, 447, 618n.255, 619n.1, 622n.60
  molecular biology (*see* genetics; phylogenetics)
  morphology, 288, 309, 314, 327–28, 333,
    383–85, 399
  mycology, 192–93, 403–4
  ontogeny (*see* ontogeny)
  palæontology, 309, 384
  pathology, 408–9
  phycology, 337, 344–45, 400, 403–5
  physiology, 309, 322–23, 408–9, 571n.29
  plant biogeography, 323
  protistology, 397–99, 407–9, 421
  protozoology, 400–3, 409
  psychology, 182–83, 182f, 261, 331, 411
  zoology, 161, 172, 187, 198–207, 208–9,
    240, 248, 293–97, 324, 325–26, 368, 381,
    385–86, 397, 408–9
scientific journals
  Archiv für Protistenkunde, 408–9, 411–12
  Centralblatt für Bakteriologie,
    Parasitenkunde und
    Infektionskrankheiten, 619n.259
  Isis (Oken), 318–19
  Journal of Bacteriology, 619n.259
  Journal of Biophysical and Biochemical
    Cytology, 628n.149
  Journal of Cell Biology, 628n.149
  Journal of General Microbiology, 621n.39
  Journal of Genetics, 630n.3
  Journal of Pathology and Bacteriology,
    619n.259
  Philosophical Transactions [*of the Royal*
    Society], 216, 232, 541n.2, 619n.267
  Protist, 618n.258
  Zeitschrift für wissenschaftliche Zoologie,
    329

826 INDEX OF SUBJECTS

sefirot, 108–9, 489n.135
ships
    HMS *Beagle*, 369
    HMS *Challenger*, 389
    HMS *Cyclops*, 389
    HMS *Endeavour*, 358
signatures (occult), 146–47, 153
Silk Road, 19, 25
snails
  as animal closest to plants, 105, 312
  aquatic and marine snails, 51, 54, 55, 167,
    201–2, 229, 517n.138
  terrestrial snails, 99–100, 105, 120, 167, 230,
    326
  unspecified snails, 54, 84, 95, 105–6, 351,
    370
Sophia (personification of wisdom), 139
sorcery, witchcraft, 3, 9, 10, 147, 148, 178, 179,
    406, 459n.63, 480n.54
soul
  as air or breath, 11, 25–26, 67, 76
  of angels, 99, 105, 212
  animate / animal soul, 36–37, 41, 64, 84,
    99, 100, 101, 104, 127, 158, 188, 205, 289,
    502n.178
  in Aquinas, 128
  in Aristotle, 35–37, 87, 188, 208
  association with stars (*see* sun, stars, and
    planets)
  as blood, 61, 78, 80
  cell-soul (Haeckel), 396
  in Chrysippus, 44
  continuity of, 12
  of crystals, minerals, atoms, electricity, 396
  in Democritus, 29
  faculties / powers of, 35–37, 68, 69–70, 71,
    72–73, 79–80, 89, 97, 99–100, 101–2, 115,
    120–21, 125, 130–31, 149, 181, 188–89,
    193, 202, 476n.173, 477n.176, 511n.194
  in al-Fārābī, 98
  in Galen, 54
  of gold, 99
  in Haeckel, 396
  human soul, 11–12, 32, 33, 44, 59–60, 64,
    66, 94, 96, 98, 100, 101, 102, 115, 116,
    152, 189, 471n.67, 473n.100, 489n.136
  in Iamblichus, 66
  immortality of, 11, 25, 33–34, 48–49, 70–71,
    117, 471n.68, 471n.69
  inherent in matter, 236
  intellective soul, 126, 149–50, 193, 219
  irrational soul, 70–71
  in al-Kindī, 97

  in Leibniz, 213, 214
  localized in parts of animal, 27, 32–33, 119–
    20, 471n.67
  localized in parts of plant, 123, 193
  in Maimonides, 108
  migration of souls (*see* metempsychosis)
  motion of (*see* at motion)
  in Neoplatonism, 73
  in Origen, 78
  origin of souls, 225
  in Philoponus, 70–71
  in Plato, 31–33, 59, 61, 78
  in Plotinus, 63–64
  preexistence of, 186
  of protists, 614n.139
  in Pythagoras, 11
  rational soul, 36, 61, 70–71, 82–83, 84, 99,
    101, 122–23, 127, 475n.140, 477n.178
  regeneration of hydra, 236
  reincarnation, 12, 30, 461n.93, 462–
    63n.103, 495n.152
  sensible, sensitive, 38, 122–23, 128, 129,
    150, 165
  in Socrates, 30–31
  in the Stoics, 44
  superadditive, 36, 73, 81–83, 115, 122–23,
    158, 188, 193, 230, 309, 331, 366–67,
    374–75, 377
  in Themistius, 67–68
  in Varro, 493–94n.102
  vegetative / vegetable soul, 36, 37, 42, 64,
    70–71, 95, 99, 101, 123, 127, 176, 188, 193
  world soul, 32, 73, 122, 316, 471n.65,
    483n.7, 493–94n.102
  of zoophytes, 70
species (biological)
  accidental (Maimonides), 108
  arising by hybridization, 251
  concept problematic for proto-organisms
    (Carus), 332–33
  defined biologically / genealogically (Ray),
    209, 245–46
  defined using single or multiple differentiæ,
    37, 188–89, 193–94, 209
  fixity of, 12, 20, 121, 165, 251, 260, 268, 315,
    324–25, 344–45, 534n.50, 565n.37
  in Jussieu , 260
  in Kützing, 348–49
  in Linnæus, 251, 268
  perfectibility / progress of, 240, 303
  as real / existing in nature, 251, 298, 411
  transmutation among, 12, 15, 20–21, 143,
    164–65, 333, 346

INDEX OF SUBJECTS   827

spermatozoa. *See* microscopic beings
sponges
  ancestral type (Haeckel), 390
  as animals, 41, 82, 100–1, 102–3, 195, 199,
    202, 208, 246–47, 262–63, 289, 295, 299–
    300, 306–7, 356–57, 368–69, 391–92,
    418*f*, 419*f*, 430, 624n.80
  in Aristotle, 39, 40, 55, 171, 195, 524n.47
  blood in, 51, 56, 207
  in Bory, 308–9, 333–34
  in Chatton, 414
  as contractile / sensate, 41, 51, 56, 70, 89,
    100–1, 125, 126, 144, 171, 184, 205, 243
  in Dioscorides, 53
  in Donati, 232
  as fish, 115–16, 124–25
  in Galen, 54
  gastraea theory (Haeckel), 390–92,
    615n.178
  germ-layers of, 390
  in Haeckel, 382, 386, 390–92
  in Hogg, 370–71, 372
  as intermediate beings, 51, 126, 140–41,
    144, 376–77 (*see also* as zoophytes)
  as irritable / not irritable, 371
  in Kützing, 348
  larvae of, 390
  in Linnæus, 249, 616n.198
  in Nieremberg, 140–41, 183, 184
  as non-living, 51
  in Plato, 34
  in Pliny, 51
  as polyparies, 338, 604–5n.158
  as protists, 418–19, 420*f*
  as protozoa, 367, 608n.23, 608n.25
  as resisting capture, 40, 51, 56, 79–80,
    535n.83
  roots of, 87, 125
  sexuality of, 51
  sponge-cutting, sponge-diving, 34, 51, 56
  Sponge-kingdom, 376–77
  spontaneous generation of, 337
  in Theophrastus, 43, 52
  as vegetables, 39, 192, 196, 203–4, 205, 219–
    20, 228, 244, 403, 588n.11
  in Wilson & Cassin, 375
  as zoophytes (or similar), 69–70, 95, 111,
    156, 171, 173, 176, 179, 195–96, 200–2,
    208–9, 232, 261–62, 575n.125
spontaneous generation, 9, 45, 49, 77, 82–83,
    92, 96, 108, 119, 170, 206, 278, 295, 297–
    98, 337, 338, 340, 348, 352–53, 360–61,
    368, 380, 391, 556n.29, 573n.79

autogony (Haeckel), 384–85, 389, 395–96,
    406–7
  disproved by Spallanzani, 216, 219, 237
  of infusoria, 273, 295, 305, 337, 380 (*see also*
    at microscopic beings)
  two stages of (Nägeli), 404
sun, stars, and planets
  analogized with elements or metals, 143,
    225
  associated with / watching over individual
    souls / beings, 32, 93, 109, 140, 183,
    210–11
  as composed of fire and air, 148–49
  creation of, 81, 88, 92, 94, 103, 120, 463n.4,
    464n.24, 489n.135, 514n.66
  as gods, 8, 32, 94, 486n.70
  inhabitants of, 211–12, 226–27
  as living / ensouled beings, 8, 23, 26, 78,
    184, 493–94n.102, 511n.201, 514n.62
  metamorphoses into, 9
  occult correspondences / signatures /
    symbolism, 140, 146–47, 153, 154–55,
    158, 225, 318, 320
  star-kingdoms, 376–77
  as totems, 6
  worship of, 81, 140
  See also cosmos; macrocosm-microcosm
symbiosis
  endosymbiosis, 463n.107, 552n.100
  eukaryotes predisposed to, 445–46
  examples among plants and animals, 423,
    424–25, 429–30, 442, 445–46
  first use of term, 423
  See also eukaryogenesis; lichens
sympathetic medicine, 151–52
systematics. *See* classification: schools of

Tabula smaragdina. See *Emerald tablet*
Theophrastus: works / corpus, 48
  De causis plantarum, 42–43, 189
  De lapidibus, 52
  Historia plantarum, 42–43, 52, 403
theurgy, 53–54, 59–60, 64, 66, 73, 92, 108–9,
    138
totemism, 6, 23
transformation. *See* alchemy; embryo;
    evolution; lycanthropy; metamorphosis;
    ontogeny; spontaneous generation;
    transmutation; transubstantiation
  genetic (*see* at genetics)
transmutation, 9–10, 12–13, 33, 110, 159,
    324–25, 422, 461n.91, 617n.223
  of metals (*see* alchemy)

828 INDEX OF SUBJECTS

transubstantiation, 9, 12–13, 115, 130, 169
Tree of Porphyry. *See* logical division

ultrastructure. *See* cellular ultrastructure
unicorn. *See* at legendary / mythical creatures
universities (selected; some names simplified)
  Bergakademie Freiburg, 322–23
  Colegio Imperial de Madrid, 183
  Collège de France, 298
  Imperial Moscow University, 281
  Imperial Novorossiysk University (Odesa),
    389
  King's College London, 369, 597n.4
  Københavns Universitet, 630n.4
  Lunds Universitet, 344–45
  Sorbonne (Université de Paris), 169, 175,
    414, 415, 510n.186
  Universidad Alcalá, 183
  Università di [Modena e] Reggio Emilia,
    350–51
  Università di Bologna, 165, 190–91
  Università di Ferrara, 134
  Università di Firenze, 512n.2
  Università di Pavia, 160
  Università di Pisa, 193
  Università di Padova (Padua), 180, 190–91
  Universität Berlin, 325, 578n.3, 620n.24
  Universität Bonn, 565n.44
  Universität Dorpat (Tartu), 273
  Universität Erfurt, 134–35
  Universität Erlangen, 565n.44
  Universität Göttingen, 221, 322–23,
    559n.93, 582n.86
  Universität Graz, 591n.65
  Universität Halle, 311–12, 594n.129
  Universität Jena, 311–12, 318–19, 322–23,
    327–28, 379, 381–82, 389, 540n.175,
    610n.75, 622n.48
  Universität Köln, 160
  Universität Marburg, 622n.48
  Universität Vienna, 591n.65
  Universität Würzburg, 311
  Universität Zürich, 352, 582n.78
  Université Catholique de Louvain, 606n.192
  Université de Dôle, 160
  Université de Genève, 341
  Université de Lille, 625n.97
  Université de Montpellier, 178, 190–91
  Université de Paris, 115, 118, 122–23, 128–
    29, 130, 131, 527n.8
  Université de Strasbourg, 414
  Université libre de Bruxelles, 397
  University College London, 597n.4
  University of Cambridge, 118, 228, 246–47,
    355, 554–55n.172

  University of Edinburgh, 355–56, 357, 358
  Univerisity of Illinois, 435–36
  University of Oxford, 118, 130, 355, 356
Urschleim, 389

vegetables (plants)
  algal filaments as basis of, 345–46
  analogies with animals, 15, 18–19, 25–26,
    30, 37–38, 43, 62, 120, 123, 228–29, 236,
    238, 255, 260, 272, 282–83, 296, 301,
    302–3, 320, 340, 342, 360, 382, 586n.165
  as animals, 68–69, 236
  aquatic (class of), 244, 259, 260
  archetype, 313, 314, 317, 345–46, 384,
    579n.26
  chemical properties, 231, 307, 336, 351, 352,
    353, 370, 371, 381–82, 622–23n.64
  chlorophyll in, 351, 353, 403–4, 414, 422–
    23, 622–23n.64, 627n.144
  definition / delineation, 196–97, 249, 371,
    372, 375
  as deities, 7
  developmental form-stages (Kützing), 348
  as "fixed animals", 159
  as "fleeting minerals", 159
  floral parts as animals, 360
  as framework or rough draft of animal, 289
  intellect, intelligence of, 42
  as inverted animals, 37, 138, 318
  as irritable / not irritable, 258–59, 261, 289,
    296, 297, 601n.76
  as living or lifeless, 22, 27, 32–33, 35, 36, 42,
    49, 67, 70, 88–89, 91, 101, 102–4, 123,
    126, 127–28, 165, 177–78, 256, 302–3,
    320, 358, 476n.166, 582n.76
  metamorphosis (Agardh, Gaillon, Kützing),
    344–46, 347–48, 349–50
  morphology / fine structure distinguishes
    from animals, 380–81, 382
  nutrition, 35–36, 37–38, 42, 61, 68, 69–70,
    72–73, 83, 98, 99, 103, 108, 150, 188, 193–
    94, 196, 331, 360, 374–75, 382, 417–18
  parasitic / mistletoe, 99, 179, 191, 228–29
  as perceptive, 20
  photosynthesis, 338, 342, 351, 417–18
  as (polar) opposites to animals, 66–67, 188,
    271–72, 280–81, 314–15, 318, 319
  pollen theory (Schleiden), 342, 344
  as polyparies, 338, 339–40, 344, 360
  prolepsis (Linnæus), 251–52, 253, 579n.24
  reproduction, seeds, spores, 193, 197, 229,
    233, 244–46, 249, 251, 254–55, 258–59,
    336–37, 341, 342, 344, 352, 358–59, 403
  as sensate / insensate, 32–33, 42, 44, 65, 66,
    69–70, 83, 88, 110, 113, 117, 129, 132,

152, 165, 171, 250–51, 291, 308, 360, 371, 374–75

soul of (*see* at soul)

as stage in animal or human ontogeny, 38, 100, 132

stirps distinguished from *planta*, 531n.5

subterranean, 322–23

transformed into animals (Kützing), 349, 590n.51

trees as the most-perfect plants, 43, 52, 128, 192–93, 194, 195, 196, 228–29, 291–92, 314–15, 322–23, 368–69, 374*f*

unicellular, 353, 380, 577n.156, 596n.184

Urpflanze (Goethe), 313–14

Urpflanze (Haeckel), 396, 399–400, 614n.147

as variants on a prototype, 290–91, 303–4, 369–70

as womb of animal-world, 343

vegetables: classification

Cesalpino as first plant systematist, 193

Cryptogamia, 249–50, 260, 335, 336, 403

emphasis on families (Magnol, Adanson), 244–45, 259

emphasis on genera (Tournefort), 245

emphasis on species (Ray), 245–46, 260

form-series in, 403–4, 405–6, 411, 412, 617n.214

fructification (characters of), 193, 249, 260

genealogical, 244–46, 251

grasses / herbs / bushes / shrubs / trees, 15, 43, 49, 52, 53, 95–96, 150, 189, 191, 192, 193, 197, 228–29, 495n.140, 528n.39

monocots versus dicots, 192, 282–83, 284–85, 287, 297, 368–69

multiple criteria in, 193, 195, 244–45, 259, 260, 534n.64

perfect versus imperfect (Ray), 246–47

Ramist trees in, 192, 195, 534n.65

presence / absence of seeds, 105, 106–7, 194, 197

Protophyta, 372, 392, 393–95, 396–97, 399–400

sexual system (Linnæus), 249, 251, 256, 259, 260, 355–56, 357–59, 598n.19

system of Jussieu, 260, 358–59

Thalassiophyta, 336, 430

Vegetable / Plant Kingdom, 103–4, 106, 138, 154, 206–7, 208, 237–38, 244, 249, 252–53, 255, 259, 263, 278, 282–83, 284, 287, 290–91, 305, 306, 308, 316, 320, 347, 348, 355–56, 368, 369, 370, 372, 373, 373*f*, 374, 375, 376, 380–81, 383, 385–86, 391–92, 393, 396–97, 398*f*, 404–5, 407–8, 408*f*, 415, 417, 418–21, 418*f*, 419*f*, 420*f*,

424, 430, 455n.3, 466n.47, 520n.209, 540n.167, 580n.40, 590n.55, 602n.96, 606n.192, 609n.61, 630n.198, 634n.102

first called a kingdom, 2, 156–57, 206–7, 208

True and Doubtful subkingdoms (Blainville), 306

vegetables (selected genera / taxa)

algae (*see* marine plants / seaweeds; microorganisms)

cotton plant, 165–66

Epipetrum, 39

ferns, 192, 194, 195, 249, 259, 268–69, 308, 322, 336–37, 342, 344, 353

fungi (*see* fungi)

horsetails, 192, 194–95, 196, 308, 322

lichens (*see* lichens)

mandrake, 7–8, 57, 147

mosses, 99, 120, 191, 192, 194, 195, 197, 228–29, 232, 233, 238, 244, 246–47, 249, 255, 259, 268–69, 308, 315–16, 322, 336–37, 342, 343–44, 345–46, 347–48, 349–50, 352, 353, 539n.152, 594n.138, 621n.36

palms (*see* palm tree)

sensitive plant / *Mimosa*, 43, 171, 238, 244, 362, 530n.83

Vestiges of the natural history of Creation, 285–86, 364, 368–69, 377, 606n.185, 606n.189

via negativa, 85–86

Victorian Age (Britain), natural histories in, A1, 375–77

viriditas (in Hildegard), 120

viruses, viroids

of archaea, 441, 444

as components of cells (Altenburg), 426

in Enderlein, 413–14

genome size / gene content of, 444

high-level taxa of, 416, 426, 444, 624n.80, 624n.81

as living / non-living, 250, 444

as polyphyletic, 444

Ultrabiontia, 634n.102

as vectors for DNA / lateral genetic transfer, 439, 444

Wheel of Fortune, 162

witchcraft, witches. *See* at sorcery

worms

as animals, 17–18, 48, 51, 79–80, 99–100, 105, 120, 124–25, 201, 203–4, 230, 234–35, 246–47, 249–50, 268–69, 270, 272, 283–84, 291–92, 293, 295, 296–98, 299, 325, 356–57, 362, 386–87, 388, 456n.19, 458n.49

cane worm (snail?), 99–100, 501n.167

830 INDEX OF SUBJECTS

worms (*cont.*)
classification of, 99–100, 101, 364–65
diverse beings called worms, 124–25, 208–9, 255–56
fungi germinate into worms, 253–54, 255
infusoria as worms, 211–12, 221, 352, 405 (*see also* Linnæan class, below)
as intermediary or linking beings, 233, 268
larval stages of insects, 214, 547n.118
Linnæan class Vermes, 249–50, 261, 293–94, 299–300, 303
metamorphoses, 168, 172, 196
regeneration, 237
spontaneous generation of, 45, 167, 170, 295, 297–98, 337, 380, 556n.29
Taenia (tapeworm), 247–48, 252
as vegetal animals, 99–100
"worm to man", 210
as zoophytes, 207, 239, 247–48, 263, 299–300

zodiac, 7, 88, 140, 144, 158
Zoophyta, zoophytes
as agents of disease, 249–50, 254
in alchemical literature, 156
animal-plants distinguished from plant-animals, 185, 296, 322, 551n.93
as animals, 67–68, 150, 156, 181, 182–83, 182*f*, 199, 201–2, 203–4, 205, 206, 219–20, 235, 238–39, 247, 250–51, 250*t*, 256, 261, 263, 270, 290, 299, 301, 302–3, 312, 318, 356–57, 364–65, 366, 391, 573n.92
as "beginning to show life", 236
classified / analogized with insects, 170–71, 200–1, 202, 219–21, 231–32, 238–39, 312, 351, 355, 360, 536–37n.106, 545n.75, 570n.12
compared / contrasted with lithophytes, 253, 258, 261, 262–63, 563n.5
coral polypary as a zoophyte, 232
date palm as animal plant, 99
as "diverse things with a common disposition", 208
exemplifying threefold order in nature, 95–96, 150–51, 151*f*
false idea / no such beings, 219–20, 238–40, 296
as a genus in a logical structure, 69–70, 71–73, 88–89, 104, 179, 182–83, 182*f*
as a grade, level or step, 70, 152, 158, 239, 258, 263, 291, 551n.93
as harmonizing / unifying nature, 156
as having a middle, median or double nature, 40–41, 66, 69–70, 89, 140–41,

173, 174–75, 176–77, 181, 185–86, 187, 194, 201–2, 237–38, 263, 276, 291, 357, 484n.36
as imperfect animals, 150, 156, 181, 182–83, 182*f*, 247, 356–57
infusion-animalcules as zoophytes, 249–50, 253, 255, 257, 263, 299–300, 303, 307, 365, 424–25
as irritable / not irritable, 219, 318
as a kingdom or other high-level taxon, 208, 322, 364–65
as knowable apart from method of logical division, 180–81
lack sensibility and irritability, 318, 371
as a link, bond, or mean between opposites, 66–67, 150–51, 151*f*
in Linnæus, 249–51, 250*t*, 252–53, 254, 255, 256, 257, 262–63
metamorphosis of, 253
misattributed to Aristotle, 37, 41, 68, 69, 70, 152, 199
movement-in-place (*see* at motion)
as neither animal nor plant (but rather a third genus), 51, 69, 70, 152, 153, 156, 195–96, 200–1, 239, 250–51, 356–57
Pflanzengeschöpfen, 312
Phytozoa (*see* high-level taxa)
as plants / vegetables, 195, 196, 312
Priestley's green matter as zoophyte, 315
as reciprocally transformable between plants and animals, 304
as sensate or insensate, 51, 56, 67, 68, 69–70, 71, 72–73, 79–80, 83, 89, 99, 102–3, 113, 121, 123–24, 127, 141–42, 144, 148, 149–50, 152, 158, 167, 171, 173, 176, 180–81, 182–83, 184–86, 200, 201–2, 205, 207, 208, 219, 228, 229, 250–51, 252–53, 268, 291, 318, 371
share analogies / affinities with animals and/or plants, 95–96, 235, 238–39, 261–62, 290
situated on branching tree, 179
terrestrial, 167, 207–8, 560n.130
three levels between plants and animals, 232
as uniting all three kingdoms, 276–77
uterine tumours as zoophytes, 167
as végéto-vital (Needham), 218–19
See also chimæric beings; high-level classification; marine invertebrates; microorganisms
Zoophyta, zoophyton (word)
in Agrippa, 153
in Arabic literature, 68, 95–96, 100, 102–3
in Budé, 175–76

in commentators, 67–68, 69–70, 71–73, 127, 150–51

in early Christian writers, 70, 79–80, 88–89, 90, 112, 113, 150–51

in early dictionaries, 173, 174–75, 176–77, 178

in early printed books, 173–74, 504n.239, 504n.240

in English, 183, 185

figurative usage, 187

first appearance (Dexippus), 66–67, 69, 150–51

in French, 177–78

as informal term, 263

in Latin translation, 45–46, 67, 68, 80, 111–13, 474n.131, 486–87n.85, 487n.86, 487n.87, 487n.88, 487n.89, 512n.228

in Neoplatonism, 66–67, 150–51

in Pico della Mirandola, 150–51, 151*f*, 152

reappearance at Council of Ferrara, 134, 173

terminology in the Ikhwān al-Ṣafā, 100

in texts translated into Latin, 67, 89, 112, 113

as "unphilosophical", 238–39

word doubtful in Sextus Empiricus, 44–46

word misattributed to / not in Aristotle, 37, 41, 68, 69, 70, 152, 194–95, 199, 200, 524n.47

word misattributed to Philo, 46, 62, 113, 173–74, 176–77, 497n.52

word not in Pliny, 51

zoospores, 341–44, 351, 380–81, 391–92